ANNOTATED INSTRUCTOR'S EDITION

Tenth Edition

Introductory Algebra

Margaret L. Lial
American River College

John Hornsby
University of New Orleans

Terry McGinnis

PEARSON

Boston Columbus Indianapolis New York San Francisco Upper Saddle River
Amsterdam Cape Town Dubai London Madrid Milan Munich Paris Montréal Toronto
Delhi Mexico City São Paulo Sydney Hong Kong Seoul Singapore Taipei Tokyo

Editorial Director	Christine Hoag
Editor in Chief	Maureen O'Connor
Executive Content Editor	Kari Heen
Content Editor	Christine Whitlock
Senior Content Editor	Lauren Morse
Assistant Editor	Rachel Haskell
Senior Managing Editor	Karen Wernholm
Senior Production Project Manager	Kathleen A. Manley
Digital Assets Manager	Marianne Groth
Supplements Production Coordinator	Kerri Consalvo
Media Producer	Stephanie Green
Software Development	Rebecca Williams, MathXL; Mary Durnwald, TestGen
Marketing Manager	Rachel Ross
Senior Author Support/Technology Specialist	Joe Vetere
Rights and Permissions Advisor	Cheryl Besenjak
Image Manager	Rachel Youdelman
Procurement Manager	Evelyn Beaton
Procurement Specialist	Debbie Rossi
Media Procurement Specialist	Ginny Michaud
Associate Director of Design	Andrea Nix
Senior Designer	Barbara Atkinson
Text Design, Production Coordination, Composition, and Illustrations	Cenveo Publisher Services/ Nesbitt Graphics, Inc.
Cover Image	*Crimson Foliage* © Lorraine Cota Manley

Library of Congress Cataloging-in-Publication Data

Lial, Margaret L.

Introductory algebra / Margaret L. Lial, John Hornsby, Terry McGinnis.—10th ed.

p. cm.

Includes index.

ISBN 978-0-321-87048-3 AIE ISBN: 978-0-321-87049-0/0-321-87049-2

1. Algebra—Textbooks. I. Hornsby, E. John. II. McGinnis, Terry. III. Title.

QA152.3.L557 2013

512.9—dc23 2012013812

1 2 3 4 5 6 7 8 9 10—CRK—16 15 14 13 12

PEARSON

www.pearsonhighered.com

ISBN 13: 978-0-321-87048-3
ISBN 10: 0-321-87048-4

To Margaret L. Lial

On March 16, 2012, the mathematics education community lost one of its most influential members with the passing of our beloved mentor, colleague, and friend Marge Lial. On that day, Marge lost her long battle with ALS. Throughout her illness, Marge showed the remarkable strength and courage that characterized her entire life.

We would like to share a few comments from among the many messages we received from friends, colleagues, and others whose lives were touched by our beloved Marge:

"What a lady"
"A remarkable person"
"Gracious to everyone"
"One of a kind"
"Truly someone special"
"A loss in the mathematical world"
"A great friend"
"Sorely missed but so fondly remembered"
"Even though our crossed path was narrow, she made an impact and I will never forget her."
"There is talent and there is Greatness. Marge was truly Great."
"Her true impact is almost more than we can imagine."

In the world of college mathematics publishing, Marge Lial was a rock star. People flocked to her, and she had a way of making everyone feel like they truly mattered. And to Marge, they did. She and Chuck Miller began writing for Scott Foresman in 1970. Just three years ago she told us that she could no longer continue because "just getting from point A to point B" had become too challenging. That's our Marge—she even gave a geometric interpretation to her illness.

It has truly been an honor and a privilege to work with Marge Lial these past twenty years. While we no longer have her wit, charm, and loving presence to guide us, so much of who we are as mathematics educators has been shaped by her influence. We will continue doing our part to make sure that the Lial name represents excellence in mathematics education. And we remember daily so many of the little ways she impacted us, including her special expressions, "Margisms" as we like to call them. She often ended emails with one of them—the single word "Onward."

We conclude with a poem penned by another of Marge's coauthors, Callie Daniels.

Your courage inspires me
Your strength...impressive
Your wit humors me
Your vision...progressive

Your determination motivates me
Your accomplishments pave my way
Your vision sketches images for me
Your influence will forever stay.

Thank you, dearest Marge.
Knowing you and working with you has been a divine gift.

Onward.

John Hornsby
Terry McGinnis

Contents

Preface

In the tenth edition of *Introductory Algebra*, we have addressed the diverse needs of today's students by creating a tightly coordinated text and technology package that includes integrated activities to help students improve their study skills, an attractive design, updated applications and graphs, helpful features, and careful explanations of concepts. We have also expanded the supplements and study aids. We have revamped the video series into a complete Lial Video Library with expanded video coverage and new, easier navigation. And we have added the new Lial MyWorkBook. We have also responded to the suggestions of users and reviewers and have added many new examples and exercises based on their feedback.

Students who have never studied algebra—as well as those who require further review of basic algebraic concepts before taking additional courses in mathematics, business, science, nursing, or other fields—will benefit from the text's student-oriented approach. Of particular interest to students and instructors will be the new guided solutions in margin problems and exercises, the new Concept Check exercises, and the enhanced Study Skills activities.

This text is part of a series that also includes the following books:

- *Basic College Mathematics,* Ninth Edition, by Lial, Salzman, and Hestwood
- *Essential Mathematics,* Fourth Edition, by Lial and Salzman
- *Prealgebra,* Fifth Edition, by Lial and Hestwood
- *Intermediate Algebra,* Tenth Edition, by Lial, Hornsby, and McGinnis
- *Introductory and Intermediate Algebra,* Fifth Edition, by Lial, Hornsby, and McGinnis
- *Prealgebra and Introductory Algebra,* Fourth Edition, by Lial, Hestwood, Hornsby, and McGinnis
- *Developmental Mathematics: Basic Mathematics and Algebra,* Third Edition, by Lial, Hornsby, McGinnis, Salzman, and Hestwood

WHAT'S NEW IN THIS EDITION

We are pleased to offer the following new textbook features and supplements.

▶ ***Engaging Chapter Openers*** The new Chapter Openers portray real life situations that reflect the mathematical content and are relevant to students. Each opener also includes an expanded outline of the chapter contents. (See pp. 27, 117, and 213—Chapters 1, 2, and 3.)

▶ ***Revised Exposition*** As each section of the text was revised, we paid special attention to the exposition, which has been tightened and polished. (See Section 1.3, Real Numbers and the Number Line, for example.) We believe this has improved discussions and presentations of topics.

▶ ***Guided Solutions*** Selected exercises in the margins and in the exercise sets, marked with a GS icon, step students through solutions as they learn new concepts or procedures. (See p. 119, margin problem, and p. 157, Exercises 23 and 24.)

▶ ***Concept Check Exercises*** Each section exercise set now begins with a group of these exercises, designated **CONCEPT CHECK**. They are designed to facilitate students' mathematical thinking and conceptual understanding. Many of these emphasize vocabulary. (See pp. 236 and 295.) Additional Concept Check problems are sprinkled throughout the exercise sets and ask students to apply mathematical processes and concepts or to identify **What Went Wrong?** in incorrect solutions. (See pp. 103 and 389.)

▶ ***Essential Study Skills*** Poor study skills are a major reason why students do not succeed in mathematics. Each one page long, these eleven enhanced activities provide helpful information, tips, and strategies on a variety of essential study skills, including *Reading Your Math Textbook, Tackling Your Homework, Taking Math Tests,* and *Managing Your Time*. While the activities are concentrated in the early chapters of the text, each has been designed independently to allow flexible use with individuals or small groups of students, or as a source of material for in-class discussions. (See pp. 55 and 104.)

▶ ***Helpful Teaching Tips*** All new **Teaching Tips**, located in the margins of the *Annotated Instructor's Edition,* provide helpful suggestions, emphasize common student trouble spots, and offer other pertinent information that instructors, especially those new to teaching this course, may find helpful. (See pp. 100 and 486.)

▶ ***Lial Video Library*** The Lial Video Library, available in MyMathLab and on the Video Resources DVD, provides students with a wealth of video resources to help them navigate the road to success. All video resources in the library include optional captions in English and Spanish. The Lial Video Library includes Section Lecture Videos, Solutions Clips, Quick Review Lectures, and Chapter Test Prep Videos. The Chapter Test Prep Videos are also available on YouTube (searchable using author name and book title), or by scanning the QR code on the inside back cover for easy access.

▶ ***MyWorkBook*** This new workbook provides Guided Examples and corresponding Now Try Exercises for each text objective. The extra practice exercises for every section of the text, with ample space for students to show their work, are correlated to Examples, Lecture Videos, and Exercise Solution Clips, to give students the help they need to successfully complete problems. Additionally, MyWorkBook lists the learning objectives and key vocabulary terms for every text section, along with vocabulary practice problems.

CONTENT CHANGES

The scope and sequence of topics in *Introductory Algebra* has stood the test of time and rates highly with our reviewers. Therefore, the table of contents remains intact, making the transition to the new edition easier. Specific content changes include the following:

▶ We gave the exercise sets special attention. There are approximately 820 new and updated exercises, including problems that check conceptual understanding, focus on skill development, and provide review. We also worked to improve the even-odd pairing of exercises.

▶ Real-world data in over 240 applications in the examples and exercises has been updated.

▶ We increased the emphasis on checking solutions and answers, as indicated by a new **CHECK** tag and ✓ in the exposition and examples.

▶ In preparation for Chapter 3 and the material on forms of linear equations, Section 2.5 provides a new example and corresponding exercises on solving a linear equation in two variables x and y for y.

▶ Section 2.6 includes entirely new discussion and examples on percent, percent equations, and percent applications, plus corresponding exercises.

▶ Section 3.4 on writing and graphing equations of lines provides increased development and coverage of the slope-intercept form, including two new examples.

▶ Sections 6.2–6.4 include three new examples on factoring trinomials.

- The following topics are among those that have been enhanced and/or expanded:

  Dividing real numbers involving zero (Section 1.6)
  Applying the properties of real numbers (Section 1.7)
  Solving applications involving consecutive integers and angle measures (Section 2.4)
  Using interval notation and solving linear inequalities (Section 2.7)
  Graphing linear equations in two variables using intercepts (Section 3.1)
  Solving systems of equations with decimal coefficients (Section 4.2)
  Finding numerical values of rational expressions and values for which rational expressions are undefined (Section 7.1)
  Solving quadratic equations by using the zero-factor property (Section 9.1)

HALLMARK FEATURES

We have enhanced the following popular features, each of which is designed to increase ease-of-use by students and/or instructors.

- ***Emphasis on Problem-Solving*** We introduce our six-step problem-solving method in Chapter 2 and integrate it throughout the text. The six steps, *Read, Assign a Variable, Write an Equation, Solve, State the Answer,* and *Check,* are emphasized in boldface type and repeated in examples and exercises to reinforce the problem-solving process for students. (See pages 148 and 316.) We also provide students with **Problem-Solving Hint** boxes that feature helpful problem-solving tips and strategies. (See pages 150 and 545.)

- ***Helpful Learning Objectives*** We begin each section with clearly stated, numbered objectives, and the included material is directly keyed to these objectives so that students and instructors know exactly what is covered in each section. (See pages 28 and 353.)

- ***Popular Cautions and Notes*** One of the most popular features of previous editions, we include information marked **CAUTION** and **Note** to warn students about common errors and emphasize important ideas throughout the exposition. The updated text design makes them easy to spot. (See pages 28 and 47.)

- ***Comprehensive Examples*** The new edition features a multitude of step-by-step, worked-out examples that include pedagogical color, helpful side comments, and special pointers. We give increased attention to checking example solutions—more checks, designated using a special **CHECK** tag and ✓, are included than in past editions. (See pages 120 and 387.)

- ***More Pointers*** Well received by both students and instructors in the previous edition, we incorporate more pointers in examples and discussions throughout this edition of the text. They provide students with important on-the-spot reminders and warnings about common pitfalls. (See pages 127 and 419.)

- ***Ample Margin Problems*** Margin problems, with answers immediately available at the bottom of the page, are found in every section of the text. (See pp. 30 and 345.) This expanded key feature allows students to immediately practice the material covered in the examples in preparation for the exercise sets. Many include new guided solutions.

- ***Updated Figures, Photos, and Hand-Drawn Graphs*** Today's students are more visually oriented than ever. As a result, we have made a concerted effort to include appealing mathematical figures, diagrams, tables, and graphs, including a "hand-drawn" style of graphs, whenever possible. (See pp. 230 and 658–659.) We have incorporated new depictions of well-known mathematicians as well as new photos to accompany applications in examples and exercises. (See pp. 219 and 547.)

- ***Optional Calculator Tips*** These tips, marked 🖩, offer helpful information and instruction for students using calculators in the course. (See pp. 31 and 571.)

▶ ***Relevant Real-Life Applications*** We include many new or updated applications from fields such as business, pop culture, sports, technology, and the health sciences that show the relevance of algebra to daily life. (See pp. 183 and 260.)

▶ ***Extensive and Varied Exercise Sets*** The text contains a wealth of exercises to provide students with opportunities to practice, apply, connect, review, and extend the skills they are learning. Numerous illustrations, tables, graphs, and photos help students visualize the problems they are solving. Problem types include skill building, writing, and calculator exercises, as well as applications, matching, true/false, multiple-choice, and fill-in-the-blank problems. (See pp. 226–227 and 401–402.)

In the Annotated Instructor's Edition of the text, the writing exercises are marked with an icon so that instructors may assign these problems at their discretion. Exercises suitable for calculator work are marked in both the student and instructor editions with a calculator icon. Students can watch an instructor work through the complete solution for all exercises marked with a Play Button icon on the Videos on DVD or in MyMathLab.

▶ ***Flexible Relating Concepts Exercises*** These help students tie concepts together and develop higher level problem-solving skills as they compare and contrast ideas, identify and describe patterns, and extend concepts to new situations. (See pp. 185 and 395.) These exercises, now located at the end of selected exercise sets, make great collaborative activities for pairs or small groups of students.

▶ ***Special Summary Exercises*** We include a set of these popular in-chapter exercises in every chapter beginning in Chapter 1. They provide students with the all-important ***mixed review problems*** they need to master topics and often include summaries of solution methods and/or additional examples. (See pp. 146 and 609.)

▶ ***Math in the Media*** Each of these one-page activities presents a relevant look at how mathematics is used in the media. Designed to help instructors answer the often-asked question, "When will I ever use this stuff?," these activities ask students to read and interpret data from newspaper articles, the Internet, and other familiar, real-world sources. (See pp. 116 and 414.) The activities are well-suited to collaborative work or they can be completed by individuals or used for open-ended class discussions.

▶ ***Step-by-Step Solutions to Selected Exercises*** Exercise numbers enclosed in a blue square, such as **11.**, indicate that a worked-out solution for the problem is included at the back of the text. These solutions are given for selected exercises that most commonly cause students difficulty. (See pp. S-1 through S-14.)

▶ ***Extensive Review Opportunities*** We conclude each chapter with the following important review components:

A **Chapter Summary** that features a helpful list of **Key Terms,** organized by section, **New Symbols, Test Your Word Power** vocabulary quiz (with answers immediately following), and a **Quick Review** of each section's contents, complete with additional examples (See pp. 276–279.)

A comprehensive set of **Chapter Review Exercises,** keyed to individual sections for easy student reference, as well as a set of **Mixed Review Exercises** that helps students further synthesize concepts (See pp. 280–284.)

A **Chapter Test** that students can take under test conditions to see how well they have mastered the chapter material (See pp. 285–286.)

A set of **Cumulative Review Exercises** (beginning in Chapter 2) that covers material going back to Chapter R (See pp. 287–288.)

STUDENT SUPPLEMENTS

Student's Solutions Manual

- By Jeffery A. Cole, Anoka-Ramsey Community College
- Provides detailed solutions to the odd-numbered section-level exercises and to all margin, Relating Concepts, Summary, Chapter Review, Chapter Test, and Cumulative Review Exercises

ISBNs: 0-321-84735-0, 978-0-321-84735-5

NEW MyWorkBook

- Provides Guided Examples and corresponding Now Try Exercises for each text objective
- Refers students to correlated Examples, Lecture Videos, and Exercise Solution Clips
- Includes extra practice exercises for every section of the text with ample space for students to show their work
- Lists learning objectives and key vocabulary terms for every text section, along with vocabulary practice problems

ISBNs: 0-321-87052-2, 978-0-321-87052-0

NEW Lial Video Library

The Lial Video Library, available in MyMathLab and on the Video Resources DVD, provides students with a wealth of video resources to help them navigate the road to success. All video resources in the library include optional captions in English and Spanish. The Lial Video Library includes the following resources:

- **Section Lecture Videos** offer a new navigation menu that allows students to easily focus on the key examples and exercises that they need to review in each section.
- **Solutions Clips** show an instructor working through the complete solutions to selected exercises from the text. Exercises with a solution clip are marked in the text and e-book with a Play Button icon ▶.
- **Quick Review Lectures** provide a short summary lecture of each key concept from the Quick Reviews at the end of every chapter in the text.
- **Chapter Test Prep Videos** provide step-by-step solutions to all exercises from the Chapter Tests. These videos provide guidance and support when students need it the most: the night before an exam. Chapter Test Prep Videos are also available on YouTube (searchable using author name and book title), or by scanning the QR code on the inside back cover for easy access.

INSTRUCTOR SUPPLEMENTS

Annotated Instructor's Edition

- Provides answers to all text exercises in color next to the corresponding problems
- Includes all NEW Teaching Tips located in the margins
- Identifies writing and calculator exercises

ISBNs: 0-321-87049-2, 978-0-321-87049-0

Instructor's Solutions Manual (Download only)

- By Jeffery A. Cole, Anoka-Ramsey Community College
- Provides complete solutions to all exercises in the text
- Available for download at www.pearsonhighered.com

ISBNs: 0-321-87050-6, 978-0-321-87050-6

Instructor's Resource Manual with Tests and Mini-Lectures (Download only)

- Contains a test bank with two diagnostic pretests, six free-response and two multiple-choice test forms per chapter, and two final exams
- Contains a mini-lecture for each section of the text with objectives, key examples, and teaching tips
- Includes a correlation guide from the ninth to the tenth edition and phonetic spellings for all key terms in the text
- Includes resources to help both new and adjunct faculty with course preparation and classroom management, by offering helpful teaching tips correlated to the sections of the text
- Available for download at www.pearsonhighered.com

ISBNs: 0-321-87325-4, 978-0-321-87325-5

ADDITIONAL MEDIA SUPPLEMENTS

MyMathLab®

MyMathLab® Online Course (access code required)

MyMathLab delivers **proven results** in helping individual students succeed. It provides **engaging experiences** that personalize, stimulate, and measure learning for each student. And, it comes from a **trusted partner** with educational expertise and an eye on the future.

To learn more about how MyMathLab combines proven learning applications with powerful assessment, visit **www.mymathlab.com** or contact your Pearson representative.

MyMathLab® Ready to Go Course (access code required)
These new Ready to Go courses provide students with all the same great MyMathLab features that they are used to but make it easier for instructors to get started. Each course includes pre-assigned homework and quizzes to make creating a course even simpler.

Ask your Pearson representative about the details for this particular course or to see a copy of this course.

MyMathLab® Plus/MyStatLab™ Plus
MyLabsPlus combines proven results and engaging experiences from MyMathLab® and MyStatLab™ with convenient management tools and a dedicated services team. Designed to support growing math and statistics programs, it includes additional features such as:

- **Batch Enrollment:** Schools can create the login name and password for every student and instructor, so everyone can be ready to start class on the first day. Automation of this process is also possible through integration with the school's Student Information System.
- **Login from your campus portal:** Instructors and their students can link directly from their campus portal into the MyLabsPlus courses. A Pearson service team works with the institution to create a single sign-on experience for instructors and students.
- **Advanced Reporting:** MyLabsPlus's advanced reporting enables instructors to review and analyze students' strengths and weaknesses by tracking their performance on tests, assignments, and tutorials. Administrators can review grades and assignments across all courses on a MyLabsPlus campus for a broad overview of program performance.
- **24/7 Support:** Students and instructors receive 24/7 support, 365 days a year, by email or online chat.

MyLabsPlus is available to qualified adopters. For more information, visit our website at *www.mylabsplus.com* or contact your Pearson representative.

MathXL®

MathXL® Online Course (access code required)
MathXL® is the homework and assessment engine that runs MyMathLab. (MyMathLab is MathXL plus a learning management system.)

With MathXL, instructors can:

- Create, edit, and assign online homework and tests using algorithmically-generated exercises correlated at the objective level to the textbook.
- Create and assign their own online exercises and import TestGen tests for added flexibility.
- Maintain records of all student work tracked in MathXL's online gradebook.

With MathXL, students can:

- Take chapter tests in MathXL and receive personalized study plans and/or personalized homework assignments based on their test results.
- Use the study plan and/or the homework to link directly to tutorial exercises for the objectives they need to study.
- Access supplemental animations and video clips directly from selected exercises.

MathXL is available to qualified adopters. For more information, visit our website at *www.mathxl.com,* or contact your Pearson representative.

TestGen®
TestGen® (*www.pearsoned.com/testgen*) enables instructors to build, edit, print, and administer tests using a computerized bank of questions developed to cover all the objectives of the text. TestGen is algorithmically based, allowing instructors to create multiple but equivalent versions of the same question or test with the click of a button. Instructors can also modify test bank questions or add new questions. The software and testbank are available for download from Pearson Education's online catalog.

PowerPoint® Lecture Slides

- Present key concepts and definitions from the text
- Available for download at *www.pearsonhighered.com* or in MyMathLab

ACKNOWLEDGMENTS

The comments, criticisms, and suggestions of users, nonusers, instructors, and students have positively shaped this textbook over the years, and we are most grateful for the many responses we have received. The feedback gathered for this revision of the text was particularly helpful, and we especially wish to thank the following individuals who provided invaluable suggestions for this and the previous edition:

Randall Allbritton, *Daytona State College*
Jannette Avery, *Monroe Community College*
Sarah E. Baxter, *Gloucester County College*
Linda Beattie, *Western New Mexico University*
Jean Bolyard, *Fairmont State College*
Tim C. Caldwell, *Meridian Community College*
Russell Campbell, *Fairmont State University*
Shawn Clift, *Eastern Kentucky University*
Bill Dunn, *Las Positas College*
Lucy Edwards, *Las Positas College*
Morris Elsen, *Cape Fear Community College*
J. Lloyd Harris, *Gulf Coast State College*
Terry Haynes, *Eastern Oklahoma State College*
Edith Hays, *Texas Woman's University*
Karen Heavin, *Morehead State University*
Christine Heinecke Lehmann, *Purdue University—North Central*
Elizabeth Heston, *Monroe Community College*
Sharon Jackson, *Brookhaven College*
Harriet Kiser, *Georgia Highlands College*
Valerie Lazzara, *Palm Beach State College*
Valerie H. Maley, *Cape Fear Community College*
Susan McClory, *San Jose State University*
Pam Miller, *Phoenix College*
Jeffrey Mills, *Ohio State University*
Linda J. Murphy, *Northern Essex Community College*
Celia Nippert, *Western Oklahoma State College*
Elizabeth Olgilvie, *Horry-Georgetown Technical College*
Enyinda Onunwor, *Stark State College*
Larry Pontaski, *Pueblo Community College*
Diann Robinson, *Ivy Tech State College—Lafayette*
Rachael Schettenhelm, *Southern Connecticut State University*
Jonathan Shands, *Cape Fear Community College*
Lee Ann Spahr, *Durham Technical Community College*
Carol Stewart, *Fairmont State University*
Fariheh Towfiq, *Palomar College*
Diane P. Veneziale, *Burlington County College*
Cora S. West, *Florida State College at Jacksonville*
Johanna Windmueller, *Seminole State College*
Gabriel Yimesghen, *Community College of Philadelphia*

Our sincere thanks go to the dedicated individuals at Pearson who have worked hard to make this revision a success: Maureen O'Connor, Kathy Manley, Barbara Atkinson, Michelle Renda, Rachel Ross, Kari Heen, Christine Whitlock, Lauren Morse, Stephanie Green, and Rachel Haskell.

Abby Tanenbaum did an excellent job updating the real-data applications, as well as helping us with manuscript preparation. We are also grateful to Marilyn Dwyer and Kathy Diamond of Cenveo/Nesbitt Graphics for their excellent production work; David Abel, for supplying his copyediting expertise; Beth Anderson, for her fine photo research; Lucie Haskins, for producing a useful index; Lisa Collette for checking the index; Jeff Cole, for writing the solutions manuals; and Janis Cimperman, Chris Heeren, Paul Lorczak, and Sarah Sponholz for timely accuracy checking of the manuscript and page proofs.

As an author team, we are committed to providing the best possible text and supplements package to help students succeed and instructors teach. As we continue to work toward this goal, we would welcome any comments or suggestions you might have via e-mail to *math@pearson.com*.

John Hornsby
Terry McGinnis

R Prealgebra Review

R.1 Fractions

OBJECTIVES

1. Identify prime numbers.
2. Write numbers in prime factored form.
3. Write fractions in lowest terms.
4. Convert between improper fractions and mixed numbers.
5. Multiply and divide fractions.
6. Add and subtract fractions.
7. Solve applied problems that involve fractions.
8. Interpret data in a circle graph.

The numbers used most often in everyday life are the **whole numbers,**

$$0, 1, 2, 3, 4, 5, \ldots$$

and **fractions,** such as

$$\frac{1}{3}, \quad \frac{5}{4}, \quad \text{and} \quad \frac{11}{12}.$$

The parts of a fraction are named as follows.

$$\text{Fraction bar} \rightarrow \frac{4}{7} \begin{matrix} \leftarrow \text{Numerator} \\ \leftarrow \text{Denominator} \end{matrix}$$

The fraction bar represents division $\left(\frac{a}{b} = a \div b\right)$.

A fraction is classified as being either a **proper fraction** or an **improper fraction.**

Proper fractions	$\frac{1}{5}, \frac{2}{7}, \frac{9}{10}, \frac{23}{25}$	Numerator is **less than** denominator. Value is less than 1.
Improper fractions	$\frac{3}{2}, \frac{5}{5}, \frac{11}{7}, \frac{28}{4}$	Numerator is **greater than or equal to** denominator. Value is greater than or equal to 1.

OBJECTIVE 1 **Identify prime numbers.** In work with fractions, we will need to write the numerators and denominators as products. A **product** is the answer to a multiplication problem. When 12 is written as the product 2×6, for example, 2 and 6 are **factors** of 12. Other factors of 12 are 1, 3, 4, and 12. A whole number is **prime** if it has exactly two different factors (itself and 1).

2, 3, 5, 7, 11, 13, 17, 19, 23, 29, 31, 37 First dozen prime numbers

A whole number greater than 1 that is not prime is a **composite number.**

4, 6, 8, 9, 10, 12, 14, 15, 16, 18, 20, 21 First dozen composite numbers

By agreement, the number 1 is neither prime nor composite.

Teaching Tip

Mention that if 1 were not excluded from the prime numbers, then *composite numbers* (those which are composed of prime factors) could be factored into primes in more than one way. This would violate the important fact that there is only one way to factor a composite number into primes.

EXAMPLE 1 **Distinguishing between Prime and Composite Numbers**

Decide whether each number is *prime* or *composite*.

(a) 33 Since 33 has factors of 3 and 11, as well as 1 and 33, it is composite.

(b) 43 Since there are no numbers other than 1 and 43 itself that divide *evenly* into 43, the number 43 is prime.

(c) 9832 Since 9832 can be divided by 2, giving 2×4916, it is composite.

◀ **Work Problem 1 at the Side.**

OBJECTIVE 2 **Write numbers in prime factored form.** We factor a number by writing it as the product of two or more numbers.

Multiplication	Factoring
$6 \cdot 3 = 18$	$18 = 6 \cdot 3$
↑ ↑ ↑	↑ ↑ ↑
Factors Product	Product Factors

Factoring is the reverse of multiplying two numbers to get the product.

In algebra, a dot $\cdot$ is used instead of the $\times$ symbol to indicate multiplication because $\times$ may be confused with the letter x. A composite number written using factors that are all prime numbers is in **prime factored form.**

EXAMPLE 2 **Writing Numbers in Prime Factored Form**

Write each number in prime factored form.

(a) 35 We factor 35 using the prime factors 5 and 7 as $35 = 5 \cdot 7$.

(b) 24 We use a factor tree, as shown below. The prime factors are circled.

Divide by the least prime factor of 24, which is 2. $24 = 2 \cdot 12$

Divide 12 by 2 to find two factors of 12. $24 = 2 \cdot 2 \cdot 6$

Now factor 6 as $2 \cdot 3$. $24 = \underbrace{2 \cdot 2 \cdot 2 \cdot 3}$

All factors are prime.

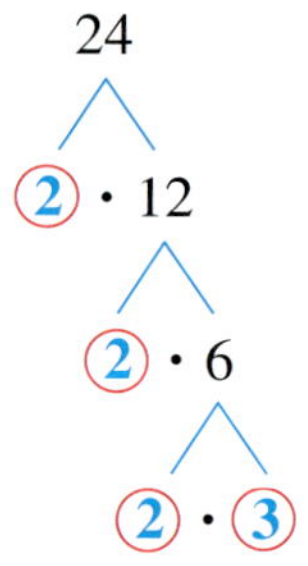

◀ **Work Problem 2 at the Side.**

Note

When factoring, we need not start with the least prime factor. No matter which prime factor we start with, we will *always* obtain the same prime factorization. Verify this in **Example 2(b)** by starting with 3 instead of 2.

OBJECTIVE 3 **Write fractions in lowest terms.** A fraction is in **lowest terms** when the numerator and denominator have no factors in common (other than 1). The following properties are useful.

Properties of 1

Any nonzero number divided by itself is equal to 1. For example, $\frac{3}{3} = 1$.

Any number multiplied by 1 remains the same. For example, $7 \cdot 1 = 7$.

1 Tell whether each number is *prime* or *composite*.

(a) 12

(b) 13

(c) 27

(d) 59

(e) 1806

2 Write each number in prime factored form.

(a) 70

(b) 72

(c) 693

(d) 97

Answers

1. **(a)** composite **(b)** prime **(c)** composite **(d)** prime **(e)** composite

2. **(a)** $2 \cdot 5 \cdot 7$ **(b)** $2 \cdot 2 \cdot 2 \cdot 3 \cdot 3$ **(c)** $3 \cdot 3 \cdot 7 \cdot 11$ **(d)** 97 is prime.

Writing a Fraction in Lowest Terms

Step 1 Write the numerator and denominator in factored form.

Step 2 Replace each pair of factors common to the numerator and denominator with 1.

Step 3 Multiply the remaining factors in the numerator and in the denominator.

(This procedure is sometimes called **"simplifying the fraction."**)

EXAMPLE 3 Writing Fractions in Lowest Terms

Write each fraction in lowest terms.

(a) $\frac{10}{15} = \frac{2 \cdot 5}{3 \cdot 5} = \frac{2}{3} \cdot \frac{5}{5} = \frac{2}{3} \cdot 1 = \frac{2}{3}$ Use the first property of 1 to replace $\frac{5}{5}$ with 1.

(b) $\frac{15}{45}$

By inspection, the greatest common factor of 15 and 45 is 15.

$$\frac{15}{45} = \frac{15}{3 \cdot 15} = \frac{1}{3 \cdot 1} = \frac{1}{3}$$ Remember to write 1 in the numerator.

If the greatest common factor is not obvious, factor the numerator and denominator into prime factors.

$$\frac{15}{45} = \frac{3 \cdot 5}{3 \cdot 3 \cdot 5} = \frac{1 \cdot 1}{3 \cdot 1 \cdot 1} = \frac{1}{3}$$ The same answer results.

(c) $\frac{150}{200} = \frac{3 \cdot 50}{4 \cdot 50} = \frac{3}{4} \cdot 1 = \frac{3}{4}$ $\frac{50}{50} = 1$

Note

When writing a fraction in lowest terms, look for the greatest common factor in the numerator and the denominator. If none is obvious, factor the numerator and the denominator into prime factors. *Any* common factor can be used and the fraction can be simplified in stages.

$$\frac{150}{200} = \frac{15 \cdot 10}{20 \cdot 10} = \frac{3 \cdot 5 \cdot 10}{4 \cdot 5 \cdot 10} = \frac{3}{4}$$ Example 3(c)

Work Problem 3 at the Side. ▶

OBJECTIVE 4 Convert between improper fractions and mixed numbers.

A **mixed number** is a single number that represents the sum of a natural number and a proper fraction.

$$\text{Mixed number} \longrightarrow 5\frac{3}{4} = 5 + \frac{3}{4}$$

Any improper fraction whose value is not a whole number can be rewritten as a mixed number, and any mixed number can be rewritten as an improper fraction.

3 Write each fraction in lowest terms.

(a) $\frac{8}{14}$

(b) $\frac{35}{42}$

(c) $\frac{72}{120}$

Teaching Tip

Students may ask, "Where does the 1 come from?" Remind them that when the numerator and the denominator are both divided by 15 (or 3 or 5), each result is 1. The phrase "cancel out" can be misleading.

Answers

3. (a) $\frac{4}{7}$ **(b)** $\frac{5}{6}$ **(c)** $\frac{3}{5}$

4 Write $\frac{92}{5}$ as a mixed number.

5 Write $11\frac{2}{3}$ as an improper fraction.

6 Find each product, and write it in lowest terms.

(a) $\frac{5}{8} \cdot \frac{2}{10}$

(b) $\frac{1}{10} \cdot \frac{12}{5}$

(c) $\frac{7}{9} \cdot \frac{12}{14}$

(d) $3\frac{1}{3} \cdot 1\frac{3}{4}$

Answers

4. $18\frac{2}{5}$

5. $\frac{35}{3}$

6. (a) $\frac{1}{8}$ (b) $\frac{6}{25}$ (c) $\frac{2}{3}$ (d) $\frac{35}{6}$, or $5\frac{5}{6}$

EXAMPLE 4 Converting an Improper Fraction to a Mixed Number

Write $\frac{59}{8}$ as a mixed number.

We divide the numerator of the improper fraction by the denominator.

	7	← Quotient
Denominator of fraction (divisor) →	$8\overline{)59}$	← Numerator of fraction (dividend)
	$\underline{56}$	
	3	← Remainder

$$\frac{59}{8} = 7\frac{3}{8}$$

◀ Work Problem **4** at the Side.

EXAMPLE 5 Converting a Mixed Number to an Improper Fraction

Write $6\frac{4}{7}$ as an improper fraction.

We multiply the denominator of the fraction by the whole number and add the numerator to get the numerator of the improper fraction.

$$7 \cdot 6 + 4 = 42 + 4 = 46$$

The denominator of the improper fraction is the same as the denominator in the mixed number, which is 7 here. Thus, $6\frac{4}{7} = \frac{46}{7}$.

◀ Work Problem **5** at the Side.

OBJECTIVE 5 Multiply and divide fractions.

Multiplying Fractions

To multiply two fractions, multiply the numerators to get the numerator of the product, and multiply the denominators to get the denominator of the product. ***The product should be written in lowest terms.***

EXAMPLE 6 Multiplying Fractions

Find each product, and write it in lowest terms.

(a) $\frac{3}{8} \cdot \frac{4}{9}$

$= \frac{3 \cdot 4}{8 \cdot 9}$ Multiply numerators. Multiply denominators.

$= \frac{3 \cdot 4}{2 \cdot 4 \cdot 3 \cdot 3}$ Factor the denominator.

$= \frac{1}{2 \cdot 3}$ $\frac{3}{3} = 1$ and $\frac{4}{4} = 1$; Remember to write 1 in the numerator.

$= \frac{1}{6}$ Write in lowest terms.

(b) $2\frac{1}{3} \cdot 5\frac{1}{2}$

$= \frac{7}{3} \cdot \frac{11}{2}$ Write each mixed number as an improper fraction.

$= \frac{77}{6}$, or $12\frac{5}{6}$ Multiply numerators and denominators. Write as a mixed number.

◀ Work Problem **6** at the Side.

Two fractions are **reciprocals** of each other if their product is 1. See the table in the margin. Because division is the opposite (or inverse) of multiplication, we use reciprocals to divide fractions.

Number	Reciprocal
$\frac{3}{4}$	$\frac{4}{3}$
$\frac{11}{7}$	$\frac{7}{11}$
$\frac{1}{5}$	5, or $\frac{5}{1}$
9, or $\frac{9}{1}$	$\frac{1}{9}$

A number and its reciprocal have a product of 1. For example,

$$\frac{3}{4} \cdot \frac{4}{3} = \frac{12}{12} = 1.$$

Dividing Fractions

To divide two fractions, multiply the first fraction by the reciprocal of the second. The result or **quotient** should be written in lowest terms.

As an example of why this works, we know that

$$20 \div 10 = 2 \quad \text{and also that} \quad 20 \cdot \frac{1}{10} = 2.$$

EXAMPLE 7 **Dividing Fractions**

Find each quotient, and write it in lowest terms.

(a) $\frac{3}{4} \div \frac{8}{5}$

$= \frac{3}{4} \cdot \frac{5}{8}$ Multiply by the reciprocal of the second fraction.

$= \frac{3 \cdot 5}{4 \cdot 8}$ Multiply numerators. Multiply denominators.

$= \frac{15}{32}$

(b) $\frac{3}{4} \div \frac{5}{8}$

$= \frac{3}{4} \cdot \frac{8}{5}$ Multiply by the reciprocal.

$= \frac{3 \cdot 4 \cdot 2}{4 \cdot 5}$ Multiply and factor.

$= \frac{6}{5}$, or $1\frac{1}{5}$

(c) $\frac{5}{8} \div 10$

$= \frac{5}{8} \cdot \frac{1}{10}$ Multiply by the reciprocal.

$= \frac{5 \cdot 1}{8 \cdot 2 \cdot 5}$ Multiply and factor.

$= \frac{1}{16}$ Remember to write 1 in the numerator.

(d) $1\frac{2}{3} \div 4\frac{1}{2}$

$= \frac{5}{3} \div \frac{9}{2}$ Write as improper fractions.

$= \frac{5}{3} \cdot \frac{2}{9}$ Multiply by the reciprocal of the second fraction.

$= \frac{10}{27}$ Multiply numerators and denominators.

Work Problem **7** at the Side. ▶

7 Find each quotient, and write it in lowest terms.

(a) $\frac{3}{10} \div \frac{2}{7}$

(b) $\frac{3}{4} \div \frac{7}{16}$

(c) $\frac{4}{3} \div 6$

(d) $3\frac{1}{4} \div 1\frac{2}{5}$

Answers

7. (a) $\frac{21}{20}$, or $1\frac{1}{20}$ **(b)** $\frac{12}{7}$, or $1\frac{5}{7}$ **(c)** $\frac{2}{9}$ **(d)** $\frac{65}{28}$, or $2\frac{9}{28}$

8 Add. Write sums in lowest terms.

(a) $\frac{3}{5} + \frac{4}{5}$

(b) $\frac{5}{14} + \frac{3}{14}$

Answers

8. **(a)** $\frac{7}{5}$, or $1\frac{2}{5}$ **(b)** $\frac{4}{7}$

OBJECTIVE 6 Add and subtract fractions. The result of adding two numbers is the **sum** of the numbers. For example, since $2 + 3 = 5$, the sum of 2 and 3 is 5.

Adding Fractions

To find the sum of two fractions with the *same* denominator, add their numerators and ***keep the same denominator.***

EXAMPLE 8 Adding Fractions with the Same Denominator

Add. Write sums in lowest terms.

(a) $\frac{3}{7} + \frac{2}{7}$

$= \frac{3+2}{7}$, or $\frac{5}{7}$ — Add numerators. Keep the same denominator.

(b) $\frac{2}{10} + \frac{3}{10}$

$= \frac{2+3}{10}$ — Add numerators. Keep the same denominator.

$= \frac{5}{10}$, or $\frac{1}{2}$ — Write in lowest terms.

◀ Work Problem 8 at the Side.

If the fractions to be added do not have the same denominator, we must first rewrite them with a common denominator. For example, to rewrite $\frac{3}{4}$ as a fraction with a denominator of 32, think as follows.

$$\frac{3}{4} = \frac{?}{32}$$

We must find the number that can be multiplied by 4 to give 32. Since $4 \cdot 8 = 32$, by the second property of 1, we multiply the numerator and the denominator by 8.

$$\frac{3}{4} = \frac{3}{4} \cdot 1 = \frac{3}{4} \cdot \frac{8}{8} = \frac{3 \cdot 8}{4 \cdot 8} = \frac{24}{32}$$

$\frac{3}{4}$ and $\frac{24}{32}$ are equivalent fractions.

This process is the reverse of writing a fraction in lowest terms.

Finding the Least Common Denominator (LCD)

Step 1 Factor all denominators to prime factored form.

Step 2 The LCD is the product of every (different) factor that appears in any of the factored denominators. If a factor is repeated, use the greatest number of repeats as factors of the LCD.

Step 3 Write each fraction with the LCD as the denominator, using the second property of 1.

EXAMPLE 9 **Adding Fractions with Different Denominators**

Add. Write sums in lowest terms.

(a) $\frac{4}{15} + \frac{5}{9}$

Step 1 To find the LCD, factor the denominators to prime factored form.

$$15 = 5 \cdot 3 \quad \text{and} \quad 9 = 3 \cdot 3$$

3 is a factor of both denominators.

Step 2

$$\text{LCD} = 5 \cdot 3 \cdot 3 = 45$$

(15 = 5 · 3 and 9 = 3 · 3)

In this example, the LCD needs one factor of 5 and two factors of 3 because the second denominator has two factors of 3.

Step 3 Now we can use the second property of 1 to write each fraction with 45 as the denominator.

$$\frac{4}{15} = \frac{4}{15} \cdot \frac{3}{3} = \frac{12}{45} \quad \text{and} \quad \frac{5}{9} = \frac{5}{9} \cdot \frac{5}{5} = \frac{25}{45}$$

At this stage, the fractions are *not* in lowest terms.

Now add the two equivalent fractions to get the sum.

$$\frac{4}{15} + \frac{5}{9}$$

$$= \frac{12}{45} + \frac{25}{45} \quad \text{Use a common denominator.}$$

$$= \frac{37}{45} \quad \text{The sum is in lowest terms.}$$

(b) $3\frac{1}{2} + 2\frac{3}{4}$

Method 1

$$3\frac{1}{2} + 2\frac{3}{4}$$

$$= \frac{7}{2} + \frac{11}{4} \quad \text{Write each mixed number as an improper fraction.}$$

$$= \frac{14}{4} + \frac{11}{4} \quad \text{Find a common denominator. The LCD is 4.}$$

Think: $\frac{7 \cdot 2}{2 \cdot 2} = \frac{14}{4}$

$$= \frac{25}{4}, \text{ or } 6\frac{1}{4} \quad \text{Add. Write as a mixed number.}$$

Method 2

$$3\frac{1}{2} = 3\frac{2}{4}$$
$$+\ 2\frac{3}{4} = 2\frac{3}{4}$$

Write $3\frac{1}{2}$ as $3\frac{2}{4}$. Then add vertically. Add the whole numbers and the fractions separately.

$$5\frac{5}{4} = 5 + 1\frac{1}{4} = 6\frac{1}{4}, \text{ or } \frac{25}{4}$$

The same answer results.

Work Problem 9 at the Side. ▶

9 Add. Write sums in lowest terms.

(a) $\frac{7}{30} + \frac{2}{45}$

(b) $\frac{17}{10} + \frac{8}{27}$

(c) $2\frac{1}{8} + 1\frac{2}{3}$

(d) $132\frac{4}{5} + 28\frac{3}{4}$

Teaching Tip

Warn students that $\frac{4}{15} + \frac{5}{9}$ is NOT EQUAL to $\frac{9}{24}$.

Answers

9. (a) $\frac{5}{18}$ **(b)** $\frac{539}{270}$, or $1\frac{269}{270}$ **(c)** $\frac{91}{24}$, or $3\frac{19}{24}$ **(d)** $161\frac{11}{20}$

10 Subtract.

(a) $\frac{9}{11} - \frac{3}{11}$

(b) $\frac{13}{15} - \frac{5}{6}$

(c) $2\frac{3}{8} - 1\frac{1}{2}$

(d) $50\frac{1}{4} - 32\frac{2}{3}$

The **difference** between two numbers is found by subtracting the numbers.

Subtracting Fractions

To find the difference between two fractions with the *same* denominator, subtract their numerators and ***keep the same denominator.***

If the fractions have *different* denominators, write them with a common denominator first.

EXAMPLE 10 Subtracting Fractions

Subtract. Write differences in lowest terms.

(a) $\frac{15}{8} - \frac{3}{8}$

$= \frac{15 - 3}{8}$ Subtract numerators. Keep the same denominator.

$= \frac{12}{8}$

$= \frac{3}{2}$, or $1\frac{1}{2}$ Write in lowest terms or as a mixed number.

(b) $\frac{15}{16} - \frac{4}{9}$

$= \frac{15}{16} \cdot \frac{9}{9} - \frac{4}{9} \cdot \frac{16}{16}$ Since 16 and 9 have no common factors greater than 1, the LCD is $16 \cdot 9 = 144$.

$= \frac{135}{144} - \frac{64}{144}$ Write equivalent fractions.

$= \frac{71}{144}$ Subtract numerators. Keep the common denominator.

(c) $4\frac{1}{2} - 1\frac{3}{4}$

Method 1 $4\frac{1}{2} - 1\frac{3}{4}$

$= \frac{9}{2} - \frac{7}{4}$ Write each mixed number as an improper fraction.

Think: $\frac{9 \cdot 2}{2 \cdot 2} = \frac{18}{4}$

$= \frac{18}{4} - \frac{7}{4}$ Find a common denominator. The LCD is 4.

$= \frac{11}{4}$, or $2\frac{3}{4}$ Subtract. Write as a mixed number.

Method 2

$4\frac{1}{2} = 4\frac{2}{4} = 3\frac{6}{4}$

$-\ 1\frac{3}{4} = 1\frac{3}{4} = 1\frac{3}{4}$

$2\frac{3}{4}$, or $\frac{11}{4}$

The LCD is 4. $4\frac{2}{4} = 3 + 1 + \frac{2}{4} = 3 + \frac{4}{4} + \frac{2}{4} = 3\frac{6}{4}$

The same answer results.

◀ Work Problem **10** at the Side.

Answers

10. **(a)** $\frac{6}{11}$ **(b)** $\frac{1}{30}$ **(c)** $\frac{7}{8}$ **(d)** $17\frac{7}{12}$

OBJECTIVE 7 Solve applied problems that involve fractions.

EXAMPLE 11 Solving an Applied Problem with Fractions

The diagram in **Figure 1** appears in the book *Woodworker's 39 Sure-Fire Projects*. Find the height of the desk to the top of the writing surface.

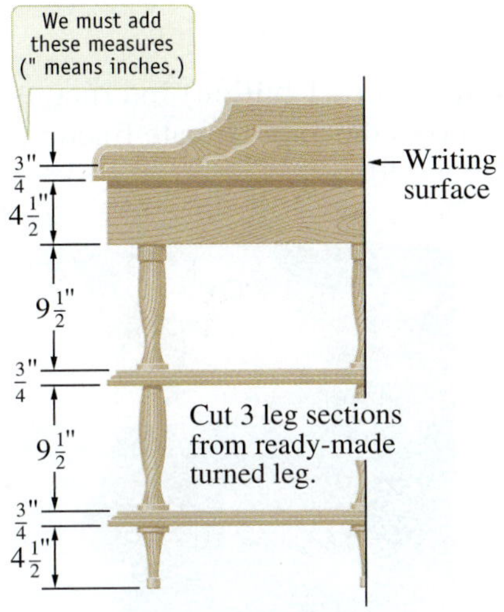

Figure 1

$$\begin{aligned}
\frac{3}{4} &\rightarrow \frac{3}{4}\\
4\frac{1}{2} &= 4\frac{2}{4}\\
9\frac{1}{2} &= 9\frac{2}{4}\\
\frac{3}{4} &\rightarrow \frac{3}{4}\\
9\frac{1}{2} &= 9\frac{2}{4}\\
\frac{3}{4} &\rightarrow \frac{3}{4}\\
+\ 4\frac{1}{2} &= 4\frac{2}{4}\\
\hline
&\ \ 26\frac{17}{4}
\end{aligned}$$

Use Method 2 from **Example 9(b).** The common denominator is 4.

Because $\frac{17}{4}$ is an improper fraction, this is not the final answer.

Since $\frac{17}{4} = 4\frac{1}{4}$, $26\frac{17}{4} = 26 + 4\frac{1}{4} = 30\frac{1}{4}$. The height is $30\frac{1}{4}$ in.

Work Problem 11 at the Side. ▶

EXAMPLE 12 Solving an Applied Problem with Fractions

An upholsterer needs $2\frac{1}{4}$ yd from a bolt of fabric to cover a chair. How many chairs can be covered with $23\frac{2}{3}$ yd of fabric?

To better understand the problem, we replace the fractions with whole numbers. Suppose each chair requires 2 yd, and we have 24 yd of fabric. Dividing 24 by 2 gives 12, the number of chairs that can be covered. To solve the original problem, we must divide $23\frac{2}{3}$ by $2\frac{1}{4}$.

$$23\frac{2}{3} \div 2\frac{1}{4}$$

$$= \frac{71}{3} \div \frac{9}{4} \quad \text{Convert the mixed numbers to improper fractions.}$$

$$= \frac{71}{3} \cdot \frac{4}{9} \quad \text{Multiply by the reciprocal.}$$

$$= \frac{284}{27}, \text{ or } 10\frac{14}{27} \quad \text{Multiply numerators and multiply denominators. Write as a mixed number.}$$

Thus, 10 chairs can be covered, with some fabric left over.

Work Problem 12 at the Side. ▶

11 Solve the problem.

To make a three-piece outfit from the same fabric, Wei Jen needs $1\frac{1}{4}$ yd for the blouse, $1\frac{2}{3}$ yd for the skirt, and $2\frac{1}{2}$ yd for the jacket. How much fabric does she need?

Teaching Tip

At times, the fractions in an application can be a distraction. It is helpful for some students to read problems through, omitting the fractional parts of the numbers, and decide which operations are appropriate to the solution. Then they can do the work using the original quantities.

12 Solve the problem.

A gallon of paint covers 500 ft^2. (ft^2 means square feet.) To paint his house, Tram needs enough paint to cover 4200 ft^2. How many gallons of paint should he buy?

Answers

11. $5\frac{5}{12}$ yd

12. $8\frac{2}{5}$ gal are needed, so he should buy 9 gal.

OBJECTIVE 8 **Interpret data in a circle graph.** In a **circle graph,** or **pie chart,** a circle is used to indicate the total of all the data categories represented. The circle is divided into *sectors*, or wedges, whose sizes show the relative magnitudes of the categories. The sum of all the fractional parts must be 1 (for 1 whole circle).

EXAMPLE 13 **Using a Circle Graph to Interpret Information**

In March 2011, there were about 2100 million (2.1 billion) Internet users worldwide. The circle graph in **Figure 2** shows the approximate fractions of these users living in various regions of the world.

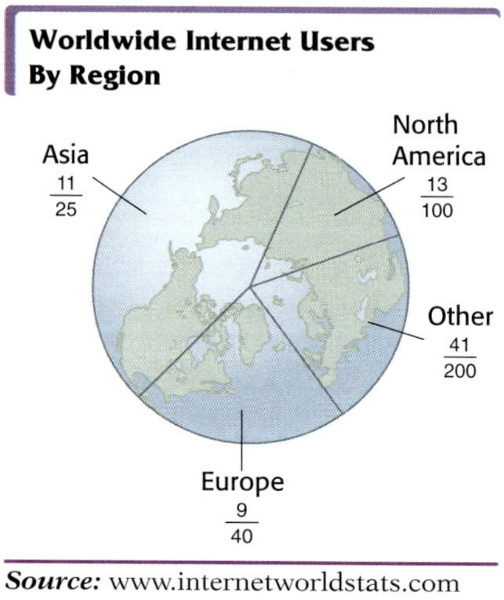

Source: www.internetworldstats.com

Figure 2

(a) Which region had the largest share of Internet users in March 2011? What was that share?

In the circle graph, the sector for Asia is the largest, so Asia had the largest share of Internet users, $\frac{11}{25}$.

(b) Estimate the number of Internet users in North America.

A share of $\frac{13}{100}$ can be rounded to $\frac{10}{100}$, or $\frac{1}{10}$, and the total number of Internet users, 2100 million, can be rounded to 2000 million (2 billion).

$$\frac{1}{10} \cdot 2000 = 200 \text{ million}$$ Multiply to estimate the number of Internet users in North America.

(c) How many actual Internet users were there in North America?

$$\frac{13}{100} \cdot 2100$$ Multiply the actual fraction from the graph for North America by the number of users.

$$= \frac{13}{100} \cdot \frac{2100}{1}$$ $a = \frac{a}{1}$ for all a.

$$= \frac{27{,}300}{100}$$ Multiply numerators and multiply denominators.

$$= 273$$ Divide.

Thus, 273 million, or 273,000,000 people in North America used the Internet.

◀ **Work Problem 13 at the Side.**

---

13 Refer to the circle graph in **Figure 2.**

(a) Which region had the second-largest number of Internet users in March 2011?

(b) Estimate the number of Internet users in Europe.

(c) How many actual Internet users were there in Europe?

Answers

13. **(a)** Europe **(b)** 500 million **(c)** 472.5 million

R.1 Exercises

FOR EXTRA HELP

Download the MyDashBoard App

MyMathLab®

CONCEPT CHECK *Decide whether each statement is* true *or* false. *If it is* false, *say why.*

1. In the fraction $\frac{3}{7}$, 3 is the numerator and 7 is the denominator.
   true

2. The mixed number equivalent of $\frac{41}{5}$ is $8\frac{1}{5}$.
   true

3. The fraction $\frac{7}{7}$ is proper.
   false; This is an improper fraction. Its value is 1.

4. The number 1 is prime.
   false; The number 1 is neither prime nor composite.

5. The fraction $\frac{17}{51}$ is in lowest terms.
   false; The fraction $\frac{17}{51}$ can be simplified to $\frac{1}{3}$.

6. The reciprocal of $\frac{8}{2}$ is $\frac{4}{1}$.
   false; The reciprocal of $\frac{8}{2} = 4$ is $\frac{2}{8} = \frac{1}{4}$.

7. The product of 8 and 2 is 10.
   false; *Product* indicates multiplication, so the product of 8 and 2 is 16.

8. The difference between 12 and 2 is 6.
   false; *Difference* indicates subtraction, so the difference between 12 and 2 is 10.

Identify each number as prime *or* composite. ***See Example 1.***

9. 19
   prime

10. 99
    composite

11. 52
    composite

12. 61
    prime

13. 2468
    composite

14. 3125
    composite

15. 97
    prime

16. 83
    prime

Write each number in prime factored form. ***See Example 2.***

17. 30
    $2 \cdot 3 \cdot 5$

18. 40
    $2 \cdot 2 \cdot 2 \cdot 5$

19. 252
    $2 \cdot 2 \cdot 3 \cdot 3 \cdot 7$

20. 168
    $2 \cdot 2 \cdot 2 \cdot 3 \cdot 7$

21. 124
    $2 \cdot 2 \cdot 31$

22. 165
    $3 \cdot 5 \cdot 11$

23. 29
    29

24. 31
    31

Write each fraction in lowest terms. ***See Example 3.***

25. $\frac{8}{16}$
    $\frac{1}{2}$

26. $\frac{4}{12}$
    $\frac{1}{3}$

27. $\frac{15}{18}$
    $\frac{5}{6}$

28. $\frac{16}{20}$
    $\frac{4}{5}$

29. $\frac{15}{75}$
    $\frac{1}{5}$

30. $\frac{24}{64}$
    $\frac{3}{8}$

31. $\frac{144}{120}$
    $\frac{6}{5}$

32. $\frac{132}{77}$
    $\frac{12}{7}$

Write each improper fraction as a mixed number. **See Example 4.**

33. $\frac{12}{7}$ $1\frac{5}{7}$

34. $\frac{28}{5}$ $5\frac{3}{5}$

35. $\frac{77}{12}$ $6\frac{5}{12}$

36. $\frac{101}{15}$ $6\frac{11}{15}$

37. $\frac{83}{11}$ $7\frac{6}{11}$

38. $\frac{67}{13}$ $5\frac{2}{13}$

Write each mixed number as an improper fraction. **See Example 5.**

39. $2\frac{3}{5}$ $\frac{13}{5}$

40. $5\frac{6}{7}$ $\frac{41}{7}$

41. $10\frac{3}{8}$ $\frac{83}{8}$

42. $12\frac{2}{3}$ $\frac{38}{3}$

43. $10\frac{1}{5}$ $\frac{51}{5}$

44. $18\frac{1}{6}$ $\frac{109}{6}$

45. CONCEPT CHECK For the fractions $\frac{p}{q}$ and $\frac{r}{s}$, which can serve as a common denominator? A

A. $q \cdot s$ **B.** $q + s$ **C.** $p \cdot r$ **D.** $p + r$

46. CONCEPT CHECK Which fraction is *not* equal to $\frac{5}{9}$? C

A. $\frac{15}{27}$ **B.** $\frac{30}{54}$ **C.** $\frac{40}{74}$ **D.** $\frac{55}{99}$

Find each product or quotient, and write it in lowest terms. **See Examples 6 and 7.**

47. $\frac{4}{5} \cdot \frac{6}{7}$ $\frac{24}{35}$

48. $\frac{5}{9} \cdot \frac{10}{7}$ $\frac{50}{63}$

49. $\frac{1}{10} \cdot \frac{12}{5}$ $\frac{6}{25}$

50. $\frac{6}{11} \cdot \frac{2}{3}$ $\frac{4}{11}$

51. $\frac{15}{4} \cdot \frac{8}{25}$ $\frac{6}{5}$, or $1\frac{1}{5}$

52. $\frac{4}{7} \cdot \frac{21}{8}$ $\frac{3}{2}$, or $1\frac{1}{2}$

53. $2\frac{2}{3} \cdot 5\frac{4}{5}$ $\frac{232}{15}$, or $15\frac{7}{15}$

54. $3\frac{3}{5} \cdot 7\frac{1}{6}$ $\frac{129}{5}$, or $25\frac{4}{5}$

55. $\frac{5}{4} \div \frac{3}{8}$ $\frac{10}{3}$, or $3\frac{1}{3}$

56. $\frac{7}{6} \div \frac{9}{10}$ $\frac{35}{27}$, or $1\frac{8}{27}$

57. $\frac{32}{5} \div \frac{8}{15}$ 12

58. $\frac{24}{7} \div \frac{6}{21}$ 12

59. $\frac{3}{4} \div 12$ $\frac{1}{16}$

60. $\frac{2}{5} \div 30$ $\frac{1}{75}$

61. $2\frac{5}{8} \div 1\frac{15}{32}$ $\frac{84}{47}$, or $1\frac{37}{47}$

62. $2\frac{3}{10} \div 7\frac{4}{5}$ $\frac{23}{78}$

Find each sum or difference, and write it in lowest terms. ***See Examples 8–10.***

63. $\frac{7}{12}+\frac{1}{12}$

$\frac{2}{3}$

64. $\frac{3}{16}+\frac{5}{16}$

$\frac{1}{2}$

65. $\frac{5}{9}+\frac{1}{3}$

$\frac{8}{9}$

66. $\frac{4}{15}+\frac{1}{5}$

$\frac{7}{15}$

67. $3\frac{1}{8}+\frac{1}{4}$

$\frac{27}{8}$, or $3\frac{3}{8}$

68. $5\frac{3}{4}+\frac{2}{3}$

$\frac{77}{12}$, or $6\frac{5}{12}$

69. $\frac{7}{12}-\frac{1}{9}$

$\frac{17}{36}$

70. $\frac{11}{16}-\frac{1}{12}$

$\frac{29}{48}$

71. $6\frac{1}{4}-5\frac{1}{3}$

$\frac{11}{12}$

72. $8\frac{4}{5}-7\frac{4}{9}$

$\frac{61}{45}$, or $1\frac{16}{45}$

73. $\frac{5}{3}+\frac{1}{6}-\frac{1}{2}$

$\frac{4}{3}$, or $1\frac{1}{3}$

74. $\frac{7}{15}+\frac{1}{6}-\frac{1}{10}$

$\frac{8}{15}$

Solve each applied problem. ***See Examples 11 and 12.***

Use the chart to answer the questions in Exercises 75 and 76.

Microwave		Stove Top		
Servings	**1**	**1**	**4**	**6**
Water	$\frac{3}{4}$ cup	1 cup	3 cups	4 cups
Grits	3 Tbsp	3 Tbsp	$\frac{3}{4}$ cup	1 cup
Salt (optional)	Dash	Dash	$\frac{1}{4}$ tsp	$\frac{1}{2}$ tsp

Source: Package of Quaker Quick Grits.

75. How many cups of water would be needed for eight microwave servings of Quaker Quick Grits?

6 cups

76. How many teaspoons of salt would be needed for five stove top servings? (*Hint:* 5 is halfway between 4 and 6.)

$\frac{3}{8}$ tsp

77. A piece of property has an irregular shape, with five sides as shown in the figure. Find the total distance around the piece of property. This is the **perimeter** of the figure.

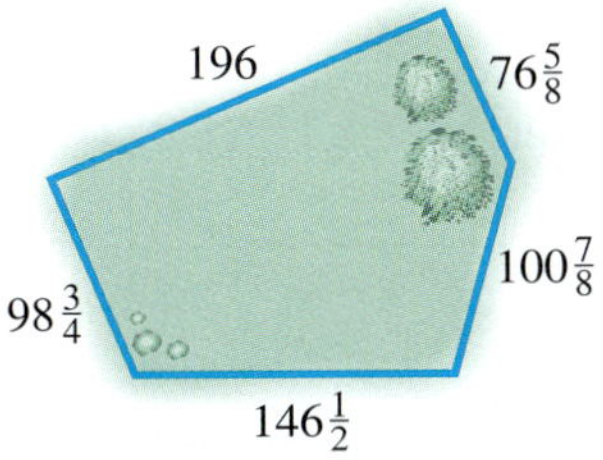

Measurements in feet

$618\frac{3}{4}$ ft

78. A triangle has sides of lengths $5\frac{1}{4}$ ft, $7\frac{1}{2}$ ft, and $10\frac{1}{8}$ ft. Find the perimeter of the triangle. **See Exercise 77.**

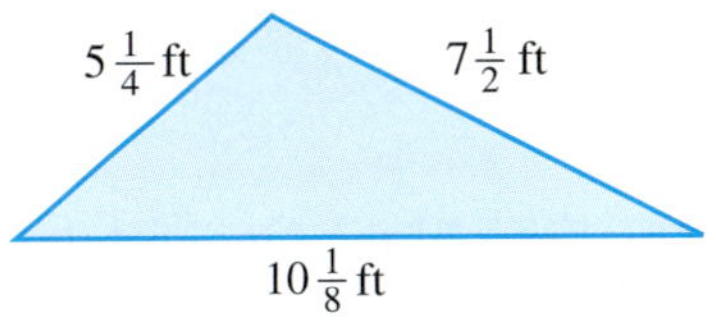

$22\frac{7}{8}$ ft

79. A hardware store sells a 40-piece socket wrench set. The measure of the largest socket is $\frac{3}{4}$ in., while the measure of the smallest socket is $\frac{3}{16}$ in. What is the difference between these measures?

$\frac{9}{16}$ in.

80. Two sockets in a socket wrench set have measures of $\frac{9}{16}$ in. and $\frac{3}{8}$ in. What is the difference between these two measures?

$\frac{3}{16}$ in.

81. Under current standards, adopted in 2001, most of the holes (called "eyes") in Grade A Swiss cheese must have a diameter between $\frac{3}{8}$ in. and $\frac{13}{16}$ in. How much less is $\frac{3}{8}$ in. than $\frac{13}{16}$ in.? (*Source*: U.S. Department of Agriculture.)

$\frac{7}{16}$ in.

82. The Pride Golf Tee Company, the only U.S. manufacturer of wooden golf tees, has created the Professional Tee System. Two lengths of tees are the ProLength Max and the Shortee, as shown in the figure. How much longer is the ProLength Max than the Shortee? (*Source: The Gazette.*)

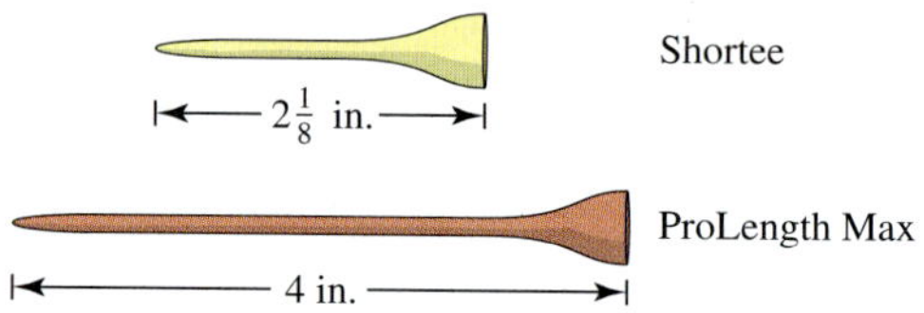

$1\frac{7}{8}$ in.

83. It takes $2\frac{3}{8}$ yd from a bolt of fabric to make a costume for a school play. How much fabric would be needed for seven costumes?

$16\frac{5}{8}$ yd

84. A cookie recipe calls for $2\frac{2}{3}$ cups of sugar. How much sugar would be needed to make four batches of cookies?

$10\frac{2}{3}$ cups

85. A cake recipe calls for $1\frac{3}{4}$ cups of sugar. A caterer has $15\frac{1}{2}$ cups of sugar on hand. How many cakes can he make?

8 cakes (There will be some sugar left over.)

86. Carlotta Valdez needs $2\frac{1}{4}$ yd of fabric to cover a chair. How many chairs can she cover with $23\frac{2}{3}$ yd of fabric?

10 chairs (There will be some fabric left over.)

Approximately 37.5 *million people living in the United States in 2010 were born in other countries. The circle graph gives the fractional number from each region of birth for these people. Use the graph to answer each question.* ***See Example 13.***

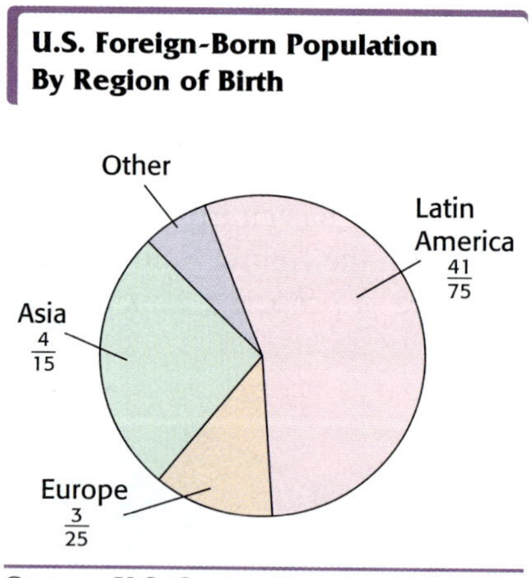

Source: U.S. Census Bureau.

87. What fractional part of the foreign-born population was from other regions?

$\frac{1}{15}$

88. What fractional part of the foreign-born population was from Latin America or Asia?

$\frac{61}{75}$

89. How many people (in millions) were born in Europe?

about $4\frac{1}{2}$ million, or 4,500,000

90. How many more people (in millions) were born in Latin America than in Asia?

about $10\frac{1}{2}$ million, or 10,500,000

R.2 Decimals and Percents

OBJECTIVES

1. Write decimals as fractions.
2. Add and subtract decimals.
3. Multiply and divide decimals.
4. Write fractions as decimals.
5. Convert percents to decimals and decimals to percents.
6. Use fraction, decimal, and percent equivalents.

Fractions are one way to represent parts of a whole. Another way is with a **decimal fraction** or **decimal,** a number written with a decimal point, such as 9.4. Each digit in a decimal number has a place value, as shown below.

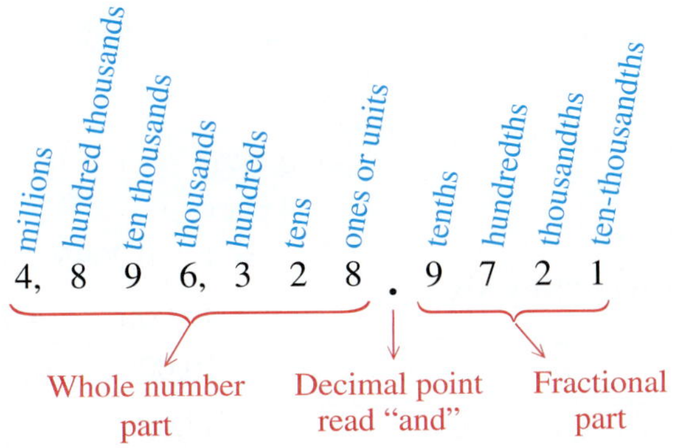

Each successive place value is ten times greater than the place value to its right and is one-tenth as great as the place value to its left.

Prices are often written as decimals. The price \$14.75 means 14 dollars and 75 cents, or 14 dollars and $\frac{75}{100}$ of a dollar.

OBJECTIVE 1 Write decimals as fractions. Place value is used to write a decimal number as a fraction. For example, since the last digit (that is, the digit farthest to the right) of 0.67 is in the *hundredths* place,

$$0.67 = \frac{67}{\mathbf{100}}.$$

Similarly, $0.9 = \frac{9}{10}$ and $0.25 = \frac{25}{100}$. Digits to the left of the decimal point indicate whole numbers, so 12.342 is the sum of 12 and 0.342.

$$12.342$$

$$= 12 + 0.342 \qquad \text{Write as the sum of a whole number and a decimal.}$$

$$= 12 + \frac{342}{\mathbf{1000}} \qquad \text{Write the decimal as a fraction.}$$

$$= \frac{12{,}000}{\mathbf{1000}} + \frac{342}{\mathbf{1000}} \qquad \text{Write with the common denominator 1000.}$$

$$= \frac{12{,}342}{\mathbf{1000}} \qquad \text{Add the numerators and keep the same denominator.}$$

These examples suggest the following rule.

Converting a Decimal to a Fraction

Read the name of the decimal using the correct place value. Write it in fraction form just as we read it. The denominator will be a **power of 10,** a number like 10, 100, 1000, and so on.

The same thing is accomplished by counting the number of digits to the right of the decimal point, and then writing the given number without a decimal point over a denominator of 1 followed by that number of zeros.

1 Write each decimal as a fraction. Do not write in lowest terms.

(a) 0.8

(b) 0.431

(c) 20.58

2 Add or subtract as indicated.

(a)
$$\begin{array}{r} 68.9 \\ 42.72 \\ +\ 8.973 \\ \hline \end{array}$$

(b)
$$\begin{array}{r} 32.5 \\ -\ 21.72 \\ \hline \end{array}$$

(c) 42.83 + 71 + 3.074

(d) 351.8 − 2.706

Answers

1. (a) $\frac{8}{10}$ (b) $\frac{431}{1000}$ (c) $\frac{2058}{100}$

2. (a) 120.593 (b) 10.78 (c) 116.904 (d) 349.094

EXAMPLE 1 Writing Decimals as Fractions

Write each decimal as a fraction. Do not write in lowest terms.

(a) 0.95

We read 0.95 as "ninety-five hundredths," so the fraction form is $\frac{95}{100}$. Using the shortcut method, since there are two places to the right of the decimal point, there will be two zeros in the denominator.

$$0.95 = \frac{95}{100}$$

2 places 2 zeros

(b) $0.056 = \frac{56}{1000}$

3 places 3 zeros

We read 0.056 as "fifty-six thousandths."

(c) 4.2095

4 places

$= 4 + 0.2095$

$= \frac{40{,}000}{10{,}000} + \frac{2095}{10{,}000}$ The LCD is 10,000.

$= \frac{42{,}095}{10{,}000}$

4 zeros

◀ Work Problem **1** at the Side.

OBJECTIVE 2 Add and subtract decimals.

EXAMPLE 2 Adding and Subtracting Decimals

Add or subtract as indicated.

(a) 6.92 + 14.8 + 3.217

Place the digits of the numbers in columns, so that tenths are in one column, hundredths in another column, and so on.

Be sure to line up decimal points.

$$\begin{array}{r} 6.92 \\ 14.8 \\ +\ 3.217 \\ \hline 24.937 \end{array}$$

Decimal points are aligned.

To avoid errors, attach zeros to make all the numbers the same length.

$$\begin{array}{r} 6.92 \\ 14.8 \\ +\ 3.217 \\ \hline \end{array} \quad \text{becomes} \quad \begin{array}{r} 6.920 \\ 14.800 \\ +\ 3.217 \\ \hline 24.937 \end{array}$$

Attach zeros.

(b) 47.6 − 32.509

$$\begin{array}{r} 47.6 \\ -\ 32.509 \\ \hline \end{array} \quad \text{becomes} \quad \begin{array}{r} 47.600 \\ -\ 32.509 \\ \hline 15.091 \end{array}$$

Write the numbers in columns, attaching zeros to 47.6.

(c) 3 − 0.253

$$\begin{array}{r} 3.000 \\ -\ 0.253 \\ \hline 2.747 \end{array}$$

A whole number is assumed to have the decimal point at the right of the number. Write 3 as 3.000.

◀ Work Problem **2** at the Side.

OBJECTIVE 3 Multiply and divide decimals.

Multiplying Decimals

Step 1 Ignore the decimal points and multiply as if the numbers were whole numbers.

Step 2 Add the number of **decimal places** (digits to the *right* of the decimal point) in each number being multiplied. Place the decimal point in the answer that many digits from the right.

EXAMPLE 3 Multiplying Decimals

Multiply.

(a) 29.3×4.52

$$\begin{array}{r} 29.3 \\ \times\ \ 4.52 \\ \hline 586 \\ 1465 \\ 1172 \\ \hline 132.436 \end{array}$$

1 decimal place in first number
2 decimal places in second number
$1 + 2 = 3$
3 decimal places in answer

(b) 7.003×55.8

$$\begin{array}{r} 7.003 \\ \times\ \ 55.8 \\ \hline 56024 \\ 35015 \\ 35015 \\ \hline 390.7674 \end{array}$$

3 decimal places
1 decimal place
$3 + 1 = 4$
4 decimal places

(c) 31.42×65

$$\begin{array}{r} 31.42 \\ \times\ \ \ \ 65 \\ \hline 15710 \\ 18852 \\ \hline 2042.30 \end{array}$$

2 decimal places
0 decimal places
$2 + 0 = 2$
2 decimal places

The final 0 can be dropped and the result can be expressed as 2042.3.

Work Problem 3 at the Side. ▶

To divide decimals, follow these steps.

Dividing Decimals

Step 1 Change the **divisor** (the number we are dividing *by*) into a whole number by moving the decimal point as many places as necessary to the right.

Step 2 Move the decimal point in the **dividend** (the number we are dividing *into*) to the right by the same number of places.

Step 3 Move the decimal point straight up and then divide as with whole numbers.

$$\begin{array}{r} 5 \\ 25\overline{)125} \end{array}$$

5 ⟵ Quotient
Divisor ⟶ 25
125 ↑ Dividend

Remember this terminology for the parts of a division problem.

---

3 Multiply.

(a) 2.13×0.05

(b) 9.32×1.4

(c) 300.2×0.052

(d) $42{,}001 \times 0.012$

Answers

3. **(a)** 0.1065 **(b)** 13.048 **(c)** 15.6104 **(d)** 504.012

4 Divide.

(a) $14.9\overline{)451.47}$

(b) $0.37\overline{)5.476}$

(c) $375.1 \div 3.001$
(Round the answer to two decimal places.)

Answers

4. (a) 30.3 **(b)** 14.8
(c) 124.99 (rounded)

EXAMPLE 4 Dividing Decimals

Divide.

(a) $233.45 \div 11.5$

Write the problem as follows.

$$11.5\overline{)233.45}$$

To change 11.5 into a whole number, move the decimal point one place to the right. Move the decimal point in 233.45 the *same* number of places to the right, to get 2334.5.

$$11.5\overline{)233.45}$$ Move each decimal point one place to the right.

To see why this works, write the division in fraction form and multiply by $\frac{10}{10}$, or 1.

$$\frac{233.45}{11.5} \cdot \frac{10}{10} = \frac{2334.5}{115}$$

The result is the same as when we moved the decimal point one place to the right in the divisor and the dividend.

Move the decimal point straight up and divide as with whole numbers.

$$\begin{array}{r} 20.3 \\ 115\overline{)2334.5} \\ \underline{230} \\ 345 \\ \underline{345} \\ 0 \end{array}$$ Move the decimal point straight up.

In the second step of the division, 115 does not divide into 34, so we used zero as a placeholder in the quotient.

(b) $73.85\overline{)1852.882}$ (Round the answer to two decimal places.)

Move the decimal point two places to the right in 73.85, to get 7385. Do the same thing with 1852.882, to get 185288.2.

$$73.85\overline{)1852.882}$$

Move the decimal point straight up and divide as with whole numbers.

$$\begin{array}{r} 25.089 \\ 7385\overline{)185288.200} \\ \underline{14770} \\ 37588 \\ \underline{36925} \\ 66320 \\ \underline{59080} \\ 72400 \\ \underline{66465} \\ 5935 \end{array}$$

We carried out the division to three decimal places so that we could round to two decimal places, obtaining the quotient 25.09.

◀ Work Problem **4** at the Side.

A shortcut can be used when multiplying or dividing by powers of 10.

Multiplying or Dividing by Powers of 10

To ***multiply*** by a power of 10, ***move the decimal point to the right*** as many places as the number of zeros.

To ***divide*** by a power of 10, ***move the decimal point to the left*** as many places as the number of zeros.

In both cases, insert 0s as placeholders if necessary.

EXAMPLE 5 Multiplying and Dividing by Powers of 10

Multiply or divide as indicated.

(a) 48.731×100
$= 48.73\,1$ or 4873.1

Move the decimal point two places to the right because 100 has two zeros.

(b) $48.7 \div 1000$
$= 048.7$ or 0.0487

Move the decimal point three places to the left because 1000 has three zeros. Insert a zero in front of the 4 to do this.

Work Problem 5 at the Side. ▶

To avoid misplacing the decimal point, check your work by estimating the answer. ***To get a quick estimate, round numbers so that only the first digit is not zero,*** using the rule for rounding. For more accurate estimates, numbers could be rounded to the first two or three nonzero digits.

Rule for Rounding

If the digit to become 0 or be dropped is 5 or more, round up by adding 1 to the final digit to be kept.

If the digit to become 0 or be dropped is 4 or less, do not round up.

For example, to estimate the answer to **Example 2(a)** from earlier in this section $(6.92 + 14.8 + 3.217)$, round as follows.

6.92 to **7**, 14.8 to **10**, and 3.217 to **3**
(5 or more; 4 or less; 4 or less)

Since $7 + 10 + 3 = 20$, our earlier answer of 24.937 is reasonable. In **Example 4(a),** round 233.45 to 200 and 11.5 to 10. Since $200 \div 10 = 20$, the answer of 20.3 is reasonable.

OBJECTIVE 4 Write fractions as decimals.

Writing a Fraction as a Decimal

Because a fraction bar indicates division, write a fraction as a decimal by dividing the numerator by the denominator.

5 Multiply or divide as indicated.

(a) 294.72×10

(b) 19.5×1000

(c) $4.793 \div 100$

(d) $960.1 \div 10$

Answers

5. **(a)** 2947.2 **(b)** 19,500 **(c)** 0.04793 **(d)** 96.01

6 Convert to decimals. For repeating decimals, write the answer two ways: using the bar notation and rounding to the nearest thousandth.

(a) $\frac{2}{9}$

(b) $\frac{17}{20}$

(c) $\frac{1}{11}$

Answers

6. **(a)** $0.\overline{2}$, 0.222 **(b)** 0.85 **(c)** $0.\overline{09}$, 0.091

EXAMPLE 6 Writing Fractions as Decimals

Write each fraction as a decimal.

(a) $\frac{19}{8}$

$$\begin{array}{r} 2.375 \\ 8\overline{)19.000} \\ \underline{16} \\ 3\,0 \\ \underline{2\,4} \\ 60 \\ \underline{56} \\ 40 \\ \underline{40} \\ 0 \end{array}$$

Divide 19 by 8. Add a decimal point and as many 0s as necessary.

$$\frac{19}{8} = 2.375$$

(b) $\frac{2}{3}$

$$\begin{array}{r} 0.6666\ldots \\ 3\overline{)2.0000\ldots} \\ \underline{1\,8} \\ 20 \\ \underline{18} \\ 20 \\ \underline{18} \\ 20 \end{array}$$

$$\frac{2}{3} = 0.6666\ldots$$

The remainder in the division in part (b) is never 0. Because 2 is always left after the subtraction, this quotient is a **repeating decimal.** A convenient notation for a repeating decimal is a bar over the digit (or digits) that repeats.

$$\frac{2}{3} = 0.6666\ldots, \quad \text{or} \quad 0.\overline{6}$$

We often round repeating decimals to as many places as needed.

$$\frac{2}{3} \approx 0.667$$ An approximation to the nearest thousandth ($\approx$ means "approximately equal to")

CAUTION

When rounding, be careful to distinguish between *thousandths* and *thousands* or between *hundredths* and *hundreds*, and so on.

◀ Work Problem **6** at the Side.

OBJECTIVE 5 Convert percents to decimals and decimals to percents. An important application of decimals is in work with percents. The word **percent** means "per one hundred." Percent is written with the symbol %. ***One percent means "one per one hundred," or "one one-hundredth."***

$$\mathbf{1\% = 0.01, \quad \text{or} \quad 1\% = \frac{1}{100}}$$

EXAMPLE 7 Converting Percents and Decimals

(a) Write 73% as a decimal.

Since $1\% = 0.01$, we convert as follows.

$$73\% = 73 \cdot 1\% = 73 \times 0.01 = 0.73$$

Also, 73% can be written as a decimal using the fraction form $1\% = \frac{1}{100}$.

$$73\% = 73 \cdot 1\% = 73 \cdot \frac{1}{100} = \frac{73}{100} = 0.73$$

Continued on Next Page

(b) Write 125% as a decimal.

$$125\% = 125 \cdot 1\% = 125 \times 0.01 = 1.25$$

A percent greater than 100 represents a number greater than 1.

(c) Write $3\frac{1}{2}\%$ as a decimal.
First write the fractional part as a decimal.

$$3\frac{1}{2}\% = (3 + 0.5)\% = 3.5\%$$

Now change the percent to decimal form.

$$3.5\% = 3.5 \times 0.01 = 0.035 \qquad 1\% = 0.01$$

(d) Write 0.32 as a percent.
Since 0.32 means 32 hundredths, write 0.32 as 32×0.01. Finally, replace 0.01 with 1%.

$$0.32 = 32 \times 0.01 = 32 \times 1\% = 32\%$$

(e) Write 2.63 as a percent.

$$2.63 = 263 \times 0.01 = 263 \times 1\% = 263\%$$

A number greater than 1 is more than 100%.

Note

A quick way to change from a percent to a decimal is to move the decimal point two places to the left and drop the % symbol. To change from a decimal to a percent, move the decimal point two places to the right and attach the percent symbol.

Decimal ← Divide by 100. / Move 2 places left. ← **Percent**

Decimal → Multiply by 100. / Move 2 places right. → **Percent**

EXAMPLE 8 **Converting Percents and Decimals by Moving the Decimal Point**

Convert each percent to a decimal and each decimal to a percent.

(a) 45% = 0.45 **(b)** 250% = 2.50

(c) 0.57 = 57% **(d)** 1.5 = 1.50 = 150%

(e) 0.327 = 32.7% **(f)** 0.007 = 0.007 = 0.7%

Work Problem **7** at the Side. ▶

In this book, we use 0 in the ones place for decimals between 0 and 1, such as 0.5 and 0.72. Some calculators, including graphing calculators, do *not* show 0 in the ones place. Either way is correct.

7 Convert as indicated.

(a) 23% to a decimal

Teaching Tip
Students are often confused by percents that are greater than 100%. Emphasize that 100% = 1 and that percents greater than 100% can be used to represent numbers greater than 1.

(b) 310% to a decimal

(c) 0.71 to a percent

(d) 1.32 to a percent

Teaching Tip
Many students are confused by percents less than 1%. They may not distinguish between 0.3 and 0.3%. Have students read these two numbers aloud to highlight how they differ.

(e) 0.685 to a percent

(f) 0.006 to a percent

Answers

7. **(a)** 0.23 **(b)** 3.10 **(c)** 71% **(d)** 132% **(e)** 68.5% **(f)** 0.6%

8 Solve each problem.

(a) Chris found a pair of jeans, regularly priced at \$50, on sale at 20% off. If he buys the jeans on sale, how much will he save? What is the sale price?

(b) Miguel was earning \$7.50 per hour on his job and received a $33\frac{1}{3}\%$ raise. How much was his hourly raise? What was his new hourly pay?

Answers

8. **(a)** \$10; \$40 **(b)** \$2.50; \$10.00

OBJECTIVE 6 Use fraction, decimal, and percent equivalents. The fraction $\frac{1}{2}$ can be written as the decimal 0.5, or as the percent 50%. ***These are three different ways of writing the same number.***

FRACTION, DECIMAL, AND PERCENT EQUIVALENTS

Fraction (or whole number)	Decimal	Percent
$\frac{1}{100}$	0.01	1%
$\frac{1}{50}$	0.02	2%
$\frac{1}{20}$	0.05	5%
$\frac{1}{10}$	0.1	10%
$\frac{1}{8}$	0.125	$12\frac{1}{2}\%$, or 12.5%
$\frac{1}{6}$	$0.1\overline{6}$	$16\frac{2}{3}\%$, or $16.\overline{6}\%$
$\frac{1}{5}$	0.2	20%
$\frac{1}{4}$	0.25	25%
$\frac{1}{3}$	$0.\overline{3}$	$33\frac{1}{3}\%$, or $33.\overline{3}\%$
$\frac{1}{2}$	0.5	50%
$\frac{2}{3}$	$0.\overline{6}$	$66\frac{2}{3}\%$, or $66.\overline{6}\%$
$\frac{3}{4}$	0.75	75%
1	1.0	100%

EXAMPLE 9 Using Percent and Fraction Equivalents

Solve each problem.

(a) Marissa bought a jacket with a regular price of \$80, on sale at 25% off. How much money will she save? What is the sale price?

From the table, $25\% = \frac{1}{4}$, and $\frac{1}{4}$ of \$80 is **\$20**, so she will save \$20.

\$80 − **\$20** = \$60 Original price − discount = sale price

(b) Caleb pays \$600 per month in rent. If his landlord raises the rent by 15%, how much more will he pay each month? What will his new rent be?

$$15\% = 10\% + 5\%$$

We know that $10\% = \frac{1}{10}$. Thus, 10%, or $\frac{1}{10}$, of \$600 is **\$60**. Also, 5% is half of 10%, so since 10% of \$600 is \$60, it follows that 5% of \$600 is half of \$60, or **\$30**. Caleb's new monthly rent will be \$600 plus these increases.

\$600 + **\$60** + **\$30** = \$690 New monthly rent

(\$60 ← 10%, \$30 ← 5%; 10% + 5% = 15%)

He will pay \$60 + \$30 = \$90 more each month.

◄ Work Problem **8** at the Side.

R.2 Exercises

FOR EXTRA HELP

Download the MyDashBoard App

MyMathLab®

CONCEPT CHECK *In Exercises 1–8, provide the correct response.*

1. In the decimal 367.9412, name the digit that has each place value.

(a) tens 6 **(b)** tenths 9 **(c)** thousandths 1

(d) ones or units 7 **(e)** hundredths 4

2. Write a number that has 5 in the thousands place, 0 in the tenths place, and 4 in the ten-thousandths place.
Answers will vary. One example is 5243.0164.

3. For the decimal number 46.249, round to the place value indicated.

(a) hundredths 46.25 **(b)** tenths 46.2

(c) ones or units 46 **(d)** tens 50

4. Round each decimal to the nearest thousandth.

(a) $0.\overline{8}$ 0.889 **(b)** $0.\overline{5}$ 0.556

(c) 0.9762 0.976 **(d)** 0.8642 0.864

5. For the sum 35.89 + 24.1, which is the best estimate? C

A. 40 **B.** 50 **C.** 60 **D.** 70

6. For the difference 109.83 − 32.4, which is the best estimate? D

A. 40 **B.** 50 **C.** 60 **D.** 70

7. For the product 84.9 × 98.3, which is the best estimate? B

A. 7000 **B.** 8000 **C.** 80,000 **D.** 70,000

8. For the quotient 9845.3 ÷ 97.2, which is the best estimate? C

A. 10 **B.** 1000 **C.** 100 **D.** 10,000

Write each decimal as a fraction. Do not write in lowest terms. ***See Example 1.***

9. 0.4 $\frac{4}{10}$ **10.** 0.6 $\frac{6}{10}$ **11.** 0.64 $\frac{64}{100}$ **12.** 0.82 $\frac{82}{100}$

13. 0.138 $\frac{138}{1000}$ **14.** 0.104 $\frac{104}{1000}$ **15.** 3.805 $\frac{3805}{1000}$ **16.** 5.166 $\frac{5166}{1000}$

Add or subtract as indicated. Make sure that your answer is reasonable by estimating first. Give your estimate, and then give the exact answer. ***See Example 2*** *and the rules for rounding in this section.*

17. 25.32 + 109.2 + 8.574
139; 143.094

18. 90.527 + 32.43 + 589.83 + 399.327
1120; 1112.114

19. 28.73 − 3.12
27; 25.61

20. 46.88 − 13.45
40; 33.43

21. 43.5 − 28.17
10; 15.33

22. 345.1 − 56.31
240; 288.79

23. 32.56 + 47.356 + 1.8
82; 81.716

24. 75.2 + 123.96 + 3.897
184; 203.057

25. 18 − 2.789
17; 15.211

26. 29 − 8.582
21; 20.418

Multiply or divide as indicated. Make sure that your answer is reasonable by estimating first. Give your estimate, and then give the exact answer. ***See Examples 3–5*** *and the rules for rounding in this section.*

27. 0.2 × 0.03
0.006; 0.006

28. 0.07 × 0.004
0.00028; 0.00028

29. 12.8 × 9.1
90; 116.48

30. 34.04 × 0.56
18; 19.0624

31. 78.65 ÷ 11
8; 7.15

32. 73.36 ÷ 14
7; 5.24

33. 19.967 ÷ 9.74
2; 2.05

34. 44.4788 ÷ 5.27
8; 8.44

35. 57.116×100 6000; 5711.6

36. 0.094×1000 100; 94

37. $1.62 \div 10$ 0.2; 0.162

38. $24.03 \div 100$ 0.2; 0.2403

Write each fraction as a decimal. For repeating decimals, write the answer two ways: using bar notation and rounding to the nearest thousandth. ***See Example 6.***

39. $\frac{1}{8}$ 0.125

40. $\frac{7}{8}$ 0.875

41. $\frac{5}{4}$ 1.25

42. $\frac{9}{5}$ 1.8

43. $\frac{5}{9}$ $0.\overline{5}$; 0.556

44. $\frac{8}{9}$ $0.\overline{8}$; 0.889

45. $\frac{1}{6}$ $0.1\overline{6}$; 0.167

46. $\frac{5}{6}$ $0.8\overline{3}$; 0.833

47. In your own words, explain how to convert a decimal to a percent.
To convert a decimal to a percent, move the decimal point two places to the right and attach a percent symbol (%).

48. In your own words, explain how to convert a percent to a decimal.
To convert a percent to a decimal, move the decimal point two places to the left and drop the percent symbol (%).

Convert each percent to a decimal. ***See Examples 7(a)–(c), 8(a), and 8(b).***

49. 54% 0.54

50. 39% 0.39

51. 117% 1.17

52. 189% 1.89

53. 2.4% 0.024

54. 3.1% 0.031

55. $6\frac{1}{4}\%$ 0.0625

56. $5\frac{1}{2}\%$ 0.055

57. 0.8% 0.008

58. 0.9% 0.009

Convert each decimal to a percent. ***See Examples 7(d), 7(e), and 8(c)–(f).***

59. 0.73 73%

60. 0.83 83%

61. 0.004 0.4%

62. 0.005 0.5%

63. 1.28 128%

64. 2.35 235%

65. 0.3 30%

66. 0.6 60%

67. 6 600%

68. 10 1000%

One method of converting a fraction to a percent is to first convert the fraction to a decimal, as shown in ***Example 6,*** *and then convert the decimal to a percent, as shown in* ***Examples 7 and 8.*** *Convert each fraction to a percent in this way.*

69. $\frac{4}{5}$ 80%

70. $\frac{3}{25}$ 12%

71. $\frac{2}{11}$ $18.\overline{18}\%$

72. $\frac{4}{9}$ $44.\overline{4}\%$

73. $\frac{7}{4}$ 175%

74. $\frac{11}{8}$ 137.5%

75. $\frac{13}{6}$ $216.\overline{6}\%$

76. $\frac{31}{9}$ $344.\overline{4}\%$

Solve each problem mentally. ***See Example 9.***

77. Liam, Shannon, Chris, and Kayla went out for dinner together. The total cost of their meals was \$75. If they decide to leave a 20% tip, how much should the tip be? \$15

78. Where Lan lives, a sales tax of 5% is added to all purchases. If he buys a Blu-ray player that costs \$80, how much will he pay for the Blu-ray player, including the tax? \$84

79. Nicole took an algebra test that was worth 50 points. If she earned 35 points on the test, what percent did she score? 70%

80. In a survey of 2400 eligible voters, 400 said that they planned to vote in an upcoming election. What percent of those surveyed planned to vote?
$16\frac{2}{3}\%$, or $16.\overline{6}\%$

Study Skills

USING YOUR MATH TEXTBOOK

Your textbook is a valuable resource. You will learn more if you fully make use of the features it offers.

General Features

Locate each feature, and complete the blanks as indicated.

- **Table of Contents** This is located at the front of the text. Find it and mark the chapters and sections you will cover, as noted on your course syllabus.
- **Answer Section** At the back of the book, answers to odd-numbered section exercises are provided. Answers to ALL Concept Check, writing, Relating Concepts, summary, chapter review, test, and cumulative review exercises are given. Tab this section so you can refer to it when doing homework.
- **Solution Section** At the back of the book, step-by-step solutions are provided to selected exercises that have a blue screen over the exercise number, such as **31**. In Section 1.1, Exercise _____ indicates a solution in the Solution Section. This solution is located on page _____.
- **List of Formulas** Inside the back cover of the text is a helpful list of geometric formulas, along with review information on triangles and angles. Use these for reference throughout the course. The formula for the volume of a cube is _______.

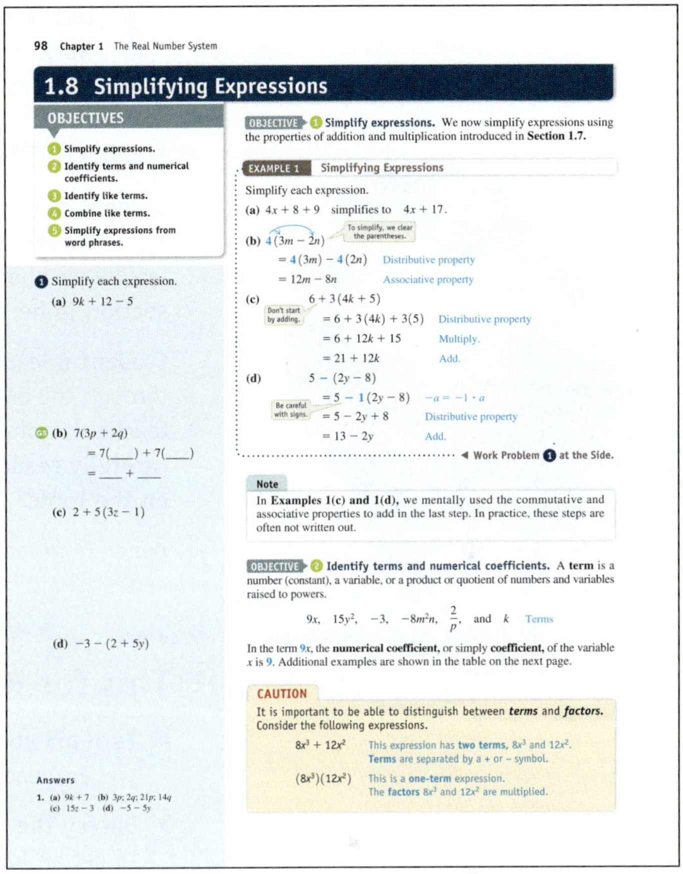

98 Chapter 1 The Real Number System

1.8 Simplifying Expressions

OBJECTIVES

1. Simplify expressions.
2. Identify terms and numerical coefficients.
3. Identify like terms.
4. Combine like terms.
5. Simplify expressions from word phrases.

OBJECTIVE 1 Simplify expressions. We now simplify expressions using the properties of addition and multiplication introduced in **Section 1.7.**

EXAMPLE 1 Simplifying Expressions

Simplify each expression.

(a) $4x + 8 + 9$ simplifies to $4x + 17$.

(b) $4(3m - 2n)$ — To simplify, we clear the parentheses.

$= 4(3m) - 4(2n)$ Distributive property

$= 12m - 8n$ Associative property

(c) $6 + 3(4k + 5)$ — Don't start by adding.

$= 6 + 3(4k) + 3(5)$ Distributive property

$= 6 + 12k + 15$ Multiply.

$= 21 + 12k$ Add.

(d) $5 - (2y - 8)$

$= 5 - 1(2y - 8)$ $-a = -1 \cdot a$

$= 5 - 2y + 8$ Distributive property — Be careful with signs.

$= 13 - 2y$ Add.

◀ Work Problem 1 at the Side.

1 Simplify each expression.

(a) $9k + 12 - 5$

(b) $7(3p + 2q)$
$= 7(\_\_\_) + 7(\_\_\_)$
$= \_\_\_ + \_\_\_$

(c) $2 + 5(3z - 1)$

(d) $-3 - (2 + 5y)$

Note

In **Examples 1(c) and 1(d),** we mentally used the commutative and associative properties to add in the last step. In practice, these steps are often not written out.

OBJECTIVE 2 Identify terms and numerical coefficients. A **term** is a number (constant), a variable, or a product or quotient of numbers and variables raised to powers.

$$9x, \quad 15y^2, \quad -3, \quad -8m^2n, \quad \frac{2}{p}, \quad \text{and} \quad k \qquad \text{Terms}$$

In the term $9x$, the **numerical coefficient,** or simply **coefficient,** of the variable x is 9. Additional examples are shown in the table on the next page.

CAUTION

It is important to be able to distinguish between ***terms*** and ***factors.*** Consider the following expressions.

$8x^3 + 12x^2$ This expression has **two terms,** $8x^3$ and $12x^2$. **Terms** are separated by a + or – symbol.

$(8x^3)(12x^2)$ This is a **one-term** expression. The **factors** $8x^3$ and $12x^2$ are multiplied.

Answers

1. (a) $9k + 7$ (b) $3p$; $2q$; $21p$; $14q$ (c) $15z - 3$ (d) $-5 - 5y$

Specific Features

- **Objectives** The objectives are listed at the beginning of each section and again within the section as the corresponding material is presented. Once you finish a section, ask yourself if you have accomplished them.
- **Margin Problems** These exercises allow you to immediately practice the material covered in the examples and prepare you for the exercise set. Check your results using the answers at the bottom of the page.
- **Pointers** These small shaded balloons provide on-the-spot warnings and reminders, point out key steps, and give other helpful tips.
- **Cautions** These provide warnings about common errors that students often make or trouble spots to avoid.
- **Notes** These provide additional explanations or emphasize important ideas.
- **Problem-Solving Hints** These give helpful tips or strategies to use when you work applications.
- **Calculator Tips** Marked with a calculator icon, these tips provide helpful information about using a calculator. A calculator icon beside an exercise is a recommendation to solve the exercise using a calculator.

Complete each of the following.

Write Objective 5 from Section 3.1. ________________

Write the answer to Section 1.7, Margin Problem 1(a). ____________

Write the information given by a pointer on page 30. ____________

List a page number for each of the following.

A Caution in Section 2.3 ________

A Note in Section 3.3 ________

A Problem-Solving Hint in Section 2.4 ________

A Calculator Tip in Section 1.1 ________

Study Skills

READING YOUR MATH TEXTBOOK

Take time to read each section and its examples before doing your homework. You will learn more and be better prepared to work the exercises your instructor assigns.

Approaches to Reading Your Math Textbook

Student A learns best by listening to her teacher explain things. She "gets it" when she sees the instructor work problems. She previews the section before the lecture, so she knows generally what to expect. **Student A carefully reads the section in her text *AFTER* she hears the classroom lecture on the topic.**

Student B learns best by reading on his own. He reads the section and works through the examples before coming to class. That way, he knows what the teacher is going to talk about and what questions he wants to ask. **Student B carefully reads the section in his text *BEFORE* he hears the classroom lecture on the topic.**

Which reading approach works best for you—that of Student A or Student B? ______

Tips for Reading Your Math Textbook

- **Turn off your cell phone.** You will be able to concentrate more fully on what you are reading.
- **Survey the material.** Take a few minutes to glance over the assigned material to get an idea of the "big picture." Look at the list of objectives to see what you will be learning.
- **Read slowly.** Read only one section—or even part of a section—at a sitting, with paper and pencil in hand.
- **Pay special attention to important information given in colored boxes or set in boldface type.** Highlight any additional information you find especially helpful.
- **Study the examples carefully.** Pay particular attention to the blue side comments and any pointers.
- **Do the margin problems in the workspace provided or on separate paper as you go.** These mirror the examples and prepare you for the exercise set. Check your answers with those given at the bottom of the page.
- **Make study cards as you read.** Make cards for new vocabulary, rules, procedures, formulas, and sample problems.
- **Mark anything you don't understand.** ***ASK QUESTIONS*** in class—everyone will benefit. Follow up with your instructor, as needed.

Mark two or three reading tips to try this week.

1 The Real Number System

Positive and *negative numbers*, shown here to indicate gains and losses, are examples of *real numbers,* the subject of this chapter.

1.1 Exponents, Order of Operations, and Inequality

OBJECTIVES

1. Use exponents.
2. Use the rules for order of operations.
3. Use more than one grouping symbol.
4. Know the meanings of $\neq$, $<$, $>$, $\leq$, and $\geq$.
5. Translate word statements to symbols.
6. Write statements that change the direction of inequality symbols.

OBJECTIVE 1 **Use exponents.** In **Chapter R,** we factored a number as the product of its prime factors. For example,

81 can be written as $3 \cdot 3 \cdot 3 \cdot 3$, $\quad$ $\cdot$ indicates multiplication.

where the factor 3 appears four times. Repeated factors are written in an abbreviated form by using an *exponent.*

$$\underbrace{3 \cdot 3 \cdot 3 \cdot 3}_{\text{4 factors of 3}} = 3^4$$

Exponent (4); Base (3)

The number 4 is the **exponent,** or **power,** and 3 is the **base** in the **exponential expression** 3^4. The exponent tells how many times the base is used as a factor. We read 3^4 as **"3 to the fourth power,"** or simply **"3 to the fourth."** ***A number raised to the first power is simply that number.*** For example,

$$6^1 = 6 \quad \text{and} \quad (2.5)^1 = 2.5. \qquad \text{In general, } a^1 = a.$$

EXAMPLE 1 **Evaluating Exponential Expressions**

Find the value of each exponential expression.

(a) 5^2 means $5 \cdot 5$, which equals 25.

5 is used as a factor 2 times.

Read 5^2 as "5 to the second power" or, more commonly, "5 squared."

(b) 6^3 means $6 \cdot 6 \cdot 6$, which equals 216.

6 is used as a factor 3 times.

Read 6^3 as "6 to the third power" or, more commonly, "6 cubed."

(c) 2^5 means $2 \cdot 2 \cdot 2 \cdot 2 \cdot 2$, which equals 32. $\quad$ 2 is used as a factor 5 times.

Read 2^5 as "2 to the fifth power."

(d) $\left(\frac{2}{3}\right)^3$ means $\frac{2}{3} \cdot \frac{2}{3} \cdot \frac{2}{3}$, which equals $\frac{8}{27}$. $\quad$ $\frac{2}{3}$ is used as a factor 3 times.

(e) $(0.3)^2$ means $0.3(0.3)$, which equals 0.09. $\quad$ 0.3 is used as a factor 2 times.

◀ Work Problem **1** at the Side.

CAUTION

Squaring, or raising a number to the second power, is not the same as doubling the number. **For example,**

$$3^2 \quad \text{means} \quad 3 \cdot 3, \quad \textit{not} \quad 2 \cdot 3.$$

Thus $3^2 = 9$, not 6. Similarly, cubing, or raising a number to the third power, does *not* mean tripling the number.

---

1 Find the value of each exponential expression.

GS **(a)** 6^2 means ____ $\cdot$ ____, which equals ____.

(b) 3^5

GS **(c)** $\left(\frac{3}{4}\right)^2$ means ____ $\cdot$ ____, which equals ____.

(d) $\left(\frac{1}{2}\right)^4$

GS **(e)** $(0.4)^3$ means ________, which equals ____.

Answers

1. (a) 6; 6; 36 **(b)** 243 **(c)** $\frac{3}{4}$; $\frac{3}{4}$; $\frac{9}{16}$ **(d)** $\frac{1}{16}$ **(e)** 0.4(0.4)(0.4); 0.064

OBJECTIVE 2 **Use the rules for order of operations.** When a problem involves more than one operation, we often use **grouping symbols.** If no grouping symbols are used, we apply the rules for order of operations.

Consider the expression $5 + 2 \cdot 3$. To show that the multiplication should be performed before the addition, we use parentheses to group $2 \cdot 3$.

$$5 + (2 \cdot 3) \quad \text{equals} \quad 5 + 6, \quad \text{or} \quad 11.$$

If addition is to be performed first, the parentheses should group $5 + 2$.

$$(5 + 2) \cdot 3 \quad \text{equals} \quad 7 \cdot 3, \quad \text{or} \quad 21.$$

Other grouping symbols are brackets [ ], braces { }, and fraction bars. (For example, in $\frac{8-2}{3}$, the expression $8 - 2$ is considered to be grouped in the numerator.)

To work problems with more than one operation, use the following rules for **order of operations.** This order is used by most calculators and computers.

Order of Operations

If grouping symbols are present, simplify within them, innermost first (and above and below fraction bars separately), in the following order.

Step 1 Apply all **exponents.**

Step 2 Do any **multiplications** or **divisions** in the order in which they occur, working from left to right.

Step 3 Do any **additions** or **subtractions** in the order in which they occur, working from left to right.

If no grouping symbols are present, start with Step 1.

Teaching Tip
Using a memory device such as "Please Excuse My Dear Aunt Sally" helps students remember the rules for order of operations: Parentheses, Exponents, Multiply, Divide, Add, Subtract.

They must remember that multiplications and divisions should be done from left to right in the order they appear. Likewise, additions and subtractions are done from left to right, in the order they appear.

Another way to show multiplication besides with a raised dot is with parentheses.

$$3(7) \text{ means } 3 \cdot 7, \quad \text{or} \quad 21.$$

$$3(4 + 5) \quad \text{means} \quad \text{"3 times the sum of 4 and 5."}$$

When simplifying $3(4 + 5)$, the sum in parentheses must be found first, then the product.

EXAMPLE 2 Using the Rules for Order of Operations

Find the value of each expression.

(a) $24 - 12 \div 3$ (Be careful. Divide first.)

$= 24 - 4$ Divide.

$= 20$ Subtract.

(b) $9(6 + 11)$

$= 9(17)$ Work inside parentheses.

$= 153$ Multiply.

(c) $6 \cdot 8 + 5 \cdot 2$

$= 48 + 10$ Multiply, working from left to right.

$= 58$ Add.

Continued on Next Page

Teaching Tip
Have students label the order in which they will evaluate the operations in each expression in Example 2 before actually carrying out the operations. For example,

$4 + 5 \cdot 6$
② ①

and $6 \cdot 8 + 5 \cdot 2.$
① ③ ②

Teaching Tip
Show why

$$100 \div 2 \times 50 = 2500.$$

The common incorrect answer is 1.

2 Label the order in which each expression should be evaluated. Then find the value of each expression.

(a) $7 + 3 \cdot 8$

② ①

$= 7 + \_\_\_\_$

$= \_\_\_\_$

(b) $2 \cdot 9 + 7 \cdot 3$

○ ○ ○

$= \_\_\_\_ + \_\_\_\_$

$= \_\_\_\_$

(c) $7 \cdot 6 - 3(8 + 1)$

○ ○○ ○

(d) $2 + 3^2 - 5$

○○○

3 Find the value of each expression.

(a) $9[(4 + 8) - 3]$

(b) $\dfrac{2(7 + 8) + 2}{3 \cdot 5 + 1}$

Answers

2. **(a)** 24; 31 **(b)** ①, ③, ②; 18; 21; 39
(c) ②, ④, ③, ①; 15
(d) ②, ①, ③; 6

3. **(a)** 81 **(b)** 2

(d) $2(5 + 6) + 7 \cdot 3$

Start here.

$= 2(11) + 7 \cdot 3$ Work inside parentheses.

$= 22 + 21$ Multiply.

$= 43$ Add.

(e) $9 + 2^3 - 5$ ($2^3 = 2 \cdot 2 \cdot 2$, not $2 \cdot 3$)

$= 9 + 8 - 5$ Apply the exponent.

$= 12$ Add, and then subtract.

◀ Work Problem **2** at the Side.

OBJECTIVE 3 **Use more than one grouping symbol.** In an expression such as $2(8 + 3(6 + 5))$, we often use brackets, [ ], in place of the outer pair of parentheses.

EXAMPLE 3 Using Brackets and Fraction Bars as Grouping Symbols

Find the value of each expression.

Start here.

(a) $2[8 + 3(6 + 5)]$

$= 2[8 + 3(11)]$ Add inside parentheses.

$= 2[8 + 33]$ Multiply inside brackets.

$= 2[41]$ Add inside brackets.

$= 82$ Multiply.

(b) $\dfrac{4(5 + 3) + 3}{2(3) - 1}$ Simplify the numerator and denominator separately.

$= \dfrac{4(8) + 3}{2(3) - 1}$ Work inside parentheses.

$= \dfrac{32 + 3}{6 - 1}$ Multiply.

$= \dfrac{35}{5}$ Add and subtract.

$= 7$ Divide.

◀ Work Problem **3** at the Side.

Note

The expression $\frac{4(5 + 3) + 3}{2(3) - 1}$ in **Example 3(b)** can be written as a quotient.

$$[4(5 + 3) + 3] \div [2(3) - 1]$$

The fraction bar "groups" the numerator and denominator separately.

Calculator Tip

Calculators follow the order of operations. Try some of the examples to see that your calculator gives the same answers. Use the parentheses keys to insert parentheses where they are needed, such as around the numerator and the denominator in **Example 3(b).**

OBJECTIVE 4 Know the meanings of $\neq$, $<$, $>$, $\leq$, and $\geq$. So far, we have used only the symbols of arithmetic, such as $+$, $-$, $\cdot$, and $\div$ and the equality symbol $=$. The equality symbol with a slash through it means "is *not* equal to."

$$7 \neq 8 \quad \text{7 is not equal to 8.}$$

If two numbers are not equal, then one of the numbers must be less than the other. The symbol $<$ represents "is less than."

$$7 < 8 \quad \text{7 is less than 8.}$$

The symbol $>$ means "is greater than."

$$8 > 2 \quad \text{8 is greater than 2.}$$

To keep the meanings of the symbols $<$ and $>$ clear, remember that the symbol always points to the lesser number.

Lesser number $\longrightarrow$ **8** < 15

$15 >$ **8** $\longleftarrow$ Lesser number

Work Problem 4 at the Side. ▶

The symbol $\leq$ means "is less than or equal to."

$$5 \leq 9 \quad \text{5 is less than or equal to 9.}$$

If either the $<$ part or the $=$ part is true, then the inequality $\leq$ is true. The statement $5 \leq 9$ is true. Also, $8 \leq 8$ is true because $8 = 8$ is true.

The symbol $\geq$ means "is greater than or equal to."

$$9 \geq 5 \quad \text{9 is greater than or equal to 5.}$$

EXAMPLE 4 Using the Symbols $\leq$ and $\geq$

Determine whether each statement is *true* or *false*.

(a) $15 \leq 20$ The statement $15 \leq 20$ is true because $15 < 20$.

(b) $12 \geq 12$ Since $12 = 12$, this statement is true.

(c) $15 \leq 20 \cdot 2$ The statement $15 \leq 20 \cdot 2$ is true, because $15 < 40$.

(d) $\dfrac{6}{15} \geq \dfrac{2}{3}$

$$\frac{6}{15} \geq \frac{10}{15} \quad \text{Find a common denominator.}$$

The statements $\frac{6}{15} > \frac{10}{15}$ and $\frac{6}{15} = \frac{10}{15}$ are false. Because at least one of them is false, $\frac{6}{15} \geq \frac{2}{3}$ is also false.

Work Problem 5 at the Side. ▶

4 Write each statement in words. Then decide whether it is *true* or *false*.

(a) $7 < 5$

(b) $12 > 6$

(c) $4 \neq 10$

5 Determine whether each statement is *true* or *false*.

(a) $30 \leq 40$

(b) $25 \geq 10$

(c) $40 \leq 10$

(d) $21 \leq 21$

(e) $9 \cdot 3 \geq 28$

(f) $\dfrac{4}{7} \leq \dfrac{5}{8}$

Answers

4. **(a)** Seven is less than five. False
(b) Twelve is greater than six. True
(c) Four is not equal to ten. True

5. **(a)** true **(b)** true **(c)** false
(d) true **(e)** false **(f)** true

OBJECTIVE 5 Translate word statements to symbols.

EXAMPLE 5 Translating from Words to Symbols

Write each word statement in symbols.

(a) Twelve **is equal to** ten **plus** two. $12 = 10 + 2$

(b) Nine **is less than** ten. $9 < 10$
Compare this with "9 less than 10," which is written $10 - 9$.

(c) Fifteen **is not equal to** eighteen. $15 \neq 18$

(d) Seven **is greater than** four. $7 > 4$

(e) Thirteen **is less than or equal to** forty. $13 \leq 40$

(f) Six **is greater than or equal to** six. $6 \geq 6$

◀ **Work Problem 6 at the Side.**

OBJECTIVE 6 Write statements that change the direction of inequality symbols. Any statement with $<$ can be converted to one with $>$, and any statement with $>$ can be converted to one with $<$. ***We do this by reversing both the order of the numbers and the direction of the symbol.***

Interchange numbers.

$6 < 10$ becomes $10 > 6$.

Reverse symbol.

EXAMPLE 6 Converting between Inequality Symbols

Parts (a)–(c) show the same statements written in two equally correct ways. In each inequality, the symbol points toward the lesser number.

(a) $5 > 2$, $2 < 5$ **(b)** $3 \leq 8$, $8 \geq 3$ **(c)** $12 \geq 5$, $5 \leq 12$

◀ **Work Problem 7 at the Side.**

Symbol	Meaning	Example
$=$	Is equal to	$0.5 = \frac{1}{2}$ means 0.5 is equal to $\frac{1}{2}$.
$\neq$	Is not equal to	$3 \neq 7$ means 3 is not equal to 7.
$<$	Is less than	$6 < 10$ means 6 is less than 10.
$>$	Is greater than	$15 > 14$ means 15 is greater than 14.
$\leq$	Is less than or equal to	$4 \leq 8$ means 4 is less than or equal to 8.
$\geq$	Is greater than or equal to	$1 \geq 0$ means 1 is greater than or equal to 0.

CAUTION

Equality and inequality symbols are used to write mathematical ***sentences,*** while operation symbols ($+$, $-$, $\cdot$, and $\div$) are used to write mathematical ***expressions*** that represent a number. Compare the following.

Sentence: $4 < 10$ ← Gives the relationship between 4 and 10

Expression: $4 + 10$ ← Tells how to operate on 4 and 10 to get 14

---

6 Write each word statement in symbols.

(a) Nine is equal to eleven minus two.

(b) Seventeen is less than thirty.

(c) Eight is not equal to ten.

(d) Fourteen is greater than twelve.

(e) Thirty is less than or equal to fifty.

(f) Two is greater than or equal to two.

7 Write each statement as another true statement with the inequality symbol reversed.

(a) $8 < 10$

(b) $3 > 1$

(c) $9 \leq 15$

(d) $6 \geq 2$

Answers

6. (a) $9 = 11 - 2$ **(b)** $17 < 30$
(c) $8 \neq 10$ **(d)** $14 > 12$
(e) $30 \leq 50$ **(f)** $2 \geq 2$

7. (a) $10 > 8$ **(b)** $1 < 3$ **(c)** $15 \geq 9$
(d) $2 \leq 6$

1.1 Exercises

FOR EXTRA HELP

Download the MyDashBoard App

MyMathLab®

CONCEPT CHECK *Decide whether each statement is* true *or* false. *If it is* false, *explain why.*

1. An exponent tells how many times its base is used as a factor.
   true

2. Some grouping symbols are +, −, •, and ÷.
   false; +, −, •, and ÷ are operation symbols.

3. When evaluated, $4 + 3(8 - 2)$ is equal to 42.
   false; Using the rules for order of operations gives $4 + 3(8 - 2) = 4 + 3(6) = 4 + 18 = 22$.

4. $3^3 = 9$
   false; $3^3 = 3 \cdot 3 \cdot 3 = 27$

5. The statement "4 is 12 less than 16" is interpreted $4 = 12 - 16$.
   false; The correct translation is $4 = 16 - 12$.

6. The statement "6 is 4 less than 10" is interpreted $6 < 10 - 4$.
   false; The correct translation is $6 = 10 - 4$.

7. **CONCEPT CHECK** When evaluating $(4^2 + 3^3)^4$, what is the *last* exponent that would be applied? Why?
   The 4 would be applied last because we work first inside the parentheses.

8. **CONCEPT CHECK** Which are not grouping symbols—parentheses, brackets, fraction bars, exponents?
   exponents

Find the value of each exponential expression. ***See Example 1.***

9. 7^2 49
10. 4^2 16
11. 12^2 144
12. 14^2 196
13. 4^3 64
14. 5^3 125
15. 10^3 1000
16. 11^3 1331
17. 3^4 81
18. 6^4 1296
19. 4^5 1024
20. 3^5 243
21. $\left(\frac{2}{3}\right)^4$ $\frac{16}{81}$
22. $\left(\frac{3}{4}\right)^3$ $\frac{27}{64}$
23. $(0.04)^3$ 0.000064
24. $(0.05)^4$ 0.00000625

Find the value of each expression. ***See Examples 2 and 3.***

25. $13 + 9 \cdot 5$ (2) (1)
    $= 13 + \underline{45}$
    $= \underline{58}$

26. $11 + 7 \cdot 6$ (2) (1)
    $= 11 + \underline{42}$
    $= \underline{53}$

27. $20 - 4 \cdot 3 + 5$ (2) (1) (3)
    $= 20 - \underline{12} + 5$
    $= \underline{8} + 5$
    $= \underline{13}$

28. $18 - 7 \cdot 2 + 6$ (2) (1) (3)
    $= 18 - \underline{14} + 6$
    $= \underline{4} + 6$
    $= \underline{10}$

29. $9 \cdot 5 - 13$ (1) (2) 32
30. $7 \cdot 6 - 11$ (1) (2) 31
31. $18 - 2 + 3$ (1) (2) 19
32. $22 - 8 + 9$ (1) (2) 23
33. $\frac{1}{4} \cdot \frac{2}{3} + \frac{2}{5} \cdot \frac{11}{3}$ $\frac{49}{30}$
34. $\frac{9}{4} \cdot \frac{2}{3} + \frac{4}{5} \cdot \frac{5}{3}$ $\frac{17}{6}$
35. $9 \cdot 4 - 8 \cdot 3$ 12
36. $11 \cdot 4 + 10 \cdot 3$ 74
37. $2.5(1.9) + 4.3(7.3)$ 36.14
38. $4.3(1.2) + 2.1(8.5)$ 23.01
39. $10 + 40 \div 5 \cdot 2$ 26
40. $12 + 35 \div 7 \cdot 3$ 27
41. $18 - 2(3 + 4)$ 4
42. $30 - 3(4 + 2)$ 12

43. $5[3+4(2^2)]$ 95

44. $4^2[(13+4)-8]$ 144

45. $\left(\frac{3}{2}\right)^2\left[\left(11+\frac{1}{3}\right)-6\right]$ 12

46. $6\left[\frac{3}{4}+8\left(\frac{1}{2}\right)^3\right]$ $\frac{21}{2}$

47. $\frac{8+6(3^2-1)}{3\cdot 2-2}$ 14

48. $\frac{8+2(8^2-4)}{4\cdot 3-10}$ 64

49. $\frac{4(7+2)+8(8-3)}{6(4-2)-2^2}$ $\frac{19}{2}$

50. $\frac{6(5+1)-9(1+1)}{5(8-4)-2^3}$ $\frac{3}{2}$

CONCEPT CHECK *Insert one pair of parentheses so that the left side of each equation is equal to the right side.*

51. $3\cdot 6+4\cdot 2=60$
$3\cdot(6+4)\cdot 2=60$

52. $2\cdot 8-1\cdot 3=42$
$2\cdot(8-1)\cdot 3=42$

53. $10-7-3=6$
$10-(7-3)=6$

54. $15-10-2=7$
$15-(10-2)=7$

55. $8+2^2=100$
$(8+2)^2=100$

56. $4+2^2=36$
$(4+2)^2=36$

Tell whether each statement is true *or* false. *In Exercises 59–68, first simplify each expression involving an operation.* ***See Example 4.***

57. $8\geq 17$ false

58. $10\geq 41$ false

59. $17\leq 18-1$ true

60. $12\geq 10+2$ true

61. $6\cdot 8+6\cdot 6\geq 0$ true

62. $4\cdot 20-16\cdot 5\geq 0$ true

63. $6[5+3(4+2)]\leq 70$ false

64. $6[2+3(2+5)]\leq 135$ false

65. $\frac{9(7-1)-8\cdot 2}{4(6-1)}>3$ false

66. $\frac{2(5+3)+2\cdot 2}{2(4-1)}>1$ true

67. $8\leq 4^2-2^2$ true

68. $10^2-8^2>6^2$ false

Write each word statement in symbols. ***See Example 5.***

69. Fifteen is equal to five plus ten. $15=5+10$

70. Twelve is equal to twenty minus eight. $12=20-8$

71. Nine is greater than five minus four. $9>5-4$

72. Ten is greater than six plus one. $10>6+1$

73. Sixteen is not equal to nineteen. $16\neq 19$

74. Three is not equal to four. $3\neq 4$

75. Two is less than or equal to three. $2\leq 3$

76. Five is less than or equal to nine. $5\leq 9$

Write each statement in words and decide whether it is true *or* false. *(Hint: To compare fractions, write them with the same denominator.)*

77. $7<19$
Seven is less than nineteen; true

78. $9<10$
Nine is less than ten; true

79. $\frac{1}{3}\neq\frac{3}{10}$
One-third is not equal to three-tenths; true

80. $\frac{10}{7}\neq\frac{3}{2}$
Ten-sevenths is not equal to three-halves; true

81. $8\geq 11$
Eight is greater than or equal to eleven; false

82. $4\leq 2$
Four is less than or equal to two; false

Write each statement as another true statement with the inequality symbol reversed. ***See Example 6.***

83. $5 < 30$ $30 > 5$

84. $8 > 4$ $4 < 8$

85. $12 \geq 3$ $3 \leq 12$

86. $25 \leq 41$ $41 \geq 25$

87. $2.5 \geq 1.3$ $1.3 \leq 2.5$

88. $4.1 \leq 5.3$ $5.3 \geq 4.1$

89. $\frac{4}{5} > \frac{3}{4}$ $\frac{3}{4} < \frac{4}{5}$

90. $\frac{8}{3} < \frac{11}{4}$ $\frac{11}{4} > \frac{8}{3}$

One way to measure a person's cardiofitness is to calculate how many METs, or metabolic units, he or she can reach at peak exertion. One MET is the amount of energy used when sitting quietly. To calculate ideal METs, we can use the following expressions.

$$14.7 - \text{age} \cdot 0.13 \quad \text{For women}$$

$$14.7 - \text{age} \cdot 0.11 \quad \text{For men}$$

(*Source: New England Journal of Medicine.*)

91. A 40-yr-old woman wishes to calculate her ideal MET.

(a) Write the expression, using her age. $14.7 - 40 \cdot 0.13$

(b) Calculate her ideal MET. (*Hint:* Use the rules for order of operations.) 9.5

(c) Researchers recommend that a person reach approximately 85% of his or her MET when exercising. Calculate 85% of the ideal MET from part (b). Then refer to the following table. What activity can the woman do that is approximately this value? 8.075; walking (5 mph)

Activity	METs	Activity	METs
Golf (with cart)	2.5	Skiing (water or downhill)	6.8
Walking (3 mph)	3.3	Swimming	7.0
Mowing lawn (power)	4.5	Walking (5 mph)	8.0
Ballroom or square dancing	5.5	Jogging	10.2
Cycling	5.7	Skipping rope	12.0

Source: Harvard School of Public Health.

(d) Repeat parts (a)–(c) for a 55-yr-old man.
$14.7 - 55 \cdot 0.11$; 8.65; 7.3525, swimming

92. Repeat parts (a)–(c) of **Exercise 91** for your age and gender. For yourself 5 yr from now.
Answers will vary.

The table shows the number of pupils per teacher in U.S. public schools in selected states during the 2009–2010 school year. Use this table to answer the questions in Exercises 93–96.

State	Pupils per Teacher
Alaska	15.3
Texas	14.7
California	22.2
Wyoming	12.7
Maine	11.8
Idaho	18.4
Missouri	14.1

Source: National Center for Education Statistics.

93. Which states had a number greater than 14.1?
Alaska, Texas, California, Idaho

94. Which states had a number that was at most 14.7?
Texas, Wyoming, Maine, Missouri

95. Which states had a number not less than 14.1?
Alaska, Texas, California, Idaho, Missouri

96. Which states had a number greater than 22.2?
none

Study Skills

TAKING LECTURE NOTES

Study the set of sample math notes given here.

- **Include the date and title** of the day's lecture topic.
- **Include definitions,** written here in parentheses—don't trust your memory.
- **Skip lines and write neatly** to make reading easier.
- **Emphasize direction words** (like *simplify*) with their explanations.
- **Mark important concepts with stars, underlining, etc.**
- **Use two columns,** which allows an example and its explanation to be close together.
- **Use brackets and arrows** to clearly show steps, related material, etc.

January 12

Exponents

Exponents used to show repeated multiplication.

$3 \cdot 3 \cdot 3 \cdot 3$ *can be written* 3^4 — *exponent (how many times it's multiplied)*; *base (the number being multiplied)*

Read 3^2 *as 3 to the 2nd power or 3 squared*

3^3 *as 3 to the 3rd power or 3 cubed*

3^4 *as 3 to the 4th power*

etc.

Simplifying an expression with exponents → *actually do the repeated multiplication*

2^3 *means* $2 \cdot 2 \cdot 2$ *and* $2 \cdot 2 \cdot 2 = 8$

☆ *Careful!* 5^2 *means* $5 \cdot 5$ NOT $5 \cdot 2$; *so* $5^2 = 5 \cdot 5 = 25$ BUT $5^2 \neq 10$

Example	*Explanation*
Simplify $2^4 \cdot 3^2$	*Exponents mean multiplication.*
$2 \cdot 2 \cdot 2 \cdot 2 \cdot 3 \cdot 3$	*Use 2 as a factor 4 times.* *Use 3 as a factor 2 times.*
$16 \cdot 9$	$2 \cdot 2 \cdot 2 \cdot 2$ *is 16*, $3 \cdot 3$ *is 9* → $16 \cdot 9$ *is 144*
144	*Simplified result is 144 (no exponents left)*

Now Try This

With a partner or in a small group, compare lecture notes.

1. **What are you doing to show main points in your notes (such as boxing, using stars, etc.)?** ______________________
2. **In what ways do you set off explanations from worked problems and subpoints (such as indenting, using arrows, circling, etc.)?** ______________________
3. **What new ideas did you learn by examining your classmates' notes?** ______________________
4. **What new techniques will you try in your notes?** ______________________

1.2 Variables, Expressions, and Equations

A **variable** is a symbol, usually a letter, used to represent an unknown number. A **constant** is a fixed, unchanging number.

x, y, or z Variables | 5, $\frac{3}{4}$, $8\frac{1}{2}$, 10.8 Constants

An **algebraic expression** is a collection of constants, variables, operation symbols, and grouping symbols, such as parentheses, square brackets, or fraction bars.

$x + 5$, $2m - 9$, $8p^2 + 6(p - 2)$ Algebraic expressions

$2m$ means $2 \cdot m$, the product of 2 and m.

$6(p - 2)$ means the product of 6 and $p - 2$.

OBJECTIVES

1. Evaluate algebraic expressions, given values for the variables.
2. Translate word phrases to algebraic expressions.
3. Identify solutions of equations.
4. Translate sentences to equations.
5. Distinguish between *equations* and *expressions*.

OBJECTIVE 1 Evaluate algebraic expressions, given values for the variables. An algebraic expression can have different numerical values for different values of the variables.

EXAMPLE 1 Evaluating Expressions

Find the value of each algebraic expression for $m = 5$ and then for $m = 9$.

(a) $8m$

$= 8 \cdot 5$ Let $m = 5$.

$= 40$ Multiply.

$8m$

$= 8 \cdot 9$ Let $m = 9$.

$= 72$ Multiply.

(b) $3m^2$

$= 3 \cdot 5^2$ Let $m = 5$. ($5^2 = 5 \cdot 5$)

$= 3 \cdot 25$ Square 5.

$= 75$ Multiply.

$3m^2$

$= 3 \cdot 9^2$ Let $m = 9$. ($9^2 = 9 \cdot 9$)

$= 3 \cdot 81$ Square 9.

$= 243$ Multiply.

Work Problem 1 at the Side. ▶

CAUTION

$3m^2$ means $3 \cdot m^2$, ***not*** $3m \cdot 3m$. See **Example 1(b).**

Unless parentheses are used, the exponent refers only to the variable or number just before it. We use parentheses to write $3m \cdot 3m$ with exponents as $(3m)^2$.

EXAMPLE 2 Evaluating Expressions

Find the value of each expression for $x = 5$ and $y = 3$.

(a) $2x + 5y$

$= 2 \cdot 5 + 5 \cdot 3$ Replace x with 5 and y with 3.

$= 10 + 15$ Multiply.

$= 25$ Add.

We could write this with parentheses as $2(5) + 5(3)$.

Follow the rules for order of operations.

Continued on Next Page

1 Find the value of each expression for $p = 3$.

GS **(a)** $6p$

$= 6 \cdot$ ____

$=$ ____

GS **(b)** $p + 12$

$=$ ____ $+ 12$

$=$ ____

GS **(c)** $5p^2$

$= 5 \cdot$ ____^{2}

$= 5 \cdot$ ____

$=$ ____

(d) $16p$

(e) $2p^3$

Teaching Tip

Mention that a variable in mathematics is analogous to a pronoun in English. For example, in "He went shopping." he can refer to any male shopper, while in $x + 3$, x can be replaced by any number.

Answers

1. **(a)** 3; 18 **(b)** 3; 15 **(c)** 3; 9; 45 **(d)** 48 **(e)** 54

2 Find the value of each expression for $x = 6$ and $y = 9$.

GS **(a)** $4x + 7y$

$= 4 \cdot \_\_\_\_ + 7 \cdot \_\_\_\_$

$= \_\_\_\_ + \_\_\_\_$

$= \_\_\_\_$

(b) $\dfrac{4x - 2y}{x + 1}$

(c) $2x^2 + y^2$

Answers

2. (a) 6; 9; 24; 63; 87 **(b)** $\frac{6}{7}$ **(c)** 153

(b) $\dfrac{9x - 8y}{2x - y}$

$= \dfrac{9 \cdot 5 - 8 \cdot 3}{2 \cdot 5 - 3}$ Replace x with 5 and y with 3.

$= \dfrac{45 - 24}{10 - 3}$ Multiply.

$= \dfrac{21}{7}$ Subtract.

$= 3$ Divide.

(c) $x^2 - 2y^2$

$= 5^2 - 2 \cdot 3^2$ Replace x with 5 and y with 3. ($3^2 = 3 \cdot 3$; $5^2 = 5 \cdot 5$)

$= 25 - 2 \cdot 9$ Apply the exponents.

$= 25 - 18$ Multiply.

$= 7$ Subtract.

◀ **Work Problem 2 at the Side.**

OBJECTIVE 2 Translate word phrases to algebraic expressions.

Problem-Solving Hint

Sometimes variables must be used to change word phrases into algebraic expressions. This process will be important in later chapters when we solve applied problems.

EXAMPLE 3 Using Variables to Write Word Phrases as Algebraic Expressions

Write each word phrase as an algebraic expression, using x as the variable.

(a) The **sum** of a number and 9

$x + 9$, or $9 + x$ "Sum" is the answer to an addition problem.

(b) 7 **minus** a number

$7 - x$ "Minus" indicates subtraction.

$x - 7$ is incorrect. We cannot subtract in either order and get the same result.

(c) A number **subtracted from 12**

$12 - x$ Be careful with order.

Compare this result with "12 subtracted from a number," which is $x - 12$.

(d) The **product** of 11 and a number

$11 \cdot x$, or $11x$

Continued on Next Page

(e) 5 **divided by** a number

$$5 \div x, \quad \text{or} \quad \frac{5}{x}$$

$\frac{x}{5}$ is *not* correct here.

(f) The **product of** 2 and the **difference** between a number and 8

We are multiplying 2 times "something." This "something" is the difference between a number and 8, written $x - 8$. We use parentheses around this difference.

$$2 \cdot (x - 8), \quad \text{or} \quad 2(x - 8)$$

$8 - x$, which means the difference between 8 and a number, is not correct.

Work Problem **3** at the Side. ▶

OBJECTIVE 3 Identify solutions of equations. An **equation** is a statement that two expressions are equal. ***An equation always includes the equality symbol, =.***

$$\left.\begin{array}{lll} x + 4 = 11, & 2y = 16, & 4p + 1 = 25 - p, \\ \frac{3}{4}x + \frac{1}{2} = 0, & z^2 = 4, & 4(m - 0.5) = 2m \end{array}\right\} \text{Equations}$$

To **solve** an equation, we must find all values of the variable that make the equation true. Such values of the variable are the **solutions** of the equation.

EXAMPLE 4 Deciding Whether a Number Is a Solution of an Equation

Decide whether the given number is a solution of the equation.

(a) $5p + 1 = 36$; 7

$$\begin{aligned} 5p + 1 &= 36 \\ 5 \cdot 7 + 1 &\stackrel{?}{=} 36 && \text{Let } p = 7. \\ 35 + 1 &\stackrel{?}{=} 36 && \text{Multiply.} \\ 36 &= 36 \ \checkmark && \text{True—the left side of the equation equals the right side.} \end{aligned}$$

Be careful. Multiply first.

The number 7 is a solution of the equation.

(b) $9m - 6 = 32$; $\frac{14}{3}$

$$\begin{aligned} 9m - 6 &= 32 \\ 9 \cdot \frac{14}{3} - 6 &\stackrel{?}{=} 32 && \text{Let } m = \tfrac{14}{3}. \\ 42 - 6 &\stackrel{?}{=} 32 && \text{Multiply.} \\ 36 &= 32 && \text{False—the left side does } \textit{not} \text{ equal the right side.} \end{aligned}$$

The number $\frac{14}{3}$ is not a solution of the equation.

Work Problem **4** at the Side. ▶

3 Write each word phrase as an algebraic expression. Use x as the variable.

(a) The sum of 5 and a number

(b) A number minus 4

(c) A number subtracted from 48

(d) The product of 6 and a number

(e) 9 multiplied by the sum of a number and 5

4 Decide whether the given number is a solution of the equation.

(a) $p - 1 = 3$; 2

(b) $2k + 3 = 15$; 7

(c) $7p - 11 = 5$; $\frac{16}{7}$

Answers

3. **(a)** $5 + x$ **(b)** $x - 4$ **(c)** $48 - x$ **(d)** $6x$ **(e)** $9(x + 5)$

4. **(a)** no **(b)** no **(c)** yes

5 Write each word sentence as an equation. Use x as the variable.

GS **(a)** Three times the sum of a number and 13 is 19.

____ (____ + ____) = ____

(b) Five times a number subtracted from 21 is 15.

(c) Five less than six times a number is equal to nineteen.

6 Decide whether each is an *equation* or an *expression*.

(a) $2x + 5y - 7$

(b) $\dfrac{3x - 1}{5}$

(c) $2x + 5 = 7$

(d) $\dfrac{x}{x - 3} = 4x$

Answers

5. **(a)** 3; x; 13; 19 **(b)** $21 - 5x = 15$
(c) $6x - 5 = 19$

6. **(a)** expression **(b)** expression
(c) equation **(d)** equation

OBJECTIVE 4 **Translate sentences to equations.** Sentences given in words are translated as equations.

EXAMPLE 5 **Translating Sentences to Equations**

Write each word sentence as an equation. Use x as the variable.

(a) Twice the sum of a number and four is six.

"Twice" means two times. The word *is* suggests equals. With x representing the number, translate as follows.

Twice	the sum of a number and four	is	six.
↓	↓	↓	↓
$2\ \cdot$	$(x + 4)$	$=$	6

The word *is* translates as =.

Use x for the unknown number.

$$2(x + 4) = 6$$

(b) Nine more than five times a number is 49.

Start with $5x$, and then add 9 to it.	The word *is* translates as =.	
↓	↓	
$5x + 9$	$=$	49

$5 \cdot x = 5x$

$$5x + 9 = 49$$

(c) Seven less than three times a number is equal to eleven.

Three times a number	less	seven	is equal to	eleven.
↓	↓	↓	↓	↓
$3x$	$-$	7	$=$	11

$$3x - 7 = 11$$

◀ Work Problem 5 at the Side.

OBJECTIVE 5 **Distinguish between equations and expressions.** Students often have trouble distinguishing between equations and expressions. ***An equation is a sentence—it has something on the left side, an = symbol, and something on the right side. An expression is a phrase that represents a number.***

$$\underbrace{4x + 5}_{\text{Left side}} = \underbrace{9}_{\text{Right side}} \qquad 4x + 5$$

Equation (points to =) Expression (points to $4x + 5$)

EXAMPLE 6 **Distinguishing between Equations and Expressions**

Decide whether each is an *equation* or an *expression*.

(a) $2x - 3$ Ask, "Is there an equality symbol?" The answer is no, so this is an expression.

(b) $2x - 3 = 8$ Because there is an equality symbol with something on either side of it, this is an equation.

(c) $5x^2 + 2y^2$ There is no equality symbol. This is an expression.

◀ Work Problem 6 at the Side.

1.2 Exercises

FOR EXTRA HELP

Download the MyDashBoard App

MyMathLab®

CONCEPT CHECK *Choose the letter(s) of the correct response.*

1. The expression $8x^2$ means ___B___.

A. $8 \cdot x \cdot 2$ **B.** $8 \cdot x \cdot x$ **C.** $8 + x^2$ **D.** $8x^2 \cdot 8x^2$

2. If $x = 2$ and $y = 1$, then the value of xy is ___C___.

A. $\frac{1}{2}$ **B.** 1 **C.** 2 **D.** 3

3. The sum of 15 and a number x is represented by ___A___.

A. $15 + x$ **B.** $15 - x$ **C.** $x - 15$ **D.** $15x$

4. Which of the following are expressions? **B, C**

A. $6x = 7$ **B.** $6x + 7$ **C.** $6x - 7$ **D.** $6x - 7 = 0$

CONCEPT CHECK *Complete each statement.*

5. For $x = 3$, the value of $x + 8$ is ___11___.

6. For $x = 1$ and $y = 2$, the value of $5xy$ is ___10___.

7. "The sum of 13 and x" is represented by the expression ___$13 + x$___. For $x = 3$, the value of this expression is ___16___.

8. $2x + 6$ is an (*equation* / <u>*expression*</u>), while $2x + 6 = 8$ is an (<u>*equation*</u> / *expression*).

In Exercises 9–12, give a short explanation.

9. If the words *more than* in **Example 5(b)** were changed to *less than,* how would the equation be changed?

The equation would be $5x - 9 = 49$.

10. Why is $2x^3$ not the same as $2x \cdot 2x \cdot 2x$? Explain, using an exponent to write $2x \cdot 2x \cdot 2x$.

$2x^3 = 2 \cdot x \cdot x \cdot x$, while $2x \cdot 2x \cdot 2x = (2x)^3$.

11. There are many pairs of values of x and y for which $2x + y$ can be evaluated. Name two such pairs and then evaluate the expression.

Answers will vary. Two such pairs are $x = 0, y = 6$ and $x = 1, y = 4$. For each pair, the value of $2x + y$ is 6.

12. When evaluating the expression $4x^2$ for $x = 3$, explain why 3 must be squared *before* multiplying by 4.

The exponent 2 applies only to the base x. The exponential must be evaluated before the product.

Find the value of each expression for **(a)** $x = 4$ *and* **(b)** $x = 6$. ***See Example 1.***

13. $4x^2$
(a) 64 **(b)** 144

14. $5x^2$
(a) 80 **(b)** 180

15. $\frac{3x - 5}{2x}$
(a) $\frac{7}{8}$ **(b)** $\frac{13}{12}$

16. $\frac{4x - 1}{3x}$
(a) $\frac{5}{4}$ **(b)** $\frac{23}{18}$

17. $\frac{6.459x}{2.7}$ (to the nearest thousandth)
(a) 9.569 **(b)** 14.353

18. $\frac{0.74x^2}{0.85}$ (to the nearest thousandth)
(a) 13.929 **(b)** 31.341

19. $3x^2 + x$
(a) 52 **(b)** 114

20. $2x + x^2$
(a) 24 **(b)** 48

Find the value of each expression for **(a)** $x = 2$ *and* $y = 1$ *and* **(b)** $x = 1$ *and* $y = 5$. ***See Example 2.***

21. $3(x + 2y)$
(a) 12 **(b)** 33

22. $2(2x + y)$
(a) 10 **(b)** 14

23. $x + \frac{4}{y}$
(a) 6 **(b)** $\frac{9}{5}$

24. $y + \frac{8}{x}$
(a) 5 **(b)** 13

25. $\frac{x}{2} + \frac{y}{3}$
(a) $\frac{4}{3}$ **(b)** $\frac{13}{6}$

26. $\frac{x}{5} + \frac{y}{4}$
(a) $\frac{13}{20}$ **(b)** $\frac{29}{20}$

27. $\frac{2x + 4y - 6}{5y + 2}$
(a) $\frac{2}{7}$ **(b)** $\frac{16}{27}$

28. $\frac{4x + 3y - 1}{2x + y}$
(a) 2 **(b)** $\frac{18}{7}$

29. $2y^2 + 5x$
(a) 12 **(b)** 55

30. $6x^2 + 4y$
(a) 28 **(b)** 26

31. $\frac{3x + y^2}{2x + 3y}$
(a) 1 **(b)** $\frac{28}{17}$

32. $\frac{x^2 + 1}{4x + 5y}$
(a) $\frac{5}{13}$ **(b)** $\frac{2}{29}$

33. $0.841x^2 + 0.32y^2$
(a) 3.684 **(b)** 8.841

34. $0.941x^2 + 0.2y^2$
(a) 3.964 **(b)** 5.941

Write each word phrase as an algebraic expression, using x as the variable. ***See Example 3.***

35. Twelve times a number

$12x$

36. Thirteen added to a number

$x + 13$

37. Two subtracted from a number

$x - 2$

38. Eight subtracted from a number

$x - 8$

39. One-third of a number, subtracted from seven

$7 - \frac{1}{3}x$

40. One-fifth of a number, subtracted from fourteen

$14 - \frac{1}{5}x$

41. The difference between twice a number and 6

$2x - 6$

42. The difference between 6 and half a number

$6 - \frac{1}{2}x$

43. 12 divided by the sum of a number and 3

$\frac{12}{x + 3}$

44. The difference between a number and 5, divided by 12

$\frac{x - 5}{12}$

45. The product of 6 and four less than a number

$6(x - 4)$

46. The product of 9 and five more than a number

$9(x + 5)$

47. Suppose that the directions on a quiz read, "Evaluate each equation for $x = 3$." How can you justify politely correcting the person who wrote these directions?

An equation cannot be *evaluated*. An equation that contains a variable can be *solved* to find the value or values of the variable that make it a true statement.

48. Suppose that the directions on a test read, "Solve the following expressions." How can you justify politely correcting the person who wrote these directions?

An expression cannot be *solved*. It indicates a series of operations to be performed. An equation can be solved.

Decide whether the given number is a solution of the equation. ***See Example 4.***

49. Is 7 a solution of $x - 5 = 12$?

no

50. Is 10 a solution of $x + 6 = 15$?

no

51. Is 1 a solution of $5x + 2 = 7$?

yes

52. Is 1 a solution of $3x + 5 = 8$?

yes

53. Is $\frac{1}{5}$ a solution of $6x + 4x + 9 = 11$?

yes

54. Is $\frac{12}{5}$ a solution of $2x + 3x + 8 = 20$?

yes

55. Is 3 a solution of $2y + 3(y - 2) = 14$?

no

56. Is 2 a solution of $6x + 2(x + 3) = 14$?

no

57. Is $\frac{1}{3}$ a solution of $\frac{z+4}{2-z} = \frac{13}{5}$?

yes

58. Is $\frac{13}{4}$ a solution of $\frac{x+6}{x-2} = \frac{37}{5}$?

yes

59. Is 4.3 a solution of $3r^2 - 2 = 53.47$?

yes

60. Is 3.7 a solution of $2x^2 + 1 = 28.38$?

yes

Write each word sentence as an equation. Use x as the variable. ***See Example 5.***

61. The sum of a number and 8 is 18.

$x + 8 = 18$

62. A number minus three equals 1.

$x - 3 = 1$

63. Five more than twice a number is 5.

$2x + 5 = 5$

64. The product of 2 and the sum of a number and 5 is 14.

$2(x + 5) = 14$

65. Sixteen minus three-fourths of a number is 13.

$16 - \frac{3}{4}x = 13$

66. The sum of six-fifths of a number and 2 is 14.

$\frac{6}{5}x + 2 = 14$

67. Three times a number is equal to 8 more than twice the number.

$3x = 2x + 8$

68. Twelve divided by a number equals $\frac{1}{3}$ times that number.

$\frac{12}{x} = \frac{1}{3}x$

Identify each as an expression *or an* equation. ***See Example 6.***

69. $3x + 2(x - 4)$

expression

70. $5y - (3y + 6)$

expression

71. $7t + 2(t + 1) = 4$

equation

72. $9r + 3(r - 4) = 2$

equation

73. $x + y = 9$

equation

74. $x + y - 9$

expression

75. $\frac{3x - 8}{2}$

expression

76. $\frac{4x + 3}{2} = 11$

equation

Relating Concepts (Exercises 77–80) For Individual or Group Work

A ***mathematical model*** *is an equation that describes the relationship between two quantities. For example, the life expectancy of Americans at birth can be approximated by the equation*

$$y = 0.174x - 270,$$

where x is a year between 1950 and 2008 and y is age in years. (Source: Centers for Disease Control and Prevention.)

Use this model to approximate life expectancy (to the nearest year) in each of the following years.

77. 1950

69 yr

78. 1975

74 yr

79. 1995

77 yr

80. 2008

79 yr

Study Skills

TACKLING YOUR HOMEWORK

You are ready to do your homework ***AFTER*** you have read the corresponding textbook section and worked through the examples and margin problems.

Homework Tips

- **Survey the exercise set.** Take a few minutes to glance over the problems that your instructor has assigned to get a general idea of the types of exercises you will be working. Skim directions, and note any references to section examples.
- **Work problems neatly.** Use pencil and write legibly, so others can read your work. Skip lines between steps. Clearly separate problems from each other.
- **Show all your work.** It is tempting to take shortcuts. Include ALL steps.
- **Check your work frequently to make sure you are on the right track.** It is hard to unlearn a mistake. For all odd-numbered problems and other selected exercises, answers are given in the back of the book.
- **If you have trouble with a problem, refer to the corresponding worked example in the section.** The exercise directions will often reference specific examples to review. Pay attention to every line of the worked example to see how to get from step to step.
- **If you are having trouble with an even-numbered problem, work the corresponding odd-numbered problem.** Check your answer in the back of the book, and apply the same steps to work the even-numbered problem.
- **Does the problem or a similar problem have a blue screen around the problem number, such as 11.?** If it does, refer to the worked-out solution in the selected solutions section at the back of the book. Study this solution.
- **Do some homework problems every day.** This is a good habit, even if your math class does not meet each day.
- **Mark any problems you don't understand.** Ask your instructor about them.

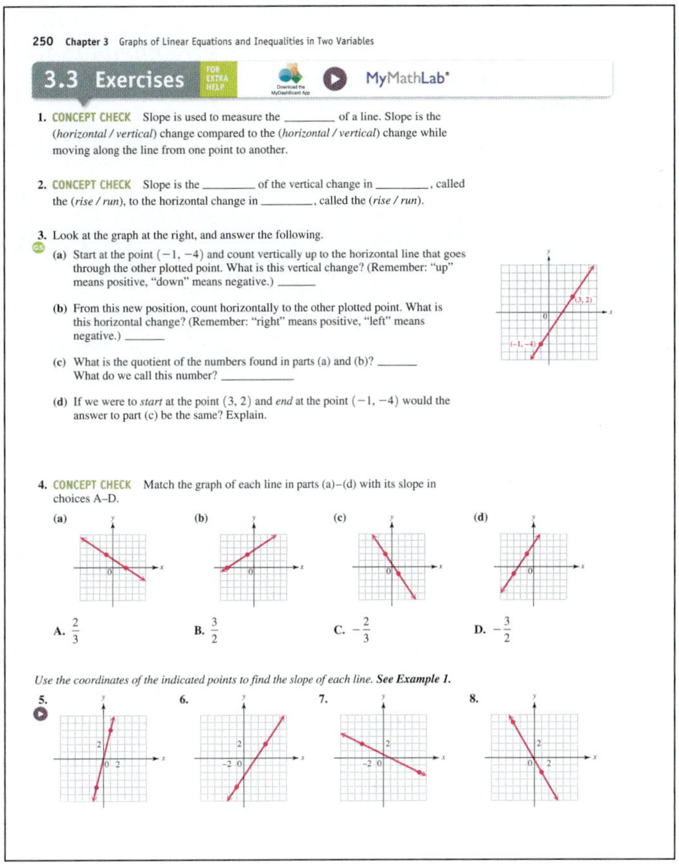

250 Chapter 3 Graphs of Linear Equations and Inequalities in Two Variables

3.3 Exercises FOR EXTRA HELP — Download the MyDashBoard App — MyMathLab®

1. CONCEPT CHECK Slope is used to measure the ________ of a line. Slope is the (*horizontal* / *vertical*) change compared to the (*horizontal* / *vertical*) change while moving along the line from one point to another.

2. CONCEPT CHECK Slope is the ________ of the vertical change in ________, called the (*rise* / *run*), to the horizontal change in ________, called the (*rise* / *run*).

3. Look at the graph at the right, and answer the following.
   (a) Start at the point $(-1, -4)$ and count vertically up to the horizontal line that goes through the other plotted point. What is this vertical change? (Remember: "up" means positive, "down" means negative.) ______
   (b) From this new position, count horizontally to the other plotted point. What is this horizontal change? (Remember: "right" means positive, "left" means negative.) ______
   (c) What is the quotient of the numbers found in parts (a) and (b)? ______ What do we call this number? ________
   (d) If we were to *start* at the point $(3, 2)$ and *end* at the point $(-1, -4)$ would the answer to part (c) be the same? Explain.

4. CONCEPT CHECK Match the graph of each line in parts (a)–(d) with its slope in choices A–D.
   (a) (b) (c) (d)
   A. $\frac{2}{3}$ B. $\frac{3}{2}$ C. $-\frac{2}{3}$ D. $-\frac{3}{2}$

Use the coordinates of the indicated points to find the slope of each line. ***See Example 1.***

5. 6. 7. 8.

Now Try This

Think through and answer each of the following.

1. **What are your biggest homework concerns?**

   ____

   ____

2. **List one or more of the homework tips to try.**

   ____

   ____

1.3 Real Numbers and the Number Line

OBJECTIVES

1. Classify numbers and graph them on number lines.
2. Tell which of two real numbers is less than the other.
3. Find the additive inverse of a real number.
4. Find the absolute value of a real number.

A **set** is a collection of objects. In mathematics, these objects are usually numbers. The objects that belong to the set are its **elements.** They are written between braces.

$\{1, 2, 3, 4, 5\}$ The set of numbers 1, 2, 3, 4, and 5

OBJECTIVE 1 Classify numbers and graph them on number lines. The set of numbers used for counting is the *natural numbers*. The set of *whole numbers* includes 0 with the natural numbers.

Natural Numbers and Whole Numbers

$\{1, 2, 3, 4, 5, \ldots\}$ is the set of **natural numbers** (or **counting numbers**).

$\{0, 1, 2, 3, 4, 5, \ldots\}$ is the set of **whole numbers.**

Teaching Tip

The "arrowhead" is used to indicate the positive direction on a number line.

We can represent numbers on a **number line** like the one in **Figure 1.**

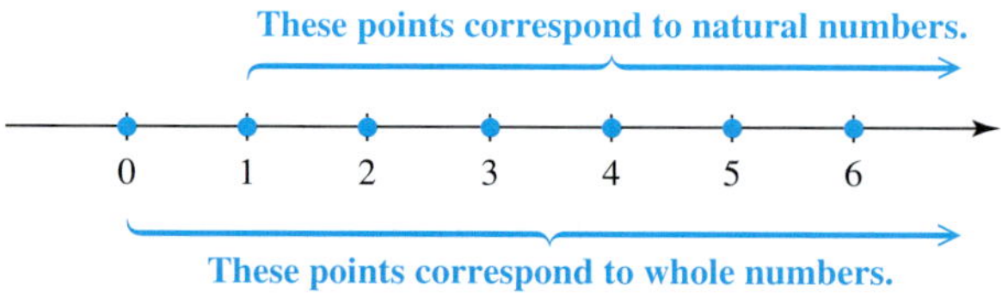

Figure 1

To draw a number line, choose any point on the line and label it 0. Then choose any point to the right of 0 and label it 1. Use the distance between 0 and 1 as the scale to locate, and then label, other points.

The natural numbers are located to the right of 0 on the number line. For each natural number, we can place a corresponding number to the left of 0, labeling the points -1, -2, -3, and so on, as shown in **Figure 2.** Each is the **opposite,** or **negative,** of a natural number. The natural numbers, their opposites, and 0 form the set of *integers*.

Integers

$\{\ldots, -3, -2, -1, 0, 1, 2, 3, \ldots\}$ is the set of **integers.**

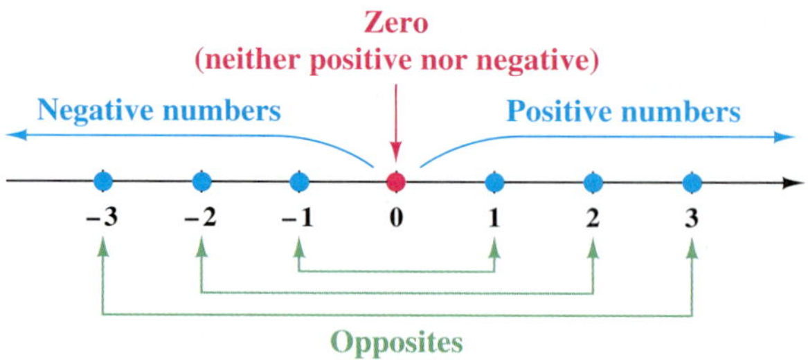

The points correspond to integers.

Figure 2

Positive numbers and *negative numbers* are **signed numbers.**

EXAMPLE 1 Using Negative Numbers in Applications

Use an integer to express the boldface italic number in each application.

(a) The lowest Fahrenheit temperature ever recorded in meteorological records was ***129***° below zero at Vostok, Antarctica, on July 21, 1983. (*Source: World Almanac and Book of Facts.*)
Use -129 because "below zero" indicates a negative number.

(b) The shore surrounding the Dead Sea is ***1348*** ft below sea level. (*Source: World Almanac and Book of Facts.*)
Again, "below sea level" indicates a negative number, -1348.

Work Problem 1 at the Side. ▶

Fractions, reviewed in **Chapter R,** are *rational numbers.*

Rational Numbers

$\{x \mid x$ is a quotient of two integers, with denominator not $0\}$ is the set of **rational numbers.**

(Read the part in the braces as "the set of all numbers x such that x is a quotient of two integers, with denominator not 0.")

Note

The set symbolism used in the definition of rational numbers,

$$\{x \mid x \textbf{ has a certain property}\},$$

is called **set-builder notation.** This notation is convenient to use when it is not possible to list all the elements of a set.

Since any number that can be written as the quotient of two integers (that is, as a fraction) is a rational number, ***all integers, mixed numbers, terminating (or ending) decimals, and repeating decimals are rational.*** The table gives examples.

Rational Number	Equivalent Quotient of Two Integers
-5	$\frac{-5}{1}$ (means $-5 \div 1$)
$1\frac{3}{4}$	$\frac{7}{4}$ (means $7 \div 4$)
0.23 (terminating decimal)	$\frac{23}{100}$ (means $23 \div 100$)
0.3333…, or $0.\overline{3}$ (repeating decimal)	$\frac{1}{3}$ (means $1 \div 3$)
4.7	$\frac{47}{10}$ (means $47 \div 10$)

To **graph** a number, we place a dot on the number line at the point that corresponds to the number. The number is the **coordinate** of the point. Think of the graph of a set of numbers as a picture of the set.

1 Use an integer to express the boldface italic number(s) in each application.

(a) Erin discovers that she has spent $***53*** more than she has in her checking account.

(b) The record high Fahrenheit temperature in the United States was ***134***° in Death Valley, California, on July 10, 1913. (*Source: World Almanac and Book of Facts.*)

(c) A football team gained ***5*** yd, then lost ***10*** yd on the next play.

Teaching Tip
A *rational* number can be written as a *ratio* (or quotient) of two integers.

Teaching Tip
Remind students that a terminating decimal is one that ends. A repeating decimal is one that repeats in a block of digits according to a pattern.

Answers

1. (a) -53 **(b)** 134 **(c)** $5, -10$

2 Graph each number on the number line.

$$-3, \quad \frac{17}{8}, \quad -2.75, \quad 1\frac{1}{2}, \quad -\frac{3}{4}$$

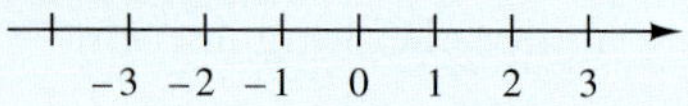

EXAMPLE 2 Graphing Numbers

Graph each number on a number line.

$$-\frac{3}{2}, \quad -\frac{2}{3}, \quad \frac{1}{2}, \quad 1\frac{1}{3}, \quad \frac{23}{8}, \quad 3\frac{1}{4}$$

To locate the improper fractions on the number line, write them as mixed numbers or decimals. The graph is shown in **Figure 3.**

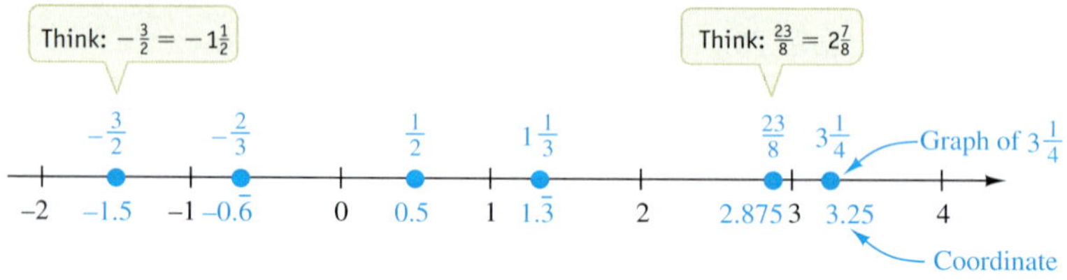

Figure 3

◀ **Work Problem 2 at the Side.**

The square root of 2, written $\sqrt{2}$, cannot be written as a quotient of two integers. Because of this, $\sqrt{2}$ is an *irrational number.* (See **Figure 4.**)

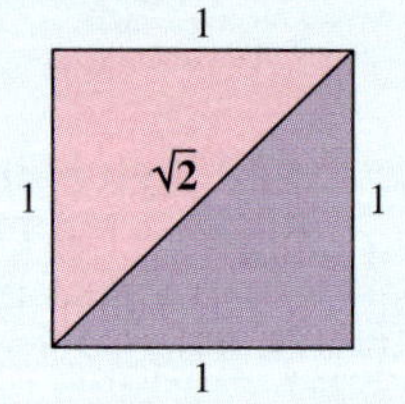

This square has diagonal of length $\sqrt{2}$. The number $\sqrt{2}$ is an irrational number.

Figure 4

Irrational Numbers

$\{x \mid x$ is a nonrational number represented by a point on the number line$\}$ is the set of **irrational numbers.**

The decimal form of an irrational number neither terminates nor repeats.

Both rational and irrational numbers can be represented by points on the number line. Together, they form the **real numbers.** See **Figure 5.**

Real Numbers

$\{x \mid x$ is a rational or an irrational number$\}$ is the set of **real numbers.**

Teaching Tip

Distinguishing between the sets of numbers is difficult for students. Consider working Exercises 21–38 in the Section 1.3 Exercises together.

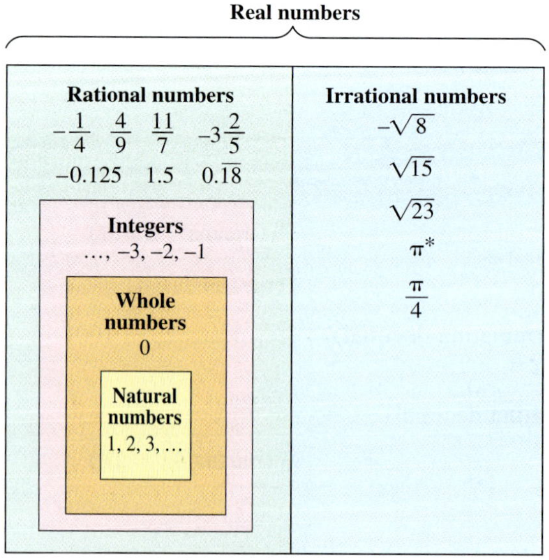

Figure 5

*The value of π (pi) is approximately 3.141592654. The decimal digits continue forever with no repeated pattern.

Answer

2. 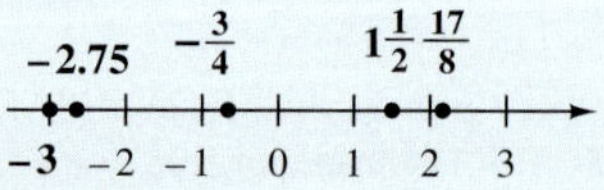

EXAMPLE 3 **Determining Whether a Number Belongs to a Set**

List the numbers in the following set that belong to each set of numbers.

$$\left\{-5,\ -\frac{2}{3},\ 0,\ 0.\overline{6},\ \sqrt{2},\ 3\frac{1}{4},\ 5,\ 5.8\right\}$$

(a) Natural numbers: 5

(b) Whole numbers: 0 and 5
The whole numbers consist of the natural (counting) numbers and 0.

(c) Integers: -5, 0, and 5

(d) Rational numbers: $-5, -\frac{2}{3}, 0, 0.\overline{6}\left(\text{or } \frac{2}{3}\right), 3\frac{1}{4}\left(\text{or } \frac{13}{4}\right), 5$, and $5.8\left(\text{or } \frac{58}{10}\right)$
Each of these numbers can be written as the quotient of two integers.

(e) Irrational numbers: $\sqrt{2}$

(f) Real numbers: All the numbers in the set are real numbers.

Work Problem 3 at the Side. ▶

OBJECTIVE 2 **Tell which of two real numbers is less than the other.** Given any two positive integers, you probably can tell which number is less than the other. Positive numbers decrease as the corresponding points on the number line go to the left. For example, $8 < 12$ because 8 is to the left of 12 on the number line. This ordering is extended to all real numbers by definition.

Ordering of Real Numbers

For any two real numbers a and b, ***a* is less than *b*** if a is to the left of b on a number line.

***a* is to the left of *b*,**
$a < b$.

We can also say that, for any real numbers a and b, ***a* is *greater than* *b*** if a is to the right of b on the number line. Any negative number is less than 0, and any negative number is less than any positive number. Also, 0 is less than any positive number.

EXAMPLE 4 **Determining the Order of Real Numbers**

Is the statement $-3 < -1$ *true* or *false*?

Locate -3 and -1 on a number line, as shown in **Figure 6.** Because -3 is to the left of -1 on the number line, -3 is less than -1. The statement $-3 < -1$ is true.

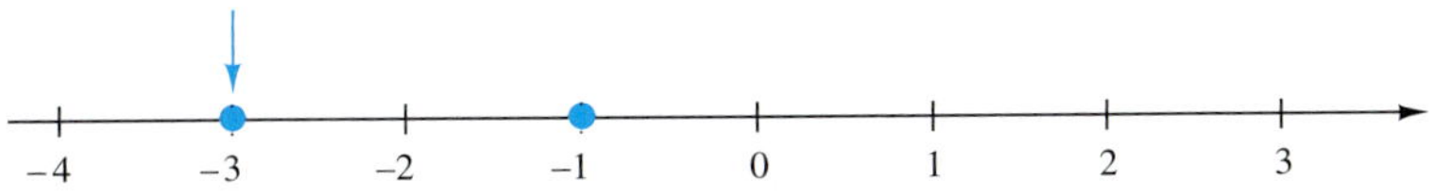

Figure 6

Work Problem 4 at the Side. ▶

3 List the numbers in the following set that belong to each set of numbers.

$$\left\{-7,\ -\frac{4}{5},\ 0,\ \sqrt{3},\ 2.7,\ 13\right\}$$

(a) Whole numbers

(b) Integers

(c) Rational numbers

(d) Irrational numbers

4 Tell whether each statement is *true* or *false*.

GS **(a)** $-2 < 4$
-2 is to the (*left* / *right*) of 4 on a number line, so the statement is (*true* / *false*).

GS **(b)** $6 > -3$
6 is to the (*left* / *right*) of -3 on a number line, so the statement is (*true* / *false*).

(c) $-9 < -12$

(d) $-4 \geq -1$

(e) $-6 \leq 0$

Answers

3. **(a)** 0, 13 **(b)** -7, 0, 13
(c) $-7, -\frac{4}{5}, 0, 2.7, 13$ **(d)** $\sqrt{3}$

4. **(a)** left; true **(b)** right; true **(c)** false
(d) false **(e)** true

5 Find the additive inverse of each number.

(a) 15

(b) -9

(c) -12

(d) $\frac{1}{2}$

(e) 0

OBJECTIVE 3 Find the additive inverse of a real number. By a property of the real numbers, for any real number x (except 0), there is exactly one number on a number line the same distance from 0 as x, but on the *opposite* side of 0. See **Figure 7.** Such pairs of numbers are *additive inverses,* or *opposites*, of each other.

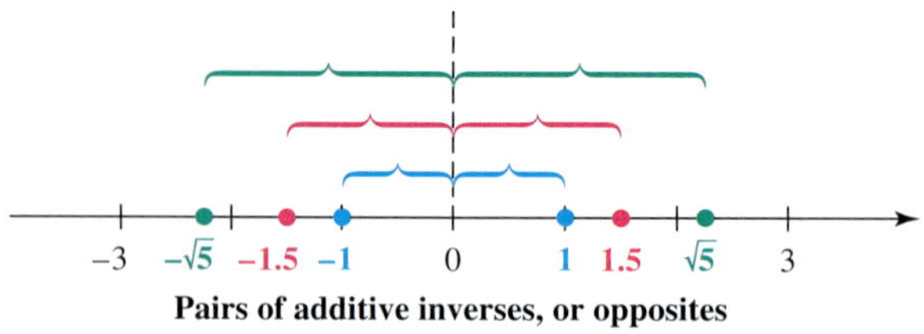

Figure 7

Additive Inverse

The **additive inverse** of a number x is the number that is the same distance from 0 on a number line as x, but on the *opposite* side of 0. (The number 0 is its own additive inverse.)

We indicate the additive inverse of a number by writing the symbol $-$ in front of the number. For example, the additive inverse of 7 is -7 (read "negative 7"). We could write the additive inverse of -4 as $-(-4)$, but we know that 4 is the opposite of -4. Since a number can have only one additive inverse, $-(-4)$ and 4 must represent the same number, so

$$-(-4) = 4.$$

This idea can be generalized.

Double Negative Rule

For any real number a, $\quad \mathbf{-(-a) = a.}$

The following table shows several numbers and their additive inverses.

Number	Additive Inverse
-4	$-(-4)$, or 4
0	0
5	-5
$-\frac{2}{3}$	$\frac{2}{3}$
0.52	-0.52

The table suggests the following rule.

The additive inverse of a nonzero number is found by changing the sign of the number.

◀ **Work Problem 5 at the Side.**

Answers

5. (a) -15 **(b)** 9 **(c)** 12 **(d)** $-\frac{1}{2}$ **(e)** 0

OBJECTIVE 4 Find the absolute value of a real number. Because additive inverses are the same distance from 0 on a number line, a number and its additive inverse have the same *absolute value*. The **absolute value** of a real number x, written $|x|$ and read **"the absolute value of x,"** can be defined as the distance between 0 and the number on a number line.

$|2| = 2$ The distance between 2 and 0 on a number line is 2 units.

$|-2| = 2$ The distance between -2 and 0 on a number line is also 2 units.

Distance is a physical measurement, which is never negative. ***Therefore, the absolute value of a number is never negative.***

Absolute Value

For any real number x,

$$|x| = \begin{cases} x & \text{if } x \geq 0 \\ -x & \text{if } x < 0. \end{cases}$$

By this definition, if x is a positive number or 0, then its absolute value is x itself. For example, since 8 is a positive number,

$$|8| = 8 \quad \text{and} \quad |0| = 0.$$

If x is a negative number, then its absolute value is the additive inverse of x.

$|-8| = -(-8) = 8$ The additive inverse of -8 is 8.

EXAMPLE 5 Finding the Absolute Value

Simplify by finding the absolute value.

(a) $|0| = 0$ **(b)** $|5| = 5$ **(c)** $|-5| = -(-5) = 5$

(d) $-|5| = -(5) = -5$ **(e)** $-|-5| = -(5) = -5$

(f) $|8-2| = |6| = 6$ **(g)** $-|8-2| = -|6| = -6$

Parts (f) and (g) show that absolute value bars are grouping symbols. We perform any operations inside absolute value symbols *before* finding the absolute value.

Work Problem 6 at the Side. ▶

---

6 Simplify by finding the absolute value.

(a) $|-6|$

Teaching Tip

The definition of absolute value usually requires several numerical examples before it is understood. When we say that the absolute value of a number is never negative, this does not mean that it is always positive, since $|0| = 0$ and 0 is neither positive nor negative.

(b) $|9|$

(c) $-|15|$

GS **(d)** $-|-9|$

$= -$ ____

$=$ ____

(e) $|9-4|$

GS **(f)** $-|32-2|$

$= -|$ ____ $|$

$=$ ____

Answers

6. (a) 6 **(b)** 9 **(c)** -15
(d) 9; -9 **(e)** 5 **(f)** 30; -30

1.3 Exercises

FOR EXTRA HELP

Download the MyDashBoard App

MyMathLab®

CONCEPT CHECK *Complete each statement.*

1. The number __0__ is a whole number, but not a natural number.

2. The natural numbers, their additive inverses, and 0 form the set of __integers__.

3. The additive inverse of every negative number is a (*negative* / *positive*) number.

4. If x and y are real numbers with $x > y$, then x lies to the (*left* / *right*) of y on a number line.

5. A rational number is the __quotient__ of two integers, with the __denominator__ not equal to 0.

6. Decimals that neither terminate nor repeat are __irrational__ numbers.

Use an integer to express each boldface italic number representing a change in the following applications. ***See Example 1.***

7. Between July 1, 2008, and July 1, 2009, the population of the United States increased by approximately ***2,632,000.*** (*Source:* U.S. Census Bureau.)

2,632,000

8. From 2010 to 2011, the mean SAT critical reading score for Illinois students increased by ***14,*** while the mathematics score increased by ***17.*** (*Source:* The College Board.)

14; 17

9. From 2005 to 2009, the paid circulation of daily newspapers in the United States went from 53,345 thousand to 46,278 thousand, representing a decrease of ***7067*** thousand. (*Source: Editor and Publisher International Yearbook,* 2010.)

−7067

10. In 1935, there were 15,295 banks in the United States. By 2010, the number was 7821, representing a decrease of ***7474*** banks. (*Source:* Federal Deposit Insurance Corporation.)

−7474

Graph each group of numbers on a number line. ***See Example 2.***

11. $0, 3, -5, -6$

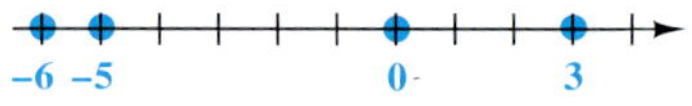

12. $2, 6, -2, -1$

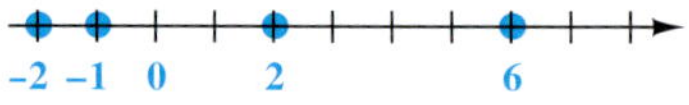

13. $-2, -6, -4, 3, 4$

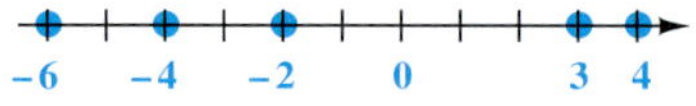

14. $-5, -3, -2, 0, 4$

15. $\frac{1}{4}, 2\frac{1}{2}, -3\frac{4}{5}, -4, -\frac{13}{8}$

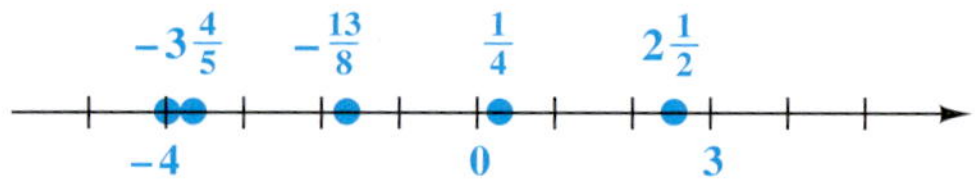

16. $5\frac{1}{4}, \frac{41}{9}, -2\frac{1}{3}, 0, -3\frac{2}{5}$

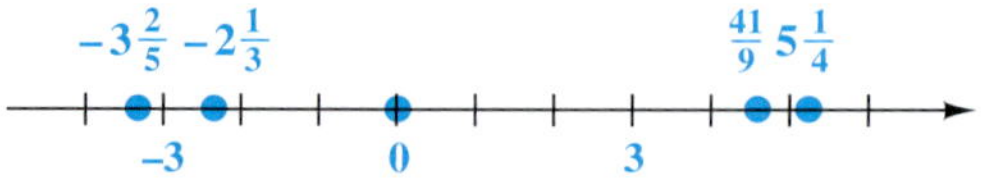

List all numbers from each set that are ***(a)*** *natural numbers,* ***(b)*** *whole numbers,* ***(c)*** *integers,* ***(d)*** *rational numbers,* ***(e)*** *irrational numbers,* ***(f)*** *real numbers.* ***See Example 3.***

17. $\left\{-9, -\sqrt{7}, -1\frac{1}{4}, -\frac{3}{5}, 0, \sqrt{5}, 3, 5.9, 7\right\}$

(a) 3, 7 **(b)** 0, 3, 7 **(c)** −9, 0, 3, 7

(d) $-9, -1\frac{1}{4}, -\frac{3}{5}, 0, 3, 5.9, 7$

(e) $-\sqrt{7}, \sqrt{5}$ **(f)** All are real numbers.

18. $\left\{-5.3, -5, -\sqrt{3}, -1, -\frac{1}{9}, 0, 1.2, 4, \sqrt{12}\right\}$

(a) 4 **(b)** 0, 4 **(c)** −5, −1, 0, 4

(d) $-5.3, -5, -1, -\frac{1}{9}, 0, 1.2, 4$

(e) $-\sqrt{3}, \sqrt{12}$ **(f)** All are real numbers.

19. $\left\{\frac{7}{9}, -2.\overline{3}, \sqrt{3}, 0, -8\frac{3}{4}, 11, -6, \pi\right\}$

(a) 11 (b) 0, 11 (c) 0, 11, −6

(d) $\frac{7}{9}, -2.\overline{3}, 0, -8\frac{3}{4}, 11, -6$

(e) $\sqrt{3}, \pi$ (f) All are real numbers.

20. $\left\{1\frac{5}{8}, -0.\overline{4}, \sqrt{6}, 9, -12, 0, \sqrt{10}, 0.026\right\}$

(a) 9 (b) 9, 0 (c) 9, −12, 0

(d) $1\frac{5}{8}, -0.\overline{4}, 9, -12, 0, 0.026$

(e) $\sqrt{6}, \sqrt{10}$ (f) All are real numbers.

CONCEPT CHECK *In Exercises 21–26, give a number that satisfies the given condition.*

21. An integer between 3.6 and 4.6

4

22. A rational number between 2.8 and 2.9

One example is 2.85. There are others.

23. A whole number that is not positive and is less than 1

0

24. A whole number greater than 3.5

One example is 4. There are others.

25. An irrational number that is between $\sqrt{12}$ and $\sqrt{14}$

One example is $\sqrt{13}$. There are others.

26. A real number that is neither negative nor positive

0

CONCEPT CHECK *In Exercises 27–32, decide whether each statement is* true *or* false.

27. Every natural number is positive.

true

28. Every whole number is positive.

false

29. Every integer is a rational number.

true

30. Every rational number is a real number.

true

31. Some numbers are both rational and irrational.

false

32. Every terminating decimal is a rational number.

true

CONCEPT CHECK *Give three numbers between −6 and 6 that satisfy each given condition.* In Exercises 33–38, answers will vary.

33. Positive real numbers but not integers

$\frac{1}{2}, \frac{5}{8}, 1\frac{3}{4}$

34. Real numbers but not positive numbers

$-1, -\frac{3}{4}, -5$

35. Real numbers but not whole numbers

$-3\frac{1}{2}, -\frac{2}{3}, \frac{3}{7}$

36. Rational numbers but not integers

$\frac{1}{2}, -\frac{2}{3}, \frac{2}{7}$

37. Real numbers but not rational numbers

$\sqrt{5}, \pi, -\sqrt{3}$

38. Rational numbers but not negative numbers

$\frac{2}{3}, \frac{5}{6}, \frac{5}{2}$

Select the lesser number in each pair. **See Example 4.**

39. $-11, -4$

-11

40. $-9, -16$

-16

41. $-21, 1$

-21

42. $-57, 3$

-57

43. $0, -100$

-100

44. $-215, 0$

-215

45. $-\frac{2}{3}, -\frac{1}{4}$

$-\frac{2}{3}$

46. $-\frac{3}{8}, -\frac{9}{16}$

$-\frac{9}{16}$

Decide whether each statement is true *or* false. **See Example 4.**

47. $8 < -16$

false

48. $12 < -24$

false

49. $-3 < -2$

true

50. $-10 < -9$

true

For each number, (a) find its additive inverse and (b) find its absolute value.

51. -2
(a) 2 (b) 2

52. -8
(a) 8 (b) 8

53. 6
(a) -6 (b) 6

54. 11
(a) -11 (b) 11

55. $-\frac{3}{4}$
(a) $\frac{3}{4}$ (b) $\frac{3}{4}$

56. $-\frac{1}{3}$
(a) $\frac{1}{3}$ (b) $\frac{1}{3}$

57. 4.95
(a) -4.95 (b) 4.95

58. 0.007
(a) -0.007 (b) 0.007

59. CONCEPT CHECK Match each expression in Column I with its value in Column II. Choices in Column II may be used once, more than once, or not at all.

I	II
(a) $\lvert -9 \rvert$ A	A. 9
(b) $-(-9)$ A	B. -9
(c) $-\lvert -9 \rvert$ B	C. Neither A nor B
(d) $-\lvert -(-9) \rvert$ B	D. Both A and B

60. CONCEPT CHECK Fill in the blanks with the correct values: The opposite of -5 is __5__, while the absolute value of -5 is __5__. The additive inverse of -5 is __5__, while the additive inverse of the absolute value of -5 is __-5__.

Simplify by finding the absolute value. ***See Example 5.***

61. $\lvert -7 \rvert$
7

62. $\lvert -3 \rvert$
3

63. $-\lvert 12 \rvert$
-12

64. $-\lvert 23 \rvert$
-23

65. $-\left\lvert -\frac{2}{3} \right\rvert$ $-\frac{2}{3}$

66. $-\left\lvert -\frac{4}{5} \right\rvert$ $-\frac{4}{5}$

67. $\lvert 13 - 4 \rvert$
9

68. $\lvert 8 - 7 \rvert$
1

Decide whether each statement is true *or* false. ***See Examples 4 and 5.***

69. $\lvert -8 \rvert < 7$
false

70. $\lvert -6 \rvert \geq -\lvert 6 \rvert$
true

71. $4 \leq \lvert 4 \rvert$
true

72. $-\lvert -3 \rvert > 2$
false

The table shows the change in the Consumer Price Index (CPI) for selected categories of goods and services from 2008 to 2009 and from 2009 to 2010. Use the table to answer Exercises 73–76.

Category	Change from 2008 to 2009	Change from 2009 to 2010
Apparel	1.2	-0.6
Food	3.9	1.6
Energy	-43.6	18.3
Medical care	11.5	12.8
Transportation	-16.2	14.1

Source: U.S. Bureau of Labor Statistics.

73. Which category for which period represents the greatest decrease?
Energy, 2008 to 2009

74. Which category for which period represents the least change?
Apparel, 2009 to 2010

75. Which has lesser absolute value, the change for Transportation from 2008 to 2009 or from 2009 to 2010?
2009 to 2010

76. Which has greater absolute value, the change for Energy from 2008 to 2009 or from 2009 to 2010?
2008 to 2009

Study Skills

USING STUDY CARDS

You may have used "flash cards" in other classes. In math, "study cards" can help you remember terms and definitions, procedures, and concepts. Use study cards to

- Help you understand and learn the material;
- Quickly review when you have a few minutes;
- Review before a quiz or test.

One of the advantages of study cards is that you learn while you are making them.

Vocabulary Cards

Put the word and a page reference on the front of the card. On the back, write the definition, an example, any related words, and a sample problem (if appropriate).

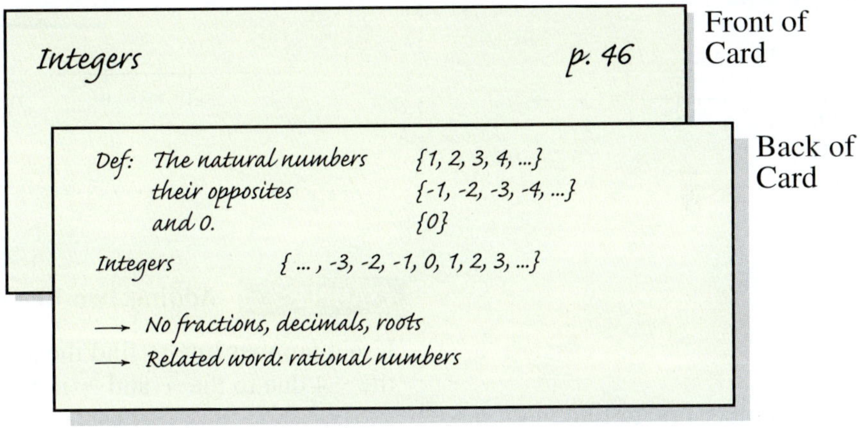

Procedure ("Steps") Cards

Write the name of the procedure on the front of the card. Then write each step in words. On the back of the card, put an example showing each step.

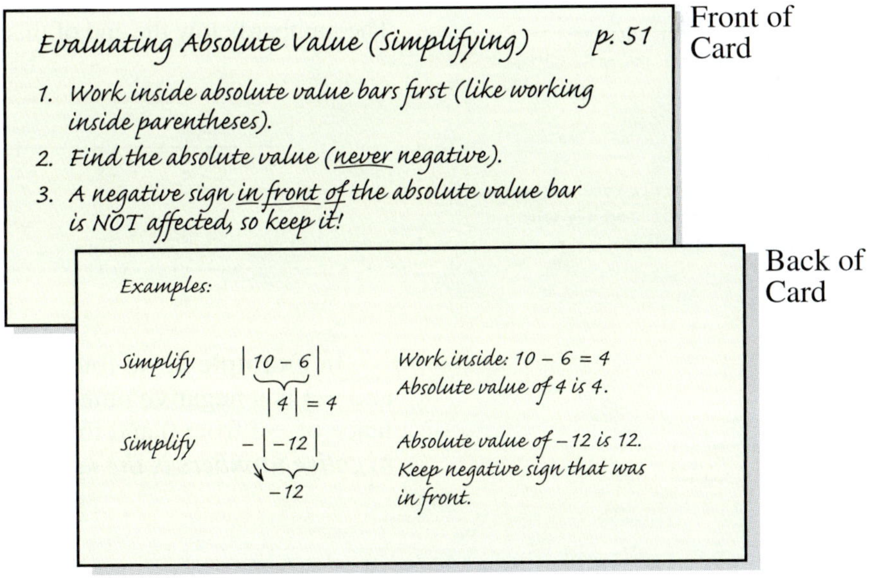

Make a vocabulary card and a procedure card for material you are learning now.

1.4 Adding Real Numbers

OBJECTIVES

1. Add two numbers with the same sign.
2. Add numbers with different signs.
3. Add mentally.
4. Use the rules for order of operations with real numbers.
5. Translate words and phrases that indicate addition.

1 Use a number line to find the sum $1 + 4$.

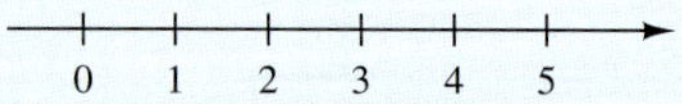

2 Use a number line to find the sum $-2 + (-5)$.

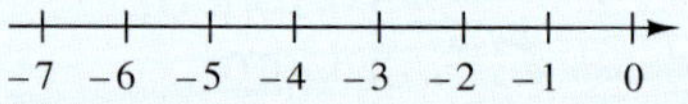

Teaching Tip
To avoid confusion, two operation symbols should not be written successively without a parenthesis between them.

Answers

1. $1 + 4 = 5$

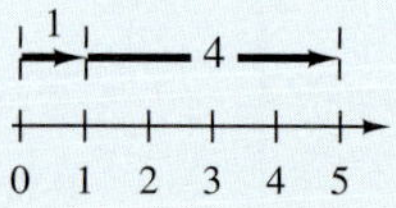

2. $-2 + (-5) = -7$

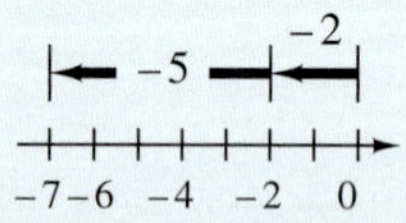

OBJECTIVE **1** **Add two numbers with the same sign.** Recall that the answer to an addition problem is the **sum.** A number line can be used to add real numbers.

EXAMPLE 1 **Adding Two Positive Numbers on a Number Line**

Use a number line to find the sum $2 + 3$.

Step 1 Start at 0 and draw an arrow 2 units to the *right*. See **Figure 8.**

Step 2 From the right end of that arrow, draw another arrow 3 units to the right.

The number below the end of this second arrow is 5, so $2 + 3 = 5$.

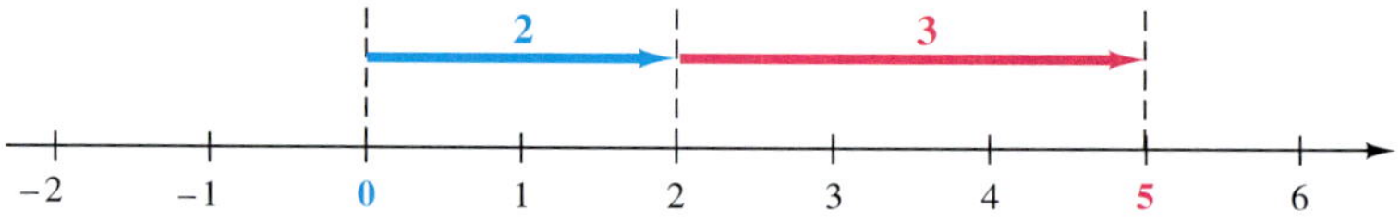

Figure 8

◀ Work Problem **1** at the Side.

EXAMPLE 2 **Adding Two Negative Numbers on a Number Line**

Use a number line to find the sum $-2 + (-4)$. (We put parentheses around the -4 due to the $+$ and $-$ next to each other.)

Step 1 Start at 0 and draw an arrow 2 units to the *left*. See **Figure 9**.

Step 2 From the left end of the first arrow, draw a second arrow 4 units to the *left* to represent the addition of a *negative* number.

The number below the end of this second arrow is -6, so $-2 + (-4) = -6$.

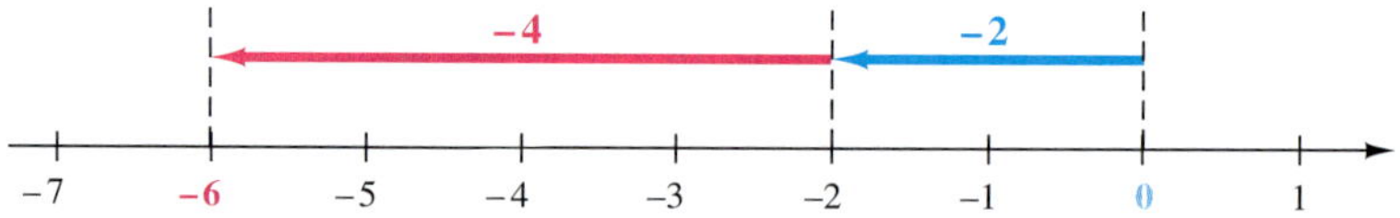

Figure 9

◀ Work Problem **2** at the Side.

In **Example 2,** we found that the sum of the two negative numbers -2 and -4 is a negative number whose distance from 0 is the sum of the distance of -2 from 0 and the distance of -4 from 0. ***That is, the sum of two negative numbers is the opposite of the sum of their absolute values.***

$$\begin{aligned}&-2 + (-4)\\&= -(|-2| + |-4|)\\&= -(2 + 4)\\&= -6\end{aligned}$$

Adding Real Numbers with the Same Sign

To add two numbers with the same sign, add the absolute values of the numbers. Give the result the same sign as the numbers being added.

Example: $-4 + (-3) = -7$

EXAMPLE 3 Adding Two Negative Numbers

Find each sum.

(a) $-2 + (-9)$
$= -11$

(b) $-8 + (-12)$
$= -20$

(c) $-15 + (-3)$
$= -18$

The sum of two negative numbers is negative.

Work Problem 3 at the Side. ▶

OBJECTIVE 2 Add numbers with different signs.

EXAMPLE 4 Adding Numbers with Different Signs

Use the number line to find the sum $-2 + 5$.

Step 1 Start at 0 and draw an arrow 2 units to the left. See **Figure 10.**

Step 2 From the left end of this arrow, draw a second arrow 5 units to the right.

The number below the end of the second arrow is 3, so $-2 + 5 = 3$.

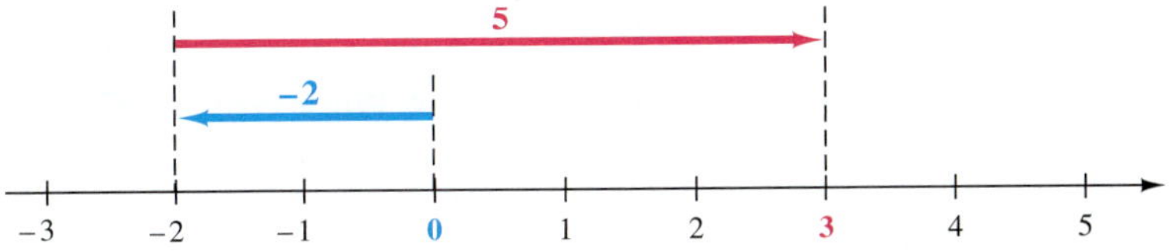

Figure 10

Work Problem 4 at the Side. ▶

Adding Real Numbers with Different Signs

To add two numbers with different signs, find the absolute values of the numbers. Subtract the lesser absolute value from the greater. Give the answer the same sign as the number with the greater absolute value.

Example: $-12 + 6 = -6$

EXAMPLE 5 Adding Numbers with Different Signs

Find the sum $-12 + 5$.

Find the absolute value of each number.

$$|-12| = 12 \quad \text{and} \quad |5| = 5$$

Then find the difference between these absolute values: $12 - 5 = 7$. The sum will be negative, since $|-12| > |5|$.

$$-12 + 5 = -7$$

Work Problem 5 at the Side. ▶

3 Find each sum.

(a) $-7 + (-3)$

(b) $-12 + (-18)$

(c) $-15 + (-4)$

4 Use a number line to find each sum.

(a) $6 + (-3)$

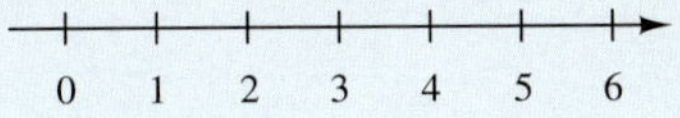

(b) $-5 + 1$

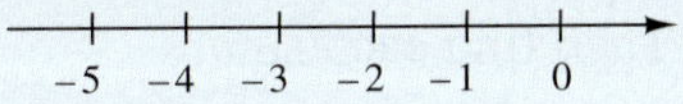

5 Find each sum.

(a) $-24 + 11$

(b) $30 + (-8)$

Answers

3. **(a)** -10 **(b)** -30 **(c)** -19

4. **(a)** $6 + (-3) = 3$

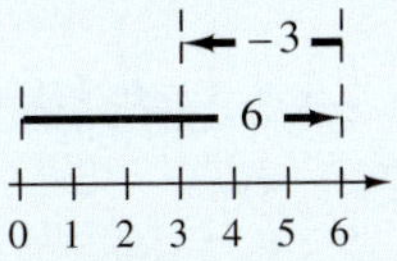

(b) $-5 + 1 = -4$

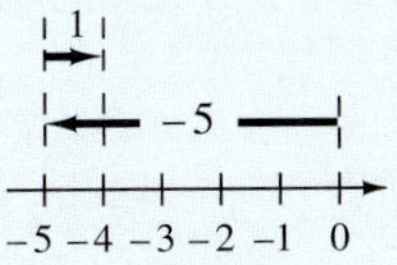

5. **(a)** -13 **(b)** 22

6 Check each answer by adding mentally. If necessary, use a number line.

(a) $-8 + 2 = -6$

(b) $-15 + 4 = -11$

(c) $17 + (-10) = 7$

(d) $\frac{3}{4} + \left(-1\frac{3}{8}\right) = -\frac{5}{8}$

(e) $-9.5 + 3.8 = -5.7$

(f) $37 + (-37) = 0$

7 GS Label the order in which each expression should be evaluated. Then find each sum.

(a) $2 + [7 + (-3)]$
② ①
$= 2 +$ ____
$=$ ____

(b) $6 + [(-2 + 5) + 7]$
○ ○ ○
$= 6 + [$____ $+ 7]$
$= 6 +$ ____
$=$ ____

(c) $-9 + [-4 + (-8 + 6)]$
○ ○ ○

Answers

6. All are correct.

7. **(a)** 4; 6 **(b)** ③, ①, ②; 3; 10; 16
(c) ③, ②, ①; −15

Calculator Tip

The (−) or (+/−) key is used to input a negative number in some scientific calculators.

OBJECTIVE 3 Add mentally.

EXAMPLE 6 Adding a Positive Number and a Negative Number

Check each answer by adding mentally. If necessary, use a number line.

(a) $7 + (-4)$
$= 3$

(b) $-8 + 12$
$= 4$

(c) $-\frac{1}{2} + \frac{1}{8}$
$= -\frac{4}{8} + \frac{1}{8}$
$= -\frac{3}{8}$

(d) $\frac{5}{6} + \left(-1\frac{1}{3}\right)$
$= \frac{5}{6} + \left(-\frac{4}{3}\right)$
$= \frac{5}{6} + \left(-\frac{8}{6}\right)$
$= -\frac{3}{6}$, or $-\frac{1}{2}$

(e) $-4.6 + 8.1$
$= 3.5$

(f) $-16 + 16$
$= 0$

(g) $42 + (-42)$
$= 0$

◀ Work Problem **6** at the Side.

Adding Signed Numbers

Same sign Add the absolute values of the numbers. Give the sum the same sign as the numbers being added.

Different signs Find the absolute values of the numbers, and subtract the lesser absolute value from the greater. Give the answer the sign of the number having the greater absolute value.

OBJECTIVE 4 Use the rules for order of operations with real numbers. When a problem involves square brackets, [ ], we do the calculations inside the brackets until a single number is obtained.

EXAMPLE 7 Adding with Brackets

Find each sum.

(a) $-3 + [4 + (-8)]$ (Start here.)
$= -3 + (-4)$
$= -7$

(b) $8 + [(-2 + 6) + (-3)]$
$= 8 + [4 + (-3)]$
$= 8 + 1$
$= 9$

Follow the order of operations.

◀ Work Problem **7** at the Side.

OBJECTIVE 5 Translate words and phrases that indicate addition. Problem solving requires translating words and phrases into symbols. We began this process in **Section 1.1.** The table lists key words and phrases that indicate addition.

Word or Phrase Indicating Addition	Example	Numerical Expression and Simplification
Sum of	The *sum of* -3 and 4	$-3 + 4$, which equals 1
Added to	5 *added to* -8	$-8 + 5$, which equals -3
More than	12 *more than* -5	$(-5) + 12$, which equals 7
Increased by	-6 *increased by* 13	$-6 + 13$, which equals 7
Plus	3 *plus* 14	$3 + 14$, which equals 17

EXAMPLE 8 Translating Words and Phrases (Addition)

Write a numerical expression for each phrase, and simplify the expression.

(a) The sum of -8 and 4 and 6

$-8 + 4 + 6$ simplifies to $-4 + 6$, which equals 2.

Add in order from left to right.

(b) 3 more than -5, increased by 12

$(-5 + 3) + 12$ simplifies to $-2 + 12$, which equals 10.

Here we *simplified* each expression by performing the operations.

Work Problem 8 at the Side. ▶

In applications, gains may be interpreted as positive numbers and losses as negative numbers.

EXAMPLE 9 Interpreting Gains and Losses

A football team gained 3 yd on first down, lost 12 yd on second down, and then gained 13 yd on third down. How many yards did the team gain or lose altogether on these plays?

$3 + (-12) + 13$ Gains plus losses

$= [3 + (-12)] + 13$ Add from left to right.

$= (-9) + 13$

$= 4$ Add.

The team gained 4 yd altogether on these plays.

Work Problem 9 at the Side. ▶

8 Write a numerical expression for each phrase, and simplify the expression.

(a) 4 more than -12

Teaching Tip
Point out that the word "and" in the first example in the table does not mean addition, but indicates where the addition symbol goes.

(b) The sum of 6 and -7

(c) -12 added to -31

(d) 7 increased by the sum of 8 and -3

9 Solve the problem.

A football team lost 8 yd on first down, lost 5 yd on second down, and then gained 7 yd on third down. How many yards did the team gain or lose altogether on these plays?

Answers

8. **(a)** $-12 + 4$; -8 **(b)** $6 + (-7)$; -1
(c) $-31 + (-12)$; -43
(d) $7 + [8 + (-3)]$; 12

9. The team lost 6 yd.

1.4 Exercises

FOR EXTRA HELP

Download the MyDashBoard App

MyMathLab®

CONCEPT CHECK *Complete each of the following.*

1. The sum of two negative numbers will always be a (*positive* / <u>*negative*</u>) number. Give a number-line illustration using the sum $-2 + (-3) =$ <u>-5</u>.

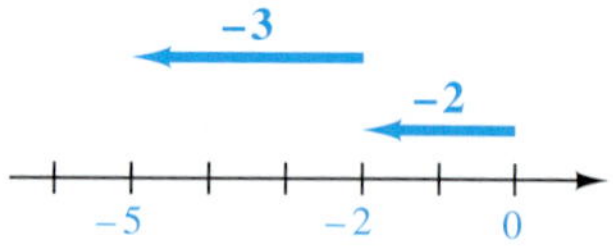

2. The sum of a number and its opposite will always be <u>zero (0)</u>.

3. When adding a positive number and a negative number, where the negative number has the greater absolute value, the sum will be a (*positive* / <u>*negative*</u>) number. Give a number-line illustration using the sum $-4 + 2 =$ <u>-2</u>.

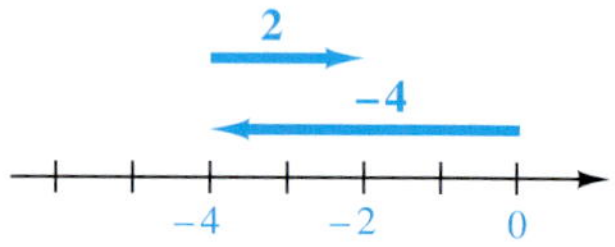

4. To simplify the expression $8 + [-2 + (-3 + 5)]$, one should begin by adding <u>-3</u> and <u>5</u>, according to the rules for order of operations.

CONCEPT CHECK *By the rules for order of operations, what is the first step you would use to simplify each expression?*

5. $4[3(-2 + 5) - 1]$ Add -2 and 5.

6. $[-4 + 7(-6 + 2)]$ Add -6 and 2.

7. $9 + ([-1 + (-3)] + 5)$ Add -1 and -3.

8. $[(-8 + 4) + (-6)] + 5$ Add -8 and 4.

Find each sum. ***See Examples 1–7.***

9. $6 + (-4)$ 2

10. $8 + (-5)$ 3

11. $12 + (-15)$ -3

12. $4 + (-8)$ -4

13. $-7 + (-3)$ -10

14. $-11 + (-4)$ -15

15. $-10 + (-3)$ -13

16. $-16 + (-7)$ -23

17. $-12.4 + (-3.5)$ -15.9

18. $-21.3 + (-2.5)$ -23.8

19. $10 + [-3 + (-2)]$
$= 10 + (\underline{-5})$
$= \underline{5}$

20. $13 + [-4 + (-5)]$
$= 13 + (\underline{-9})$
$= \underline{4}$

21. $5 + [14 + (-6)]$ 13

22. $7 + [3 + (-14)]$ -4

23. $-3 + [5 + (-2)]$ 0

24. $-7 + [10 + (-3)]$ 0

25. $-8 + [3 + (-1) + (-2)]$ -8

26. $-7 + [5 + (-8) + 3]$ -7

27. $\frac{9}{10}+\left(-\frac{3}{5}\right)$ $\frac{3}{10}$

28. $\frac{5}{8}+\left(-\frac{17}{12}\right)$ $-\frac{19}{24}$

29. $-\frac{1}{6}+\frac{2}{3}$ $\frac{1}{2}$

30. $-\frac{6}{25}+\frac{19}{20}$ $\frac{71}{100}$

31. $2\frac{1}{2}+\left(-3\frac{1}{4}\right)$ $-\frac{3}{4}$

32. $6\frac{1}{2}+\left(-4\frac{3}{8}\right)$ $\frac{17}{8}$, or $2\frac{1}{8}$

33. $7.8+(-9.4)$ -1.6

34. $14.7+(-10.1)$ 4.6

35. $-7.1+[3.3+(-4.9)]$ -8.7

36. $-9.5+[-6.8+(-1.3)]$ -17.6

37. $[-8+(-3)]+[-7+(-7)]$
$= \underline{-11} + (\underline{-14})$
$= \underline{-25}$

38. $[-5+(-4)]+[9+(-2)]$
$= \underline{-9} + \underline{7}$
$= \underline{-2}$

39. $\left(-\frac{1}{2}+0.25\right)+\left(-\frac{3}{4}+0.75\right)$ $-\frac{1}{4}$, or -0.25

40. $\left(-\frac{3}{2}+0.75\right)+\left(-\frac{1}{2}+2.25\right)$ 1

Perform each operation, and then determine whether the statement is true *or* false. *Try to do all work mentally.* ***See Examples 6 and 7.***

41. $-11+13=13+(-11)$ true

42. $16+(-9)=-9+16$ true

43. $-10+6+7=-3$ false

44. $-12+8+5=-1$ false

45. $\frac{7}{3}+\left(-\frac{1}{3}\right)+\left(-\frac{6}{3}\right)=0$ true

46. $-\frac{3}{2}+1+\frac{1}{2}=0$ true

47. $|-8+10|=-8+(-10)$ false

48. $|-4+6|=-4+(-6)$ false

49. $2\frac{1}{5}+\left(-\frac{6}{11}\right)=-\frac{6}{11}+2\frac{1}{5}$ true

50. $-1\frac{1}{2}+\frac{5}{8}=\frac{5}{8}+\left(-1\frac{1}{2}\right)$ true

51. $-7+[-5+(-3)]=[(-7)+(-5)]+3$ false

52. $6+[-2+(-5)]=[(-4)+(-2)]+5$ true

Write a numerical expression for each phrase, and simplify the expression. ***See Example 8.***

53. The sum of -5 and 12 and 6
$-5+12+6$; 13

54. The sum of -3 and 5 and -12
$-3+5+(-12)$; -10

55. 14 added to the sum of -19 and -4
$[-19+(-4)]+14$; -9

56. -2 added to the sum of -18 and 11
$(-18+11)+(-2)$; -9

57. The sum of −4 and −10, increased by 12
$[-4+(-10)]+12; -2$

58. The sum of −7 and −13, increased by 14
$[-7+(-13)]+14; -6$

59. $\frac{2}{7}$ more than the sum of $\frac{5}{7}$ and $-\frac{9}{7}$
$\left[\frac{5}{7}+\left(-\frac{9}{7}\right)\right]+\frac{2}{7}; -\frac{2}{7}$

60. 0.85 more than the sum of −1.25 and −4.75
$[-1.25+(-4.75)]+0.85; -5.15$

Solve each problem. ***See Example 9.***

61. Nathaniel owed his older sister Jenna $24 for his share of the bill when they took their mother out to dinner for her birthday. He later borrowed $38 from his younger sister Ilana to buy two DVDs. What positive or negative number represents Nathaniel's financial situation with his siblings? −$62

62. Bonika's checking account balance is $54.00. She then takes a gamble by writing a check for $89.00. What is her new balance? (Write the balance as a signed number.) −$35.00

63. The surface, or rim, of a canyon is at altitude 0. On a hike down into the canyon, a party of hikers stops for a rest at 130 m below the surface. They then descend another 54 m. What is their new altitude? (Write the altitude as a signed number.) −184 m

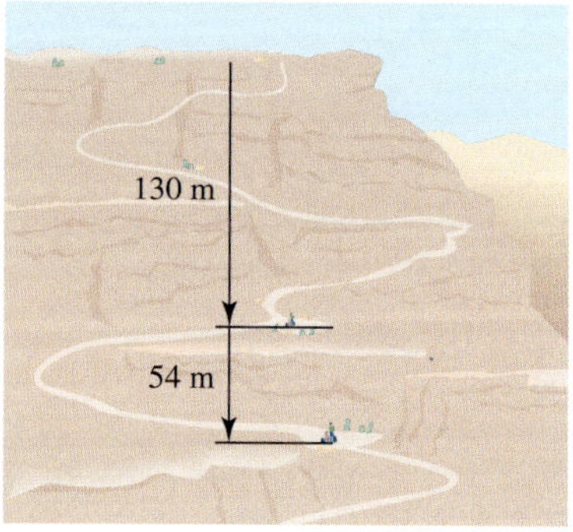

64. A pilot announces to the passengers that the current altitude of their plane is 34,000 ft. Because of some unexpected turbulence, the pilot is forced to descend 2100 ft. What is the new altitude of the plane? (Write the altitude as a signed number.) 31,900 ft

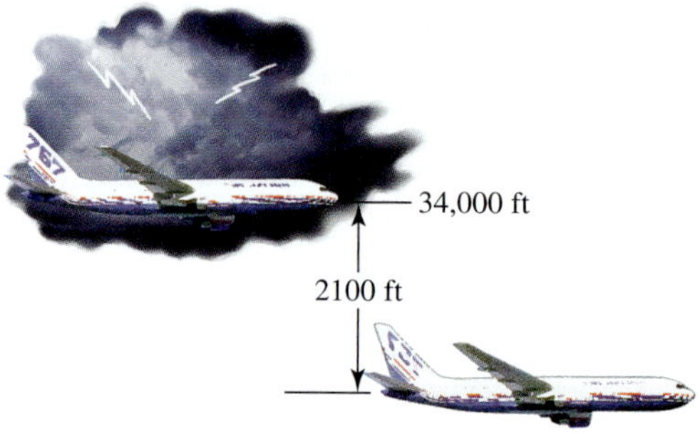

65. J. D. Patin enjoys playing Triominoes every Wednesday night. Last Wednesday, on four successive turns, his scores were −19, 28, −5, and 13. What was his final score for the four turns? 17

66. Gay Aguillard also enjoys playing Triominoes. On five successive turns, her scores were −13, 15, −12, 24, and 14. What was her total score for the five turns? 28

67. On three consecutive passes, a quarterback passed for a gain of 6 yd, was sacked for a loss of 12 yd, and passed for a gain of 43 yd. What positive or negative number represents the total net yardage for the plays? 37 yd

68. On a series of three consecutive running plays, a running back gained 4 yd, lost 3 yd, and lost 2 yd. What positive or negative number represents his total net yardage for the series of plays? −1 yd

69. The lowest temperature ever recorded in Arkansas was -29°F. The highest temperature ever recorded there was 149°F more than the lowest. What was this highest temperature? (*Source: National Climatic Data Center.*) **120°F**

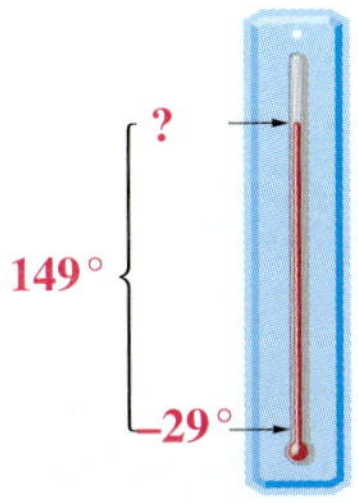

70. On January 23, 1943, the temperature rose 49°F in two minutes in Spearfish, South Dakota. If the starting temperature was -4°F, what was the temperature two minutes later? **45°F**

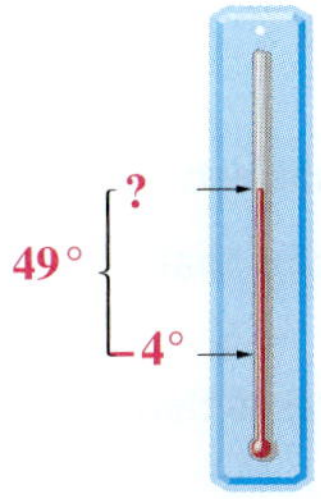

71. Dana Weightman owes \$153 to a credit card company. She makes a \$14 purchase with the card, and then pays \$60 on the account. What is her current balance as a signed number? **−\$107**

72. A female polar bear weighed 660 lb when she entered her winter den. She lost 45 lb during each of the first two months of hibernation, and another 205 lb before leaving the den with her two cubs in March. How much did she weigh when she left the den? **365 lb**

73. Based on the 2010 Census, New York lost 2 seats in the U.S. House of Representatives. Illinois, Iowa, Missouri, and Massachusetts were among the states that each lost 1 seat. Write a signed number that represents the total change in number of seats for these five states. (*Source:* U.S. Census Bureau.) **−6**

74. Based on the 2010 Census, Ohio and New York lost the most seats in the U.S. House of Representatives, each losing 2 seats. The states gaining the most seats were Texas with 4 and Florida with 2. Write a signed number that represents the algebraic sum of these changes. (*Source:* U.S. Census Bureau.) **+2**

1.5 Subtracting Real Numbers

OBJECTIVES

1. Subtract two numbers on a number line.
2. Use the definition of subtraction.
3. Work subtraction problems that involve brackets.
4. Translate words and phrases that indicate subtraction.

OBJECTIVE 1 Subtract two numbers on a number line. Recall that the answer to a subtraction problem is a **difference.** In the subtraction $x - y$, x is the **minuend** and y is the **subtrahend.**

$$x \quad - \quad y \quad = \quad z$$

Minuend (↑ x) Subtrahend (↑ y) Difference (↑ z)

EXAMPLE 1 Subtracting Numbers on a Number Line

Use a number line to find the difference $7 - 4$.

Step 1 Start at 0 and draw an arrow 7 units to the *right*. See **Figure 11.**

Step 2 From the right end of the first arrow, draw a second arrow 4 units to the *left* to represent the subtraction of a positive number.

The number below the end of this second arrow is 3, so $7 - 4 = 3$.

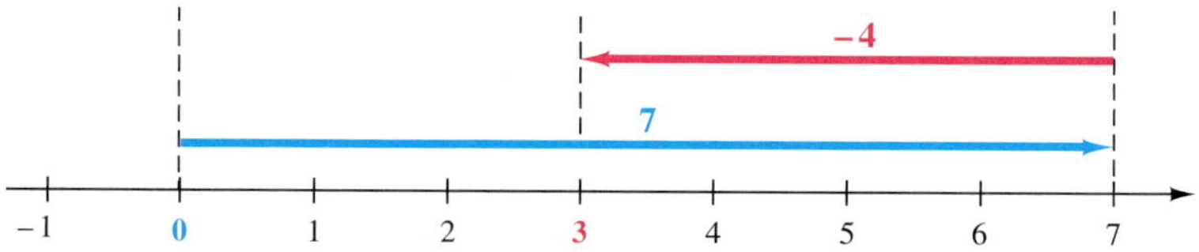

Figure 11

◀ **Work Problem 1 at the Side.**

1 Use a number line to find each difference.

(a) $5 - 1$

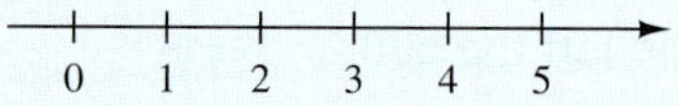

(b) $6 - 2$

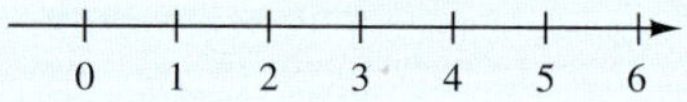

OBJECTIVE 2 Use the definition of subtraction. The procedure used in **Example 1** to find $7 - 4$ is the same procedure for finding $7 + (-4)$.

$$7 - 4 \quad \text{is equal to} \quad 7 + (-4).$$

This shows that *subtracting* a positive number from a larger positive number is the same as *adding* the opposite of the smaller number to the larger.

Definition of Subtraction

For any real numbers a and b,

$$a - b \quad \text{is defined as} \quad a + (-b).$$

To subtract b from a, add the additive inverse (or opposite) of b to a. In words, change the subtrahend to its opposite and add.

Example: $4 - 9$

$= 4 + (-9)$

$= -5$

Subtracting Signed Numbers

Step 1 Change the subtraction symbol to an addition symbol, and change the sign of the subtrahend.

Step 2 Add, as in **Section 1.4.**

Answers

1. (a) $5 - 1 = 4$

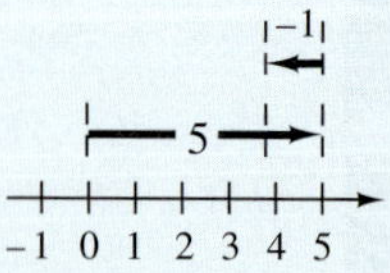

(b) $6 - 2 = 4$

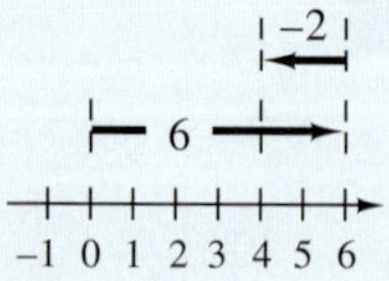

EXAMPLE 2 **Using the Definition of Subtraction**

Subtract.

(No change; Change − to +; Opposite of 3)

(a) $12 - 3 = 12 + (-3) = 9$

(b) $5 - 7 = 5 + (-7) = -2$ (−7 has the greater absolute value, so the sum is negative.)

(c) $-8 - 15 = -8 + (-15) = -23$

(No change; Change − to +; Opposite of −5)

(d) $-3 - (-5) = -3 + (5) = 2$

(e) $\frac{3}{8} - \left(-\frac{4}{5}\right)$

$= \frac{15}{40} - \left(-\frac{32}{40}\right)$ Find a common denominator.

$= \frac{15}{40} + \frac{32}{40}$ Use the definition of subtraction.

$= \frac{47}{40}$ Add the fractions.

Work Problem 2 at the Side. ▶

Uses of the Symbol −

We use the symbol − for three purposes.

1. It can represent ***subtraction,*** as in $9 - 5 = 4$.
2. It can represent ***negative numbers,*** such as −10, −2, and −3.
3. It can represent ***the opposite (or additive inverse) of a number,*** as in "the opposite (or additive inverse) of 8 is −8."

We may see more than one use in the same problem, such as $-6 - (-9)$, where −9 is subtracted from −6. The meaning of the symbol depends on its position in the algebraic expression.

OBJECTIVE 3 **Work subtraction problems that involve brackets.** As before, first perform any operations inside the parentheses and brackets.

EXAMPLE 3 **Subtracting with Grouping Symbols**

Perform each operation.

(a) $-6 - [2 - (8 + 3)]$ (Work from the inside out.)

$= -6 - [2 - 11]$ Add inside the parentheses.

$= -6 - [2 + (-11)]$ Definition of subtraction

$= -6 - (-9)$ Add.

$= -6 + 9$ Definition of subtraction

$= 3$ Add.

Continued on Next Page

2 Subtract.

GS **(a)** $6 - 10$

$= 6 + (\_\_\_)$

$= \_\_\_$

(b) $-2 - 4$

GS **(c)** $3 - (-5)$

$= 3 + \_\_\_$

$= \_\_\_$

(d) $-8 - (-12)$

(e) $\frac{5}{4} - \left(-\frac{3}{7}\right)$

Teaching Tip

Tell students to read **Example 2(d)** as "negative three minus negative five."

Teaching Tip

Emphasize the uses of the symbol − with students.

Answers

2. **(a)** −10; −4 **(b)** −6 **(c)** 5; 8 **(d)** 4 **(e)** $\frac{47}{28}$

3 Label the order in which each **GS** expression should be evaluated. Then perform each operation.

(a) $2 - [(-3) - (4 + 6)]$

◯ ◯ ◯

(b) $[(5 - 7) + 3] - 8$

◯ ◯ ◯

(c) $6 - [(-1 - 4) - 2]$

◯ ◯ ◯

(b) $5 - \left[\left(-\frac{1}{3} - \frac{1}{2}\right) - (4 - 1)\right]$ Work from the inside out.

$= 5 - \left[\left(-\frac{1}{3} + \left(-\frac{1}{2}\right)\right) - 3\right]$ Start within each set of parentheses inside the brackets.

$= 5 - \left[\left(-\frac{5}{6}\right) - 3\right]$ Use 6 as the common denominator; $-\frac{1}{3} + \left(-\frac{1}{2}\right) = -\frac{2}{6} + \left(-\frac{3}{6}\right) = -\frac{5}{6}$

$= 5 - \left[\left(-\frac{5}{6}\right) + (-3)\right]$ Definition of subtraction

$= 5 - \left[\left(-\frac{5}{6}\right) + \left(-\frac{18}{6}\right)\right]$ Use 6 as the common denominator; $-\frac{3}{1} \cdot \frac{6}{6} = -\frac{18}{6}$

$= 5 - \left(-\frac{23}{6}\right)$ Add inside the brackets.

$= 5 + \frac{23}{6}$ Definition of subtraction

$= \frac{30}{6} + \frac{23}{6}$ Write 5 as an improper fraction.

$= \frac{53}{6}$ Add fractions.

◀ **Work Problem 3 at the Side.**

OBJECTIVE 4 Translate words and phrases that indicate subtraction. The table lists words and phrases that indicate subtraction in problem solving.

Word, Phrase, or Sentence Indicating Subtraction	Example	Numerical Expression and Simplification
Difference between	The ***difference between*** −3 and −8	−3 − (−8) simplifies to −3 + 8, which equals 5
Subtracted from*	12 ***subtracted from*** 18	18 − 12, which equals 6
From . . . , subtract	***From*** 12, ***subtract*** 8.	12 − 8 simplifies to 12 + (−8), which equals 4
Less	6 ***less*** 5	6 − 5, which equals 1
Less than*	6 ***less than*** 5	5 − 6 simplifies to 5 + (−6), which equals −1
Decreased by	9 ***decreased by*** −4	9 − (−4) simplifies to 9 + 4, which equals 13
Minus	8 ***minus*** 5	8 − 5, which equals 3

*Be careful with order when translating.

CAUTION

When subtracting two numbers, be careful to write them in the correct order, because, in general,

$x - y \neq y - x.$ For example, $5 - 3 \neq 3 - 5$.

Think carefully before interpreting an expression involving subtraction.

Answers

3. (a) ③, ②, ①; 15

(b) ①, ②, ③; −7

(c) ③, ①, ②; 13

EXAMPLE 4 Translating Words and Phrases (Subtraction)

Write a numerical expression for each phrase, and simplify the expression.

(a) The **difference between** -8 and 5

When "difference between" is used, write the numbers in the order they are given.

$$-8 - 5 \quad \text{simplifies to} \quad -8 + (-5), \quad \text{which equals} \quad -13.$$

(b) 4 **subtracted from** the sum of 8 and -3

Here the operation of addition is also used, as indicated by the word *sum*. First, add 8 and -3. Next, subtract 4 from this sum. The expression is

$$[8 + (-3)] - 4 \quad \text{simplifies to} \quad 5 - 4, \quad \text{which equals} \quad 1.$$

(c) 4 **less than** -6

Here 4 must be taken *from* -6, so write -6 first.

Be careful with order.

$$-6 - 4 \quad \text{simplifies to} \quad -6 + (-4), \quad \text{which equals} \quad -10.$$

Notice that "4 less than -6" differs from "4 *is less than* -6." The statement "4 is less than -6" is symbolized as $4 < -6$ (which is a false statement).

(d) 8, **decreased by** 5 **less than** 12

First, write "5 less than 12" as $12 - 5$. Next, subtract $12 - 5$ from 8.

$$8 - (12 - 5) \quad \text{simplifies to} \quad 8 - 7, \quad \text{which equals} \quad 1.$$

Work Problem **4** at the Side. ▶

EXAMPLE 5 Solving an Applied Problem Involving Subtraction

The record high temperature of 134°F in the United States was recorded at Death Valley, California, in 1913. The record low was -80°F, at Prospect Creek, Alaska, in 1971. See **Figure 12.** What is the difference between these highest and lowest temperatures? (*Source:* National Climatic Data Center.)

We must subtract the lowest temperature from the highest temperature.

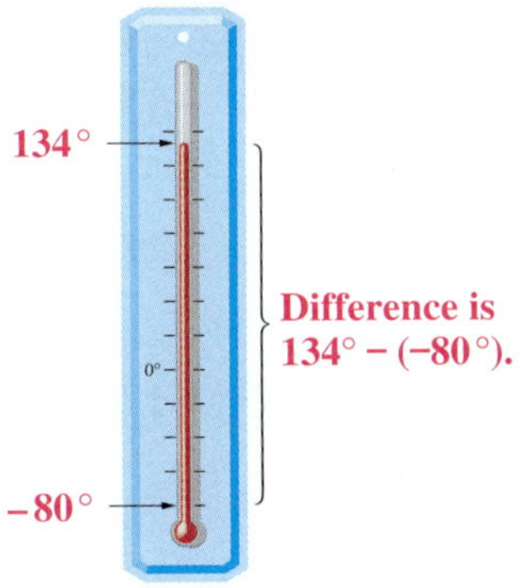

Figure 12

Order of numbers matters in subtraction.

$$\begin{aligned} &134 - (-80) \\ &= 134 + 80 \quad \text{Use the definition of subtraction.} \\ &= 214 \quad \text{Add.} \end{aligned}$$

The difference between the two temperatures is 214°F.

Work Problem **5** at the Side. ▶

---

4 Write a numerical expression for each phrase, and simplify the expression.

(a) The difference between -5 and -12

(b) -2 subtracted from the sum of 4 and -4

(c) 7 less than -2

(d) 9, decreased by 10 less than 7

5 Solve the problem.

The highest elevation in Argentina is Mt. Aconcagua, which is 6960 m above sea level. The lowest point in Argentina is the Valdes Peninsula, 40 m below sea level. Find the difference between the highest and lowest elevations.

Answers

4. (a) $-5 - (-12)$; 7
   (b) $[4 + (-4)] - (-2)$; 2
   (c) $-2 - 7$; -9
   (d) $9 - (7 - 10)$; 12
5. 7000 m

1.5 Exercises

FOR EXTRA HELP

Download the MyDashBoard App

MyMathLab®

CONCEPT CHECK *Fill in each blank with the correct response.*

1. By the definition of subtraction, in order to perform the subtraction $-6-(-8)$, we must add the opposite of ___-8___ to ___-6___ to get ___2___.

2. By the rules for order of operations, to simplify $8-[3-(-4-5)]$, the first step is to subtract ___5___ from ___-4___.

3. "The difference between 7 and 12" translates as ___$7-12$___, while "the difference between 12 and 7" translates as ___$12-7$___.

4. To subtract b from a, add the ___additive___ ___inverse___, or ___opposite___, of b to a.

5. $-8-4$ can be rewritten as $-8+($___-4___$)$.

6. $-19-22$ can be rewritten as $-19+($___-22___$)$.

Perform the indicated operations. **See Examples 1–3.**

7. $-7-3$
   $=-7+($___-3___$)$
   $=$ ___-10___

8. $-12-5$
   $=-12+($___-5___$)$
   $=$ ___-17___

9. $-10-6$ -16

10. $-13-16$ -29

11. $7-(-4)$
   $=7+$ ___4___
   $=$ ___11___

12. $9-(-6)$
   $=9+$ ___6___
   $=$ ___15___

13. $6-(-13)$ 19

14. $13-(-3)$ 16

15. $-7-(-3)$ -4

16. $-8-(-6)$ -2

17. $3-(4-6)$ 5

18. $6-(7-14)$ 13

19. $-3-(6-9)$ 0

20. $-4-(5-12)$ 3

21. $\frac{1}{2}-\left(-\frac{1}{4}\right)$ $\frac{3}{4}$

22. $\frac{1}{3}-\left(-\frac{4}{3}\right)$ $\frac{5}{3}$

23. $-\frac{3}{4}-\frac{5}{8}$ $-\frac{11}{8}$

24. $-\frac{5}{6}-\frac{1}{2}$ $-\frac{4}{3}$

25. $\frac{5}{8}-\left(-\frac{1}{2}-\frac{3}{4}\right)$ $\frac{15}{8}$

26. $\frac{9}{10}-\left(-\frac{1}{8}-\frac{3}{10}\right)$ $\frac{53}{40}$

27. $4.4-(-9.2)$ 13.6

28. $6.7-(-12.6)$ 19.3

29. $-7.4-4.5$ -11.9

30. $-5.4-9.6$ -15.0

31. $8-(-3)-9+6$ 8

32. $12-(-8)-25+9$ 4

33. $-5.2-(8.4-10.8)$ -2.8

34. $-9.6-(3.5-12.6)$ -0.5

35. $[(-3.1)-4.5]-(0.8-2.1)$ -6.3

36. $[(-7.8)-9.3]-(0.6-3.5)$ -14.2

37. $-12-[(9-2)-(-6-3)]$ -28

38. $-4-[(6-9)-(-7-4)]$ -12

39. $\left(-\frac{3}{8}-\frac{2}{3}\right)-\left(-\frac{9}{8}-3\right)$ $\frac{37}{12}$

40. $\left(-\frac{3}{4}-\frac{5}{2}\right)-\left(-\frac{1}{8}-1\right)$ $-\frac{17}{8}$

41. $[-12.25-(8.34+3.57)]-17.88$ -42.04

42. $[-34.99+(6.59-12.25)]-8.33$ -48.98

Write a numerical expression for each phrase and simplify. **See Example 4.**

43. The difference between 4 and -8
$4-(-8)$; 12

44. The difference between 7 and -14
$7-(-14)$; 21

45. 8 less than -2
$-2-8$; -10

46. 9 less than -13
$-13-9$; -22

47. The sum of 9 and -4, decreased by 7
$[9+(-4)]-7$; -2

48. The sum of 12 and -7, decreased by 14
$[12+(-7)]-14$; -9

49. 12 less than the difference between 8 and -5
$[8-(-5)]-12$; 1

50. 19 less than the difference between 9 and -2
$[9-(-2)]-19$; -8

Solve each problem. **See Example 5.**

51. The lowest temperature ever recorded in Illinois was -36°F on January 5, 1999. The lowest temperature ever recorded in Utah was on February 1, 1985, and was 33°F lower than Illinois's record low. What is the record low temperature for Utah? (*Source:* National Climatic Data Center.)
-69°F

52. The lowest temperature ever recorded in South Carolina was -19°F on January 21, 1985. The lowest temperature ever recorded in Wisconsin was 36° lower than South Carolina's record low. What is the record low temperature in Wisconsin? (*Source:* National Climatic Data Center.)
-55°F

53. The top of Mount Whitney, visible from Death Valley, has an altitude of 14,494 ft above sea level. The bottom of Death Valley is 282 ft below sea level. Using 0 as sea level, find the difference between these two elevations. (*Source: World Almanac and Book of Facts.*)
14,776 ft

54. The height of Mount Pumasillo in Peru is 20,492 ft. The depth of the Peru-Chile Trench in the Pacific Ocean is 26,457 ft below sea level. Find the difference between these two elevations. (*Source: World Almanac and Book of Facts.*)
46,949 ft

55. Samir owed his brother \$10. He later borrowed \$70. What positive or negative number represents his present financial status?
−\$80

56. Francesca has \$15 in her purse, and Emilio has a debt of \$12. Find the difference between these amounts.
\$27

57. A chemist is running an experiment under precise conditions. At first, she runs it at −174.6°F. She then lowers the temperature by 2.3°F. What is the new temperature for the experiment?
−176.9°F

58. At 2:00 A.M., a plant worker found that a dial reading was 7.904. At 3:00 A.M., she found the reading to be −3.291. Find the difference between these two readings.
11.195

59. In August, Kari Heen began with a checking account balance of \$904.89. Her checks and deposits for August are given below.

Checks	Deposits
\$35.84	\$85.00
\$26.14	\$120.76
\$3.12	

Assuming no other transactions, what was her account balance at the end of August?
\$1045.55

60. In September, Derek Bowen began with a checking account balance of \$904.89. His checks and deposits for September are given below.

Checks	Deposits
\$41.29	\$80.59
\$13.66	\$276.13
\$84.40	

Assuming no other transactions, what was his account balance at the end of September?
\$1122.26

61. A certain Greek mathematician was born in 426 B.C. His father was born 43 years earlier. In what year was his father born?
469 B.C.

62. A certain Roman philosopher was born in 325 B.C. Her mother was born 35 years earlier. In what year was her mother born?
360 B.C.

63. Kim Falgout owes $870.00 on her MasterCard account. She returns two items costing $35.90 and $150.00 and receives credits for these on the account. Next, she makes a purchase of $82.50, and then two more purchases of $10.00 each. She makes a payment of $500.00. She then incurs a finance charge of $37.23. How much does she still owe?

$323.83

64. Charles Vosburg owes $679.00 on his Visa account. He returns three items costing $36.89, $29.40, and $113.55 and receives credits for these on the account. Next, he makes purchases of $135.78 and $412.88, and two purchases of $20.00 each. He makes a payment of $400. He then incurs a finance charge of $24.57. How much does he still owe?

$712.39

65. José Martinez enjoys diving in Lake Okoboji. He dives to 34 ft below the surface of the lake. His partner, Sean O'Malley, dives to 40 ft below the surface, but then ascends 20 ft. What is the vertical distance between José and Sean?

14 ft

66. Rhonda Alessi also enjoys diving. She dives to 12 ft below the surface of False River. Her sister, Sandy, dives to 20 ft below the surface, but then ascends 10 ft. What is the vertical distance between Rhonda and Sandy?

2 ft

67. The height of Mt. Foraker is 17,400 ft, while the depth of the Java Trench is 23,376 ft. What is the vertical distance between the top of Mt. Foraker and the bottom of the Java Trench? (*Source: World Almanac and Book of Facts.*)

40,776 ft

68. The height of Mt. Wilson in Colorado is 14,246 ft, while the depth of the Cayman Trench is 24,721 ft. What is the vertical distance between the top of Mt. Wilson and the bottom of the Cayman Trench? (*Source: World Almanac and Book of Facts.*)

38,967 ft

69. In 2005, Americans saved −0.5% of their after-tax incomes, the first negative personal savings rate since 1933. In the fourth quarter of 2011, they saved 3.9%. Find the difference between the two amounts for 2011 and 2005.

4.4%

70. Refer to **Exercise 69.** How is it possible that Americans had a negative personal savings rate in 2005?

Americans spent more money than they earned, which means that they had to dip into savings or borrow money.

71. In 2000, the federal budget had a surplus of \$236 billion. In 2010, the federal budget had a deficit of \$1294 billion. Find the difference between the amounts for 2000 and 2010. (*Source:* U.S. Office of Management and Budget.)

\$1530 billion

72. In 1998, undergraduate college students had an average (mean) credit card balance of \$1879. The average balance increased \$869 by 2000, then dropped \$579 by 2004, and then increased \$1004 by 2008. What was the average credit card balance of undergraduate college students in 2008? (*Source:* Sallie Mae.)

\$3173

Median sales prices for existing single-family homes in the United States for the years 2005 through 2009 are shown in the table. Complete the table, determining the change from one year to the next by subtraction.

	Year	Median Sales Price	Change from Previous Year
	2005	\$219,900	——
73.	2006	\$221,900	**\$2900**
74.	2007	\$219,000	**−\$2900**
75.	2008	\$198,100	**−\$20,900**
76.	2009	\$175,500	**−\$22,600**

Source: National Association of Realtors.

The table shows the heights of some selected mountains. Use the information given to answer Exercises 77 and 78.

77. How much higher is Mt. Wilson than Pikes Peak?

136 ft

78. If Mt. Wilson and Pikes Peak were stacked one on top of the other, how much higher would they be than Mt. Foraker?

10,956 ft

Mountain	Height (in feet)
Foraker	17,400
Wilson	14,246
Pikes Peak	14,110

Source: *World Almanac and Book of Facts.*

CONCEPT CHECK *In Exercises 79–84, suppose that x represents a positive number and y represents a negative number. Determine whether the given expression must represent a positive number or a negative number.*

79. $y - x$

negative

80. $x - y$

positive

81. $x + |y|$

positive

82. $y - |x|$

negative

83. $|x| + |y|$

positive

84. $-(|x| + |y|)$

negative

1.6 Multiplying and Dividing Real Numbers

OBJECTIVES

1. Find the product of a positive number and a negative number.
2. Find the product of two negative numbers.
3. Use the reciprocal of a number to apply the definition of division.
4. Use the rules for order of operations when multiplying and dividing signed numbers.
5. Evaluate expressions involving variables.
6. Translate words and phrases involving multiplication and division.
7. Translate simple sentences into equations.

The result of multiplication is the **product.** We know that the product of two positive numbers is positive. We also know that the product of 0 and any positive number is 0, so we extend that property to all real numbers.

Multiplication Property of 0

For any real number a, the following hold.

$$\mathbf{a \cdot 0 = 0} \quad \text{and} \quad \mathbf{0 \cdot a = 0}$$

OBJECTIVE 1 Find the product of a positive number and a negative number. Observe the following pattern.

$$3 \cdot 5 = 15$$
$$3 \cdot 4 = 12$$
$$3 \cdot 3 = 9$$
$$3 \cdot 2 = 6$$
$$3 \cdot 1 = 3$$
$$3 \cdot 0 = 0$$
$$3 \cdot (-1) = ?$$

The products decrease by 3.

What should $3(-1)$ equal? Since multiplication can also be considered repeated addition, the product $3(-1)$ represents the sum

$$-1 + (-1) + (-1), \quad \text{which equals} \quad -3,$$

so the product should be -3, which fits the pattern. Also, $3(-2)$ represents the sum

$$-2 + (-2) + (-2), \quad \text{which equals} \quad -6.$$

Work Problem 1 at the Side. ▶

These results suggest the following rule.

Multiplying Numbers with Different Signs

The product of a positive number and a negative number is negative.

Examples: $6(-3) = -18$ and $-3(6) = -18$

EXAMPLE 1 Multiplying a Positive Number and a Negative Number

Find each product using the multiplication rule given in the box.

(a) $8(-5)$
$= -(8 \cdot 5)$
$= -40$

(b) $-9\left(\frac{1}{3}\right)$
$= -3$

(c) $-6.2(4.1)$
$= -25.42$

Work Problem 2 at the Side. ▶

1 Find each product by finding the sum of three numbers.

(a) $3(-3)$

(b) $3(-4)$

(c) $3(-5)$

2 Find each product.

(a) $2(-6)$

(b) $7(-8)$

(c) $-9(2)$

(d) $-16\left(\frac{5}{32}\right)$

(e) $4.56(-10)$

Answers

1. (a) -9 (b) -12 (c) -15
2. (a) -12 (b) -56 (c) -18 (d) $-\frac{5}{2}$ (e) -45.6

3 Find each product.

(a) $-5(-6)$

(b) $-7(-3)$

(c) $-8(-5)$

(d) $-11(-2)$

(e) $-17(3)(-7)$

(f) $-41(2)(-13)$

Answers

3. (a) 30 (b) 21 (c) 40 (d) 22
(e) 357 (f) 1066

OBJECTIVE 2 Find the product of two negative numbers. The product of two positive numbers is positive, and the product of a positive number and a negative number is negative. What about the product of two negative numbers? Look at another pattern.

$$-5(4) = -20$$
$$-5(3) = -15$$
$$-5(2) = -10$$
$$-5(1) = -5$$
$$-5(0) = 0$$
$$-5(-1) = ?$$

The products increase by 5.

The numbers in color on the left side of the equality symbols decrease by 1 for each step down the list. The products on the right increase by 5 for each step down the list. To maintain this pattern, $-5(-1)$ should be 5 more than $-5(0)$, or 5 more than 0, so

$$-5(-1) = 5.$$

The pattern continues with

$$-5(-2) = 10$$
$$-5(-3) = 15$$
$$-5(-4) = 20$$
$$-5(-5) = 25, \quad \text{and so on.}$$

This pattern suggests the next rule.

Multiplying Two Negative Numbers

The product of two negative numbers is positive.

Example: $-5(-4) = 20$

EXAMPLE 2 Multiplying Two Negative Numbers

Find each product using the multiplication rule given in the box.

(a) $-9(-2)$
$= 18$

(b) $-6(-12)$
$= 72$

(c) $-2(4)(-1)$
$= -8(-1)$
$= 8$

(d) $3(-5)(-2)$
$= -15(-2)$
$= 30$

◀ Work Problem **3** at the Side.

Multiplying Signed Numbers

The product of two numbers having the *same* sign is *positive*.

The product of two numbers having *different* signs is *negative*.

OBJECTIVE 3 Use the reciprocal of a number to apply the definition of division. Recall that the result of division is the **quotient.** The *quotient* of two numbers involves multiplying by the *reciprocal* of the second number, which is the *divisor.*

Reciprocals

Pairs of numbers whose product is 1 are **reciprocals, or multiplicative inverses,** of each other.

The following table shows several numbers and their reciprocals.

Number	Reciprocal
4	$\frac{1}{4}$
-5	$\frac{1}{-5}$, or $-\frac{1}{5}$
0.3, or $\frac{3}{10}$	$\frac{10}{3}$
$-\frac{5}{8}$	$-\frac{8}{5}$

A number and its reciprocal have a product of 1. For example,

$$4 \cdot \frac{1}{4} = \frac{4}{4}, \text{ or } 1.$$

0 has no reciprocal because the product of 0 and any number is 0, and cannot be 1.

Work Problem 4 at the Side. ▶

The quotient of a and b is the product of a and the reciprocal of b.

Definition of Division

The quotient $\frac{a}{b}$ of real numbers a and b, with $b \neq 0$, is

$$\frac{a}{b} = a \cdot \frac{1}{b}.$$

Example: $\frac{8}{-4}$ means $8\left(-\frac{1}{4}\right)$, which equals -2.

Since division is defined in terms of multiplication, the rules for multiplying signed numbers also apply to dividing them.

EXAMPLE 3 Using the Definition of Division

Find each quotient using the definition of division.

(a) $\frac{12}{3}$

$= 12 \cdot \frac{1}{3}$ $\quad \frac{a}{b} = a \cdot \frac{1}{b}$

$= 4$

(b) $\frac{5(-2)}{2}$

$= -10 \cdot \frac{1}{2}$

$= -5$

(c) $\frac{-1.47}{-7}$

$= -1.47 \cdot \left(-\frac{1}{7}\right)$

$= 0.21$

(d) $-\frac{2}{3} \div \left(-\frac{5}{4}\right)$

$= -\frac{2}{3} \cdot \left(-\frac{4}{5}\right)$

$= \frac{8}{15}$

Work Problem 5 at the Side. ▶

---

4 Complete the table.

	Number	Reciprocal
(a)	6	____
(b)	-2	____
(c)	$\frac{2}{3}$	____
(d)	$-\frac{1}{4}$	____
(e)	0.75	____
(f)	0	____

5 Find each quotient.

(a) $\frac{42}{7}$

(b) $\frac{-36}{(-2)(-3)}$

(c) $\frac{-12.56}{-0.4}$

(d) $\frac{10}{7} \div \left(-\frac{24}{5}\right)$

Answers

4. **(a)** $\frac{1}{6}$ **(b)** $\frac{1}{-2}$, or $-\frac{1}{2}$ **(c)** $\frac{3}{2}$ **(d)** -4 **(e)** $\frac{4}{3}$ **(f)** none

5. **(a)** 6 **(b)** -6 **(c)** 31.4 **(d)** $-\frac{25}{84}$

6 Decide whether each quotient is undefined or equal to 0.

(a) $\frac{-3}{0}$

Teaching Tip

Stress the distinction between expressions such as $\frac{0}{3}$ and $\frac{3}{0}$. (Division is *undefined* when 0 is *underneath*.)

Also tell students that $\frac{0}{0}$ is undefined, or indeterminate.

(b) $\frac{0}{-53}$

7 Find each quotient.

(a) $\frac{-8}{-2}$

Teaching Tip

Show students that a calculator returns an error message when $\frac{1}{0}$ is entered.

(b) $\frac{-16.4}{2.05}$

(c) $\frac{1}{4} \div \left(-\frac{2}{3}\right)$

(d) $\frac{12}{-4}$

Answers

6. (a) undefined (b) 0

7. (a) 4 (b) −8 (c) $-\frac{3}{8}$ (d) −3

In **Example 3(a),** $\frac{12}{3} = 4$, since $4 \cdot 3 = 12$. Multiply to check a division problem.

This relationship between multiplication and division allows us to investigate division involving 0. Consider the quotient $\frac{0}{3}$.

$$\frac{0}{3} = 0, \quad \text{since} \quad 0 \cdot 3 = 0.$$

Now consider $\frac{3}{0}$.

$$\frac{3}{0} = ?$$

We need to find a number that when multiplied by 0 will equal 3, that is, $? \cdot 0 = 3$. *No* real number satisfies this equation, since the product of any real number and 0 must be 0. ***Thus, division by 0 is undefined.***

Division Involving 0

For any real number x, with $x \neq 0$,

$$\mathbf{\frac{0}{x} = 0} \quad \text{and} \quad \mathbf{\frac{x}{0}} \textbf{ is undefined.}$$

Examples: $\frac{0}{-10} = 0$ and $\frac{-10}{0}$ is undefined.

◄ **Work Problem 6 at the Side.**

When dividing fractions, multiplying by the reciprocal of the divisor works well. However it is often easier to divide in the usual way, and then determine the sign of the answer.

Dividing Signed Numbers

The quotient of two numbers having the *same* sign is *positive.*

The quotient of two numbers having *different* signs is *negative.*

Examples: $\frac{-15}{-5} = 3$, $\frac{15}{-5} = -3$, and $\frac{-15}{5} = -3$

EXAMPLE 4 Dividing Signed Numbers

Find each quotient.

(a) $\frac{8}{-2} = -4$ **(b)** $\frac{-10}{2} = -5$ **(c)** $\frac{-4.5}{-0.09} = 50$

(d) $-\frac{1}{8} \div \left(-\frac{3}{4}\right)$

$= -\frac{1}{8} \cdot \left(-\frac{4}{3}\right)$ Multiply by the reciprocal of the divisor.

$= \frac{1}{6}$ Write in lowest terms.

◄ **Work Problem 7 at the Side.**

From the definitions of multiplication and division of real numbers,

$$\frac{-40}{8} = -5 \quad \text{and} \quad \frac{40}{-8} = -5, \quad \text{so} \quad \frac{-40}{8} = \frac{40}{-8}.$$

Based on this example, the quotient of a positive number and a negative number can be written in any of the following three forms.

Equivalent Forms

For any positive real numbers a and b, the following equivalences hold.

$$\frac{-a}{b} = \frac{a}{-b} = -\frac{a}{b}$$

Similarly, the quotient of two negative numbers can be expressed as the quotient of two positive numbers.

Equivalent Forms

For any positive real numbers a and b, the following equivalence holds.

$$\frac{-a}{-b} = \frac{a}{b}$$

OBJECTIVE 4 Use the rules for order of operations when multiplying and dividing signed numbers.

EXAMPLE 5 Using the Rules for Order of Operations

Simplify.

(a) $-9(2) - (-3)(2)$

$= -18 - (-6)$ Multiply.

$= -18 + 6$ Definition of subtraction

$= -12$ Add.

(b) $-6(-2) - 3(-4)$

$= 12 - (-12)$ Multiply.

$= 12 + 12$ Definition of subtraction

$= 24$ Add.

(c) $\dfrac{5(-2) - 3(4)}{2(1 - 6)}$

$= \dfrac{-10 - 12}{2(-5)}$ Simplify the numerator and denominator separately.

$= \dfrac{-22}{-10}$ Subtract in the numerator. Multiply in the denominator.

$= \dfrac{11}{5}$ Write in lowest terms.

Work Problem 8 at the Side. ▶

8 Label the order in which each expression should be evaluated. Then perform the indicated operations.

GS **(a)** $-3(4) - 2(6)$

○ ○○

$= \_\_\_ - \_\_\_$

$= \_\_\_$

GS **(b)** $-8[-1 - (-4)(-5)]$

○ ○ ○

$= -8[-1 - \_\_\_]$

$= -8[\_\_\_]$

$= \_\_\_$

(c) $\dfrac{6(-4) - 2(5)}{3(2 - 7)}$

(d) $\dfrac{-6(-8) + 3(9)}{-2[4 - (-3)]}$

Answers

8. **(a)** (1), (3), (2); -12; 12; -24

(b) (3), (2), (1); 20; -21; 168

(c) $\dfrac{34}{15}$ **(d)** $-\dfrac{75}{14}$

9 Evaluate each expression.

GS **(a)** $2x - 7(y+1)$,
for $x = -4$ and $y = 3$

$2x - 7(y+1)$

$= 2(\_\_\_\_) - 7(\_\_\_\_ + 1)$

$= \_\_\_\_ - 7(\_\_\_\_)$

$= \_\_\_\_ - \_\_\_\_$

$= \_\_\_\_$

(b) $2x^2 - 4y^2$,
for $x = -2$ and $y = -3$

Teaching Tip

At this point, review the rules for order of operations.

(c) $\dfrac{4x - 2y}{-3x}$,
for $x = 2$ and $y = -1$

(d) $\left(\frac{2}{5}x - \frac{5}{6}y\right)\left(-\frac{1}{3}z\right)$,
for $x = 10$, $y = 6$, and $z = -9$

Answers

9. **(a)** -4; 3; -8; 4; -8; 28; -36
(b) -28 **(c)** $-\frac{5}{3}$ **(d)** -3

OBJECTIVE 5 Evaluate expressions involving variables. To *evaluate* an expression means to find its *value*.

EXAMPLE 6 Evaluating Expressions for Numerical Values

Evaluate each expression for $x = -1$, $y = -2$, and $m = -3$.

(a) $(3x + 4y)(-2m)$

Use parentheses around substituted negative values to avoid errors.

$= [3(-1) + 4(-2)][-2(-3)]$ Substitute the given values for the variables.

$= [-3 + (-8)][6]$ Multiply.

$= [-11]6$ Add inside the brackets.

$= -66$ Multiply.

(b) $2x^2 - 3y^2$

$= 2(-1)^2 - 3(-2)^2$ Substitute.

Think: $(-2)^2 = -2(-2)$

Think: $(-1)^2 = -1(-1)$

$= 2(1) - 3(4)$ Apply the exponents.

$= 2 - 12$ Multiply.

$= -10$ Subtract.

(c) $\dfrac{4y^2 + x}{m}$

$= \dfrac{4(-2)^2 + (-1)}{-3}$ Substitute.

$= \dfrac{4(4) + (-1)}{-3}$ Apply the exponent.

$= \dfrac{16 + (-1)}{-3}$ Multiply.

$= \dfrac{15}{-3}$, or -5 Add, and then divide.

(d) $\left(\frac{3}{4}x + \frac{5}{8}y\right)\left(-\frac{1}{2}m\right)$

$= \left[\frac{3}{4}(-1) + \frac{5}{8}(-2)\right] \cdot \left[-\frac{1}{2}(-3)\right]$ Substitute.

$= \left[-\frac{3}{4} + \left(-\frac{5}{4}\right)\right] \cdot \left[\frac{3}{2}\right]$ Multiply inside the brackets.

$= \left[-\frac{8}{4}\right] \cdot \left[\frac{3}{2}\right]$ Add inside the brackets.

$= (-2)\left(\frac{3}{2}\right)$ Divide.

$= -3$ Multiply.

◀ Work Problem **9** at the Side.

OBJECTIVE 6 Translate words and phrases involving multiplication and division. The table gives words and phrases that indicate multiplication.

Word or Phrase Indicating Multiplication	Example	Numerical Expression and Simplification
Product of	The *product of* -5 and -2	$-5(-2)$, which equals 10
Times	13 *times* -4	$13(-4)$, which equals -52
Twice (meaning "2 times")	*Twice* 6	$2(6)$, which equals 12
Of (used with fractions)	$\frac{1}{2}$ *of* 10	$\frac{1}{2}(10)$, which equals 5
Percent of	12*% of* -16	$0.12(-16)$, which equals -1.92
As much as	$\frac{2}{3}$ *as much as* 30	$\frac{2}{3}(30)$, which equals 20

EXAMPLE 7 Translating Words and Phrases (Multiplication)

Write a numerical expression for each phrase, and simplify the expression.

(a) The **product of** 12 and the sum of 3 and -6

$12[3 + (-6)]$ simplifies to $12[-3]$, which equals -36.

(b) **Twice** the difference between 8 and -4

$2[8 - (-4)]$ simplifies to $2[12]$, which equals 24.

(c) Two-thirds **of** the sum of -5 and -3

$\frac{2}{3}[-5 + (-3)]$ simplifies to $\frac{2}{3}[-8]$, which equals $-\frac{16}{3}$.

(d) 15**% of** the difference between 14 and -2

$0.15[14 - (-2)]$ simplifies to $0.15[16]$, which equals 2.4.

Remember that 15% = 0.15.

(e) **Double** the product of 3 and 4

$2 \cdot (3 \cdot 4)$ simplifies to $2(12)$, which equals 24.

Work Problem 10 at the Side. ▶

The word *quotient* refers to the answer in a division problem. In algebra, a quotient is usually represented with a fraction bar. The table gives some phrases associated with division.

Phrase Indicating Division	Example	Numerical Expression and Simplification
Quotient of	The *quotient of* -24 and 3	$\frac{-24}{3}$, which equals -8
Divided by	-16 *divided by* -4	$\frac{-16}{-4}$, which equals 4
Ratio of	The *ratio of* 2 to 3	$\frac{2}{3}$

When translating a phrase involving division, we write the first number named as the numerator and the second as the denominator.

10 Write a numerical expression for each phrase, and simplify the expression.

(a) The product of 6 and the sum of -5 and -4

(b) Three times the difference between 4 and -6

(c) Three-fifths of the sum of 2 and -7

(d) 20% of the sum of 9 and -4

(e) Triple the product of 5 and 6

Answers

10. (a) $6[(-5) + (-4)]$; -54
(b) $3[4 - (-6)]$; 30
(c) $\frac{3}{5}[2 + (-7)]$; -3
(d) $0.20[9 + (-4)]$; 1
(e) $3(5 \cdot 6)$; 90

11 Write a numerical expression for each phrase, and simplify the expression.

(a) The quotient of 20 and the sum of 8 and -3

(b) The product of -9 and 2, divided by the difference between 5 and -1

12 Write each sentence in symbols, using x to represent the number.

(a) Twice a number is -6.

(b) The difference between -8 and a number is -11.

(c) The sum of 5 and a number is 8.

(d) The quotient of a number and -2 is 6.

Answers

11. **(a)** $\frac{20}{8+(-3)}$; 4 **(b)** $\frac{-9(2)}{5-(-1)}$; -3

12. **(a)** $2x = -6$ **(b)** $-8 - x = -11$ **(c)** $5 + x = 8$ **(d)** $\frac{x}{-2} = 6$

EXAMPLE 8 Translating Words and Phrases (Division)

Write a numerical expression for each phrase, and simplify the expression.

(a) The **quotient of** 14 and the sum of -9 and 2

$$\frac{14}{-9+2} \text{ simplifies to } \frac{14}{-7}, \text{ which equals } -2.$$

"Quotient" indicates division.

(b) The product of 5 and -6, **divided by** the difference between -7 and 8

$$\frac{5(-6)}{-7-8} \text{ simplifies to } \frac{-30}{-15}, \text{ which equals } 2.$$

◀ Work Problem **11** at the Side.

OBJECTIVE 7 Translate simple sentences into equations. We can use words and phrases to translate sentences into equations.

EXAMPLE 9 Translating Sentences into Equations

Write each sentence in symbols, using x to represent the number.

(a) Three **times** a number **is** -18.

The word *times* indicates multiplication. The word *is* translates as $=$.

$$3 \cdot x = -18, \quad \text{or} \quad 3x = -18 \qquad 3 \cdot x = 3x$$

(b) The **sum** of a number and 9 **is** 12.

$$x + 9 = 12$$

(c) The **difference between** a number and 5 **is** 0.

$$x - 5 = 0$$

(d) The **quotient of** 24 and a number **is** -2.

$$\frac{24}{x} = -2$$

◀ Work Problem **12** at the Side.

CAUTION

In **Examples 7 and 8,** the *phrases* translate as *expressions*, while in **Example 9,** the *sentences* translate as *equations*. ***An expression is a phrase. An equation is a sentence with something on the left side, an = symbol, and something on the right side.***

$$\frac{5(-6)}{-7-8} \qquad 3x = -18$$

Expression ↑ Equation ↑

1.6 Exercises

FOR EXTRA HELP

Download the MyDashBoard App

MyMathLab®

CONCEPT CHECK *Fill in each blank with one of the following:*

greater than 0, less than 0, *or* equal to 0.

1. The product or the quotient of two numbers with the same sign is **greater than 0**.
2. The product or the quotient of two numbers with different signs is **less than 0**.
3. If three negative numbers are multiplied together, the product is **less than 0**.
4. If two negative numbers are multiplied and then their product is divided by a negative number, the result is **less than 0**.
5. If a negative number is squared and the result is added to a positive number, the final answer is **greater than 0**.
6. The reciprocal of a negative number is **less than 0**.

Find each product. ***See Examples 1 and 2.***

7. $-7(4)$ **-28**
8. $-8(5)$ **-40**
9. $-5(-6)$ **30**
10. $-4(-20)$ **80**
11. $-8(0)$ **0**
12. $0(-12)$ **0**
13. $-\frac{3}{8}\left(-\frac{20}{9}\right)$ **$\frac{5}{6}$**
14. $-\frac{5}{4}\left(-\frac{6}{25}\right)$ **$\frac{3}{10}$**
15. $-6.8(0.35)$ **-2.38**
16. $-4.6(0.24)$ **-1.104**
17. $-6\left(-\frac{1}{4}\right)$ **$\frac{3}{2}$**
18. $-8\left(-\frac{1}{2}\right)$ **4**

Find each quotient. ***See Examples 3 and 4*** *and the discussion of division involving 0.*

19. $\frac{-15}{5}$ **-3**
20. $\frac{-18}{6}$ **-3**
21. $\frac{20}{-10}$ **-2**
22. $\frac{28}{-4}$ **-7**
23. $\frac{-160}{-10}$ **16**
24. $\frac{-260}{-20}$ **13**
25. $\frac{0}{-3}$ **0**
26. $\frac{0}{-5}$ **0**
27. $\frac{-10.252}{0}$ **undefined**
28. $\frac{-29.584}{0}$ **undefined**
29. $\left(-\frac{3}{4}\right)\div\left(-\frac{1}{2}\right)$ **$\frac{3}{2}$**
30. $\left(-\frac{3}{16}\right)\div\left(-\frac{5}{8}\right)$ **$\frac{3}{10}$**

31. CONCEPT CHECK Which expression is undefined?

A. $\frac{5-5}{5+5}$ **B.** $\frac{5+5}{5+5}$ **C.** $\frac{5-5}{5-5}$ **D.** $\frac{5-5}{5}$ C

32. CONCEPT CHECK Which expression is undefined?

A. $13 \div 0$ **B.** $13 \div 13$ **C.** $0 \div 13$ **D.** $13 \cdot 0$ A

Perform each indicated operation. ***See Example 5.***

33. $\frac{-5(-6)}{9-(-1)}$

$= \frac{30}{10}$

$= 3$

34. $\frac{-12(-5)}{7-(-5)}$

$= \frac{60}{12}$

$= 5$

35. $\frac{-21(3)}{-3-6}$ 7

36. $\frac{-40(3)}{-2-3}$ 24

37. $\frac{-10(2)+6(2)}{-3-(-1)}$ 4

38. $\frac{8(-1)+6(-2)}{-6-(-1)}$ 4

39. $\frac{-27(-2)-(-12)(-2)}{-2(3)-2(2)}$ −3

40. $\frac{-13(-4)-(-8)(-2)}{(-10)(2)-4(-2)}$ −3

41. $\frac{3^2-4^2}{7(-8+9)}$ −1

42. $\frac{5^2-7^2}{2(3+3)}$ −2

43. $\frac{4(2^3-5)-5(-3^3+21)}{3[6-(-2)]}$ $\frac{7}{4}$

44. $\frac{-3(-2^4+10)+4(2^5-12)}{-2[8-(-7)]}$ $-\frac{49}{15}$

Evaluate each expression for $x = 6$, $y = -4$, *and* $a = 3$. ***See Example 6.***

45. $6x - 5y + 4a$ 68

46. $5x - 2y + 3a$ 47

47. $(5x - 2y)(-2a)$ −228

48. $(2x + y)(3a)$ 72

49. $\left(\frac{5}{6}x + \frac{3}{2}y\right)\left(-\frac{1}{3}a\right)$ 1

50. $\left(\frac{1}{3}x - \frac{4}{5}y\right)\left(-\frac{1}{5}a\right)$ $-\frac{78}{25}$

51. $(6 - x)(5 + y)(3 + a)$ 0

52. $(-5 + x)(-3 + y)(3 - a)$ 0

53. $5x - 4a^2$ −6

54. $-2y^2 + 3a$ −23

55. $\frac{xy + 9a}{x + y - 2}$ undefined

56. $\frac{2y^2 - x}{a - 3}$ undefined

Write a numerical expression for each phrase and simplify. ***See Examples 7 and 8.***

57. The product of 4 and -7, added to -12

$-12 + 4(-7)$; -40

58. The product of -9 and 2, added to 9

$9 + (-9)(2)$; -9

59. Twice the product of -8 and 2, subtracted from -1

$-1 - 2(-8)(2)$; 31

60. Twice the product of -1 and 6, subtracted from -4

$-4 - 2(-1)(6)$; 8

61. The product of -3 and the difference between 3 and -7

$-3[3 - (-7)]$; -30

62. The product of 12 and the difference between 9 and -8

$12[9 - (-8)]$; 204

63. Three-tenths of the sum of -2 and -28

$\frac{3}{10}[-2 + (-28)]$; -9

64. Four-fifths of the sum of -8 and -2

$\frac{4}{5}[-8 + (-2)]$; -8

65. The quotient of -20 and the sum of -8 and -2

$\frac{-20}{-8 + (-2)}$; 2

66. The quotient of -12 and the sum of -5 and -1

$\frac{-12}{-5 + (-1)}$; 2

67. The sum of -18 and -6, divided by the product of 2 and -4

$\frac{-18 + (-6)}{2(-4)}$; 3

68. The sum of 15 and -3, divided by the product of 4 and -3

$\frac{15 + (-3)}{4(-3)}$; -1

69. The product of $-\frac{2}{3}$ and $-\frac{1}{5}$, divided by $\frac{1}{7}$

$\frac{-\frac{2}{3}\left(-\frac{1}{5}\right)}{\frac{1}{7}}$; $\frac{14}{15}$

70. The product of $-\frac{1}{2}$ and $\frac{3}{4}$, divided by $-\frac{2}{3}$

$\frac{-\frac{1}{2}\left(\frac{3}{4}\right)}{-\frac{2}{3}}$; $\frac{9}{16}$

Write each sentence in symbols, using x to represent the number. ***See Example 9.***

71. Nine times a number is -36.

$9x = -36$

72. Seven times a number is -42.

$7x = -42$

73. The quotient of a number and 4 is -1.

$\frac{x}{4} = -1$

74. The quotient of a number and 3 is -3.

$\frac{x}{3} = -3$

75. $\frac{9}{11}$ less than a number is 5.

$x - \frac{9}{11} = 5$

76. $\frac{1}{2}$ less than a number is 2.

$x - \frac{1}{2} = 2$

77. When 6 is divided by a number, the result is -3.

$\frac{6}{x} = -3$

78. When 15 is divided by a number, the result is -5.

$\frac{15}{x} = -5$

In 2011, the following question and expression appeared on boxes of Swiss Miss Chocolate: On average, how many mini-marshmallows are in one serving?

$$3 + 2 \times 4 \div 2 - 3 \times 7 - 4 + 47$$

79. The box gave 92 as the answer. What is the *correct* answer? (*Source*: Swiss Miss Chocolate box.)

29

80. Explain the error that somebody at the company made in calculating the answer.

The incorrect answer, 92, was obtained by performing all of the operations in order from left to right rather than following the rules for order of operations. The multiplications and divisions need to be done in order, before the additions and subtractions.

Relating Concepts (Exercises 81–86) For Individual or Group Work

To find the ***average*** *of a group of numbers, we add the numbers and then divide this sum by the number of terms added.* ***Work Exercises 81–84 in order,*** *to find the average of* 23, 18, 13, -4, *and* -8. *Then find the averages in* ***Exercises 85 and 86.***

81. Find the sum of the given group of numbers.

42

82. How many numbers are in the group?

5

83. Divide your answer for **Exercise 81** by your answer for **Exercise 82.** Give the quotient as a mixed number.

$8\frac{2}{5}$

84. What is the average of the given group of numbers?

$8\frac{2}{5}$

85. What is the average of all integers between -10 and 14, including both -10 and 14?

2

86. What is the average of all integers between -15 and -10, including both -15 and -10?

$-12\frac{1}{2}$

Summary Exercises *Performing Operations with Real Numbers*

Summary Exercises provide mixed practice to better prepare students for tests.

Operations with Signed Numbers

Addition

Same sign Add the absolute values of the numbers. The sum has the same sign as the numbers.

Different signs Find the absolute values of the numbers, and subtract the lesser absolute value from the greater. Give the sum the sign of the number having the greater absolute value.

Subtraction

Add the opposite of the subtrahend to the minuend.

Multiplication and Division

Same sign The product or quotient of two numbers with the same sign is positive.

Different signs The product or quotient of two numbers with different signs is negative.

Division by 0 is undefined.

Perform each indicated operation.

1. $14 - 3 \cdot 10$
−16

2. $-3(8) - 4(-7)$
4

3. $(3 - 8)(-2) - 10$
0

4. $-6(7 - 3)$
−24

5. $7 - (-3)(2 - 10)$
−17

6. $-4[(-2)(6) - 7]$
76

7. $(-4)(7) - (-5)(2)$
−18

8. $-5[-4 - (-2)(-7)]$
90

9. $40 - (-2)[8 - 9]$
38

10. $\dfrac{5(-4)}{-7 - (-2)}$
4

11. $\dfrac{-3 - (-9 + 1)}{-7 - (-6)}$
−5

12. $\dfrac{5(-8 + 3)}{13(-2) + (-7)(-3)}$
5

13. $\dfrac{6^2 - 8}{-2(2) + 4(-1)}$
$-\frac{7}{2}$, or $-3\frac{1}{2}$

14. $\dfrac{16(-8 + 5)}{15(-3) + (-7 - 4)(-3)}$
4

15. $\dfrac{9(-6) - 3(8)}{4(-7) + (-2)(-11)}$
13

16. $\dfrac{2^2 + 4^2}{5^2 - 3^2}$

$\frac{5}{4}$, or $1\frac{1}{4}$

17. $\dfrac{(2+4)^2}{(5-3)^2}$

9

18. $\dfrac{4^3 - 3^3}{-5(-4+2)}$

$\frac{37}{10}$, or $3\frac{7}{10}$

19. $\dfrac{-9(-6) + (-2)(27)}{3(8-9)}$

0

20. $|-4(9)| - |-11|$

25

21. $\dfrac{6(-10+3)}{15(-2) - 3(-9)}$

14

22. $\dfrac{(-9)^2 - 9^2}{3^2 - 5^2}$

0

23. $\dfrac{(-10)^2 + 10^2}{-10(5)}$

-4

24. $-\frac{3}{4} \div \left(-\frac{5}{8}\right)$

$\frac{6}{5}$, or $1\frac{1}{5}$

25. $\frac{1}{2} \div \left(-\frac{1}{2}\right)$

-1

26. $\dfrac{8^2 - 12}{(-5)^2 + 2(6)}$

$\frac{52}{37}$, or $1\frac{15}{37}$

27. $\left[\frac{5}{8} - \left(-\frac{1}{16}\right)\right] + \frac{3}{8}$

$\frac{17}{16}$, or $1\frac{1}{16}$

28. $\left(\frac{1}{2} - \frac{1}{3}\right) - \frac{5}{6}$

$-\frac{2}{3}$

29. $-0.9(-3.7)$

3.33

30. $-5.1(-0.2)$

1.02

31. $-3^2 - 2^2$

-13

32. $|-2(3) + 4| - |-2|$

0

33. $40 - (-2)[-5 - 3]$

24

Evaluate each expression for $x = -2$, $y = 3$, and $a = 4$.

34. $-x + y - 3a$

-7

35. $(x+6)^3 - y^3$

37

36. $(x - y) - (a - 2y)$

-3

37. $\left(\frac{1}{2}x + \frac{2}{3}y\right)\left(-\frac{1}{4}a\right)$

-1

38. $\dfrac{2x + 3y}{a - xy}$

$\frac{1}{2}$

39. $\dfrac{x^2 - y^2}{x^2 + y^2}$

$-\frac{5}{13}$

40. $-x^2 + 3y$

5

41. $\dfrac{-x + 2y}{2x + a}$

undefined

42. $\dfrac{2x + a}{-x + 2y}$

0

1.7 Properties of Real Numbers

OBJECTIVES

1. Use the commutative properties.
2. Use the associative properties.
3. Use the identity properties.
4. Use the inverse properties.
5. Use the distributive property.

In the basic properties covered in this section, a, b, and c represent real numbers.

OBJECTIVE 1 Use the commutative properties. The word *commute* means to go back and forth. Many people commute to work or to school. If you travel from home to work and follow the same route from work to home, you travel the same distance each time.

The **commutative properties** say that if two numbers are added or multiplied in either order, the result is the same.

Commutative Properties

$a + b = b + a$ Addition

$ab = ba$ Multiplication

EXAMPLE 1 Using the Commutative Properties

Use a commutative property to complete each statement.

(a) $-8 + 5 = 5 + \underline{\ ?\ }$

$-8 + 5 = 5 + (-8)$ Commutative property of addition

Notice that the "order" changed.

(b) $(-2)7 = \underline{\ ?\ }(-2)$

$-2(7) = 7(-2)$ Commutative property of multiplication

Work Problem 1 at the Side. ▶

1 Complete each statement. Use a commutative property.

(a) $x + 9 = 9 + \_\_\_\_$

(b) $-12(4) = \_\_\_\_(-12)$

(c) $5x = x \cdot \_\_\_\_$

OBJECTIVE 2 Use the associative properties. When we *associate* one object with another, we think of those objects as being grouped together.

The **associative properties** say that when we add or multiply three numbers, we can group the first two together or the last two together and get the same answer.

Associative Properties

$(a + b) + c = a + (b + c)$ Addition

$(ab)c = a(bc)$ Multiplication

EXAMPLE 2 Using the Associative Properties

Use an associative property to complete each statement.

(a) $-8 + (1 + 4) = (-8 + \underline{\ ?\ }) + 4$

$-8 + (1 + 4) = (-8 + 1) + 4$ Associative property of addition

The "order" is the same. The "grouping" changed.

(b) $[2 \cdot (-7)] \cdot 6 = 2 \cdot \underline{\ ?\ }$

$[2 \cdot (-7)] \cdot 6 = 2 \cdot [(-7) \cdot 6]$ Associate property of multiplication

Work Problem 2 at the Side. ▶

2 Complete each statement. Use an associative property.

(a) $(9 + 10) + (-3)$

$= 9 + [\_\_\_\_ + (-3)]$

(b) $-5 + (2 + 8)$

$= (\_\_\_\_\_\_) + 8$

(c) $10 \cdot [-8 \cdot (-3)]$

$= \_\_\_\_\_\_$

Answers

1. **(a)** x **(b)** 4 **(c)** 5

2. **(a)** 10 **(b)** $-5 + 2$ **(c)** $[10 \cdot (-8)] \cdot (-3)$

3 Decide whether each statement is an example of a commutative property, an associative property, or both.

(a) $2 \cdot (4 \cdot 6) = (2 \cdot 4) \cdot 6$

(b) $(2 \cdot 4) \cdot 6 = (4 \cdot 2) \cdot 6$

(c) $(2 + 4) + 6 = 4 + (2 + 6)$

4 Find each sum or product.

(a) $5 + 18 + 29 + 31 + 12$

(b) $5(37)(20)$

Answers

3. **(a)** associative **(b)** commutative **(c)** both
4. **(a)** 95 **(b)** 3700

By the associative property, the sum (or product) of three numbers will be the same no matter how the numbers are "associated" in groups. Parentheses can be left out if a problem contains only addition (or multiplication). For example,

$$(-1 + 2) + 3 \quad \text{and} \quad -1 + (2 + 3) \quad \text{can be written as} \quad -1 + 2 + 3.$$

EXAMPLE 3 **Distinguishing between Properties**

Decide whether each statement is an example of a commutative property, an associative property, or both.

(a) $(2 + 4) + 5 = 2 + (4 + 5)$

The order of the three numbers is the same on both sides of the equality symbol. The only change is in the *grouping*, or association, of the numbers. This is an example of the associative property.

(b) $6 \cdot (3 \cdot 10) = 6 \cdot (10 \cdot 3)$

The same numbers, 3 and 10, are grouped on each side. On the left, the 3 appears first, but on the right, the 10 appears first. Since the only change involves the *order* of the numbers, this is an example of the commutative property.

(c) $(8 + 1) + 7 = 8 + (7 + 1)$

Both the order and the grouping are changed. On the left, the order of the three numbers is 8, 1, and 7. On the right, it is 8, 7, and 1. On the left, the 8 and 1 are grouped. On the right, the 7 and 1 are grouped. Therefore, *both* properties are used.

◀ **Work Problem 3 at the Side.**

Example 4 illustrates that we can sometimes use the commutative and associative properties to rearrange and regroup numbers to simplify calculations.

EXAMPLE 4 **Using the Commutative and Associative Properties**

Find each sum or product.

(a) $23 + 41 + 2 + 9 + 25$

$= (41 + 9) + (23 + 2) + 25$

$= 50 + 25 + 25$ Use the commutative and associative properties.

$= 100$

(b) $25(69)(4)$

$= 25(4)(69)$

$= 100(69)$

$= 6900$

◀ **Work Problem 4 at the Side.**

OBJECTIVE 3 **Use the identity properties.** If a child wears a costume on Halloween, the child's appearance is changed, but his or her *identity* is unchanged. The identity of a real number is left unchanged when identity properties are applied.

The **identity properties** say that the sum of 0 and any number equals that number, and the product of 1 and any number equals that number.

Identity Properties

$a + 0 = a$	and	$0 + a = a$	Addition
$a \cdot 1 = a$	and	$1 \cdot a = a$	Multiplication

The number 0 leaves the identity, or value, of any real number unchanged by addition, so 0 is the **identity element for addition**, or the **additive identity.** Since multiplication by 1 leaves any real number unchanged, 1 is the **identity element for multiplication**, or the **multiplicative identity.**

EXAMPLE 5 Using the Identity Properties

Use an identity property to complete each statement.

(a) $-3 + \underline{\ ?\ } = -3$

$-3 + \mathbf{0} = -3$

Identity property for addition

(b) $\underline{\ ?\ } \cdot \frac{1}{2} = \frac{1}{2}$

$\mathbf{1} \cdot \frac{1}{2} = \frac{1}{2}$

Identity property for multiplication

Work Problem 5 at the Side. ▶

We use the identity property for multiplication to write fractions in lowest terms and to find common denominators.

EXAMPLE 6 Using the Identity Property for Multiplication

In part (a), write in lowest terms. In part (b), perform the operation.

(a) $\frac{49}{35}$

$= \frac{7 \cdot 7}{5 \cdot 7}$ Factor.

$= \frac{7}{5} \cdot \frac{7}{7}$ Write as a product.

$= \frac{7}{5} \cdot 1$ Property of 1

$= \frac{7}{5}$ Identity property

(b) $\frac{3}{4} + \frac{5}{24}$

$= \frac{3}{4} \cdot 1 + \frac{5}{24}$ Identity property

$= \frac{3}{4} \cdot \frac{6}{6} + \frac{5}{24}$ Use $1 = \frac{6}{6}$ to get a common denominator.

$= \frac{18}{24} + \frac{5}{24}$ Multiply.

$= \frac{23}{24}$ Add.

Work Problem 6 at the Side. ▶

OBJECTIVE 4 Use the inverse properties. Each day before you go to work or school, you probably put on your shoes. Before you go to sleep at night, you probably take them off. These operations from everyday life are examples of *inverse* operations.

The **inverse properties** of addition and multiplication lead to the additive and multiplicative identities, respectively. Recall that $-a$ is the **additive inverse,** or **opposite,** of a and $\frac{1}{a}$ is the **multiplicative inverse,** or **reciprocal,** of the nonzero number a. The sum of the numbers a and $-a$ is 0, and the product of the nonzero numbers a and $\frac{1}{a}$ is 1.

Inverse Properties

$a + (-a) = 0$ and $-a + a = 0$ Addition

$a \cdot \frac{1}{a} = 1$ and $\frac{1}{a} \cdot a = 1 \quad (a \neq 0)$ Multiplication

---

5 Use an identity property to complete each statement.

(a) $9 + 0 = \underline{\qquad}$

(b) $\underline{\qquad} + (-7) = -7$

(c) $\underline{\qquad} \cdot 1 = 5$

6 In part (a), write in lowest terms. In part (b), perform the operation.

(a) $\frac{85}{105}$

(b) $\frac{9}{10} - \frac{53}{50}$

Answers

5. (a) 9 **(b)** 0 **(c)** 5

6. (a) $\frac{17}{21}$ **(b)** $-\frac{4}{25}$

7 Complete each statement so that it is an example of either an identity property or an inverse property. Tell which property is used.

(a) $-6 + \_\_\_\_ = 0$

(b) $\frac{4}{3} \cdot \_\_\_\_ = 1$

(c) $-\frac{1}{9} \cdot (\_\_\_\_) = 1$

(d) $275 + \_\_\_\_ = 275$

(e) $-0.75 + \frac{3}{4} = \_\_\_\_$

(f) $0.2(5) = \_\_\_\_$

Teaching Tip

The identity and inverse properties are often confused. Cover them with several examples.

EXAMPLE 7 Using the Inverse Properties

Use an inverse property to complete each statement.

(a) $\underline{\ ?\ } + \frac{1}{2} = 0$

$-\frac{1}{2} + \frac{1}{2} = 0$

(b) $4 + \underline{\ ?\ } = 0$

$4 + (-4) = 0$

(c) $-0.75 + \frac{3}{4} = \underline{\ ?\ }$

$-0.75 + \frac{3}{4} = 0$

The inverse property for addition is used in parts (a)–(c).

(d) $\underline{\ ?\ } \cdot \frac{5}{2} = 1$

$\frac{2}{5} \cdot \frac{5}{2} = 1$

(e) $-5(\underline{\ ?\ }) = 1$

$-5\left(-\frac{1}{5}\right) = 1$

(f) $4(0.25) = \underline{\ ?\ }$

$4(0.25) = 1$

The inverse property for multiplication is used in parts (d)–(f).

◀ **Work Problem 7 at the Side.**

OBJECTIVE 5 Use the distributive property. The word *distribute* means "to give out from one to several." Consider the following expressions.

$$2(5+8) \quad \text{equals} \quad 2(13), \quad \text{or} \quad 26.$$

$$2(5) + 2(8) \quad \text{equals} \quad 10 + 16, \quad \text{or} \quad 26.$$

Since both expressions equal 26,

$$2(5 + 8) = 2(5) + 2(8).$$

This result is an example of the *distributive property of multiplication with respect to addition,* the only property involving *both* addition and multiplication. With this property, a product can be changed to a sum. This idea is illustrated by the divided rectangle in **Figure 13.**

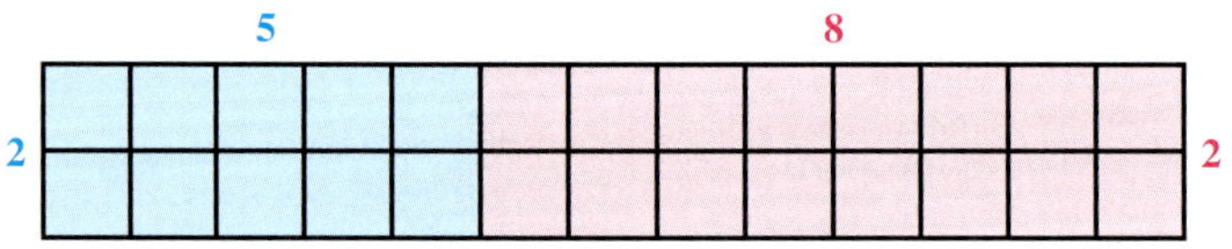

The area of the left part is 2(5) = 10.
The area of the right part is 2(8) = 16.
The total area is 2(5 + 8) = 26 or the total area is
2(5) + 2(8) = 10 + 16 = 26.
Thus, 2(5 + 8) = 2(5) + 2(8).

Figure 13

The **distributive property** says that multiplying a number a by a sum of numbers $b + c$ gives the same result as multiplying a by b and a by c and then adding the two products.

Distributive Property

$$a(b + c) = ab + ac \quad \text{and} \quad (b + c)a = ba + ca$$

As the arrows show, the a outside the parentheses is "distributed" over the b and c inside.

Answers

7. **(a)** 6; inverse **(b)** $\frac{3}{4}$; inverse **(c)** -9; inverse **(d)** 0; identity **(e)** 0; inverse **(f)** 1; inverse

The distributive property is also valid for subtraction.

$$a(b - c) = ab - ac \quad \text{and} \quad (b - c)a = ba - ca$$

The distributive property also can be extended to more than two numbers.

$$a(b + c + d) = ab + ac + ad$$

The distributive property can also be written "in reverse."

$$ab + ac = a(b + c)$$

We will use this form in the next section to combine like terms.

EXAMPLE 8 **Using the Distributive Property**

Use the distributive property to rewrite each expression.

(a) $5(9 + 6)$

$= 5 \cdot 9 + 5 \cdot 6$ Distributive property

$= 45 + 30$ Multiply.

$= 75$ Add.

Multiply first.

(b) $4(x + 5 + y)$

$= 4x + 4 \cdot 5 + 4y$ Distributive property

$= 4x + 20 + 4y$ Multiply.

(c) $-2(x + 3)$

$= -2x + (-2)(3)$ Distributive property

$= -2x + (-6)$ Multiply.

$= -2x - 6$ Definition of subtraction

(d) $3(k - 9)$

Be careful here.

$= 3[k + (-9)]$ Definition of subtraction

$= 3k + 3(-9)$ Distributive property

$= 3k - 27$ Multiply.

(e) $8(3r + 11t + 5z)$

$= 8(3r) + 8(11t) + 8(5z)$ Distributive property

$= (8 \cdot 3)r + (8 \cdot 11)t + (8 \cdot 5)z$ Associative property

$= 24r + 88t + 40z$ Multiply.

(f) $6 \cdot 8 + 6 \cdot 2$

$= 6(8 + 2)$ Distributive property in reverse

$= 6(10)$ Add.

$= 60$ Multiply.

(g) $4x - 4m$

$= 4(x - m)$ Distributive property in reverse

Work Problem **8** at the Side. ▶

8 Use the distributive property to rewrite each expression.

GS **(a)** $2(p + 5)$

$= 2 \cdot \_\_\_\_ + 2 \cdot \_\_\_\_$

$= \_\_\_\_ + \_\_\_\_$

(b) $-4(y + 7)$

(c) $5(m - 4)$

(d) $7(2y + 7k - 9m)$

GS **(e)** $9 \cdot k + 9 \cdot 5$

$= \_\_\_\_(\_\_\_\_ + \_\_\_\_)$

(f) $3a - 3b$

Answers

8. **(a)** p; 5; $2p$; 10 **(b)** $-4y - 28$ **(c)** $5m - 20$ **(d)** $14y + 49k - 63m$ **(e)** 9; k; 5 **(f)** $3(a - b)$

9 Write each expression without parentheses.

GS **(a)** $-(3k-5)$

$= \_\_\_\_ \cdot (3k-5)$

$= \_\_\_\_ \cdot 3k + (-1)(\_\_\_\_)$

$= \_\_\_\_ + \_\_\_\_$

(b) $-(2-r)$

(c) $-(-5y+8)$

Teaching Tip

Ask students to tell in their own words what the properties say about numbers. For example, the identity property for addition can be stated as follows: "If 0 is added to any number, the result is the number you started with."

(d) $-(-z+4)$

(e) $-(-t-4u+5v)$

Answers

9. (a) -1; -1; -5; $-3k$; 5 **(b)** $-2+r$
(c) $5y-8$ **(d)** $z-4$
(e) $t+4u-5v$

The symbol $-a$ may be interpreted as $-1 \cdot a$. Using this result and the distributive property, we can remove (or clear) parentheses from some expressions.

EXAMPLE 9 Using the Distributive Property to Remove (Clear) Parentheses

Write each expression without parentheses.

(a) $-(2y+3)$

$= -1 \cdot (2y+3)$ $-a = -1 \cdot a$

$= -1 \cdot 2y + (-1) \cdot 3$ Distributive property

$= -2y - 3$ Multiply.

The − symbol indicates a factor of −1.

(b) $-(-9w-2)$

$= -1(-9w-2)$

$= -1(-9w) - 1(-2)$

$= 9w + 2$

We can also interpret the negative sign in front of the parentheses to mean the *opposite* of each of the terms within the parentheses.

(c) $-(-x-3y+6z)$

$= -1(-1x-3y+6z)$ Be careful with signs.

$= -1(-1x) - 1(-3y) - 1(6z)$ Distributive property

$= x + 3y - 6z$ $-1(-1x) = 1x = x$

◀ Work Problem 9 at the Side.

Here is a summary of the basic properties of real numbers.

Properties of Addition and Multiplication

For any real numbers a, b, and c, the following properties hold.

Commutative properties $\quad a+b=b+a \qquad ab=ba$

Associative properties $\quad (a+b)+c=a+(b+c)$

$(ab)c = a(bc)$

Identity properties There is a real number 0 such that

$a+0=a$ and $0+a=a$.

There is a real number 1 such that

$a \cdot 1 = a$ and $1 \cdot a = a$.

Inverse properties For each real number a, there is a single real number $-a$ such that

$a+(-a)=0$ and $(-a)+a=0$.

For each nonzero real number a, there is a single real number $\frac{1}{a}$ such that

$a \cdot \frac{1}{a} = 1$ and $\frac{1}{a} \cdot a = 1$.

Distributive property $\quad a(b+c)=ab+ac$

$(b+c)a = ba+ca$

1.7 Exercises

FOR EXTRA HELP

Download the MyDashBoard App

MyMathLab®

1. CONCEPT CHECK Match each item in Column I with the correct choice(s) from Column II. Choices may be used once, more than once, or not at all.

I

(a) Identity element for addition B

(b) Identity element for multiplication F

(c) Additive inverse of a C

(d) Multiplicative inverse, or reciprocal, of the nonzero number a I

(e) The number that is its own additive inverse B

(f) The two numbers that are their own multiplicative inverses D, F

(g) The only number that has no multiplicative inverse B

(h) An example of the associative property A

(i) An example of the commutative property G

(j) An example of the distributive property H

II

A. $(5 \cdot 4) \cdot 3 = 5 \cdot (4 \cdot 3)$

B. 0

C. $-a$

D. -1

E. $5 \cdot 4 \cdot 3 = 60$

F. 1

G. $(5 \cdot 4) \cdot 3 = 3 \cdot (5 \cdot 4)$

H. $5(4 + 3) = 5 \cdot 4 + 5 \cdot 3$

I. $\frac{1}{a}$

2. CONCEPT CHECK Fill in the blanks: The commutative property allows us to change the __order__ of the terms in a sum or the factors in a product. The associative property allows us to change the __grouping__ of the terms in a sum or the factors in a product.

CONCEPT CHECK *Tell whether or not the following everyday activities are commutative.*

3. Washing your face and brushing your teeth yes

4. Putting on your left sock and putting on your right sock yes

5. Preparing a meal and eating a meal no

6. Starting a car and driving away in a car no

7. Putting on your socks and putting on your shoes no

8. Getting undressed and taking a shower no

9. CONCEPT CHECK Use parentheses to show how the associative property can be used to give two different meanings to the phrase "foreign sales clerk."
(foreign sales) clerk; foreign (sales clerk)

10. CONCEPT CHECK Use parentheses to show how the associative property can be used to give two different meanings to the phrase "defective merchandise counter."
(defective merchandise) counter; defective (merchandise counter)

Use the commutative or the associative property to complete each statement. State which property is used. ***See Examples 1 and 2.***

11. $-15 + 9 = 9 + \_\_\_\_$ -15; commutative property

12. $6 + (-2) = -2 + \_\_\_\_$ 6; commutative property

13. $-8 \cdot 3 = \_\_\_\_ \cdot (-8)$ 3; commutative property

14. $-12 \cdot 4 = 4 \cdot \_\_\_\_$ -12; commutative property

15. $(3 + 6) + 7 = 3 + (\_\_\_\_ + 7)$ 6; associative property

16. $(-2 + 3) + 6 = -2 + (\_\_\_\_ + 6)$ 3; associative property

17. $7 \cdot (2 \cdot 5) = (\_\_\_\_ \cdot 2) \cdot 5$ 7; associative property

18. $8 \cdot (6 \cdot 4) = (8 \cdot \_\_\_\_) \cdot 4$ 6; associative property

19. CONCEPT CHECK Evaluate $25 - (6 - 2)$ and evaluate $(25 - 6) - 2$. Do you think subtraction is associative?

Subtraction is not associative.

20. CONCEPT CHECK Evaluate $180 \div (15 \div 3)$ and evaluate $(180 \div 15) \div 3$. Do you think division is associative?

Division is not associative.

21. CONCEPT CHECK Complete the table and the statements beside it.

Number	Additive Inverse	Multiplicative Inverse
5	-5	$\frac{1}{5}$
-10	10	$-\frac{1}{10}$
$-\frac{1}{2}$	$\frac{1}{2}$	-2
$\frac{3}{8}$	$-\frac{3}{8}$	$\frac{8}{3}$
x	$-x$	$\frac{1}{x}$ $(x \neq 0)$
$-y$	y	$-\frac{1}{y}$ $(y \neq 0)$

In general, a number and its additive inverse have (*the same* / *opposite*) signs.

A number and its multiplicative inverse have (*the same* / *opposite*) signs.

22. CONCEPT CHECK The following conversation actually took place between one of the authors of this book and his son, Jack, when Jack was 4 years old.

DADDY: "Jack, what is $3 + 0$?"
JACK: "3."
DADDY: "Jack, what is $4 + 0$?"
JACK: "4. And Daddy, *string* plus zero equals *string*!"

What property of addition did Jack recognize? **identity property**

Decide whether each statement is an example of a commutative, associative, identity, or inverse property, or of the distributive property. ***See Examples 1, 2, 3, and 5–8.***

23. $\frac{2}{3}(-4) = -4\left(\frac{2}{3}\right)$

commutative property

24. $6\left(-\frac{5}{6}\right) = \left(-\frac{5}{6}\right)6$

commutative property

25. $-6 + (12 + 7) = (-6 + 12) + 7$

associative property

26. $(-8 + 13) + 2 = -8 + (13 + 2)$

associative property

27. $-6 + 6 = 0$

inverse property

28. $12 + (-12) = 0$

inverse property

29. $\left(\frac{2}{3}\right)\left(\frac{3}{2}\right) = 1$

inverse property

30. $\left(\frac{5}{8}\right)\left(\frac{8}{5}\right) = 1$

inverse property

31. $2.34 \cdot 1 = 2.34$

identity property

32. $-8.456 \cdot 1 = -8.456$

identity property

33. $(4 + 17) + 3 = 3 + (4 + 17)$

commutative property

34. $(-8 + 4) + (-12) = -12 + (-8 + 4)$

commutative property

35. $6(x+y)=6x+6y$
distributive property

36. $14(t+s)=14t+14s$
distributive property

37. $-\frac{5}{9}=-\frac{5}{9}\cdot\frac{3}{3}=-\frac{15}{27}$
identity property

38. $\frac{13}{12}=\frac{13}{12}\cdot\frac{7}{7}=\frac{91}{84}$
identity property

39. $5(2x)+5(3y)=5(2x+3y)$
distributive property

40. $3(5t)-3(7r)=3(5t-7r)$
distributive property

41. **CONCEPT CHECK** Suppose that a student simplifies the expression $-3(4-6)$ as shown.

$$\begin{aligned}&-3(4-6)\\&=-3(4)-3(6)\\&=-12-18\\&=-30\end{aligned}$$

What Went Wrong? Work the problem correctly.

The expression following the first equality symbol should be $-3(4)-3(-6)$.
$$\begin{aligned}&-3(4-6)\\&=-3(4)-3(-6)\\&=-12+18\\&=6\end{aligned}$$

42. Explain how the procedure of changing $\frac{3}{4}$ to $\frac{9}{12}$ requires the use of the multiplicative identity element, 1.

We must multiply $\frac{3}{4}$ by 1 in the form of a fraction, $\frac{3}{3}$.
$$\frac{3}{4}\cdot\frac{3}{3}=\frac{9}{12}$$

Write a new expression that is equal to the given expression, using the given property. Then simplify the new expression if possible. ***See Examples 1, 2, 5, 7, and 8.***

43. $r+7$; commutative
$7+r$

44. $t+9$; commutative
$9+t$

45. $s+0$; identity
s

46. $w+0$; identity
w

47. $-6(x+7)$; distributive
$-6x+(-6)7$; $-6x-42$

48. $-5(y+2)$; distributive
$-5y+(-5)2$; $-5y-10$

49. $(w+5)+(-3)$; associative
$w+[5+(-3)]$; $w+2$

50. $(b+8)+(-10)$; associative
$b+[8+(-10)]$; $b-2$

Use the properties of this section to perform the operations. ***See Example 4.***

51. $50(67)2$
6700

52. $26+8-26+12$
20

53. $-\frac{3}{8}+\frac{2}{5}+\frac{8}{5}+\frac{3}{8}$
2

54. $-\frac{5}{12}-\frac{3}{7}+\frac{5}{12}+\frac{10}{7}$
1

55. $\left(-\frac{8}{5}\right)(0.77)\left(-\frac{5}{8}\right)$
0.77

56. $\frac{9}{7}(-0.38)\left(\frac{7}{9}\right)$
-0.38

Use the distributive property to rewrite each expression. Simplify if possible.
See Example 8.

57. $4(t+3)$
$4t + 12$

58. $5(w+4)$
$5w + 20$

59. $-8(r+3)$
$-8r - 24$

60. $-11(x+4)$
$-11x - 44$

61. $-5(y-4)$
$= -5(\underline{\ y\ }) - 5(\underline{-4})$
$= \underline{-5y + 20}$

62. $-9(g-4)$
$= -9(\underline{\ g\ }) - 9(\underline{-4})$
$= \underline{-9g + 36}$

63. $-\frac{4}{3}(12y + 15z)$
$-16y - 20z$

64. $-\frac{2}{5}(10b + 20a)$
$-4b - 8a$

65. $8 \cdot z + 8 \cdot w$
$8(z + w)$

66. $4 \cdot s + 4 \cdot r$
$4(s + r)$

67. $5 \cdot 3 + 5 \cdot 17$
$5(3 + 17)$; 100

68. $15 \cdot 6 + 5 \cdot 6$
$(15 + 5)6$; 120

69. $7(2v) + 7(5r)$
$7(2v + 5r)$

70. $13(5w) + 13(4p)$
$13(5w + 4p)$

71. $8(3r + 4s - 5y)$
$24r + 32s - 40y$

72. $2(5u - 3v + 7w)$
$10u - 6v + 14w$

73. $-3(8x + 3y + 4z)$
$-24x - 9y - 12z$

74. $-5(2x - 5y + 6z)$
$-10x + 25y - 30z$

Use the distributive property to write each expression without parentheses.
See Example 9.

75. $-(4t + 5m)$
$-4t - 5m$

76. $-(9x + 12y)$
$-9x - 12y$

77. $-(-5c - 4d)$
$5c + 4d$

78. $-(-13x - 15y)$
$13x + 15y$

79. $-(-3q + 5r - 8s)$
$3q - 5r + 8s$

80. $-(-4z + 5w - 9y)$
$4z - 5w + 9y$

Study Skills

REVIEWING A CHAPTER

Your textbook provides material to help you prepare for quizzes or tests in this course. Refer to the **Chapter Summary** as you read through the following techniques.

Chapter Reviewing Techniques

- **Review the Key Terms.** Make a study card for each. Include the given definition, an example, a sketch (if appropriate), and a section or page reference.
- **Take the Test Your Word Power quiz** to check your understanding of new vocabulary. The answers immediately follow.
- **Read the Quick Review.** Pay special attention to the headings. Study the explanations and examples given for each concept. Try to think about the whole chapter.
- **Reread your lecture notes.** Focus on what your instructor has emphasized in class, and review that material in your text.
- **Work the Review Exercises.** They are grouped by section.
  - ✓ Pay attention to direction words, such as *simplify*, *solve*, and *estimate.*
  - ✓ After you've done each section of exercises, check your answers in the answer section.
  - ✓ Are your answers exact and complete? Did you include the correct labels, such as \$, cm^2, ft, etc.?
  - ✓ Make study cards for difficult problems.
- **Work the Mixed Review Exercises.** They are in mixed-up order. Check your answers in the answer section.
- **Take the Chapter Test under test conditions.**
  - ✓ Time yourself.
  - ✓ Use a calculator or notes (if your instructor permits them on tests).
  - ✓ Take the test in one sitting.
  - ✓ Show all your work.
  - ✓ Check your answers in the back of the book.

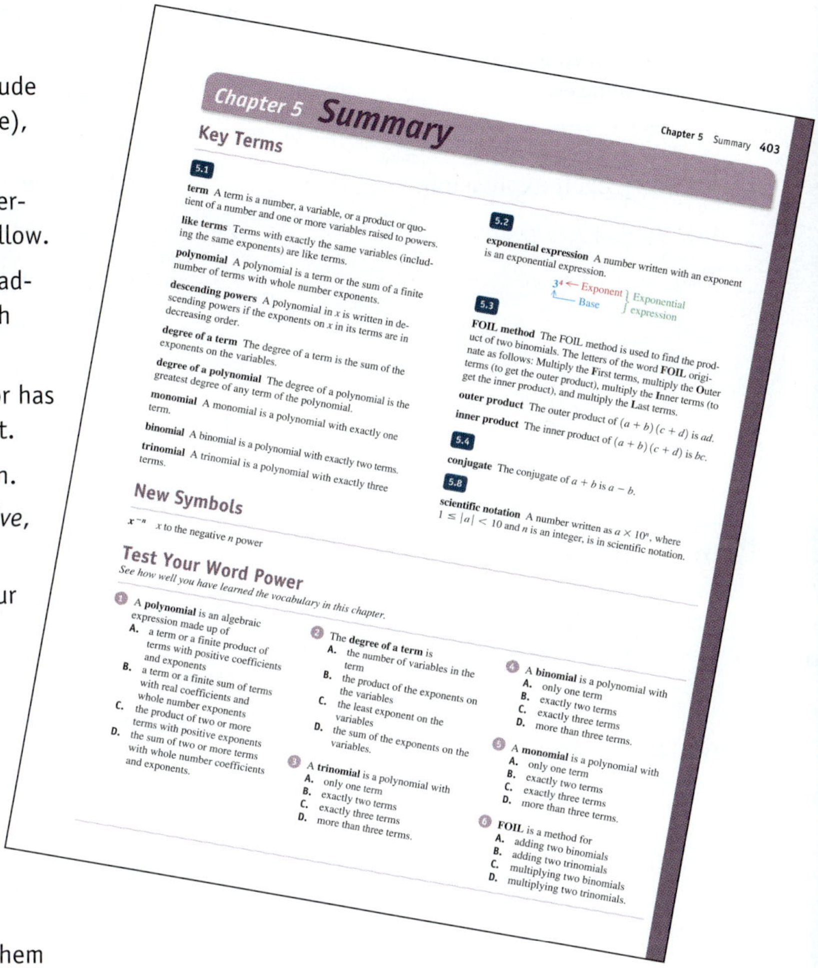

Chapter 5 Summary

Chapter 5 Summary 403

Key Terms

5.1

term A term is a number, a variable, or a product or quotient of a number and one or more variables raised to powers.

like terms Terms with exactly the same variables (including the same exponents) are like terms.

polynomial A polynomial is a term or the sum of a finite number of terms with whole number exponents.

descending powers A polynomial in x is written in descending powers if the exponents on x in its terms are in decreasing order.

degree of a term The degree of a term is the sum of the exponents on the variables.

degree of a polynomial The degree of a polynomial is the greatest degree of any term of the polynomial.

monomial A monomial is a polynomial with exactly one term.

binomial A binomial is a polynomial with exactly two terms.

trinomial A trinomial is a polynomial with exactly three terms.

5.2

exponential expression A number written with an exponent is an exponential expression.

3^4 ← Exponent; Base; Exponential expression

5.3

FOIL method The FOIL method is used to find the product of two binomials. The letters of the word **FOIL** originate as follows: Multiply the **F**irst terms, multiply the **O**uter terms (to get the outer product), multiply the **I**nner terms (to get the inner product), and multiply the **L**ast terms.

outer product The outer product of $(a + b)(c + d)$ is ad.

inner product The inner product of $(a + b)(c + d)$ is bc.

5.4

conjugate The conjugate of $a + b$ is $a - b$.

5.8

scientific notation A number written as $a \times 10^n$, where $1 \leq |a| < 10$ and n is an integer, is in scientific notation.

New Symbols

x^{-n} x to the negative n power

Test Your Word Power

See how well you have learned the vocabulary in this chapter.

1. A **polynomial** is an algebraic expression made up of
   A. a term or a finite product of terms with positive coefficients and exponents
   B. a term or a finite sum of terms with real coefficients and whole number exponents
   C. the product of two or more terms with positive exponents
   D. the sum of two or more terms with whole number coefficients and exponents.
2. The **degree of a term** is
   A. the number of variables in the term
   B. the product of the exponents on the variables
   C. the least exponent on the variables
   D. the sum of the exponents on the variables.
3. A **trinomial** is a polynomial with
   A. only one term
   B. exactly two terms
   C. exactly three terms
   D. more than three terms.
4. A **binomial** is a polynomial with
   A. only one term
   B. exactly two terms
   C. exactly three terms
   D. more than three terms.
5. A **monomial** is a polynomial with
   A. only one term
   B. exactly two terms
   C. exactly three terms
   D. more than three terms.
6. **FOIL** is a method for
   A. adding two binomials
   B. adding two trinomials
   C. multiplying two binomials
   D. multiplying two trinomials.

Reviewing a chapter will take some time. Avoid rushing through your review in one night. Use the suggestions over a few days or evenings to better understand the material and remember it longer.

Follow these reviewing techniques for your next test. After the test, evaluate how they worked for you. What will you do differently when reviewing for your next test?

1.8 Simplifying Expressions

OBJECTIVES

1. Simplify expressions.
2. Identify terms and numerical coefficients.
3. Identify like terms.
4. Combine like terms.
5. Simplify expressions from word phrases.

1 Simplify each expression.

(a) $9k + 12 - 5$

GS **(b)** $7(3p + 2q)$

$= 7(\_\_\_\_) + 7(\_\_\_\_)$

$= \_\_\_\_ + \_\_\_\_$

(c) $2 + 5(3z - 1)$

(d) $-3 - (2 + 5y)$

Answers

1. (a) $9k + 7$ **(b)** $3p$; $2q$; $21p$; $14q$
(c) $15z - 3$ **(d)** $-5 - 5y$

OBJECTIVE 1 Simplify expressions. We now simplify expressions using the properties of addition and multiplication introduced in **Section 1.7.**

EXAMPLE 1 Simplifying Expressions

Simplify each expression.

(a) $4x + 8 + 9$ simplifies to $4x + 17$.

(b) $4(3m - 2n)$ — To simplify, we clear the parentheses.

$= 4(3m) - 4(2n)$ Distributive property

$= 12m - 8n$ Associative property

(c) $6 + 3(4k + 5)$ — Don't start by adding.

$= 6 + 3(4k) + 3(5)$ Distributive property

$= 6 + 12k + 15$ Multiply.

$= 21 + 12k$ Add.

(d) $5 - (2y - 8)$

$= 5 - 1(2y - 8)$ $\quad -a = -1 \cdot a$ — Be careful with signs.

$= 5 - 2y + 8$ Distributive property

$= 13 - 2y$ Add.

◀ Work Problem **1** at the Side.

Note

In **Examples 1(c) and 1(d),** we mentally used the commutative and associative properties to add in the last step. In practice, these steps are often not written out.

OBJECTIVE 2 Identify terms and numerical coefficients. A **term** is a number (constant), a variable, or a product or quotient of numbers and variables raised to powers.

$$9x, \quad 15y^2, \quad -3, \quad -8m^2n, \quad \frac{2}{p}, \quad \text{and} \quad k \qquad \text{Terms}$$

In the term $9x$, the **numerical coefficient,** or simply **coefficient,** of the variable x is 9. Additional examples are shown in the table on the next page.

CAUTION

It is important to be able to distinguish between ***terms*** and ***factors.*** Consider the following expressions.

$8x^3 + 12x^2$ — This expression has **two terms,** $8x^3$ and $12x^2$. **Terms** are separated by a + or − symbol.

$(8x^3)(12x^2)$ — This is a **one-term** expression. The **factors** $8x^3$ and $12x^2$ are multiplied.

Term	Numerical Coefficient
$-7y$	-7
$34r^3$	34
$-26x^5yz^4$	-26
$-k$, or $-1 \cdot k$	-1
r, or $1r$	1
$\frac{3x}{8} = \frac{3}{8}x$	$\frac{3}{8}$

Work Problem 2 at the Side. ▶

OBJECTIVE 3 Identify like terms. Terms with exactly the same variables that have the same exponents are **like terms.** Here are some examples.

Like Terms			Unlike Terms			
$9t$	and	$4t$	$4y$	and	$7t$	Different variables
$6x^2$	and	$-5x^2$	$17x$	and	$-8x^2$	Different exponents
$-2pq$	and	$11pq$	$4xy^2$	and	$4xy$	Different exponents
$3x^2y$	and	$5x^2y$	$-7wz^3$	and	$2xz^3$	Different variables

Work Problem 3 at the Side. ▶

OBJECTIVE 4 Combine like terms. Recall the distributive property.

$x(y+z) = xy + xz$ can also be written "in reverse" as $xy + xz = x(y+z)$.

This last form, which may be used to find the sum or difference of like terms, provides justification for **combining like terms.**

EXAMPLE 2 Combining Like Terms

Combine like terms in each expression.

(a) $-9m + 5m$
$= (-9+5)m$
$= -4m$

(b) $6r + 3r + 2r$
$= (6+3+2)r$
$= 11r$

(c) $4x + x$
$= 4x + 1x$ $x = 1x$
$= (4+1)x$
$= 5x$

(d) $16y^2 - 9y^2$
$= (16-9)y^2$
$= 7y^2$

(e) $32y + 10y^2$ These unlike terms cannot be combined.

Work Problem 4 at the Side. ▶

Simplifying an Expression

An expression has been simplified when the following conditions have been met.

- All grouping symbols have been removed.
- All like terms have been combined.
- Operations have been performed, when possible.

2 Give the numerical coefficient of each term.

(a) $15q$

(b) $-2m^3$

(c) $-18m^7q^4$

(d) $-r$

(e) $\frac{5x}{4}$

3 Identify each pair of terms as *like* or *unlike*.

(a) $9x, 4x$

(b) $-8y^3, 12y^2$

(c) $5x^2y^4, 5x^4y^2$

(d) $7x^2y^4, -7x^2y^4$

(e) $13kt, 4tk$

4 Combine like terms.

(a) $4k + 7k$

(b) $4r - r$

(c) $5z + 9z - 4z$

(d) $8p + 8p^2$

Answers

2. **(a)** 15 **(b)** -2 **(c)** -18 **(d)** -1 **(e)** $\frac{5}{4}$

3. **(a)** like **(b)** unlike **(c)** unlike **(d)** like **(e)** like

4. **(a)** $11k$ **(b)** $3r$ **(c)** $10z$ **(d)** cannot be combined

5 Simplify.

GS **(a)** $10p + 3(5 + 2p)$

$= 10p + \_\_\_\_(5) + 3(\_\_\_\_)$

$= 10p + \_\_\_\_ + \_\_\_\_$

$= \_\_\_\_\_\_$

(b) $7z - 2 - (1 + z)$

Teaching Tip

Constants are like terms and may be combined.

(c) $-(3k^2 + 5k) + 7(k^2 - 4k)$

Teaching Tip

In Example 3(c), point out that

$$-(2 - r)$$

can be thought of as the opposite of $(2 - r)$, or $-2 + r$.

(d) $-\frac{3}{4}(x - 8) - \frac{1}{3}x$

6 Translate each phrase into a mathematical expression and simplify.

(a) Three times a number, subtracted from the sum of the number and 8

(b) Twice a number added to the sum of 6 and the number

Answers

5. **(a)** 3; $2p$; 15; $6p$; $16p + 15$ **(b)** $6z - 3$
(c) $4k^2 - 33k$ **(d)** $-\frac{13}{12}x + 6$

6. **(a)** $(x + 8) - 3x$; $-2x + 8$
(b) $(6 + x) + 2x$; $3x + 6$

CAUTION

Remember that only like terms may be combined.

EXAMPLE 3 Simplifying Expressions Involving Like Terms

Simplify each expression.

(a) $14y + 2(6 + 3y)$

$= 14y + 2(6) + 2(3y)$ Distributive property

$= 14y + 12 + 6y$ Multiply.

$= 20y + 12$ Combine like terms.

(b) $9k - 6 - 3(2 - 5k)$ Be careful with signs.

$= 9k - 6 - 3(2) - 3(-5k)$ Distributive property

$= 9k - 6 - 6 + 15k$ Multiply.

$= 24k - 12$ Combine like terms.

(c) $-(2 - r) + 10r$

$= -1(2 - r) + 10r$ $-(2 - r) = -1(2 - r)$

$= -1(2) - 1(-r) + 10r$ Distributive property

Be careful with signs.

$= -2 + r + 10r$ Multiply.

$= -2 + 11r$ Combine like terms.

(d) $-\frac{2}{3}(x - 6) - \frac{1}{6}x$

$= -\frac{2}{3}x - \frac{2}{3}(-6) - \frac{1}{6}x$ Distributive property

$= -\frac{2}{3}x + 4 - \frac{1}{6}x$ Multiply.

$= -\frac{4}{6}x + 4 - \frac{1}{6}x$ Get a common denominator.

$= -\frac{5}{6}x + 4$ Combine like terms,

◀ Work Problem 5 at the Side.

OBJECTIVE 5 Simplify expressions from word phrases.

EXAMPLE 4 Translating Words into a Mathematical Expression

Translate the phrase into a mathematical expression and simplify.

The sum of 9, five times a number, four times the number, and six times the number

The word "sum" indicates that the terms should be added. Use x for the number.

$9 + 5x + 4x + 6x$ simplifies to $9 + 15x$. Combine like terms.

This is an expression to be simplified, *not* an equation to be solved.

◀ Work Problem 6 at the Side.

1.8 Exercises

FOR EXTRA HELP

Download the MyDashBoard App

MyMathLab®

CONCEPT CHECK *In Exercises 1–4, choose the letter of the correct response.*

1. Which expression is a simplified form of $-(6x - 3)$? **B**

A. $-6x - 3$ **B.** $-6x + 3$

C. $6x - 3$ **D.** $6x + 3$

2. Which is an example of a pair of like terms? **C**

A. $6t, 6w$ **B.** $-8x^2y, 9xy^2$

C. $5ry, 6yr$ **D.** $-5x^2, 2x^3$

3. Which is an example of a term with numerical coefficient 5? **A**

A. $5x^3y^7$ **B.** x^5

C. $\frac{x}{5}$ **D.** 5^2xy^3

4. Which is a correct translation for "six times a number, subtracted from the product of eleven and the number" (if x represents the number)? **B**

A. $6x - 11x$ **B.** $11x - 6x$

C. $(11 + x) - 6x$ **D.** $6x - (11 + x)$

Simplify each expression. ***See Examples 1 and 2.***

5. $3x + 12x$ **15x**

6. $4y + 9y$ **13y**

7. $12b + b$ **13b**

8. $19x + x$ **20x**

9. $4r + 19 - 8$ **4r + 11**

10. $7t + 18 - 4$ **7t + 14**

11. $5 + 2(x - 3y)$ **5 + 2x − 6y**

12. $8 + 3(s - 6t)$ **8 + 3s − 18t**

13. $-2 - (5 - 3p)$ **−7 + 3p**

14. $-10 - (7 - 14r)$ **−17 + 14r**

15. $-\frac{5}{6}(y + 12) - \frac{1}{2}y$ $-\frac{4}{3}y - 10$

16. $-\frac{2}{3}(w + 15) - \frac{1}{4}w$ $-\frac{11}{12}w - 10$

Give the numerical coefficient. ***See Objective 2.***

17. $-12k$ **−12**

18. $-23y$ **−23**

19. $5m^2$ **5**

20. $-3n^6$ **−3**

21. xw **1**

22. pq **1**

23. $-x$ **−1**

24. $-t$ **−1**

25. 74 **74**

26. 98 **98**

27. $-\frac{3}{8}x$ $-\frac{3}{8}$

28. $-\frac{5}{4}z$ $-\frac{5}{4}$

29. $\frac{x}{2}$ $\frac{1}{2}$

30. $\frac{x}{6}$ $\frac{1}{6}$

31. $\frac{2x}{5}$ $\frac{2}{5}$

32. $\frac{8x}{9}$ $\frac{8}{9}$

33. $-1.28r^2$ **−1.28**

34. $-2.985t^3$ **−2.985**

Identify each group of terms as like *or* unlike. ***See Objective 3.***

35. $8r, -13r$ **like**

36. $-7a, 12a$ **like**

37. $5z^4, 9z^3$ **unlike**

38. $8x^5, -10x^3$ **unlike**

39. $4, 9, -24$ **like**

40. $7, 17, -83$ **like**

41. x, y **unlike**

42. t, s **unlike**

Simplify each expression. ***See Examples 2 and 3.***

43. $-5 - 2(x - 3)$

$= -5 - 2(\underline{\ x\ }) - 2(\underline{\ -3\ })$

$= -5 - \underline{\ 2x\ } + \underline{\ 6\ }$

$= \underline{\ 1 - 2x\ }$

44. $-8 - 3(2x + 4)$

$= -8 - 3(\underline{\ 2x\ }) - 3(\underline{\ 4\ })$

$= -8 - \underline{\ 6x\ } - \underline{\ 12\ }$

$= \underline{\ -20 - 6x\ }$

45. $-\frac{4}{3} + 2t + \frac{1}{3}t - 8 - \frac{8}{3}t$ $\quad -\frac{1}{3}t - \frac{28}{3}$

46. $-\frac{5}{6} + 8x + \frac{1}{6}x - 7 - \frac{7}{6}$ $\quad \frac{49}{6}x - 9$

47. $-5.3r + 4.9 - (2r + 0.7) + 3.2r$ $\quad -4.1r + 4.2$

48. $2.7b + 5.8 - (3b + 0.5) - 4.4b$ $\quad -4.7b + 5.3$

49. $2y^2 - 7y^3 - 4y^2 + 10y^3$ $\quad -2y^2 + 3y^3$

50. $9x^4 - 7x^6 + 12x^4 + 14x^6$ $\quad 21x^4 + 7x^6$

51. $13p + 4(4 - 8p)$ $\quad -19p + 16$

52. $5x + 3(7 - 2x)$ $\quad -x + 21$

53. $\frac{1}{2}(2x + 4) - \frac{1}{3}(9x - 6)$ $\quad -2x + 4$

54. $\frac{1}{4}(8x + 16) - \frac{1}{5}(20x - 15)$ $\quad -2x + 7$

55. $-\frac{2}{3}(5x + 7) - \frac{1}{3}(4x + 8)$ $\quad -\frac{14}{3}x - \frac{22}{3}$

56. $-\frac{3}{4}(7x + 9) - \frac{1}{4}(5x + 7)$ $\quad -\frac{13}{2}x - \frac{17}{2}$

57. $-\frac{4}{3}(y - 12) - \frac{1}{6}y$ $\quad -\frac{3}{2}y + 16$

58. $-\frac{7}{5}(t - 15) - \frac{1}{2}t$ $\quad -\frac{19}{10}t + 21$

59. $-5(5y - 9) + 3(3y + 6)$ $\quad -16y + 63$

60. $-3(2t + 4) + 8(2t - 4)$ $\quad 10t - 44$

61. $-3(2r - 3) + 2(5r + 3)$ $\quad 4r + 15$

62. $-4(5y - 7) + 3(2y - 5)$ $\quad -14y + 13$

63. $8(2k - 1) - (4k - 3)$ $\quad 12k - 5$

64. $6(3p - 2) - (5p + 1)$ $\quad 13p - 13$

65. $-2(-3k + 2) - (5k - 6) - 3k - 5$ $\quad -2k - 3$

66. $-2(3r - 4) - (6 - r) + 2r - 5$ $\quad -3r - 3$

67. $-4(-3x + 3) - (6x - 4) - 2x + 1$ $\quad 4x - 7$

68. $-5(8x + 2) - (5x - 3) - 3x + 17$ $\quad -48x + 10$

69. $-7.5(2y + 4) - 2.9(3y - 6)$ $\quad -23.7y - 12.6$

70. $8.4(6t - 6) + 2.4(9 - 3t)$ $\quad 43.2t - 28.8$

Write each phrase as a mathematical expression. Use x to represent the number. Combine like terms when possible. ***See Example 4.***

71. Five times a number, added to the sum of the number and three

$(x + 3) + 5x$; $6x + 3$

72. Six times a number, added to the sum of the number and six

$(x + 6) + 6x$; $7x + 6$

73. A number multiplied by -7, subtracted from the sum of 13 and six times the number

$(13 + 6x) - (-7x)$; $13 + 13x$

74. A number multiplied by 5, subtracted from the sum of 14 and eight times the number

$(14 + 8x) - 5x$; $14 + 3x$

75. Six times a number added to -4, subtracted from twice the sum of three times the number and 4

$2(3x + 4) - (-4 + 6x)$; 12

76. Nine times a number added to 6, subtracted from triple the sum of 12 and 8 times the number

$3(12 + 8x) - (6 + 9x)$; $30 + 15x$

77. Write the expression $9x - (x + 2)$ using words, as in **Exercises 71–76.**

Wording may vary. One example is "the difference between 9 times a number and the sum of the number and 2."

78. Write the expression $2(3x + 5) - 2(x + 4)$ using words, as in **Exercises 71–76.**

Wording may vary. One example is "the difference between twice the sum of three times a number and 5, and twice the sum of the number and 4."

79. CONCEPT CHECK A student simplified the expression $7x - 2(3 - 2x)$ as shown.

$$\begin{aligned} &7x - 2(3 - 2x) \\ &= 7x - 2(3) - 2(2x) \\ &= 7x - 6 - 4x \\ &= 3x - 6 \end{aligned}$$

What Went Wrong? Find the correct simplified answer.

The student made a sign error when applying the distributive property.

$$\begin{aligned} &7x - 2(3 - 2x) \\ &= 7x - 2(3) - 2(-2x) \\ &= 7x - 6 + 4x \\ &= 11x - 6 \end{aligned}$$

80. CONCEPT CHECK A student simplified the expression $3 + 2(4x - 5)$ as shown.

$$\begin{aligned} &3 + 2(4x - 5) \\ &= 5(4x - 5) \\ &= 5(4x) + 5(-5) \\ &= 20x - 25 \end{aligned}$$

What Went Wrong? Find the correct simplified answer.

The student incorrectly started by adding $3 + 2$. As the first step, 2 must be multiplied by $4x - 5$.

$$\begin{aligned} &3 + 2(4x - 5) \\ &= 3 + 2(4x) + 2(-5) \\ &= 3 + 8x - 10 \\ &= 8x - 7 \end{aligned}$$

Relating Concepts (Exercises 81–84) For Individual or Group Work

A manufacturer has fixed costs of \$1000 *to produce widgets. Each widget costs* \$5 *to make. The fixed cost to produce gadgets is* \$750, *and each gadget costs* \$3 *to make.* ***Work Exercises 81–84 in order.***

81. Write an expression for the cost to make x widgets. (*Hint:* The cost will be the sum of the fixed cost and the cost per item times the number of items.)

$1000 + 5x$ (dollars)

82. Write an expression for the cost to make y gadgets.

$750 + 3y$ (dollars)

83. Write an expression for the total cost to make x widgets and y gadgets.

$1000 + 5x + 750 + 3y$ (dollars)

84. Simplify the expression you wrote in **Exercise 83.**

$1750 + 5x + 3y$ (dollars)

Study Skills
TAKING MATH TESTS

Techniques To Improve Your Test Score	Comments
Come prepared with a pencil, eraser, paper, and calculator, if allowed.	Working in pencil lets you erase, keeping your work neat and readable.
Scan the entire test, note the point values of different problems, and plan your time accordingly.	To do 20 problems in 50 minutes, allow $50 \div 20 = 2.5$ minutes per problem. Spend less time on the easier problems.
Do a "knowledge dump" when you get the test. Write important notes, such as formulas, to yourself in a corner of the test.	Writing down tips and things that you've memorized at the beginning allows you to relax later.
Read directions carefully, and circle any significant words. When you finish a problem, read the directions again to make sure you did what was asked.	Pay attention to announcements written on the board or made by your instructor. Ask if you don't understand.
Show all your work. Many teachers give partial credit if some steps are correct, even if the final answer is wrong. ***Write neatly.***	If your teacher can't read your writing, you won't get credit for it. If you need more space to work, ask to use extra paper.
Write down anything that might help solve a problem: a formula, a diagram, etc. If you can't get it, circle the problem and come back to it later. Do *not* erase anything you wrote down.	If you know even a little bit about the problem, write it down. The answer may come to you as you work on it, or you may get partial credit. Don't spend too long on any one problem.
If you can't solve a problem, make a guess. Do not change it unless you find an obvious mistake.	Have a good reason for changing an answer. Your first guess is often your best bet.
Check that the answer to an application problem is reasonable and makes sense. Read the problem again to make sure you've answered the question.	Use common sense. Can the father really be seven years old? Would a month's rent be $32,140? Label your answer: $, years, inches, etc.
Check for careless errors. Rework the problem without looking at your previous work. Compare the two answers.	Reworking the problem from the beginning forces you to rethink it. If possible, use a different method to solve the problem.

Mark several tips to try when you take your next math test. After the test, evaluate how they worked for you. What will you do differently when taking your next test?

Chapter 1 Summary

Key Terms

1.1

exponent An exponent, or **power,** is a number that indicates how many times a factor is repeated.

base The base is the number that is a repeated factor when written with an exponent.

exponential expression A number written with an exponent is an exponential expression.

1.2

variable A variable is a symbol, usually a letter, used to represent an unknown number.

constant A constant is a fixed, unchanging number.

algebraic expression An algebraic expression is a collection of numbers (constants), variables, operation symbols, and grouping symbols.

equation An equation is a statement that says two expressions are equal.

solution A solution of an equation is any value of the variable that makes the equation true.

1.3

natural numbers The set of natural numbers is $\{1, 2, 3, 4, \ldots\}$.

whole numbers The set of whole numbers is $\{0, 1, 2, 3, 4, \ldots\}$.

number line A number line shows the ordering of real numbers on a line.

additive inverse The additive inverse of a number a is the number that is the same distance from 0 on the number line as a, but on the opposite side of 0. This number is also called the opposite of a.

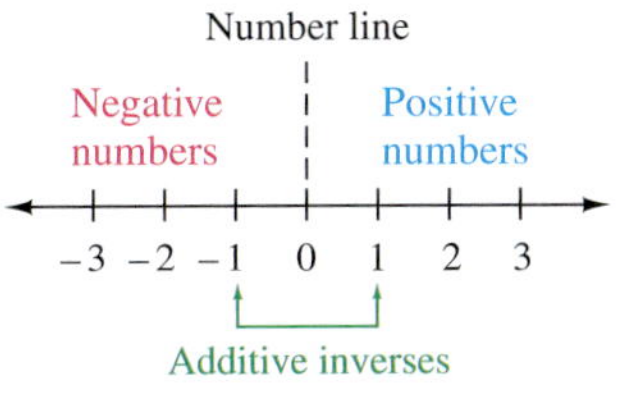

integers The set of integers is $\{\ldots, -3, -2, -1, 0, 1, 2, 3, \ldots\}$.

negative number A negative number is located to the *left* of 0 on the number line.

positive number A positive number is located to the *right* of 0 on the number line.

signed numbers Positive numbers and negative numbers are signed numbers.

rational numbers A rational number is a number that can be written as the quotient of two integers, with denominator not 0.

set-builder notation Set-builder notation uses a variable and a description to describe a set. It is often used to describe sets whose elements cannot easily be listed.

coordinate The number that corresponds to a point on the number line is the coordinate of that point.

irrational numbers An irrational number is a real number that is not a rational number.

real numbers Real numbers are numbers that can be represented by points on the number line (that is, all rational and irrational numbers).

absolute value The absolute value of a number is the distance between 0 and the number on a number line.

1.4

sum The answer to an addition problem is the sum.

1.5

minuend In the operation $a - b$, a is the minuend.

subtrahend In the operation $a - b$, b is the subtrahend.

difference The answer to a subtraction problem is the difference.

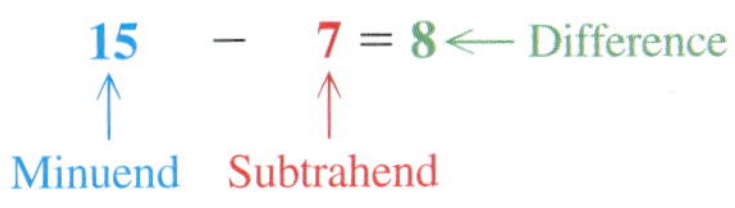

1.6

product The answer to a multiplication problem is the product.

quotient The answer to a division problem is the quotient.

reciprocal Pairs of numbers whose product is 1 are reciprocals, or **multiplicative inverses,** of each other.

(continued)

1.7

identity element for addition When the identity element for addition, which is 0, is added to a number, the number is unchanged.

identity element for multiplication When a number is multiplied by the identity element for multiplication, which is 1, the number is unchanged.

1.8

term A term is a number (constant), a variable, or a product or quotient of a number and one or more variables raised to powers.

Numerical coefficient
↓
$\underbrace{-7x^2}_{\text{Term}}$

numerical coefficient The numerical factor of a term is its numerical coefficient, or **coefficient.**

like terms Terms with exactly the same variables (including the same exponents) are like terms.

New Symbols

a^n	n factors of a	$a(b)$, $(a)b$, $(a)(b)$, $a \cdot b$, or ab	a times b
$=$	is equal to	$\frac{a}{b}$, a/b, or $a \div b$	a divided by b
$\neq$	is not equal to	$\{\ \}$	set braces
$<$	is less than	$\{x \mid x$ **has a certain property**$\}$	set-builder notation
$\leq$	is less than or equal to	$[\ ]$	square brackets
$>$	is greater than	$\lvert x \rvert$	absolute value of x
$\geq$	is greater than or equal to	$-x$	additive inverse, or opposite, of x
		$\frac{1}{x}$	multiplicative inverse, or reciprocal, of x (where $x \neq 0$)

Test Your Word Power

See how well you have learned the vocabulary in this chapter.

1 A **product** is
- **A.** the answer in an addition problem
- **B.** the answer in a multiplication problem
- **C.** one of two or more numbers that are added to get another number
- **D.** one of two or more numbers that are multiplied to get another number.

2 An **exponent** is
- **A.** a symbol that tells how many numbers are being multiplied
- **B.** a number raised to a power
- **C.** a number that tells how many times a factor is repeated
- **D.** one of two or more numbers that are multiplied.

3 A **variable** is
- **A.** a symbol used to represent an unknown number
- **B.** a value that makes an equation true
- **C.** a solution of an equation
- **D.** the answer in a division problem.

4 An **integer** is
- **A.** a positive or negative number
- **B.** a natural number, its opposite, or zero
- **C.** any number that can be graphed on a number line
- **D.** the quotient of two numbers.

5 A **coordinate** is
- **A.** the number that corresponds to a point on a number line
- **B.** the graph of a number
- **C.** any point on a number line
- **D.** the distance from 0 on a number line.

6 The **absolute value** of a number is
- **A.** the graph of the number
- **B.** the reciprocal of the number
- **C.** the opposite of the number
- **D.** the distance between 0 and the number on a number line.

7 A **term** is
- **A.** a numerical factor
- **B.** a number, a variable, or a product or quotient of numbers and variables raised to powers
- **C.** one of several variables with the same exponents
- **D.** a sum of numbers and variables raised to powers.

8 The **subtrahend** in $a - b = c$ is
- **A.** a
- **B.** b
- **C.** c
- **D.** $a - b$.

Answers to Test Your Word Power

1. B; *Example:* The product of 2 and 5, or 2 times 5, is 10.
2. C; *Example:* In 2^3, the number 3 is the exponent (or power), so 2 is a factor three times; $2^3 = 2 \cdot 2 \cdot 2 = 8$.
3. A; *Examples:* x, y, z
4. B; *Examples:* $-9, 0, 6$
5. A; *Example:* The point graphed three units to the right of 0 on a number line has coordinate 3.
6. D; *Examples:* $|2| = 2$ and $|-2| = 2$
7. B; *Examples:* $6, \frac{x}{2}, -4ab^2$
8. B; *Example:* In $5 - 3 = 2$, 5 is the minuend, 3 is the subtrahend, and 2 is the difference.

Quick Review

Concepts | **Examples**

1.1 Exponents, Order of Operations, and Inequality

Order of Operations
Simplify within any parentheses or brackets and above and below fraction bars, using the following steps.

Step 1 Apply all exponents.

Step 2 Multiply or divide from left to right.

Step 3 Add or subtract from left to right.

$$\frac{9(2+6)}{2} - 2(2^3 + 3)$$

$$= \frac{9(8)}{2} - 2(8+3) \quad \text{Add inside parentheses. Apply the exponent.}$$

$$= 36 - 2(8+3) \quad \text{Multiply and divide.}$$

$$= 36 - 2(11) \quad \text{Add inside parentheses.}$$

$$= 36 - 22 \quad \text{Multiply.}$$

$$= 14 \quad \text{Subtract.}$$

1.2 Variables, Expressions, and Equations

To evaluate an expression means to find its value. Evaluate an expression with a variable by substituting a given number for the variable.

Evaluate $2x + y^2$ for $x = 3$ and $y = -4$.

$$2x + y^2$$

$$= 2(3) + (-4)^2 \quad \text{Substitute.}$$

$$= 6 + 16 \quad \text{Multiply and apply the exponent.}$$

$$= 22 \quad \text{Add.}$$

Values of a variable that make an equation true are solutions of the equation.

Is 2 a solution of $5x + 3 = 18$?

$$5(2) + 3 \stackrel{?}{=} 18 \quad \text{Let } x = 2.$$

$$13 = 18 \quad \text{False}$$

2 is not a solution.

1.3 Real Numbers and the Number Line

Ordering Real Numbers
a is less than b if a is to the left of b on a number line.

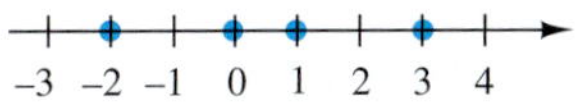

$-2 < 3$ $3 > 0$ $1 < 3$

The additive inverse, or opposite, of a is $-a$.

$-(5) = -5$ $-(-7) = 7$ $-0 = 0$

The absolute value of a, written $|a|$, is the distance between a and 0 on a number line.

$|13| = 13$ $|0| = 0$ $|-5| = 5$

1.4 Adding Real Numbers

To add two numbers with the *same sign*, add their absolute values. The sum has that same sign.

To add two numbers with *different signs*, subtract the lesser absolute value from the greater. The sum has the sign of the number with greater absolute value.

$$9 + 4 = 13$$

$$-8 + (-5) = -13$$

$$7 + (-12) = -5$$

$$-5 + 13 = 8$$

Concepts	Examples
1.5 Subtracting Real Numbers	
For any real numbers a and b, $a - b$ is defined as $a + (-b)$.	$5 - (-2) = 5 + 2 = 7$ $-3 - 4 = -3 + (-4) = -7$ $-2 - (-6) = -2 + 6 = 4$
1.6 Multiplying and Dividing Real Numbers	
The product (or quotient) of two numbers having the *same sign* is *positive*.	$6 \cdot 5 = 30$ $\quad -7(-8) = 56$ $\quad \frac{-24}{-6} = 4$
The product (or quotient) of two numbers having *different signs* is *negative*.	$-6(5) = -30$ $\quad 6(-5) = -30$
To divide a by b, multiply a by the reciprocal of b.	$-18 \div 9 = \frac{-18}{9} = -2$ $49 \div (-7) = \frac{49}{-7} = -7$ $10 \div \frac{2}{3} = 10 \cdot \frac{3}{2} = 15$
0 divided by a nonzero number is 0. Division by 0 is undefined.	$\frac{0}{5} = 0$ $\quad \frac{5}{0}$ is undefined.
1.7 Properties of Real Numbers	
Commutative Properties $a + b = b + a$ $ab = ba$	$7 + (-1) = -1 + 7$ $5(-3) = (-3)5$
Associative Properties $(a + b) + c = a + (b + c)$ $(ab)c = a(bc)$	$(3 + 4) + 8 = 3 + (4 + 8)$ $[-2(6)]4 = -2[6(4)]$
Identity Properties $a + 0 = a$ $\quad 0 + a = a$ $a \cdot 1 = a$ $\quad 1 \cdot a = a$	$-7 + 0 = -7$ $\quad 0 + (-7) = -7$ $9 \cdot 1 = 9$ $\quad 1 \cdot 9 = 9$
Inverse Properties $a + (-a) = 0$ $\quad -a + a = 0$ $a \cdot \frac{1}{a} = 1$ $\quad \frac{1}{a} \cdot a = 1$ $\quad (a \neq 0)$	$7 + (-7) = 0$ $\quad -7 + 7 = 0$ $-2\left(-\frac{1}{2}\right) = 1$ $\quad -\frac{1}{2}(-2) = 1$
Distributive Properties $a(b + c) = ab + ac$ $(b + c)a = ba + ca$ $a(b - c) = ab - ac$	$5(4 + 2) = 5(4) + 5(2)$ $(4 + 2)5 = 4(5) + 2(5)$ $9(5 - 4) = 9(5) - 9(4)$
1.8 Simplifying Expressions	
Only like terms may be combined. We use the distributive property.	$4(3 + 2x) - 6(5 - x)$ $= 12 + 8x - 30 + 6x$ $= 14x - 18$

Chapter 1 Review Exercises

If you need help with any of these Review Exercises, look in the section indicated.

1.1 *Find the value of each exponential expression.*

1. 5^4 625

2. $(0.03)^4$ 0.00000081

3. 0.21^3 0.009261

4. $\left(\frac{5}{2}\right)^3$ $\frac{125}{8}$

Find the value of each expression.

5. $8 \cdot 5 - 13$ 27

6. $5[4^2 + 3(2^3)]$ 200

7. $\frac{7(3^2 - 5)}{16 - 2 \cdot 6}$ 7

8. $\frac{5(6^2 - 2^4)}{3 \cdot 5 + 10}$ 4

Write each word sentence in symbols.

9. Thirteen is less than seventeen.
$13 < 17$

10. Five plus two is not equal to ten.
$5 + 2 \neq 10$

11. Write $6 < 15$ in words.
Six is less than fifteen.

12. Construct a false statement that involves addition on the left side, the symbol $\geq$, and division on the right side.
Answers will vary. One example is $2 + 5 \geq \frac{16}{2}$.

1.2 *Evaluate each expression for $x = 6$ and $y = 3$.*

13. $2x + 6y$ 30

14. $4(3x - y)$ 60

15. $\frac{x}{3} + 4y$ 14

16. $\frac{x^2 + 3}{3y - x}$ 13

Change each word phrase to an algebraic expression. Use x to represent the number.

17. Six added to a number $x + 6$

18. A number subtracted from eight $8 - x$

19. Nine subtracted from six times a number $6x - 9$

20. Three-fifths of a number added to 12 $12 + \frac{3}{5}x$

Decide whether the given number is a solution of the equation.

21. $5x + 3(x + 2) = 22$; 2 yes

22. $\frac{x + 5}{3x} = 1$; 6 no

Change each word sentence to an equation. Use x to represent the number.

23. Six less than twice a number is 10. $2x - 6 = 10$

24. The product of a number and 4 is 8. $4x = 8$

Identify each of the following as either an equation *or an* expression.

25. $5r - 8(r + 7) = 2$ equation

26. $2y + (5y - 9) + 2$ expression

1.3 *Graph each group of numbers on a number line.*

27. $-4, -\frac{1}{2}, 0, 2.5, 5$

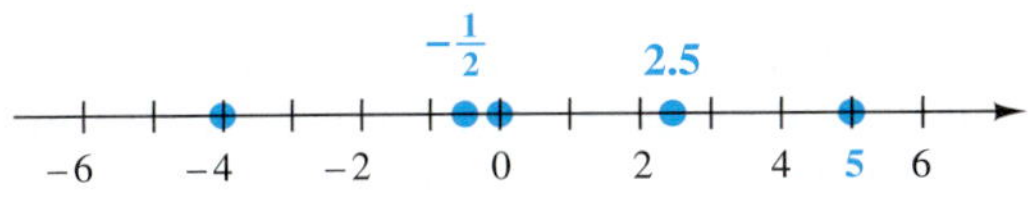

28. $-3\frac{1}{4}, \frac{14}{5}, -1\frac{1}{8}, \frac{5}{6}$

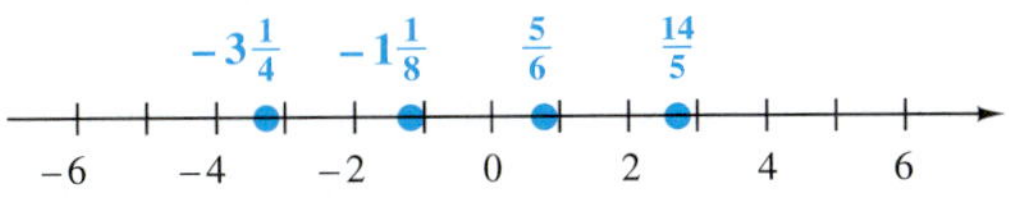

Select the lesser number in each pair.

29. $-10, 5$ -10

30. $-8, -9$ -9

31. $-\frac{2}{3}, -\frac{3}{4}$ $-\frac{3}{4}$

32. $0, -|23|$ $-|23|$

Decide whether each statement is true *or* false.

33. $12 > -13$ true

34. $0 > -5$ true

35. $-9 < -7$ true

36. $-13 > -13$ false

Simplify by finding the absolute value.

37. $-|3|$ -3

38. $-|-19|$ -19

39. $-|9-2|$ -7

40. $|15-6|$ 9

1.4 *Find each sum.*

41. $-10 + 4$ -6

42. $14 + (-18)$ -4

43. $-8 + (-9)$ -17

44. $\frac{4}{9} + \left(-\frac{5}{4}\right)$ $-\frac{29}{36}$

45. $[-6 + (-8) + 8] + [9 + (-13)]$ -10

46. $(-4 + 7) + (-11 + 3) + (-15 + 1)$ -19

Write a numerical expression for each phrase, and simplify the expression.

47. 19 added to the sum of -31 and 12

$(-31 + 12) + 19; 0$

48. 13 more than the sum of -4 and -8

$[-4 + (-8)] + 13; 1$

Solve each problem.

49. Like many people, Otis Taylor neglects to keep up his checkbook balance. When he finally balanced his account, he found that the balance was $-\$23.75$, so he deposited \$50.00. What is his new balance?

\$26.25

50. The low temperature in Yellowknife, in the Canadian Northwest Territories, one January day was $-26°$F. It rose 16° to its high that day. What was the high temperature?

$-10°$F

1.5 *Find each difference.*

51. $-7 - 4$ -11

52. $-12 - (-11)$ -1

53. $5 - (-2)$ 7

54. $-\frac{3}{7} - \frac{4}{5}$ $-\frac{43}{35}$

55. $2.56 - (-7.75)$

10.31

56. $(-10 - 4) - (-2)$

-12

57. $(-3 + 4) - (-1)$

2

58. $|5 - 9| - |-3 + 6|$

1

Write a numerical expression for each phrase, and simplify the expression.

59. The difference between -4 and -6

$-4 - (-6); 2$

60. Five less than the sum of 4 and -8

$[4 + (-8)] - 5; -9$

61. The difference between 18 and -23, decreased by 15

$[18 - (-23)] - 15; 26$

62. Nineteen, decreased by 12 less than -7

$19 - (-7 - 12); 38$

Solve each problem.

63. The quaterback for the Chicago Bears passed for a gain of 8 yd, was sacked for a loss of 12 yd, and then threw a 42 yd touchdown pass. What positive or negative number represents the total net yardage for the plays?

38

64. On Tuesday, January 10, 2012, the Dow Jones Industrial Average closed at 12,462.47, up 69.78 from the previous day. What was the closing price on Monday, January 9, 2012? (*Source: The New York Times.*)

12,392.69

The table shows the number of people naturalized in the United States (that is, made citizens of the United States) for the years 2003 through 2010. In Exercises 65–68, use a signed number to represent the change in the number of people naturalized for each time period.

Year	Number of People (in thousands)
2003	463
2004	337
2005	604
2006	702
2007	1383
2008	526
2009	570
2010	710

Source: U.S. Department of Homeland Security.

65. 2003–2004

-126 thousand

66. 2004–2005

267 thousand

67. 2006–2007

681 thousand

68. 2007–2008

-857 thousand

1.6 *Perform the indicated operations.*

69. $(-12)(-3)$

36

70. $15(-7)$

-105

71. $-\frac{4}{3}\left(-\frac{3}{8}\right)$

$\frac{1}{2}$

72. $-4.8(-2.1)$

10.08

73. $5(8-12)$

-20

74. $(5-7)(8-3)$

-10

75. $2(-6)-(-4)(-3)$

-24

76. $3(-10)-5$

-35

77. $\frac{-36}{-9}$

4

78. $\frac{220}{-11}$

-20

79. $-\frac{1}{2} \div \frac{2}{3}$

$-\frac{3}{4}$

80. $-33.9 \div (-3)$

11.3

81. $\frac{-5(3)-1}{8-4(-2)}$

-1

82. $\frac{5(-2)-3(4)}{-2[3-(-2)]+10}$

undefined

83. $\frac{10^2-5^2}{8^2+3^2-(-2)}$

1

84. $\frac{4^2-8 \cdot 2}{(-1.2)^2-(-0.56)}$

0

Evaluate each expression for $x=-5$, $y=4$, and $z=-3$.

85. $6x-4z$

-18

86. $5x+y-z$

-18

87. $5x^2$

125

88. $z^2(3x-8y)$

-423

Write a numerical expression for each phrase, and simplify the expression.

89. Nine less than the product of -4 and 5

$-4(5) - 9; -29$

90. Five-sixths of the sum of 12 and -6

$\frac{5}{6}[12 + (-6)]; 5$

91. The quotient of 12 and the sum of 8 and -4

$\frac{12}{8 + (-4)}; 3$

92. The product of -20 and 12, divided by the difference between 15 and -15

$\frac{-20(12)}{15 - (-15)}; -8$

Translate each sentence to an equation, using x to represent the number.

93. The quotient of a number and the sum of the number and 5 is -2.

$\frac{x}{x + 5} = -2$

94. 3 less than 8 times a number is -7.

$8x - 3 = -7$

1.7 *Decide whether each statement is an example of a commutative, associative, identity, or inverse property, or of the distributive property.*

95. $6 + 0 = 6$

identity property

96. $5 \cdot 1 = 5$

identity property

97. $-\frac{2}{3}\left(-\frac{3}{2}\right) = 1$

inverse property

98. $17 + (-17) = 0$

inverse property

99. $3x + 3y = 3(x + y)$

distributive property

100. $w(xy) = (wx)y$

associative property

101. $5 + (-9 + 2) = [5 + (-9)] + 2$

associative property

102. $(1 + 2) + 3 = 3 + (1 + 2)$

commutative property

Use the distributive property to rewrite each expression. Simplify if possible.

103. $7y + y$

$(7 + 1)y; 8y$

104. $-12(4 - t)$

$-12 \cdot 4 - 12(-t); -48 + 12t$

105. $3(2s) + 3(4y)$

$3(2s + 4y); 6s + 12y$

106. $-(-4r + 5s)$

$-1(-4r) + (-1)(5s); 4r - 5s$

1.8 *Use the distributive property as necessary and combine like terms.*

107. $16p^2 - 8p^2 + 9p^2$

$17p^2$

108. $4r^2 - 3r + 10r + 12r^2$

$16r^2 + 7r$

109. $-8(5k - 6) + 3(7k + 2)$

$-19k + 54$

110. $2s - (-3s + 6)$

$5s - 6$

111. $-7(2t - 4) - 4(3t + 8) - 19(t + 1)$

$-45t - 23$

112. $3.6t^2 + 9t - 8.1(6t^2 + 4t)$

$-45t^2 - 23.4t$

*Mixed Review Exercises**

Perform the indicated operations.

113. $\frac{15}{2} \cdot \left(-\frac{4}{5}\right)$ -6

114. $\left(-\frac{5}{6}\right)^2$ $\frac{25}{36}$

115. $-|(-7)(-4)| - (-2)$ -26

116. $\frac{6(-4) + 2(-12)}{5(-3) + (-3)}$ $\frac{8}{3}$

117. $\frac{3}{8} - \frac{5}{12}$ $-\frac{1}{24}$

118. $\frac{12^2 + 2^2 - 8}{10^2 - (-4)(-15)}$ $\frac{7}{2}$

119. $\frac{8^2 + 6^2}{7^2 + 1^2}$ 2

120. $-16(-3.5) - 7.2(-3)$ 77.6

121. $2\frac{5}{6} - 4\frac{1}{3}$ $-1\frac{1}{2}$

122. $-8 + [(-4 + 17) - (-3 - 3)]$ 11

123. $-\frac{12}{5} \div \frac{9}{7}$ $-\frac{28}{15}$

124. $(-8 - 3) - 5(2 - 9)$ 24

125. $[-7 + (-2) - (-3)] + [8 + (-13)]$ -11

126. $[(-2) + 7 - (-5)] + [-4 - (-10)]$ 16

Solve each problem.

127. The highest temperature ever recorded in Iowa was 118°F at Keokuk on July 20, 1934. The lowest temperature ever recorded in the state was at Elkader on February 3, 1996, and was 165° lower than the highest temperature. What is the record low temperature for Iowa? (*Source:* National Climatic Data Center.)

-47°F

128. Humpback whales love to heave their 45-ton bodies out of the water. This is called *breaching*. Chantelle, a researcher based on the island of Maui, noticed that one of her favorite whales, "Pineapple," breached 15 ft above the surface of the ocean while her mate cruised 12 ft below the surface. What is the difference between these two heights?

27 ft

*The order of exercises in this final group does not correspond to the order in which topics occur in the chapter. This random ordering should help you prepare for the chapter test in yet another way.

Chapter 1 Test

CHAPTER Test Prep VIDEO *The Chapter Test Prep Videos with test solutions are available on DVD, in MyMathLab, and on YouTube—search "LialIntroductoryAlg" and click on "Channels."*

Decide whether each statement is true *or* false.

1. $4[-20 + 7(-2)] \leq -135$

true

2. $\left(\frac{1}{2}\right)^2 + \left(\frac{2}{3}\right)^2 = \left(\frac{1}{2} + \frac{2}{3}\right)^2$

false

3. Graph the numbers -1, -3, $|-4|$, and $|-1|$ on the number line.

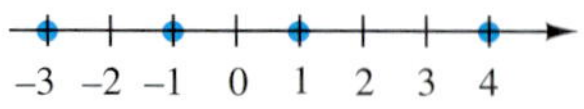

Select the lesser number from each pair.

4. $6, -|-8|$

$-|-8|$ (or -8)

5. $-0.742, -1.277$

-1.277

6. Write in symbols: The quotient of -6 and the sum of 2 and -8. Simplify the expression.

$\frac{-6}{2 + (-8)}$; 1

7. If a and b are both negative, is $\frac{a + b}{a \cdot b}$ positive or negative?

negative

Perform the indicated operations whenever possible.

8. $-2 - (5 - 17) + (-6)$

4

9. $-5\frac{1}{2} + 2\frac{2}{3}$

$-2\frac{5}{6}$

10. $4^2 + (-8) - (2^3 - 6)$

6

11. $-6.2 - [-7.1 + (2.0 - 3.1)]$

2

12. $(-5)(-12) + 4(-4) + (-8)^2$

108

13. $\frac{-7 - |-6 + 2|}{-5 - (-4)}$

11

14. $\frac{30(-1 - 2)}{-9[3 - (-2)] - 12(-2)}$

$\frac{30}{7}$

In Exercises 15 and 16, evaluate each expression for $x = -2$ and $y = 4$.

15. $3x - 4y^2$

-70

16. $\frac{5x + 7y}{3(x + y)}$

3

Solve each problem.

17. The highest Fahrenheit temperature ever recorded in Idaho was 118°F, while the lowest was −60°F. What is the difference between these highest and lowest temperatures? (*Source: World Almanac and Book of Facts.*)
178°F

18. For a certain system of rating relief pitchers, 3 points are awarded for a save, 3 points are awarded for a win, 2 points are subtracted for a loss, and 2 points are subtracted for a blown save. If a pitcher has 4 saves, 3 wins, 2 losses, and 1 blown save, how many points does he have?
15

19. For 2009, the U.S. federal government collected \$2.10 trillion in revenues, but spent \$3.52 trillion. Write the federal budget deficit as a signed number. (*Source: The Gazette.*)
−\$1.42 trillion

Match each statement in Column I with the property it illustrates in Column II.

I	II
20. $3x + 0 = 3x$ **D**	**A.** Commutative property
21. $(5 + 2) + 8 = 8 + (5 + 2)$ **A**	**B.** Associative property
22. $-3(x + y) = -3x + (-3y)$ **E**	**C.** Inverse property
23. $-5 + (3 + 2) = (-5 + 3) + 2$ **B**	**D.** Identity property
24. $-\frac{5}{3}\left(-\frac{3}{5}\right) = 1$ **C**	**E.** Distributive property

Simplify.

25. $8x + 4x - 6x + x + 14x$
21x

26. $-2(3x^2 + 4) - 3(x^2 + 2x)$
$-9x^2 - 6x - 8$

27. Which properties are used to show that $-(3x + 1) = -3x - 1$?
identity and distributive properties

28. Consider the expression $-6[5 + (-2)]$.

(a) Evaluate it by first working within the brackets. **−18**

(b) Evaluate it by using the distributive property. **−18**

(c) Why must the answers in parts (a) and (b) be the same?
The distributive property tells us that the two methods produce equal results.

Math in the Media

THE MAGIC NUMBER IN SPORTS

The climax of any sports season is the playoffs. Baseball fans eagerly debate predictions of which team will win the pennant for their division. The *magic number* for each first-place team is often reported in media outlets. The **magic number** (sometimes called the **elimination number**) is the combined number of wins by the first-place team and losses by the second-place team that would clinch the title for the first-place team.

American League

East	W	L	PCT	GB
New York	87	57	.604	—
Boston	85	60	.586	2.5
Tampa Bay	80	64	.556	7.0
Toronto	73	73	.500	15.0
Baltimore	58	86	.403	29.0
Central	**W**	**L**	**PCT**	**GB**
Detroit	83	62	.572	—
Chicago	73	71	.507	9.5
Cleveland	71	72	.497	11.0
Kansas City	61	86	.415	23.0
Minnesota	59	86	.407	24.0
West	**W**	**L**	**PCT**	**GB**
Texas	82	64	.562	—
Los Angeles	80	65	.552	1.5
Oakland	66	79	.455	15.5
Seattle	61	84	.421	20.5

Source: mlb.com

To calculate the magic number, consider the following conditions.

The number of wins for the first-place team (W_1) *plus the magic number* (M) *is one more than the sum of the number of wins to date* (W_2) *and the number of games remaining in the season* (N_2) *for the second-place team.*

1. First, use the variable definitions to write an equation involving the magic number. Second, solve the equation for the magic number and write a formula for it.

$W_1 + M = W_2 + N_2 + 1;$

$M = W_2 + N_2 + 1 - W_1$

2. The American League standings on September 10, 2011, are shown above. There were 162 regulation games in the 2011 season. Find the magic number for each team. The number of games remaining in the season for the second-place team is calculated as

$$N_2 = 162 - (W_2 + L_2),$$

where L_2 represents the number of losses for the second-place team.

(a) AL East: New York vs Boston

Magic Number 16

(b) AL Central: Detroit vs Chicago

Magic Number 9

(c) AL West: Texas vs Los Angeles

Magic Number 16

3. Try to calculate the magic number for Seattle vs Texas. (Treat Seattle as if it were the second-place team.) How can we interpret the result?

$M = -3$; A negative magic number indicates that Seattle has been eliminated from winning the division.

2 Equations, Inequalities, and Applications

Solving *linear equations*, the subject of this chapter, can be thought of in terms of the concept of balance.

2.1 The Addition Property of Equality

OBJECTIVES

1. Identify linear equations.
2. Use the addition property of equality.
3. Simplify, and then use the addition property of equality.

Teaching Tip
Students may confuse expressions and equations. Keep reminding them, as we will, of the difference.

An *equation* is a statement asserting that two algebraic expressions are equal. In this chapter, we *solve* equations.

CAUTION

Remember that an equation always includes an equality symbol.

Equation	Expression
Left side → $x - 5 = 2$ ← Right side	$x - 5$
An equation can be solved.	An expression **cannot** be solved. (It can often be *evaluated* or *simplified*.)

OBJECTIVE 1 Identify linear equations.

Linear Equation in One Variable

A **linear equation in one variable** can be written in the form

$$Ax + B = C,$$

where A, B, and C are real numbers, with $A \neq 0$.

$4x + 9 = 0$, $2x - 3 = 5$, and $x = 7$ Linear equations

$x^2 + 2x = 5$, $x^3 - x = 6$, and $|2x + 6| = 0$ *Non*linear equations

Teaching Tip
Mention that other letters besides x can be used for variables in equations.

Recall that a **solution** of an equation is a number that makes the equation true when it replaces the variable. An equation is solved by finding its **solution set,** the set of all solutions. Equations that have exactly the same solution sets are **equivalent equations.** A linear equation in x is solved by using a series of steps to produce a simpler equivalent equation of the form

$$x = \text{a number} \quad \text{or} \quad \text{a number} = x.$$

OBJECTIVE 2 Use the addition property of equality. In the equation $x - 5 = 2$, both $x - 5$ and 2 represent the same number because that is the meaning of the equality symbol. To solve the equation, we change the left side from $x - 5$ to just x, as follows.

Teaching Tip
Emphasize that the goal here is to isolate x.

$x - 5 = 2$ Given equation

$x - 5 + 5 = 2 + 5$ Add 5 to *each* side to keep them equal.

$x + 0 = 7$ Additive inverse property

$x = 7$ Additive identity property

(Add 5. It is the opposite (additive inverse) of −5, and −5 + 5 = 0.)

The solution is 7. We check by replacing x with 7 in the original equation.

CHECK

$x - 5 = 2$ Original equation

$7 - 5 \stackrel{?}{=} 2$ Let $x = 7$.

$2 = 2$ ✓ True (The left side equals the right side.)

Since the final equation is true, 7 checks as the solution and $\{7\}$ is the solution set. We write a solution set using set braces $\{\ \}$.

To solve the equation $x - 5 = 2$, we added the same number, 5, to each side. The **addition property of equality** justifies this step.

Addition Property of Equality

If A, B, and C represent real numbers, then the equations

$$\boldsymbol{A = B} \quad \text{and} \quad \boldsymbol{A + C = B + C}$$

are equivalent equations.

In words, we can add the same number to each side of an equation without changing the solution set.

In this property, any quantity that represents a real number C can be added to each side of an equation to obtain an equivalent equation.

Note

Equations can be thought of in terms of a balance. Thus, adding the *same* quantity to each side does not affect the balance. See **Figure 1.**

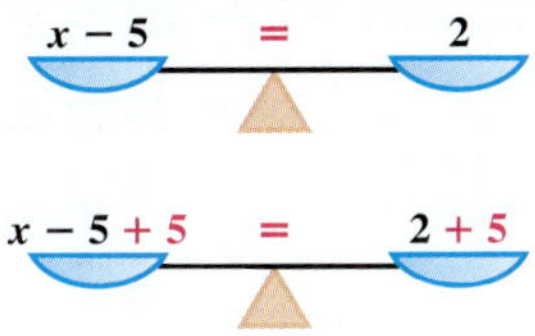

Figure 1

EXAMPLE 1 Using the Addition Property of Equality

Solve $x - 16 = 7$.

Our goal is to get an equivalent equation of the form **x = a number.**

$$x - 16 = 7$$

$$x - 16 + 16 = 7 + 16 \qquad \text{Add 16 to each side.}$$

$$x = 23 \qquad \text{Combine like terms.}$$

CHECK $\quad x - 16 = 7 \qquad$ Original equation

$$23 - 16 \stackrel{?}{=} 7 \qquad \text{Substitute 23 for } x.$$

$$7 = 7 \checkmark \qquad \text{True}$$

7 is *not* the solution.

Since a true statement results, 23 is the solution and $\{23\}$ is the solution set.

Work Problem **1** at the Side. ▶

EXAMPLE 2 Using the Addition Property of Equality

Solve $x - 2.9 = -6.4$.

Our goal is to isolate x.

$$x - 2.9 = -6.4$$

$$x - 2.9 + 2.9 = -6.4 + 2.9 \qquad \text{Add 2.9 to each side.}$$

$$x = -3.5$$

CHECK $\quad x - 2.9 = -6.4 \qquad$ Original equation

$$-3.5 - 2.9 \stackrel{?}{=} -6.4 \qquad \text{Let } x = -3.5.$$

$$-6.4 = -6.4 \checkmark \qquad \text{True}$$

Since a true statement results, the solution set is $\{-3.5\}$.

Work Problem **2** at the Side. ▶

1 Solve.

GS **(a)** Fill in the blanks.

$$x - 12 = 9$$

$$x - 12 + \_\_\_\_ = 9 + \_\_\_\_$$

$$x = \_\_\_\_$$

CHECK $\quad x - 12 = 9$

$$\_\_\_\_ - 12 \stackrel{?}{=} 9$$

$$\_\_\_\_ = 9 \quad (\textit{True / False})$$

The solution set is ____.

(b) $x - 25 = -18$

(c) $x - 8 = -13$

Teaching Tip

In Example 1, point out that we combined the steps that change $x - 16 + 16$ to $x + 0$ and $x + 0$ to x. Emphasize that the final line of the check does *not* give the answer, only a confirmation that the answer found is correct.

2 Solve.

(a) $x - 3.7 = -8.1$

(b) $a - 4.1 = 6.3$

Answers

1. (a) 12; 12; 21; 21; 9; True; $\{21\}$ **(b)** $\{7\}$ **(c)** $\{-5\}$

2. (a) $\{-4.4\}$ **(b)** $\{10.4\}$

3 Solve.

GS **(a)** $-3 = a + 2$

Our goal is to isolate ____.

On each side of the equation, (*add* / *subtract*) 2.

Now solve.

(b) $22 = -16 + r$

Answers

3. **(a)** *a*; subtract; $\{-5\}$ **(b)** $\{38\}$

In **Section 1.5,** subtraction was defined as addition of the opposite. Thus, we can also use the following rule when solving an equation.

> **The same number may be *subtracted* from each side of an equation without changing the solution set.**

For example, to solve the equation $x + 4 = 10$, we *subtract* 4 from each side, which is the same as adding -4. The result is $x = 6$.

EXAMPLE 3 Using the Addition Property of Equality

Solve $-7 = x + 22$.

Here the variable x is on the right side of the equation.

$$-7 = x + 22$$

The variable can be isolated on *either* side.

$$-7 - 22 = x + 22 - 22 \qquad \text{Subtract 22 from each side.}$$

$$-29 = x, \quad \text{or} \quad x = -29 \qquad \text{Rewrite; a number} = x, \text{ or } x = \text{a number.}$$

CHECK

$$-7 = x + 22 \qquad \text{Original equation}$$

$$-7 \stackrel{?}{=} -29 + 22 \qquad \text{Let } x = -29.$$

$$-7 = -7 \ ✓ \qquad \text{True}$$

The check confirms that the solution set is $\{-29\}$.

◀ Work Problem **3** at the Side.

Note

In **Example 3,** what happens if we subtract $-7 - 22$ incorrectly, obtaining $x = -15$ (instead of $x = -29$) as the last line of the solution? A check should indicate an error.

CHECK

$$-7 = x + 22 \qquad \text{Original equation from Example 3}$$

$$-7 \stackrel{?}{=} -15 + 22 \qquad \text{Let } x = -15.$$

$$-7 = 7 \qquad \text{False}$$

The left side does *not* equal the right side.

The false statement indicates that -15 is *not* a solution of the equation. If this happens, rework the problem.

EXAMPLE 4 Subtracting a Variable Term

Solve $\frac{3}{5}k + 15 = \frac{8}{5}k$.

$$\frac{3}{5}k + 15 = \frac{8}{5}k$$

We must get the terms with k on the same side of the = symbol.

$$\frac{3}{5}k + 15 - \frac{3}{5}k = \frac{8}{5}k - \frac{3}{5}k \qquad \text{Subtract } \tfrac{3}{5}k \text{ from each side.}$$

From now on we will skip this step.

$$15 = 1k \qquad \tfrac{3}{5}k - \tfrac{3}{5}k = 0;\ \tfrac{8}{5}k - \tfrac{3}{5}k = \tfrac{5}{5}k = 1k$$

$$15 = k \qquad \text{Multiplicative identity property}$$

Check by substituting 15 in the original equation. The solution set is $\{15\}$.

What happens if we solve the equation in **Example 4** by first subtracting $\frac{8}{5}k$ from each side?

$$\frac{3}{5}k + 15 = \frac{8}{5}k \quad \text{Original equation from Example 4}$$

$$\frac{3}{5}k + 15 - \frac{8}{5}k = \frac{8}{5}k - \frac{8}{5}k \quad \text{Subtract } \tfrac{8}{5}k \text{ from each side.}$$

$$15 - k = 0 \quad \tfrac{3}{5}k - \tfrac{8}{5}k = -\tfrac{5}{5}k = -1k = -k;\ \tfrac{8}{5}k - \tfrac{8}{5}k = 0$$

$$15 - k - 15 = 0 - 15 \quad \text{Subtract 15 from each side.}$$

$$-k = -15 \quad \text{Combine like terms; additive inverse}$$

This result gives the value of $-k$, but not of k itself. However, it does say that the additive inverse of k is -15, which means that k must be 15.

$$k = 15 \quad \text{Same result as in Example 4}$$

(This result can also be justified using the multiplication property of equality, covered in **Section 2.2.**) We can make the following generalization.

If a is a number and $-x = a$, then $x = -a$.

Work Problem 4 at the Side. ▶

EXAMPLE 5 Using the Addition Property of Equality Twice

Solve $8 - 6p = -7p + 5$.

We must get all terms with variables on the same side of the equation and all terms without variables on the other side of the equation.

$$8 - 6p = -7p + 5$$

$$8 - 6p + 7p = -7p + 5 + 7p \quad \text{Add } 7p \text{ to each side.}$$

$$8 + p = 5 \quad \text{Combine like terms.}$$

$$8 + p - 8 = 5 - 8 \quad \text{Subtract 8 from each side.}$$

$$p = -3 \quad \text{Combine like terms.}$$

CHECK Substitute -3 for p in the original equation.

$$8 - 6p = -7p + 5 \quad \text{Original equation}$$

$$8 - 6(-3) \stackrel{?}{=} -7(-3) + 5 \quad \text{Let } p = -3.$$

Use parentheses when substituting to avoid errors.

$$8 + 18 \stackrel{?}{=} 21 + 5 \quad \text{Multiply.}$$

$$26 = 26 \ ✓ \quad \text{True}$$

The check results in a true statement, so the solution set is $\{-3\}$.

Work Problem 5 at the Side. ▶

Note

There are often several equally correct ways to solve an equation. In **Example 5,** we could begin by adding $6p$, instead of $7p$, to each side. Combining like terms and subtracting 5 from each side gives $3 = -p$. (Try this.) If $3 = -p$, then $-3 = p$, and the variable has been isolated on the right side of equation. The same solution results.

4 **(a)** Solve $5m + 4 = 6m$.

(b) Solve $\frac{7}{2}m + 1 = \frac{9}{2}m$.

(c) What is the solution set of $-x = 6$?

(d) What is the solution set of $-x = -12$?

5 Solve.

GS **(a)** $10 - a = -2a + 9$

$10 - a + \_\_\_\_ = -2a + 9 + 2a$

$10 + a = 9$

$10 + a - \_\_\_\_ = 9 - \_\_\_\_$

$a = \_\_\_\_$

The solution set is ____.

(b) $6x - 8 = 12 + 5x$

(c) $8t + 6 = 6 + 7t$

Answers

4. **(a)** $\{4\}$ **(b)** $\{1\}$ **(c)** $\{-6\}$ **(d)** $\{12\}$

5. **(a)** $2a$; 10; 10; -1; $\{-1\}$
(b) $\{20\}$ **(c)** $\{0\}$

OBJECTIVE 3 **Simplify, and then use the addition property of equality.**

EXAMPLE 6 **Combining Like Terms When Solving**

Solve $3t - 12 + t + 2 = 5 + 3t + 2$.

$$3t - 12 + t + 2 = 5 + 3t + 2 \qquad \text{Combine like terms on each side.}$$
$$4t - 10 = 7 + 3t$$
$$4t - 10 - 3t = 7 + 3t - 3t \qquad \text{Subtract } 3t \text{ from each side.}$$
$$t - 10 = 7 \qquad \text{Combine like terms.}$$
$$t - 10 + 10 = 7 + 10 \qquad \text{Add 10 to each side.}$$
$$t = 17 \qquad \text{Combine like terms.}$$

CHECK
$$3t - 12 + t + 2 = 5 + 3t + 2 \qquad \text{Original equation}$$
$$3(17) - 12 + 17 + 2 \stackrel{?}{=} 5 + 3(17) + 2 \qquad \text{Let } t = 17.$$
$$51 - 12 + 17 + 2 \stackrel{?}{=} 5 + 51 + 2 \qquad \text{Multiply.}$$
$$58 = 58 \ ✓ \qquad \text{True}$$

The check results in a true statement, so the solution set is $\{17\}$.

◀ **Work Problem 6 at the Side.**

CAUTION

The final line of the check does not give the solution to the equation. It gives a confirmation that the solution found is correct.

EXAMPLE 7 **Using the Distributive Property When Solving**

Solve $3(2 + 5x) - (1 + 14x) = 6$.

$$3(2 + 5x) - (1 + 14x) = 6$$

Be sure to distribute to *all* terms within the parentheses.

$$3(2 + 5x) - 1(1 + 14x) = 6 \qquad -(1 + 14x) = -1(1 + 14x)$$
$$3(2) + 3(5x) - 1(1) - 1(14x) = 6 \qquad \text{Distributive property}$$
$$6 + 15x - 1 - 14x = 6 \qquad \text{Multiply.}$$

Be careful here.

$$x + 5 = 6 \qquad \text{Combine like terms.}$$
$$x + 5 - 5 = 6 - 5 \qquad \text{Subtract 5 from each side.}$$
$$x = 1 \qquad \text{Combine like terms.}$$

Check by substituting 1 for x in the original equation. The solution set is $\{1\}$.

◀ **Work Problem 7 at the Side.**

CAUTION

Be careful to apply the distributive property correctly in a problem like that in **Example 7**, or a sign error may result.

---

6 Solve.

(a) $4x + 6 + 2x - 3 = 9 + 5x - 4$

(b) $9r + 4r + 6 - 2 = 9r + 4 + 3r$

7 Solve.

(a) $4(r + 1) - (3r + 5) = 1$

Teaching Tip

Many students make errors when a subtraction sign appears in front of parentheses, as in Example 7. Warn them about such errors.

GS **(b)** $2(5 + 2m) - 3(m - 4) = 29$

$2(5) + 2(2m) -$ ____ m

$-$ ____ (____) $= 29$

$10 + 4m -$ ____ $+$ ____ $= 29$

Now complete the solution.

Answers

6. **(a)** $\{2\}$ **(b)** $\{0\}$

7. **(a)** $\{2\}$ **(b)** $3; 3; -4; 3m; 12; \{7\}$

2.1 Exercises

FOR EXTRA HELP

Download the MyDashBoard App

MyMathLab®

CONCEPT CHECK *In Exercises 1–4, fill in each blank with the correct response.*

1. An equation includes a(n) __equality__ symbol, while a(n) __expression__ does not.

2. A(n) __linear__ equation in one variable can be written in the form $Ax + B$ __=__ C.

3. Equations that have exactly the same solution set are __equivalent__ equations.

4. The addition property of equality states that the same expression may be __added to__ or __subtracted from__ each side of an equation without changing the solution set.

5. **CONCEPT CHECK** Substitute to determine which number is a solution of the equation $2x - 5 = 3x$.

A. 0 **B.** 5 **C.** -5 **D.** -1

C

6. **CONCEPT CHECK** Which of the following are *not* linear equations in one variable?

A. $x^2 - 5x + 6 = 0$ **B.** $x^3 = x$

C. $3x - 4 = 0$ **D.** $7x - 6x = 3 + 9x$

A, B

7. **CONCEPT CHECK** Decide whether each is an *expression* or an *equation*. If it is an expression, simplify it. If it is an equation, solve it.

(a) $5x + 8 - 4x + 7$ expression; $x + 15$

(b) $-6m + 12 + 7m - 5$ expression; $m + 7$

(c) $5x + 8 - 4x = 7$ equation; $\{-1\}$

(d) $-6m + 12 + 7m = -5$ equation; $\{-17\}$

8. Explain how to check a solution of an equation.

Replace the variable(s) in the original equation with the proposed solution. A true statement will result if the proposed solution is correct.

Solve each equation, and check your solution. ***See Examples 1–5.***

9. $x - 4 = 8$ $\{12\}$

10. $x - 8 = 9$ $\{17\}$

11. $x - 5 = -8$ $\{-3\}$

12. $x - 7 = -9$ $\{-2\}$

13. $r + 9 = 13$ $\{4\}$

14. $t + 6 = 10$ $\{4\}$

15. $x + 26 = 17$ $\{-9\}$

16. $x + 45 = 24$ $\{-21\}$

17. $x + \frac{1}{4} = -\frac{1}{2}$ $\left\{-\frac{3}{4}\right\}$

18. $x + \frac{2}{3} = -\frac{1}{6}$ $\left\{-\frac{5}{6}\right\}$

19. $x - 8.4 = -2.1$ $\{6.3\}$

20. $z - 15.5 = -5.1$ $\{10.4\}$

21. $t + 12.3 = -4.6$ $\{-16.9\}$

22. $x + 21.5 = -13.4$ $\{-34.9\}$

23. $7 + r = -3$ $\{-10\}$

24. $8 + k = -4$ $\{-12\}$

25. $2 = p + 15$ $\{-13\}$

26. $3 = z + 17$ $\{-14\}$

27. $-\frac{1}{3} = x - \frac{3}{5}$ $\left\{\frac{4}{15}\right\}$

28. $-\frac{1}{4} = x - \frac{2}{3}$ $\left\{\frac{5}{12}\right\}$

29. $3x = 2x + 7$ $\{7\}$

30. $5x = 4x + 9$ $\{9\}$

31. $8x - 3 = 9x$ $\{-3\}$

32. $4x - 1 = 5x$ $\{-1\}$

33. $10x + 4 = 9x$ $\{-4\}$

34. $8t + 5 = 7t$ $\{-5\}$

35. $\frac{9}{7}r - 3 = \frac{2}{7}r$ $\{3\}$

36. $\frac{8}{5}w - 6 = \frac{3}{5}w$ $\{6\}$

37. $5.6x + 2 = 4.6x$ $\{-2\}$

38. $9.1x + 5 = 8.1x$ $\{-5\}$

39. $3p + 6 = 10 + 2p$ $\{4\}$

40. $8x + 4 = -6 + 7x$ $\{-10\}$

41. $5 - x = -2x - 11$ $\{-16\}$

42. $3 - 8x = -9x - 1$ $\{-4\}$

43. $-4z + 7 = -5z + 9$ $\{2\}$

44. $-6q + 3 = -7q + 10$ $\{7\}$

Solve each equation, and check your solution. ***See Examples 6 and 7.***

45. $3x + 6 - 10 = 2x - 2$ $\{2\}$

46. $8k - 4 + 6 = 7k + 1$ $\{-1\}$

47. $6x + 5 + 7x + 3 = 12x + 4$ $\{-4\}$

48. $4x - 3 - 8x + 1 = -5x + 9$ $\{11\}$

49. $10x + 5x + 7 - 8 = 12x + 3 + 2x$ $\{4\}$

50. $7p + 4p + 13 - 7 = 7p + 9 + 3p$ $\{3\}$

51. $5.2q - 4.6 - 7.1q = -0.9q - 4.6$ $\{0\}$

52. $-4.0x + 2.7 - 1.6x = -4.6x + 2.7$ $\{0\}$

53. $\frac{5}{7}x + \frac{1}{3} = \frac{2}{5} - \frac{2}{7}x + \frac{2}{5}$ $\left\{\frac{7}{15}\right\}$

54. $\frac{6}{7}s - \frac{3}{4} = \frac{4}{5} - \frac{1}{7}s + \frac{1}{6}$ $\left\{\frac{103}{60}\right\}$

55. $(5x + 6) - (3 + 4x) = 10$ $\{7\}$

56. $(8r - 3) - (7r + 1) = -6$ $\{-2\}$

57. $2(p + 5) - (9 + p) = -3$ $\{-4\}$

58. $4(k - 6) - (3k + 2) = -5$ $\{21\}$

59. $-6(2x + 1) + (13x - 7) = 0$ $\{13\}$

60. $-5(3w - 3) + (16w + 1) = 0$ $\{-16\}$

61. $10(-2x + 1) = -19(x + 1)$ $\{29\}$

62. $2(-3r + 2) = -5(r - 3)$ $\{-11\}$

63. $-2(8p + 2) - 3(2 - 7p) = 2(4 + 2p)$ $\{18\}$

64. $4(3 - z) - 5(1 - 2z) = 7(3 + z)$ $\{-14\}$

65. CONCEPT CHECK Write an equation that requires the use of the addition property of equality, in which 6 must be added to each side to solve the equation and the solution is a negative number.

Answers will vary. One example is $x - 6 = -8$.

66. CONCEPT CHECK Write an equation that requires the use of the addition property of equality, in which $\frac{1}{2}$ must be subtracted from each side and the solution is a positive number.

Answers will vary. One example is $x + \frac{1}{2} = 1$.

Study Skills

MANAGING YOUR TIME

Many college students juggle a difficult schedule and multiple responsibilities, including school, work, and family demands.

Time Management Tips

- **Read the syllabus for each class.** Understand class policies, such as attendance, late homework, and make-up tests. Find out how you are graded.
- **Make a semester or quarter calendar.** Put test dates and major due dates for *all* your classes on the *same* calendar. Try using a different color pen for each class.
- **Make a weekly schedule.** After you fill in your classes and other regular responsibilities, block off some study periods. Aim for 2 hours of study for each 1 hour in class.
- **Make "to-do" lists.** Number tasks in order of importance. Cross off tasks as you complete them.
- **Break big assignments into smaller chunks.** Make deadlines for each smaller chunk so that you stay on schedule.
- **Choose a regular study time and place** (such as the campus library). Routine helps.
- **Keep distractions to a minimum.** Get the most out of the time you have set aside for studying by limiting interruptions. Turn off your cell phone. Take a break from social media. Avoid studying in front of the TV.
- **Take breaks when studying.** Do not try to study for hours at a time. Take a 10-minute break each hour or so.
- **Ask for help when you need it.** Talk with your instructor during office hours. Make use of the learning center, tutoring center, counseling office, or other resources available at your school.

Now Try This

Think through and answer each question.

1. How many hours do you have available for studying this week?

   ____

2. Which two or three of the above suggestions will you try this week to improve your time management?

   ____

3. Once the week is over, evaluate how these suggestions worked. What will you do differently next week?

   ____

2.2 The Multiplication Property of Equality

OBJECTIVES

1. Use the multiplication property of equality.
2. Simplify, and then use the multiplication property of equality.

1 Check that 5 is the solution
GS of $3x = 15$.

$$3x = 15$$

$$3(\_\_\_\_) \stackrel{?}{=} 15$$

$$\_\_\_\_ = 15 \quad (\textit{True / False})$$

The solution set is ____.

Teaching Tip

Remind students of the equation and balance analogy from Section 2.1. If we double the weight on one side of the balance, what do we have to do to the other side to maintain the balance?

Answer

1. 5; 15; True; $\{5\}$

OBJECTIVE 1 Use the multiplication property of equality. The addition property of equality is not enough to solve some equations.

$$3x + 2 = 17$$

$$3x + 2 - 2 = 17 - 2 \qquad \text{Subtract 2 from each side.}$$

$$3x = 15 \qquad \text{Combine like terms.}$$

The coefficient of x here is 3, not 1 as desired. The **multiplication property of equality** is needed to change $3x = 15$ to an equation of the form

$$x = \text{a number.}$$

Since $3x = 15$, both $3x$ and 15 must represent the same number. Multiplying both $3x$ and 15 by the same number will result in an equivalent equation.

Multiplication Property of Equality

If A, B, and C (where $C \neq 0$) represent real numbers, then the equations

$$A = B \quad \text{and} \quad AC = BC$$

are equivalent equations.

In words, we can multiply each side of an equation by the same nonzero number without changing the solution set.

In $3x = 15$, we must change $3x$ to $1x$, or x. To do this, we multiply each side of the equation by $\frac{1}{3}$, the reciprocal of 3, because $\frac{1}{3} \cdot 3 = \frac{3}{3} = 1$.

$$3x = 15$$

$$\frac{1}{3}(3x) = \frac{1}{3}(15) \qquad \text{Multiply each side by } \tfrac{1}{3}.$$

$$\left(\frac{1}{3} \cdot 3\right)x = \frac{1}{3}(15) \qquad \text{Associative property}$$

$$1x = 5 \qquad \text{Multiplicative inverse property}$$

$$x = 5 \qquad \text{Multiplicative identity property}$$

The product of a number and its reciprocal is 1.

The solution is 5. We can check this result in the original equation.

◀ Work Problem **1** at the Side.

Just as the addition property of equality permits *subtracting* the same number from each side of an equation, the multiplication property of equality permits *dividing* each side of an equation by the same nonzero number.

$$3x = 15$$

$$\frac{3x}{3} = \frac{15}{3} \qquad \text{Divide each side by 3.}$$

$$x = 5 \qquad \text{Same result as above}$$

We can divide each side of an equation by the same nonzero number without changing the solution. ***Do not, however, divide each side by a variable, since the variable might be equal to 0.***

Note

In practice, it is usually easier to multiply on each side if the coefficient of the variable is a fraction, and divide on each side if the coefficient is an integer or a decimal. For example, to solve

$\frac{3}{4}x = 12$, it is easier to multiply by $\frac{4}{3}$ than to divide by $\frac{3}{4}$.

On the other hand, to solve

$5x = 20$, it is easier to divide by 5 than to multiply by $\frac{1}{5}$.

EXAMPLE 1 Dividing Each Side of an Equation by a Nonzero Number

Solve $5x = 60$.

$$5x = 60$$ Our goal is to isolate x.

$$\frac{5x}{5} = \frac{60}{5}$$ Divide each side by 5, the coefficient of x.

Dividing by 5 is the same as multiplying by $\frac{1}{5}$.

$$x = 12$$ $\frac{5x}{5} = \frac{5}{5}x = 1x = x$

CHECK Substitute 12 for x in the original equation.

$$5x = 60$$ Original equation

$$5(12) \stackrel{?}{=} 60$$ Let $x = 12$.

$$60 = 60 \checkmark$$ True

60 is *not* the solution.

Since a true statement results, the solution set is $\{12\}$.

Work Problem 2 at the Side. ▶

EXAMPLE 2 Using the Multiplication Property of Equality

Solve $-25p = 30$.

$$-25p = 30$$

$$\frac{-25p}{-25} = \frac{30}{-25}$$ Divide each side by -25, the coefficient of p.

$$p = -\frac{30}{25}$$ $\frac{a}{-b} = -\frac{a}{b}$

$$p = -\frac{6}{5}$$ Write in lowest terms.

CHECK

$$-25p = 30$$ Original equation

$$\frac{-25}{1}\left(-\frac{6}{5}\right) \stackrel{?}{=} 30$$ Let $p = -\frac{6}{5}$.

$$30 = 30 \checkmark$$ True

The check confirms that the solution set is $\{-\frac{6}{5}\}$.

Work Problem 3 at the Side. ▶

---

2 Solve.

GS **(a)** Fill in the blanks.

$$7x = 91$$

$$\frac{7x}{\_\_\_\_} = \frac{91}{\_\_\_\_}$$

$$x = \_\_\_\_$$

CHECK

$$7x = 91$$

$$7(\_\_\_\_) \stackrel{?}{=} 91$$

$$\_\_\_\_ = 91 \quad (True / False)$$

The solution set is ____.

(b) $3r = -12$

(c) $15x = 75$

3 Solve.

(a) $2m = 15$

(b) $-6x = 14$

(c) $10z = -45$

Answers

2. **(a)** 7; 7; 13; 13; 91; True; $\{13\}$
(b) $\{-4\}$ **(c)** $\{5\}$

3. **(a)** $\left\{\frac{15}{2}\right\}$ **(b)** $\left\{-\frac{7}{3}\right\}$ **(c)** $\left\{-\frac{9}{2}\right\}$

4 Solve.

(a) $-0.7m = -5.04$

(b) $-63.75 = 12.5k$

EXAMPLE 3 Using the Multiplication Property of Equality

Solve $6.09 = 2.1x$.

$$6.09 = 2.1x$$ Isolate x on the right.

$$\frac{6.09}{2.1} = \frac{2.1x}{2.1}$$ Divide each side by 2.1.

$$2.9 = x, \quad \text{or} \quad x = 2.9$$ Use a calculator if necessary.

Check to confirm that the solution set is $\{2.9\}$.

◀ Work Problem **4** at the Side.

5 Solve.

GS **(a)** $\frac{x}{5} = 5$

What is the coefficient of x here? ____ By what number should we multiply each side? ____ Now solve.

(b) $\frac{p}{4} = -6$

EXAMPLE 4 Using the Multiplication Property of Equality

Solve $\frac{x}{4} = 3$.

$$\frac{x}{4} = 3$$

$$\frac{1}{4}x = 3$$ $\frac{x}{4} = \frac{1x}{4} = \frac{1}{4}x$

$$4 \cdot \frac{1}{4}x = 4 \cdot 3$$ Multiply each side by 4, the reciprocal of $\frac{1}{4}$.

$$x = 12$$ $4 \cdot \frac{1}{4}x = 1x = x$ Multiplicative inverse property; multiplicative identity property

CHECK

$$\frac{x}{4} = 3$$ Original equation

$$\frac{12}{4} \stackrel{?}{=} 3$$ Let $x = 12$.

$$3 = 3 \checkmark$$ True

Since a true statement results, the solution set is $\{12\}$.

◀ Work Problem **5** at the Side.

6 Solve.

GS **(a)** $-\frac{5}{6}t = -15$

By what number should we multiply each side? ____ Now solve.

(b) $\frac{3}{5}k = -21$

EXAMPLE 5 Using the Multiplication Property of Equality

Solve $\frac{3}{4}h = 6$.

$$\frac{3}{4}h = 6$$

$$\frac{4}{3} \cdot \frac{3}{4}h = \frac{4}{3} \cdot 6$$ Multiply each side by $\frac{4}{3}$, the reciprocal of $\frac{3}{4}$.

$$1 \cdot h = \frac{4}{3} \cdot \frac{6}{1}$$ Multiplicative inverse property

$$h = 8$$ Multiplicative identity property; multiply fractions.

Check to confirm that the solution set is $\{8\}$.

◀ Work Problem **6** at the Side.

Answers

4. **(a)** $\{7.2\}$ **(b)** $\{-5.1\}$

5. **(a)** $\frac{1}{5}$; 5; $\{25\}$ **(b)** $\{-24\}$

6. **(a)** $-\frac{6}{5}$; $\{18\}$ **(b)** $\{-35\}$

In **Section 2.1,** we obtained the equation $-k = -15$ in our alternative solution to **Example 4.** We reasoned that since this equation says that the additive inverse (or opposite) of k is -15, then k must equal 15. We can also use the multiplication property of equality to obtain the same result.

EXAMPLE 6 Using the Multiplication Property of Equality When the Coefficient of the Variable Is -1

Solve $-k = -15$.

$-k = -15$	
$-1 \cdot k = -15$	$-k = -1 \cdot k$
$-1(-1 \cdot k) = -1(-15)$	Multiply each side by -1.
$[-1(-1)] \cdot k = 15$	Associative property; multiply.
$1 \cdot k = 15$	Multiplicative inverse property
$k = 15$	Multiplicative identity property

CHECK

$-k = -15$	Original equation
$-(15) \stackrel{?}{=} -15$	Let $k = 15$.
$-15 = -15$ ✓	True

The solution, 15, checks, so the solution set is $\{15\}$.

Work Problem 7 at the Side. ▶

OBJECTIVE 2 Simplify, and then use the multiplication property of equality.

EXAMPLE 7 Combining Like Terms When Solving

Solve $5m + 6m = 33$.

$5m + 6m = 33$	
$11m = 33$	Combine like terms.
$\frac{11m}{11} = \frac{33}{11}$	Divide each side by 11.
$m = 3$	Multiplicative identity property

CHECK

$5m + 6m = 33$	Original equation
$5(3) + 6(3) \stackrel{?}{=} 33$	Let $m = 3$.
$15 + 18 \stackrel{?}{=} 33$	Multiply.
$33 = 33$ ✓	True

Since a true statement results, the solution set is $\{3\}$.

Work Problem 8 at the Side. ▶

7 Solve.

(a) $-m = 2$

(b) $-p = -7$

8 Solve.

(a) $7m - 5m = -12$

(b) $4r - 9r = 20$

Answers

7. **(a)** $\{-2\}$ **(b)** $\{7\}$

8. **(a)** $\{-6\}$ **(b)** $\{-4\}$

2.2 Exercises

FOR EXTRA HELP

Download the MyDashBoard App

MyMathLab®

1. CONCEPT CHECK Tell whether you would use the addition or multiplication property of equality to solve each equation. *Do not actually solve.*

(a) $3x = 12$ (b) $3 + x = 12$

(c) $-x = 4$ (d) $-12 = 6 + x$

(a) and (c): multiplication property of equality
(b) and (d): addition property of equality

2. CONCEPT CHECK Which equation does *not* require the use of the multiplication property of equality?

A. $3x - 5x = 6$ B. $-\frac{1}{4}x = 12$

C. $5x - 4x = 7$ D. $\frac{x}{3} = -2$

C

CONCEPT CHECK *By what number is it necessary to multiply each side of each equatio n in order to isolate x on the left side? Do not actually solve.*

3. $\frac{2}{3}x = 8$ $\frac{3}{2}$

4. $\frac{4}{5}x = 6$ $\frac{5}{4}$

5. $\frac{x}{10} = 3$ 10

6. $\frac{x}{100} = 8$ 100

7. $-\frac{9}{2}x = -4$ $-\frac{2}{9}$

8. $-\frac{8}{3}x = -11$ $-\frac{3}{8}$

9. $-x = 0.36$ -1

10. $-x = 0.29$ -1

CONCEPT CHECK *By what number is it necessary to divide each side of each equation in order to isolate x on the left side? Do not actually solve.*

11. $6x = 5$ 6

12. $7x = 10$ 7

13. $-4x = 13$ -4

14. $-13x = 6$ -13

15. $0.12x = 48$ 0.12

16. $0.21x = 63$ 0.21

17. $-x = 23$ -1

18. $-x = 49$ -1

19. CONCEPT CHECK In the solution of a linear equation, the next-to-the-last step reads "$-x = -\frac{3}{4}$." Which of the following would be the solution of this equation?

A. $-\frac{3}{4}$ B. $\frac{3}{4}$ C. -1 D. $\frac{4}{3}$ B

20. CONCEPT CHECK Which of the following is the solution of the equation

$$-x = -24?$$

A. 24 B. -24 C. 1 D. -1 A

Solve each equation, and check your solution. ***See Examples 1–7.***

21. $5x = 30$ $\{6\}$

22. $7x = 56$ $\{8\}$

23. $2m = 15$ $\left\{\frac{15}{2}\right\}$

24. $3m = 10$ $\left\{\frac{10}{3}\right\}$

25. $3a = -15$ $\{-5\}$

26. $5k = -70$ $\{-14\}$

27. $10t = -36$ $\left\{-\frac{18}{5}\right\}$

28. $4s = -34$ $\left\{-\frac{17}{2}\right\}$

29. $-6x = -72$ $\{12\}$

30. $-8x = -64$ $\{8\}$

31. GS

$2r = 0$

$\frac{2r}{\underline{2}} = \frac{0}{\underline{2}}$

$r = \underline{0}$

Solution set: $\underline{\{0\}}$

32. GS

$5x = 0$

$\frac{5x}{\underline{5}} = \frac{0}{\underline{5}}$

$x = \underline{0}$

Solution set: $\underline{\{0\}}$

33. $-x = 12$
$\{-12\}$

34. $-t = 14$
$\{-14\}$

35. $0.2t = 8$
$\{40\}$

36. $0.9x = 18$
$\{20\}$

37. $-2.1m = 25.62$
$\{-12.2\}$

38. $-3.9a = 31.2$
$\{-8\}$

39. $\frac{1}{4}x = -12$
$\{-48\}$

40. $\frac{1}{5}p = -3$
$\{-15\}$

41. $\frac{z}{6} = 12$
$\{72\}$

42. $\frac{x}{5} = 15$
$\{75\}$

43. $\frac{x}{7} = -5$
$\{-35\}$

44. $\frac{k}{8} = -3$
$\{-24\}$

45. $\frac{2}{7}p = 4$
$\{14\}$

46. $\frac{3}{8}x = 9$
$\{24\}$

47. $-\frac{7}{9}c = \frac{3}{5}$
$\left\{-\frac{27}{35}\right\}$

48. $-\frac{5}{6}d = \frac{4}{9}$
$\left\{-\frac{8}{15}\right\}$

49. $4x + 3x = 21$
$\{3\}$

50. $9x + 2x = 121$
$\{11\}$

51. $3r - 5r = 10$
$\{-5\}$

52. $9p - 13p = 24$
$\{-6\}$

53. $\frac{2}{5}x - \frac{3}{10}x = 2$
$\{20\}$

54. $\frac{2}{3}x - \frac{5}{9}x = 4$
$\{36\}$

55. $5m + 6m - 2m = 63$
$\{7\}$

56. $11r - 5r + 6r = 168$
$\{14\}$

57. $x + x - 3x = 12$
(*Hint:* What is the coefficient of each x? __1__)
$\{-12\}$

58. $z - 3z + z = -16$
(*Hint:* What is the coefficient of each z? __1__)
$\{16\}$

59. $-6x + 4x - 7x = 0$
$\{0\}$

60. $-5x + 4x - 8x = 0$
$\{0\}$

61. $0.9w - 0.5w + 0.1w = -3$
$\{-6\}$

62. $0.5x - 0.6x + 0.3x = -1$
$\{-5\}$

63. **CONCEPT CHECK** Write an equation that requires the use of the multiplication property of equality, where each side must be multiplied by $\frac{2}{3}$ and the solution is a negative number.
Answers will vary. One example is $\frac{3}{2}x = -6$.

64. **CONCEPT CHECK** Write an equation that requires the use of the multiplication property of equality, where each side must be divided by 100 and the solution is not an integer.
Answers will vary. One example is $100x = 17$.

Write an equation using the information given in the problem. Use x as the variable. Then solve the equation.

65. When a number is multiplied by -4, the result is 10. Find the number.
$-4x = 10; -\frac{5}{2}$

66. When a number is multiplied by 4, the result is 6. Find the number.
$4x = 6; \frac{3}{2}$

67. When a number is divided by -5, the result is 2. Find the number.
$\frac{x}{-5} = 2; -10$

68. If twice a number is divided by 5, the result is 4. Find the number.
$\frac{2x}{5} = 4; 10$

Study Skills
ANALYZING YOUR TEST RESULTS

An exam is a learning opportunity—learn from your mistakes. After a test is returned, do the following:

- **Note what you got wrong and why you had points deducted.**
- **Figure out how to solve the problems you missed.** Check your textbook or notes, or ask your instructor. Rework the problems correctly.
- **Keep all quizzes and tests that are returned to you.** Use them to study for future tests and the final exam.

Typical Reasons for Errors on Math Tests

These are test taking errors. They are easy to correct if you read carefully, show all your work, proofread, and double-check units and labels.

1. You read the directions wrong.
2. You read the question wrong or skipped over something.
3. You made a computation error.
4. You made a careless error. (For example, you incorrectly copied a correct answer onto a separate answer sheet.)
5. Your answer is not complete.
6. You labeled your answer wrong. (For example, you labeled an answer "ft" instead of "ft^2".)
7. You didn't show your work.

These are test preparation errors. You must practice the kinds of problems that you will see on tests.

8. You didn't understand a concept.
9. You were unable to set up the problem (in an application).
10. You were unable to apply a procedure.

Below are sample charts for tracking your test taking progress. Use them to find out if you tend to make certain kinds of errors on tests. Check the appropriate box when you've made an error in a particular category.

Test Taking Errors

Test	Read directions wrong	Read question wrong	Computation error	Not exact or accurate	Not complete	Labeled wrong	Didn't show work
1							
2							
3							

Test Preparation Errors

Test	Didn't understand concept	Didn't set up problem correctly	Couldn't apply concept to new situation
1			
2			
3			

What will you do to avoid these kinds of errors on your next test?

2.3 More on Solving Linear Equations

In this section, we solve linear equations using *both* properties of equality introduced in **Sections 2.1 and 2.2.**

Work Problem 1 at the Side. ▶

OBJECTIVE 1 **Learn and use the four steps for solving a linear equation.** We use the following method.

Solving a Linear Equation

Step 1 **Simplify each side separately.** Clear (eliminate) parentheses, fractions, and decimals, using the distributive property as needed, and combine like terms.

Step 2 **Isolate the variable term on one side.** Use the addition property so that the variable term is on one side of the equation and a number is on the other.

Step 3 **Isolate the variable.** Use the multiplication property to get the equation in the form $x =$ a number, or a number $= x$. (Other letters may be used for the variable.)

Step 4 **Check.** Substitute the proposed solution into the original equation to see if a true statement results. If not, rework the problem.

Remember that when we solve an equation, our primary goal is to isolate the variable on one side of the equation.

EXAMPLE 1 Using Both Properties of Equality to Solve a Linear Equation

Solve $-6x + 5 = 17$.

Step 1 There are no parentheses, fractions, or decimals in this equation, so this step is not necessary.

(Our goal is to isolate x.) $-6x + 5 = 17$

Step 2 $-6x + 5 - 5 = 17 - 5$ Subtract 5 from each side.

$-6x = 12$ Combine like terms.

Step 3 $\frac{-6x}{-6} = \frac{12}{-6}$ Divide each side by -6.

$x = -2$

Step 4 Check by substituting -2 for x in the original equation.

CHECK

$-6x + 5 = 17$ Original equation

$-6(-2) + 5 \stackrel{?}{=} 17$ Let $x = -2$.

$12 + 5 \stackrel{?}{=} 17$ Multiply.

(17 is *not* the solution.) $17 = 17$ ✓ True

The solution, -2, checks, so the solution set is $\{-2\}$.

Work Problem 2 at the Side. ▶

OBJECTIVES

1. Learn and use the four steps for solving a linear equation.
2. Solve equations that have no solution or infinitely many solutions.
3. Solve equations with fractions or decimals as coefficients.
4. Write expressions for two related unknown quantities.

1 As a review, tell whether you would use the addition or multiplication property of equality to solve each equation. *Do not actually solve.*

(a) $7 + x = -9$

(b) $-13x = 26$

(c) $-x = \frac{3}{4}$

(d) $-12 = x - 4$

2 Solve.

(a) $-5p + 4 = 19$

(b) $7 + 2m = -3$

Answers

1. **(a)** and **(d):** addition property of equality **(b)** and **(c):** multiplication property of equality
2. **(a)** $\{-3\}$ **(b)** $\{-5\}$

3 Solve.

GS (a) $5 - 8k = 2k - 5$

Begin with Step 2.

$$5 - 8k - 2k = 2k - 5 - \_\_\_\_$$

$$5 - \_\_\_\_ = -5$$

$$5 - 10k - \_\_\_\_ = -5 - \_\_\_\_$$

$$-10k = -10$$

$$\frac{-10k}{\_\_\_\_} = \frac{-10}{\_\_\_\_}$$

$$k = \_\_\_\_$$

Check by substituting ____ for k in the original equation.

The solution set is ____.

(b) $2q + 3 = 4q - 9$

(c) $6x + 7 = -8 + 3x$

Answers

3. **(a)** $2k$; $10k$; 5; 5; -10; -10; 1; 1; $\{1\}$
(b) $\{6\}$ **(c)** $\{-5\}$

EXAMPLE 2 Using Both Properties of Equality to Solve a Linear Equation

Solve $3x + 2 = 5x - 8$.

Step 1 There are no parentheses, fractions, or decimals in the equation, so we begin with Step 2.

$$3x + 2 = 5x - 8$$ Our goal is to isolate x.

Step 2	$3x + 2 - 5x = 5x - 8 - 5x$	Subtract $5x$ from each side.
	$-2x + 2 = -8$	Combine like terms.
	$-2x + 2 - 2 = -8 - 2$	Subtract 2 from each side.
	$-2x = -10$	Combine like terms.
Step 3	$\frac{-2x}{-2} = \frac{-10}{-2}$	Divide each side by -2.
	$x = 5$	

Step 4 Check by substituting 5 for x in the original equation.

CHECK	$3x + 2 = 5x - 8$	Original equation
	$3(5) + 2 \stackrel{?}{=} 5(5) - 8$	Let $x = 5$.
	$15 + 2 \stackrel{?}{=} 25 - 8$	Multiply.
	$17 = 17$ ✓	True

The solution, 5, checks, so the solution set is $\{5\}$.

Note

Remember that a variable can be isolated on either side of an equation. In **Example 2,** x will be isolated on the right if we begin by subtracting $3x$, instead of $5x$, from each side of the equation.

$3x + 2 = 5x - 8$	Equation from **Example 2**
$3x + 2 - 3x = 5x - 8 - 3x$	Subtract $3x$ from each side.
$2 = 2x - 8$	Combine like terms.
$2 + 8 = 2x - 8 + 8$	Add 8 to each side.
$10 = 2x$	Combine like terms.
$\frac{10}{2} = \frac{2x}{2}$	Divide each side by 2.
$5 = x$	The same solution results.

There are often several equally correct ways to solve an equation.

◀ Work Problem **3** at the Side.

EXAMPLE 3 Solving a Linear Equation

Solve $4(k-3)-k=k-6$.

Step 1 Clear the parentheses using the distributive property.

$$4(k-3)-k=k-6$$

$$4(k)+4(-3)-k=k-6 \quad \text{Distributive property}$$

$$4k-12-k=k-6 \quad \text{Multiply.}$$

$$3k-12=k-6 \quad \text{Combine like terms.}$$

Step 2

$$3k-12-k=k-6-k \quad \text{Subtract } k.$$

$$2k-12=-6 \quad \text{Combine like terms.}$$

$$2k-12+12=-6+12 \quad \text{Add 12.}$$

$$2k=6 \quad \text{Combine like terms.}$$

Step 3

$$\frac{2k}{2}=\frac{6}{2} \quad \text{Divide by 2.}$$

$$k=3$$

Step 4 Check by substituting 3 for k in the original equation.

CHECK

$$4(k-3)-k=k-6 \quad \text{Original equation}$$

$$4(3-3)-3\stackrel{?}{=}3-6 \quad \text{Let } k=3.$$

$$4(0)-3\stackrel{?}{=}3-6 \quad \text{Work inside the parentheses.}$$

$$-3=-3 \ \checkmark \quad \text{True}$$

The solution, 3, checks, so the solution set is $\{3\}$.

Work Problem 4 at the Side. ▶

EXAMPLE 4 Solving a Linear Equation

Solve $8a-(3+2a)=3a+1$.

Step 1

$$8a-(3+2a)=3a+1$$

$$8a-1(3+2a)=3a+1 \quad \text{Multiplicative identity property}$$

$$8a-3-2a=3a+1 \quad \text{Distributive property}$$

Be careful with signs.

$$6a-3=3a+1 \quad \text{Combine like terms.}$$

Step 2

$$6a-3-3a=3a+1-3a \quad \text{Subtract } 3a.$$

$$3a-3=1 \quad \text{Combine like terms.}$$

$$3a-3+3=1+3 \quad \text{Add 3.}$$

$$3a=4 \quad \text{Combine like terms.}$$

Step 3

$$\frac{3a}{3}=\frac{4}{3} \quad \text{Divide by 3.}$$

$$a=\frac{4}{3}$$

Step 4 Check that the solution set is $\{\frac{4}{3}\}$.

---

4 Solve.

GS **(a)** $7(p-2)+p=2p+4$

Step 1 ____ the parentheses.

$$\_\_\_\_(p)+7(\_\_\_\_)+p=2p+4$$

$$\_\_\_\_-\_\_\_\_+p=2p+4$$

$$\_\_\_\_-14=2p+4$$

Now complete the solution. Give the solution set.

(b) $11+3(x+1)=5x+16$

Answers

4. (a) Clear; 7; −2; 7p; 14; 8p; The solution set is $\{3\}$.

(b) $\{-1\}$

5 Solve.

(a) $7m - (2m - 9) = 39$

(b) $5x - (x + 9) = x - 4$

6 Solve.

(a) $2(4 + 3r) = 3(r + 1) + 11$

(b) $2 - 3(2 + 6z)$
$= 4(z + 1) - 8$

CAUTION

In an expression such as $8a - (3 + 2a)$ in **Example 4,** the $-$ sign acts like a factor of -1 and affects the sign of *every* term within the parentheses.

$$8a - (3 + 2a) \quad \leftarrow \text{Left side of the equation in Example 4}$$
$$= 8a - 1(3 + 2a)$$
$$= 8a + (-1)(3 + 2a)$$
$$= 8a - 3 - 2a$$

Change to $-$ in *both* terms.

◀ **Work Problem 5 at the Side.**

EXAMPLE 5 Solving a Linear Equation

Solve $4(4 - 3x) = 32 - 8(x + 2)$.

Do *not* subtract 8 from 32 here.

Be careful with signs.

Step 1	$4(4 - 3x) = 32 - 8(x + 2)$	
	$16 - 12x = 32 - 8x - 16$	Distributive property
	$16 - 12x = 16 - 8x$	Combine like terms.
Step 2	$16 - 12x + 8x = 16 - 8x + 8x$	Add $8x$.
	$16 - 4x = 16$	Combine like terms.
	$16 - 4x - 16 = 16 - 16$	Subtract 16.
	$-4x = 0$	Combine like terms.
Step 3	$\frac{-4x}{-4} = \frac{0}{-4}$	Divide by -4.
	$x = 0$	
Step 4 CHECK	$4(4 - 3x) = 32 - 8(x + 2)$	Original equation
	$4[4 - 3(0)] \stackrel{?}{=} 32 - 8(0 + 2)$	Let $x = 0$.
	$4(4 - 0) \stackrel{?}{=} 32 - 8(2)$	Work inside the brackets and parentheses.
	$4(4) \stackrel{?}{=} 32 - 16$	Subtract and multiply.
	$16 = 16$ ✓	True

Since the solution 0 checks, the solution set is $\{0\}$.

◀ **Work Problem 6 at the Side.**

OBJECTIVE 2 Solve equations that have no solution or infinitely many solutions. Each equation so far has had exactly one solution. An equation with exactly one solution is a **conditional equation** because it is only true under certain conditions. Some equations have no solution or infinitely many solutions.

Answers

5. (a) $\{6\}$ (b) $\left\{\frac{5}{3}\right\}$

6. (a) $\{2\}$ (b) $\{0\}$

EXAMPLE 6 Solving an Equation That Has Infinitely Many Solutions

Solve $5x - 15 = 5(x - 3)$.

$$5x - 15 = 5(x - 3)$$

$$5x - 15 = 5x - 15 \quad \text{Distributive property}$$

$$5x - 15 - 5x = 5x - 15 - 5x \quad \text{Subtract } 5x.$$

Notice that the variable "disappeared."

$$-15 = -15 \quad \text{Combine like terms.}$$

$$-15 + 15 = -15 + 15 \quad \text{Add 15.}$$

$$0 = 0 \quad \text{True}$$

Solution set: {all real numbers}

Since the last statement $(0 = 0)$ is true, *any* real number is a solution. We could have predicted this from the second line in the solution,

$$5x - 15 = 5x - 15 \longleftarrow \text{This is true for } any \text{ value of } x.$$

Try several values for x in the original equation to see that they all satisfy it.

An equation with both sides exactly the same, like $0 = 0$, is an **identity.** An identity is true for all replacements of the variables. As shown above, we write the solution set as **{all real numbers}.**

CAUTION

In **Example 6,** do not write $\{0\}$ as the solution set of the equation. While 0 is a solution, there are infinitely many other solutions. ***For $\{0\}$ to be the solution set, the last line must include a variable, such as x, and read $x = 0$ (as in Example 5), not $0 = 0$.***

EXAMPLE 7 Solving an Equation That Has No Solution

Solve $2x + 3(x + 1) = 5x + 4$.

$$2x + 3(x + 1) = 5x + 4$$

$$2x + 3x + 3 = 5x + 4 \quad \text{Distributive property}$$

$$5x + 3 = 5x + 4 \quad \text{Combine like terms.}$$

$$5x + 3 - 5x = 5x + 4 - 5x \quad \text{Subtract } 5x.$$

Again, the variable "disappeared."

$$3 = 4 \quad \text{False}$$

There is no solution. Solution set: $\emptyset$

A false statement $(3 = 4)$ results. A **contradiction** is an equation that has no solution. Its solution set is the **empty set,** or **null set,** symbolized **$\emptyset$.**

Work Problem 7 at the Side. ▶

CAUTION

Do not write $\{\emptyset\}$ to represent the empty set.

OBJECTIVE 3 **Solve equations with fractions or decimals as coefficients.** To avoid messy computations, we clear an equation of fractions by multiplying each side by the least common denominator (LCD) of all the fractions in the equation. Doing this will give an equation with only *integer* coefficients.

7 Solve.

(a) $2(x - 6) = 2x - 12$

(b) $3x + 6(x + 1) = 9x - 4$

(c) $-4x + 12 = 3 - 4(x - 3)$

Teaching Tip

Stress the difference between a conditional equation with solution set {0}, as in Example 5, and an identity with solution set {all real numbers}, as in Example 6.

Teaching Tip

For some classes, it may be desirable to cover solving equations with fractions or decimals at a second meeting.

Answers

7. **(a)** {all real numbers} **(b)** $\emptyset$ **(c)** $\emptyset$

8 Solve.

(a) $\frac{1}{4}x - 4 = \frac{3}{2}x + \frac{3}{4}x$

Teaching Tip

Emphasize that *each* term of the equation must be multiplied by the LCD. It is best to identify the number of terms in the equation *before* multiplying.

(b) $\frac{1}{2}x + \frac{5}{8}x = \frac{3}{4}x - 6$

Answers

8. **(a)** $\{-2\}$ **(b)** $\{-16\}$

EXAMPLE 8 **Solving an Equation with Fractions as Coefficients**

Solve $\frac{2}{3}x - \frac{1}{2}x = -\frac{1}{6}x - 2$.

Step 1 $\frac{2}{3}x - \frac{1}{2}x = -\frac{1}{6}x - 2$

Pay particular attention here.

$6\left(\frac{2}{3}x - \frac{1}{2}x\right) = 6\left(-\frac{1}{6}x - 2\right)$ Multiply each side by 6, the LCD.

$6\left(\frac{2}{3}x\right) + 6\left(-\frac{1}{2}x\right) = 6\left(-\frac{1}{6}x\right) + 6(-2)$ Distributive property; multiply *each* term inside the parentheses by 6.

The fractions have been cleared.

$4x - 3x = -x - 12$ Multiply.

$x = -x - 12$ Combine like terms.

Step 2 $x + x = -x - 12 + x$ Add x.

$2x = -12$ Combine like terms.

Step 3 $\frac{2x}{2} = \frac{-12}{2}$ Divide by 2.

$x = -6$

Step 4 **CHECK** $\frac{2}{3}x - \frac{1}{2}x = -\frac{1}{6}x - 2$ Original equation

$\frac{2}{3}(-6) - \frac{1}{2}(-6) \stackrel{?}{=} -\frac{1}{6}(-6) - 2$ Let $x = -6$.

$-4 + 3 \stackrel{?}{=} 1 - 2$ Multiply.

$-1 = -1$ ✓ True

The solution, -6, checks. The solution set is $\{-6\}$.

◀ Work Problem **8** at the Side.

CAUTION

When clearing an equation of fractions, be sure to multiply every term on each side of the equation by the LCD.

EXAMPLE 9 **Solving an Equation with Fractions as Coefficients**

Solve $\frac{1}{3}(x + 5) - \frac{3}{5}(x + 2) = 1$.

Step 1 $\frac{1}{3}(x + 5) - \frac{3}{5}(x + 2) = 1$

$15\left[\frac{1}{3}(x + 5) - \frac{3}{5}(x + 2)\right] = 15(1)$ To clear the fractions, multiply by 15, the LCD.

$15\left[\frac{1}{3}(x + 5)\right] + 15\left[-\frac{3}{5}(x + 2)\right] = 15(1)$ Distributive property

$5(x + 5) - 9(x + 2) = 15$ Multiply.

$15\left[\frac{1}{3}(x+5)\right] = 15 \cdot \frac{1}{3} \cdot (x+5) = 5(x+5)$

$5x + 25 - 9x - 18 = 15$ Distributive property

$-4x + 7 = 15$ Combine like terms.

Continued on Next Page

Step 2 $-4x + 7 - 7 = 15 - 7$ Subtract 7.

$-4x = 8$ Combine like terms.

Step 3 $\frac{-4x}{-4} = \frac{8}{-4}$ Divide by -4.

$x = -2$

Step 4 Check to confirm that $\{-2\}$ is the solution set.

Work Problem 9 at the Side. ▶

CAUTION

Be sure you understand how to multiply by the LCD to clear an equation of fractions. ***Study Step 1 in Examples 8 and 9 carefully.***

EXAMPLE 10 Solving an Equation with Decimals as Coefficients

Solve $0.1t + 0.05(20 - t) = 0.09(20)$.

Step 1 The decimals here are expressed as tenths (0.1) and hundredths $(0.05$ and $0.09)$. We choose the least exponent on 10 needed to eliminate the decimal points. In this equation, we use $10^2 = 100$.

$$0.1t + 0.05(20 - t) = 0.09(20)$$

$$0.10t + 0.05(20 - t) = 0.09(20) \quad 0.1 = 0.10$$

$$100[0.10t + 0.05(20 - t)] = 100[0.09(20)] \quad \text{Multiply by 100.}$$

$$100(0.10t) + 100[0.05(20 - t)] = 100[0.09(20)] \quad \text{Distributive property}$$

$$10t + 5(20 - t) = 9(20) \quad \text{Multiply.}$$

$$10t + 5(20) + 5(-t) = 180 \quad \text{Distributive property}$$

$$10t + 100 - 5t = 180 \quad \text{Multiply.}$$

$$5t + 100 = 180 \quad \text{Combine like terms.}$$

Step 2 $5t + 100 - 100 = 180 - 100$ Subtract 100.

$5t = 80$ Combine like terms.

Step 3 $\frac{5t}{5} = \frac{80}{5}$ Divide by 5.

$t = 16$

Step 4 Check to confirm that $\{16\}$ is the solution set.

Work Problem 10 at the Side. ▶

Note

In **Example 10,** multiplying by 100 is accomplished by moving the decimal point two places to the right.

$$0.10t + 0.05(20 - t) = 0.09(20)$$

$$10t + 5(20 - t) = 9(20) \quad \text{Multiply by 100.}$$

9 Solve.

$$\frac{1}{4}(x + 3) - \frac{2}{3}(x + 1) = -2$$

10 Solve.

$$0.06(100 - x) + 0.04x = 0.05(92)$$

Answers

9. $\{5\}$
10. $\{70\}$

11 Perform each translation.

(a) Two numbers have a sum of 36. One of the numbers is represented by r. Write an expression for the other number.

(b) Two numbers have a product of 18. One of the numbers is represented by x. Write an expression for the other number.

Answers

11. **(a)** $36 - r$ **(b)** $\frac{18}{x}$

The table summarizes the solution sets of the equations in this section.

Type of Equation	Final Equation in Solution	Number of Solutions	Solution Set
Conditional **(See Examples 1–5, 8–10.)**	$x =$ a number	One	{a number}
Identity **(See Example 6.)**	A true statement with no variable, such as $0 = 0$	Infinite	{all real numbers}
Contradiction **(See Example 7.)**	A false statement with no variable, such as $3 = 4$	None	$\emptyset$

OBJECTIVE 4 Write expressions for two related unknown quantities.

Problem-Solving Hint

When we solve applied problems, we must often write *expressions* to relate unknown quantities. We then use these expressions to write the *equation* needed to solve the application. The next example provides preparation for doing this in **Section 2.4.**

EXAMPLE 11 Translating Phrases into Algebraic Expressions

Perform each translation.

(a) Two numbers have a sum of 23. If one of the numbers is represented by x, find an expression for the other number.

First, suppose that the sum of two numbers is 23, and one of the numbers is **10**. To find the other number, we would subtract **10** from 23.

$$23 - \mathbf{10} \leftarrow \text{This gives 13 as the other number.}$$

Instead of using **10** as one of the numbers, we use x. The other number would be obtained in the same way—by subtracting x from 23.

$$23 - x$$

$x - 23$ is *not* correct. Subtraction is *not* commutative.

To check, we find the sum of the two numbers.

$$x + (23 - x) = 23, \quad \text{as required.}$$

(b) Two numbers have a product of 24. If one of the numbers is represented by x, find an expression for the other number.

Suppose that one of the numbers is 4. To find the other number, we would divide 24 by 4.

$$\frac{24}{4} \leftarrow \text{This gives 6 as the other number. The product } 6 \cdot 4 \text{ is 24.}$$

In the same way, if x is one of the numbers, then we divide 24 by x to find the other number.

$$\frac{24}{x} \leftarrow \text{The other number}$$

◀ Work Problem 11 at the Side.

2.3 Exercises

FOR EXTRA HELP

Download the MyDashBoard App

MyMathLab®

CONCEPT CHECK *Based on the methods of this section, fill in each blank to indicate what we should do first to solve the given equation. Do not actually solve.*

1. $7x + 8 = 1$

Use the __addition__ property of equality to __subtract__ 8 from each side.

2. $7x - 5x + 15 = 8 + x$

On the __left__ side, combine __like__ terms.

3. $3(2t - 4) = 20 - 2t$

Use the __distributive__ property to clear __parentheses__ on the left side of the equation.

4. $\frac{3}{4}z = -15$

Use the __multiplication__ property of equality to multiply each side by __$\frac{4}{3}$__ to obtain z on the left.

5. $\frac{2}{3}x - \frac{1}{6} = \frac{3}{2}x + 1$

Clear __fractions__ by multiplying by __6__, the LCD.

6. $0.9x + 0.3(x + 12) = 6$

Clear __decimals__ by multiplying each side by __10__ to obtain $9x$ as the first term on the left.

7. CONCEPT CHECK Suppose that when solving three linear equations, we obtain the final results shown in parts (a)–(c). Fill in the blanks in parts (a)–(c), and then match each result with the solution set in choices A–C for the *original* equation.

(a) $6 = 6$ (The original equation is a(n) __identity__.) B

(b) $x = 0$ (The original equation is a(n) __conditional__ equation.) A

(c) $-5 = 0$ (The original equation is a(n) __contradiction__.) C

A. $\{0\}$

B. {all real numbers}

C. $\emptyset$

CONCEPT CHECK *In Exercises 8–10, give the letter of the correct choice.*

8. Which linear equation does *not* have all real numbers as solutions?

A. $5x = 4x + x$ **B.** $2(x + 6) = 2x + 12$ **C.** $\frac{1}{2}x = 0.5x$ **D.** $3x = 2x$ D

9. The expression $12\left[\frac{1}{6}(x + 2) - \frac{2}{3}(x + 1)\right]$ is equivalent to which of the following?

A. $-6x - 4$ **B.** $2x - 8$ **C.** $-6x + 1$ **D.** $-6x + 4$ A

10. The expression $100[0.03(x - 10)]$ is equivalent to which of the following?

A. $0.03x - 0.3$ **B.** $3x - 3$ **C.** $3x - 10$ **D.** $3x - 30$ D

Solve each equation, and check your solution. ***See Examples 1–7.***

11. $3x + 2 = 14$ $\{4\}$

12. $4x + 3 = 27$ $\{6\}$

13. $-5z - 4 = 21$ $\{-5\}$

14. $-7w - 4 = 10$ $\{-2\}$

15. $4p - 5 = 2p$
$\left\{\frac{5}{2}\right\}$

16. $6q - 2 = 3q$
$\left\{\frac{2}{3}\right\}$

17. $5m + 8 = 7 + 3m$
$\left\{-\frac{1}{2}\right\}$

18. $4r + 2 = r - 6$
$\left\{-\frac{8}{3}\right\}$

19. $10p + 6 = 12p - 4$
$\{5\}$

20. $-5x + 8 = -3x + 10$
$\{-1\}$

21. $7r - 5r + 2 = 5r - r$
$\{1\}$

22. $9p - 4p + 6 = 7p - p$
$\{6\}$

23. $x + 3 = -(2x + 2)$
$\left\{-\frac{5}{3}\right\}$

24. $2x + 1 = -(x + 3)$
$\left\{-\frac{4}{3}\right\}$

25. $3(4x - 2) + 5x = 30 - x$
$\{2\}$

26. $5(2m + 3) - 4m = 8m + 27$
$\{-6\}$

27. $-2p + 7 = 3 - (5p + 1)$
$\left\{-\frac{5}{3}\right\}$

28. $4x + 9 = 3 - (x - 2)$
$\left\{-\frac{4}{5}\right\}$

29. $11x - 5(x + 2) = 6x + 5$
$\emptyset$

30. $6x - 4(x + 1) = 2x + 4$
$\emptyset$

31. $-(8x - 2) + 5x - 6 = -4$
$\{0\}$

32. $-(7x - 5) + 4x - 7 = -2$
$\{0\}$

33. $4(2x - 1) = -6(x + 3)$
$\{-1\}$

34. $6(3w + 5) = 2(10w + 10)$
$\{5\}$

35. $6(4x - 1) = 12(2x + 3)$
$\emptyset$

36. $6(2x + 8) = 4(3x - 6)$
$\emptyset$

37. $3(2x - 4) = 6(x - 2)$
{all real numbers}

38. $3(6 - 4x) = 2(-6x + 9)$
{all real numbers}

39. $24 - 4(7 - 2t) = 4(t - 1)$
$\{0\}$

40. $8 - 2(2 - x) = 4(x + 1)$
$\{0\}$

Solve each equation, and check your solution. ***See Examples 8–10.***

41. $-\frac{2}{7}r + 2r = \frac{1}{2}r + \frac{17}{2}$

(*Hint:* Clear the fractions by multiplying each side of the equation by the LCD, __14__.)
$\{7\}$

42. $\frac{3}{5}t - \frac{1}{10}t = t - \frac{5}{2}$

(*Hint:* Clear the fractions by multiplying each side of the equation by the LCD, __10__.)
$\{5\}$

43. $\frac{3}{4}x - \frac{1}{3}x + 5 = \frac{5}{6}x$
$\{12\}$

44. $\frac{1}{5}x - \frac{2}{3}x - 2 = -\frac{2}{5}x$
$\{-30\}$

45. $\frac{1}{7}(3x + 2) - \frac{1}{5}(x + 4) = 2$
$\{11\}$

46. $\frac{1}{4}(3x - 1) + \frac{1}{6}(x + 3) = 3$
$\{3\}$

47. $\frac{1}{9}(x + 18) + \frac{1}{3}(2x + 3) = x + 3$

$\{0\}$

48. $-\frac{1}{4}(x - 12) + \frac{1}{2}(x + 2) = x + 4$

$\{0\}$

49. $-\frac{5}{6}q - \left(q - \frac{1}{2}\right) = \frac{1}{4}(q + 1)$

$\left\{\frac{3}{25}\right\}$

50. $\frac{2}{3}k - \left(k + \frac{1}{4}\right) = \frac{1}{12}(k + 4)$

$\left\{-\frac{7}{5}\right\}$

51. $0.3(30) + 0.15x = 0.2(30 + x)$

(*Hint:* Clear the decimals by multiplying each side of the equation by 100.)

$\{60\}$

52. $0.2(60) + 0.05x = 0.1(60 + x)$

(*Hint:* Clear the decimals by multiplying each side of the equation by 100.)

$\{120\}$

53. $0.92x + 0.98(12 - x) = 0.96(12)$

$\{4\}$

54. $1.00x + 0.05(12 - x) = 0.10(63)$

$\{6\}$

55. $0.02(5000) + 0.03x = 0.025(5000 + x)$

$\{5000\}$

56. $0.06(10{,}000) + 0.08x = 0.072(10{,}000 + x)$

$\{15{,}000\}$

Solve each equation, and check your solution. ***See Examples 1–10.***

57. $-3(5z + 24) + 2 = 2(3 - 2z) - 4$

$\left\{-\frac{72}{11}\right\}$

58. $-2(2s - 4) - 8 = -3(4s + 4) - 1$

$\left\{-\frac{13}{8}\right\}$

59. $-(6k - 5) - (-5k + 8) = -3$

$\{0\}$

60. $-(4x + 2) - (-3x - 5) = 3$

$\{0\}$

61. $8(t - 3) + 4t = 6(2t + 1) - 10$

$\emptyset$

62. $9(v + 1) - 3v = 2(3v + 1) - 8$

$\emptyset$

63. $4(x+3)=2(2x+8)-4$

{all real numbers}

64. $4(x+8)=2(2x+6)+20$

{all real numbers}

65. $\frac{1}{3}(x+3)+\frac{1}{6}(x-6)=x+3$

$\{-6\}$

66. $\frac{1}{2}(x+2)+\frac{3}{4}(x+4)=x+5$

$\{4\}$

67. $0.3(x+15)+0.4(x+25)=25$

$\{15\}$

68. $0.1(x+80)+0.2x=14$

$\{20\}$

Write the answer to each problem in terms of the variable. **See Example 11.**

69. Two numbers have a sum of 12. One of the numbers is q. What expression represents the other number?

$12-q$

70. Two numbers have a sum of 26. One of the numbers is r. What expression represents the other number?

$26-r$

71. The product of two numbers is 9. One of the numbers is z. What expression represents the other number?

$\frac{9}{z}$

72. The product of two numbers is 13. One of the numbers is k. What expression represents the other number?

$\frac{13}{k}$

73. A football player gained x yards rushing. On the next down, he gained 29 yd. What expression represents the number of yards he gained altogether?

$x+29$

74. A football player gained y yards on a punt return. On the next return, he gained 25 yd. What expression represents the number of yards he gained altogether?

$y+25$

75. Monica is a years old. What expression represents her age 12 yr from now? 2 yr ago?

$a+12$; $a-2$

76. Chandler is b years old. What expression represents his age 3 yr ago? 5 yr from now?

$b-3$; $b+5$

77. Tom has r quarters. Express the value of the quarters in cents.

$25r$

78. Jean has y dimes. Express the value of the dimes in cents.

$10y$

79. A bank teller has t dollars, all in \$5 bills. What expression represents the number of \$5 bills the teller has?

$\frac{t}{5}$

80. A clerk has v dollars, all in \$10 bills. What expression represents the number of \$10 bills the clerk has?

$\frac{v}{10}$

81. A plane ticket costs x dollars for an adult and y dollars for a child. Find an expression that represents the total cost for 3 adults and 2 children.

$3x+2y$

82. A concert ticket costs p dollars for an adult and q dollars for a child. Find an expression that represents the total cost for 4 adults and 6 children.

$4p+6q$

Study Skills

USING STUDY CARDS REVISITED

Two additional types of study cards follow. As with vocabulary and procedure cards introduced earlier, use tough problem and practice quiz cards to do the following.

- To help you understand and learn the material
- To quickly review when you have a few minutes
- To review before a quiz or test

Tough Problem Cards

When you are doing your homework and encounter a "difficult" problem, write the procedure to work the problem on the front of a card in words. Include special notes or tips (like what *not* to do). On the back of the card, work an example. Show all steps, and label what you are doing.

Front of Card

When solving a linear equation, be careful when clearing parentheses if there is a minus sign in front. p. 135

$6x - (x + 3)$

The minus sign acts like −1, so change the sign of every term inside the parentheses.

Back of Card

Solve.

	$6x - (x + 3) = 7$	
	$6x - 1(x + 3) = 7$	
Change both signs.	$6x - x - 3 = 7$	*Distributive property*
	$5x - 3 = 7$	*Combine like terms.*
	$5x - 3 + 3 = 7 + 3$	*Add 3.*
	$5x = 10$	*Combine like terms.*
	$\frac{5x}{5} = \frac{10}{5}$	*Divide by 5.*
Solution set: {2}	$x = 2$	

Practice Quiz Cards

Write a problem with direction words (like *solve, simplify*) on the front of a card, and work the problem on the back. Make one for each type of problem you learn.

Front of Card

Solve $4(3x - 4) = 2(6x - 9) + 2.$ p. 137

Back of Card

	$4(3x - 4) = 2(6x - 9) + 2$	
	$12x - 16 = 12x - 18 + 2$	*Distributive property*
	$12x - 16 = 12x - 16$	*Combine like terms.*
	$12x - 16 + 16 = 12x - 16 + 16$	*Add 16.*
	$12x = 12x$	*Combine like terms.*
	$12x - 12x = 12x - 12x$	*Subtract 12x.*
When both sides of an equation are the same, it is called an identity.	$0 = 0$	*True*

Any real number will work, so the solution set is {all real numbers} (not just {0}).

Make a tough problem card and a practice quiz card for material you are learning now.

Summary Exercises *Applying Methods for Solving Linear Equations*

Summary Exercises provide mixed practice to better prepare students for tests.

These exercises provide mixed practice in solving all the types of linear equations introduced in **Sections 2.1–2.3.** Refer back to the examples in these sections to review the various solution methods.

Solve each equation, and check your solution.

1. $a + 2 = -3$
$\{-5\}$

2. $2m + 8 = 16$
$\{4\}$

3. $16.5k = -84.15$
$\{-5.1\}$

4. $-x = -25$
$\{25\}$

5. $\frac{4}{5}x = -20$
$\{-25\}$

6. $9x - 7x = -12$
$\{-6\}$

7. $5x - 9 = 4(x - 3)$
$\{-3\}$

8. $\frac{a}{-2} = 8$
$\{-16\}$

9. $5x - 9 = 3(x - 3)$
$\{0\}$

10. $\frac{2}{3}x + 8 = \frac{1}{4}x$
$\left\{-\frac{96}{5}\right\}$

11. $4x + 2(3 - 2x) = 6$
{all real numbers}

12. $x - 16.2 = 7.5$
$\{23.7\}$

13. $-3(t - 5) + 2(7 + 2t) = 36$
$\{7\}$

14. $2 - (m + 4) = 3m - 2$
$\{0\}$

15. $0.08x + 0.06(x + 9) = 1.24$
$\{5\}$

16. $-0.3x + 2.1(x - 4) = -6.6$
$\{1\}$

17. $-x = 16$
$\{-16\}$

18. $3(m + 5) - 1 + 2m = 5(m + 2)$
$\emptyset$

19. $10m - (5m - 9) = 39$

$\{6\}$

20. $7(p - 2) + p = 2(p + 2)$

$\{3\}$

21. $-2t + 5t - 9 = 3(t - 4) - 5$

$\emptyset$

22. $-9z = -21$

$\left\{\frac{7}{3}\right\}$

23. $0.02(50) + 0.08r = 0.04(50 + r)$

$\{25\}$

24. $2.3x + 13.7 = 1.3x + 2.9$

$\{-10.8\}$

25. $2(3 + 7x) - (1 + 15x) = 2$

$\{3\}$

26. $6q - 9 = 12 + 3q$

$\{7\}$

27. $2(5 + 3x) = 3(x + 1) + 13$

$\{2\}$

28. $r + 9 + 7r = 4(3 + 2r) - 3$

{all real numbers}

29. $\frac{5}{6}x + \frac{1}{3} = 2x + \frac{2}{3}$

$\left\{-\frac{2}{7}\right\}$

30. $0.06x + 0.09(15 - x) = 0.07(15)$

$\{10\}$

31. $\frac{3}{4}(a - 2) - \frac{1}{3}(5 - 2a) = -2$

$\left\{\frac{14}{17}\right\}$

32. $2 - (m + 4) = 3m + 8$

$\left\{-\frac{5}{2}\right\}$

33. $5.2x - 4.6 - 7.1x = -2.1 - 1.9x - 2.5$

{all real numbers}

34. $9(2m - 3) - 4(5 + 3m) = 5(4 + m) - 3$

$\{64\}$

2.4 An Introduction to Applications of Linear Equations

OBJECTIVES

1. Learn the six steps for solving applied problems.
2. Solve problems involving unknown numbers.
3. Solve problems involving sums of quantities.
4. Solve problems involving consecutive integers.
5. Solve problems involving complementary and supplementary angles.

Teaching Tip

Mention how the nature of the applied problem restricts the set of possible solutions. For example, an answer such as -33 medals or $25\frac{1}{2}$ medals should be recognized as inappropriate in **Example 2.**

OBJECTIVE 1 **Learn the six steps for solving applied problems.** While there is not one specific method, we suggest the following.

Solving an Applied Problem

Step 1 **Read** the problem, several times if necessary. What information is given? What is to be found?

Step 2 **Assign a variable** to represent the unknown value. Use a sketch, diagram, or table, as needed. Express any other unknown values in terms of the variable.

Step 3 **Write an equation** using the variable expression(s).

Step 4 **Solve** the equation.

Step 5 **State the answer.** Label it appropriately. Does the answer seem reasonable?

Step 6 **Check** the answer in the words of the *original* problem.

OBJECTIVE 2 **Solve problems involving unknown numbers.**

EXAMPLE 1 **Finding the Value of an Unknown Number**

The product of 4, and a number decreased by 7, is 100. What is the number?

Step 1 **Read** the problem carefully. We are asked to find a number.

Step 2 **Assign a variable** to represent the unknown quantity.

Let $x =$ the number.

Step 3 **Write an equation.**

Writing a "word equation" is often helpful.

The product of 4,	and	a number	decreased by	7,	is	100.
↓		↓	↓	↓	↓	↓
$4 \cdot$		$(x$	$-$	$7)$	$=$	100

Note the careful use of parentheses.

Is, are, was, and *were* translate as =.

Because of the commas in the given problem, writing the equation as $4x - 7 = 100$ is *incorrect*. The equation $4x - 7 = 100$ corresponds to the statement "The product of 4 and a number, decreased by 7, is 100."

Step 4 **Solve** the equation.

$$4(x - 7) = 100 \quad \text{Equation from Step 3}$$

$$4x - 28 = 100 \quad \text{Distributive property}$$

$$4x - 28 + 28 = 100 + 28 \quad \text{Add 28.}$$

$$4x = 128 \quad \text{Combine like terms.}$$

$$\frac{4x}{4} = \frac{128}{4} \quad \text{Divide by 4.}$$

$$x = 32$$

Continued on Next Page

Step 5 **State the answer.** The number is 32.

Step 6 **Check.** The number 32 decreased by 7 is 25. The product of 4 and 25 is 100, as required. The answer, 32, is correct.

Work Problem **1** at the Side. ▶

OBJECTIVE 3 Solve problems involving sums of quantities.

Problem-Solving Hint

In general, to solve problems involving sums of quantities, choose a variable to represent one of the unknowns. ***Then represent the other quantity in terms of the same variable.*** (See **Example 11** in **Section 2.3.**)

EXAMPLE 2 Finding Numbers of Olympic Medals

At the 2010 Winter Olympics in Vancouver, Canada, the United States won 14 more medals than Norway. The two countries won a total of 60 medals. How many medals did each country win? (*Source: World Almanac and Book of Facts.*)

Step 1 **Read** the problem. We are given the total number of medals and asked to find the number each country won.

Step 2 **Assign a variable.**

Let x = the number of medals Norway won.

Then $x + 14$ = the number of medals the United States won.

Step 3 **Write an equation.**

The total	is	the number of medals Norway won	plus	the number of medals the U.S. won.
↓	↓	↓	↓	↓
60	$=$	x	$+$	$(x + 14)$

Step 4 **Solve** the equation.

$$60 = 2x + 14 \quad \text{Combine like terms.}$$

$$60 - 14 = 2x + 14 - 14 \quad \text{Subtract 14.}$$

$$46 = 2x \quad \text{Combine like terms.}$$

$$\frac{46}{2} = \frac{2x}{2} \quad \text{Divide by 2.}$$

$$23 = x, \quad \text{or} \quad x = 23$$

Step 5 **State the answer.** The variable x represents the number of medals Norway won, so Norway won 23 medals. Then the number of medals the United States won is

$$x + 14 = 23 + 14, \quad \text{or} \quad 37.$$

Step 6 **Check.** Since the United States won 37 medals and Norway won 23, the total number of medals was $37 + 23 = 60$. Because $37 - 23 = 14$, the United States won 14 more medals than Norway. This information agrees with what is given in the problem, so the answer checks.

---

1 Solve each problem.

GS **(a)** If 5 is added to a number, the result is 7 less than 3 times the number. Find the number.

Step 1
We must find a ________.

Step 2
Let x = the ________.

Step 3

If 5 is added to a number,	the result is	7 less than 3 times the number.
↓	↓	↓
________	=	________

Complete Steps 4–6 to solve the problem. Give the answer.

(b) If 5 is added to the product of 9 and a number, the result is 19 less than the number. Find the number.

Answers

1. **(a)** number; number; $x + 5$ (or $5 + x$); $3x - 7$; The number is 6.
(b) -3

2 Solve each problem.

GS **(a)** The 150-member Iowa legislature includes 18 fewer Democrats than Republicans. (No other parties are represented.) How many Democrats and Republicans are there in the legislature? (*Source*: www.legis.iowa.gov)

Step 1

We must find the number of Democrats and ________.

Step 2

Let $x =$ the number of Republicans.

Then ________ = the number of ________.

Complete Steps 3–6 to solve the problem. Give the equation and the answer.

(b) At the 2010 Winter Olympics, Germany won 19 more medals than China. The two countries won a total of 41 medals. How many medals did each country win? (*Source: World Almanac and Book of Facts.*)

3 Solve the problem.

At a club meeting, each member brought two nonmembers. If a total of 27 people attended, how many were members and how many were nonmembers?

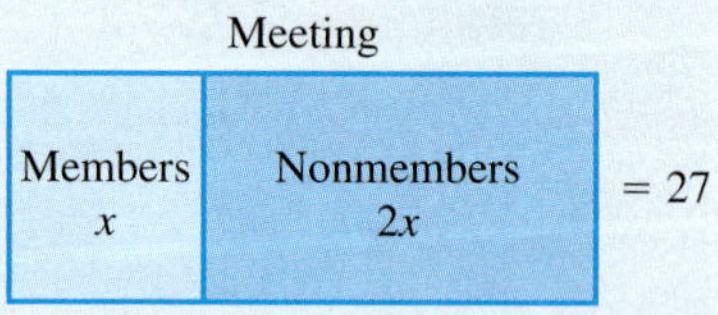

Answers

2. (a) Republicans; $x - 18$; Democrats; $x + (x - 18) = 150$; Republicans: 84; Democrats: 66
(b) Germany: 30; China: 11
3. members: 9; nonmembers: 18

Problem-Solving Hint

The problem in **Example 2** could also be solved by letting x represent the number of medals the United States won. Then $x - 14$ would represent the number of medals Norway won. The equation would be different.

$$60 = x + (x - 14) \quad \text{Alternative equation for Example 2}$$

The solution of this equation is 37, which is the number of U.S. medals. The number of Norwegian medals would be $37 - 14 = 23$. ***The answers are the same,*** whichever approach is used, even though the equation and its solution are different.

◀ Work Problem **2** at the Side.

EXAMPLE 3 Analyzing a Gasoline/Oil Mixture

A lawn trimmer uses a mixture of gasoline and oil. The mixture contains 16 oz of gasoline for each 1 ounce of oil. If the tank holds 68 oz of the mixture, how many ounces of oil and how many ounces of gasoline does it require when it is full?

Step 1 **Read** the problem. We must find how many ounces of oil and gasoline are needed to fill the tank.

Step 2 **Assign a variable.**

Let $x =$ the number of ounces of oil required.

Then $16x =$ the number of ounces of gasoline required.

A diagram like the following is sometimes helpful.

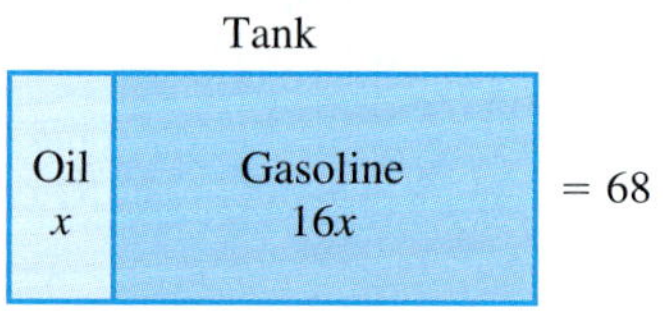

Step 3 **Write an equation.**

Amount of gasoline	plus	amount of oil	is	total amount in tank.
↓	↓	↓	↓	↓
$16x$	$+$	x	$=$	68

Step 4 **Solve.**

$$17x = 68 \quad \text{Combine like terms.}$$

$$\frac{17x}{17} = \frac{68}{17} \quad \text{Divide by 17.}$$

$$x = 4$$

Step 5 **State the answer.** When full, the lawn trimmer requires 4 oz of oil and $16x = 16(4)$ oz, or 64 oz of gasoline.

Step 6 **Check.** Since $4 + 64 = 68$, and $64 = 16(4)$, the answer checks.

◀ Work Problem **3** at the Side.

Problem-Solving Hint

Sometimes we must find three unknown quantities. ***When the three unknowns are compared in pairs, let the variable represent the unknown found in both pairs.***

EXAMPLE 4 **Dividing a Board into Pieces**

A project calls for three pieces of wood. The longest piece must be twice the length of the middle-sized piece. The shortest piece must be 10 in. shorter than the middle-sized piece. If a board 70 in. long is to be used, how long must each piece be?

Step 1 **Read** the problem. Three lengths must be found.

Step 2 **Assign a variable.** Since the middle-sized piece appears in both pairs of comparisons, let x represent the length, in inches, of the middle-sized piece.

Let $x =$ the length of the middle-sized piece.

Then $2x =$ the length of the longest piece,

and $x - 10 =$ the length of the shortest piece.

A sketch is helpful here. See **Figure 2.**

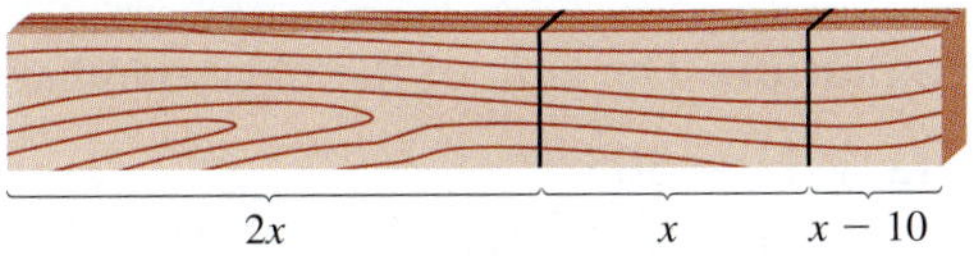

Figure 2

Step 3 **Write an equation.**

Longest	plus	middle-sized	plus	shortest	is	total length.
↓	↓	↓	↓	↓	↓	↓
$2x$	$+$	x	$+$	$(x - 10)$	$=$	70

Step 4 **Solve.**

$$4x - 10 = 70 \quad \text{Combine like terms.}$$

$$4x - 10 + 10 = 70 + 10 \quad \text{Add 10.}$$

$$4x = 80 \quad \text{Combine like terms.}$$

$$\frac{4x}{4} = \frac{80}{4} \quad \text{Divide by 4.}$$

$$x = 20$$

Step 5 **State the answer.** The middle-sized piece is 20 in. long, the longest piece is $2(20) = 40$ in. long, and the shortest piece is $20 - 10 = 10$ in. long.

Step 6 **Check.** The sum of the lengths is 70 in. All conditions of the problem are satisfied.

Work Problem 4 at the Side. ▶

4 Solve each problem.

(a) Over a 6-hr period, a basketball player spent twice as much time lifting weights as practicing free throws and 2 hr longer watching game films than practicing free throws. How many hours did he spend on each task?

Steps 1 and 2
The unknown found in both pairs of comparisons is the time spent ______.

Let $x =$ the time spent practicing free throws.

Then ____ = the time spent lifting weights, and

____ = the time spent watching game films.

Step 3
Write an equation.

______________ = 6

Complete Steps 4–6 to solve the problem. Give the answer.

(b) A piece of pipe is 50 in. long. It is cut into three pieces. The longest piece is 10 in. longer than the middle-sized piece, and the shortest piece measures 5 in. less than the middle-sized piece. Find the lengths of the three pieces.

Answers

4. (a) practicing free throws; $2x$; $x + 2$; $x + 2x + (x + 2)$; practicing free throws: 1 hr; lifting weights: 2 hr; watching game films: 3 hr

(b) longest: 25 in.; middle: 15 in.; shortest: 10 in.

5 Solve the problem.

Two pages that face each other in this book have a sum of 569. What are the page numbers?

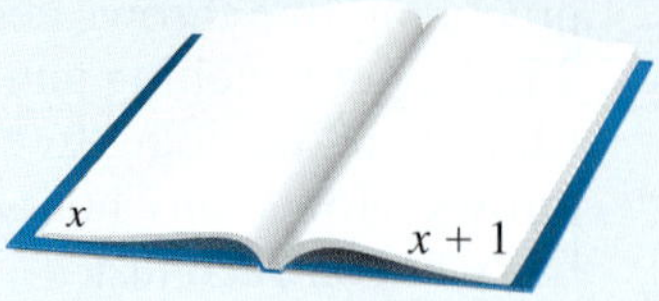

OBJECTIVE 4 Solve problems involving consecutive integers. Two integers that differ by 1 are **consecutive integers.** For example, 3 and 4, 6 and 7, and −2 and −1 are pairs of consecutive integers. See **Figure 3.**

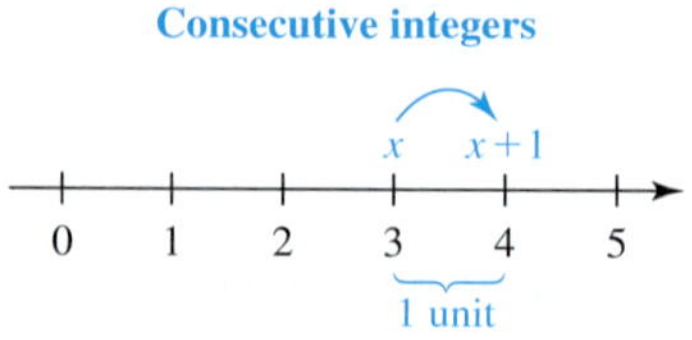

Figure 3

In general, if x represents an integer, then x + 1 represents the next greater consecutive integer.

EXAMPLE 5 Finding Consecutive Integers

Two pages that face each other in this book have 305 as the sum of their page numbers. What are the page numbers?

Step 1 **Read** the problem. Because the two pages face each other, they must have page numbers that are consecutive integers.

Step 2 **Assign a variable.**

Let $x =$ the lesser page number.

Then $x + 1 =$ the greater page number.

Step 3 **Write an equation.** The sum of the page numbers is 305.

$$x + (x + 1) = 305$$

Step 4 **Solve.**

$2x + 1 = 305$ Combine like terms.

$2x = 304$ Subtract 1.

$x = 152$ Divide by 2.

Step 5 **State the answer.** The lesser page number is 152, and the greater is $152 + 1 = 153$. (Your book is opened to these two pages.)

Step 6 **Check.** The sum of 152 and 153 is 305. The answer is correct.

◀ **Work Problem 5 at the Side.**

Teaching Tip
In Step 3 of **Example 5,** stress that it is a good idea to use parentheses around $x + 1$ (even though they are not necessary here).

Consecutive *even* integers, such as 2 and 4, and 8 and 10, differ by 2. Similarly, **consecutive *odd* integers,** such as 1 and 3, and 9 and 11, also differ by 2. See **Figure 4.**

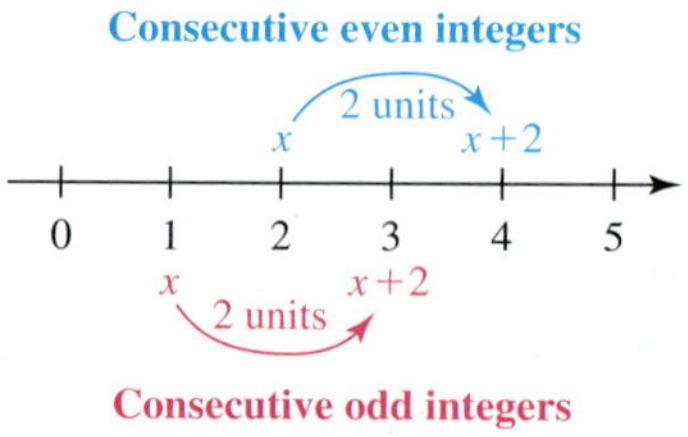

Figure 4

In general, if x represents an even (or odd) integer, then x + 2 represents the next greater consecutive even (or odd) integer, respectively.

Answer

5. 284, 285

In this book, we list consecutive integers in increasing order.

Problem-Solving Hint

When solving consecutive integer problems, if $x =$ the lesser integer, then the following apply.

For two consecutive integers, use	$x, \quad x + 1.$
For two consecutive *even* integers, use	$x, \quad x + 2.$
For two consecutive *odd* integers, use	$x, \quad x + 2.$

EXAMPLE 6 **Finding Consecutive Odd Integers**

If the lesser of two consecutive odd integers is doubled, the result is 7 more than the greater of the two integers. Find the two integers.

Step 1 **Read** the problem. We must find two consecutive odd integers.

Step 2 **Assign a variable.**

Let $x =$ the lesser consecutive odd integer.

Then $x + 2 =$ the greater consecutive odd integer.

Step 3 **Write an equation.**

If the lesser is doubled,	the result is	7	more than	the greater.
↓	↓	↓	↓	↓
$2x$	$=$	7	$+$	$x + 2$

Step 4 **Solve.**

$2x = 9 + x$ Combine like terms.

$x = 9$ Subtract x.

Step 5 **State the answer.** The lesser integer is 9. The greater is $9 + 2 = 11$.

Step 6 **Check.** When 9 is doubled, we get 18, which is 7 more than the greater odd integer, 11. The answers are correct.

Work Problem 6 at the Side. ▶

OBJECTIVE 5 **Solve problems involving complementary and supplementary angles.** An angle can be measured by a unit called the degree (°), which is $\frac{1}{360}$ of a complete rotation. Two angles whose sum is 90° are *complements* of each other, or **complementary.** An angle that measures 90° is a **right angle.** Two angles whose sum is 180° are *supplements* of each other, or **supplementary.** One angle *supplements* the other to form a **straight angle** of 180°. See **Figure 5.**

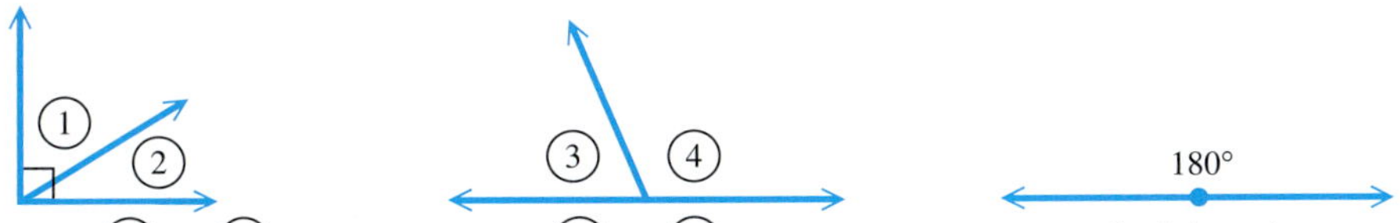

Angles (1) and (2) are complementary. They form a right angle, indicated by ⌉.

Angles (3) and (4) are supplementary. They form a straight angle.

Straight angle

Figure 5

6 Solve each problem.

(a) The sum of two consecutive even integers is 254. Find the integers.

Step 1
We must find two consecutive ______.

Step 2
Let $x =$ the lesser of the two ______ even integers.

Then ______ = the greater of the two consecutive even integers.

Step 3

The lesser consecutive even integer	+	the greater consecutive even integer	is	the total.
↓	↓	↓	↓	↓
______	+	______	=	______

Complete Steps 4–6 to solve the problem. Give the answer.

(b) Find two consecutive odd integers such that the sum of twice the lesser and three times the greater is 191.

Teaching Tip
Based on preference and/or time constraints, Objective 5 could be covered with Section 2.5.

Answers

6. **(a)** even integers; consecutive; $x + 2$; x; $x + 2$; 254; 126 and 128
(b) 37, 39

7 Fill in the blank below each figure. Then solve.

(a) Find the complement of an angle that measures 26°.

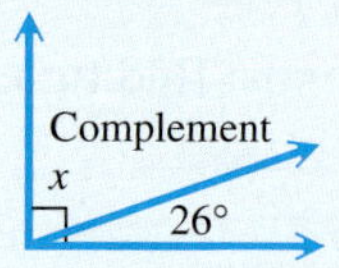

$x + 26 =$ _______

(b) Find the supplement of an angle that measures 92°.

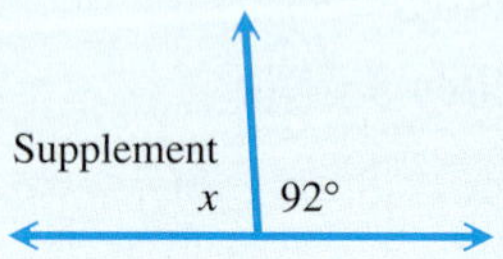

$x + 92 =$ _______

8 Solve each problem. Give the equation and the answer. (Let $x =$ the degree measure of the angle.)

(a) Find the measure of an angle whose complement is eight times its measure.

(b) Find the measure of an angle whose supplement is twice its measure.

Answers

7. **(a)** 90; 64° **(b)** 180; 88°

8. **(a)** $90 - x = 8x$; 10°

(b) $180 - x = 2x$; 60°

Problem-Solving Hint

If x represents the degree measure of an angle, then

$\mathbf{90 - x}$ represents the degree measure of its complement,

and $\mathbf{180 - x}$ represents the degree measure of its supplement.

◀ **Work Problem 7 at the Side.**

EXAMPLE 7 Finding the Measure of an Angle

Find the measure of an angle whose complement is five times its measure.

Step 1 **Read** the problem. We must find the measure of an angle, given information about the measure of its complement.

Step 2 **Assign a variable.**

Let $x =$ the degree measure of the angle.

Then $90 - x =$ the degree measure of its complement.

Step 3 **Write an equation.**

Measure of the complement	is	5 times the measure of the angle.
↓	↓	↓
$90 - x$	$=$	$5x$

Step 4 **Solve.**

$$90 - x + x = 5x + x \quad \text{Add } x.$$

$$90 = 6x \quad \text{Combine like terms.}$$

$$\frac{90}{6} = \frac{6x}{6} \quad \text{Divide by 6.}$$

$$15 = x, \quad \text{or} \quad x = 15$$

Step 5 **State the answer.** The measure of the angle is 15°.

Step 6 **Check.** If the angle measures 15°, then its complement measures 90° − 15° = 75°, which is equal to five times 15°, as required.

◀ **Work Problem 8 at the Side.**

EXAMPLE 8 Finding the Measure of an Angle

Find the measure of an angle whose supplement is 10° more than twice its complement.

Step 1 **Read** the problem. We must find the measure of an angle, given information about its complement and its supplement.

Step 2 **Assign a variable.**

Let $x =$ the degree measure of the angle.

Then $90 - x =$ the degree measure of its complement,

and $180 - x =$ the degree measure of its supplement.

Continued on Next Page

We can visualize this information using a sketch. See **Figure 6.**

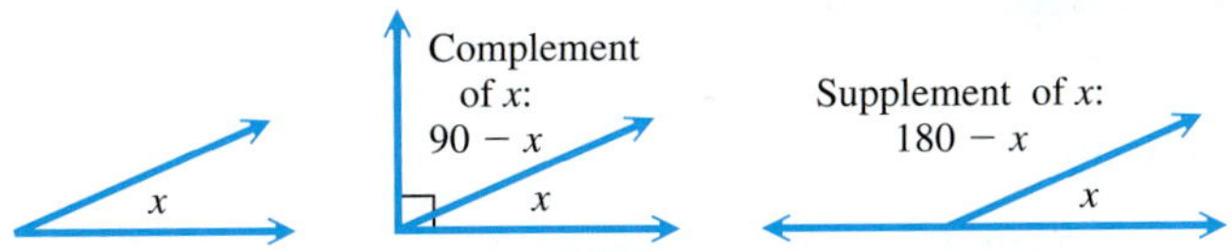

Figure 6

Step 3 **Write an equation.**

Supplement	is	10	more than	twice	its complement.
↓	↓	↓	↓	↓	↓
$180 - x$	$=$	10	$+$	2	$\cdot\ (90 - x)$

Be sure to use parentheses here.

Step 4 **Solve.**

$$180 - x = 10 + 2(90 - x)$$

$$180 - x = 10 + 180 - 2x \quad \text{Distributive property}$$

$$180 - x = 190 - 2x \quad \text{Combine like terms.}$$

$$180 - x + 2x = 190 - 2x + 2x \quad \text{Add } 2x.$$

$$180 + x = 190 \quad \text{Combine like terms.}$$

$$180 + x - 180 = 190 - 180 \quad \text{Subtract 180.}$$

$$x = 10$$

Step 5 **State the answer.** The measure of the angle is 10°.

Step 6 **Check.** The complement of 10° is 80° and the supplement of 10° is 170°. Also, 170° is equal to 10° more than twice 80° (that is, $170 = 10 + 2(80)$ is true). Therefore, the answer is correct.

Work Problem 9 at the Side. ▶

9 Solve each problem.

GS **(a)** Find the measure of an angle such that twice its complement is 30° less than its supplement.

Let x = the degree measure of the angle.

Then ________ = the degree measure of its complement,

and ________ = the degree measure of its supplement.

Complete the solution. Give the equation and the answer.

(b) Find the measure of an angle whose supplement is 46° less than three times its complement.

Answers

9. **(a)** $90 - x$; $180 - x$; $2(90 - x) = (180 - x) - 30$; 30°
(b) 22°

2.4 Exercises

FOR EXTRA HELP

Download the MyDashBoard App

MyMathLab®

1. Give the six steps introduced in this section for solving application problems.

Step 1: Read the problem carefully.

Step 2: Assign a variable to represent the unknown.

Step 3: Write an equation.

Step 4: Solve the equation.

Step 5: State the answer.

Step 6: Check the answer.

2. **CONCEPT CHECK** List some of the words that translate as "=" when writing an equation to solve an applied problem.

Some examples are *is, are, was,* and *were.*

CONCEPT CHECK *In Exercises 3–6, which choice would* **not** *be a reasonable answer? Justify your response.*

3. A problem requires finding the number of cars on a dealer's lot.

A. 0 **B.** 45 **C.** 1 **D.** $6\frac{1}{2}$

D; There cannot be a fractional number of cars.

4. A problem requires finding the number of hours a light bulb is on during a day.

A. 0 **B.** 4.5 **C.** 13 **D.** 25

D; A day cannot have more than 24 hr.

5. A problem requires finding the distance traveled in miles.

A. -10 **B.** 1.8 **C.** $10\frac{1}{2}$ **D.** 50

A; Distance cannot be negative.

6. A problem requires finding the time in minutes.

A. 0 **B.** 10.5 **C.** -5 **D.** 90

C; Time cannot be negative.

CONCEPT CHECK *Fill in each blank with the correct response.*

7. Consecutive integers differ by _______, such as 15 and _______, and -8 and _______.

1; 16 (or 14); -7 (or -9)

8. Consecutive odd integers are _______ integers that differ by _______, such as _______ and 13. Consecutive even integers are _______ integers that differ by _______, such as 12 and _______.

odd; 2; 11 (or 15); even; 2; 10 (or 14)

9. Two angles whose measures sum to 90° are _______ angles. Two angles whose measures sum to 180° are _______ angles.

complementary; supplementary

10. A right angle has measure _______. A straight angle has measure _______.

90°; 180°

CONCEPT CHECK *Answer each question.*

11. Is there an angle that is equal to its supplement? Is there an angle that is equal to its complement? If the answer is yes to either question, give the measure of the angle.

yes, 90°; yes, 45°

12. If x represents an integer, how can you express the next *smaller* consecutive integer in terms of x? The next *smaller* even integer?

$x - 1$; $x - 2$

Solve each problem. In each case, give the equation using x as the variable, and give the answer. ***See Example 1.***

13. The product of 8, and a number increased by 6, is 104. What is the number? $8 \cdot (x + 6) = 104; 7$

14. The product of 5, and 3 more than twice a number, is 85. What is the number? $5 \cdot (2x + 3) = 85; 7$

15. If 2 is added to five times a number, the result is equal to 5 more than four times the number. Find the number.
$5x + 2 = 4x + 5; 3$

16. If four times a number is added to 8, the result is three times the number, added to 5. Find the number.
$8 + 4x = 5 + 3x; -3$

17. Two less than three times a number is equal to 14 more than five times the number. What is the number?
$3x - 2 = 5x + 14; -8$

18. Nine more than five times a number is equal to 3 less than seven times the number. What is the number?
$5x + 9 = 7x - 3; 6$

19. If 2 is subtracted from a number and this difference is tripled, the result is 6 more than the number. Find the number. $3(x - 2) = x + 6; 6$

20. If 3 is added to a number and this sum is doubled, the result is 2 more than the number. Find the number.
$2(x + 3) = x + 2; -4$

21. The sum of three times a number and 7 more than the number is the same as the difference between −11 and twice the number. What is the number?
$3x + (x + 7) = -11 - 2x; -3$

22. If 4 is added to twice a number and this sum is multiplied by 2, the result is the same as if the number is multiplied by 3 and 4 is added to the product. What is the number? $2(2x + 4) = 3x + 4; -4$

GS *In Exercises 23 and 24, complete the six problem-solving steps to solve each problem.* ***See Example 2.***

23. Pennsylvania and New York were among the states with the most remaining drive-in movie screens in 2012. Pennsylvania had 3 more screens than New York, and there were 59 screens total in the two states. How many drive-in movie screens remained in each state? (*Source:* www.Drive-Ins.com)

Step 1 **Read.** What are we asked to find?
We are asked to find the number of drive-in movie screens in the two states.

Step 2 **Assign a variable.** Let $x =$ the number of screens in New York.
Then $x + 3 =$ the number of screens in Pennsylvania.

Step 3 **Write an equation.**
$x + x + 3 = 59$

Step 4 **Solve** the equation.
$x = 28$

Step 5 **State the answer.** New York had 28 screens. Pennsylvania had $28 + 3 = 31$ screens.

Step 6 **Check.** The number of screens in Pennsylvania was 3 more than the number of screens in New York. The total number of screens was $28 + 31 = 59$.

24. Two of the most watched episodes in television were the final episodes of *M*A*S*H* and *Cheers*. The total number of viewers for these two episodes was about 92 million, with 8 million more people watching the *M*A*S*H* episode than the *Cheers* episode. How many people watched each episode? (*Source:* Nielsen Media Research.)

Step 1 **Read.** What are we asked to find?
We are asked to find the number of people in millions who watched each episode.

Step 2 **Assign a variable.** Let $x =$ the number of people who watched the *Cheers* episode.
Then $x + 8 =$ the number of people in millions who watched the *M*A*S*H* episode.

Step 3 **Write an equation.**
$x + x + 8 = 92$

Step 4 **Solve** the equation.
$x = 42$

Step 5 **State the answer.** 42 million people watched the *Cheers* episode. $42 + 8 = 50$ million watched the *M*A*S*H** episode.

Step 6 **Check.** The number of people who watched the *M*A*S*H** episode was 8 million more than the number who watched the *Cheers* episode. The total number of viewers in millions was $42 + 50 = 92$.

Solve each problem. ***See Example 2.***

25. The total number of Democrats and Republicans in the U.S. House of Representatives during the 112th session (2011–2012) was 435. There were 49 more Republicans than Democrats. How many members of each party were there? (*Source: World Almanac and Book of Facts.*)

Democrats: 193; Republicans: 242

26. During the 112th session, the U.S. Senate had a total of 98 Democrats and Republicans. There were 4 more Democrats than Republicans. How many Democrats and Republicans were there in the Senate? (*Source: World Almanac and Book of Facts.*)

Democrats: 51; Republicans: 47

27. Bon Jovi and Roger Waters had the two top-grossing North American concert tours in 2010, together generating \$197.7 million in ticket sales. If Roger Waters took in \$18.7 million less than Bon Jovi, how much did each tour generate? (*Source:* Pollstar.)

Bon Jovi: \$108.2 million; Roger Waters: \$89.5 million

28. In the United States in 2010, Honda Accord sales were 30 thousand less than Toyota Camry sales, and 596 thousand of these two cars were sold. How many of each make of car were sold? (*Source:* www.wikipedia.org)

Camry: 313 thousand; Accord: 283 thousand

29. In the 2010–2011 NBA regular season, the Los Angeles Lakers won 7 more than twice as many games as they lost. The Lakers played 82 games. How many wins and losses did the team have? (*Source:* www.nba.com)

wins: 57; losses: 25

30. In the 2011 regular MLB season, the Detroit Tigers won 39 fewer than twice as many games as they lost. The Tigers played 162 regular season games. How many wins and losses did the team have? (*Source:* www.mlb.com)

wins: 95; losses: 67

31. A one-cup serving of orange juice contains 3 mg less than four times the amount of vitamin C as a one-cup serving of pineapple juice. Servings of the two juices contain a total of 122 mg of vitamin C. How many milligrams of vitamin C are in a serving of each type of juice? (*Source:* U.S. Agriculture Department.)

orange: 97 mg; pineapple: 25 mg

32. A one-cup serving of pineapple juice has 9 more than three times as many calories as a one-cup serving of tomato juice. Servings of the two juices contain a total of 173 calories. How many calories are in a serving of each type of juice? (*Source:* U.S. Agriculture Department.)

tomato: 41 calories; pineapple: 132 calories

Solve each problem. ***See Example 3.***

33. U.S. five-cent coins are made from a combination of nickel and copper. For every 1 lb of nickel, 3 lb of copper are used. How many pounds of copper would be needed to make 560 lb of five-cent coins? (*Source:* The United States Mint.)

420 lb

34. The recipe for a special whole-grain bread calls for 1 oz of rye flour for every 4 oz of whole-wheat flour. How many ounces of each kind of flour should be used to make a loaf of bread weighing 32 oz?

whole wheat: 25.6 oz; rye: 6.4 oz

35. A medication contains 9 mg of active ingredients for every 1 mg of inert ingredients. How much of each kind of ingredient would be contained in a single 250-mg caplet?

active: 225 mg; inert: 25 mg

36. A recipe for salad dressing uses 2 oz of olive oil for each 1 oz of red wine vinegar. If 42 oz of salad dressing are needed, how many ounces of each ingredient should be used?

olive oil: 28 oz; red wine vinegar: 14 oz

37. The value of a "Mint State-63" (uncirculated) 1950 Jefferson nickel minted at Denver is $\frac{4}{3}$ the value of a similar condition 1944 nickel minted at Philadelphia. Together, the value of the two coins is \$28.00. What is the value of each coin? (*Source:* Yeoman, R., *A Guide Book of United States Coins.*)

1950 Denver nickel: \$16.00; 1944 Philadelphia nickel: \$12.00

38. In one day, a store sold $\frac{8}{5}$ as many DVDs as Blu-ray discs. The total number of DVDs and Blu-ray discs sold that day was 273. How many DVDs were sold?

168 DVDs

39. The world's largest taco contained approximately 1 kg of onion for every 6.6 kg of grilled steak. The total weight of these two ingredients was 617.6 kg. To the nearest tenth of a kilogram, how many kilograms of each ingredient were used? (*Source: Guinness World Records.*)

onions: 81.3 kg; grilled steak: 536.3 kg

40. As of mid-2011, the combined population of China and India was estimated at 2.5 billion. If there were about 88% as many people living in India as China, what was the population of each country, to the nearest tenth of a billion? (*Source:* U.S. Census Bureau.)

China: 1.3 billion; India: 1.2 billion

Solve each problem. ***See Example 4.***

41. An office manager booked 55 airline tickets. He booked 7 more tickets on American Airlines than United Airlines. On Southwest Airlines, he booked 4 more than twice as many tickets as on United. How many tickets did he book on each airline?

American: 18; United: 11; Southwest: 26

42. A mathematics textbook editor spent 7.5 hr making telephone calls, writing e-mails, and attending meetings. She spent twice as much time attending meetings as making telephone calls and 0.5 hr longer writing e-mails than making telephone calls. How many hours did she spend on each task?

telephone calls: 1.75 hr; e-mails: 2.25 hr; meetings: 3.5 hr

43. Nagaraj Nanjappa has a party-length submarine sandwich 59 in. long. He wants to cut it into three pieces so that the middle piece is 5 in. longer than the shortest piece and the shortest piece is 9 in. shorter than the longest piece. How long should the three pieces be?

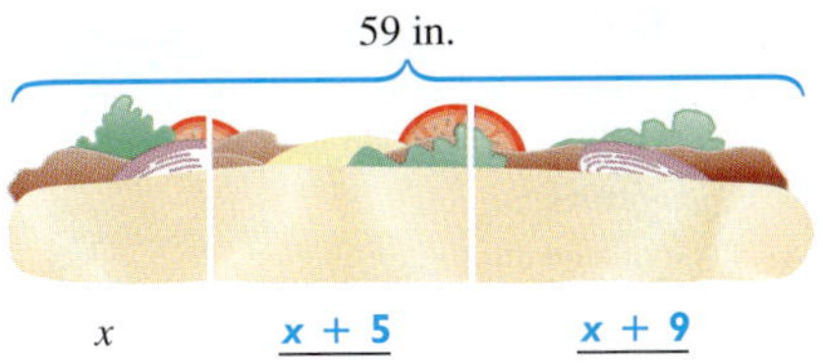

shortest piece: 15 in.; middle piece: 20 in.; longest piece: 24 in.

44. A three-foot-long deli sandwich must be split into three pieces so that the middle piece is twice as long as the shortest piece and the shortest piece is 8 in. shorter than the longest piece. How long should the three pieces be? (*Hint:* How many inches are in 3 ft?)

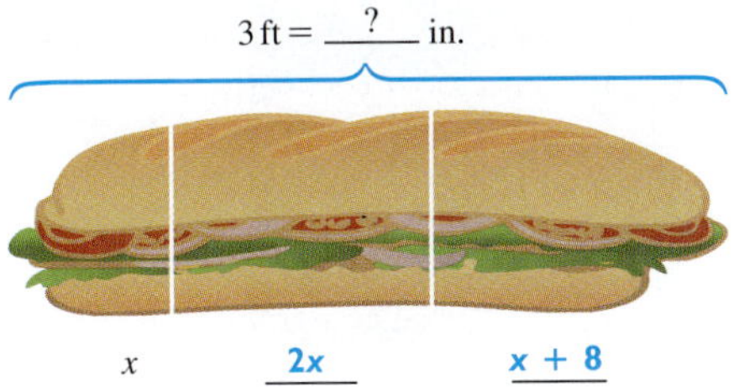

shortest piece: 7 in.; middle piece: 14 in.; longest piece: 15 in.

45. The United States earned 104 medals at the 2012 Summer Olympics. The number of silver medals earned was the same as the number of bronze medals. The number of gold medals was 17 more than the number of silver medals. How many of each kind of medal did the United States earn? (*Source:* www.london2012.com)

gold: 46; silver: 29; bronze: 29

46. China earned 88 medals at the 2012 Summer Olympics. The number of gold medals earned was 15 more than the number of bronze medals. The number of silver medals earned was 4 more than the number of bronze medals. How many of each kind of medal did China earn? (*Source:* www.london2012.com)

gold: 38; silver: 27; bronze: 23

47. Venus is 31.2 million mi farther from the sun than Mercury, while Earth is 57 million mi farther from the sun than Mercury. If the total of the distances from these three planets to the sun is 196.2 million mi, how far away from the sun is Mercury? (All distances given here are mean (*average*) distances.) (*Source: The New York Times Almanac.*)

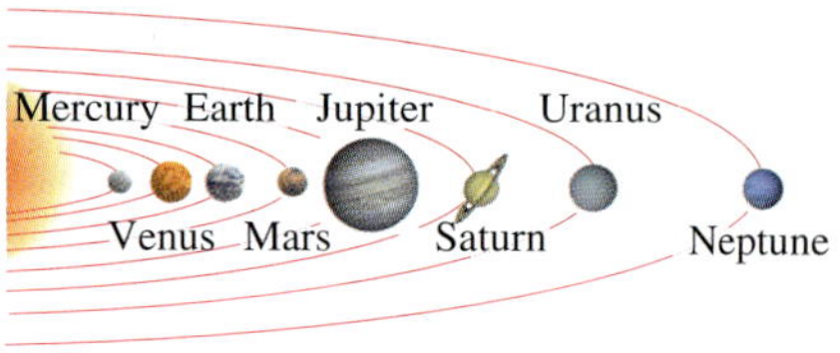

36 million mi

48. Saturn, Jupiter, and Uranus have a total of 153 known satellites (moons). Jupiter has 2 more satellites than Saturn, and Uranus has 35 fewer satellites than Saturn. How many known satellites does Uranus have? (*Source:* http://solarsystem.nasa.gov)

27

49. The sum of the measures of the angles of any triangle is 180°. In triangle *ABC*, angles *A* and *B* have the same measure, while the measure of angle *C* is 60° greater than each of *A* and *B*. What are the measures of the three angles?

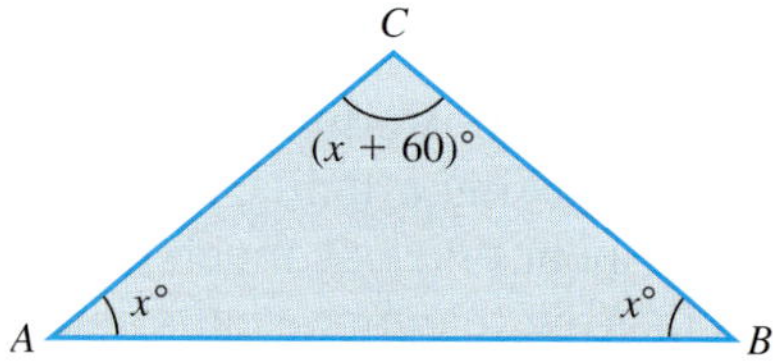

A and *B*: 40°; *C*: 100°

50. In triangle *ABC*, the measure of angle *A* is 141° more than the measure of angle *B*. The measure of angle *B* is the same as the measure of angle *C*. Find the measure of each angle. (*Hint:* See **Exercise 49.**)

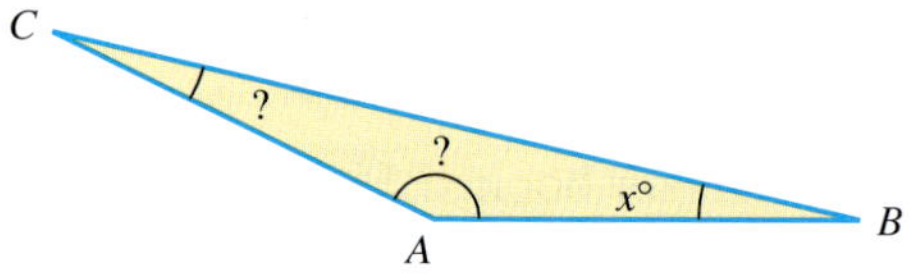

A: 154°; *B* and *C*: 13°

Solve each problem. ***See Examples 5 and 6.***

51. The numbers on two consecutively numbered gym lockers have a sum of 137. What are the locker numbers? 68, 69

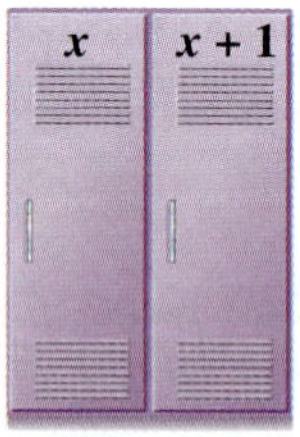

52. The sum of two consecutive check numbers is 357. Find the numbers. 178, 179

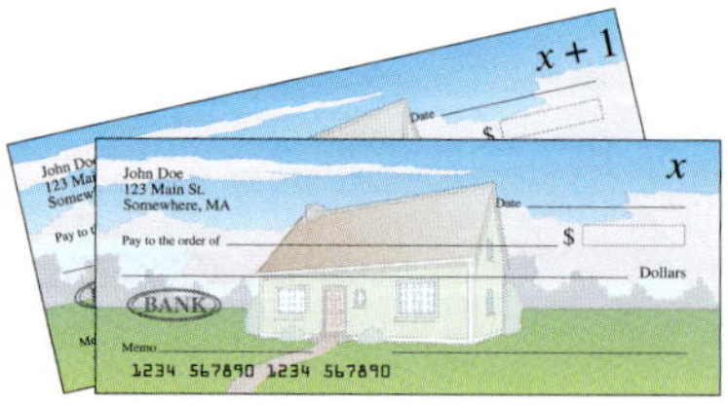

53. Two pages that are back-to-back in this book have 293 as the sum of their page numbers. What are the page numbers? 146, 147

54. Two hotel rooms have room numbers that are consecutive integers. The sum of the numbers is 515. What are the two room numbers? 257, 258

55. Find two consecutive even integers such that the lesser added to three times the greater gives a sum of 46. 10, 12

56. Find two consecutive even integers such that six times the lesser added to the greater gives a sum of 86. 12, 14

57. Find two consecutive odd integers such that 59 more than the lesser is four times the greater. 17, 19

58. Find two consecutive odd integers such that twice the greater is 17 more than the lesser. 13, 15

59. When the lesser of two consecutive integers is added to three times the greater, the result is 43. Find the integers. 10, 11

60. If five times the lesser of two consecutive integers is added to three times the greater, the result is 59. Find the integers. 7, 8

61. If the sum of three consecutive even integers is 60, what is the first of the three even integers? (*Hint:* If x and $x + 2$ represent the first two consecutive even integers, how would we represent the third consecutive even integer?) 18

62. If the sum of three consecutive odd integers is 69, what is the third of the three odd integers? (*Hint:* If x and $x + 2$ represent the first two consecutive odd integers, how would we represent the third consecutive odd integer?) 25

Solve each problem. ***See Examples 7 and 8.***

63. Find the measure of an angle whose complement is four times its measure. 18°

64. Find the measure of an angle whose complement is five times its measure. 15°

65. Find the measure of an angle whose supplement is eight times its measure. 20°

66. Find the measure of an angle whose supplement is three times its measure. 45°

67. Find the measure of an angle whose supplement measures 39° more than twice its complement. 39°

68. Find the measure of an angle whose supplement measures 38° less than three times its complement. 26°

69. Find the measure of an angle such that the difference between the measures of its supplement and three times its complement is 10°. 50°

70. Find the measure of an angle such that the sum of the measures of its complement and its supplement is 160°. 55°

2.5 Formulas and Additional Applications from Geometry

OBJECTIVES

1. Solve a formula for one variable, given the values of the other variables.
2. Use a formula to solve an applied problem.
3. Solve problems involving vertical angles and straight angles.
4. Solve a formula for a specified variable.

A **formula** is an equation in which variables are used to describe a relationship. For example, formulas exist for finding perimeters and areas of geometric figures, for calculating money earned on bank savings, and for converting among measurements.

$$P = 4s, \quad A = \pi r^2, \quad I = prt, \quad F = \frac{9}{5}C + 32 \qquad \text{Formulas}$$

Many of the formulas used in this book are given inside the back cover.

OBJECTIVE 1 Solve a formula for one variable, given the values of the other variables. In **Example 1,** we use the idea of *area.* The **area** of a plane (two-dimensional) geometric figure is a measure of the surface covered by the figure. Area is measured in square units.

EXAMPLE 1 Using Formulas to Evaluate Variables

Find the value of the remaining variable in each formula.

(a) $A = LW$; $A = 64, L = 10$

This formula gives the area of a rectangle. See **Figure 7.** Substitute the given values for A and L into the formula.

$A = LW$ (Solve for W.)

$64 = 10W$ Let $A = 64$ and $L = 10$.

$\frac{64}{10} = \frac{10W}{10}$ Divide by 10.

$6.4 = W$

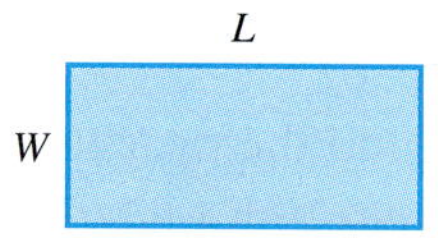

Rectangle
$A = LW$

Figure 7

The width is **6.4**. Since $10(6.4) = 64$, the given area, the answer checks.

(b) $A = \frac{1}{2}h(b + B)$; $A = 210, B = 27, h = 10$

This formula gives the area of a trapezoid. See **Figure 8.**

$A = \frac{1}{2}h(b + B)$ (Solve for b.)

$210 = \frac{1}{2}(10)(b + 27)$ Let $A = 210, h = 10, B = 27$.

$210 = 5(b + 27)$ Multiply $\frac{1}{2}(10)$.

$210 = 5b + 135$ Distributive property

$210 - 135 = 5b + 135 - 135$ Subtract 135.

$75 = 5b$ Combine like terms.

$\frac{75}{5} = \frac{5b}{5}$ Divide by 5.

$15 = b$

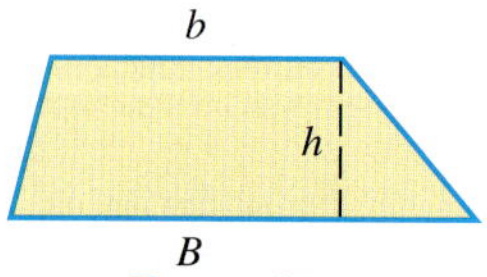

Trapezoid
$A = \frac{1}{2}h(b + B)$

Figure 8

The length of the shorter parallel side, b, is **15**. Since $\frac{1}{2}(10)(15 + 27) = 210$, the given area, the answer checks.

◄ **Work Problem 1 at the Side.**

---

1 Find the value of the remaining variable in each formula.

(a) $A = bh$
(area of a parallelogram);
$A = 96, h = 8$

Given values for A and h, solve for the value of ____.

GS **(b)** $I = prt$ (simple interest);
$I = 246, r = 0.06$ (that is, 6%), $t = 2$

$I = prt$

____ $= p($____$)($____$)$

$246 =$ ____ p

$\frac{246}{\text{\_\_\_\_}} = \frac{0.12p}{0.12}$

____ $= p$

(c) $P = 2L + 2W$
(perimeter of a rectangle);
$P = 126, W = 25$

Answers

1. (a) b; $b = 12$
(b) 246; 0.06; 2; 0.12; 0.12; 2050
(c) $L = 38$

OBJECTIVE 2 **Use a formula to solve an applied problem.** **Examples 2 and 3** use the idea of *perimeter*. The **perimeter** of a plane (two-dimensional) geometric figure is the distance around the figure. For a polygon (e.g., a rectangle, square, or triangle), it is the sum of the lengths of its sides.

EXAMPLE 2 Finding the Dimensions of a Rectangular Yard

Kari Heen's backyard is in the shape of a rectangle. The length is 5 m less than twice the width, and the perimeter is 80 m. Find the dimensions of the yard.

Step 1 **Read** the problem. We must find the dimensions of the yard.

Step 2 **Assign a variable.** Let $W =$ the width of the lot, in meters. Since the length is 5 m less than twice the width, the length is given by $L = 2W - 5$. See **Figure 9.**

Step 3 **Write an equation.** Use the formula for the perimeter of a rectangle.

$$P = 2L + 2W$$

Perimeter = 2 · Length + 2 · Width

$$80 = 2(2W - 5) + 2W \qquad \text{Substitute } 2W - 5 \text{ for length } L.$$

Step 4 **Solve.**

$$80 = 4W - 10 + 2W \qquad \text{Distributive property}$$

$$80 = 6W - 10 \qquad \text{Combine like terms.}$$

$$80 + 10 = 6W - 10 + 10 \qquad \text{Add 10.}$$

$$90 = 6W \qquad \text{Combine like terms.}$$

$$\frac{90}{6} = \frac{6W}{6} \qquad \text{Divide by 6.}$$

$$15 = W$$

Step 5 **State the answer.** The width is 15 m. The length is $2(15) - 5 = 25$ m.

Step 6 **Check.** If the width of the yard is 15 m and the length is 25 m, the perimeter is $2(25) + 2(15) = 50 + 30 = 80$ m, as required.

Work Problem 2 at the Side. ▶

EXAMPLE 3 Finding the Dimensions of a Triangle

The longest side of a triangle is 3 ft longer than the shortest side. The medium side is 1 ft longer than the shortest side. If the perimeter of the triangle is 16 ft, what are the lengths of the three sides?

Step 1 **Read** the problem. We are given the perimeter of a triangle and must find the lengths of the three sides.

Step 2 **Assign a variable.** The shortest side is mentioned in each pair of comparisons.

Let $s =$ the length of the shortest side, in feet.

Then $s + 1 =$ the length of the medium side, in feet,

and $s + 3 =$ the length of the longest side, in feet.

It is a good idea to draw a sketch. See **Figure 10.**

Continued on Next Page

Teaching Tip

Mention to students that many of the formulas used in this section and in the exercises can be found inside the back cover of this book.

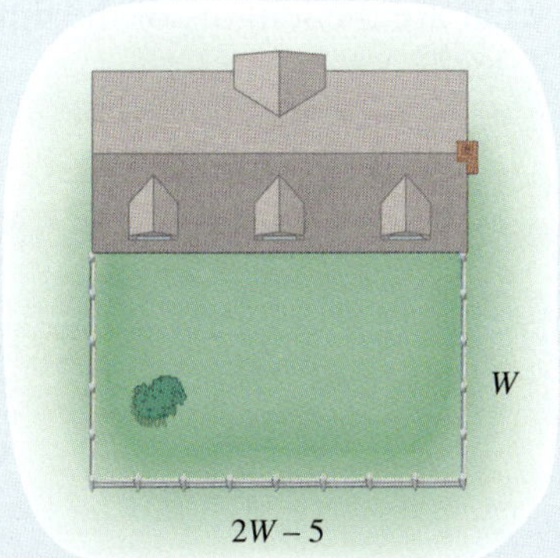

Figure 9

2 Solve the problem.

A farmer has 800 m of fencing material to enclose a rectangular field. The width of the field is 175 m. Find the length of the field.

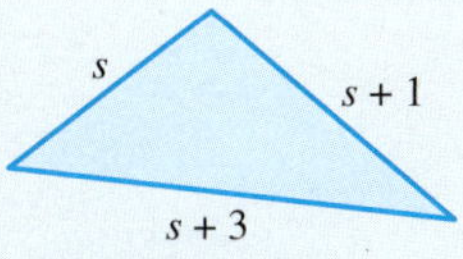

Figure 10

Answer

2. 225 m

3 Solve the problem.

The longest side of a triangle is 1 in. longer than the medium side. The medium side is 5 in. longer than the shortest side. If the perimeter is 32 in., what are the lengths of the three sides?

Step 3 **Write an equation.** Use the formula for the perimeter of a triangle.

$P = a + b + c$ Perimeter of a triangle

$16 = s + (s + 1) + (s + 3)$ Substitute.

Step 4 **Solve.** $16 = 3s + 4$ Combine like terms.

$12 = 3s$ Subtract 4.

$4 = s$ Divide by 3.

Step 5 **State the answer.** Since s represents the length of the shortest side, its measure is 4 ft.

$s + 1 = 4 + 1 = 5$ ft Length of the medium side

$s + 3 = 4 + 3 = 7$ ft Length of the longest side

Step 6 **Check.** The medium side, 5 ft, is 1 ft longer than the shortest side, and the longest side, 7 ft, is 3 ft longer than the shortest side. Furthermore, the perimeter is

$4 + 5 + 7 = 16$ ft, as required.

◀ **Work Problem 3 at the Side.**

EXAMPLE 4 Finding the Height of a Triangular Sail

The area of a triangular sail of a sailboat is 126 ft^2. (Recall that ft^2 means "square feet.") The base of the sail is 12 ft. Find the height of the sail.

Step 1 **Read** the problem. We must find the height of the triangular sail.

Step 2 **Assign a variable.** Let h = the height of the sail, in feet. See **Figure 11.**

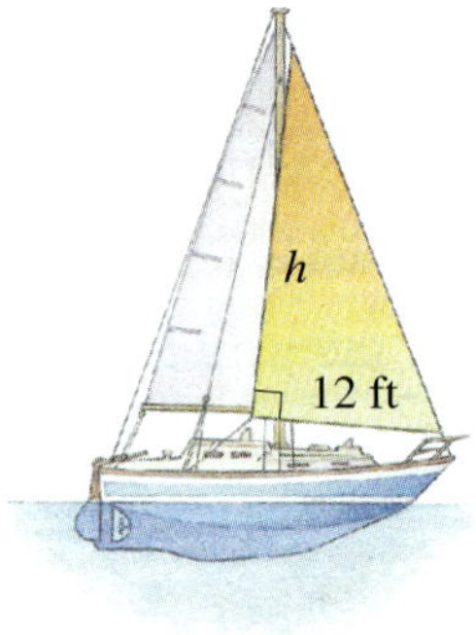

Figure 11

Step 3 **Write an equation.** Use the formula for the area of a triangle.

$A = \frac{1}{2}bh$ A is the area, b is the base, and h is the height.

$126 = \frac{1}{2}(12)h$ Substitute $A = 126$, $b = 12$.

Step 4 **Solve.** $126 = 6h$ Multiply.

$21 = h$ Divide by 6.

Step 5 **State the answer.** The height of the sail is 21 ft.

Step 6 **Check** to see that the values $A = 126$, $b = 12$, and $h = 21$ satisfy the formula for the area of a triangle.

◀ **Work Problem 4 at the Side.**

4 Solve the problem.

The area of a triangle is 120 m^2. The height is 24 m. Find the length of the base of the triangle.

Answers

3. 7 in.; 12 in.; 13 in.
4. 10 m

OBJECTIVE 3 Solve problems involving vertical angles and straight angles. **Figure 12** shows two intersecting lines forming angles that are numbered ①, ②, ③, and ④. Angles ① and ③ lie "opposite" each other. They are **vertical angles.** Another pair of vertical angles is ② and ④. ***Vertical angles have equal measures.***

Now look at angles ① and ②. When their measures are added, we get 180°, the measure of a **straight angle.** There are three other angle pairs that form straight angles: ② and ③, ③ and ④, and ① and ④.

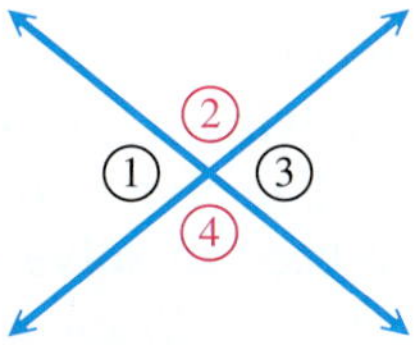

Figure 12

EXAMPLE 5 Finding Angle Measures

Refer to the appropriate figure in each part.

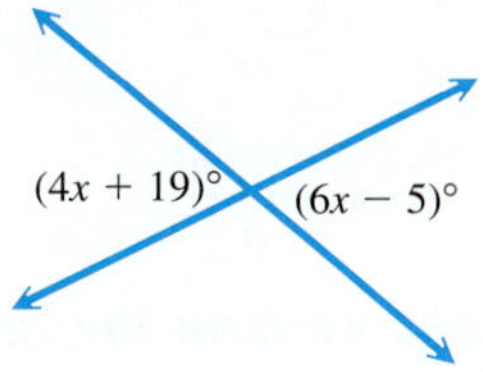

Figure 13

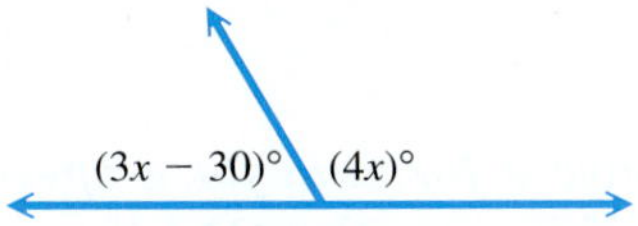

Figure 14

(a) Find the measure of each marked angle in **Figure 13.**
Since the marked angles are vertical angles, they have equal measures.

$$4x + 19 = 6x - 5 \quad \text{Set } 4x + 19 \text{ equal to } 6x - 5.$$
$$19 = 2x - 5 \quad \text{Subtract } 4x.$$
$$24 = 2x \quad \text{Add } 5.$$
$$12 = x \quad \text{Divide by } 2.$$

This is *not* the answer.

Since $x = 12$, one angle has measure $4(12) + 19 = 67$ degrees. The other has the same measure, since $6(12) - 5 = 67$ as well. Each angle measures 67°.

(b) Find the measure of each marked angle in **Figure 14.**
The measures of the marked angles must add to 180° because together they form a straight angle. (They are also *supplements* of each other.)

$$(3x - 30) + 4x = 180 \quad \text{Supplementary angles sum to } 180°.$$
$$7x - 30 = 180 \quad \text{Combine like terms.}$$
$$7x = 210 \quad \text{Add } 30.$$
$$x = 30 \quad \text{Divide by } 7.$$

Don't stop here!

Replace x with 30 in the measure of each marked angle.

$$3x - 30 = 3(30) - 30 = 90 - 30 = 60$$
$$4x = 4(30) = 120$$

The measures of the angles add to 180°, as required.

The two angle measures are 60° and 120°.

CAUTION

In **Example 5,** the answer is *not* the value of x. ***Remember to substitute the value of the variable into the expression given for each angle.***

Work Problem 5 at the Side. ▶

---

5 Find the measure of each marked angle.

(a)

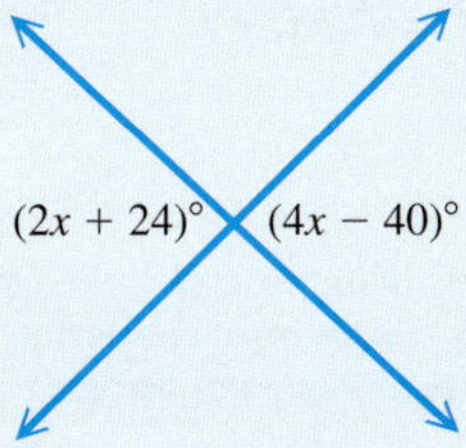

(b)

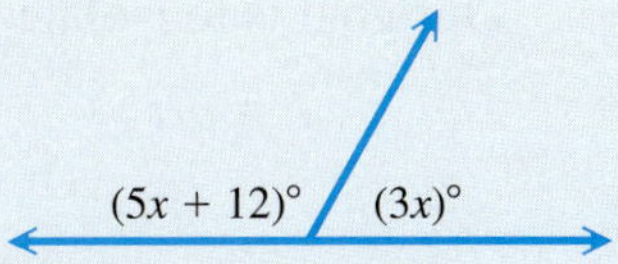

GS **(c)**

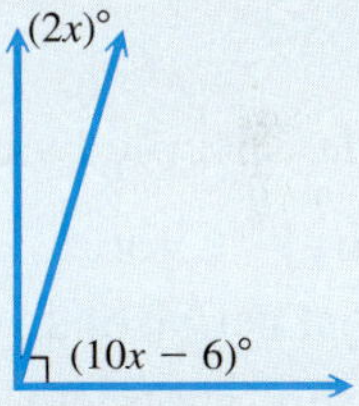

Because of the ⌐ symbol, the marked angles are ________ angles and have a sum of ____. The equation to use is

________ = 90.

Solve this equation to find that x = ____. Then substitute to find the measure of each angle.

$$2x = 2(\_\_\_\_)$$
$$= \_\_\_\_$$

$$10x - 6 = 10(\_\_\_\_) - 6$$
$$= \_\_\_\_$$

The two angles measure ________.

Answers

5. (a) Both measure 88°. **(b)** 117° and 63° **(c)** complementary; 90°; $2x + (10x - 6)$; 8; 8; 16; 8; 74; 16° and 74°

6 Solve each formula for the specified variable.

(a) $W = Fd$ for F

Teaching Tip

Stress that solving for a specified variable uses exactly the same steps as solving any of the previous linear equations.

GS **(b)** $I = prt$ for t

Our goal is to isolate ____.

$$I = prt$$

$$\frac{I}{\_\_\_\_} = \frac{prt}{\_\_\_\_}$$

$$\_\_\_\_ = t$$

7 Solve for the specified variable.

GS **(a)** $Ax + By = C$ for A

Our goal is to isolate ____. To do this, subtract ____ from each side. Then ____ each side by ____.

Show these steps and write the formula solved for A.

(b) $Ax + By = C$ for B

Teaching Tip

We use color for the specified variable. Encourage students to circle or underline the variable they are solving for.

Answers

6. **(a)** $F = \frac{W}{d}$ **(b)** t; pr; pr; $\frac{I}{pr}$

7. **(a)** A; By; divide; x; $A = \frac{C - By}{x}$

(b) $B = \frac{C - Ax}{y}$

OBJECTIVE 4 **Solve a formula for a specified variable.** Sometimes we want to rewrite a formula in terms of a *different* variable in the formula. For example, consider $A = LW$, the formula for the area of a rectangle.

How can we rewrite $A = LW$ in terms of W?

The process whereby we do this is called **solving for a specified variable,** or **solving a literal equation.**

To solve a formula for a specified variable, we use the *same* steps that we used to solve an equation with just one variable. Consider the parallel reasoning to solve each of the following for x.

$3x + 4 = 13$		$ax + b = c$	
$3x + 4 - 4 = 13 - 4$	Subtract 4.	$ax + b - b = c - b$	Subtract b.
$3x = 9$		$ax = c - b$	
$\frac{3x}{3} = \frac{9}{3}$	Divide by 3.	$\frac{ax}{a} = \frac{c - b}{a}$	Divide by a.
$x = 3$		$x = \frac{c - b}{a}$	

When we solve a formula for a specified variable, we treat the specified variable as if it were the ONLY variable in the equation, and treat the other variables as if they were numbers.

EXAMPLE 6 **Solving for a Specified Variable**

Solve $A = LW$ for W.

Think of undoing what has been done to W. Since W is multiplied by L, undo the multiplication by dividing each side of $A = LW$ by L.

$$A = LW$$ (Our goal is to isolate W.)

$$\frac{A}{L} = \frac{LW}{L} \quad \text{Divide by } L.$$

$$\frac{A}{L} = W, \quad \text{or} \quad W = \frac{A}{L} \qquad \frac{LW}{L} = \frac{L}{L} \cdot W = 1 \cdot W = W$$

◀ Work Problem **6** at the Side.

EXAMPLE 7 **Solving for a Specified Variable**

Solve $P = 2L + 2W$ for L.

$$P = 2L + 2W$$ (Our goal is to isolate L.)

$$P - 2W = 2L + 2W - 2W \quad \text{Subtract } 2W.$$

$$P - 2W = 2L \quad \text{Combine like terms.}$$

$$\frac{P - 2W}{2} = \frac{2L}{2} \quad \text{Divide by 2.}$$

$$\frac{P - 2W}{2} = L, \quad \text{or} \quad L = \frac{P - 2W}{2} \qquad \frac{2L}{2} = \frac{2}{2} \cdot L = 1 \cdot L = L$$

◀ Work Problem **7** at the Side.

EXAMPLE 8 Solving for a Specified Variable

Solve $F = \frac{9}{5}C + 32$ for C.

Our goal is to isolate C.

$$F = \frac{9}{5}C + 32$$ This is the formula for converting temperatures from Celsius to Fahrenheit.

$$F - 32 = \frac{9}{5}C + 32 - 32$$ Subtract 32.

$$F - 32 = \frac{9}{5}C$$

Be sure to use parentheses.

$$\frac{5}{9}(F - 32) = \frac{5}{9} \cdot \frac{9}{5}C$$ Multiply by $\frac{5}{9}$.

$$\frac{5}{9}(F - 32) = C$$ $\frac{5}{9} \cdot \frac{9}{5}C = 1C = C$

$$C = \frac{5}{9}(F - 32)$$ This is the formula for converting temperatures from Fahrenheit to Celsius.

Work Problem 8 at the Side. ▶

EXAMPLE 9 Solving for a Specified Variable

Solve each equation for y.

(a) $2x - y = 7$ — Our goal is to isolate y.

$$2x - y - 2x = 7 - 2x$$ Subtract $2x$.

$$-y = 7 - 2x$$ Combine like terms.

$$-1(-y) = -1(7 - 2x)$$ Multiply by -1.

$$y = -7 + 2x$$ Multiply; distributive property

$$y = 2x - 7$$ $-a + b = b - a$

We could have added y and subtracted 7 from each side of the equation to isolate y on the right, giving $2x - 7 = y$, a different form of the same result. There is often more than one way to solve for a specified variable.

(b) $-3x + 2y = 6$

$$-3x + 2y + 3x = 6 + 3x$$ Add $3x$.

$$2y = 3x + 6$$ Combine like terms; commutative property

$$\frac{2y}{2} = \frac{3x + 6}{2}$$ Divide by 2.

Be careful here.

$$y = \frac{3x}{2} + \frac{6}{2}$$ $\frac{a+b}{c} = \frac{a}{c} + \frac{b}{c}$

$\frac{3x}{2} = \frac{3}{2} \cdot \frac{x}{1} = \frac{3}{2}x$

$$y = \frac{3}{2}x + 3$$ Simplify.

Although we could have given our answer as $y = \frac{3x + 6}{2}$, in preparation for our work in **Chapter 3,** we simplified further as shown in the last two lines of the solution.

Work Problem 9 at the Side. ▶

8 Solve each formula for the specified variable.

(a) $A = p + prt$ for t

(b) $x = u + zs$ for z

9 Solve each equation for y.

(a) $5x + y = 3$

(b) $x - 2y = 8$

Teaching Tip

Mention to students that the concepts of Example 9 will be used extensively in Chapter 3.

Answers

8. (a) $t = \frac{A - p}{pr}$ (b) $z = \frac{x - u}{s}$

9. (a) $y = -5x + 3$ (b) $y = \frac{1}{2}x - 4$

2.5 Exercises

FOR EXTRA HELP

Download the MyDashBoard App

MyMathLab®

CONCEPT CHECK *Give a one-sentence definition of each term.*

1. Perimeter of a plane geometric figure

The perimeter of a plane geometric figure is the distance around the figure.

2. Area of a plane geometric figure

The area of a plane geometric figure is the measure of the surface covered or enclosed by the figure.

CONCEPT CHECK *Decide whether perimeter or area would be used to solve a problem concerning the measure of the quantity.*

3. Sod for a lawn

area

4. Carpeting for a bedroom

area

5. Baseboards for a living room

perimeter

6. Fencing for a yard

perimeter

7. Fertilizer for a garden

area

8. Tile for a bathroom

area

9. Determining the cost of planting rye grass in a lawn for the winter

area

10. Determining the cost of replacing a linoleum floor with a wood floor

area

In the following exercises a formula is given, along with the values of all but one of the variables in the formula. Find the value of the variable that is not given. In Exercises 21–24, use 3.14 as an approximation for π (pi). **See Example 1.**

11. $P = 2L + 2W$ (perimeter of a rectangle); $L = 8$, $W = 5$

$P = 26$

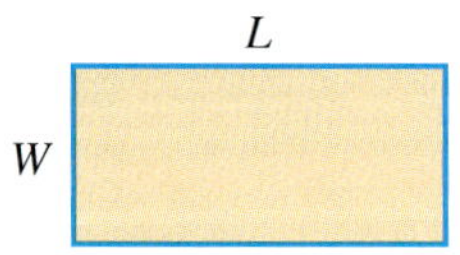

12. $P = 2L + 2W$; $L = 6$, $W = 4$

$P = 20$

13. $A = \frac{1}{2}bh$ (area of a triangle); $b = 8$, $h = 16$

$A = 64$

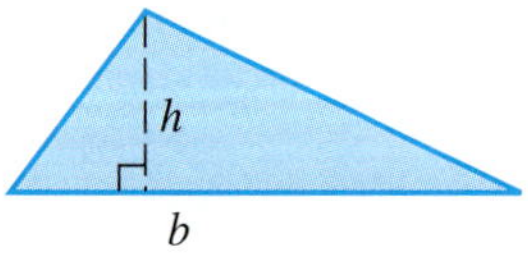

14. $A = \frac{1}{2}bh$; $b = 10$, $h = 14$

$A = 70$

15. $P = a + b + c$ (perimeter of a triangle); $P = 12$, $a = 3$, $c = 5$

$b = 4$

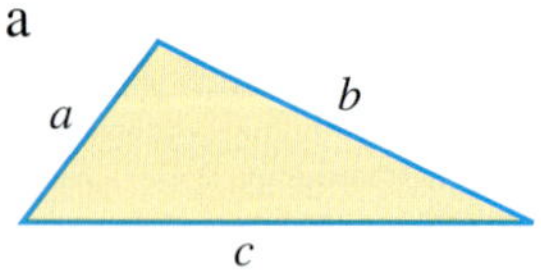

16. $P = a + b + c$; $P = 15$, $a = 3$, $b = 7$

$c = 5$

17. $d = rt$ (distance formula); $d = 252$, $r = 45$

$t = 5.6$

18. $d = rt$; $d = 100$, $t = 2.5$

$r = 40$

19. $A = \frac{1}{2}h(b + B)$ (area of a trapezoid); $A = 91$, $b = 12$, $B = 14$

$h = 7$

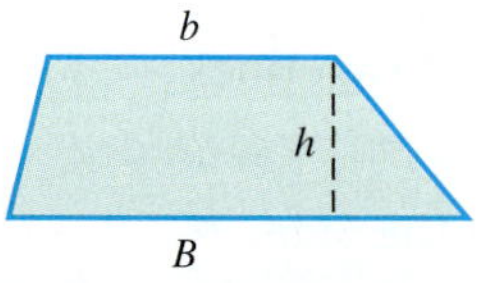

20. $A = \frac{1}{2}h(b + B)$; $A = 75$, $b = 19$, $B = 31$

$h = 3$

21. $C = 2\pi r$ (circumference of a circle); $C = 16.328$

$r = 2.6$

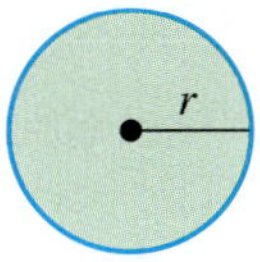

22. $C = 2\pi r$; $C = 8.164$

$r = 1.3$

23. $A = \pi r^2$ (area of a circle); $r = 4$

$A = 50.24$

24. $A = \pi r^2$; $r = 12$

$A = 452.16$

The ***volume*** *of a three-dimensional object is a measure of the space occupied by the object. For example, we would need to know the volume of a gasoline tank in order to know how many gallons of gasoline it would take to completely fill the tank. Volume is measured in cubic units.*

In Exercises 25–30, a formula for the volume (V) of a three-dimensional object is given, along with values for the other variables. Evaluate V. (Use 3.14 as an approximation for π.) ***See Example 1.***

25. $V = LWH$ (volume of a rectangular box); $L = 10$, $W = 5$, $H = 3$

$V = 150$

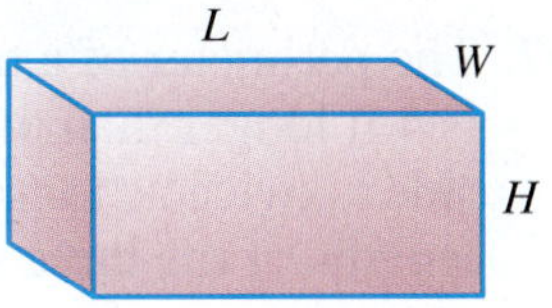

26. $V = LWH$; $L = 12$, $W = 8$, $H = 4$

$V = 384$

27. $V = \frac{1}{3}Bh$ (volume of a pyramid); $B = 12$, $h = 13$

$V = 52$

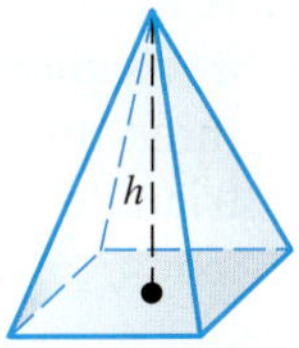

28. $V = \frac{1}{3}Bh$; $B = 36$, $h = 4$

$V = 48$

29. $V = \frac{4}{3}\pi r^3$ (volume of a sphere); $r = 12$

$V = 7234.56$

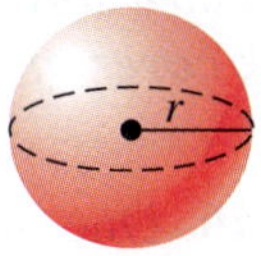

30. $V = \frac{4}{3}\pi r^3$; $r = 6$

$V = 904.32$

Simple interest I in dollars is calculated using the following formula.

$$I = prt \quad \text{Simple interest formula}$$

Here, p represents the principal, or amount, in dollars that is invested or borrowed, r represents the annual interest rate, expressed as a percent, and t represents time, in years.

In Exercises 31–36, find the value of the remaining variable in the simple interest formula. ***See Example 1.*** *(Hint: Write percents as decimals. See* ***Section R.2*** *to review how to do this if necessary.)*

31. $p = \$7500$, $r = 4\%$, $t = 2$ yr

$I = \$600$

32. $p = \$3600$, $r = 3\%$, $t = 4$ yr

$I = \$432$

33. $I = \$33$, $r = 2\%$, $t = 3$ yr

$p = \$550$

34. $I = \$270$, $r = 5\%$, $t = 6$ yr

$p = \$900$

35. $I = \$180$, $p = \$4800$, $r = 2.5\%$

(*Hint:* 2.5% written as a decimal is 0.025.)

$t = 1.5$ yr

36. $I = \$162$, $p = \$2400$, $r = 1.5\%$

(*Hint:* 1.5% written as a decimal is 0.015.)

$t = 4.5$ yr

Solve each perimeter problem. ***See Examples 2 and 3.***

37. The length of a rectangle is 9 in. more than the width. The perimeter is 54 in. Find the length and the width of the rectangle.

length: 18 in.; width: 9 in.

38. The width of a rectangle is 3 ft less than the length. The perimeter is 62 ft. Find the length and the width of the rectangle.

length: 17 ft; width: 14 ft

39. The perimeter of a rectangle is 36 m. The length is 2 m more than three times the width. Find the length and the width of the rectangle.

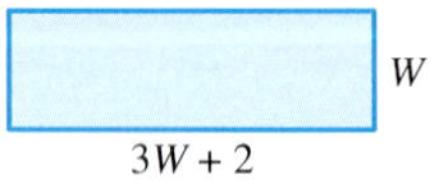

length: 14 m; width: 4 m

40. The perimeter of a rectangle is 36 yd. The width is 18 yd less than twice the length. Find the length and the width of the rectangle.

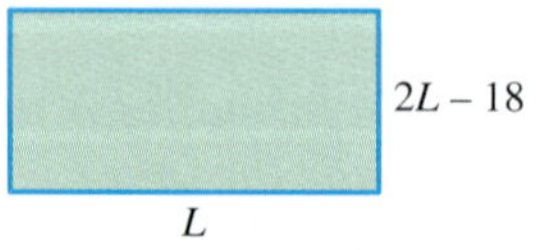

length: 12 yd; width: 6 yd

41. The longest side of a triangle is 3 in. longer than the shortest side. The medium side is 2 in. longer than the shortest side. If the perimeter of the triangle is 20 in., what are the lengths of the three sides?

shortest: 5 in.; medium: 7 in.; longest: 8 in.

42. The perimeter of a triangle is 28 ft. The medium side is 4 ft longer than the shortest side, while the longest side is twice as long as the shortest side. What are the lengths of the three sides?

shortest: 6 ft; medium: 10 ft; longest: 12 ft

43. Two sides of a triangle have the same length. The third side measures 4 m less than twice that length. The perimeter of the triangle is 24 m. Find the lengths of the three sides.

two equal sides: 7 m; third side: 10 m

44. A triangle is such that its medium side is twice as long as its shortest side and its longest side is 7 yd less than three times its shortest side. The perimeter of the triangle is 47 yd. What are the lengths of the three sides?

shortest 9 yd; medium: 18 yd; longest: 20 yd

Use a formula to write an equation for each application, and then solve. (Use 3.14 *as an approximation for* π*.)* ***Formulas are found inside the back cover of this book. See Examples 2–4.***

45. One of the largest fashion catalogues in the world was published in Hamburg, Germany. Each of the 212 pages in the catalogue measured 1.2 m by 1.5 m. What was the perimeter of a page? What was the area? (*Source: Guinness World Records.*)

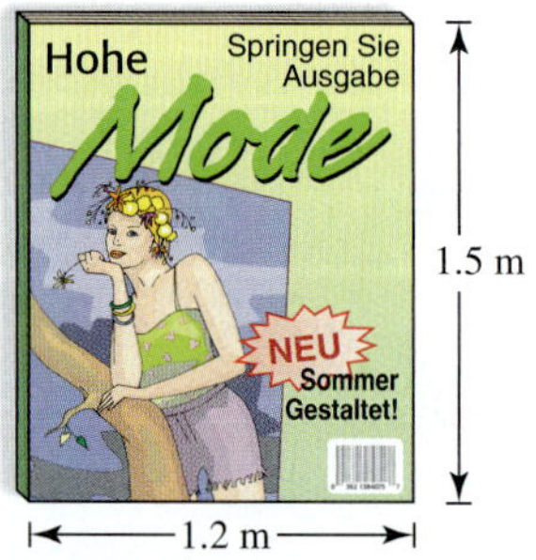

perimeter: 5.4 m; area: 1.8 m^2

46. One of the world's largest mandalas (sand paintings) measures 12.24 m by 12.24 m. What is the perimeter of the sand painting? To the nearest hundredth of a square meter, what is the area? (*Source: Guinness World Records.*)

perimeter: 48.96 m; area: 149.82 m^2

47. The area of a triangular road sign is 70 ft^2. If the base of the sign measures 14 ft, what is the height of the sign?

10 ft

48. The area of a triangular advertising banner is 96 ft^2. If the height of the banner measures 12 ft, find the measure of the base.

16 ft

49. A prehistoric ceremonial site dating to about 3000 B.C. was discovered at Stanton Drew in southwestern England. The site, which is larger than Stonehenge, is a nearly perfect circle, consisting of nine concentric rings that probably held upright wooden posts. Around this timber temple is a wide, encircling ditch enclosing an area with a diameter of 443 ft. Find this enclosed area to the nearest thousand square feet. (*Source: Archaeology.*)

about 154,000 ft^2

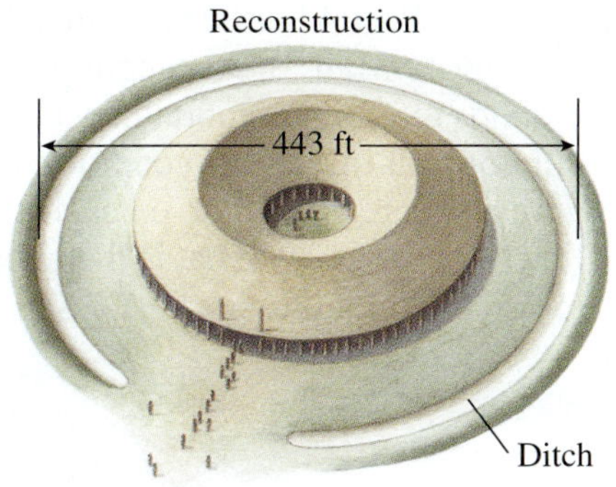

50. The Rogers Centre in Toronto, Canada, is the first stadium with a hard-shell, retractable roof. The steel dome is 630 ft in diameter. To the nearest foot, what is the circumference of this dome? (*Source:* www.ballparks.com)

about 1978 ft

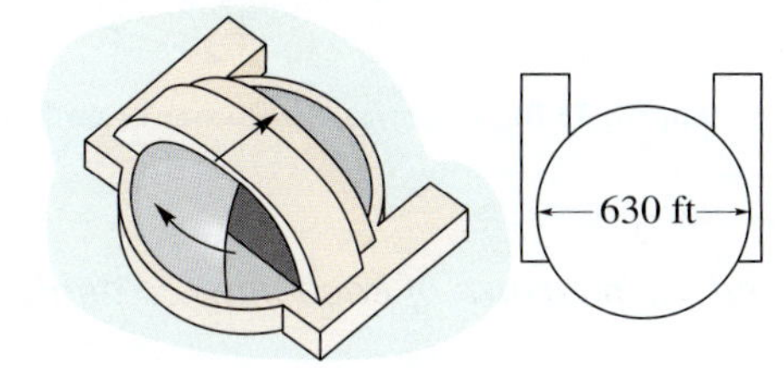

51. One of the largest drums ever constructed was made from Japanese cedar and cowhide, with radius 7.87 ft. What was the area of the circular face of the drum? What was the circumference of the drum? Round your answers to the nearest hundredth. (*Source: Guinness World Records.*)

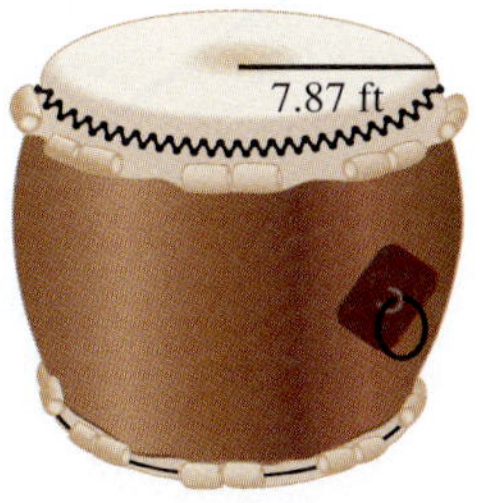

194.48 ft^2; 49.42 ft

52. A drum played at the Royal Festival Hall in London had radius 6.5 ft. What was the area of the circular face of the drum? What was the circumference of the drum? (*Source: Guinness World Records.*)

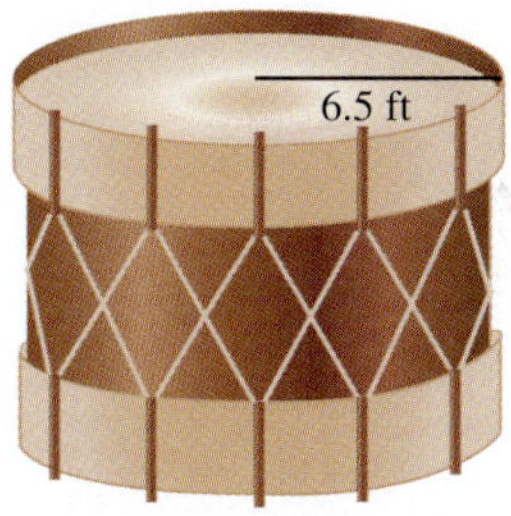

132.665 ft^2; 40.82 ft

53. The survey plat depicted here shows two lots that form a trapezoid. The measures of the parallel sides are 115.80 ft and 171.00 ft. The height of the trapezoid is 165.97 ft. Find the combined area of the two lots. Round your answer to the nearest hundredth of a square foot.

23,800.10 ft^2

54. Lot A in the figure is in the shape of a trapezoid. The parallel sides measure 26.84 ft and 82.05 ft. The height of the trapezoid is 165.97 ft. Find the area of Lot A. Round your answer to the nearest hundredth of a square foot.

9036.24 ft^2

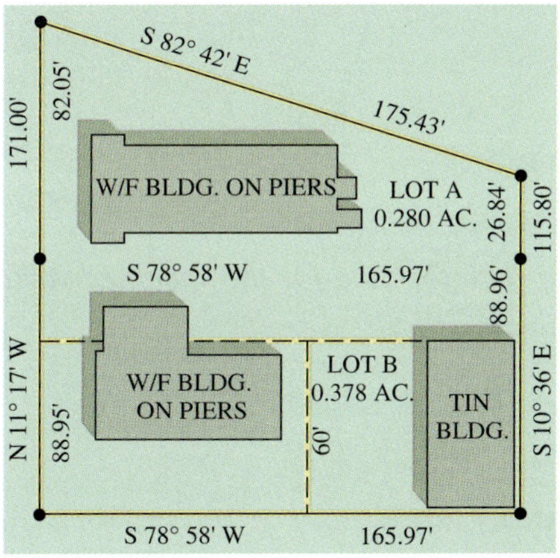

Source: Property survey in New Roads, Louisiana.

55. The U.S. Postal Service requires that any box sent by Priority Mail® have length plus girth (distance around) totaling no more than 108 in. The maximum volume that meets this condition is contained by a box with a square end 18 in. on each side. What is the length of the box? What is the maximum volume? (*Source:* United States Postal Service.)

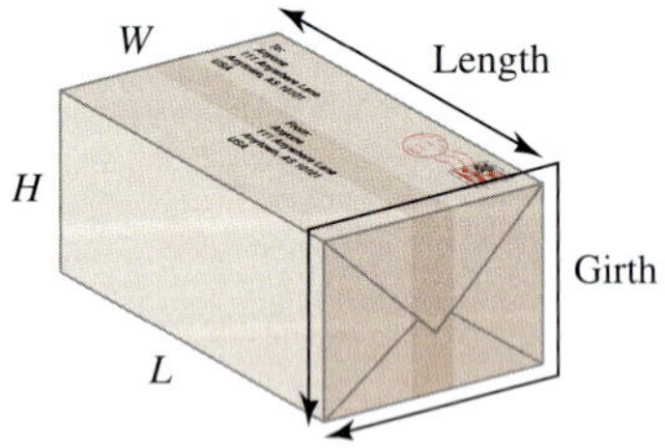

length: 36 in.; maximum volume: 11,664 in.^{3}

56. One of the world's largest sandwiches, made by Wild Woody's Chill and Grill in Roseville, Michigan, was 12 ft long, 12 ft wide, and $17\frac{1}{2}$ in. $(1\frac{11}{24}$ ft) thick. What was the volume of the sandwich? (*Source: Guinness World Records.*)

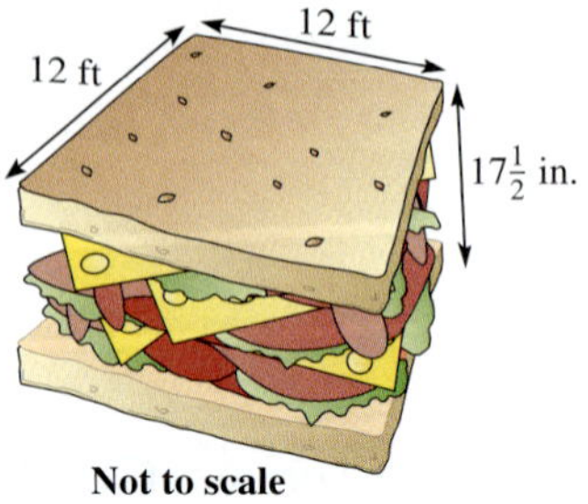

210 ft^3

Find the measure of each marked angle. ***See Example 5.***

57.

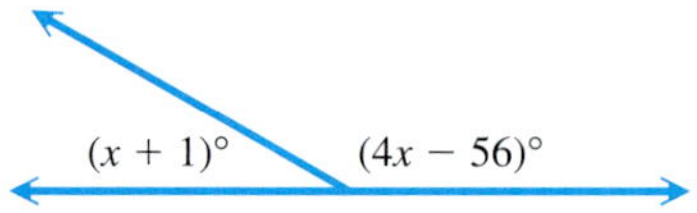

48°, 132°

58.

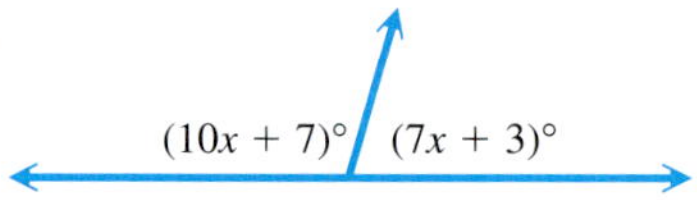

107°, 73°

59.

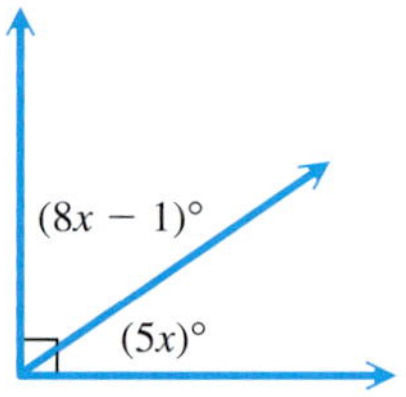

55°, 35°

60.

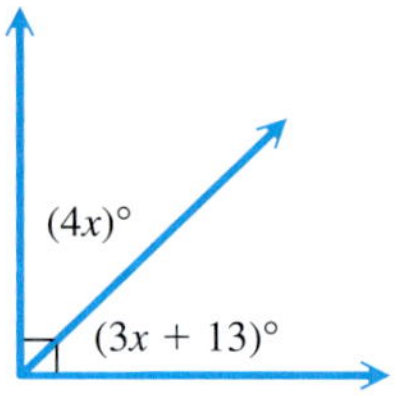

46°, 44°

61.

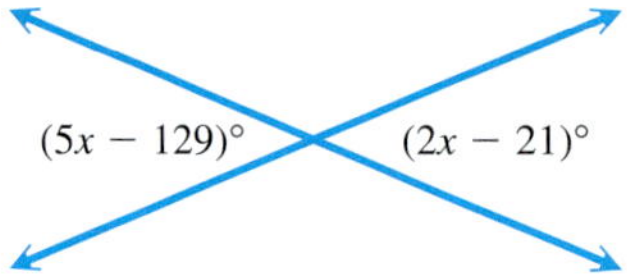

51°, 51°

62.

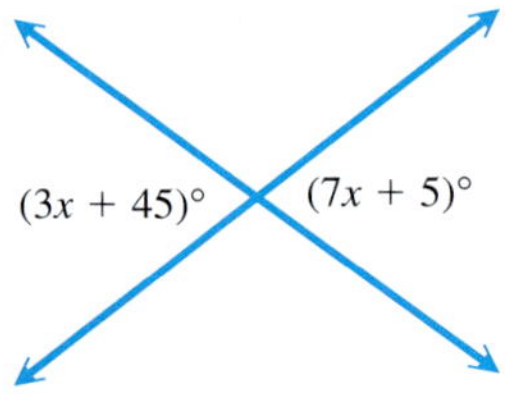

75°, 75°

63.

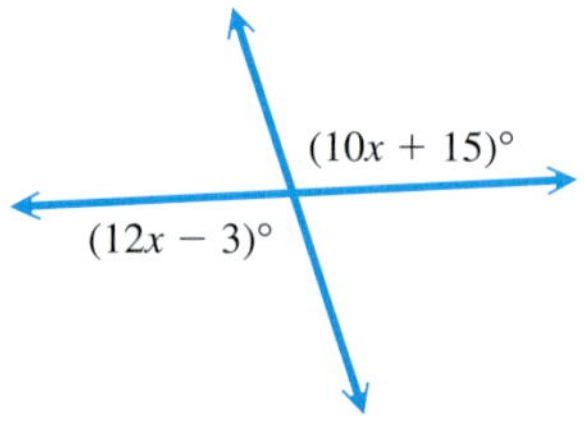

105°, 105°

64.

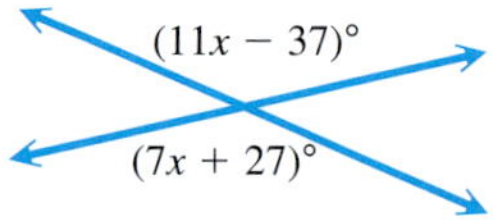

139°, 139°

Solve each formula for the specified variable. ***See Examples 6–8.***

65. $d = rt$ for t

$t = \frac{d}{r}$

66. $d = rt$ for r

$r = \frac{d}{t}$

67. $V = LWH$ for H

$H = \frac{V}{LW}$

68. $V = LWH$ for L

$L = \frac{V}{WH}$

69. $P = a + b + c$ for b

$b = P - a - c$

70. $P = a + b + c$ for c

$c = P - a - b$

71. $C = 2\pi r$ for r

$r = \frac{C}{2\pi}$

72. $C = \pi d$ for d

$d = \frac{C}{\pi}$

73. $I = prt$ for r

$r = \frac{I}{pt}$

74. $I = prt$ for p

$p = \frac{I}{rt}$

75. $A = \frac{1}{2}bh$ for h

$h = \frac{2A}{b}$

76. $A = \frac{1}{2}bh$ for b

$b = \frac{2A}{h}$

77. $V = \frac{1}{3}\pi r^2 h$ for h

$h = \frac{3V}{\pi r^2}$

78. $V = \pi r^2 h$ for h

$h = \frac{V}{\pi r^2}$

79. $P = 2L + 2W$ for W

$W = \frac{P - 2L}{2}$

80. $A = p + prt$ for r

$r = \frac{A - p}{pt}$

81. $y = mx + b$ for m

$m = \frac{y - b}{x}$

82. $y = mx + b$ for x

$x = \frac{y - b}{m}$

83. $Ax + By = C$ for y

$y = \frac{C - Ax}{B}$

84. $Ax + By = C$ for x

$x = \frac{C - By}{A}$

85. $M = C(1 + r)$ for r

$r = \frac{M - C}{C}$

86. $C = \frac{5}{9}(F - 32)$ for F

$F = \frac{9}{5}C + 32$

87. $P = 2(a + b)$ for a

$a = \frac{P - 2b}{2}$

88. $P = 2(a + b)$ for b

$b = \frac{P - 2a}{2}$

Solve each equation for y. ***See Example 9.***

We give one possible answer for Exercises 89–96. There are other correct forms.

89. $6x + y = 4$

$y = -6x + 4$

90. $3x + y = 6$

$y = -3x + 6$

91. $5x - y = 2$

$y = 5x - 2$

92. $4x - y = 1$

$y = 4x - 1$

93. $-3x + 5y = -15$

$y = \frac{3}{5}x - 3$

94. $-2x + 3y = -9$

$y = \frac{2}{3}x - 3$

95. $x - 3y = 12$

$y = \frac{1}{3}x - 4$

96. $x - 5y = 10$

$y = \frac{1}{5}x - 2$

2.6 Ratio, Proportion, and Percent

OBJECTIVES

1. Write ratios.
2. Solve proportions.
3. Solve applied problems using proportions.
4. Find percents and percentages.

OBJECTIVE 1 Write ratios. A **ratio** is a comparison of two quantities using a quotient.

Ratio

The ratio of the number a to the number b (where $b \neq 0$) is written as follows.

$$a \text{ to } b, \quad a:b, \quad \text{or} \quad \frac{a}{b}$$

Writing a ratio as a quotient $\frac{a}{b}$ is most common in algebra.

EXAMPLE 1 Writing Word Phrases as Ratios

Write a ratio for each word phrase.

(a) 5 hr to 3 hr $\quad \frac{5 \text{ hr}}{3 \text{ hr}} = \frac{5}{3}$

(b) 6 hr to 3 days

First convert 3 days to hours.

$$3 \text{ days} = 3 \cdot 24 = 72 \text{ hr} \quad 1 \text{ day} = 24 \text{ hr}$$

Now write the ratio using the common unit of measure, hours.

$$\frac{6 \text{ hr}}{3 \text{ days}} = \frac{6 \text{ hr}}{72 \text{ hr}} = \frac{6}{72}, \quad \text{or} \quad \frac{1}{12} \quad \text{Write in lowest terms.}$$

◀ Work Problem 1 at the Side.

1 Write each ratio.

(a) 9 women to 5 women

GS **(b)** 4 in. to 1 ft

First convert 1 ft to ____ in.

Then write a ratio.

(c) 8 months to 2 yr

An application of ratios is in *unit pricing,* to see which size of an item offered in different sizes produces the best price per unit.

EXAMPLE 2 Finding Price per Unit

A Jewel-Osco supermarket charges the following prices for a jar of extra crunchy peanut butter.

PEANUT BUTTER

Size	Price
18 oz	\$3.49
28 oz	\$4.99
40 oz	\$6.79

Which size is the best buy? That is, which size has the lowest unit price?

Continued on Next Page

Answers

1. (a) $\frac{9}{5}$ (b) 12; $\frac{4}{12}$, or $\frac{1}{3}$ (c) $\frac{8}{24}$, or $\frac{1}{3}$

To find the best buy, write ratios comparing the price for each size jar to the number of units (ounces) per jar.

Size	Unit Price (dollars per ounce)
18 oz	$\frac{\$3.49}{18} = \$0.194$
28 oz	$\frac{\$4.99}{28} = \$0.178$
40 oz	$\frac{\$6.79}{40} = \$0.170$ ← The best buy

(Results are rounded to the nearest thousandth.)

Because the 40-oz size produces the lowest unit price, it is the best buy. Buying the largest size does not always provide the best buy, although it often does, as in this case.

Work Problem **2** at the Side. ▶

OBJECTIVE 2 Solve proportions. A ratio is used to compare two numbers or amounts. A **proportion** says that two ratios are equal. For example, the proportion

$$\frac{3}{4} = \frac{15}{20}$$

A proportion is a special type of equation.

says that the ratios $\frac{3}{4}$ and $\frac{15}{20}$ are equal. In the proportion

$$\frac{a}{b} = \frac{c}{d} \quad (\text{where } b, d \neq 0),$$

a, b, c, and d are the **terms** of the proportion. The a and d terms are called the **extremes,** and the b and c terms are called the **means.** We read the proportion $\frac{a}{b} = \frac{c}{d}$ as **"*a* is to *b* as *c* is to *d*."** Multiplying each side of this proportion by the common denominator, bd, gives the following.

$$bd \cdot \frac{a}{b} = bd \cdot \frac{c}{d} \quad \text{Multiply each side by } bd.$$

$$\frac{b}{b}(d \cdot a) = \frac{d}{d}(b \cdot c) \quad \text{Associative and commutative properties}$$

$$ad = bc \quad \text{Commutative and identity properties}$$

We can also find the products ad and bc by multiplying diagonally.

$$ad = bc$$

$$\frac{a}{b} = \frac{c}{d}$$

For this reason, ad and bc are called **cross products of the proportion.**

Cross Products of a Proportion

If $\frac{a}{b} = \frac{c}{d}$, then the cross products ad and bc are equal—***the product of the extremes equals the product of the means.***

Also, if $ad = bc$, then $\frac{a}{b} = \frac{c}{d}$ (where $b, d \neq 0$).

---

2 Solve each problem.

GS **(a)** A supermarket charges the following prices for a popular brand of pork and beans.

PORK AND BEANS

Size	Price
8 oz	\$0.76
28 oz	\$1.00
53 oz	\$1.99

Calculate the unit price to the nearest thousandth of an ounce for each size.

8 oz: $\frac{\$0.76}{\_\_\_\_} = \_\_\_\_$

28 oz: $\frac{\_\_\_\_}{28} = \_\_\_\_$

53 oz: $\frac{\_\_\_\_}{\_\_\_\_} = \_\_\_\_$

Which size is the best buy? What is the unit price for that size?

(b) A supermarket charges the following prices for a certain brand of liquid detergent.

LAUNDRY DETERGENT

Size	Price
75 oz	\$8.94
100 oz	\$13.97
150 oz	\$19.97

Which size is the best buy? What is the unit price to the nearest thousandth for that size?

Teaching Tip

Some students erroneously try to "reduce" across the equality symbol when working with proportions. Give examples to show why this cannot be done.

Answers

2. (a) 8; \$0.095; \$1.00; \$0.036; \$1.99; 53; \$0.038; 28 oz; \$0.036 per oz
(b) 75 oz; \$0.119 per oz

3 Solve each proposition.

GS **(a)** $\frac{y}{6} = \frac{35}{42}$

$y \cdot$ ____ $=$ ____ $\cdot\, 35$

____ $y =$ ____

$y =$ ____

The solution set is ____.

(b) $\frac{a}{24} = \frac{15}{16}$

Teaching Tip

Caution students that finding cross products is *not* appropriate for solving equations such as

$$\frac{x}{3} + \frac{5}{4} = \frac{1}{2}$$

or for adding fractions such as

$$\frac{2}{3} + \frac{1}{9}.$$

4 Solve each equation.

(a) $\frac{z}{2} = \frac{z+1}{3}$

(b) $\frac{p+3}{3} = \frac{p-5}{4}$

Answers

3. **(a)** 42; 6; 42; 210; 5; $\{5\}$ **(b)** $\left\{\frac{45}{2}\right\}$

4. **(a)** $\{2\}$ **(b)** $\{-27\}$

Note

If $\frac{a}{c} = \frac{b}{d}$, then $ad = cb$, or $ad = bc$. This means that the two proportions are equivalent, and the proportion

$$\frac{a}{b} = \frac{c}{d} \quad \text{can also be written as} \quad \frac{a}{c} = \frac{b}{d}.$$

Sometimes one form is more convenient to work with than the other.

Four numbers are used in a proportion. If any three of these numbers are known, the fourth can be found.

EXAMPLE 3 Finding an Unknown in a Proportion

Solve the proportion $\frac{5}{9} = \frac{x}{63}$.

$$\frac{5}{9} = \frac{x}{63} \qquad \text{Solve for } x.$$

$5 \cdot 63 = 9 \cdot x$ Cross products must be equal.

$315 = 9x$ Multiply.

$35 = x$ Divide by 9.

Check by substituting 35 for x in the proportion. The solution set is $\{35\}$.

◄ **Work Problem 3 at the Side.**

CAUTION

The cross product method cannot be used directly if there is more than one term on either side of the equality symbol.

EXAMPLE 4 Solving an Equation by Using Cross Products

Solve the equation $\frac{m-2}{5} = \frac{m+1}{3}$.

$$\frac{m-2}{5} = \frac{m+1}{3}$$

Be sure to use parentheses.

$3(m-2) = 5(m+1)$ Find the cross products.

$3m - 6 = 5m + 5$ Distributive property

$3m = 5m + 11$ Add 6.

$-2m = 11$ Subtract $5m$.

$m = -\frac{11}{2}$ Divide by -2.

A check confirms that the solution is $-\frac{11}{2}$, so the solution set is $\left\{-\frac{11}{2}\right\}$.

◄ **Work Problem 4 at the Side.**

Note

When we set cross products equal to each other, we are really multiplying each ratio in the proportion by a common denominator.

OBJECTIVE 3 Solve applied problems using proportions.

EXAMPLE 5 Applying Proportions

After Lee Ann Spahr pumped 5.0 gal of gasoline, the display showing the price read \$23.20. When she finished pumping the gasoline, the price display read \$67.28. How many gallons did she pump?

To solve this problem, set up a proportion, with prices in the numerators and gallons in the denominators. Let $x =$ the number of gallons pumped.

$$\frac{\$23.20}{5.0} = \frac{\$67.28}{x} \qquad \begin{matrix}\leftarrow \text{Price} \\ \leftarrow \text{Gallons}\end{matrix}$$

Be sure that numerators represent the *same* quantities and denominators represent the *same* quantities.

$23.20x = 5.0(67.28)$ Cross products

$23.20x = 336.40$ Multiply.

$x = 14.5$ Divide by 23.20.

She pumped a total of 14.5 gal. Check this answer. Notice that the way the proportion was set up uses the fact that the unit price is the same, no matter how many gallons are purchased.

Calculator Tip

Using a calculator to perform the arithmetic in **Example 5** reduces the possibility of errors.

Work Problem 5 at the Side. ▶

OBJECTIVE 4 Find percents and percentages. ***A percent is a ratio where the second number is always 100.***

50% represents the ratio of 50 to 100, that is, $\frac{50}{100}$, or, **0.50**.

27% represents the ratio of 27 to 100, that is, $\frac{27}{100}$, or, **0.27**.

125% represents the ratio of 125 to 100, that is, $\frac{125}{100}$, or, **1.25**.

5% represents the ratio of 5 to 100, that is, $\frac{5}{100}$, or, **0.05**.

Recall from **Section R.2** that ***the decimal point can be moved two places to the left and the percent symbol dropped to change a percent to a decimal.***

Calculator Tip

Many calculators have a percent key that does this automatically.

We can solve a percent problem involving $x\%$ by writing it as a proportion.

$$\frac{\textbf{\textit{amount}}}{\textbf{\textit{base}}} = \frac{x}{100}$$

The amount, or **percentage,** is compared to the **base** (the whole amount). We can also write this proportion as follows.

$$\frac{\text{amount}}{\text{base}} = \text{percent (as a decimal)}$$ $\frac{x}{100}$ is equivalent to percent as a decimal.

or **amount = percent (as a decimal) · base** Basic percent equation

5 Solve each problem.

(a) Twelve gal of gasoline costs \$57.48. To the nearest cent, how much would 16.5 gal of the same fuel cost?

(b) Eight quarts of oil cost \$14.00. How much do 5 qt of oil cost?

Answers

5. (a) \$79.04 **(b)** \$8.75

6 Solve each problem.

(a) What is 20% of 70?

Teaching Tip

Stress that the decimal form of a percent is used in the percent equation.

(b) 40% of what number is 130?

(c) 121 is what percent of 484?

Answers

6. **(a)** 14 **(b)** 325 **(c)** 25%

EXAMPLE 6 Solving Percent Equations

Solve each problem.

(a) What is 15% of 600?

Let n = the number. The word *of* indicates multiplication.

What / is / 15% / of / 600? — Translate each word or phrase to write the equation.

$$n = 0.15 \cdot 600 \quad \text{Write the percent equation.}$$

Write 15% as a decimal.

$$n = \mathbf{90} \quad \text{Multiply.}$$

Thus, **90** is 15% of 600.

(b) 32% of what number is 64?

32% / of / what number / is / 64?

$$0.32 \cdot n = 64 \quad \text{Write the percent equation.}$$

Write 32% as a decimal.

$$n = \frac{64}{0.32} \quad \text{Divide by 0.32.}$$

$$n = \mathbf{200} \quad \text{Simplify. Use a calculator.}$$

32% of **200** is 64.

(c) 90 is what percent of 360?

90 / is / what percent / of / 360?

$$90 = p \cdot 360 \quad \text{Write the percent equation.}$$

$$\frac{90}{360} = p \quad \text{Divide by 360.}$$

$$\mathbf{0.25} = p, \quad \text{or} \quad \mathbf{25\%} = p \quad \text{Simplify. Write 0.25 as a percent.}$$

Thus, 90 is **25%** of 360.

◀ Work Problem **6** at the Side.

EXAMPLE 7 Solving Applied Percent Problems

Solve each problem.

(a) A DVD with a regular price of \$18 is on sale this week at 22% off. Find the amount of the discount and the sale price of the disc.

The discount is 22% of 18, so we must find the number that is 22% of 18.

What number / is / 22% / of / 18?

$$n = 0.22 \cdot 18 \quad \text{Write the percent equation.}$$

$$n = 3.96 \quad \text{Multiply.}$$

The discount is \$3.96, so the sale price is found by subtracting.

$$\$18.00 - \$3.96 = \$14.04 \quad \text{Original price} - \text{discount} = \text{sale price}$$

Continued on Next Page

(b) A newspaper ad offered a set of tires at a sale price of \$258. The regular price was \$300. What percent of the regular price was the savings?

The savings amounted to \$300 − \$258 = \$42. We can now restate the problem: What percent of 300 is 42?

What percent	of	300	is	42?	
↓	↓	↓	↓	↓	
p	·	300	=	42	Write the percent equation.

$$p = \frac{42}{300} \quad \text{Divide by 300.}$$

$$p = 0.14, \quad \text{or} \quad 14\% \quad \text{Simplify. Write 0.14 as a percent.}$$

The sale price represents a 14% savings.

Work Problem **7** at the Side. ▶

7 Solve each problem.

(a) A television set with a regular price of \$950 is on sale at 25% off. Find the amount of the discount and the sale price of the set.

(b) A winter coat is on a clearance sale for \$48. The regular price is \$120. What percent of the regular price is the savings?

GS **(c)** Mark scored 34 points on a test, which was 85% of the possible points. How many possible points were on the test?

Write the percent equation. Give the percent as a decimal.

34	was	85%	of	what number?
↓	↓	↓	↓	↓
34	____	____	____	n

Complete the solution, and give the answer to the problem.

Answers

7. **(a)** \$237.50; \$712.50 **(b)** 60%
(c) =; 0.85; · ; 40 points

2.6 Exercises

FOR EXTRA HELP

Download the MyDashBoard App

MyMathLab®

1. CONCEPT CHECK Ratios are used to _______ two numbers or quantities. Which of the following indicate the ratio of a to b?

   A. $\frac{a}{b}$ B. $\frac{b}{a}$ C. $a \cdot b$ D. $a : b$

   compare; A, D

2. CONCEPT CHECK A proportion says that two _______ are equal. The equation $\frac{a}{b} = \frac{c}{d}$ (where $b, d, \neq 0$) is a _______, where ad and bc are the _______ .

   ratios; proportion; cross products

3. CONCEPT CHECK Match each ratio in Column I with the ratio equivalent to it in Column II.

I	II
(a) 75 to 100 C	A. 80 to 100
(b) 5 to 4 D	B. 50 to 100
(c) $\frac{1}{2}$ B	C. 3 to 4
(d) 4 to 5 A	D. 15 to 12

4. CONCEPT CHECK Which of the following represent a ratio of 4 days to 2 weeks?

   A. $\frac{4}{2}$ B. $\frac{4}{7}$ C. $\frac{4}{14}$

   D. $\frac{2}{1}$ E. $\frac{2}{7}$ F. $\frac{1}{2}$

   C, E

Write a ratio for each word phrase. Write fractions in lowest terms. ***See Example 1.***

5. 60 ft to 70 ft $\frac{6}{7}$
6. 30 mi to 40 mi $\frac{3}{4}$
7. 72 dollars to 220 dollars $\frac{18}{55}$
8. 80 people to 120 people $\frac{2}{3}$
9. 30 in. to 8 ft $\frac{5}{16}$
10. 8 ft to 20 yd $\frac{2}{15}$
11. 16 min to 1 hr $\frac{4}{15}$
12. 24 min to 2 hr $\frac{1}{5}$
13. 2 yd to 60 in. $\frac{6}{5}$
14. 3 days to 40 hr $\frac{9}{5}$

Find the best buy for each item. Give the unit price to the nearest thousandth for that size. ***See Example 2.*** (*Source:* Jewel-Osco and HyVee.)

15. GRANULATED SUGAR

Size	Price
4 lb	$3.29
10 lb	$7.49

10 lb; $0.749

16. APPLESAUCE

Size	Price
23 oz	$1.99
48 oz	$3.49

48 oz; $0.073

17. ORANGE JUICE

Size	Price
64 oz	$2.99
89 oz	$4.79
128 oz	$6.49

64 oz; $0.047

18. SALAD DRESSING

Size	Price
8 oz	$1.69
16 oz	$1.97
36 oz	$5.99

16 oz; $0.123

19. MAPLE SYRUP

Size	Price
8.5 oz	$5.79
12.5 oz	$7.99
32 oz	$16.99

32 oz; $0.531

20. MOUTHWASH

Size	Price
16.9 oz	$3.39
33.8 oz	$3.49
50.7 oz	$5.29

33.8 oz; $0.103

21. TOMATO KETCHUP

Size	Price
32 oz	$1.79
36 oz	$2.69
40 oz	$2.49
64 oz	$4.38

32 oz; $0.056

22. GRAPE JELLY

Size	Price
1 lb	$0.79
2 lb	$1.49
5 lb	$3.59
20 lb	$12.99

20 lb; $0.650

Solve each equation. ***See Examples 3 and 4.***

23. $\frac{k}{4} = \frac{175}{20}$ {35}

24. $\frac{x}{6} = \frac{18}{4}$ {27}

25. $\frac{49}{56} = \frac{z}{8}$ {7}

26. $\frac{20}{100} = \frac{z}{80}$ {16}

27. $\frac{x}{4} = \frac{12}{30}$ $\left\{\frac{8}{5}\right\}$

28. $\frac{x}{6} = \frac{5}{21}$ $\left\{\frac{10}{7}\right\}$

29. $\frac{z}{4} = \frac{z+1}{6}$ {2}

30. $\frac{m}{5} = \frac{m-2}{2}$ $\left\{\frac{10}{3}\right\}$

31. $\frac{3y-2}{5} = \frac{6y-5}{11}$ {−1}

32. $\frac{2r+8}{4} = \frac{3r-9}{3}$ {10}

33. $\frac{5k+1}{6} = \frac{3k-2}{3}$ {5}

34. $\frac{x+4}{6} = \frac{x+10}{8}$ {14}

35. $\frac{2p+7}{3} = \frac{p-1}{4}$ $\left\{-\frac{31}{5}\right\}$

36. $\frac{3m-2}{5} = \frac{4-m}{3}$ $\left\{\frac{13}{7}\right\}$

Solve each problem. ***See Example 5.***

37. If 16 candy bars cost \$20.00, how much do 24 candy bars cost? **\$30.00**

38. If 12 ring tones cost \$30.00, how much do 8 ring tones cost? **\$20.00**

39. If 6 gal of premium gasoline cost \$22.74, how much would it cost to completely fill a 15-gal tank? **\$56.85**

40. If sales tax on a \$16.00 DVD is \$1.32, how much would the sales tax be on a \$120.00 DVD player? **\$9.90**

41. Biologists tagged 500 fish in Grand Bay. At a later date, they found 7 tagged fish in a sample of 700. Estimate the total number of fish in Grand Bay to the nearest hundred. **50,000 fish**

42. Researchers at West Okoboji Lake tagged 840 fish. A later sample of 1000 fish contained 18 that were tagged. Approximate the fish population in West Okoboji Lake to the nearest hundred. **46,700 fish**

43. The distance between Kansas City, Missouri, and Denver is 600 mi. On a certain wall map, this is represented by a length of 2.4 ft. On the map, how many feet would there be between Memphis and Philadelphia, two cities that are actually 1000 mi apart? **4 ft**

44. The distance between Singapore and Tokyo is 3300 mi. On a certain wall map, this distance is represented by a length of 11 in. The actual distance between Mexico City and Cairo is 7700 mi. How far apart are they on the same map? **$25\frac{2}{3}$ in.**

45. In 2009, there were 87 licensed drivers for every 100 U.S. residents of driving age (16 and over). How many of the 241 million residents of driving age were licensed drivers? (*Source:* Bureau of Transportation Statistics.)

Complete the proportion, solve it, and give the answer.

$$\frac{87}{100} = \frac{x}{241} \quad \begin{matrix} \leftarrow \text{Drivers (in millions)} \\ \leftarrow \text{Residents (in millions)} \end{matrix}$$

209.67 million drivers

46. In 2009, there were 7 licensed drivers for every 10 registered vehicles in California. There were 33.8 million registered vehicles. How many licensed drivers were there? (*Source:* Bureau of Transportation Statistics.)

Complete the proportion (using *x* as the variable), solve it, and give the answer.

$$\begin{matrix} \text{Drivers} \rightarrow \\ \text{Vehicles} \rightarrow \end{matrix} \frac{7}{10} = \frac{x}{33.8}$$

23.66 million drivers

47. According to the directions on the label of a bottle of Armstrong® Concentrated Floor Cleaner, for routine cleaning, $\frac{1}{4}$ cup of cleaner should be mixed with 1 gal of warm water. How much cleaner should be mixed with $10\frac{1}{2}$ gal of water?

$2\frac{5}{8}$ **cups**

48. The directions on the bottle mentioned in **Exercise 47** also specify that, for extra-strength cleaning, $\frac{1}{2}$ cup of cleaner should be used for each 1 gal of water. For extra-strength cleaning, how much cleaner should be mixed with $15\frac{1}{2}$ gal of water?

$7\frac{3}{4}$ **cups**

49. On January 3, 2012, the exchange rate between euros and U.S. dollars was 1 euro to \$1.3051. Ashley went to Rome and exchanged her U.S. currency for euros, receiving 300 euros. How much in U.S. dollars did she exchange? (*Source:* www.xe.com/ucc)

\$391.53

50. On the same date as in **Exercise 49,** 12 U.S. dollars could be exchanged for 163.8 Mexican pesos. At that exchange rate, how many pesos could Carlos obtain for \$100? (*Source:* www.xe.com/ucc)

1365 pesos

Two triangles are ***similar*** *if they have the same shape (but not necessarily the same size). Similar triangles have sides that are proportional. The figure shows two similar triangles. Notice that the ratios of the corresponding sides all equal* $\frac{3}{2}$.

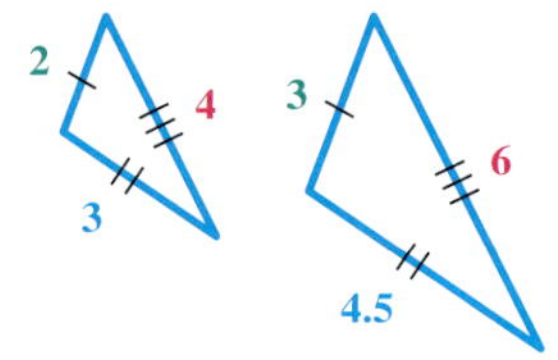

$$\frac{3}{2} = \frac{3}{2} \qquad \frac{4.5}{3} = \frac{3}{2} \qquad \frac{6}{4} = \frac{3}{2}$$

If we know that two triangles are similar, we can set up a proportion to solve for the length of an unknown side.

In Exercises 51–56, find the lengths x and y as needed in each pair of similar triangles.

51. GS

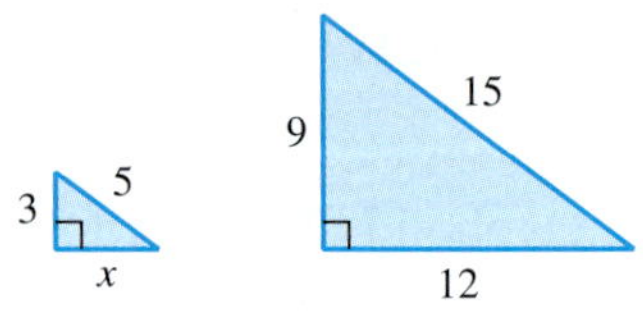

Complete the proportion and then solve for x.

$$\frac{3}{\underline{\ 9\ }} = \frac{x}{12}$$

$x = 4$

52. GS

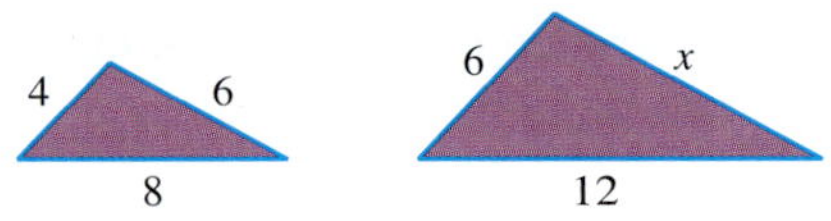

Complete the proportion and then solve for x.

$$\frac{x}{6} = \frac{12}{\underline{\ 8\ }}$$

$x = 9$

53.

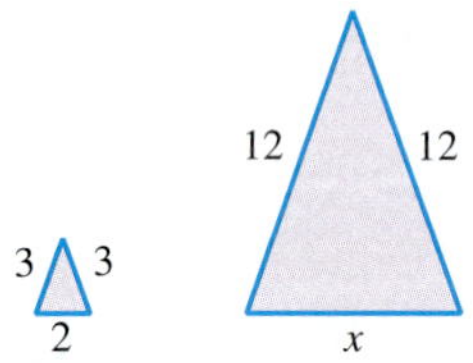

$x = 8$

54.

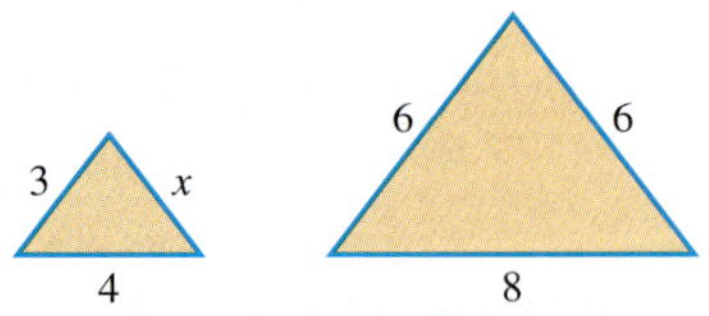

$x = 3$

55.

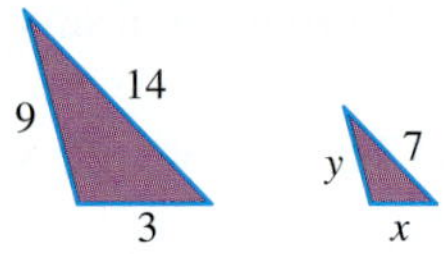

$x = 1.5$; $y = 4.5$

56.

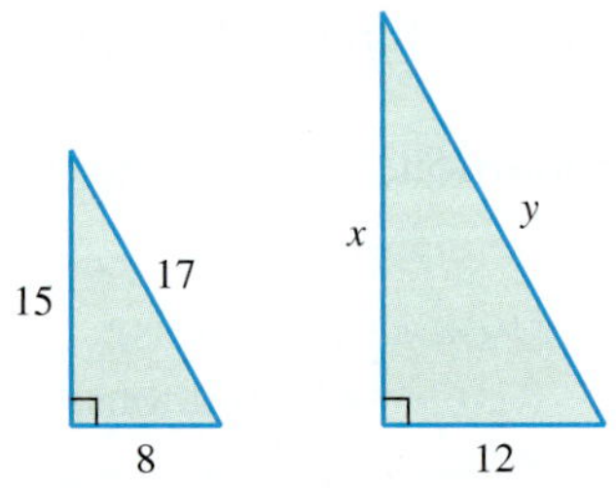

$x = 22.5$; $y = 25.5$

GS *Use the information in each problem to complete the diagram of similar triangles. Then write a proportion and solve the problem.*

57. An enlarged version of the chair used by George Washington at the Constitutional Convention casts a shadow 18 ft long at the same time a vertical pole 12 ft high casts a shadow 4 ft long. How tall is the chair?

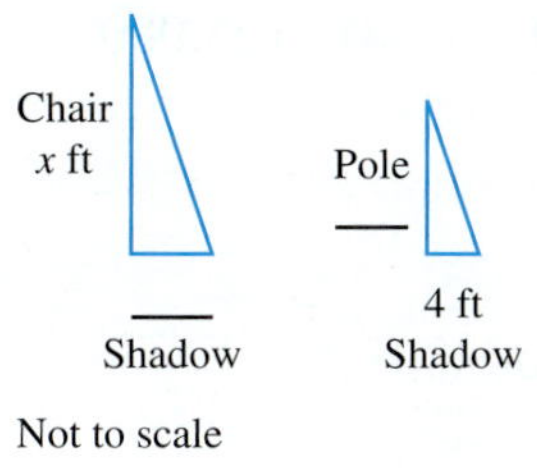

18 ft; 12 ft; One proportion is $\frac{x}{12} = \frac{18}{4}$; 54 ft

58. One of the tallest candles ever constructed was exhibited at the 1897 Stockholm Exhibition. If it cast a shadow 5 ft long at the same time a vertical pole 32 ft high cast a shadow 2 ft long, how tall was the candle?

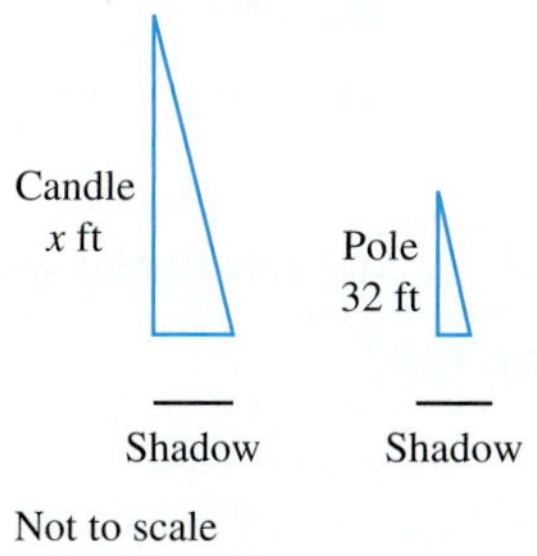

5 ft; 2 ft; One proportion is $\frac{x}{32} = \frac{5}{2}$; 80 ft

GS *Children are often given antibiotics in liquid form, called an oral suspension. Pharmacists make up these suspensions by mixing the medication in powder form with water. They use proportions to calculate the volume of the suspension for the amount of medication that has been prescribed. In Exercises 59 and 60, do each of the following.*

(a) *Find the total amount of medication in milligrams to be given over the full course of treatment.*

(b) *Write a proportion that can be solved to find the total volume of the liquid suspension that the pharmacist will prepare. Use x as the variable.*

(c) *Solve the proportion to determine the total volume of the oral suspension.*

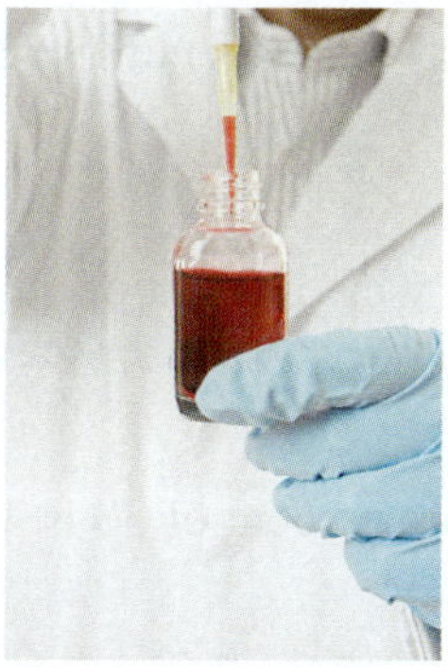

59. Amoxil is a common antibiotic used to treat a variety of infections. Logan's pediatric nurse practitioner has prescribed 375 mg of amoxil a day for 7 days to treat his ear infection. The pharmacist uses 125 mg of amoxil in each 5 mL of the suspension. (*Source:* www.drugs.com)

(a) 2625 mg (b) $\frac{125 \text{ mg}}{5 \text{ mL}} = \frac{2625 \text{ mg}}{x \text{ mL}}$ (c) 105 mL

60. An amoxil oral suspension can also be made by using 250 mg for each 5 mL of suspension. Ava's pediatrician prescribed 900 mg a day for 10 days to treat her bronchitis. (*Source:* www.drugs.com)

(a) 9000 mg (b) $\frac{250 \text{ mg}}{5 \text{ mL}} = \frac{9000 \text{ mg}}{x \text{ mL}}$ (c) 180 mL

The Consumer Price Index (CPI) provides a means of determining the purchasing power of the U.S. dollar from one year to the next. Using the period from 1982 to 1984 as a measure of 100.0, the CPI for selected years from 1998 to 2010 is shown in the table. To use the CPI to predict a price in a particular year, we set up a proportion and compare it with a known price in another year.

$$\frac{\text{price in year } A}{\text{index in year } A} = \frac{\text{price in year } B}{\text{index in year } B}$$

Use the CPI figures in the table to find the amount that would be charged for using the same amount of electricity that cost \$225 in 1998. Give answers to the nearest dollar.

Year	Consumer Price Index
1998	163.0
2000	172.2
2002	179.9
2004	188.9
2006	201.6
2008	215.3
2010	218.1

Source: U.S. Bureau of Labor Statistics.

61. in 2000
\$238

62. in 2002
\$248

63. in 2006
\$278

64. in 2010
\$301

CONCEPT CHECK *Use mental techniques to answer each question.*

65. The 2010 U.S. Census gave the population of Alabama as 4,780,000, where 26.0% were African-Americans. What is the best estimate of the African-American population in Alabama? (*Source:* U.S. Census Bureau.) C

A. 500,000 **B.** 750,000

C. 1,250,000 **D.** 1,500,000

66. The same census gave the population of New Mexico as 2,059,000, with 46.3% being Hispanic. What is the best estimate of the Hispanic population in New Mexico? (*Source:* U.S. Census Bureau.) A

A. 950,000 **B.** 95,000

C. 1,250,000 **D.** 125,000

Solve each problem. ***See Examples 6 and 7.***

67. What is 14% of 780?
109.2

68. What is 26% of 480?
124.8

69. 42% of what number is 294?
700

70. 18% of what number is 108?
600

71. 120% of what number is 510?
425

72. 140% of what number is 315?
225

73. 4 is what percent of 50?
8%

74. 8 is what percent of 64?
12.5%

75. What percent of 30 is 36?
120%

76. What percent of 48 is 96?
200%

77. Find the discount on a leather recliner with a regular price of $795 if the recliner is 15% off. What is the sale price of the recliner?
$119.25; $675.75

78. A laptop computer with a regular price of $597 is on sale at 20% off. Find the amount of the discount and the sale price of the computer.
$119.40; $477.60

79. Clayton earned 48 points on a 60-point geometry project. What percent of the total points did he earn?
80%

80. On a 75-point algebra test, Grady scored 63 points. What percent of the total points did he score?
84%

81. Anna saved $1950, which was 65% of the amount she needed for a used car. What was the total amount she needed for the car?
$3000

82. Bryn had $525, which was 70% of the total amount she needed for a deposit on an apartment. What was the total deposit she needed?
$750

Work each percent problem. Round all money amounts to the nearest dollar and percents to the nearest tenth, as needed. ***See Examples 5–7.***

83. A family of four with a monthly income of $3800 plans to spend 8% of this amount on entertainment. How much will be spent on entertainment?
$304

84. Quinhon Dac Ho earns $3200 per month. He saves 12% of this amount. How much does he save?
$384

85. In 2010, the U.S. civilian labor force consisted of 153,889,000 persons. Of this total, 14,825,000 were unemployed. What was the percent of unemployment? (*Source:* U.S. Bureau of Labor Statistics.)

The number unemployed	was	what percent	of	the total civilian labor force?
↓	↓	↓	↓	↓
14,825,000	=	p	·	153,889,000

9.6%

86. In 2010, the U.S. civilian labor force consisted of 124,073,000 persons. Of this total, 14,715,000 were union members. What percent were union members? (*Source:* U.S. Bureau of Labor Statistics.)

What percent	of	the total U.S. labor force	was	the number of union members?
↓	↓	↓	↓	↓
p	·	124,073,000	=	14,715,000

11.9%

87. As of February 2011, U.S. households owned 377,410,000 pets. Of these, 78,200,000 were dogs. What percent of the pets were *not* dogs? (*Source:* American Pet Product Manufacturers Association.)

79.3%

88. In the fall of 2010, total public and private school enrollment in the United States was 62.4 million. Of this total, 10.9% of the students were enrolled in private schools. How many students were enrolled in *public* schools? (*Source:* National Center for Education Statistics.)

55.6 million

89. The 1916 dime minted in Denver is quite rare. The 1979 edition of *A Guide Book of United States Coins* listed its value in Extremely Fine condition as \$625. The 2012 value had increased to \$6200. What was the percent increase in the value of this coin? (*Hint:* First subtract to find the increase in value. Then write a percent equation that uses this increase.) 892%

90. Here is a common business problem: If the sales tax rate is 6.5% and we have collected \$3400 in sales tax, how much were sales? (*Hint:* To solve using a percent equation, ask "6.5% of what number is \$3400?" Write 6.5% as a decimal.) \$52,308

Relating Concepts (Exercises 91–94) For Individual or Group Work

Work Exercises 91–94 in order. *The steps justify the method of solving a proportion by cross products.*

91. What is the LCD of the fractions in the following equation?

$$\frac{x}{6} = \frac{2}{5}$$ 30

92. Solve the equation in **Exercise 91** as follows.

(a) Multiply each side by the LCD. What equation results? $5x = 12$

(b) Solve the equation from part (a) by dividing each side by the coefficient of x. $\left\{\frac{12}{5}\right\}$

93. Solve the equation in **Exercise 91** using cross products.

$\left\{\frac{12}{5}\right\}$

94. Compare your answers from **Exercises 92(b) and 93.** What do you notice?

Both methods give the same solution set.

Summary Exercises *Applying Problem-Solving Techniques*

Summary Exercises provide mixed practice to better prepare students for tests.

The following problems are of the various types discussed in this chapter. Solve each problem.

1. On an algebra test, the highest grade was 42 points more than the lowest grade. The sum of the two grades was 138. Find the lowest grade.

48

2. Find the measure of an angle whose measure is 70° more than its complement.

80°

3. The perimeter of a certain square is seven times the length of a side, decreased by 12. Find the length of a side.

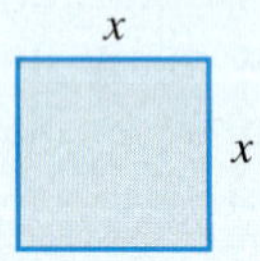

4

4. Find the measures of the marked angles.

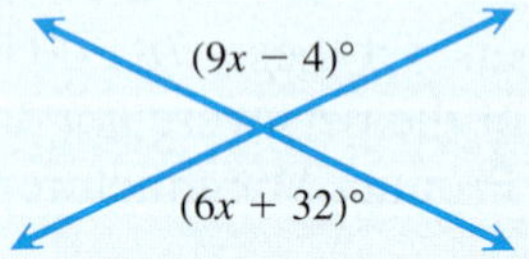

104°, 104°

5. If 2 is added to five times a number, the result is equal to 5 more than four times the number. Find the number.

3

6. Find two consecutive even integers such that four times the greater added to the lesser is 98.

18, 20

7. Find the measures of the marked angles.

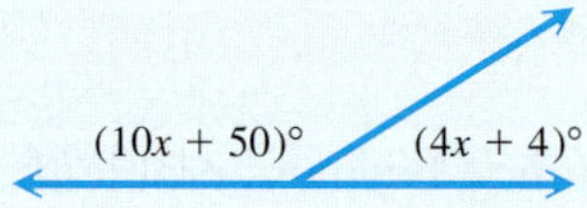

140°, 40°

8. A store has 39 qt of milk, some in pint cartons and some in quart cartons. There are six times as many quart cartons as pint cartons. How many quart cartons are there? (*Hint:* 1 qt = 2 pt.)

36 quart cartons

9. Grant Wood painted his most famous work, *American Gothic,* in 1930 on composition board with perimeter 108.44 in. If the width of the painting is 29.88 in., find the length. Then find the area. Round your answers to the nearest hundredth. (*Source: The Gazette.*)

24.34 in.; 727.28 in.^{2}

10. The interest in 1 yr on a deposit of \$11,000 was \$682. What percent interest was paid?

6.2%

11. Soccer teams Barcelona and Real Madrid along with baseball's New York Yankees were the best-paid global sports teams during the 2009–2010 season. Barcelona's average pay was \$0.5 million more than Real Madrid's. The Yankees average pay was \$0.6 million less than Real Madrid's. Average pay for the three teams totaled \$22.1 million. What was the average pay for each team? (*Source:* www.sportingintelligence.com)

Barcelona: \$7.9 million; Real Madrid: \$7.4 million; New York Yankees: \$6.8 million

12. A fully inflated professional basketball has a circumference of 78 cm. What is the radius of a circular cross section through the center of the ball? (Use 3.14 as the approximation for π.) Round your answer to the nearest hundredth.

12.42 cm

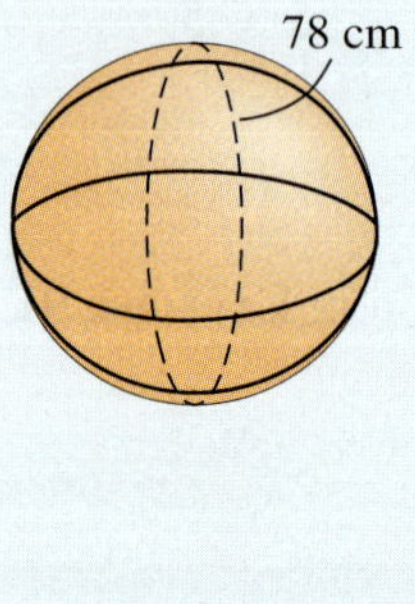

13. A music player that normally sells for $90 is on sale for $75. What is the percent discount on the player?

$16\frac{2}{3}\%$

14. Two slices of bacon contain 85 calories. How many calories are there in twelve slices of bacon?

510 calories

15. Athletes in vigorous training programs can eat 50 calories per day for every 2.2 lb of body weight. To the nearest hundred, how many calories can a 175 lb athlete consume per day? (*Source: The Gazette.*)

4000 calories

16. In the 2011 Masters Golf Tournament in Augusta, Georgia, Adam Scott finished with a score of 2 more than the tournament winner, Charl Schwartzel. The sum of their scores was 550. Find their scores. (*Source:* www.golf.about.com)

Charl Schwartzel: 274; Adam Scott: 276

17. Find the best buy (based on price per unit). Give the unit price to the nearest thousandth for that size. (*Source:* HyVee.)

SPAGHETTI SAUCE

Size	Price
14 oz	$1.79
24 oz	$1.77
48 oz	$3.65

24 oz; $0.074

18. In the 2010 Winter Olympics in Vancouver, Canada, athletes from the host country earned 26 medals. Seven of every 13 medals were gold. How many gold medals did the Canadian team earn? (*Source: World Almanac and Book of Facts.*)

14 gold medals

19. The average cost of a traditional Thanksgiving dinner for 10 people, featuring turkey, stuffing, cranberries, pumpkin pie, and trimmings, was $43.47 in 2010. This price increased 13.2% in 2011. What was the price of this traditional Thanksgiving dinner in 2011? Round your answer to the nearest cent. (*Source:* American Farm Bureau.)

$49.21

20. Refer to **Exercise 19.** The cost of a traditional Thanksgiving dinner in 1986, the year this information was first collected, was $28.74. What percent increase is the 2011 cost over the 1986 cost? Round your answer to the nearest tenth of a percent. (*Source:* American Farm Bureau.)

71.2%

2.7 Solving Linear Inequalities

OBJECTIVES

1. Graph intervals on a number line.
2. Use the addition property of inequality.
3. Use the multiplication property of inequality.
4. Solve linear inequalities by using both properties of inequality.
5. Use inequalities to solve applied problems.
6. Solve linear inequalities with three parts.

An **inequality** relates algebraic expressions using the symbols

$<$ "is less than," $\leq$ "is less than or equal to,"

$>$ "is greater than," $\geq$ "is greater than or equal to."

In each case, the interpretation is based on reading the symbol from left to right.

Linear Inequality in One Variable

A **linear inequality in one variable** can be written in the form

$$Ax + B < C, \quad Ax + B \leq C, \quad Ax + B > C, \quad \text{or} \quad Ax + B \geq C,$$

where A, B, and C represent real numbers, and $A \neq 0$.

Some examples of linear inequalities in one variable follow.

$$x + 5 < 2, \quad z - \frac{3}{4} \geq 5, \quad \text{and} \quad 2k + 5 \leq 10 \qquad \text{Linear inequalities}$$

We solve a linear inequality by finding all of its real number solutions. For example, the set

$$\{x \mid x \leq 2\}$$ Set-builder notation **(Section 1.3)**

The set of | all x | such that | x is less than or equal to 2

includes *all real numbers* that are less than or equal to 2, not just the *integers* less than or equal to 2.

Teaching Tip
Mention that some texts use a solid circle • rather than a square bracket to indicate that an endpoint is included in a number line graph. An open circle ○ may be used to indicate noninclusion, rather than a parenthesis.

OBJECTIVE 1 **Graph intervals on a number line.** Graphing is a good way to show the solution set of an inequality. To graph all real numbers belonging to the set $\{x \mid x \leq 2\}$, we place a square bracket at 2 on a number line and draw an arrow extending from the bracket to the left (since all numbers *less than* 2 are also part of the graph). See **Figure 15.**

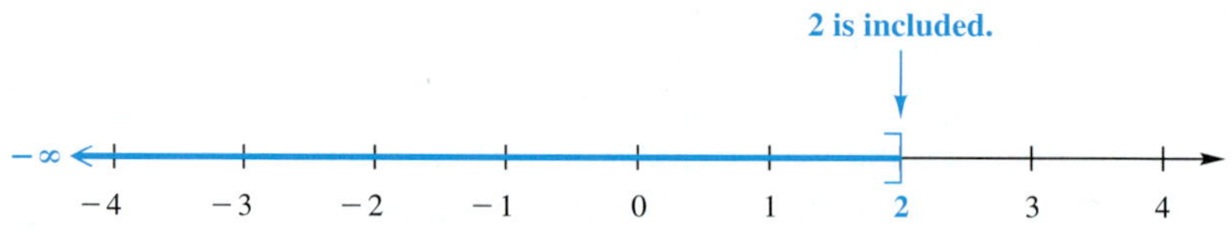

Figure 15 Graph of the interval $(-\infty, 2]$

Teaching Tip
Stress when to use a square bracket and when to use a parenthesis. Mention that brackets and parentheses open in the direction of the shading.

The set of numbers less than or equal to 2 is an example of an **interval** on the number line. We can write this interval using **interval notation,** which uses the **infinity symbols** ∞ or $-\infty$.

$$(-\infty, 2] \qquad \text{Interval notation}$$

The negative infinity symbol $-\infty$ here does not indicate a number, but shows that the interval includes *all* real numbers less than 2. Again, the square bracket indicates that 2 is part of the solution.

EXAMPLE 1 Graphing an Interval on a Number Line

Graph $x > -5$.

The statement $x > -5$ says that x can represent any value greater than -5 but cannot equal -5, written $(-5, \infty)$. We graph this interval by placing a parenthesis at -5 and drawing an arrow to the right, as in **Figure 16.** The parenthesis at -5 indicates that -5 is *not* part of the graph.

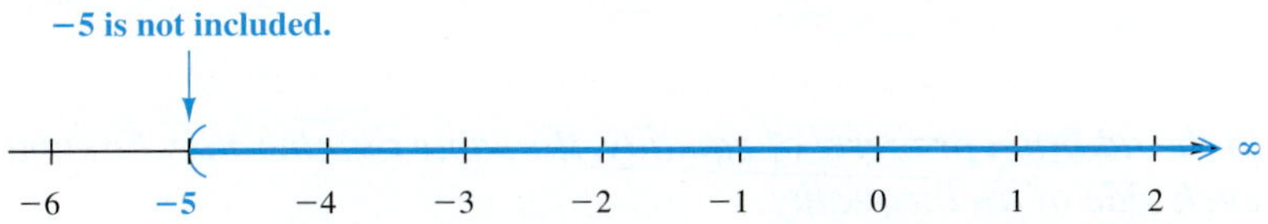

Figure 16 Graph of the interval $(-5, \infty)$

Work Problem 1 at the Side. ▶

Keep the following important concepts regarding interval notation in mind.

1. A parenthesis indicates that an endpoint is *not included* in a solution set.
2. A bracket indicates that an endpoint is *included* in a solution set.
3. A parenthesis is *always* used next to an infinity symbol, $-\infty$ or ∞.
4. The set of all real numbers is written in interval notation as $(-\infty, \infty)$.

EXAMPLE 2 Graphing an Interval on a Number Line

Graph $3 > x$.

The statement $3 > x$ means the same as $x < 3$. ***The inequality symbol continues to point to the lesser value.*** The graph of $x < 3$, written in interval notation as $(-\infty, 3)$, is shown in **Figure 17.**

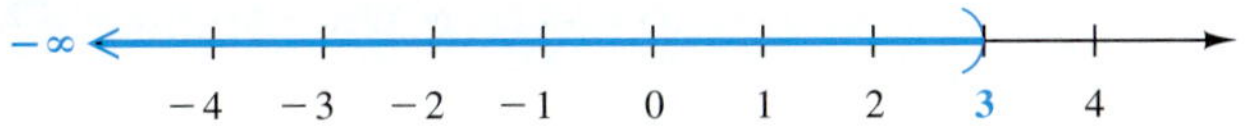

Figure 17 Graph of the interval $(-\infty, 3)$

Work Problem 2 at the Side. ▶

The table summarizes solution sets of linear inequalities.

Set-Builder Notation	Interval Notation	Graph
$\{x \mid x < a\}$	$(-\infty, a)$	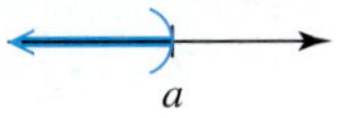
$\{x \mid x \leq a\}$	$(-\infty, a]$	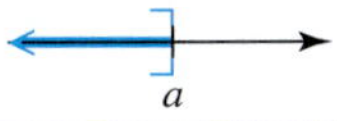
$\{x \mid x > a\}$	(a, ∞)	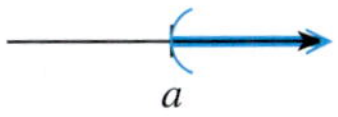
$\{x \mid x \geq a\}$	$[a, \infty)$	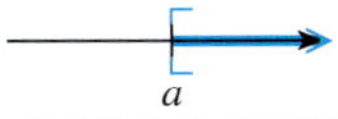
$\{x \mid x \text{ is a real number}\}$	$(-\infty, \infty)$	

OBJECTIVE 2 Use the addition property of inequality. Consider the true inequality $2 < 5$. If 4 is added to each side, the result is

$$2 + 4 < 5 + 4 \quad \text{Add 4.}$$

$$6 < 9, \quad \text{True}$$

also a true sentence. This suggests the **addition property of inequality.**

1 Write each inequality in interval notation, and graph the interval.

(a) $x \leq 3$

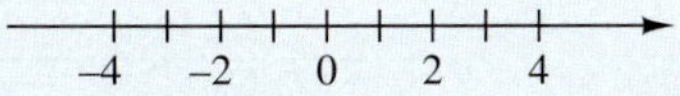

(b) $x > -4$

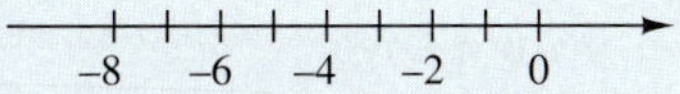

(c) $x \leq -\frac{3}{4}$

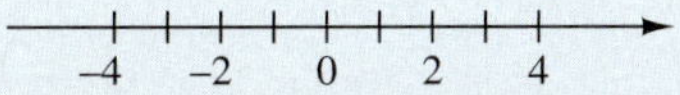

2 Write each inequality in interval notation and graph the interval.

(a) $-4 \geq x$

Teaching Tip

Emphasize that $-\infty$ and ∞ are *never* next to a square bracket.

(b) $0 < x$

Teaching Tip

If interval notation is new to your students, consider working **Exercises 1–18** in the exercise set together as a class.

Answers

1. (a) $(-\infty, 3]$

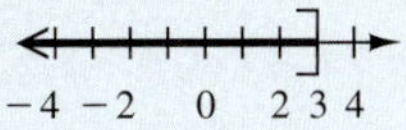

(b) $(-4, \infty)$

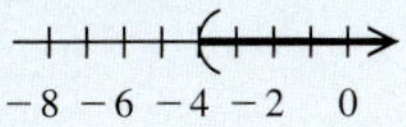

(c) $\left(-\infty, -\frac{3}{4}\right]$

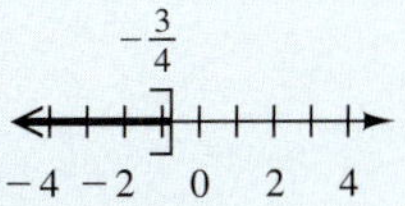

2. (a) $(-\infty, -4]$

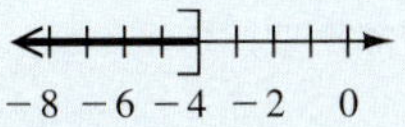

(b) $(0, \infty)$

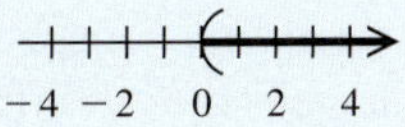

Addition Property of Inequality

For any real numbers A, B, and C, the inequalities

$$A < B \quad \text{and} \quad A + C < B + C$$

have exactly the same solutions.

In words, the same number may be added to each side of an inequality without changing the solutions.

As with the addition property of equality, the same number may be subtracted from each side of an inequality.

EXAMPLE 3 **Using the Addition Property of Inequality**

Solve $7 + 3k \geq 2k - 5$, and graph the solution set.

$$7 + 3k \geq 2k - 5$$

$$7 + 3k - 2k \geq 2k - 5 - 2k \qquad \text{Subtract } 2k.$$

$$7 + k \geq -5 \qquad \text{Combine like terms.}$$

$$7 + k - 7 \geq -5 - 7 \qquad \text{Subtract 7.}$$

$$k \geq -12 \qquad \text{Combine like terms.}$$

The solution set is $[-12, \infty)$. Its graph is shown in **Figure 18.**

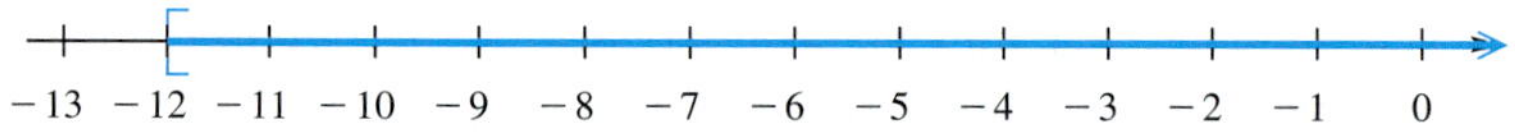

Figure 18

◀ **Work Problem 3 at the Side.**

Note

Because an inequality has many solutions, we cannot check all of them by substitution as we did with the single solution of an equation. To check the solutions in the **Example 3** solution set $[-12, \infty)$, we use a multistep process. First, we substitute -12 for k in the related *equation*.

CHECK

$$7 + 3k = 2k - 5 \qquad \text{Related equation}$$

$$7 + 3(-12) \stackrel{?}{=} 2(-12) - 5 \qquad \text{Let } k = -12.$$

$$7 - 36 \stackrel{?}{=} -24 - 5 \qquad \text{Multiply.}$$

$$-29 = -29 \ \checkmark \qquad \text{True}$$

A true statement results, so -12 is indeed the "boundary" point. Now we test a number other than -12 from the interval $[-12, \infty)$. We choose 0 since it is easy to substitute.

CHECK

$$7 + 3k \geq 2k - 5 \qquad \text{Original inequality}$$

$$7 + 3(0) \stackrel{?}{\geq} 2(0) - 5 \qquad \text{Let } k = 0.$$

$$7 \geq -5 \ \checkmark \qquad \text{True}$$

(0 is easy to substitute.)

Again, a true statement results, so the checks confirm that solutions to the inequality are in the interval $[-12, \infty)$. Any number "outside" the interval $[-12, \infty)$, that is, any number in $(-\infty, -12)$, will give a false statement when tested. (Try this.)

3 Solve each inequality, and graph the solution set.

(a) $-1 + 8r < 7r + 2$

(b) $5 + 5x \geq 4x + 3$

Answers

3. (a) $(-\infty, 3)$

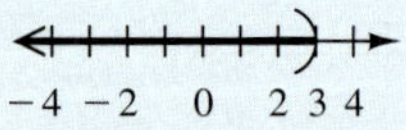

(b) $[-2, \infty)$

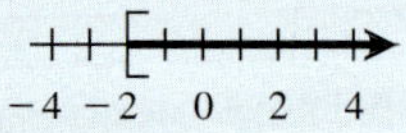

OBJECTIVE 3 **Use the multiplication property of inequality.** Consider the true inequality $3 < 7$. Multiply each side by the positive number 2.

$$3 < 7$$

$$2(3) < 2(7) \quad \text{Multiply by 2.}$$

$$6 < 14 \quad \text{True}$$

Now multiply each side of $3 < 7$ by the negative number -5.

$$3 < 7$$

$$-5(3) < -5(7) \quad \text{Multiply by } -5.$$

$$-15 < -35 \quad \text{False}$$

To get a true statement when multiplying each side by -5, ***we must reverse the direction of the inequality symbol.***

$$3 < 7$$

$$-5(3) > -5(7) \quad \text{Multiply by } -5. \text{ Reverse the direction of the symbol.}$$

$$-15 > -35 \quad \text{True}$$

Work Problem 4 at the Side. ▶

These examples suggest the **multiplication property of inequality.**

Multiplication Property of Inequality

Let A, B, and C (where $C \neq 0$) be real numbers.

1. If C is ***positive,*** then the inequalities

$$A < B \quad \text{and} \quad AC < BC$$

have exactly the same solutions.

2. If C is ***negative,*** then the inequalities

$$A < B \quad \text{and} \quad AC > BC$$

have exactly the same solutions.

In words, each side of an inequality may be multiplied by the same positive number without changing the solutions. ***If the multiplier is negative, we must reverse the direction of the inequality symbol.***

As with the multiplication property of equality, the same nonzero number may be divided into each side.

Note the following differences for positive and negative numbers.

1. When each side of an inequality is multiplied or divided by a ***positive number,*** the direction of the inequality symbol *does not change.*

2. When each side of an inequality is multiplied or divided by a ***negative number, reverse the direction of the inequality symbol.***

4 Work each of the following.

(a) Multiply each side of

$$-3 < 7$$

by 2 and then by -5. Reverse the direction of the inequality symbol if necessary to make a true statement.

(b) Multiply each side of

$$3 > -7$$

by 2 and then by -5. Reverse the direction of the inequality symbol if necessary to make a true statement.

(c) Multiply each side of

$$-7 < -3$$

by 2 and then by -5. Reverse the direction of the inequality symbol if necessary to make a true statement.

Teaching Tip

Multiplying each side of an inequality by a variable quantity should be avoided, since it is not known whether the variable quantity is positive or negative.

Answers

4. (a) $-6 < 14$; $15 > -35$
(b) $6 > -14$; $-15 < 35$
(c) $-14 < -6$; $35 > 15$

5 Solve each inequality. Graph the solution set.

GS **(a)** $9x < -18$

$$\frac{9x}{9} \;(< / >)\; \frac{-18}{\_\_\_\_}$$

$$x \;(< / >)\; \_\_\_\_$$

The solution set is ____.

GS **(b)** $-2r > -12$

$$\frac{-2r}{\_\_\_\_} \;(< / >)\; \frac{-12}{\_\_\_\_}$$

$$r \;(< / >)\; \_\_\_\_$$

The solution set is ____.

(c) $-5p \leq 0$

Teaching Tip

Remind students to reverse the direction of the inequality symbol when multiplying or dividing by a negative number.

Teaching Tip

The steps for solving an inequality parallel those for solving an equation, with the additional requirement of reversal of the direction of the inequality symbol when multiplying or dividing by a negative number.

Answers

5. (a) $<$; 9; $<$; -2; $(-\infty, -2)$

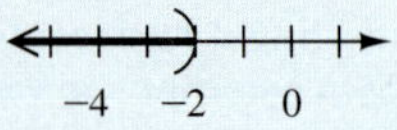

(b) -2; $<$; -2; $<$; 6; $(-\infty, 6)$

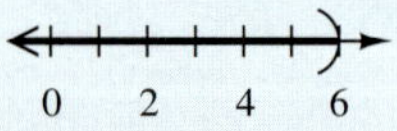

(c) $[0, \infty)$

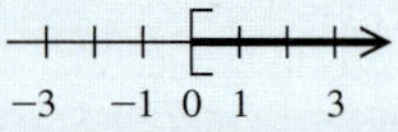

EXAMPLE 4 **Using the Multiplication Property of Inequality**

Solve $3r < -18$, and graph the solution set.

We divide each side by 3, a positive number, so the direction of the inequality symbol *does not* change. ***(It does not matter that the number on the right side of the inequality is negative.)***

$$3r < -18$$

$$\frac{3r}{3} < \frac{-18}{3} \qquad \text{Divide by 3.}$$

3 is *positive*. Do NOT reverse the direction of the symbol.

$$r < -6$$

The graph of the solution set, $(-\infty, -6)$, is shown in **Figure 19.**

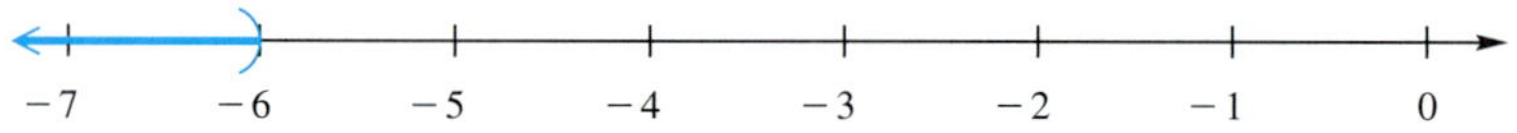

Figure 19

EXAMPLE 5 **Using the Multiplication Property of Inequality**

Solve $-4t \geq 8$, and graph the solution set.

Here each side of the inequality must be divided by -4, a negative number, which *does* require changing the direction of the inequality symbol.

$$-4t \geq 8$$

$$\frac{-4t}{-4} \leq \frac{8}{-4} \qquad \text{Divide by } -4. \text{ Reverse the symbol.}$$

-4 is *negative*. Change $\geq$ to $\leq$.

$$t \leq -2$$

The solution set, $(-\infty, -2]$, is graphed in **Figure 20.**

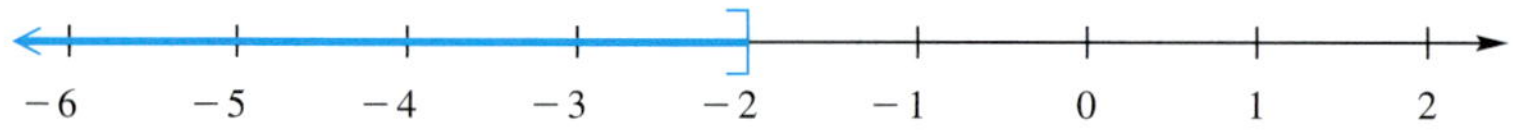

Figure 20

◀ **Work Problem 5 at the Side.**

OBJECTIVE 4 **Solve linear inequalities by using both properties of inequality.** The steps to solve a linear inequality are summarized below.

Solving a Linear Inequality

Step 1 **Simplify each side separately.** Use the distributive property to clear parentheses and combine like terms on each side as needed.

Step 2 **Isolate the variable term on one side.** Use the addition property of inequality to get all terms with variables on one side of the inequality and all numbers on the other side.

Step 3 **Isolate the variable.** Use the multiplication property of inequality to write the inequality as $x <$ a number or $x >$ a number.

Remember: Reverse the direction of the inequality symbol only when multiplying or dividing each side of an inequality by a negative number.

EXAMPLE 6 Solving a Linear Inequality

Solve $3x + 2 - 5 > -x + 7 + 2x$. Graph the solution set.

Step 1 Combine like terms and simplify.

$$3x + 2 - 5 > -x + 7 + 2x$$
$$3x - 3 > x + 7$$

Step 2 Use the addition property of inequality.

$$3x - 3 - x > x + 7 - x \quad \text{Subtract } x.$$
$$2x - 3 > 7 \quad \text{Combine like terms.}$$
$$2x - 3 + 3 > 7 + 3 \quad \text{Add 3.}$$
$$2x > 10 \quad \text{Combine like terms.}$$

Step 3 Use the multiplication property of inequality.

$$\frac{2x}{2} > \frac{10}{2} \quad \text{Divide by 2.}$$

Because 2 is positive, keep the symbol $>$.

$$x > 5$$

The solution set is $(5, \infty)$. Its graph is shown in **Figure 21.**

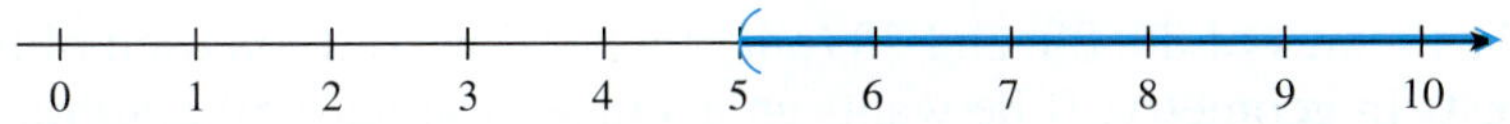

Figure 21

Work Problem 6 at the Side. ▶

EXAMPLE 7 Solving a Linear Inequality

Solve $5(k - 3) - 7k \geq 4(k - 3) + 9$. Graph the solution set.

Step 1

$$5(k - 3) - 7k \geq 4(k - 3) + 9$$

Start by clearing parentheses.

$$5k - 15 - 7k \geq 4k - 12 + 9 \quad \text{Distributive property}$$
$$-2k - 15 \geq 4k - 3 \quad \text{Combine like terms.}$$

Step 2

$$-2k - 15 - 4k \geq 4k - 3 - 4k \quad \text{Subtract } 4k.$$
$$-6k - 15 \geq -3 \quad \text{Combine like terms.}$$
$$-6k - 15 + 15 \geq -3 + 15 \quad \text{Add 15.}$$
$$-6k \geq 12 \quad \text{Combine like terms.}$$

Step 3

$$\frac{-6k}{-6} \leq \frac{12}{-6} \quad \text{Divide by } -6. \text{ Reverse the symbol.}$$

Because -6 is negative, change $\geq$ to $\leq$.

$$k \leq -2$$

The solution set is $(-\infty, -2]$. Its graph is shown in **Figure 22.**

Figure 22

Work Problem 7 at the Side. ▶

6 Solve.

$$7x - 6 + 1 \geq 5x - x + 2$$

Graph the solution set.

7 Solve.

$$-15 - (2x + 1) \geq 4(x - 1) - 3x$$

Graph the solution set.

Answers

6. $\left[\frac{7}{3}, \infty\right)$

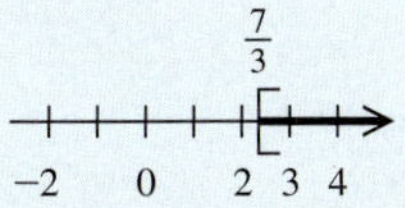

7. $(-\infty, -4]$

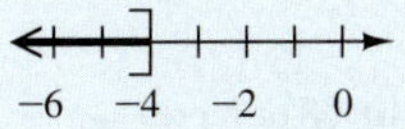

8 Translate each statement into an inequality, using x as the variable.

(a) The total cost is less than \$10.

(b) Chicago received at most 5 in. of snow.

(c) The car's speed exceeded 60 mph.

(d) You must be at least 18 yr old to vote.

9 Solve each problem.

(a) Kristine has grades of 98 and 85 on her first two tests in algebra. If she wants an average of at least 92 after her third test, what score must she make on that test?

(b) Maggie has scores of 98, 86, and 88 on her first three tests in algebra. If she wants an average of at least 90 after her fourth test, what score must she make on her fourth test?

Answers

8. **(a)** $x < 10$ **(b)** $x \le 5$ **(c)** $x > 60$ **(d)** $x \ge 18$

9. **(a)** 93 or more **(b)** 88 or more

OBJECTIVE 5 Use inequalities to solve applied problems. The table gives some common phrases that suggest inequality.

Phrase/Word	Example	Inequality
Is greater than	A number *is greater than* 4	$x > 4$
Is less than	A number *is less than* -12	$x < -12$
Exceeds	A number *exceeds* 3.5	$x > 3.5$
Is at least	A number *is at least* 6	$x \ge 6$
Is at most	A number *is at most* 8	$x \le 8$

◀ Work Problem **8** at the Side.

The next example uses the idea of finding the average of a number of scores. ***In general, to find the average of n numbers, add the numbers and divide by n.*** We use the six problem-solving steps from **Section 2.4,** changing Step 3 to "Write an inequality."

EXAMPLE 8 Finding an Average Test Score

Brent has scores of 86, 88, and 78 (out of a possible 100) on each of his first three tests in geometry. If he wants an average of at least 80 after his fourth test, what are the possible scores he can make on that test?

Step 1 **Read** the problem again.

Step 2 **Assign a variable.**

Let x = Brent's score on his fourth test.

Step 3 **Write an inequality.**

Average ↓ is at least ↓ 80. ↓

$$\frac{86 + 88 + 78 + x}{4} \ge 80 \qquad \text{To find his average after four tests, add the test scores and divide by 4.}$$

Step 4 **Solve.**

$$\frac{252 + x}{4} \ge 80 \qquad \text{Add the known scores.}$$

$$4\left(\frac{252 + x}{4}\right) \ge 4(80) \qquad \text{Multiply by 4.}$$

$$252 + x \ge 320$$

$$252 + x - 252 \ge 320 - 252 \qquad \text{Subtract 252.}$$

$$x \ge 68 \qquad \text{Combine like terms.}$$

Step 5 **State the answer.** He must score 68 or more on the fourth test to have an average of *at least* 80.

Step 6 **Check.** $\dfrac{86 + 88 + 78 + 68}{4} = \dfrac{320}{4} = 80$

To complete the check, also show that any number greater than 68 (but less than or equal to 100) makes the average greater than 80.

◀ Work Problem **9** at the Side.

OBJECTIVE 6 Solve linear inequalities with three parts. An inequality that says that one number is *between* two other numbers is a **three-part inequality.** For example,

$$-3 < 5 < 7 \quad \text{says that} \quad 5 \quad \text{is between} \quad -3 \text{ and } 7.$$

EXAMPLE 9 Graphing a Three-Part Inequality

Write the inequality $-3 \le x < 2$ in interval notation, and graph the interval.

The statement is read "-3 is less than or equal to x *and* x is less than 2." We want the set of numbers that are *between* -3 and 2, with -3 included and 2 excluded. In interval notation, we write $[-3, 2)$, using a square bracket at -3 because -3 is part of the graph and a parenthesis at 2 because 2 is not part of the graph. See **Figure 23.**

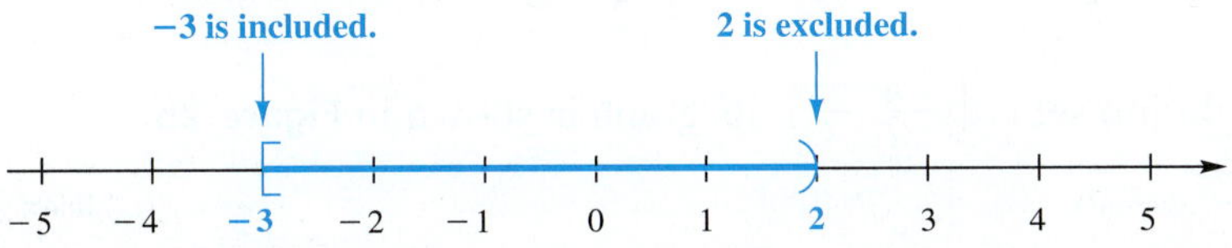

Figure 23 Graph of the interval $[-3, 2)$

Work Problem 10 at the Side. ▶

The three-part inequality $3 < x + 2 < 8$ says that $x + 2$ is between 3 and 8. We solve this inequality as follows.

$$3 - 2 < x + 2 - 2 < 8 - 2 \qquad \text{Subtract 2 from } \textit{each} \text{ part.}$$

$$1 < x < 6$$

The idea is to get the inequality in the form

a number $< x <$ another number.

CAUTION

Three-part inequalities are written so that the symbols point in the same direction and both point toward the lesser number. It would be *wrong* to write an inequality as $8 < x + 2 < 3$, since this would imply that $8 < 3$, a false statement.

EXAMPLE 10 Solving a Three-Part Inequality

Solve each inequality, and graph the solution set.

(a)
$$4 \le 3x - 5 < 10$$

$$4 + 5 \le 3x - 5 + 5 < 10 + 5 \qquad \text{Add 5 to each part.}$$

$$9 \le 3x < 15$$

$$\frac{9}{3} \le \frac{3x}{3} < \frac{15}{3} \qquad \text{Divide each part by 3.}$$

Remember to divide all *three* parts by 3.

$$3 \le x < 5$$

The solution set is $[3, 5)$. Its graph is shown in **Figure 24.**

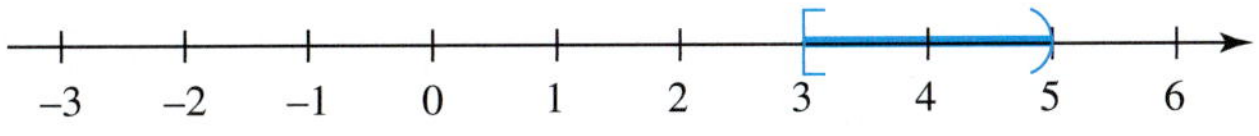

Figure 24

Continued on Next Page

10 Write each inequality in interval notation and graph the interval.

(a) $-7 < x < -2$

(b) $-6 < x \le -4$

Teaching Tip

Remind students to work with all three parts of the inequality.

Answers

10. (a) $(-7, -2)$

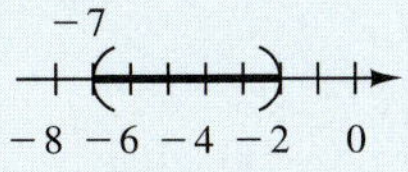

(b) $(-6, -4]$

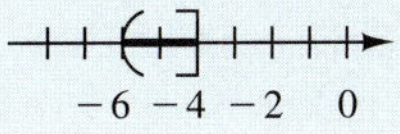

11 Solve each inequality, and graph the solution set.

(a) $2 \leq 3x - 1 \leq 8$

(b) $-4 \leq \frac{3}{2}x - 1 \leq 0$

(b)

$$-4 \leq \frac{2}{3}m - 1 < 8$$ Work with all three parts at the same time.

$$3(-4) \leq 3\left(\frac{2}{3}m - 1\right) < 3(8)$$ Multiply each part by 3 to clear the fraction.

$$-12 \leq 2m - 3 < 24$$ Multiply. Use the distributive property.

$$-12 + 3 \leq 2m - 3 + 3 < 24 + 3$$ Add 3 to each part.

$$-9 \leq 2m < 27$$

$$\frac{-9}{2} \leq \frac{2m}{2} < \frac{27}{2}$$ Divide each part by 2.

$$-\frac{9}{2} \leq m < \frac{27}{2}$$

The solution set is $\left[-\frac{9}{2}, \frac{27}{2}\right)$. Its graph is shown in **Figure 25.**

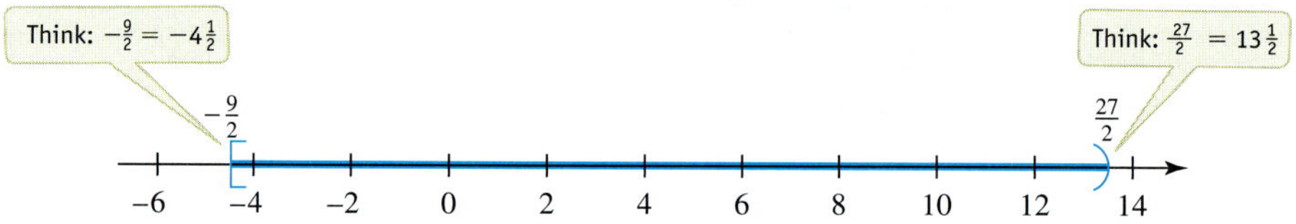

Figure 25

◀ **Work Problem 11 at the Side.**

Note

The inequality in **Example 10(b),** $-4 \leq \frac{2}{3}m - 1 < 8$, can also be solved by first adding 1 to each part and then multiplying each part by $\frac{3}{2}$. Try this.

The table summarizes solution sets of three-part inequalities.

Set-Builder Notation	Interval Notation	Graph
$\{x \mid a < x < b\}$	(a, b)	
$\{x \mid a < x \leq b\}$	$(a, b]$	
$\{x \mid a \leq x < b\}$	$[a, b)$	
$\{x \mid a \leq x \leq b\}$	$[a, b]$	

Answers

11. (a) $[1, 3]$

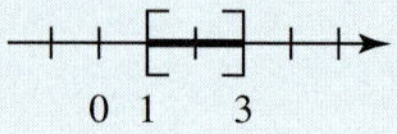

(b) $\left[-2, \frac{2}{3}\right]$

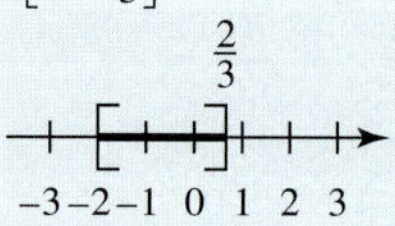

2.7 Exercises

FOR EXTRA HELP

Download the MyDashBoard App

MyMathLab®

CONCEPT CHECK *Work each problem.*

1. When graphing an inequality, use a parenthesis if the inequality symbol is __< (or >)__ or __> (or <)__. Use a square bracket if the inequality symbol is __≤ (or ≥)__ or __≥ (or ≤)__.

2. *True* or *false*? In interval notation, a square bracket is sometimes used next to an infinity symbol.

false

3. In interval notation, the set $\{x \mid x > 0\}$ is written __$(0, \infty)$__.

4. In interval notation, the set of all real numbers is written __$(-\infty, \infty)$__.

CONCEPT CHECK *Write an inequality using the variable x that corresponds to each graph of solutions on a number line.*

5.

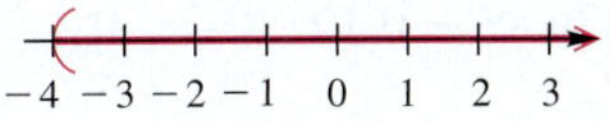

$x > -4$

6.

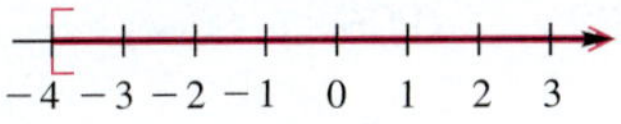

$x \geq -4$

7.

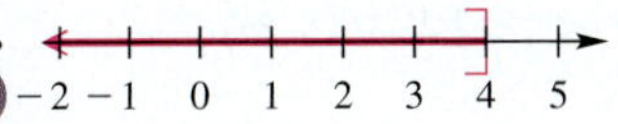

$x \leq 4$

8.

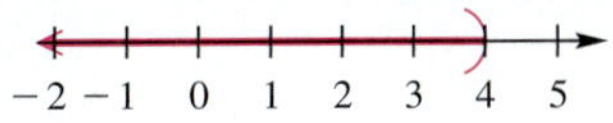

$x < 4$

9.

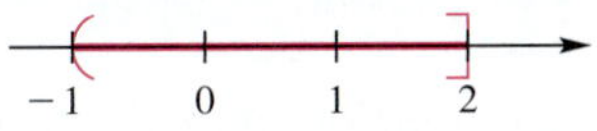

$-1 < x \leq 2$

10.

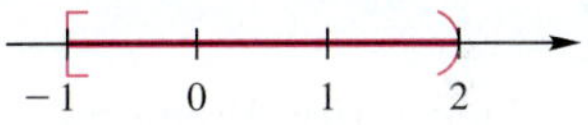

$-1 \leq x < 2$

Write each inequality in interval notation, and graph the interval. ***See Examples 1, 2, and 9.***

11. $k \leq 4$

$(-\infty, 4]$

12. $r \leq -10$

$(-\infty, -10]$

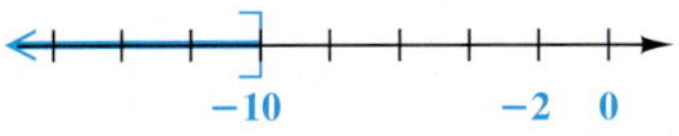

13. $x > -3$

$(-3, \infty)$

14. $x > 3$

$(3, \infty)$

15. $8 \leq x \leq 10$

$[8, 10]$

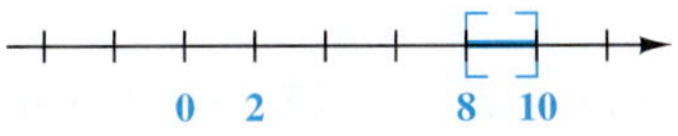

16. $3 \leq x \leq 5$

$[3, 5]$

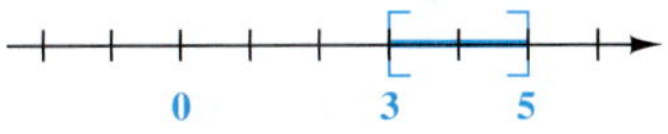

17. $0 < x \leq 10$

$(0, 10]$

18. $-3 \leq x < 5$

$[-3, 5)$

Solve each inequality. Write the solution set in interval notation, and graph it. ***See Example 3.***

19. $z - 8 \geq -7$

$[1, \infty)$

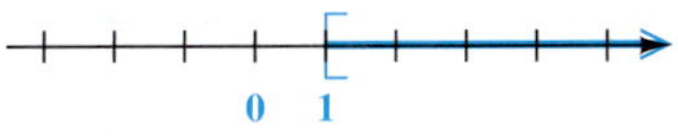

20. $p - 3 \geq -11$

$[-8, \infty)$

21. $2k + 3 \geq k + 8$

$[5, \infty)$

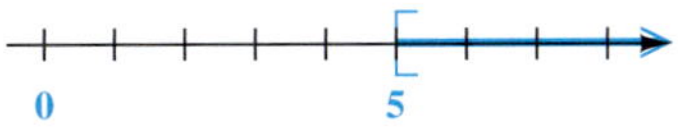

22. $3x + 7 \geq 2x + 11$

$[4, \infty)$

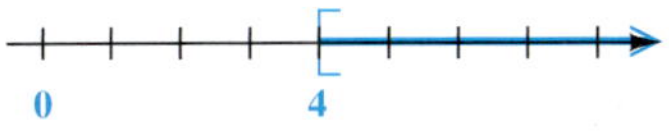

23. $3n + 5 < 2n - 1$

$(-\infty, -6)$

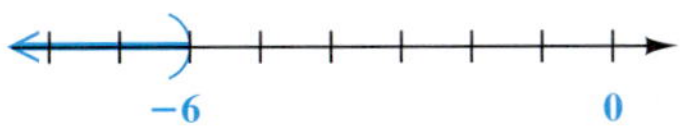

24. $5x - 2 < 4x - 5$

$(-\infty, -3)$

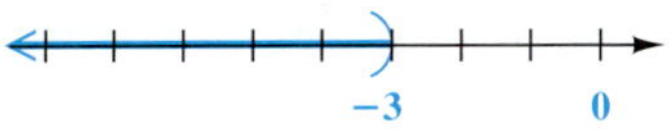

25. Under what conditions must the inequality symbol be reversed when using the multiplication property of inequality?

It must be reversed when multiplying or dividing by a negative number.

26. Explain the steps you would use to solve the inequality $-5x > 20$.

Divide by -5 and reverse the direction of the inequality symbol to get $x < -4$.

Solve each inequality. Write the solution set in interval notation, and graph it. ***See Examples 4 and 5.***

27. $3x < 18$ $(-\infty, 6)$

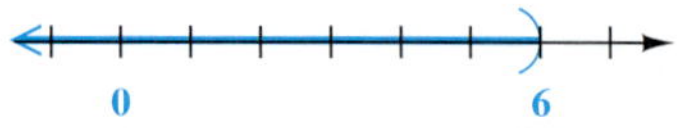

28. $5x < 35$ $(-\infty, 7)$

29. $2x \geq -20$ $[-10, \infty)$

30. $6m \geq -24$ $[-4, \infty)$

31. $-8t > 24$ $(-\infty, -3)$

32. $-7x > 49$ $(-\infty, -7)$

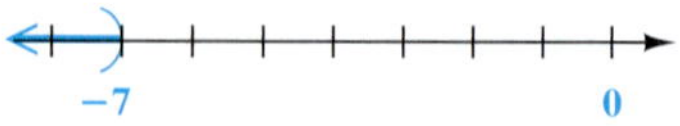

33. $-x \geq 0$ $(-\infty, 0]$

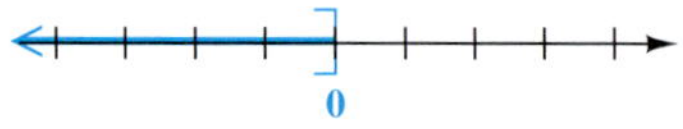

34. $-k < 0$ $(0, \infty)$

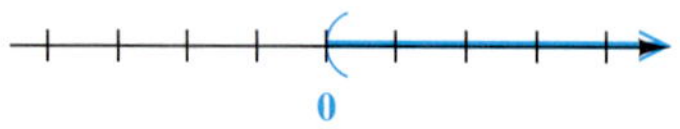

35. $-\frac{3}{4}r < -15$ $(20, \infty)$

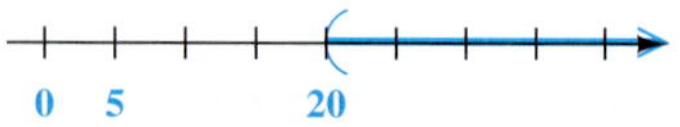

36. $-\frac{7}{8}t < -14$ $(16, \infty)$

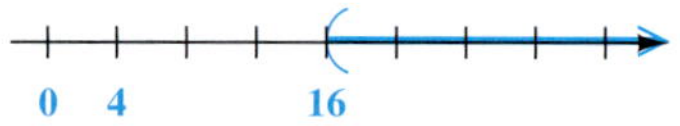

37. $-0.02x \leq 0.06$ $[-3, \infty)$

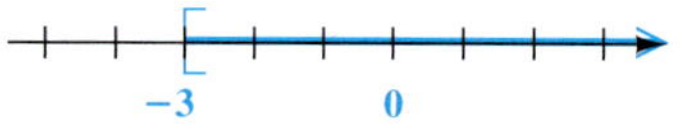

38. $-0.03v \geq -0.12$ $(-\infty, 4]$

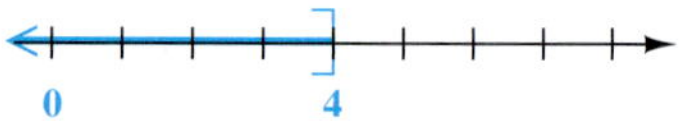

Solve each inequality. Write the inequality in interval notation, and graph it. ***See Examples 3–7.***

39. $8x + 9 \leq -15$ $(-\infty, -3]$

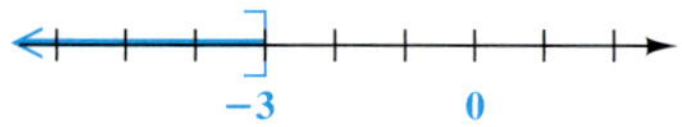

40. $6x + 7 \leq -17$ $(-\infty, -4]$

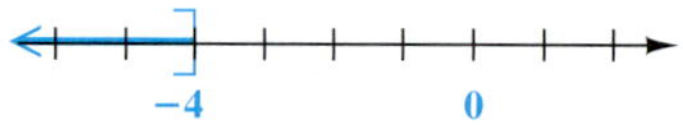

41. $-4x - 3 < 1$ $(-1, \infty)$

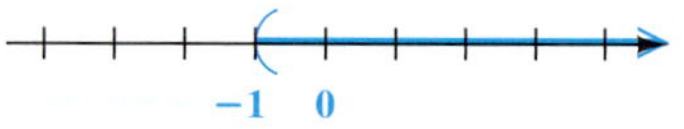

42. $-5x - 4 < 6$ $(-2, \infty)$

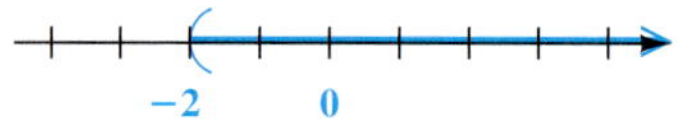

43. $5r + 1 \geq 3r - 9$ $[-5, \infty)$

44. $6t + 3 < 3t + 12$ $(-\infty, 3)$

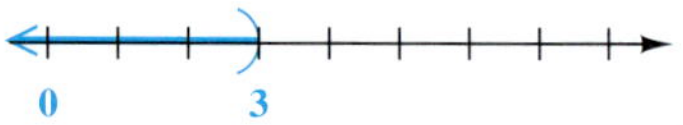

45. $6x + 3 + x < 2 + 4x + 4$

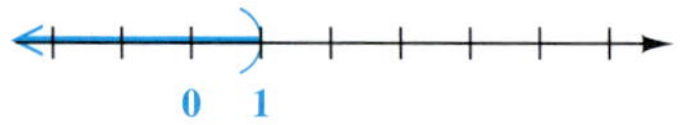

$(-\infty, 1)$

46. $-4w + 12 + 9w \geq w + 9 + w$

$[-1, \infty)$

47. $-x + 4 + 7x \leq -2 + 3x + 6$

$(-\infty, 0]$

48. $14x - 6 + 7x > 4 + 10x - 10$

$(0, \infty)$

49. $5(t - 1) > 3(t - 2)$ $\left(-\frac{1}{2}, \infty\right)$

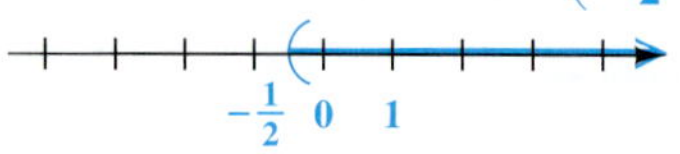

50. $7(m - 2) < 4(m - 4)$ $\left(-\infty, -\frac{2}{3}\right)$

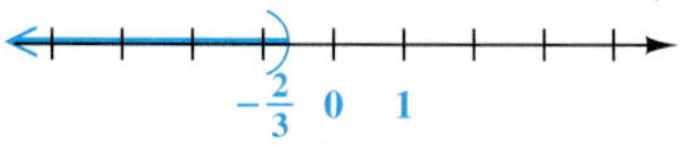

51. $5(x + 3) - 6x \leq 3(2x + 1) - 4x$ $[4, \infty)$

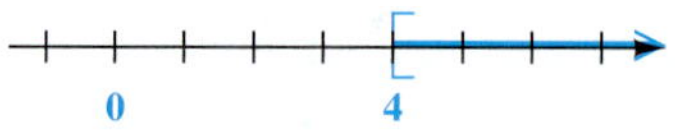

52. $2(x - 5) + 3x < 4(x - 6) + 1$ $(-\infty, -13)$

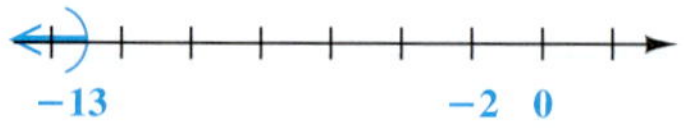

53. $\frac{2}{3}(p + 3) > \frac{5}{6}(p - 4)$ $(-\infty, 32)$

54. $\frac{7}{9}(n - 4) \leq \frac{4}{3}(n + 5)$ $\left[-\frac{88}{5}, \infty\right)$

55. $4x - (6x + 1) \leq 8x + 2(x - 3)$ $\left[\frac{5}{12}, \infty\right)$

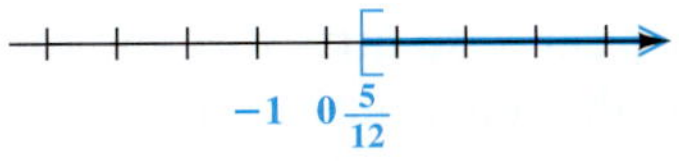

56. $2x - (4x + 3) < 6x + 3(x + 4)$ $\left(-\frac{15}{11}, \infty\right)$

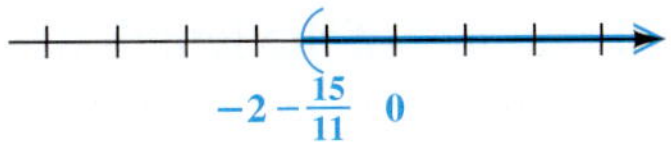

57. $5(2k + 3) - 2(k - 8) > 3(2k + 4) + k - 2$

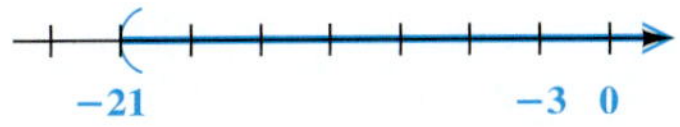

$(-21, \infty)$

58. $2(3z - 5) + 4(z + 6) \geq 2(3z + 2) + 3z - 15$

$[-25, \infty)$

CONCEPT CHECK *Translate each statement into an inequality. Use x as the variable.*

59. You must be at least 16 yr old to drive.
$x \geq 16$

60. Less than 1 in. of rain fell.
$x < 1$

61. Denver received more than 8 in. of snow.
$x > 8$

62. A full-time student must take at least 12 credits.
$x \geq 12$

63. Tracy could spend at most \$20 on a gift.
$x \leq 20$

64. The wind speed exceeded 40 mph.
$x > 40$

Solve each problem. ***See Example 8.***

65. John Douglas has grades of 84 and 98 on his first two history tests. What must he score on his third test so that his average is at least 90?
88 or more

66. Elizabeth Gainey has scores of 74 and 82 on her first two algebra tests. What must she score on her third test so that her average is at least 80?
84 or more

67. A student has scores of 87, 84, 95, and 79 on four quizzes. What must she score on the fifth quiz to have an average of at least 85?
80 or more

68. Another student has scores of 82, 93, 94, and 86 on four quizzes. What must he score on the fifth quiz to have an average of at least 90?
95 or more

69. When 2 is added to the difference between six times a number and 5, the result is greater than 13 added to 5 times the number. Find all such numbers.
all numbers greater than 16

70. When 8 is subtracted from the sum of three times a number and 6, the result is less than 4 more than the number. Find all such numbers.
all numbers less than 3

71. The formula for converting Celsius temperature to Fahrenheit is $F = \frac{9}{5}C + 32$. The Fahrenheit temperature of Providence, Rhode Island, has never exceeded 104°. How would you describe this using Celsius temperature?
It has never exceeded 40°C.

72. The formula for converting Fahrenheit temperature to Celsius is $C = \frac{5}{9}(F - 32)$. If the Celsius temperature on a certain day in San Diego, California, is never more than 25°, how would you describe the corresponding Fahrenheit temperature?
It is never more than 77°F.

73. For what values of x would the rectangle have perimeter of at least 400?

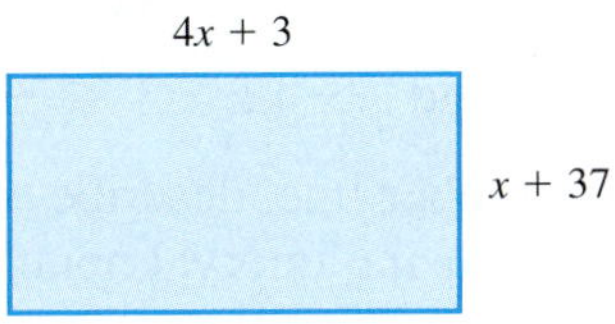

32 or greater

74. For what values of x would the triangle have perimeter of at least 72?

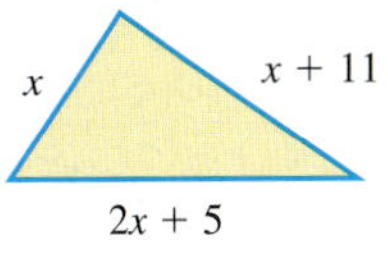

14 or greater

75. An international phone call costs \$2.00, plus \$0.30 per minute or fractional part of a minute. If x represents the number of minutes of the length of the call, then $2 + 0.30x$ represents the cost of the call. If Jorge has \$5.60 to spend on a call, what is the maximum total time he can use the phone?
12 min

76. At the Speedy Gas 'n Go, a car wash costs \$4.50, and gasoline is selling for \$3.40 per gal. Terri Hoelker has \$43.60 to spend, and her car is so dirty that she must have it washed. What is the maximum number of gallons of gasoline that she can purchase?
11.5 gal

Solve each inequality. Write the solution set in interval notation, and graph it. ***See Example 10.***

77. $-5 \le 2x - 3 \le 9$ $[-1, 6]$

78. $-7 \le 3x - 4 \le 8$ $[-1, 4]$

79. $10 < 7p + 3 < 24$ $(1, 3)$

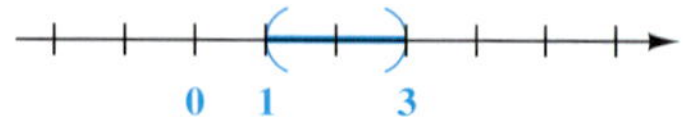

80. $-8 \le 3r - 1 \le -1$ $\left[-\frac{7}{3}, 0\right]$

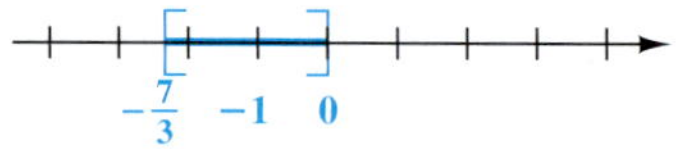

81. $-12 < -1 + 6m \le -5$ $\left(-\frac{11}{6}, -\frac{2}{3}\right]$

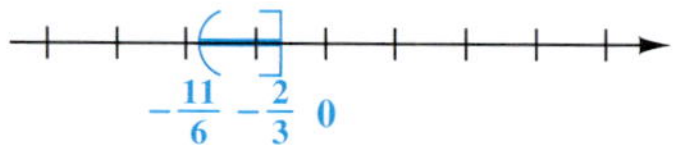

82. $-14 \le 1 + 5q < 3$ $\left[-3, \frac{2}{5}\right)$

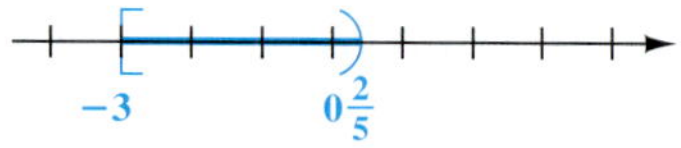

83. $-12 \le \frac{1}{2}z + 1 \le 4$ $[-26, 6]$

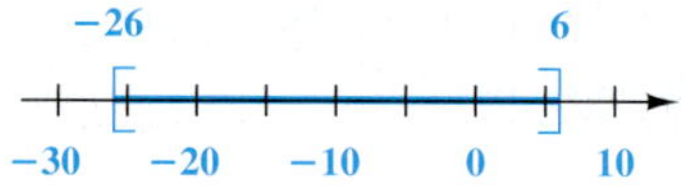

84. $-6 \le 3 + \frac{1}{3}x \le 5$ $[-27, 6]$

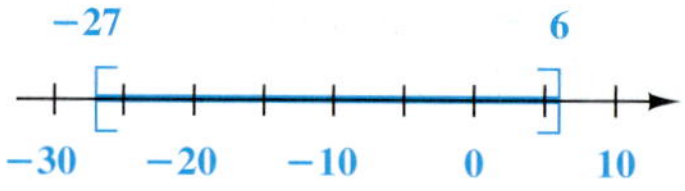

85. $1 \le \frac{2}{3}p + 3 \le 7$ $[-3, 6]$

86. $2 < \frac{3}{4}x + 6 < 12$ $\left(-\frac{16}{3}, 8\right)$

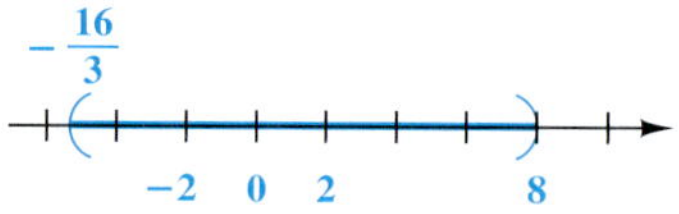

Relating Concepts (Exercises 87–90) For Individual or Group Work

Work Exercises 87–90 in order, *to see the connection between the solution of an equation and the solutions of the corresponding inequalities.*

In Exercises 87 and 88, solve and graph the solutions.

87. $3x + 2 = 14$ $\{4\}$

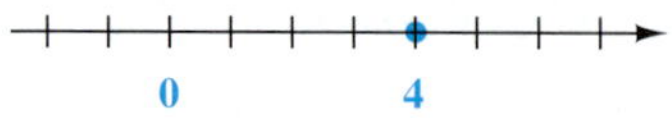

88. (a) $3x + 2 < 14$ $(-\infty, 4)$

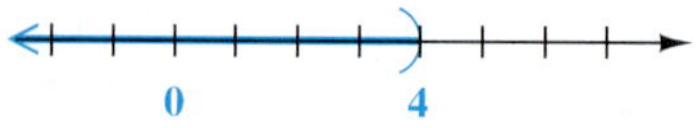

(b) $3x + 2 > 14$ $(4, \infty)$

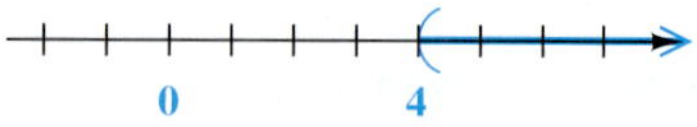

89. Now graph all the solutions together on the following number line.

Describe the graph.
It is the set of all real numbers.

90. Based on the results from **Exercises 87–89,** if we were to graph the solutions of

$$-4x + 3 = -1, \quad -4x + 3 > -1,$$
$$\text{and} \quad -4x + 3 < -1$$

on the same number line, describe the graph. Give this solution set using interval notation.
The graph would be the set of all real numbers; $(-\infty, \infty)$

Chapter 2 Summary

Key Terms

2.1

linear equation A linear equation in one variable is an equation that can be written in the form $Ax + B = C$, where A, B, and C are real numbers, with $A \neq 0$.

solution set The set of all solutions of an equation is its solution set.

equivalent equations Equations that have exactly the same solution sets are equivalent equations.

2.3

conditional equation A conditional equation is an equation that is true for some values of the variable and false for others.

identity An identity is an equation that is true for all values of the variable.

contradiction A contradiction is an equation with no solution.

2.4

consecutive integers Two integers that differ by 1 are consecutive integers.

consecutive even (or odd) integers Two even (or odd) integers that differ by 2 are consecutive even (or odd) integers.

complementary angles Two angles whose measures have a sum of 90° are complementary angles.

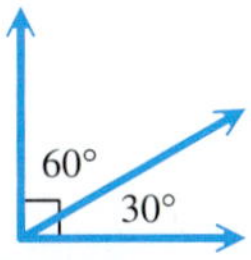

right angle A right angle measures 90°.

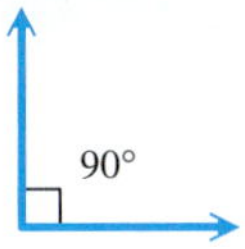

supplementary angles Two angles whose measures have a sum of 180° are supplementary angles.

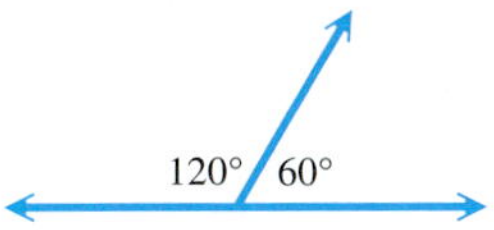

straight angle A straight angle measures 180°.

2.5

formula A formula is an equation in which variables are used to describe a relationship.

area The area of a plane geometric figure is a measure of the surface covered by the figure.

perimeter The perimeter of a plane geometric figure is the distance around the figure.

vertical angles Vertical angles are angles formed by intersecting lines. They have the same measure. (In the figure, ① and ③ are vertical angles, as are ② and ④.)

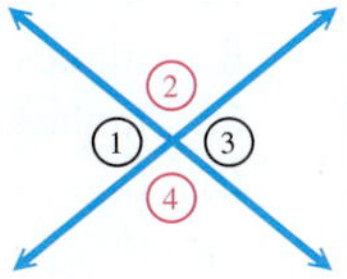

2.6

ratio A ratio is a comparison of two quantities using a quotient.

proportion A proportion is a statement that two ratios are equal.

cross products of a proportion The method of cross products provides a way of determining whether a proportion is true.

$$\frac{a}{b} = \frac{c}{d}$$

ad and *bc* are cross products.

terms In the proportion $\frac{a}{b} = \frac{c}{d}$, a, b, c, and d are the terms. The a and d terms are the **extremes,** and the b and c terms are the **means.**

2.7

inequality Inequalities are statements relating algebraic expressions using $<$, $\leq$, $>$, or $\geq$.

linear inequality A linear inequality in one variable can be written in the form $Ax + B < C$, $Ax + B \leq C$, $Ax + B > C$, or $Ax + B \geq C$, where A, B, and C are real numbers, with $A \neq 0$.

interval An interval is a portion of a number line.

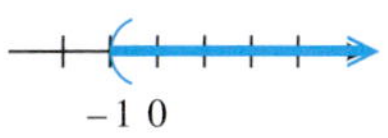

The interval $(-1, \infty)$

interval notation Interval notation is a special notation that uses parentheses ( ) and/or brackets [ ] to describe an interval on a number line.

three-part inequality An inequality that says that one number is between two other numbers is a three-part inequality.

New Symbols

$\emptyset$	empty (null) set	a to b, $a:b$, or $\frac{a}{b}$	the ratio of a to b	∞	infinity
$1°$	one degree	(a, b)	interval notation for $a < x < b$	$-\infty$	negative infinity
⅂	right angle	$[a, b]$	interval notation for $a \le x \le b$	$(-\infty, \infty)$	set of all real numbers

Test Your Word Power

See how well you have learned the vocabulary in this chapter.

1. A **solution** of an equation is a number that
   A. makes an expression undefined
   B. makes the equation false
   C. makes the equation true
   D. makes an expression equal to 0.

2. **Complementary angles** are angles
   A. formed by two parallel lines
   B. whose sum is 90°
   C. whose sum is 180°
   D. formed by perpendicular lines.

3. **Supplementary angles** are angles
   A. formed by two parallel lines
   B. whose sum is 90°
   C. whose sum is 180°
   D. formed by perpendicular lines.

4. A **ratio**
   A. compares two quantities using a quotient
   B. says that two quotients are equal
   C. is a product of two quantities
   D. is a difference between two quantities.

5. A **proportion**
   A. compares two quantities using a quotient
   B. says that two ratios are equal
   C. is a product of two quantities
   D. is a difference between two quantities.

Answers to Test Your Word Power

1. C; *Example:* 8 is the solution of $2x + 5 = 21$.

2. B; *Example:* Angles with measures 35° and 55° are complementary angles.

3. C; *Example:* Angles with measures 112° and 68° are supplementary angles.

4. A; *Example:* $\frac{7 \text{ in.}}{12 \text{ in.}} = \frac{7}{12}$

5. B; *Example:* $\frac{2}{3} = \frac{8}{12}$

Quick Review

Concepts	*Examples*
2.1 The Addition Property of Equality The same number may be added to (or subtracted from) each side of an equation without changing the solution set.	Solve. $x - 6 = 12$ $x - 6 + 6 = 12 + 6$ Add 6. $x = 18$ Combine like terms. Solution set: $\{18\}$
2.2 The Multiplication Property of Equality Each side of an equation may be multiplied (or divided) by the same nonzero number without changing the solution set.	Solve. $\frac{3}{4}x = -9$ $\frac{4}{3} \cdot \frac{3}{4}x = \frac{4}{3}(-9)$ Multiply by $\frac{4}{3}$. $x = -12$ Solution set: $\{-12\}$

Concepts	*Examples*
2.3 More on Solving Linear Equations	
Step 1 **Simplify each side separately.**	Solve. $2x + 2(x + 1) = 14 + x$ $2x + 2x + 2 = 14 + x$ Distributive property $4x + 2 = 14 + x$ Combine like terms.
Step 2 **Isolate the variable term on one side.**	$4x + 2 - x - 2 = 14 + x - x - 2$ Subtract x. Subtract 2. $3x = 12$ Combine like terms.
Step 3 **Isolate the variable.**	$\frac{3x}{3} = \frac{12}{3}$ Divide by 3. $x = 4$
Step 4 **Check.**	CHECK $2(4) + 2(4 + 1) \stackrel{?}{=} 14 + 4$ Let $x = 4$. $18 = 18$ ✓ True Solution set: $\{4\}$
2.4 An Introduction to Applications of Linear Equations	
Step 1 **Read.**	One number is 5 more than another. Their sum is 21. What are the numbers? We are looking for two numbers.
Step 2 **Assign a variable.**	Let $x =$ the lesser number. Then $x + 5 =$ the greater number.
Step 3 **Write an equation.**	$x + (x + 5) = 21$
Step 4 **Solve the equation.**	$2x + 5 = 21$ Combine like terms. $2x = 16$ Subtract 5. $x = 8$ Divide by 2.
Step 5 **State the answer.**	The numbers are 8 and 13.
Step 6 **Check.**	13 is 5 more than 8, and $8 + 13 = 21$. The answer checks.
2.5 Formulas and Additional Applications from Geometry	
To find the value of one of the variables in a formula, given values for the others, substitute the known values into the formula.	Find L if $A = LW$, given that $A = 24$ and $W = 3$. $24 = L \cdot 3$ $A = 24, W = 3$ $\frac{24}{3} = \frac{L \cdot 3}{3}$ Divide by 3. $8 = L$
To solve a formula for one of the variables, isolate that variable by treating the other variables as numbers and using the steps for solving equations.	Solve $P = 2a + 2b$ for b. $P - 2a = 2a + 2b - 2a$ Subtract $2a$. $P - 2a = 2b$ Combine like terms. $\frac{P - 2a}{2} = \frac{2b}{2}$ Divide by 2. $\frac{P - 2a}{2} = b$, or $b = \frac{P - 2a}{2}$

Concepts	Examples

2.6 Ratio, Proportion, and Percent

To write a ratio, express quantities in the same units.

4 ft to 8 in. can be written **48 in.** to 8 in., which is the ratio $\frac{48}{8}$, or $\frac{6}{1}$.

To solve a proportion, use the method of cross products.

Solve. $\frac{x}{12} = \frac{35}{60}$

$$60x = 12 \cdot 35 \quad \text{Cross products}$$
$$60x = 420 \quad \text{Multiply.}$$
$$x = 7 \quad \text{Divide by 60.}$$

Solution set: $\{7\}$

To solve a percent problem, use the percent equation.

amount = percent (as a decimal) · base

65 is what percent of 325?

$$65 = p \cdot 325$$
$$\frac{65}{325} = p$$
$$\mathbf{0.2} = p, \quad \text{or} \quad \mathbf{20\%} = p$$

65 is **20%** of 325.

2.7 Solving Linear Inequalities

Step 1 **Simplify each side separately.**

Step 2 **Isolate the variable term on one side.**

Step 3 **Isolate the variable.**

Be sure to reverse the direction of the inequality symbol when multiplying or dividing by a negative number.

Solve and graph the solution set.

$$3(1 - x) + 5 - 2x > 9 - 6$$
$$3 - 3x + 5 - 2x > 9 - 6 \quad \text{Distributive property}$$
$$8 - 5x > 3 \quad \text{Combine like terms.}$$
$$8 - 5x - 8 > 3 - 8 \quad \text{Subtract 8.}$$
$$-5x > -5 \quad \text{Combine like terms.}$$
$$\frac{-5x}{-5} < \frac{-5}{-5} \quad \text{Divide by } -5. \text{ Change } > \text{ to } <.$$
$$x < 1$$

Solution set: $(-\infty, 1)$

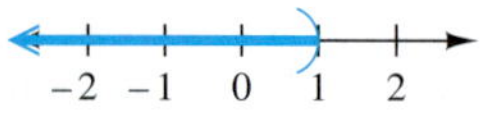

To solve a three-part inequality such as

$$4 < 2x + 6 < 8,$$

work with all three parts at the same time.

Solve. $4 < 2x + 6 < 8$

$$4 - 6 < 2x + 6 - 6 < 8 - 6 \quad \text{Subtract 6.}$$
$$-2 < 2x < 2$$
$$\frac{-2}{2} < \frac{2x}{2} < \frac{2}{2} \quad \text{Divide by 2.}$$
$$-1 < x < 1$$

Solution set: $(-1, 1)$

Chapter 2 Review Exercises

2.1–2.3 *Solve each equation. Check the solution.*

1. $x - 7 = 2$
$\{9\}$

2. $4r - 6 = 10$
$\{4\}$

3. $5x + 8 = 4x + 2$
$\{-6\}$

4. $8t = 7t + \frac{3}{2}$
$\left\{\frac{3}{2}\right\}$

5. $(4r - 8) - (3r + 12) = 0$
$\{20\}$

6. $7(2x + 1) = 6(2x - 9)$
$\left\{-\frac{61}{2}\right\}$

7. $-\frac{6}{5}y = -18$
$\{15\}$

8. $\frac{1}{2}r - \frac{1}{6}r + 3 = 2 + \frac{1}{6}r + 1$
$\{0\}$

9. $3x - (-2x + 6) = 4(x - 4) + x$
$\emptyset$

10. $0.10(x + 80) + 0.20x = 8 + 0.30x$
{all real numbers}

2.4 *Solve each problem.*

11. If 7 is added to five times a number, the result is equal to three times the number. Find the number.
$-\frac{7}{2}$

12. If 4 is subtracted from twice a number, the result is 36. Find the number.
20

13. The land area of Hawaii is 5213 mi^2 greater than that of Rhode Island. Together, the areas total 7637 mi^2. What is the area of each state?
Hawaii: 6425 mi^2; Rhode Island: 1212 mi^2

14. The height of Seven Falls in Colorado is $\frac{5}{2}$ the height (in feet) of Twin Falls in Idaho. The sum of the heights is 420 ft. Find the height of each.
Seven Falls: 300 ft; Twin Falls: 120 ft

15. The supplement of an angle measures 10 times the measure of its complement. What is the measure of the angle (in degrees)?
80°

16. Find two consecutive odd integers such that when the lesser is added to twice the greater, the result is 24 more than the greater integer.
11, 13

2.5 *A formula is given in each exercise, along with the values for all but one of the variables. Find the value of the variable that is not given. (For Exercises 19 and 20, use 3.14 as an approximation for π.)*

17. $A = \frac{1}{2}bh$; $A = 44$, $b = 8$
$h = 11$

18. $A = \frac{1}{2}h(b + B)$; $b = 3$, $B = 4$, $h = 8$
$A = 28$

19. $C = 2\pi r$; $C = 29.83$
$r = 4.75$

20. $V = \frac{4}{3}\pi r^3$; $r = 9$
$V = 3052.08$

Solve each formula for the specified variable.

21. $A = bh$ for h $\quad h = \frac{A}{b}$

22. $A = \frac{1}{2}h(b + B)$ for h $\quad h = \frac{2A}{b + B}$

23. Solve each equation for y. We give one possible answer in each case. There are other correct forms.

(a) $x + y = 11$

$y = -x + 11$

(b) $3x - 2y = 12$

$y = \frac{3}{2}x - 6$

Find the measure of each marked angle.

24.

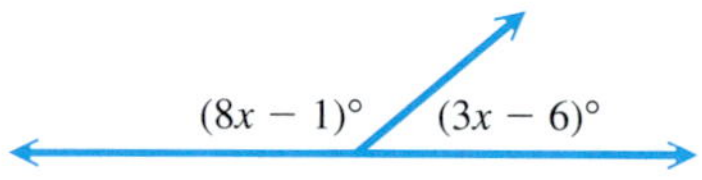

135°, 45°

25.

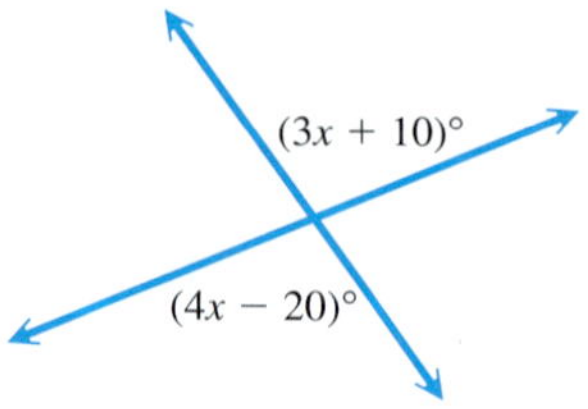

100°, 100°

Solve each application of geometry.

26. A cinema screen in Indonesia has length 92.75 ft and width 70.5 ft. What is the perimeter? What is the area? (*Source: Guinness World Records.*)

perimeter: 326.5 ft; area: 6538.875 ft^2

27. A Montezuma cypress in Mexico is 137 ft tall and has a circumference of about 146.9 ft. What is the diameter of the tree? What is its radius? Use 3.14 as an approximation for π. Round answers to the nearest hundredth. (*Source: Guinness World Records.*)

diameter: 46.78 ft; radius: 23.39 ft

2.6 *Write a ratio for each word phrase, giving fractions in lowest terms.*

28. 60 cm to 40 cm $\frac{3}{2}$

29. 5 days to 2 weeks $\frac{5}{14}$

30. 90 in. to 10 ft $\frac{3}{4}$

Solve each equation.

31. $\frac{p}{21} = \frac{5}{30}$ $\left\{\frac{7}{2}\right\}$

32. $\frac{5+x}{3} = \frac{2-x}{6}$ $\left\{-\frac{8}{3}\right\}$

33. $\frac{y}{5} = \frac{6y-5}{11}$ $\left\{\frac{25}{19}\right\}$

34. Explain how 40% can be expressed as a ratio of two whole numbers.

40% means $\frac{40}{100}$, or $\frac{2}{5}$. It is the same as the ratio 2 to 5.

Solve each problem.

35. If 2 lb of fertilizer will cover 150 ft^2 of lawn, how many pounds would be needed to cover 500 ft^2?

$6\frac{2}{3}$ lb

36. If 8 oz of medicine must be mixed with 20 oz of water, how many ounces of medicine must be mixed with 90 oz of water?

36 oz

37. The distance between two cities on a road map is 32 cm. The two cities are actually 150 km apart. The distance on the map between two other cities is 80 cm. How far apart are these cities?

375 km

38. Find the best buy. Give the unit price to the nearest thousandth for that size. (*Source:* Jewel-Osco.)

18 oz; $0.249

CEREAL

Size	Price
9 oz	$3.49
14 oz	$3.99
18 oz	$4.49

39. What is 8% of 75? 6

40. What percent of 12 is 21? 175%

41. 6 is what percent of 18? $33\frac{1}{3}\%$

42. 36% of what number is 900? 2500

43. Nicholas paid \$28,790, including sales tax, for his 2012 Prius v in the base model. The sales tax rate where he lives is 6%. What was the actual price of the car to the nearest dollar? (*Source:* www.toyota.com)

\$27,160

44. Maureen, from the mathematics editorial division of Pearson Education, took the mathematics faculty from a community college out to dinner. The bill was \$304.75. Maureen added a 15% tip, and paid for the meal with her corporate credit card. What was the total price she paid to the nearest cent?

\$350.46

2.7 *Write each inequality in interval notation, and graph it.*

45. $p \geq -4$

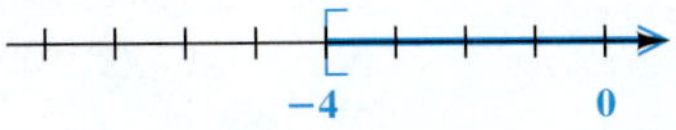

$[-4, \infty)$

46. $x < 7$

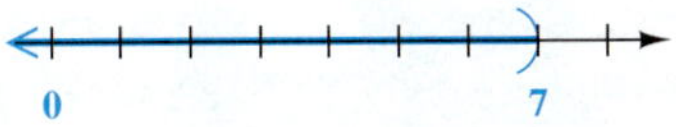

$(-\infty, 7)$

47. $-5 \leq k < 6$

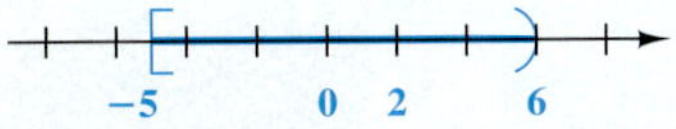

$[-5, 6)$

48. $r \geq \frac{1}{2}$

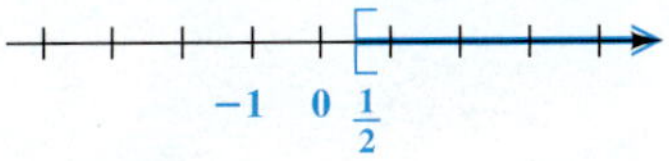

$\left[\frac{1}{2}, \infty\right)$

Solve each inequality. Write the solution set in interval notation, and graph it.

49. $x + 6 \geq 3$

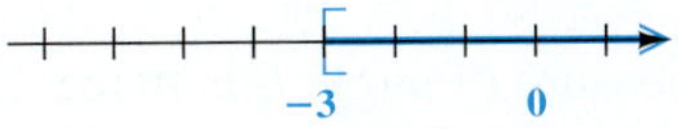

$[-3, \infty)$

50. $5t < 4t + 2$

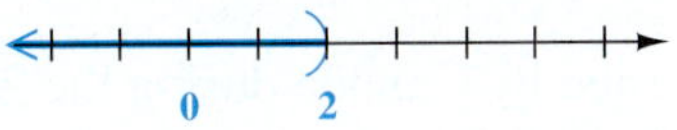

$(-\infty, 2)$

51. $-6x \leq -18$

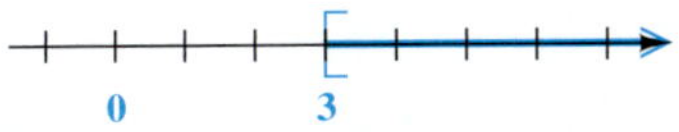

$[3, \infty)$

52. $8(k - 5) - (2 + 7k) \geq 4$

$[46, \infty)$

53. $4x - 3x > 10 - 4x + 7x$

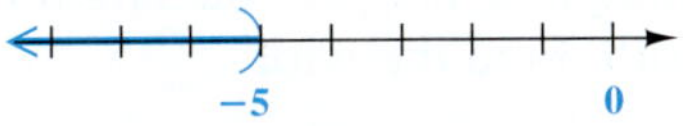

$(-\infty, -5)$

54. $3(2w + 5) + 4(8 + 3w) < 5(3w + 2) + 2w$

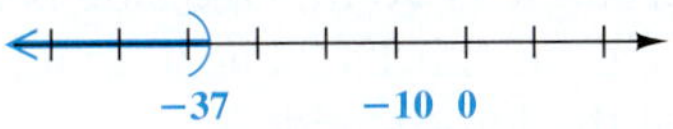

$(-\infty, -37)$

55. $-3 \leq 2x + 1 < 4$

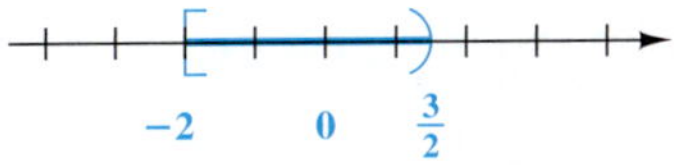

$\left[-2, \frac{3}{2}\right)$

56. $8 < 3x + 5 \leq 20$

$(1, 5]$

57. Justin Sudak has grades of 94 and 88 on his first two calculus tests. What possible scores on a third test will give him an average of at least 90?

88 or more

58. If nine times a number is added to 6, the result is at most 3. Find all such numbers.

all numbers less than or equal to $-\frac{1}{3}$

Mixed Review Exercises

Solve.

59. $\frac{x}{7} = \frac{x - 5}{2}$ $\{7\}$

60. $d = 2r$ for r $r = \frac{d}{2}$

61. $-2x > -4$ $(-\infty, 2)$

62. $2k - 5 = 4k + 13$ $\{-9\}$

63. $0.05x + 0.02x = 4.9$ $\{70\}$

64. $2 - 3(t - 5) = 4 + t$ $\left\{\frac{13}{4}\right\}$

65. $9x - (7x + 2) = 3x + (2 - x)$ $\emptyset$

66. $\frac{1}{3}s + \frac{1}{2}s + 7 = \frac{5}{6}s + 5 + 2$ {all real numbers}

67. In 2010, the top-selling frozen pizza brands, DiGiorno and Red Baron, together had sales of $937.5 million. Red Baron's sales were $399.9 million less than DiGiorno's. What were sales in millions for each brand? (*Source:* www.aibonline.org)

DiGiorno: $668.7 million; Red Baron: $268.8 million

68. On a world globe, the distance between Capetown and Bangkok, two cities that are actually 10,080 km apart, is 12.4 in. The actual distance between Moscow and Berlin is 1610 km. How far apart are Moscow and Berlin on this globe, to the nearest inch?

2.0 in.

69. Of the 26 medals earned by Canada during the 2010 Winter Olympic games, there were two times as many gold as silver medals and 9 more gold than bronze medals. How many of each medal did Canada earn? (*Source: World Almanac and Book of Facts.*)

gold: 14; silver: 7; bronze: 5

70. In triangle *DEF*, the measure of angle *E* is twice the measure of angle *D*. Angle *F* has measure 18° less than six times the measure of angle *D*. Find the measure of each angle.

***D*: 22°; *E*: 44°; *F*: 114°**

71. The perimeter of a triangle is 96 m. One side is twice as long as another, and the third side is 30 m long. What is the length of the longest side?

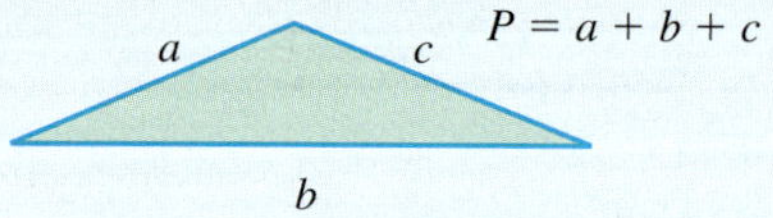

44 m

72. The perimeter of a rectangle is 288 ft. The length is 4 ft longer than the width. Find the width.

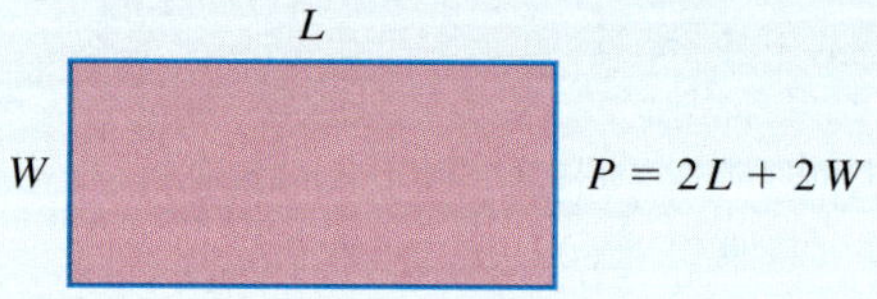

70 ft

73. Find the best buy. Give the unit price to the nearest thousandth for that size. (*Source:* Jewel-Osco.)

LAUNDRY DETERGENT

Size	Price
50 oz	$3.99
100 oz	$7.29
160 oz	$9.99

160 oz; $0.062

74. Find the measure of each marked angle.

24°, 66°

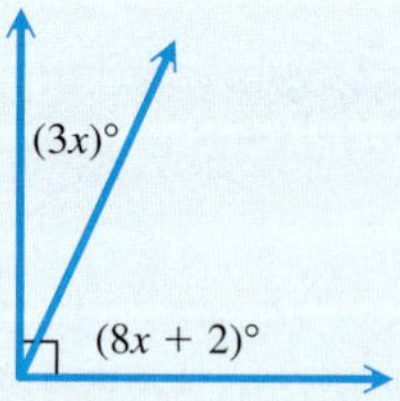

75. Latarsha has grades of 82 and 96 on her first two English tests. What must she make on her third test so that her average will be at least 90?

92 or more

76. If nine pairs of jeans cost $355.50, find the cost of five pairs. (Assume all are equally priced.)

$197.50

Chapter 2 Test

CHAPTER Test Prep VIDEO — *The Chapter Test Prep Videos with test solutions are available on DVD, in MyMathLab, and on YouTube—search "LialIntroductoryAlg" and click on "Channels."*

Solve each equation, and check your solution.

1. $3x - 7 = 11$

$\{6\}$

2. $5x + 9 = 7x + 21$

$\{-6\}$

3. $2 - 3(x - 5) = 3 + (x + 1)$

$\left\{\frac{13}{4}\right\}$

4. $2.3x + 13.7 = 1.3x + 2.9$

$\{-10.8\}$

5. $-\frac{4}{7}x = -12$

$\{21\}$

6. $-8(2x + 4) = -4(4x + 8)$

{all real numbers}

7. $0.06(x + 20) + 0.08(x - 10) = 4.6$

$\{30\}$

8. $7 - (m - 4) = -3m + 2(m + 1)$

$\emptyset$

Solve each problem.

9. The Chicago Bulls finished with the best record for the 2010–2011 NBA regular season. They won 2 more than three times as many games as they lost. They played 82 games. How many games did they win and lose? (*Source:* www.nba.com)

wins: 62; losses: 20

10. Three islands in the Hawaiian island chain are Hawaii (the Big Island), Maui, and Kauai. Together, their areas total 5300 mi^2. The island of Hawaii is 3293 mi^2 larger than the island of Maui, and Maui is 177 mi^2 larger than Kauai. What is the area of each island?

Hawaii: 4021 mi^2; Maui: 728 mi^2; Kauai: 551 mi^2

11. If the lesser of two consecutive even integers is tripled, the result is 20 more than twice the greater integer. Find the two integers.

24, 26

12. Find the measure of an angle if its supplement measures 10° more than three times its complement.

50°

13. The formula for the perimeter of a rectangle is $P = 2L + 2W$.

(a) Solve for W.

$W = \frac{P - 2L}{2}$

(b) If $P = 116$ and $L = 40$, find the value of W.

18

14. Solve the following equation for y.

$$5x - 4y = 8$$

$y = \frac{5}{4}x - 2$ (Other correct answers are possible.)

Find the measure of each marked angle.

15.

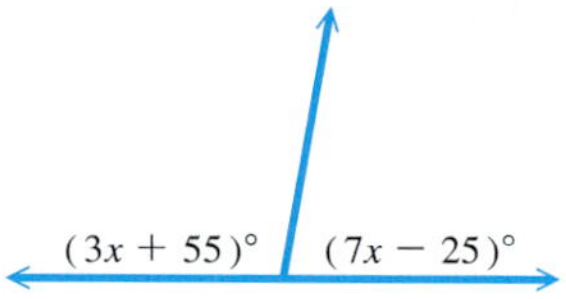

100°, 80°

16.

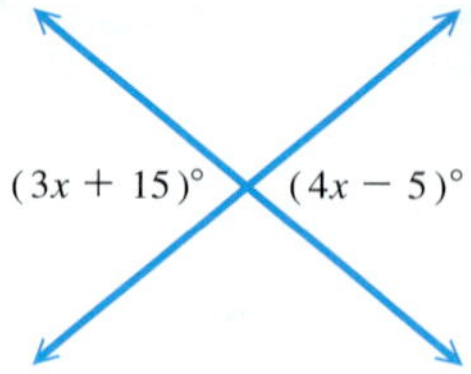

75°, 75°

Solve each equation.

17. $\frac{z}{8} = \frac{12}{16}$

{6}

18. $\frac{x+5}{3} = \frac{x-3}{4}$

{−29}

Solve each problem.

19. Find the best buy. Give the unit price to the nearest thousandth for that size. (*Source:* Jewel-Osco.)

PROCESSED CHEESE SLICES

Size	Price
8 oz	\$2.99
16 oz	\$3.99
48 oz	\$14.69

16 oz; \$0.249

20. The distance between Milwaukee and Boston is 1050 mi. On a certain map, this distance is represented by 42 in. On the same map, Seattle and Cincinnati are 92 in. apart. What is the actual distance between Seattle and Cincinnati?

2300 mi

21. An eReader regularly priced at \$149 is on sale at 18% off. Find the amount of the discount and the sale price of the eReader.

\$26.82; \$122.18

22. What percent of 65 is 26?

40%

23. Write an inequality involving x that describes the numbers graphed.

(a)

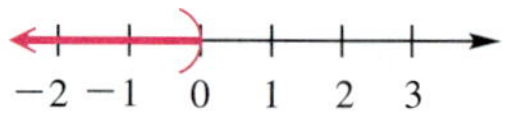

$x < 0$

(b)

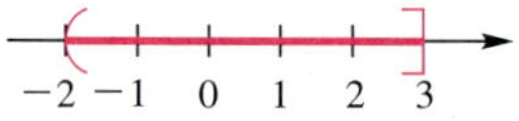

$-2 < x \le 3$

Solve each inequality. Write the solution set in interval notation, and graph it.

24. $-3x > -33$ $(-\infty, 11)$

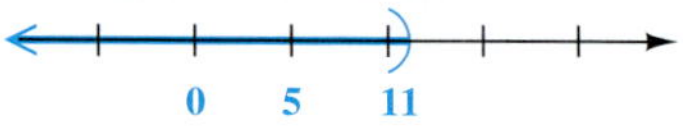

25. $-0.04x \le 0.12$ $[-3, \infty)$

26. $-4x + 2(x - 3) \ge 4x - (3 + 5x) - 7$ $(-\infty, 4]$

27. $-10 < 3x - 4 \le 14$ $(-2, 6]$

28. Shania Johnson has scores of 76 and 81 on her first two algebra tests. If she wants an average of at least 80 after her third test, what score must she make on her third test?

83 or more

Chapters R–2 Cumulative Review Exercises

Write each fraction in lowest terms.

1. $\frac{15}{40}$ $\frac{3}{8}$

2. $\frac{108}{144}$ $\frac{3}{4}$

Perform the indicated operations.

3. $\frac{5}{6}+\frac{1}{4}+\frac{7}{15}$ $\frac{31}{20}$

4. $16\frac{7}{8}-3\frac{1}{10}$ $\frac{551}{40}$, or $13\frac{31}{40}$

5. $\frac{9}{8}\cdot\frac{16}{3}$ 6

6. $\frac{3}{4}\div\frac{5}{8}$ $\frac{6}{5}$

7. $4.8+12.5+16.73$ 34.03

8. $56.3-28.99$ 27.31

9. $67.8(0.45)$ 30.51

10. $236.46\div 4.2$ 56.3

11. In making dresses, Earth Works uses $\frac{5}{8}$ yd of trim per dress. How many yards of trim would be used to make 56 dresses?
35 yd

12. A cook wants to increase a recipe for Quaker Quick Grits that serves 4 to make enough for 10 people. The recipe calls for 3 cups of water. How much water will be needed to serve 10?
$7\frac{1}{2}$ cups

13. First published in 1953, the digest-sized *TV Guide* changed to a full-sized magazine in 2005. See the figure. The new magazine is 3 in. wider than the old guide. What is the difference in their heights? (*Source: TV Guide.*)
$3\frac{3}{8}$ in.

14. A small business owner bought 3 business laptop computers for \$529.99, \$599.99, and \$629.99 and 3 ergonomic office chairs for \$279.99 each. What was the final bill (without tax)? (*Source:* www.staples.com)
\$2599.94

Tell whether each inequality is true *or* false.

15. $\frac{8(7)-5(6+2)}{3\cdot 5+1}\geq 1$ true

16. $\frac{4(9+3)-8(4)}{2+3-3}\geq 2$ true

Perform the indicated operations.

17. $-11+20+(-2)$ 7

18. $13+(-19)+7$ 1

19. $9-(-4)$ 13

20. $-2(-5)(-4)$ -40

21. $\frac{4\cdot 9}{-3}$ -12

22. $\frac{8}{7-7}$ undefined

23. $(-5+8)+(-2-7)$ -6

24. $(-7-1)(-4)+(-4)$ 28

25. $\frac{-3-(-5)}{1-(-1)}$ 1

26. $\frac{6(-4)-(-2)(12)}{3^2+7^2}$ 0

27. $\frac{(-3)^2-(-4)(2^4)}{5\cdot 2-(-2)^3}$ $\frac{73}{18}$

28. $\frac{-2(5^3)-6}{4^2+2(-5)+(-2)}$ -64

Find the value of each expression for $x = -2$, $y = -4$, and $z = 3$.

29. $xz^3 - 5y^2$ $\quad -134$

30. $\dfrac{xz - y^3}{-4z}$ $\quad -\dfrac{29}{6}$

Name the property illustrated by each equation.

31. $7(k + m) = 7k + 7m$
distributive property

32. $3 + (5 + 2) = 3 + (2 + 5)$
commutative property

33. $7 + (-7) = 0$
inverse property

34. $3.5(1) = 3.5$
identity property

Simplify each expression.

35. $4p - 6 + 3p - 8$ $\quad 7p - 14$

36. $-4(k + 2) + 3(2k - 1)$ $\quad 2k - 11$

Solve each equation, and check the solution.

37. $2r - 6 = 8$ $\quad \{7\}$

38. $2(p - 1) = 3p + 2$ $\quad \{-4\}$

39. $4 - 5(a + 2) = 3(a + 1) - 1$ $\quad \{-1\}$

40. $2 - 6(z + 1) = 4(z - 2) + 10$
$\left\{-\frac{3}{5}\right\}$

41. $-(m - 1) = 3 - 2m$
$\{2\}$

42. $\dfrac{x - 2}{3} = \dfrac{2x + 1}{5}$
$\{-13\}$

43. $\dfrac{2x + 3}{5} = \dfrac{x - 4}{2}$
$\{26\}$

44. $\dfrac{2}{3}x + \dfrac{3}{4}x = -17$
$\{-12\}$

Solve each formula for the indicated variable.

45. $P = a + b + c + B$ for c
$c = P - a - b - B$

46. $P = 4s$ for s
$s = \dfrac{P}{4}$

Solve each inequality. Write the solution set in interval notation, and graph it.

47. $-5z \geq 4z - 18$

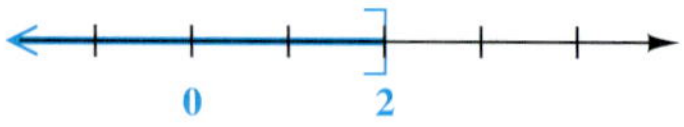

$(-\infty, 2]$

48. $6(r - 1) + 2(3r - 5) < -4$

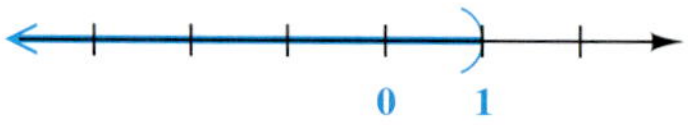

$(-\infty, 1)$

Solve each problem.

49. Abby Tanenbaum bought textbooks at the college bookstore for \$244.33, including 6% sales tax. What did the books cost before tax? \$230.50

50. A used car has a price of \$9500. For trading in her old car, Shannon d'Hemecourt will get 25% off. Find the price of the car with the trade-in. \$7125

51. The perimeter of a rectangle is 98 cm. The width is 19 cm. Find the length. 30 cm

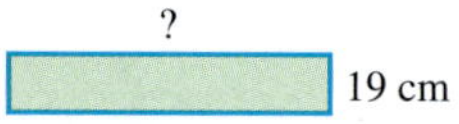

52. The area of a triangle is 104 in.^{2}. The base is 13 in. Find the height.
16 in.

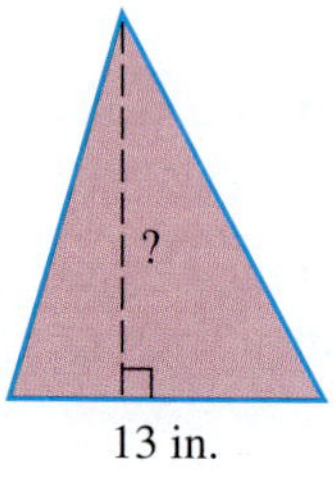

3 Graphs of Linear Equations and Inequalities in Two Variables

We determine location on a map using *coordinates*, a concept that is based on a *rectangular coordinate system*, one of the topics of this chapter.

3.1 Linear Equations in Two Variables; The Rectangular Coordinate System

OBJECTIVES

1. Interpret graphs.
2. Write a solution as an ordered pair.
3. Decide whether a given ordered pair is a solution of a given equation.
4. Complete ordered pairs for a given equation.
5. Complete a table of values.
6. Plot ordered pairs.

As we saw in **Section R.1,** circle graphs (pie charts) provide a convenient way to organize and communicate information. Along with *bar graphs* and *line graphs*, they can be used to analyze data, make predictions, or simply to entertain us.

OBJECTIVE 1 Interpret graphs. A **bar graph** is used to show comparisons. It consists of a series of bars (or simulations of bars) arranged either vertically or horizontally. In a bar graph, values from two categories are paired with each other.

EXAMPLE 1 Interpreting a Bar Graph

The bar graph in **Figure 1** shows annual per-capita spending on health care in the United States for the years 2004 through 2009.

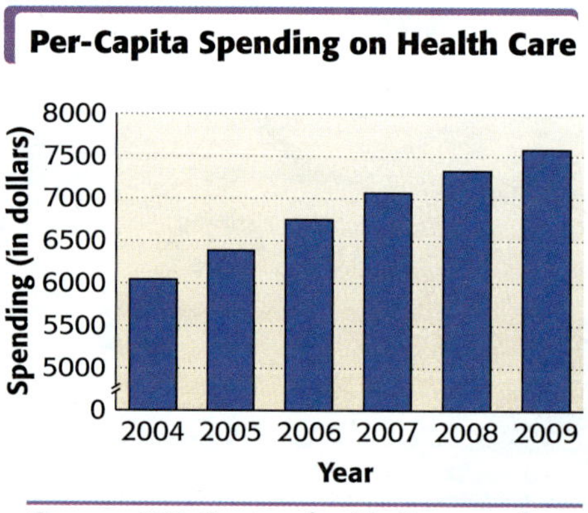

Source: U.S. Centers for Medicare and Medicaid Services.

Figure 1

(a) In what years was per-capita health care spending greater than $7000?

Locate 7000 on the vertical axis and follow the line across to the right. Three years—2007, 2008, and 2009—have bars that extend above the line for 7000, so per-capita health care spending was greater than $7000 in those years.

(b) Estimate per-capita health care spending in 2004 and 2008.

Locate the top of the bar for 2004 and move horizontally across to the vertical scale to see that it is about 6000. Per-capita health care spending for 2004 was about $6000.

Similarly, follow the top of the bar for 2008 across to the vertical scale to see that it lies a little more than halfway between 7000 and 7500, so per-capita health care spending in 2008 was about $7300.

(c) Describe the change in per-capita spending as the years progressed.

As the years progressed, per-capita spending on health care increased steadily, from about $6000 in 2004 to about $7500 in 2009.

◀ Work Problem 1 at the Side.

1 Refer to the bar graph in **Figure 1.**

(a) Which years had per-capita health care spending less than $7000?

(b) Estimate per-capita health care spending in 2006 and 2007.

(c) Describe the change in per-capita health care spending from 2006 to 2007.

Answers

1. (a) 2004, 2005, 2006
(b) 2006: about $6750; 2007: about $7000
(c) Per-capita health care spending increased by about $250 from 2006 to 2007.

A **line graph** is used to show changes or trends in data over time. To form a line graph, we connect a series of points representing data with line segments.

EXAMPLE 2 Interpreting a Line Graph

The line graph in **Figure 2** shows average prices of a gallon of regular unleaded gasoline in the United States for the years 2004 through 2011.

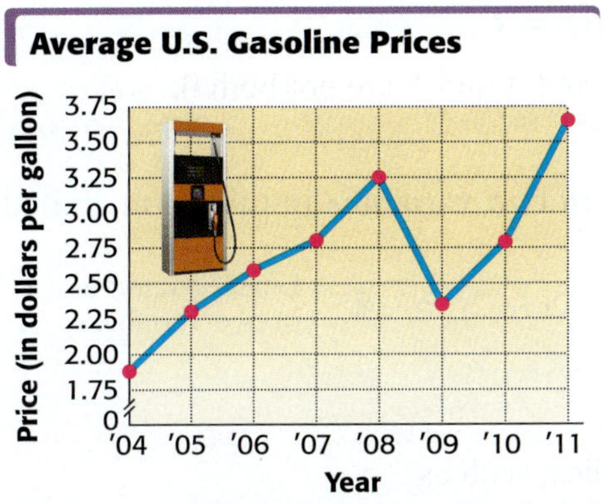

Source: U.S. Department of Energy.

Figure 2

(a) Between which years did the average price of a gallon of gasoline decrease?

The line between 2008 and 2009 falls, so the average price of a gallon of gasoline decreased from 2008 to 2009.

(b) What was the general trend in the average price of a gallon of gasoline from 2004 through 2008?

The line graph rises from 2004 to 2008, so the average price of a gallon of gasoline increased over those years.

(c) Estimate the average price of a gallon of gasoline in 2004 and 2008. About how much did the price increase between 2004 and 2008?

Move up from 2004 on the horizontal scale to the point plotted for 2004. This point is about halfway between the lines on the vertical scale for \$1.75 and \$2.00. Halfway between \$1.75 and \$2.00 would be about \$1.88. So, it cost about \$1.88 for a gallon of gasoline in 2004.

Similarly, locate the point plotted for 2008. Moving across to the vertical scale, the graph indicates that the price for a gallon of gasoline in 2008 was about \$3.25.

Between 2004 and 2008, the average price of a gallon of gasoline increased by about

$$\$3.25 - \$1.88 = \$1.37.$$

Work Problem **2** at the Side. ▶

The line graph in **Figure 2** relates years to average prices for a gallon of gasoline. We can also represent these two related quantities using a table of data, as shown in the margin. In table form, we can see more precise data rather than estimating it. Trends in the data are easier to see from the graph, which gives a "picture" of the data.

2 Refer to the line graph in **Figure 2.**

(a) Estimate the average price of a gallon of regular unleaded gasoline in 2005.

(b) About how much did the average price of a gallon of gasoline increase from 2005 to 2008?

(c) About how much did the average price of a gallon of gasoline decrease from 2008 to 2009?

Year	Average Price (in dollars per gallon)
2004	1.88
2005	2.30
2006	2.59
2007	2.80
2008	3.25
2009	2.35
2010	2.79
2011	3.65

Source: U.S. Department of Energy.

Answers

2. **(a)** about \$2.30 **(b)** about \$0.95 **(c)** about \$0.90

3 Write each solution as an ordered pair.

GS **(a)** $x = 5$ and $y = 7$

x-value y-value
↓ ↓
(____, ____)

GS **(b)** $y = 6$ and $x = -1$

x-value y-value
↓ ↓
(____, ____)

(c) $y = 4$ and $x = -3$

(d) $x = \frac{2}{3}$ and $y = -12$

Teaching Tip

Ask students for other ordered pairs that satisfy the equation $y = 4x + 5$. Make the point that an infinite number of such pairs will satisfy a linear equation in two variables.

(e) $y = 1.5$ and $x = -2.4$

(f) $x = 0$ and $y = 0$

Answers

3. (a) 5; 7 (b) -1; 6 (c) $(-3, 4)$
(d) $\left(\frac{2}{3}, -12\right)$ (e) $(-2.4, 1.5)$
(f) $(0, 0)$

We can extend these ideas to the subject of this chapter, *linear equations in **two** variables.* A linear equation in two variables, one for each of the quantities being related, can be used to represent the data in a table or graph. ***The graph of a linear equation in two variables is a line.***

Linear Equation in Two Variables

A **linear equation in two variables** is an equation that can be written in the form

$$Ax + By = C,$$

where A, B, and C are real numbers and A and B are not both 0.

Some examples of linear equations in two variables in this form, called *standard form*, follow.

$$3x + 4y = 9, \quad x - y = 0, \quad \text{and} \quad x + 2y = -8$$ Linear equations in two variables

Note

Other linear equations in two variables, such as

$$y = 4x + 5 \quad \text{and} \quad 3x = 7 - 2y,$$

are not written in standard form, but could be algebraically rewritten in this form. We discuss the forms of linear equations in **Section 3.4.**

OBJECTIVE 2 **Write a solution as an ordered pair.** Recall from **Section 1.2** that a *solution* of an equation is a number that makes the equation true when it replaces the variable. For example, the linear equation in *one* variable $x - 2 = 5$ has solution 7, since replacing x with 7 gives a true statement.

A solution of a linear equation in two variables requires two numbers, one for each variable. For example, a true statement results when we replace x with 2 and y with 13 in the equation $y = 4x + 5$ since

$$13 = 4(2) + 5. \quad \text{Let } x = 2 \text{ and } y = 13.$$

The pair of numbers $x = 2$ and $y = 13$ gives one solution of the equation $y = 4x + 5$. The phrase "$x = 2$ and $y = 13$" is abbreviated

x-value ↓ ↓ y-value
(2, 13)
Ordered pair

with the x-value, 2, and the y-value, 13, given as a pair of numbers written inside parentheses. ***The x-value is always given first.*** A pair of numbers such as (2, 13) is called an **ordered pair.**

CAUTION

The ordered pairs (2, 13) and (13, 2) are *not* the same. In the first pair, $x = 2$ and $y = 13$. In the second pair, $x = 13$ and $y = 2$. ***The order in which the numbers are written in an ordered pair is important.*** For two ordered pairs to be equal, their x-values must be equal and their y-values must be equal.

◀ Work Problem 3 at the Side.

OBJECTIVE 3 Decide whether a given ordered pair is a solution of a given equation. We substitute the x- and y-values of an ordered pair into a linear equation in two variables to see whether the ordered pair is a solution.

EXAMPLE 3 Deciding Whether Ordered Pairs Are Solutions

Decide whether each ordered pair is a solution of the equation $2x + 3y = 12$.

(a) $(3, 2)$

Substitute 3 for x and 2 for y in the given equation $2x + 3y = 12$.

$$2x + 3y = 12$$
$$2(3) + 3(2) \stackrel{?}{=} 12 \qquad \text{Let } x = 3 \text{ and } y = 2.$$
$$6 + 6 \stackrel{?}{=} 12 \qquad \text{Multiply.}$$
$$12 = 12 \checkmark \qquad \text{True}$$

This result is true, so $(3, 2)$ is a solution of $2x + 3y = 12$.

(b) $(-2, -7)$

$$2x + 3y = 12$$
$$2(-2) + 3(-7) \stackrel{?}{=} 12 \qquad \text{Let } x = -2 \text{ and } y = -7.$$
$$-4 + (-21) \stackrel{?}{=} 12 \qquad \text{Multiply.}$$
$$-25 = 12 \qquad \text{False}$$

Use parentheses to avoid errors.

This result is false, so $(-2, -7)$ is *not* a solution of $2x + 3y = 12$.

Work Problem 4 at the Side. ▶

OBJECTIVE 4 Complete ordered pairs for a given equation. We substitute a number for one variable to find the value of the other variable.

EXAMPLE 4 Completing Ordered Pairs

Complete each ordered pair for the equation $y = 4x + 5$.

(a) $(7, \_\_\_)$ The x-value always comes first.

In this ordered pair, $x = 7$. Replace x with 7 in the given equation.

$$y = 4x + 5$$
$$y = 4(7) + 5 \qquad \text{Let } x = 7.$$
$$y = 28 + 5 \qquad \text{Multiply.}$$
$$y = 33 \qquad \text{Add.}$$

Solve for the value of y.

The ordered pair is $(7, 33)$.

(b) $(\_\_\_, -3)$

In this ordered pair, $y = -3$. Replace y with -3 in the given equation.

$$y = 4x + 5$$
$$-3 = 4x + 5 \qquad \text{Let } y = -3.$$
$$-8 = 4x \qquad \text{Subtract 5 from each side.}$$
$$-2 = x \qquad \text{Divide each side by 4.}$$

Solve for the value of x.

The ordered pair is $(-2, -3)$.

Work Problem 5 at the Side. ▶

4 Decide whether each ordered pair is a solution of the equation $5x + 2y = 20$.

GS **(a)** $(0, 10)$

In this ordered pair, $x = \_\_\_\_$ and $y = \_\_\_\_$.

$$5x + 2y = 20$$
$$5(\_\_\_\_) + 2(\_\_\_\_) \stackrel{?}{=} 20$$
$$\_\_\_\_ + 20 \stackrel{?}{=} 20$$
$$\_\_\_\_ = 20$$

Is $(0, 10)$ a solution? (yes / no)

(b) $(2, -5)$

(c) $(3, 2)$

(d) $(-4, 20)$

5 Complete each ordered pair for the equation $y = 2x - 9$.

GS **(a)** $(5, \_\_\_\_)$

In this ordered pair, $x = \_\_\_\_$.

We must find the corresponding value of $\_\_\_\_$.

$$y = 2x - 9$$
$$y = 2(\_\_\_\_) - 9$$
$$y = \_\_\_\_ - 9$$
$$y = \_\_\_\_$$

The ordered pair is $\_\_\_\_$.

(b) $(2, \_\_\_\_)$

(c) $(\_\_\_\_, 7)$

(d) $(\_\_\_\_, -13)$

Answers

4. **(a)** 0; 10; 0; 10; 0; 20; yes
(b) no **(c)** no **(d)** yes

5. **(a)** 5; y; 5; 10; 1; $(5, 1)$
(b) $(2, -5)$ **(c)** $(8, 7)$ **(d)** $(-2, -13)$

6 Complete the table of values for each equation. Write the results as ordered pairs.

GS **(a)** $2x - 3y = 12$

x	y
0	
	0
3	
	-3

From the first row of the table, let $x =$ ____.

$$2x - 3y = 12$$
$$2(\_\_\_\_) - 3y = 12$$
$$\_\_\_\_ - 3y = 12$$
$$-3y = 12$$
$$\frac{-3y}{\_\_\_\_} = \frac{12}{\_\_\_\_}$$
$$y = \_\_\_\_$$

Write ____ for y in the first row of the table. Repeat this process to complete the rest of the table.

(b) $x = -1$

x	y
	-4
	0
-1	2

(c) $y = 4$

x	y
-3	4
2	
5	

Answers

6. (a) 0; 0; 0; -3; -3; -4; -4

x	y
0	-4
6	0
3	-2
$\frac{3}{2}$	-3

$(0, -4), (6, 0), (3, -2), \left(\frac{3}{2}, -3\right)$

(b)

x	y
-1	-4
-1	0
-1	2

$(-1, -4), (-1, 0), (-1, 2)$

(c)

x	y
-3	4
2	4
5	4

$(-3, 4), (2, 4), (5, 4)$

OBJECTIVE 5 Complete a table of values. Ordered pairs are often displayed in a **table of values.** Although we usually write tables of values vertically, they may be written horizontally.

EXAMPLE 5 Completing Tables of Values

Complete the table of values for each equation. Then write the results as ordered pairs.

(a) $x - 2y = 8$

x	y	Ordered Pairs
2		$(2, \_\_\_\_)$
10		$(10, \_\_\_\_)$
	0	$(\_\_\_\_, 0)$
	-2	$(\_\_\_\_, -2)$

From the first row of the table, let $x = 2$ in the equation. From the second row of the table, let $x = 10$.

If	$x = 2$,	If	$x = 10$,
then	$x - 2y = 8$	then	$x - 2y = 8$
becomes	$2 - 2y = 8$	becomes	$10 - 2y = 8$
	$-2y = 6$		$-2y = -2$
	$y = -3$.		$y = 1$.

The first two ordered pairs are $(2, -3)$ and $(10, 1)$. From the third and fourth rows of the table, let $y = 0$ and $y = -2$, respectively.

If	$y = 0$,	If	$y = -2$,
then	$x - 2y = 8$	then	$x - 2y = 8$
becomes	$x - 2(0) = 8$	becomes	$x - 2(-2) = 8$
	$x - 0 = 8$		$x + 4 = 8$
	$x = 8$.		$x = 4$.

The last two ordered pairs are $(8, 0)$ and $(4, -2)$. The completed table of values and corresponding ordered pairs follow.

Write y-values in the second column.

Write x-values in the first column.

x	y	Ordered Pairs
2	-3	$(2, -3)$
10	1	$(10, 1)$
8	0	$(8, 0)$
4	-2	$(4, -2)$

Each ordered pair is a solution of the given equation $x - 2y = 8$.

(b) $x = 5$

x	y
	-2
	6
	3

The given equation is $x = 5$. No matter which value of y is chosen, the value of x is *always* 5.

x	y	Ordered Pairs
5	-2	$(5, -2)$
5	6	$(5, 6)$
5	3	$(5, 3)$

◀ **Work Problem 6 at the Side.**

Note

We can think of $x = 5$ in **Example 5(b)** as an equation in two variables by rewriting $x = 5$ as

$$x + 0y = 5.$$

This form shows that, for any value of y, the value of x is 5. Similarly, $y = 4$ in **Margin Problem 6(c)** on the preceding page is the same as

$$0x + y = 4.$$

OBJECTIVE 6 Plot ordered pairs. In **Section 2.3,** we saw that linear equations in *one* variable have either one, zero, or an infinite number of real number solutions. These solutions can be graphed on *one* number line. For example, the linear equation in one variable $x - 2 = 5$ has solution 7, which is graphed on the number line in **Figure 3.**

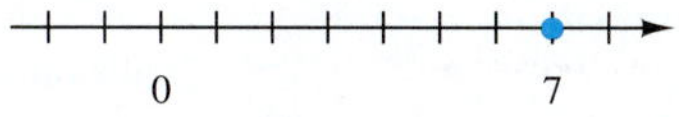

Figure 3

Every linear equation in *two* variables has an infinite number of ordered pairs as solutions. To graph these solutions, represented as the ordered pairs (x, y), we need *two* number lines, one for each variable. These two number lines are drawn at right angles as in **Figure 4.** The horizontal number line is the ***x*-axis**, and the vertical line is the ***y*-axis.** Together, the x-axis and y-axis form a **rectangular coordinate system.** The point at which the x-axis and y-axis intersect is the **origin.**

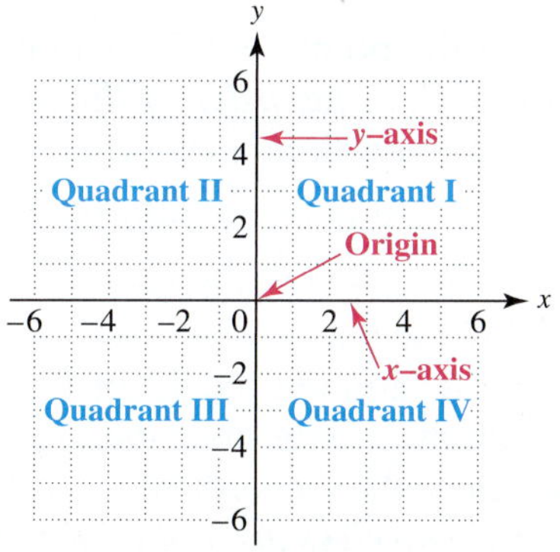

Rectangular coordinate system

Figure 4

The rectangular coordinate system is divided into four regions, or **quadrants.** These quadrants are numbered counterclockwise. See **Figure 4.**

Work Problem 7 at the Side. ▶

The x-axis and y-axis determine a **plane**—a flat surface (as illustrated by a sheet of paper). By referring to the two axes, every point in the plane can be associated with an ordered pair. The numbers in the ordered pair are the **coordinates** of the point.

Note

In a plane, *both* numbers in the ordered pair are needed to locate a point. The ordered pair is a name for the point.

7 Name the quadrant in which each point in the figure is located.

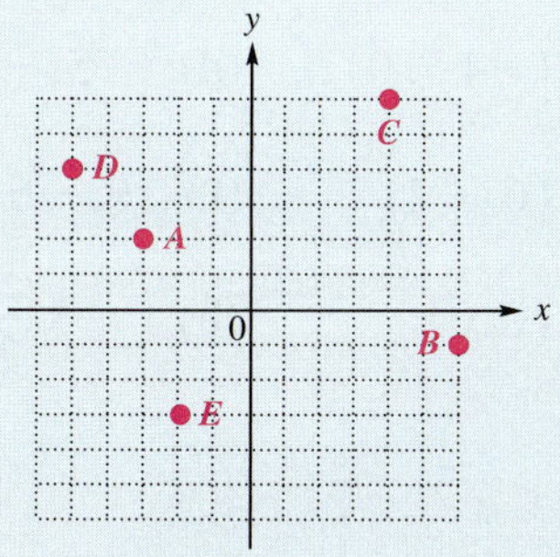

Teaching Tip

Students will find it helpful to have graph paper while studying this chapter.

René Descartes (1596–1650)

The rectangular coordinate system is also called the **Cartesian coordinate system,** in honor of René Descartes, the French mathematician credited with its invention.

Answer

7. *A*: II; *B*: IV; *C*: I; *D*: II; *E*: III

8 Plot the given points in a coordinate system.

(a) $(3, 5)$ **(b)** $(-2, 6)$

(c) $(-4.5, 0)$ **(d)** $(-5, -2)$

(e) $(6, -2)$ **(f)** $(0, -6)$

(g) $(0, 0)$ **(h)** $\left(-3, \frac{5}{2}\right)$

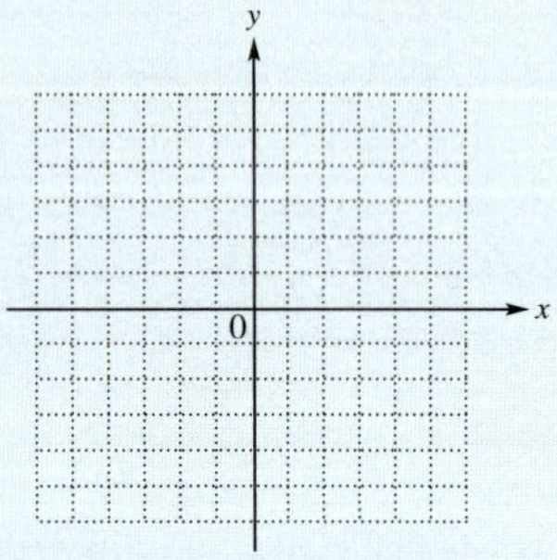

Indicate the points that lie in each quadrant.

quadrant I: ________

quadrant II: ________

quadrant III: ________

quadrant IV: ________

no quadrant: ________

Teaching Tip

Relate placing a point in the coordinate plane to finding a town on a road map to help students visualize the process.

Answers

8.

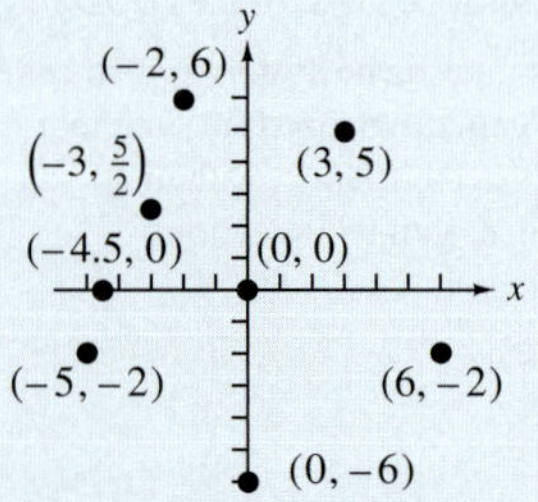

quadrant I: $(3, 5)$

quadrant II: $(-2, 6)$, $\left(-3, \frac{5}{2}\right)$

quadrant III: $(-5, -2)$

quadrant IV: $(6, -2)$

no quadrant: $(-4.5, 0)$, $(0, 0)$, $(0, -6)$

EXAMPLE 6 Plotting Ordered Pairs

Plot the given points in a coordinate system.

(a) $(2, 3)$ **(b)** $(-1, -4)$ **(c)** $(-2, 3)$ **(d)** $(3, -2)$ **(e)** $\left(\frac{3}{2}, 2\right)$

(f) $(4, -3.75)$ **(g)** $(5, 0)$ **(h)** $(0, -3)$ **(i)** $(0, 0)$

The point $(2, 3)$ from part (a) is **plotted** (graphed) in **Figure 5.** The other points are plotted in **Figure 6.** In each case, we begin at the origin.

Step 1 Move right or left the number of units that corresponds to the x-coordinate in the ordered pair—***right if the x-coordinate is positive or left if it is negative.***

Step 2 Then turn and move up or down the number of units that corresponds to the y-coordinate—***up if the y-coordinate is positive or down if it is negative.***

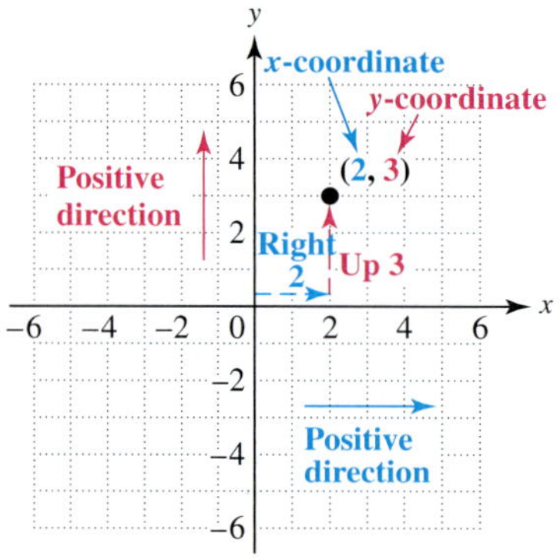

Figure 5

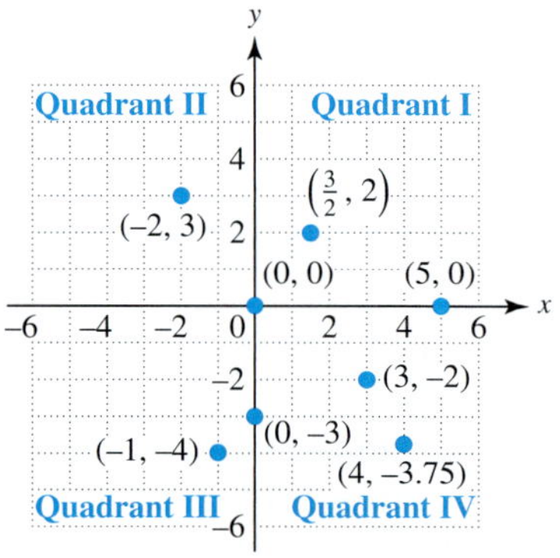

Figure 6

Notice in **Figure 6** that the point $(-2, 3)$ is in quadrant II, whereas the point $(3, -2)$ is in quadrant IV. ***The order of the coordinates is important. The x-coordinate is always given first in an ordered pair.***

To plot the point $\left(\frac{3}{2}, 2\right)$, think of the improper fraction $\frac{3}{2}$ as the mixed number $1\frac{1}{2}$ and move $\frac{3}{2}$ $\left(\text{or } 1\frac{1}{2}\right)$ units to the right along the x-axis. Then turn and go 2 units up, parallel to the y-axis. The point $(4, -3.75)$ is plotted similarly, by approximating the location of the decimal y-coordinate.

The point $(5, 0)$ lies on the x-axis since the y-coordinate is 0. The point $(0, -3)$ lies on the y-axis since the x-coordinate is 0. The point $(0, 0)$ is at the origin. ***Points on the axes themselves are not in any quadrant.***

◀ **Work Problem 8 at the Side.**

We can use a linear equation in two variables to mathematically describe, or *model*, certain real-life situations, as shown in the next example.

EXAMPLE 7 Using a Linear Model

The annual number of twin births in the United States from 2001 through 2007 can be approximated by the linear equation

Number of twin births → y; Year → x

$$y = 2.929x - 5738,$$

which relates x, the year, and y, the number of twin births in thousands. (*Source:* National Center for Health Statistics.)

Continued on Next Page

(a) Complete the table of values for the given linear equation.

x (Year)	y (Number of Twin Births, in thousands)
2001	
2004	
2007	

To find y for $x = 2001$, substitute into the equation.

$$y = 2.929x - 5738$$

$$y = 2.929(2001) - 5738 \quad \text{Let } x = 2001.$$

$$y \approx 123 \quad \text{Use a calculator.}$$

≈ means "is approximately equal to."

In 2001, there were about 123 thousand (or 123,000) twin births.

Work Problem 9 at the Side. ▶

Including the results from **Margin Problem 9** gives the completed table that follows.

x (Year)	y (Number of Twin Births, in thousands)	Ordered Pairs (x, y)
2001	123	→ (2001, 123)
2004	132	→ (2004, 132)
2007	141	→ (2007, 141)

Here each year x is paired with a number of twin births y (in thousands).

(b) Graph the ordered pairs found in part (a).
See **Figure 7.** A graph of ordered pairs of data is a **scatter diagram.**

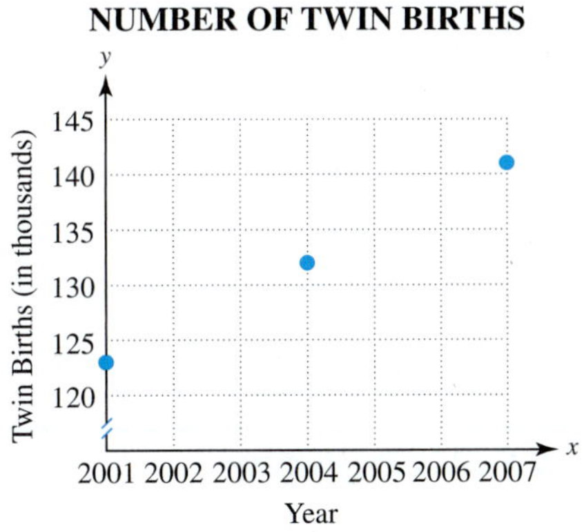

Figure 7

Notice the axis labels and scales. Each square represents 1 unit in the horizontal direction and 5 units in the vertical direction. We show a break in the y-axis, to indicate the jump from 0 to 120.

A scatter diagram can indicate whether two quantities are related. In **Figure 7,** the plotted points could be connected to approximate a straight ***line***, so the variables x (year) and y (number of twin births) have a ***line***ar relationship. The increase in the number of twin births is also reflected.

CAUTION

The equation in **Example 7** is valid only for the years 2001 through 2007. ***Do not assume that it would provide reliable data for other years.***

9 Refer to the linear equation in **Example 7.** Round to the nearest whole number.

GS **(a)** Find the y-value for $x = 2004$.

$y = 2.929x - 5738$

$y = 2.929(\_\_\_\_) - \_\_\_\_$

$y \approx \_\_\_\_$

(b) Find the y-value for $x = 2007$. Interpret your result.

Teaching Tip

Mention that it is a good idea, if possible, to use the same scale on the x-axis and the y-axis when graphing in a rectangular coordinate system. Talking about scale now is good preparation for Section 3.3 on the slope of a line.

Answers

9. (a) 2004; 5738; 132
(b) 141; In 2007, there were about 141 thousand (or 141,000) twin births in the United States.

3.1 Exercises

FOR EXTRA HELP

Download the MyDashBoard App

MyMathLab®

CONCEPT CHECK *Complete each statement.*

1. The symbol (x, y) (*does* / *does not*) represent an ordered pair, while the symbols $[x, y]$ and $\{x, y\}$ (*do* / *do not*) represent ordered pairs.

2. The point whose graph has coordinates $(-4, 2)$ is in quadrant **II**.

3. The point whose graph has coordinates $(0, 5)$ lies on the **y**-axis.

4. The ordered pair $(4, \underline{3})$ is a solution of the equation $y = 3$.

5. The ordered pair $(\underline{6}, -2)$ is a solution of the equation $x = 6$.

6. The ordered pair $(3, 2)$ is a solution of the equation $2x - 5y = \underline{-4}$.

7. Do $(4, -1)$ and $(-1, 4)$ represent the same ordered pair? Explain.

No. For two ordered pairs to be equal, their x-values must be equal and their y-values must be equal. Here we have $4 \neq -1$ and $-1 \neq 4$.

8. Explain why it would be easier to find the corresponding y-value for $x = \frac{1}{3}$ in the equation $y = 6x + 2$ than it would be for $x = \frac{1}{7}$.

Substituting $\frac{1}{3}$ for x in $y = 6x + 2$ gives $y = 6\left(\frac{1}{3}\right) + 2$. Here $6\left(\frac{1}{3}\right) = 2$, so $y = 2 + 2$, or 4. Because $6\left(\frac{1}{7}\right) = \frac{6}{7}$, calculating y for $x = \frac{1}{7}$ requires working with fractions.

CONCEPT CHECK *Fill in each blank with the word* positive *or the word* negative.

The point with coordinates (x, y) is in

9. quadrant III if x is **negative** and y is **negative**.

10. quadrant II if x is **negative** and y is **positive**.

11. quadrant IV if x is **positive** and y is **negative**.

12. quadrant I if x is **positive** and y is **positive**.

The bar graph shows total U.S. milk production in billions of pounds for the years 2004 through 2010. Use the bar graph to work Exercises 13–16. ***See Example 1.***

13. In what years was U.S. milk production greater than 185 billion pounds?

2007, 2008, 2009, 2010

14. In what years was U.S. milk production about the same?

2008 and 2009

15. Estimate U.S. milk production in 2004 and 2010.

2004: about 171 billion lb; 2010: about 193 billion lb

16. Describe the change in U.S. milk production from 2004 to 2010.

U.S. milk production increased about 22 billion lb during these years, from about 171 billion lb to about 193 billion lb.

U.S. Milk Production

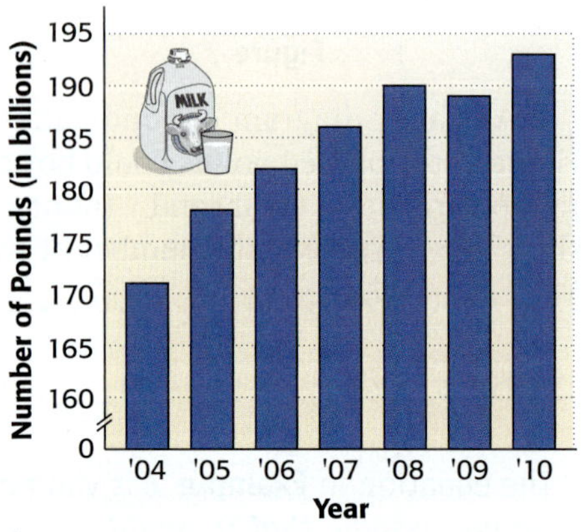

Source: U.S. Department of Agriculture.

The line graph shows the number of new and used passenger cars (in millions) imported into the United States over the years 2005 through 2010. Use the line graph to work Exercises 17–20. ***See Example 2.***

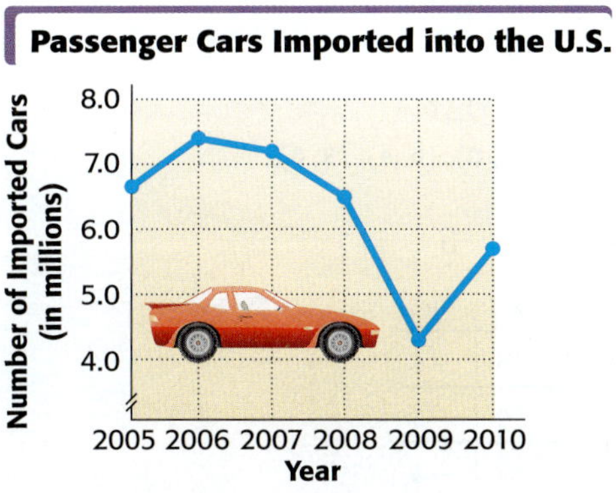

Source: U.S. Census Bureau.

17. Over which two consecutive years did the number of imported cars increase the most? About how much was this increase?

2009 to 2010; about 1.5 million

18. Estimate the number of cars imported during 2007, 2008, and 2009.

2007: 7.2 million; 2008: 6.5 million; 2009: 4.3 million

19. Describe the trend in car imports from 2006 to 2009.

The number of cars imported each year was decreasing.

20. During which year(s) were fewer than 6 millions cars imported into the United States?

2009, 2010

Decide whether each ordered pair is a solution of the given equation. ***See Example 3.***

21. $x + y = 9;\ (0, 9)$
yes

22. $x + y = 8;\ (0, 8)$
yes

23. $2x - y = 6;\ (4, 2)$
yes

24. $2x + y = 5;\ (3, -1)$
yes

25. $4x - 3y = 6;\ (2, 1)$
no

26. $5x - 3y = 15;\ (5, 2)$
no

27. $y = \frac{2}{3}x;\ (-6, -4)$
yes

28. $y = -\frac{1}{4}x;\ (-8, 2)$
yes

29. $x = -6;\ (5, -6)$
no

30. $x = -9;\ (8, -9)$
no

31. $y = 2;\ (2, 4)$
no

32. $y = 7;\ (7, -2)$
no

Complete each ordered pair for the equation $y = 2x + 7$. ***See Example 4.***

33. (2, __11__)

34. (0, __7__)

35. (__$-\frac{7}{2}$__, 0)

36. (__−5__, −3)

Complete each ordered pair for the equation $y = -4x - 4$. ***See Example 4.***

37. (0, __−4__)

38. (__−1__, 0)

39. (__−5__, 16)

40. (__−7__, 24)

Complete each table of values. Write the results as ordered pairs. ***See Example 5.***

41. $2x + 3y = 12$

x	y
0	4
6	0
−6	8

(0, 4), (6, 0), (−6, 8)

42. $4x + 3y = 24$

x	y
0	8
6	0
3	4

(0, 8), (6, 0), (3, 4)

43. $3x - 5y = -15$

x	y
0	3
−5	0
−15	−6

(0, 3), (−5, 0), (−15, −6)

44. $4x - 9y = -36$

x	y
−9	0
0	**4**
9	8

(−9, 0), (0, 4), (9, 8)

45. $x = -9$

x	y
−9	6
−9	2
−9	−3

(−9, 6), (−9, 2), (−9, −3)

46. $x = 12$

x	y
12	3
12	8
12	0

(12, 3), (12, 8), (12, 0)

47. $y = -6$

x	y
8	**−6**
4	**−6**
−2	**−6**

(8, −6), (4, −6), (−2, −6)

48. $y = -10$

x	y
4	**−10**
0	**−10**
−4	**−10**

(4, −10), (0, −10), (−4, −10)

49. $x - 8 = 0$

x	y
8	8
8	3
8	0

(8, 8), (8, 3), (8, 0)

50. $x + 4 = 0$

x	y
−4	4
−4	0
−4	−4

(−4, 4), (−4, 0), (−4, −4)

51. $y + 2 = 0$

x	y
9	**−2**
2	**−2**
0	**−2**

(9, −2), (2, −2), (0, −2)

52. $y + 1 = 0$

x	y
1	**−1**
0	**−1**
−1	**−1**

(1, −1), (0, −1), (−1, −1)

Give the ordered pairs for the points labeled A–F in the figure. (All coordinates are integers.) Tell the quadrant in which each point is located. ***See Example 6.***

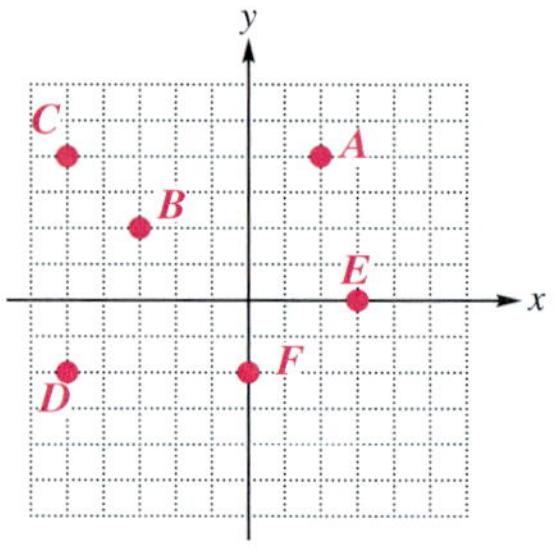

53. *A*
(2, 4); I

54. *B*
(−3, 2); II

55. *C*
(−5, 4); II

56. *D*
(−5, −2); III

57. *E*
(3, 0); no quadrant

58. *F*
(0, −2); no quadrant

59. A point (x, y) has the property that $xy < 0$. In which quadrant(s) must the point lie? Explain.

If $xy < 0$, then either $x < 0$ and $y > 0$ or $x > 0$ and $y < 0$. If $x < 0$ and $y > 0$, then the point lies in quadrant II. If $x > 0$ and $y < 0$, then the point lies in quadrant IV.

60. A point (x, y) has the property that $xy > 0$. In which quadrant(s) must the point lie? Explain.

If $xy > 0$, then either $x > 0$ and $y > 0$ or $x < 0$ and $y < 0$. If $x > 0$ and $y > 0$, then the point lies in quadrant I. If $x < 0$ and $y < 0$, then the point lies in quadrant III.

Plot each ordered pair on the rectangular coordinate system provided. ***See Example 6.***

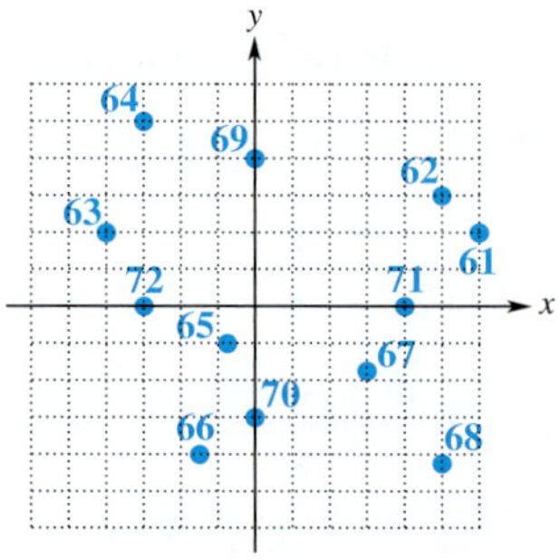

61. $(6, 2)$

62. $(5, 3)$

63. $(-4, 2)$

64. $(-3, 5)$

65. $\left(-\frac{4}{5}, -1\right)$

66. $\left(-\frac{3}{2}, -4\right)$

67. $(3, -1.75)$

68. $(5, -4.25)$

69. $(0, 4)$

70. $(0, -3)$

71. $(4, 0)$

72. $(-3, 0)$

Complete each table of values, and then plot the ordered pairs. ***See Examples 5 and 6.***

73. $x - 2y = 6$

x	y
0	-3
6	0
2	-2
4	-1

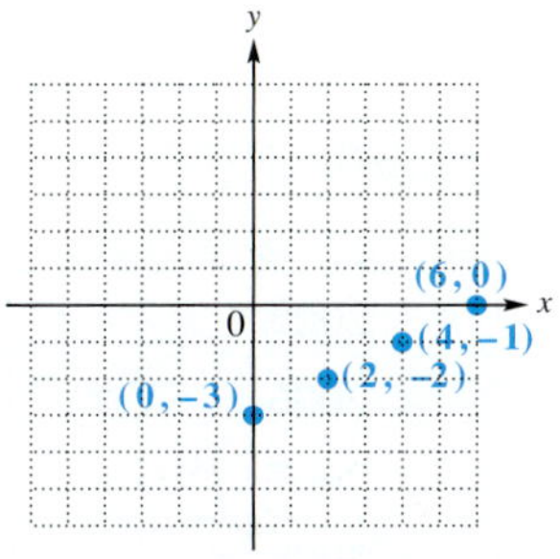

74. $2x - y = 4$

x	y
0	-4
2	0
1	-2
-1	-6

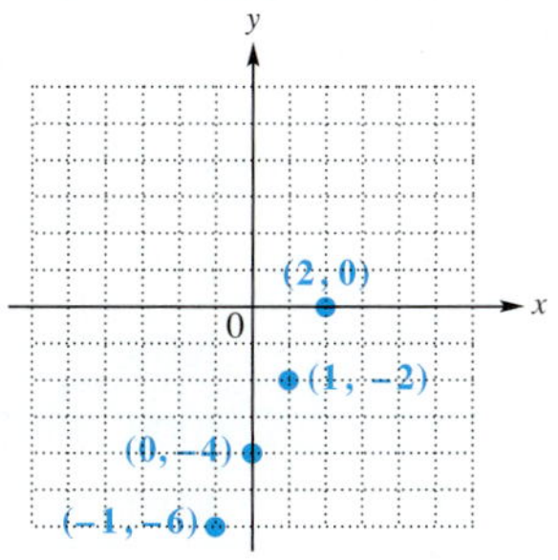

75. $3x - 4y = 12$

x	y
0	-3
4	0
-4	-6
$-\frac{4}{3}$	-4

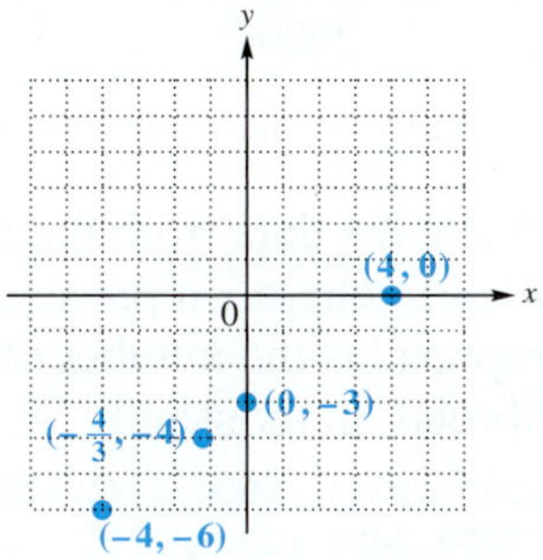

76. $2x - 5y = 10$

x	y
0	-2
5	0
-5	-4
$-\frac{5}{2}$	-3

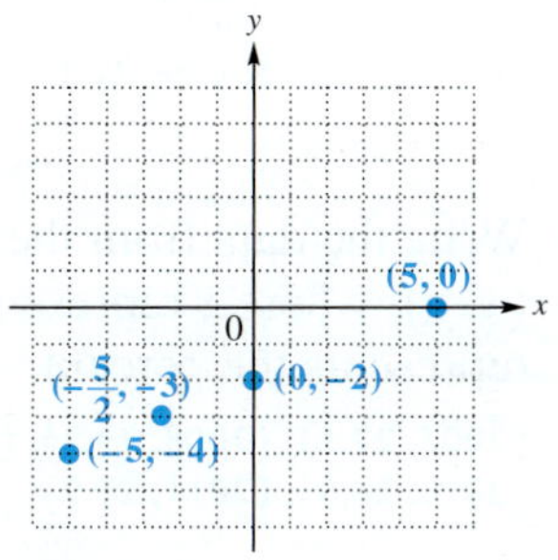

77. $y + 4 = 0$

x	y
0	-4
5	-4
-2	-4
-3	-4

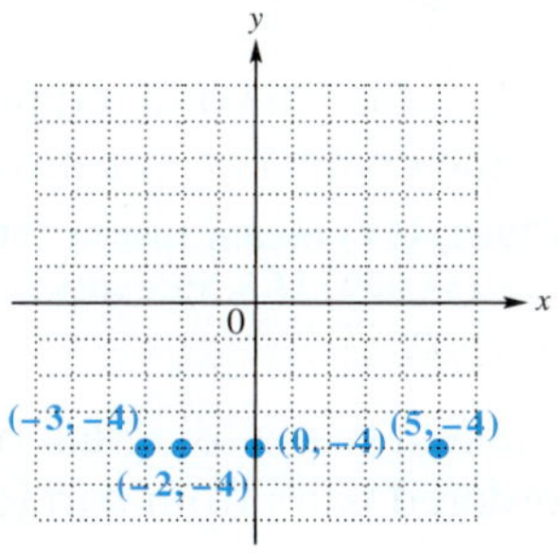

78. $x - 5 = 0$

x	y
5	1
5	0
5	6
5	-4

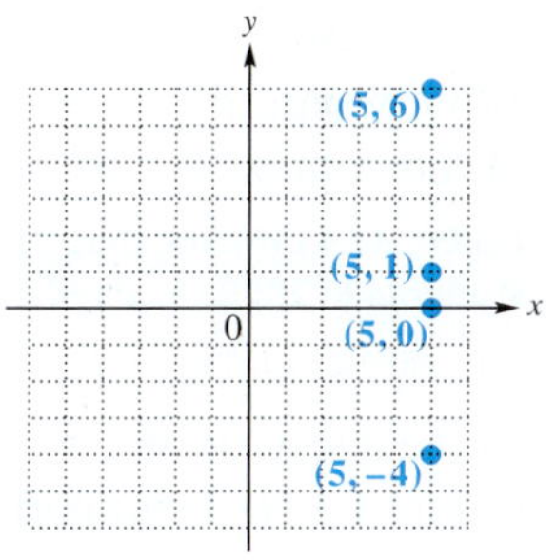

79. Describe the pattern indicated by the plotted points in **Exercises 73–78.**

The points in each graph appear to lie on a straight line.

80. Answer each question.

(a) A line through the plotted points in **Exercise 77** would be horizontal. What do you notice about the y-coordinates of the ordered pairs?

They are the same (all -4).

(b) A line through the plotted points in **Exercise 78** would be vertical. What do you notice about the x-coordinates of the ordered pairs?

They are the same (all 5).

Work each problem. ***See Example 7.***

81. Suppose that it costs a flat fee of \$20 plus \$5 per day to rent a pressure washer. Therefore, the cost y in dollars to rent the pressure washer for x days is given by

$$y = 5x + 20.$$

Express each of the following as an ordered pair.

(a) When the washer is rented for 5 days, the cost is \$45. (*Hint:* What does x represent? What does y represent?)

(5, 45)

(b) We paid \$50 when we returned the washer, so we must have rented it for 6 days.

(6, 50)

82. Suppose that it costs \$5000 to start up a business selling snow cones. Furthermore, it costs \$0.50 per cone in labor, ice, syrup, and overhead. Then the cost y in dollars to make x snow cones is given by

$$y = 0.50x + 5000.$$

Express each of the following as an ordered pair.

(a) When 100 snow cones are made, the cost is \$5050. (*Hint:* What does x represent? What does y represent?)

(100, 5050)

(b) When the cost is \$6000, the number of snow cones made is 2000.

(2000, 6000)

83. The table shows the rate (in percent) at which 2-year college students (public) complete a degree within 3 years.

Year	Percent
2007	27.1
2008	29.3
2009	28.3
2010	28.0
2011	26.9

Source: ACT.

(a) Write the data from the table as ordered pairs (x, y), where x represents the year and y represents the percent.

(2007, 27.1), (2008, 29.3), (2009, 28.3), (2010, 28.0), (2011, 26.9)

(b) What would the ordered pair (2000, 32.4) mean in the context of this problem?

(2000, 32.4) means that 32.4 percent of 2-year college students in 2000 received a degree within 3 years.

(c) Make a scatter diagram of the data using the ordered pairs from part (a). **See the graph.**

2-YEAR COLLEGE STUDENTS COMPLETING A DEGREE WITHIN 3 YEARS

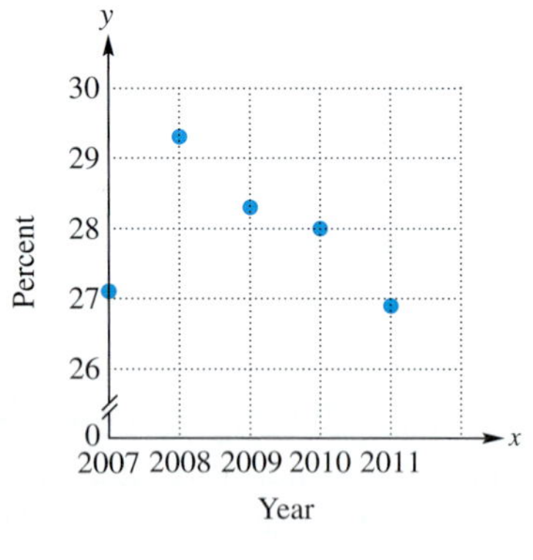

(d) Describe the pattern indicated by the points on the scatter diagram. What is happening to the rates at which 2-year college students complete a degree within 3 years?

With the exception of the point for 2007, the points lie approximately on a straight line. Rates at which 2-year college students complete a degree within 3 years were generally decreasing.

84. The table shows the number of U.S. students studying abroad (in thousands) for recent academic years.

Academic Year	Number of Students (in thousands)
2005	224
2006	242
2007	262
2008	260
2009	271

Source: Institute of International Education.

(a) Write the data from the table as ordered pairs (x, y), where x represents the year and y represents the number of U.S. students studying abroad, in thousands.

(2005, 224), (2006, 242), (2007, 262), (2008, 260), (2009, 271)

(b) What does the ordered pair (2000, 154) mean in the context of this problem?

(2000, 154) means that in 2000, about 154 thousand (or 154,000) U.S. students studied abroad.

(c) Make a scatter diagram of the data using the ordered pairs from part (a). **See the graph.**

U.S. STUDENTS STUDYING ABROAD

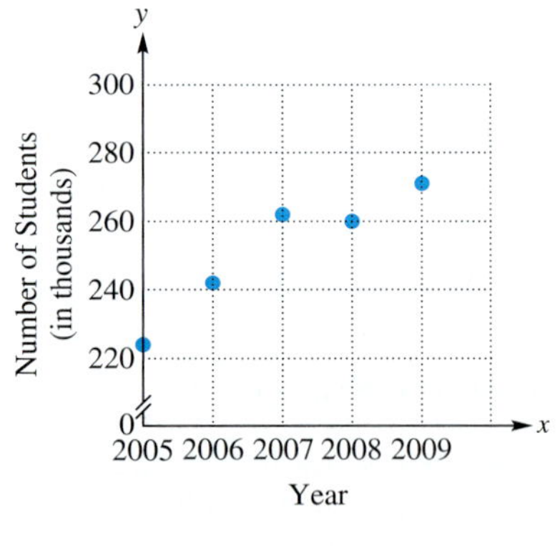

(d) Describe the pattern indicated by the points on the scatter diagram. What is the trend in the number of U.S. students studying abroad?

The points for 2005, 2006, and 2007 lie approximately on a straight line. The number of U.S. students studying abroad has been increasing year by year, except for a small dip in 2008.

85. The maximum benefit for the heart from exercising occurs if the heart rate is in the target heart rate zone. The lower limit of this target zone can be approximated by the linear equation

$$y = -0.5x + 108,$$

where x represents age and y represents heartbeats per minute. (*Source:* www.fitresource.com)

(a) Complete the table of values for this linear equation.

Age	Heartbeats (per minute)
20	**98**
40	**88**
60	**78**
80	**68**

See the table.

(b) Write the data from the table of values as ordered pairs.

(20, 98), (40, 88), (60, 78), (80, 68)

(c) Make a scatter diagram of the data. Do the points lie in a linear pattern?

TARGET HEART RATE ZONE
(Lower Limit)

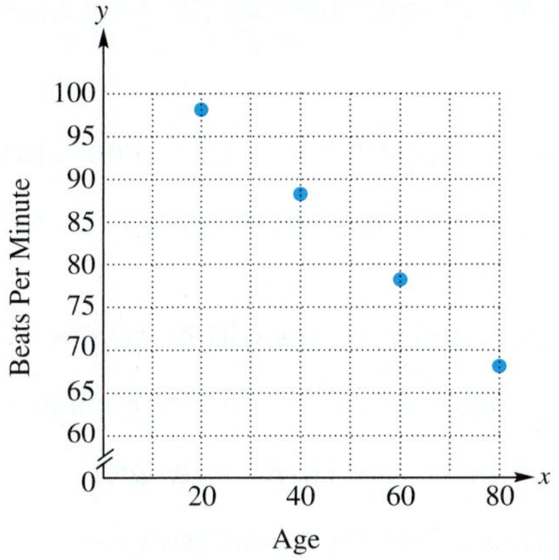

See the graph; yes

86. (See **Exercise 85.**) The upper limit of the target heart rate zone can be approximated by the linear equation

$$y = -0.8x + 173,$$

where x represents age and y represents heartbeats per minute. (*Source:* www.fitresource.com)

(a) Complete the table of values for this linear equation.

Age	Heartbeats (per minute)
20	**157**
40	**141**
60	**125**
80	**109**

See the table.

(b) Write the data from the table of values as ordered pairs.

(20, 157), (40, 141), (60, 125), (80, 109)

(c) Make a scatter diagram of the data. Describe the pattern indicated by the data.

TARGET HEART RATE ZONE
(Upper Limit)

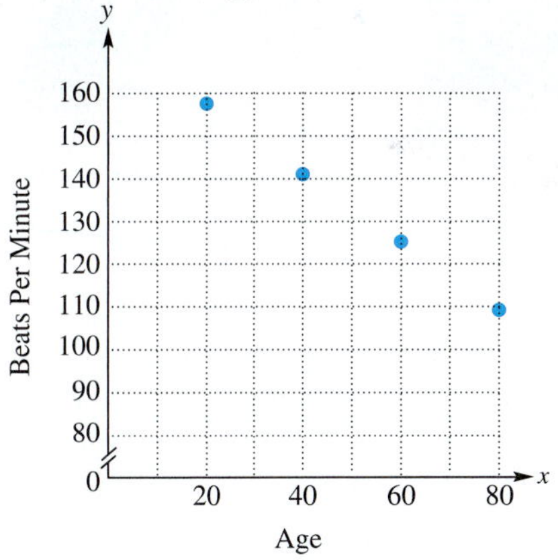

See the graph; The points lie in a linear pattern.

87. See **Exercises 85 and 86.** What is the target heart rate zone for age 20? Age 40?

between 98 and 157 beats per minute;
between 88 and 141 beats per minute

88. See **Exercises 85 and 86.** What is the target heart rate zone for age 60? Age 80?

between 78 and 125 beats per minute;
between 68 and 109 beats per minute

Study Skills

PREPARING FOR YOUR MATH FINAL EXAM

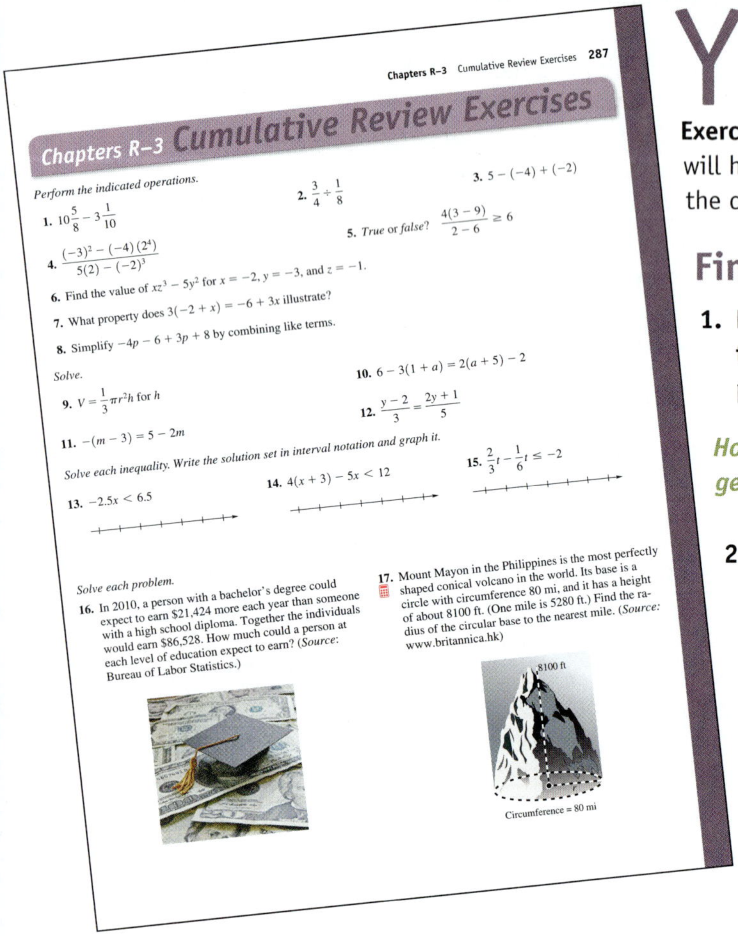

Chapters R–3 Cumulative Review Exercises 287

Chapters R–3 Cumulative Review Exercises

Perform the indicated operations.

1. $10\frac{5}{8} - 3\frac{1}{10}$
2. $\frac{3}{4} \div \frac{1}{8}$
3. $5 - (-4) + (-2)$
4. $\frac{(-3)^2 - (-4)(2^4)}{5(2) - (-2)^3}$
5. *True* or *false*? $\frac{4(3-9)}{2-6} \geq 6$
6. Find the value of $xz^3 - 5y^2$ for $x = -2$, $y = -3$, and $z = -1$.
7. What property does $3(-2 + x) = -6 + 3x$ illustrate?
8. Simplify $-4p - 6 + 3p + 8$ by combining like terms.

Solve.

9. $V = \frac{1}{3}\pi r^2 h$ for h
10. $6 - 3(1 + a) = 2(a + 5) - 2$
11. $-(m - 3) = 5 - 2m$
12. $\frac{y-2}{3} = \frac{2y+1}{5}$

Solve each inequality. Write the solution set in interval notation and graph it.

13. $-2.5x < 6.5$
14. $4(x + 3) - 5x < 12$
15. $\frac{2}{3}t - \frac{1}{6}t \leq -2$

Solve each problem.

16. In 2010, a person with a bachelor's degree could expect to earn \$21,424 more each year than someone with a high school diploma. Together the individuals would earn \$86,528. How much could a person at each level of education expect to earn? (*Source:* Bureau of Labor Statistics.)
17. Mount Mayon in the Philippines is the most perfectly shaped conical volcano in the world. Its base is a circle with circumference 80 mi, and it has a height of about 8100 ft. (One mile is 5280 ft.) Find the radius of the circular base to the nearest mile. (*Source:* www.britannica.hk)

8100 ft

Circumference = 80 mi

Your math final exam is likely to be a comprehensive exam, which means it will cover material from the entire term. **One way to prepare for it now is by working a set of Cumulative Review Exercises** each time your class finishes a chapter. This continual review will help you remember concepts and procedures as you progress through the course.

Final Exam Preparation Suggestions

1. **Figure out the grade you need to earn on the final exam to get the course grade you want.** Check your course syllabus for grading policies, or ask your instructor if you are not sure.

How many points do you need to earn on your math final exam to get the grade you want?

2. **Create a final exam week plan.** Set priorities that allow you to spend extra time studying. This may mean making adjustments, in advance, in your work schedule or enlisting extra help with family responsibilities.

What adjustments do you need to make for final exam week? List two or three here.

3. **Use the following suggestions to guide your studying.**

- **Begin reviewing several days before the final exam.** DON'T wait until the last minute.
- **Know exactly which chapters and sections will be covered.**
- **Divide up the chapters.** Decide how much to review each day.
- **Keep returned quizzes and tests. Use them to review.**
- **Practice all types of problems. Use the Cumulative Review Exercises** at the end of each chapter in your textbook beginning in Chapter 2. All answers are given in the answer section.
- **Review or rewrite your notes** to create summaries of key information.
- **Make study cards for all types of problems.** Carry the cards with you, and review them whenever you have a few minutes.
- **Take plenty of short breaks as you study to reduce physical and mental stress.** Exercising, listening to music, and enjoying a favorite activity are effective stress busters.

Finally, *DON'T* stay up all night the night before an exam—*get a good night's sleep.*

Which of these suggestions will you use as you study for your math final exam?

3.2 Graphing Linear Equations in Two Variables

OBJECTIVES

1. **Graph linear equations by plotting ordered pairs.**
2. **Find intercepts.**
3. **Graph linear equations of the form $Ax + By = 0$.**
4. **Graph linear equations of the form $y = k$ or $x = k$.**
5. **Use a linear equation to model data.**

OBJECTIVE 1 **Graph linear equations by plotting ordered pairs.** There are infinitely many ordered pairs that satisfy an equation in two variables. We find these ordered-pair solutions by choosing as many values of x (or y) as we wish and then completing each ordered pair.

For example, consider the equation $x + 2y = 7$. If we choose $x = 1$, then we can substitute to find the corresponding value of y.

$$x + 2y = 7 \quad \text{Given equation}$$
$$1 + 2y = 7 \quad \text{Let } x = 1.$$
$$2y = 6 \quad \text{Subtract 1.}$$
$$y = 3 \quad \text{Divide by 2.}$$

If $x = 1$, then $y = 3$, and the ordered pair $(1, 3)$ is a solution of the equation.

$$1 + 2(3) = 7 \checkmark \quad (1, 3) \text{ is a solution.}$$

Work Problem **1** at the Side. ▶

Figure 8 shows a graph of all the ordered-pair solutions found above and in **Margin Problem 1** for $x + 2y = 7$.

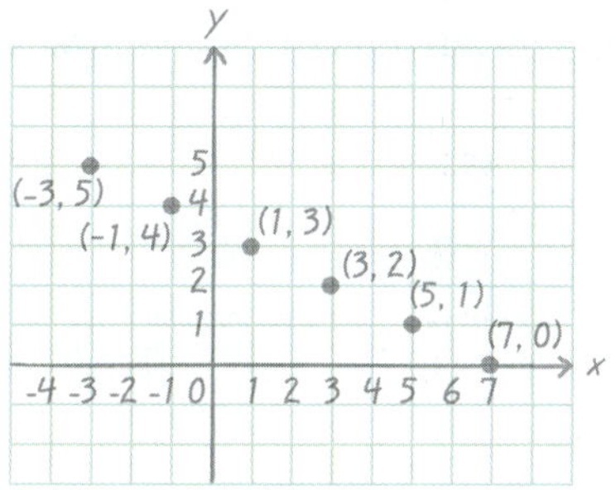

Figure 8

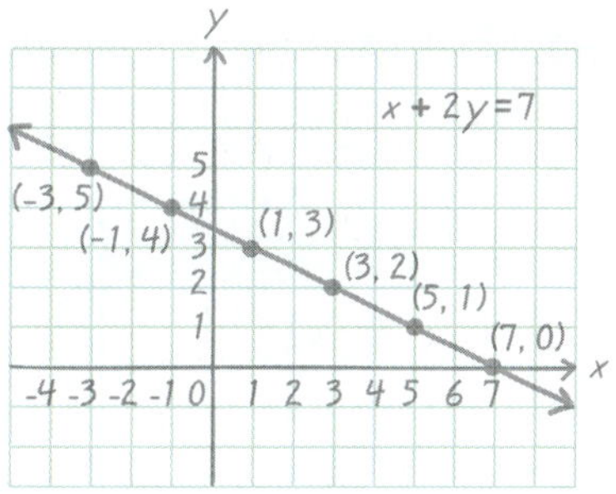

Figure 9

Notice that the points plotted in **Figure 8** all appear to lie on a straight line, as shown in **Figure 9.** In fact, the following is true.

Every point on the line represents a solution of the equation $x + 2y = 7$, and every solution of the equation corresponds to a point on the line.

This line gives a "picture" of all the solutions of the equation $x + 2y = 7$. The line extends indefinitely in both directions, as suggested by the arrowhead on each end, and is the **graph** of the equation $x + 2y = 7$. The process of plotting the ordered pairs and drawing the line through the corresponding points is called **graphing.**

The preceding discussion can be generalized.

Graph of a Linear Equation

The graph of any linear equation in two variables is a straight line.

Notice that the word ***line*** appears in the term "***line***ar equation."

1 Complete each ordered pair to find additional solutions of the equation $x + 2y = 7$.

GS **(a)** $(-3, \_\_\_\_)$

Substitute ____ for x in the equation and solve for ____.

(b) $(-1, \_\_\_\_)$

(c) $(3, \_\_\_\_)$

(d) $(5, \_\_\_\_)$

(e) $(7, \_\_\_\_)$

Answers

1. (a) -3; y; $(-3, 5)$ **(b)** $(-1, 4)$ **(c)** $(3, 2)$ **(d)** $(5, 1)$ **(e)** $(7, 0)$

EXAMPLE 1 Graphing a Linear Equation

Graph $4x - 5y = 20$.

At least two different points are needed to draw the graph. First let $x = 0$ and then let $y = 0$ to determine two ordered pairs.

$$4x - 5y = 20 \qquad \text{0 is easy to substitute.}$$
$$4(0) - 5y = 20$$
$$0 - 5y = 20 \qquad \text{Multiply.}$$
$$-5y = 20 \qquad \text{Subtract.}$$
$$y = -4 \qquad \text{Divide by } -5.$$

$$4x - 5y = 20 \qquad \text{0 is easy to substitute.}$$
$$4x - 5(0) = 20$$
$$4x - 0 = 20 \qquad \text{Multiply.}$$
$$4x = 20 \qquad \text{Subtract.}$$
$$x = 5 \qquad \text{Divide by 4.}$$

The ordered pairs are $(0, -4)$ and $(5, 0)$. Find a third ordered pair (as a check) by choosing a number *other than 0* for x or y. We choose $y = 2$.

$$4x - 5y = 20 \qquad \text{We arbitrarily let } y = 2. \text{ Other numbers could be used for } y, \text{ or for } x, \text{ instead.}$$
$$4x - 5(2) = 20$$
$$4x - 10 = 20 \qquad \text{Multiply.}$$
$$4x = 30 \qquad \text{Add 10.}$$
$$x = \frac{30}{4}, \text{ or } \frac{15}{2} \qquad \text{Divide by 4. Write in lowest terms.}$$

This gives the ordered pair $(\frac{15}{2}, 2)$, or $(7\frac{1}{2}, 2)$. We plot the three ordered pairs $(0, -4)$, $(5, 0)$, and $(7\frac{1}{2}, 2)$, and draw a line through them. See **Figure 10.**

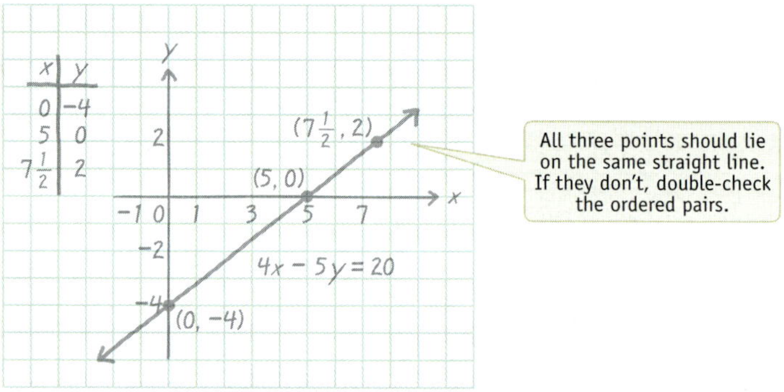

Figure 10

◀ Work Problem **2** at the Side.

OBJECTIVE 2 Find intercepts. In **Figure 10,** the graph crosses, or intersects, the y-axis at $(0, -4)$ and the x-axis at $(5, 0)$. For this reason, $(0, -4)$ is called the ***y*-intercept,** and $(5, 0)$ is called the ***x*-intercept** of the graph. The intercepts are particularly useful for graphing linear equations.

Finding Intercepts

To find the x-intercept, let $y = 0$ in the given equation and solve for x. Then $(x, 0)$ is the x-intercept.

To find the y-intercept, let $x = 0$ in the given equation and solve for y. Then $(0, y)$ is the y-intercept.

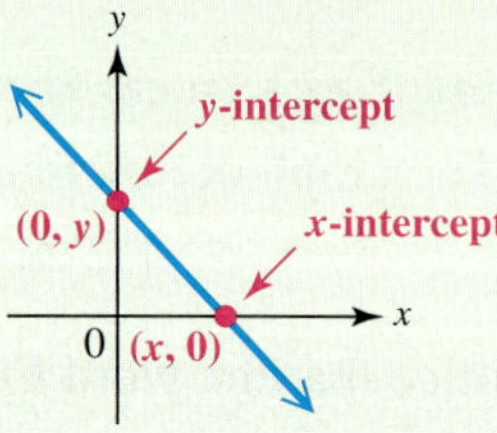

2 Graph the linear equation.

GS $x + y = 6$

x	y
0	
	0
2	

From the first row of the table, let $x =$ ____.

$$x + y = 6$$
$$\_\_\_\_ + y = 6$$
$$y = \_\_\_\_$$

Write ____ for the y-value in the first row of the table.

The first ordered pair is ____.

Repeat this process to complete the table. Then plot the corresponding points, and draw the ____ through them.

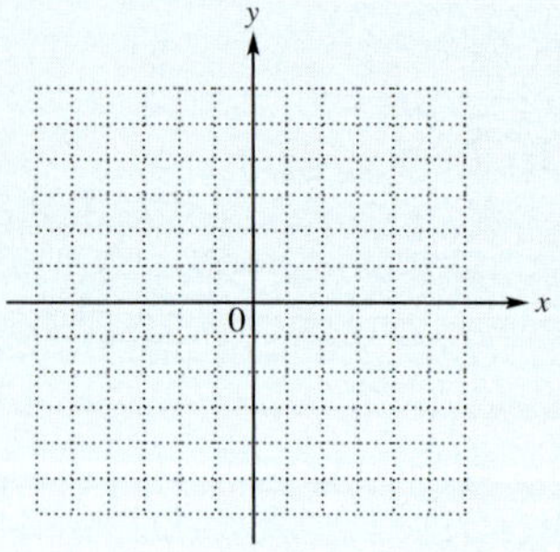

Answer

2. 0; 0; 6; 6; (0, 6); line

x	y
0	6
6	0
2	4

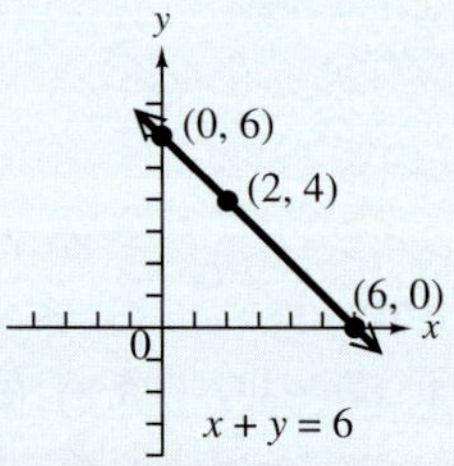

EXAMPLE 2 Graphing a Linear Equation Using Intercepts

Graph $2x + y = 4$ using intercepts.

To find the y-intercept, let $x = 0$.

$$2x + y = 4$$
$$2(0) + y = 4 \quad \text{Let } x = 0.$$
$$0 + y = 4 \quad \text{Multiply.}$$
$$y = 4 \quad y\text{-intercept is } (0, 4).$$

To find the x-intercept, let $y = 0$.

$$2x + y = 4$$
$$2x + 0 = 4 \quad \text{Let } y = 0.$$
$$2x = 4 \quad \text{Add.}$$
$$x = 2 \quad x\text{-intercept is } (2, 0).$$

The intercepts are $(0, 4)$ and $(2, 0)$. To find a third point (as a check), we let $x = 1$.

$$2x + y = 4$$
$$2(1) + y = 4 \quad \text{Let } x = 1.$$
$$2 + y = 4 \quad \text{Multiply.}$$
$$y = 2 \quad \text{Subtract 2.}$$

This gives the ordered pair $(1, 2)$. The graph, with the two intercepts in red, is shown in **Figure 11.**

	x	y
y-intercept ⟶	0	4
x-intercept ⟶	2	0
	1	2

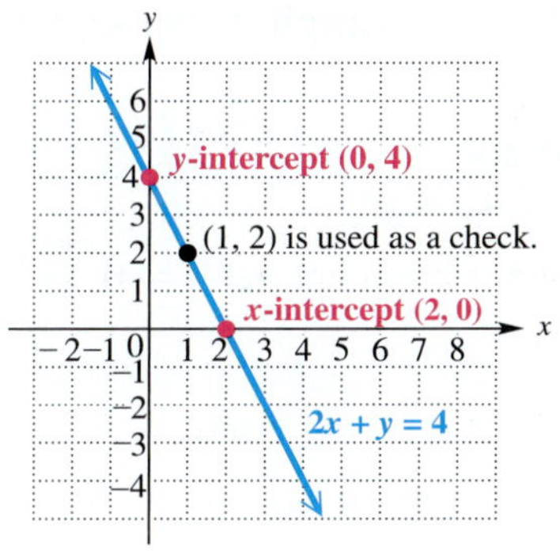

Figure 11

Work Problem **3** at the Side. ▶

3 Graph the linear equation using intercepts. (Be sure to get a third point as a check.)

GS

$$5x + 2y = 10$$

Find the y-intercept by letting $x =$ ____ in the equation.

The y-intercept is ____. Find the x-intercept by letting $y =$ ____ in the equation.

The x-intercept is ____.

Find a third point with $x = 4$ as a check, and graph the equation.

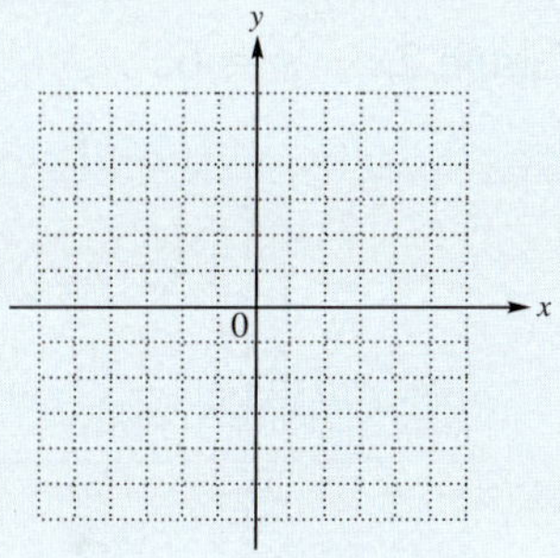

Teaching Tip

Point out that, in general, any point on the y-axis has x-coordinate 0 and any point on the x-axis has y-coordinate 0.

EXAMPLE 3 Graphing a Linear Equation Using Intercepts

Graph $y = -\frac{3}{2}x + 3$.

Although this linear equation is not in standard form $Ax + By = C$, it *could* be written in that form. The intercepts can be found by first letting $x = 0$ and then letting $y = 0$.

$$y = -\frac{3}{2}x + 3$$
$$y = -\frac{3}{2}(0) + 3 \quad \text{Let } x = 0.$$
$$y = 0 + 3 \quad \text{Multiply.}$$
$$y = 3 \quad \text{Add.}$$

$$y = -\frac{3}{2}x + 3$$
$$0 = -\frac{3}{2}x + 3 \quad \text{Let } y = 0.$$
$$\frac{3}{2}x = 3 \quad \text{Add } \tfrac{3}{2}x.$$
$$x = 2 \quad \text{Multiply by } \tfrac{2}{3}.$$

This gives the intercepts $(0, 3)$ and $(2, 0)$.

Continued on Next Page

Answer

3. 0; (0, 5); 0; (2, 0)

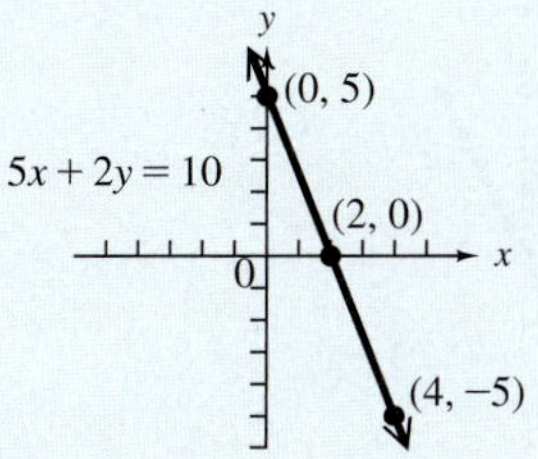

4 Graph the linear equation using intercepts.

$$y = \frac{2}{3}x - 2$$

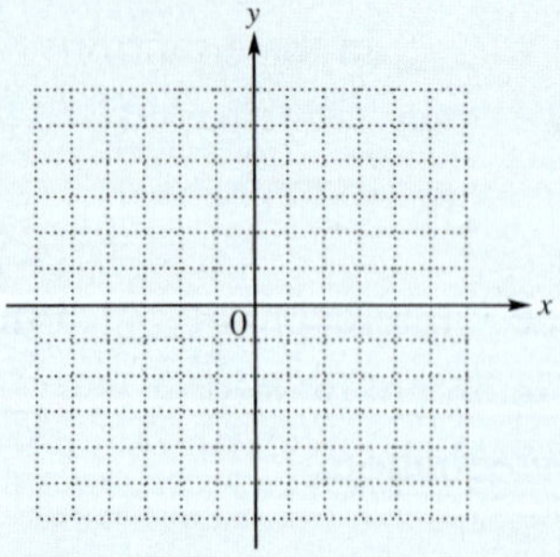

To find a third point, we let $x = -2$.

$$y = -\frac{3}{2}\boldsymbol{x} + 3$$

Choosing a multiple of 2 makes multiplying by $-\frac{3}{2}$ easier.

$$y = -\frac{3}{2}(\boldsymbol{-2}) + 3 \quad \text{Let } x = -2.$$

$$y = 3 + 3 \quad \text{Multiply.}$$

$$\boldsymbol{y = 6} \quad \text{Add.}$$

This gives the ordered pair $(-2, 6)$. We plot this ordered pair and both intercepts and draw a line through them, as shown in **Figure 12.**

x	y
0	3
2	0
−2	6

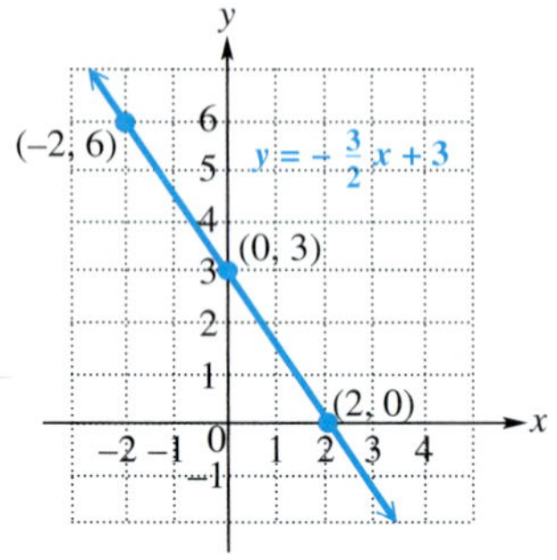

Figure 12

◀ **Work Problem 4 at the Side.**

5 Graph $2x - y = 0$.

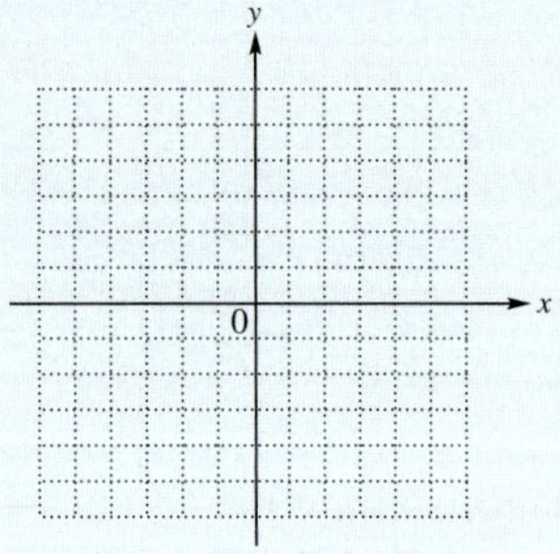

OBJECTIVE 3 Graph linear equations of the form $Ax + By = 0$.

EXAMPLE 4 Graphing an Equation with *x*- and *y*-Intercepts (0, 0)

Graph $x - 3y = 0$.

To find the y-intercept, let $x = 0$.

$$\boldsymbol{x} - 3y = 0$$

$$\boldsymbol{0} - 3y = 0 \quad \text{Let } x = 0.$$

$$-3y = 0 \quad \text{Subtract.}$$

$$y = 0 \quad y\text{-intercept is } (0, 0).$$

To find the x-intercept, let $y = 0$.

$$x - 3\boldsymbol{y} = 0$$

$$x - 3(\boldsymbol{0}) = 0 \quad \text{Let } y = 0.$$

$$x - 0 = 0 \quad \text{Multiply.}$$

$$x = 0 \quad x\text{-intercept is } (0, 0).$$

The x- and y-intercepts are the *same* point, $(0, 0)$. We must select *two other values* for x or y to find two other points. We choose $x = 6$ and $x = -6$.

$$\boldsymbol{x} - 3y = 0$$

$$\boldsymbol{6} - 3y = 0 \quad \text{Let } x = 6.$$

$$-3y = -6 \quad \text{Subtract 6.}$$

$$y = 2 \quad \text{Gives } (6, 2)$$

$$\boldsymbol{x} - 3y = 0$$

$$\boldsymbol{-6} - 3y = 0 \quad \text{Let } x = -6.$$

$$-3y = 6 \quad \text{Add 6.}$$

$$y = -2 \quad \text{Gives } (-6, -2)$$

We use $(-6, -2)$, $(0, 0)$, and $(6, 2)$ to draw the graph in **Figure 13.**

x	y
0	0
6	2
−6	−2

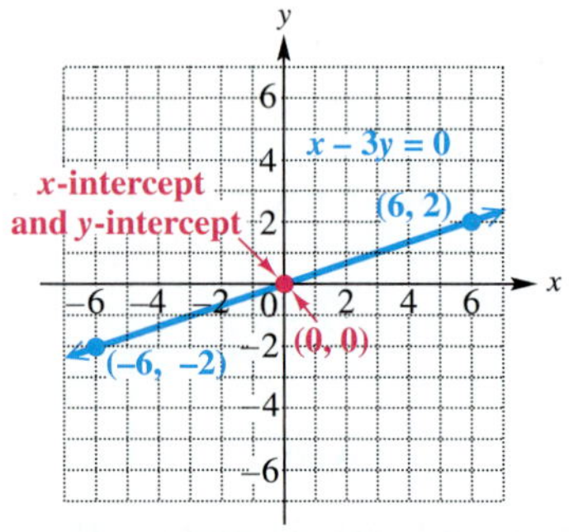

Figure 13

◀ **Work Problem 5 at the Side.**

Answers

4.

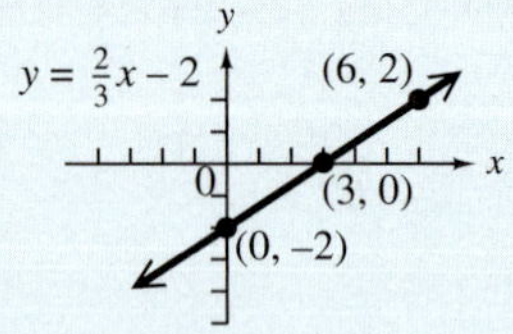

5.

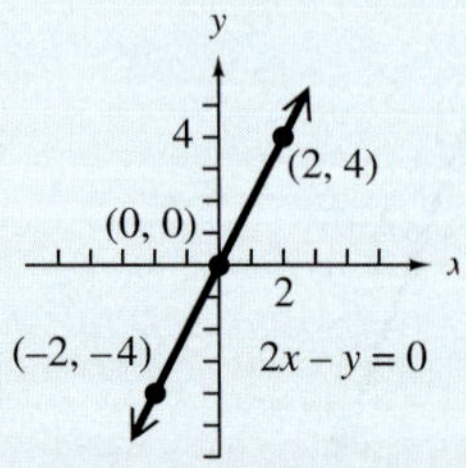

Line through the Origin

The graph of a linear equation of the form

$$Ax + By = 0,$$

where A and B are nonzero real numbers, passes through the origin $(0, 0)$.

OBJECTIVE 4 **Graph linear equations of the form $y = k$ or $x = k$.** Consider the following linear equations.

$y = -4$ can be written as $0x + y = -4$.

$x = 3$ can be written as $x + 0y = 3$.

When the coefficient of x or y is 0, the graph is a horizontal or vertical line.

EXAMPLE 5 **Graphing a Horizontal Line ($y = k$)**

Graph $y = -4$.

For any value of x, y is always equal to -4. Three ordered pairs that satisfy the equation are shown in the table of values. Drawing a line through these points gives the **horizontal line** shown in **Figure 14**. The y-intercept is $(0, -4)$. There is no x-intercept.

x can be *any* real number.

x	y
-2	-4
0	-4
3	-4

y must be -4.

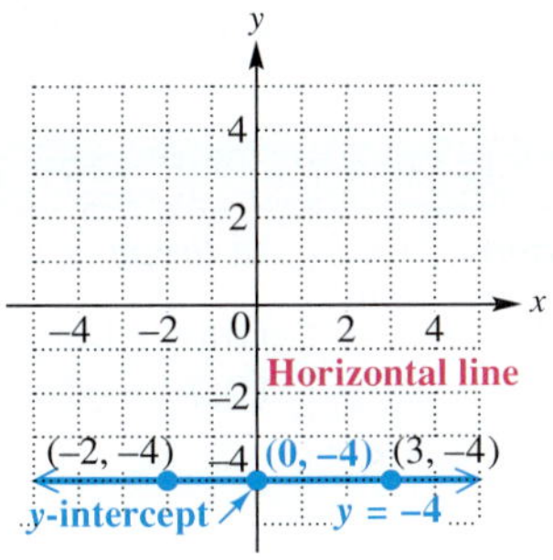

Figure 14

Work Problem 6 at the Side. ▶

EXAMPLE 6 **Graphing a Vertical Line ($x = k$)**

Graph $x - 3 = 0$.

First we add 3 to each side of the equation $x - 3 = 0$ to get the equivalent equation $x = 3$. All ordered-pair solutions of this equation have x-coordinate 3. Any number can be used for y. We show three ordered pairs that satisfy the equation in the table of values. The graph is the **vertical line** shown in **Figure 15**. The x-intercept is $(3, 0)$. There is no y-intercept.

y can be *any* real number.

x	y
3	3
3	0
3	-2

x must be 3.

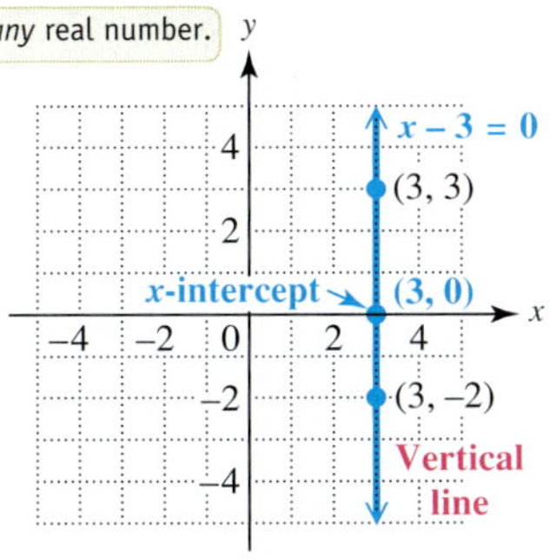

Figure 15

Work Problem 7 at the Side. ▶

6 Graph $y = -5$.

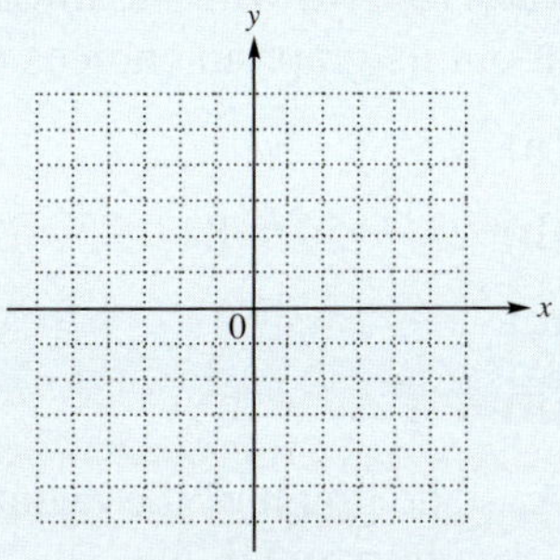

This is the graph of a ________ line. There is no x-intercept.

7 Graph $x + 4 = 6$.

GS First subtract ____ from each side to obtain the equivalent equation $x =$ ____. Now graph this equation.

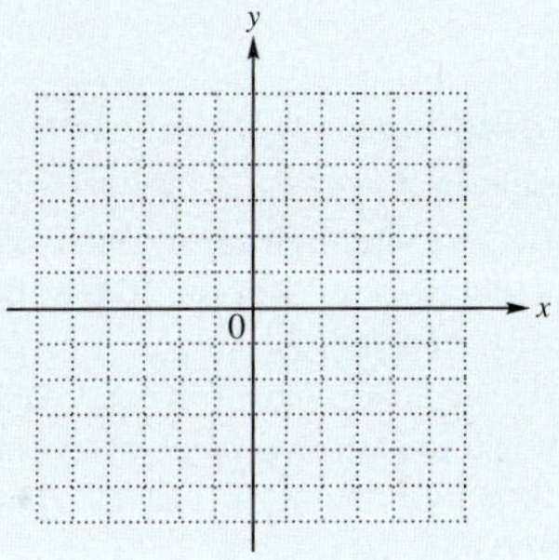

This is the graph of a ________ line. There is no y-intercept.

Answers

6.

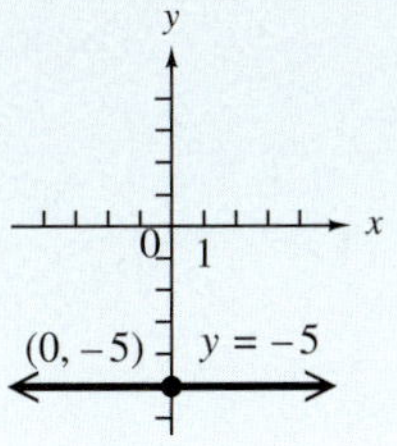

horizontal

7. 4; 2

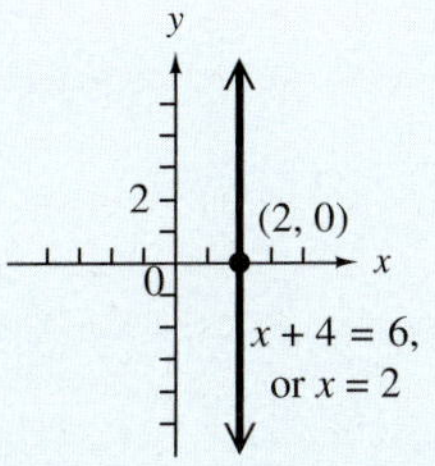

vertical

8 Match each linear equation in parts (a)–(d) with the information about its graph in choices A–D.

(a) $x = 5$

(b) $2x - 5y = 8$

(c) $y - 2 = 3$

(d) $x + 4y = 0$

A. The graph of the equation is a horizontal line.

B. The graph of the equation passes through the origin.

C. The graph of the equation is a vertical line.

D. The graph of the equation passes through $(9, 2)$.

Teaching Tip

Helpful hints:

1. An equation with only the variable x will always intersect the x-axis and thus be *vertical*.
2. An equation with only the variable y will always intersect the y-axis and thus be *horizontal*.

Answers

8. **(a)** C **(b)** D **(c)** A **(d)** B

From **Examples 5 and 6,** we make the following observations.

Horizontal and Vertical Lines

The graph of the linear equation $y = k$, where k is a real number, is the **horizontal line** with y-intercept $(0, k)$. There is no x-intercept (unless the horizontal line is the x-axis itself).

The graph of the linear equation $x = k$, where k is a real number, is the **vertical line** with x-intercept $(k, 0)$. There is no y-intercept (unless the vertical line is the y-axis itself).

The x-axis is the horizontal line given by the equation $y = 0$, and the y-axis is the vertical line given by the equation $x = 0$.

CAUTION

The equations of horizontal and vertical lines are often confused with each other. The graph of $y = k$ is parallel to the x-axis and the graph of $x = k$ is parallel to the y-axis (for $k \neq 0$).

A summary of the forms of linear equations from this section follows.

Graphing a Linear Equation

Equation	*To Graph*	*Example*
$y = k$	Draw a horizontal line, through $(0, k)$.	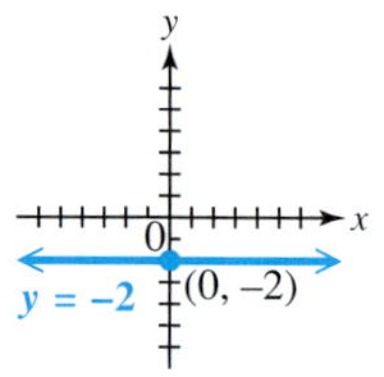
$x = k$	Draw a vertical line, through $(k, 0)$.	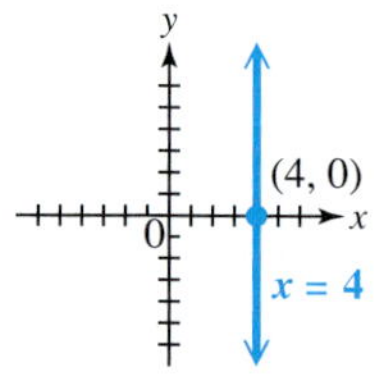
$Ax + By = 0$	The graph passes through $(0, 0)$. To find additional points that lie on the graph, choose any value for x or y, except 0.	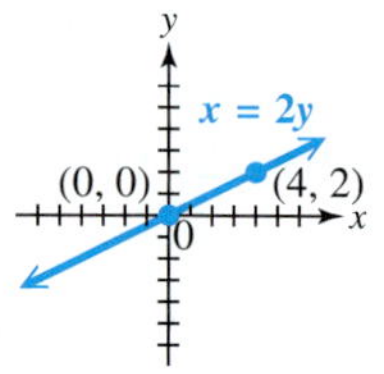
$Ax + By = C$ (where $A, B, C \neq 0$)	Find any two points on the line. A good choice is to find the intercepts. Let $x = 0$, and find the corresponding value of y. Then let $y = 0$, and find x. As a check, get a third point by choosing a value for x or y that has not yet been used.	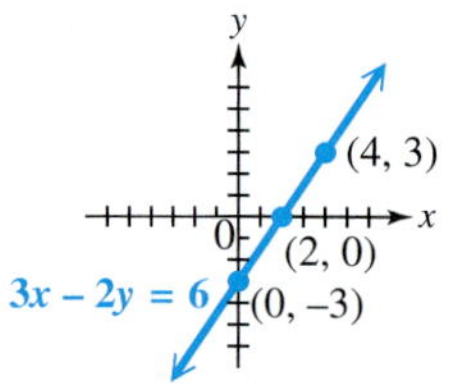

◀ Work Problem **8** at the Side.

OBJECTIVE 5 Use a linear equation to model data.

EXAMPLE 7 Using a Linear Equation to Model Credit Card Debt

The amount of credit card debt y in billions of dollars in the United States from 2000 through 2008 can be modeled by the linear equation

$$y = 32.0x + 684,$$

where $x = 0$ represents the year 2000, $x = 1$ represents 2001, and so on. (*Source:* The Nilson Report.)

(a) Use the equation to approximate credit card debt in the years 2000, 2004, and 2008.

Substitute the appropriate value for each year x to find credit card debt in that year.

For 2000: $y = 32.0(0) + 684$ Replace x with 0.

$y = 684$ billion dollars Multiply, and then add.

For 2004: $y = 32.0(4) + 684$ $2004 - 2000 = 4$

$y = 812$ billion dollars Replace x with 4.

For 2008: $y = 32.0(8) + 684$ $2008 - 2000 = 8$

$y = 940$ billion dollars Replace x with 8.

(b) Write the information from part (a) as three ordered pairs, and use them to graph the given linear equation.

Since x represents the year and y represents the debt, the ordered pairs are (0, 684), (4, 812), and (8, 940). See **Figure 16.** (Arrowheads are not included with the graphed line, since the data are for the years 2000 to 2008 only—that is, from $x = 0$ to $x = 8$.)

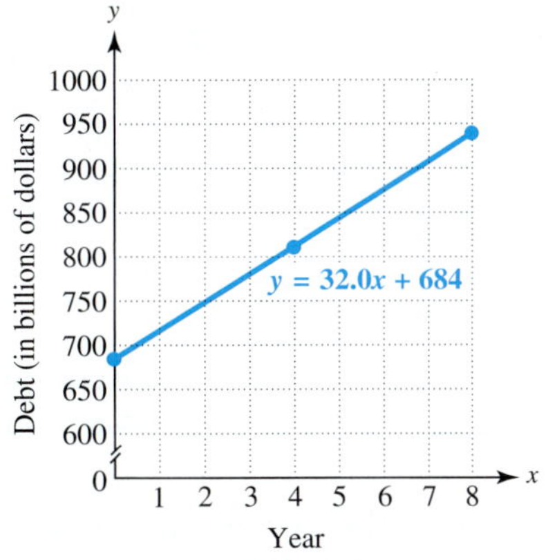

Figure 16

(c) Use the graph and the equation to approximate credit card debt in 2002.

For 2002, $x = 2$. On the graph, find 2 on the horizontal axis, move up to the graphed line and then across to the vertical axis. It appears that credit card debt in 2002 was about 750 billion dollars.

To use the equation, substitute 2 for x.

$y = 32.0x + 684$ Given linear equation

$y = 32.0(2) + 684$ Let $x = 2$.

$y = 748$ billion dollars Multiply, and then add.

This result for 2002 is close to our estimate of 750 billion dollars from the graph.

Work Problem 9 at the Side. ▶

9 Use **(a)** the graph and **(b)** the equation in **Example 7** to approximate credit card debt in 2006.

Answers

9. **(a)** about 875 billion dollars
(b) 876 billion dollars

3.2 Exercises

FOR EXTRA HELP

CONCEPT CHECK *Fill in each blank with the correct response.*

1. A linear equation in two variables can be written in the form $Ax +$ __By__ $=$ __C__, where A, B, and C are real numbers and A and B are not both __0__.

2. The graph of any linear equation in two variables is a straight __line__. Every point on the line represents a __solution__ of the equation.

3. **CONCEPT CHECK** Match the information about each graph in Column I with the correct linear equation in Column II.

I	II
(a) The graph of the equation has y-intercept $(0, -4)$. A	**A.** $3x + y = -4$
(b) The graph of the equation has $(0, 0)$ as x-intercept and y-intercept. C	**B.** $x - 4 = 0$
(c) The graph of the equation does not have an x-intercept. D	**C.** $y = 4x$
(d) The graph of the equation has x-intercept $(4, 0)$. B	**D.** $y = 4$

4. **CONCEPT CHECK** Which of these equations have a graph with only one intercept? A, C, D

A. $x + 8 = 0$ **B.** $x - y = 3$ **C.** $x + y = 0$ **D.** $y = 4$

CONCEPT CHECK *Identify the intercepts of each graph. (All coordinates are integers.)*

5.
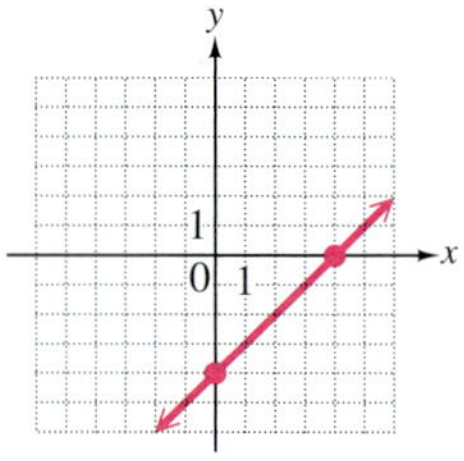

x-intercept: $(4, 0)$; y-intercept: $(0, -4)$

6.
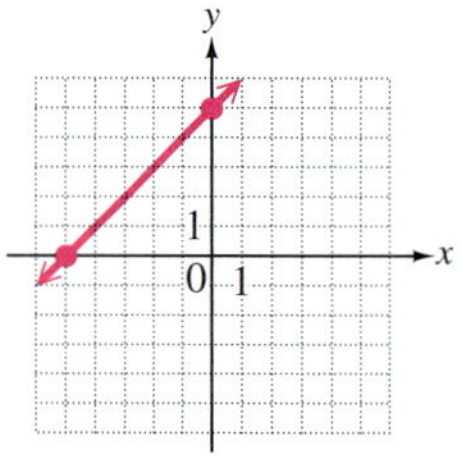

x-intercept: $(-5, 0)$; y-intercept: $(0, 5)$

7.
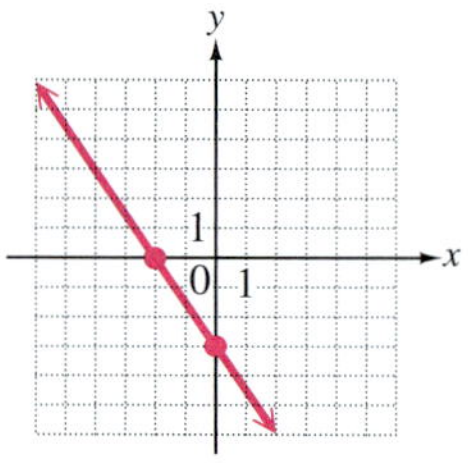

x-intercept: $(-2, 0)$; y-intercept: $(0, -3)$

8.
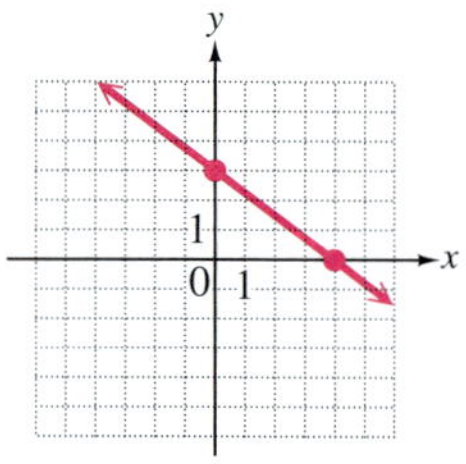

x-intercept: $(4, 0)$; y-intercept: $(0, 3)$

Complete the given ordered pairs for each equation. Then graph each equation by plotting the points and drawing the line through them. ***See Examples 1 and 2.***

9. $x + y = 5$

$(0,$ __5__$)$, $($__5__$, 0)$, $(2,$ __3__$)$

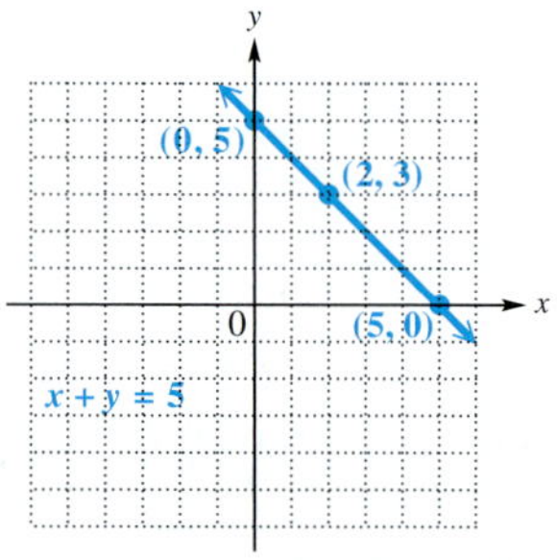

10. $x - y = 2$

$(0,$ __−2__$)$, $($__2__$, 0)$, $(5,$ __3__$)$

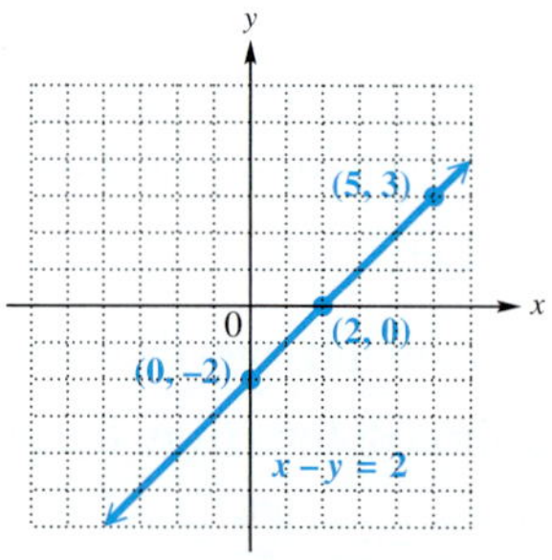

11. $y = \frac{2}{3}x + 1$

$(0,$ __1__$)$, $(3,$ __3__$)$, $(-3,$ __−1__$)$

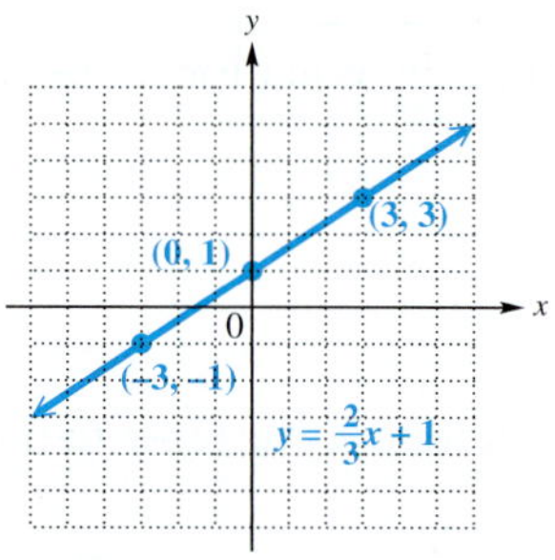

12. $y = -\frac{3}{4}x + 2$

$(0, \underline{2}), (4, \underline{-1}), (-4, \underline{5})$

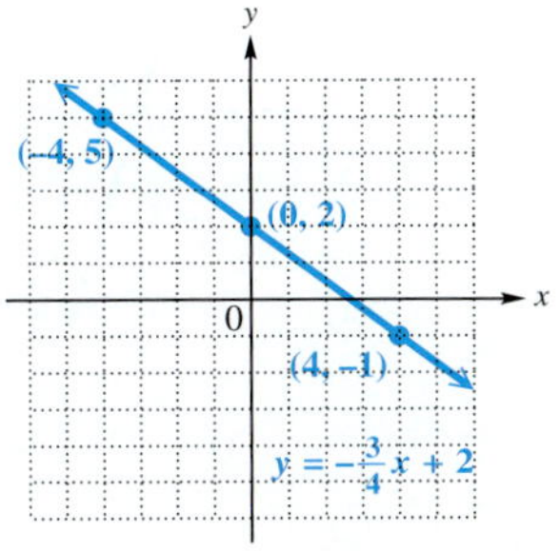

13. $3x = -y - 6$

$(0, \underline{-6}), (\underline{-2}, 0), \left(-\frac{1}{3}, \underline{-5}\right)$

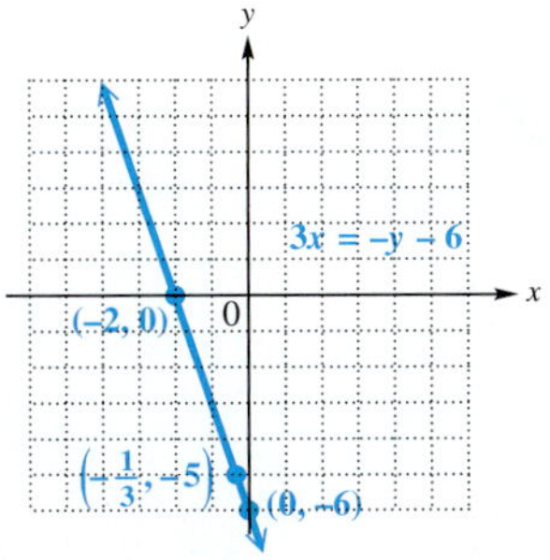

14. $x = 2y + 3$

$(\underline{3}, 0), (0, \underline{-\frac{3}{2}}), \left(\underline{4}, \frac{1}{2}\right)$

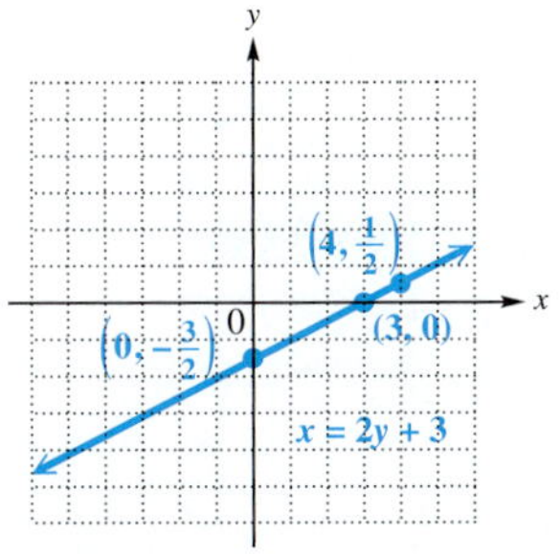

Find the intercepts for the graph of each equation. ***See Example 2.***

15. $x - y = 8$

GS

To find the y-intercept,

let $x = \underline{0}$.

The y-intercept is $\underline{(0, -8)}$.

To find the x-intercept,

let $y = \underline{0}$.

The x-intercept is $\underline{(8, 0)}$.

16. $x - y = 10$

GS

To find the y-intercept,

let $x = \underline{0}$.

The y-intercept is $\underline{(0, -10)}$.

To find the x-intercept,

let $y = \underline{0}$.

The x-intercept is $\underline{(10, 0)}$.

17. $2x - 3y = 24$

y-intercept: $\underline{(0, -8)}$

x-intercept: $\underline{(12, 0)}$

18. $-3x + 8y = 48$

y-intercept: $\underline{(0, 6)}$

x-intercept: $\underline{(-16, 0)}$

19. $x + 6y = 0$

y-intercept: $\underline{(0, 0)}$

x-intercept: $\underline{(0, 0)}$

20. $3x - y = 0$

y-intercept: $\underline{(0, 0)}$

x-intercept: $\underline{(0, 0)}$

Graph each linear equation using intercepts. ***See Examples 1–6.***

21. $y = x - 2$

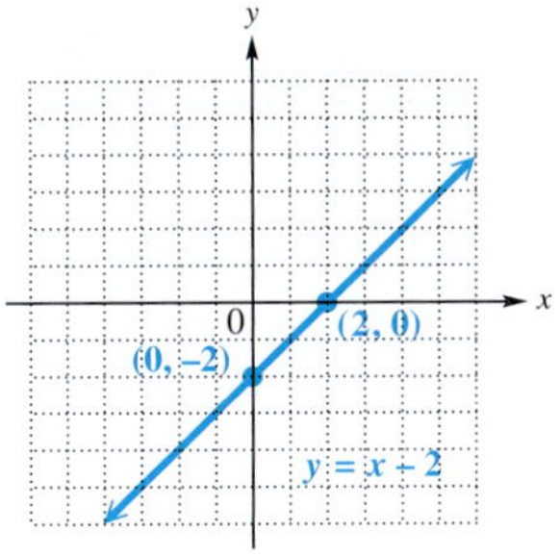

22. $y = -x + 6$

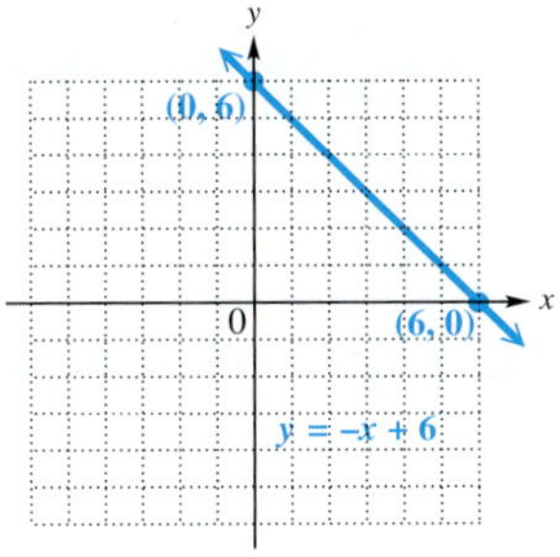

23. $x - y = 4$

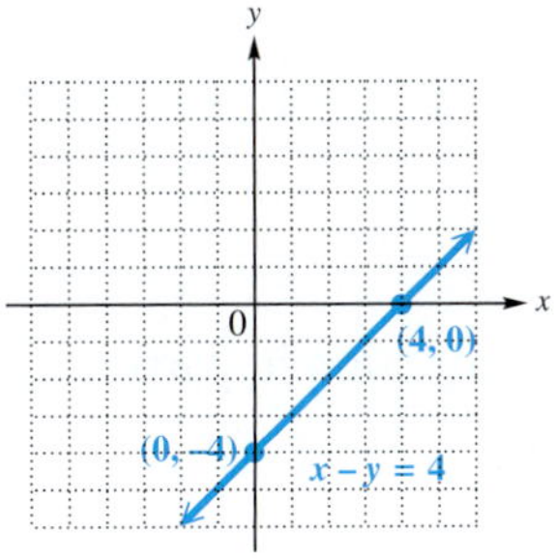

24. $x - y = 5$

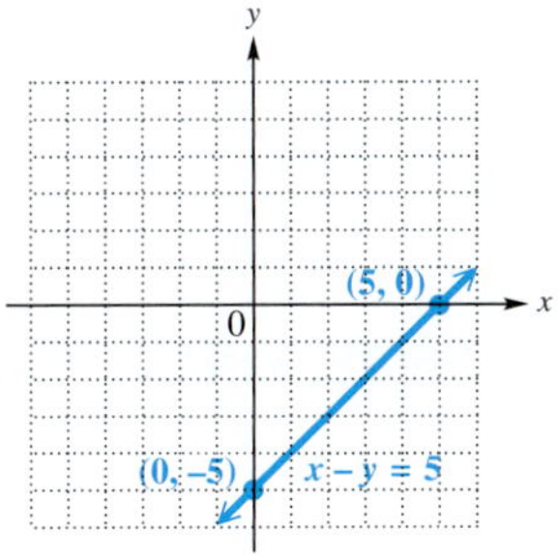

25. $2x + y = 6$

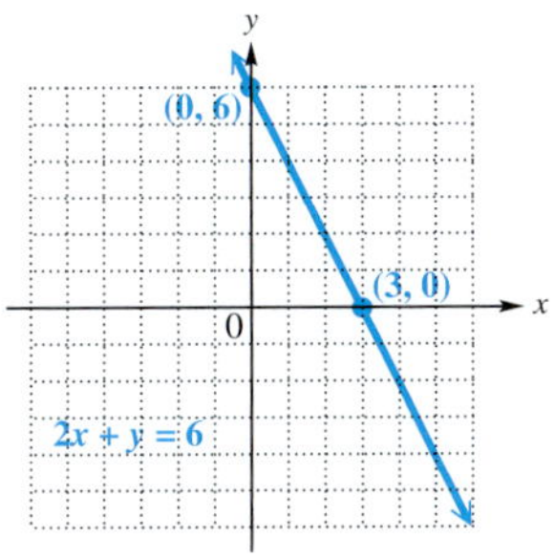

26. $-3x + y = -6$

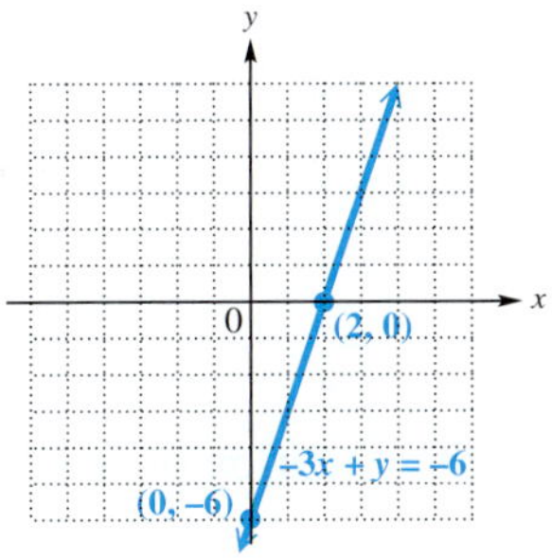

27. $y = 2x - 5$

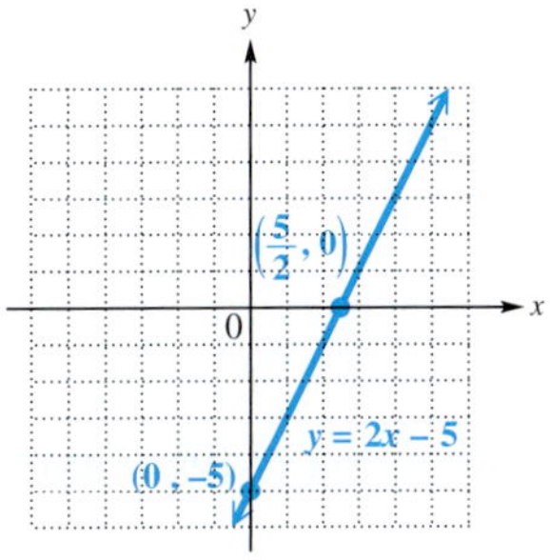

28. $y = 4x + 3$

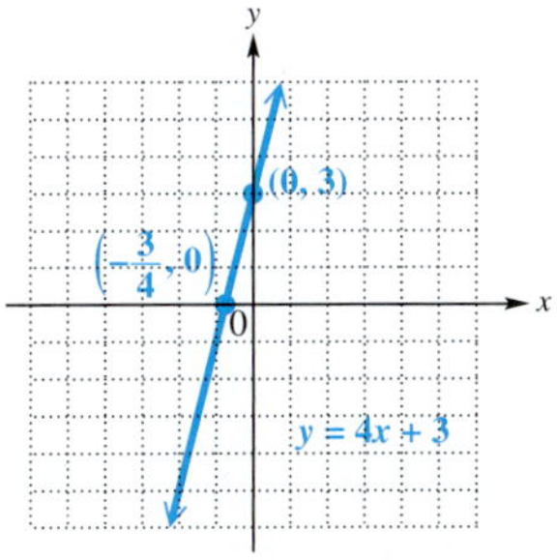

29. $3x + 7y = 14$

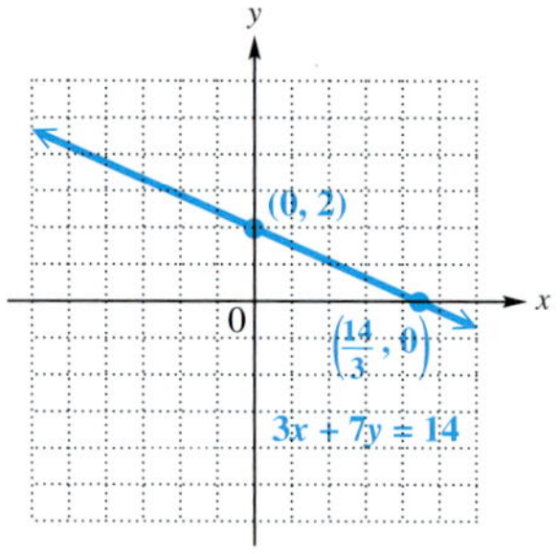

30. $6x - 5y = 18$

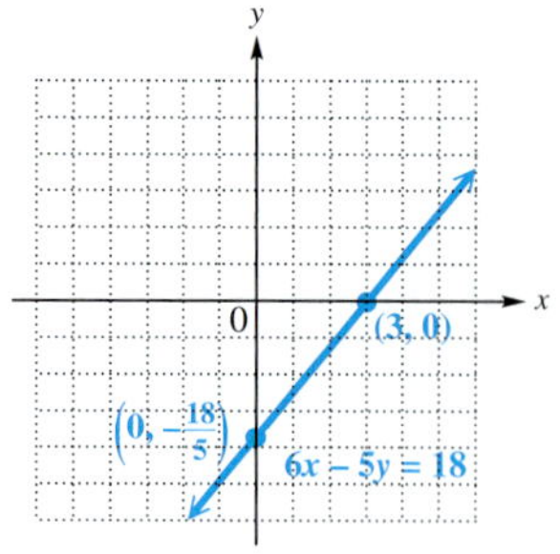

31. $y - 2x = 0$

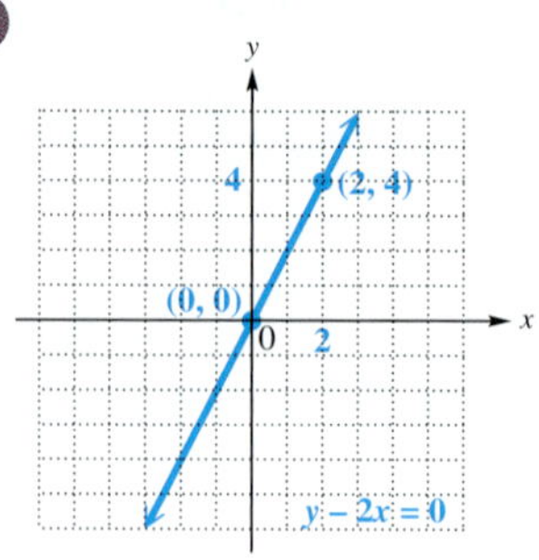

32. $y + 3x = 0$

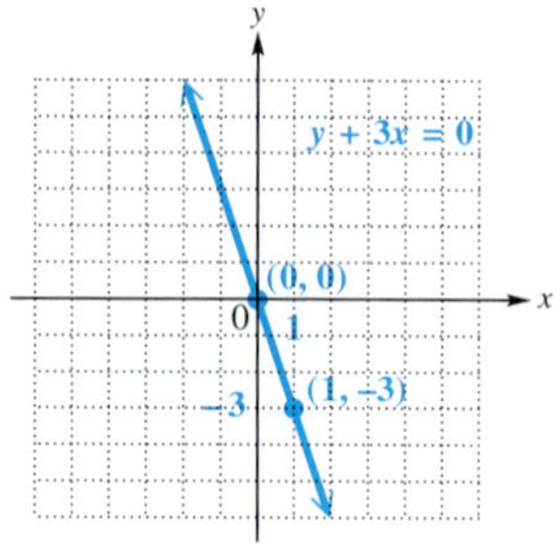

33. $y = -6x$

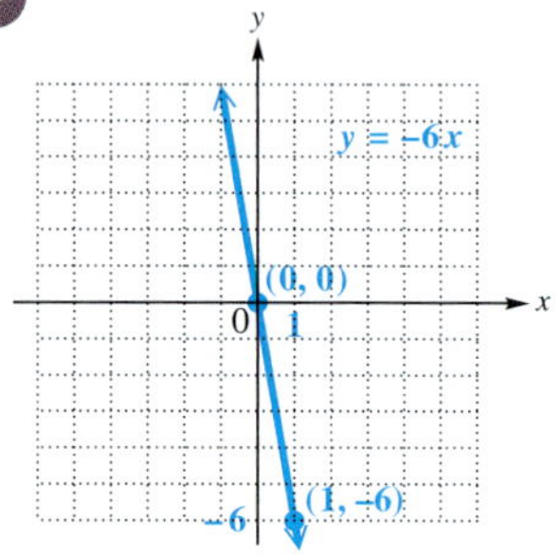

34. $y = 4x$

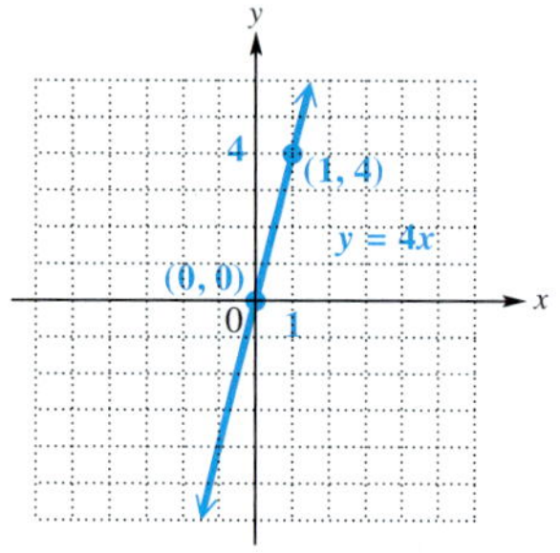

35. $x = -2$

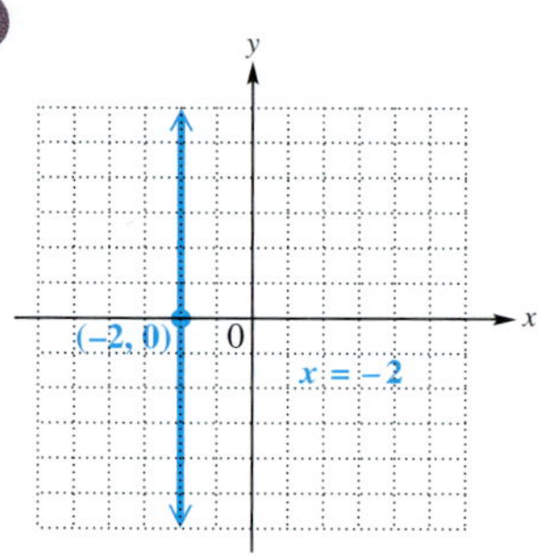

36. $x = 4$

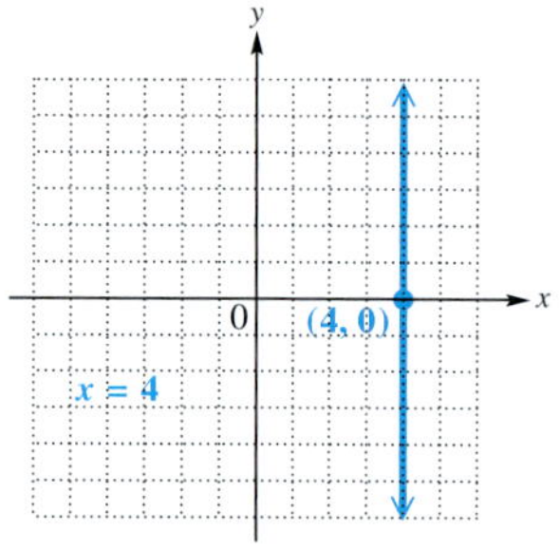

37. $y - 3 = 0$

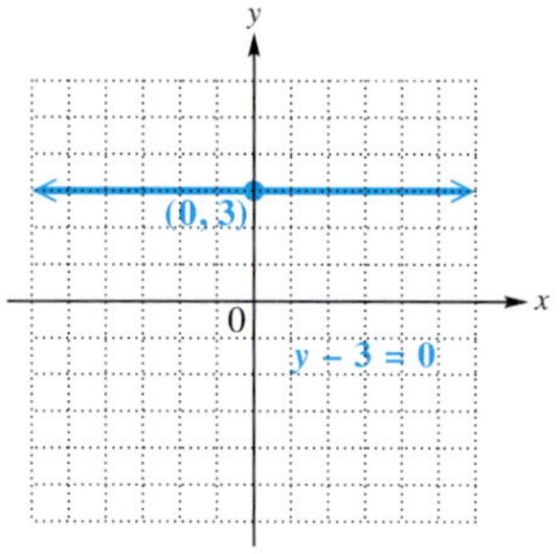

38. $y + 1 = 0$

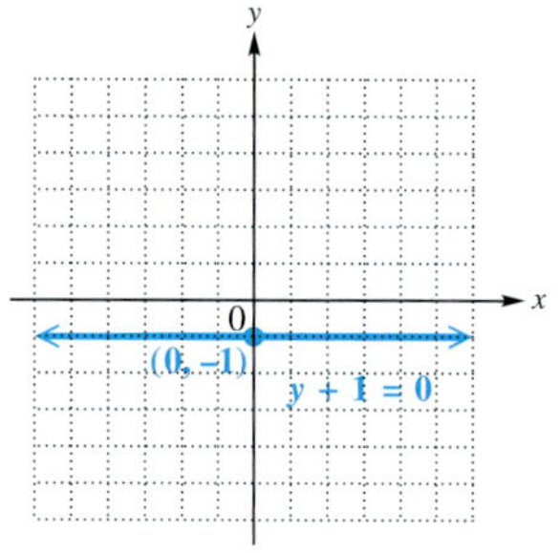

39. $-3y = 15$

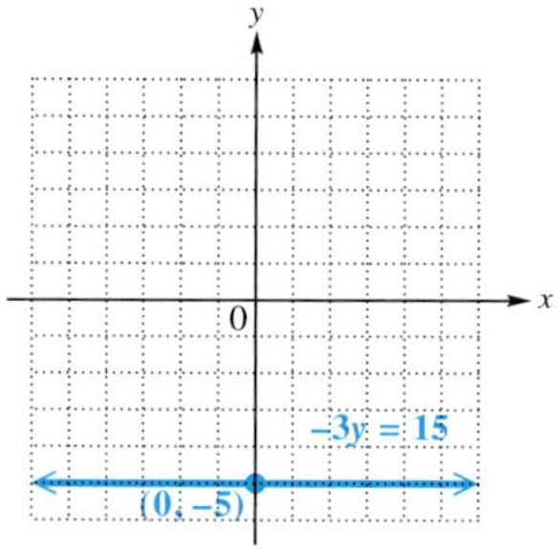

40. $-2y = 12$

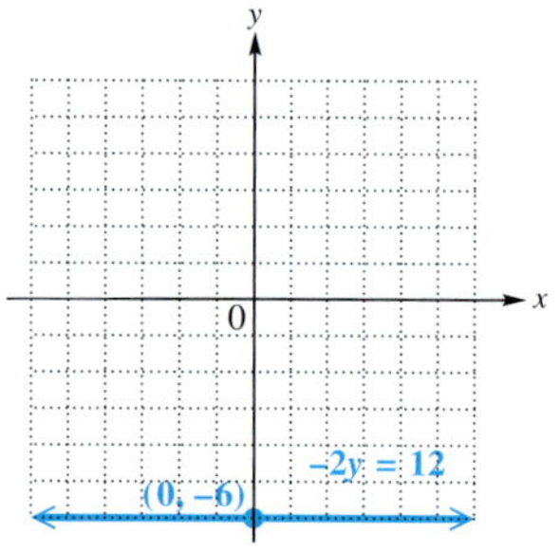

41. CONCEPT CHECK Match each equation in parts (a)–(d) with its graph in choices A–D.

(a) $x = -2$ D **(b)** $y = -2$ C **(c)** $x = 2$ B **(d)** $y = 2$ A

A.

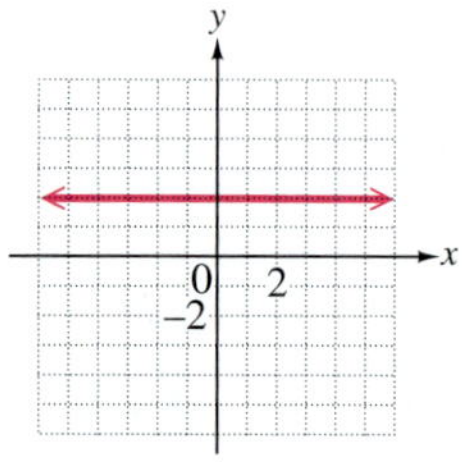

B.

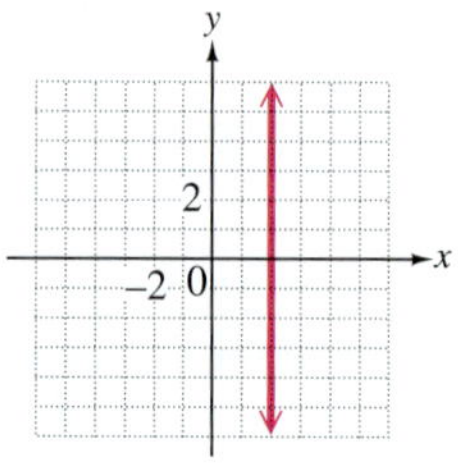

C.

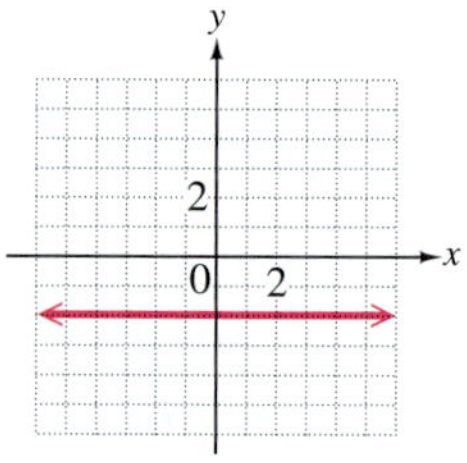

D.

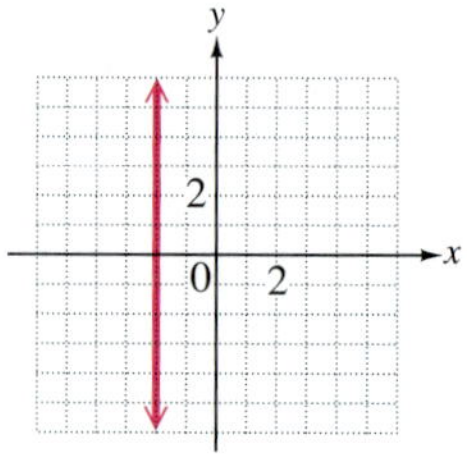

42. CONCEPT CHECK What is the equation of the *x*-axis? What is the equation of the *y*-axis?

$y = 0$; $x = 0$

In Exercises 43–46, describe what the graph of each linear equation will look like on the coordinate plane. (Hint: Rewrite the equation if necessary so that it is in a more recognizable form.) **In Exercises 43–46, descriptions may vary.**

43. $3x = y - 9$

The graph is a line with x-intercept $(-3, 0)$ and y-intercept $(0, 9)$.

44. $x - 10 = 1$

The graph is a vertical line with x-intercept $(11, 0)$.

45. $3y = -6$

The graph is a horizontal line with y-intercept $(0, -2)$.

46. $2x = 4y$

The graph passes through the origin $(0, 0)$ and the points $(2, 1)$ and $(4, 2)$.

Solve each problem. ***See Example 7.***

47. The height y of a woman is related to the length of her radius bone x and is approximated by the linear equation

$$y = 3.9x + 73.5,$$

where both x and y are in centimeters.

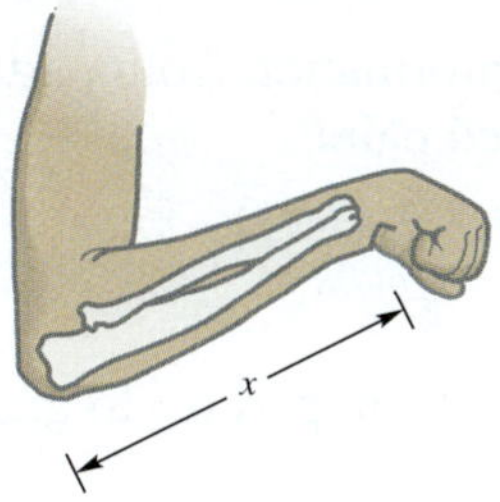

(a) Use the equation to find the approximate heights of women with radius bones of lengths 20 cm, 22 cm, and 26 cm.

151.5 cm, 159.3 cm, 174.9 cm

(b) Write the information from part (a) as three ordered pairs.

(20, 151.5), (22, 159.3), (26, 174.9)

(c) Graph the equation using the data from part (b).

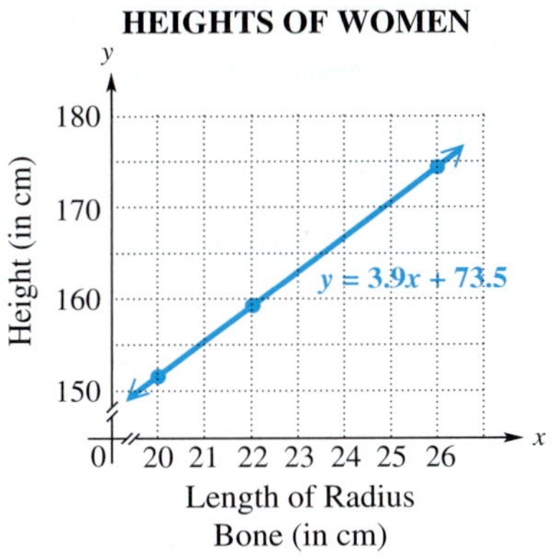

See the graph.

(d) Use the graph to estimate the length of the radius bone in a woman who is 167 cm tall. Then use the equation to find the length of this radius bone to the nearest centimeter. (*Hint:* Substitute for y in the equation.)

24 cm; 24 cm

48. The weight y (in pounds) of a man taller than 60 in. can be roughly approximated by the linear equation

$$y = 5.5x - 220,$$

where x is the height of the man in inches.

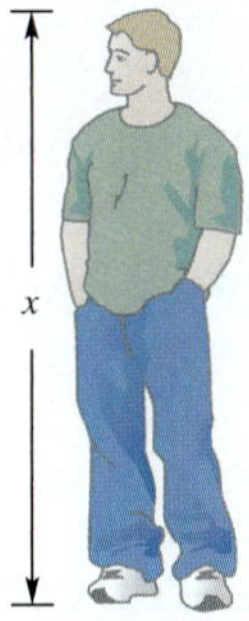

(a) Use the equation to approximate the weights of men whose heights are 62 in., 66 in., and 72 in.

121 lb, 143 lb, 176 lb

(b) Write the information from part (a) as three ordered pairs.

(62, 121), (66, 143), (72, 176)

(c) Graph the equation using the data from part (b).

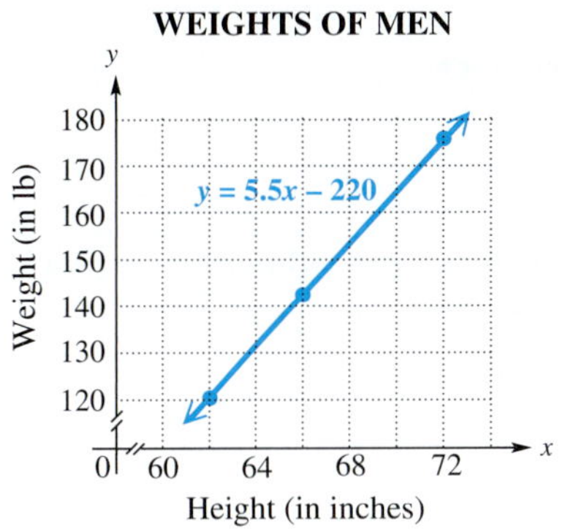

See the graph.

(d) Use the graph to estimate the height of a man who weighs 155 lb. Then use the equation to find the height of this man to the nearest inch. (*Hint:* Substitute for y in the equation.)

68 in.; 68 in.

49. As a fundraiser, a school club is selling posters. The printer charges a \$25 set-up fee, plus \$0.75 for each poster. The cost y in dollars to print x posters is given by the linear equation

$$y = 0.75x + 25.$$

(a) What is the cost y in dollars to print 50 posters? To print 100 posters?

\$62.50; \$100

(b) Find the number of posters x if the printer billed the club for costs of \$175.

200

(c) Write the information from parts (a) and (b) as three ordered pairs.

(50, 62.50), (100, 100), (200, 175)

(d) Use the data from part (c) to graph the equation.

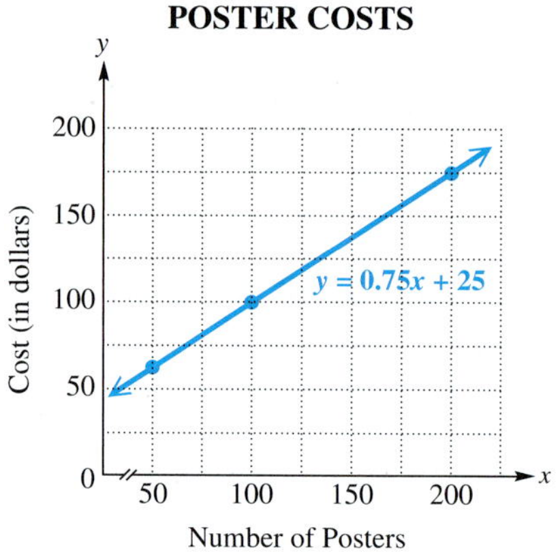

See the graph.

50. A gas station is selling gasoline for \$4.50 per gallon and charges \$7 for a car wash. The cost y in dollars for x gallons of gasoline and a car wash is given by the linear equation

$$y = 4.50x + 7.$$

(a) What is the cost y in dollars for 9 gallons of gasoline and a car wash? For 4 gallons of gasoline and a car wash?

\$47.50; \$25

(b) Find the number of gallons of gasoline x if the cost for the gasoline and a car wash is \$43.

8 gal

(c) Write the information from parts (a) and (b) as three ordered pairs.

(9, 47.50), (4, 25), (8, 43)

(d) Use the data from part (c) to graph the equation.

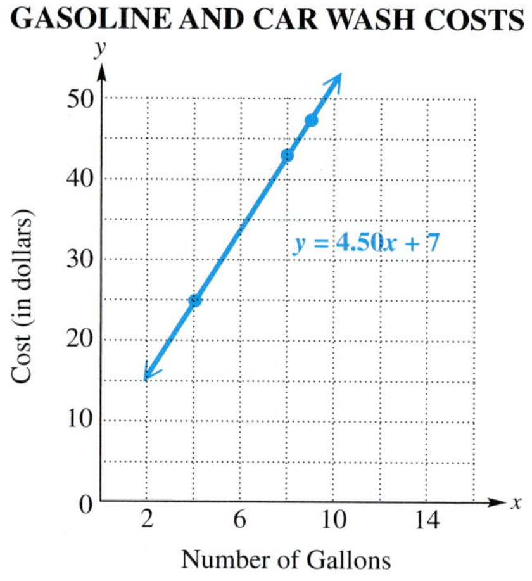

See the graph.

51. The graph shows the value of a certain sport-utility vehicle over the first 5 yr of ownership. Use the graph to do the following.

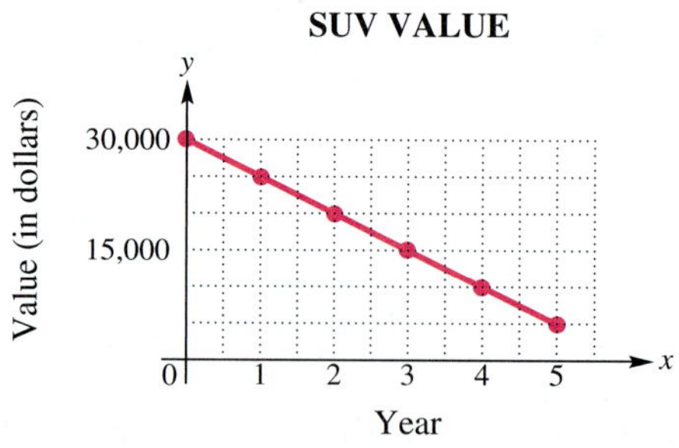

(a) Determine the initial value of the SUV.

\$30,000

(b) Find the **depreciation** (loss in value) from the original value after the first 3 yr.

\$15,000

(c) What is the annual or yearly depreciation in each of the first 5 yr?

\$5000

(d) What does the ordered pair (5, 5000) mean in the context of this problem?

After 5 yr, the SUV has a value of \$5000.

52. Demand for an item is often closely related to its price. As price increases, demand decreases, and as price decreases, demand increases. Suppose demand for a video game is 2000 units when the price is \$40, and demand is 2500 units when the price is \$30.

(a) Let x be the price and y be the demand for the game. Graph the two given pairs of prices and demands.

VIDEO GAME PRICE/DEMAND

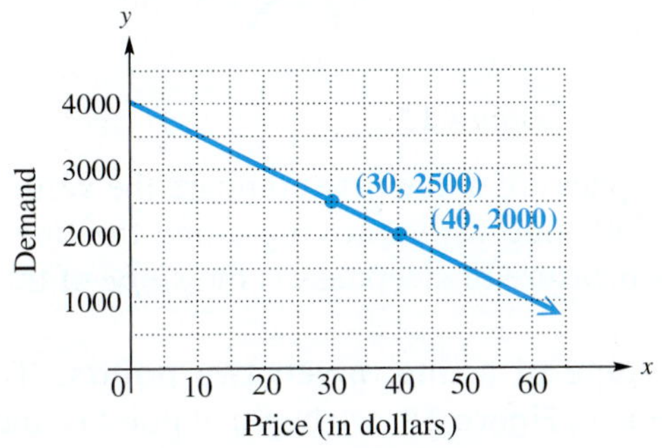

See the indicated points on the graph.

(b) Assume the relationship is linear. Draw a line through the two points from part (a). From your graph, estimate the demand if the price drops to \$20.

See the graph; 3000 units

(c) Use the graph to estimate the price if the demand is 3500 units.

\$10

(d) Write the prices and demands from parts (b) and (c) as ordered pairs.

(20, 3000), (10, 3500)

53. U.S. per capita consumption of cheese increased for the years 1980 through 2009 as shown in the graph. If $x = 0$ represents 1980, $x = 5$ represents 1985, and so on, per capita consumption y in pounds can be modeled by the linear equation

$$y = 0.5249x + 18.54.$$

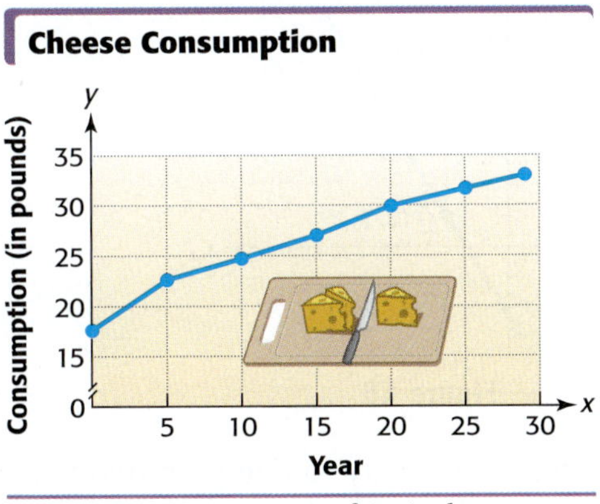

Source: U.S. Department of Agriculture.

(a) Use the equation to approximate cheese consumption in 1990, 2000, and 2009 to the nearest tenth.

1990: 23.8 lb; 2000: 29.0 lb; 2009: 33.8 lb

(b) Use the graph to estimate cheese consumption for the same years.

1990: 25 lb; 2000: 30 lb; 2009: 33 lb

(c) How do the approximations using the equation compare to the estimates from the graph?

The values are quite close.

54. The number of U.S. marathon finishers y (in thousands) from 1990 through 2010 are shown in the graph and modeled by the linear equation

$$y = 13.36x + 220.8,$$

where $x = 0$ corresponds to 1990, $x = 5$ corresponds to 1995, and so on.

Source: Running U.S.A.

(a) Use the equation to approximate the number of U.S. marathon finishers in 1990, 2000, and 2010 to the nearest thousand.

1990: 221 thousand; 2000: 354 thousand; 2010: 488 thousand

(b) Use the graph to estimate the number of U.S. marathon finishers for the same years.

1990: 225 thousand; 2000: 350 thousand; 2010: 500 thousand

(c) How do the approximations using the equation compare to the estimates from the graph?

The approximations using the equation are fairly close to the graph estimates for 1990 and 2000. The approximation using the equation is low for 2010.

3.3 The Slope of a Line

OBJECTIVES

1. Find the slope of a line, given two points.
2. Find the slope from the equation of a line.
3. Use slope to determine whether two lines are parallel, perpendicular, or neither.

An important characteristic of the lines we graphed in the previous section is their slant or "steepness," as viewed from *left to right*. See **Figure 17.**

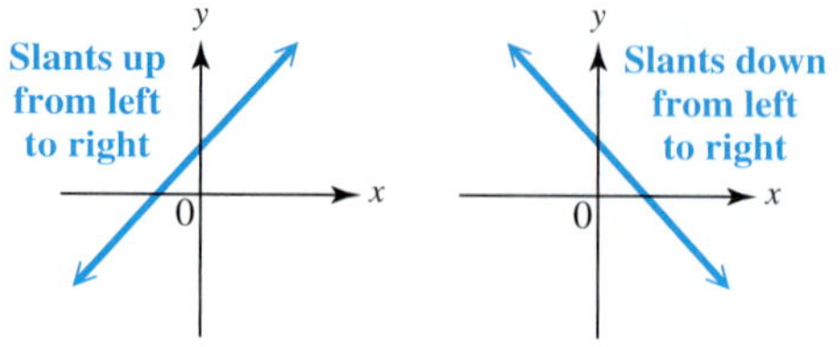

Figure 17

One way to measure the steepness of a line is to compare the vertical change in the line to the horizontal change while moving along the line from one fixed point to another. This measure of steepness is the *slope* of the line.

OBJECTIVE 1 Find the slope of a line, given two points. To find the steepness, or slope, of the line in **Figure 18,** we begin at point Q and move to point P. The vertical change, or **rise,** is the change in the y-values, which is the difference

$$6 - 1 = 5 \text{ units.}$$

The horizontal change, or **run,** from Q to P is the change in the x-values, which is the difference

$$5 - 2 = 3 \text{ units.}$$

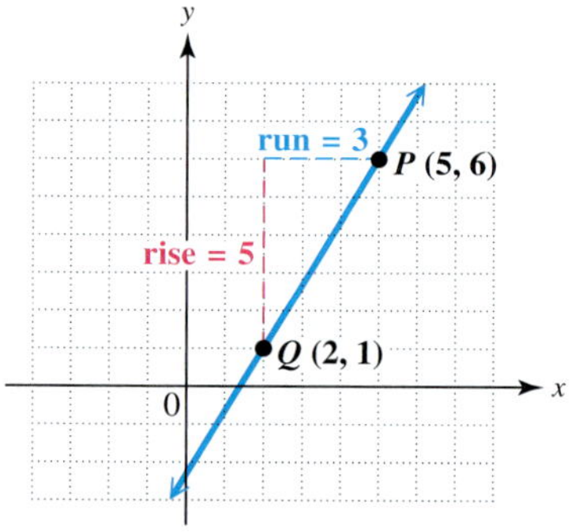

Figure 18

Remember from **Section 2.6** that one way to compare two numbers is by using a ratio. **Slope** is the ratio of the vertical change in y to the horizontal change in x. The line in **Figure 18** has

$$\text{slope} = \frac{\text{vertical change in } y \text{ (rise)}}{\text{horizontal change in } x \text{ (run)}} = \frac{5}{3}.$$

To confirm this ratio, we can count grid squares. We start at point Q in **Figure 18** and count *up* 5 grid squares to find the vertical change (rise). To find the horizontal change (run) and arrive at point P, we count to the *right* 3 grid squares. The slope is $\frac{5}{3}$, as found above.

Slope is a single number that allows us to determine the direction in which a line is slanting from left to right, as well as how much slant there is to the line.

Teaching Tip

When students count grid squares, make sure they understand that

- **upward moves are positive,**
- **downward moves are negative,**
- **rightward moves are positive,**
- **leftward moves are negative.**

EXAMPLE 1 Finding the Slope of a Line

Find the slope of the line in **Figure 19.**

We use the two points shown on the line. The vertical change is the difference in the y-values, or $-1 - 3 = -4$, and the horizontal change is the difference in the x-values, or $6 - 2 = 4$. Thus, the line has

$$\text{slope} = \frac{\text{change in } y \text{ (rise)}}{\text{change in } x \text{ (run)}} = \frac{-4}{4}, \text{ or } -1.$$

Counting grid squares, we begin at point P and count *down* 4 grid squares. Because we counted down, we write the vertical change as a negative number, -4 here. Then we count to the *right* 4 grid squares to reach point Q. The slope is $\frac{-4}{4}$, or -1.

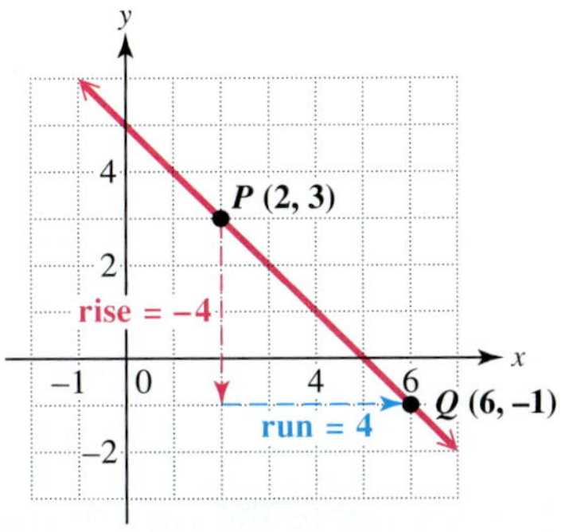

Figure 19

Work Problem 1 at the Side. ▶

Note

The slope of a line is the same for any two points on the line. To see this, refer to **Figure 19.** Find two points, say $(3, 2)$ and $(5, 0)$, which also lie on the line. If we start at $(3, 2)$ and count *down* 2 units and then to the *right* 2 units, we arrive at $(5, 0)$. The slope is $\frac{-2}{2}$, or -1, the same slope we found in **Example 1.**

The concept of slope is used in many everyday situations. See **Figure 20.** A highway with a 10%, or $\frac{1}{10}$, grade (or slope) rises 1 m for every 10 m horizontally. Architects specify the pitch of a roof by using slope. A $\frac{5}{12}$ roof means that the roof rises 5 ft vertically for every 12 ft that it runs horizontally. The slope of a stairwell indicates the ratio of the vertical rise to the horizontal run. The slope of the stairwell is $\frac{8}{12}$, or $\frac{2}{3}$.

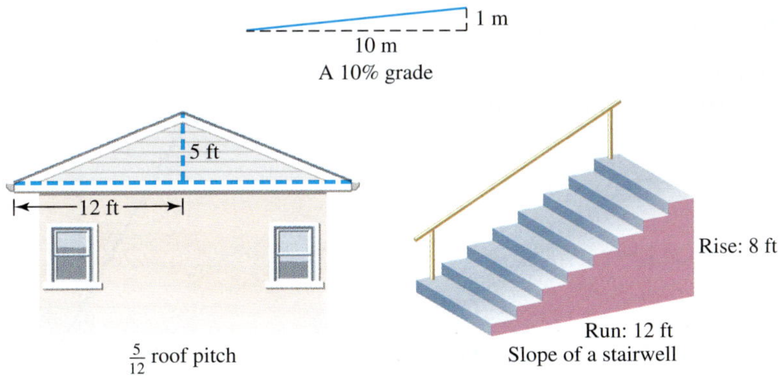

Figure 20

We can generalize the preceding discussion and find the slope of a line through two nonspecific points (x_1, y_1) and (x_2, y_2). This notation is called **subscript notation.** Read x_1 as **"x-sub-one"** and x_2 as **"x-sub-two."** See **Figure 21** on the next page.

1 Find the slope of each line.

GS **(a)**

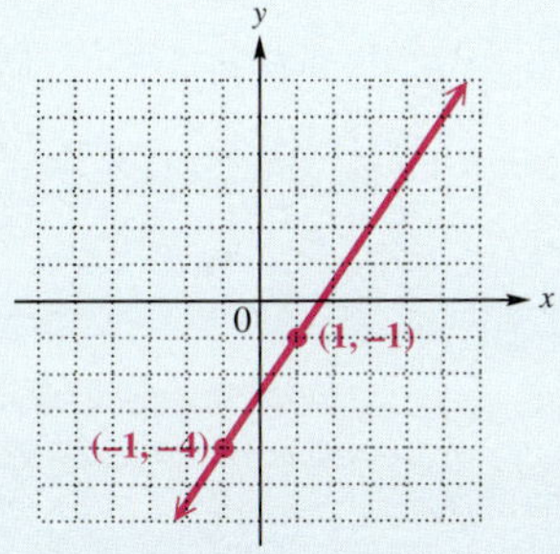

Begin at $(-1, -4)$.

Count (*up* / *down*) ____ units.

Then count to the (*right* / *left*) ____ units.

$$\text{slope} = \frac{\text{vertical change in } y}{\text{horizontal change in } x} = \frac{\_\_\_\_}{\_\_\_\_}$$

The slope is ____.

(b)

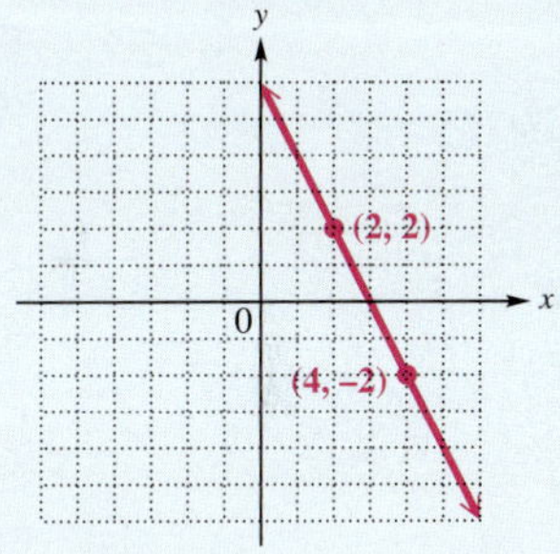

Answers

1. **(a)** up; 3; right; 2; 3; 2; $\frac{3}{2}$
(b) -2

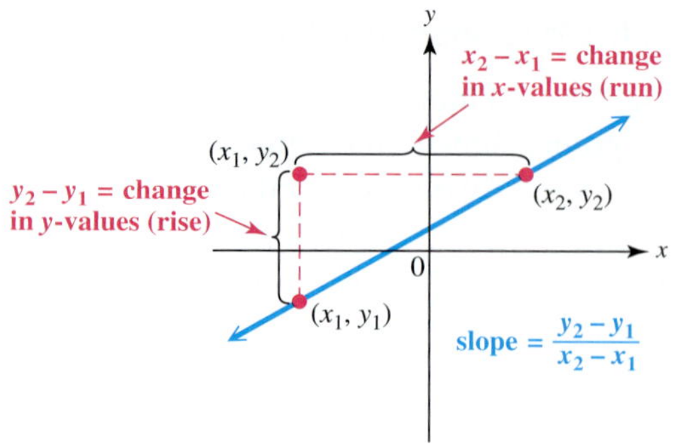

Figure 21

Moving along the line from the point (x_1, y_1) to the point (x_2, y_2), we see that y changes by $y_2 - y_1$ units. This is the vertical change (rise). Similarly, x changes by $x_2 - x_1$ units, which is the horizontal change (run). The slope of the line is the ratio of $y_2 - y_1$ to $x_2 - x_1$.

Teaching Tip
Stress the difference between x_2 and x^2. Students sometimes confuse these. Also explain the difference in how they are read.

Note

Subscript notation is used to identify a point. It does *not* indicate any operation. Note the difference between x_2, a nonspecific value, and x^2, which means $x \cdot x$. Read x_2 as "x-sub-two," *not* "x squared."

Slope Formula

The **slope *m*** of the line passing through the points (x_1, y_1) and (x_2, y_2) is defined as follows. (Traditionally, the letter m represents slope.)

$$m = \frac{\textbf{change in } y}{\textbf{change in } x} = \frac{y_2 - y_1}{x_2 - x_1} \qquad (\text{where } x_1 \neq x_2)$$

The slope gives the change in y for each unit of change in x.

Teaching Tip
Remind students that

$$-\frac{x}{y} = \frac{-x}{y} = \frac{x}{-y}$$

and

$$\frac{-x}{-y} = \frac{x}{y}.$$

EXAMPLE 2 Finding Slopes of Lines Given Two Points

Find the slope of each line.

(a) The line passing through $(-4, 7)$ and $(1, -2)$

Label the given points, and then apply the slope formula.

$$\begin{matrix} (x_1, y_1) & & (x_2, y_2) \\ \downarrow\ \downarrow & & \downarrow\ \downarrow \\ (-4, 7) & \text{and} & (1, -2) \end{matrix}$$

$$\text{slope } m = \frac{y_2 - y_1}{x_2 - x_1} = \frac{-2 - 7}{1 - (-4)} \quad \text{Substitute carefully.}$$

$$= \frac{-9}{5}, \quad \text{or} \quad -\frac{9}{5}$$

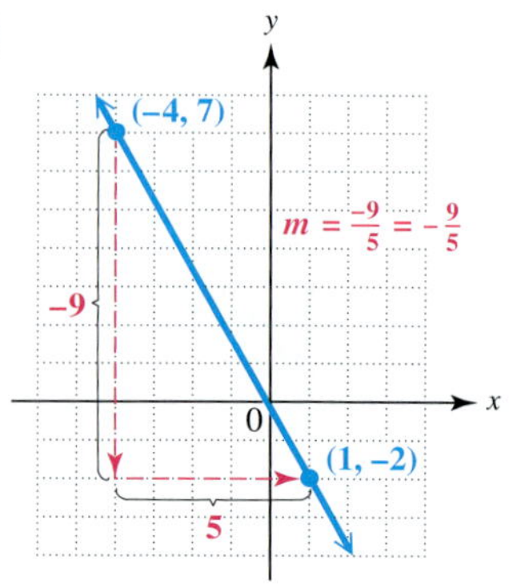

Figure 22

Count grid squares in **Figure 22** to confirm that the slope is $-\frac{9}{5}$.

Continued on Next Page

(b) The line passing through $(-9, -2)$ and $(12, 5)$

Label the points, and then apply the slope formula.

$$\overset{(x_1,\ y_1)}{(-9,\ -2)} \quad \text{and} \quad \overset{(x_2,\ y_2)}{(12,\ 5)}$$

$$\text{slope } m = \frac{y_2 - y_1}{x_2 - x_1} = \frac{5 - (-2)}{12 - (-9)}$$

$$= \frac{7}{21}, \text{ or } \frac{1}{3} \quad \text{Write in lowest terms.}$$

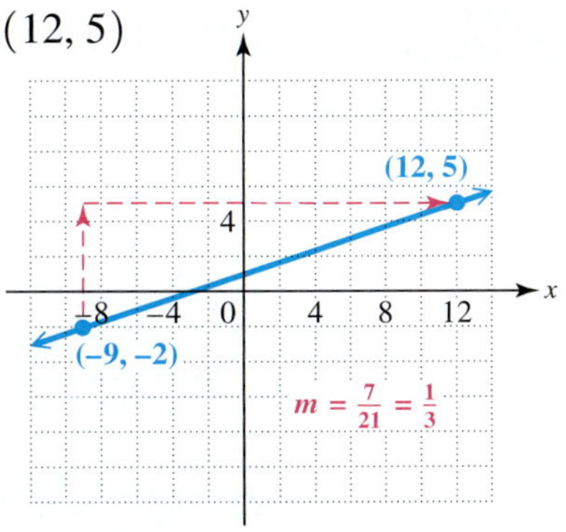

Figure 23

Confirm this calculation using **Figure 23.**

The same slope is obtained if we label the points in reverse order. ***It makes no difference which point is identified as (x_1, y_1) or (x_2, y_2).***

$$\overset{(x_2,\ y_2)}{(-9,\ -2)} \quad \text{and} \quad \overset{(x_1,\ y_1)}{(12,\ 5)}$$

$$\text{slope } m = \frac{y_2 - y_1}{x_2 - x_1} = \frac{-2 - 5}{-9 - 12} \quad \text{Substitute.}$$

$$= \frac{-7}{-21}, \text{ or } \frac{1}{3} \quad \text{The same slope results.}$$

Work Problem 2 at the Side. ▶

The lines in **Figures 22 and 23** suggest the following generalization.

Orientation of Lines with Positive and Negative Slopes

A line with positive slope rises (slants up) from left to right.

A line with negative slope falls (slants down) from left to right.

EXAMPLE 3 Finding the Slope of a Horizontal Line

Find the slope of the line passing through $(-5, 4)$ and $(2, 4)$.

$$\overset{(x_1,\ y_1)}{(-5,\ 4)} \quad \text{and} \quad \overset{(x_2,\ y_2)}{(2,\ 4)} \quad \text{Label the points.}$$

$$m = \frac{y_2 - y_1}{x_2 - x_1} = \frac{4 - 4}{2 - (-5)} \quad \text{Substitute in the slope formula.}$$

$$= \frac{0}{7}, \text{ or } 0 \quad \text{Slope 0}$$

As shown in **Figure 24,** the line through these two points is horizontal, with equation $y = 4$. ***All horizontal lines have slope 0,*** since the difference in their y-values is always 0.

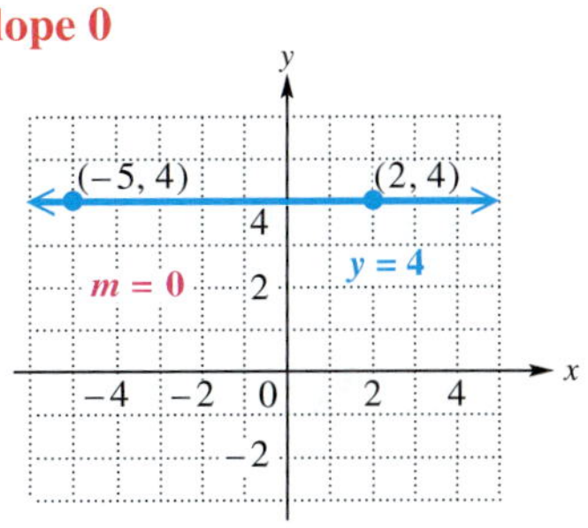

Figure 24

---

2 Find the slope of each line.

GS (a) The line passing through $(6, -2)$ and $(5, 4)$

Label the points.

$$\overset{(x_1,\ y_1)}{(6,\ -2)} \quad \text{and} \quad \overset{(\_\_\_,\ \_\_\_)}{(5,\ 4)}$$

$$\text{slope } m = \frac{y_2 - y_1}{x_2 - x_1}$$

$$= \frac{\_\_\_ - (\_\_\_)}{\_\_\_ - 6}$$

$$= \frac{6}{\_\_\_}, \text{ or } \_\_\_$$

(b) The line passing through $(-3, 5)$ and $(-4, -7)$

Teaching Tip

Emphasize starting with the x- and y-values of the *same* point and subtracting the x- and y-values of the other point.

(c) The line passing through $(6, -8)$ and $(-2, 4)$

(Find this slope in two different ways as in **Example 2(b).**)

Teaching Tip

Emphasize that students must follow the graph of a line *from left to right* to determine whether its slope is positive or negative.

Answers

2. (a) x_2; y_2; 4; -2; 5; -1; -6

(b) 12 **(c)** $-\frac{3}{2}$; $-\frac{3}{2}$

3 Find the slope of each line.

(a) The line passing through $(2, 5)$ and $(-1, 5)$

(b) The line passing through $(3, 1)$ and $(3, -4)$

EXAMPLE 4 **Finding the Slope of a Vertical Line**

Find the slope of the line passing through $(6, 2)$ and $(6, -4)$.

$$\begin{matrix} (x_1, y_1) & & (x_2, y_2) \\ \downarrow\ \downarrow & & \downarrow\ \downarrow \\ (6, 2) & \text{and} & (6, -4) \end{matrix} \qquad \text{Label the points.}$$

$$m = \frac{y_2 - y_1}{x_2 - x_1} = \frac{-4 - 2}{6 - 6} \qquad \text{Substitute in the slope formula.}$$

$$= \frac{-6}{0} \qquad \textbf{Undefined slope}$$

Because division by 0 is undefined, this line has undefined slope. (This is why the slope formula at the beginning of this section had the restriction $x_1 \neq x_2$.) The graph in **Figure 25** shows that this line is vertical, with equation $x = 6$. All points on a vertical line have the same x-value, so ***the slope of any vertical line is undefined.***

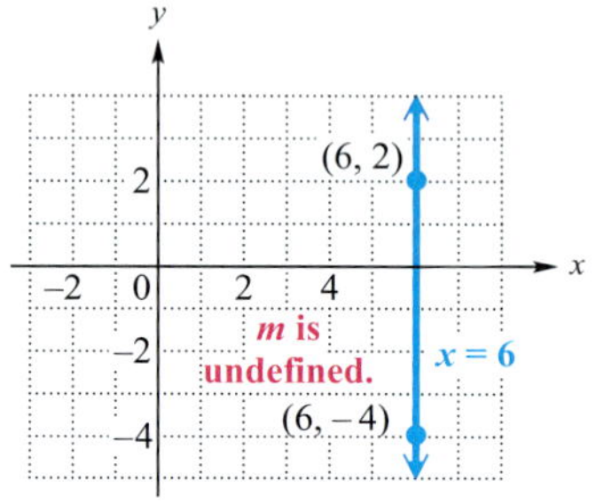

Figure 25

◀ **Work Problem 3 at the Side.**

Slopes of Horizontal and Vertical Lines

Horizontal lines, which have equations of the form $y = k$ (where k is a constant (number)), have **slope 0.**

Vertical lines, which have equations of the form $x = k$ (where k is a constant (number)), have **undefined slope.**

◀ **Work Problem 4 at the Side.**

4 Find the slope of each line.

GS **(a)** The line with equation $y = -1$

This is the equation of a (*horizontal* / *vertical*) line. It has (0 / *undefined*) slope.

(b) The line with equation $x - 4 = 0$

Rewrite this equation in an equivalent form.

$$x - 4 = 0$$

$$x = \_\_\_\_$$

This is the equation of a (*horizontal* / *vertical*) line. It has (0 / *undefined*) slope.

Figure 26 summarizes the four cases for slopes of lines.

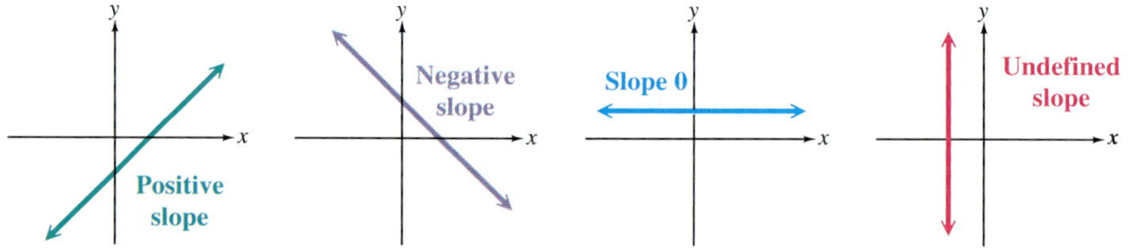

Slopes of lines

Figure 26

OBJECTIVE 2 **Find the slope from the equation of a line.** Consider the linear equation $y = -3x + 5$. We can find the slope of the line using any two points on the line. We get these two points by first choosing two different values of x and then finding the corresponding values of y. We arbitrarily choose $x = -2$ and $x = 4$.

$y = -3x + 5$		$y = -3x + 5$	
$y = -3(-2) + 5$	Let $x = -2$.	$y = -3(4) + 5$	Let $x = 4$.
$y = 6 + 5$	Multiply.	$y = -12 + 5$	Multiply.
$y = 11$	Add.	$y = -7$	Add.

The ordered pairs are $(-2, 11)$ and $(4, -7)$.

Teaching Tip

Summarize the four cases for slopes of lines as in Figure 26. Emphasize that *0 slope* is not the same as *undefined slope*.

Answers

3. **(a)** 0 **(b)** undefined
4. **(a)** horizontal; 0
(b) 4; vertical; undefined

Now we apply the slope formula using the two points $(-2, 11)$ and $(4, -7)$.

$$m = \frac{-7 - 11}{4 - (-2)} = \frac{-18}{6} = -3$$

The slope, $m = -3$, is the same number as the coefficient of x in the equation $y = -3x + 5$. It can be shown that this always happens, *as long as the equation is solved for y*. This fact is used to find the slope of a line from its equation.

Finding the Slope of a Line from Its Equation

Step 1 Solve the equation for y. (See **Section 2.5.**)

Step 2 The slope is given by the coefficient of x.

EXAMPLE 5 **Finding Slopes from Equations**

Find the slope of each line.

(a) $2x - 5y = 4$

Step 1 Solve the equation for y.

$2x - 5y = 4$ (Isolate y on one side.)

$-5y = 4 - 2x$ Subtract $2x$.

$-5y = -2x + 4$ Commutative property

$y = \frac{2}{5}x - \frac{4}{5}$ Divide *each* term by -5. ($\frac{-2x}{-5} = \frac{-2}{-5}x = \frac{2}{5}x$)

↑ Slope

Step 2 The slope is given by the coefficient of x, so the slope is $\frac{2}{5}$.

(b) $8x + 4y = 1$ (Solve for y.)

$4y = 1 - 8x$ Subtract $8x$.

$4y = -8x + 1$ Commutative property

$y = -2x + \frac{1}{4}$ Divide *each* term by 4.

The slope of this line is given by the coefficient of x, which is -2.

(c) $3y + x = -3$ (We leave out the step showing the commutative property.)

$3y = -x - 3$ Subtract x.

$y = \frac{-x}{3} - 1$ Divide *each* term by 3.

$y = -\frac{1}{3}x - 1$ $\frac{-x}{3} = \frac{-1x}{3} = -\frac{1}{3}x$ (The slope is $-\frac{1}{3}$, *not* $\frac{-x}{3}$ or $-\frac{x}{3}$.)

The coefficient of x is $-\frac{1}{3}$, so the slope of this line is $-\frac{1}{3}$.

Work Problem 5 at the Side. ▶

5 Find the slope of each line.

GS **(a)** $y = -\frac{7}{2}x + 1$

Since this equation is already solved for y, Step 1 is not needed. The slope, which is given by the ________ of ____, can be read directly from the equation. The slope is ____.

GS **(b)** $4y = 4x - 3$

Solve for y here by dividing each term by ____ to get the equation

$y =$ ________.

The coefficient of x is ____, so the slope is ____.

(c) $3x + 2y = 9$

(d) $5y - x = 10$

Teaching Tip

Using the equation $2x + 3y = 12$, demonstrate that the slope is the same whether we determine it from two ordered pairs or we solve for y and identify the coefficient of x.

Answers

5. **(a)** coefficient; x; $-\frac{7}{2}$ **(b)** 4; $x - \frac{3}{4}$; 1; 1 **(c)** $-\frac{3}{2}$ **(d)** $\frac{1}{5}$

OBJECTIVE 3 Use slope to determine whether two lines are parallel, perpendicular, or neither. Two lines in a plane that never intersect are **parallel.** We use slopes to tell whether two lines are parallel.

Figure 27 shows the graphs of $x + 2y = 4$ and $x + 2y = -6$. These lines appear to be parallel. We solve each equation for y to find the slope.

$x + 2y = 4$		$x + 2y = -6$	
$2y = -x + 4$	Subtract x.	$2y = -x - 6$	Subtract x.
$y = \frac{-x}{2} + 2$	Divide by 2.	$y = \frac{-x}{2} - 3$	Divide by 2.
$y = -\frac{1}{2}x + 2$ ($\frac{-x}{2} = \frac{-1x}{2} = -\frac{1}{2}x$) ↑ Slope		$y = -\frac{1}{2}x - 3$ (The slope is $-\frac{1}{2}$, *not* $-\frac{x}{2}$.) ↑ Slope	

Each line has slope $-\frac{1}{2}$. ***Nonvertical parallel lines always have equal slopes.***

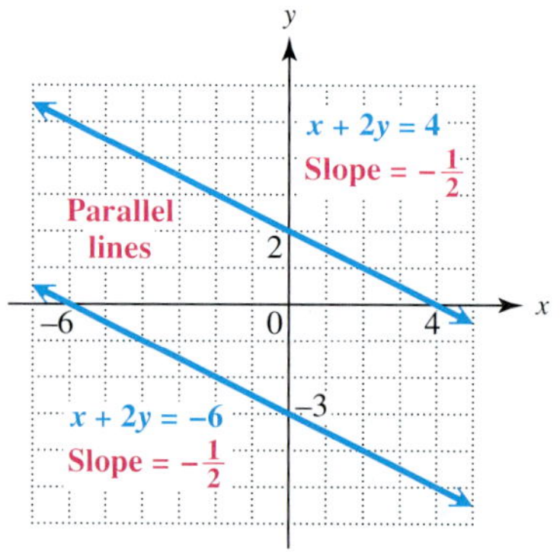

Figure 27

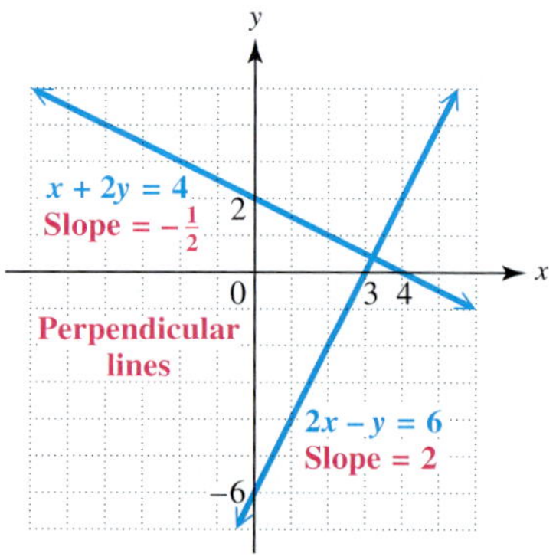

Figure 28

Figure 28 shows the graphs of $x + 2y = 4$ and $2x - y = 6$. These lines appear to be **perpendicular** (that is, they intersect at a 90° angle). As shown on the left above, solving $x + 2y = 4$ for y gives $y = -\frac{1}{2}x + 2$, with slope $-\frac{1}{2}$. We must solve $2x - y = 6$ for y.

$$2x - y = 6$$

$$-y = -2x + 6 \quad \text{Subtract } 2x.$$

$$y = 2x - 6 \quad \text{Multiply by } -1.$$

↑ Slope

The product of the two slopes $-\frac{1}{2}$ and 2 is

$$-\frac{1}{2}(2) = -1.$$

This condition is true in general. ***The product of the slopes of two perpendicular lines, neither of which is vertical, is always* −1.** This means that the slopes of perpendicular lines are negative (or opposite) reciprocals—if one slope is the nonzero number a, then the other is $-\frac{1}{a}$. The table in the margin shows several examples.

Teaching Tip

Stress that the slope is the coefficient of x. Encourage students to rewrite a term like $\frac{-x}{2}$ as $-\frac{1}{2}x$ so that the coefficient can be more easily recognized.

Number	Negative Reciprocal
$\frac{3}{4}$	$-\frac{4}{3}$
$\frac{1}{2}$	$-\frac{2}{1}$, or -2
-6, or $-\frac{6}{1}$	$\frac{1}{6}$
-0.4, or $-\frac{4}{10}$	$\frac{10}{4}$, or 2.5

***The product of each number and its negative reciprocal is* −1.**

Slopes of Parallel and Perpendicular Lines

Two lines with the same slope are parallel.

Two lines whose slopes have a product of -1 are perpendicular.

EXAMPLE 6 Deciding Whether Two Lines Are Parallel or Perpendicular

Decide whether each pair of lines is *parallel, perpendicular,* or *neither.*

(a) $x + 2y = 7$ and $-2x + y = 3$

Find the slope of each line by first solving each equation for y.

$$x + 2y = 7$$
$$2y = -x + 7 \quad \text{Subtract } x.$$
$$y = -\frac{1}{2}x + \frac{7}{2} \quad \text{Divide by 2.}$$

The slope is $-\frac{1}{2}$.

$$-2x + y = 3$$
$$y = 2x + 3 \quad \text{Add } 2x.$$

The slope is 2.

Because the slopes are not equal, the lines are not parallel. Check the product of the slopes: $-\frac{1}{2}(2) = -1$. The two lines are perpendicular because the product of their slopes is -1. See **Figure 29.**

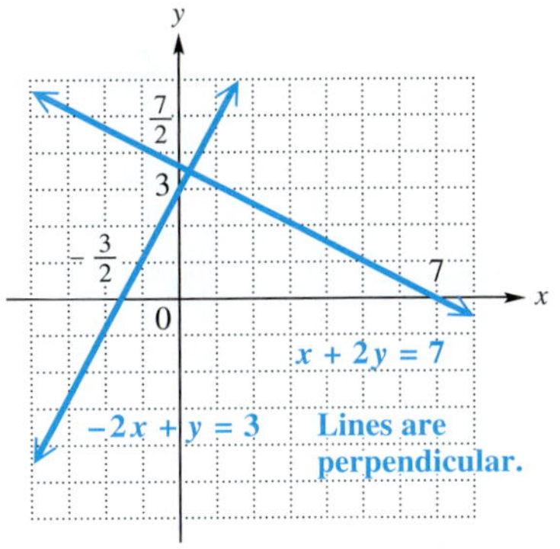

Figure 29

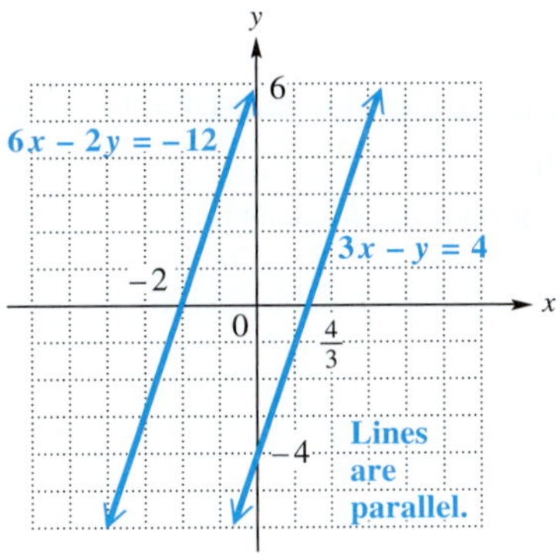

Figure 30

(b) $3x - y = 4$ and $6x - 2y = -12$ $\xrightarrow{\text{Solve each equation for } y.}$ $y = 3x - 4$ and $y = 3x + 6$

Both lines have slope 3, so the lines are parallel. See **Figure 30.**

(c) $4x + 3y = 6$ and $2x - y = 5$ $\xrightarrow{\text{Solve each equation for } y.}$ $y = -\frac{4}{3}x + 2$ and $y = 2x - 5$

The slopes are $-\frac{4}{3}$ and 2. Because $-\frac{4}{3} \neq 2$ and $-\frac{4}{3}(2) \neq -1$, these lines are neither parallel nor perpendicular.

(d) $5x - y = 1$ and $x - 5y = -10$ $\xrightarrow{\text{Solve each equation for } y.}$ $y = 5x - 1$ and $y = \frac{1}{5}x + 2$

$5(\frac{1}{5}) = 1$, *not* -1.

Here the slopes are 5 and $\frac{1}{5}$. The lines are not parallel, nor are they perpendicular.

Work Problem 6 at the Side. ▶

6 Decide whether each pair of lines is *parallel, perpendicular,* or *neither.*

(a) $x + y = 6$
$x + y = 1$

(b) $3x - y = 4$
$x + 3y = 9$

(c) $2x - y = 5$
$2x + y = 3$

(d) $3x - 7y = 35$
$7x - 3y = -6$

Teaching Tip

Discuss the graphical meaning of *neither:* The lines must intersect, but not at a right angle.

Answers

6. (a) parallel **(b)** perpendicular **(c)** neither **(d)** neither

3.3 Exercises

FOR EXTRA HELP

Download the MyDashBoard App

MyMathLab®

1. CONCEPT CHECK Slope is used to measure the ___steepness___ of a line. Slope is the (*horizontal* / *vertical*) change compared to the (*horizontal* / *vertical*) change while moving along the line from one point to another.

2. CONCEPT CHECK Slope is the ___ratio___ of the vertical change in ___y___, called the (*rise* / *run*), to the horizontal change in ___x___, called the (*rise* / *run*).

3. Look at the graph at the right, and answer the following.

GS

(a) Start at the point $(-1, -4)$ and count vertically up to the horizontal line that goes through the other plotted point. What is this vertical change? (Remember: "up" means positive, "down" means negative.) ___6___

(b) From this new position, count horizontally to the other plotted point. What is this horizontal change? (Remember: "right" means positive, "left" means negative.) ___4___

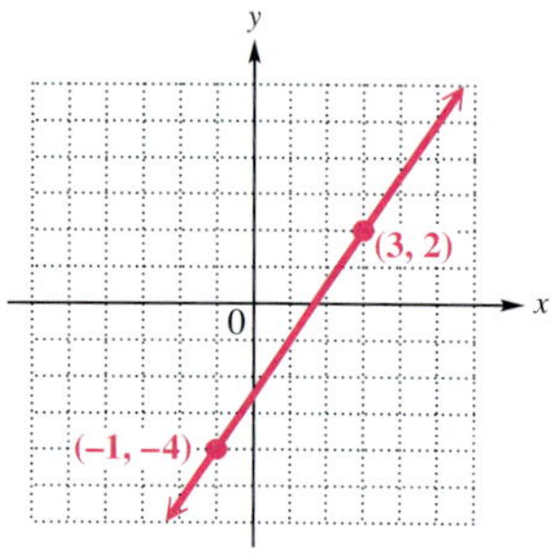

(c) What is the quotient of the numbers found in parts (a) and (b)? ___$\frac{6}{4}$, or $\frac{3}{2}$___
What do we call this number? ___slope of the line___

(d) If we were to *start* at the point $(3, 2)$ and *end* at the point $(-1, -4)$ would the answer to part (c) be the same? Explain.

Yes, it doesn't matter which point we start with. The slope would be expressed as the quotient of -6 and -4, which simplifies to $\frac{3}{2}$.

4. CONCEPT CHECK Match the graph of each line in parts (a)–(d) with its slope in choices A–D.

(a) C

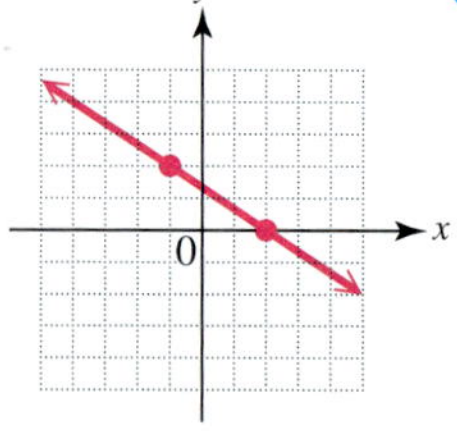

(b) A

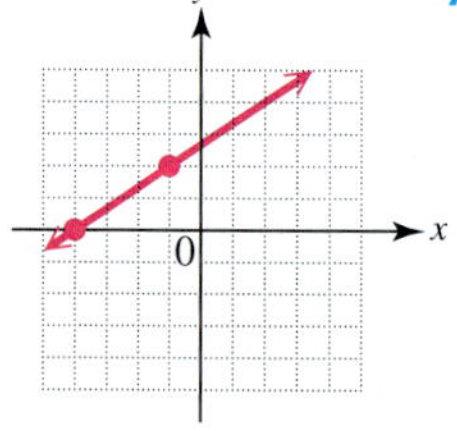

(c) D

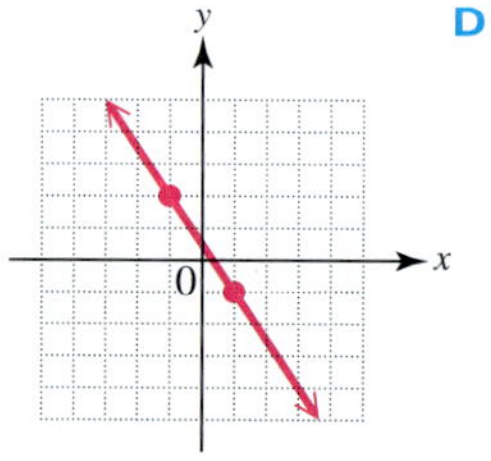

(d) B

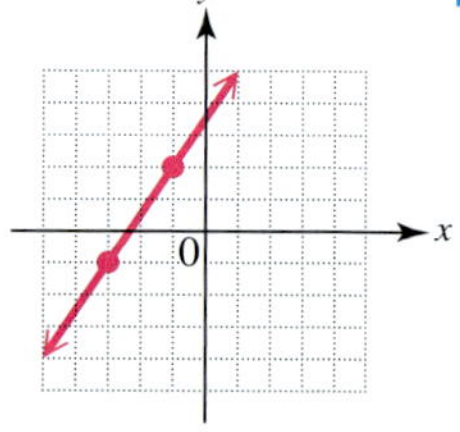

A. $\frac{2}{3}$ **B.** $\frac{3}{2}$ **C.** $-\frac{2}{3}$ **D.** $-\frac{3}{2}$

Use the coordinates of the indicated points to find the slope of each line. ***See Example 1.***

5. 4

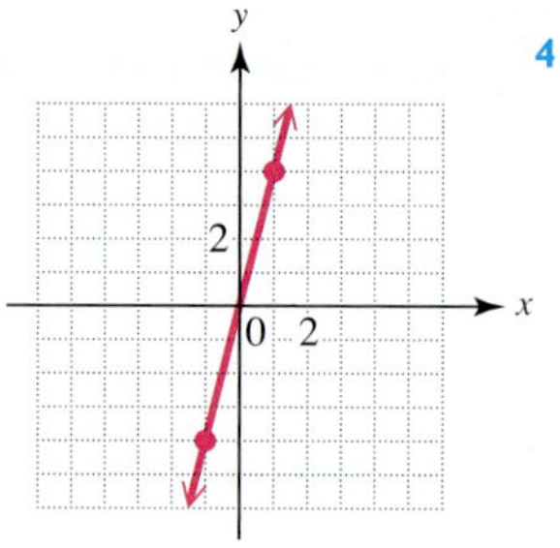

6. $\frac{3}{2}$

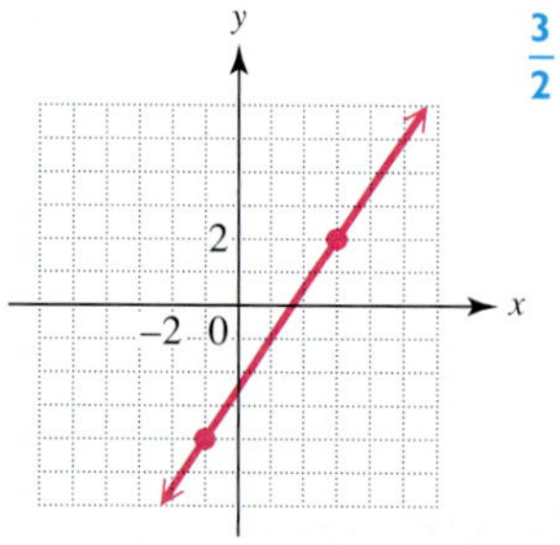

7. $-\frac{1}{2}$

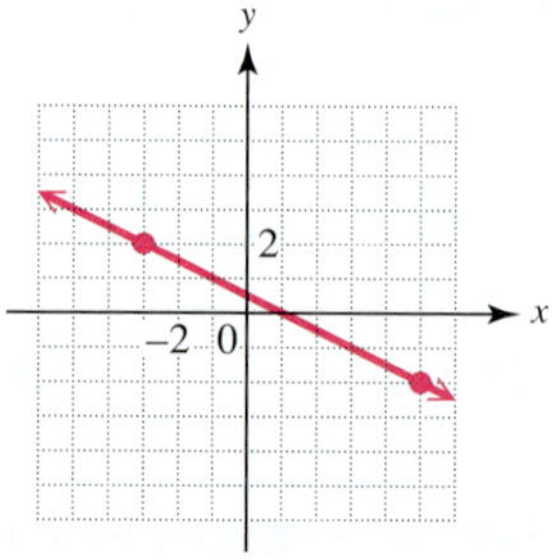

8. $-\frac{7}{4}$

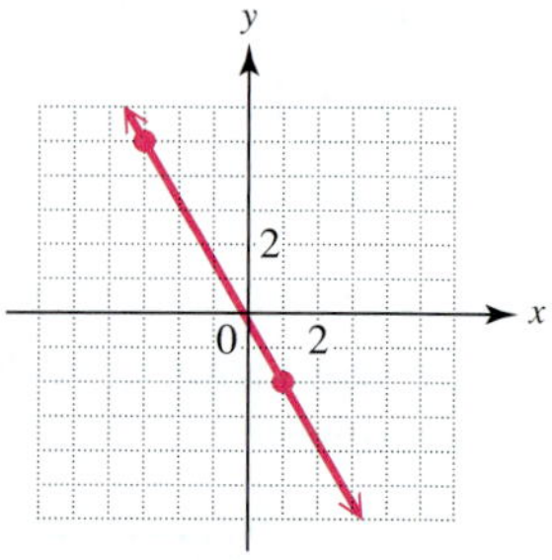

9. CONCEPT CHECK On the given axes, sketch the graph of a straight line with the indicated slope.

(a) Negative **(b)** Positive **(c)** Undefined **(d)** Zero

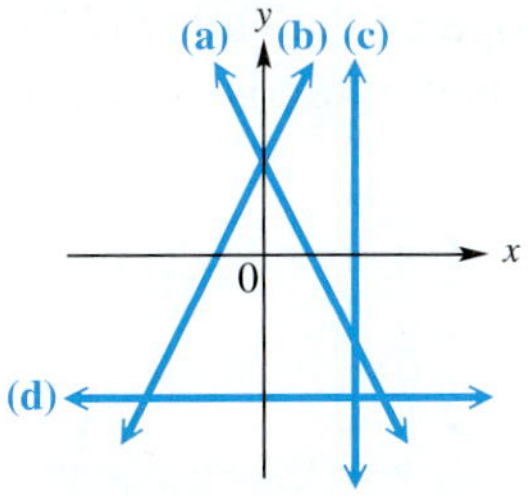

Answers will vary.

10. CONCEPT CHECK Decide whether the line with the given slope m *rises from left to right, falls from left to right,* is *horizontal,* or is *vertical.*

(a) $m = -4$ falls from left to right

(b) $m = 0$ horizontal

(c) m is undefined. vertical

(d) $m = \frac{3}{7}$ rises from left to right

CONCEPT CHECK *The figure at the right shows a line that has a positive slope (because it rises from left to right) and a positive y-value for the y-intercept (because it intersects the y-axis above the origin).*

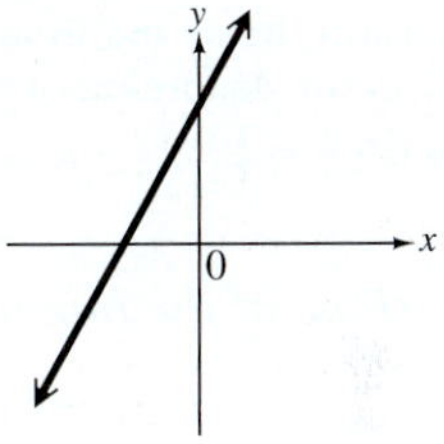

*For each figure in Exercises 11–16, decide whether **(a)** the slope is* positive, negative, *or* 0 *and whether **(b)** the y-value of the y-intercept is* positive, negative, *or* 0.

11. (a) negative **(b)** 0

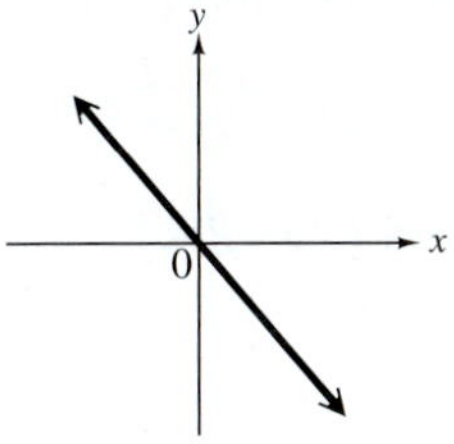

12. (a) negative **(b)** negative

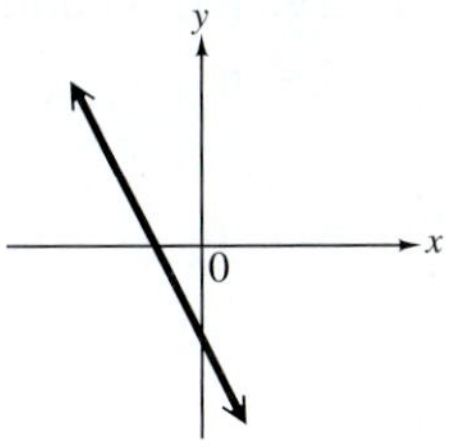

13. (a) positive **(b)** negative

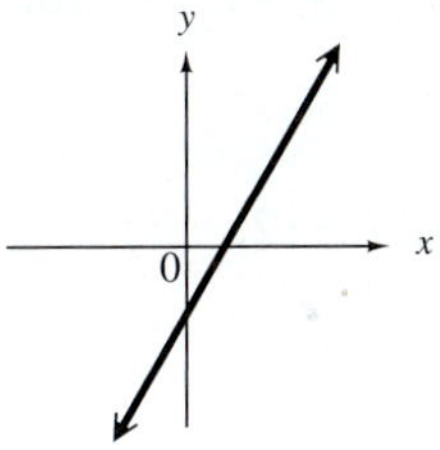

14. (a) positive **(b)** 0

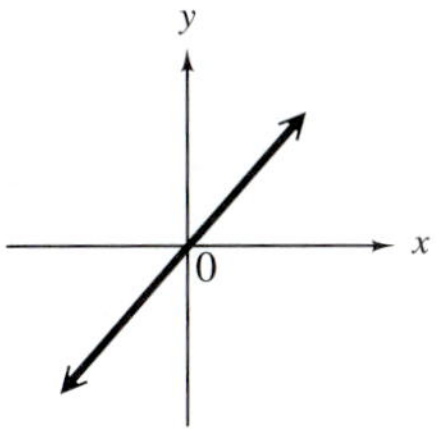

15. (a) 0 **(b)** negative

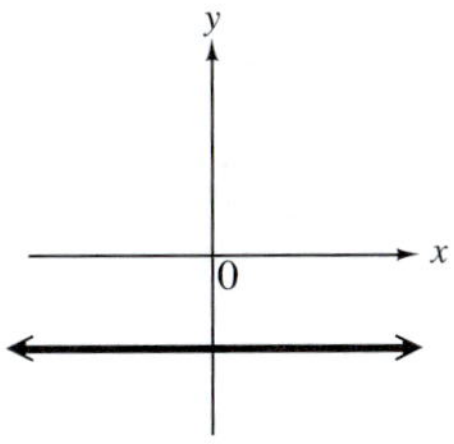

16. (a) 0 **(b)** positive

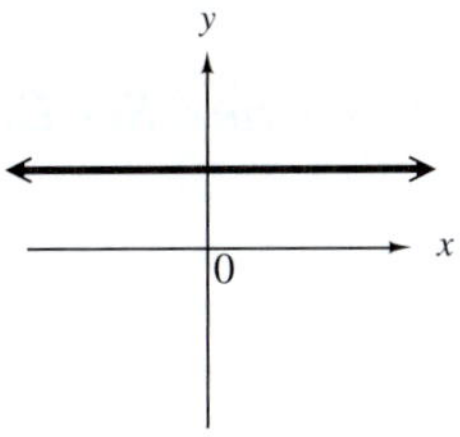

CONCEPT CHECK *Answer each question.*

17. What is the slope (or grade) of this hill?

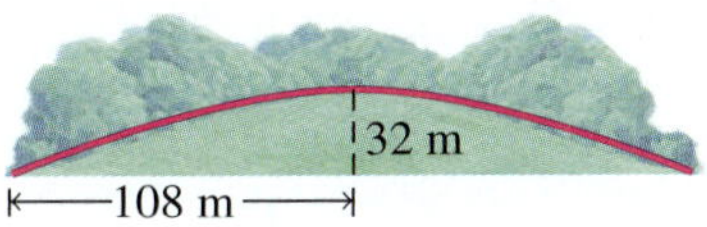

$\frac{8}{27}$

18. What is the slope (or pitch) of this roof?

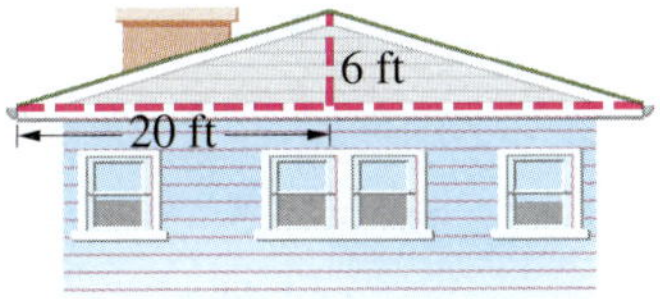

$\frac{3}{10}$

19. What is the slope of the slide? (*Hint:* The slide *drops* 8 ft vertically as it extends 12 ft horizontally.)

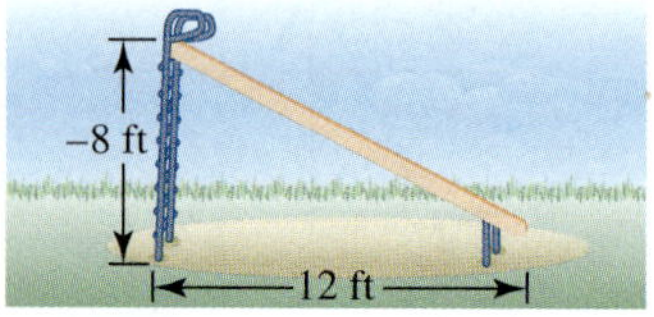

$-\frac{2}{3}$

20. What is the slope (or grade) of this ski slope? (*Hint:* The ski slope *drops* 25 ft vertically as it extends 100 ft horizontally.)

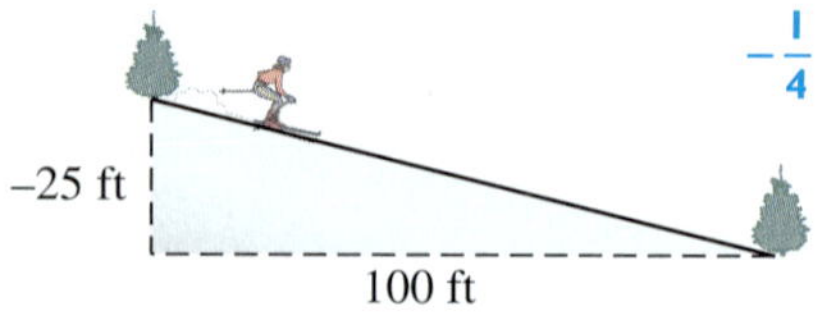

$-\frac{1}{4}$

21. CONCEPT CHECK A student found the slope of the line through the points $(2, 5)$ and $(-1, 3)$ as follows.

$$\frac{3-5}{2-(-1)} = \frac{-2}{3}, \quad \text{or} \quad -\frac{2}{3}$$

What Went Wrong? Give the correct slope.

Because the student found the difference $3 - 5 = -2$ in the numerator, he or she should have subtracted in the same order in the denominator to get $-1 - 2 = -3$. The correct slope is $\frac{-2}{-3} = \frac{2}{3}$.

22. CONCEPT CHECK A student found the slope of the line through the points $(-2, 4)$ and $(6, -1)$ as follows.

$$\frac{6-(-2)}{-1-4} = \frac{8}{-5}, \quad \text{or} \quad -\frac{8}{5}$$

What Went Wrong? Give the correct slope.

Slope is defined as $\frac{\text{change in } y}{\text{change in } x}$, but the student found $\frac{\text{change in } x}{\text{change in } y}$. The correct slope is $\frac{-5}{8} = -\frac{5}{8}$.

Find the slope of the line passing through each pair of points. **See Examples 2–4.**

23. $(1, -2)$ and $(-3, -7)$ $\frac{5}{4}$

24. $(4, -1)$ and $(-2, -8)$ $\frac{7}{6}$

25. $(0, 3)$ and $(-2, 0)$ $\frac{3}{2}$

26. $(-8, 0)$ and $(0, -5)$ $-\frac{5}{8}$

27. $(-2, 4)$ and $(-3, 7)$ -3

28. $(-4, -5)$ and $(-5, -8)$ 3

29. $(4, 3)$ and $(-6, 3)$ 0

30. $(6, -5)$ and $(-12, -5)$ 0

31. $(-12, 3)$ and $(-12, -7)$ undefined

32. $(-8, 6)$ and $(-8, -1)$ undefined

33. $(4.8, 2.5)$ and $(3.6, 2.2)$ $\frac{1}{4}$

34. $(3.1, 2.6)$ and $(1.6, 2.1)$ $\frac{1}{3}$

35. $\left(-\frac{7}{5}, \frac{3}{10}\right)$ and $\left(\frac{1}{5}, -\frac{1}{2}\right)$ $-\frac{1}{2}$

36. $\left(-\frac{4}{3}, \frac{1}{2}\right)$ and $\left(\frac{1}{3}, -\frac{5}{6}\right)$ $-\frac{4}{5}$

Find the slope of each line. **See Example 5.**

37. $y = 5x + 12$ 5

38. $y = 2x - 3$ 2

39. $4y = x + 1$ $\frac{1}{4}$

40. $2y = -x + 4$ $-\frac{1}{2}$

41. $3x - 2y = 3$ $\frac{3}{2}$

42. $-6x + 4y = 4$ $\frac{3}{2}$

43. $-2y - 3x = 5$ $-\frac{3}{2}$

44. $-4y - 3x = 2$ $-\frac{3}{4}$

45. $y = 6$ 0

46. $y = 4$ 0

47. $x = -2$ undefined

48. $x = 5$ undefined

49. $x - y = 0$ 1

50. $x + y = 0$ -1

For each pair of equations, give the slope of each line, and then determine whether the two lines are parallel, perpendicular, *or* neither. ***See Example 6.***

51. $-4x + 3y = 4$
$-8x + 6y = 0$
$\frac{4}{3}$; $\frac{4}{3}$; parallel

52. $2x + 5y = 4$
$4x + 10y = 1$
$-\frac{2}{5}$; $-\frac{2}{5}$; parallel

53. $5x - 3y = -2$
$3x - 5y = -8$
$\frac{5}{3}$; $\frac{3}{5}$; neither

54. $8x - 9y = 6$
$8x + 6y = -5$
$\frac{8}{9}$; $-\frac{4}{3}$; neither

55. $3x - 5y = -1$
$5x + 3y = 2$
$\frac{3}{5}$; $-\frac{5}{3}$; perpendicular

56. $3x - 2y = 6$
$2x + 3y = 3$
$\frac{3}{2}$; $-\frac{2}{3}$; perpendicular

Relating Concepts (Exercises 57–62) For Individual or Group Work

Figure A *gives the percent of first-time full-time freshmen at 4-year colleges and universities who planned to major in the Biological Sciences.* **Figure B** *shows the percent of the same group of students who planned to major in Business.*

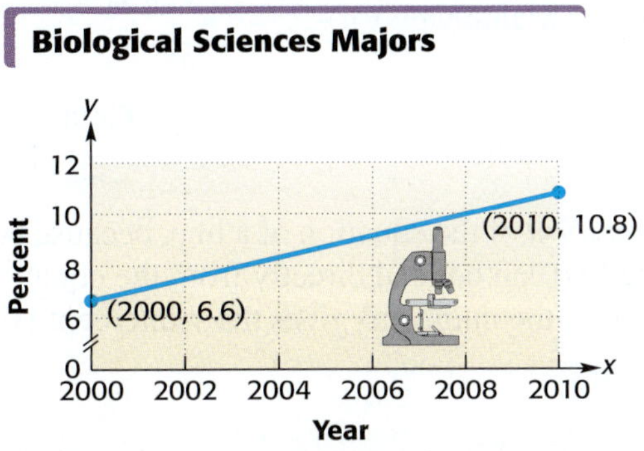

Source: Higher Education Research Institute, UCLA.

Figure A

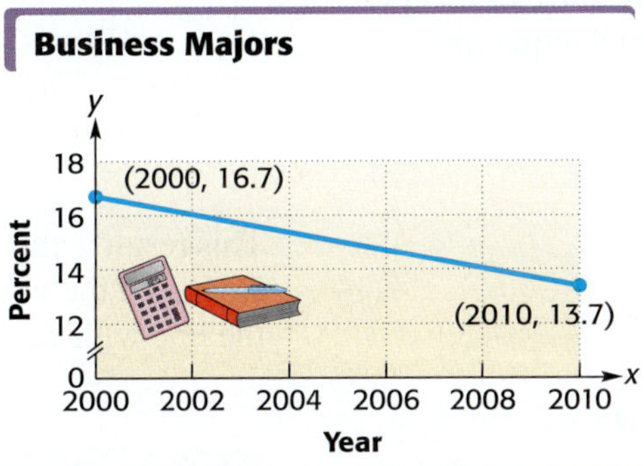

Source: Higher Education Research Institute, UCLA.

Figure B

Work Exercises 57–62 in order.

57. Use the given ordered pairs to find the slope of the line in **Figure A.** 0.42

58. The slope of the line in **Figure A** is (*positive / negative*). This means that during the period represented, the percent of freshmen planning to major in the Biological Sciences (*increased / decreased*).

59. The slope of a line represents the *rate of change*. Based on **Figure A,** what was the increase in the percent of freshmen *per year* who planned to major in the Biological Sciences during the period shown? 0.42%

60. Use the given ordered pairs to find the slope of the line in **Figure B.** −0.3

61. The slope of the line in **Figure B** is (*positive / negative*). This means that during the period represented, the percent of freshmen planning to major in Business (*increased / decreased*).

62. Based on **Figure B,** what was the decrease in the percent of freshmen *per year* who planned to major in Business? 0.3%

3.4 Writing and Graphing Equations of Lines

OBJECTIVES

1. Use the slope-intercept form of the equation of a line.
2. Graph a line by using its slope and a point on the line.
3. Write an equation of a line by using its slope and any point on the line.
4. Write an equation of a line by using two points on the line.
5. Write an equation of a line that fits a data set.

OBJECTIVE 1 Use the slope-intercept form of the equation of a line. In **Section 3.3,** we found the slope (steepness) of a line by solving the equation of the line for y. In that form, the slope is the coefficient of x. For example, the graph of the equation

$$y = 2x + 3 \quad \text{has slope} \quad 2.$$

What does the number 3 represent? To find out, suppose a line has slope m and y-intercept $(0, b)$. We can find an equation of this line by choosing another point (x, y) on the line, as shown in **Figure 31.** Then we use the slope formula.

$$m = \frac{y - b}{x - 0} \quad \begin{matrix}\leftarrow \text{Change in } y\text{-values} \\ \leftarrow \text{Change in } x\text{-values}\end{matrix}$$

$$m = \frac{y - b}{x} \quad \text{Subtract in the denominator.}$$

$$mx = y - b \quad \text{Multiply by } x.$$

$$mx + b = y \quad \text{Add } b.$$

$$y = mx + b \quad \text{Rewrite.}$$

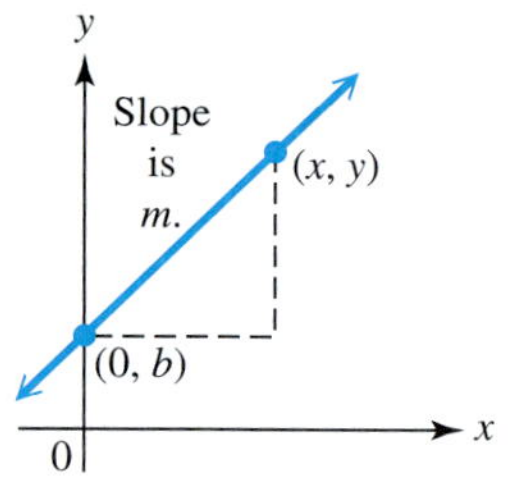

Figure 31

This result is the *slope-intercept form* of the equation of a line, because both the slope and the y-intercept of the line can be read directly from the equation. For the line with equation $y = 2x + 3$, the number 3 gives the y-intercept $(0, 3)$.

Slope-Intercept Form

The **slope-intercept form** of the equation of a line with slope m and y-intercept $(0, b)$ is given as follows.

$$\mathbf{y = mx + b}$$

Slope ↑ m; ↑ $(0, b)$ is the y-intercept.

The intercept given is the y-intercept.

EXAMPLE 1 Identifying Slopes and y-Intercepts

Identify the slope and y-intercept of the line with each equation.

(a) $y = -4x + 1$

Slope: -4; y-intercept $(0, 1)$

(b) $y = x - 8$ can be written as $y = 1x + (-8)$.

Slope: 1; y-intercept $(0, -8)$

(c) $y = 6x$ can be written as $y = 6x + 0$.

Slope: 6; y-intercept $(0, 0)$

(d) $y = \frac{x}{4} - \frac{3}{4}$ can be written as $y = \frac{1}{4}x + \left(-\frac{3}{4}\right)$.

Slope: $\frac{1}{4}$; y-intercept $\left(0, -\frac{3}{4}\right)$

◄ Work Problem 1 at the Side.

1 Identify the slope and y-intercept of the line with each equation.

(a) $y = 2x - 6$

(b) $y = -\frac{3}{5}x - 9$

(c) $y = -\frac{x}{3} + \frac{7}{3}$

(d) $y = -x$

Answers

1. (a) slope: 2; y-intercept: $(0, -6)$
(b) slope: $-\frac{3}{5}$; y-intercept: $(0, -9)$
(c) slope: $-\frac{1}{3}$; y-intercept: $\left(0, \frac{7}{3}\right)$
(d) slope: -1; y-intercept: $(0, 0)$

EXAMPLE 2 **Writing an Equation of a Line**

Write an equation of the line with slope $\frac{2}{3}$ and y-intercept $(0, -1)$.

Here, $m = \frac{2}{3}$ and $b = -1$, so we can write the following equation.

Slope → m; y-intercept $(0, b)$ → b

$$y = mx + b \quad \text{Slope-intercept form}$$

$$y = \frac{2}{3}x + (-1), \quad \text{or} \quad y = \frac{2}{3}x - 1 \quad \text{Substitute for } m \text{ and } b.$$

Work Problem 2 at the Side. ▶

OBJECTIVE 2 **Graph a line by using its slope and a point on the line.** We can use the slope and y-intercept to graph a line.

Graphing a Line by Using Its Slope and *y*-Intercept

Step 1 Write the equation in slope-intercept form $y = mx + b$, if necessary, by solving for y.

Step 2 Identify the y-intercept. Graph the point $(0, b)$.

Step 3 Identify slope m of the line. Use the geometric interpretation of slope ("rise over run") to find another point on the graph by counting from the y-intercept.

Step 4 Join the two points with a line to obtain the graph.

EXAMPLE 3 **Graphing Lines by Using Slopes and *y*-Intercepts**

Graph each equation by using the slope and y-intercept.

(a) $y = \frac{2}{3}x - 1$

Step 1 The equation is in slope-intercept form.

$$y = \frac{2}{3}x - 1$$

Slope ↑ (the $\frac{2}{3}$); Value of b in y-intercept $(0, b)$ ↑ (the -1)

Step 2 The y-intercept is $(0, -1)$. Graph this point. See **Figure 32.**

Step 3 The slope is $\frac{2}{3}$. By definition,

$$\text{slope } m = \frac{\text{change in } y \text{ (rise)}}{\text{change in } x \text{ (run)}} = \frac{2}{3}.$$

From the y-intercept, count up 2 units and to the right 3 units to obtain the point $(3, 1)$.

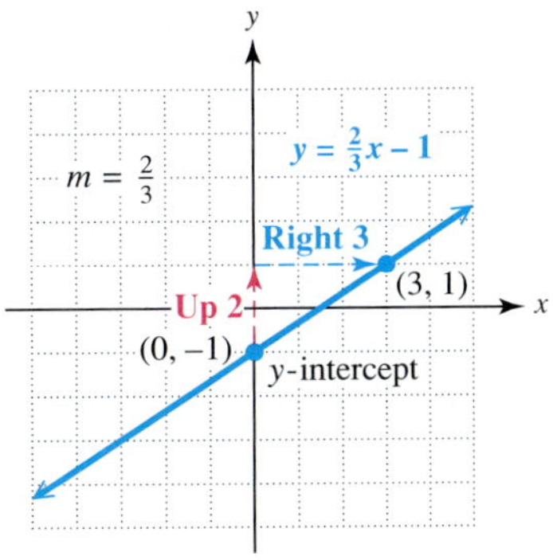

Figure 32

Step 4 Draw the line through the points $(0, -1)$ and $(3, 1)$ to obtain the graph in **Figure 32.**

Continued on Next Page

2 Write an equation of the line with the given slope and y-intercept.

(a) slope $\frac{1}{2}$; y-intercept $(0, -4)$

(b) slope -1; y-intercept $(0, 8)$

(c) slope 3; y-intercept $(0, 0)$

(d) slope 0; y-intercept $(0, 2)$

(e) slope 1; y-intercept $(0, 0.75)$

Teaching Tip

Emphasize that the y-intercept is a point with coordinates $(0, b)$. Students may confuse a statement like $b = 3$, indicating y-intercept $(0, 3)$, with $y = 3$, the equation of a horizontal line.

Answers

2. (a) $y = \frac{1}{2}x - 4$ **(b)** $y = -x + 8$ **(c)** $y = 3x$ **(d)** $y = 2$ **(e)** $y = x + 0.75$

3 Graph $3x - 4y = 8$ by using the slope and y-intercept.

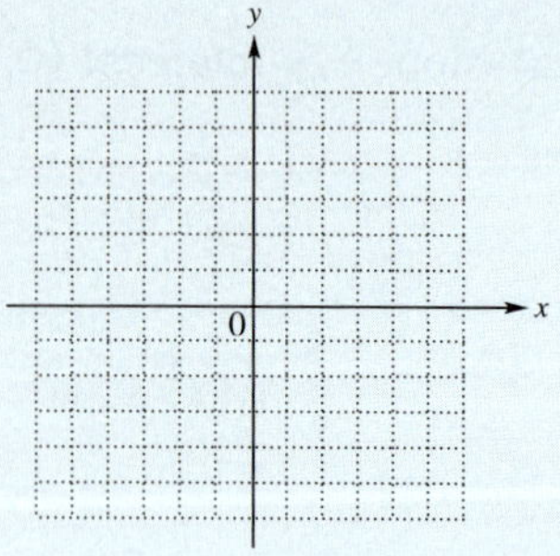

4 Graph the line passing through the point $(2, -3)$, with slope $-\frac{1}{3}$.

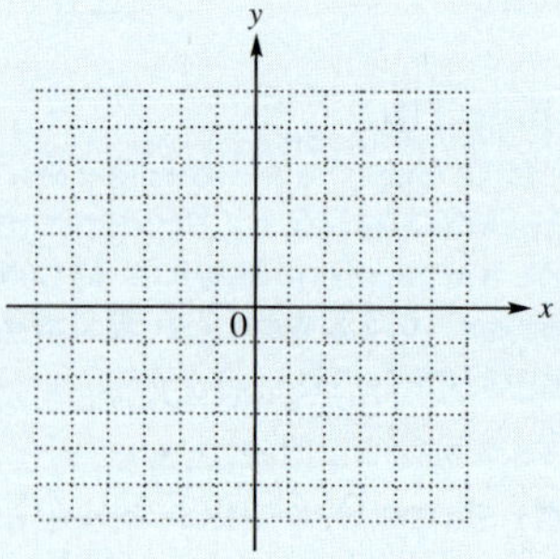

Teaching Tip

In Example 4, emphasize that the count 4 units down and 1 unit to the right starts at the given point, *not* at the origin.

Answers

3.

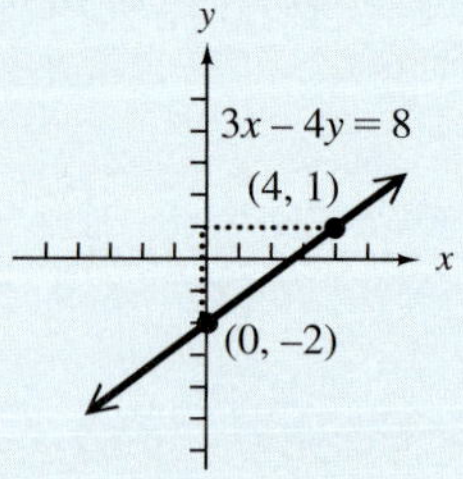

4.

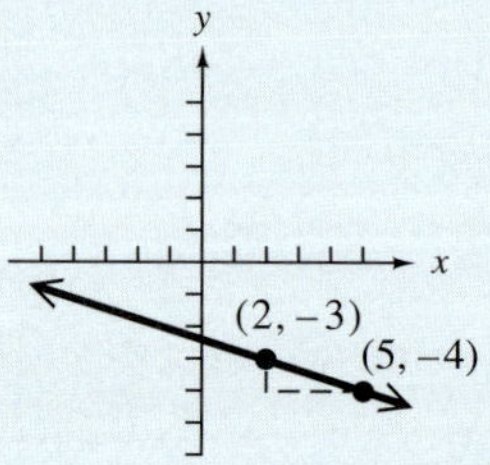

(b) $3x + 4y = 8$

Step 1 Solve for y to write the equation in slope-intercept form.

$$3x + 4y = 8$$

Isolate y on one side.

$$4y = -3x + 8 \qquad \text{Subtract } 3x.$$

$$\text{Slope-intercept form} \longrightarrow y = -\frac{3}{4}x + 2 \qquad \text{Divide } each \text{ term by 4.}$$

Step 2 The y-intercept is $(0, 2)$. Graph this point. See **Figure 33.**

Step 3 The slope is $-\frac{3}{4}$, which can be written as either $\frac{-3}{4}$ or $\frac{3}{-4}$. We use $\frac{-3}{4}$ here.

$$m = \frac{\text{change in } y \text{ (rise)}}{\text{change in } x \text{ (run)}} = \frac{-3}{4}$$

From the y-intercept, count *down* 3 units (because of the negative sign) and to the right 4 units, to obtain the point $(4, -1)$.

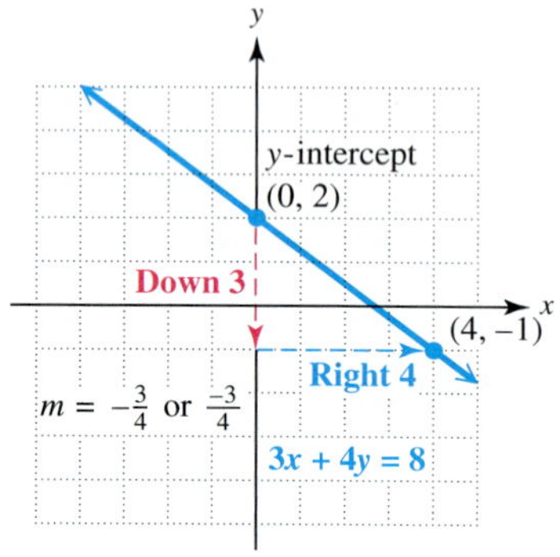

Figure 33

Step 4 Draw the line through the two points $(0, 2)$ and $(4, -1)$ to obtain the graph in **Figure 33.**

◀ **Work Problem 3 at the Side.**

Note

In Step 3 of **Example 3(b),** we could use $\frac{3}{-4}$ for the slope. From the y-intercept, count up 3 units and to the *left* 4 units (because of the negative sign) to obtain the point $(-4, 5)$. Verify that this produces the same line.

EXAMPLE 4 Graphing a Line by Using Its Slope and a Point

Graph the line passing through the point $(-2, 3)$, with slope -4.

First, locate the point $(-2, 3)$. See **Figure 34.** Then write the slope -4 as

$$\text{slope } m = \frac{\text{change in } y}{\text{change in } x} = -4 = \frac{-4}{1}.$$

Locate another point on the line by counting *down* 4 units from $(-2, 3)$ and then to the right 1 unit. Finally, draw the line through this new point P and the given point $(-2, 3)$. See **Figure 34.**

We could have written the slope as $\frac{4}{-1}$ instead. In this case, we would move up 4 units from $(-2, 3)$ and then to the *left* 1 unit. Verify that this produces the same line.

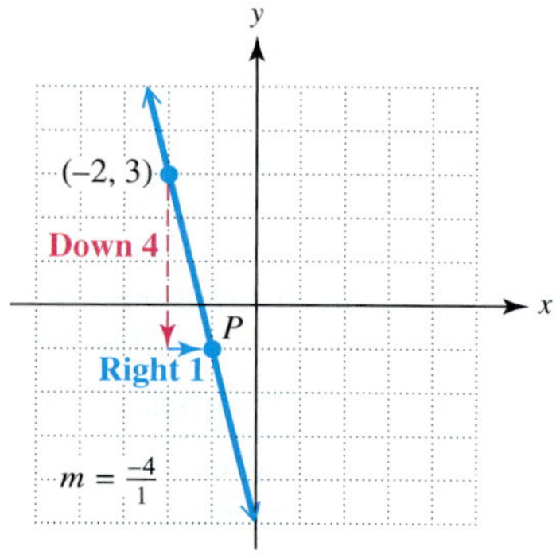

Figure 34

◀ **Work Problem 4 at the Side.**

OBJECTIVE 3 Write an equation of a line by using its slope and any point on the line. We can use the slope-intercept form to do this.

EXAMPLE 5 Using the Slope-Intercept Form to Write an Equation of a Line

Write an equation, in slope-intercept form, of the line having slope 4 passing through the point $(2, 5)$.

Since the line passes through the point $(2, 5)$, we can substitute $x = 2$, $y = 5$, and the given slope $m = 4$ into $y = mx + b$ and solve for b.

$y = mx + b$ — Slope-intercept form

$5 = 4(2) + b$ — Let $x = 2$, $y = 5$, and $m = 4$.

$5 = 8 + b$ — Multiply.

$-3 = b$ — Subtract 8.

($(0, b)$ is the y-intercept. Don't stop here.)

Now substitute the values of m and b into slope-intercept form.

$y = mx + b$ — Slope-intercept form

$y = 4x - 3$ — Let $m = 4$ and $b = -3$.

Work Problem 5 at the Side. ▶

Note

Slope-intercept form is an especially useful form for a linear equation because of the information we can determine from it. It is the form used by graphing calculators and the one that describes *a linear function.*

There is another form that can be used to write the equation of a line. To develop this form, let m represent the slope of a line and let (x_1, y_1) represent a given point on the line. Let (x, y) represent any other point on the line. See **Figure 35.**

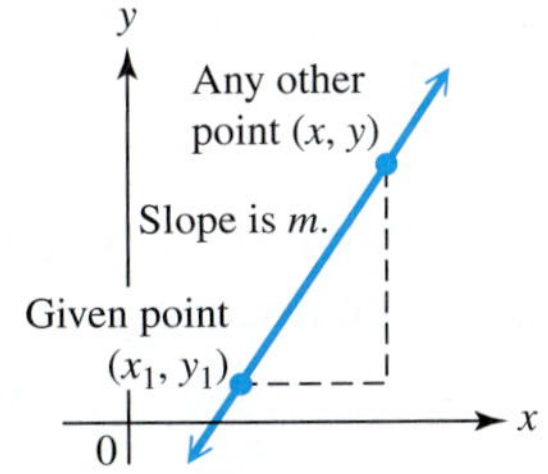

Figure 35

$$m = \frac{y - y_1}{x - x_1} \quad \text{Definition of slope}$$

$$m(x - x_1) = y - y_1 \quad \text{Multiply each side by } x - x_1.$$

$$y - y_1 = m(x - x_1) \quad \text{Rewrite.}$$

This result is the **point-slope form** of the equation of a line.

Point-Slope Form

The point-slope form of the equation of a line with slope m passing through the point (x_1, y_1) is given as follows.

$$\mathbf{y - y_1 = m(x - x_1)}$$

(Slope: m; Given point: (x_1, y_1))

---

5 Write an equation, in slope-intercept form, of each line.

GS **(a)** The line having slope -2 passing through the point $(-1, 4)$

Substitute values and solve for b.

$y = mx + b$

____ = ____(____) + b

$4 =$ ____ $+ b$

____ $= b$

An equation of the line is $y =$ ________.

(b) The line having slope 3 passing through the point $(-2, 1)$

Teaching Tip

For some classes, it may be desirable to cover the point-slope form in a second class session.

Answers

5. (a) 4; -2; -1; 2; 2; $-2x + 2$
(b) $y = 3x + 7$

6 Write an equation of each line. Give the final answer in slope-intercept form.

GS **(a)** The line passing through $(-1, 3)$, with slope -2

$$y - y_1 = m(x - x_1)$$
$$y - \_\_\_ = \_\_\_[x - (\_\_\_)]$$
$$y - 3 = -2(x + \_\_\_)$$
$$y - 3 = -2x - \_\_\_$$
$$y = \_\_\_\_$$

(b) The line passing through $(5, 2)$, with slope $-\frac{1}{3}$

Answers

6. (a) $3; -2; -1; 1; 2; -2x + 1$

(b) $y = -\frac{1}{3}x + \frac{11}{3}$

EXAMPLE 6 Using Point-Slope Form to Write Equations

Write an equation of each line. Give the final answer in slope-intercept form.

(a) The line passing through $(-2, 4)$, with slope -3

The given point is $(-2, 4)$ so $x_1 = -2$ and $y_1 = 4$. Also, $m = -3$. Substitute these values into the point-slope form.

Only y_1, m, and x_1 are replaced with numbers.

$y - y_1 = m(x - x_1)$	Point-slope form
$y - 4 = -3[x - (-2)]$	Let $x_1 = -2$, $y_1 = 4$, $m = -3$.
$y - 4 = -3(x + 2)$	Definition of subtraction
$y - 4 = -3x - 6$	Distributive property
$y = -3x - 2$	Add 4.

(b) The line passing through $(4, 2)$, with slope $\frac{3}{5}$

$y - y_1 = m(x - x_1)$	Point-slope form
$y - 2 = \frac{3}{5}(x - 4)$	Let $x_1 = 4$, $y_1 = 2$, $m = \frac{3}{5}$.
$y - 2 = \frac{3}{5}x - \frac{12}{5}$	Distributive property
$y = \frac{3}{5}x - \frac{12}{5} + \frac{10}{5}$	Add $2 = \frac{10}{5}$.
$y = \frac{3}{5}x - \frac{2}{5}$	Combine like terms.

We did not clear fractions after the substitution step because we want the equation in slope-intercept form—that is, solved for y.

◄ **Work Problem 6 at the Side.**

OBJECTIVE 4 Write an equation of a line by using two points on the line. Many of the linear equations in **Sections 3.1–3.3** were given in the form

$$Ax + By = C,$$

called **standard form,** where A, B, and C are real numbers and A and B are not both 0. In most cases, A, B, and C are rational numbers. For consistency in this book, we give answers so that A, B, and C are integers with greatest common factor 1 and $A \geq 0$. (If $A = 0$, then we give $B > 0$.)

Note

The definition of standard form is not the same in all texts. A linear equation can be written in many different, yet equally correct, ways. For example,

$$3x + 4y = 12, \quad 6x + 8y = 24, \quad \text{and} \quad -9x - 12y = -36$$

all represent the same set of ordered pairs. When giving answers, $3x + 4y = 12$ is preferable to the other forms because the greatest common factor of 3, 4, and 12 is 1 and $A \geq 0$.

EXAMPLE 7 **Writing an Equation of a Line by Using Two Points**

Write an equation of the line passing through the points $(3, 4)$ and $(-2, 5)$. Give the final answer in slope-intercept form and then in standard form.

First, find the slope of the line.

$$\overset{(x_1, y_1)}{(3, 4)} \quad \text{and} \quad \overset{(x_2, y_2)}{(-2, 5)}$$ Label the points.

$$\text{slope } m = \frac{y_2 - y_1}{x_2 - x_1} = \frac{5-4}{-2-3}$$ Apply the slope formula.

$$= \frac{1}{-5}, \quad \text{or} \quad -\frac{1}{5}$$ Simplify the fraction.

Now use (x_1, y_1), here $(3, 4)$, and either slope-intercept or point-slope form.

$$y - y_1 = m(x - x_1)$$ We choose point-slope form.

$$y - 4 = -\frac{1}{5}(x - 3)$$ Let $x_1 = 3$, $y_1 = 4$, $m = -\frac{1}{5}$.

$$y - 4 = -\frac{1}{5}x + \frac{3}{5}$$ Distributive property

$$y = -\frac{1}{5}x + \frac{3}{5} + \frac{20}{5}$$ Add $4 = \frac{20}{5}$ to each side.

Slope-intercept form ⟶ $y = -\frac{1}{5}x + \frac{23}{5}$ Combine like terms.

$$5y = -x + 23$$ Multiply by 5 to clear fractions.

Standard form ⟶ $x + 5y = 23$ Add x.

The same result would be found using $(-2, 5)$ for (x_1, y_1).

Work Problem **7** at the Side. ▶

Summary of Forms of Linear Equations

Equation	*Description*	*Example*
$x = k$	**Vertical line** Slope is undefined. x-intercept is $(k, 0)$.	$x = 3$
$y = k$	**Horizontal line** Slope is 0. y-intercept is $(0, k)$.	$y = 3$
$y = mx + b$	**Slope-intercept form** Slope is m. y-intercept is $(0, b)$.	$y = \frac{3}{2}x - 6$
$y - y_1 = m(x - x_1)$	**Point-slope form** Slope is m. Line passes through (x_1, y_1).	$y + 3 = \frac{3}{2}(x - 2)$
$Ax + By = C$	**Standard form** Slope is $-\frac{A}{B}$. x-intercept is $\left(\frac{C}{A}, 0\right)$. y-intercept is $\left(0, \frac{C}{B}\right)$.	$3x - 2y = 12$

7 Write an equation of the line passing through each pair of points. Give the final answer in slope-intercept form and then in standard form.

(a) $(2, 5)$ and $(-1, 6)$

Teaching Tip

Show students that the same result would be found in Example 7 by substituting the slope and either given point in slope-intercept form $y = mx + b$ and then solving for b.

(b) $(-3, 1)$ and $(2, 4)$

Teaching Tip

Have students rewrite the equations for slope-intercept form and point-slope form from the summary box in standard form to see that the three equations are different, but equivalent, forms of the same equation,

$$3x - 2y = 12.$$

Answers

7. (a) $y = -\frac{1}{3}x + \frac{17}{3}$; $x + 3y = 17$

(b) $y = \frac{3}{5}x + \frac{14}{5}$; $3x - 5y = -14$

OBJECTIVE 5 Write an equation of a line that fits a data set. If a given set of data fits a linear pattern—that is, its graph consists of points lying close to a straight line—we can write a linear equation that models the data.

Year	Cost (in dollars)
1	3766
3	4645
5	5491
7	6185
9	7050
11	8244

Source: The College Board.

8 Use the points $(1, 3766)$ and $(9, 7050)$ to find an equation in slope-intercept form that approximates the data of **Example 8.** How well does this equation approximate the cost in 2007?

EXAMPLE 8 Writing an Equation of a Line That Describes Data

The table in the margin lists the average annual cost (in dollars) of tuition and fees for instate students at public 4-year colleges and universities for selected years. Year 1 represents 2001, year 3 represents 2003, and so on. Plot the data and find an equation that approximates it.

Letting y represent the cost in year x, we plot the data as shown in **Figure 36.**

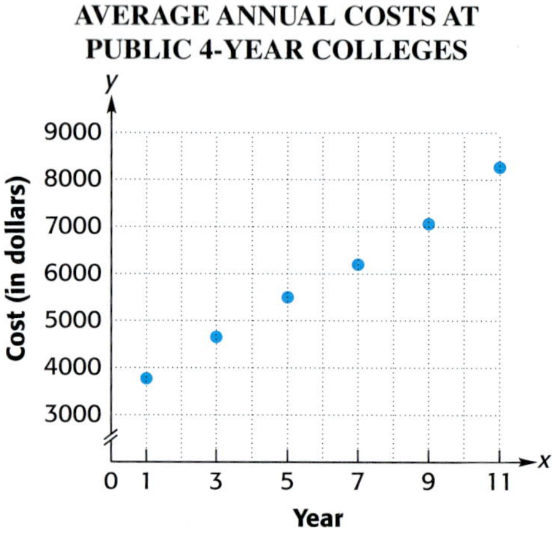

Figure 36

The points appear to lie approximately in a straight line. To find an equation of the line, we choose the ordered pairs $(5, 5491)$ and $(7, 6185)$ from the table and determine the slope of the line through these points.

$$m = \frac{y_2 - y_1}{x_2 - x_1} = \frac{6185 - 5491}{7 - 5} = 347$$ Let $(7, 6185) = (x_2, y_2)$ and $(5, 5491) = (x_1, y_1)$.

The slope, 347, is positive, indicating that tuition and fees *increased* \$347 each year. Use the slope and the point $(5, 5491)$ in the slope-intercept form.

$y = mx + b$	Slope-intercept form
$5491 = 347(5) + b$	Substitute for x, y, and m.
$5491 = 1735 + b$	Multiply.
$3756 = b$	Subtract 1735.

Solve for b, the y-value of the y-intercept.

Thus, $m = 347$ and $b = 3756$, so an equation of the line is

$$y = 347x + 3756.$$

To see how well this equation approximates the ordered pairs in the data table, let $x = 3$ (for 2003) and find y.

$y = 347x + 3756$	Equation of the line
$y = 347(3) + 3756$	Substitute 3 for x.
$y = 4797$	Multiply, and then add.

The corresponding value in the table for $x = 3$ is 4645, so the equation approximates the data reasonably well.

◀ Work Problem 8 at the Side.

Teaching Tip

Perhaps discuss what constitutes a good fit or approximation. Point out in Example 8 that if we had chosen two different data points, we would have obtained a slightly different equation.

Answer

8. $y = 410.5x + 3355.5$;
The equation gives $y = 6229$ when $x = 7$, which is a little high but a reasonable approximation.

3.4 Exercises

CONCEPT CHECK *Work each problem.*

1. Match each description of a line in Column I with the correct equation in Column II.

I		II
(a) Slope -2, passes through the point $(4, 1)$	D	**A.** $y = 4x$
(b) Slope -2, y-intercept $(0, 1)$	C	**B.** $y = \frac{1}{4}x$
(c) Passes through the points $(0, 0)$ and $(4, 1)$	B	**C.** $y = -2x + 1$
(d) Passes through the points $(0, 0)$ and $(1, 4)$	A	**D.** $y - 1 = -2(x - 4)$

2. Which equations are equivalent to $2x - 3y = 6$? **A, B, D**

A. $y = \frac{2}{3}x - 2$ **B.** $-2x + 3y = -6$ **C.** $y = -\frac{3}{2}x + 3$ **D.** $y - 2 = \frac{2}{3}(x - 6)$

3. Match each equation with the graph that would most closely resemble its graph.

(a) $y = x + 3$ **C** **(b)** $y = -x + 3$ **B** **(c)** $y = x - 3$ **A** **(d)** $y = -x - 3$ **D**

A.

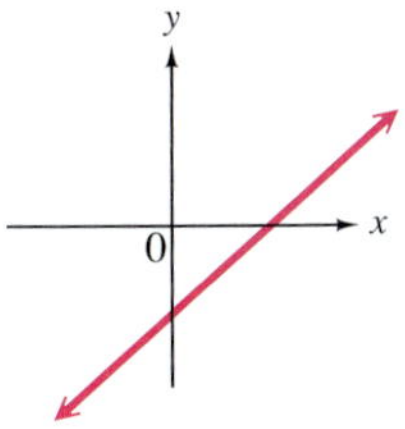

B.

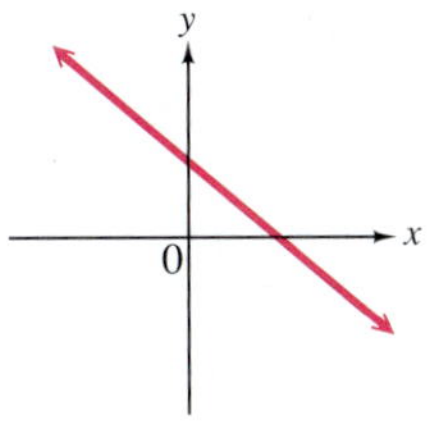

C.

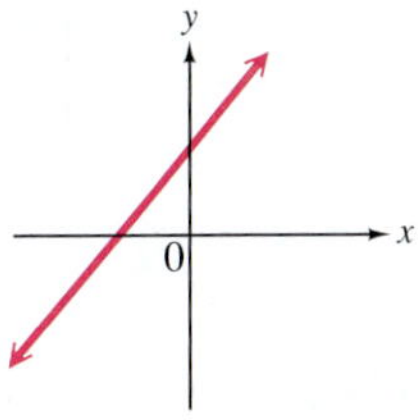

D.

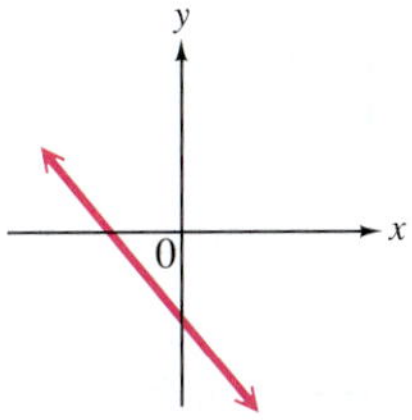

4. (a) What is the common name given to the vertical line whose x-intercept is the origin?
the y-axis

(b) What is the common name given to the line with slope 0 whose y-intercept is the origin?
the x-axis

Identify the slope and y-intercept of the line with each equation. ***See Example 1.***

5. $y = \frac{5}{2}x - 4$
slope: $\frac{5}{2}$; y-intercept: $(0, -4)$

6. $y = \frac{7}{3}x - 6$
slope: $\frac{7}{3}$; y-intercept: $(0, -6)$

7. $y = -x + 9$
slope: -1; y-intercept: $(0, 9)$

8. $y = x + 1$
slope: 1; y-intercept: $(0, 1)$

9. $y = \frac{x}{5} - \frac{3}{10}$
slope: $\frac{1}{5}$; y-intercept: $\left(0, -\frac{3}{10}\right)$

10. $y = \frac{x}{7} - \frac{5}{14}$
slope: $\frac{1}{7}$; y-intercept: $\left(0, -\frac{5}{14}\right)$

Write the equation of the line with the given slope and y-intercept. ***See Example 2.***

11. slope 4, y-intercept $(0, -3)$
$y = 4x - 3$

12. slope -5, y-intercept $(0, 6)$
$y = -5x + 6$

13. $m = -1$, y-intercept $(0, -7)$
$y = -x - 7$

14. $m = 1$, y-intercept $(0, -9)$
$y = x - 9$

15. slope 0, y-intercept $(0, 3)$

$y = 3$

16. slope 0, y-intercept $(0, -4)$

$y = -4$

17. undefined slope, y-intercept $(0, -2)$

$x = 0$

18. undefined slope, y-intercept $(0, 5)$

$x = 0$

CONCEPT CHECK *Use the geometric interpretation of slope (rise divided by run, from* ***Section 3.3****) to find the slope of each line. Then, by identifying the y-intercept from the graph, write the slope-intercept form of the equation of the line.*

19.

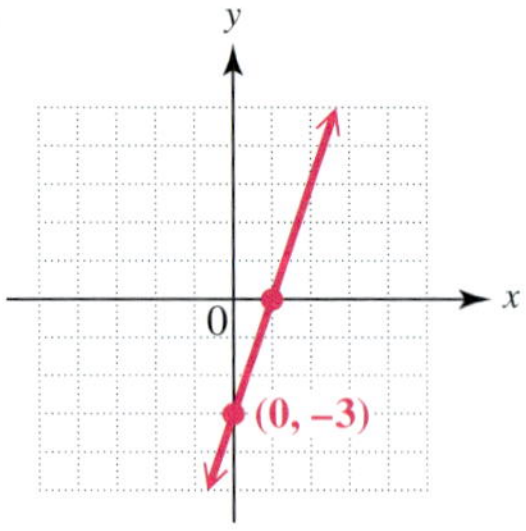

$y = 3x - 3$

20.

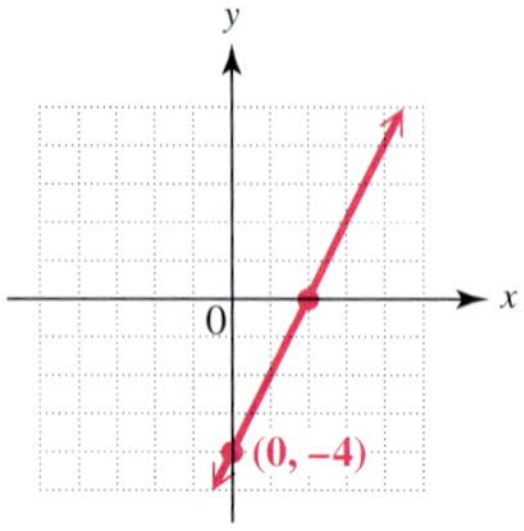

$y = 2x - 4$

21.

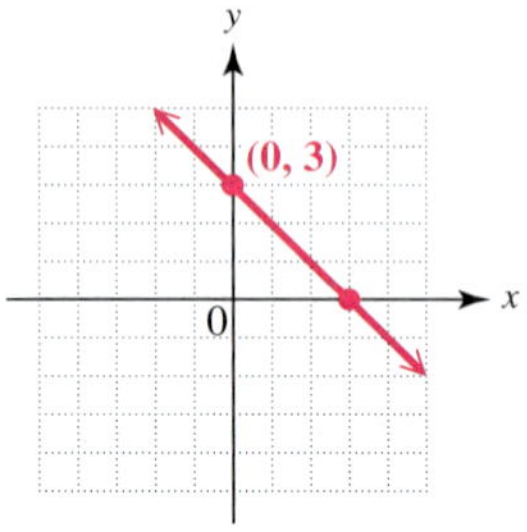

$y = -x + 3$

22.

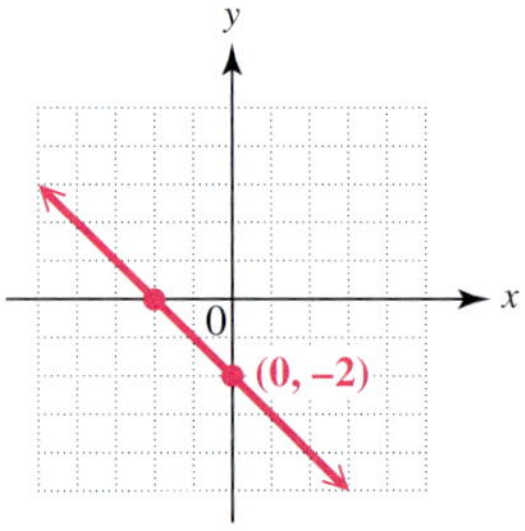

$y = -x - 2$

23.

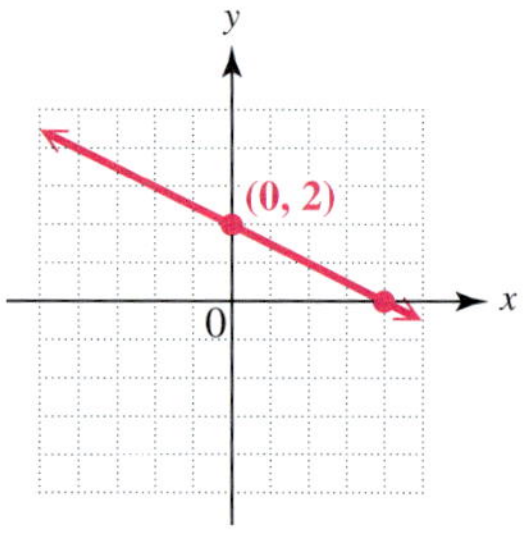

$y = -\frac{1}{2}x + 2$

24.

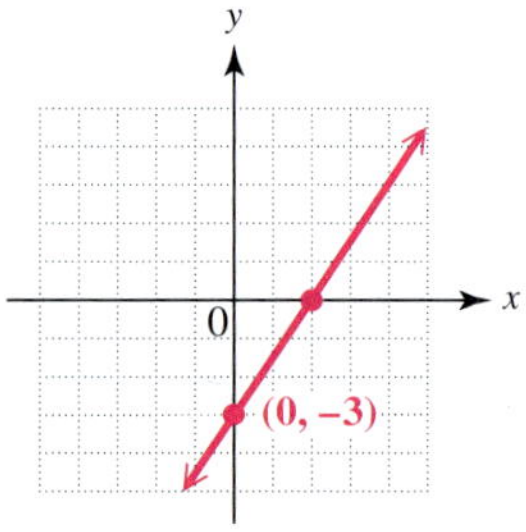

$y = \frac{3}{2}x - 3$

Graph each equation by identifying the slope and y-intercept, and using their definitions to find two points on the line. ***See Example 3.***

25. $y = 3x + 2$

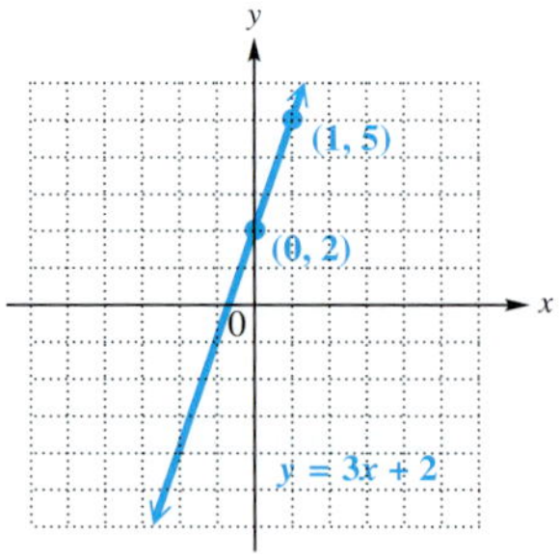

26. $y = 4x - 4$

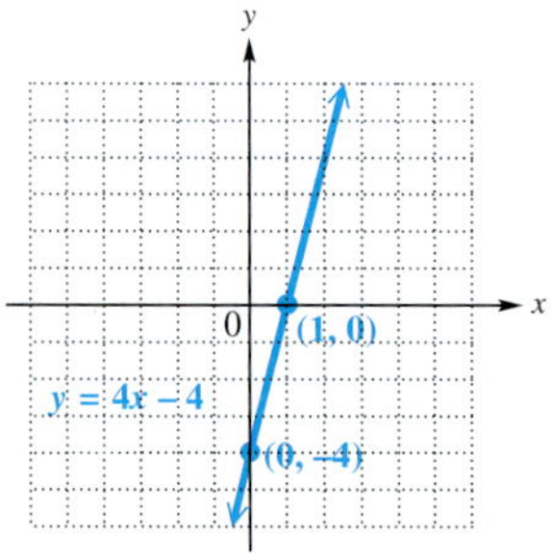

27. $y = \frac{3}{4}x - 1$

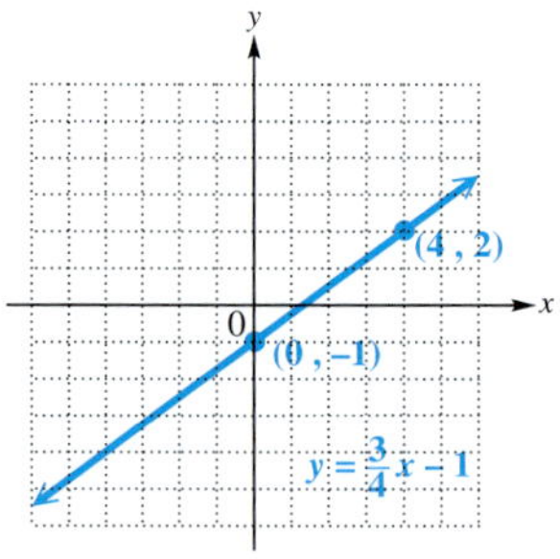

28. $y = \frac{3}{2}x + 2$

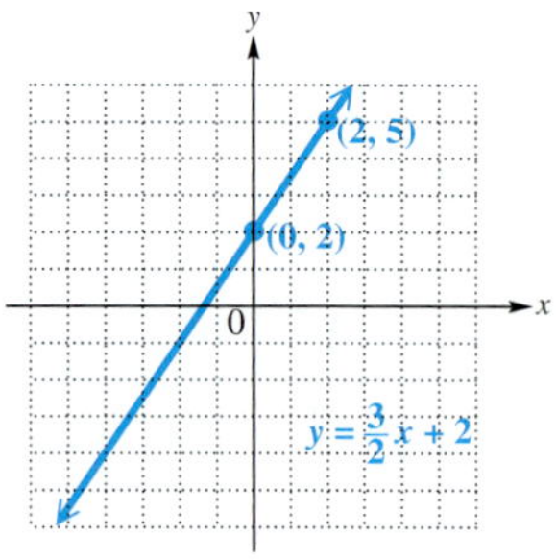

29. $2x + y = -5$

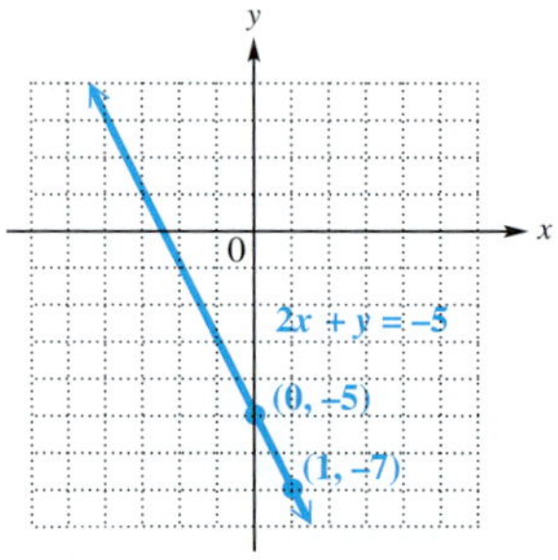

30. $3x + y = -2$

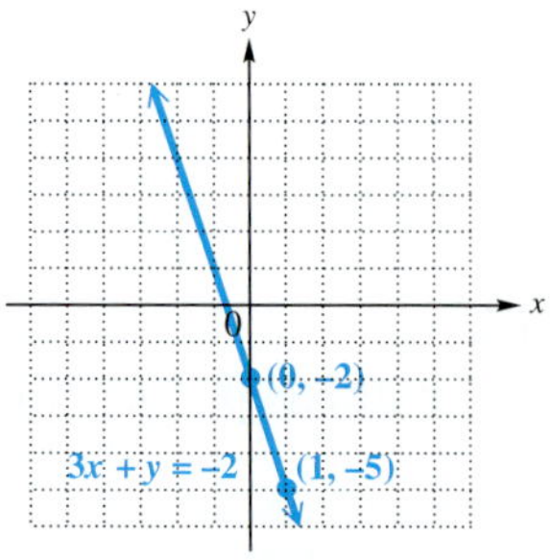

31. $x + 2y = 4$

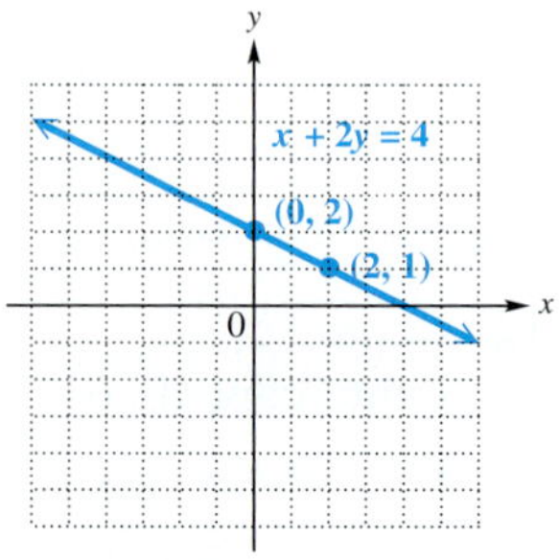

32. $x + 3y = 12$

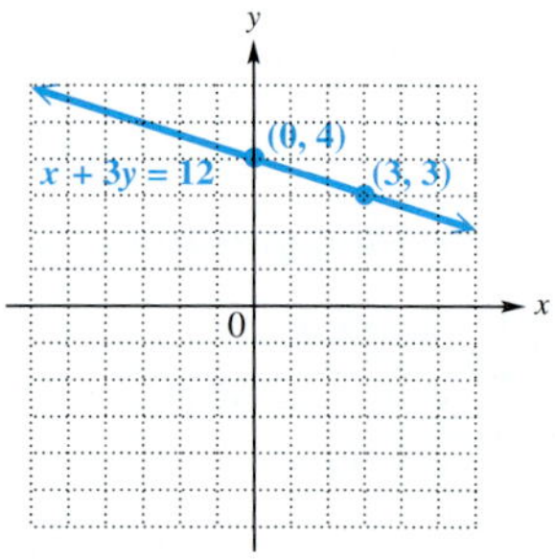

Graph each line passing through the given point and having the given slope. (In Exercises 37–40, recall the types of lines having slope 0 and undefined slope.) Give the slope-intercept form of the equation of the line if possible. ***See Examples 4 and 5.***

33. $(-2, 3)$, $m = \frac{1}{2}$

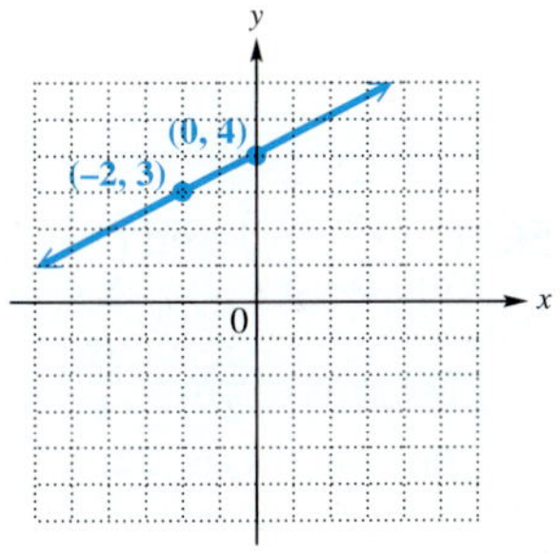

$y = \frac{1}{2}x + 4$

34. $(-4, -1)$, $m = \frac{3}{4}$

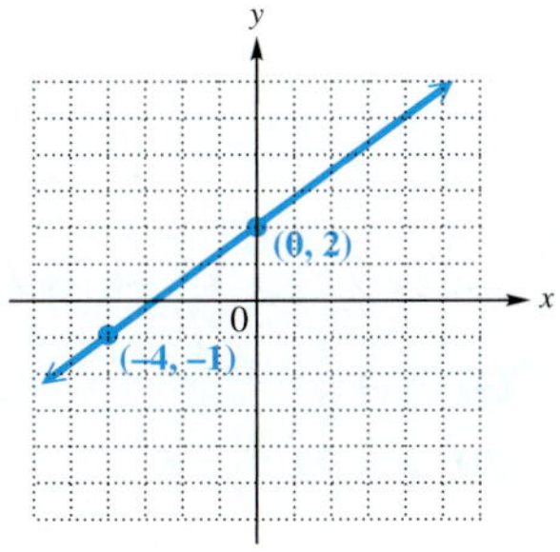

$y = \frac{3}{4}x + 2$

35. $(1, -5)$, $m = -\frac{2}{5}$

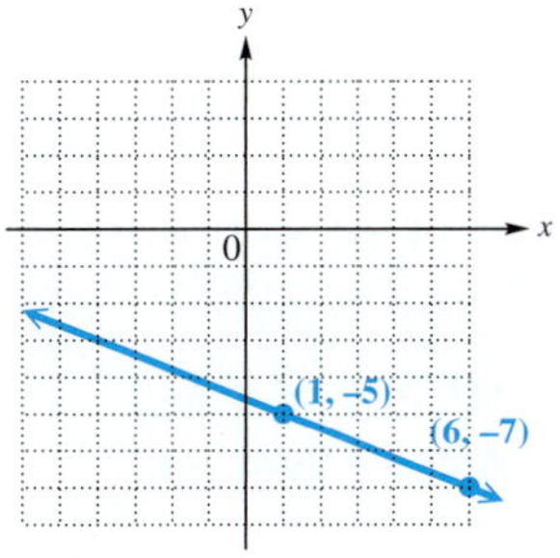

$y = -\frac{2}{5}x - \frac{23}{5}$

36. $(2, -1)$, $m = -\frac{1}{3}$

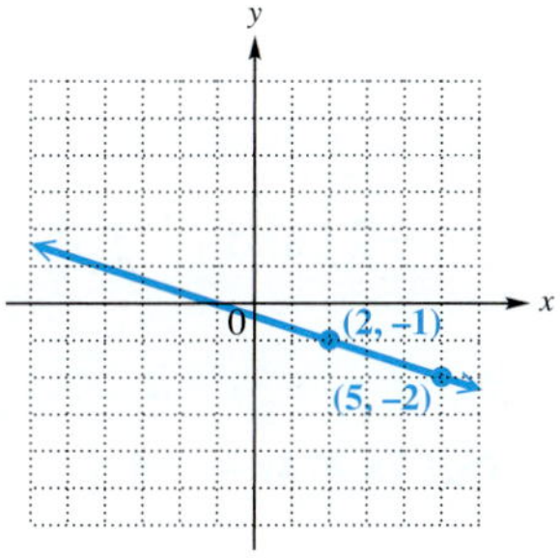

$y = -\frac{1}{3}x - \frac{1}{3}$

37. $(3, 2)$, $m = 0$

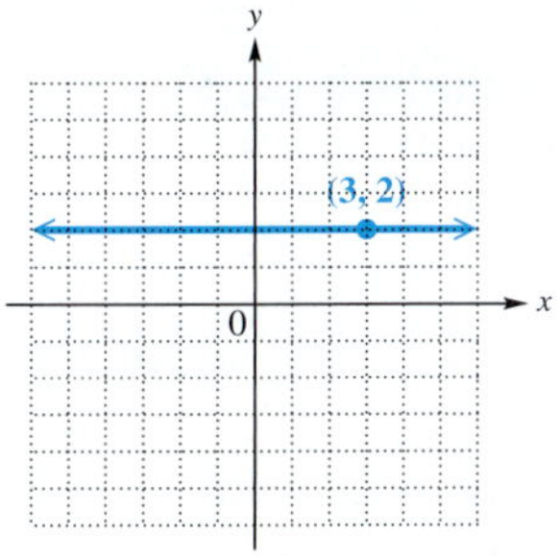

$y = 2$

38. $(-2, 3)$, $m = 0$

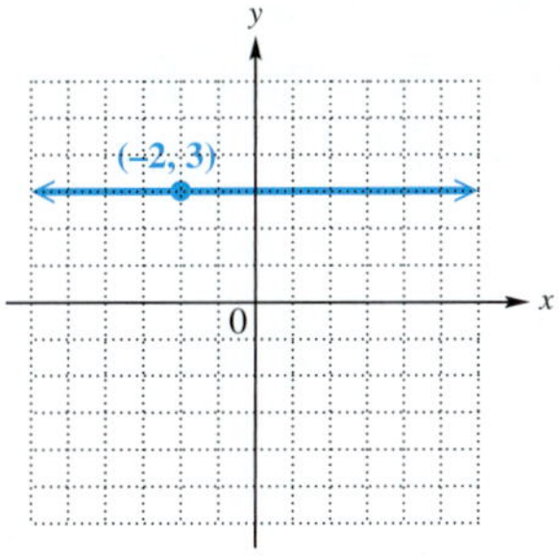

$y = 3$

39. $(3, -2)$, undefined slope

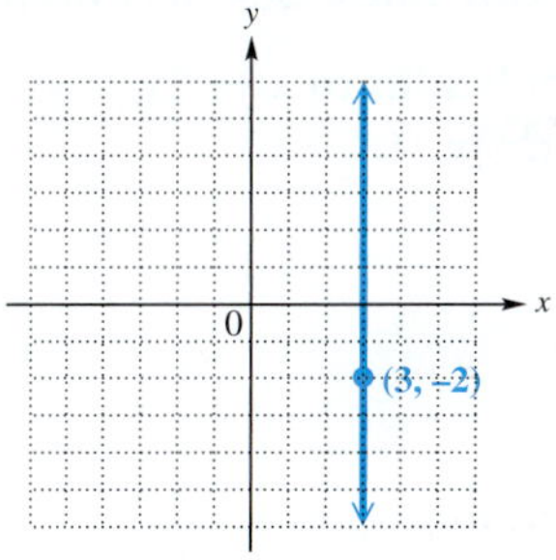

$x = 3$ (no slope-intercept form)

40. $(2, 4)$, undefined slope

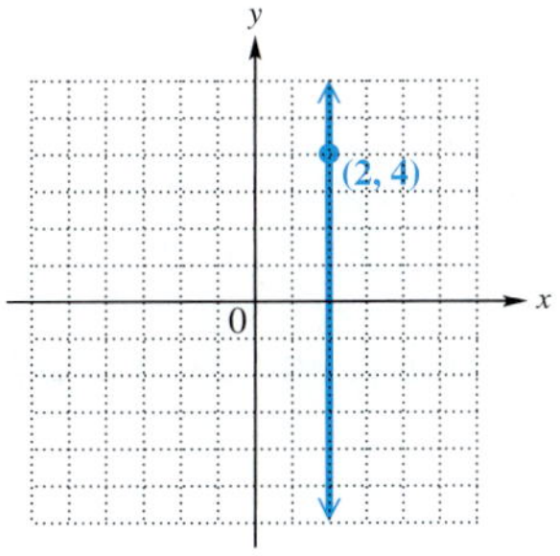

$x = 2$ (no slope-intercept form)

41. $(0, 0)$, $m = \frac{2}{3}$

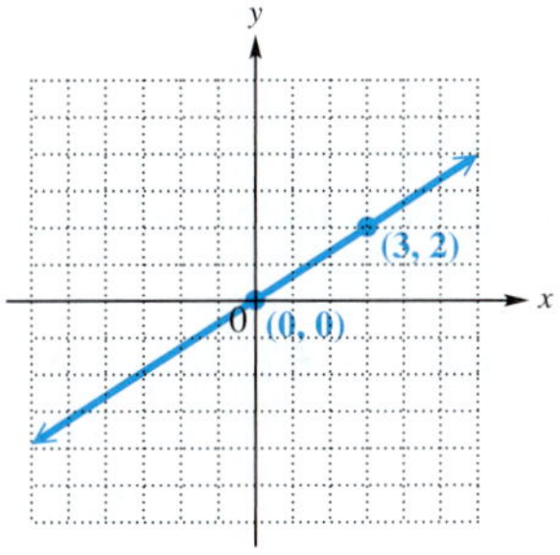

$y = \frac{2}{3}x$

42. $(0, 0)$, $m = \frac{5}{2}$

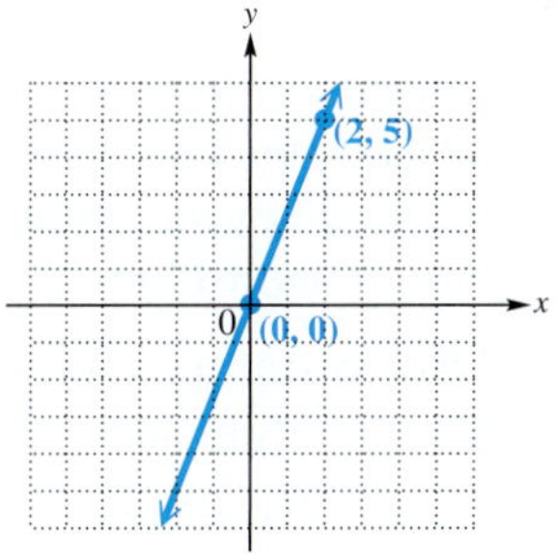

$y = \frac{5}{2}x$

Write an equation of the line passing through the given point and having the given slope. Give the final answer in slope-intercept form. ***See Examples 5 and 6.***

43. $(4, 1)$, $m = 2$

$y = 2x - 7$

44. $(2, 7)$, $m = 3$

$y = 3x + 1$

45. $(3, -10)$, $m = -2$

$y = -2x - 4$

46. $(2, -5)$, $m = -4$

$y = -4x + 3$

47. $(-2, 5)$, $m = \frac{2}{3}$

$y = \frac{2}{3}x + \frac{19}{3}$

48. $(-4, 1)$, $m = \frac{3}{4}$

$y = \frac{3}{4}x + 4$

*Write an equation of the line passing through each pair of points. Give the final answer in **(a)** slope-intercept form and **(b)** standard form. **See Example 7.***

49. $(8, 5)$ and $(9, 6)$

(a) $y = x - 3$

(b) $x - y = 3$

50. $(4, 10)$ and $(6, 12)$

(a) $y = x + 6$

(b) $x - y = -6$

51. $(-1, -7)$ and $(-8, -2)$

(a) $y = -\frac{5}{7}x - \frac{54}{7}$

(b) $5x + 7y = -54$

52. $(-2, -1)$ and $(3, -4)$

(a) $y = -\frac{3}{5}x - \frac{11}{5}$

(b) $3x + 5y = -11$

53. $(0, -2)$ and $(-3, 0)$

(a) $y = -\frac{2}{3}x - 2$

(b) $2x + 3y = -6$

54. $(-4, 0)$ and $(0, 2)$

(a) $y = \frac{1}{2}x + 2$

(b) $x - 2y = -4$

55. $\left(\frac{1}{2}, \frac{3}{2}\right)$ and $\left(-\frac{1}{4}, \frac{5}{4}\right)$

(a) $y = \frac{1}{3}x + \frac{4}{3}$

(b) $x - 3y = -4$

56. $\left(-\frac{2}{3}, \frac{8}{3}\right)$ and $\left(\frac{1}{3}, \frac{7}{3}\right)$

(a) $y = -\frac{1}{3}x + \frac{22}{9}$

(b) $3x + 9y = 22$

*Write an equation of the line satisfying the given conditions. Give the final answer in slope-intercept form. (Hint: Recall the relationships among slopes of parallel and perpendicular lines. See **Section 3.3.**)*

57. Perpendicular to $x - 2y = 7$, y-intercept $(0, -3)$

$y = -2x - 3$

58. Parallel to $5x = 2y + 10$, y-intercept $(0, 4)$

$y = \frac{5}{2}x + 4$

59. Passing through $(2, 3)$, parallel to $4x - y = -2$

$y = 4x - 5$

60. Passing through $(4, 2)$, perpendicular to $x - 3y = 7$

$y = -3x + 14$

61. Passing through $(2, -3)$, parallel to $3x = 4y + 5$

$y = \frac{3}{4}x - \frac{9}{2}$

62. Passing through $(-1, 4)$, perpendicular to $2x + 3y = 8$

$y = \frac{3}{2}x + \frac{11}{2}$

*The cost y to produce x items is, in some cases, expressed as $y = mx + b$. The number b gives the **fixed cost** (the cost that is the same no matter how many items are produced), and the number m gives the **variable cost** (the cost to produce an additional item). Use this information to work Exercises 63 and 64.*

63. It costs \$400 to start up a business selling campaign buttons. Each button costs \$0.25 to produce.

(a) What is the fixed cost?

\$400

(b) What is the variable cost?

\$0.25

(c) Write the cost equation.

$y = 0.25x + 400$

(d) What will be the cost to produce 100 campaign buttons, based on the cost equation?

\$425

(e) How many campaign buttons will be produced if total cost is \$775?

1500

64. It costs \$2000 to purchase a copier, and each copy costs \$0.02 to make.

(a) What is the fixed cost?

\$2000

(b) What is the variable cost?

\$0.02

(c) Write the cost equation.

$y = 0.02x + 2000$

(d) What will be the cost to produce 10,000 copies, based on the cost equation?

\$2200

(e) How many copies will be produced if total cost is \$2600?

30,000

Solve each problem. ***See Example 8.***

65. The table lists the average annual cost y (in dollars) of tuition and fees at 2-year public colleges for selected years x, where $x = 1$ represents 2007, $x = 2$ represents 2008, and so on.

Year x	Cost y (in dollars)
1	2294
2	2372
3	2558
4	2727
5	2963

Source: The College Board.

(a) Write five ordered pairs for the data.

(1, 2294), (2, 2372), (3, 2558), (4, 2727), (5, 2963)

(b) Plot the ordered pairs. Do the points lie approximately in a straight line?

AVERAGE ANNUAL COSTS AT 2-YEAR COLLEGES

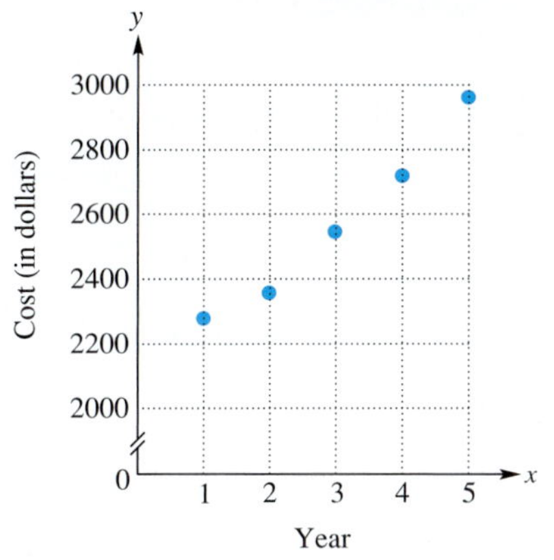

See the graph; yes

(c) Use the ordered pairs $(2, 2372)$ and $(5, 2963)$ to find the equation of a line that approximates the data. Write the equation in slope-intercept form.

$y = 197x + 1978$

(d) Use the equation from part (c) to estimate the average annual cost at 2-year colleges in 2012 to the nearest dollar. (*Hint:* What is the value of x for 2012?)

$3160

66. The table gives heavy-metal nuclear waste (in thousands of metric tons) from spent reactor fuel awaiting permanent storage. (*Source: Scientific American.*)

Year x	Waste y (in thousands of tons)
1995	32
2000	42
2010*	61
2020*	76

*Estimates by the U.S. Department of Energy.

Let $x = 0$ represent 1995, $x = 5$ represent 2000 (since $2000 - 1995 = 5$), and so on.

(a) For 1995, the ordered pair is $(0, 32)$. Write ordered pairs for the data for the other years given in the table.

(5, 42), (15, 61), (25, 76)

(b) Plot the ordered pairs (x, y). Do the points lie approximately in a straight line?

HEAVY-METAL NUCLEAR WASTE AWAITING STORAGE

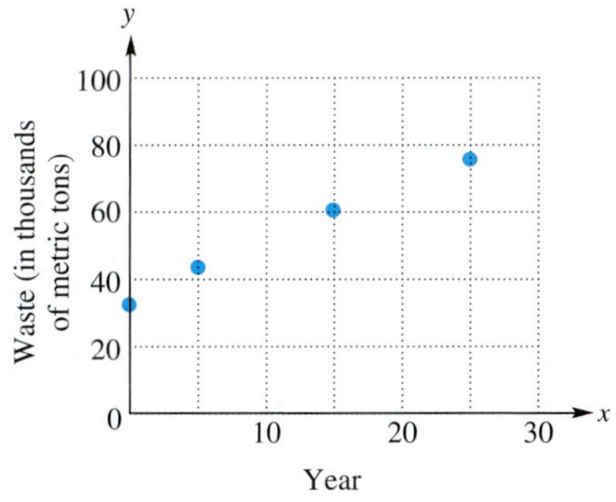

See the graph; yes

(c) Use the ordered pairs $(0, 32)$ and $(25, 76)$ to find the equation of a line that approximates the data. Write the equation in slope-intercept form.

$y = 1.76x + 32$

(d) Use the equation from part (c) to estimate the amount of nuclear waste in 2015. (*Hint:* What is the value of x for 2015?)

67.2 thousand (or 67,200) metric tons

The points on the graph show the number of colleges that teamed up with banks to issue student ID cards which doubled as debit cards from 2002 through 2007. The graph of a linear equation that models the data is also shown.

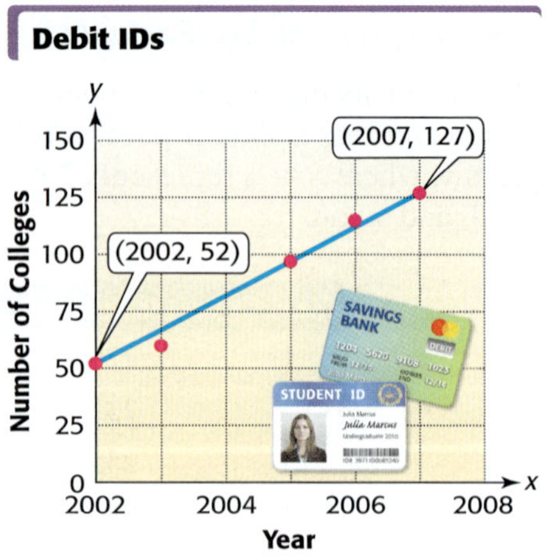

Source: CR80News.
* Data for 2004 unavailable.

67. Use the ordered pairs shown on the graph to write an equation of the line that models the data. Give the equation in slope-intercept form.

$y = 15x - 29{,}978$

68. Use the equation from **Exercise 67** to estimate the number of colleges that teamed up with banks to offer debit IDs in 2004, the year with unavailable data.

82 colleges

Relating Concepts (Exercises 69–76) For Individual or Group Work

If we think of ordered pairs of the form (C, F), then the two most common methods of measuring temperature, Celsius and Fahrenheit, can be related as follows:

When $C = 0, F = 32,$ *and when* $C = 100, F = 212.$

Work Exercises 69–76 in order.

69. Write two ordered pairs relating these two temperature scales.

$(0, 32); (100, 212)$

70. Find the slope of the line through the two points.

$\frac{9}{5}$

71. Use the point-slope form to find an equation of the line. (Your variables should be C and F rather than x and y.)

$F - 32 = \frac{9}{5}(C - 0)$

72. Write an equation for F in terms of C.

$F = \frac{9}{5}C + 32$

73. Use the equation from **Exercise 72** to write an equation for C in terms of F.

$C = \frac{5}{9}(F - 32)$

74. Use the equation from **Exercise 72** to find the Fahrenheit temperature when $C = 30$. 86°

75. Use the equation from **Exercise 73** to find the Celsius temperature when $F = 50$. 10°

76. For what temperature is $F = C$? (Use the photo shown at the right to confirm your answer.) −40°

Summary Exercises *Applying Graphing and Equation-Writing Techniques for Lines*

Summary Exercises provide mixed practice to better prepare students for tests.

1. CONCEPT CHECK Match the description of a line in Column I with the correct equation in Column II.

I	II
(a) Slope -0.5, $b = -2$ **B**	**A.** $y = -\frac{1}{2}x$
(b) x-intercept $(4, 0)$, y-intercept $(0, 2)$ **D**	**B.** $y = -\frac{1}{2}x - 2$
(c) Passes through $(4, -2)$ and $(0, 0)$ **A**	**C.** $x - 2y = 2$
(d) $m = \frac{1}{2}$, passes through $(-2, -2)$ **C**	**D.** $x + 2y = 4$
	E. $x = 2y$

2. CONCEPT CHECK Which equations are equivalent to $2x + 5y = 20$?

A. $y = -\dfrac{2}{5}x + 4$ **B.** $y - 2 = -\dfrac{2}{5}(x - 5)$ **C.** $y = \dfrac{5}{2}x - 4$ **D.** $2x = 5y - 20$

A, B

Graph each line, using the given information or equation.

3. $m = 1$, $b = -2$

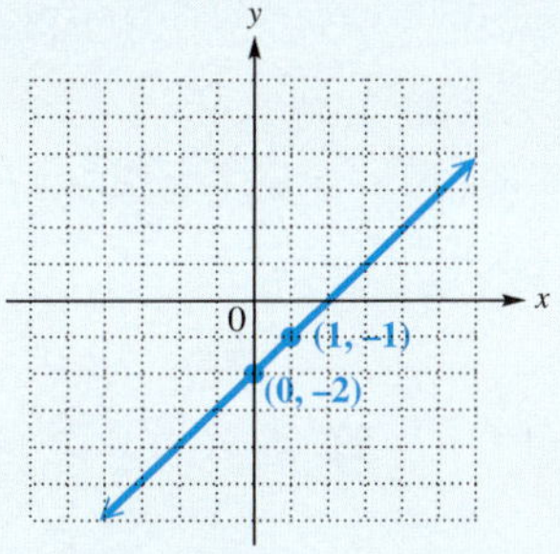

4. $m = 1$, y-intercept $(0, -4)$

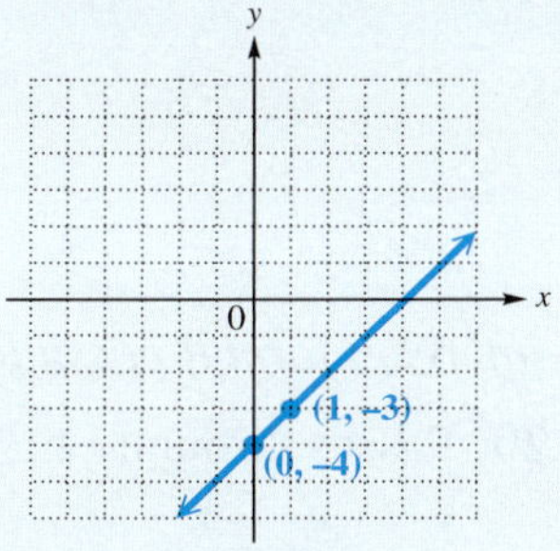

5. $y = -2x + 6$

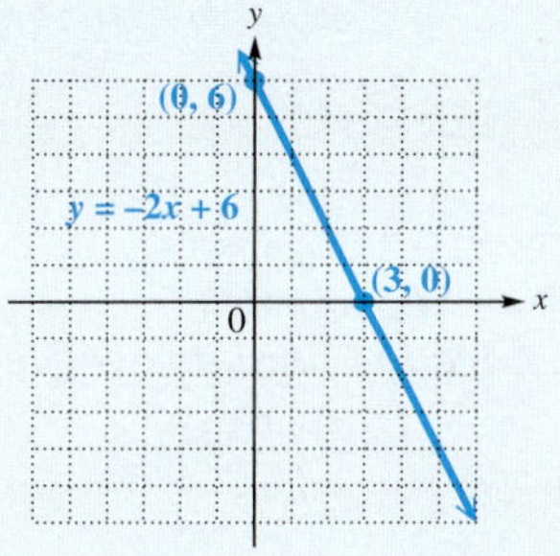

6. $x + 4 = 0$

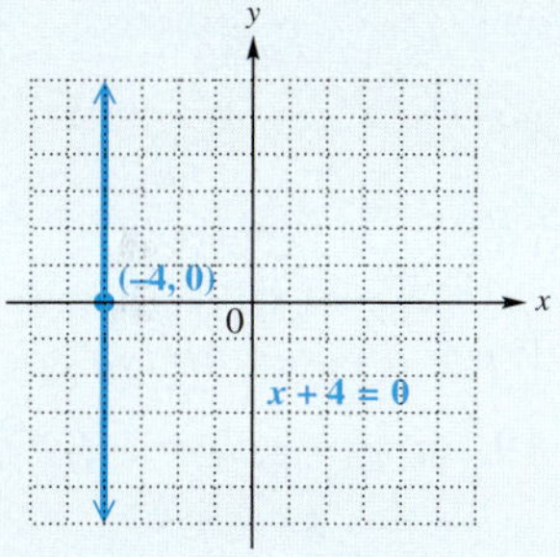

7. $m = -\frac{2}{3}$, passes through $(3, -4)$

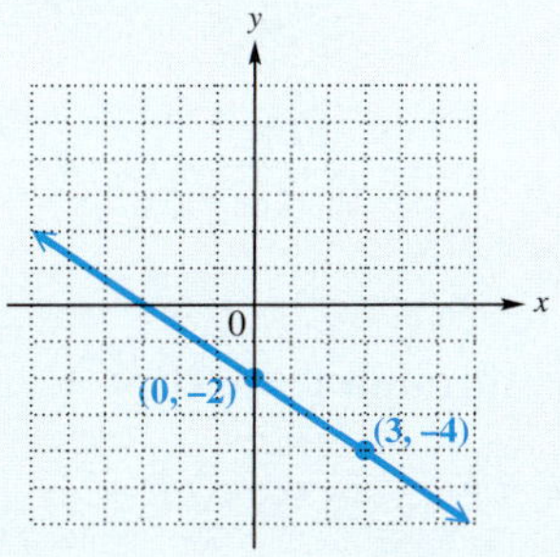

8. $y = -\dfrac{1}{2}x + 2$

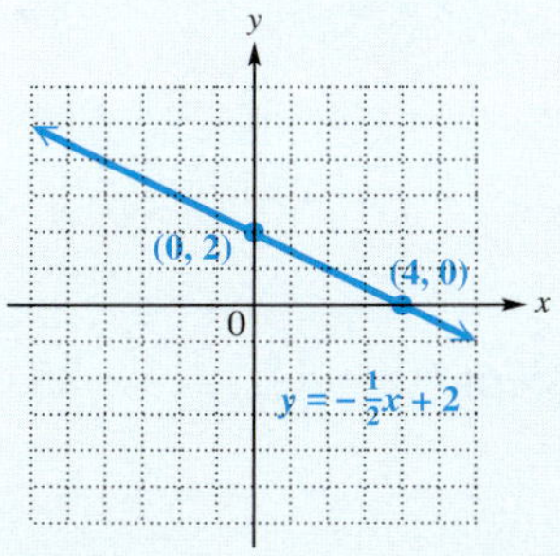

9. $y - 4 = -9$

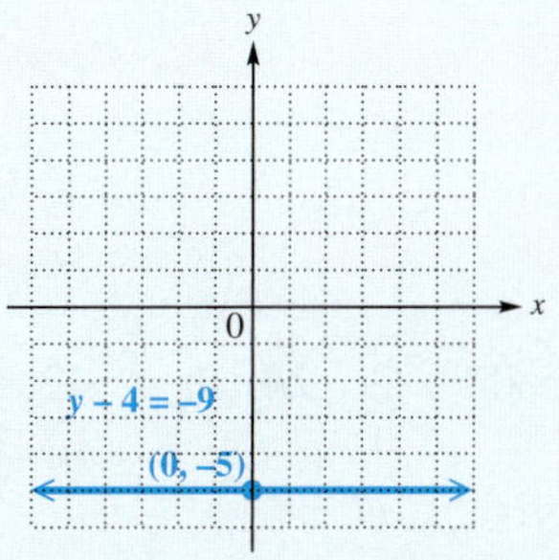

10. $m = -\frac{3}{4}$, passes through $(4, -4)$

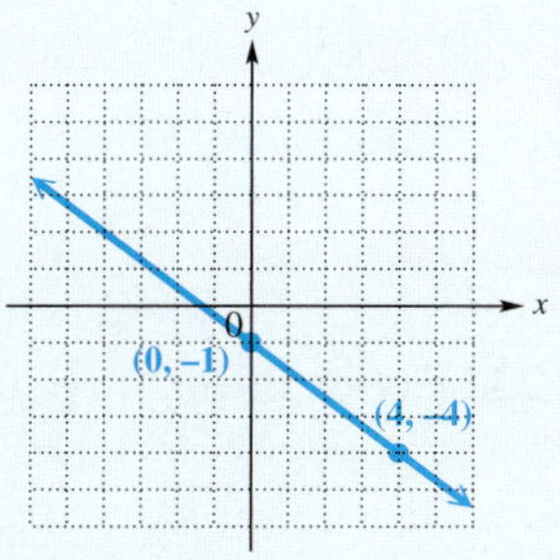

11. Undefined slope, passes through $(3.5, 0)$

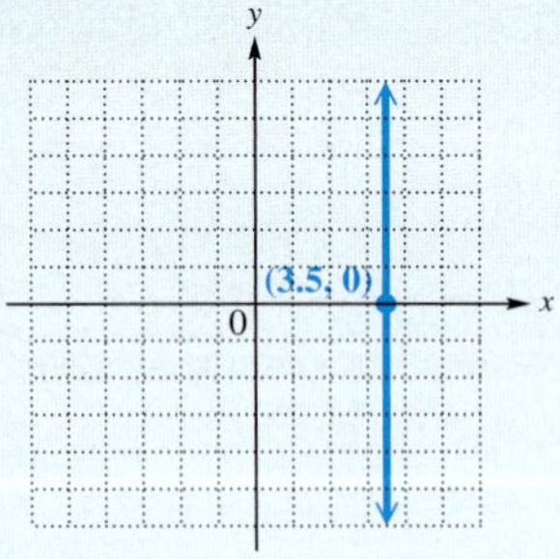

12. Slope $-\frac{1}{5}$, passes through $(0, 0)$

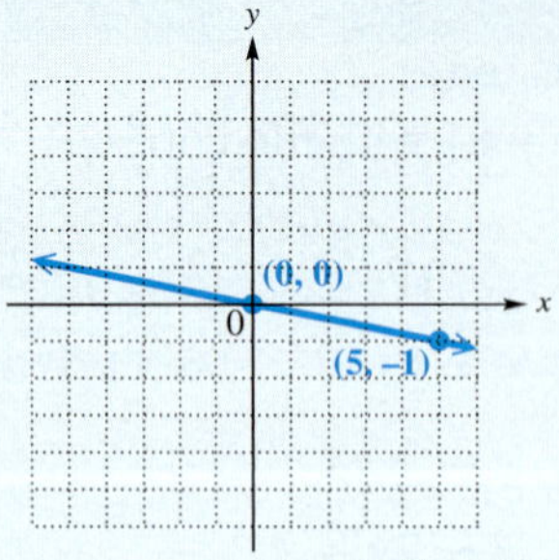

13. $4x - 5y = 20$

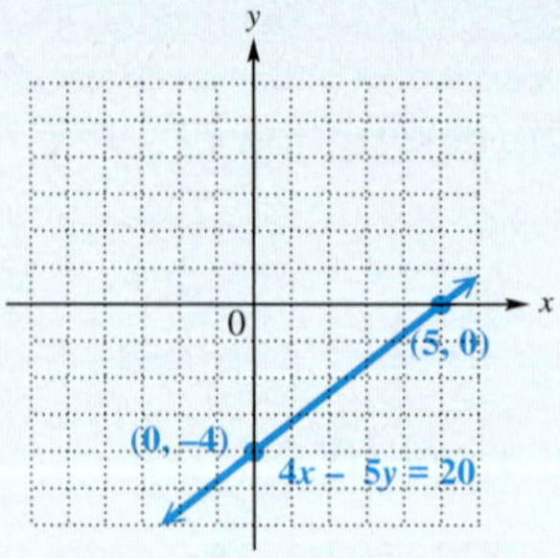

14. $6x - 5y = 30$

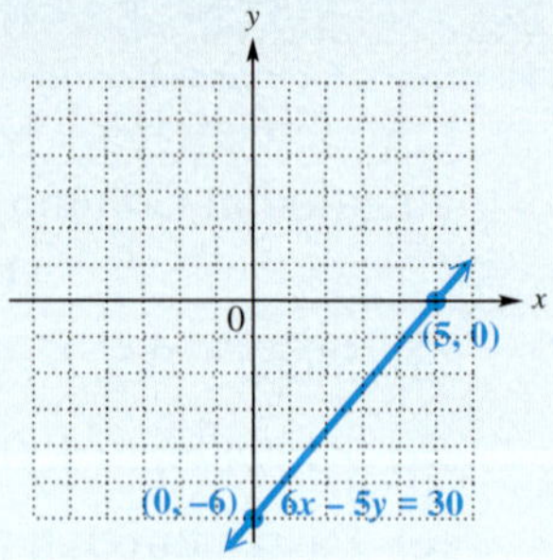

15. $x - 4y = 0$

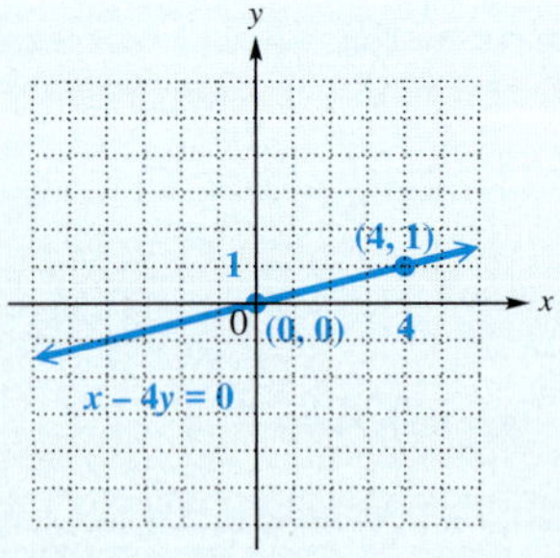

16. $m = 0$, passes through $\left(0, \frac{3}{2}\right)$

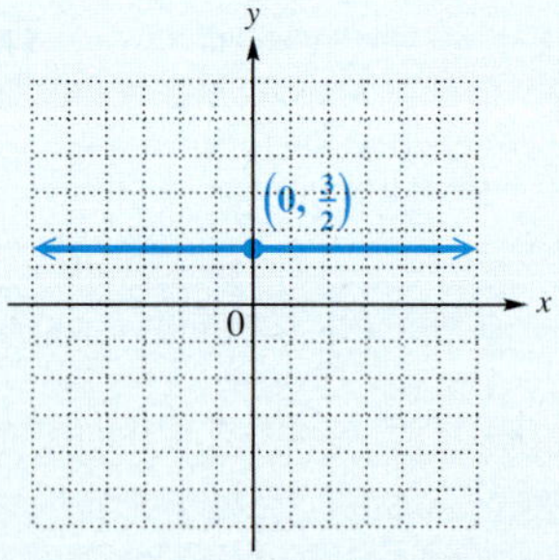

17. $3y = 12 - 2x$

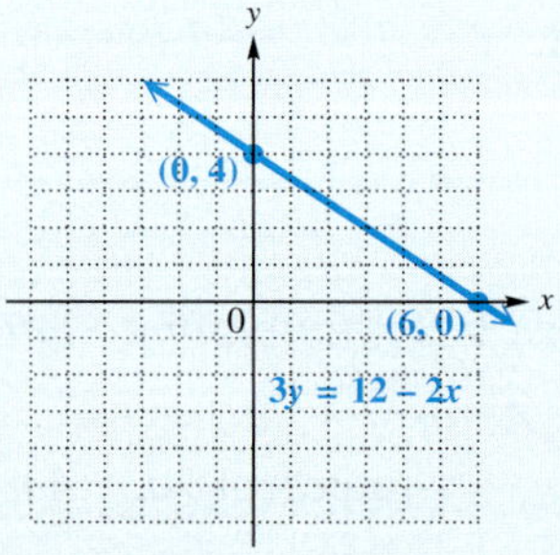

18. $8x = 6y + 24$

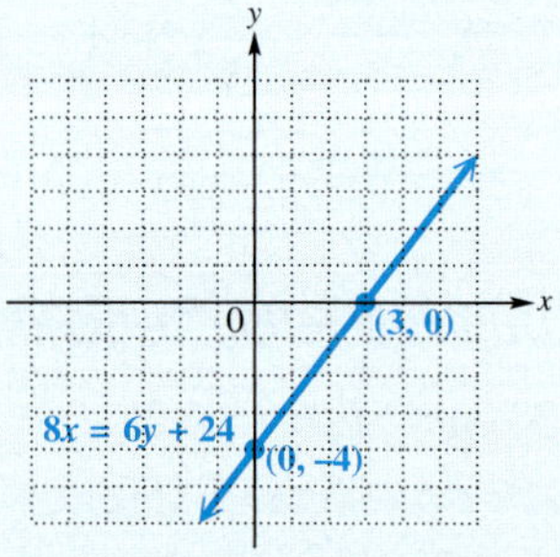

Write an equation of each line. Give the final answer in slope-intercept form if possible.

19. $m = -3, b = -6$

$y = -3x - 6$

20. Passes through $(1, -7)$ and $(-2, 5)$

$y = -4x - 3$

21. Passes through $(0, 0)$ and $(5, 3)$

$y = \frac{3}{5}x$

22. Passes through $(0, 0)$, undefined slope

$x = 0$

23. Passes through $(0, 0)$, $m = 0$

$y = 0$

24. $m = -2$, y-intercept $(0, -4)$

$y = -2x - 4$

25. $m = \frac{5}{3}$, through $(-3, 0)$

$y = \frac{5}{3}x + 5$

3.5 Graphing Linear Inequalities in Two Variables

OBJECTIVES

1. Graph linear inequalities in two variables.
2. Graph an inequality with a boundary line through the origin.

In **Section 3.2** we graphed linear equations, such as

$$2x + 3y = 6.$$

Now we extend this work to include *linear inequalities in two variables,* such as

$$2x + 3y \leq 6.$$

(Recall that $\leq$ is read "is less than or equal to.")

Linear Inequality in Two Variables

An inequality that can be written in the form

$$Ax + By < C, \quad Ax + By > C, \quad Ax + By \leq C, \quad \text{or} \quad Ax + By \geq C,$$

where A, B, and C are real numbers and A and B are not both 0, is a **linear inequality in two variables.**

OBJECTIVE 1 **Graph linear inequalities in two variables.** Consider the graph in **Figure 37.**

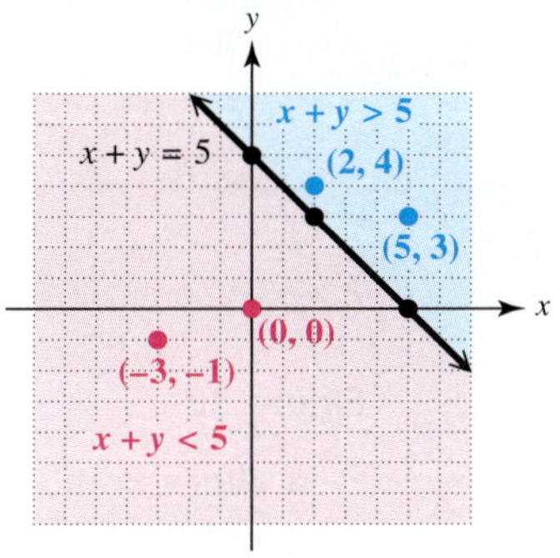

Figure 37

The graph of the line $x + y = 5$ in **Figure 37** divides the points in the rectangular coordinate system into three sets.

1. Those points that lie on the line itself and satisfy the equation $x + y = 5$ (like $(0, 5)$, $(2, 3)$, and $(5, 0)$)
2. Those points that lie in the region above the line and satisfy the inequality $x + y > 5$ (like $(5, 3)$ and $(2, 4)$)
3. Those points that lie in the region below the line and satisfy the inequality $x + y < 5$ (like $(0, 0)$ and $(-3, -1)$)

The graph of the line $x + y = 5$ is the **boundary line** for the inequalities

$$x + y > 5 \quad \text{and} \quad x + y < 5.$$

Graphs of linear inequalities in two variables are regions in the real number plane that may or may not include boundary lines.

Teaching Tip
Stress the fact that *equality* is the boundary between *is less than* and *is greater than.*

1 Use $(0, 0)$ as a test point, and **GS** shade the appropriate region for the linear inequality.

$$3x + 4y \leq 12$$

$$3(\_\_\_\_) + 4(\_\_\_\_) \overset{?}{\leq} 12$$

$$\_\_\_\_ \overset{?}{\leq} 12$$ (*True* / *False*)

Shade the region of the graph that (*includes* / *does not include*) the test point $(0, 0)$.

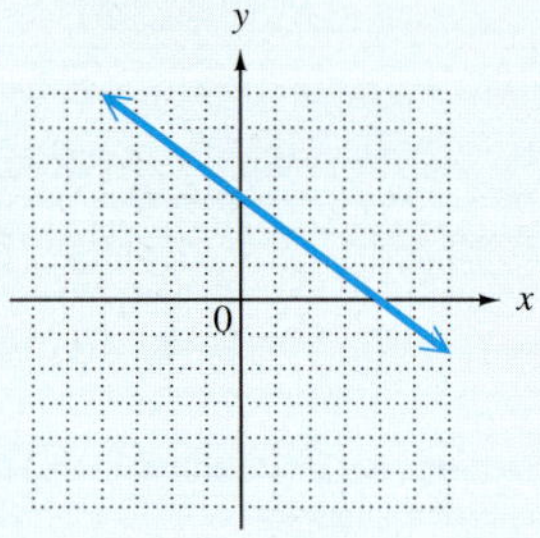

2 Graph $4x - 5y \leq 20$.

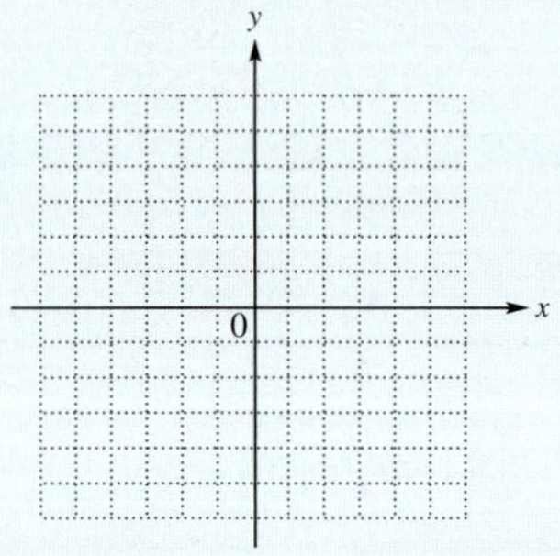

EXAMPLE 1 **Graphing a Linear Inequality**

Graph $2x + 3y \leq 6$.

The inequality $2x + 3y \leq 6$ means that

$$2x + 3y < 6 \quad \text{or} \quad 2x + 3y = 6.$$

We begin by graphing the equation $2x + 3y = 6$, a line with intercepts $(0, 2)$ and $(3, 0)$ as shown in **Figure 38.** This boundary line divides the plane into two regions, one of which satisfies the inequality. To find the correct region, we choose a test point *not* on the boundary line and substitute it into the given inequality to see whether the resulting statement is true or false. The point $(0, 0)$ is a convenient choice.

$2x + 3y \leq 6$	Original inequality
$2(0) + 3(0) \overset{?}{\leq} 6$	Let $x = 0$ and $y = 0$.
$0 + 0 \overset{?}{\leq} 6$	Multiply.
$0 \leq 6$	True

Use (0, 0) as a test point.

Since the last statement is true, we shade the region that includes the test point $(0, 0)$. See **Figure 38.** The shaded region, along with the boundary line, is the desired graph.

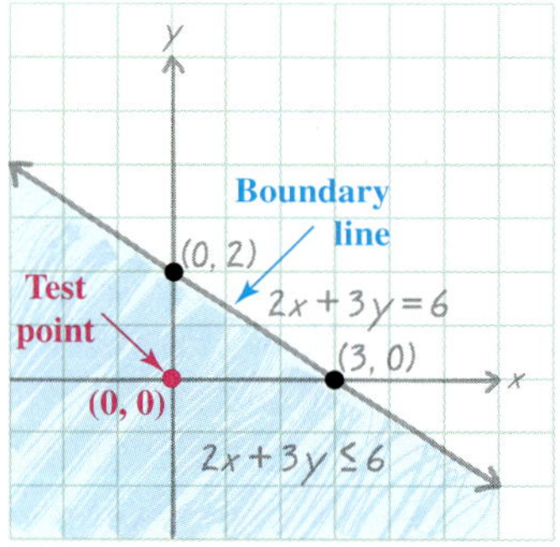

Figure 38

◀ **Work Problems 1 and 2 at the Side.**

Note

Alternatively in **Example 1,** we can find the required region by solving the given inequality for y.

$2x + 3y \leq 6$	Inequality from **Example 1**
$3y \leq -2x + 6$	Subtract $2x$.
$y \leq -\frac{2}{3}x + 2$	Divide each term by 3.

Ordered pairs in which y is equal to $-\frac{2}{3}x + 2$ are on the boundary line, so pairs in which y *is less than* $-\frac{2}{3}x + 2$ will be *below* that line. (As we move *down* vertically, the y-values *decrease*.) This gives the same region that we shaded in **Figure 38.** (Ordered pairs in which y *is greater than* $-\frac{2}{3}x + 2$ will be *above* the boundary line.)

Answers

1. 0; 0; 0; True; includes

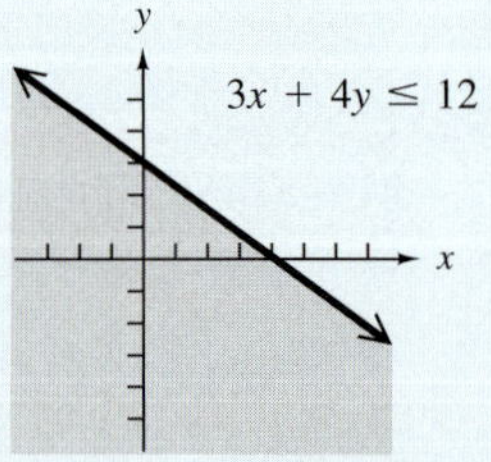

2.

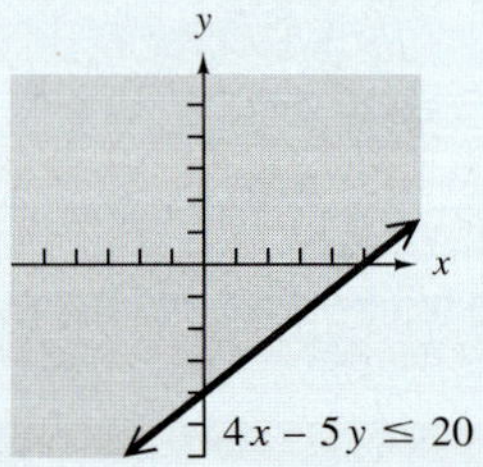

EXAMPLE 2 Graphing a Linear Inequality

Graph $x - y > 5$.

This inequality does *not* involve equality. Therefore, the points on the line $x - y = 5$ do *not* belong to the graph. However, the line still serves as a boundary for two regions, one of which satisfies the inequality.

To graph the inequality, first graph the equation $x - y = 5$. Use a *dashed line* to show that the points on the line are *not* solutions of the inequality $x - y > 5$. See **Figure 39.** Then choose a test point to see which region satisfies the inequality.

$x - y > 5$ Original inequality

$0 - 0 \overset{?}{>} 5$ Let $x = 0$ and $y = 0$.

$0 > 5$ False

(0, 0) is a convenient test point.

Since $0 > 5$ is false, the graph of the inequality is the region that *does not* contain $(0, 0)$. Shade the *other* region, as shown in **Figure 39.**

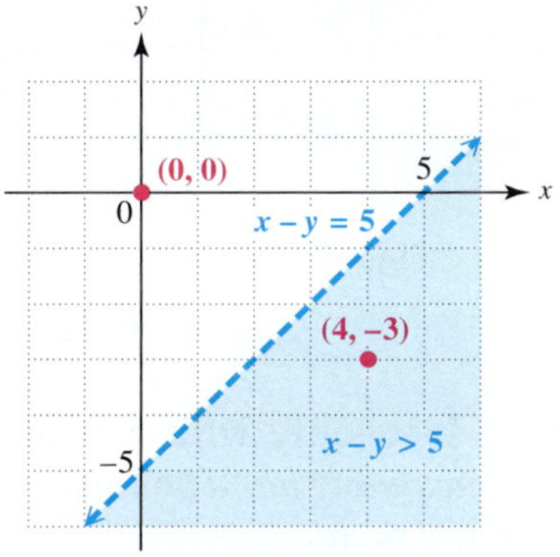

Figure 39

CHECK To check that the proper region is shaded, we test a point in the shaded region. For example, we use $(4, -3)$.

$x - y > 5$

$4 - (-3) \overset{?}{>} 5$ Let $x = 4$ and $y = -3$.

$7 > 5$ ✓ True

Use parentheses to avoid errors.

This verifies that the correct region is shaded in **Figure 39.**

Work Problems 3 and 4 at the Side. ▶

A summary of the steps used to graph a linear inequality in two variables follows.

Graphing a Linear Inequality in Two Variables

Step 1 **Graph the boundary.** Graph the line that is the boundary of the region. Use the methods of **Section 3.2.** Draw a solid line if the inequality involves $\leq$ or $\geq$. Draw a dashed line if the inequality involves $<$ or $>$.

Step 2 **Shade the appropriate side.** Use any point not on the line as a test point. Substitute for x and y in the *inequality*. If a true statement results, shade the region containing the test point. If a false statement results, shade the other region.

3 Use $(0, 0)$ as a test point and shade the appropriate region for the linear inequality.

$$3x + 5y > 15$$

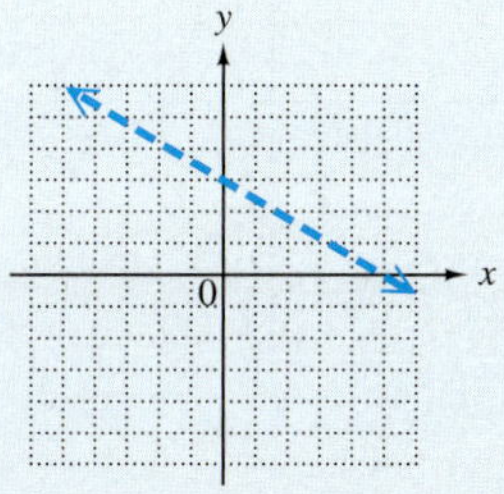

Teaching Tip

Having students select two test points, one on either side of the boundary line as in Example 2, provides an additional check that they have shaded the correct region. Point out that one of the points will give a true statement and the other a false one.

4 Graph $x + 2y > 6$.

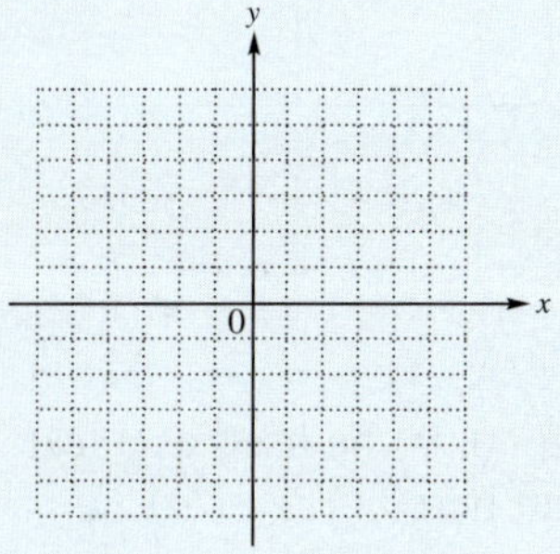

Teaching Tip

In Example 2, to emphasize that the points on the boundary line are *not* solutions of the inequality, select an ordered pair on the line. Substitute it into the inequality to show that a false statement results.

Answers

3.

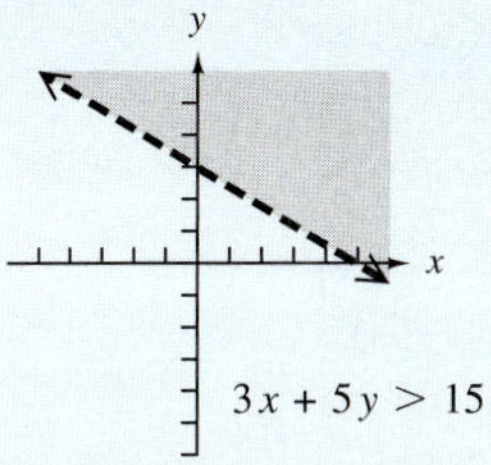

4.

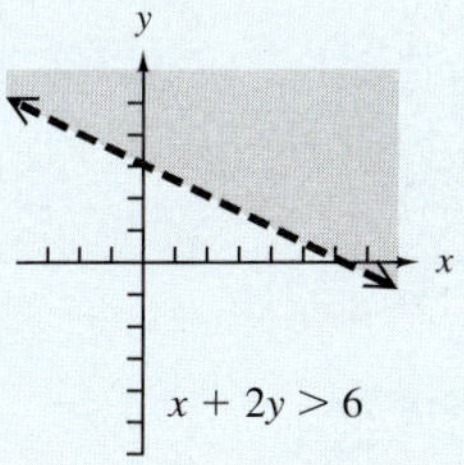

EXAMPLE 3 Graphing a Linear Inequality with a Vertical Boundary Line

Graph $x < 3$.

First graph $x = 3$, a vertical line through the point $(3, 0)$. Use a dashed line, and choose $(0, 0)$ as a test point.

$$x < 3 \quad \text{Original inequality}$$
$$0 \overset{?}{<} 3 \quad \text{Let } x = 0.$$
$$0 < 3 \quad \text{True}$$

Because $0 < 3$ is true, we shade the region containing $(0, 0)$, as in **Figure 40.**

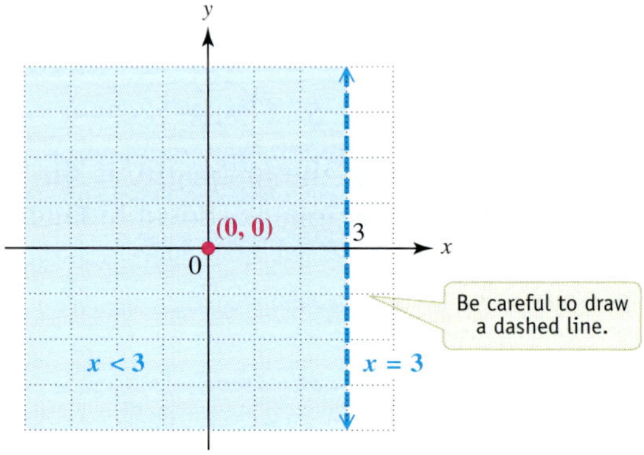

Figure 40

◀ **Work Problem 5 at the Side.**

OBJECTIVE 2 Graph an inequality with a boundary line through the origin. ***If the graph of an inequality has a boundary line through the origin, (0, 0) cannot be used as a test point.***

EXAMPLE 4 Graphing a Linear Inequality with a Boundary Line through the Origin

Graph $x \leq 2y$.

We graph $x = 2y$ using a solid line through $(0, 0)$, $(6, 3)$, and $(4, 2)$. Because $(0, 0)$ is *on* the line $x = 2y$, it cannot be used as a test point. Instead, we choose a test point *off* the line, say $(1, 3)$.

$$x \leq 2y \quad \text{Original inequality}$$
$$1 \overset{?}{\leq} 2(3) \quad \text{Let } x = 1 \text{ and } y = 3.$$
$$1 \leq 6 \quad \text{True}$$

Because $1 \leq 6$ is true, we shade the side of the graph containing the test point $(1, 3)$. See **Figure 41.**

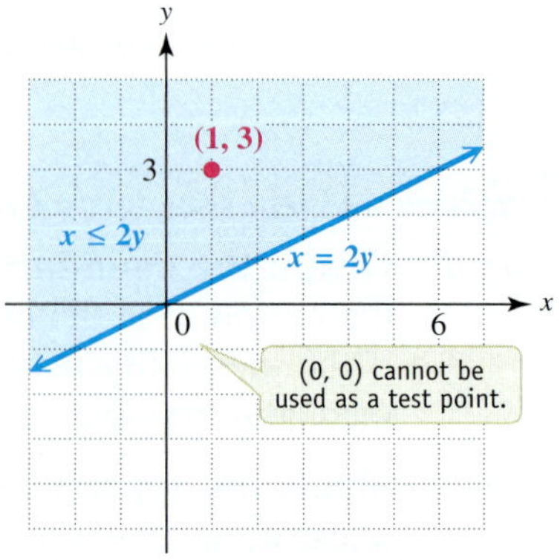

Figure 41

◀ **Work Problem 6 at the Side.**

5 Graph $y < 4$.

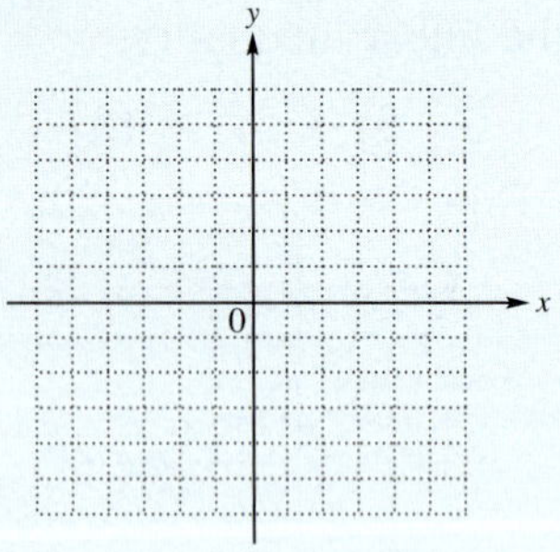

6 Graph $x \geq -3y$.

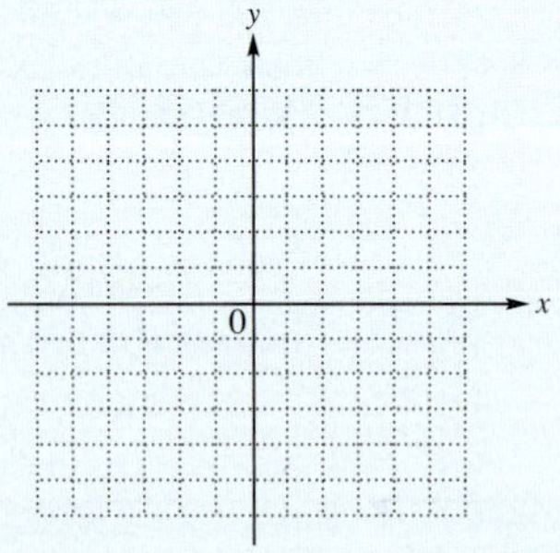

Can $(0, 0)$ be used as a test point here? (*Yes* / *No*)

Answers

5.

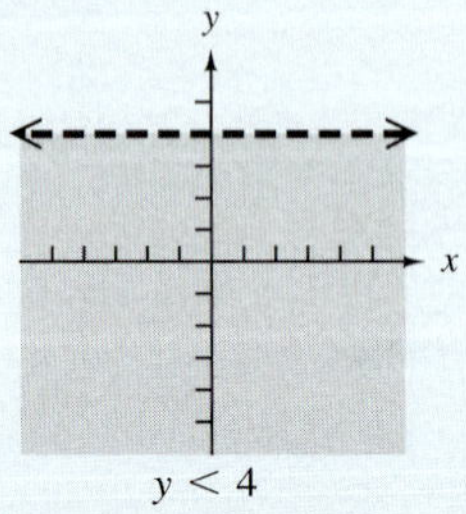

6. No

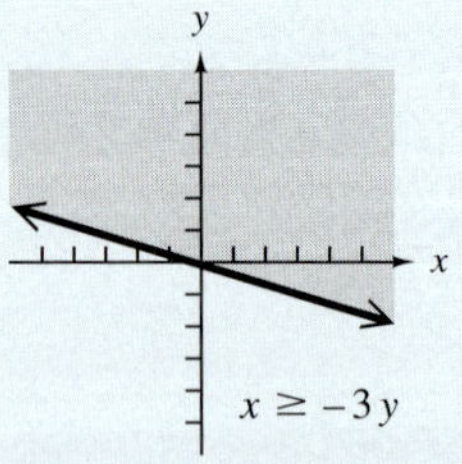

3.5 Exercises

FOR EXTRA HELP

CONCEPT CHECK *The statements in Exercises 1–4 were taken from various web sites. Each includes a phrase that can be symbolized with one of the inequality symbols* $<$, $\leq$, $>$, *or* $\geq$. *Give the inequality symbol for the boldfaced words.*

1. In 2009, carbon dioxide emissions from the consumption of fossil fuels were **more than** 5400 million metric tons in the United States and **more than** 7700 million metric tons in China. (*Source:* www.eia.gov)

   $>$, $>$

2. The number of motor vehicle deaths in the United States in 2009 was 3615 **less than** the number in 2008. (*Source:* www.nhtsa.gov)

   $<$

3. As of January 2012, American Airlines passengers were allowed one carry-on bag, with dimensions (length + width + height) of **at most** 45 in. (*Source:* www.aa.com)

   $\leq$

4. As of January 2012, the usage charge on many of AT&T's calling plans was **at least** \$9.99 per month. (*Source:* www.consumer.att.com)

   $\geq$

CONCEPT CHECK *Decide whether each statement is* true *or* false. *If false, explain why.*

5. The point $(4, 0)$ lies on the graph of $3x - 4y < 12$.

   false; The point $(4, 0)$ lies on the boundary line $3x - 4y = 12$, which is *not* part of the graph because the symbol $<$ does not involve equality.

6. The point $(4, 0)$ lies on the graph of $3x - 4y \leq 12$.

   true

7. The point $(0, 0)$ can be used as a test point to determine which region to shade when graphing the linear inequality $x + 4y > 0$.

   false; Because $(0, 0)$ is on the boundary line $x + 4y = 0$, it cannot be used as a test point. Use a test point *off* the line.

8. When graphing the linear inequality $3x + 2y \geq 12$, use a dashed line for the boundary line.

   false; Use a solid line for the boundary line because of the equality portion of the $\geq$ symbol.

9. Both points $(4, 1)$ and $(0, 0)$ lie on the graph of $3x - 2y \geq 0$.

   true

10. The graph of $y > x$ does not contain points in quadrant IV.

    true

In Exercises 11–16, the straight-line boundary has been drawn. Complete each graph by shading the correct region. ***See Examples 1–4.***

11. $x + y \geq 4$

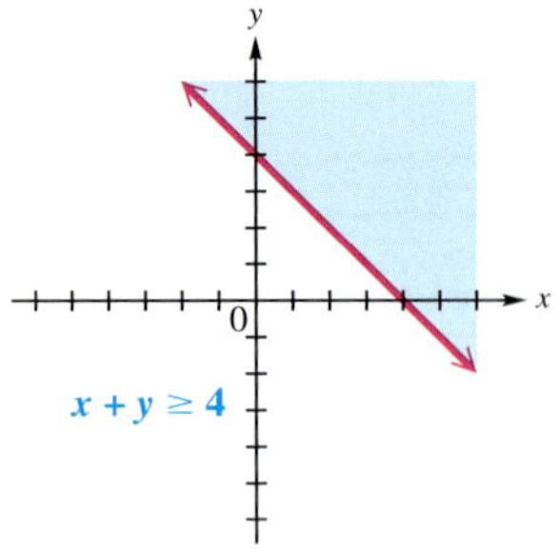

12. $x + y \leq 2$

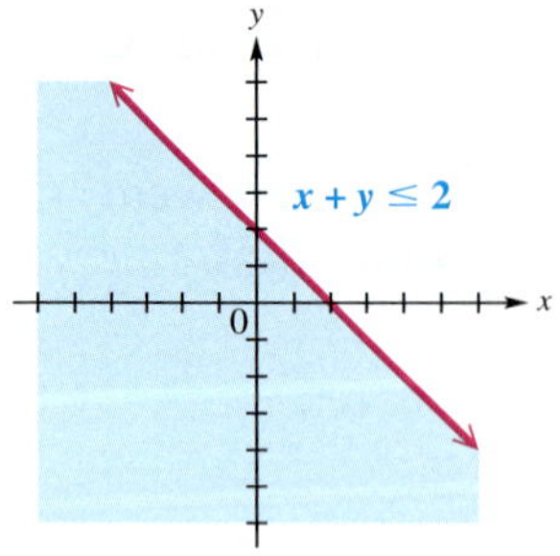

13. $x + 2y \geq 7$

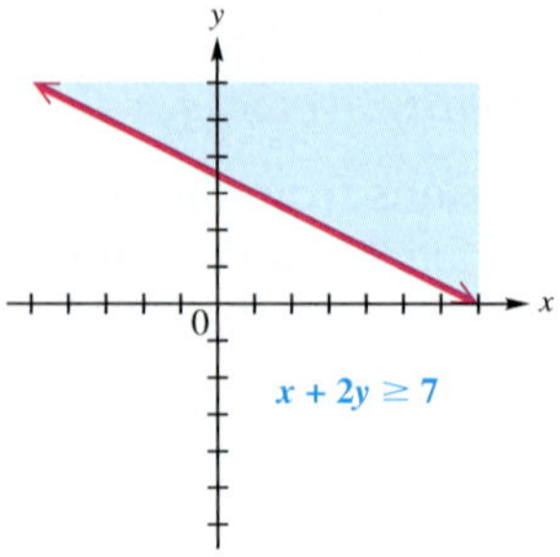

14. $2x + y \geq 5$

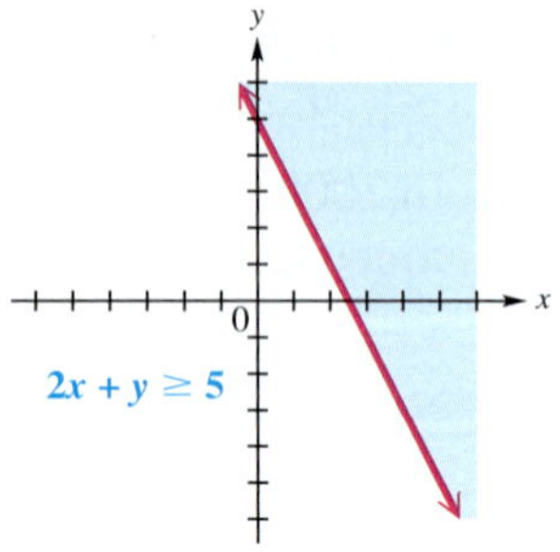

15. $-3x + 4y > 12$

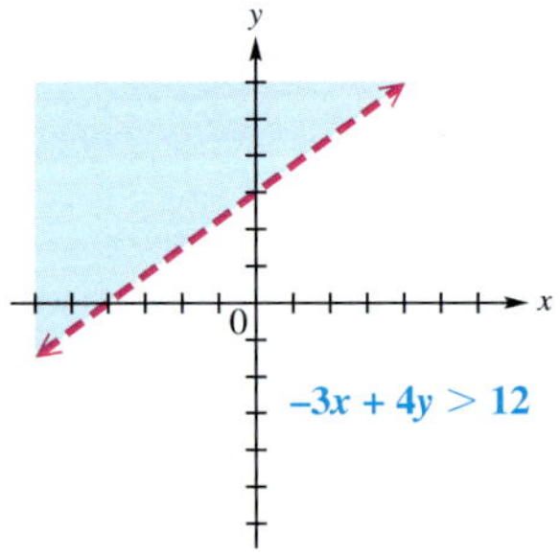

16. $4x - 5y < 20$

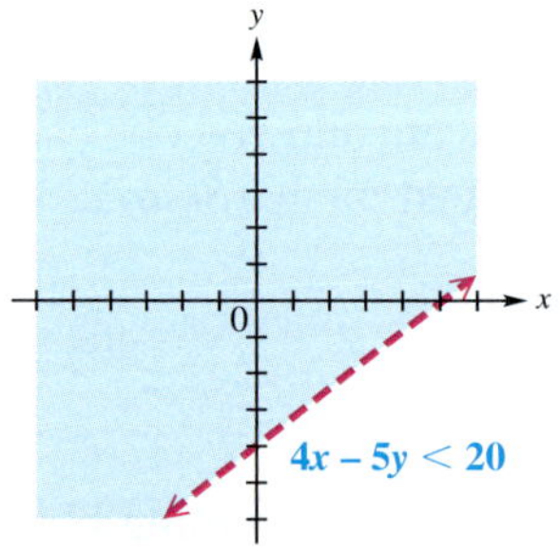

17. $x > 4$

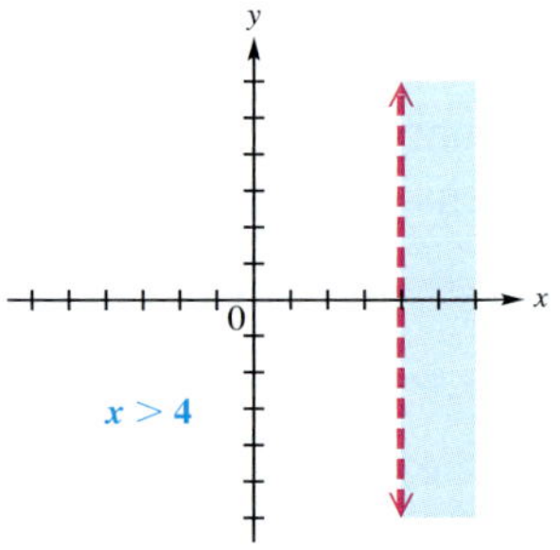

18. $y < -1$

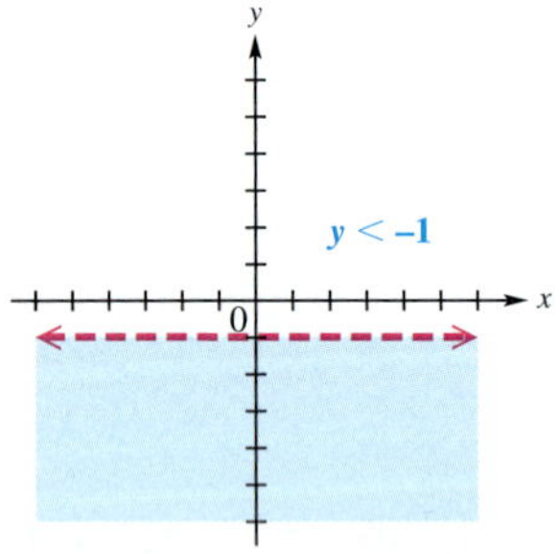

19. $x \geq -y$

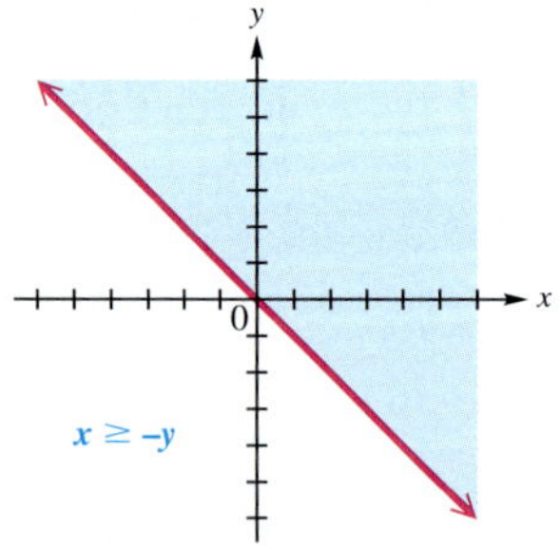

20. $x > 3y$

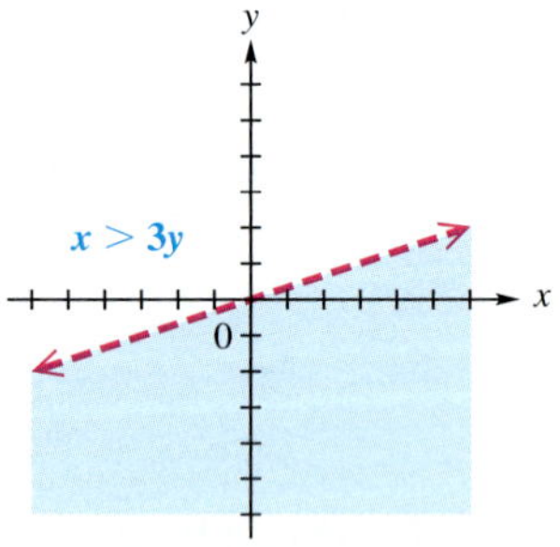

21. Explain how to determine whether to use a dashed line or a solid line when graphing a linear inequality in two variables.

Use a dashed line if the symbol is $<$ or $>$. Use a solid line if the symbol is $\leq$ or $\geq$.

22. Explain why the point $(0, 0)$ is not an appropriate choice for a test point when graphing an inequality whose boundary goes through the origin.

A test point cannot lie on the boundary line. It must lie on one side of the boundary.

Graph each linear inequality. ***See Examples 1–4.***

23. $x + y \le 5$

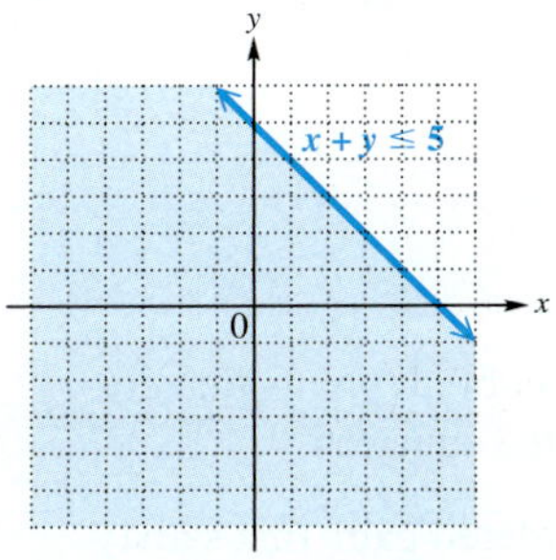

24. $x + y \ge 3$

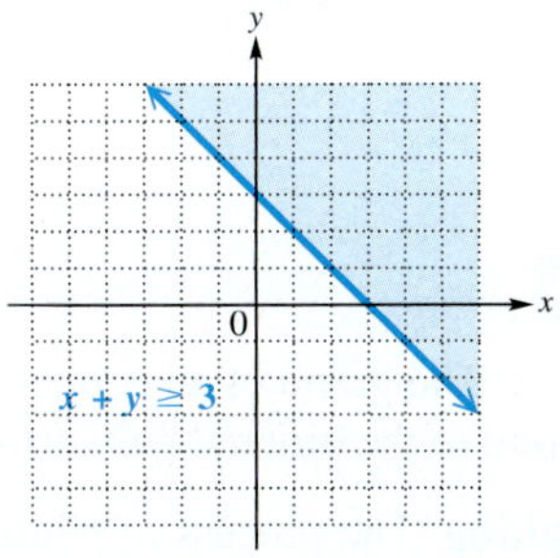

25. $x + 2y < 4$

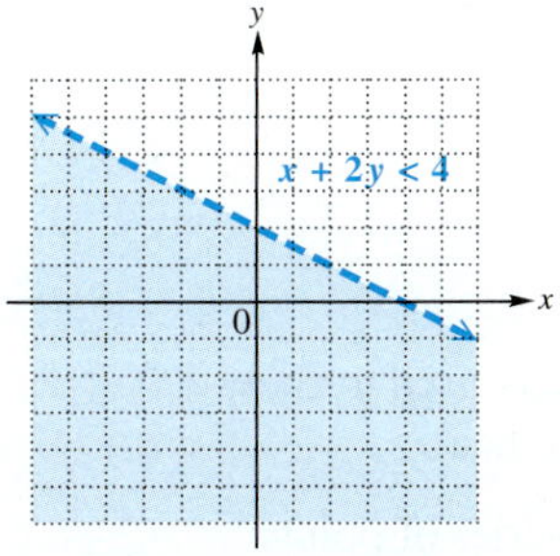

26. $x + 3y > 6$

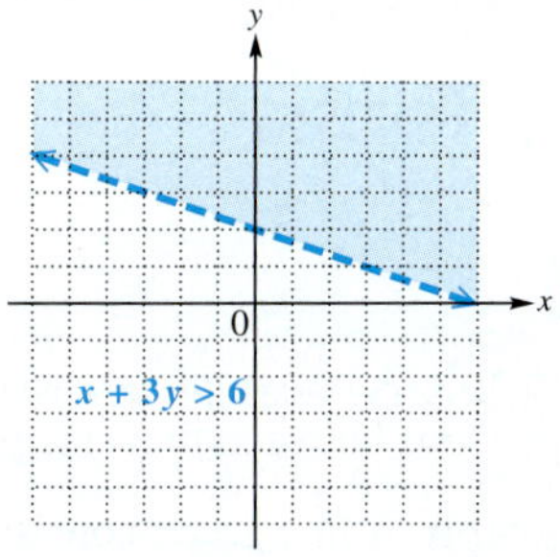

27. $2x + 6 > -3y$

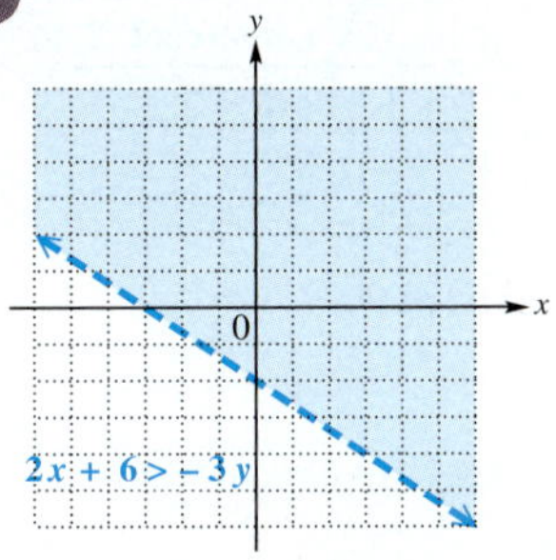

28. $-4y > 3x - 12$

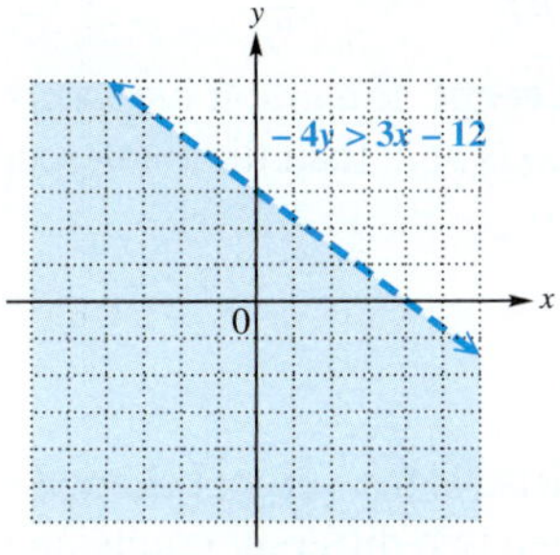

29. $y \ge 2x + 1$

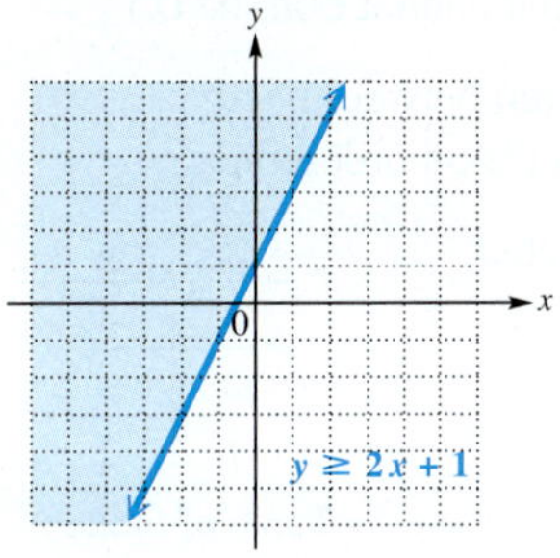

30. $y < -3x + 1$

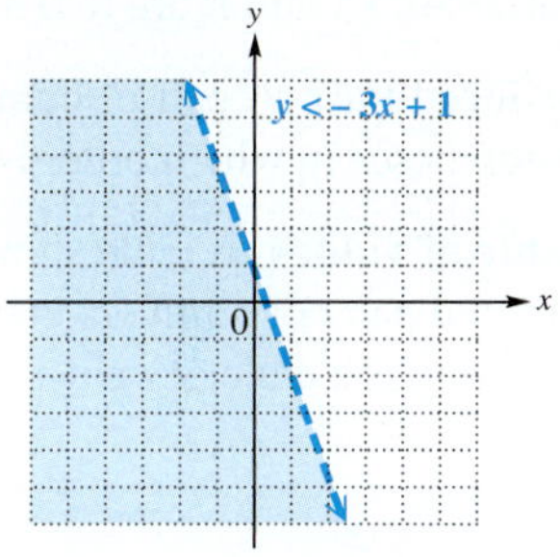

31. $x \le -2$

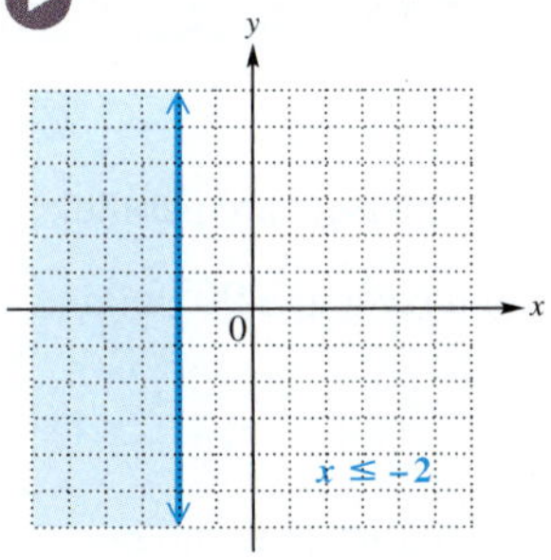

32. $x \ge 1$

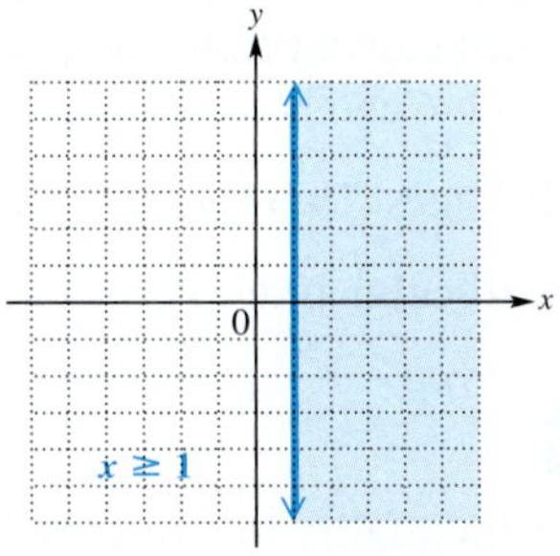

33. $y < 5$

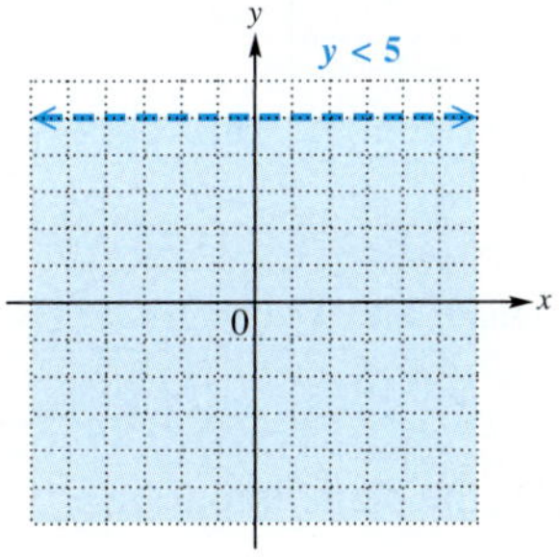

34. $y < -3$

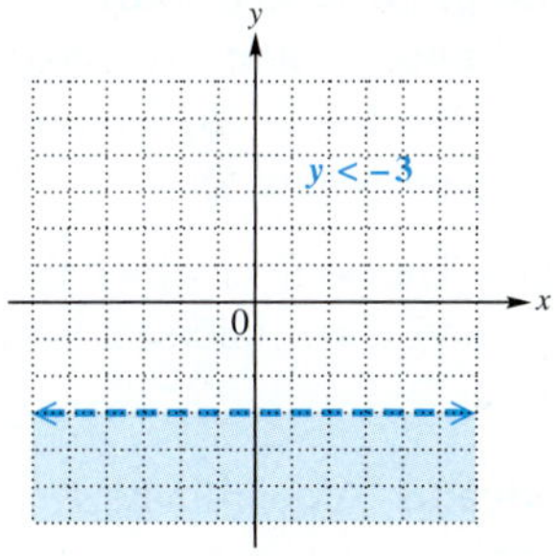

35. $y \ge 4x$

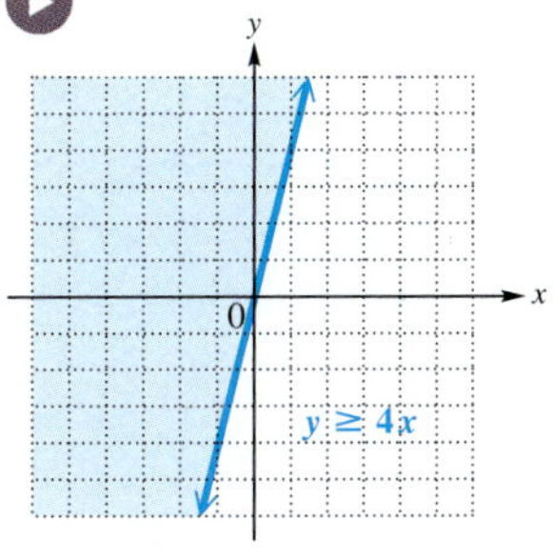

36. $y \le 2x$

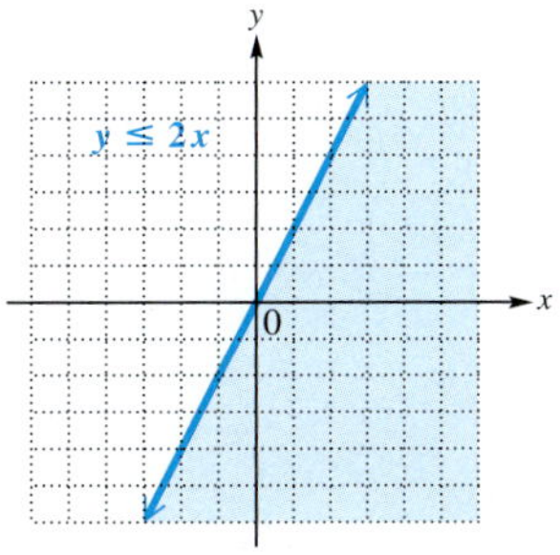

37. $x < -2y$

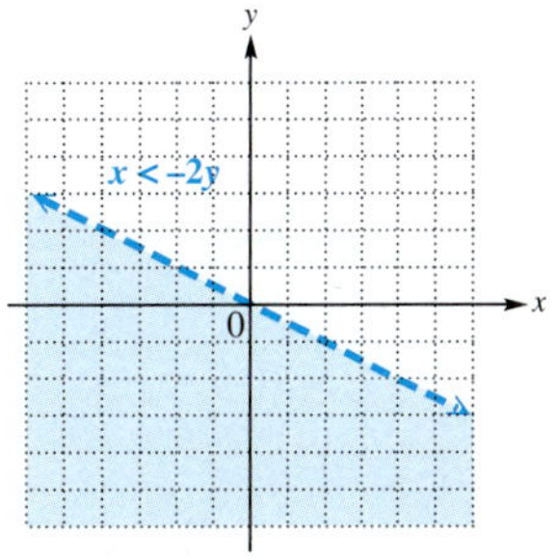

38. $x > -5y$

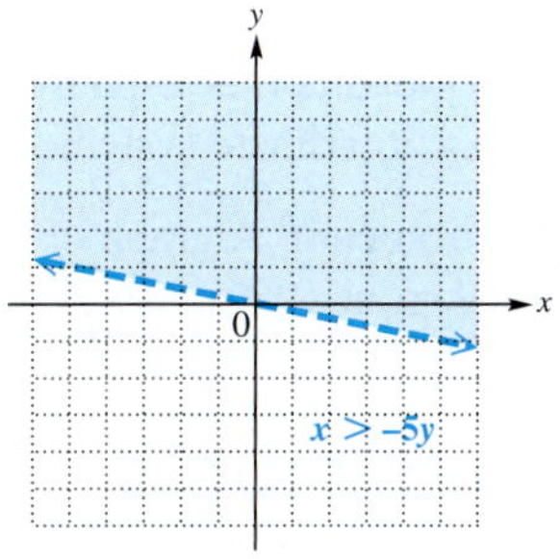

Chapter 3 Summary

Key Terms

3.1

bar graph A bar graph is a series of bars used to show comparisons between two categories of data.

line graph A line graph consists of a series of points that are connected with line segments and is used to show changes or trends in data.

linear equation in two variables An equation that can be written in the form $Ax + By = C$ is a linear equation in two variables. (A and B are real numbers that cannot both be 0.)

ordered pair A pair of numbers written between parentheses in which order is important is an ordered pair.

table of values A table showing selected ordered pairs of numbers that satisfy an equation is a table of values.

x	y
0	4
2	0
1	2

Table of values for $2x + y = 4$

***x*-axis** The horizontal axis in a coordinate system is the x-axis.

***y*-axis** The vertical axis in a coordinate system is the y-axis.

rectangular (Cartesian) coordinate system An x-axis and y-axis at right angles form a coordinate system.

origin The point at which the x-axis and y-axis intersect is the origin.

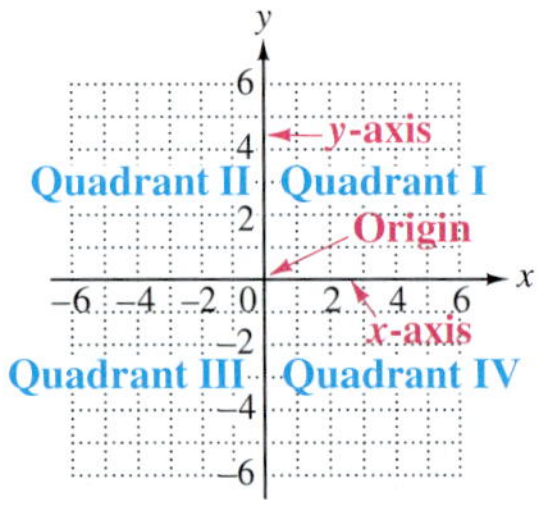

quadrants A coordinate system divides the plane into four regions or quadrants.

plane A flat surface determined by two intersecting lines is a plane.

coordinates The numbers in an ordered pair are the coordinates of the corresponding point.

plot To plot an ordered pair is to find the corresponding point on a coordinate system.

scatter diagram A graph of ordered pairs of data is a scatter diagram.

3.2

graph The graph of an equation is the set of all points that correspond to the ordered pairs that satisfy the equation.

graphing The process of plotting the ordered pairs that satisfy a linear equation and drawing a line through them is called graphing.

***y*-intercept** If a graph intersects the y-axis at k, then the y-intercept is $(0, k)$.

***x*-intercept** If a graph intersects the x-axis at k, then the x-intercept is $(k, 0)$.

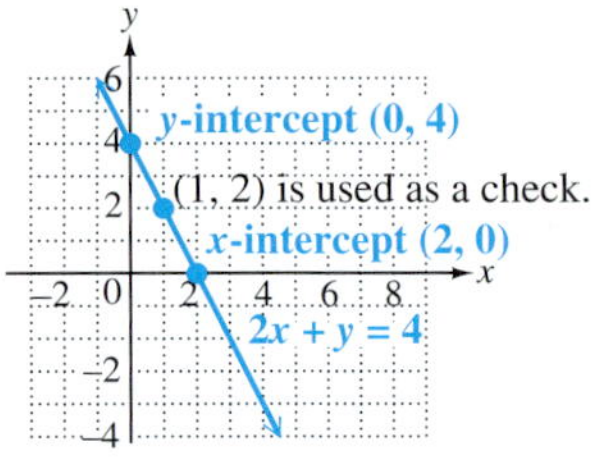

Graph of $2x + y = 4$

3.3

rise Rise is the vertical change between two different points on a line.

run Run is the horizontal change between two different points on a line.

slope The slope of a line is the ratio of the change in y compared to the change in x when moving along the line from one point to another.

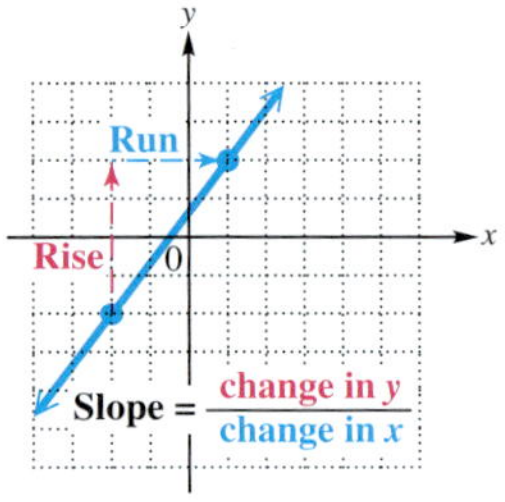

parallel lines Two lines in a plane that never intersect are parallel.

perpendicular lines Perpendicular lines intersect at a 90° angle.

3.5

linear inequality in two variables An inequality that can be written in the form $Ax + By < C$, $Ax + By > C$, $Ax + By \leq C$, or $Ax + By \geq C$ is a linear inequality in two variables.

boundary line In the graph of a linear inequality, the boundary line separates the region that satisfies the inequality from the region that does not satisfy the inequality.

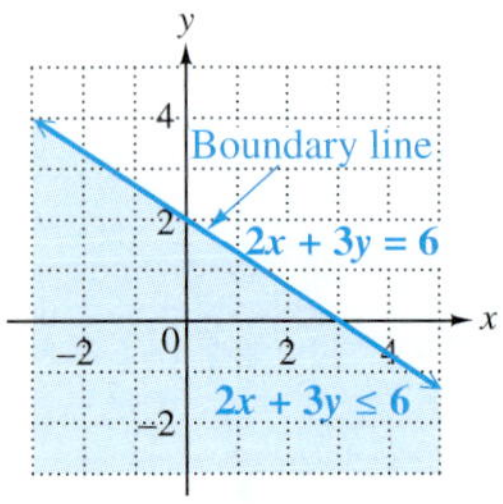

New Symbols

(x, y) ordered pair

(x_1, y_1) subscript notation; x-sub-one, y-sub-one

m slope

Test Your Word Power

See how well you have learned the vocabulary in this chapter.

1. An **ordered pair** is a pair of numbers written
   - **A.** in numerical order between brackets
   - **B.** between parentheses or brackets
   - **C.** between parentheses in which order is important
   - **D.** between parentheses in which order does not matter.

2. The **coordinates** of a point are
   - **A.** the numbers in the corresponding ordered pair
   - **B.** the solution of an equation
   - **C.** the values of the x- and y-intercepts
   - **D.** the graph of the point.

3. An **intercept** is
   - **A.** the point where the x-axis and y-axis intersect
   - **B.** a pair of numbers written in parentheses in which order is important
   - **C.** one of the four regions determined by a rectangular coordinate system
   - **D.** the point where a graph intersects the x-axis or the y-axis.

4. The **slope** of a line is
   - **A.** the measure of the run over the rise of the line
   - **B.** the distance between two points on the line
   - **C.** the ratio of the change in y to the change in x along the line
   - **D.** the horizontal change compared to the vertical change of two points on the line.

5. Two lines in a plane are **parallel** if
   - **A.** they represent the same line
   - **B.** they never intersect
   - **C.** they intersect at a 90° angle
   - **D.** one has a positive slope and one has a negative slope.

6. Two lines in a plane are **perpendicular** if
   - **A.** they represent the same line
   - **B.** they never intersect
   - **C.** they intersect at a 90° angle
   - **D.** one has a positive slope and one has a negative slope.

Answers to Test Your Word Power

1. C; *Examples:* $(0, 3)$, $(3, 8)$, $(4, 0)$

2. A; *Example:* The point associated with the ordered pair $(1, 2)$ has x-coordinate 1 and y-coordinate 2.

3. D; *Example:* The graph of the equation $4x - 3y = 12$ has x-intercept $(3, 0)$ and y-intercept $(0, -4)$.

4. C; *Example:* The line through $(3, 6)$ and $(5, 4)$ has slope $\frac{4-6}{5-3} = \frac{-2}{2} = -1$.

5. B; *Example:* See **Figure 27** in **Section 3.3.**

6. C; *Example:* See **Figure 28** in **Section 3.3.**

Quick Review

Concepts	Examples
3.1 Linear Equations in Two Variables; The Rectangular Coordinate System	
An ordered pair is a solution of an equation if it makes the equation a true statement.	Is $(2, -5)$ or $(0, -6)$ a solution of $4x - 3y = 18$? $4(2) - 3(-5) \stackrel{?}{=} 18$ $8 + 15 \stackrel{?}{=} 18$ $23 = 18$ False $(2, -5)$ is not a solution. $4(0) - 3(-6) \stackrel{?}{=} 18$ $0 + 18 \stackrel{?}{=} 18$ $18 = 18$ ✓ True $(0, -6)$ is a solution.
If a value of either variable in an equation is given, the value of the other variable can be found by substitution.	Complete the ordered pair $(0, \_\_\_\_)$ for $3x = y + 4$. $3(0) = y + 4$ Let $x = 0$. $0 = y + 4$ Multiply. $-4 = y$ Subtract 4. The ordered pair is $(0, -4)$.

(continued)

Concepts	Examples

3.1 Linear Equations in Two Variables; The Rectangular Coordinate System *(continued)*

To plot the ordered pair $(-3, 4)$, start at the origin, move 3 units to the left, and from there move 4 units up.

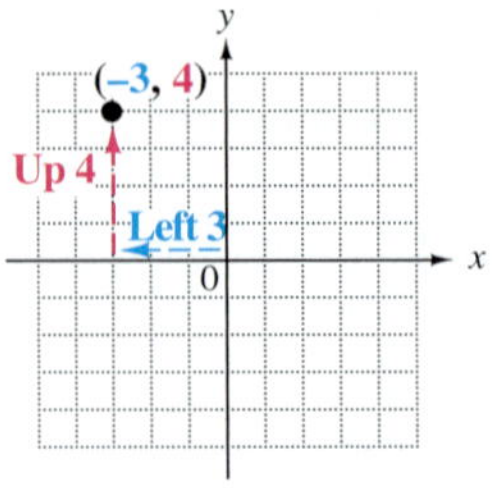

3.2 Graphing Linear Equations in Two Variables

To graph a linear equation, follow these steps.

Step 1 Find at least two ordered pairs that are solutions of the equation. (The intercepts are good choices.)

Step 2 Plot the corresponding points.

Step 3 Draw a straight line through the points.

Graph $x - 2y = 4$.

x	y
0	-2
4	0
-2	-3

Intercepts (first two rows)

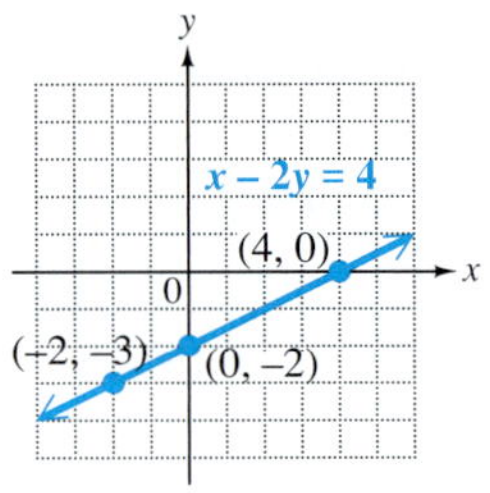

The graph of $Ax + By = 0$ passes through the origin. In this case, find and plot at least one other point that satisfies the equation. Then draw the line through these points.

The graph of $y = k$ is a horizontal line through $(0, k)$.

The graph of $x = k$ is a vertical line through $(k, 0)$.

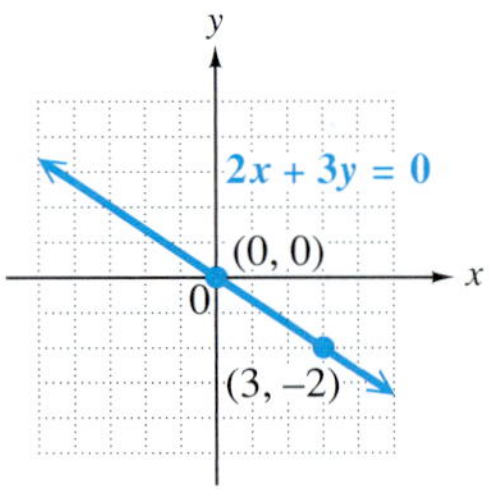

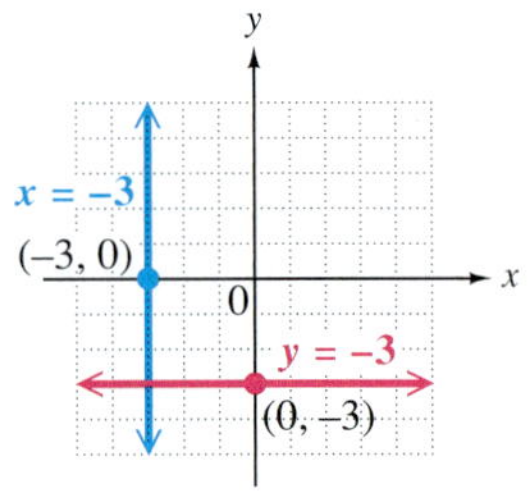

3.3 The Slope of a Line

The slope m of the line passing through the points (x_1, y_1) and (x_2, y_2) is defined as follows.

$$m = \frac{\textbf{change in } y}{\textbf{change in } x} = \frac{y_2 - y_1}{x_2 - x_1} \quad (\text{where } x_1 \neq x_2)$$

The line through $(-2, 3)$ and $(4, -5)$ has slope

$$m = \frac{-5 - 3}{4 - (-2)} = \frac{-8}{6} = -\frac{4}{3}.$$

Horizontal lines have slope 0.

Vertical lines have undefined slope.

The line $y = -2$ has slope **0**.

The line $x = 4$ has **undefined slope**.

To find the slope of a line from its equation, solve for y. The slope is the coefficient of x.

Find the slope of the graph of $3x - 4y = 12$.

$$-4y = -3x + 12 \quad \text{Add } -3x.$$

$$y = \frac{3}{4}x - 3 \quad \text{Divide by } -4.$$

↑ Slope (the coefficient $\frac{3}{4}$)

Parallel lines have the same slope.

The lines $y = 3x - 1$ and $y = 3x + 4$ are parallel because both have slope **3**.

The slopes of perpendicular lines are negative reciprocals (that is, their product is -1).

The lines $y = -3x - 1$ and $y = \frac{1}{3}x + 4$ are perpendicular because their slopes are -3 and $\frac{1}{3}$, and $-3\left(\frac{1}{3}\right) = -1$.

Concepts	Examples

3.4 Writing and Graphing Equations of Lines

Slope-Intercept Form
$y = mx + b$
m is the slope.
$(0, b)$ is the y-intercept.

Write an equation in slope-intercept form of the line with slope 2 and y-intercept $(0, -5)$.

$$y = 2x - 5$$

Point-Slope Form
$y - y_1 = m(x - x_1)$
m is the slope.
(x_1, y_1) is a point on the line.

Write an equation of the line with slope $-\frac{1}{2}$ through $(-4, 5)$.

$$y - y_1 = m(x - x_1) \quad \text{Point-slope form}$$
$$y - 5 = -\frac{1}{2}[x - (-4)] \quad \text{Substitute.}$$
$$y - 5 = -\frac{1}{2}(x + 4) \quad \text{Definition of subtraction}$$
$$y - 5 = -\frac{1}{2}x - 2 \quad \text{Distributive property}$$
$$y = -\frac{1}{2}x + 3 \quad \text{Add 5.}$$

Standard Form
$Ax + By = C$
A, B, and C are real numbers and A and B are not both 0.

The equation $y = -\frac{1}{2}x + 3$ is written in standard form as

$x + 2y = 6$. $\quad A = 1, B = 2,$ and $C = 6.$

3.5 Graphing Linear Inequalities in Two Variables

Step 1 Graph the line that is the boundary of the region. Draw a solid line if the inequality is $\leq$ or $\geq$. Draw a dashed line if the inequality is $<$ or $>$.

Step 2 Use any point not on the line as a test point. Substitute for x and y in the inequality. If the result is true, shade the region containing the test point. If the result is false, shade the other region.

Graph $2x + y \leq 5$.

Graph the line $2x + y = 5$. Make it solid because the symbol $\leq$ includes equality.

Use $(0, 0)$ as a test point.

$$2(0) + 0 \overset{?}{\leq} 5$$
$$0 \leq 5 \quad \text{True}$$

Shade the side of the line containing $(0, 0)$.

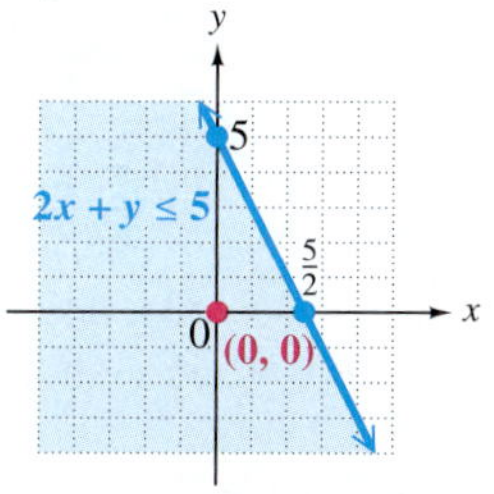

Chapter 3 Review Exercises

3.1 *The percents of first-year college students at two-year public institutions who returned for a second year for the years 2006 through 2011 are shown in the graph.*

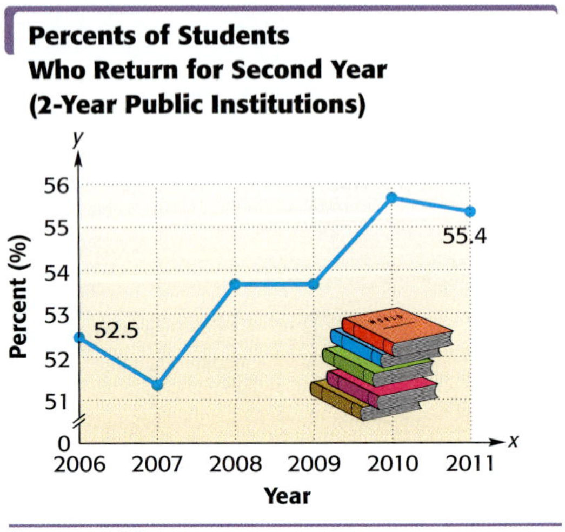

Source: ACT.

1. Write two ordered pairs of the form (year, percent) for the data shown in the graph for the years 2006 and 2011.

(2006, 52.5), (2011, 55.4)

2. What does the ordered pair (2000, 52.0) mean in the context of these problems?

In the year 2000, 52.0% of first-year college students at two-year public institutions returned for a second year.

3. Over which two consecutive years did the percent stay the same? Estimate this percent.

2008 to 2009; about 53.7%

4. Over which pairs of consecutive years did the percent increase? Estimate the increases.

2007 to 2008; 2009 to 2010; about 2% increase each time

Complete the given ordered pairs for each equation.

5. $y = 3x + 2$; $(-1, \underline{\ -1\ })$, $(0, \underline{\ 2\ })$, $(\underline{\ 1\ }, 5)$

6. $4x + 3y = 6$; $(0, \underline{\ 2\ })$, $(\underline{\ \frac{3}{2}\ }, 0)$, $(-2, \underline{\ \frac{14}{3}\ })$

7. $x = 3y$; $(0, \underline{\ 0\ })$, $(8, \underline{\ \frac{8}{3}\ })$, $(\underline{\ -9\ }, -3)$

8. $x - 7 = 0$; $(\underline{\ 7\ }, -3)$, $(\underline{\ 7\ }, 0)$, $(\underline{\ 7\ }, 5)$

Decide whether each ordered pair is a solution of the given equation.

9. $x + y = 7$; $(2, 5)$

yes

10. $2x + y = 5$; $(-1, 3)$

no

11. $3x - y = 4$; $\left(\frac{1}{3}, -3\right)$

yes

12. $x = -1$; $(0, -1)$

no

Name the quadrant in which each point lies. Then plot each ordered pair on the given coordinate system.

13. $(2, 3)$

quadrant I

14. $(-4, 2)$

quadrant II

15. $(3, 0)$

no quadrant

16. $(0, -6)$

no quadrant

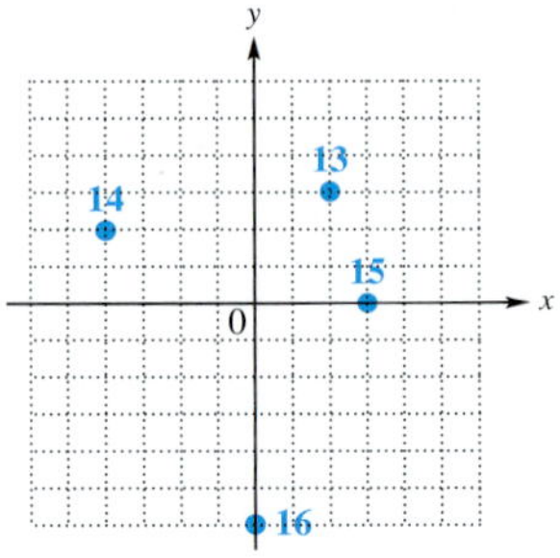

17. If $x > 0$ and $y < 0$, in what quadrant(s) must (x, y) lie? Explain.

x is positive in quadrants I and IV; y is negative in quadrants III and IV. Thus, if x is positive and y is negative, (x, y) must lie in quadrant IV.

18. On what axis does the point $(k, 0)$ lie for any real value of k? The point $(0, k)$? Explain.

In the ordered pair $(k, 0)$, the y-value is 0, so the point lies on the x-axis. In the ordered pair $(0, k)$, the x-value is 0, so the point lies on the y-axis.

3.2 *Graph each linear equation using intercepts.*

19. $2x - y = 3$

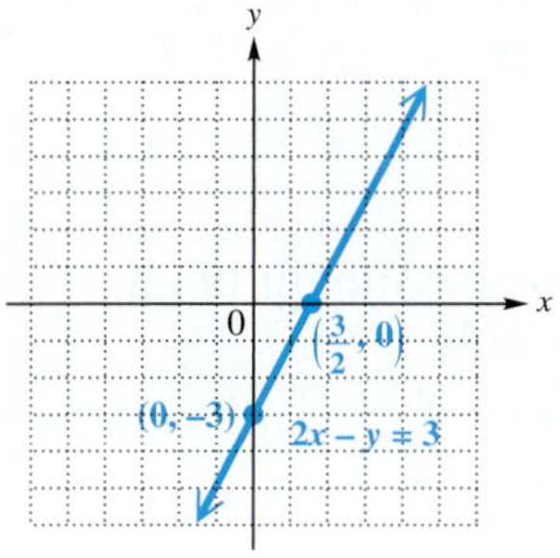

20. $x + 2y = -4$

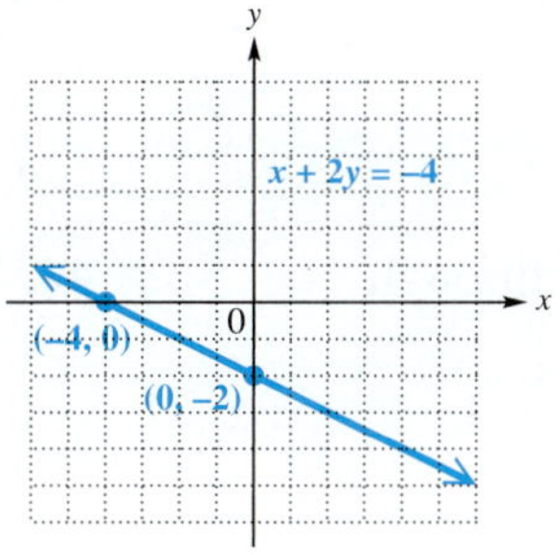

21. $x + y = 0$

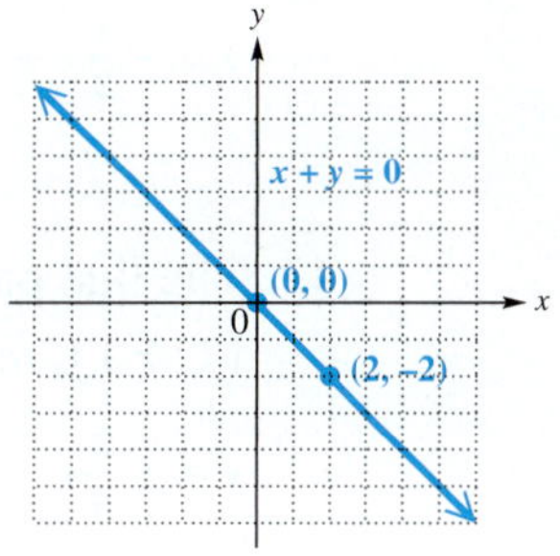

22. $y = 3x + 4$

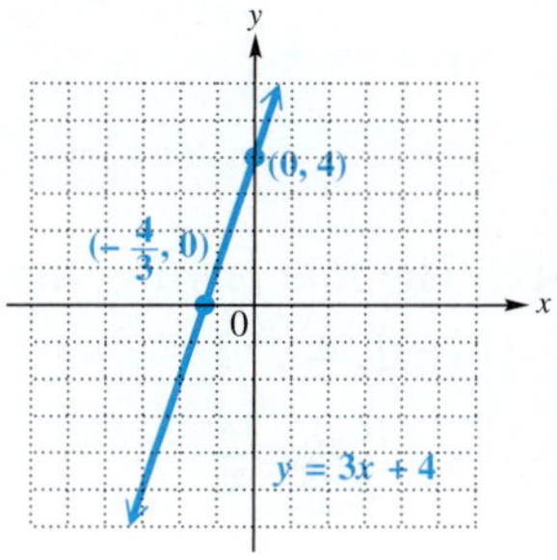

3.3 *Find the slope of each line.*

23. The line passing through $(2, 3)$ and $(-4, 6)$

$-\frac{1}{2}$

24. The line passing through $(0, 0)$ and $(-3, 2)$

$-\frac{2}{3}$

25. The line passing through $(0, 6)$ and $(1, 6)$

0

26. The line passing through $(2, 5)$ and $(2, 8)$

undefined

27. $y = 3x - 4$

3

28. $y = \frac{2}{3}x + 1$

$\frac{2}{3}$

29.

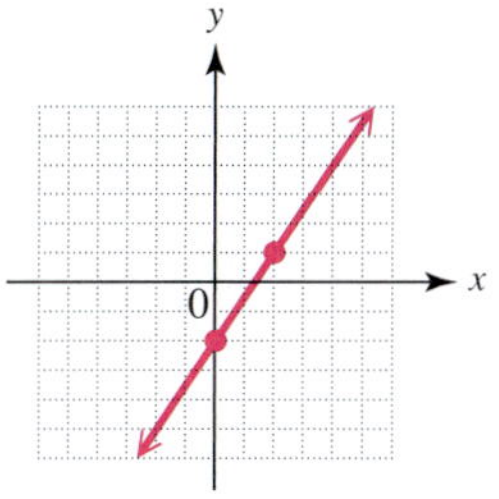

$\frac{3}{2}$

30.

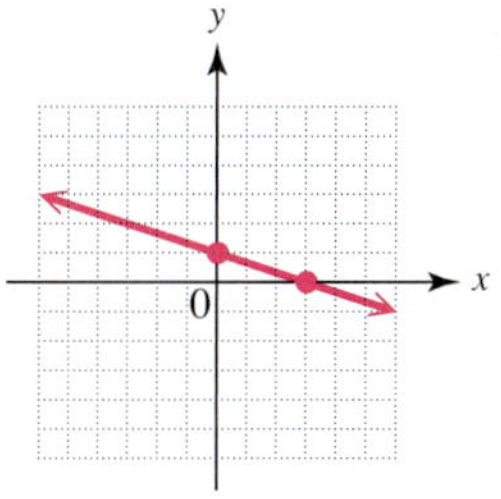

$-\frac{1}{3}$

31. The line having these points $\frac{3}{2}$

x	y
0	1
2	4
6	10

32. $x = 0$

undefined

33. $y = 4$

0

34. (a) A line parallel to the graph of $y = 2x + 3$ **2**

(b) A line perpendicular to the graph of $y = -3x + 3$

$\frac{1}{3}$

Decide whether each pair of lines is parallel, perpendicular, *or* neither.

35. $3x + 2y = 6$
$6x + 4y = 8$
parallel

36. $x - 3y = 1$
$3x + y = 4$
perpendicular

37. $x - 2y = 8$
$x + 2y = 8$
neither

38. CONCEPT CHECK What is the slope of a line perpendicular to a line with undefined slope? 0

3.4 *Write an equation of each line. Give the final answer in slope-intercept form (if possible).*

39. $m = -1,\ b = \frac{2}{3}$

$y = -x + \frac{2}{3}$

40. The line in **Exercise 30**

$y = -\frac{1}{3}x + 1$

41. The line passing through $(4, -3),\ m = 1$

$y = x - 7$

42. The line passing through $(-1, 4),\ m = \frac{2}{3}$

$y = \frac{2}{3}x + \frac{14}{3}$

43. The line passing through $(1, -1),\ m = -\frac{3}{4}$

$y = -\frac{3}{4}x - \frac{1}{4}$

44. The line passing through $(2, 1)$ and $(-2, 2)$

$y = -\frac{1}{4}x + \frac{3}{2}$

45. The line passing through $(-4, 1)$, slope 0

$y = 1$

46. The line passing through $\left(\frac{1}{3}, -\frac{3}{4}\right)$, undefined slope

$x = \frac{1}{3}$

47. Consider the linear equation $x + 3y = 15$.

(a) Write it in the form $y = mx + b$.

$y = -\frac{1}{3}x + 5$

(b) What is the slope? What is the *y*-intercept?

slope: $-\frac{1}{3}$; *y*-intercept: $(0, 5)$

(c) Use the slope and *y*-intercept to graph the line. Indicate two points on the graph.

See the graph.

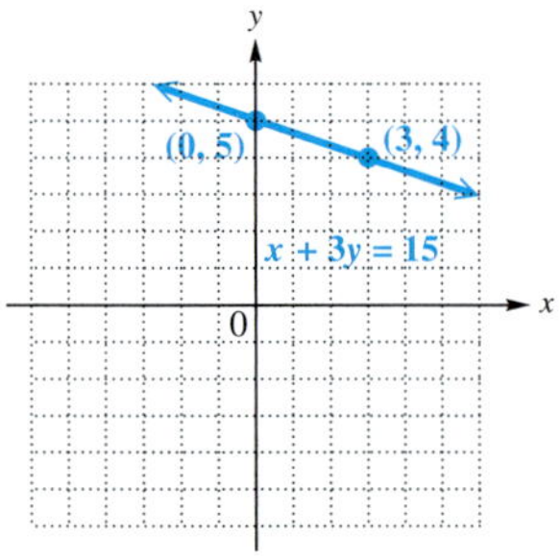

3.5 *Graph each linear inequality.*

48. $3x + 5y > 9$

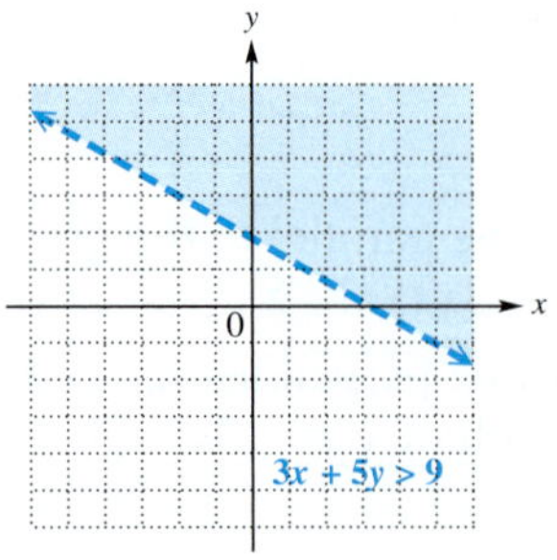

49. $2x - 3y > -6$

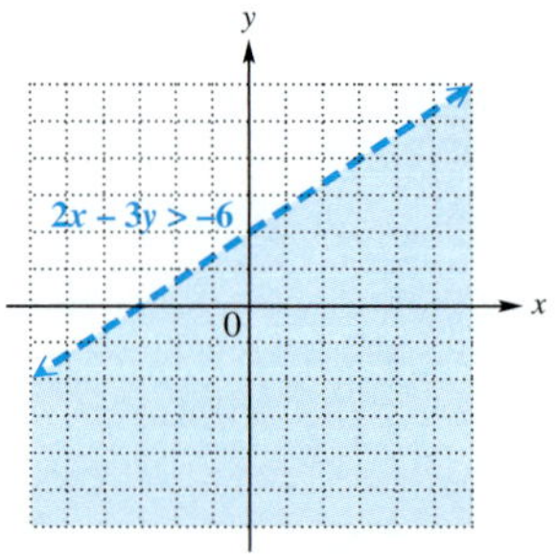

50. $x \geq -4$

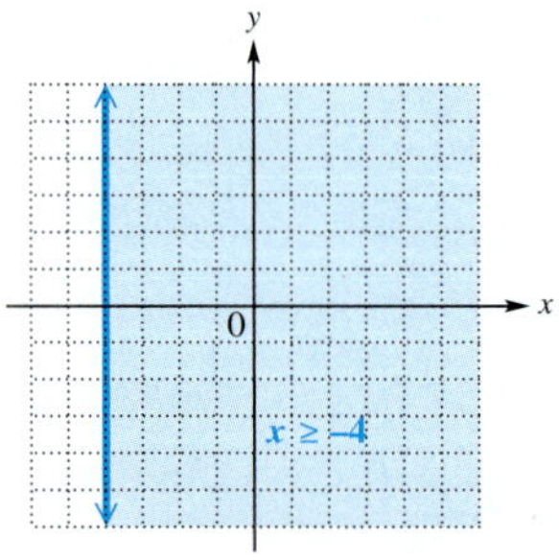

Mixed Review Exercises

CONCEPT CHECK *In Exercises 51–56, match each statement to the appropriate graph or graphs in choices A–D. Graphs may be used more than once.*

A.

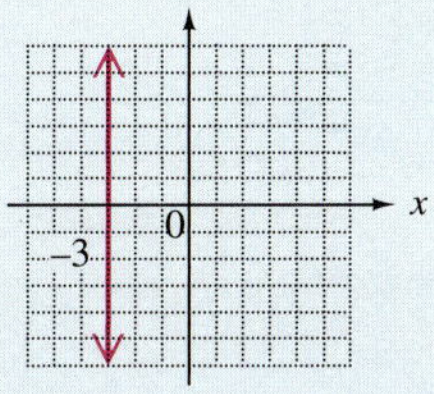

B.

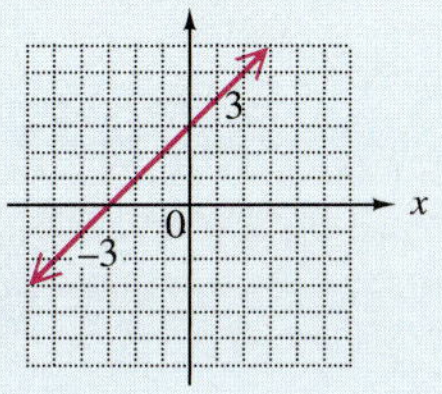

C.

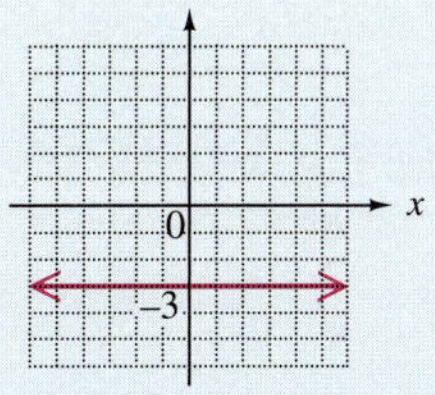

D.

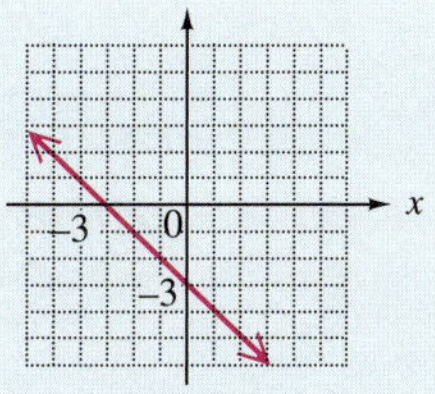

51. The line shown in the graph has undefined slope.

A

52. The graph of the equation has y-intercept $(0, -3)$.

C, D

53. The graph of the equation has x-intercept $(-3, 0)$.

A, B, D

54. The line shown in the graph has negative slope.

D

55. The graph is that of the equation $y = -3$.

C

56. The line shown in the graph has slope 1.

B

Find the intercepts and the slope of each line. Then use them to graph the line.

57. $y = -2x - 5$

y-intercept: $(0, -5)$

x-intercept: $\left(-\frac{5}{2}, 0\right)$

slope: -2

58. $x + 3y = 0$

y-intercept: $(0, 0)$

x-intercept: $(0, 0)$

slope: $-\frac{1}{3}$

59. $y - 5 = 0$

y-intercept: $(0, 5)$

x-intercept: none

slope: 0

60. $x + 4 = 3$

y-intercept: none

x-intercept: $(-1, 0)$

slope: undefined

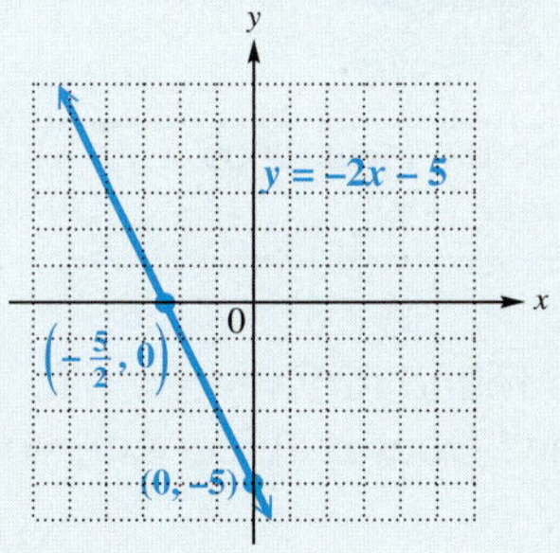

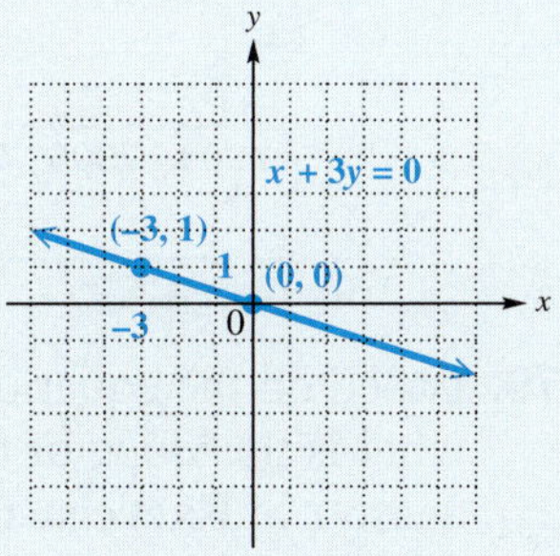

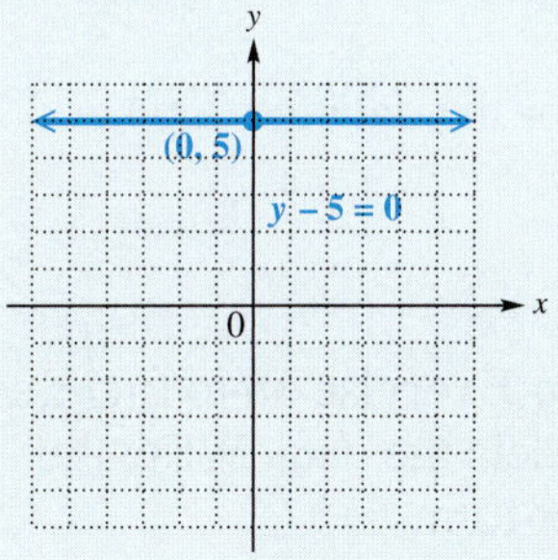

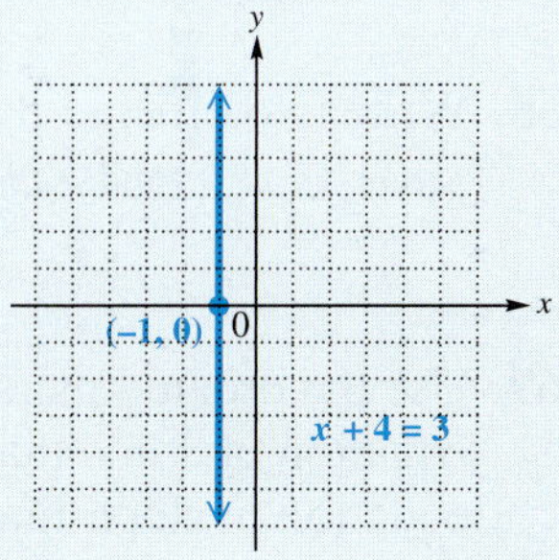

*Write an equation of each line. Give the final answer in **(a)** slope-intercept form and **(b)** standard form.*

61. $m = -\frac{1}{4}, b = -\frac{5}{4}$

(a) $y = -\frac{1}{4}x - \frac{5}{4}$

(b) $x + 4y = -5$

62. The line passing through $(8, 6)$, $m = -3$

(a) $y = -3x + 30$

(b) $3x + y = 30$

63. The line passing through $(3, -5)$ and $(-4, -1)$

(a) $y = -\frac{4}{7}x - \frac{23}{7}$

(b) $4x + 7y = -23$

Graph each inequality.

64. $y < -4x$

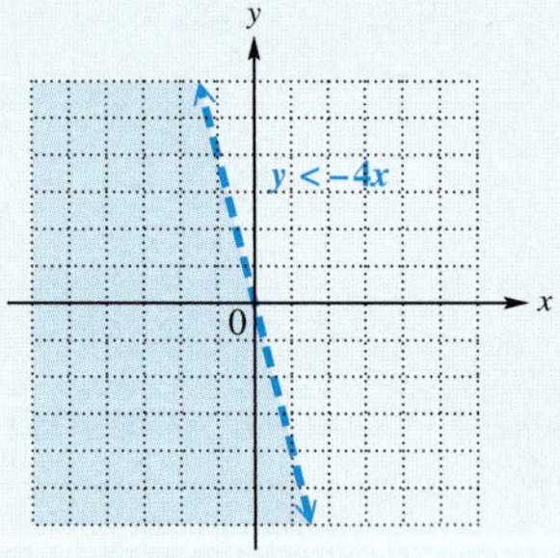

65. $x - 2y \leq 6$

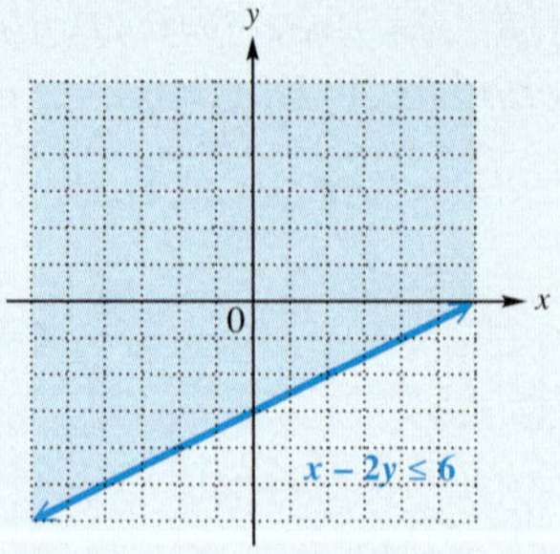

Relating Concepts (Exercises 66–72) *For Individual or Group Work*

The percents of four-year college students in public schools who earned a degree within five years of entry between 2004 and 2009 are shown in the graph. Use the graph to ***work Exercises 66–72 in order.***

66. What was the increase in percent from 2004 to 2009?

1.7%

Percents of Students Graduating Within 5 Years (Public Institutions)

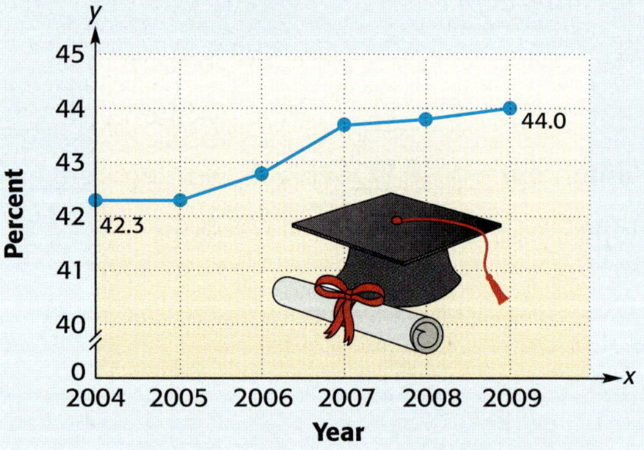

Source: ACT.

67. Since the points of the graph lie approximately in a linear pattern, a straight line can be used to model the data. Will this line have positive or negative slope? Explain.

Since the graph rises from left to right, the slope is positive.

68. Write two ordered pairs for the data for 2004 and 2009.

(2004, 42.3), (2009, 44.0)

69. Use the ordered pairs from **Exercise 68** to find the equation of a line that models the data. Write the equation in slope-intercept form.

$y = 0.34x - 639.06$

70. Based on the equation you found in **Exercise 69,** what is the slope of the line? Does it agree with your answer in **Exercise 67?**

0.34; yes

71. Use the equation from **Exercise 69** to approximate the percents for 2005 through 2008 to the nearest tenth, and complete the table.

Year	Percent
2005	42.6
2006	43.0
2007	43.3
2008	43.7

72. Use the equation from **Exercise 69** to estimate the percent for 2010. Can we be sure that this estimate is accurate?

44.3%; No. The equation is based on data only from 2004 through 2009.

Chapter 3 Test

Chapter Test Prep Video

The Chapter Test Prep Videos with test solutions are available on DVD, in MyMathLab, and on YouTube—search "LialIntroductoryAlg" and click on "Channels."

The line graph shows the overall unemployment rate in the U.S. civilian labor force for the years 2003 through 2010. Use the graph to work **Exercises 1–3.**

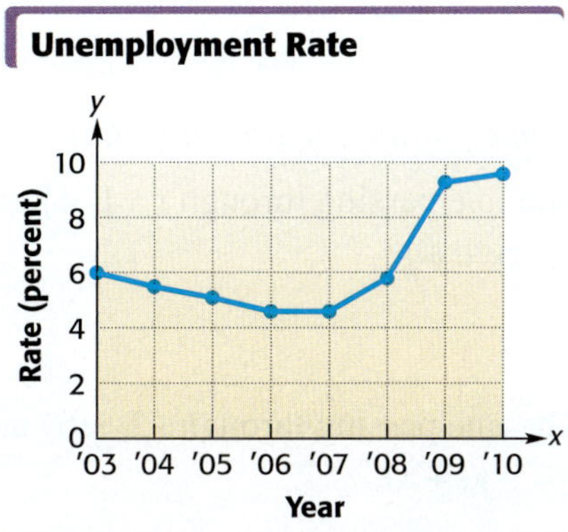

Source: Bureau of Labor Statistics.

1. Between which pairs of consecutive years did the unemployment rate decrease?

between 2003 and 2004, 2004 and 2005, and 2005 and 2006

2. What was the general trend in the unemployment rate between 2007 and 2010?

The unemployment rate was increasing.

3. Estimate the overall unemployment rate in 2008 and 2009. About how much did the unemployment rate increase between 2008 and 2009?

2008: 5.8%; 2009: 9.3%; increase: 3.5%

Graph each linear equation. Give the intercepts.

4. $3x + y = 6$

y-intercept: **(0, 6)**

x-intercept: **(2, 0)**

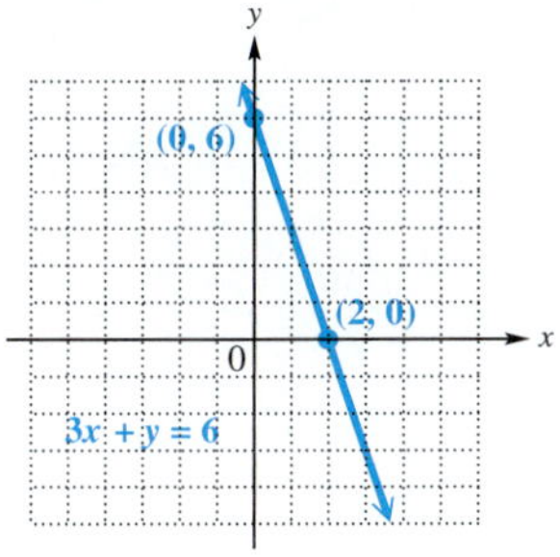

5. $y - 2x = 0$

y-intercept: **(0, 0)**

x-intercept: **(0, 0)**

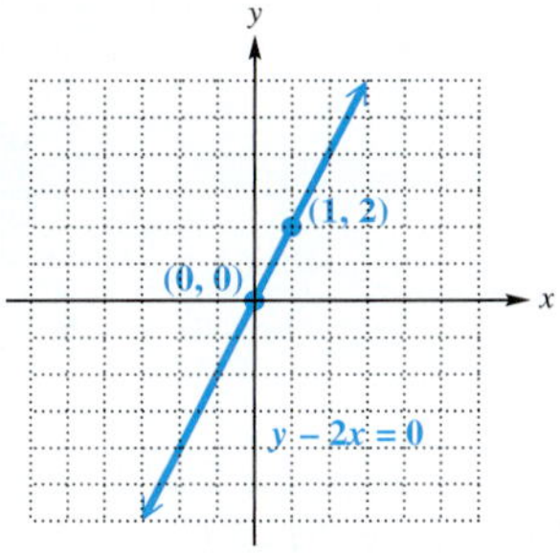

6. $x + 3 = 0$

y-intercept: **none**

x-intercept: **(−3, 0)**

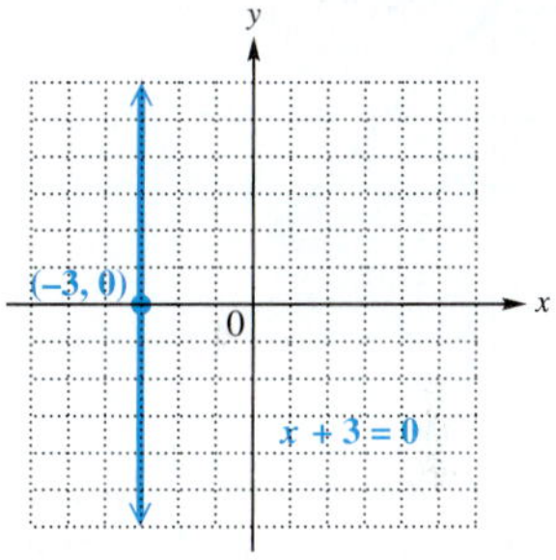

7. $y = 1$

y-intercept: **(0, 1)**

x-intercept: **none**

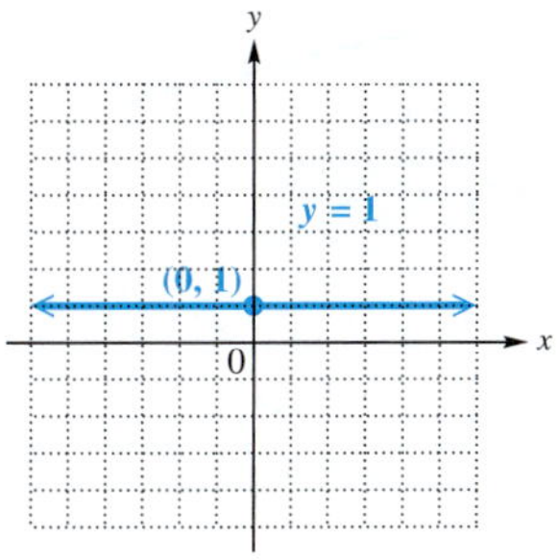

8. $x - y = 4$

y-intercept: **(0, −4)**

x-intercept: **(4, 0)**

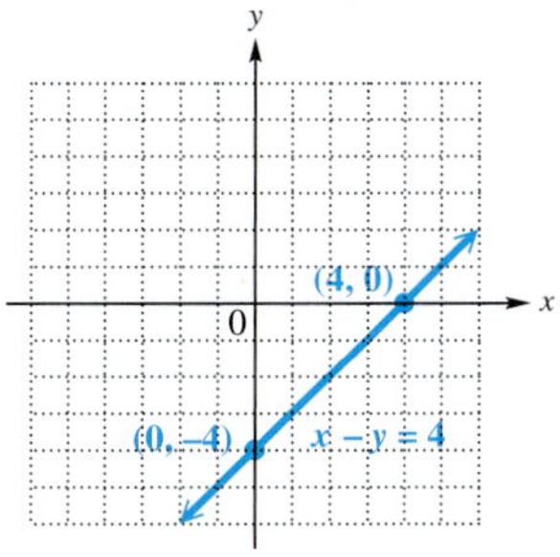

Find the slope of each line.

9. The line passing through $(-4, 6)$ and $(-1, -2)$

$-\frac{8}{3}$

10. $2x + y = 10$

-2

11. $x + 12 = 0$

undefined

12. $\frac{5}{2}$

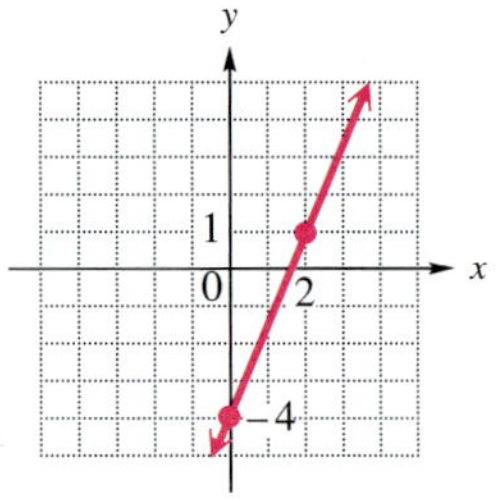

13. A line parallel to the graph of $y - 4 = 6$ 0

Write an equation for each line. Give the final answer in slope-intercept form.

14. The line passing through $(-1, 4)$, $m = 2$

$y = 2x + 6$

15. The line in **Exercise 12**

$y = \frac{5}{2}x - 4$

16. The line passing through $(2, -6)$ and $(1, 3)$

$y = -9x + 12$

17. x-intercept: $(3, 0)$; y-intercept: $\left(0, \frac{9}{2}\right)$

$y = -\frac{3}{2}x + \frac{9}{2}$

Graph each linear inequality.

18. $x + y \leq 3$

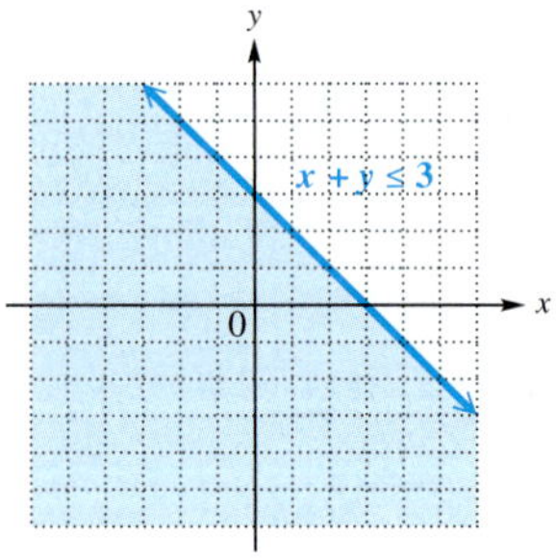

19. $3x - y > 0$

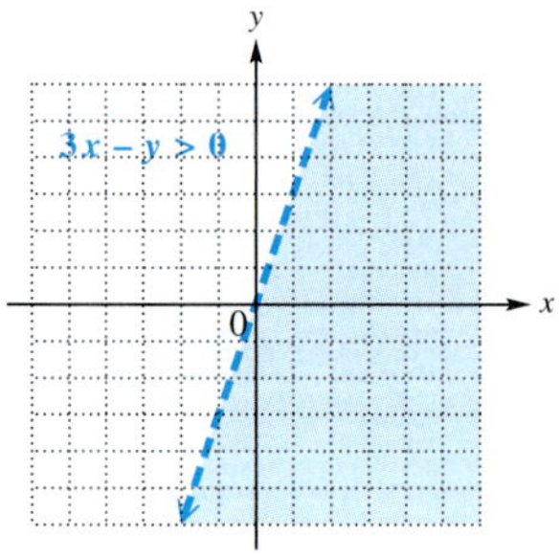

The graph shows worldwide snowmobile sales from 2000 through 2011, where 2000 corresponds to $x = 0$. Use the graph to work Exercises 20–24.

20. Is the slope of the line in the graph positive or negative? Explain.

The slope is negative since sales are decreasing.

Worldwide Snowmobile Sales

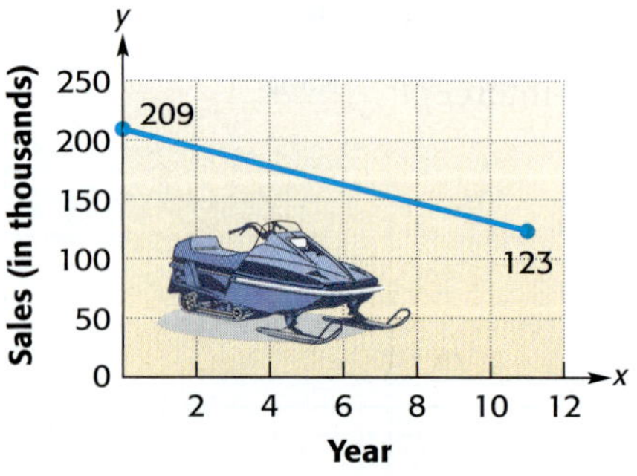

Source: www.snowmobile.org

21. Write two ordered pairs for the data points shown in the graph. Use them to find the slope of the line to the nearest tenth.

(0, 209), (11, 123); −7.8

22. Use the ordered pairs and slope from **Exercise 21** to find an equation of a line that models the data. Write the equation in slope-intercept form.

$y = -7.8x + 209$

23. Use the equation from **Exercise 22** to approximate worldwide snowmobile sales for 2007. How does your answer compare to the actual sales of 160.3 thousand?

154.4 thousand; The equation gives a number that is lower than the actual sales.

24. What does the ordered pair (11, 123) mean in the context of this problem?

In 2011, worldwide snowmobile sales were 123 thousand.

Chapters R–3 Cumulative Review Exercises

Perform the indicated operations.

1. $10\frac{5}{8} - 3\frac{1}{10}$ $\frac{301}{40}$, or $7\frac{21}{40}$

2. $\frac{3}{4} \div \frac{1}{8}$ 6

3. $5 - (-4) + (-2)$ 7

4. $\frac{(-3)^2 - (-4)(2^4)}{5(2) - (-2)^3}$ $\frac{73}{18}$, or $4\frac{1}{18}$

5. *True* or *false*? $\frac{4(3-9)}{2-6} \geq 6$ true

6. Find the value of $xz^3 - 5y^2$ for $x = -2$, $y = -3$, and $z = -1$. -43

7. What property does $3(-2 + x) = -6 + 3x$ illustrate? distributive property

8. Simplify $-4p - 6 + 3p + 8$ by combining like terms. $-p + 2$

Solve.

9. $V = \frac{1}{3}\pi r^2 h$ for h $h = \frac{3V}{\pi r^2}$

10. $6 - 3(1 + a) = 2(a + 5) - 2$ $\{-1\}$

11. $-(m - 3) = 5 - 2m$ $\{2\}$

12. $\frac{y-2}{3} = \frac{2y+1}{5}$ $\{-13\}$

Solve each inequality. Write the solution set in interval notation and graph it.

13. $-2.5x < 6.5$

$(-2.6, \infty)$

14. $4(x + 3) - 5x < 12$

$(0, \infty)$

15. $\frac{2}{3}t - \frac{1}{6}t \leq -2$

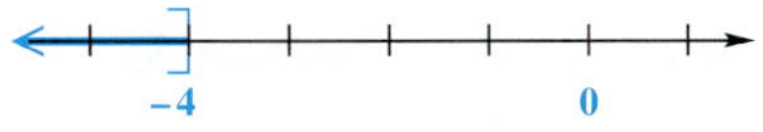

$(-\infty, -4]$

Solve each problem.

16. In 2010, a person with a bachelor's degree could expect to earn \$21,424 more each year than someone with a high school diploma. Together the individuals would earn \$86,528. How much could a person at each level of education expect to earn? (*Source*: Bureau of Labor Statistics.)

high school diploma: \$32,552; bachelor's degree: \$53,976

17. Mount Mayon in the Philippines is the most perfectly shaped conical volcano in the world. Its base is a circle with circumference 80 mi, and it has a height of about 8100 ft. (One mile is 5280 ft.) Find the radius of the circular base to the nearest mile. (*Source:* www.britannica.hk)

Circumference = 80 mi

13 mi

18. The winning times in seconds for the women's 1000 m speed skating event in the Winter Olympics for the years 1960 through 2010 can be closely approximated by the linear equation

$$y = -0.4336x + 94.30,$$

where x is the number of years since 1960. That is, $x = 4$ represents 1964, $x = 8$ represents 1968, and so on. (*Source: World Almanac and Book of Facts.*)

(a) Use this equation to complete the table of values. Round times to the nearest hundredth of a second.

x	y
20	**85.63**
38	**77.82**
42	**76.09**

(b) Write ordered pairs for the data given in the table. **(20, 85.63), (38, 77.82), (42, 76.09)**

(c) In the context of this problem, what does the ordered pair (32, 80.42) mean?

In 1992, the winning time was approximately 80.42 sec.

19. In 2010, approximately 60 million people from other countries visited the United States. The circle graph shows the distribution of these international visitors by their home countries or regions.

International Travelers to the United States, 2010

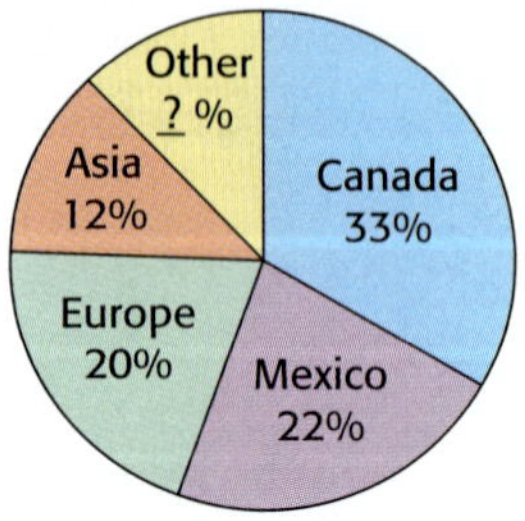

Source: U.S. Department of Commerce.

(a) About how many travelers visited the United States from Canada? **19.8 million**

(b) About how many travelers visited the United States from Mexico? **13.2 million**

(c) About how many travelers visited the United States from places other than Canada, Mexico, Europe, or Asia? **7.8 million**

Consider the linear equation $-3x + 4y = 12$. *Find the following.*

20. The x- and y-intercepts

(−4, 0); (0, 3)

21. The slope

$\frac{3}{4}$

22. The graph

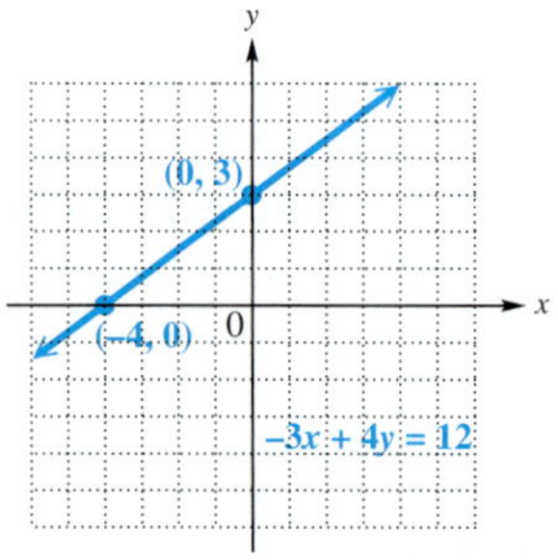

23. The y-value of the point having x-value 4

6

24. Are the lines with equations $x + 5y = -6$ and $y = 5x - 8$ *parallel, perpendicular,* or *neither?*

perpendicular

Write an equation of each line. Give the final answer in slope-intercept form if possible.

25. The line passing through $(2, -5)$ with slope 3

$y = 3x - 11$

26. The line passing through $(0, 4)$ and $(2, 4)$

$y = 4$

4 Systems of Linear Equations and Inequalities

The point of intersection of two lines can be found using a *system of linear equations*, the subject of this chapter.

4.1 Solving Systems of Linear Equations by Graphing

OBJECTIVES

1. Decide whether a given ordered pair is a solution of a system.
2. Solve linear systems by graphing.
3. Solve special systems by graphing.
4. Identify special systems without graphing.

A **system of linear equations**, also called a **linear system,** consists of two or more linear equations with the same variables.

$$\begin{aligned} 2x + 3y &= 4 \\ 3x - y &= -5 \end{aligned} \qquad \begin{aligned} x + 3y &= 1 \\ -y &= 4 - 2x \end{aligned} \qquad \begin{aligned} x - y &= 1 \\ y &= 3 \end{aligned} \qquad \text{Linear systems}$$

In the system on the right, think of $y = 3$ as an equation in two variables by writing it as $\mathbf{0}x + y = 3$.

OBJECTIVE 1 Decide whether a given ordered pair is a solution of a system. A **solution of a system** of linear equations is an ordered pair that makes both equations true at the same time. A solution of an equation is said to *satisfy* the equation.

EXAMPLE 1 Determining Whether an Ordered Pair Is a Solution

Determine whether the ordered pair $(4, -3)$ is a solution of each system.

(a) $x + 4y = -8$

$3x + 2y = 6$

To decide whether $(4, -3)$ is a solution of the system, substitute 4 for x and -3 for y in each equation.

$x + 4y = -8$		$3x + 2y = 6$	
$4 + 4(-3) \stackrel{?}{=} -8$	Substitute.	$3(4) + 2(-3) \stackrel{?}{=} 6$	Substitute.
$4 + (-12) \stackrel{?}{=} -8$	Multiply.	$12 + (-6) \stackrel{?}{=} 6$	Multiply.
$-8 = -8$ ✓	True	$6 = 6$ ✓	True

Because $(4, -3)$ satisfies *both* equations, it is a solution of the system.

(b) $2x + 5y = -7$

$3x + 4y = 2$

Again, substitute 4 for x and -3 for y in each equation.

$2x + 5y = -7$		$3x + 4y = 2$	
$2(4) + 5(-3) \stackrel{?}{=} -7$	Substitute.	$3(4) + 4(-3) \stackrel{?}{=} 2$	Substitute.
$8 + (-15) \stackrel{?}{=} -7$	Multiply.	$12 + (-12) \stackrel{?}{=} 2$	Multiply.
$-7 = -7$ ✓	True	$0 = 2$	False

The ordered pair $(4, -3)$ is *not* a solution of this system because it does not satisfy the second equation.

◀ Work Problem 1 at the Side.

OBJECTIVE 2 Solve linear systems by graphing. The set of all ordered pairs that are solutions of a system is its **solution set.** One way to find the solution set of a system of two linear equations is to graph both equations on the same axes. Any intersection point would be on both lines and would therefore be a solution of *both* equations. ***Thus, the coordinates of any point where the lines intersect give a solution of the system.***

1 Determine whether the given ordered pair is a solution of the system.

GS **(a)** $(2, 5)$

$3x - 2y = -4$

$5x + y = 15$

$3x - 2y = -4$

$3(\_\_\_) - 2(\_\_\_) \stackrel{?}{=} -4$

$5x + y = 15$

$5(2) + \_\_\_ \stackrel{?}{=} \_\_\_$

$(2, 5)$ (*is / is not*) a solution.

(b) $(1, -2)$

$x - 3y = 7$

$4x + y = 5$

$(1, -2)$ (*is / is not*) a solution.

Answers

1. **(a)** 2; 5; 5; 15; is **(b)** is not

The graph in **Figure 1** shows that the solution of the system in **Example 1(a)** is the intersection point $(4, -3)$. ***The solution (point of intersection) is always written as an ordered pair.*** Because *two different* straight lines can intersect at no more than one point, there can never be more than one solution for such a system.

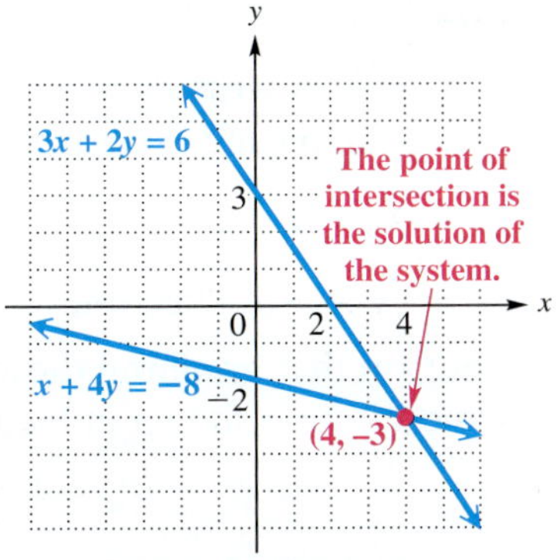

Figure 1

EXAMPLE 2 Solving a System by Graphing

Solve the system of equations by graphing both equations on the same axes.

$$2x + 3y = 4$$
$$3x - y = -5$$

We graph these two equations by plotting several points for each line. Recall from **Section 3.2** that the intercepts are often convenient choices. As review, we show finding the intercepts for $2x + 3y = 4$.

To find the y-intercept, let $x = 0$.

$$2x + 3y = 4$$
$$2(0) + 3y = 4 \quad \text{Let } x = 0.$$
$$3y = 4$$
$$y = \frac{4}{3} \quad \text{Divide by 3.}$$

To find the x-intercept, let $y = 0$.

$$2x + 3y = 4$$
$$2x + 3(0) = 4 \quad \text{Let } y = 0.$$
$$2x = 4$$
$$x = 2 \quad \text{Divide by 2.}$$

The tables show the intercepts and a check point for each graph.

$2x + 3y = 4$

x	y	
0	$\frac{4}{3}$	← y-intercept
2	0	← x-intercept
−2	$\frac{8}{3}$	

Find a third ordered pair as a check.

$3x - y = -5$

x	y	
0	5	← y-intercept
$-\frac{5}{3}$	0	← x-intercept
−2	−1	

The lines in **Figure 2** suggest that the graphs intersect at the point $(-1, 2)$. We check by substituting -1 for x and 2 for y in *both* equations.

CHECK

$$2x + 3y = 4$$
$$2(-1) + 3(2) \stackrel{?}{=} 4$$
$$4 = 4 \quad ✓ \text{ True}$$
$$3x - y = -5$$
$$3(-1) - 2 \stackrel{?}{=} -5$$
$$-5 = -5 \quad ✓ \text{ True}$$

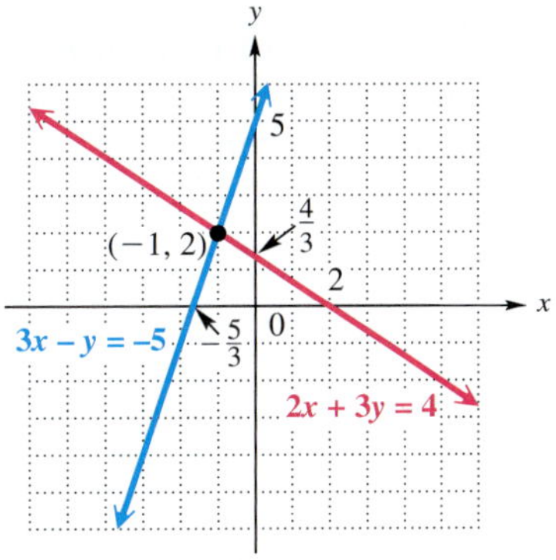

Figure 2

Because $(-1, 2)$ satisfies both equations, the solution set of this system is $\{(-1, 2)\}$.

Work Problem **2** at the Side. ▶

2 Solve each system of equations by graphing both equations on the same axes. Check each solution.

GS **(a)** $5x - 3y = 9$

$x + 2y = 7$

One of the lines is already graphed. Complete the table, and graph the other line.

$5x - 3y = 9$

x	y
0	−3
$\frac{9}{5}$	0
3	2

$x + 2y = 7$

x	y
0	___
___	0
−1	___

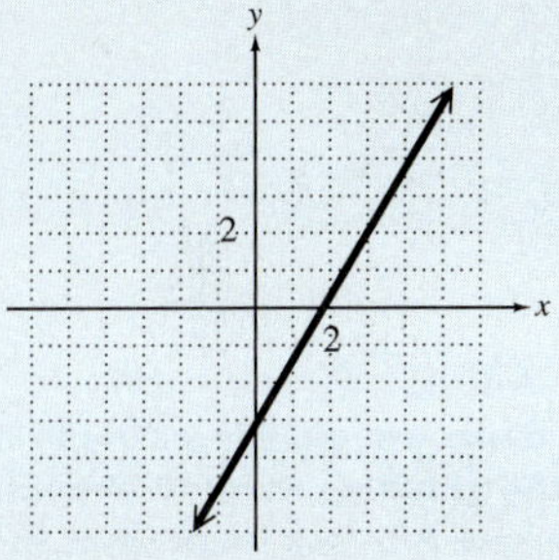

The point of intersection of the lines is the ordered pair ________, so the solution set of the system of equations is ________.

(b) $x + y = 4$

$2x - y = -1$

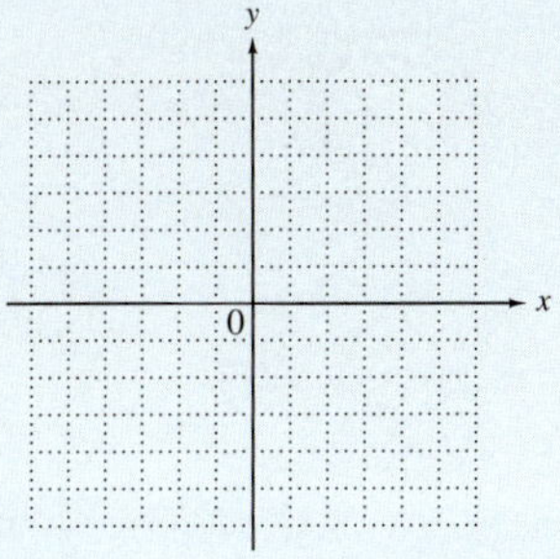

Teaching Tip

Review how to graph lines by determining *x*- and *y*-intercepts (where possible) from Section 3.2. Students *must* use a straightedge to solve systems of equations graphically.

Answers

2. **(a)** $\frac{7}{2}$; 7; 4; (3, 2); $\{(3, 2)\}$ **(b)** $\{(1, 3)\}$

Note

We can also write each equation in a system in slope-intercept form and use the slope and y-intercept to graph each line. For **Example 2,**

$2x + 3y = 4$ becomes $y = -\frac{2}{3}x + \frac{4}{3}$. y-intercept $\left(0, \frac{4}{3}\right)$; slope $-\frac{2}{3}$

$3x - y = -5$ becomes $y = 3x + 5$. y-intercept $(0, 5)$; slope 3, or $\frac{3}{1}$

Confirm that graphing these equations results in the same lines and the same solution shown in **Figure 2** on the preceding page.

Solving a Linear System by Graphing

Step 1 **Graph each equation** of the system on the same coordinate axes.

Step 2 **Find the coordinates of the point of intersection** of the graphs if possible, and write it as an ordered pair. This is the solution of the system.

Step 3 **Check** the ordered-pair solution in *both* of the original equations. If it satisfies *both* equations, write the solution set.

Teaching Tip

Students will quickly learn the drawbacks of the graphing method. A good oral exercise is to ask them what some of these drawbacks are.

CAUTION

With the graphing method, it may not be possible to determine from the graph the exact coordinates of the point that represents the solution, particularly if these coordinates are not integers. The graphing method does, however, show geometrically how solutions are found and is useful when approximate answers will suffice.

OBJECTIVE 3 Solve special systems by graphing. The graphs of the equations in a system may not intersect at all or may be the same line.

EXAMPLE 3 Solving Special Systems

Solve each system by graphing.

(a) $2x + y = 2$

$2x + y = 8$

The graphs of these lines are shown in **Figure 3.** The two lines are parallel and have no points in common. For such a system, there is no solution. We write the solution set as $\emptyset$.

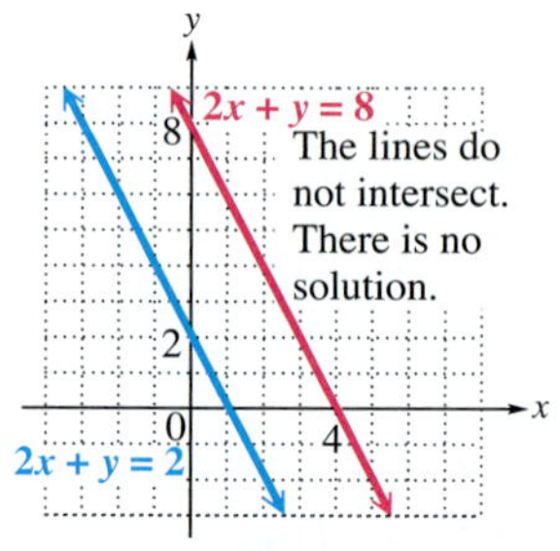

Figure 3

Continued on Next Page

(b) $2x + 5y = 1$

$6x + 15y = 3$

The graphs of these two equations are the same line. See **Figure 4.** We can obtain the second equation by multiplying each side of the first equation by 3. In this case, every point on the line is a solution of the system, and the solution set contains an infinite number of ordered pairs, each of which satisfies both equations of the system.

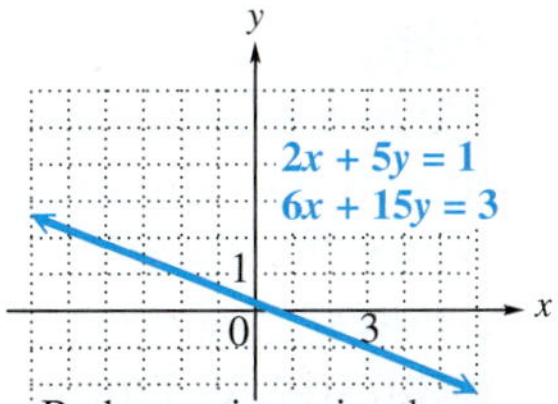

Both equations give the same graph. There is an infinite number of solutions.

Figure 4

We write the solution set as

$$\{(x, y) \mid 2x + 5y = 1\},$$

This is the first equation in the system. See the Note on the next page.

read "the set of ordered pairs (x, y) such that $2x + 5y = 1$." Recall from **Section 1.3** that this notation is called **set-builder notation.**

Work Problem 3 at the Side. ▶

The system in **Example 2** has exactly one solution. A system with at least one solution is a **consistent system.** A system of equations with no solution, such as the one in **Example 3(a),** is an **inconsistent system.**

The equations in **Example 2** are **independent equations** with different graphs. The equations of the system in **Example 3(b)** have the same graph and are equivalent. Because they are different forms of the same equation, these equations are **dependent equations.**

Three Cases for Solutions of Linear Systems with Two Variables

Case 1 The graphs intersect at exactly one point, which gives the (single) ordered-pair solution of the system. The **system is consistent** and the **equations are independent.** See **Figure 5(a).**

Case 2 The graphs are parallel lines, so there is no solution and the solution set is $\emptyset$. The **system is inconsistent** and the **equations are independent.** See **Figure 5(b).**

Case 3 The graphs are the same line. There is an infinite number of solutions, and the solution set is written in set-builder notation as

$$\{(x, y) \mid \underline{\qquad\qquad}\},$$

where one of the equations follows the | symbol. The **system is consistent** and the **equations are dependent.** See **Figure 5(c).**

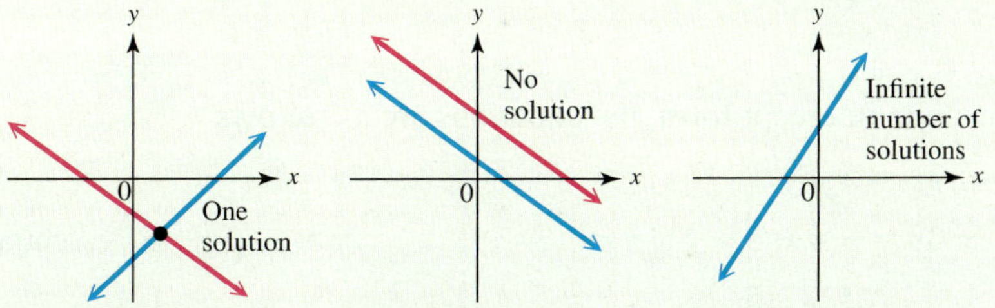

(a) The system is consistent. The equations are independent.

(b) The system is inconsistent. The equations are independent.

(c) The system is consistent. The equations are dependent.

Figure 5

3 Solve each system of equations by graphing both equations on the same axes.

GS **(a)** $3x - y = 4$

$6x - 2y = 12$

One of the lines is already graphed.

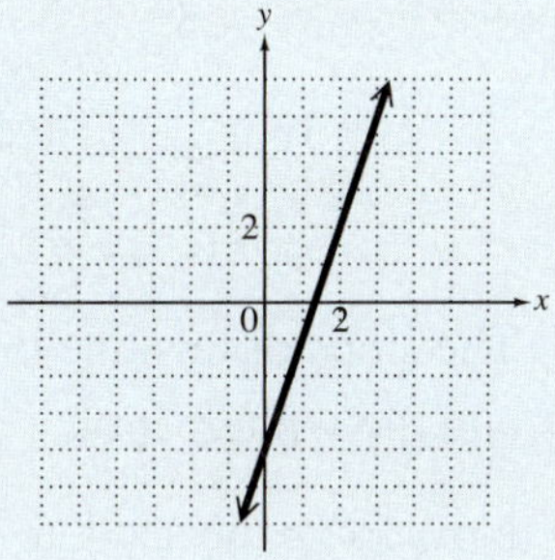

(b) $x - 3y = -2$

$2x - 6y = -4$

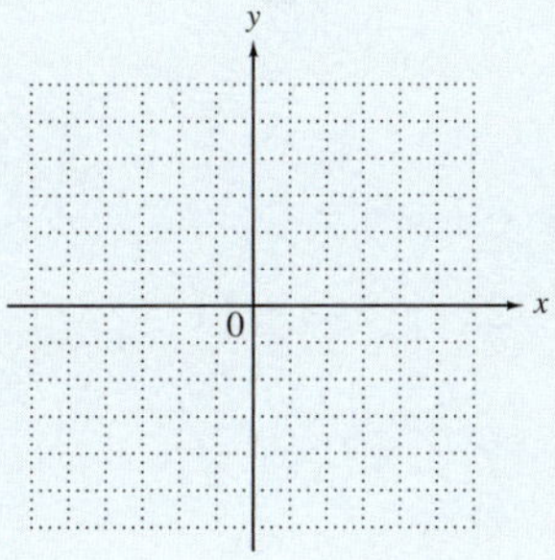

Answers

3. **(a)** $\emptyset$ **(b)** $\{(x, y) \mid x - 3y = -2\}$

Note

When a system has an infinite number of solutions, as in Case 3, either equation of the system can be used to write the solution set. ***We prefer to use the equation in standard form with integer coefficients having greatest common factor 1.*** If neither of the given equations is in this form, use an *equivalent* equation that is in standard form with integer coefficients having greatest common factor 1.

OBJECTIVE 4 Identify special systems without graphing.

EXAMPLE 4 Identifying the Three Cases by Using Slopes

Describe each system without graphing. State the number of solutions.

(a) $3x + 2y = 6$

$-2y = 3x - 5$

Write each equation in slope-intercept form, $y = mx + b$.

$3x + 2y = 6$ (Solve for y.)		$-2y = 3x - 5$ (Solve for y.)	
$2y = -3x + 6$	Subtract $3x$.	$y = -\frac{3}{2}x + \frac{5}{2}$	Divide *each* term by -2.
$y = -\frac{3}{2}x + 3$	Divide *each* term by 2.		

Both equations have slope $-\frac{3}{2}$ but they have different y-intercepts, $(0, 3)$ and $\left(0, \frac{5}{2}\right)$. Lines with the same slope are parallel (**Section 3.3**), so these equations have graphs that are parallel lines. Thus, the system has no solution.

(b) $2x - y = 4$

$x = \frac{y}{2} + 2$

Again, write each equation in slope-intercept form.

$2x - y = 4$		$x = \frac{y}{2} + 2$	
$-y = -2x + 4$	Subtract $2x$.	$\frac{y}{2} + 2 = x$	Interchange sides.
$\mathbf{y = 2x - 4}$	Multiply by -1.	$\mathbf{y = 2x - 4}$	Subtract 2. Multiply by 2.

The equations are exactly the same—their graphs are the same line. Thus, the system has an infinite number of solutions.

(c) $x - 3y = 5$

$2x + y = 8$

In slope-intercept form, the equations are as follows.

$x - 3y = 5$		$2x + y = 8$	
$-3y = -x + 5$	Subtract x.	$y = -2x + 8$	Subtract $2x$.
$y = \frac{1}{3}x - \frac{5}{3}$	Divide by -3.		

The graphs of these equations are neither parallel nor the same line, since the slopes are different. This system has exactly one solution.

◀ **Work Problem 4 at the Side.**

4 Describe each system without graphing. State the number of solutions.

GS **(a)** $2x - 3y = 5$

$3y = 2x - 7$

Solve the first equation for ____.

$2x - 3y = 5$

$-3y =$ ____ $+ 5$

$y =$ ____ $-$ ____

The slope of this line is ____, and the y-intercept is ____.

Now solve the second equation for y, and complete the solution.

Teaching Tip

Emphasize the Note at the top of this page with students. Work Exercise 8 in the exercise set as a class to reinforce the Note.

(b) $-x + 3y = 2$

$2x - 6y = -4$

(c) $6x + y = 3$

$2x - y = -11$

Answers

4. (a) y; $-2x$; $\frac{2}{3}x$; $\frac{5}{3}$; $\frac{2}{3}$; $\left(0, -\frac{5}{3}\right)$
The equations represent parallel lines. The system has no solution.

(b) The equations represent the same line. The system has an infinite number of solutions.

(c) The equations represent lines that are neither parallel nor the same line. The system has exactly one solution.

4.1 Exercises

FOR EXTRA HELP

Download the MyDashBoard App

MyMathLab®

CONCEPT CHECK *Complete each statement. The following terms may be used once, more than once, or not at all.*

consistent	system of linear equations	inconsistent	solution
ordered pair	independent	linear equation	dependent

1. A(n) __system of linear equations__ consists of two or more linear equations with the (*same* / *different*) variables.

2. A solution of a system of linear equations is a(n) __ordered pair__ that makes all equations of the system (*true* / *false*) at the same time.

3. The equations of two parallel lines form a(n) __inconsistent__ system that has (*one* / *no* / *infinitely many*) solution(s). The equations are __independent__ because their graphs are different.

4. If the graphs of a linear system intersect in one point, the point of intersection is the __solution__ of the system. The system is __consistent__ and the equations are independent.

5. If two equations of a linear system have the same graph, the equations are __dependent__. The system is __consistent__ and has (*one* / *no* / *infinitely many*) solution(s).

6. If a linear system is inconsistent, the graphs of the two equations are (*intersecting* / *parallel* / *the same*) line(s). The system has no __solution__.

CONCEPT CHECK *Work each problem.*

7. A student determined that the ordered pair $(1, -2)$ is a solution of the following system. His reasoning was that the ordered pair satisfies the first equation $x + y = -1$, since $1 + (-2) = -1$.
What Went Wrong?

$$x + y = -1$$
$$2x + y = 4$$

It is not a solution of the system because it is not a solution of the second equation, $2x + y = 4$.

8. The following system has infinitely many solutions. Write its solution set, using set-builder notation.

$$6x - 4y = 8$$
$$3x - 2y = 4$$

$\{(x, y) | 3x - 2y = 4\}$

9. Which ordered pair could be a solution of the system graphed? Why is it the only valid choice?

A. $(2, 2)$
B. $(-2, 2)$
C. $(-2, -2)$
D. $(2, -2)$

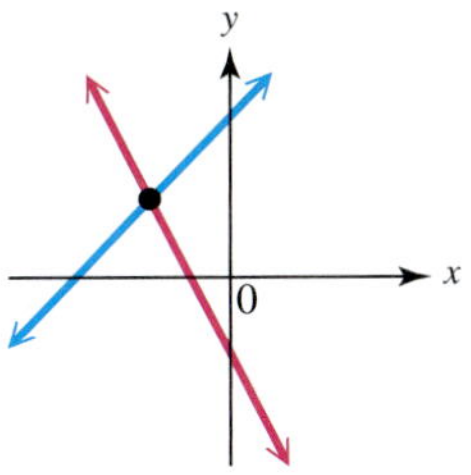

B, because the ordered pair must be in quadrant II.

10. Which ordered pair could be a solution of the system graphed? Why is it the only valid choice?

A. $(2, 0)$
B. $(0, 2)$
C. $(-2, 0)$
D. $(0, -2)$

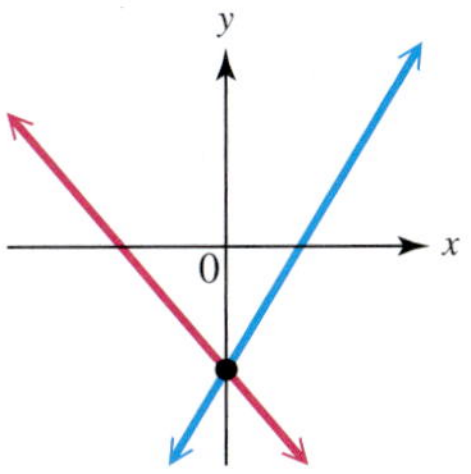

D, because the ordered pair must be on the y-axis, with $y < 0$.

Decide whether the given ordered pair is a solution of the given system. ***See Example 1.***

11. $(2, -3)$
$$x + y = -1$$
$$2x + 5y = 19$$
no

12. $(4, 3)$
$$x + 2y = 10$$
$$3x + 5y = 3$$
no

13. $(-1, -3)$
$$3x + 5y = -18$$
$$4x + 2y = -10$$
yes

14. $(-9, -2)$
$$2x - 5y = -8$$
$$3x + 6y = -39$$
yes

15. $(7, -2)$
$4x = 26 - y$
$3x = 29 + 4y$
yes

16. $(9, 1)$
$2x = 23 - 5y$
$3x = 24 + 3y$
yes

17. $(6, -8)$
$-2y = x + 10$
$3y = 2x + 30$
no

18. $(-5, 2)$
$5y = 3x + 20$
$3y = -2x - 4$
no

Solve each system of equations by graphing. If the system is inconsistent or the equations are dependent, say so. ***See Examples 2 and 3.*** **We show the graphs here only for Exercises 19–24.**

19. $x - y = 2$
$x + y = 6$ $\{(4, 2)\}$

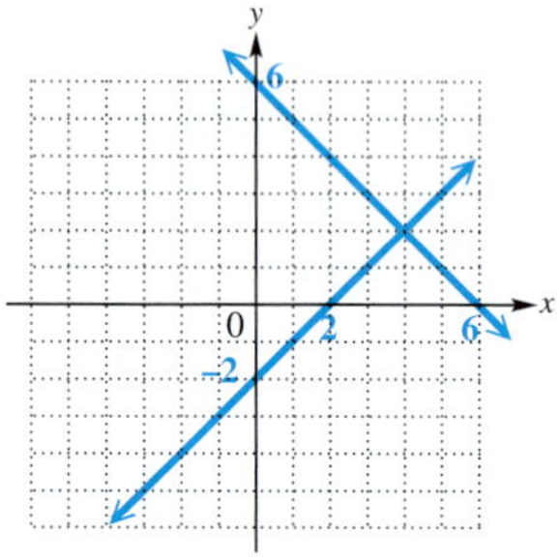

20. $x - y = 3$
$x + y = -1$ $\{(1, -2)\}$

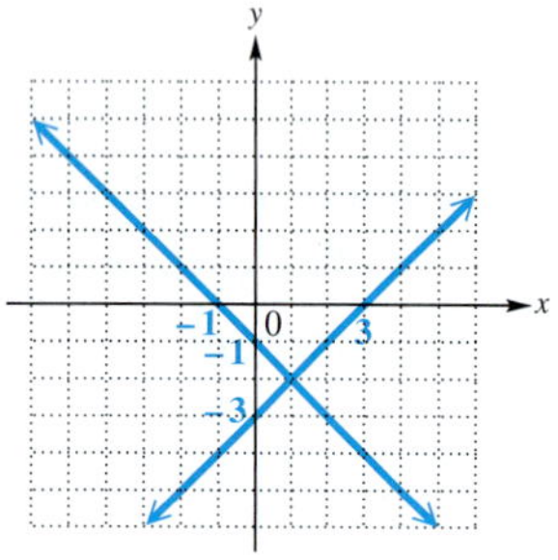

21. $x + y = 4$
$y - x = 4$ $\{(0, 4)\}$

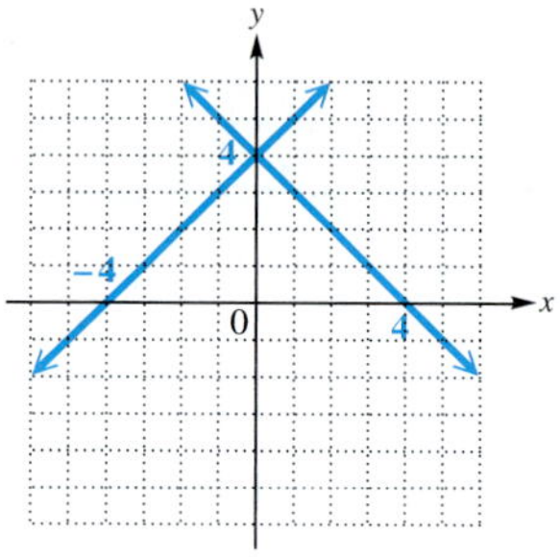

22. $x + y = -5$
$x - y = 5$ $\{(0, -5)\}$

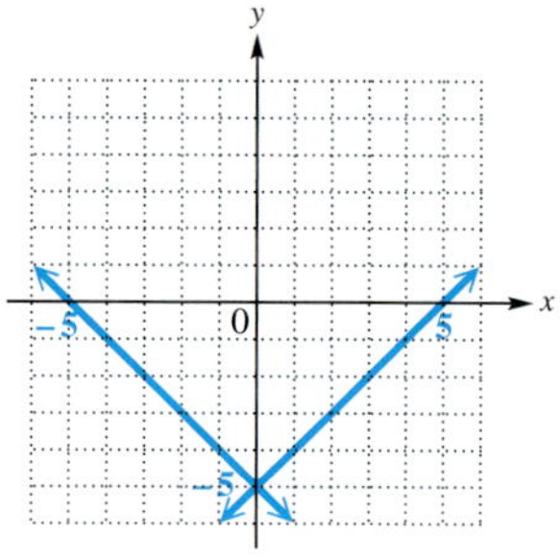

23. $x - 2y = 6$
$x + 2y = 2$ $\{(4, -1)\}$

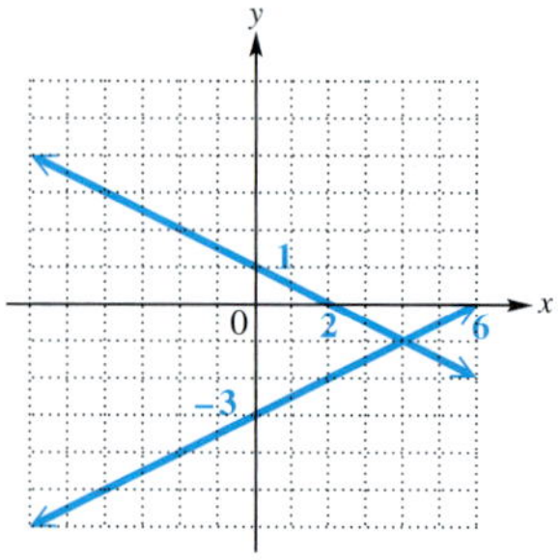

24. $2x - y = 4$
$4x + y = 2$ $\{(1, -2)\}$

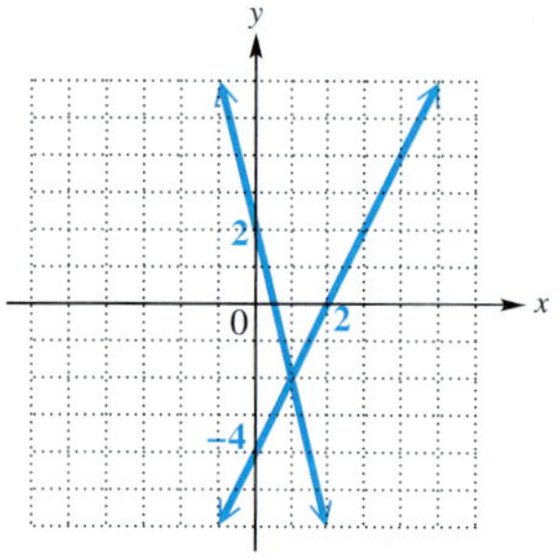

25. $3x - 2y = -3$
$-3x - y = -6$ $\{(1, 3)\}$

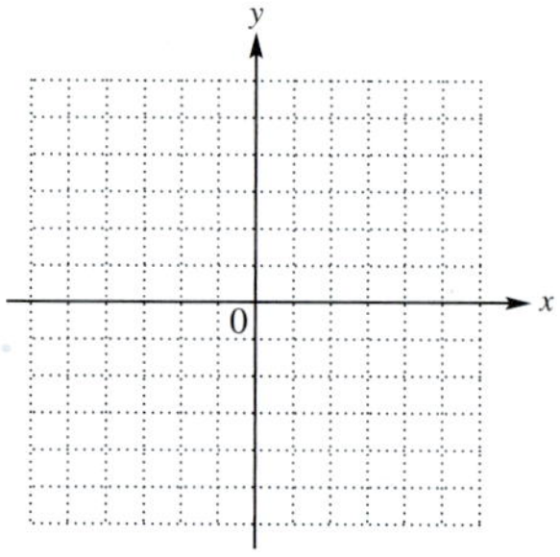

26. $2x - y = 4$
$2x + 3y = 12$ $\{(3, 2)\}$

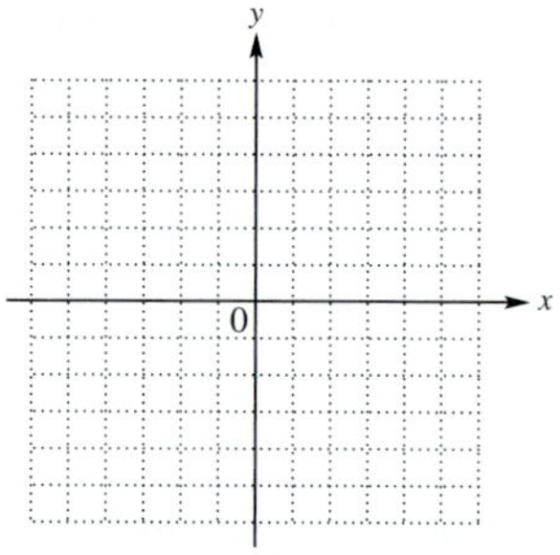

27. $2x - 3y = -6$
$y = -3x + 2$

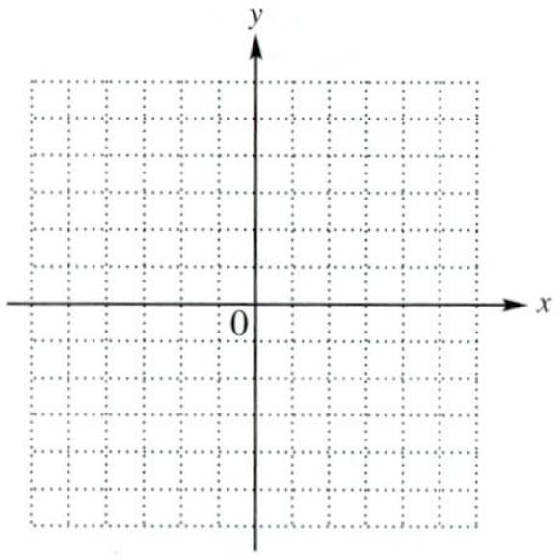

$\{(0, 2)\}$

28. $-3x + y = -3$
$y = x - 3$

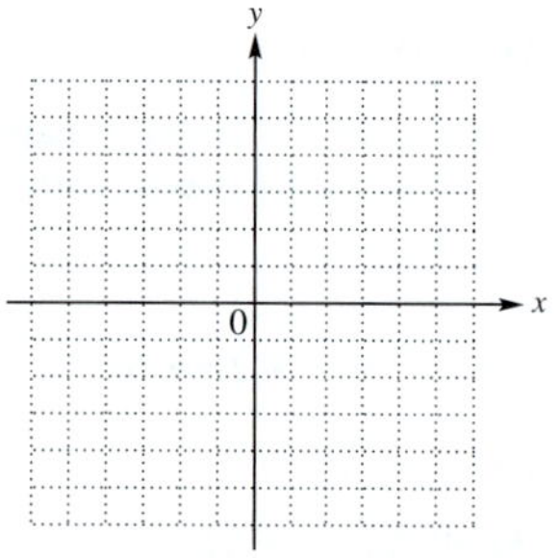

$\{(0, -3)\}$

29. $x + 2y = 6$
$2x + 4y = 8$

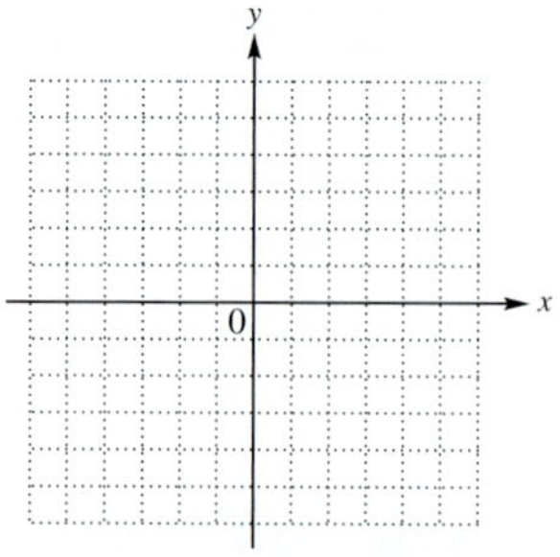

$\emptyset$ (inconsistent system)

30. $2x - y = 6$
$6x - 3y = 12$

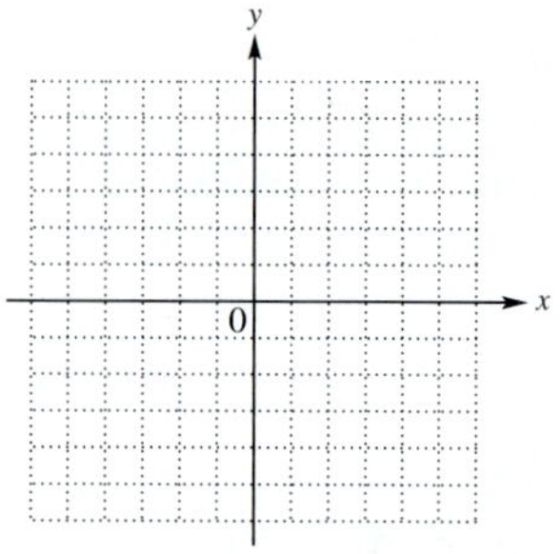

$\emptyset$ (inconsistent system)

31. $5x - 3y = 2$

$10x - 6y = 4$

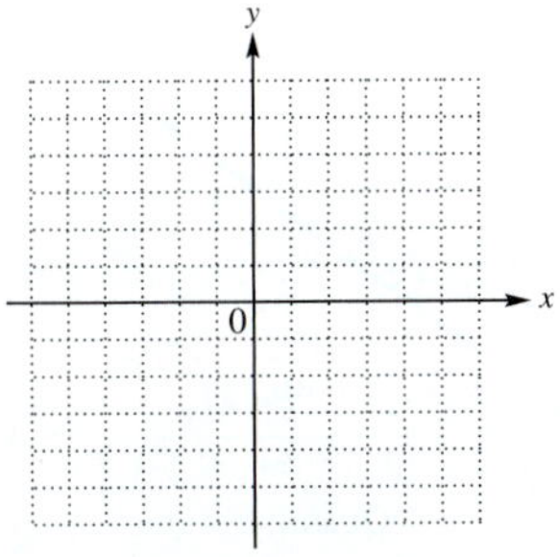

$\{(x, y) \mid 5x - 3y = 2\}$
(dependent equations)

32. $2x - 5y = 8$

$4x - 10y = 16$

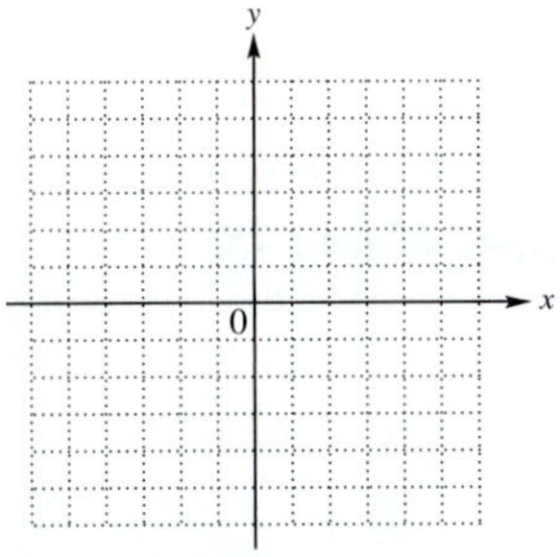

$\{(x, y) \mid 2x - 5y = 8\}$
(dependent equations)

33. $3x - 4y = 24$

$y = -\frac{3}{2}x + 3$

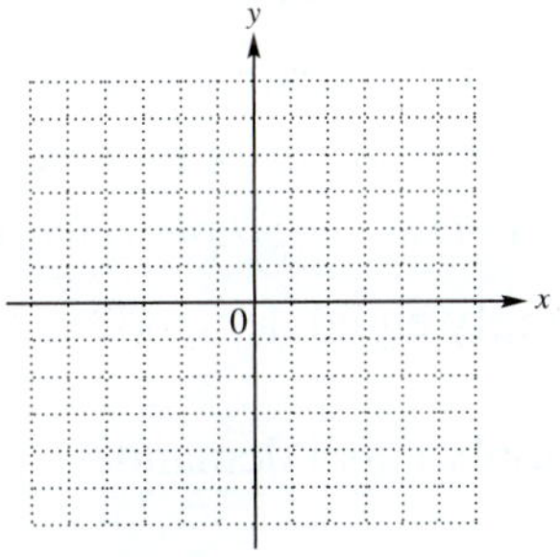

$\{(4, -3)\}$

34. $3x - 2y = 12$

$y = -4x + 5$

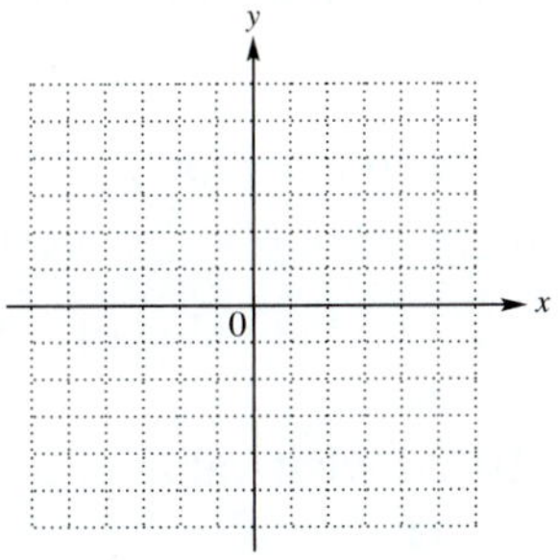

$\{(2, -3)\}$

35. $4x - 2y = 8$

$2x = y + 4$

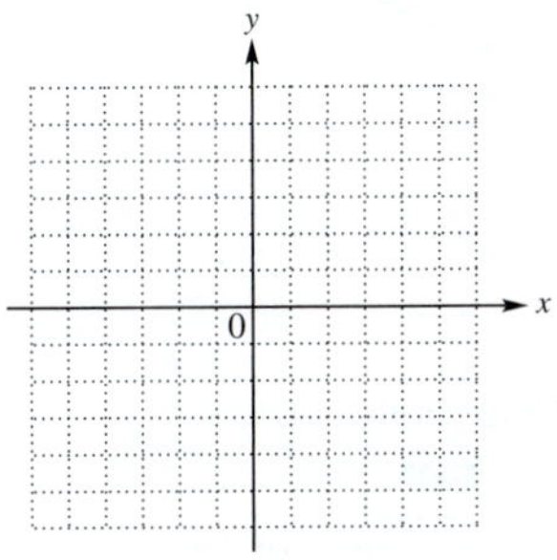

$\{(x, y) \mid 2x - y = 4\}$
(dependent equations)

36. $3x = 5 - y$

$6x + 2y = 10$

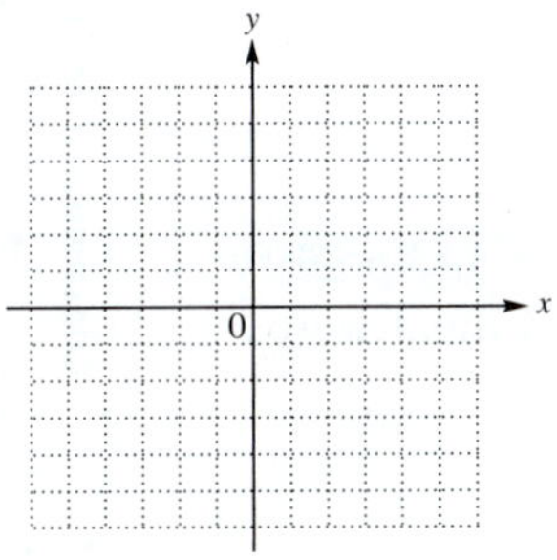

$\{(x, y) \mid 3x + y = 5\}$
(dependent equations)

37. $3x = y + 5$

$6x - 5 = 2y$

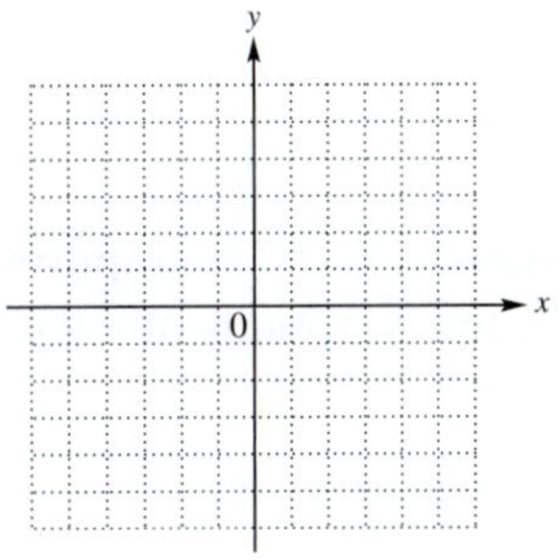

$\emptyset$ **(inconsistent system)**

38. $2x = y - 4$

$4x - 2y = -4$

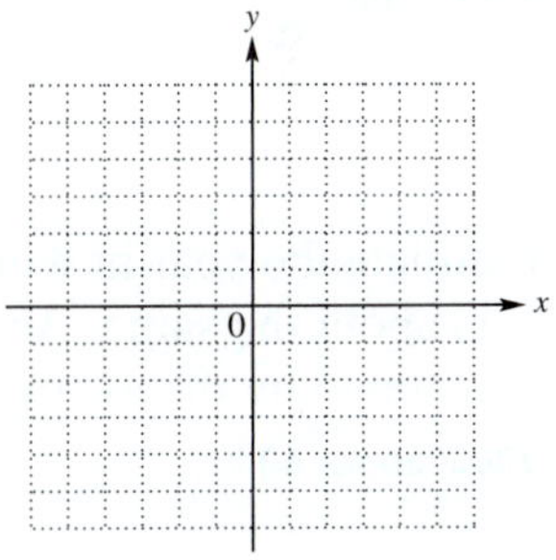

$\emptyset$ **(inconsistent system)**

Without graphing, answer the following questions for each linear system. ***See Example 4.***

(a) *Is the system inconsistent, are the equations dependent, or neither?*
(b) *Is the graph a pair of intersecting lines, a pair of parallel lines, or one line?*
(c) *Does the system have one solution, no solution, or an infinite number of solutions?*

39. $y - x = -5$

$x + y = 1$

(a) neither
(b) intersecting lines
(c) one solution

40. $2x + y = 6$

$x - 3y = -4$

(a) neither
(b) intersecting lines
(c) one solution

41. $x + 2y = 0$

$4y = -2x$

(a) dependent
(b) one line
(c) infinite number of solutions

42. $4x - 6y = 10$

$-6x + 9y = -15$

(a) dependent
(b) one line
(c) infinite number of solutions

43. $5x + 4y = 7$

$10x + 8y = 4$

(a) inconsistent
(b) parallel lines
(c) no solution

44. $y = 3x$

$y + 3 = 3x$

(a) inconsistent
(b) parallel lines
(c) no solution

45. $x - 3y = 5$

$2x + y = 8$

(a) neither
(b) intersecting lines
(c) one solution

46. $2x + 3y = 12$

$2x - y = 4$

(a) neither
(b) intersecting lines
(c) one solution

*Economics deals with **supply and demand.** Typically, as the price of an item increases, the demand for the item decreases while the supply increases. If supply and demand can be described by straight-line equations, the point at which the lines intersect determines the **equilibrium supply** and **equilibrium demand.***

The price per unit, p, and the demand, x, for a particular aluminum siding are related by the linear equation $p = 60 - \frac{3}{4}x$, while the price and the supply are related by the linear equation $p = \frac{3}{4}x$. See the figure. Use the graph to work Exercises 47 and 48.

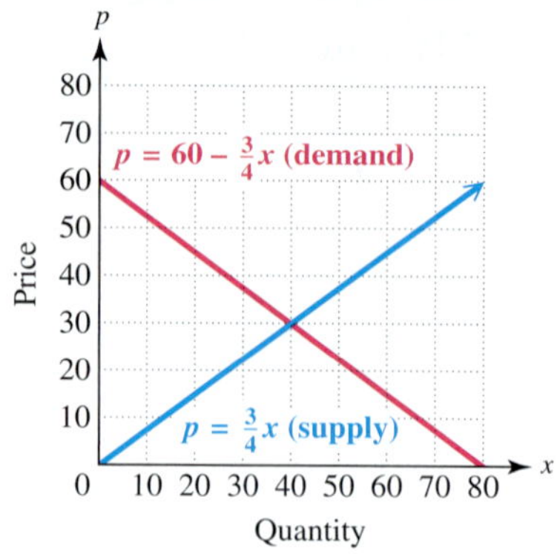

47. **(a)** At what value of x does supply equal demand?

40

(b) At what value of p does supply equal demand?

30

48. When $x > 40$, does demand exceed supply or does supply exceed demand?

Supply exceeds demand.

The per-capita consumption of beef and poultry in the United States in selected years is shown in the graph. Use the graph to work Exercises 49 and 50.

U.S. Per-Capita Meat Consumption

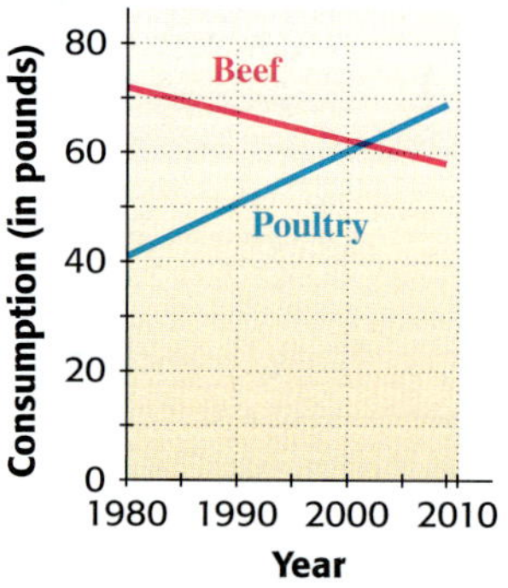

Source: U.S. Department of Agr

49. For which years did per-capita consumption of beef exceed that of poultry?

1980–2000

50. Estimate the year in which the per-capita consumption of beef and poultry was about the same. About how many pounds of each kind of meat were consumed by the average American in that year?

2000; about 60 lb

The graph shows sales of music CDs and digital downloads of single songs (in millions) in the United States over selected years. Use the graph to work Exercises 51–54.

Music Going Digital

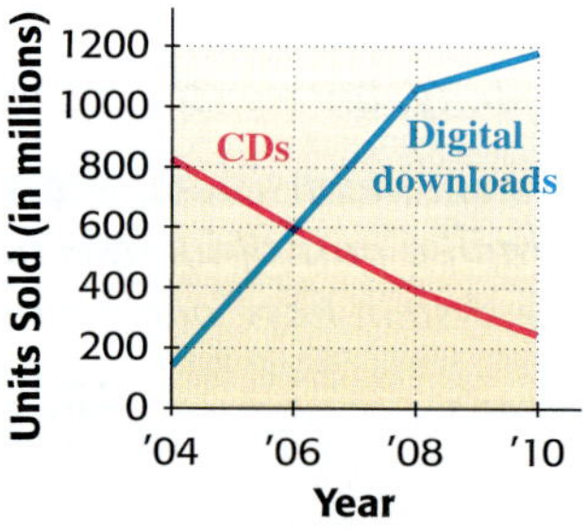

Source: Recording Industry Association of America.

51. In what year did Americans purchase about the same number of CDs as single digital downloads? How many units was this?

2006; about 600 (million) units

52. Express the point of intersection of the two graphs as an ordered pair of the form (year, units in millions).

(2006, 600)

53. Describe the trend in sales of music CDs over the years 2004 to 2010. If a straight line were used to approximate its graph, would the line have positive, negative, or zero slope? Explain.

The slope would be negative. Sales of CDs were decreasing during this period.

54. If a straight line were used to approximate the graph of sales of digital downloads over the years 2004 to 2010, would the line have positive, negative, or zero slope? Explain.

The slope would be positive. Sales of digital downloads were increasing during this period.

4.2 Solving Systems of Linear Equations by Substitution

OBJECTIVES

1. Solve linear systems by substitution.
2. Solve special systems by substitution.
3. Solve linear systems with fractions and decimals.

OBJECTIVE 1 Solve linear systems by substitution.

Work Problem 1 at the Side. ▶

As we see in **Margin Problem 1,** graphing to solve a system of equations has a serious drawback: It is difficult to accurately find a solution such as $\left(\frac{11}{3}, -\frac{4}{9}\right)$ from a graph.

As a result, there are algebraic methods for solving systems of equations. The **substitution method,** which gets its name from the fact that an expression in one variable is *substituted* for the other variable, is one such method.

EXAMPLE 1 Using the Substitution Method

Solve the system by the substitution method.

$$3x + 5y = 26 \quad (1)$$
$$y = 2x \quad (2)$$

We number the equations for reference in our discussion.

Equation (2) is already solved for y. This equation says that $y = 2x$, so we substitute $2x$ for y in equation (1).

$3x + 5y = 26$	(1)
$3x + 5(2x) = 26$	Let $y = 2x$.
$3x + 10x = 26$	Multiply.
$13x = 26$	Combine like terms.
$x = 2$	Divide by 13.

Don't stop here.

Now we can find the value of y by substituting 2 for x in either equation. We choose equation (2) since the substitution is easier.

$y = 2x$	(2)
$y = 2(2)$	Let $x = 2$.
$y = 4$	Multiply.

We check the solution $(2, 4)$ by substituting 2 for x and 4 for y in *both* equations.

CHECK

$3x + 5y = 26$	(1)	$y = 2x$	(2)
$3(2) + 5(4) \stackrel{?}{=} 26$	Substitute.	$4 \stackrel{?}{=} 2(2)$	Substitute.
$6 + 20 \stackrel{?}{=} 26$	Multiply.	$4 = 4$ ✓	True
$26 = 26$ ✓	True		

Since $(2, 4)$ satisfies both equations, the solution set of the system is $\{(2, 4)\}$.

Work Problem 2 at the Side. ▶

CAUTION

A system is not completely solved until values for both x and y are found. Write the solution set as a set containing an ordered pair.

1 Solve the system by graphing.

$$2x + 3y = 6$$
$$x - 3y = 5$$

Can you determine the answer? Why or why not?

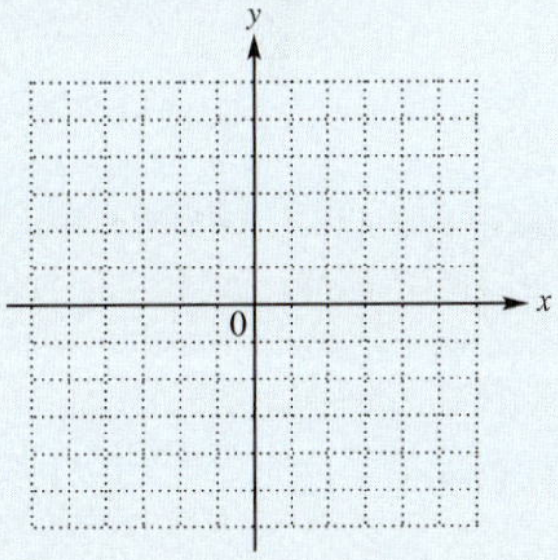

2 (GS) Fill in the blanks to solve by the substitution method. Check the solution.

$$3x + 5y = 69$$
$$y = 4x$$

$3x + 5(\_\_\_\_) = 69$

$3x + \_\_\_\_ = 69$

$\_\_\_\_ = 69$

$x = \_\_\_\_$

$y = 4(\_\_\_\_) = \_\_\_\_$

The solution set is ____.

Answers

1. The answer cannot be determined from the graph because it is too difficult to read the exact coordinates.
2. $4x$; $20x$; $23x$; 3; 3; 12; $\{(3, 12)\}$

3 Solve each system by the substitution method. Check each solution.

(a) $2x + 7y = -12$

$x = 3 - 2y$

(b) $x = y - 3$

$4x + 9y = 1$

Teaching Tip

Emphasize that the solution set found by the substitution method will be the same as the solution set found by graphing. Have students graph the system in Example 1 or Example 2 to see this.

Teaching Tip

Mention that the substitution method is also essential in solving some nonlinear systems of equations, a topic covered in intermediate and college algebra courses.

Answers

3. **(a)** $\{(15, -6)\}$ **(b)** $\{(-2, 1)\}$

EXAMPLE 2 Using the Substitution Method

Solve the system by the substitution method.

$$2x + 5y = 7 \quad (1)$$

$$x = -1 - y \quad (2)$$

Equation (2) gives x in terms of y. We substitute $-1 - y$ for x in equation (1).

$$2x + 5y = 7 \quad (1)$$

$$2(-1 - y) + 5y = 7 \quad \text{Let } x = -1 - y.$$

Distribute 2 to *both* -1 and $-y$.

$$-2 - 2y + 5y = 7 \quad \text{Distributive property}$$

$$-2 + 3y = 7 \quad \text{Combine like terms.}$$

$$3y = 9 \quad \text{Add 2.}$$

$$y = 3 \quad \text{Divide by 3.}$$

To find x, substitute 3 for y in equation (2).

$$x = -1 - y \quad (2)$$

$$x = -1 - 3 \quad \text{Let } y = 3.$$

$$x = -4 \quad \text{Subtract.}$$

Write the x-coordinate first.

We check the solution $(-4, 3)$ in *both* equations.

CHECK

$2x + 5y = 7$	(1)	$x = -1 - y$	(2)
$2(-4) + 5(3) \stackrel{?}{=} 7$	Substitute.	$-4 \stackrel{?}{=} -1 - 3$	Substitute.
$-8 + 15 \stackrel{?}{=} 7$	Multiply.	$-4 = -4$ ✓	True
$7 = 7$ ✓ True			

Both results are true, so the solution set of the system is $\{(-4, 3)\}$.

◀ **Work Problem 3 at the Side.**

CAUTION

Even though we found y first in **Example 2, *the x-coordinate is always written first in the ordered-pair solution of a system.***

To solve a system by substitution, follow these steps.

Solving a Linear System by Substitution

Step 1 **Solve one equation for either variable.** If one of the variables has coefficient 1 or -1, choose it, since it usually makes the substitution easier.

Step 2 **Substitute** for that variable in the other equation. The result should be an equation with just one variable.

Step 3 **Solve** the equation from Step 2.

Step 4 **Substitute** the result from Step 3 into the equation from Step 1 to find the value of the other variable.

Step 5 **Check** the ordered-pair solution in both of the *original* equations. Then write the solution set.

EXAMPLE 3 Using the Substitution Method

Solve the system by the substitution method.

$$2x = 4 - y \quad (1)$$
$$5x + 3y = 10 \quad (2)$$

Step 1 We must solve one of the equations for either x or y. Because the coefficient of y in equation (1) is -1, we avoid fractions by choosing this equation and solving for y.

$$2x = 4 - y \quad (1)$$
$$y + 2x = 4 \quad \text{Add } y.$$
$$y = -2x + 4 \quad \text{Subtract } 2x.$$

Step 2 Now substitute $-2x + 4$ for y in equation (2).

$$5x + 3y = 10 \quad (2)$$
$$5x + 3(-2x + 4) = 10 \quad \text{Let } y = -2x + 4.$$

Step 3 Solve the equation from Step 2.

$$5x - 6x + 12 = 10 \quad \text{Distributive property}$$

Distribute 3 to *both* $-2x$ and 4.

$$-x + 12 = 10 \quad \text{Combine like terms.}$$
$$-x = -2 \quad \text{Subtract 12.}$$
$$x = 2 \quad \text{Multiply by } -1.$$

Step 4 Equation (1) solved for y is $y = -2x + 4$. Substitute 2 for x.

$$y = -2(2) + 4 \quad \text{Substitute in (1).}$$
$$y = 0 \quad \text{Multiply, and then add.}$$

Step 5 Check that $(2, 0)$ is the solution.

CHECK

$2x = 4 - y$	(1)	$5x + 3y = 10$	(2)
$2(2) \stackrel{?}{=} 4 - 0$	Substitute.	$5(2) + 3(0) \stackrel{?}{=} 10$	Substitute.
$4 = 4$ ✓	True	$10 = 10$ ✓	True

Since both results are true, the solution set of the system is $\{(2, 0)\}$.

Work Problem 4 at the Side. ▶

OBJECTIVE 2 **Solve special systems by substitution.**

EXAMPLE 4 Solving an Inconsistent System by Substitution

Use substitution to solve the system.

$$x = 5 - 2y \quad (1)$$
$$2x + 4y = 6 \quad (2)$$

Because equation (1) is already solved for x, we substitute $5 - 2y$ for x in equation (2).

$$2x + 4y = 6 \quad (2)$$
$$2(5 - 2y) + 4y = 6 \quad \text{Let } x = 5 - 2y.$$
$$10 - 4y + 4y = 6 \quad \text{Distributive property}$$
$$10 = 6 \quad \text{False}$$

Continued on Next Page

4 Solve each system by the substitution method. Check each solution.

GS **(a)** Fill in the blanks to solve.

$$x + 4y = -1$$
$$2x - 5y = 11$$

Solve the first equation for x.

$$x = -1 - \_\_\_\_$$

Substitute into the second equation to find y.

$$2(\_\_\_\_) - 5y = 11$$
$$-2 - 8y - 5y = 11$$
$$-2 - \_\_\_\_ y = 11$$
$$\_\_\_\_ y = 13$$
$$y = \_\_\_\_$$

Find x.

$$x = -1 - 4y$$
$$x = -1 - 4(\_\_\_\_)$$
$$x = \_\_\_\_$$

The solution set is ____.

(b) $2x + 5y = 4$
$x + y = -1$

Teaching Tip

As an additional classroom example, solve the system in Margin Problem 1 at the beginning of this section with students.

$$2x + 3y = 6$$
$$x - 3y = 5$$

Answer: $\left\{\left(\frac{11}{3}, -\frac{4}{9}\right)\right\}$

Answers

4. **(a)** $4y$; $-1 - 4y$; 13; -13; -1; -1; 3; $\{(3, -1)\}$
(b) $\{(-3, 2)\}$

5 Use substitution to solve the system.

$$8x - 2y = 1$$
$$y = 4x - 8$$

6 Solve each system by substitution.

(a) $7x - 6y = 10$
$-14x + 20 = -12y$

(b) $8x - y = 4$
$y = 8x + 4$

Teaching Tip

Explain to students that once a true statement (identity) is obtained, even if it is not 0 = 0, we can stop. For instance, in *Example 5* we might also obtain the true statement

$$-12 = -12.$$

Answers

5. ∅
6. (a) $\{(x, y) \mid 7x - 6y = 10\}$ **(b)** ∅

A false result, here $10 = 6$, means that the equations in the system

$$x = 5 - 2y$$
$$2x + 4y = 6$$

have graphs that are parallel lines. The system is inconsistent and has no solution, so the solution set is ∅. See **Figure 6.**

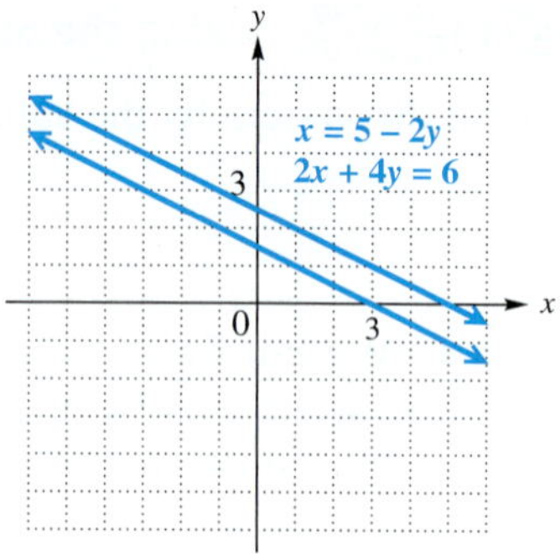

Figure 6

◄ **Work Problem 5 at the Side.**

CAUTION

It is a common error to give "false" as the answer to an inconsistent system. The correct response is ∅.

EXAMPLE 5 Solving a System with Dependent Equations by Substitution

Solve the system by the substitution method.

$$3x - y = 4 \quad (1)$$
$$-9x + 3y = -12 \quad (2)$$

Begin by solving equation (1) for y to get $y = 3x - 4$. Substitute $3x - 4$ for y in equation (2) and solve the resulting equation.

$$-9x + 3y = -12 \quad (2)$$
$$-9x + 3(3x - 4) = -12 \quad \text{Let } y = 3x - 4.$$
$$-9x + 9x - 12 = -12 \quad \text{Distributive property}$$
$$0 = 0 \quad \text{Add 12. Combine like terms.}$$

This true result means that every solution of one equation is also a solution of the other, so the system has an infinite number of solutions—all the ordered pairs corresponding to points that lie on the common graph. The solution set is

$$\{(x, y) \mid 3x - y = 4\}.$$

A graph of the equations of this system is shown in **Figure 7.**

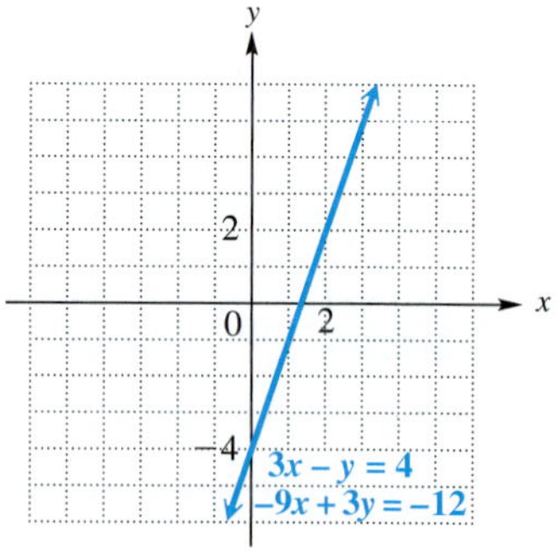

Figure 7

◄ **Work Problem 6 at the Side.**

CAUTION

It is a common error to give "true" as the solution of a system of dependent equations. Write the solution set in set-builder notation using the equation in the system (or an equivalent equation) that is in standard form, with integer coefficients having greatest common factor 1.

OBJECTIVE 3 Solve linear systems with fractions and decimals.

EXAMPLE 6 Using the Substitution Method with Fractions

Solve the system by the substitution method.

$$3x + \frac{1}{4}y = 2 \quad (1)$$

$$\frac{1}{2}x + \frac{3}{4}y = -\frac{5}{2} \quad (2)$$

Clear equation (1) of fractions by multiplying each side by 4.

$$4\left(3x + \frac{1}{4}y\right) = 4(2) \quad \text{Multiply by 4, the common denominator.}$$

$$4(3x) + 4\left(\frac{1}{4}y\right) = 4(2) \quad \text{Distributive property}$$

$$12x + y = 8 \quad (3)$$

Now clear equation (2) of fractions by multiplying each side by 4.

$$\frac{1}{2}x + \frac{3}{4}y = -\frac{5}{2} \quad (2)$$

$$4\left(\frac{1}{2}x + \frac{3}{4}y\right) = 4\left(-\frac{5}{2}\right) \quad \text{Multiply by 4, the common denominator.}$$

$$4\left(\frac{1}{2}x\right) + 4\left(\frac{3}{4}y\right) = 4\left(-\frac{5}{2}\right) \quad \text{Distributive property.}$$

$$2x + 3y = -10 \quad (4)$$

The given system of equations has been simplified to an equivalent system.

$$12x + y = 8 \quad (3)$$

$$2x + 3y = -10 \quad (4)$$

To solve this system by substitution, solve equation (3) for y.

$$12x + y = 8 \quad (3)$$

$$y = -12x + 8 \quad \text{Subtract } 12x.$$

Now substitute this result for y in equation (4).

$$2x + 3y = -10 \quad (4)$$

$$2x + 3(-12x + 8) = -10 \quad \text{Let } y = -12x + 8.$$

$$2x - 36x + 24 = -10 \quad \text{Distributive property}$$

$$-34x = -34 \quad \text{Combine like terms. Subtract 24.}$$

$$x = 1 \quad \text{Divide by } -34.$$

Distribute 3 to both $-12x$ and 8.

Substitute 1 for x in $y = -12x + 8$ (equation (3) solved for y).

$$y = -12(1) + 8 = -4$$

Check $(1, -4)$ in both of the original equations. The solution set of the system is $\{(1, -4)\}$.

Work Problem 7 at the Side. ▶

---

7 Solve each system by the substitution method.

GS **(a)** $\frac{2}{3}x + \frac{1}{2}y = 6 \quad (1)$

$\frac{1}{2}x - \frac{3}{4}y = 0 \quad (2)$

First clear all fractions. The LCD for the denominators 3 and 2 in (1) is ____.

$$\_\_\_\left(\frac{2}{3}x + \frac{1}{2}y\right) = \_\_\_ \quad (6)$$

$$\_\_\_\left(\frac{2}{3}x\right) + 6(\_\_\_) = 36$$

$$\_\_\_ + 3y = 36$$

Clear fractions in (2) by multiplying both sides by ____, the LCD of denominators 2 and 4.

Complete the solution of the system. The solution set is ____.

(b) $x + \frac{1}{2}y = \frac{1}{2}$

$\frac{1}{6}x - \frac{1}{3}y = \frac{4}{3}$

Answers

7. (a) 6; 6; 6; 6; $\frac{1}{2}y$; $4x$; 4; $\{(6, 4)\}$

(b) $\{(2, -3)\}$

8 Solve the system by the substitution method.

$$0.2x + 0.3y = 0.5$$
$$0.3x - 0.1y = 1.3$$

EXAMPLE 7 Using the Substitution Method with Decimals

Solve the system by the substitution method.

$$0.5x + 2.4y = 4.2 \quad (1)$$
$$-0.1x + 1.5y = 5.1 \quad (2)$$

Clear each equation of decimals, as in **Section 2.3,** by multiplying by 10.

$$10(0.5x + 2.4y) = 10(4.2) \quad \text{Multiply equation (1) by 10.}$$
$$10(0.5x) + 10(2.4y) = 10(4.2) \quad \text{Distributive property}$$
$$5x + 24y = 42 \quad (3)$$

$$10(-0.1x + 1.5y) = 10(5.1) \quad \text{Multiply equation (2) by 10.}$$
$$10(-0.1x) + 10(1.5y) = 10(5.1) \quad \text{Distributive property}$$
$$-x + 15y = 51 \quad (4)$$

$10(-0.1x) = -1x = -x$

Now solve the equivalent system of equations by substitution.

$$5x + 24y = 42 \quad (3)$$
$$-x + 15y = 51 \quad (4)$$

Equation (4) can be solved for x.

$$x = 15y - 51 \quad \text{Equation (4) solved for } x$$

Substitute this result for x in equation (3).

$$5x + 24y = 42 \quad (3)$$
$$5(15y - 51) + 24y = 42 \quad \text{Let } x = 15y - 51.$$
$$75y - 255 + 24y = 42 \quad \text{Distributive property}$$
$$99y = 297 \quad \text{Combine like terms. Add 255.}$$
$$y = 3 \quad \text{Divide by 99.}$$

Since equation (4) solved for x is $x = 15y - 51$, substitute 3 for y.

$$x = 15(3) - 51 = -6$$

Check $(-6, 3)$ in both of the original equations. The solution set is $\{(-6, 3)\}$.

◀ Work Problem **8** at the Side.

Answer

8. $\{(4, -1)\}$

4.2 Exercises

FOR EXTRA HELP

Download the MyDashBoard App

MyMathLab®

CONCEPT CHECK *Work each problem.*

1. A student solves the following system and finds that $x = 3$, which is correct. The student gives $\{3\}$ as the solution set. **What Went Wrong?**

$$5x - y = 15$$
$$7x + y = 21$$

The student must find the value of y and write the solution as an ordered pair. The solution set is $\{(3, 0)\}$.

2. A student solves the following system and obtains the statement $0 = 0$. The student gives the solution set as $\{(0, 0)\}$. **What Went Wrong?**

$$x + y = 4$$
$$2x + 2y = 8$$

The true result $0 = 0$ means that the system has an infinite number of solutions. The solution set is $\{(x, y) \mid x + y = 4\}$.

3. When we use the substitution method, how can we tell that a system has no solution?

A false statement, such as $0 = 3$, occurs.

4. When we use the substitution method, how can we tell that a system has an infinite number of solutions?

A true statement, such as $0 = 0$, occurs.

Solve each system by the substitution method. Check each solution. **See Examples 1–5.**

5. $x + y = 12$
$y = 3x$
$\{(3, 9)\}$

6. $x + 3y = -28$
$y = -5x$
$\{(2, -10)\}$

7. $3x + 2y = 27$
$x = y + 4$
$\{(7, 3)\}$

8. $4x + 3y = -5$
$x = y - 3$
$\{(-2, 1)\}$

9. $3x + 5y = 14$
$x - 2y = -10$
$\{(-2, 4)\}$

10. $5x + 2y = -1$
$2x - y = -13$
$\{(-3, 7)\}$

11. $3x + 4 = -y$
$2x + y = 0$
$\{(-4, 8)\}$

12. $2x - 5 = -y$
$x + 3y = 0$
$\{(3, -1)\}$

13. $7x + 4y = 13$
$x + y = 1$
$\{(3, -2)\}$

14. $3x - 2y = 19$
$x + y = 8$
$\{(7, 1)\}$

15. $3x - y = 5$
$y = 3x - 5$
$\{(x, y) \mid 3x - y = 5\}$

16. $4x - y = -3$
$y = 4x + 3$
$\{(x, y) \mid 4x - y = -3\}$

17. $2x + 8y = 3$
$x = 8 - 4y$
$\emptyset$

18. $2x + 10y = 3$
$x = 1 - 5y$
$\emptyset$

19. $2y = 4x + 24$
$2x - y = -12$
$\{(x, y) \mid 2x - y = -12\}$

20. $2y = 14 - 6x$
$3x + y = 7$
$\{(x, y) | 3x + y = 7\}$

21. $6x - 8y = 6$
$2y = -2 + 3x$
$\left\{\left(\frac{1}{3}, -\frac{1}{2}\right)\right\}$

22. $3x + 2y = 6$
$6x = 8 + 4y$
$\left\{\left(\frac{5}{3}, \frac{1}{2}\right)\right\}$

Solve each system by the substitution method. Check each solution. ***See Examples 6 and 7.***

23. $\frac{1}{5}x + \frac{2}{3}y = -\frac{8}{5}$
$3x - y = 9$
$\{(2, -3)\}$

24. $\frac{1}{3}x - \frac{1}{2}y = -\frac{2}{3}$
$4x + y = 6$
$\{(1, 2)\}$

25. $\frac{1}{2}x + \frac{1}{3}y = -\frac{1}{3}$
$\frac{1}{2}x + 2y = -7$
$\{(2, -4)\}$

26. $\frac{1}{6}x + \frac{1}{6}y = 1$
$-\frac{1}{2}x - \frac{1}{3}y = -5$
$\{(18, -12)\}$

27. $\frac{x}{5} + 2y = \frac{16}{5}$
$\frac{3x}{5} + \frac{y}{2} = -\frac{7}{5}$
$\{(-4, 2)\}$

28. $\frac{x}{2} + \frac{y}{3} = \frac{7}{6}$
$\frac{x}{4} - \frac{3y}{2} = \frac{9}{4}$
$\{(3, -1)\}$

29. $0.1x + 0.9y = -2$
$0.5x - 0.2y = 4.1$
$\{(7, -3)\}$

30. $0.2x - 1.3y = -3.2$
$-0.1x + 2.7y = 9.8$
$\{(10, 4)\}$

31. $0.8x - 0.1y = 1.3$
$2.2x + 1.5y = 8.9$
$\{(2, 3)\}$

32. $0.3x - 0.1y = 2.1$
$0.6x + 0.3y = -0.3$
$\{(4, -9)\}$

Relating Concepts (Exercises 33–36) For Individual or Group Work

A system of linear equations can be used to model the cost and the revenue of a business. ***Work Exercises 33–36 in order.***

33. Suppose that to start a business manufacturing and selling bicycles, it costs \$5000. Each bicycle will cost \$400 to manufacture. Explain why the linear equation

$$y_1 = 400x + 5000 \quad (y_1 \text{ in dollars})$$

gives the *total* cost to manufacture x bicycles.

To find the total cost, multiply the number of bicycles (x) by the cost per bicycle (400 dollars) and add the fixed cost (5000 dollars). Thus, $y_1 = 400x + 5000$ gives this total cost (in dollars).

34. We decide to sell each bicycle for \$600. Write an equation using y_2 (in dollars) to express the revenue when we sell x bicycles.

$y_2 = 600x$

35. Form a system from the two equations in **Exercises 33 and 34.** Then solve the system, assuming $y_1 = y_2$, that is, cost = revenue.

$y_1 = 400x + 5000$
$y_2 = 600x$;
solution set: $\{(25, 15{,}000)\}$

36. The value of x in **Exercise 35** is the number of bicycles it takes to *break even.* Fill in the blanks: When __25__ bicycles are sold, the break-even point is reached. At that point, we have spent __15,000__ dollars and taken in __15,000__ dollars.

4.3 Solving Systems of Linear Equations by Elimination

OBJECTIVES

1. Solve linear systems by elimination.
2. Multiply when using the elimination method.
3. Use an alternative method to find the second value in a solution.
4. Solve special systems by elimination.

OBJECTIVE 1 Solve linear systems by elimination. Recall that adding the same quantity to each side of an equation results in equal sums.

$$\text{If} \quad A = B, \quad \text{then} \quad A + C = B + C.$$

We can take this addition a step further. Adding *equal* quantities, rather than the *same* quantity, to each side of an equation also results in equal sums.

$$\text{If} \quad A = B \quad \text{and} \quad C = D, \quad \text{then} \quad A + C = B + D.$$

Using the addition property of equality to solve systems is called the **elimination method.**

EXAMPLE 1 Using the Elimination Method

Use the elimination method to solve the system.

$$x + y = 5 \quad (1)$$
$$x - y = 3 \quad (2)$$

Each equation in this system is a statement of equality, so the sum of the left sides equals the sum of the right sides. Adding vertically in this way gives the following.

$$\begin{aligned} x + y &= 5 && (1) \\ x - y &= 3 && (2) \\ \hline 2x \quad &= 8 && \text{Add left sides and add right sides.} \\ x &= 4 && \text{Divide by 2.} \end{aligned}$$

Notice that y has been eliminated. The result, $x = 4$, gives the x-value of the ordered-pair solution of the given system. To find the y-value of the solution, substitute 4 for x in either of the two equations of the system. We choose equation (1).

$$\begin{aligned} x + y &= 5 && (1) \\ 4 + y &= 5 && \text{Let } x = 4. \\ y &= 1 && \text{Subtract 4.} \end{aligned}$$

Check the solution, (4, 1), by substituting 4 for x and 1 for y in both equations of the given system.

CHECK

$$\begin{aligned} x + y &= 5 && (1) \\ 4 + 1 &\stackrel{?}{=} 5 && \text{Substitute.} \\ 5 &= 5 \ \checkmark && \text{True} \end{aligned}$$

$$\begin{aligned} x - y &= 3 && (2) \\ 4 - 1 &\stackrel{?}{=} 3 && \text{Substitute.} \\ 3 &= 3 \ \checkmark && \text{True} \end{aligned}$$

Since *both* results are true, the solution set of the system is $\{(4, 1)\}$.

Work Problem 1 at the Side. ▶

CAUTION

A system is not completely solved until values for both x and y are found. Do not stop after finding the value of only one variable. Remember to write the solution set as a set containing an ordered pair.

1 Solve each system by the elimination method. Check each solution.

GS **(a)** Fill in the blanks to solve the following system.

$$\begin{aligned} x + y &= 8 && (1) \\ x - y &= 2 && (2) \\ \hline \_\_\_\_ &= \_\_\_\_ && \text{Add.} \\ x &= \_\_\_\_ \end{aligned}$$

Find y.

$$\begin{aligned} x - y &= 2 && (2) \\ \_\_\_\_ - y &= 2 \\ -y &= \_\_\_\_ \\ y &= \_\_\_\_ \end{aligned}$$

The solution set is ____.

(b) $3x - y = 7$
$2x + y = 3$

Answers

1. **(a)** $2x$; 10; 5; 5; -3; 3; $\{(5, 3)\}$
(b) $\{(2, -1)\}$

Teaching Tip
Remind students that the solution of a system of linear equations is the point of intersection of the two lines and that it could be obtained using the graphing method.

You might have students solve the system in Example 1 by graphing and/or substitution to see that the solution set is the same.

With the elimination method, the idea is to *eliminate* one of the variables. ***To do this, one pair of variable terms in the two equations must have coefficients that are opposites.***

Solving a Linear System by Elimination

Step 1 **Write both equations in standard form** $Ax + By = C$.

Step 2 **Transform so that the coefficients of one pair of variable terms are opposites.** Multiply one or both equations by appropriate numbers so that the sum of the coefficients of either the x- or y-terms is 0.

Step 3 **Add** the new equations to eliminate a variable. The sum should be an equation with just one variable.

Step 4 **Solve** the equation from Step 3 for the remaining variable.

Step 5 **Substitute** the result from Step 4 into *either* of the original equations, and solve for the other variable.

Step 6 **Check** the ordered-pair solution in *both* of the *original* equations. Then write the solution set.

It does not matter which variable is eliminated first. Usually we choose the one that is more convenient to work with.

Teaching Tip
To make up an additional system to work as a classroom example, use the form

$$ax + by = c$$
$$dx + ey = f,$$

where a, b, c, d, e, and f are consecutive integers. The solution set will always be $\{(-1, 2)\}$.

EXAMPLE 2 **Using the Elimination Method**

Solve the system.

$$y + 11 = 2x \quad (1)$$
$$5x = y + 26 \quad (2)$$

Step 1 Write both equations in standard form $Ax + By = C$.

$$-2x + y = -11 \quad \text{Subtract } 2x \text{ and } 11 \text{ in equation (1).}$$
$$5x - y = 26 \quad \text{Subtract } y \text{ in equation (2).}$$

Step 2 Because the coefficients of y are 1 and -1, adding will eliminate y. It is not necessary to multiply either equation by a number.

Step 3 Add the two equations.

$$\begin{aligned} -2x + y &= -11 \\ 5x - y &= 26 \\ \hline 3x \quad &= 15 \end{aligned} \quad \text{Add in columns.}$$

Step 4 Solve.

$$x = 5 \quad \text{Divide by 3.}$$

Step 5 Find the value of y by substituting 5 for x in either of the original equations.

$$y + 11 = 2x \quad (1)$$
$$y + 11 = 2(5) \quad \text{Let } x = 5.$$
$$y + 11 = 10 \quad \text{Multiply.}$$
$$y = -1 \quad \text{Subtract 11.}$$

Continued on Next Page

Step 6 Check the ordered-pair solution $(5, -1)$ by substituting $x = 5$ and $y = -1$ into both of the original equations.

CHECK

$y + 11 = 2x$	(1)	$5x = y + 26$	(2)
$-1 + 11 \stackrel{?}{=} 2(5)$	Substitute.	$5(5) \stackrel{?}{=} -1 + 26$	Substitute.
$10 = 10$ ✓	True	$25 = 25$ ✓	True

Since $(5, -1)$ is a solution of *both* equations, the solution set is $\{(5, -1)\}$.

Work Problem **2** at the Side. ▶

OBJECTIVE 2 **Multiply when using the elimination method.** Sometimes we need to multiply each side of one or both equations in a system by some number so that adding the equations will eliminate a variable.

EXAMPLE 3 **Using the Elimination Method**

Solve the system.

$$2x + 3y = -15 \quad (1)$$
$$5x + 2y = 1 \quad (2)$$

Adding the two equations gives $7x + 5y = -14$, which does not eliminate either variable. However, we can multiply each equation by a suitable number so that the coefficients of one of the two variables are opposites. For example, to eliminate x, multiply each side of equation (1) by 5, and each side of equation (2) by -2.

$$10x + 15y = -75 \quad \text{Multiply equation (1) by 5.}$$
$$-10x - 4y = -2 \quad \text{Multiply equation (2) by } -2.$$
$$11y = -77 \quad \text{Add.}$$
$$y = -7 \quad \text{Divide by 11.}$$

The coefficients of x are opposites.

Substituting -7 for y in either equation (1) or (2) gives $x = 3$. Check that the solution set of the system is $\{(3, -7)\}$.

Work Problem **3** at the Side. ▶

OBJECTIVE 3 **Use an alternative method to find the second value in a solution.** Sometimes it is easier to find the value of the second variable in a solution by using the elimination method twice.

EXAMPLE 4 **Finding the Second Value Using an Alternative Method**

Solve the system.

$$4x = 9 - 3y \quad (1)$$
$$5x - 2y = 8 \quad (2)$$

Write equation (1) in standard form by adding $3y$ to each side.

$$4x + 3y = 9 \quad (3)$$
$$5x - 2y = 8 \quad (4)$$

One way to proceed is to eliminate y by multiplying each side of equation (3) by 2 and each side of equation (2) by 3, and then adding.

Continued on Next Page

2 Solve each system by the elimination method. Check each solution.

(a) $2x - y = 2$
$4x + y = 10$

(b) $8x - 5y = 32$
$4x + 5y = 4$

3 **(a)** Solve the system in **Example 3** by first eliminating the variable y. Check the solution.

(b) Solve the system, and check the solution.

$$6x + 7y = 4$$
$$5x + 8y = -1$$

Answers

2. **(a)** $\{(2, 2)\}$ **(b)** $\left\{\left(3, -\frac{8}{5}\right)\right\}$

3. **(a)** $\{(3, -7)\}$ **(b)** $\{(3, -2)\}$

4 Solve each system of equations.

(a) $5x = 7 + 2y$

$5y = 5 - 3x$

(b) $3y = 8 + 4x$

$6x = 9 - 2y$

Teaching Tip

Show how to find the value of y in Example 4 by substituting $\frac{42}{23}$ for x in one of the original equations so that students can see the advantage of the method used in this example.

5 Solve each system by the elimination method.

(a) $4x + 3y = 10$

$2x + \frac{3}{2}y = 12$

(b) $4x - 6y = 10$

$-10x + 15y = -25$

Teaching Tip

Refer students to Section 4.2, Example 5, where the system from Example 5(b) here was solved by substitution.

Answers

4. (a) $\left\{\left(\frac{45}{31}, \frac{4}{31}\right)\right\}$ (b) $\left\{\left(\frac{11}{26}, \frac{42}{13}\right)\right\}$

5. (a) $\emptyset$
(b) $\{(x, y) \mid 2x - 3y = 5\}$
(*Note:* To write the solution set, we divided each term of the equation $4x - 6y = 10$ by 2 so that the coefficients would have greatest common factor 1.)

$$\begin{aligned} 8x + 6y &= 18 && \text{Multiply equation (3) } 4x + 3y = 9 \text{ by 2.} \\ 15x - 6y &= 24 && \text{Multiply equation (2) } 5x - 2y = 8 \text{ by 3.} \\ \hline 23x &= 42 && \text{Add.} \\ x &= \frac{42}{23} && \text{Divide by 23.} \end{aligned}$$

The coefficients of y are opposites.

Substituting $\frac{42}{23}$ for x in one of the given equations would give y, but the arithmetic would be complicated. Instead, solve for y by starting again with the original equations in standard form (equations (3) and (2)) and eliminating x.

$$\begin{aligned} 20x + 15y &= 45 && \text{Multiply equation (3) } 4x + 3y = 9 \text{ by 5.} \\ -20x + 8y &= -32 && \text{Multiply equation (2) } 5x - 2y = 8 \text{ by } -4. \\ \hline 23y &= 13 && \text{Add.} \\ y &= \frac{13}{23} && \text{Divide by 23.} \end{aligned}$$

The coefficients of x are opposites.

Check that the solution set is $\left\{\left(\frac{42}{23}, \frac{13}{23}\right)\right\}$.

◄ Work Problem **4** at the Side.

Note

When the value of the first variable is a fraction, the method used in **Example 4** helps avoid arithmetic errors. This method could be used to solve any system.

OBJECTIVE 4 Solve special systems by elimination.

EXAMPLE 5 Solve Special Systems Using the Elimination Method

Solve each system by the elimination method.

(a)
$$\begin{aligned} 2x + 4y &= 5 && (1) \\ 4x + 8y &= -9 && (2) \end{aligned}$$

Multiply each side of equation (1) by -2. Then add the two equations.

$$\begin{aligned} -4x - 8y &= -10 && \text{Multiply equation (1) by } -2. \\ 4x + 8y &= -9 && (2) \\ \hline 0 &= -19 && \text{False} \end{aligned}$$

The false statement $0 = -19$ indicates that the system has solution set $\emptyset$.

(b)
$$\begin{aligned} 3x - y &= 4 && (1) \\ -9x + 3y &= -12 && (2) \end{aligned}$$

Multiply each side of equation (1) by 3. Then add the two equations.

$$\begin{aligned} 9x - 3y &= 12 && \text{Multiply equation (1) by 3.} \\ -9x + 3y &= -12 && (2) \\ \hline 0 &= 0 && \text{True} \end{aligned}$$

A true statement occurs when the equations are equivalent. This indicates that every solution of one equation is also a solution of the other. The solution set is $\{(x, y) \mid 3x - y = 4\}$.

◄ Work Problem **5** at the Side.

4.3 Exercises

FOR EXTRA HELP

Download the MyDashBoard App

MyMathLab®

CONCEPT CHECK *In Exercises 1–4, answer* true *or* false *for each statement. If false, tell why.*

1. The ordered pair (0, 0) *must* be a solution of a system in the following form.

$$Ax + By = 0$$
$$Cx + Dy = 0$$

true

2. To eliminate the y-terms in the following system, we should multiply equation (2) by 3 and then add the result to equation (1).

$$2x + 12y = 7 \quad (1)$$
$$3x + 4y = 1 \quad (2)$$

false; Multiply by -3.

3. The following system has solution set $\emptyset$.

$$x + y = 1$$
$$x + y = 2$$

true

4. The ordered pair $(4, -5)$ cannot be a solution of a system that contains the following equation.

$$5x - 4y = 0$$

true

Solve each system by the elimination method. Check each solution. ***See Examples 1 and 2.***

5. $x + y = 2$
$2x - y = -5$
$\{(-1, 3)\}$

6. $3x - y = -12$
$x + y = 4$
$\{(-2, 6)\}$

7. $2x + y = -5$
$x - y = 2$
$\{(-1, -3)\}$

8. $2x + y = -15$
$-x - y = 10$
$\{(-5, -5)\}$

9. $3x + 2y = 0$
$-3x - y = 3$
$\{(-2, 3)\}$

10. $5x - y = 5$
$-5x + 2y = 0$
$\{(2, 5)\}$

11. $6x - y = -1$
$5y = 17 + 6x$
$\left\{\left(\frac{1}{2}, 4\right)\right\}$

12. $y = 9 - 6x$
$-6x + 3y = 15$
$\left\{\left(\frac{1}{2}, 6\right)\right\}$

Solve each system by the elimination method. Check each solution. ***See Examples 3–5.***

13. $2x - y = 12$
$3x + 2y = -3$
$\{(3, -6)\}$

14. $x + y = 3$
$-3x + 2y = -19$
$\{(5, -2)\}$

15. $x + 3y = 19$
$2x - y = 10$
$\{(7, 4)\}$

16. $4x - 3y = -19$
$2x + y = 13$
$\{(2, 9)\}$

17. $x + 4y = 16$
$3x + 5y = 20$
$\{(0, 4)\}$

18. $2x + y = 8$
$5x - 2y = -16$
$\{(0, 8)\}$

19. $5x - 3y = -20$
$-3x + 6y = 12$
$\{(-4, 0)\}$

20. $4x + 3y = -28$
$5x - 6y = -35$
$\{(-7, 0)\}$

21. $2x - 8y = 0$
$4x + 5y = 0$
$\{(0, 0)\}$

22. $3x - 15y = 0$
$6x + 10y = 0$
$\{(0, 0)\}$

23. $x + y = 7$
$x + y = -3$
$\varnothing$

24. $x - y = 4$
$x - y = -3$
$\varnothing$

25. $-x + 3y = 4$
$-2x + 6y = 8$
$\{(x, y) \mid x - 3y = -4\}$

26. $6x - 2y = 24$
$-3x + y = -12$
$\{(x, y) \mid 3x - y = 12\}$

27. $2x + 3y = 21$
$5x - 2y = -14$
$\{(0, 7)\}$

28. $5x + 4y = 12$
$3x + 5y = 15$
$\{(0, 3)\}$

29. $3x - 7 = -5y$
$5x + 4y = -10$
$\{(-6, 5)\}$

30. $2x + 3y = 13$
$6 + 2y = -5x$
$\{(-4, 7)\}$

31. $2x + 3y = 0$
$4x + 12 = 9y$
$\left\{\left(-\frac{6}{5}, \frac{4}{5}\right)\right\}$

32. $4x - 3y = -2$
$5x + 3 = 2y$
$\left\{\left(-\frac{5}{7}, -\frac{2}{7}\right)\right\}$

33. $24x + 12y = -7$
$16x - 17 = 18y$
$\left\{\left(\frac{1}{8}, -\frac{5}{6}\right)\right\}$

34. $9x + 4y = -3$
$6x + 7 = -6y$
$\left\{\left(\frac{1}{3}, -\frac{3}{2}\right)\right\}$

35. $5x - 2y = 3$
$10x - 4y = 5$
$\varnothing$

36. $3x - 5y = 1$
$6x - 10y = 4$
$\varnothing$

37. $6x + 3y = 0$
$-18x - 9y = 0$
$\{(x, y) \mid 2x + y = 0\}$

38. $3x - 5y = 0$
$9x - 15y = 0$
$\{(x, y) \mid 3x - 5y = 0\}$

39. $3x = 3 + 2y$
$-\frac{4}{3}x + y = \frac{1}{3}$
$\{(11, 15)\}$

40. $3x = 27 + 2y$
$x - \frac{7}{2}y = -25$
$\{(17, 12)\}$

41. $\frac{1}{5}x + y = \frac{6}{5}$

$\frac{1}{10}x + \frac{1}{3}y = \frac{5}{6}$

$\left\{\left(13, -\frac{7}{5}\right)\right\}$

42. $\frac{1}{3}x + \frac{1}{2}y = \frac{13}{6}$

$\frac{1}{2}x - \frac{1}{4}y = -\frac{3}{4}$

$\left\{\left(\frac{1}{2}, 4\right)\right\}$

43. $2.4x + 1.7y = 7.6$

$1.2x - 0.5y = 9.2$

$\{(6, -4)\}$

44. $0.5x + 3.4y = 13$

$1.5x - 2.6y = -25$

$\{(-8, 5)\}$

Relating Concepts (Exercises 45–48) For Individual or Group Work

The graph shows average U.S. movie theater ticket prices from 2001 through 2010. In 2001, the average ticket price was \$5.66, *as represented on the graph by the point* $P(2001, 5.66)$. *In 2010, the average ticket price was* \$7.89, *as represented on the graph by the point* $Q(2010, 7.89)$. ***Work Exercises 45–48 in order.***

Average Movie Ticket Price

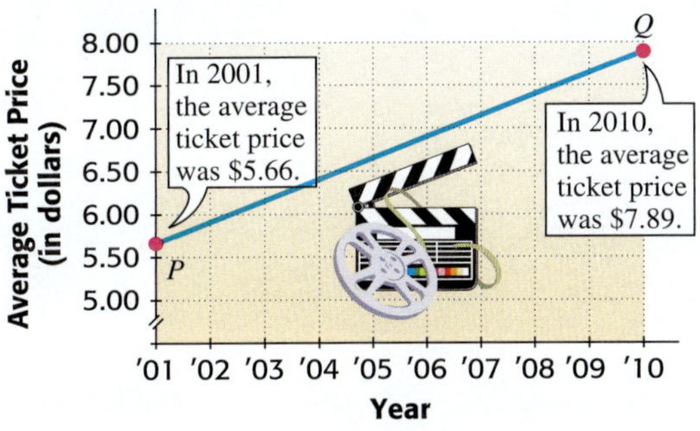

Source: Motion Picture Association of America.

45. The line segment has an equation that can be written in the form $y = ax + b$. Using the coordinates of point P with $x = 2001$ and $y = 5.66$, write an equation in the variables a and b. $5.66 = 2001a + b$

46. Using the coordinates of point Q with $x = 2010$ and $y = 7.89$, write a second equation in the variables a and b. $7.89 = 2010a + b$

47. Write the system of equations formed from the two equations in **Exercises 45 and 46.** Solve the system, giving the values of a and b to three decimal places. (*Hint*: Eliminate b using the elimination method.)

$2001a + b = 5.66$

$2010a + b = 7.89$; solution set: $\{(0.248, -490.590)\}$

48. **(a)** What is the equation of the segment PQ?

$y = 0.248x - 490.590$

(b) Let $x = 2008$ in the equation from part (a), and solve for y (to two decimal places). How does the result compare with the actual figure of \$7.18?

\$7.39; This is a bit more than the actual figure.

Summary Exercises *Applying Techniques for Solving Systems of Linear Equations*

Summary Exercises provide mixed practice to better prepare students for tests.

Use the following guidelines to help decide whether to use substitution or elimination to solve a system of linear equations.

Guidelines for Choosing a Method to Solve a System of Linear Equations

1. If one of the equations of the system is already solved for one of the variables, as indicated by the arrows in the following systems, the substitution method is the better choice.

$$\begin{aligned} 3x + 4y &= 9 \\ \rightarrow \mathbf{y} &= 2x - 6 \end{aligned} \qquad \text{and} \qquad \begin{aligned} \rightarrow \mathbf{x} &= 3y - 7 \\ -5x + 3y &= 9 \end{aligned}$$

2. If both equations are in standard $Ax + By = C$ form and none of the variables has coefficient -1 or 1, as in the following system, the elimination method is the better choice.

$$\begin{aligned} 4x - 11y &= 3 \\ -2x + 3y &= 4 \end{aligned}$$

3. If one or both of the equations are in standard form and the coefficient of one of the variables is -1 or 1, as indicated by the arrows in the following systems, either method is appropriate.

$$\begin{aligned} \rightarrow 3x + \mathbf{y} &= -2 \\ -5x + 2y &= 4 \end{aligned} \qquad \text{and} \qquad \begin{aligned} 3x - 2y &= 8 \\ \rightarrow \mathbf{-x} + 3y &= -4 \end{aligned}$$

CONCEPT CHECK *Use the preceding guidelines to solve each problem.*

1. To minimize the amount of work required, tell whether you would use the substitution or elimination method to solve each system, and why. *Do not actually solve.*

(a) $3x + 2y = 18$

$y = 3x$

Use substitution since the second equation is solved for y.

(b) $3x + y = -7$

$x - y = -5$

Use elimination since the coefficients of the y-terms are opposites.

(c) $3x - 2y = 0$

$9x + 8y = 7$

Use elimination since the equations are in standard form with no coefficients of 1 or −1. Solving by substitution would involve fractions.

2. Which system would be easier to solve using the substitution method? Why?

System A: $\begin{aligned} 5x - 3y &= 7 \\ 2x + 8y &= 3 \end{aligned}$ *System B*: $\begin{aligned} 7x + 2y &= 4 \\ y &= -3x + 1 \end{aligned}$

System B is easier to solve by substitution because the second equation is already solved for y.

*In Exercises 3 and 4, **(a)** solve the system by the elimination method, **(b)** solve the system by the substitution method, and **(c)** tell which method you prefer for that particular system and why.*

3. $4x - 3y = -8$
$x + 3y = 13$
(a) $\{(1, 4)\}$
(b) $\{(1, 4)\}$
(c) Answers will vary.

4. $2x + 5y = 0$
$x = -3y + 1$
(a) $\{(-5, 2)\}$
(b) $\{(-5, 2)\}$
(c) Answers will vary.

*Solve each system by the method of your choice. (For Exercises 5–7, see your answers for **Exercise 1**.)*

5. $3x + 2y = 18$
$y = 3x$
$\{(2, 6)\}$

6. $3x + y = -7$
$x - y = -5$
$\{(-3, 2)\}$

7. $3x - 2y = 0$
$9x + 8y = 7$
$\left\{\left(\frac{1}{3}, \frac{1}{2}\right)\right\}$

8. $x + y = 7$
$x = -3 - y$
$\emptyset$

9. $5x - 4y = 15$
$-3x + 6y = -9$
$\{(3, 0)\}$

10. $4x + 2y = 3$
$y = -x$
$\left\{\left(\frac{3}{2}, -\frac{3}{2}\right)\right\}$

11. $3x = 7 - y$
$2y = 14 - 6x$
$\{(x, y) \mid 3x + y = 7\}$

12. $3x - 5y = 7$
$2x + 3y = 30$
$\{(9, 4)\}$

13. $5x = 7 + 2y$
$5y = 5 - 3x$
$\left\{\left(\frac{45}{31}, \frac{4}{31}\right)\right\}$

14. $4x + 3y = 1$
$3x + 2y = 2$
$\{(4, -5)\}$

15. $2x - 3y = 7$
$-4x + 6y = 14$
$\emptyset$

16. $7x - 4y = 0$
$3x = 2y$
$\{(0, 0)\}$

17. $6x + 5y = 13$
$3x + 3y = 4$
$\left\{\left(\frac{19}{3}, -5\right)\right\}$

18. $x - 3y = 7$
$4x + y = 5$
$\left\{\left(\frac{22}{13}, -\frac{23}{13}\right)\right\}$

Solve each system by any method. First clear all fractions or decimals.

19. $\frac{1}{4}x - \frac{1}{5}y = 9$
$y = 5x$
$\{(-12, -60)\}$

20. $\frac{1}{5}x + \frac{2}{3}y = -\frac{8}{5}$
$x - 5y = 17$
$\{(2, -3)\}$

21. $\frac{1}{6}x + \frac{1}{6}y = 2$
$-\frac{1}{2}x - \frac{1}{3}y = -8$
$\{(24, -12)\}$

22. $\frac{x}{5} + 2y = \frac{8}{5}$
$\frac{3x}{5} + \frac{y}{2} = -\frac{7}{10}$
$\{(-2, 1)\}$

23. $\frac{x}{5} + y = 6$
$\frac{x}{10} + \frac{y}{3} = \frac{5}{6}$
$\{(-35, 13)\}$

24. $\frac{2}{5}x + \frac{4}{3}y = -8$
$\frac{7}{10}x - \frac{2}{9}y = 9$
$\{(10, -9)\}$

25. $0.2x + 0.3y = 1.0$
$-0.3x + 0.1y = 1.8$
$\{(-4, 6)\}$

26. $0.3x - 0.2y = 0.9$
$0.2x - 0.3y = 0.1$
$\{(5, 3)\}$

4.4 Applications of Linear Systems

OBJECTIVES

1. Solve problems about unknown numbers.
2. Solve problems about quantities and their costs.
3. Solve problems about mixtures.
4. Solve problems about distance, rate (or speed), and time.

Teaching Tip

Mention to students that most of the problems in this section could also be solved by using only one variable. Demonstrate this with **Example 1.**

We modify the six-step problem-solving method from **Section 2.4** slightly to allow for two variables and two equations.

Solving an Applied Problem with Two Variables

Step 1 **Read** the problem carefully. What information is given? What is to be found?

Step 2 **Assign variables** to represent the unknown values. Use a sketch, diagram, or table, as needed. Write down what each variable represents.

Step 3 **Write two equations** using both variables.

Step 4 **Solve** the system of two equations.

Step 5 **State the answer.** Label it appropriately. Does it seem reasonable?

Step 6 **Check** the answer in the words of the *original* problem.

OBJECTIVE 1 Solve problems about unknown numbers.

EXAMPLE 1 Solving a Problem about Two Unknown Numbers

In 2011, consumer sales of sports equipment were \$307 million more for snow skiing than for snowboarding. Together, total equipment sales for these two sports were \$931 million. What were equipment sales for each sport? (*Source:* National Sporting Goods Association.)

Step 1 **Read** the problem carefully. We must find equipment sales (in millions of dollars) for snow skiing and for snowboarding. We know how much more equipment sales were for snow skiing than for snowboarding. Also, we know the total sales.

Step 2 **Assign variables.**

Let x = equipment sales for skiing (in millions of dollars),

and y = equipment sales for snowboarding (in millions of dollars).

Step 3 **Write two equations.**

$x = 307 + y$ Equipment sales for skiing were \$307 million more than equipment sales for snowboarding. (1)

$x + y = 931$ Total sales were \$931 million. (2)

Step 4 **Solve** the system from Step 3. We use the substitution method since the first equation is already solved for x.

$x + y = 931$ (2)

$(307 + y) + y = 931$ Let $x = 307 + y$.

$307 + 2y = 931$ Combine like terms.

$2y = 624$ Subtract 307.

Don't stop here. $y = 312$ Divide by 2.

Continued on Next Page

To find the value of x, we substitute 312 for y in either equation (1) or (2).

$$x = 307 + 312 \quad \text{Let } y = 312 \text{ in equation (1).}$$

$$x = 619 \quad \text{Add.}$$

Step 5 **State the answer.** Equipment sales for skiing were $619 million, and equipment sales for snowboarding were $312 million.

Step 6 **Check** the answer in the original problem. Since

$$619 = 307 + 312 \quad \text{and} \quad 619 + 312 = 931,$$

the answer satisfies the information given in the problem.

Work Problem 1 at the Side. ▶

CAUTION

If an applied problem asks for *two* values as in **Example 1,** be sure to give both of them in your answer.

OBJECTIVE ▶ 2 Solve problems about quantities and their costs.

EXAMPLE 2 Solving a Problem about Quantities and Costs

For a production of *Jersey Boys* at the August Wilson Theatre in New York, main floor tickets cost $127 and rear mezzanine tickets cost $97. The members of a club spent a total of $3150 for 30 tickets. How many tickets of each kind did they buy? (*Source*: www.broadway.com)

Step 1 **Read** the problem several times.

Step 2 **Assign variables.**

Let x = the number of main floor tickets,

and y = the number of rear mezzanine tickets.

Summarize the information given in the problem in a table. The entries in the first two rows of the Total Value column were found by multiplying Number of Tickets by Price per Ticket.

	Number of Tickets	Price per Ticket (in dollars)	Total Value (in dollars)
Main Floor	x	127	$127x$
Mezzanine	y	97	$97y$
Total	30	XXXXXX	3150

Step 3 **Write two equations.**

$$x + y = 30 \quad \text{Total number of tickets was 30.} \quad (1)$$

$$127x + 97y = 3150 \quad \text{Total value of tickets was \$3150.} \quad (2)$$

Step 4 **Solve** the system formed in Step 3 using the elimination method.

$$x + y = 30 \quad (1)$$

$$127x + 97y = 3150 \quad (2)$$

Continued on Next Page

1 Solve each problem.

GS **(a)** In 2011, consumer sales of sports equipment were $26 million less for tennis than for archery. Together, total equipment sales for these two sports were $876 million. What were equipment sales for each sport? (*Source*: National Sporting Goods Association.)

Step 1
We must find ________.

Step 2
Let x = equipment sales for tennis (in millions of dollars), and y = equipment sales for ________ (in millions of dollars).

Step 3
Write two equations.

x = ________ (1)
(See the first sentence in the problem.)

________ (2)
(See the second sentence in the problem.)

Complete Steps 4–6 to solve the problem. Give the answer.

(b) Two of the most popular movies of 2011 were *Fast Five* and *Cars 2*. Together, their domestic gross was about $401 million. *Cars 2* grossed $18 million less than *Fast Five*. How much did each movie gross? (*Source:* www.boxofficemojo.com)

Answers

1. **(a)** equipment sales for each sport; archery; $y - 26$; $x + y = 876$; equipment sales for tennis: $425 million; equipment sales for archery: $451 million
(b) *Fast Five:* $209.5 million; *Cars 2:* $191.5 million

2 GS For a production of *The Lion King* playing at the Minskoff Theatre in New York, orchestra tickets cost \$129 and rear mezzanine tickets cost \$87. If a group of 18 people attended the show and spent a total of \$1776 for their tickets, how many of each kind of ticket did they buy? (*Source:* www.broadway.com)

(a) Complete the table.

	Number of Tickets	Price per Ticket (dollars)	Total Value (dollars)
Orchestra	x	____	____
Mezzanine	y	____	____
Total	____	XXX	____

(b) Write a system of equations.

(c) Use the system of equations to solve the problem. Check your answer in the words of the original problem.

Answers

2. (a)

	Number of Tickets	Price per Ticket (dollars)	Total Value (dollars)
Orchestra	x	129	$129x$
Mezzanine	y	87	$87y$
Total	18	XXX	1776

(b) $x + y = 18$
$129x + 87y = 1776$

(c) orchestra: 5; mezzanine: 13

$$\begin{aligned} -97x - 97y &= -2910 && \text{Multiply equation (1) by } -97. \\ 127x + 97y &= 3150 && (2) \\ \hline 30x &= 240 && \text{Add.} \end{aligned}$$

Main floor tickets $\longrightarrow x = \mathbf{8}$ Divide by 30.

Substitute 8 for x in equation (1).

$$\begin{aligned} \mathbf{x} + y &= 30 && (1) \\ \mathbf{8} + y &= 30 && \text{Let } x = 8. \end{aligned}$$

Mezzanine tickets $\longrightarrow y = \mathbf{22}$ Subtract 8.

Step 5 **State the answer.** The club members bought 8 main floor tickets and 22 rear mezzanine tickets.

Step 6 **Check.** The sum of 8 and 22 is 30, so the total number of tickets is correct. Since 8 tickets were purchased at \$127 each and 22 at \$97 each, the total spent on all the tickets is

$$\$127(8) + \$97(22) = \$3150, \quad \text{as required.}$$

◀ Work Problem **2** at the Side.

OBJECTIVE 3 Solve problems about mixtures. In **Section 2.6,** we solved percent problems using one variable. Mixture problems that involve percent can be solved using a system of two equations in two variables.

EXAMPLE 3 Solving a Mixture Problem Involving Percent

A pharmacist needs 100 L of 50% alcohol solution. She has a 30% alcohol solution and an 80% alcohol solution, which she can mix. How many liters of each will be required to make the 100 L of a 50% alcohol solution?

Step 1 **Read** the problem. Note the percent of each solution and of the mixture.

Step 2 **Assign variables.**

Let x = the number of liters of 30% alcohol needed,

and y = the number of liters of 80% alcohol needed.

Liters of Mixture	Percent (as a decimal)	Liters of Pure Alcohol
x	0.30	$0.30x$
y	0.80	$0.80y$
100	0.50	$0.50(100)$

Summarize the information in a table. Percents are written as decimals.

Figure 8 gives an idea of what is happening in this problem.

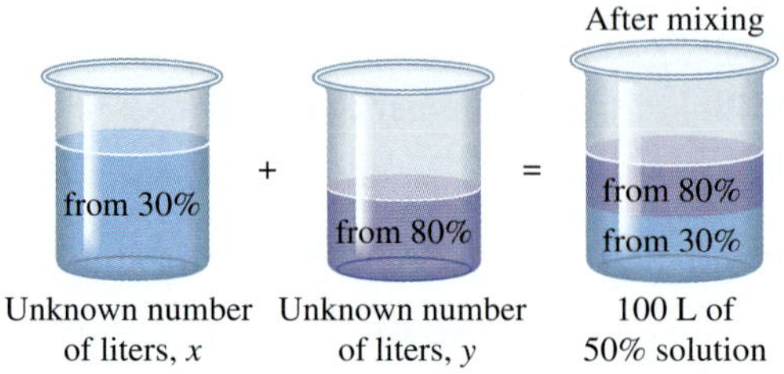

Figure 8

Continued on Next Page

Step 3 **Write two equations.** The total number of liters in the final mixture will be 100, which gives one equation.

$$x + y = 100$$

To find the amount of pure alcohol in each mixture, multiply the number of liters by the concentration. (Refer to the table.) The amount of pure alcohol in the 30% solution added to the amount of pure alcohol in the 80% solution will equal the amount of pure alcohol in the final 50% solution. This gives a second equation.

$$0.30x + 0.80y = 0.50(100)$$

These two equations form a system.

Be sure to write two equations.

$$x + y = 100 \quad (1)$$
$$0.30x + 0.80y = 50 \quad (2) \quad 0.50(100) = 50$$

Step 4 **Solve** the system using the substitution method. Solving equation (1) for x gives $x = 100 - y$. Substitute $100 - y$ for x in equation (2).

$$0.30x + 0.80y = 50 \quad (2)$$
$$0.30(100 - y) + 0.80y = 50 \quad \text{Let } x = 100 - y.$$
$$30 - 0.30y + 0.80y = 50 \quad \text{Distributive property}$$
$$30 + 0.50y = 50 \quad \text{Combine like terms.}$$
$$0.50y = 20 \quad \text{Subtract 30.}$$
$$\text{Liters of 80\% solution} \longrightarrow y = 40 \quad \text{Divide by 0.50.}$$

Equation (1) solved for x is $x = 100 - y$. Substitute 40 for y to find the value of x.

$$x = 100 - 40 \quad \text{Let } y = 40.$$
$$\text{Liters of 30\% solution} \longrightarrow x = 60 \quad \text{Subtract.}$$

Step 5 **State the answer.** The pharmacist should use 60 L of the 30% solution and 40 L of the 80% solution.

Step 6 **Check** the answer in the original problem. Since

$$60 + 40 = 100 \quad \text{and} \quad 0.30(60) + 0.80(40) = 50,$$

this mixture will give the 100 L of 50% solution, as required.

Work Problems 3 and 4 at the Side. ▶

Note

In Example 3, we could have used the elimination method. Also, we could have cleared decimals by multiplying each side of equation (2) by 10.

OBJECTIVE 4 Solve problems about distance, rate (or speed), and time. If an automobile travels at an average rate of 50 mph for 2 hr, then it travels

$$50 \times 2 = 100 \text{ mi.}$$

This is an example of the basic relationship between distance, rate, and time,

distance = rate × time, given by the formula $d = rt$.

Work Problem 5 at the Side. ▶

3 (GS) How many liters of a 25% alcohol solution must be mixed with a 12% solution to get 13 L of a 15% solution?

(a) Complete the table.

Liters	Percent (as a decimal)	Liters of Pure Alcohol
x	0.25	$0.25x$
y	0.12	______
13	0.15	______

(b) Write a system of equations, and use it to solve the problem.

4 Solve the problem.

Joe needs 60 milliliters (mL) of 20% acid solution for a chemistry experiment. The lab has on hand only 10% and 25% solutions. How much of each should he mix to get the desired amount of 20% solution?

5 Solve using the distance formula $d = rt$.

A small plane traveled from Stockholm, Sweden, to Oslo, Norway, averaging 244 km per hr. The trip took 1.7 hr. To the nearest kilometer, what is the distance between the two cities?

Answers

3. (a)

Liters	Percent (as a decimal)	Liters of Pure Alcohol
x	0.25	$0.25x$
y	0.12	$0.12y$
13	0.15	$0.15(13)$

(b) $x + y = 13$
$0.25x + 0.12y = 0.15(13)$;
3 L of 25%; 10 L of 12%

4. 20 mL of 10%; 40 mL of 25%

5. 415 km

6 Two cars that were 450 mi apart **GS** traveled toward each other. They met after 5 hr. If one car traveled twice as fast as the other, what were their rates?

(a) Complete this table.

	r	t	d
Faster Car	x	5	___
Slower Car	y	5	___

(b) Write a system of equations, and use it to solve the problem.

7 Solve the problem.

From a truck stop, two trucks travel in opposite directions on a straight highway. In 3 hr they are 405 mi apart. Find the rate of each truck if one travels 5 mph faster than the other.

Answers

6. (a)

	r	t	d
Faster Car	x	5	$5x$
Slower Car	y	5	$5y$

(b) $5x + 5y = 450$
$x = 2y$;
faster car: 60 mph; slower car: 30 mph

7. faster truck: 70 mph; slower truck: 65 mph

EXAMPLE 4 Solving a Problem about Distance, Rate, and Time

Two executives in cities 400 mi apart drive to a business meeting at a location on the line between their cities. They meet after 4 hr. Find the rate (speed) of each car if one car travels 20 mph faster than the other.

Step 1 **Read** the problem carefully.

Step 2 **Assign variables.**

Let x = the rate of the faster car,

and y = the rate of the slower car.

Make a table using the formula $d = rt$, and draw a sketch. See **Figure 9.**

	r	t	d
Faster Car	x	4	$x \cdot 4$, or $4x$
Slower Car	y	4	$y \cdot 4$, or $4y$

Since each car travels for 4 hr, the time t for each car is 4. Find d, using $d = rt$ (or $rt = d$).

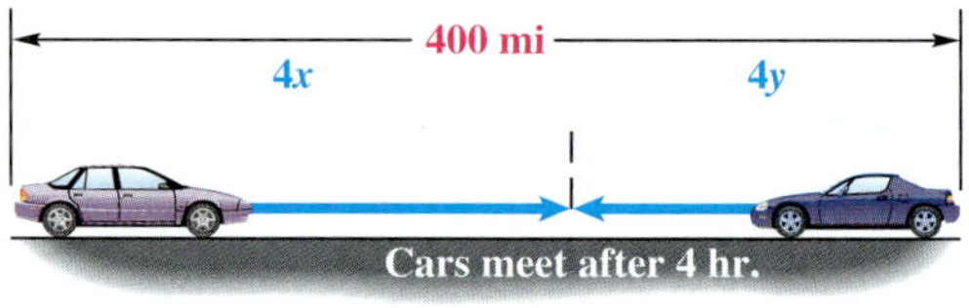

Figure 9

Step 3 **Write two equations.** The total distance traveled by both cars is 400 mi, which gives equation (1). The faster car goes 20 mph faster than the slower car, which gives equation (2).

$$4x + 4y = 400 \quad (1)$$
$$x = 20 + y \quad (2)$$

Step 4 **Solve** the system of equations by substitution. Replace x with $20 + y$ in equation (1) and then solve for y.

$$4x + 4y = 400 \quad (1)$$
$$4(20 + y) + 4y = 400 \quad \text{Let } x = 20 + y.$$
$$80 + 4y + 4y = 400 \quad \text{Distributive property}$$
$$80 + 8y = 400 \quad \text{Combine like terms.}$$
$$8y = 320 \quad \text{Subtract 80.}$$
$$\text{Slower car} \rightarrow y = 40 \quad \text{Divide by 8.}$$

To find x, substitute 40 for y in equation (2), $x = 20 + y$.

$$x = 20 + 40 \quad \text{Let } y = 40.$$
$$\text{Faster car} \rightarrow x = 60 \quad \text{Add.}$$

Step 5 **State the answer.** The rates of the cars are 40 mph and 60 mph.

Step 6 **Check** the answer. Since each car travels for 4 hr, total distance is

$$4(60) + 4(40) = 240 + 160 = 400 \text{ mi}, \quad \text{as required.}$$

◀ Work Problems **6** and **7** at the Side.

CAUTION

Be careful. ***When you use two variables to solve a problem, you must write two equations.***

EXAMPLE 5 Solving a Problem about Distance, Rate, and Time

A plane flies 560 mi in 1.75 hr traveling with the wind. The return trip against the same wind takes the plane 2 hr. Find the rate (speed) of the plane and the wind speed.

Step 1 **Read** the problem several times.

Step 2 **Assign variables.**

Let $x =$ the rate of the plane,

and $y =$ the wind speed.

When the plane is traveling *with* the wind, the wind "pushes" the plane. In this case, the rate (speed) of the plane is the *sum* of the rate of the plane and the wind speed, $(x + y)$ mph. See **Figure 10.**

When the plane is traveling *against* the wind, the wind "slows" the plane down. In this case, the rate (speed) of the plane is the *difference* between the rate of the plane and the wind speed, $(x - y)$ mph. Again, see **Figure 10.**

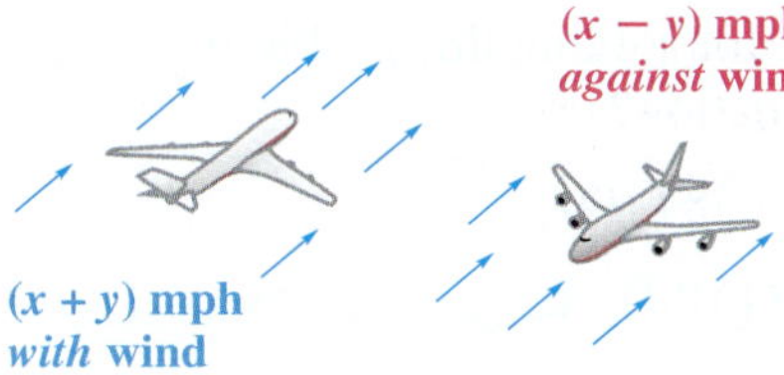

Figure 10

	r	t	d
With Wind	$x + y$	1.75	560
Against Wind	$x - y$	2	560

Summarize this information in a table. Use the formula $d = rt$ (or $rt = d$).

Step 3 **Write two equations.** Refer to the table to do this.

$$(x + y)1.75 = 560 \xrightarrow{\text{Divide by 1.75.}} x + y = 320 \quad (1)$$

$$(x - y)2 = 560 \xrightarrow{\text{Divide by 2.}} x - y = 280 \quad (2)$$

Step 4 **Solve** the system of equations (1) and (2) using elimination.

$$\begin{aligned} x + y &= 320 \quad (1) \\ x - y &= 280 \quad (2) \\ \hline 2x \quad &= 600 \quad \text{Add.} \\ x &= 300 \quad \text{Divide by 2.} \end{aligned}$$

Since $x + y = 320$ and $x = 300$, it follows that $y = 20$.

Step 5 **State the answer.** The rate of the plane is 300 mph, and the wind speed is 20 mph.

Step 6 **Check.** The answer seems reasonable, and true statements result when the values are substituted into the equations of the system.

Work Problem 8 at the Side. ▶

8 Solve each problem.

GS **(a)** In 1 hr, Gigi can row 2 mi against the current or 10 mi with the current. Find the rate of the current and Gigi's rate in still water.

Steps 1 and 2

Let $x =$ the rate of the current,

and $y =$ _______ in still water.

Then her rate *against* the current is (_______) mph,

and her rate with the current is (_______) mph.

Complete the table.

	r	t	d
Against Current	___	___	___
With Current	___	___	___

Step 3

Write two equations.

$(y - x) \cdot 1 =$ ____ (1)

(See the first row of the table.)

______________ (2)

(See the second row of the table.)

Complete Steps 4–6 to solve the problem. Give the answer.

(b) In 1 hr, a boat travels 15 mph with the current and 9 mph against the current. Find the rate of the boat and the rate of the current.

Answers

8. (a) Gigi's rate; $y - x$; $y + x$

	r	t	d
Against Current	$y - x$	1	2
With Current	$y + x$	1	10

2; $(y + x) \cdot 1 = 10$;
rate of the current: 4 mph;
Gigi's rate: 6 mph

(b) rate of the boat: 12 mph;
rate of the current: 3 mph

4.4 Exercises

FOR EXTRA HELP

Download the MyDashBoard App

MyMathLab®

CONCEPT CHECK *Choose the correct response in Exercises 1–8.*

1. Which expression represents the monetary value of x 20-dollar bills?

A. $\frac{x}{20}$ dollars **B.** $\frac{20}{x}$ dollars **C.** $(20 + x)$ dollars **D.** $20x$ dollars D

2. Which expression represents the cost of t pounds of candy that sells for \$1.95 per lb?

A. $\$1.95t$ **B.** $\frac{\$1.95}{t}$ **C.** $\frac{t}{\$1.95}$ **D.** $\$1.95 + t$ A

3. Which expression represents the amount of interest earned on d dollars invested at an interest rate of 2% for 1 yr?

A. $2d$ dollars **B.** $0.02d$ dollars **C.** $0.2d$ dollars **D.** $200d$ dollars B

4. Suppose that x liters of a 40% acid solution are mixed with y liters of a 35% solution to obtain 100 L of a 38% solution. One equation in a system for solving this problem is $x + y = 100$. Which one of the following is the other equation?

A. $0.35x + 0.40y = 0.38(100)$ **B.** $0.40x + 0.35y = 0.38(100)$

C. $35x + 40y = 38$ **D.** $40x + 35y = 0.38(100)$ B

5. According to *Natural History* magazine, the speed of a cheetah is 70 mph. If a cheetah runs for x hours, how many miles does the cheetah cover?

A. $(70 + x)$ miles **B.** $(70 - x)$ miles **C.** $\frac{70}{x}$ miles **D.** $70x$ miles D

6. How far does a car travel in 2.5 hr if it travels at an average rate of x miles per hour?

A. $(x + 2.5)$ miles **B.** $\frac{2.5}{x}$ miles **C.** $\frac{x}{2.5}$ miles **D.** $2.5x$ miles D

7. What is the rate of a plane that travels at 560 mph *with* a wind of r mph?

A. $\frac{r}{560}$ mph **B.** $(560 - r)$ mph **C.** $(560 + r)$ mph **D.** $(r - 560)$ mph C

8. What is the rate of a plane that travels at 560 mph *against* a wind of r mph?

A. $(560 + r)$ mph **B.** $\frac{560}{r}$ mph **C.** $(560 - r)$ mph **D.** $(r - 560)$ mph C

GS *In Exercises 9 and 10, refer to the six-step problem-solving method. Fill in the blanks for Steps 2 and 3, and then complete the solution by applying Steps 4–6.*

9. The sum of two numbers is 98 and the difference between them is 48. Find the two numbers.

Step 1 **Read** the problem carefully.

Step 2 **Assign variables.**

Let $x =$ the first number

and $y =$ the second number.

Step 3 **Write two equations.**

First equation: $x + y = 98$

Second equation: $x - y = 48$

The two numbers are 73 and 25.

10. The sum of two numbers is 201 and the difference between them is 11. Find the two numbers.

Step 1 **Read** the problem carefully.

Step 2 **Assign variables.**

Let $x =$ the first number

and $y =$ the second number.

Step 3 **Write two equations.**

First equation: $x + y = 201$

Second equation: $x - y = 11$

The two numbers are 106 and 95.

Solve each problem using a system of equations. ***See Example 1.***

11. Two of the longest-running shows on Broadway are *The Phantom of the Opera* and *Cats*. As of August 21, 2011, there had been a total of 17,288 performances of the two shows, with 2318 more performances of *The Phantom of the Opera* than *Cats*. How many performances were there of each show? (*Source:* The Broadway League.)

The Phantom of the Opera: 9803; *Cats*: 7485

12. During Broadway runs of *A Chorus Line* and *Beauty and the Beast,* there have been 676 fewer performances of *Beauty and the Beast* than of *A Chorus Line*. There were a total of 11,598 performances of the two shows. How many performances were there of each show? (*Source:* The Broadway League.)

A Chorus Line: 6137; *Beauty and the Beast*: 5461

13. The two domestic top-grossing movies of 2011 were *Harry Potter and the Deathly Hallows Part 2* and *Transformers: Dark of the Moon*. The Transformers movie grossed $29 million less than the Harry Potter movie, and together the two films took in $733 million. How much did each of these movies earn? (*Source:* www.boxofficemojo.com)

Harry Potter and the Deathly Hallows Part 2: $381 million; *Transformers: Dark of the Moon*: $352 million

14. During their opening weekends, the movies *Harry Potter and the Deathly Hallows Part 2* and *Transformers: Dark of the Moon* grossed a total of $267 million, with the Harry Potter movie grossing $71 million more than the Transformers movie. How much did each of these movies earn during their opening weekends? (*Source:* www.boxofficemojo.com)

Harry Potter and the Deathly Hallows Part 2: $169 million; *Transformers: Dark of the Moon*: $98 million

15. The Terminal Tower in Cleveland, Ohio, is 239 ft shorter than the Key Tower, also in Cleveland. The total of the heights of the two buildings is 1655 ft. Find the heights of the buildings. (*Source: World Almanac and Book of Facts.*)

Terminal Tower: 708 ft; Key Tower: 947 ft

16. The total of the heights of the Chase Tower and the One America Tower, both in Indianapolis, Indiana, is 1234 ft. The Chase Tower is 168 ft taller than the One America Tower. Find the heights of the two buildings. (*Source: World Almanac and Book of Facts.*)

Chase Tower: 701 ft; One America Tower: 533 ft

17. GS An official playing field (including end zones) for the Indoor Football League has length 38 yd longer than its width. The perimeter of the rectangular field is 188 yd. Find the length and width of the field. (*Source:* Indoor Football League.)

Steps 1 and 2 **Read** carefully, and **assign** variables.
Let $x =$ the length (in yards),
and $y =$ the width (in yards).

Step 3 **Write two equations.**
Equation (1): See the first sentence in the problem. Express the length in terms of the width.

$$x = 38 + y \qquad (1)$$

Equation (2): See the second sentence in the problem. Perimeter of a rectangle equals twice the length plus twice the width.

$$2x + 2y = 188 \qquad (2)$$

Now complete the solution.

length: 66 yd; width: 28 yd

18. GS Pickleball is a combination of badminton, tennis, and ping pong. The perimeter of the rectangular-shaped court is 128 ft. The width is 24 ft shorter than the length. Find the length and width of the court. (*Source:* www.sportsknowhow.com)

Steps 1 and 2 **Read** carefully, and **assign** variables.
Let $x =$ the length (in feet),
and $y =$ the width (in feet).

Step 3 **Write two equations.**
Equation (1): Perimeter of a rectangle equals twice the length plus twice the width.

$$2x + 2y = 128 \qquad (1)$$

Equation (2): See the third sentence in the problem. Express the width in terms of the length.

$$y = x - 24 \qquad (2)$$

Now complete the solution.

length: 44 ft; width: 20 ft

Suppose that x units of a product cost C dollars to manufacture and earn revenue of R dollars. The value of x, where the expressions for C and R are equal, is the ***break-even quantity,*** *the number of units that produce 0 profit.*

In Exercises 19 and 20, ***(a)*** *find the break-even quantity, and* ***(b)*** *decide whether the product should be produced based on if it will earn a profit. (Profit equals revenue minus cost.)*

19. $C = 85x + 900$; $R = 105x$;
No more than 38 units can be sold.

(a) 45 units

(b) Do not produce—the product will lead to a loss.

20. $C = 105x + 6000$; $R = 255x$;
No more than 400 units can be sold.

(a) 40 units

(b) Produce—the product will earn a profit.

For each problem, complete any tables. Then solve the problem using a system of equations. ***See Example 2.***

21. A motel clerk counts his \$1 and \$10 bills at the end of a day. He finds that he has a total of 74 bills having a combined monetary value of \$326. Find the number of bills of each denomination that he has.

Number of Bills	Denomination of Bill (in dollars)	Total Value (in dollars)
x	1	1x, or x
y	10	10y
74	XXXXXX	326

46 ones; 28 tens

22. Carly is a bank teller. At the end of a day, she has a total of 69 \$5 and \$10 bills. The total value of the money is \$590. How many of each denomination does she have?

Number of Bills	Denomination of Bill (in dollars)	Total Value (in dollars)
x	5	$5x$
y	10	10y
69	XXXXXX	590

20 fives; 49 tens

23. Tracy Sudak bought each of her seven nephews a DVD of *Moneyball* or a Blu-ray disc of *The Concert for George*. Each DVD cost \$16.99. Each Blu-ray disc cost \$22.42. Tracy spent a total of \$129.79. How many of each did she buy? (*Source:* www.amazon.com)

5 DVDs of *Moneyball*; 2 Blu-ray discs of *The Concert for George*

24. Terry Wong bought each of his five nieces a DVD of *Dolphin Tale* or a Blu-ray disc of *Treasure Buddies*. Each DVD cost \$14.99. Each Blu-ray disc cost \$24.99. Terry spent a total of \$84.95. How many of each did he buy? (*Source:* www.amazon.com)

4 DVDs of *Dolphin Tale*; 1 Blu-ray disc of *Treasure Buddies*

25. GS Maria Lopez has twice as much money invested at 5% simple annual interest as she does at 4%. If her yearly income from these two investments is \$350, how much does she have invested at each rate?

Amount Invested (in dollars)	Rate of Interest	Interest for One Year (in dollars)
x	5%, or 0.05	$0.05x$
y	**4%, or 0.04**	**0.04y**
XXXXX	XXXXX	350

Equation (1): $x =$ **2y**

Equation (2): $0.05x +$ **0.04y** $=$ **350**

\$2500 at 4%; \$5000 at 5%

26. GS Charles Miller invested in two accounts, one paying 3% simple annual interest and the other paying 2% interest. He earned a total of \$880 interest. If he invested three times as much in the 3% account as in the 2% account, how much did he invest at each rate?

Amount Invested (in dollars)	Rate of Interest	Interest for One Year (in dollars)
x	**3%, or 0.03**	**0.03x**
y	**2%, or 0.02**	**0.02y**
XXXXX	XXXX	**880**

Equation (1): **0.03x** + **0.02y** $= 880$

Equation (2): $x =$ **3y**

\$24,000 at 3%; \$8000 at 2%

27. The two top-grossing North American concert tours in 2011 were U2 and Taylor Swift. Based on average ticket prices, it cost a total of \$912 to buy six tickets for U2 and five tickets for Taylor Swift. Three tickets for U2 and four tickets for Taylor Swift cost \$564. How much did an average ticket cost for each tour? (*Source:* Pollstar.)

U2: \$92; Taylor Swift: \$72

28. Two other popular North American concert tours in 2011 were Kenny Chesney and Lady Gaga. Based on average ticket prices, it cost a total of \$875 to buy eight tickets for Kenny Chesney and three tickets for Lady Gaga. Four tickets for Kenny Chesney and five tickets for Lady Gaga cost \$777. How much did an average ticket cost for each tour? (*Source:* Pollstar.)

Kenny Chesney: \$73; Lady Gaga: \$97

For each problem, complete any tables. Then solve the problem using a system of equations. ***See Example 3.***

29. A 40% dye solution is to be mixed with a 70% dye solution to get 120 L of a 50% solution. How many liters of the 40% and 70% solutions will be needed?

Liters of Solution	Percent (as a decimal)	Liters of Pure Dye
x	0.40	**0.40x**
y	0.70	**0.70y**
120	0.50	**0.50(120), or 60**

80 L of 40% solution; 40 L of 70% solution

30. A 90% antifreeze solution is to be mixed with a 75% solution to make 120 L of a 78% solution. How many liters of the 90% and 75% solutions will be used?

Liters of Solution	Percent (as a decimal)	Liters of Pure Antifreeze
x	0.90	**0.90x**
y	0.75	**0.75y**
120	0.78	**0.78(120), or 93.6**

24 L of 90% solution; 96 L of 75% solution

31. Ahmad Hashemi wishes to mix coffee worth \$6 per lb with coffee worth \$3 per lb to get 90 lb of a mixture worth \$4 per lb. How many pounds of the \$6 and the \$3 coffees will be needed?

Number of Pounds	Dollars per Pound	Cost (in dollars)
x	6	6x
y	3	3y
90	4	4(90), or 360

30 lb at \$6 per lb; 60 lb at \$3 per lb

32. Mariana Coanda wishes to blend candy selling for \$1.20 per lb with candy selling for \$1.80 per lb to get a mixture that will be sold for \$1.40 per lb. How many pounds of the \$1.20 and the \$1.80 candies should be used to get 45 lb of the mixture?

Number of Pounds	Dollars per Pound	Cost (in dollars)
x	1.20	1.20x
y	1.80	1.80y
45	1.40	1.40(45), or 63

30 lb at \$1.20 per lb; 15 lb at \$1.80 per lb

33. How many pounds of nuts selling for \$6 per lb and raisins selling for \$3 per lb should Kelli Hammer combine to obtain 60 lb of a trail mix selling for \$5 per lb?

nuts: 40 lb; raisins: 20 lb

34. Callie Daniels is preparing cheese trays. She uses some cheeses that sell for \$8 per lb and others that sell for \$12 per lb. How many pounds of each cheese should she use in order for the mixed cheeses on the trays to weigh a total of 56 lb and sell for \$10.50 per lb?

21 lb at \$8 per lb; 35 lb at \$12 per lb

For each problem, complete any tables. Then solve the problem using a system of equations. ***See Examples 4 and 5.***

35. Two trains start from towns 495 mi apart and travel toward each other on parallel tracks. They pass each other 4.5 hr later. If one train travels 10 mph faster than the other, find the rate of each train.

	r	t	d
Train 1	x	4.5	4.5x
Train 2	y	4.5	4.5y

Equation (1): $4.5x +$ 4.5y $=$ 495

Equation (2): $x =$ 10 $+$ y

60 mph; 50 mph

36. Two trains that are 495 mi apart travel toward each other. They pass each other 5 hr later. If one train travels half as fast as the other, what are their rates?

	r	t	d
Train 1	x	5	5x
Train 2	y	5	$5y$

Equation (1): 5x $+ 5y =$ 495

Equation (2): $x =$ $\frac{1}{2}$ y, or 2x $= y$

66 mph; 33 mph

37. RAGBRAI®, the Des Moines **R**egister's **A**nnual **G**reat **B**icycle **R**ide **A**cross **I**owa, is the longest and oldest touring bicycle ride in the world. Suppose a cyclist began the 471-mi ride on July 22, 2012, in western Iowa at the same time that a car traveling toward it left eastern Iowa. If the bicycle and the car met after 7.5 hr and the car traveled 35.8 mph faster than the bicycle, find the average rate of each. (*Source:* www.ragbrai.com)

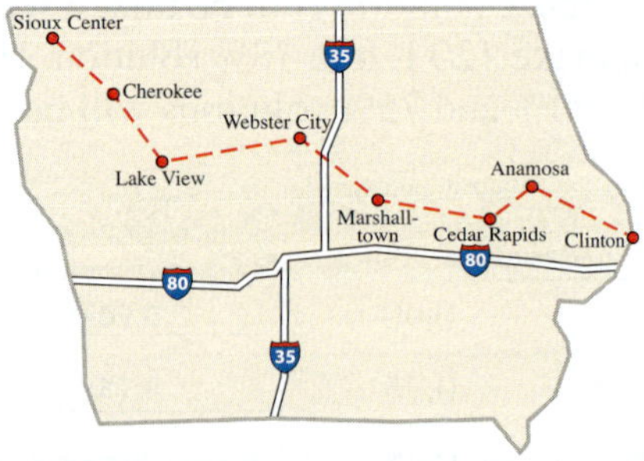

bicycle: 13.5 mph; car: 49.3 mph

38. In 2010, Atlanta's Hartsfield Airport was the nation's busiest. Suppose two planes leave the airport at the same time, one traveling east and the other traveling west. If the planes are 2100 mi apart after 2 hr and one plane travels 50 mph faster than the other, find the rate of each plane. (*Source:* Airports Council International.)

500 mph; 550 mph

39. Toledo and Cincinnati are 200 mi apart. A car leaves Toledo traveling toward Cincinnati, and another car leaves Cincinnati at the same time, traveling toward Toledo. The car leaving Toledo averages 15 mph faster than the other, and they meet after 1 hr, 36 min. What are the rates of the cars?

car leaving Cincinnati: 55 mph; car leaving Toledo: 70 mph

40. Kansas City and Denver are 600 mi apart. Two cars start from these cities, traveling toward each other. They meet after 6 hr. Find the rate of each car if one travels 30 mph slower than the other.

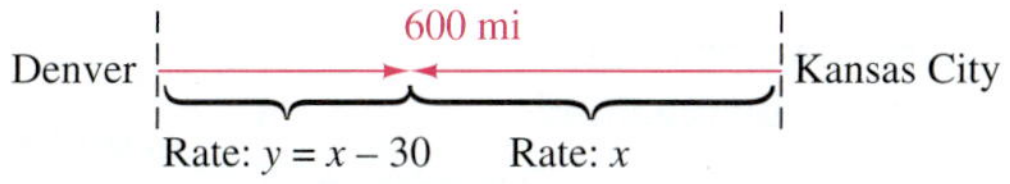

35 mph; 65 mph

41. At the beginning of a bicycle ride for charity, Roberto and Juana are 30 mi apart. If they leave at the same time and ride in the same direction, Roberto overtakes Juana in 6 hr. If they ride toward each other, they meet in 1 hr. What are their rates?

Roberto: 17.5 mph; Juana: 12.5 mph

42. Mr. Abbot left Farmersville in a plane at noon to travel to Exeter. Mr. Baker left Exeter in his automobile at 2 P.M. to travel to Farmersville. It is 400 mi from Exeter to Farmersville. If the sum of their rates was 120 mph, and if they crossed paths at 4 P.M., find the rate of each.

Abbot: 80 mph; Baker: 40 mph

43. A boat takes 3 hr to go 24 mi upstream. It can go 36 mi downstream in the same time. Find the rate of the current and the rate of the boat in still water.

Let $x =$ the rate of the boat in still water and $y =$ the rate of the current.

	r	*t*	*d*
Downstream	$x + y$	**3**	36
Upstream	$x - y$	**3**	**24**

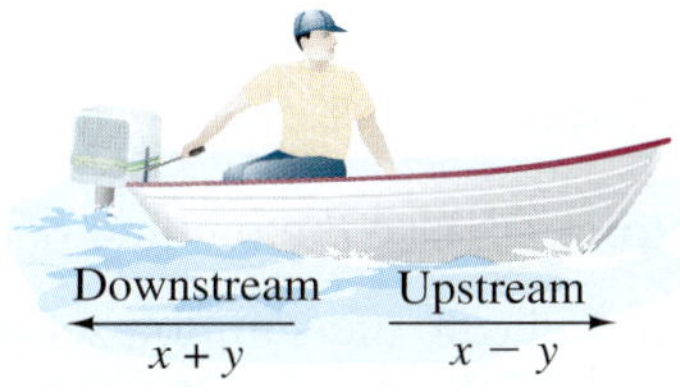

boat: 10 mph; current: 2 mph

44. It takes a boat $1\frac{1}{2}$ hr to go 12 mi downstream, and 6 hr to return. Find the rate of the boat in still water and the rate of the current.

Let $x =$ the rate of the boat in still water and $y =$ the rate of the current.

	r	*t*	*d*
Downstream	$x + y$	$\frac{3}{2}$	**12**
Upstream	$x - y$	6	**12**

boat: 5 mph; current: 3 mph

45. If a plane can travel 440 mph against the wind and 500 mph with the wind, find the wind speed and the rate of the plane in still air.

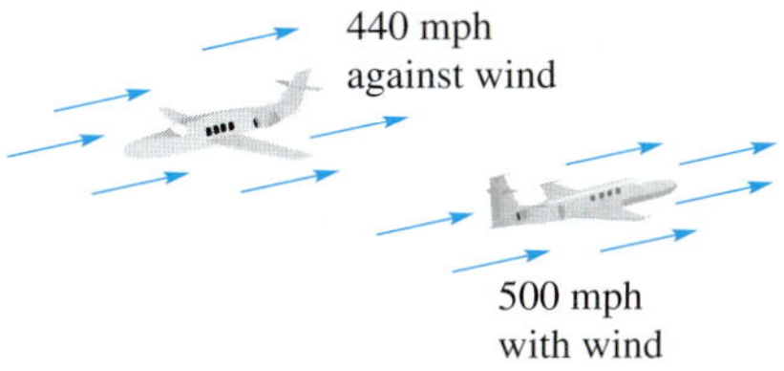

plane: 470 mph; wind: 30 mph

46. A small plane travels 200 mph with the wind and 120 mph against it. Find the wind speed and the rate of the plane in still air.

plane: 160 mph; wind: 40 mph

4.5 Solving Systems of Linear Inequalities

OBJECTIVE

1 Solve systems of linear inequalities by graphing.

We graphed the solutions of a linear inequality in **Section 3.5.** For example, to graph the solutions of $x + 3y > 12$, we first graph the boundary line $x + 3y = 12$ by finding and plotting a few ordered pairs that satisfy the equation. (The *x*- and *y*-intercepts are good choices.) Because the $>$ symbol does not include equality, the points on the line do *not* satisfy the inequality, and we graph it using a dashed line. To decide which region includes the points that are solutions, we choose a test point not on the line.

$$x + 3y > 12 \quad \text{Original inequality}$$

$$0 + 3(0) \overset{?}{>} 12 \quad \text{Let } x = 0 \text{ and } y = 0.$$

$$0 > 12 \quad \text{False}$$

(0, 0) is a convenient test point.

This false result indicates that the solutions are those points on the side of the line that does *not* include (0, 0), as shown in **Figure 11.**

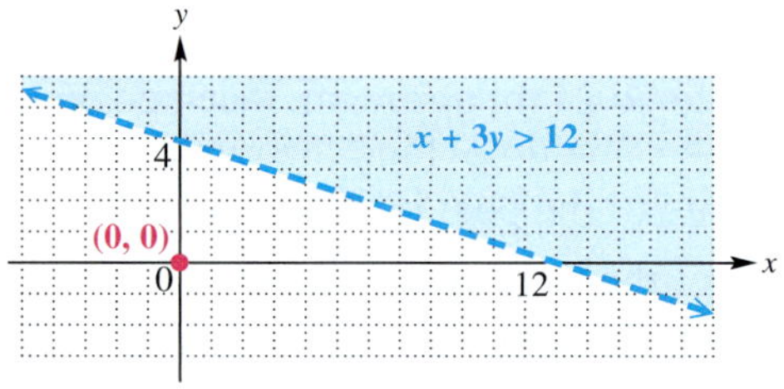

Figure 11

OBJECTIVE 1 Solve systems of linear inequalities by graphing. A **system of linear inequalities** consists of two or more linear inequalities. The **solution set of a system of linear inequalities** includes all points that make all inequalities of the system true at the same time.

Solving a System of Linear Inequalities

Step 1 **Graph each linear inequality.** Use the method of **Section 3.5.**

Step 2 **Choose the intersection.** Indicate the solution set of the system by shading the intersection of the graphs (the region where the graphs overlap).

EXAMPLE 1 Solving a System of Two Linear Inequalities

Graph the solution set of the system.

$$3x + 2y \le 6$$

$$2x - 5y \ge 10$$

Step 1 To graph $3x + 2y \le 6$, graph the solid boundary line $3x + 2y = 6$ using the intercepts $(0, 3)$ and $(2, 0)$. Determine the region to shade.

$$3x + 2y \le 6$$

$$3(0) + 2(0) \overset{?}{\le} 6 \quad \text{Use } (0, 0) \text{ as a test point.}$$

$$0 \le 6 \quad \text{True}$$

Shade the region containing $(0, 0)$. See **Figure 12(a)** on the next page.

Continued on Next Page

Now graph $2x - 5y \geq 10$ with solid boundary line $2x - 5y = 10$ using the intercepts $(0, -2)$ and $(5, 0)$. Determine the region to shade.

$$2x - 5y \geq 10$$

$$2(0) - 5(0) \overset{?}{\geq} 10 \quad \text{Use } (0, 0) \text{ as a test point.}$$

$$0 \geq 10 \quad \text{False}$$

Shade the region that does *not* contain $(0, 0)$. See **Figure 12(b).**

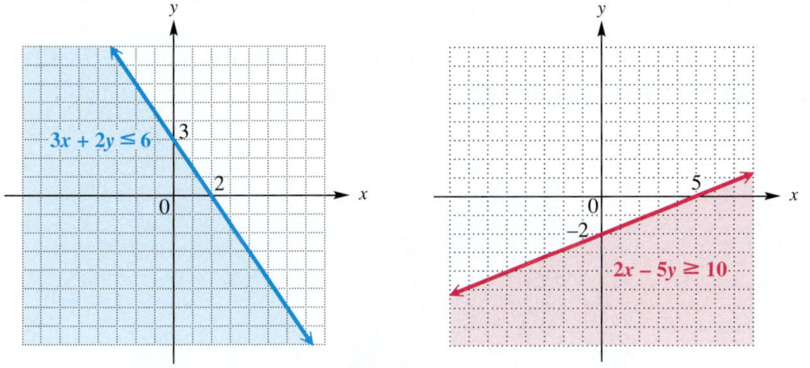

(a) (b)

Figure 12

Step 2 The solution set of this system includes all points in the intersection (overlap) of the graphs of the two inequalities. As shown in **Figure 13,** this intersection is the gray shaded region and portions of the two boundary lines that surround it.

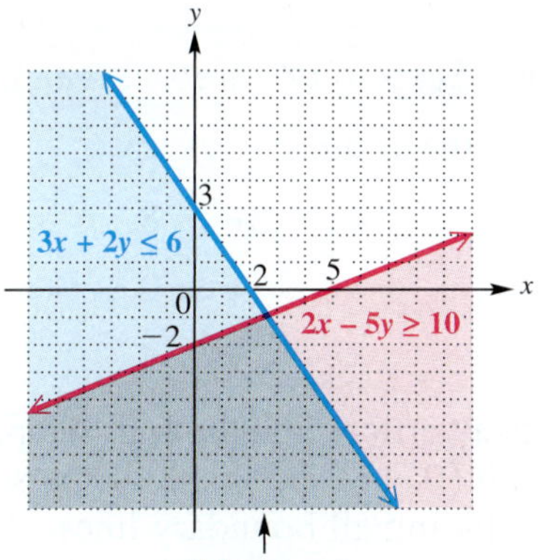

Figure 13

CHECK To confirm that the correct region of intersection is shaded, select a test point in the gray shaded region, say $(0, -4)$, and substitute it into *both* inequalities of the given system to make sure that true statements result. Try this. (Using an ordered pair that has one coordinate 0 makes the substitution easier.) ✓

Work Problem 1 at the Side. ▶

Note

We usually do all the work on one set of axes. In the remaining examples, only one graph is shown. Be sure that the region of the final solution set is clearly indicated.

1 Graph the solution set of each system.

GS **(a)** $x - 2y \leq 8$

$3x + y \geq 6$

To get you started, the graphs of $x - 2y = 8$ and $3x + y = 6$ are shown.

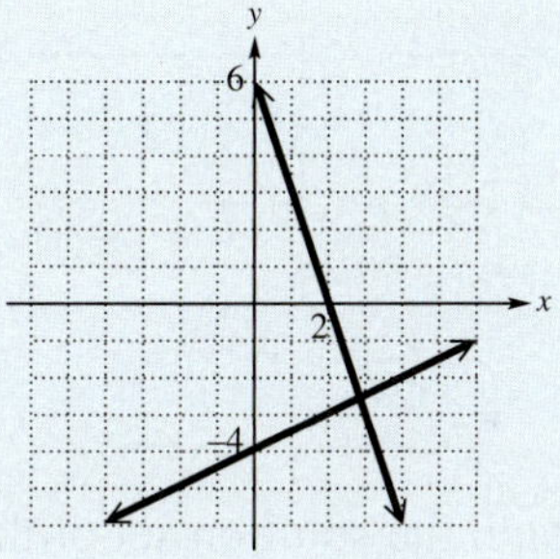

(b) $4x - 2y \leq 8$

$x + 3y \geq 3$

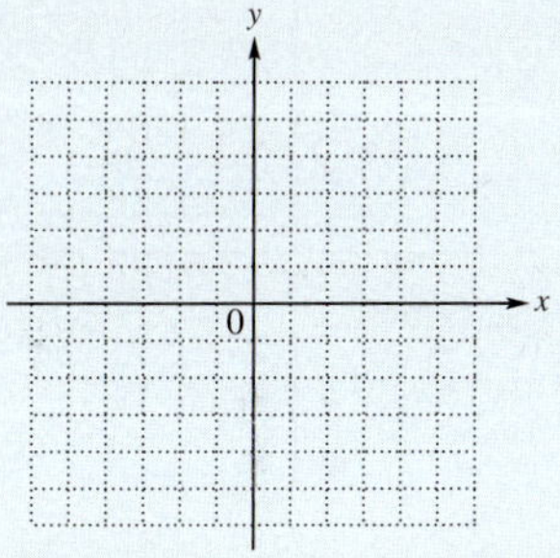

Answers

1. (a)

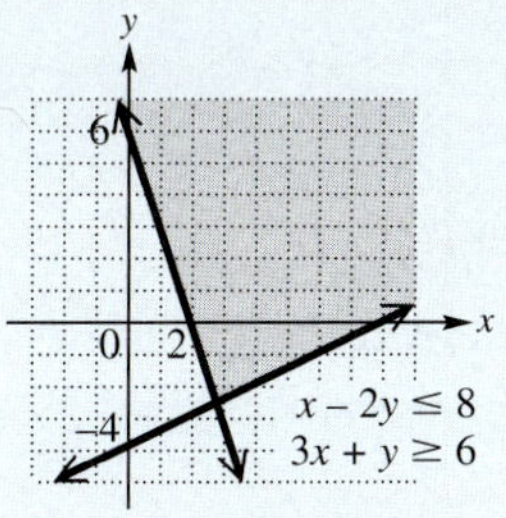

(b)

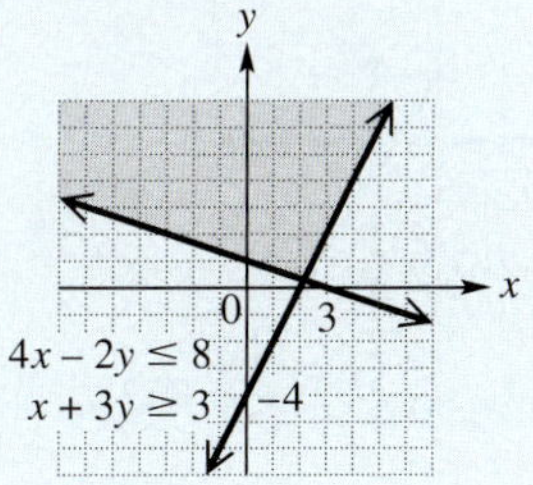

2 Graph the solution set of the system.

$$x + 2y < 0$$
$$3x - 4y < 12$$

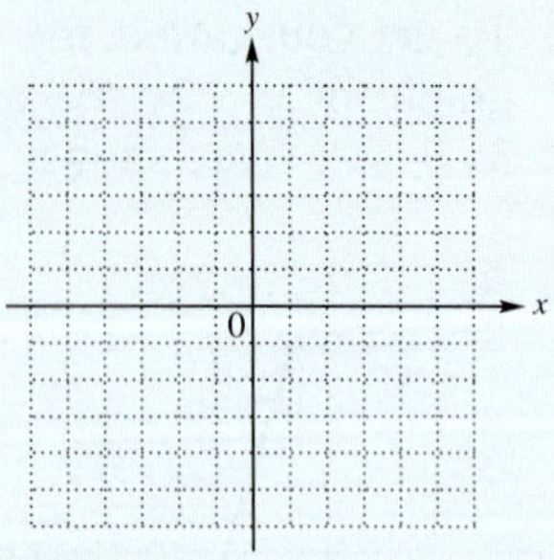

3 Graph the solution set of the system.

$$3x + 2y \le 12$$
$$x \le 2$$
$$y \le 4$$

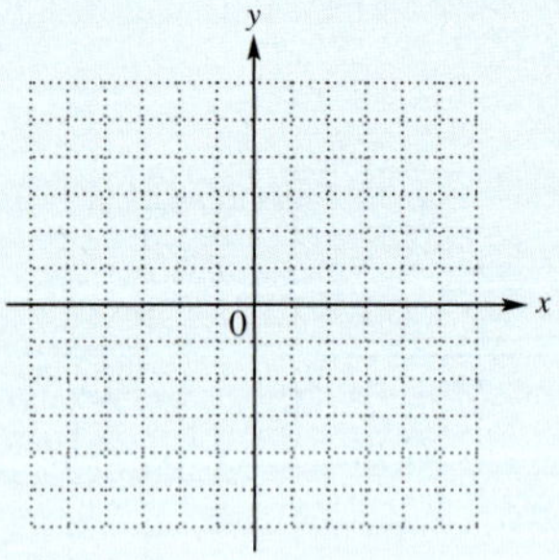

EXAMPLE 2 Solving a System of Two Linear Inequalities

Graph the solution set of the system.

$$x - y > 5$$
$$2x + y < 2$$

Figure 14 shows the graphs of both $x - y > 5$ and $2x + y < 2$. Dashed lines show that the graphs of the inequalities do not include their boundary lines. Use $(0, 0)$ as a test point to determine the region to shade for each inequality.

The solution set of the system is the region with the darkest shading. The solution set does not include either boundary line.

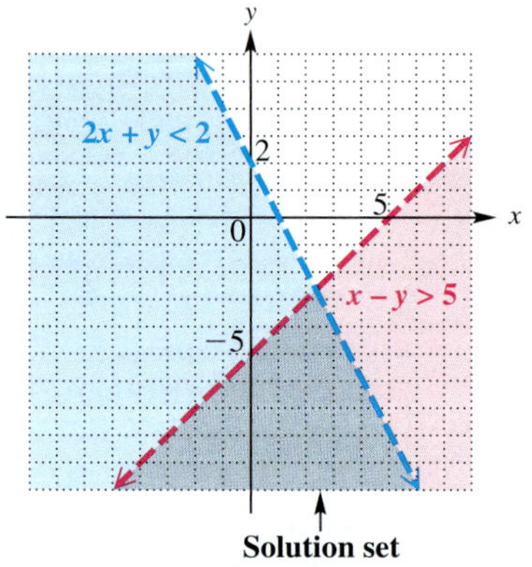

Figure 14

◀ Work Problem **2** at the Side.

EXAMPLE 3 Solving a System of Three Linear Inequalities

Graph the solution set of the system.

$$4x - 3y \le 8$$
$$x \ge 2$$
$$y \le 4$$

Recall that $x = 2$ is a vertical line through the point $(2, 0)$, and $y = 4$ is a horizontal line through $(0, 4)$. The graph of the solution set is the shaded region in **Figure 15,** including all boundary lines. (Here, use $(3, 2)$ as a test point to confirm that the correct region is shaded.)

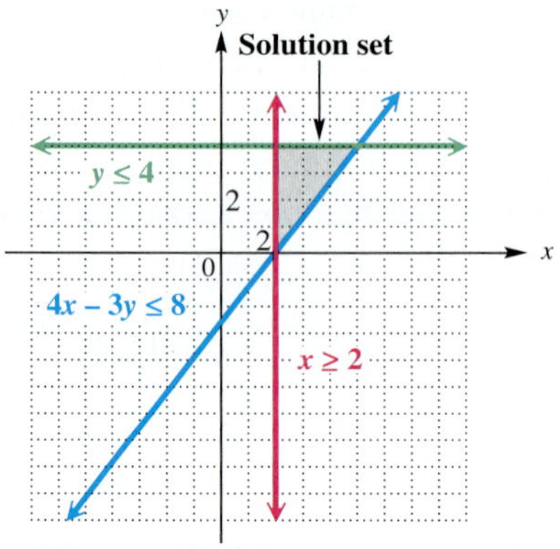

Figure 15

◀ Work Problem **3** at the Side.

Answers

2.

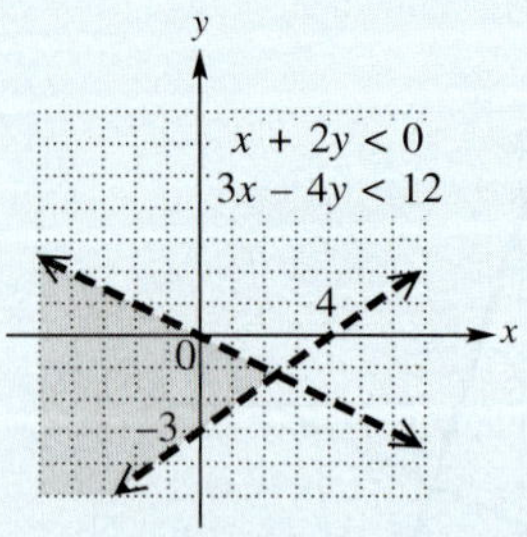

3.

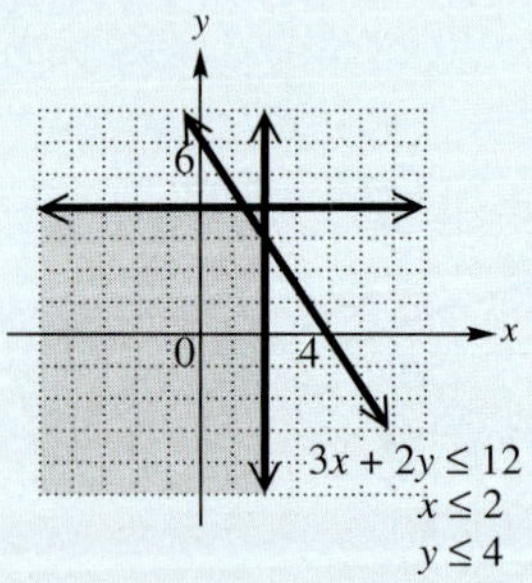

4.5 Exercises

FOR EXTRA HELP

Download the MyDashBoard App

MyMathLab®

CONCEPT CHECK *Match each system of inequalities with the correct graph from choices A–D.*

1. $x \geq 5$
$y \leq -3$ C

2. $x \leq 5$
$y \geq -3$ A

3. $x > 5$
$y < -3$ B

4. $x < 5$
$y > -3$ D

A.

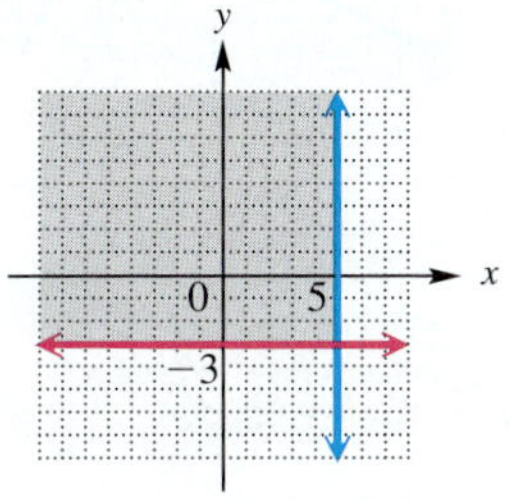

B.

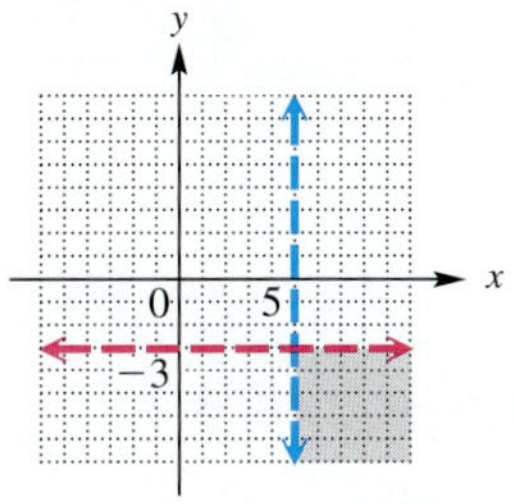

C.

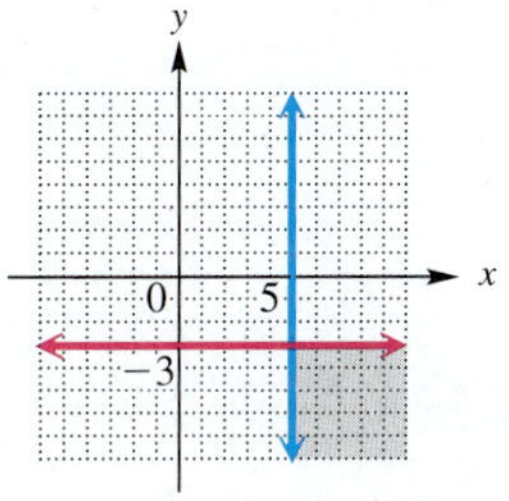

D.

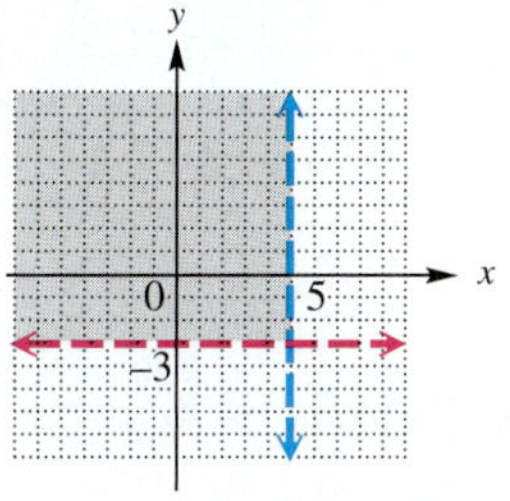

CONCEPT CHECK *In Exercises 5–8, shade the solution set of each system of inequalities. Boundary lines are already graphed.*

5. $x - 3y \leq 6$
$x \geq -4$

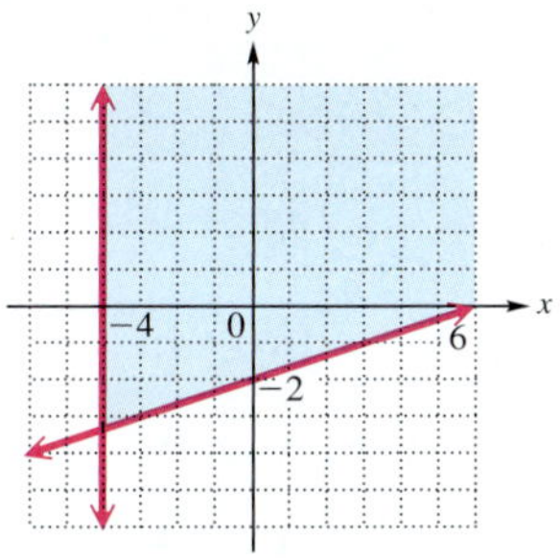

6. $x - 2y \geq 4$
$x \leq -2$

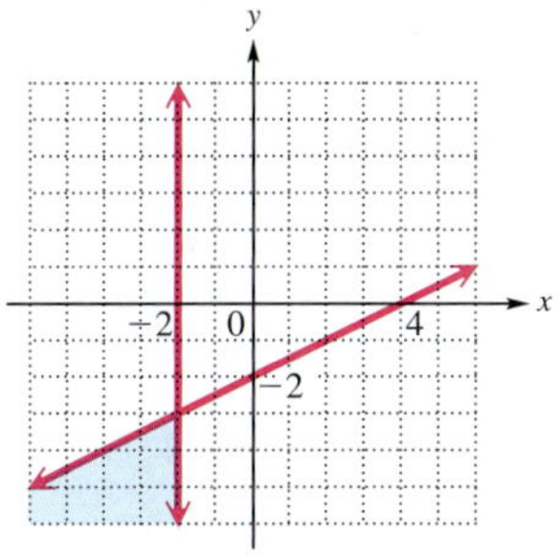

7. $x + y > 4$
$5x - 3y < 15$

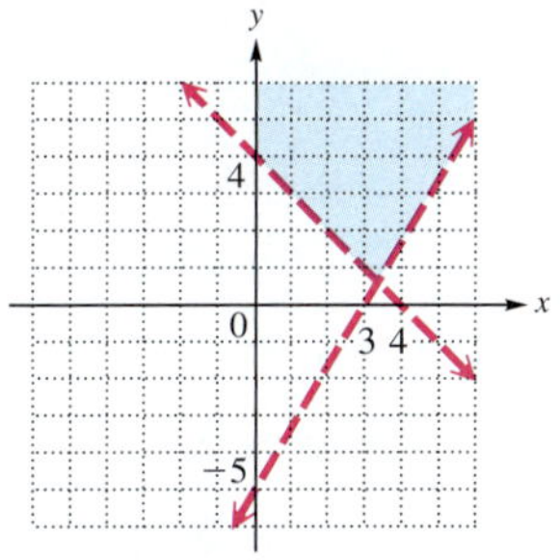

8. $3x - 2y > 12$
$4x + 3y < 12$

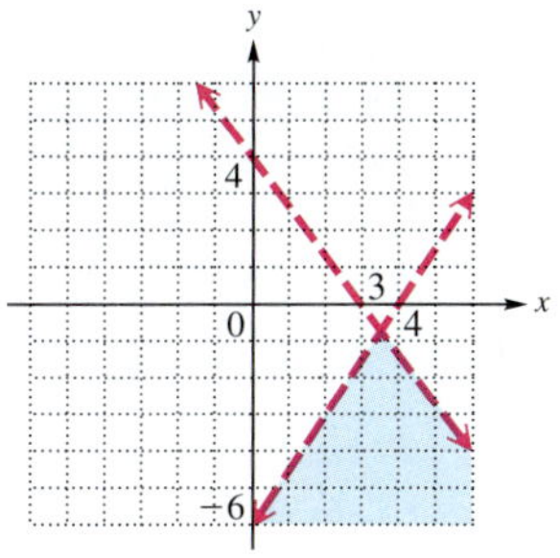

Graph the solution set of each system of linear inequalities. ***See Examples 1–3.***

9. $x + y \leq 6$
$x - y \geq 1$

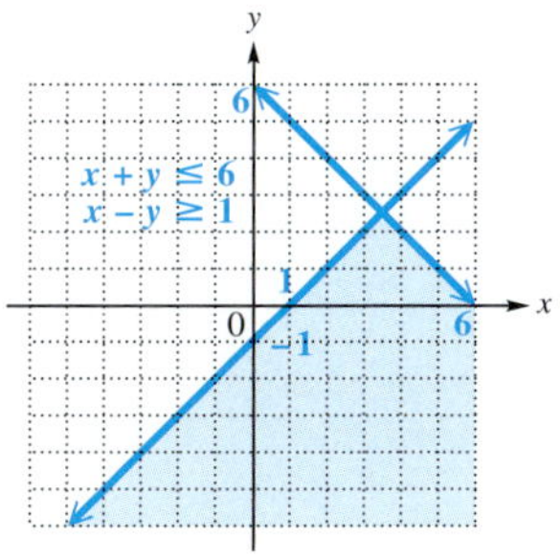

10. $x + y \leq 2$
$x - y \geq 3$

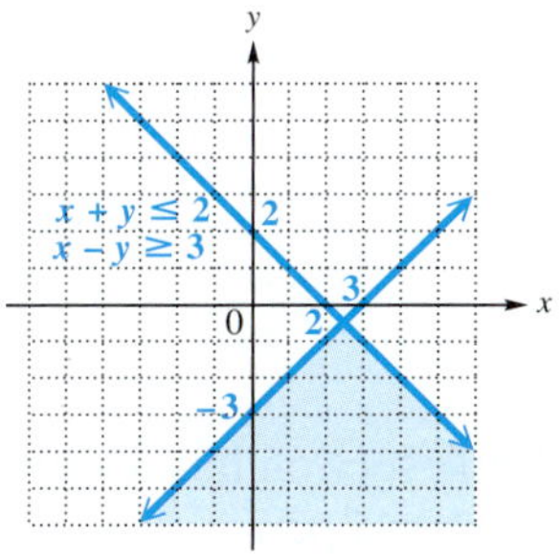

11. $4x + 5y \geq 20$
$x - 2y \leq 5$

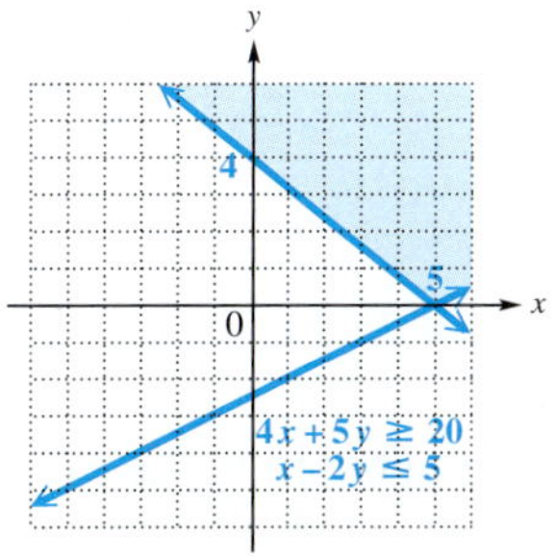

12. $x + 4y \leq 8$
$2x - y \geq 4$

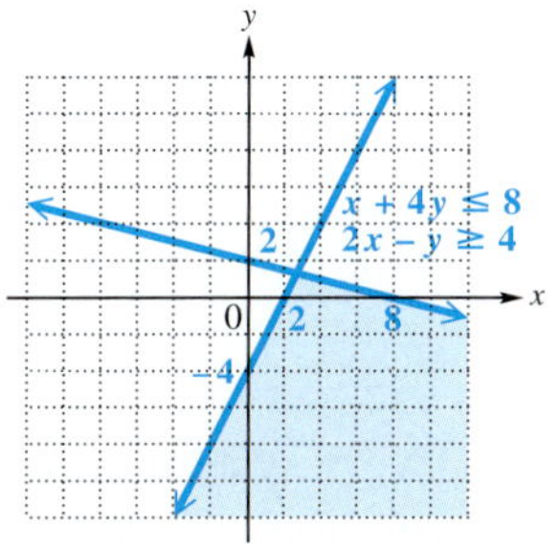

13. $2x + 3y < 6$
$x - y < 5$

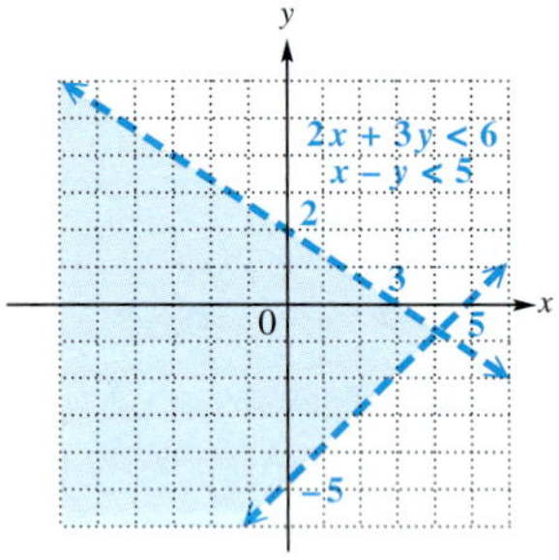

14. $x + 2y < 4$
$x - y < -1$

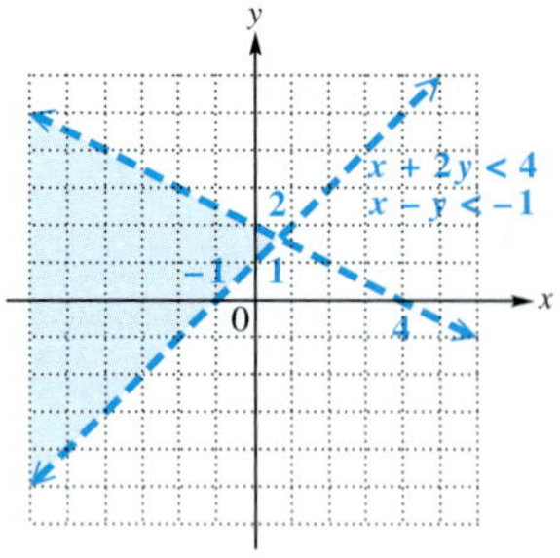

15. $y \leq 2x - 5$
$x < 3y + 2$

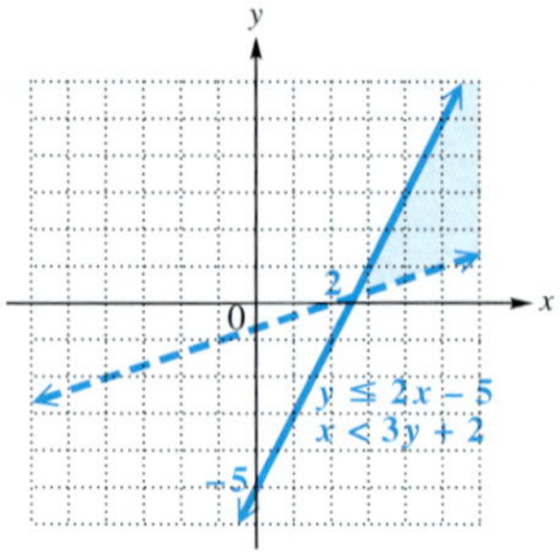

16. $x \geq 2y + 6$
$y > -2x + 4$

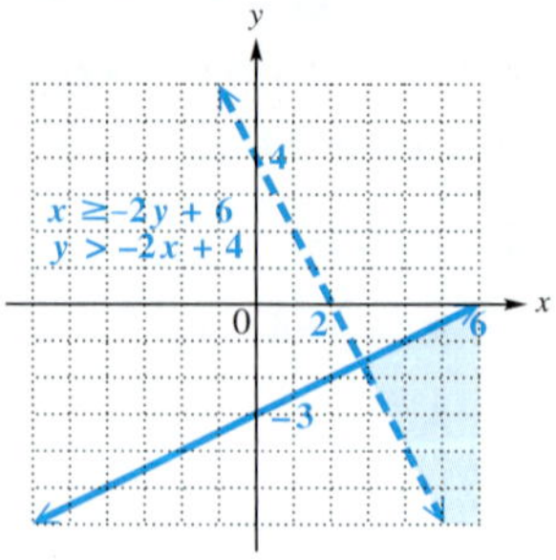

17. $4x + 3y < 6$
$x - 2y > 4$

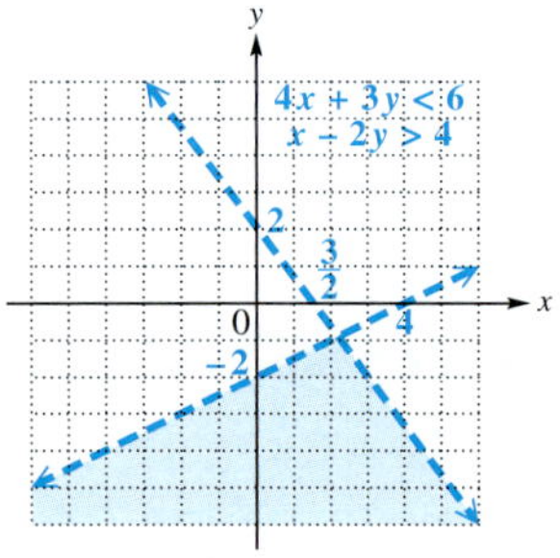

18. $3x + y > 4$
$x + 2y < 2$

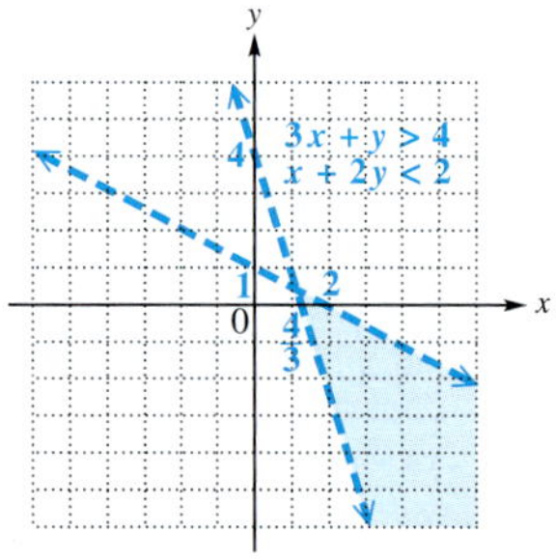

19. $x \leq 2y + 3$
$x + y < 0$

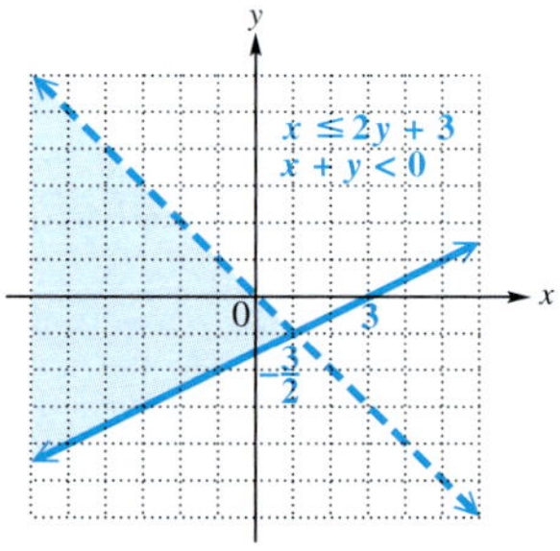

20. $x \leq 4y + 3$
$x + y > 0$

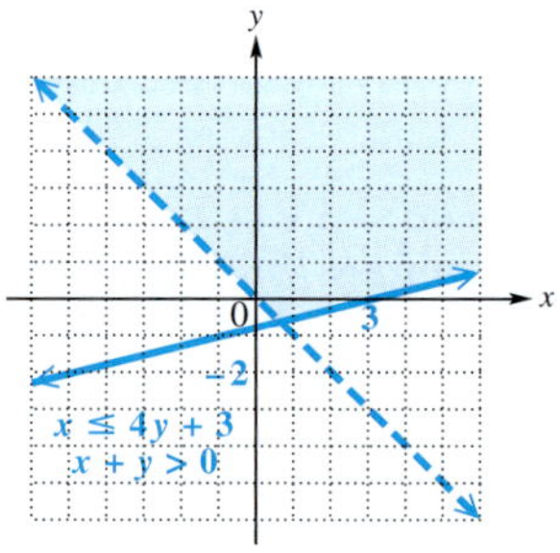

21. $4x + 5y < 8$
$y > -2$
$x > -4$

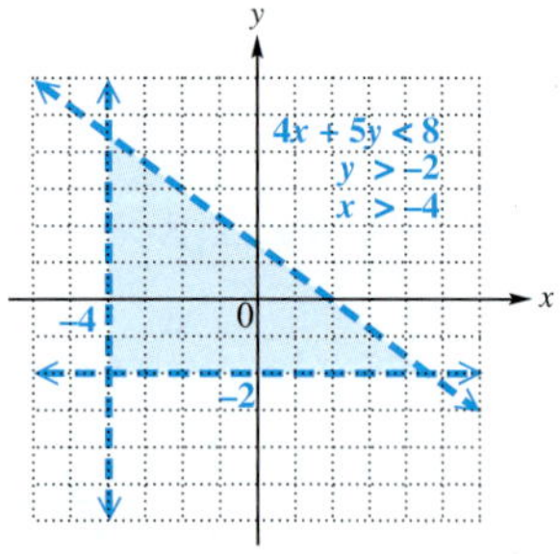

22. $x + y \geq -3$
$x - y \leq 3$
$y \leq 3$

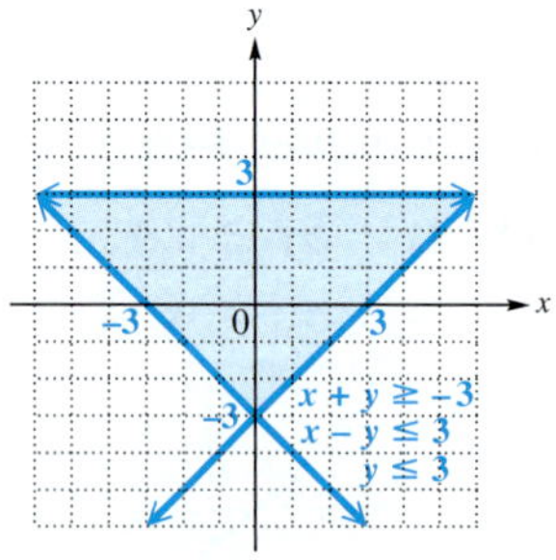

23. $3x - 2y \geq 6$
$x + y \leq 4$
$x \geq 0$
$y \geq -4$

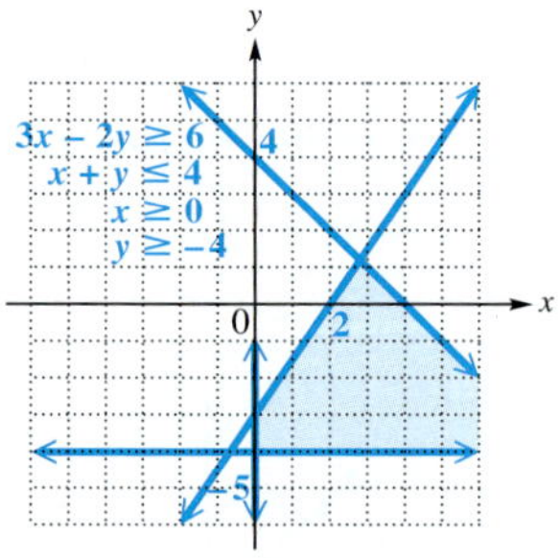

24. $2x - 3y < 6$
$x + y > 3$
$x < 4$
$y < 4$

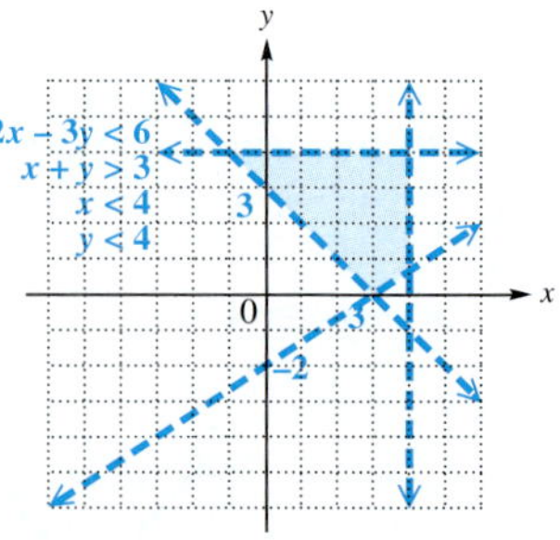

25. Every system of inequalities illustrated in the examples of this section has infinitely many solutions. Explain why this is so. Does this mean that *any* ordered pair is a solution?

There are infinitely many points in any of the shaded regions. There are, however, points that are not in the shaded regions, and thus there are ordered pairs that are not solutions.

Chapter 4 Summary

Key Terms

4.1

system of linear equations A system of linear equations (or **linear system**) consists of two or more linear equations with the same variables.

solution of a system A solution of a system of linear equations is an ordered pair that makes all the equations of the system true at the same time.

solution set of a system The set of all ordered pairs that are solutions of a system is its solution set.

consistent system A system of equations with at least one solution is a consistent system.

inconsistent system An inconsistent system is a system of equations with no solution.

independent equations Equations of a system that have different graphs are independent equations.

dependent equations Equations of a system that have the same graph (because they are different forms of the same equation) are dependent equations.

4.5

system of linear inequalities A system of linear inequalities contains two or more linear inequalities (and no other kinds of inequalities).

solution set of a system of linear inequalities The solution set of a system of linear inequalities includes all ordered pairs that make all inequalities of the system true at the same time.

Test Your Word Power

See how well you have learned the vocabulary in this chapter.

1. A **system of linear equations** consists of
   - **A.** at least two linear equations with different variables
   - **B.** two or more linear equations that have an infinite number of solutions
   - **C.** two or more linear equations with the same variables
   - **D.** two or more linear inequalities.

2. A **solution of a system** of linear equations is
   - **A.** an ordered pair that makes one equation of the system true
   - **B.** an ordered pair that makes all the equations of the system true at the same time
   - **C.** any ordered pair that makes one or the other or both equations of the system true
   - **D.** the set of values that make all the equations of the system false.

3. A **consistent system** is a system of equations
   - **A.** with at least one solution
   - **B.** with no solution
   - **C.** with an infinite number of solutions
   - **D.** that have the same graph.

4. An **inconsistent system** is a system of equations
   - **A.** with one solution
   - **B.** with no solution
   - **C.** with an infinite number of solutions
   - **D.** that have the same graph.

5. **Dependent equations**
   - **A.** have different graphs
   - **B.** have no solution
   - **C.** have one solution
   - **D.** are different forms of the same equation.

Answers to Test Your Word Power

1. C; *Example:* $2x + y = 7$
$3x - y = 3$

2. B; *Example:* The ordered pair $(2, 3)$ satisfies both equations of the system in the Answer 1 example, so it is a solution of the system.

3. A; *Example:* The system in the Answer 1 example is consistent. The graphs of the equations intersect at exactly one point, in this case the solution $(2, 3)$.

4. B; *Example:* The equations of two parallel lines make up an inconsistent system. Their graphs never intersect, so there is no solution of the system.

5. D; *Example:* The equations $4x - y = 8$ and $8x - 2y = 16$ are dependent because their graphs are the same line.

Quick Review

Concepts	Examples
4.1 Solving Systems of Linear Equations by Graphing	
An ordered pair is a solution of a system if it makes all equations of the system true at the same time.	Is $(4, -1)$ a solution of the system $\begin{matrix} x + y = 3 \\ 2x - y = 9 \end{matrix}$? Because $4 + (-1) = 3$ and $2(4) - (-1) = 9$ are both true, $(4, -1)$ is a solution.
To solve a linear system by graphing, follow these steps. ***Step 1*** Graph each equation of the system on the same axes. ***Step 2*** Find the coordinates of the point of intersection. ***Step 3*** Check. Write the solution set. If the graphs of the equations do not intersect (that is, the lines are parallel), then the system has *no solution* and the *solution set is* $\emptyset$. If the graphs of the equations are the same line, then the system has an *infinite number of solutions*. Use set-builder notation to write the solution set as $\{(x, y) \mid \_\_\_\_\_\_\_\_\}$, where a form of the equation is written on the blank.	Solve the system by graphing. $x + y = 5$ $2x - y = 4$ 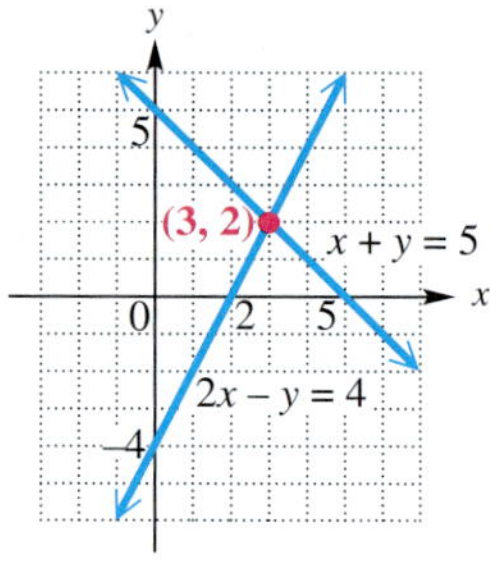 The ordered pair $(3, 2)$ satisfies both equations, so $\{(3, 2)\}$ is the solution set.
4.2 Solving Systems of Linear Equations by Substitution	
Step 1 Solve one equation for either variable.	Solve by substitution. $x + 2y = -5$ (1) $y = -2x - 1$ (2) Equation (2) is already solved for y.
Step 2 Substitute for that variable in the other equation to get an equation in one variable. ***Step 3*** Solve the equation from Step 2.	Substitute $-2x - 1$ for y in equation (1). $x + 2(-2x - 1) = -5$ Let $y = -2x - 1$ in (1). $x - 4x - 2 = -5$ Distributive property $-3x - 2 = -5$ Combine like terms. $-3x = -3$ Add 2. $x = 1$ Divide by -3.
Step 4 Substitute the result into the equation from Step 1 to get the value of the other variable.	To find y, let $x = 1$ in equation (2). $y = -2(1) - 1$ Let $x = 1$. $y = -3$ Multiply, and then subtract.
Step 5 Check. Write the solution set.	The solution $(1, -3)$ checks, so $\{(1, -3)\}$ is the solution set.

Concepts	Examples

4.3 Solving Systems of Linear Equations by Elimination

Step 1 Write both equations in standard form $Ax + By = C$.

Step 2 Multiply one or both equations by appropriate numbers as needed so that the sum of the coefficients of either the x- or y-terms is 0.

Step 3 Add the equations to get an equation with only one variable (or no variable).

Step 4 Solve the equation from Step 3.

Step 5 Substitute the solution from Step 4 into either of the original equations to find the value of the remaining variable.

Step 6 Check. Write the solution set.

Solve by elimination.

$$x + 3y = 7 \quad (1)$$
$$3x - y = 1 \quad (2)$$

Multiply equation (1) by -3 to eliminate the x-terms.

$$\begin{aligned} -3x - 9y &= -21 && \text{Multiply (1) by } -3. \\ 3x - y &= 1 && (2) \\ \hline -10y &= -20 && \text{Add.} \\ y &= 2 && \text{Divide by } -10. \end{aligned}$$

Substitute to get the value of x.

$$\begin{aligned} x + 3(2) &= 7 && \text{Let } y = 2 \text{ in (1).} \\ x + 6 &= 7 && \text{Multiply.} \\ x &= 1 && \text{Subtract 6.} \end{aligned}$$

Since $1 + 3(2) = 7$ and $3(1) - 2 = 1$, the solution $(1, 2)$ checks. The solution set is $\{(1, 2)\}$.

4.4 Applications of Linear Systems

Use the modified six-step method.

Step 1 **Read** the problem carefully.

Step 2 **Assign variables** for each unknown value. Use diagrams or tables as needed.

Step 3 **Write two equations** using both variables.

Step 4 **Solve** the system.

Step 5 **State the answer.**

Step 6 **Check** the answer in the words of the original problem.

The sum of two numbers is 30. Their difference is 6. Find the numbers.

Let x represent one number.

Let y represent the other number.

$$\begin{aligned} x + y &= 30 && (1) \\ x - y &= 6 && (2) \\ \hline 2x \quad &= 36 && \text{Add.} \\ x &= 18 && \text{Divide by 2.} \end{aligned}$$

Let $x = 18$ in equation (1): $18 + y = 30$. Solve to get $y = 12$. The numbers are 18 and 12.

The sum of 18 and 12 is 30, and the difference between 18 and 12 is 6, so the answer checks.

4.5 Solving Systems of Linear Inequalities

To solve a system of linear inequalities, follow these steps.

Step 1 Graph each inequality on the same axes. (This was explained in **Section 3.5.**)

Step 2 Choose the intersection. The solution set of the system is formed by the overlap of the regions of the two graphs.

The shaded region is the solution of the following system.

$$2x + 4y \geq 5$$
$$x \geq 1$$

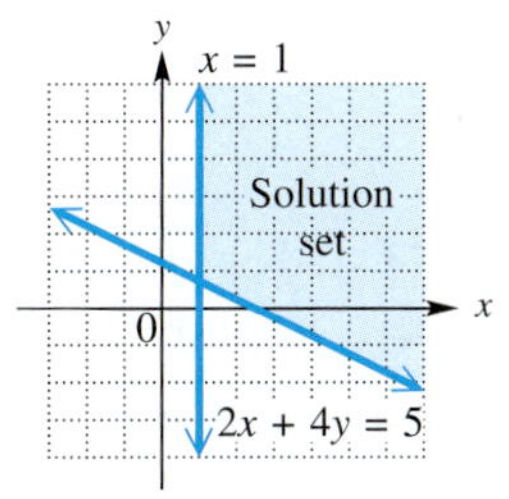

Chapter 4 Review Exercises

4.1 *Decide whether the given ordered pair is a solution of the given system.*

1. $(3, 4)$

$4x - 2y = 4$

$5x + y = 19$ **yes**

2. $(-5, 2)$

$x - 4y = -13$

$2x + 3y = 4$ **no**

Solve each system by graphing. **We do not show the graphs for Exercises 3–6.**

3. $x + y = 4$

$2x - y = 5$

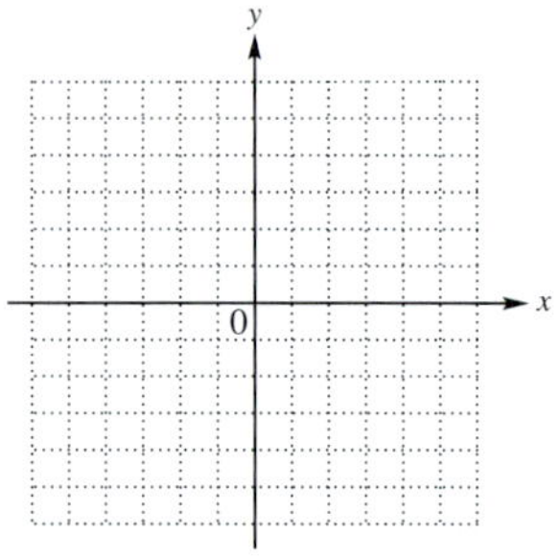

$\{(3, 1)\}$

4. $x - 2y = 4$

$2x + y = -2$

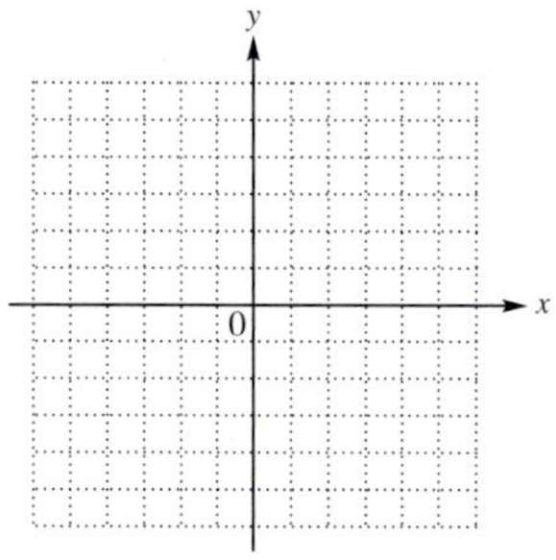

$\{(0, -2)\}$

5. $x - 2 = 2y$

$2x - 4y = 4$

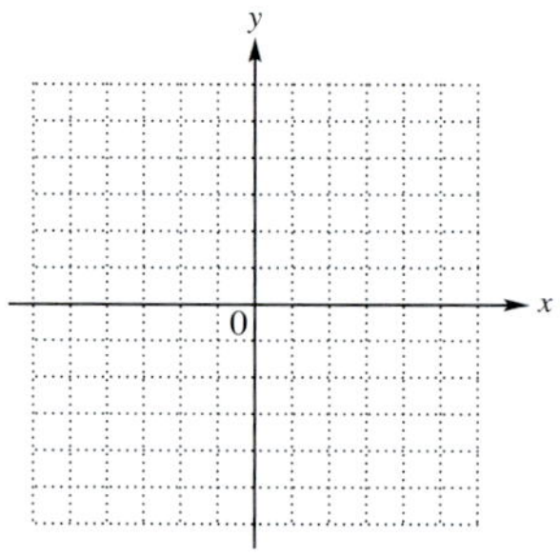

$\{(x, y) \mid x - 2y = 2\}$

6. $2x + 4 = 2y$

$y - x = -3$

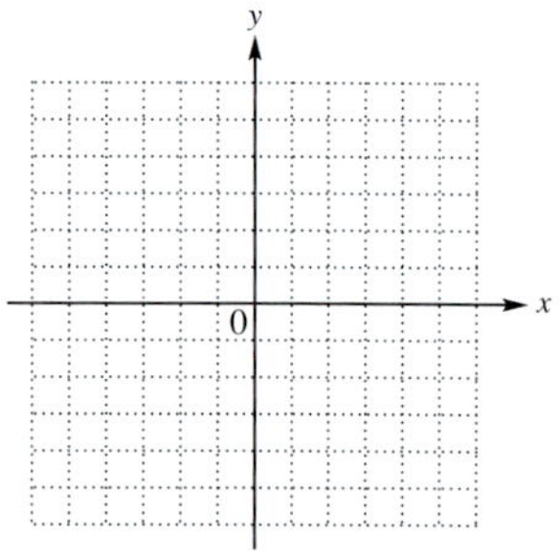

$\emptyset$

4.2 *Solve each system by the substitution method.*

7. $3x + y = 7$

$x = 2y$

$\{(2, 1)\}$

8. $2x - 5y = -19$

$y = x + 2$

$\{(3, 5)\}$

9. $4x + 5y = 44$

$x + 2 = 2y$

$\{(6, 4)\}$

10. $5x + 15y = 3$

$x + 3y = 2$

$\emptyset$

4.3 *Solve each system by the elimination method.*

11. $2x - y = 13$

$x + y = 8$

$\{(7, 1)\}$

12. $3x - y = -13$

$x - 2y = -1$

$\{(-5, -2)\}$

13. $-4x + 3y = 25$

$6x - 5y = -39$

$\{(-4, 3)\}$

14. $3x - 4y = 9$

$6x - 8y = 18$

$\{(x, y) \mid 3x - 4y = 9\}$

4.1–4.3 *Solve each system by any method.*

15. $x - 2y = 5$

$y = x - 7$

$\{(9, 2)\}$

16. $5x - 3y = 11$

$2y = x - 4$

$\left\{\left(\frac{10}{7}, -\frac{9}{7}\right)\right\}$

17. $\frac{x}{2} + \frac{y}{3} = 7$

$\frac{x}{4} + \frac{2y}{3} = 8$

$\{(8, 9)\}$

18. $\frac{3x}{4} - \frac{y}{3} = \frac{7}{6}$

$\frac{x}{2} + \frac{2y}{3} = \frac{5}{3}$

$\{(2, 1)\}$

19. $0.2x + 1.2y = -1$

$0.1x + 0.3y = 0.1$

$\{(7, -2)\}$

20. $0.1x + y = 1.6$

$0.6x + 0.5y = -1.4$

$\{(-4, 2)\}$

4.4 *For each problem, complete any tables. Then solve the problem using a system of equations.*

21. The two leading pizza chains in the United States are Pizza Hut and Domino's. In 2010, Pizza Hut had 2613 more locations than Domino's, and together the two chains had 12,471 locations. How many locations did each chain have? (*Source: PMQ Pizza Magazine.*)

Pizza Hut: 7542 locations; Domino's: 4929 locations

22. Together, the average paid circulation for *Reader's Digest* and *People* magazines in 2011 was 9.3 million. The circulation for *People* was 2.1 million less than that of *Reader's Digest*. What were the circulation figures for each magazine? (*Source:* Audit Bureau of Circulations.)

***Reader's Digest*: 5.7 million; *People*: 3.6 million**

23. Candy that sells for \$1.30 per lb is to be mixed with candy selling for \$0.90 per lb to get 100 lb of a mix that will sell for \$1 per lb. How much of each type should be used?

Number of Pounds	Cost per Pound (in dollars)	Total Value (in dollars)
x	1.30	$1.30x$
y	**0.90**	**0.90y**
100	1.00	**100(1), or 100**

25 lb of \$1.30 candy; 75 lb of \$0.90 candy

24. A cashier has 20 bills, all of which are \$10 or \$20 bills. The total value of the money is \$330. How many of each type does the cashier have?

Number of Bills	Denomination of Bills (in dollars)	Total Value (in dollars)
x	10	**10x**
y	**20**	$20y$
20	XXXXXX	330

13 twenties; 7 tens

25. The perimeter of a rectangle is 90 m. Its length is $1\frac{1}{2}$ times its width. Find the length and width of the rectangle.

length: 27 m; width: 18 m

26. A certain plane flying with the wind travels 540 mi in 2 hr. Later, flying against the same wind, the plane travels 690 mi in 3 hr. Find the rate of the plane in still air and the wind speed.

plane: 250 mph; wind: 20 mph

27. After taxes, Ms. Cesar's game show winnings were \$18,000. She invested part of it at 3% annual simple interest and the rest at 4%. Her interest income for the first year was \$650. How much did she invest at each rate?

Amount Invested (in dollars)	Percent (as a decimal)	Interest (in dollars)
x	0.03	**0.03x**
y	**0.04**	**0.04y**
18,000	XXXXX	**650**

\$7000 at 3%; \$11,000 at 4%

28. A 40% antifreeze solution is to be mixed with a 70% solution to get 90 L of a 50% solution. How many liters of the 40% and 70% solutions will be needed?

Number of Liters	Percent (as a decimal)	Amount of Pure Antifreeze
x	0.40	**0.40x**
y	**0.70**	**0.70y**
90	0.50	**0.50(90), or 45**

60 L of 40% solution; 30 L of 70% solution

4.5 **CONCEPT CHECK** *Answer each question.*

29. Which system of linear inequalities is graphed in the figure?

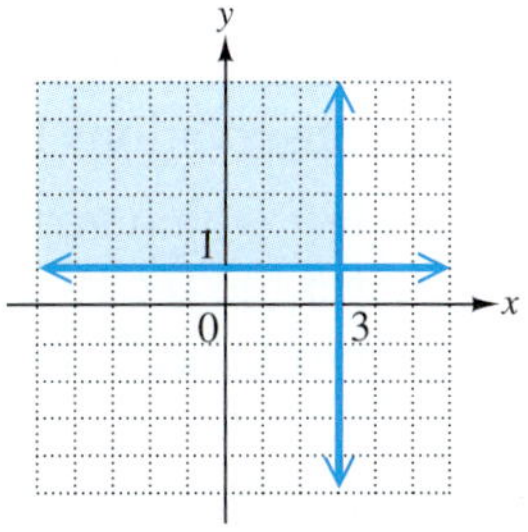

A. $x \leq 3$, $y \leq 1$ **B.** $x \leq 3$, $y \geq 1$ **C.** $x \geq 3$, $y \leq 1$ **D.** $x \geq 3$, $y \geq 1$

B

30. Which system of inequalities has no solution? (Do not graph.)

A. $x \geq 4$, $y \leq 3$ **B.** $x + y > 4$, $x + y < 3$ **C.** $x > 2$, $y < 1$ **D.** $x + y < 4$, $x - y < 3$

B

Graph the solution set of each system of linear inequalities.

31. $x + y \geq 2$
$x - y \leq 4$

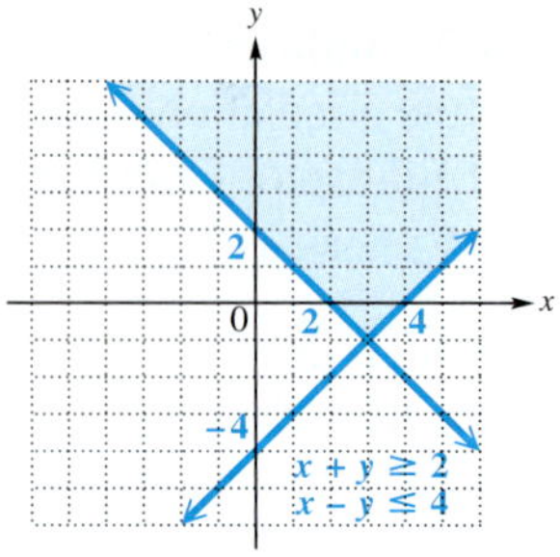

32. $y \geq 2x$
$2x + 3y \leq 6$

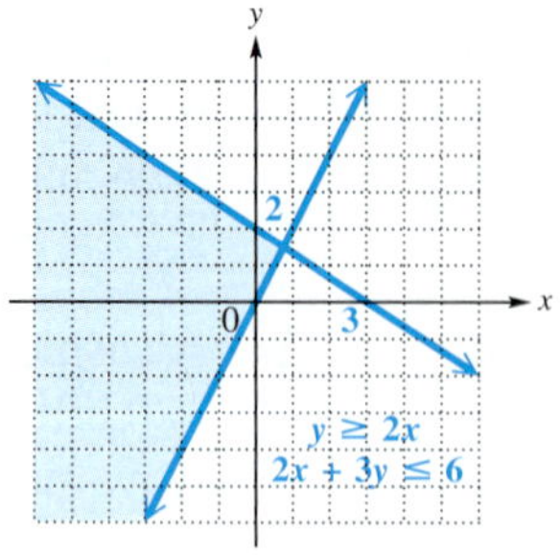

33. $x + y < 3$
$2x > y$

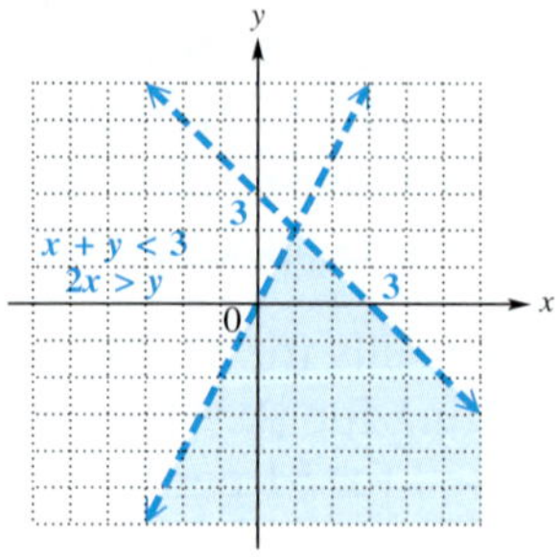

Mixed Review Exercises

Solve.

34. $3x + 4y = 6$
$4x - 5y = 8$ **{(2, 0)}**

35. $\frac{3x}{2} + \frac{y}{5} = -3$
$4x + \frac{y}{3} = -11$ **{(−4, 15)}**

36. $x + 6y = 3$
$2x + 12y = 2$ **∅**

37. $x + y < 5$
$x - y \geq 2$

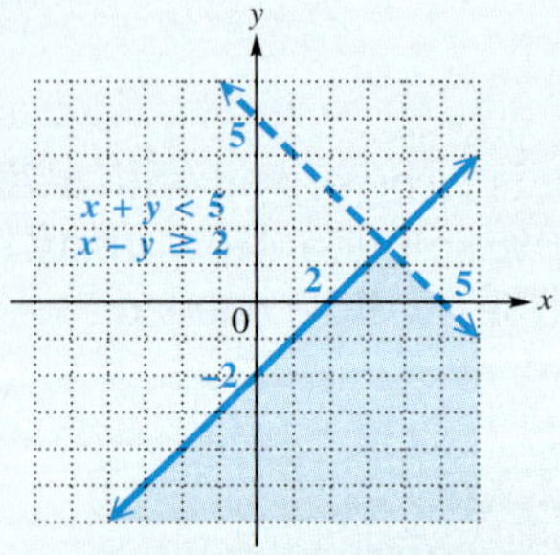

38. $y \leq 2x$
$x + 2y > 4$

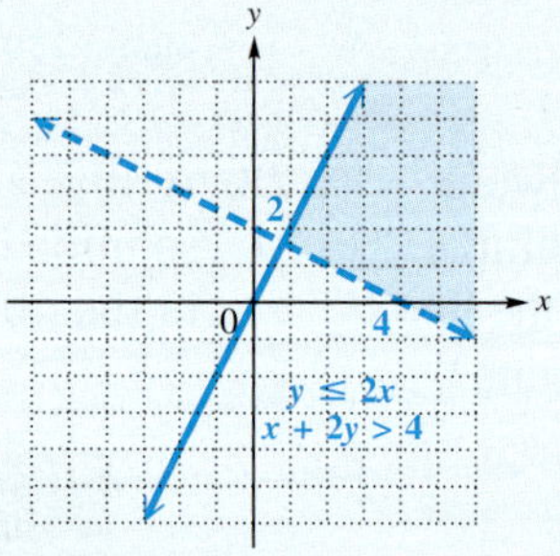

39. $y < -4x$
$y < -2$

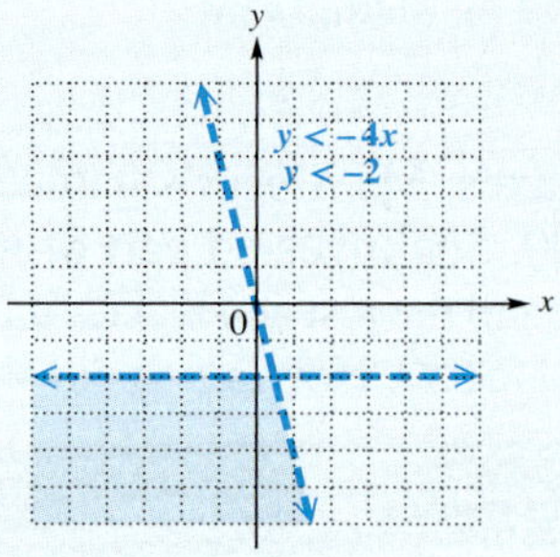

40. The perimeter of an isosceles triangle is 29 in. One side of the triangle is 5 in. longer than each of the two equal sides. Find the lengths of the sides of the triangle.

8 in., 8 in., and 13 in.

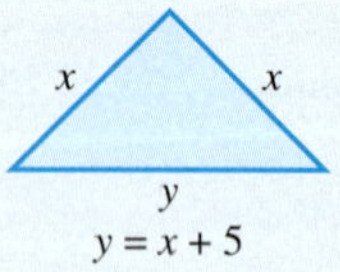

41. In Super Bowl XLVI, the New York Giants beat the New England Patriots by 4 points, and the winning score was 13 points less than twice the losing score. What was the final score of the game? (*Source:* NFL.)

Giants: 21; Patriots: 17

42. Eboni Perkins compared the monthly payments she would incur for two types of mortgages: fixed-rate and variable-rate. Her observations led to the following graph.

(a) For which years would the monthly payment be more for the fixed-rate mortgage than for the variable-rate mortgage? **years 0–6**

(b) In what year would the payments be the same, and what would those payments be? **year 6; about $650**

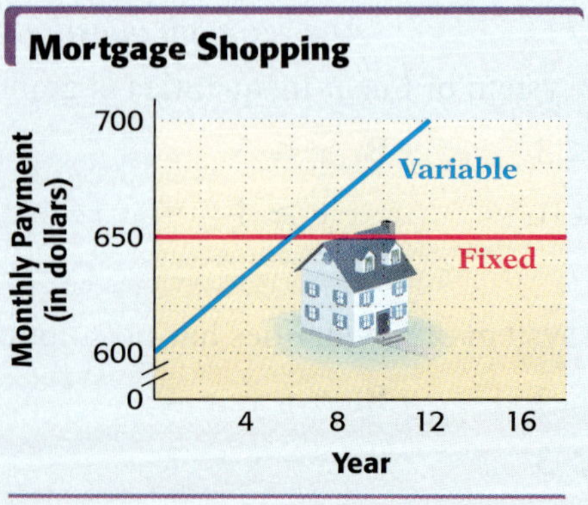

Chapter 4 Test

CHAPTER Test Prep VIDEO

The Chapter Test Prep Videos with test solutions are available on DVD, in MyMathLab, and on YouTube—search "LialIntroductoryAlg" and click on "Channels."

1. Decide whether each ordered pair is a solution of the system.

$$2x + y = -3$$
$$x - y = -9$$

(a) $(1, -5)$ **(b)** $(1, 10)$ **(c)** $(-4, 5)$

no **no** **yes**

2. Solve the system by graphing.

$$2x + y = 1$$
$$3x - y = 9$$

{(2, −3)}

We do not show the graph.

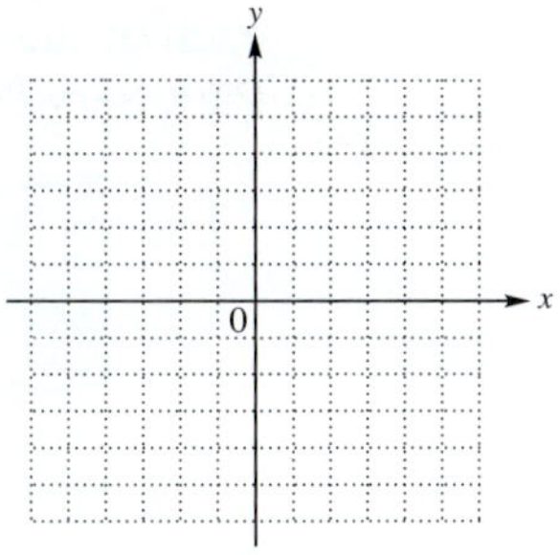

3. Suppose that the graph of a system of two linear equations consists of lines that have the same slope but different y-intercepts. How many solutions does the system have?

It has no solution.

Solve each system by the substitution method.

4. $2x + y = -4$
$x = y + 7$
{(1, −6)}

5. $4x + 3y = -35$
$x + y = 0$
{(−35, 35)}

Solve each system by the elimination method.

6. $2x - y = 4$
$3x + y = 21$
{(5, 6)}

7. $4x + 2y = 2$
$5x + 4y = 7$
{(−1, 3)}

8. $3x + 4y = 9$
$2x + 5y = 13$
{(−1, 3)}

9. $6x - 5y = 0$
$-2x + 3y = 0$
{(0, 0)}

10. $4x + 5y = 2$
$-8x - 10y = 6$
∅

Solve each system by any method.

11. $3x = 6 + y$

$6x - 2y = 12$

$\{(x, y) \mid 3x - y = 6\}$

12. $\frac{6}{5}x - \frac{1}{3}y = -20$

$-\frac{2}{3}x + \frac{1}{6}y = 11$

$\{(-15, 6)\}$

Solve each problem.

13. The distance between Memphis and Atlanta is 782 mi less than the distance between Minneapolis and Houston. Together, the two distances total 1570 mi. How far is it between Memphis and Atlanta? How far is it between Minneapolis and Houston? (*Source: Rand McNally Road Atlas.*)

Memphis and Atlanta: 394 mi;
Minneapolis and Houston: 1176 mi

14. In 2010, a total of 7.8 million people visited the Statue of Liberty and the National World War II Memorial, two popular tourist attractions. The Statue of Liberty had 0.2 million fewer visitors than the National World War II Memorial. How many visitors did each of these attractions have? (*Source:* National Park Service, Department of the Interior.)

Statue of Liberty: 3.8 million;
National World War II Memorial: 4.0 million

15. A 15% solution of alcohol is to be mixed with a 40% solution to get 50 L of a final mixture that is 30% alcohol. How much of each of the original solutions should be used?

Liters of Solution	Percent (as a decimal)	Liters of Pure Alcohol

20 L of 15% solution; 30 L of 40% solution

16. Two cars leave from Perham, Minnesota, and travel in the same direction. One car travels $1\frac{1}{3}$ times as fast as the other. After 3 hr they are 45 mi apart. What are the rates of the cars?

	r	t	d
Faster Car			
Slower Car			

slower car: 45 mph; faster car: 60 mph

Graph the solution set of each system of inequalities.

17. $2x + 7y \leq 14$

$x - y \geq 1$

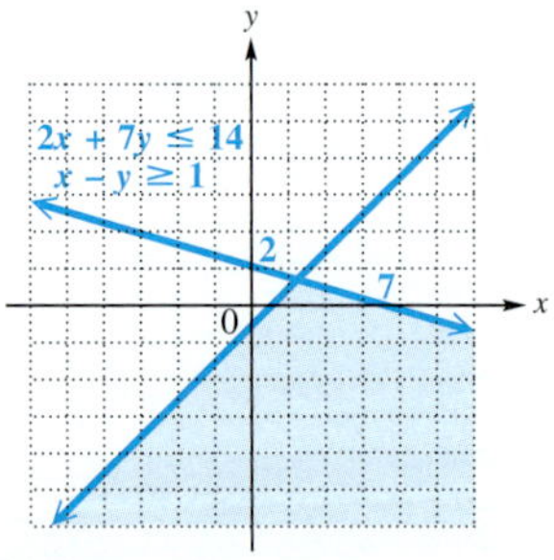

18. $2x - y > 6$

$4y + 12 \geq -3x$

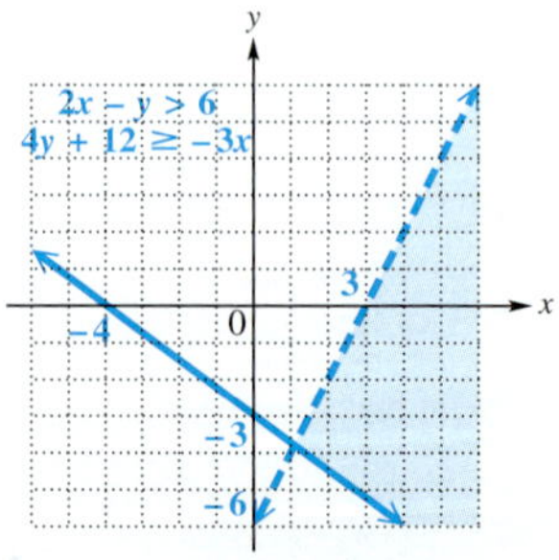

Chapters R–4 Cumulative Review Exercises

1. List all integer factors of 40.

$-1, 1, -2, 2, -4, 4, -5, 5, -8, 8, -10, 10, -20, 20, -40, 40$

2. Find the value of the expression for $x = 1$ and $y = 5$.

$$\frac{3x^2 + 2y^2}{10y + 3}$$

1

Name the property that justifies each statement.

3. $5 + (-4) = (-4) + 5$

commutative property

4. $r(s - k) = rs - rk$

distributive property

5. $-\frac{2}{3} + \frac{2}{3} = 0$

inverse property

6. Evaluate $-2 + 6[3 - (4 - 9)]$.

46

7. Solve the formula $P = \frac{kT}{V}$ for T.

$T = \frac{PV}{k}$

Solve each linear equation.

8. $2 - 3(6x + 2) = 4(x + 1) + 18$

$\left\{-\frac{13}{11}\right\}$

9. $\frac{3}{2}\left(\frac{1}{3}x + 4\right) = 6\left(\frac{1}{4} + x\right)$

$\left\{\frac{9}{11}\right\}$

Solve each linear inequality. Write the solution set in interval notation.

10. $-\frac{5}{6}x < 15$

$(-18, \infty)$

11. $-8 < 2x + 3$

$\left(-\frac{11}{2}, \infty\right)$

12. The third generation iPad tablet computer, released in March 2012, has a perimeter of 33.62 in., and its width is 2.19 in. less than its length. What are its dimensions? (*Source:* www.apple.com)

length: 9.50 in.; width: 7.31 in.

Graph each linear equation.

13. $x - y = 4$

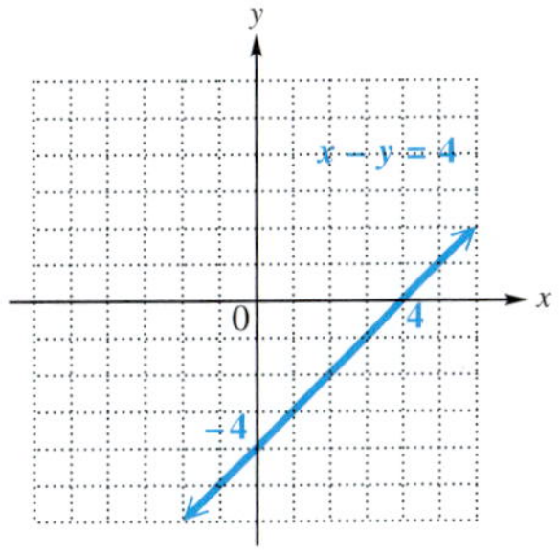

14. $3x + y = 6$

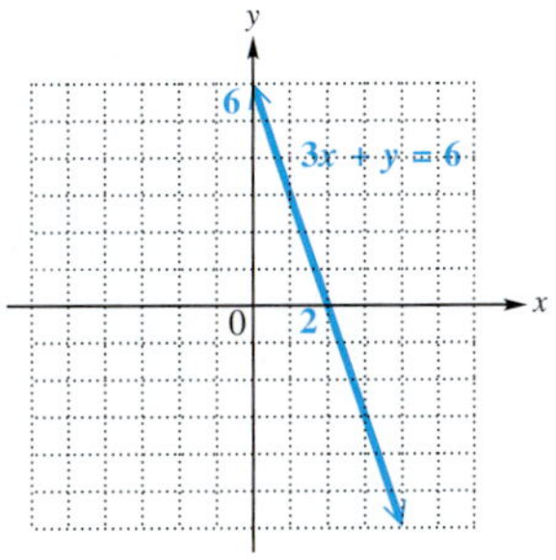

Find the slope of each line.

15. Through $(-5, 6)$ and $(1, -2)$

$-\frac{4}{3}$

16. Perpendicular to the line $y = 4x - 3$

$-\frac{1}{4}$

Write an equation in slope-intercept form for each line.

17. Through $(-4, 1)$ with slope $\frac{1}{2}$

$y = \frac{1}{2}x + 3$

18. Through the points $(1, 3)$ and $(-2, -3)$

$y = 2x + 1$

19. **(a)** Write an equation of the vertical line through $(9, -2)$. $x = 9$

(b) Write an equation of the horizontal line through $(4, -1)$. $y = -1$

Solve each system by any method.

20. $2x - y = -8$

$x + 2y = 11$

$\{(-1, 6)\}$

21. $4x + 5y = -8$

$3x + 4y = -7$

$\{(3, -4)\}$

22. $3x + 4y = 2$

$6x + 8y = 1$

$\emptyset$

Use a system of equations to solve each problem.

23. Admission prices at a high school football game were \$6 for adults and \$2 for children. The total value of the tickets sold was \$2528, and 454 tickets were sold. How many adults and how many children attended the game?

Kind of Ticket	Number Sold	Cost of Each (in dollars)	Total Value (in dollars)
Adult	x	6	$6x$
Child	y	2	$2y$
Total	454	XXXXXX	2528

405 adults, 49 children

24. The perimeter of a triangle is 53 in. If two sides are of equal length, and the third side measures 4 in. less than each of the equal sides, what are the lengths of the three sides?

19 in., 19 in., 15 in.

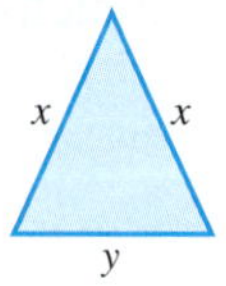

25. Graph the solution set of the system.

$x + 2y \leq 12$

$2x - y \leq 8$

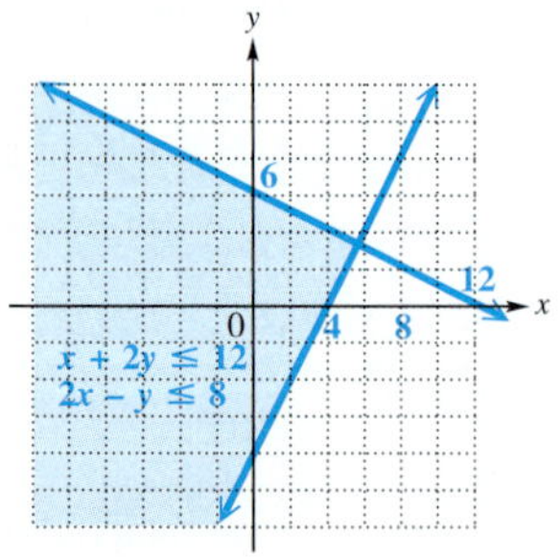

5 Exponents and Polynomials

Numbers such as the one shown here are often expressed using *scientific notation*, a method of expressing them using a power of 10 as a factor: $15{,}680{,}758{,}917{,}289 = 1.5680758917289 \times 10^{13}$.

5.1 Adding and Subtracting Polynomials

OBJECTIVES

1. Review combining like terms.
2. Know the vocabulary for polynomials.
3. Evaluate polynomials.
4. Add polynomials.
5. Subtract polynomials.
6. Add and subtract polynomials with more than one variable.

Recall from **Section 1.8** that in an algebraic expression such as

$$4x^3 + 6x^2 + 5x + 8,$$

the quantities that are added, $4x^3$, $6x^2$, $5x$, and 8, are **terms.** In the term $4x^3$, the number 4 is the **numerical coefficient,** or simply the **coefficient,** of x^3. In the same way, 6 is the coefficient of x^2 in the term $6x^2$, and 5 is the coefficient of x in the term $5x$. The ***constant*** term 8 can be thought of as $8 \cdot 1 = 8x^0$, so 8 is the coefficient in the term 8. Other examples are given in the table at the side.

Term	Numerical Coefficient
$-7y$	-7
$34r^3$	34
$-26x^5yz^4$	-26
$-k = -1k$	-1
$r = 1r$	1
$\frac{3x}{8} = \frac{3}{8}x$	$\frac{3}{8}$
$\frac{x}{3} = \frac{1x}{3} = \frac{1}{3}x$	$\frac{1}{3}$

OBJECTIVE 1 Review combining like terms. Recall from **Section 1.8** that **like terms** have exactly the same combination of variables, with the same exponents on the variables. ***Only the coefficients may differ.***

Examples of like terms			Examples of unlike terms		
$19m^5$	and	$14m^5$	$7x$	and	$7y$
$-37y^9$	and	y^9	z^4	and	z
$3pq$	and	$-2pq$	$2pq$	and	$2p$
$2xy^2$	and	$-xy^2$	$-4xy^2$	and	$5x^2y$

Using the distributive property, we combine, or add, like terms by adding their coefficients.

EXAMPLE 1 Combining Like Terms

Simplify each expression by combining like terms.

(a) $-4x^3 + 6x^3$

$= (-4 + 6)x^3$

$= 2x^3$

(b) $8rs - 13rs + 9rs$

$= (8 - 13 + 9)rs$

$= 4rs$

(c) $9x^6 - 14x^6 + x^6$ ($x^6 = 1x^6$)

$= (9 - 14 + 1)x^6$

$= -4x^6$

(d) $y + \frac{2}{3}y$

$= 1y + \frac{2}{3}y$ ($y = 1y$)

$= \left(\frac{3}{3} + \frac{2}{3}\right)y$

$= \frac{5}{3}y$

(e) $12m^2 + 5m + 4m^2$

$= (12 + 4)m^2 + 5m$

$= 16m^2 + 5m$

(f) $5u + 11v$

These are unlike terms. They cannot be combined.

◀ Work Problem 1 at the Side.

CAUTION

In **Example 1(e),** we cannot combine $16m^2$ and $5m$. These two terms are unlike because the exponents on the variables are different. ***Unlike terms have different variables or different exponents on the same variables.***

1 Combine like terms.

(a) $5x^4 + 7x^4$

(b) $9pq + 3pq - 2pq$

(c) $r^2 + 3r + 5r^2$

(d) $x + \frac{1}{2}x$

(e) $8t + 6w$

(f) $3x^4 - 3x^2$

Answers

1. (a) $12x^4$ (b) $10pq$ (c) $6r^2 + 3r$ (d) $\frac{3}{2}x$
(e) These are unlike terms. They cannot be combined.
(f) These are unlike terms. They cannot be combined.

OBJECTIVE 2 Know the vocabulary for polynomials. A **polynomial in x** is a term or the sum of a finite number of terms of the form ax^n, for any real number a and any whole number n. For example,

$$16x^8 - 7x^6 + 5x^4 - 3x^2 + 4 \quad \text{Polynomial in } x \text{ (The 4 can be written as } 4x^0.)$$

is a polynomial in x. It is written in **descending powers,** because the exponents on x decrease from left to right. By contrast, the expression

$$2x^3 - x^2 + \frac{4}{x} \quad \text{Not a polynomial}$$

is not a polynomial in x. A variable appears in a denominator.

Note

We can define *polynomial* using any variable and not just x, as in **Example 1(d).** Polynomials may have terms with more than one variable, as in **Example 1(b).**

Work Problem 2 at the Side. ▶

The **degree of a term** is the sum of the exponents on the variables. The **degree of a polynomial** is the greatest degree of any nonzero term of the polynomial. The table gives several examples.

Term	Degree	Polynomial	Degree
$3x^4$	4	$3x^4 - 5x^2 + 6$	4
$5x$, or $5x^1$	1	$5x + 7$	1
-7, or $-7x^0$	0	$x^2y + xy - 5y^2$	3
$2x^2y$, or $2x^2y^1$	$2 + 1 = 3$	$x^5 + 3x^6$	6

A polynomial with only one term is a **monomial.** (*Mono*- means "one," as in *mono*rail.) A polynomial with exactly two terms is a **binomial.** (*Bi*- means "two," as in *bi*cycle.) A polynomial with exactly three terms is a **trinomial.** (*Tri*- means "three," as in *tri*angle.)

$$9m, \quad -6y^5, \quad a^2, \quad \text{and} \quad 6 \qquad \text{Monomials}$$

$$-9x^4 + 9x^3, \quad 8m^2 + 6m, \quad \text{and} \quad 3m^5 - 9m^2 \qquad \text{Binomials}$$

$$9m^3 - 4m^2 + 6, \quad \frac{19}{3}y^2 + \frac{8}{3}y + 5, \quad \text{and} \quad -3m^5 - 9m^2 + 2 \qquad \text{Trinomials}$$

EXAMPLE 2 Classifying Polynomials

Simplify each polynomial if possible. Then give the degree and tell whether the polynomial is a *monomial*, a *binomial*, a *trinomial*, or *none of these*.

(a) $2x^3 + 5$ We cannot simplify further. This is a binomial of degree 3.

(b) $4x - 5x + 2x$

$= x$ Combine like terms to simplify.

The degree is 1 (since $x = x^1$). The simplified polynomial is a monomial.

Work Problem 3 at the Side. ▶

2 Choose all descriptions that apply for each of the expressions in parts (a)–(d).

A. Polynomial

B. Polynomial written in descending powers

C. Not a polynomial

(a) $3m^5 + 5m^2 - 2m + 1$

(b) $2p^4 + p^6$

(c) $\frac{1}{x} + 2x^2 + 3$

(d) $x - 3$

3 Simplify each polynomial if possible. Then give the degree and tell whether the polynomial is a *monomial,* a *binomial,* a *trinomial,* or *none of these.*

(a) $3x^2 + 2x - 4$

(b) $x^3 + 4x^3$

(c) $x^8 - x^7 + 2x^8$

Answers

2. (a) A and B **(b)** A **(c)** C **(d)** A and B

3. (a) degree 2; trinomial

(b) degree 3; monomial (simplify to $5x^3$)

(c) degree 8; binomial (simplify to $3x^8 - x^7$)

4 Find the value of $2x^3 + 8x - 6$ in each case.

GS **(a)** For $x = -1$

$$2x^3 + 8x - 6$$
$$= 2(\_\_)^3 + 8(\_\_) - 6$$
$$= 2(\_\_) + 8(-1) - 6$$
$$= \_\_\_ - \_\_\_ - 6$$
$$= \_\_\_$$

(b) For $x = 4$

(c) For $x = -2$

5 Add each pair of polynomials.

(a) $4x^3 - 3x^2 + 2x$ and $6x^3 + 2x^2 - 3x$

(b) $x^2 - 2x + 5$ and $4x^2 - 2$

Answers

4. **(a)** -1; -1; -1; -2; 8; -16 **(b)** 154 **(c)** -38

5. **(a)** $10x^3 - x^2 - x$ **(b)** $5x^2 - 2x + 3$

OBJECTIVE 3 **Evaluate polynomials.** When we evaluate an expression, we find its value. A polynomial usually represents different numbers for different values of the variable.

EXAMPLE 3 Evaluating a Polynomial

Find the value of $3x^4 + 5x^3 - 4x - 4$ for **(a)** $x = -2$ and for **(b)** $x = 3$.

(a) $3x^4 + 5x^3 - 4x - 4$

$= 3(-2)^4 + 5(-2)^3 - 4(-2) - 4$ Substitute -2 for x.

$= 3(16) + 5(-8) - 4(-2) - 4$ Apply the exponents.

(Use parentheses to avoid errors.)

$= 48 - 40 + 8 - 4$ Multiply.

$= 12$ Add and subtract.

(b) $3x^4 + 5x^3 - 4x - 4$

$= 3(3)^4 + 5(3)^3 - 4(3) - 4$ Let $x = 3$.

$= 3(81) + 5(27) - 4(3) - 4$ Apply the exponents.

$= 243 + 135 - 12 - 4$ Multiply.

$= 362$ Add and subtract.

◀ Work Problem **4** at the Side.

OBJECTIVE 4 **Add polynomials.**

Adding Polynomials

To add two polynomials, add like terms.

EXAMPLE 4 Adding Polynomials Vertically

(a) Add $6x^3 - 4x^2 + 3$ and $-2x^3 + 7x^2 - 5$.

$$\begin{array}{r} 6x^3 - 4x^2 + 3 \\ \underline{-2x^3 + 7x^2 - 5} \end{array}$$ Write like terms in columns.

Now add, column by column.

(Combine the coefficients only. Do *not* add the exponents.)

$$\begin{array}{ccc} 6x^3 & -4x^2 & 3 \\ \underline{-2x^3} & \underline{7x^2} & \underline{-5} \\ 4x^3 & 3x^2 & -2 \end{array}$$

Add the three sums together to obtain the answer.

$$4x^3 + 3x^2 + (-2) = 4x^3 + 3x^2 - 2$$

(b) Add $2x^2 - 4x + 3$ and $x^3 + 5x$.
Write like terms in columns and add column by column.

$$\begin{array}{r} 2x^2 - 4x + 3 \\ \underline{x^3 \qquad\quad + 5x \qquad} \\ x^3 + 2x^2 + x + 3 \end{array}$$

(Leave spaces for missing terms.)

◀ Work Problem **5** at the Side.

The polynomials in **Example 4** also could be added horizontally.

EXAMPLE 5 Adding Polynomials Horizontally

Find each sum.

(a) Add $6x^3 - 4x^2 + 3$ and $-2x^3 + 7x^2 - 5$.

$(6x^3 - 4x^2 + 3) + (-2x^3 + 7x^2 - 5) = 4x^3 + 3x^2 - 2$ Combine like terms. Same answer found in **Example 4(a)**

(b) $(2x^2 - 4x + 3) + (x^3 + 5x)$

$= x^3 + 2x^2 - 4x + 5x + 3$ Commutative property

$= x^3 + 2x^2 + x + 3$ Combine like terms.

Work Problem 6 at the Side. ▶

OBJECTIVE 5 Subtract polynomials. In **Section 1.5,** the difference $x - y$ was defined as $x + (-y)$. (We find the difference $x - y$ by adding x and the opposite of y.)

$7 - 2$ is equivalent to $7 + (-2)$, which equals 5.

$-8 - (-2)$ is equivalent to $-8 + 2$, which equals -6.

A similar method is used to subtract polynomials.

Subtracting Polynomials

To subtract two polynomials, change all the signs of the subtrahend (second polynomial) and add the result to the minuend (first polynomial).

EXAMPLE 6 Subtracting Polynomials Horizontally

Perform each subtraction.

(a) $(5x - 2) - (3x - 8)$

$= (5x - 2) + [-(3x - 8)]$ Definition of subtraction

$= (5x - 2) + [-1(3x - 8)]$ $-a = -1a$

$= (5x - 2) + (-3x + 8)$ Distributive property

$= 2x + 6$ Combine like terms.

(b) Subtract $6x^3 - 4x^2 + 2$ from $11x^3 + 2x^2 - 8$.

$(11x^3 + 2x^2 - 8) - (6x^3 - 4x^2 + 2)$ Be careful to write the problem in the correct order.

$= (11x^3 + 2x^2 - 8) + (-6x^3 + 4x^2 - 2)$

$= 5x^3 + 6x^2 - 10$ Answer

CHECK To check a subtraction problem, use the following fact.

If $a - b = c$, then $a = b + c$.

Here, add $6x^3 - 4x^2 + 2$ and $5x^3 + 6x^2 - 10$.

$(6x^3 - 4x^2 + 2) + (5x^3 + 6x^2 - 10)$

$= 11x^3 + 2x^2 - 8$ ✓

Work Problem 7 at the Side. ▶

6 Find each sum.

(a) $(2x^4 - 6x^2 + 7)$
$+ (-3x^4 + 5x^2 + 2)$

(b) $(3x^2 + 4x + 2)$
$+ (6x^3 - 5x - 7)$

7 Subtract, and check your answers by addition.

GS **(a)** $(14y^3 - 6y^2 + 2y - 5)$
$- (2y^3 - 7y^2 - 4y + 6)$
$= (\_\_\_\_\_\_\_\_)$
$+ (\_\_\_\_\_\_\_\_)$
$= \_\_\_\_\_\_\_\_$

(b) Subtract

$\left(-\frac{3}{2}y^2 + \frac{4}{3}y + 6\right)$

from $\left(\frac{7}{2}y^2 - \frac{11}{3}y + 8\right)$.

Answers

6. **(a)** $-x^4 - x^2 + 9$
(b) $6x^3 + 3x^2 - x - 5$

7. **(a)** $14y^3 - 6y^2 + 2y - 5$; $-2y^3 + 7y^2 + 4y - 6$; $12y^3 + y^2 + 6y - 11$
(b) $5y^2 - 5y + 2$

Subtraction also can be done in columns. We use vertical subtraction in **Section 5.7** when we study polynomial division.

EXAMPLE 7 Subtracting Polynomials Vertically

Subtract by columns: $(14y^3 - 6y^2 + 2y - 5) - (2y^3 - 7y^2 - 4y + 6)$.

$$\begin{array}{r} 14y^3 - 6y^2 + 2y - 5 \\ \underline{2y^3 - 7y^2 - 4y + 6} \end{array}$$ Arrange like terms in columns.

Change all signs in the second row, and then add.

$$\begin{array}{r} 14y^3 - 6y^2 + 2y - 5 \\ \underline{-2y^3 + 7y^2 + 4y - 6} \\ 12y^3 + y^2 + 6y - 11 \end{array}$$

Change signs.

Add.

◀ Work Problem 8 at the Side.

EXAMPLE 8 Adding and Subtracting More Than Two Polynomials

Perform the indicated operations to simplify the expression

$$(4 - x + 3x^2) - (2 - 3x + 5x^2) + (8 + 2x - 4x^2).$$

Rewrite, changing the subtraction to adding the opposite.

$(4 - x + 3x^2) - (2 - 3x + 5x^2) + (8 + 2x - 4x^2)$

$= (4 - x + 3x^2) + (-2 + 3x - 5x^2) + (8 + 2x - 4x^2)$

$= (2 + 2x - 2x^2) + (8 + 2x - 4x^2)$ Combine like terms.

$= 10 + 4x - 6x^2$ Combine like terms.

◀ Work Problem 9 at the Side.

OBJECTIVE 6 Add and subtract polynomials with more than one variable. Polynomials in more than one variable are added and subtracted by combining like terms, just as with single-variable polynomials.

EXAMPLE 9 Adding and Subtracting Multivariable Polynomials

Add or subtract as indicated.

(a) $(4a + 2ab - b) + (3a - ab + b)$

$= 4a + 2ab - b + 3a - ab + b$

$= 7a + ab$ Combine like terms.

(b) $(2x^2y + 3xy + y^2) - (3x^2y - xy - 2y^2)$

$= 2x^2y + 3xy + y^2 - 3x^2y + xy + 2y^2$

$= -x^2y + 4xy + 3y^2$

Be careful with signs.

◀ Work Problem 10 at the Side.

---

8 Subtract by columns.

$(4y^3 - 16y^2 + 2y)$
$- (12y^3 - 9y^2 + 16)$

9 Perform the indicated operations.

$(6p^4 - 8p^3 + 2p - 1)$
$- (-7p^4 + 6p^2 - 12)$
$+ (p^4 - 3p + 8)$

10 Add or subtract.

(a) $(3mn + 2m - 4n)$
$+ (-mn + 4m + n)$

(b) $(5p^2q^2 - 4p^2 + 2q)$
$- (2p^2q^2 - p^2 - 3q)$

Answers

8. $-8y^3 - 7y^2 + 2y - 16$
9. $14p^4 - 8p^3 - 6p^2 - p + 19$
10. (a) $2mn + 6m - 3n$
(b) $3p^2q^2 - 3p^2 + 5q$

5.1 Exercises

FOR EXTRA HELP

Download the MyDashBoard App

MyMathLab®

CONCEPT CHECK *Complete each statement.*

1. In the term $7x^5$, the coefficient is __7__ and the exponent is __5__.

2. The expression $5x^3 - 4x^2$ has (*one* / *two* / *three*) term(s). (two)

3. The degree of the term $-4x^8$ is __8__.

4. The polynomial $4x^2 - y^2$ (*is* / *is not*) an example of a trinomial. (is not)

5. When $x^2 + 10$ is evaluated for $x = 4$, the result is __26__.

6. __$5x^9$__ is an example of a monomial with coefficient 5, in the variable x, having degree 9.

For each polynomial, determine the number of terms, and give the coefficient of each term. ***See Objective 2.***

7. $6x^4$ 1; 6

8. $-9y^5$ 1; -9

9. t^4 1; 1

10. s^7 1; 1

11. $\frac{x}{5}$ 1; $\frac{1}{5}$

12. $\frac{z}{8}$ 1; $\frac{1}{8}$

13. $-19r^2 - r$ 2; -19, -1

14. $2y^3 - y$ 2; 2, -1

15. $x - 8x^2 + \frac{2}{3}x^3$ 3; 1, -8, $\frac{2}{3}$

16. $v - 2v^3 + \frac{3}{4}v^2$ 3; 1, -2, $\frac{3}{4}$

In each polynomial, combine like terms whenever possible. Write the result with descending powers. ***See Example 1.***

17. $-3m^5 + 5m^5$ $2m^5$

18. $-4y^3 + 3y^3$ $-y^3$

19. $2r^5 + (-3r^5)$ $-r^5$

20. $-19y^2 + 9y^2$ $-10y^2$

21. $\frac{1}{2}x^4 + \frac{1}{6}x^4$ $\frac{2}{3}x^4$

22. $\frac{3}{10}x^6 + \frac{1}{5}x^6$ $\frac{1}{2}x^6$

23. $-0.5m^2 + 0.2m^5$ cannot be simplified; $0.2m^5 - 0.5m^2$

24. $-0.9y + 0.9y^2$ cannot be simplified; $0.9y^2 - 0.9y$

25. $-3x^5 + 2x^5 - 4x^5$ $-5x^5$

26. $6x^3 - 8x^3 + 9x^3$ $7x^3$

27. $-4p^7 + 8p^7 + 5p^9$ $5p^9 + 4p^7$

28. $-3a^8 + 4a^8 - 3a^2$ $a^8 - 3a^2$

29. $-4y^2 + 3y^2 - 2y^2 + y^2$ $-2y^2$

30. $3r^5 - 8r^5 + r^5 + 2r^5$ $-2r^5$

For each polynomial, first simplify, if possible, and write it with descending powers. Then give the degree of the resulting polynomial, and tell whether it is a monomial, *a* binomial, *a* trinomial, *or* none of these. ***See Example 2.***

31. $6x^4 - 9x$ already simplified; 4; binomial

32. $7t^3 - 3t$ already simplified; 3; binomial

33. $5m^4 - 3m^2 + 6m^5 - 7m^3$ already simplified; $6m^5 + 5m^4 - 7m^3 - 3m^2$; 5; none of these

34. $6p^5 + 4p^3 - 8p^4 + 10p^2$

already simplified; $6p^5 - 8p^4 + 4p^3 + 10p^2$; 5; none of these

35. $\frac{5}{3}x^4 - \frac{2}{3}x^4 + \frac{1}{3}x^2 - 4$

$x^4 + \frac{1}{3}x^2 - 4$; 4; trinomial

36. $\frac{4}{5}r^6 + \frac{1}{5}r^6 - r^4 + \frac{2}{5}r$

$r^6 - r^4 + \frac{2}{5}r$; 6; trinomial

37. $0.8x^4 - 0.3x^4 - 0.5x^4 + 7$

7; 0; monomial

38. $1.2t^3 - 0.9t^3 - 0.3t^3 + 9$

9; 0; monomial

39. $2.5x^2 + 0.5x + x^2 - x - 2x^2$

$1.5x^2 - 0.5x$; 2; binomial

40. $8.3y - 9.2y^3 - 2.6y + 4.8y^3 - 6.7y^2$

$-4.4y^3 - 6.7y^2 + 5.7y$; 3; trinomial

Find the value of each polynomial for **(a)** $x = 2$ *and for* **(b)** $x = -1$. ***See Example 3.***

41. $-2x + 3$

(a) -1 (b) 5

42. $5x - 4$

(a) 6 (b) -9

43. $2x^2 + 5x + 1$

(a) 19 (b) -2

44. $-3x^2 + 14x - 2$

(a) 14 (b) -19

45. $2x^5 - 4x^4 + 5x^3 - x^2$

(a) 36 (b) -12

46. $x^4 - 6x^3 + x^2 + 1$

(a) -27 (b) 9

47. $-4x^5 + x^2$

(a) -124 (b) 5

48. $2x^6 - 4x$

(a) 120 (b) 6

Add or subtract as indicated. ***See Examples 4 and 7.***

49. Add.

$$\begin{array}{r} 3m^2 + 5m \\ \underline{2m^2 - 2m} \end{array}$$

$5m^2 + 3m$

50. Add.

$$\begin{array}{r} 4a^3 - 4a^2 \\ \underline{6a^3 + 5a^2} \end{array}$$

$10a^3 + a^2$

51. Subtract.

$$\begin{array}{r} 12x^4 - x^2 \\ \underline{8x^4 + 3x^2} \end{array}$$

$4x^4 - 4x^2$

52. Subtract.

$$\begin{array}{r} 13y^5 - y^3 \\ \underline{7y^5 + 5y^3} \end{array}$$

$6y^5 - 6y^3$

53. Add.

$$\begin{array}{r} \frac{2}{3}x^2 + \frac{1}{5}x + \frac{1}{6} \\ \underline{\frac{1}{2}x^2 - \frac{1}{3}x + \frac{2}{3}} \end{array}$$

$\frac{7}{6}x^2 - \frac{2}{15}x + \frac{5}{6}$

54. Add.

$$\begin{array}{r} \frac{4}{7}y^2 - \frac{1}{5}y + \frac{7}{9} \\ \underline{\frac{1}{3}y^2 - \frac{1}{3}y + \frac{2}{5}} \end{array}$$

$\frac{19}{21}y^2 - \frac{8}{15}y + \frac{53}{45}$

55. Subtract.

$$\begin{array}{r} 12m^3 - 8m^2 + 6m + 7 \\ \underline{5m^2 \qquad\quad - 4} \end{array}$$

$12m^3 - 13m^2 + 6m + 11$

56. Subtract.

$$\begin{array}{r} 5a^4 - 3a^3 + 2a^2 - a + 6 \\ \underline{-6a^4 \qquad\quad - a^2 + a - 1} \end{array}$$

$11a^4 - 3a^3 + 3a^2 - 2a + 7$

57. Subtract.

$$\begin{array}{r} 4.3x^3 - 6.1x^2 - 3.0x - 5 \\ \underline{1.4x^3 - 2.6x^2 - 1.5x + 4} \end{array}$$

$2.9x^3 - 3.5x^2 - 1.5x - 9$

58. Subtract.

$$\begin{array}{r} 8.7z^3 + 4.2z^2 - 9.0z - 7 \\ \underline{5.9z^3 - 6.6z^2 + 3.5z - 4} \end{array}$$

$2.8z^3 + 10.8z^2 - 12.5z - 3$

Perform the indicated operations. ***See Examples 5, 6, and 8.***

59. $(2r^2 + 3r - 12) + (6r^2 + 2r)$

$8r^2 + 5r - 12$

60. $(3r^2 + 5r - 6) + (2r - 5r^2)$

$-2r^2 + 7r - 6$

61. $(8m^2 - 7m) - (3m^2 + 7m - 6)$

$5m^2 - 14m + 6$

62. $(x^2 + x) - (3x^2 + 2x - 1)$

$-2x^2 - x + 1$

63. $(16x^3 - x^2 + 3x) + (-12x^3 + 3x^2 + 2x)$

$4x^3 + 2x^2 + 5x$

64. $(-2b^6 + 3b^4 - b^2) + (b^6 + 2b^4 + 2b^2)$

$-b^6 + 5b^4 + b^2$

65. $(7y^4 + 3y^2 + 2y) - (18y^5 - 5y^3 + y)$

$-18y^5 + 7y^4 + 5y^3 + 3y^2 + y$

66. $(8t^5 + 3t^3 + 5t) - (19t^4 - 6t^2 + t)$

$8t^5 - 19t^4 + 3t^3 + 6t^2 + 4t$

67. $[(8m^2 + 4m - 7) - (2m^3 - 5m + 2)] - (m^2 + m)$

$-2m^3 + 7m^2 + 8m - 9$

68. $[(9b^3 - 4b^2 + 3b + 2) - (-2b^3 + b)] - (8b^3 + 6b + 4)$

$3b^3 - 4b^2 - 4b - 2$

69. Subtract $9x^2 - 3x + 7$ from $-2x^2 - 6x + 4$.

$-11x^2 - 3x - 3$

70. Subtract $-5w^3 + 5w^2 - 7$ from $6w^3 + 8w + 5$.

$11w^3 - 5w^2 + 8w + 12$

Find a polynomial that represents the perimeter of each square, rectangle, or triangle.

71.

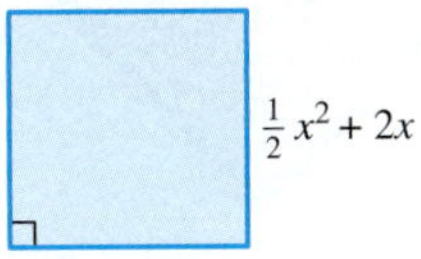

$2x^2 + 8x$

72.

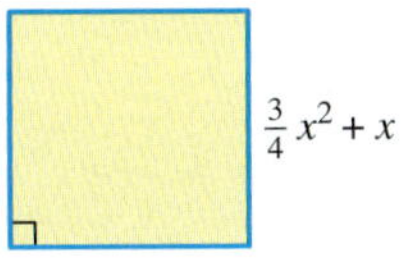

$3x^2 + 4x$

73.

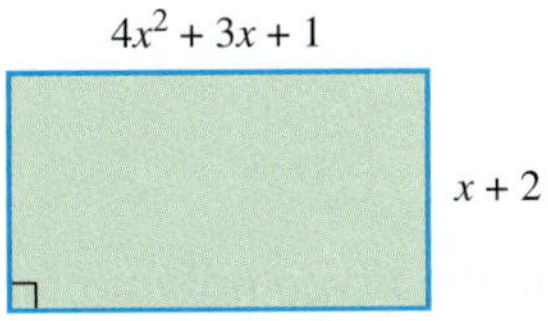

$8x^2 + 8x + 6$

74.

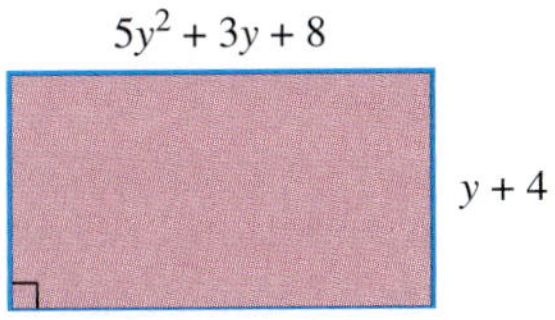

$10y^2 + 8y + 24$

75.

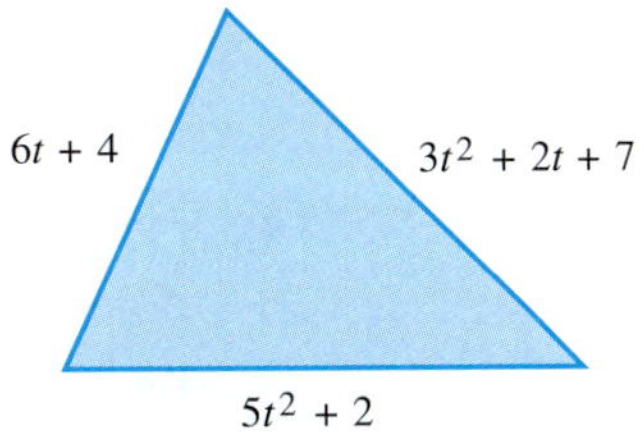

$8t^2 + 8t + 13$

76.

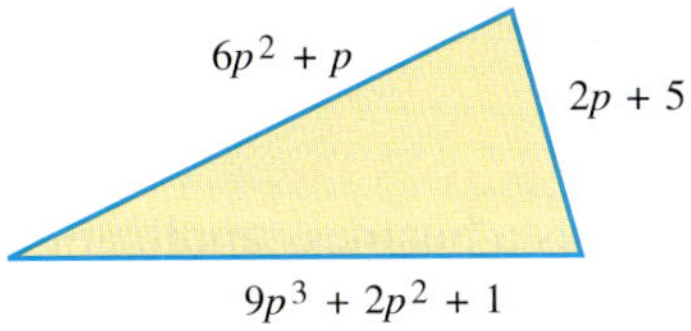

$9p^3 + 8p^2 + 3p + 6$

Add or subtract as indicated. ***See Example 9.***

77. $(9a^2b - 3a^2 + 2b) + (4a^2b - 4a^2 - 3b)$

$13a^2b - 7a^2 - b$

78. $(4xy^3 - 3x + y) + (5xy^3 + 13x - 4y)$

$9xy^3 + 10x - 3y$

79. $(2c^4d + 3c^2d^2 - 4d^2) - (c^4d + 8c^2d^2 - 5d^2)$

$c^4d - 5c^2d^2 + d^2$

80. $(3k^2h^3 + 5kh + 6k^3h^2) - (2k^2h^3 - 9kh + k^3h^2)$

$k^2h^3 + 14kh + 5k^3h^2$

81. Subtract.

$$\begin{array}{r} 9m^3n - 5m^2n^2 + 4mn^2 \\ \underline{-3m^3n + 6m^2n^2 + 8mn^2} \end{array}$$

$12m^3n - 11m^2n^2 - 4mn^2$

82. Subtract.

$$\begin{array}{r} 12r^5t + 11r^4t^2 - 7r^3t^3 \\ \underline{-8r^5t + 10r^4t^2 + 3r^3t^3} \end{array}$$

$20r^5t + r^4t^2 - 10r^3t^3$

*Find **(a)** a polynomial that represents the perimeter of each triangle and **(b)** the measures of the angles of the triangle. (Hint: In part (b), the sum of the measures of the angles of any triangle is* 180°.*)*

83.

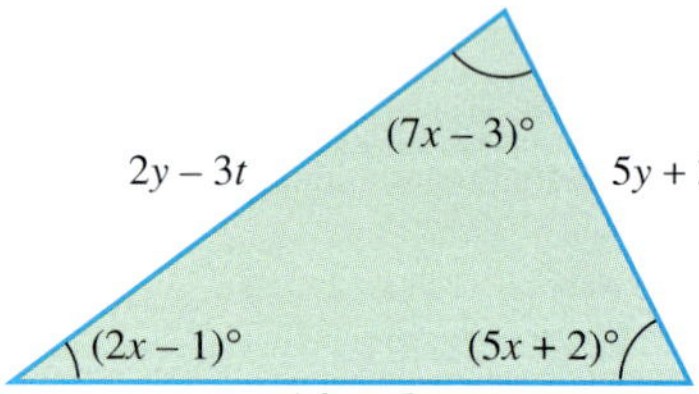

(a) $23y + 5t$ (b) 25°, 67°, 88°

84.

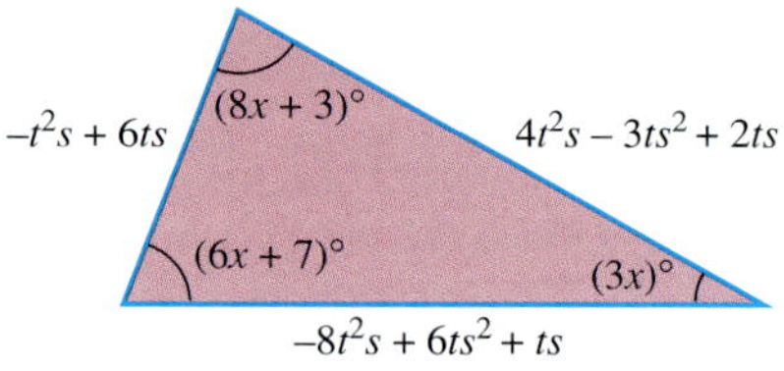

(a) $-5t^2s + 3ts^2 + 9ts$ (b) 30°, 67°, 83°

Relating Concepts (Exercises 85–88) For Individual or Group Work

A polynomial can model the distance in feet that a car going approximately 68 *mph will skid in t seconds. If we let D represent this distance, then*

$$D = 100t - 13t^2.$$

*Each time we evaluate this polynomial for a value of t, we get one and only one output value D. This idea is basic to the concept of a **function,** an important concept in mathematics. Exercises 85–88 illustrate this idea with this polynomial and two others.* ***Work them in order.***

85. Evaluate the given polynomial for $t = 5$. Use the result to fill in the blanks:

In ___5___ seconds, the car will skid ___175___ feet.

86. Use the polynomial equation $D = 100t - 13t^2$ to find the distance the car will skid in 1 sec. Write an ordered pair of the form (t, D).

87 ft; (1, 87)

87. (This is a business application of functions.) If it costs \$15 plus \$2 per day to rent a chain saw, the binomial $2x + 15$ gives the cost in dollars to rent the chain saw for x days. Evaluate this polynomial for $x = 6$. Use the result to fill in the blanks:

If the saw is rented for ___6___ days, the cost is ___\$27___.

88. (This is a physics application of functions.) If an object is projected upward under certain conditions, its height in feet is given by the trinomial $-16t^2 + 60t + 80$, where t is in seconds. Evaluate this trinomial for $t = 2.5$, and then use the result to fill in the blanks:

If ___2.5___ seconds have elapsed, the height of the object is ___130___ feet.

5.2 The Product Rule and Power Rules for Exponents

OBJECTIVES

1. Use exponents.
2. Use the product rule for exponents.
3. Use the rule $(a^m)^n = a^{mn}$.
4. Use the rule $(ab)^m = a^m b^m$.
5. Use the rule $\left(\frac{a}{b}\right)^m = \frac{a^m}{b^m}$.
6. Use combinations of the rules for exponents.
7. Use the rules for exponents in a geometry problem.

OBJECTIVE 1 **Use exponents.** In **Section 1.1,** we used exponents to write repeated products. In the expression 5^2, the number 5 is the **base** and 2 is the **exponent,** or **power.** The expression 5^2 is an **exponential expression.** Although we do not usually write the exponent when it is 1, in general,

$$a^1 = a, \quad \text{for any quantity } a.$$

EXAMPLE 1 Using Exponents

Write $3 \cdot 3 \cdot 3 \cdot 3 \cdot 3$ in exponential form and evaluate.

Since 3 occurs as a factor five times, the base is 3 and the exponent is 5. The exponential expression is 3^5, read "3 to the fifth power," or simply "3 to the fifth."

$$\underbrace{3 \cdot 3 \cdot 3 \cdot 3 \cdot 3}_{5 \text{ factors of } 3} \quad \text{means} \quad 3^5, \quad \text{which equals} \quad 243.$$

Work Problem 1 at the Side. ▶

EXAMPLE 2 Evaluating Exponential Expressions

Name the base and the exponent of each expression. Then evaluate.

Expression	Base	Exponent	Value
(a) 5^4	5	4	$5 \cdot 5 \cdot 5 \cdot 5$, which equals 625
(b) -5^4	5	4	$-1 \cdot (5 \cdot 5 \cdot 5 \cdot 5)$, which equals -625
(c) $(-5)^4$	-5	4	$(-5)(-5)(-5)(-5)$, which equals 625

CAUTION

Compare **Examples 2(b) and 2(c).** In -5^4, the absence of parentheses shows that the exponent 4 applies only to the base 5, not -5. In $(-5)^4$, the parentheses show that the exponent 4 applies to the base -5. In summary, $-a^n$ and $(-a)^n$ are not necessarily the same.

Expression	Base	Exponent	Example
$-a^n$	a	n	$-3^2 = -(3 \cdot 3) = -9$
$(-a)^n$	$-a$	n	$(-3)^2 = (-3)(-3) = 9$

Work Problem 2 at the Side. ▶

OBJECTIVE 2 **Use the product rule for exponents.** To develop the product rule, we use the definition of an exponent.

$$2^4 \cdot 2^3$$
$$= \underbrace{(2 \cdot 2 \cdot 2 \cdot 2)}_{4 \text{ factors}}\underbrace{(2 \cdot 2 \cdot 2)}_{3 \text{ factors}}$$
$$= \underbrace{2 \cdot 2 \cdot 2 \cdot 2 \cdot 2 \cdot 2 \cdot 2}_{4 + 3 = 7 \text{ factors}}$$
$$= 2^7$$

1 Write each product in exponential form and evaluate.

(a) $2 \cdot 2 \cdot 2 \cdot 2$

(b) $(-3)(-3)(-3)$

2 Name the base and the exponent of each expression. Then evaluate.

(a) 2^5 **(b)** -2^5

(c) $(-2)^5$ **(d)** 4^2

(e) -4^2 **(f)** $(-4)^2$

Answers

1. (a) 2^4; 16 (b) $(-3)^3$; -27
2. (a) 2; 5; 32 (b) 2; 5; -32 (c) -2; 5; -32 (d) 4; 2; 16 (e) 4; 2; -16 (f) -4; 2; 16

3 Simplify by using the product rule, if possible.

GS **(a)** $8^2 \cdot 8^5$

$= 8^{\_\_+\_\_}$

$= \_\_\_\_$

(b) $(-7)^5(-7)^3$

Teaching Tip

Be aware of the common student error.

$6^2 \cdot 6^3 = 36^5$ INCORRECT

Stress keeping the same base and adding the exponents.

GS **(c)** $y^3 \cdot y$

$= y^3 \cdot y^{\_\_}$

$= y^{\_\_+\_\_}$

$= \_\_\_\_$

(d) $z^2z^5z^6$

(e) $4^2 \cdot 3^5$

(f) $6^4 + 6^2$

Answers

3. **(a)** 2; 5; 8^7 **(b)** $(-7)^8$
(c) 1; 3; 1; y^4 **(d)** z^{13}
(e) The product rule does not apply. (The product is 3888.)
(f) The product rule does not apply. (The sum is 1332.)

Also,

$$\begin{aligned} &6^2 \cdot 6^3 \\ &= (6 \cdot 6)(6 \cdot 6 \cdot 6) \\ &= 6 \cdot 6 \cdot 6 \cdot 6 \cdot 6 \\ &= 6^5. \end{aligned}$$

Generalizing from these examples, we have the following.

$2^4 \cdot 2^3$ is equal to 2^{4+3}, which equals 2^7.

$6^2 \cdot 6^3$ is equal to 6^{2+3}, which equals 6^5.

In each case, adding the exponents gives the exponent of the product, suggesting the **product rule for exponents.**

Product Rule for Exponents

For any positive integers m and n, $\boldsymbol{a^m \cdot a^n = a^{m+n}}$.
(Keep the same base and add the exponents.)

Example: $6^2 \cdot 6^5 = 6^{2+5} = 6^7$

CAUTION

Do not multiply the bases when using the product rule. ***Keep the same base and add the exponents.*** For example,

$6^2 \cdot 6^5$ is equal to 6^7, ***not*** 36^7.

EXAMPLE 3 Using the Product Rule

Use the product rule for exponents to simplify, if possible.

(a) $6^3 \cdot 6^5$

$= 6^{3+5}$ Product rule

$= 6^8$ Add the exponents.

(b) $(-4)^7(-4)^2$

$= (-4)^{7+2}$ Product rule

$= (-4)^9$ Add the exponents.

(c) $x^2 \cdot x$

$= x^2 \cdot x^1$ $a = a^1$, for all a.

$= x^{2+1}$ Product rule

$= x^3$ Add the exponents.

(d) $m^4m^3m^5$

$= m^{4+3+5}$ Product rule

$= m^{12}$ Add the exponents.

(e) $2^3 \cdot 3^2$ (Think: 2^3 means $2 \cdot 2 \cdot 2$.) (Think: 3^2 means $3 \cdot 3$.)

$= 8 \cdot 9$ Evaluate 2^3 and 3^2.

$= 72$ Multiply.

The product rule does not apply because the bases are different.

(f) $2^3 + 2^4$

$= 8 + 16$ Evaluate 2^3 and 2^4.

$= 24$ Add.

The product rule does not apply. This is a *sum,* not a *product.*

◀ **Work Problem 3 at the Side.**

EXAMPLE 4 **Using the Product Rule**

Multiply $2x^3$ and $3x^7$.

$2x^3 \cdot 3x^7$ ($2x^3 = 2 \cdot x^3$; $3x^7 = 3 \cdot x^7$)

$= (2 \cdot 3) \cdot (x^3 \cdot x^7)$ Commutative and associative properties

$= 6x^{3+7}$ Multiply; product rule

$= 6x^{10}$ Add the exponents.

Work Problem 4 at the Side. ▶

CAUTION

Be sure that you understand the difference between *adding* and *multiplying* exponential expressions. For example,

$8x^3 + 5x^3$ means $(8 + 5)x^3$, which equals $13x^3$,

but $(8x^3)(5x^3)$ means $(8 \cdot 5)x^{3+3}$, which equals $40x^6$.

OBJECTIVE 3 **Use the rule $(a^m)^n = a^{mn}$.** We can simplify an expression such as $(5^2)^4$ with the product rule for exponents, as follows.

$(5^2)^4$

$= 5^2 \cdot 5^2 \cdot 5^2 \cdot 5^2$ Definition of exponent

$= 5^{2+2+2+2}$ Product rule

$= 5^8$ Add the exponents.

Observe that $2 \cdot 4 = 8$. This example suggests **power rule (a) for exponents.**

Power Rule (a) for Exponents

For any positive integers m and n, $(a^m)^n = a^{mn}$.
(Raise a power to a power by multiplying exponents.)

Example: $(3^2)^4 = 3^{2 \cdot 4} = 3^8$

EXAMPLE 5 **Using Power Rule (a)**

Use power rule (a) for exponents to simplify.

(a) $(2^5)^3$
$= 2^{5 \cdot 3}$
$= 2^{15}$

(b) $(5^7)^2$
$= 5^{7 \cdot 2}$
$= 5^{14}$

(c) $(x^2)^5$
$= x^{2 \cdot 5}$ Power rule (a)
$= x^{10}$ Multiply.

Work Problem 4 at the Side. ▶

OBJECTIVE 4 **Use the rule $(ab)^m = a^m b^m$.** We can rewrite the expression $(4x)^3$ as shown below.

$(4x^3)$

$= (4x)(4x)(4x)$ Definition of exponent

$= 4 \cdot 4 \cdot 4 \cdot x \cdot x \cdot x$ Commutative and associative properties

$= 4^3x^3$ Definition of exponent

---

4 Multiply.

GS **(a)** $5m^2 \cdot 2m^6$

$= (5 \cdot \_\_\_\_) \cdot (m^{\_\_} \cdot m^{\_\_})$

$= \_\_\_\_ \, m^{\_\_+\_\_}$

$= \_\_\_\_$

(b) $3p^5 \cdot 9p^4$

(c) $-7p^5 \cdot (3p^8)$

5 Simplify.

GS **(a)** $(5^3)^4$

$= 5^{\_\_ \cdot \_\_}$

$= \_\_\_\_$

(b) $(6^2)^5$

(c) $(3^2)^4$

(d) $(a^6)^5$

Answers

4. **(a)** 2; 2; 6; 10; 2; 6; $10m^8$
(b) $27p^9$ **(c)** $-21p^{13}$

5. **(a)** 3; 4; 5^{12} **(b)** 6^{10} **(c)** 3^8 **(d)** a^{30}

6 Simplify.

GS **(a)** $(2ab)^4$

$= 2\_\_a\_\_b\_\_$

$= \_\_\_\_$

(b) $5(mn)^3$

(c) $(3a^2b^4)^5$

(d) $(-5m^2)^3$

Teaching Tip

In Example 6(d), point out that $(-5^6)^3 = -5^{18}$, but $(-5^3)^6 = 5^{18}$. Stress that the factor -1 must also be raised to the designated power, or a sign error could result. Mention that -1 raised to an even power is 1 and -1 raised to an odd power is -1.

Answers

6. **(a)** 4; 4; 4; $16a^4b^4$ **(b)** $5m^3n^3$ **(c)** $243a^{10}b^{20}$ **(d)** $-125m^6$

The example $(4x)^3 = 4^3x^3$ suggests **power rule (b) for exponents.**

Power Rule (b) for Exponents

For any positive integer m, $\quad (ab)^m = a^mb^m$.
(Raise a product to a power by raising each factor to the power.)

Example: $\quad (2p)^5 = 2^5p^5$

EXAMPLE 6 Using Power Rule (b)

Use power rule (b) for exponents to simplify.

(a) $(3xy)^2$

$= 3^2x^2y^2$ Power rule (b)

$= 9x^2y^2$ $\quad 3^2 = 3 \cdot 3 = 9$

(b) $9(pq)^2$

$= 9(p^2q^2)$ Power rule (b)

$= 9p^2q^2$ Multiply.

(c) $5(2m^2p^3)^4$

$= 5[2^4(m^2)^4(p^3)^4]$ Power rule (b)

$= 5(2^4m^8p^{12})$ Power rule (a)

$= 80m^8p^{12}$ $\quad 5 \cdot 2^4 = 5 \cdot 16 = 80$

(d) $(-5^6)^3$

$= (-1 \cdot 5^6)^3$ $\quad -a = -1 \cdot a$

$= (-1)^3(5^6)^3$ Power rule (b) — Raise -1 to the designated power.

$= -1 \cdot 5^{18}$ Power rule (a)

$= -5^{18}$ Multiply.

◀ Work Problem **6** at the Side.

CAUTION

Power rule (b) does not apply to a sum.

$$(4x)^2 = 4^2x^2, \quad \text{but} \quad (4 + x)^2 \neq 4^2 + x^2.$$

OBJECTIVE 5 **Use the rule $\left(\frac{a}{b}\right)^m = \frac{a^m}{b^m}$.** Since the quotient $\frac{a}{b}$ can be written as $a \cdot \frac{1}{b}$, we can use power rule (b), together with some of the properties of real numbers, to state **power rule (c) for exponents.**

Power Rule (c) for Exponents

For any positive integer m, $\quad \left(\frac{a}{b}\right)^m = \frac{a^m}{b^m}$ **(where $b \neq 0$).**

(Raise a quotient to a power by raising both the numerator and the denominator to the power.)

Example: $\quad \left(\frac{5}{3}\right)^2 = \frac{5^2}{3^2}$

EXAMPLE 7 Using Power Rule (c)

Use power rule (c) for exponents to simplify.

(a) $\left(\frac{2}{3}\right)^5$

$= \frac{2^5}{3^5}$

$= \frac{32}{243}$

(b) $\left(\frac{m}{n}\right)^4$

$= \frac{m^4}{n^4}$

(where $n \neq 0$)

(c) $\left(\frac{1}{5}\right)^4$

$= \frac{1^4}{5^4}$ Power rule (c)

$= \frac{1}{625}$ Simplify.

Note

In **Example 7(c),** we used the fact that $1^4 = 1$ since $1 \cdot 1 \cdot 1 \cdot 1 = 1$.

In general, $\boldsymbol{1^n = 1}$, ***for any integer n.***

Work Problem 7 at the Side. ▶

The rules for exponents discussed in this section should be *memorized.*

Rules for Exponents

For positive integers m and n, the following are true.

		Examples
Product rule	$a^m \cdot a^n = a^{m+n}$	$6^2 \cdot 6^5 = 6^{2+5} = 6^7$
Power rules (a)	$(a^m)^n = a^{mn}$	$(3^2)^4 = 3^{2 \cdot 4} = 3^8$
(b)	$(ab)^m = a^m b^m$	$(2p)^5 = 2^5 p^5$
(c)	$\left(\frac{a}{b}\right)^m = \frac{a^m}{b^m}$ $(b \neq 0)$	$\left(\frac{5}{3}\right)^2 = \frac{5^2}{3^2}$

OBJECTIVE 6 **Use combinations of the rules for exponents.**

EXAMPLE 8 Using Combinations of Rules

Simplify each expression.

(a) $\left(\frac{2}{3}\right)^2 \cdot 2^3$

$= \frac{2^2}{3^2} \cdot \frac{2^3}{1}$ Power rule (c)

$= \frac{2^2 \cdot 2^3}{3^2 \cdot 1}$ Multiply fractions.

$= \frac{2^{2+3}}{3^2}$ Product rule

$= \frac{2^5}{3^2}$, or $\frac{32}{9}$

(b) $(5x)^3(5x)^4$

$= (5x)^7$ Product rule

$= 5^7x^7$ Power rule (b)

Continued on Next Page

7 Simplify. Assume that all variables represent nonzero real numbers.

GS (a) $\left(\frac{5}{2}\right)^4$

$= \frac{5\_\_}{2\_\_}$

$= \_\_\_\_$

(b) $\left(\frac{p}{q}\right)^2$

(c) $\left(\frac{r}{t}\right)^3$

(d) $\left(\frac{1}{3}\right)^5$

(e) $\left(\frac{1}{x}\right)^{10}$

Answers

7. (a) 4; 4; $\frac{625}{16}$ (b) $\frac{p^2}{q^2}$ (c) $\frac{r^3}{t^3}$

(d) $\frac{1}{243}$ (e) $\frac{1}{x^{10}}$

8 Simplify.

GS **(a)** $(2m)^5(2m)^3$

$= (2m)^{—}$

$=$ ____

(b) $\left(\dfrac{5k^3}{3}\right)^2$

(c) $\left(\dfrac{1}{5}\right)^4(2x)^2$

(d) $(-3xy^2)^3(x^2y)^4$

9 Find a polynomial that represents the area of each figure.

(a)

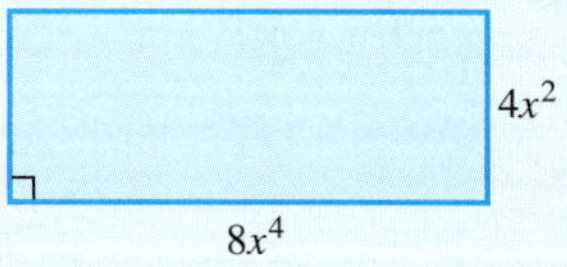

(b)

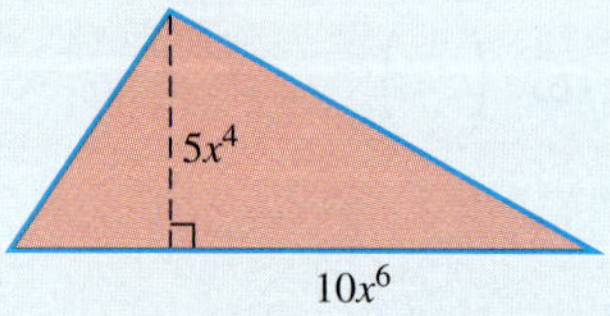

Answers

8. **(a)** 8; 2^8m^8, or $256m^8$ **(b)** $\dfrac{5^2k^6}{3^2}$, or $\dfrac{25k^6}{9}$
(c) $\dfrac{2^2x^2}{5^4}$, or $\dfrac{4x^2}{625}$
(d) $-3^3x^{11}y^{10}$, or $-27x^{11}y^{10}$

9. **(a)** $32x^6$ **(b)** $25x^{10}$

(c) $(2x^2y^3)^4(3xy^2)^3$

$= 2^4(x^2)^4(y^3)^4 \cdot 3^3x^3(y^2)^3$ Power rule (b)

$= 2^4x^8y^{12} \cdot 3^3x^3y^6$ Power rule (a)

$= 2^4 \cdot 3^3x^8x^3y^{12}y^6$ Commutative and associative properties

$= 16 \cdot 27x^{11}y^{18}$ Product rule

$= 432x^{11}y^{18}$ Multiply.

Notice that $(2x^2y^3)^4$ means $2^4x^{2\cdot4}y^{3\cdot4}$, ***not*** $(2\cdot4)\,x^{2\cdot4}y^{3\cdot4}$.

Do *not* multiply the coefficient 2 and the exponent 4.

(d) $(-x^3y)^2(-x^5y^4)^3$

$= (-1x^3y)^2(-1x^5y^4)^3$ $-a = -1 \cdot a$

Think of the negative sign in each factor as -1.

$= (-1)^2(x^3)^2(y^2) \cdot (-1)^3(x^5)^3(y^4)^3$ Power rule (b)

$= (-1)^2(x^6)(y^2) \cdot (-1)^3(x^{15})(y^{12})$ Power rule (a)

$= (-1)^5(x^{6+15})(y^{2+12})$ Product rule

$= -1x^{21}y^{14}$ Add the exponents.

$= -x^{21}y^{14}$ $-1 \cdot a = -a$

◄ Work Problem **8** at the Side.

OBJECTIVE 7 **Use the rules for exponents in a geometry problem.**

EXAMPLE 9 **Using Area Formulas**

Find a polynomial that represents the area in **(a) Figure 1** and **(b) Figure 2**.

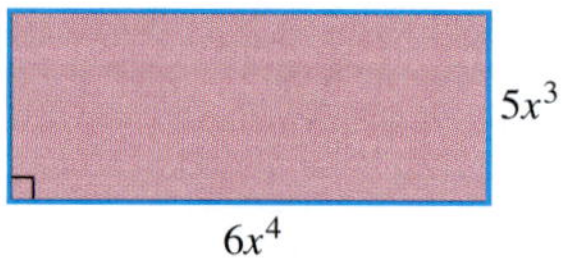

Figure 1

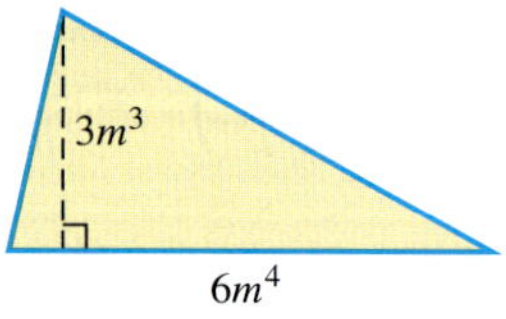

Figure 2

(a) For **Figure 1**, use the formula for the area of a rectangle, $A = LW$.

$A = (6x^4)(5x^3)$ Substitute.

$A = 6 \cdot 5 \cdot x^{4+3}$ Commutative property and product rule

$A = 30x^7$ Multiply. Add the exponents.

(b) **Figure 2** is a triangle with base $6m^4$ and height $3m^3$.

$A = \dfrac{1}{2}bh$ Area formula

$A = \dfrac{1}{2}(6m^4)(3m^3)$ Substitute.

$A = \dfrac{1}{2}(6 \cdot 3 \cdot m^{4+3})$ Commutative property and product rule

$A = 9m^7$ Multiply. Add the exponents.

◄ Work Problem **9** at the Side.

5.2 Exercises

FOR EXTRA HELP

Download the MyDashBoard App

MyMathLab®

CONCEPT CHECK *Decide whether each statement is* true *or* false.

1. $3^3 = 9$ false

2. $(-2)^4 = 2^4$ true

3. $(a^2)^3 = a^5$ false

4. $\left(\frac{1}{4}\right)^2 = \frac{1}{4^2}$ true

5. CONCEPT CHECK What exponent is understood on the base x in the expression xy^2? 1

6. CONCEPT CHECK How are the expressions 3^2, 5^3, and 7^4 read?

three squared; five cubed; seven to the fourth power

Write each expression using exponents. **See Example 1.**

7. $t \cdot t \cdot t \cdot t \cdot t \cdot t \cdot t$ t^7

8. $w \cdot w \cdot w \cdot w \cdot w \cdot w$ w^6

9. $\left(\frac{1}{2}\right)\left(\frac{1}{2}\right)\left(\frac{1}{2}\right)\left(\frac{1}{2}\right)\left(\frac{1}{2}\right)$ $\left(\frac{1}{2}\right)^5$

10. $\left(-\frac{1}{4}\right)\left(-\frac{1}{4}\right)\left(-\frac{1}{4}\right)\left(-\frac{1}{4}\right)$ $\left(-\frac{1}{4}\right)^4$

11. $(-8p)(-8p)$ $(-8p)^2$

12. $(-7x)(-7x)(-7x)$ $(-7x)^3$

Identify the base and the exponent for each exponential expression. In Exercises 13–16, also evaluate the expression. **See Example 2.**

13. 3^5 base: 3; exponent: 5; 243

14. 2^7 base: 2; exponent: 7; 128

15. $(-3)^5$ base: -3; exponent: 5; -243

16. $(-2)^7$ base: -2; exponent: 7; -128

17. $(-6x)^4$ base: $-6x$; exponent: 4

18. $(-8x)^4$ base: $-8x$; exponent: 4

19. $-6x^4$ base: x; exponent: 4

20. $-8x^4$ base: x; exponent: 4

Use the product rule for exponents to simplify each expression, if possible. If it does not apply, say so. Write each answer in exponential form. **See Examples 3 and 4.**

21. $5^2 \cdot 5^6$ 5^8

22. $3^6 \cdot 3^7$ 3^{13}

23. $4^2 \cdot 4^7 \cdot 4^3$ 4^{12}

24. $5^3 \cdot 5^8 \cdot 5^2$ 5^{13}

25. $(-7)^3(-7)^6$ $(-7)^9$

26. $(-9)^8(-9)^5$ $(-9)^{13}$

27. $t^3t^8t^{13}$ t^{24}

28. $n^5n^6n^9$ n^{20}

29. $(-8r^4)(7r^3)$ $-56r^7$

30. $(10a^7)(-4a^3)$ $-40a^{10}$

31. $(-6p^5)(-7p^5)$ $42p^{10}$

32. $(-5w^8)(-9w^8)$ $45w^{16}$

33. $3^8 + 3^9$

34. $4^{12} + 4^5$

35. $5^8 \cdot 3^8$

36. $6^3 \cdot 8^3$

In Exercises 33–36, the product rule does not apply.

Use the power rules for exponents to simplify each expression. **See Examples 5–7.**

37. $(4^3)^2$ 4^6

38. $(8^3)^6$ 8^{18}

39. $(t^4)^5$ t^{20}

40. $(y^6)^5$ y^{30}

41. $(7r)^3$ $343r^3$

42. $(11x)^4$ 11^4x^4

43. $(-5^2)^6$ 5^{12}

44. $(-9^4)^8$ 9^{32}

45. $(-8^3)^5$ -8^{15}

46. $(-7^5)^7$ -7^{35}

47. $(5xy)^5$ $5^5x^5y^5$

48. $(9pq)^6$ $9^6p^6q^6$

49. $8(qr)^3$ $8q^3r^3$

50. $4(vw)^5$ $4v^5w^5$

51. $\left(\frac{1}{2}\right)^3$ $\frac{1}{8}$

52. $\left(\frac{1}{3}\right)^5$ $\frac{1}{243}$

53. $\left(\frac{a}{b}\right)^3$ $(b \neq 0)$ $\frac{a^3}{b^3}$

54. $\left(\frac{r}{t}\right)^4$ $(t \neq 0)$ $\frac{r^4}{t^4}$

55. $\left(\frac{9}{5}\right)^8$ $\frac{9^8}{5^8}$

56. $\left(\frac{12}{7}\right)^6$ $\frac{12^6}{7^6}$

57. $(-2x^2y)^3$ $-8x^6y^3$

58. $(-5m^4p^2)^3$ $-125m^{12}p^6$

59. $(3a^3b^2)^2$ $9a^6b^4$

60. $(4x^3y^5)^4$ $256x^{12}y^{20}$

Simplify each expression. ***See Example 8.***

61. $\left(\frac{5}{2}\right)^3 \cdot \left(\frac{5}{2}\right)^2$ $\frac{5^5}{2^5}$

62. $\left(\frac{3}{4}\right)^5 \cdot \left(\frac{3}{4}\right)^6$ $\frac{3^{11}}{4^{11}}$

63. $\left(\frac{9}{8}\right)^3 \cdot 9^2$ $\frac{9^5}{8^3}$

64. $\left(\frac{8}{5}\right)^4 \cdot 8^3$ $\frac{8^7}{5^4}$

65. $(2x)^9(2x)^3$ $2^{12}x^{12}$

66. $(6y)^5(6y)^8$ $6^{13}y^{13}$

67. $(-6p)^4(-6p)$ -6^5p^5

68. $(-13q)^3(-13q)$ 13^4q^4

69. $(6x^2y^3)^5$ $6^5x^{10}y^{15}$

70. $(5r^5t^6)^7$ $5^7r^{35}t^{42}$

71. $(x^2)^3(x^3)^5$ x^{21}

72. $(y^4)^5(y^3)^5$ y^{35}

73. $(2w^2x^3y)^2(x^4y)^5$ $4w^4x^{26}y^7$

74. $(3x^4y^2z)^3(yz^4)^5$ $27x^{12}y^{11}z^{23}$

75. $(-r^4s)^2(-r^2s^3)^5$ $-r^{18}s^{17}$

76. $(-ts^6)^4(-t^3s^5)^3$ $-t^{13}s^{39}$

77. $\left(\frac{5a^2b^5}{c^6}\right)^3$ $(c \neq 0)$ $\frac{125a^6b^{15}}{c^{18}}$

78. $\left(\frac{6x^3y^9}{z^5}\right)^4$ $(z \neq 0)$ $\frac{1296x^{12}y^{36}}{z^{20}}$

79. $(-5m^3p^4q)^2(p^2q)^3$ $25m^6p^{14}q^5$

80. $(-a^4b^5)(-6a^3b^3)^2$ $-36a^{10}b^{11}$

81. $(2x^2y^3z)^4(xy^2z^3)^2$ $16x^{10}y^{16}z^{10}$

82. $(4q^2r^3s^5)^3(qr^2s^3)^4$ $64q^{10}r^{17}s^{27}$

83. CONCEPT CHECK A student simplified $(10^2)^3$ as 1000^6. **What Went Wrong?**

Using power rule (a), simplify as follows: $(10^2)^3 = 10^{2 \cdot 3} = 10^6 = 1{,}000{,}000$.

84. CONCEPT CHECK A student simplified $(3x^2y^3)^4$ as $12x^8y^{12}$. **What Went Wrong?**

The 4 is used as an *exponent* on 3. It is *not* multiplied by 3. The correct simplification is $3^4 \cdot x^8 \cdot y^{12} = 81x^8y^{12}$.

Find a polynomial that represents the area of each figure. In Exercise 88, leave π in your answer. ***See Example 9.*** *(If necessary, refer to the formulas inside the back cover. The ⌝ in the figures indicates 90° (right) angles.)*

85.

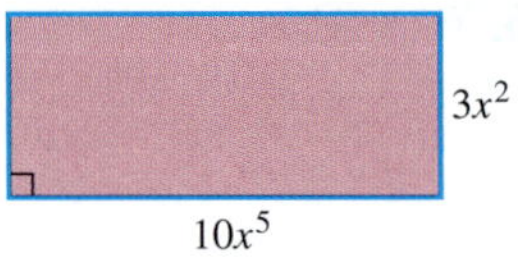

$30x^7$

86.

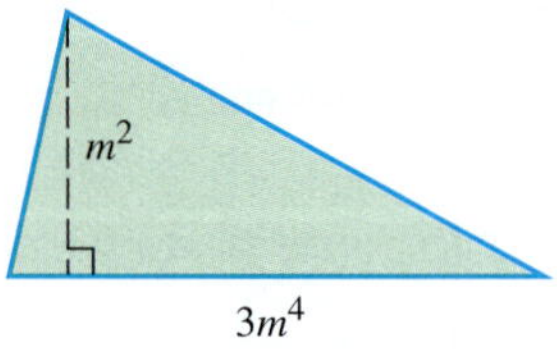

$\frac{3}{2}m^6$

87.

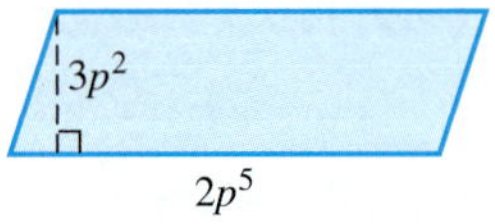

$6p^7$

88.

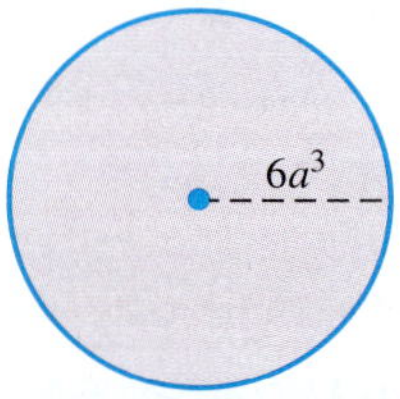

$36\pi a^6$

5.3 Multiplying Polynomials

OBJECTIVE 1 Multiply a monomial and a polynomial. In **Section 5.2,** we found the product of two monomials as follows.

$$(-8m^6)(-9n^6)$$
$$= (-8)(-9)(m^6)(n^6) \quad \text{Commutative and associative properties}$$
$$= 72m^6n^6 \quad \text{Multiply.}$$

CAUTION

Do not confuse addition of terms with multiplication of terms.

$7q^5 + 2q^5$	$(7q^5)(2q^5)$
$= (7 + 2)q^5$	$= 7 \cdot 2q^{5+5}$
$= 9q^5$	$= 14q^{10}$

To find the product of a monomial and a polynomial with more than one term, we use the distributive property and multiplication of monomials.

EXAMPLE 1 Multiplying Monomials and Polynomials

Find each product.

(a) $4x^2(3x + 5)$
$$= 4x^2(3x) + 4x^2(5) \quad \text{Distributive property}$$
$$= 12x^3 + 20x^2 \quad \text{Multiply monomials.}$$

(b) $-8m^3(4m^3 + 3m^2 + 2m - 1)$
$$= -8m^3(4m^3) + (-8m^3)(3m^2)$$
$$+ (-8m^3)(2m) + (-8m^3)(-1) \quad \text{Distributive property}$$
$$= -32m^6 - 24m^5 - 16m^4 + 8m^3 \quad \text{Multiply monomials.}$$

Work Problem 1 at the Side. ▶

OBJECTIVE 2 Multiply two polynomials. To find the product of the polynomials $x^2 + 3x + 5$ and $x - 4$, think of $x - 4$ as a single quantity and use the distributive property as follows.

$$(x^2 + 3x + 5)(x - 4)$$
$$= x^2(x - 4) + 3x(x - 4) + 5(x - 4) \quad \text{Distributive property}$$
$$= x^2(x) + x^2(-4) + 3x(x) + 3x(-4) + 5(x) + 5(-4) \quad \text{Distributive property again}$$
$$= x^3 - 4x^2 + 3x^2 - 12x + 5x - 20 \quad \text{Multiply monomials.}$$
$$= x^3 - x^2 - 7x - 20 \quad \text{Combine like terms.}$$

Multiplying Polynomials

To multiply two polynomials, multiply each term of the second polynomial by each term of the first polynomial and add the products.

OBJECTIVES

1. Multiply a monomial and a polynomial.
2. Multiply two polynomials.
3. Multiply binomials by the FOIL method.

1 Find each product.

GS **(a)** $5m^3(2m + 7)$
$$= 5m^3(\_\_\_) + 5m^3(\_\_\_)$$
$$= \_\_\_\_\_$$

(b) $2x^4(3x^2 + 2x - 5)$

(c) $-4y^2(3y^3 + 2y^2 - 4y + 8)$

Teaching Tip

Mention that polynomials are usually arranged in descending powers before multiplying.

If students have difficulty following this explanation, use the rectangular model procedure following Example 3.

Answers

1. (a) $2m$; 7; $10m^4 + 35m^3$
(b) $6x^6 + 4x^5 - 10x^4$
(c) $-12y^5 - 8y^4 + 16y^3 - 32y^2$

EXAMPLE 2 **Multiplying Two Polynomials**

Multiply $(m^2 + 5)(4m^3 - 2m^2 + 4m)$.

$(m^2 + 5)(4m^3 - 2m^2 + 4m)$ — Multiply each term of the second ploynomial by each term of the first.

$= m^2(4m^3) + m^2(-2m^2) + m^2(4m) + 5(4m^3) + 5(-2m^2) + 5(4m)$ — Distributive property

$= 4m^5 - 2m^4 + 4m^3 + 20m^3 - 10m^2 + 20m$ — Distributive property again

$= 4m^5 - 2m^4 + 24m^3 - 10m^2 + 20m$ — Combine like terms.

◀ **Work Problem 2 at the Side.**

EXAMPLE 3 **Multiplying Polynomials Vertically**

Multiply $(x^3 + 2x^2 + 4x + 1)(3x + 5)$ vertically.

$$\begin{array}{r} x^3 + 2x^2 + 4x + 1 \\ 3x + 5 \\ \hline \end{array}$$

Write the polynomials vertically.

Begin by multiplying each of the terms in the top row by 5.

$$\begin{array}{r} x^3 + 2x^2 + 4x + 1 \\ 3x + 5 \\ \hline 5x^3 + 10x^2 + 20x + 5 \end{array}$$

$5(x^3 + 2x^2 + 4x + 1)$

Now multiply each term in the top row by $3x$. Then add like terms.

$$\begin{array}{r} x^3 + 2x^2 + 4x + 1 \\ 3x + 5 \\ \hline 5x^3 + 10x^2 + 20x + 5 \\ 3x^4 + 6x^3 + 12x^2 + 3x \\ \hline 3x^4 + 11x^3 + 22x^2 + 23x + 5 \end{array}$$

This process is similar to multiplication of whole numbers.

Place *like* terms in columns so they can be added.

$3x(x^3 + 2x^2 + 4x + 1)$

Add in columns.

The product is $3x^4 + 11x^3 + 22x^2 + 23x + 5$.

◀ **Work Problem 3 at the Side.**

We can use a rectangle to model polynomial multiplication. For example, to find the product

$$(2x + 1)(3x + 2),$$

we label a rectangle with each term as shown below on the left. Then we write the product of each pair of monomials in the appropriate box as shown on the right.

	$3x$	2
$2x$		
1		

	$3x$	2
$2x$	$6x^2$	$4x$
1	$3x$	2

The product of the binomials is the sum of these four monomial products.

$$\begin{aligned} &(2x + 1)(3x + 2) \\ &= 6x^2 + 4x + 3x + 2 \\ &= 6x^2 + 7x + 2 \end{aligned}$$

Combine like terms.

◀ **Work Problem 4 at the Side.**

2 Multiply.

(a) $(m + 3)(m^2 - 2m + 1)$

(b) $(6p^2 + 2p - 4)(3p^2 - 5)$

3 Multiply vertically.

$$\begin{array}{r} 3x^2 + 4x - 5 \\ x + 4 \\ \hline \end{array}$$

4 Use the rectangle method to find each product.

GS **(a)** $(4x + 3)(x + 2)$

	x	2
$4x$	$4x^2$	
3		

(b) $(x + 5)(x^2 + 3x + 1)$

Answers

2. (a) $m^3 + m^2 - 5m + 3$
(b) $18p^4 + 6p^3 - 42p^2 - 10p + 20$

3. $3x^3 + 16x^2 + 11x - 20$

4. (a)

	x	2
$4x$	$4x^2$	$8x$
3	$3x$	6

;

$4x^2 + 11x + 6$

(b) $x^3 + 8x^2 + 16x + 5$

OBJECTIVE 3 Multiply binomials by the FOIL method. When multiplying binomials, the **FOIL method** reduces the rectangle method to a systematic approach without the rectangle. Consider this example.

$$(x + 3)(x + 5)$$

$= (x + 3)x + (x + 3)5$	Distributive property
$= x(x) + 3(x) + x(5) + 3(5)$	Distributive property again
$= x^2 + 3x + 5x + 15$	Multiply.
$= x^2 + 8x + 15$	Combine like terms.

The letters of the word FOIL refer to the positions of the terms.

$(x + 3)(x + 5)$	Multiply the **First terms:** $x(x)$.	**F**
$(x + 3)(x + 5)$	Multiply the **Outer terms:** $x(5)$. This is the **outer product.**	**O**
$(x + 3)(x + 5)$	Multiply the **Inner terms:** $3(x)$. This is the **inner product.**	**I**
$(x + 3)(x + 5)$	Multiply the **Last terms:** $3(5)$.	**L**

We add the outer product, $5x$, and the inner product, $3x$, mentally so that the three terms of the answer can be written without extra steps.

$$(x + 3)(x + 5)$$
$$= x^2 + 8x + 15$$

Multiplying Binomials by the FOIL Method

Step 1 Multiply the two **F**irst terms of the binomials to get the first term of the answer.

Step 2 Find the **O**uter product and the **I**nner product and combine them (when possible) to get the middle term of the answer.

Step 3 Multiply the two **L**ast terms of the binomials to get the last term of the answer.

F $= x^2$ **L** $= 15$

$(x + 3)(x + 5)$

I $= 3x$
O $= 5x$
$8x$ Add.

$(x + 3)(x + 5)$
$= x^2 + 8x + 15$

Add the terms found in Steps 1–3.

Work Problem 5 at the Side.

5 For the product

$$(2p - 5)(3p + 7),$$

find and simplify the following.

(a) Product of first terms

____ (____)
= ____

(b) Outer product

____ (____)
= ____

(c) Inner product

____ (____)
= ____

(d) Product of last terms

____ (____)
= ____

(e) Complete product in simplified form

____

Answers

5. (a) $2p$; $3p$; $6p^2$ **(b)** $2p$; 7; $14p$
(c) -5; $3p$; $-15p$ **(d)** -5; 7; -35
(e) $6p^2 - p - 35$

EXAMPLE 4 **Using the FOIL Method**

Use the FOIL method to find the product $(x + 8)(x - 6)$.

Step 1 **F** Multiply the **first** terms: $x(x) = x^2$.

Step 2 **O** Find the **outer** product: $x(-6) = -6x$.

I Find the **inner** product: $8(x) = 8x$.

Add the outer and inner products mentally: $-6x + 8x = 2x$.

Step 3 **L** Multiply the **last** terms: $8(-6) = -48$.

The product $(x + 8)(x - 6)$ is $x^2 + 2x - 48$. Add the terms found in Steps 1–3.

Shortcut:

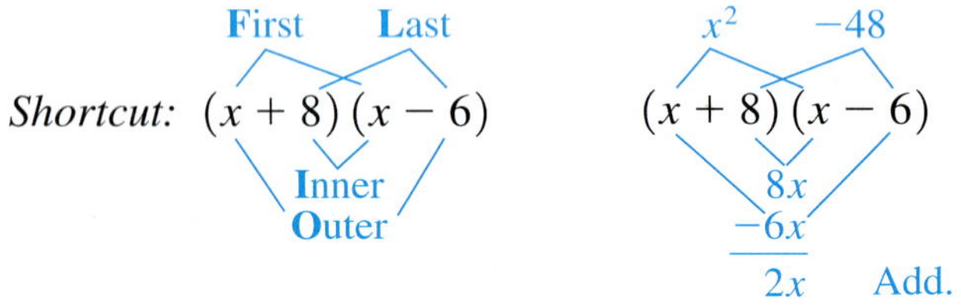

$(x + 8)(x - 6)$
$= x^2 + 2x - 48$

◀ **Work Problem 6 at the Side.**

EXAMPLE 5 **Using the FOIL Method**

Multiply $(9x - 2)(3y + 1)$.

First	$(9x - 2)(3y + 1)$	$27xy$
Outer	$(9x - 2)(3y + 1)$	$9x$
Inner	$(9x - 2)(3y + 1)$	$-6y$
Last	$(9x - 2)(3y + 1)$	-2

These unlike terms *cannot* be added.

F O I L

The product $(9x - 2)(3y + 1)$ is $27xy + 9x - 6y - 2$.

◀ **Work Problem 7 at the Side.**

EXAMPLE 6 **Using the FOIL Method**

Find each product.

(a) $(2k + 5y)(k + 3y)$

$= 2k(k) + 2k(3y) + 5y(k) + 5y(3y)$ FOIL method

$= 2k^2 + 6ky + 5ky + 15y^2$ Multiply.

$= 2k^2 + 11ky + 15y^2$ Combine like terms.

(b) $(7p + 2q)(3p - q)$

$= 21p^2 - 7pq + 6pq - 2q^2$

$= 21p^2 - pq - 2q^2$

(c) $2x^2(x - 3)(3x + 4)$

$= 2x^2(3x^2 - 5x - 12)$

$= 6x^4 - 10x^3 - 24x^2$

◀ **Work Problem 8 at the Side.**

Note

Alternatively, **Example 6(c)** can be simplified as follows.

$2x^2(x - 3)(3x + 4)$ Multiply $2x^2$ and $x - 3$ first.

$= (2x^3 - 6x^2)(3x + 4)$ Multiply that product and $3x + 4$.

$= 6x^4 - 10x^3 - 24x^2$ Same answer

6 Use the FOIL method to find each product.

GS **(a)** $(m + 4)(m - 3)$

$= m(\_\_\_) + m(\_\_\_)$
$+ 4(\_\_\_) + 4(\_\_\_)$
$= \_\_\_\_\_\_$

(b) $(y + 7)(y + 2)$

(c) $(r - 8)(r - 5)$

7 Find the product.

$(4x - 3)(2y + 5)$

8 Find each product.

GS **(a)** $(6m + 5)(m - 4)$

$= 6m(\_\_\_) + 6m(\_\_\_)$
$+ 5(\_\_\_) + 5(\_\_\_)$
$= \_\_\_\_\_\_$

(b) $(3r + 2t)(3r + 4t)$

GS **(c)** $y^2(8y + 3)(2y + 1)$

$= y^2(\_\_\_\_\_\_)$
$= \_\_\_\_\_\_$

Answers

6. **(a)** m; -3; m; -3; $m^2 + m - 12$
(b) $y^2 + 9y + 14$
(c) $r^2 - 13r + 40$

7. $8xy + 20x - 6y - 15$

8. **(a)** m; -4; m; -4; $6m^2 - 19m - 20$
(b) $9r^2 + 18rt + 8t^2$
(c) $16y^2 + 14y + 3$; $16y^4 + 14y^3 + 3y^2$

5.3 Exercises

FOR EXTRA HELP

Download the MyDashBoard App

MyMathLab®

CONCEPT CHECK *In Exercises 1 and 2, match each product in Column I with the correct polynomial in Column II.*

I	II
1. (a) $5x^3(6x^7)$ B	**A.** $125x^{21}$
(b) $-5x^7(6x^3)$ D	**B.** $30x^{10}$
(c) $(5x^7)^3$ A	**C.** $-216x^9$
(d) $(-6x^3)^3$ C	**D.** $-30x^{10}$

I	II
2. (a) $(x-5)(x+4)$ C	**A.** $x^2+9x+20$
(b) $(x+5)(x+4)$ A	**B.** $x^2-9x+20$
(c) $(x-5)(x-4)$ B	**C.** x^2-x-20
(d) $(x+5)(x-4)$ D	**D.** x^2+x-20

CONCEPT CHECK *Fill in each blank with the correct response.*

3. In multiplying a monomial by a polynomial, such as in $4x(3x^2+7x^3)=4x(3x^2)+4x(7x^3)$, the first property that is used is the __distributive__ property.

4. The FOIL method can only be used to multiply two polynomials when both polynomials are __binomials__.

Find each product. **See Section 5.2.**

5. $5p(3q^2)$ $15pq^2$

6. $4a^3(3b^2)$ $12a^3b^2$

7. $(-6m^3)(3n^2)$ $-18m^3n^2$

8. $(9r^3)(-2s^2)$ $-18r^3s^2$

9. $y^5 \cdot 9y \cdot y^4$ $9y^{10}$

10. $x^2 \cdot 3x^3 \cdot 2x$ $6x^6$

11. $(4x^3)(2x^2)(-x^5)$ $-8x^{10}$

12. $(7t^5)(3t^4)(-t^8)$ $-21t^{17}$

Find each product. **See Example 1.**

13. $-2m(3m+2)$ $-6m^2-4m$

14. $-5p(6+3p)$ $-30p-15p^2$

15. $\frac{3}{4}p(8-6p+12p^3)$ $6p-\frac{9}{2}p^2+9p^4$

16. $\frac{4}{3}x(3+2x+5x^3)$ $4x+\frac{8}{3}x^2+\frac{20}{3}x^4$

17. $2y^5(5y^4+2y+3)$ $10y^9+4y^6+6y^5$

18. $2m^4(3m^2+5m+6)$ $6m^6+10m^5+12m^4$

19. $2y^3(3y^3+2y+1)$ $6y^6+4y^4+2y^3$

20. $2m^4(3m^2+5m+1)$ $6m^6+10m^5+2m^4$

21. $-4r^3(-7r^2+8r-9)$ $28r^5-32r^4+36r^3$

22. $-9a^5(-3a^6-2a^4+8a^2)$ $27a^{11}+18a^9-72a^7$

23. $3a^2(2a^2-4ab+5b^2)$ $6a^4-12a^3b+15a^2b^2$

24. $4z^3(8z^2+5zy-3y^2)$ $32z^5+20z^4y-12z^3y^2$

25. $7m^3n^2(3m^2+2mn-n^3)$

$= 7m^3n^2(\underline{3m^2}) + 7m^3n^2(\underline{2mn}) - 7m^3n^2(\underline{n^3})$

$= \underline{21m^5n^2+14m^4n^3-7m^3n^5}$

26. $2p^2q(3p^2q^2-5p+2q^2)$

$= \underline{2p^2q}(3p^2q^2) + \underline{2p^2q}(-5p) + \underline{2p^2q}(2q^2)$

$= \underline{6p^4q^3-10p^3q+4p^2q^3}$

Find each product. **See Examples 2 and 3.**

27. $(6x + 1)(2x^2 + 4x + 1)$
$12x^3 + 26x^2 + 10x + 1$

28. $(9a + 2)(9a^2 + a + 1)$
$81a^3 + 27a^2 + 11a + 2$

29. $(2r - 1)(3r^2 + 4r - 4)$
$6r^3 + 5r^2 - 12r + 4$

30. $(9y - 2)(8y^2 - 6y + 1)$
$72y^3 - 70y^2 + 21y - 2$

31. $(4m + 3)(5m^3 - 4m^2 + m - 5)$
$20m^4 - m^3 - 8m^2 - 17m - 15$

32. $(y + 4)(3y^3 - 2y^2 + y + 3)$
$3y^4 + 10y^3 - 7y^2 + 7y + 12$

33. $(5x^2 + 2x + 1)(x^2 - 3x + 5)$
$5x^4 - 13x^3 + 20x^2 + 7x + 5$

34. $(2m^2 + m - 3)(m^2 - 4m + 5)$
$2m^4 - 7m^3 + 3m^2 + 17m - 15$

35. $(6x^4 - 4x^2 + 8x)\left(\frac{1}{2}x + 3\right)$
$3x^5 + 18x^4 - 2x^3 - 8x^2 + 24x$

36. $(8y^6 + 4y^4 - 12y^2)\left(\frac{3}{4}y^2 + 2\right)$
$6y^8 + 19y^6 - y^4 - 24y^2$

Find each product using the rectangle method shown in the text.

37. $(x + 3)(x + 4)$

	x	4
x	x^2	$4x$
3	$3x$	12

Product: $x^2 + 7x + 12$

38. $(x + 5)(x + 2)$

	x	2
x	x^2	$2x$
5	$5x$	10

Product: $x^2 + 7x + 10$

39. $(2x + 1)(x^2 + 3x + 2)$

	x^2	$3x$	2
$2x$	$2x^3$	$6x^2$	$4x$
1	x^2	$3x$	2

Product: $2x^3 + 7x^2 + 7x + 2$

40. $(x + 4)(3x^2 + 2x + 1)$

	$3x^2$	$2x$	1
x	$3x^3$	$2x^2$	x
4	$12x^2$	$8x$	4

Product: $3x^3 + 14x^2 + 9x + 4$

Find each product. Use the FOIL method. In Exercises 61–66, use the FOIL method and the distributive property. **See Examples 4–6.**

41. $(m + 7)(m + 5)$
$m^2 + 12m + 35$

42. $(x + 4)(x + 7)$
$x^2 + 11x + 28$

43. $(n - 2)(n + 3)$
$n^2 + n - 6$

44. $(r - 6)(r + 8)$
$r^2 + 2r - 48$

45. $(4r + 1)(2r - 3)$
$8r^2 - 10r - 3$

46. $(5x + 2)(2x - 7)$
$10x^2 - 31x - 14$

47. $(3x + 2)(3x - 2)$
$9x^2 - 4$

48. $(7x + 3)(7x - 3)$
$49x^2 - 9$

49. $(3q + 1)(3q + 1)$
$9q^2 + 6q + 1$

50. $(4w + 7)(4w + 7)$
$16w^2 + 56w + 49$

51. $(5x + 7)(3y - 8)$
$15xy - 40x + 21y - 56$

52. $(4x + 3)(2y - 1)$
$8xy - 4x + 6y - 3$

53. $(3t + 4s)(2t + 5s)$
$6t^2 + 23st + 20s^2$

54. $(8v + 5w)(2v + 3w)$
$16v^2 + 34vw + 15w^2$

55. $(-0.3t + 0.4)(t + 0.6)$

$-0.3t^2 + 0.22t + 0.24$

56. $(-0.5x + 0.9)(x - 0.2)$

$-0.5x^2 + x - 0.18$

57. $\left(x - \frac{2}{3}\right)\left(x + \frac{1}{4}\right)$

$x^2 - \frac{5}{12}x - \frac{1}{6}$

58. $\left(y + \frac{3}{5}\right)\left(y - \frac{1}{2}\right)$

$y^2 + \frac{1}{10}y - \frac{3}{10}$

59. $\left(-\frac{5}{4} + 2r\right)\left(-\frac{3}{4} - r\right)$

$\frac{15}{16} - \frac{1}{4}r - 2r^2$

60. $\left(-\frac{8}{3} + 3k\right)\left(-\frac{2}{3} - k\right)$

$\frac{16}{9} + \frac{2}{3}k - 3k^2$

61. $x(2x - 5)(x + 3)$

$2x^3 + x^2 - 15x$

62. $m(4m - 1)(2m + 3)$

$8m^3 + 10m^2 - 3m$

63. $3y^3(2y + 3)(y - 5)$

$6y^5 - 21y^4 - 45y^3$

64. $5t^4(t + 3)(3t - 1)$

$15t^6 + 40t^5 - 15t^4$

65. $-8r^3(5r^2 + 2)(5r^2 - 2)$

$-200r^7 + 32r^3$

66. $-5t^4(2t^4 + 1)(2t^4 - 1)$

$-20t^{12} + 5t^4$

*Find polynomials that represent **(a)** the area and **(b)** the perimeter of each square or rectangle. (If necessary, refer to the formulas inside the back cover.)*

67.

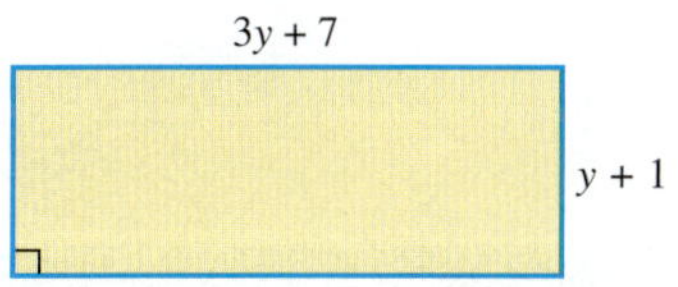

(a) $3y^2 + 10y + 7$

(b) $8y + 16$

68.

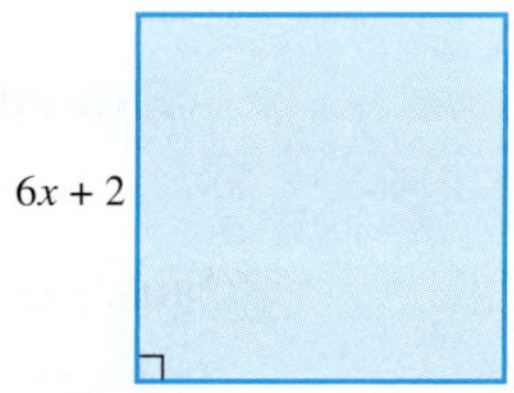

(a) $36x^2 + 24x + 4$

(b) $24x + 8$

Relating Concepts (Exercises 69–74) For Individual or Group Work

Work Exercises 69–74 in order. *(All units are in yards.)*

69. Find a polynomial that represents the area of the rectangle.

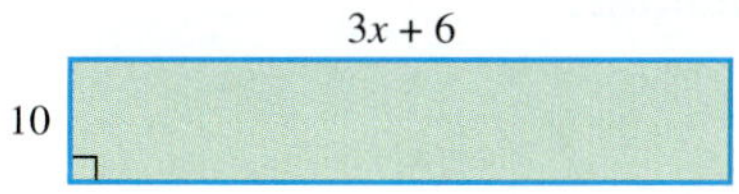

$30x + 60$

70. Suppose you know that the area of the rectangle is 600 yd^2. Use this information and the polynomial from **Exercise 69** to write an equation in x, and solve it.

$30x + 60 = 600$; $\{18\}$

71. Refer to **Exercise 70.** What are the dimensions of the rectangle?

10 yd by 60 yd

72. Use the result of **Exercise 71** to find the perimeter of the rectangle.

140 yd

73. Suppose the rectangle represents a lawn and it costs \$3.50 per square yard to lay sod on the lawn. How much will it cost to sod the entire lawn?

\$2100

74. Again, suppose the rectangle represents a lawn and it costs \$9.00 per yard to fence the lawn. How much will it cost to fence the lawn?

\$1260

5.4 Special Products

OBJECTIVES

1. Square binomials.
2. Find the product of the sum and difference of two terms.
3. Find greater powers of binomials.

OBJECTIVE 1 Square binomials. The square of a binomial can be found quickly by using the method shown in **Example 1.**

EXAMPLE 1 Squaring a Binomial

Find $(m + 3)^2$.

$(m+3)^2$ means $(m+3)(m+3)$.

$$(m + 3)(m + 3)$$

$$= m^2 + 3m + 3m + 9 \quad \text{FOIL method}$$

$$= m^2 + 6m + 9 \quad \text{Combine like terms.}$$

This result has the squares of the first and the last terms of the binomial.

$$m^2 = m^2 \quad \text{and} \quad 3^2 = 9$$

The middle term, 6m, is twice the product of the two terms of the binomial, since the outer and inner products are $m(3)$ and $3(m)$.

$$m(3) + 3(m) = 2(m)(3) = 6m$$

◀ Work Problem 1 at the Side.

Example 1 suggests the following rules.

Square of a Binomial

The square of a binomial is a trinomial consisting of

the square of the first term + twice the product of the two terms + the square of the last term.

For a and b, the following are true.

$$(a + b)^2 = a^2 + 2ab + b^2$$
$$(a - b)^2 = a^2 - 2ab + b^2$$

EXAMPLE 2 Squaring Binomials

Square each binomial.

$$(a - b)^2 = a^2 - 2 \cdot a \cdot b + b^2$$

(a) $(5z - 1)^2 = (5z)^2 - 2(5z)(1) + (1)^2$

$= 5^2z^2 - 10z + 1$ $\quad$ $(5z)^2 = 5^2z^2 = 25z^2$ by power rule (b).

$= 25z^2 - 10z + 1$

(b) $(3b + 5r)^2$

$= (3b)^2 + 2(3b)(5r) + (5r)^2$ $\quad$ $(a+b)^2 = a^2 + 2ab + b^2$

$= 9b^2 + 30br + 25r^2$ $\quad$ $(3b)^2 = 3^2b^2 = 9b^2$ and $(5r)^2 = 5^2r^2 = 25r^2$

Continued on Next Page

---

1 Consider the binomial $x + 4$.

GS **(a)** What is the first term of the binomial? ____

Square it. ____

(b) What is the last term of the binomial? ____

Square it. ____

(c) Find twice the product of the two terms of the binomial.

$2(\_\_\_\_)(\_\_\_\_) = \_\_\_\_$

(d) Use the results of parts (a)–(c) to find $(x + 4)^2$.

____

Answers

1. **(a)** x; x^2 **(b)** 4; 16 **(c)** x; 4; $8x$ **(d)** $x^2 + 8x + 16$

(c) $(2a - 9x)^2$

$= (2a)^2 - 2(2a)(9x) + (9x)^2$ $\quad (a - b)^2 = a^2 - 2ab + b^2$

$= 4a^2 - 36ax + 81x^2$ $\quad (2a)^2 = 2^2a^2 = 4a^2$ and $(9x)^2 = 9^2x^2 = 81x^2$

(d) $\left(4m + \frac{1}{2}\right)^2$

$= (4m)^2 + 2(4m)\left(\frac{1}{2}\right) + \left(\frac{1}{2}\right)^2$ $\quad (a + b)^2 = a^2 + 2ab + b^2$

$= 16m^2 + 4m + \frac{1}{4}$ $\quad (4m)^2 = 4^2m^2 = 16m^2$ and $\left(\frac{1}{2}\right)^2 = \frac{1}{4}$

(e) $x(4x - 3)^2$ — Remember the middle term.

$= x(16x^2 - 24x + 9)$ Square the binomial.

$= 16x^3 - 24x^2 + 9x$ Distributive property

In the square of a sum, all of the terms are positive, as in Examples 2(b) and (d). In the square of a difference, the middle term is negative, as in Examples 2(a) and (c).

CAUTION

A common error when squaring a binomial is to forget the middle term of the product. In general, remember the following.

$(a + b)^2 = a^2 + 2ab + b^2$, **not** $a^2 + b^2$,

and $(a - b)^2 = a^2 - 2ab + b^2$, **not** $a^2 - b^2$.

Work Problem 2 at the Side. ▶

OBJECTIVE 2 Find the product of the sum and difference of two terms. In binomial products of the form $(a + b)(a - b)$, one binomial is the sum of two terms, and the other is the difference of the *same* two terms. Consider the following.

$(x + 2)(x - 2)$

$= x^2 - 2x + 2x - 4$ FOIL method

$= x^2 - 4$ Combine like terms.

Thus, the product of $a + b$ and $a - b$ is the **difference of two squares.**

Product of the Sum and Difference of Two Terms

$$(a + b)(a - b) = a^2 - b^2$$

Note

The expressions $a + b$ and $a - b$, the sum and difference of the *same* two terms, are **conjugates.** In the example above, $x + 2$ and $x - 2$ are conjugates.

2 Square each binomial.

GS **(a)** $(t - 6)^2$

$= (\_\_\_)^2 - 2(\_\_\_)(\_\_\_) + (\_\_\_)^2$

$= \_\_\_\_\_\_\_\_$

(b) $(2m - p)^2$

GS **(c)** $(4p + 3q)^2$

$= (\_\_\_)^2 + \_\_\_(\_\_\_)3q + (\_\_\_)^2$

$= \_\_\_\_\_\_\_\_$

(d) $(5r - 6s)^2$

(e) $\left(3k - \frac{1}{2}\right)^2$

(f) $x(2x + 7)^2$

Answers

2. (a) t; t; 6; 6; $t^2 - 12t + 36$
(b) $4m^2 - 4mp + p^2$
(c) $4p$; 2; $4p$; $3q$; $16p^2 + 24pq + 9q^2$
(d) $25r^2 - 60rs + 36s^2$
(e) $9k^2 - 3k + \frac{1}{4}$
(f) $4x^3 + 28x^2 + 49x$

3 Find each product.

GS **(a)** $(y+3)(y-3)$

$= (\_\_\_\_)^2 - (\_\_\_\_)^2$

$= \_\_\_\_\_\_$

(b) $(10m+7)(10m-7)$

GS **(c)** $(7p+2q)(7p-2q)$

$= (\_\_\_\_)^2 - (\_\_\_\_)^2$

$= \_\_\_\_\_\_$

(d) $\left(3r-\frac{1}{2}\right)\left(3r+\frac{1}{2}\right)$

(e) $3x(x^3-4)(x^3+4)$

Answers

3. **(a)** y; 3; y^2-9 **(b)** $100m^2-49$
(c) $7p$; $2q$; $49p^2-4q^2$ **(d)** $9r^2-\frac{1}{4}$
(e) $3x^7-48x$

EXAMPLE 3 Finding the Product of the Sum and Difference of Two Terms

Find each product.

(a) $(x+4)(x-4)$

Use the rule for the product of the sum and difference of two terms.

$(x+4)(x-4)$

$= x^2 - 4^2$ $\quad (a+b)(a-b)=a^2-b^2$

$= x^2 - 16$ $\quad$ Square 4.

(b) $\left(\frac{2}{3}-w\right)\left(\frac{2}{3}+w\right)$

$= \left(\frac{2}{3}+w\right)\left(\frac{2}{3}-w\right)$ $\quad$ Commutative property

$= \left(\frac{2}{3}\right)^2 - w^2$ $\quad (a+b)(a-b)=a^2-b^2$

$= \frac{4}{9} - w^2$ $\quad$ Square $\frac{2}{3}$.

(c) $x(x+2)(x-2)$

$= x(x^2-4)$ $\quad$ Find the product of the sum and difference of two terms.

$= x^3 - 4x$ $\quad$ Distributive property

EXAMPLE 4 Finding the Product of the Sum and Difference of Two Terms

Find each product.

$(a \quad + b)\ (a \quad - b)$

$\downarrow \quad \downarrow \quad \downarrow \quad \downarrow$

(a) $(5m+3)(5m-3)$

Use the rule for the product of the sum and difference of two terms.

$(5m+3)(5m-3)$

$= (5m)^2 - 3^2$ $\quad (a+b)(a-b)=a^2-b^2$

$= 25m^2 - 9$ $\quad$ Apply the exponents.

Be careful to square $5m$ correctly.

(b) $(4x+y)(4x-y)$

$= (4x)^2 - y^2$

$= 16x^2 - y^2$

$(4x)^2 = 4^2x^2 = 16x^2$

(c) $\left(z-\frac{1}{4}\right)\left(z+\frac{1}{4}\right)$

$= z^2 - \left(\frac{1}{4}\right)^2$

$= z^2 - \frac{1}{16}$

(d) $2p(p^2+3)(p^2-3)$

$= 2p(p^4-9)$ $\quad$ Multiply the conjugates.

$= 2p^5 - 18p$ $\quad$ Distributive property

◀ Work Problem **3** at the Side.

The product rules of this section will be important in **Chapters 6 and 7** and should be *memorized.*

OBJECTIVE 3 Find greater powers of binomials. The methods used in the previous section and this section can be combined to find greater powers of binomials.

EXAMPLE 5 Finding Greater Powers of Binomials

Find each product.

(a) $(x+5)^3$

$= (x+5)^2(x+5)$	$a^3 = a^2 \cdot a$
$= (x^2+10x+25)(x+5)$	Square the binomial.
$= x^3+10x^2+25x+5x^2+50x+125$	Multiply polynomials.
$= x^3+15x^2+75x+125$	Combine like terms.

(b) $(2y-3)^4$

$= (2y-3)^2(2y-3)^2$	$a^4 = a^2 \cdot a^2$
$= (4y^2-12y+9)(4y^2-12y+9)$	Square each binomial.
$= 16y^4-48y^3+36y^2-48y^3+144y^2$	Multiply polynomials.
$-108y+36y^2-108y+81$	
$= 16y^4-96y^3+216y^2-216y+81$	Combine like terms.

(c) $-2r(r+2)^3$

$= -2r(r+2)(r+2)^2$	$a^3 = a \cdot a^2$
$= -2r(r+2)(r^2+4r+4)$	Square the binomial.
$= -2r(r^3+4r^2+4r+2r^2+8r+8)$	Multiply polynomials.
$= -2r(r^3+6r^2+12r+8)$	Combine like terms.
$= -2r^4-12r^3-24r^2-16r$	Distributive property

Work Problem 4 at the Side. ▶

4 Find each product.

GS **(a)** $(m+1)^3$

$= (\_\_\_\_)^2(\_\_\_\_)$

$= (\_\_\_\_\_\_)(m+1)$

$= \_\_\_\_\_\_\_\_$

(b) $(3k-2)^4$

(c) $-3x(x-4)^3$

Answers

4. (a) $m+1$; $m+1$; m^2+2m+1; m^3+3m^2+3m+1
(b) $81k^4-216k^3+216k^2-96k+16$
(c) $-3x^4+36x^3-144x^2+192x$

5.4 Exercises

FOR EXTRA HELP

Download the MyDashBoard App

MyMathLab®

CONCEPT CHECK *Fill in each blank with the correct response.*

1. The square of a binomial is a trinomial consisting of the __square__ of the first term + __twice__ the __product__ of the two terms + the __square__ of the last term.

2. In the square of a sum, all of the terms are __positive__, while in the square of a difference, the middle term is __negative__.

3. The product of the sum and difference of two terms is the __difference__ of the __squares__ of the two terms.

4. The word __conjugates__ describes two expressions that represent the sum and the difference of the same two terms.

5. Consider the binomial square $(2x + 3)^2$.

(a) What is the simplest form of the square of the first term, $(2x)^2$? $4x^2$

(b) What is the simplest form of twice the product of the two terms, $2(2x)(3)$? $12x$

(c) What is the simplest form of the square of the last term, 3^2? 9

(d) Write the final product, which is a trinomial, using your results from parts (a)–(c). $4x^2 + 12x + 9$

6. Repeat **Exercise 5** for the binomial square $(3x - 2)^2$.
(a) $9x^2$ (b) $-12x$ (c) 4 (d) $9x^2 - 12x + 4$

Square each binomial. ***See Examples 1 and 2.***

7. $(p + 2)^2$ — $p^2 + 4p + 4$

8. $(r + 5)^2$ — $r^2 + 10r + 25$

9. $(z - 5)^2$ — $z^2 - 10z + 25$

10. $(x - 3)^2$ — $x^2 - 6x + 9$

11. $\left(x - \frac{3}{4}\right)^2$ — $x^2 - \frac{3}{2}x + \frac{9}{16}$

12. $\left(y + \frac{5}{8}\right)^2$ — $y^2 + \frac{5}{4}y + \frac{25}{64}$

13. $(v + 0.4)^2$ — $v^2 + 0.8v + 0.16$

14. $(w - 0.9)^2$ — $w^2 - 1.8w + 0.81$

15. $(4x - 3)^2$ — $16x^2 - 24x + 9$

16. $(9y - 4)^2$ — $81y^2 - 72y + 16$

17. $(10z + 6)^2$ — $100z^2 + 120z + 36$

18. $(5y + 2)^2$ — $25y^2 + 20y + 4$

19. $(2p + 5q)^2$ — $4p^2 + 20pq + 25q^2$

20. $(8a - 3b)^2$ — $64a^2 - 48ab + 9b^2$

21. $(0.8t + 0.7s)^2$ — $0.64t^2 + 1.12ts + 0.49s^2$

22. $(0.7z - 0.3w)^2$ — $0.49z^2 - 0.42zw + 0.09w^2$

23. $\left(5x + \frac{2}{5}y\right)^2$ — $25x^2 + 4xy + \frac{4}{25}y^2$

24. $\left(6m - \frac{4}{5}n\right)^2$ — $36m^2 - \frac{48}{5}mn + \frac{16}{25}n^2$

25. $t(3t - 1)^2$ — $9t^3 - 6t^2 + t$

26. $x(2x + 5)^2$ — $4x^3 + 20x^2 + 25x$

27. $3t(4t + 1)^2$ — $48t^3 + 24t^2 + 3t$

28. $2x(7x - 2)^2$ — $98x^3 - 56x^2 + 8x$

29. $-(4r - 2)^2$ — $-16r^2 + 16r - 4$

30. $-(3y - 8)^2$ — $-9y^2 + 48y - 64$

31. Consider the product $(7x + 3y)(7x - 3y)$.
GS
(a) What is the simplest form of the product of the first terms, $7x(7x)$? $49x^2$

(b) Multiply the outer terms, $7x(-3y)$. Then multiply the inner terms, $3y(7x)$. Add the results. What is this sum? 0

(c) What is the simplest form of the product of the last terms, $3y(-3y)$? $-9y^2$

(d) Write the final product using your answers in parts (a) and (c). $49x^2 - 9y^2$ Why is the sum found in part (b) omitted here?

Because 0 is the identity element for addition, it is not necessary to write "+0."

32. Repeat **Exercise 31** for the product $(5x + 7y)(5x - 7y)$.
GS
(a) $25x^2$ (b) 0 (c) $-49y^2$ (d) $25x^2 - 49y^2$;
Because 0 is the identity element for addition, it is not necessary to write "+0."

Find each product. ***See Examples 3 and 4.***

33. $(q + 2)(q - 2)$
$q^2 - 4$

34. $(x + 8)(x - 8)$
$x^2 - 64$

35. $\left(r - \frac{3}{4}\right)\left(r + \frac{3}{4}\right)$
$r^2 - \frac{9}{16}$

36. $\left(q - \frac{7}{8}\right)\left(q + \frac{7}{8}\right)$
$q^2 - \frac{49}{64}$

37. $(s + 2.5)(s - 2.5)$
$s^2 - 6.25$

38. $(t + 1.4)(t - 1.4)$
$t^2 - 1.96$

39. $(2w + 5)(2w - 5)$
$4w^2 - 25$

40. $(3z + 8)(3z - 8)$
$9z^2 - 64$

41. $(10x + 3y)(10x - 3y)$
$100x^2 - 9y^2$

42. $(13r + 2z)(13r - 2z)$
$169r^2 - 4z^2$

43. $(2x^2 - 5)(2x^2 + 5)$
$4x^4 - 25$

44. $(9y^2 - 2)(9y^2 + 2)$
$81y^4 - 4$

45. $\left(7x + \frac{3}{7}\right)\left(7x - \frac{3}{7}\right)$
$49x^2 - \frac{9}{49}$

46. $\left(9y + \frac{2}{3}\right)\left(9y - \frac{2}{3}\right)$
$81y^2 - \frac{4}{9}$

47. $p(3p + 7)(3p - 7)$
$9p^3 - 49p$

48. $q(5q - 1)(5q + 1)$
$25q^3 - q$

Find each product. ***See Example 5.***

49. $(m - 5)^3$
$m^3 - 15m^2 + 75m - 125$

50. $(p + 3)^3$
$p^3 + 9p^2 + 27p + 27$

51. $(y + 2)^3$
$y^3 + 6y^2 + 12y + 8$

52. $(x - 7)^3$
$x^3 - 21x^2 + 147x - 343$

53. $(2a + 1)^3$
$8a^3 + 12a^2 + 6a + 1$

54. $(3m - 1)^3$
$27m^3 - 27m^2 + 9m - 1$

55. $(3r - 2t)^4$
$81r^4 - 216r^3t + 216r^2t^2 - 96rt^3 + 16t^4$

56. $(2z + 5y)^4$
$16z^4 + 160z^3y + 600z^2y^2 + 1000zy^3 + 625y^4$

57. $3x^2(x - 3)^3$
$3x^5 - 27x^4 + 81x^3 - 81x^2$

58. $4p^3(p + 4)^3$
$4p^6 + 48p^5 + 192p^4 + 256p^3$

59. $-8x^2y(x + y)^4$
$-8x^6y - 32x^5y^2 - 48x^4y^3 - 32x^3y^4 - 8x^2y^5$

60. $-5uv^2(u - v)^4$
$-5u^5v^2 + 20u^4v^3 - 30u^3v^4 + 20u^2v^5 - 5uv^6$

Determine a polynomial that represents the area of each figure. In Exercise 65, leave π in your answer. (If necessary, refer to the formulas inside the back cover.)

61.

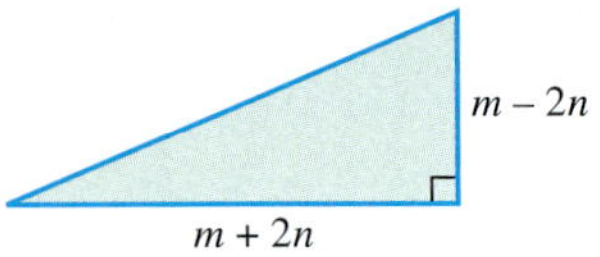

$\frac{1}{2}m^2 - 2n^2$

62.

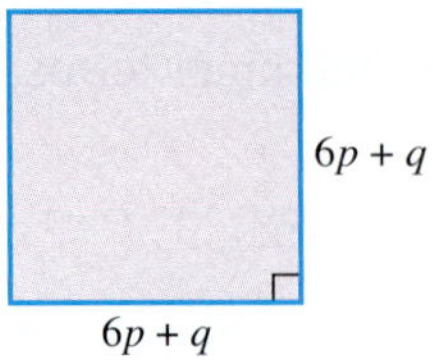

$36p^2 + 12pq + q^2$

63.

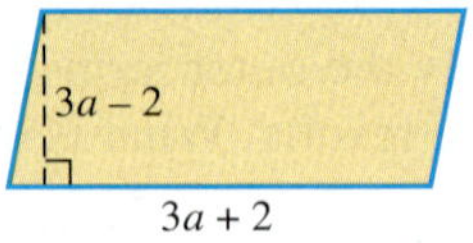

$9a^2 - 4$

64.

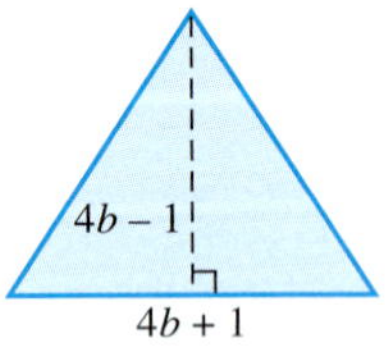

$8b^2 - \frac{1}{2}$

65.

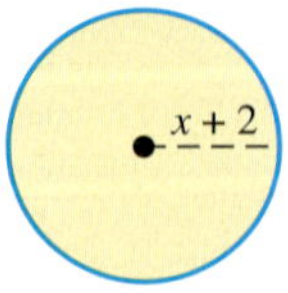

$\pi x^2 + 4\pi x + 4\pi$

66.

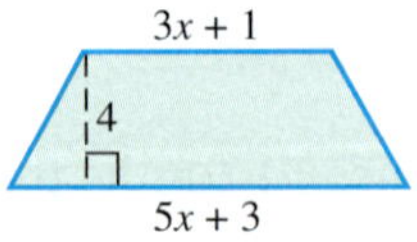

$16x + 8$

In Exercises 67 and 68, refer to the figure shown here.

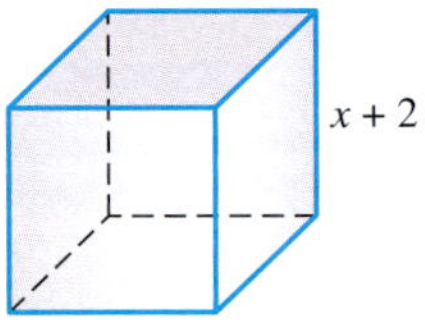

67. Find a polynomial that represents the volume of the cube (in cubic units).

$x^3 + 6x^2 + 12x + 8$

68. If the value of x is 6, what is the volume of the cube (in cubic units)? 512

Relating Concepts (Exercises 69–78) For Individual or Group Work

Special products can be illustrated by using areas of rectangles. Use the figure and ***work Exercises 69–74 in order,*** *to justify the special product $(a + b)^2 = a^2 + 2ab + b^2$.*

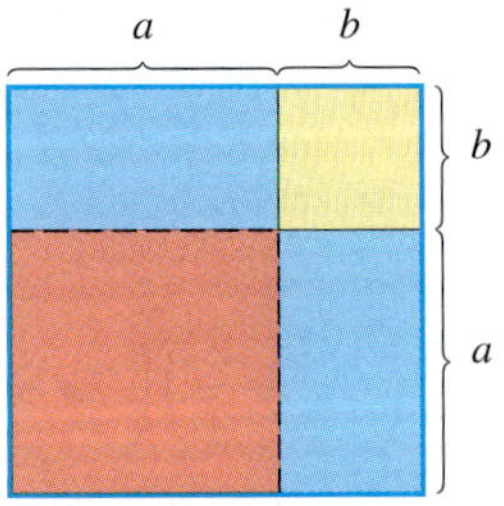

69. Express the area of the large square as the square of a binomial.

$(a + b)^2$

70. Give the monomial that represents the area of the red square.

a^2

71. Give the monomial that represents the sum of the areas of the blue rectangles.

$2ab$

72. Give the monomial that represents the area of the yellow square.

b^2

73. What is the sum of the monomials you obtained in **Exercises 70–72?**

$a^2 + 2ab + b^2$

74. Explain why the binomial square found in **Exercise 69** must equal the polynomial found in **Exercise 73.**

They both represent the area of the entire large square.

To understand how the special product $(a + b)^2 = a^2 + 2ab + b^2$ can be applied to a purely numerical problem, ***work Exercises 75–78 in order.***

75. Evaluate 35^2 using either traditional paper-and-pencil methods or a calculator.

1225

76. The number 35 can be written as $30 + 5$. Therefore, $35^2 = (30 + 5)^2$. Use the special product for squaring a binomial with $a = 30$ and $b = 5$ to write an expression for $(30 + 5)^2$. Do not simplify at this time.

$30^2 + 2(30)(5) + 5^2$

77. Use the order of operations to simplify the expression found in **Exercise 76.**

1225

78. How do the answers in **Exercises 75 and 77** compare?

They are equal.

5.5 Integer Exponents and the Quotient Rule

OBJECTIVES

1. Use 0 as an exponent.
2. Use negative numbers as exponents.
3. Use the quotient rule for exponents.
4. Use combinations of rules.

Consider the following list.

$$2^4 = 16$$
$$2^3 = 8$$
$$2^2 = 4$$

Each time we decrease the exponent by 1, the value is divided by 2 (the base). Using this pattern, we can continue the list to lesser and lesser integer exponents.

$$2^1 = 2$$
$$2^0 = 1$$
$$2^{-1} = \frac{1}{2}$$

Work Problem 1 at the Side. ▶

From the preceding list and the answers to **Margin Problem 1,** it appears that we should define 2^0 as 1 and negative exponents as reciprocals.

OBJECTIVE 1 **Use 0 as an exponent.** We want the definitions of 0 and negative exponents to satisfy the rules for exponents from **Section 5.2.** For example, if $6^0 = 1$,

$$6^0 \cdot 6^2 = 1 \cdot 6^2 = 6^2 \quad \text{and} \quad 6^0 \cdot 6^2 = 6^{0+2} = 6^2,$$

so the product rule is satisfied. The power rules are also valid for a 0 exponent. Thus, we define a 0 exponent as follows.

Zero Exponent

For any nonzero real number a, $\quad \boldsymbol{a^0 = 1.}$

Example: $\quad 17^0 = 1$

EXAMPLE 1 Using Zero Exponents

Evaluate.

(a) $60^0 = 1$

(b) $(-60)^0 = 1$

(c) $-60^0 = -(1)$, or -1

(d) $y^0 = 1 \quad (y \neq 0)$

(e) $6y^0 = 6(1)$, or $6 \quad (y \neq 0)$

(f) $(6y)^0 = 1 \quad (y \neq 0)$

Work Problem 2 at the Side. ▶

CAUTION

Look again at **Examples 1(b) and 1(c).** In $(-60)^0$, the base is -60, and since any nonzero base raised to the 0 exponent is 1, $(-60)^0 = 1$. In -60^0, which can be written $-(60)^0$, the base is 60, so $-60^0 = -1$.

1 Continue the list of exponentials using -2, -3, and -4 as exponents.

$2^{-2} =$ ____

$2^{-3} =$ ____

$2^{-4} =$ ____

2 Evaluate.

(a) 28^0

(b) $(-16)^0$

(c) -7^0

(d) $m^0 \quad (m \neq 0)$

(e) $-2p^0 \quad (p \neq 0)$

(f) $(5r)^0 \quad (r \neq 0)$

Answers

1. $2^{-2} = \frac{1}{4}$; $2^{-3} = \frac{1}{8}$; $2^{-4} = \frac{1}{16}$

2. (a) 1 **(b)** 1 **(c)** -1 **(d)** 1 **(e)** -2 **(f)** 1

OBJECTIVE 2 Use negative numbers as exponents. Review the lists at the beginning of this section and **Margin Problem 1.** Since $2^{-2} = \frac{1}{4}$ and $2^{-3} = \frac{1}{8}$, we can think that 2^{-n} should perhaps equal $\frac{1}{2^n}$. Is the product rule valid in such cases? For example, if we multiply 6^{-2} by 6^2, we get

$$6^{-2} \cdot 6^2 = 6^{-2+2} = 6^0 = 1.$$

The expression 6^{-2} behaves as if it were the reciprocal of 6^2, because their product is 1. The reciprocal of 6^2 may be written $\frac{1}{6^2}$, leading us to define 6^{-2} as $\frac{1}{6^2}$, and generalize accordingly.

Negative Exponents

For any nonzero real number a and any integer n, $\quad a^{-n} = \frac{1}{a^n}$.

Example: $3^{-2} = \frac{1}{3^2}$

By definition, a^{-n} and a^n are reciprocals, since

$$a^n \cdot a^{-n} = a^n \cdot \frac{1}{a^n} = 1.$$

Since $1^n = 1$, the definition of a^{-n} can also be written

$$a^{-n} = \frac{1}{a^n} = \frac{1^n}{a^n} = \left(\frac{1}{a}\right)^n.$$

For example, $\quad 6^{-3} = \left(\frac{1}{6}\right)^3 \quad$ and $\quad \left(\frac{1}{3}\right)^{-2} = 3^2.$

EXAMPLE 2 Using Negative Exponents

Simplify by writing with positive exponents. Assume that all variables represent nonzero real numbers.

(a) $3^{-2} = \frac{1}{3^2}$, or $\frac{1}{9}$ $\quad a^{-n} = \frac{1}{a^n}$

(b) $5^{-3} = \frac{1}{5^3}$, or $\frac{1}{125}$ $\quad a^{-n} = \frac{1}{a^n}$

(c) $\left(\frac{1}{2}\right)^{-3} = 2^3$, or 8 $\quad \frac{1}{2}$ and 2 are reciprocals.

Notice that we can change the base to its reciprocal if we also change the sign of the exponent.

(d) $\left(\frac{2}{5}\right)^{-4}$

$= \left(\frac{5}{2}\right)^4$ $\quad \frac{2}{5}$ and $\frac{5}{2}$ are reciprocals.

$= \frac{5^4}{2^4}$ $\quad$ Power rule (c)

$= \frac{625}{16}$ $\quad$ Apply the exponents.

(e) $\left(\frac{4}{3}\right)^{-5}$

$= \left(\frac{3}{4}\right)^5$ $\quad \frac{4}{3}$ and $\frac{3}{4}$ are reciprocals.

$= \frac{3^5}{4^5}$ $\quad$ Power rule (c)

$= \frac{243}{1024}$ $\quad$ Apply the exponents.

Continued on Next Page

(f) $4^{-1} - 2^{-1}$

$= \frac{1}{4} - \frac{1}{2}$ Apply the exponents.

$= \frac{1}{4} - \frac{2}{4}$ Find a common denominator.

$= -\frac{1}{4}$ Subtract.

(g) $3p^{-2}$

$= \frac{3}{1} \cdot \frac{1}{p^2}$ $a^{-n} = \frac{1}{a^n}$

$= \frac{3}{p^2}$ Multiply.

(h) $\frac{1}{x^{-4}}$

$= \frac{1^{-4}}{x^{-4}}$ $1^n = 1$, for any integer n

$= \left(\frac{1}{x}\right)^{-4}$ Power rule (c)

$= x^4$ $\frac{1}{x}$ and x are reciprocals.

(i) x^3y^{-4}

$= \frac{x^3}{1} \cdot \frac{1}{y^4}$ $a^{-n} = \frac{1}{a^n}$

$= \frac{x^3}{y^4}$ Multiply.

In general, $\frac{1}{a^{-n}} = a^n$.

CAUTION

A negative exponent does not indicate a negative number. Negative exponents lead to reciprocals.

Expression	Example	
a^{-n}	$3^{-2} = \frac{1}{3^2} = \frac{1}{9}$	Not negative
$-a^{-n}$	$-3^{-2} = -\frac{1}{3^2} = -\frac{1}{9}$	Negative

Work Problem 3 at the Side. ▶

Consider the following.

$$\frac{2^{-3}}{3^{-4}}$$

$$= \frac{\frac{1}{2^3}}{\frac{1}{3^4}} \quad \text{Definition of negative exponent}$$

$$= \frac{1}{2^3} \div \frac{1}{3^4} \quad \frac{a}{b} \text{ means } a \div b.$$

$$= \frac{1}{2^3} \cdot \frac{3^4}{1} \quad \text{To divide, multiply by the reciprocal of the divisor.}$$

$$= \frac{3^4}{2^3} \quad \text{Multiply.}$$

Therefore, $\frac{2^{-3}}{3^{-4}} = \frac{3^4}{2^3}$.

3 Simplify by writing with positive exponents. Assume that all variables represent nonzero real numbers.

(a) 4^{-3}

(b) 6^{-2}

(c) $\left(\frac{1}{4}\right)^{-2}$

(d) $\left(\frac{2}{3}\right)^{-2}$

(e) $2^{-1} + 5^{-1}$

(f) $7m^{-5}$

(g) $\frac{1}{z^{-6}}$

(h) p^2q^{-5}

Answers

3. (a) $\frac{1}{4^3}$, or $\frac{1}{64}$ **(b)** $\frac{1}{6^2}$, or $\frac{1}{36}$
(c) 4^2, or 16 **(d)** $\left(\frac{3}{2}\right)^2$, or $\frac{9}{4}$
(e) $\frac{1}{2} + \frac{1}{5} = \frac{7}{10}$ **(f)** $\frac{7}{m^5}$ **(g)** z^6 **(h)** $\frac{p^2}{q^5}$

4 Simplify. Assume that all variables represent nonzero real numbers.

GS **(a)** $\frac{7^{-1}}{5^{-4}}$

$= \frac{5\_\_}{7\_\_}$

$= \_\_\_\_$

(b) $\frac{x^{-3}}{y^{-2}}$

(c) $\frac{4h^{-5}}{m^{-2}k}$

GS **(d)** $\left(\frac{3m}{p}\right)^{-2}$

$= \left(\frac{p}{3m}\right)^{\_\_}$

$= \frac{p^{\_\_}}{3^{\_\_}m^{\_\_}}$

$= \_\_\_\_$

Teaching Tip

Mention that the final expression in the Caution is a *complex fraction.* We simplify complex fractions in **Section 7.5.**

Answers

4. **(a)** 4; 1; $\frac{625}{7}$ **(b)** $\frac{y^2}{x^3}$ **(c)** $\frac{4m^2}{h^5k}$

(d) 2; 2; 2; 2; $\frac{p^2}{9m^2}$

Changing from Negative to Positive Exponents

For any nonzero numbers a and b, and any integers m and n,

$$\frac{a^{-m}}{b^{-n}} = \frac{b^n}{a^m} \quad \text{and} \quad \left(\frac{a}{b}\right)^{-m} = \left(\frac{b}{a}\right)^m.$$

Examples: $\frac{3^{-5}}{2^{-4}} = \frac{2^4}{3^5}$ and $\left(\frac{4}{5}\right)^{-3} = \left(\frac{5}{4}\right)^3$

EXAMPLE 3 **Changing from Negative to Positive Exponents**

Simplify. Assume that all variables represent nonzero real numbers.

(a) $\frac{4^{-2}}{5^{-3}} = \frac{5^3}{4^2}$, or $\frac{125}{16}$

(b) $\frac{m^{-5}}{p^{-1}} = \frac{p^1}{m^5}$, or $\frac{p}{m^5}$

(c) $\frac{a^{-2}b}{3d^{-3}} = \frac{bd^3}{3a^2}$ Notice that b in the numerator and the coefficient 3 in the denominator are not affected.

(d) $\left(\frac{x}{2y}\right)^{-4}$

$= \left(\frac{2y}{x}\right)^4$ Negative-to-positive rule

$= \frac{2^4y^4}{x^4}$ Power rules (b) and (c)

$= \frac{16y^4}{x^4}$ Apply the exponent.

◀ Work Problem **4** at the Side.

CAUTION

Be careful. We cannot use the rule $\frac{a^{-m}}{b^{-n}} = \frac{b^n}{a^m}$ to change negative exponents to positive exponents if the exponents occur in a *sum* or *difference* of terms. For example,

$$\frac{5^{-2} + 3^{-1}}{7 - 2^{-3}} \quad \text{would be written with positive exponents as} \quad \frac{\frac{1}{5^2} + \frac{1}{3}}{7 - \frac{1}{2^3}}.$$

OBJECTIVE 3 **Use the quotient rule for exponents.** Consider a quotient of two exponential expressions with the same base.

$$\frac{6^5}{6^3} = \frac{6 \cdot 6 \cdot 6 \cdot 6 \cdot 6}{6 \cdot 6 \cdot 6} = 6^2$$

The difference between the exponents, $5 - 3 = 2$, is the exponent in the quotient. Also,

$$\frac{6^2}{6^4} = \frac{6 \cdot 6}{6 \cdot 6 \cdot 6 \cdot 6} = \frac{1}{6^2} = 6^{-2}.$$

Here, $2 - 4 = -2$. These examples suggest the **quotient rule for exponents.**

Quotient Rule for Exponents

For any nonzero real number a and any integers m and n,

$$\frac{a^m}{a^n} = a^{m-n}.$$

(Keep the same base and subtract the exponents.)

Example: $\frac{5^8}{5^4} = 5^{8-4} = 5^4$

CAUTION

A common **error** is to write $\frac{5^8}{5^4} = 1^{8-4} = 1^4$. ***This is incorrect.*** By the quotient rule, the quotient must have the *same base*, which is 5 here.

$$\frac{5^8}{5^4} = 5^{8-4} = 5^4$$

We can confirm this by writing out the factors.

$$\frac{5^8}{5^4} = \frac{5 \cdot 5 \cdot 5 \cdot 5 \cdot 5 \cdot 5 \cdot 5 \cdot 5}{5 \cdot 5 \cdot 5 \cdot 5} = 5^4$$

EXAMPLE 4 **Using the Quotient Rule**

Simplify. Assume that all variables represent nonzero real numbers.

(a) $\frac{5^8}{5^6} = 5^{8-6} = 5^2 = 25$ — Keep the same base.

(b) $\frac{4^2}{4^9} = 4^{2-9} = 4^{-7} = \frac{1}{4^7}$

(c) $\frac{5^{-3}}{5^{-7}} = 5^{-3-(-7)} = 5^4 = 625$ — Be careful with signs.

(d) $\frac{q^5}{q^{-3}} = q^{5-(-3)} = q^8$

(e) $\frac{3^2x^5}{3^4x^3}$

$= \frac{3^2}{3^4} \cdot \frac{x^5}{x^3}$

$= 3^{2-4} \cdot x^{5-3}$ Quotient rule

$= 3^{-2}x^2$ Subtract.

$= \frac{x^2}{3^2}$, or $\frac{x^2}{9}$

(f) $\frac{(m+n)^{-2}}{(m+n)^{-4}}$

$= (m+n)^{-2-(-4)}$

$= (m+n)^{-2+4}$

$= (m+n)^2 \quad (m \neq -n)$

The restriction $m \neq -n$ is necessary to prevent a denominator of 0 in the original expression. Division by 0 is undefined.

(g) $\frac{7x^{-3}y^2}{2^{-1}x^2y^{-5}}$

$= \frac{7 \cdot 2^1y^2y^5}{x^2x^3}$ Definition of negative exponent

$= \frac{14y^7}{x^5}$ Multiply; product rule

Work Problem **5** at the Side. ▶

5 Simplify. Assume that all variables represent nonzero real numbers.

GS **(a)** $\frac{5^{11}}{5^8}$

$= 5^{\_\_-\_\_}$

$= 5^{\_\_}$

$= \_\_\_\_$

(b) $\frac{4^7}{4^{10}}$

GS **(c)** $\frac{6^{-5}}{6^{-2}}$

$= 6^{\_\_-(\_\_)}$

$= 6^{\_\_}$

$= \frac{1}{6^{\_\_}}$

$= \_\_\_\_$

(d) $\frac{8^4m^9}{8^5m^{10}}$

(e) $\frac{3^{-1}(x+y)^{-3}}{2^{-2}(x+y)^{-4}} \quad (x \neq -y)$

Answers

5. **(a)** 11; 8; 3; 125 **(b)** $\frac{1}{64}$

(c) -5; -2; -3; 3; $\frac{1}{216}$

(d) $\frac{1}{8m}$ **(e)** $\frac{4}{3}(x+y)$

Teaching Tip

Stress that the product, quotient, and power rules are the same for positive and negative exponents.

The definitions and rules for exponents given in this section and **Section 5.2** are summarized here.

Definitions and Rules for Exponents

For any integers m and n, the following are true.

		Examples
Product rule	$a^m \cdot a^n = a^{m+n}$	$7^4 \cdot 7^5 = 7^{4+5} = 7^9$
Zero exponent	$a^0 = 1 \quad (a \neq 0)$	$(-3)^0 = 1$
Negative exponent	$a^{-n} = \frac{1}{a^n} \quad (a \neq 0)$	$5^{-3} = \frac{1}{5^3}$
Quotient rule	$\frac{a^m}{a^n} = a^{m-n} \quad (a \neq 0)$	$\frac{2^2}{2^5} = 2^{2-5} = 2^{-3} = \frac{1}{2^3}$
Power rules (a)	$(a^m)^n = a^{mn}$	$(4^2)^3 = 4^{2 \cdot 3} = 4^6$
(b)	$(ab)^m = a^m b^m$	$(3k)^4 = 3^4 k^4$
(c)	$\left(\frac{a}{b}\right)^m = \frac{a^m}{b^m} \quad (b \neq 0)$	$\left(\frac{2}{3}\right)^2 = \frac{2^2}{3^2}$
Negative-to-positive rules	$\frac{a^{-m}}{b^{-n}} = \frac{b^n}{a^m} \quad (a, b \neq 0)$	$\frac{2^{-4}}{5^{-3}} = \frac{5^3}{2^4}$
	$\left(\frac{a}{b}\right)^{-m} = \left(\frac{b}{a}\right)^m$	$\left(\frac{4}{7}\right)^{-2} = \left(\frac{7}{4}\right)^2$

OBJECTIVE 4 **Use combinations of rules.** We sometimes need to use more than one rule to simplify an expression.

EXAMPLE 5 **Using a Combination of Rules**

Simplify each expression. Assume that all variables represent nonzero real numbers.

(a) $\frac{(4^2)^3}{4^5}$

$= \frac{4^6}{4^5}$ Power rule (a)

$= 4^{6-5}$ Quotient rule

$= 4^1$ Subtract the exponents.

$= 4$ $4^1 = 4$

(b) $(2x)^3(2x)^2$

$= (2x)^5$ Product rule

$= 2^5 x^5$ Power rule (b)

$= 32x^5$ $2^5 = 32$

Continued on Next Page

(c) $\left(\frac{2x^3}{5}\right)^{-4}$

$= \left(\frac{5}{2x^3}\right)^4$ Negative-to-positive rule

$= \frac{5^4}{2^4x^{12}}$ Power rules (a)–(c)

$= \frac{625}{16x^{12}}$ Apply the exponents.

(d) $\left(\frac{3x^{-2}}{4^{-1}y^3}\right)^{-3}$

$= \frac{3^{-3}x^6}{4^3y^{-9}}$ Power rules (a)–(c)

$= \frac{x^6y^9}{4^3 \cdot 3^3}$ Negative-to-positive rule

$= \frac{x^6y^9}{1728}$ $4^3 \cdot 3^3 = 64 \cdot 27 = 1728$

(e) $\frac{(4m)^{-3}}{(3m)^{-4}}$

$= \frac{4^{-3}m^{-3}}{3^{-4}m^{-4}}$ Power rule (b)

$= \frac{3^4m^4}{4^3m^3}$ Negative-to-positive rule

$= \frac{3^4m^{4-3}}{4^3}$ Quotient rule

$= \frac{3^4m}{4^3}$ Subtract.

$= \frac{81m}{64}$ Apply the exponents.

Note

Since the steps can be done in several different orders, there are many equally correct ways to simplify expressions like those in **Examples 5(c) through 5(e).**

Work Problem 6 at the Side. ▶

6 Simplify each expression. Assume that all variables represent nonzero real numbers.

(a) $\frac{(3^4)^2}{3^3}$

(b) $(4x)^2(4x)^4$

(c) $\left(\frac{5}{2z^4}\right)^{-3}$

(d) $\left(\frac{6y^{-4}}{7^{-1}z^5}\right)^{-2}$

(e) $\frac{(6x)^{-1}}{(3x^2)^{-2}}$

Answers

6. **(a)** 243 **(b)** 4^6x^6, or $4096x^6$ **(c)** $\frac{8z^{12}}{125}$ **(d)** $\frac{y^8z^{10}}{1764}$ **(e)** $\frac{3x^3}{2}$

5.5 Exercises

FOR EXTRA HELP

Download the MyDashBoard App

MyMathLab®

CONCEPT CHECK *Decide whether each expression is positive, negative, or* 0.

1. $(-2)^{-3}$ negative
2. $(-3)^{-2}$ positive
3. -2^4 negative
4. -3^6 negative
5. $\left(\frac{1}{4}\right)^{-2}$ positive
6. $\left(\frac{1}{5}\right)^{-2}$ positive
7. $1 - 5^0$ 0
8. $1 - 7^0$ 0

CONCEPT CHECK *In Exercises 9 and 10, match each expression in Column I with the equivalent expression in Column II. Choices in Column II may be used once, more than once, or not at all. (In Exercise 9, $x \neq 0$.)*

	I		II
9.	**(a)** x^0	B	**A.** 0
	(b) $-x^0$	C	**B.** 1
	(c) $7x^0$	D	**C.** -1
	(d) $(7x)^0$	B	**D.** 7
	(e) $-7x^0$	E	**E.** -7
	(f) $(-7x)^0$	B	**F.** $\frac{1}{7}$

	I		II
10.	**(a)** -2^{-4}	C	**A.** 8
	(b) $(-2)^{-4}$	F	**B.** 16
	(c) 2^{-4}	F	**C.** $-\frac{1}{16}$
	(d) $\frac{1}{2^{-4}}$	B	**D.** -8
	(e) $\frac{1}{-2^{-4}}$	E	**E.** -16
	(f) $\frac{1}{(-2)^{-4}}$	B	**F.** $\frac{1}{16}$

Decide whether each expression is equal to 0, 1, *or* -1. ***See Example 1.***

11. 9^0 1
12. 5^0 1
13. $(-4)^0$ 1
14. $(-10)^0$ 1
15. -9^0 -1
16. -5^0 -1
17. $(-2)^0 - 2^0$ 0
18. $(-8)^0 - 8^0$ 0
19. $\frac{0^{10}}{10^0}$ 0
20. $\frac{0^5}{5^0}$ 0

Evaluate each expression. ***See Examples 1 and 2.***

21. $7^0 + 9^0$ 2
22. $8^0 + 6^0$ 2
23. 4^{-3} $\frac{1}{64}$
24. 5^{-4} $\frac{1}{625}$
25. $\left(\frac{1}{2}\right)^{-4}$ 16
26. $\left(\frac{1}{3}\right)^{-3}$ 27
27. $\left(\frac{6}{7}\right)^{-2}$ $\frac{49}{36}$
28. $\left(\frac{2}{3}\right)^{-3}$ $\frac{27}{8}$
29. $(-3)^{-4}$ $\frac{1}{81}$
30. $(-4)^{-3}$ $-\frac{1}{64}$
31. $5^{-1} + 3^{-1}$ $\frac{8}{15}$
32. $6^{-1} + 2^{-1}$ $\frac{2}{3}$
33. $-2^{-1} + 3^{-2}$ $-\frac{7}{18}$
34. $(-3)^{-2} + (-4)^{-1}$ $-\frac{5}{36}$

Simplify by writing each expression with positive exponents. Assume that all variables represent nonzero real numbers. ***See Examples 2–4.***

35. $\frac{9^4}{9^5}$ $\frac{1}{9}$

36. $\frac{7^3}{7^4}$ $\frac{1}{7}$

37. $\frac{6^{-3}}{6^2}$ $\frac{1}{6^5}$, or $\frac{1}{7776}$

38. $\frac{4^{-2}}{4^3}$ $\frac{1}{4^5}$, or $\frac{1}{1024}$

39. $\frac{1}{6^{-3}}$ 216

40. $\frac{1}{5^{-2}}$ 25

41. $\frac{2}{r^{-4}}$ $2r^4$

42. $\frac{3}{s^{-8}}$ $3s^8$

43. $\frac{4^{-3}}{5^{-2}}$ $\frac{25}{64}$

44. $\frac{6^{-2}}{5^{-4}}$ $\frac{625}{36}$

45. p^5q^{-8} $\frac{p^5}{q^8}$

46. $x^{-8}y^4$ $\frac{y^4}{x^8}$

47. $\frac{r^5}{r^{-4}}$ r^9

48. $\frac{a^6}{a^{-4}}$ a^{10}

49. $\frac{6^4x^8}{6^5x^3}$ $\frac{x^5}{6}$

50. $\frac{3^8y^5}{3^{10}y^2}$ $\frac{y^3}{9}$

51. $\frac{6y^3}{2y}$ $3y^2$

52. $\frac{5m^2}{m}$ $5m$

53. $\frac{3x^5}{3x^2}$ x^3

54. $\frac{10p^8}{2p^4}$ $5p^4$

55. $\frac{x^{-3}y}{4z^{-2}}$ $\frac{yz^2}{4x^3}$

56. $\frac{p^{-5}q^{-4}}{9r^{-3}}$ $\frac{r^3}{9p^5q^4}$

57. $\frac{(a+b)^{-3}}{(a+b)^{-4}}$ $a+b$

58. $\frac{(x+y)^{-8}}{(x+y)^{-9}}$ $x+y$

Simplify by writing each expression with positive exponents. Assume that all variables represent nonzero real numbers. ***See Example 5.***

59. $\frac{(7^4)^3}{7^9}$ 343

60. $\frac{(5^3)^2}{5^2}$ 625

61. $x^{-3} \cdot x^5 \cdot x^{-4}$ $\frac{1}{x^2}$

62. $y^{-8} \cdot y^5 \cdot y^{-2}$ $\frac{1}{y^5}$

63. $\frac{(3x)^{-2}}{(4x)^{-3}}$ $\frac{64x}{9}$

64. $\frac{(2y)^{-3}}{(5y)^{-4}}$ $\frac{625y}{8}$

65. $\left(\frac{x^{-1}y}{z^2}\right)^{-2}$ $\frac{x^2z^4}{y^2}$

66. $\left(\frac{p^{-4}q}{r^{-3}}\right)^{-3}$ $\frac{p^{12}}{r^9q^3}$

67. $(6x)^4(6x)^{-3}$ $6x$

68. $(10y)^9(10y)^{-8}$ $10y$

69. $\frac{(m^7n)^{-2}}{m^{-4}n^3}$ $\frac{1}{m^{10}n^5}$

70. $\frac{(m^8n^{-4})^2}{m^{-2}n^5}$ $\frac{m^{18}}{n^{13}}$

71. $\frac{5x^{-3}}{(4x)^2}$ $\frac{5}{16x^5}$

72. $\frac{-3k^5}{(2k)^2}$ $-\frac{3k^3}{4}$

73. $\left(\frac{2p^{-1}q}{3^{-1}m^2}\right)^2$ $\frac{36q^2}{m^4p^2}$

74. $\left(\frac{4xy^2}{x^{-1}y}\right)^{-2}$ $\frac{1}{16x^4y^2}$

75. CONCEPT CHECK A student simplified $\frac{16^3}{2^2}$ as shown.

$$\frac{16^3}{2^2} = \left(\frac{16}{2}\right)^{3-2} = 8^1 = 8$$

What Went Wrong? Give the correct answer.

The student attempted to use the quotient rule with unequal bases. The correct way to simplify is

$$\frac{16^3}{2^2} = \frac{(2^4)^3}{2^2} = \frac{2^{12}}{2^2} = 2^{12-2} = 2^{10} = 1024.$$

76. CONCEPT CHECK A student simplified 5^{-4} as shown.

$$5^{-4} = -5^4 = -625$$

What Went Wrong? Give the correct answer.

The student incorrectly assumed that the negative exponent indicated a negative number. The correct way to simplify is

$$5^{-4} = \frac{1}{5^4} = \frac{1}{625}.$$

Summary Exercises *Applying the Rules for Exponents*

Summary Exercises provide mixed practice to better prepare students for tests.

CONCEPT CHECK *Decide whether each expression is positive or negative.*

1. $(-5)^4$ positive

2. -5^4 negative

3. $(-2)^5$ negative

4. -2^5 negative

5. $\left(-\frac{3}{7}\right)^2$ positive

6. $\left(-\frac{3}{7}\right)^3$ negative

7. $\left(-\frac{3}{7}\right)^{-2}$ positive

8. $\left(-\frac{3}{7}\right)^{-5}$ negative

Simplify each expression. Assume that all variables represent nonzero real numbers.

9. $\left(\frac{6x^2}{5}\right)^{12}$ $\frac{6^{12}x^{24}}{5^{12}}$

10. $\left(\frac{rs^2t^3}{3t^4}\right)^6$ $\frac{r^6s^{12}}{729t^6}$

11. $(10x^2y^4)^2(10xy^2)^3$ $100{,}000x^7y^{14}$

12. $(-2ab^3c)^4(-2a^2b)^3$ $-128a^{10}b^{15}c^4$

13. $\left(\frac{9wx^3}{y^4}\right)^3$ $\frac{729w^3x^9}{y^{12}}$

14. $(4x^{-2}y^{-3})^{-2}$ $\frac{x^4y^6}{16}$

15. $\frac{c^{11}(c^2)^4}{(c^3)^3(c^2)^{-6}}$ c^{22}

16. $\left(\frac{k^4t^2}{k^2t^{-4}}\right)^{-2}$ $\frac{1}{k^4t^{12}}$

17. $5^{-1}+6^{-1}$ $\frac{11}{30}$

18. $\frac{(3y^{-1}z^3)^{-1}(3y^2)}{(y^3z^2)^{-3}}$ $y^{12}z^3$

19. $\frac{(2xy^{-1})^3}{2^3x^{-3}y^2}$ $\frac{x^6}{y^5}$

20. $-8^0+(-8)^0$ 0

21. $(z^4)^{-3}(z^{-2})^{-5}$ $\frac{1}{z^2}$

22. $\left(\frac{r^2st^5}{3r}\right)^{-2}$ $\frac{9}{r^2s^2t^{10}}$

23. $\frac{(3^{-1}x^{-3}y)^{-1}(2x^2y^{-3})^2}{(5x^{-2}y^2)^{-2}}$ $\frac{300x^3}{y^3}$

24. $\left(\frac{5x^2}{3x^{-4}}\right)^{-1}$ $\frac{3}{5x^6}$

25. $\left(\frac{-2x^{-2}}{2x^2}\right)^{-2}$ x^8

26. $\frac{(x^{-4}y^2)^3(x^2y)^{-1}}{(xy^2)^{-3}}$ $\frac{y^{11}}{x^{11}}$

27. $\dfrac{(a^{-2}b^3)^{-4}}{(a^{-3}b^2)^{-2}(ab)^{-4}}$

$\dfrac{a^6}{b^4}$

28. $(2a^{-30}b^{-29})(3a^{31}b^{30})$

$6ab$

29. $5^{-2} + 6^{-2}$

$\dfrac{61}{900}$

30. $\left(\dfrac{(x^{47}y^{23})^2}{x^{-26}y^{-42}}\right)^0$

1

31. $\left(\dfrac{7a^2b^3}{2}\right)^3$

$\dfrac{343a^6b^9}{8}$

32. $-(-12^0)$

1

33. $-(-12)^0$

-1

34. $\dfrac{0^{12}}{12^0}$

0

35. $\dfrac{(2xy^{-3})^{-2}}{(3x^{-2}y^4)^{-3}}$

$\dfrac{27y^{18}}{4x^8}$

36. $\left(\dfrac{a^2b^3c^4}{a^{-2}b^{-3}c^{-4}}\right)^{-2}$

$\dfrac{1}{a^8b^{12}c^{16}}$

37. $(6x^{-5}z^3)^{-3}$

$\dfrac{x^{15}}{216z^9}$

38. $(2p^{-2}qr^{-3})(2p)^{-4}$

$\dfrac{q}{8p^6r^3}$

39. $\dfrac{(xy)^{-3}(xy)^5}{(xy)^{-4}}$

x^6y^6

40. $42^0 - (-12)^0$

0

41. $\dfrac{(7^{-1}x^{-3})^{-2}(x^4)^{-6}}{7^{-1}x^{-3}}$

$\dfrac{343}{x^{15}}$

42. $\left(\dfrac{3^{-4}x^{-3}}{3^{-3}x^{-6}}\right)^{-2}$

$\dfrac{9}{x^6}$

43. $(5p^{-2}q)^{-3}(5pq^3)^4$

$5p^{10}q^9$

44. $8^{-1} + 6^{-1}$

$\dfrac{7}{24}$

45. $\left(\dfrac{4r^{-6}s^{-2}t}{2r^8s^{-4}t^2}\right)^{-1}$

$\dfrac{r^{14}t}{2s^2}$

46. $(13x^{-6}y)(13x^{-6}y)^{-1}$

1

47. $\dfrac{(8pq^{-2})^4}{(8p^{-2}q^{-3})^3}$

$8p^{10}q$

48. $\left(\dfrac{mn^{-2}p}{m^2np^4}\right)^{-2}\left(\dfrac{mn^{-2}p}{m^2np^4}\right)^3$

$\dfrac{1}{mn^3p^3}$

49. $-(-3^0)^0$

-1

50. $5^{-1} - 8^{-1}$

$\dfrac{3}{40}$

5.6 Dividing a Polynomial by a Monomial

OBJECTIVE

1 Divide a polynomial by a monomial.

1 Divide.

GS (a) $\dfrac{6p^4 + 18p^7}{3p^2}$

$= \dfrac{\_\_\_}{3p^2} + \dfrac{\_\_\_}{3p^2}$

$= \_\_\_\_\_$

(b) $\dfrac{12m^6 + 18m^5 + 30m^4}{6m^2}$

(c) $(18r^7 - 9r^2) \div (3r)$

Answers

1. (a) $6p^4$; $18p^7$; $2p^2 + 6p^5$
(b) $2m^4 + 3m^3 + 5m^2$ (c) $6r^6 - 3r$

OBJECTIVE 1 Divide a polynomial by a monomial. We add two fractions with a common denominator as follows.

$$\frac{a}{c} + \frac{b}{c} = \frac{a + b}{c}$$

In reverse, this statement gives a rule for dividing a polynomial by a monomial.

Dividing a Polynomial by a Monomial

To divide a polynomial by a monomial, divide each term of the polynomial by the monomial.

$$\frac{a + b}{c} = \frac{a}{c} + \frac{b}{c} \quad (\text{where } c \neq 0)$$

Examples: $\dfrac{2 + 5}{3} = \dfrac{2}{3} + \dfrac{5}{3}$ and $\dfrac{x + 3z}{2y} = \dfrac{x}{2y} + \dfrac{3z}{2y}$ $(y \neq 0)$

The parts of a division problem are named here.

$$\text{Dividend} \rightarrow \frac{12x^2 + 6x}{6x} = 2x + 1 \leftarrow \text{Quotient}$$
Divisor → $6x$

EXAMPLE 1 Dividing a Polynomial by a Monomial

Divide $5m^5 - 10m^3$ by $5m^2$.

$\dfrac{5m^5 - 10m^3}{5m^2}$

$= \dfrac{5m^5}{5m^2} - \dfrac{10m^3}{5m^2}$ Use the preceding rule, with + replaced by −.

$= m^3 - 2m$ Quotient rule

CHECK Multiply. $5m^2 \cdot (m^3 - 2m) = 5m^5 - 10m^3$ ✓

Divisor ($5m^2$), Quotient ($m^3 - 2m$), Original polynomial (Dividend) ($5m^5 - 10m^3$)

Because division by 0 is undefined, the quotient $\frac{5m^5 - 10m^3}{5m^2}$ is undefined if $m = 0$. From now on, we assume that no denominators are 0.

◀ Work Problem **1** at the Side.

EXAMPLE 2 Dividing a Polynomial by a Monomial

Divide.

$\dfrac{16a^5 - 12a^4 + 8a^2}{4a^3}$

$= \dfrac{16a^5}{4a^3} - \dfrac{12a^4}{4a^3} + \dfrac{8a^2}{4a^3}$ Divide each term by $4a^3$.

This becomes $\frac{2}{a}$, not $2a$.

$= 4a^2 - 3a + \dfrac{2}{a}$ Quotient rule

Continued on Next Page

The quotient $4a^2 - 3a + \frac{2}{a}$ is *not* a polynomial because of the expression $\frac{2}{a}$, which has a variable in the denominator. While the sum, difference, and product of two polynomials are always polynomials, the quotient of two polynomials may not be.

CHECK $4a^3\left(4a^2 - 3a + \dfrac{2}{a}\right)$ — Divisor × Quotient should equal Dividend.

$= 4a^3(4a^2) + 4a^3(-3a) + 4a^3\left(\dfrac{2}{a}\right)$ — Distributive property

$= 16a^5 - 12a^4 + 8a^2$ ✓ — Dividend

Work Problem 2 at the Side. ▶

EXAMPLE 3 Dividing a Polynomial by a Monomial with a Negative Coefficient

Divide $-7x^3 + 12x^4 - 4x$ by $-4x$.

Write the polynomial in descending powers as $12x^4 - 7x^3 - 4x$ before dividing.

Write in descending powers.

$$\frac{12x^4 - 7x^3 - 4x}{-4x}$$

$= \dfrac{12x^4}{-4x} - \dfrac{7x^3}{-4x} - \dfrac{4x}{-4x}$ — Divide each term by $-4x$.

$= -3x^3 - \dfrac{7x^2}{-4} - (-1)$ — Quotient rule

$= -3x^3 + \dfrac{7}{4}x^2 + 1$ — Be sure to include the 1 in the answer.

CHECK $-4x\left(-3x^3 + \dfrac{7}{4}x^2 + 1\right)$ — Divisor × Quotient should equal Dividend.

$= -4x(-3x^3) - 4x\left(\dfrac{7}{4}x^2\right) - 4x(1)$ — Distributive property

$= 12x^4 - 7x^3 - 4x$ ✓ — Dividend

Work Problem 3 at the Side. ▶

EXAMPLE 4 Dividing a Polynomial by a Monomial

Divide $180x^4y^{10} - 150x^3y^8 + 120x^2y^6 - 90xy^4 + 100y$ by $-30xy^2$.

$$\frac{180x^4y^{10} - 150x^3y^8 + 120x^2y^6 - 90xy^4 + 100y}{-30xy^2}$$

$$= \frac{180x^4y^{10}}{-30xy^2} - \frac{150x^3y^8}{-30xy^2} + \frac{120x^2y^6}{-30xy^2} - \frac{90xy^4}{-30xy^2} + \frac{100y}{-30xy^2}$$

$$= -6x^3y^8 + 5x^2y^6 - 4xy^4 + 3y^2 - \frac{10}{3xy}$$

Check by multiplying the quotient by the divisor.

Work Problem 4 at the Side. ▶

2 Divide.

GS **(a)** $\dfrac{20x^4 - 25x^3 + 5x}{5x^2}$

$= \dfrac{\_\_\_}{\_\_\_} - \dfrac{\_\_\_}{\_\_\_} + \dfrac{\_\_\_}{\_\_\_}$

$= \_\_\_\_\_\_$

(b) $\dfrac{50m^4 - 30m^3 + 20m}{10m^3}$

3 Divide.

(a) $\dfrac{-9y^6 + 8y^7 - 11y - 4}{y^2}$

(b) $\dfrac{-8p^4 - 6p^3 - 12p^5}{-3p^3}$

4 Divide.

$$\frac{45x^4y^3 + 30x^3y^2 - 60x^2y}{-15x^2y}$$

Answers

2. (a) The complete expression is $\dfrac{20x^4}{5x^2} - \dfrac{25x^3}{5x^2} + \dfrac{5x}{5x^2}$; $4x^2 - 5x + \dfrac{1}{x}$

(b) $5m - 3 + \dfrac{2}{m^2}$

3. (a) $8y^5 - 9y^4 - \dfrac{11}{y} - \dfrac{4}{y^2}$

(b) $4p^2 + \dfrac{8p}{3} + 2$

4. $-3x^2y^2 - 2xy + 4$

5.6 Exercises

FOR EXTRA HELP

Download the MyDashBoard App

MyMathLab®

CONCEPT CHECK *In Exercises 1–4, complete each statement.*

1. In the statement $\frac{6x^2+8}{2} = 3x^2 + 4$, $\underline{6x^2+8}$ is the dividend, $\underline{2}$ is the divisor, and $\underline{3x^2+4}$ is the quotient.

2. The expression $\frac{3x+12}{x}$ is undefined if $x =$ $\underline{0}$.

3. To check the division shown in **Exercise 1,** multiply $\underline{3x^2+4}$ by $\underline{2}$ and show that the product is $\underline{6x^2+8}$.

(The first two responses may be reversed.)

4. The expression $5x^2 - 3x + 6 + \frac{2}{x}$ (*is* / *is not*) a polynomial.

5. Explain why the division problem $\frac{16m^3 - 12m^2}{4m}$ can be performed using the method of this section, while the division problem $\frac{4m}{16m^3 - 12m^2}$ cannot.

To use the method of this section, the divisor must be a monomial. This is true of the first problem, but not the second.

6. **CONCEPT CHECK** Evaluate $\frac{5y+6}{2}$ for $y = 2$. Evaluate $5y + 3$ for $y = 2$. Is $\frac{5y+6}{2}$ equivalent to $5y + 3$?

8; 13; no

7. **CONCEPT CHECK** A polynomial in the variable x has degree 6 and is divided by a monomial in the variable x having degree 4. What is the degree of the quotient?

2

8. **CONCEPT CHECK** If $-60x^5 - 30x^4 + 20x^3$ is divided by $3x^2$, what is the sum of the coefficients of the third- and second-degree terms in the quotient?

−30

Perform each division. ***See Examples 1–4.***

9. $\frac{12m^4 - 6m^3}{6m^2}$

$2m^2 - m$

10. $\frac{35n^5 - 5n^2}{5n}$

$7n^4 - n$

11. $\frac{60x^4 - 20x^2 + 10x}{2x}$

$30x^3 - 10x + 5$

12. $\frac{120x^6 - 60x^3 + 80x^2}{2x}$

$60x^5 - 30x^2 + 40x$

13. $\frac{20m^5 - 10m^4 + 5m^2}{-5m^2}$

$-4m^3 + 2m^2 - 1$

14. $\frac{12t^5 - 6t^3 + 6t^2}{-6t^2}$

$-2t^3 + t - 1$

15. $\frac{8t^5 - 4t^3 + 4t^2}{2t}$

$4t^4 - 2t^2 + 2t$

16. $\frac{8r^4 - 4r^3 + 6r^2}{2r}$

$4r^3 - 2r^2 + 3r$

17. $\frac{4a^5 - 4a^2 + 8}{4a}$

$a^4 - a + \frac{2}{a}$

18. $\frac{5t^8 + 5t^7 + 15}{5t}$

$t^7 + t^6 + \frac{3}{t}$

19. $\frac{12x^5 - 4x^4 + 6x^3}{-6x^2}$

$-2x^3 + \frac{2x^2}{3} - x$

20. $\frac{24x^6 - 14x^5 + 32x^4}{-4x^2}$

$-6x^4 + \frac{7x^3}{2} - 8x^2$

21. $\frac{4x^2 + 20x^3 - 36x^4}{4x^2}$

$-9x^2 + 5x + 1$

22. $\dfrac{5x^2 - 30x^4 + 30x^5}{5x^2}$

$6x^3 - 6x^2 + 1$

23. $\dfrac{-3x^3 - 4x^4 + 2x}{-3x^2}$

$\dfrac{4x^2}{3} + x - \dfrac{2}{3x}$

24. $\dfrac{-8x + 6x^3 - 5x^4}{-3x^2}$

$\dfrac{5x^2}{3} - 2x + \dfrac{8}{3x}$

25. $\dfrac{27r^4 - 36r^3 - 6r^2 + 3r - 2}{3r}$

$= \dfrac{27r^4}{3r} - \dfrac{36r^3}{3r} - \dfrac{6r^2}{3r} + \dfrac{3r}{3r} - \dfrac{2}{3r}$

$= 9r^3 - 12r^2 - 2r + 1 - \dfrac{2}{3r}$

26. $\dfrac{8k^4 - 12k^3 - 2k^2 - 2k - 3}{2k}$

$= \dfrac{8k^4}{2k} - \dfrac{12k^3}{2k} - \dfrac{2k^2}{2k} - \dfrac{2k}{2k} - \dfrac{3}{2k}$

$= 4k^3 - 6k^2 - k - 1 - \dfrac{3}{2k}$

27. $\dfrac{2m^5 - 6m^4 + 8m^2}{-2m^3}$

$-m^2 + 3m - \dfrac{4}{m}$

28. $\dfrac{6r^5 - 8r^4 + 10r^2}{-2r^4}$

$-3r + 4 - \dfrac{5}{r^2}$

29. $(120x^{11} - 60x^{10} + 140x^9 - 100x^8) \div (10x^{12})$

$\dfrac{12}{x} - \dfrac{6}{x^2} + \dfrac{14}{x^3} - \dfrac{10}{x^4}$

30. $(120x^{12} - 84x^9 + 60x^8 - 36x^7) \div (12x^9)$

$10x^3 - 7 + \dfrac{5}{x} - \dfrac{3}{x^2}$

31. $(20a^4b^3 - 15a^5b^2 + 25a^3b) \div (-5a^4b)$

$-4b^2 + 3ab - \dfrac{5}{a}$

32. $(16y^5z - 8y^2z^2 + 12yz^3) \div (-4y^2z^2)$

$-\dfrac{4y^3}{z} + 2 - \dfrac{3z}{y}$

Use an area formula to answer each question. (If necessary, refer to the formulas inside the back cover.)

33. What expression represents the length of the rectangle?

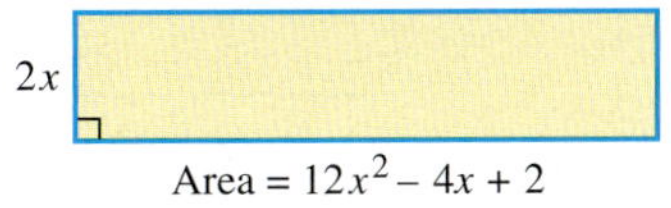

$6x - 2 + \dfrac{1}{x}$

34. What expression represents the length of the base of the triangle?

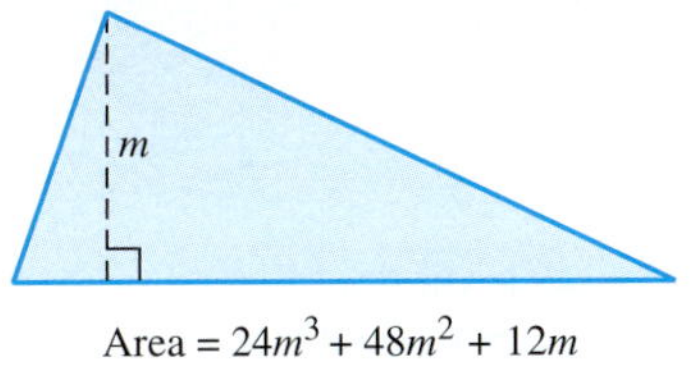

$48m^2 + 96m + 24$

35. CONCEPT CHECK What polynomial, when divided by $5x^3$, yields $3x^2 - 7x + 7$ as a quotient?

$15x^5 - 35x^4 + 35x^3$

36. CONCEPT CHECK The quotient of a certain polynomial and $-12y^3$ is $6y^3 - 5y^2 + 2y - 3 + \frac{7}{y}$. What is this polynomial?

$-72y^6 + 60y^5 - 24y^4 + 36y^3 - 84y^2$

5.7 Dividing a Polynomial by a Polynomial

OBJECTIVES

1. Divide a polynomial by a polynomial.
2. Apply division to a geometry problem.

Teaching Tip

Mention that the method used to divide a polynomial by a polynomial is similar to the method of long division used with whole numbers.

Students often make sign errors when performing the subtraction step in polynomial division. There are several strategies that can be used in the classroom to guard against such errors. One of the more popular ones is to have students circle the original signs in the bottom polynomial and then write in the opposite sign. Then the operation of addition is performed.

Teaching Tip

By the method suggested in the previous Teaching Tip, Step 5 would look like this:

$$\begin{array}{r} 4x^2 - 8x \phantom{{}-14x+15} \\ 2x+3\overline{)8x^3 - 4x^2 - 14x + 15} \\ \underline{\oplus 8x^3 \oplus 12x^2} \phantom{{}-14x+15} \\ -16x^2 - 14x \phantom{{}+15} \\ \underline{\ominus 16x^2 \ominus 24x} \phantom{{}+15} \\ 10x + 15 \end{array}$$

(In the first subtraction, the signs of $8x^3$ and $12x^2$ are marked $-$ and rewritten as $+$; in the second, the signs of $16x^2$ and $24x$ are marked $+$ and rewritten as $-$.)

OBJECTIVE 1 Divide a polynomial by a polynomial. We use a method of "long division" to divide a polynomial by a polynomial (other than a monomial). ***Both polynomials must be written in descending powers.***

Dividing Whole Numbers / Dividing Polynomials

Step 1

Dividing Whole Numbers: Divide 6696 by 27.

$$27\overline{)6696}$$

Dividing Polynomials: Divide $8x^3 - 4x^2 - 14x + 15$ by $2x + 3$.

$$2x + 3\overline{)8x^3 - 4x^2 - 14x + 15}$$

Step 2

Dividing Whole Numbers: 66 divided by 27 = **2**.

$2 \cdot 27 = $ **54**

$$\begin{array}{r} 2 \\ 27\overline{)6696} \\ \underline{54} \end{array}$$

Dividing Polynomials: $8x^3$ divided by $2x = \mathbf{4x^2}$.

$4x^2(2x + 3) = \mathbf{8x^3 + 12x^2}$

$$\begin{array}{r} 4x^2\phantom{{}-14x+15} \\ 2x+3\overline{)8x^3 - 4x^2 - 14x + 15} \\ \underline{8x^3 + 12x^2}\phantom{{}-14x+15} \end{array}$$

Step 3

Dividing Whole Numbers: Subtract. Then bring down the next digit.

$$\begin{array}{r} 2 \\ 27\overline{)6696} \\ \underline{54} \\ 129 \end{array}$$

Dividing Polynomials: Subtract. Then bring down the next term.

$$\begin{array}{r} 4x^2\phantom{{}-14x+15} \\ 2x+3\overline{)8x^3 - 4x^2 - 14x + 15} \\ \underline{8x^3 + 12x^2}\phantom{{}-14x+15} \\ -16x^2 - 14x\phantom{{}+15} \end{array}$$

(To subtract two polynomials, change the signs of the second and then add.)

Step 4

Dividing Whole Numbers: 129 divided by 27 = **4**.

$4 \cdot 27 = $ **108**

$$\begin{array}{r} 24 \\ 27\overline{)6696} \\ \underline{54} \\ 129 \\ \underline{108} \end{array}$$

Dividing Polynomials: $-16x^2$ divided by $2x = \mathbf{-8x}$.

$-8x(2x + 3) = \mathbf{-16x^2 - 24x}$

$$\begin{array}{r} 4x^2 - 8x\phantom{{}+15} \\ 2x+3\overline{)8x^3 - 4x^2 - 14x + 15} \\ \underline{8x^3 + 12x^2}\phantom{{}-14x+15} \\ -16x^2 - 14x\phantom{{}+15} \\ \underline{-16x^2 - 24x}\phantom{{}+15} \end{array}$$

Step 5

Dividing Whole Numbers: Subtract. Then bring down the next digit.

$$\begin{array}{r} 24 \\ 27\overline{)6696} \\ \underline{54} \\ 129 \\ \underline{108} \\ 216 \end{array}$$

Dividing Polynomials: Subtract. Then bring down the next term.

$$\begin{array}{r} 4x^2 - 8x\phantom{{}+15} \\ 2x+3\overline{)8x^3 - 4x^2 - 14x + 15} \\ \underline{8x^3 + 12x^2}\phantom{{}-14x+15} \\ -16x^2 - 14x\phantom{{}+15} \\ \underline{-16x^2 - 24x}\phantom{{}+15} \\ 10x + 15 \end{array}$$

(continued)

Step 6	
216 divided by 27 = **8**.	$10x$ divided by $2x = \mathbf{5}$.
$8 \cdot 27 = \mathbf{216}$	$5(2x+3) = \mathbf{10x + 15}$
$$\begin{array}{r} 248 \\ 27\overline{)6696} \\ \underline{54} \\ 129 \\ \underline{108} \\ 216 \\ \underline{216} \\ \text{Remainder} \rightarrow 0 \end{array}$$	$$\begin{array}{r} 4x^2 - 8x + 5 \\ 2x+3\overline{)8x^3 - 4x^2 - 14x + 15} \\ \underline{8x^3 + 12x^2} \\ -16x^2 - 14x \\ \underline{-16x^2 - 24x} \\ 10x + 15 \\ \underline{10x + 15} \\ \text{Remainder} \rightarrow 0 \end{array}$$
6696 divided by 27 is 248.	$8x^3 - 4x^2 - 14x + 15$ divided by $2x + 3$ is $4x^2 - 8x + 5$.
Step 7 Multiply to check.	
CHECK $27 \cdot 248 = 6696$ ✓	CHECK $(2x+3)(4x^2 - 8x + 5) = 8x^3 - 4x^2 - 14x + 15$ ✓

EXAMPLE 1 Dividing a Polynomial by a Polynomial

Divide $5x + 4x^3 - 8 - 4x^2$ by $2x - 1$.

The dividend, the first polynomial here, must be written in descending powers as $4x^3 - 4x^2 + 5x - 8$. Then divide by $2x - 1$.

$$\begin{array}{r} 2x^2 - x + 2 \\ 2x-1\overline{)4x^3 - 4x^2 + 5x - 8} \\ \underline{4x^3 - 2x^2} \\ -2x^2 + 5x \\ \underline{-2x^2 + x} \\ 4x - 8 \\ \underline{4x - 2} \\ -6 \leftarrow \text{Remainder} \end{array}$$

Write in descending powers.

To subtract, add the opposite.

Step 1 $4x^3$ divided by $2x$ is $\mathbf{2x^2}$. $2x^2(2x-1) = 4x^3 - 2x^2$

Step 2 Subtract. Bring down the next term.

Step 3 $-2x^2$ divided by $2x$ is $\mathbf{-x}$. $-x(2x-1) = -2x^2 + x$

Step 4 Subtract. Bring down the next term.

Step 5 $4x$ divided by $2x$ is **2**. $2(2x-1) = 4x - 2$

Step 6 Subtract. The remainder is **−6**. Write the remainder as the numerator of a fraction that has $2x - 1$ as its denominator. Because of the nonzero remainder, the answer is not a polynomial.

$$\underset{\text{Dividend} \rightarrow,\ \text{Divisor} \rightarrow}{\frac{4x^3 - 4x^2 + 5x - 8}{2x - 1}} = \underbrace{2x^2 - x + 2}_{\text{Quotient polynomial}} + \underbrace{\frac{-6}{2x-1}}_{\text{Fractional part of quotient}} \begin{array}{l} \leftarrow \text{Remainder} \\ \leftarrow \text{Divisor} \end{array}$$

Continued on Next Page

1 Divide.

(a) $(x^3 + x^2 + 4x - 6) \div (x - 1)$

(b) $\dfrac{p^3 - 2p^2 - 5p + 9}{p + 2}$

2 Divide.

(a) $\dfrac{r^2 - 5}{r + 4}$

(b) $(x^3 - 8) \div (x - 2)$

Teaching Tip

Explain to students how placeholders here are used in polynomial division just as in whole numbers, where 0 distinguishes 302 from 32, for example.

Answers

1. (a) $x^2 + 2x + 6$

(b) $p^2 - 4p + 3 + \dfrac{3}{p + 2}$

2. (a) $r - 4 + \dfrac{11}{r + 4}$

(b) $x^2 + 2x + 4$

Step 7 Multiply to check.

CHECK $(2x - 1)\left(2x^2 - x + 2 + \dfrac{-6}{2x - 1}\right)$ Multiply Divisor × (Quotient including the Remainder).

$$= (2x - 1)(2x^2) + (2x - 1)(-x) + (2x - 1)(2) + (2x - 1)\left(\frac{-6}{2x - 1}\right)$$

$$= 4x^3 - 2x^2 - 2x^2 + x + 4x - 2 - 6$$

$$= 4x^3 - 4x^2 + 5x - 8 \checkmark$$

◀ **Work Problem 1 at the Side.**

CAUTION

Remember to include "$+ \dfrac{\text{remainder}}{\text{divisor}}$" as part of the answer.

EXAMPLE 2 Dividing into a Polynomial with Missing Terms

Divide $x^3 - 1$ by $x - 1$.

Here the dividend, $x^3 - 1$, is missing the x^2-term and the x-term. We use **0** as the coefficient for each missing term. Thus, $x^3 - 1 = x^3 + 0x^2 + 0x - 1$.

$$\begin{array}{r} x^2 + x + 1 \\ x - 1\overline{)x^3 + 0x^2 + 0x - 1} \\ \underline{x^3 - x^2} \qquad\qquad \\ x^2 + 0x \qquad \\ \underline{x^2 - x} \qquad \\ x - 1 \\ \underline{x - 1} \\ 0 \end{array}$$

Insert placeholders for the missing terms.

The remainder is 0. The quotient is $x^2 + x + 1$.

CHECK $(x - 1)(x^2 + x + 1)$

$$= x^3 + x^2 + x - x^2 - x - 1$$

$$= x^3 - 1 \checkmark \qquad \text{Divisor} \times \text{Quotient} = \text{Dividend}$$

◀ **Work Problem 2 at the Side.**

EXAMPLE 3 Dividing by a Polynomial with Missing Terms

Divide $x^4 + 2x^3 + 2x^2 - x - 1$ by $x^2 + 1$.

Since $x^2 + 1$ has a missing x-term, write it as $x^2 + 0x + 1$.

$$\begin{array}{r} x^2 + 2x + 1 \\ x^2 + 0x + 1\overline{)x^4 + 2x^3 + 2x^2 - x - 1} \\ \underline{x^4 + 0x^3 + x^2} \qquad\qquad \\ 2x^3 + x^2 - x \qquad \\ \underline{2x^3 + 0x^2 + 2x} \qquad \\ x^2 - 3x - 1 \\ \underline{x^2 + 0x + 1} \\ -3x - 2 \end{array}$$

Insert a placeholder for the missing term.

$-3x - 2$ ← Remainder

Continued on Next Page

When the result of subtracting ($-3x - 2$, in this case) is a constant or a polynomial of degree less than the divisor $(x^2 + 0x + 1)$, that constant or polynomial is the remainder. We write the answer as follows.

$$x^2 + 2x + 1 + \frac{-3x - 2}{x^2 + 1}$$

Remember to include "$+ \frac{\text{remainder}}{\text{divisor}}$."

CHECK Show that $(x^2 + 1)\left(x^2 + 2x + 1 + \frac{-3x - 2}{x^2 + 1}\right)$ gives the original dividend, $x^4 + 2x^3 + 2x^2 - x - 1$. ✓

Work Problem 3 at the Side. ▶

EXAMPLE 4 Dividing a Polynomial When the Quotient Has Fractional Coefficients

Divide $4x^3 + 2x^2 + 3x + 2$ by $4x - 4$.

$\frac{6x^2}{4x} = \frac{3}{2}x$

$\frac{9x}{4x} = \frac{9}{4}$

$$\begin{array}{r} x^2 + \frac{3}{2}x + \frac{9}{4} \\ 4x - 4\overline{)4x^3 + 2x^2 + 3x + 2} \\ \underline{4x^3 - 4x^2 } \\ 6x^2 + 3x \\ \underline{6x^2 - 6x } \\ 9x + 2 \\ \underline{9x - 9} \\ 11 \end{array}$$

The answer is $x^2 + \frac{3}{2}x + \frac{9}{4} + \frac{11}{4x - 4}$.

Work Problem 4 at the Side. ▶

OBJECTIVE 2 Apply division to a geometry problem.

EXAMPLE 5 Using an Area Formula

The area of the rectangle in **Figure 3** is given by $(x^3 + 4x^2 + 8x + 8)$ sq. units. The width is given by $(x + 2)$ units. What is its length?

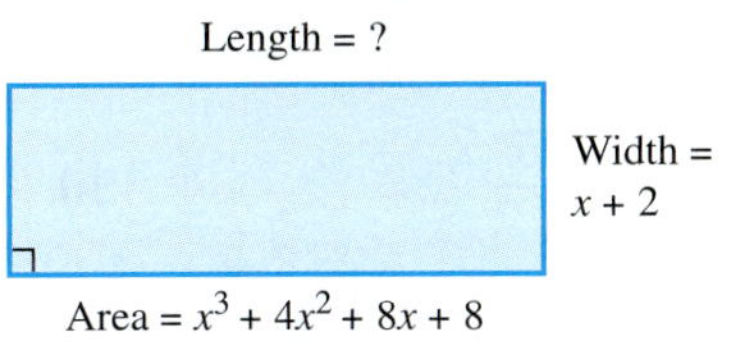

Figure 3

For a rectangle, $A = LW$. Solving for L gives $L = \frac{A}{W}$. Divide the area, $x^3 + 4x^2 + 8x + 8$, by the width, $x + 2$.

Work Problem 5 at the Side. ▶

The quotient from **Margin Problem 5,** $x^2 + 2x + 4$, represents the length of the rectangle in units.

---

3 Divide.

(a) $(2x^4 + 3x^3 - x^2 + 6x + 5) \div (x^2 - 1)$

(b)

$$\frac{2m^5 + m^4 + 6m^3 - 3m^2 - 18}{m^2 + 3}$$

4 Divide $3x^3 + 7x^2 + 7x + 10$ by $3x + 6$.

5 Divide $x^3 + 4x^2 + 8x + 8$ by $x + 2$.

Answers

3. **(a)** $2x^2 + 3x + 1 + \frac{9x + 6}{x^2 - 1}$

(b) $2m^3 + m^2 - 6$

4. $x^2 + \frac{1}{3}x + \frac{5}{3}$

5. $x^2 + 2x + 4$

5.7 Exercises

FOR EXTRA HELP

Download the MyDashBoard App

MyMathLab®

CONCEPT CHECK *Answer each question.*

1. In the division problem $(4x^4 + 2x^3 - 14x^2 + 19x + 10) \div (2x + 5) = 2x^3 - 4x^2 + 3x + 2$, which polynomial is the divisor? Which is the quotient?
The divisor is $2x + 5$. The quotient is $2x^3 - 4x^2 + 3x + 2$.

2. When dividing one polynomial by another, how do you know when to stop dividing?
Stop when the degree of the remainder is less than the degree of the divisor, or when the remainder is 0.

3. In dividing $12m^2 - 20m + 3$ by $2m - 3$, what is the first step?
Divide $12m^2$ by $2m$ to get $6m$.

4. In the division in **Exercise 3,** what is the second step?
Multiply $6m$ by $2m - 3$ to get $12m^2 - 18m$.

Perform each division. **See Example 1.**

5. $\dfrac{x^2 - x - 6}{x - 3}$
$x + 2$

6. $\dfrac{m^2 - 2m - 24}{m - 6}$
$m + 4$

7. $\dfrac{2y^2 + 9y - 35}{y + 7}$
$2y - 5$

8. $\dfrac{2y^2 + 9y + 7}{y + 1}$
$2y + 7$

9. $\dfrac{p^2 + 2p + 20}{p + 6}$
$p - 4 + \dfrac{44}{p + 6}$

10. $\dfrac{x^2 + 11x + 16}{x + 8}$
$x + 3 + \dfrac{-8}{x + 8}$

11. $(r^2 - 8r + 15) \div (r - 3)$
$r - 5$

12. $(t^2 + 2t - 35) \div (t - 5)$
$t + 7$

13. $\dfrac{4a^2 - 22a + 32}{2a + 3}$
$2a - 14 + \dfrac{74}{2a + 3}$

14. $\dfrac{9w^2 + 6w + 10}{3w - 2}$
$3w + 4 + \dfrac{18}{3w - 2}$

15. $\dfrac{8x^3 - 10x^2 - x + 3}{2x + 1}$
$4x^2 - 7x + 3$

16. $\dfrac{12t^3 - 11t^2 + 9t + 18}{4t + 3}$
$3t^2 - 5t + 6$

Perform each division. **See Examples 2–4.**

17. $\dfrac{3y^3 + y^2 + 2}{y + 1}$
$3y^2 - 2y + 2$

18. $\dfrac{2r^3 - 6r - 36}{r - 3}$
$2r^2 + 6r + 12$

19. $\dfrac{2x^3 + x + 2}{x + 1}$
$2x^2 - 2x + 3 + \dfrac{-1}{x + 1}$

20. $\dfrac{3x^3 + x + 5}{x + 1}$
$3x^2 - 3x + 4 + \dfrac{1}{x + 1}$

21. $\dfrac{3k^3 - 4k^2 - 6k + 10}{k^2 - 2}$
$3k - 4 + \dfrac{2}{k^2 - 2}$

22. $\dfrac{5z^3 - z^2 + 10z + 2}{z^2 + 2}$
$5z - 1 + \dfrac{4}{z^2 + 2}$

23. $(x^4 - x^2 - 2) \div (x^2 - 2)$
$x^2 + 1$

24. $(r^4 + 2r^2 - 3) \div (r^2 - 1)$
$r^2 + 3$

25. $\dfrac{x^4 - 1}{x^2 - 1}$
$x^2 + 1$

26. $\dfrac{y^3 + 1}{y + 1}$
$y^2 - y + 1$

27. $\dfrac{6p^4 - 15p^3 + 14p^2 - 5p + 10}{3p^2 + 1}$
$2p^2 - 5p + 4 + \dfrac{6}{3p^2 + 1}$

28. $\dfrac{6r^4 - 10r^3 - r^2 + 15r - 8}{2r^2 - 3}$
$3r^2 - 5r + 4 + \dfrac{4}{2r^2 - 3}$

29. $\dfrac{2x^5 + x^4 + 11x^3 - 8x^2 - 13x + 7}{2x^2 + x - 1}$
$x^3 + 6x - 7$

30. $\dfrac{4t^5 - 11t^4 - 6t^3 + 5t^2 - t + 3}{4t^2 + t - 3}$
$t^3 - 3t^2 - 1$

31. $(10x^3 + 13x^2 + 4x + 1) \div (5x + 5)$
$2x^2 + \frac{3}{5}x + \frac{1}{5}$

32. $(6x^3 - 19x^2 - 19x - 4) \div (2x - 8)$
$3x^2 + \frac{5}{2}x + \frac{1}{2}$

Work each problem. ***See Example 5.*** *(If necessary, refer to the formulas inside the back cover.)*

33. Give the length of the rectangle.

The area is $(5x^3 + 7x^2 - 13x - 6)$ sq. units.

$(x^2 + x - 3)$ units

34. Find the measure of the base of the parallelogram.

The area is $(2x^3 + 2x^2 - 3x - 1)$ sq. units.

$(2x^2 + 4x + 1)$ units

Relating Concepts (Exercises 35–38) For Individual or Group Work

We can find the value of a polynomial in x for a given value of x by substituting that number for x. Surprisingly, we can accomplish the same thing by division. For example, to find the value of $2x^2 - 4x + 3$ *for* $x = -3$*, we would divide* $2x^2 - 4x + 3$ *by* $x - (-3)$*. The remainder will give the value of the polynomial for* $x = -3$*.*
Work Exercises 35–38 in order.

35. Find the value of $2x^2 - 4x + 3$ for $x = -3$ by substitution. 33

36. Divide $2x^2 - 4x + 3$ by $x + 3$. Give the remainder. 33

37. Compare your answers to **Exercises 35 and 36.** What do you notice?
They are the same.

38. Choose another polynomial and evaluate it both ways for some value of the variable. Do the answers agree?
The answers should agree.

5.8 An Application of Exponents: Scientific Notation

OBJECTIVES

1. Express numbers in scientific notation.
2. Convert numbers in scientific notation to numbers without exponents.
3. Use scientific notation in calculations.

Teaching Tip

To introduce scientific notation, have students find

$$5{,}000{,}000 \times 6{,}000{,}000$$

on their calculators. Discuss the calculator results.

Teaching Tip

Point out that positive numbers in scientific notation with positive exponents on 10 are relatively "large," while those with negative exponents on 10 are relatively "small."

OBJECTIVE 1 **Express numbers in scientific notation.** Numbers occurring in science are often extremely large (such as the distance from Earth to the sun, 93,000,000 mi) or extremely small (the wavelength of yellow-green light, approximately 0.0000006 m). Because of the difficulty of working with many zeros, scientists often express such numbers with exponents, using a form called *scientific notation*.

Scientific Notation

A number is written in **scientific notation** when it is expressed in the form

$$a \times 10^n,$$

where $1 \le |a| < 10$ and n is an integer.

In **scientific notation,** there is always one nonzero digit to the left of the decimal point. This is shown in the following, where the first number is in scientific notation.

$3.19 \times 10^1 = 3.19 \times 10 = 31.9$ Decimal point moves 1 place to the right.

$3.19 \times 10^2 = 3.19 \times 100 = 319.$ Decimal point moves 2 places to the right.

$3.19 \times 10^3 = 3.19 \times 1000 = 3190.$ Decimal point moves 3 places to the right.

$3.19 \times 10^{-1} = 3.19 \times 0.1 = 0.319$ Decimal point moves 1 place to the left.

$3.19 \times 10^{-2} = 3.19 \times 0.01 = 0.0319$ Decimal point moves 2 places to the left.

$3.19 \times 10^{-3} = 3.19 \times 0.001 = 0.00319$ Decimal point moves 3 places to the left.

Note

In scientific notation, a multiplication cross (or symbol) $\times$ is commonly used.

A number in scientific notation is always written with the decimal point after the first nonzero digit and then multiplied by the appropriate power of 10. For example, 56,200 is written 5.62×10^4, since

$$56{,}200 = 5.62 \times 10{,}000 = 5.62 \times 10^4.$$

Other examples include

	42,000,000	written	4.2×10^7,
	0.000586	written	5.86×10^{-4},
and	2,000,000,000	written	2×10^9.

It is not necessary to write 2.0.

To write a positive number in scientific notation, follow the steps given on the next page. (For a negative number, follow these steps using the *absolute value* of the number. Then make the result negative.)

Writing a Positive Number in Scientific Notation

Step 1 Move the decimal point to the right of the first nonzero digit.

Step 2 Count the number of places you moved the decimal point.

Step 3 The number of places in Step 2 is the absolute value of the exponent on 10.

Step 4 The exponent on 10 is positive if the original number is greater than the number in Step 1. The exponent is negative if the original number is less than the number in Step 1. If the decimal point is not moved, the exponent is 0.

EXAMPLE 1 Using Scientific Notation

Write each number in scientific notation.

(a) 93,000,000

Move the decimal point to follow the first nonzero digit (the 9). Count the number of places the decimal point was moved.

93,000,000. ← Decimal point
7 places

The number will be written in scientific notation as 9.3×10^n. To find the value of n, first compare the original number, 93,000,000, with 9.3. Since 93,000,000 is *greater* than 9.3, we must multiply by a *positive* power of 10 so that the product 9.3×10^n will equal the larger number.

Since the decimal point was moved 7 places, and since n is positive,

$$93{,}000{,}000 = 9.3 \times 10^7.$$

(b) $63{,}200{,}000{,}000 = 6.3200000000 = 6.32 \times 10^{10}$
10 places

(c) $3.021 = 3.021 \times 10^0$

(d) 0.00462

Move the decimal point to the right of the first nonzero digit and count the number of places the decimal point was moved.

0.00462 3 places

Since 0.00462 is *less* than 4.62, the exponent must be *negative*.

$$0.00462 = 4.62 \times 10^{-3}$$

(e) $-0.0000762 = -7.62 \times 10^{-5}$
5 places — Remember the *negative* sign.

Work Problem **1** at the Side. ▶

Note

When writing a positive number in scientific notation, think as follows.

1. If the original number is "large," like 93,000,000, use a *positive* exponent on 10, since positive is greater than negative.
2. If the original number is "small," like 0.00462, use a *negative* exponent on 10, since negative is less than positive.

---

1 Write each number in scientific notation.

GS **(a)** 63,000

The first nonzero digit is ____.

The decimal point should be moved ____ places.

$63{,}000 = \_\_\_\_ \times 10^{\_\_}$

(b) 5,870,000

(c) 7.0065

GS **(d)** 0.0571

The first nonzero digit is ____.

The decimal point should be moved ____ places.

$0.0571 = \_\_\_\_ \times 10^{\_\_}$

(e) −0.00062

Answers

1. (a) 6; 4; 6.3; 4 **(b)** 5.87×10^6
(c) 7.0065×10^0 **(d)** 5; 2; 5.71; −2
(e) -6.2×10^{-4}

2 Write without exponents.

GS **(a)** 4.2×10^3

Move the decimal point ____ places to the ____.

$4.2 \times 10^3 =$ ________

(b) 8.7×10^5

GS **(c)** 6.42×10^{-3}

Move the decimal point ____ places to the ____.

$6.42 \times 10^{-3} =$ ________

(d) -5.27×10^{-1}

3 Perform each calculation. Write answers in scientific notation and also without exponents.

(a) $(2.6 \times 10^4)(2 \times 10^{-6})$

(b) $(3 \times 10^5)(5 \times 10^{-2})$

(c) $\dfrac{4.8 \times 10^2}{2.4 \times 10^{-3}}$

Answers

2. **(a)** 3; right; 4200 **(b)** 870,000 **(c)** 3; left; 0.00642 **(d)** −0.527
3. **(a)** 5.2×10^{-2}; 0.052 **(b)** 1.5×10^4; 15,000 **(c)** 2×10^5; 200,000

OBJECTIVE 2 Convert numbers in scientific notation to numbers without exponents. To convert a number written in scientific notation to a number without exponents, work in reverse.

Multiplying a positive number by a positive power of 10 will make the number greater. Multiplying by a negative power of 10 will make the number less.

EXAMPLE 2 Writing Numbers without Exponents

Write each number without exponents.

(a) 6.2×10^3

Since the exponent is positive, make 6.2 greater by moving the decimal point 3 places to the right. It is necessary to attach two 0s.

$$6.2 \times 10^3 = 6.200 = 6200$$

(b) $4.283 \times 10^5 = 4.28300 = 428{,}300$ Move 5 places to the right. Attach 0s as necessary.

(c) $-9.73 \times 10^{-2} = -09.73 = -0.0973$ Move 2 places to the left.

The exponent tells the number of places and the direction in which the decimal point is moved.

◀ Work Problem 2 at the Side.

OBJECTIVE 3 Use scientific notation in calculations. The next example uses scientific notation with products and quotients.

EXAMPLE 3 Multiplying and Dividing with Scientific Notation

Perform each calculation. Write answers in scientific notation and also without exponents.

(a)

$(7 \times 10^3)(5 \times 10^4)$	
$= (7 \times 5)(10^3 \times 10^4)$	Commutative and associative properties
$= 35 \times 10^7$	Multiply and use the product rule.
$= (3.5 \times 10^1) \times 10^7$	Write 35 in scientific notation.
$= 3.5 \times (10^1 \times 10^7)$	Associative property
$= 3.5 \times 10^8$	Product rule
$= 350{,}000{,}000$	Write without exponents.

Don't stop. This number is *not* in scientific notation, since 35 is not between 1 and 10.

(b) $\dfrac{4 \times 10^{-5}}{2 \times 10^3}$

$= \dfrac{4}{2} \times \dfrac{10^{-5}}{10^3}$	
$= 2 \times 10^{-8}$	Divide and use the quotient rule.
$= 0.00000002$	Write without exponents.

◀ Work Problem 3 at the Side.

Calculator Tip

Calculators usually have a key labeled EE or EXP for scientific notation. See your owner's manual for more information.

Note

Multiplying or dividing numbers written in scientific notation may produce an answer in the form $a \times 10^0$. Since $10^0 = 1$, $a \times 10^0 = a$. For example,

$$(8 \times 10^{-4})(5 \times 10^4) = 40 \times 10^0 = 40. \quad 10^0 = 1$$

Also, if $a = 1$, then $a \times 10^n = 10^n$. For example, we could write 1,000,000 as 10^6 instead of 1×10^6.

EXAMPLE 4 **Using Scientific Notation to Solve an Application**

A *nanometer* is a very small unit of measure that is equivalent to about 0.00000003937 in. About how much would 700,000 nanometers measure in inches? (*Source: World Almanac and Book of Facts.*)

Write each number in scientific notation, and then multiply.

$$700{,}000(0.00000003937)$$

$= (7 \times 10^5)(3.937 \times 10^{-8})$ Write in scientific notation.

$= (7 \times 3.937)(10^5 \times 10^{-8})$ Properties of real numbers

$= 27.559 \times 10^{-3}$ Multiply; product rule

Don't stop here.

$= (2.7559 \times 10^1) \times 10^{-3}$ Write 27.559 in scientific notation.

$= 2.7559 \times 10^{-2}$ Product rule

$= 0.027559$ Write without exponents.

Thus, 700,000 nanometers would measure

$$2.7559 \times 10^{-2} \text{ in.,} \quad \text{or} \quad 0.027559 \text{ in.}$$

Work Problem 4 at the Side. ▶

EXAMPLE 5 **Using Scientific Notation to Solve an Application**

In 2010, the national debt was $\$1.3562 \times 10^{13}$ (which is more than \$13 trillion). The population of the United States was approximately 309 million that year. About how much would each person have had to contribute in order to pay off the national debt? (*Source:* U.S. Department of the Treasury; U.S. Census Bureau.)

Write the population in scientific notation. Then divide to obtain the per person contribution.

$$\frac{1.3562 \times 10^{13}}{309{,}000{,}000}$$

$= \dfrac{1.3562 \times 10^{13}}{3.09 \times 10^8}$ Write 309 million in scientific notation.

$= \dfrac{1.3562}{3.09} \times 10^5$ Quotient rule

$\approx 0.4389 \times 10^5$ Divide. Round to 4 decimal places.

$= 43{,}890$ Write without exponents.

Each person would have to pay about \$43,890.

Work Problem 5 at the Side. ▶

4 The speed of light is approximately 3.0×10^5 km per sec. How far does light travel in 6.0×10^1 sec? (*Source: World Almanac and Book of Facts.*)

5 If the speed of light is approximately 3.0×10^5 km per sec, how many seconds does it take light to travel approximately 1.5×10^8 km from the Sun to Earth? (*Source: World Almanac and Book of Facts.*)

Answers

4. 1.8×10^7 km, or 18,000,000 km
5. 5×10^2 sec, or 500 sec

5.8 Exercises

FOR EXTRA HELP

Download the MyDashBoard App

MyMathLab®

CONCEPT CHECK *Match each number written in scientific notation in Column I with the correct choice from Column II. Not all choices in Column II will be used.*

I		II
1. (a) 4.6×10^{-4}	C	**A.** 46,000
(b) 4.6×10^{4}	A	**B.** 460,000
(c) 4.6×10^{5}	B	**C.** 0.00046
(d) 4.6×10^{-5}	D	**D.** 0.000046
		E. 4600

I		II
2. (a) 1×10^{9}	A	**A.** 1 billion
(b) 1×10^{6}	C	**B.** 100 million
(c) 1×10^{8}	B	**C.** 1 million
(d) 1×10^{10}	D	**D.** 10 billion
		E. 100 billion

CONCEPT CHECK *Determine whether or not the given number is written in scientific notation as defined in* ***Objective 1.*** *If it is not, write it as such.*

3. 4.56×10^3
in scientific notation

4. 7.34×10^5
in scientific notation

5. 5,600,000
not in scientific notation; 5.6×10^6

6. 34,000
not in scientific notation; 3.4×10^4

7. 0.004
not in scientific notation; 4×10^{-3}

8. 0.0007
not in scientific notation; 7×10^{-4}

9. 0.8×10^2
not in scientific notation; 8×10^1

10. 0.9×10^3
not in scientific notation; 9×10^2

11. Explain what it means for a number to be written in scientific notation.
A number is written in scientific notation if it is the product of a number whose absolute value is between 1 and 10 (inclusive of 1) and a power of 10.

12. Explain how to multiply a number by a positive power of ten. Then explain how to multiply a number by a negative power of ten.
To multiply by a positive power of 10, move the decimal point to the right as many places as the exponent on 10. With a negative power of 10, move the decimal point to the left as many places as the absolute value of the exponent on 10.

Write each number in scientific notation. ***See Example 1.***

13. 5,876,000,000
5.876×10^9

14. 9,994,000,000
9.994×10^9

15. 82,350
8.235×10^4

16. 78,330
7.833×10^4

17. 0.000007
7×10^{-6}

18. 0.0000004
4×10^{-7}

19. -0.00203
-2.03×10^{-3}

20. -0.0000578
-5.78×10^{-5}

Write each number without exponents. ***See Example 2.***

21. 7.5×10^5
750,000

22. 8.8×10^6
8,800,000

23. 5.677×10^{12}
5,677,000,000,000

24. 8.766×10^9
8,766,000,000

25. 1×10^{12}
1,000,000,000,000

26. 1×10^7
10,000,000

27. -6.21×10^0
-6.21

28. -8.56×10^0
-8.56

29. 7.8×10^{-4}
0.00078

30. 8.9×10^{-5}
0.000089

31. 5.134×10^{-9}
0.000000005134

32. 7.123×10^{-10}
0.0000000007123

Perform the indicated operations. Write the answers in scientific notation and then without exponents. ***See Example 3.***

33. $(2 \times 10^8)(3 \times 10^3)$
6×10^{11}; 600,000,000,000

34. $(3 \times 10^7)(3 \times 10^3)$
9×10^{10}; 90,000,000,000

35. $(5 \times 10^4)(3 \times 10^2)$
1.5×10^7; 15,000,000

36. $(8 \times 10^5)(2 \times 10^3)$
1.6×10^9; 1,600,000,000

37. $(4 \times 10^{-6})(2 \times 10^3)$
8×10^{-3}; 0.008

38. $(3 \times 10^{-7})(2 \times 10^2)$
6×10^{-5}; 0.00006

39. $(6 \times 10^3)(4 \times 10^{-2})$
2.4×10^2; 240

40. $(7 \times 10^5)(3 \times 10^{-4})$
2.1×10^2; 210

41. $(9 \times 10^4)(7 \times 10^{-7})$
6.3×10^{-2}; 0.063

42. $(6 \times 10^4)(8 \times 10^{-8})$
4.8×10^{-3}; 0.0048

43. $(3.15 \times 10^{-4})(2.04 \times 10^8)$
6.426×10^4; 64,260

44. $(4.92 \times 10^{-3})(2.25 \times 10^7)$
1.107×10^5; 110,700

45. $\dfrac{9 \times 10^{-5}}{3 \times 10^{-1}}$
3×10^{-4}; 0.0003

46. $\dfrac{12 \times 10^{-4}}{4 \times 10^{-3}}$
3×10^{-1}; 0.3

47. $\dfrac{8 \times 10^3}{2 \times 10^2}$
4×10^1; 40

48. $\dfrac{15 \times 10^4}{3 \times 10^3}$
5×10^1; 50

49. $\dfrac{2.6 \times 10^{-3}}{2 \times 10^2}$
1.3×10^{-5}; 0.000013

50. $\dfrac{9.5 \times 10^{-1}}{5 \times 10^3}$
1.9×10^{-4}; 0.00019

51. $\dfrac{4 \times 10^5}{8 \times 10^2}$
5×10^2; 500

52. $\dfrac{3 \times 10^9}{6 \times 10^5}$
5×10^3; 5000

53. $\dfrac{2.6 \times 10^{-3} \times 7.0 \times 10^{-1}}{2 \times 10^2 \times 3.5 \times 10^{-3}}$
2.6×10^{-3}; 0.0026

54. $\dfrac{9.5 \times 10^{-1} \times 2.4 \times 10^4}{5 \times 10^3 \times 1.2 \times 10^{-2}}$
3.8×10^2; 380

55. $\dfrac{(1.65 \times 10^8)(5.24 \times 10^{-2})}{(6 \times 10^4)(2 \times 10^7)}$
7.205×10^{-6}; 0.000007205

56. $\dfrac{(3.25 \times 10^7)(6.48 \times 10^{-4})}{(5 \times 10^2)(4 \times 10^6)}$
1.053×10^{-5}; 0.00001053

Each statement taken from a book or newspaper contains a number in boldface italic type. If the number is in scientific notation, write it without exponents. If the number is not in scientific notation, write it as such. ***See Examples 1 and 2.***

57. The muon, a close relative of the electron produced by the bombardment of cosmic rays against the upper atmosphere, has a half-life of 2 millionths of a second ($\boldsymbol{2 \times 10^{-6}}$ s). (*Source:* Hewitt, Paul G., *Conceptual Physics*.)
0.000002

58. There are 13 red balls and 39 black balls in a box. Mix them up and draw 13 out one at a time without returning any ball . . . the probability that the 13 drawings each will produce a red ball is . . . $\boldsymbol{1.6 \times 10^{-12}}$. (*Source:* Weaver, Warren, *Lady Luck*.)
0.0000000000016

59. An electron and a positron attract each other in two ways: the electromagnetic attraction of their opposite electric charges, and the gravitational attraction of their two masses. The electromagnetic attraction is

4,200,000,000,000,000,000,000,000,000,000,000,000,000,000,000

times as strong as the gravitational. (*Source:* Asimov, I., *Isaac Asimov's Book of Facts*.)
4.2×10^{42}

60. A googol is

10,000,000,000,000,000,000,000,000,000,000,000,000,000,000,000,000,000,000,000,000, 000,000,000,000,000,000,000,000,000,000,000,000,000,000,000,000.

The Web search engine Google is named after a googol. Sergey Brin, president and cofounder of Google, Inc., was a mathematics major. He chose the name Google to describe the vast reach of this search engine. (*Source: The Gazette*.)
1×10^{100}

Use scientific notation to calculate the answer to each problem. In Exercises 63–65 and 69–70, give answers without exponents. ***See Examples 4 and 5.***

61. The Double Helix Nebula, a conglomeration of dust and gas stretching across the center of the Milky Way galaxy, is 25,000 light-years from Earth. If one light-year is about 6,000,000,000,000 mi, about how many miles is the Double Helix Nebula from Earth? (*Source:* www.spitzer.caltech.edu) **1.5×10^{17} mi**

62. Pollux, one of the brightest stars in the night sky, is 33.7 light-years from Earth. If one light-year is about 6,000,000,000,000 mi (that is, 6 trillion mi), about how many miles is Pollux from Earth? (*Source: World Almanac and Book of Facts.*) **2.022×10^{14} mi**

63. In 2011, the population of the United States was about 311.6 million. To the nearest dollar, calculate how much each person in the United States would have had to contribute in order to make one person a trillionaire (that is, to give that person $1,000,000,000,000). (*Source:* U.S. Census Bureau.) **$3209**

64. In 2010, the U.S. government collected about $3767 per person in individual income taxes. If the population at that time was 309,000,000, how much did the government collect in taxes for 2010? (*Source*: *World Almanac and Book of Facts.*) **$1,164,003,000,000**

65. In 2011, Congress raised the debt limit to $\$1.64 \times 10^{13}$. When this national debt limit is reached, about how much is it for every man, woman, and child in the country? Use 311 million as the population of the United States. (*Source:* www.tradingnrg.org) **$52,733**

66. In 2010, the state of Minnesota had about 8.1×10^4 farms with an average of 3.44×10^2 acres per farm. What was the total number of acres devoted to farmland in Minnesota that year? (*Source:* U.S. Department of Agriculture.) **2.7864×10^7, or 27,864,000, acres**

67. Venus is 6.68×10^7 mi from the Sun. If light travels at a speed of 1.86×10^5 mi per sec, how long does it take light to travel from the Sun to Venus? (*Source: World Almanac and Book of Facts.*) **3.59×10^2 sec, or 359 sec**

68. The distance to Earth from Pluto is 4.58×10^9 km. *Pioneer 10* transmitted radio signals from Pluto to Earth at the speed of light, 3.00×10^5 km per sec. How long (in seconds) did it take for the signals to reach Earth? **about 1.53×10^4 sec, or 15,300 sec**

69. During the 2010–2011 season, Broadway shows grossed a total of 1.08×10^9 dollars. Total attendance for the season was 1.25×10^7. What was the average ticket price for a Broadway show? (*Source:* The Broadway League.) **$86.40**

70. In 2010, 10.6×10^9 dollars were spent to attend motion pictures in the United States and Canada. The total number of tickets sold that year totaled 1.34 billion. What was the average ticket price? (*Source:* Motion Picture Association of America.) **$7.91**

Chapter 5 Summary

Key Terms

5.1

term A term is a number, a variable, or a product or quotient of a number and one or more variables raised to powers.

like terms Terms with exactly the same variables (including the same exponents) are like terms.

polynomial A polynomial is a term or the sum of a finite number of terms with whole number exponents.

descending powers A polynomial in x is written in descending powers if the exponents on x in its terms are in decreasing order.

degree of a term The degree of a term is the sum of the exponents on the variables.

degree of a polynomial The degree of a polynomial is the greatest degree of any term of the polynomial.

monomial A monomial is a polynomial with exactly one term.

binomial A binomial is a polynomial with exactly two terms.

trinomial A trinomial is a polynomial with exactly three terms.

5.2

exponential expression A number written with an exponent is an exponential expression.

3^4 ← Exponent; Base (3); Exponential expression (3^4)

5.3

FOIL method The FOIL method is used to find the product of two binomials. The letters of the word **FOIL** originate as follows: Multiply the **F**irst terms, multiply the **O**uter terms (to get the outer product), multiply the **I**nner terms (to get the inner product), and multiply the **L**ast terms.

outer product The outer product of $(a + b)(c + d)$ is ad.

inner product The inner product of $(a + b)(c + d)$ is bc.

5.4

conjugate The conjugate of $a + b$ is $a - b$.

5.8

scientific notation A number written as $a \times 10^n$, where $1 \leq |a| < 10$ and n is an integer, is in scientific notation.

New Symbols

x^{-n} x to the negative n power

Test Your Word Power

See how well you have learned the vocabulary in this chapter.

1. A **polynomial** is an algebraic expression made up of
   A. a term or a finite product of terms with positive coefficients and exponents
   B. a term or a finite sum of terms with real coefficients and whole number exponents
   C. the product of two or more terms with positive exponents
   D. the sum of two or more terms with whole number coefficients and exponents.

2. The **degree of a term** is
   A. the number of variables in the term
   B. the product of the exponents on the variables
   C. the least exponent on the variables
   D. the sum of the exponents on the variables.

3. A **trinomial** is a polynomial with
   A. only one term
   B. exactly two terms
   C. exactly three terms
   D. more than three terms.

4. A **binomial** is a polynomial with
   A. only one term
   B. exactly two terms
   C. exactly three terms
   D. more than three terms.

5. A **monomial** is a polynomial with
   A. only one term
   B. exactly two terms
   C. exactly three terms
   D. more than three terms.

6. **FOIL** is a method for
   A. adding two binomials
   B. adding two trinomials
   C. multiplying two binomials
   D. multiplying two trinomials.

Answers to Test Your Word Power

1. B; *Example:* $5x^3 + 2x^2 - 7$
2. D; *Examples:* The term 6 has degree 0, $3x$ has degree 1, $-2x^8$ has degree 8, and $5x^2y^4$ has degree 6.
3. C; *Example:* $2a^2 - 3ab + b^2$
4. B; *Example:* $3t^3 + 5t$
5. A; *Examples:* -5 and $4xy^5$
6. C; *Example:* $(m + 4)(m - 3)$

   F O I L

   $= m(m) - 3m + 4m + 4(-3)$

   $= m^2 + m - 12$

Quick Review

Concepts	Examples

5.1 Adding and Subtracting Polynomials

Addition
Add like terms.

Add.
$$\begin{array}{r} 2x^2 + 5x - 3 \\ \underline{5x^2 - 2x + 7} \\ 7x^2 + 3x + 4 \end{array}$$

Subtraction
Change the signs of the terms in the second polynomial and add to the first polynomial.

Subtract. $(2x^2 + 5x - 3) - (5x^2 - 2x + 7)$

$= (2x^2 + 5x - 3) + (-5x^2 + 2x - 7)$

$= -3x^2 + 7x - 10$

5.2 The Product Rule and Power Rules for Exponents

For any integers m and n, the following hold.

Product rule $a^m \cdot a^n = a^{m+n}$

Power rules (a) $(a^m)^n = a^{mn}$

(b) $(ab)^m = a^m b^m$

(c) $\left(\frac{a}{b}\right)^m = \frac{a^m}{b^m}$ (where $b \neq 0$)

Perform the operations by using the rules for exponents.

$2^4 \cdot 2^5 = 2^{4+5} = 2^9$

$(3^4)^2 = 3^{4 \cdot 2} = 3^8$

$(6a)^5 = 6^5 a^5$

$\left(\frac{2}{3}\right)^4 = \frac{2^4}{3^4}$

5.3 Multiplying Polynomials

Multiply each term of the first polynomial by each term of the second polynomial. Then add like terms.

FOIL method

Step 1 Multiply the two **F**irst terms to get the first term of the answer.

Step 2 Find the **O**uter product and the **I**nner product and mentally add them, when possible, to get the middle term of the answer.

Step 3 Multiply the two **L**ast terms to get the last term of the answer.

Add the terms found in Steps 1–3.

Multiply.
$$\begin{array}{rrrrr} & 3x^3 & -\ 4x^2 & +\ 2x & -\ 7 \\ & & & 4x & +\ 3 \\ \hline & 9x^3 & -\ 12x^2 & +\ 6x & -\ 21 \\ 12x^4 & -\ 16x^3 & +\ 8x^2 & -\ 28x & \\ \hline 12x^4 & -\ 7x^3 & -\ 4x^2 & -\ 22x & -\ 21 \end{array}$$

Multiply $(2x + 3)(5x - 4)$.

$2x(5x) = 10x^2$ F

$2x(-4) + 3(5x) = 7x$ O, I

$3(-4) = -12$ L

The product is $10x^2 + 7x - 12$.

5.4 Special Products

Square of a Binomial

$$(a + b)^2 = a^2 + 2ab + b^2$$
$$(a - b)^2 = a^2 - 2ab + b^2$$

Multiply.

$(3x + 1)^2$

$= (3x)^2 + 2(3x)(1) + 1^2$

$= 9x^2 + 6x + 1$

$(2m - 5n)^2$

$= (2m)^2 - 2(2m)(5n) + (5n)^2$

$= 4m^2 - 20mn + 25n^2$

Concepts	Examples
5.4 Special Products *(continued)*	
Product of the Sum and Difference of Two Terms $$(a+b)(a-b)=a^2-b^2$$	$(4a+3)(4a-3)$ $= (4a)^2 - 3^2$ $= 16a^2 - 9$
5.5 Integer Exponents and the Quotient Rule	
If $a, b \neq 0$, for integers m and n, the following hold.	Simplify by using the rules for exponents.
Zero exponent $a^0 = 1$	$15^0 = 1$
Negative exponent $a^{-n} = \frac{1}{a^n}$	$5^{-2} = \frac{1}{5^2} = \frac{1}{25}$
Quotient rule $\frac{a^m}{a^n} = a^{m-n}$	$\frac{4^8}{4^3} = 4^{8-3} = 4^5$
Negative-to-positive rules $\frac{a^{-m}}{b^{-n}} = \frac{b^n}{a^m}$ $\left(\frac{a}{b}\right)^{-m} = \left(\frac{b}{a}\right)^m$	$\frac{6^{-2}}{7^{-3}} = \frac{7^3}{6^2}$ $\left(\frac{5}{3}\right)^{-4} = \left(\frac{3}{5}\right)^4$
5.6 Dividing a Polynomial by a Monomial	Divide.
Divide each term of the polynomial by the monomial. $$\frac{a+b}{c} = \frac{a}{c} + \frac{b}{c}$$	$\frac{4x^3 - 2x^2 + 6x - 8}{2x}$ $= \frac{4x^3}{2x} - \frac{2x^2}{2x} + \frac{6x}{2x} - \frac{8}{2x}$ Divide each term in the dividend by $2x$, the divisor. $= 2x^2 - x + 3 - \frac{4}{x}$
5.7 Dividing a Polynomial by a Polynomial	Divide.
Use "long division."	$$\begin{array}{r} 2x - 5 \\ 3x+4\overline{)6x^2 - 7x - 21} \\ \underline{6x^2 + 8x} \\ -15x - 21 \\ \underline{-15x - 20} \\ -1 \end{array}$$ $-1 \leftarrow$ Remainder. The answer is $2x - 5 + \frac{-1}{3x+4}$.
5.8 An Application of Exponents: Scientific Notation	
To write a positive number in scientific notation $$a \times 10^n,$$ move the decimal point to follow the first nonzero digit. **1.** If moving the decimal point makes the number less, n is positive. **2.** If it makes the number greater, n is negative. **3.** If the decimal point is not moved, n is 0. For a negative number, follow these steps using the *absolute value* of the number. Then make the result negative.	Write in scientific notation. $247 = 2.47 \times 10^2$ $0.0051 = 5.1 \times 10^{-3}$ Write without exponents. $3.25 \times 10^5 = 325{,}000$ $8.44 \times 10^{-6} = 0.00000844$

Chapter 5 Review Exercises

5.1 *Combine terms where possible in each polynomial. Write the answer in descending powers of the variable. Give the degree of the answer. Identify the polynomial as a* monomial, *a* binomial, *a* trinomial, *or* none of these.

1. $9m^2 + 11m^2 + 2m^2$

$22m^2$; degree 2; monomial

2. $-4p + p^3 - p^2 + 8p + 2$

$p^3 - p^2 + 4p + 2$; degree 3; none of these

3. $12a^5 - 9a^4 + 8a^3 + 2a^2 - a + 3$

already in descending powers; degree 5; none of these

4. $-7y^5 - 8y^4 - y^5 + y^4 + 9y$

$-8y^5 - 7y^4 + 9y$; degree 5; trinomial

Add or subtract as indicated.

5. Add.

$$\begin{array}{r} -2a^3 + 5a^2 \\ \underline{-3a^3 - a^2} \end{array}$$

$-5a^3 + 4a^2$

6. Add.

$$\begin{array}{r} 4r^3 - 8r^2 + 6r \\ \underline{-2r^3 + 5r^2 + 3r} \end{array}$$

$2r^3 - 3r^2 + 9r$

7. Subtract.

$$\begin{array}{r} 6y^2 - 8y + 2 \\ \underline{-5y^2 + 2y - 7} \end{array}$$

$11y^2 - 10y + 9$

8. Subtract.

$$\begin{array}{r} -12k^4 - 8k^2 + 7k - 5 \\ \underline{k^4 + 7k^2 + 11k + 1} \end{array}$$

$-13k^4 - 15k^2 - 4k - 6$

9. $(2m^3 - 8m^2 + 4) + (8m^3 + 2m^2 - 7)$

$10m^3 - 6m^2 - 3$

10. $(-5y^2 + 3y + 11) + (4y^2 - 7y + 15)$

$-y^2 - 4y + 26$

11. $(6p^2 - p - 8) - (-4p^2 + 2p + 3)$

$10p^2 - 3p - 11$

12. $(12r^4 - 7r^3 + 2r^2) - (5r^4 - 3r^3 + 2r^2 + 1)$

$7r^4 - 4r^3 - 1$

5.2 *Simplify each expression.*

13. $4^3 \cdot 4^8$

4^{11}

14. $(-5)^6(-5)^5$

-5^{11}

15. $(-8x^4)(9x^3)$

$-72x^7$

16. $(2x^2)(5x^3)(x^9)$

$10x^{14}$

17. $(19x)^5$

19^5x^5

18. $(-4y)^7$

-4^7y^7

19. $5(pt)^4$

$5p^4t^4$

20. $\left(\frac{7}{5}\right)^6$

$\frac{7^6}{5^6}$

21. $(3x^2y^3)^3$

$27x^6y^9$

22. $(t^4)^8(t^2)^5$

t^{42}

23. $(6x^2z^4)^2(x^3yz^2)^4$

$36x^{16}y^4z^{16}$

24. $\left(\frac{2m^3n}{p^2}\right)^3$

$\frac{8m^9n^3}{p^6}$

25. Find a polynomial that represents the volume of the figure. (If necessary, refer to the formulas inside the back cover.)

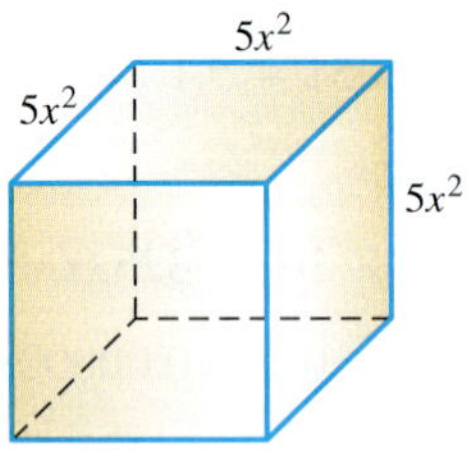

$125x^6$

26. Explain why the product rule for exponents does not apply to the expression

$$7^2 + 7^4.$$

The product rule for exponents does not apply here because we want the sum of 7^2 and 7^4, not their product.

5.3 *Find each product.*

27. $5x(2x + 14)$
$10x^2 + 70x$

28. $-3p^3(2p^2 - 5p)$
$-6p^5 + 15p^4$

29. $(3r - 2)(2r^2 + 4r - 3)$
$6r^3 + 8r^2 - 17r + 6$

30. $(2y + 3)(4y^2 - 6y + 9)$
$8y^3 + 27$

31. $(5p^2 + 3p)(p^3 - p^2 + 5)$
$5p^5 - 2p^4 - 3p^3 + 25p^2 + 15p$

32. $(x + 6)(x - 3)$
$x^2 + 3x - 18$

33. $(3k - 6)(2k + 1)$
$6k^2 - 9k - 6$

34. $(6p - 3q)(2p - 7q)$
$12p^2 - 48pq + 21q^2$

35. $(m^2 + m - 9)(2m^2 + 3m - 1)$
$2m^4 + 5m^3 - 16m^2 - 28m + 9$

5.4 *Find each product.*

36. $(a + 4)^2$
$a^2 + 8a + 16$

37. $(3p - 2)^2$
$9p^2 - 12p + 4$

38. $(2r + 5s)^2$
$4r^2 + 20rs + 25s^2$

39. $(r + 2)^3$
$r^3 + 6r^2 + 12r + 8$

40. $(2x - 1)^3$
$8x^3 - 12x^2 + 6x - 1$

41. $(2z + 7)(2z - 7)$
$4z^2 - 49$

42. $(6m - 5)(6m + 5)$
$36m^2 - 25$

43. $(5a + 6b)(5a - 6b)$
$25a^2 - 36b^2$

44. $(2x^2 + 5)(2x^2 - 5)$
$4x^4 - 25$

45. **CONCEPT CHECK** The square of a binomial leads to a polynomial with how many terms? The product of the sum and difference of two terms leads to a polynomial with how many terms?
three; two

46. Explain why $(a + b)^2$ is not equivalent to $a^2 + b^2$.
$(a + b)^2 = (a + b)(a + b) = a^2 + 2ab + b^2$. The term $2ab$ is not in $a^2 + b^2$.

5.5 *Evaluate each expression.*

47. $5^0 + 8^0$
2

48. 2^{-5}
$\frac{1}{32}$

49. $\left(\frac{6}{5}\right)^{-2}$
$\frac{25}{36}$

50. $4^{-2} - 4^{-1}$
$-\frac{3}{16}$

Simplify each expression. Assume that all variables represent nonzero numbers.

51. $\frac{6^{-3}}{6^{-5}}$ 36

52. $\frac{x^{-7}}{x^{-9}}$ x^2

53. $\frac{p^{-8}}{p^4}$ $\frac{1}{p^{12}}$

54. $\frac{r^{-2}}{r^{-6}}$ r^4

55. $(2^4)^2$ 2^8

56. $(9^3)^{-2}$ $\frac{1}{9^6}$

57. $(5^{-2})^{-4}$ 5^8

58. $(8^{-3})^4$ $\frac{1}{8^{12}}$

59. $\frac{(m^2)^3}{(m^4)^2}$ $\frac{1}{m^2}$

60. $\frac{y^4 \cdot y^{-2}}{y^{-5}}$ y^7

61. $\frac{r^9 \cdot r^{-5}}{r^{-2} \cdot r^{-7}}$ r^{13}

62. $(-5m^3)^2$ $25m^6$

63. $(2y^{-4})^{-3}$
$\frac{y^{12}}{8}$

64. $\frac{ab^{-3}}{a^4b^2}$
$\frac{1}{a^3b^5}$

65. $\frac{(6r^{-1})^2 \cdot (2r^{-4})}{r^{-5}(r^2)^{-3}}$
$72r^5$

66. $\frac{(2m^{-5}n^2)^3(3m^2)^{-1}}{m^{-2}n^{-4}(m^{-1})^2}$
$\frac{8n^{10}}{3m^{13}}$

5.6 *Perform each division.*

67. $\frac{-15y^4}{-9y^2}$

$\frac{5y^2}{3}$

68. $\frac{-12x^3y^2}{6xy}$

$-2x^2y$

69. $\frac{6y^4 - 12y^2 + 18y}{-6y}$

$-y^3 + 2y - 3$

70. $\frac{2p^3 - 6p^2 + 5p}{2p^2}$

$p - 3 + \frac{5}{2p}$

71. $(5x^{13} - 10x^{12} + 20x^7 - 35x^5) \div (-5x^4)$

$-x^9 + 2x^8 - 4x^3 + 7x$

72. $(-10m^4n^2 + 5m^3n^3 + 6m^2n^4) \div (5m^2n)$

$-2m^2n + mn^2 + \frac{6n^3}{5}$

5.7 *Perform each division.*

73. $(2r^2 + 3r - 14) \div (r - 2)$

$2r + 7$

74. $\frac{12m^2 - 11m - 10}{3m - 5}$

$4m + 3 + \frac{5}{3m - 5}$

75. $\frac{10a^3 + 5a^2 - 14a + 9}{5a^2 - 3}$

$2a + 1 + \frac{-8a + 12}{5a^2 - 3}$

76. $\frac{2k^4 + 4k^3 + 9k^2 - 8}{2k^2 + 1}$

$k^2 + 2k + 4 + \frac{-2k - 12}{2k^2 + 1}$

5.8 *Write each number in scientific notation.*

77. 48,000,000

4.8×10^7

78. 28,988,000,000

2.8988×10^{10}

79. 0.000065

6.5×10^{-5}

80. 0.0000000824

8.24×10^{-8}

Write each number without exponents.

81. 2.4×10^4

24,000

82. 7.83×10^7

78,300,000

83. 8.97×10^{-7}

0.000000897

84. 9.95×10^{-12}

0.00000000000995

Perform the indicated operations. Write the answers in scientific notation and then without exponents.

85. $(2 \times 10^{-3})(4 \times 10^5)$

8×10^2; 800

86. $\frac{8 \times 10^4}{2 \times 10^{-2}}$

4×10^6; 4,000,000

87. $\frac{12 \times 10^{-5} \times 5 \times 10^4}{4 \times 10^3 \times 6 \times 10^{-2}}$

2.5×10^{-2}; 0.025

88. $\frac{2.5 \times 10^5 \times 4.8 \times 10^{-4}}{7.5 \times 10^8 \times 1.6 \times 10^{-5}}$

1×10^{-2}; 0.01

89. A computer can perform 466,000,000 calculations per second. How many calculations can it perform per minute? Per hour?

2.796×10^{10} calculations; 1.6776×10^{12} calculations

90. In theory, there are 1×10^9 possible Social Security numbers. The population of the United States is about 3×10^8. How many Social Security numbers are available for each person? (*Source:* U.S. Census Bureau.)

about 3.3

Mixed Review Exercises

Perform each indicated operation. Assume that all variables represent nonzero real numbers.

91. $19^0 - 3^0$

0

92. $(3p)^4(3p^{-7})$

$\frac{243}{p^3}$

93. 7^{-2}

$\frac{1}{49}$

94. $(-7 + 2k)^2$

$49 - 28k + 4k^2$

95. $\frac{2y^3 + 17y^2 + 37y + 7}{2y + 7}$

$y^2 + 5y + 1$

96. $\left(\frac{6r^2s}{5}\right)^4$

$\frac{1296r^8s^4}{625}$

97. $-m^5(8m^2 + 10m + 6)$

$-8m^7 - 10m^6 - 6m^5$

98. $\left(\frac{1}{2}\right)^{-5}$

32

99. $(25x^2y^3 - 8xy^2 + 15x^3y) \div (5x)$

$5xy^3 - \frac{8y^2}{5} + 3x^2y$

100. $(6r^{-2})^{-1}$

$\frac{r^2}{6}$

101. $(2x + y)^3$

$8x^3 + 12x^2y + 6xy^2 + y^3$

102. $2^{-1} + 4^{-1}$

$\frac{3}{4}$

103. $(a + 2)(a^2 - 4a + 1)$

$a^3 - 2a^2 - 7a + 2$

104. $(5y^3 - 8y^2 + 7) - (-3y^3 + y^2 + 2)$

$8y^3 - 9y^2 + 5$

105. $(2r + 5)(5r - 2)$

$10r^2 + 21r - 10$

106. $(12a + 1)(12a - 1)$

$144a^2 - 1$

107. What polynomial represents the area of this rectangle? The perimeter?

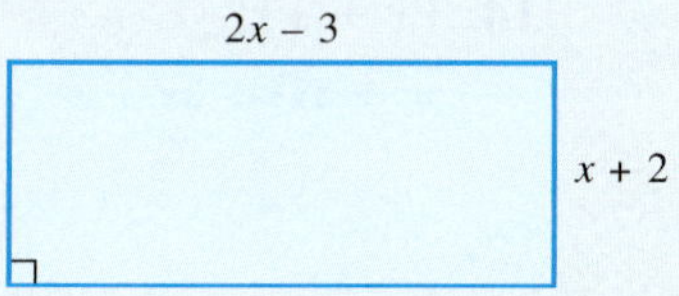

$2x^2 + x - 6;\ 6x - 2$

108. What polynomial represents the perimeter of this square? The area?

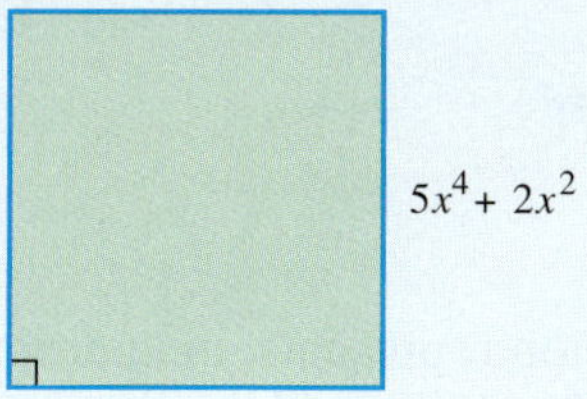

$20x^4 + 8x^2;\ 25x^8 + 20x^6 + 4x^4$

109. **CONCEPT CHECK** One of your friends in class simplified

$$\frac{6x^2 - 12x}{6} \quad \text{as} \quad x^2 - 12x.$$

What Went Wrong? Give the correct answer.

The friend wrote the second term of the quotient as $-12x$ rather than $-2x$. Here is the correct method:

$$\frac{6x^2 - 12x}{6} = \frac{6x^2}{6} - \frac{12x}{6} = x^2 - 2x.$$

110. **CONCEPT CHECK** What polynomial, when multiplied by $6m^2n$, gives the following product?

$$12m^3n^2 + 18m^6n^3 - 24m^2n^2$$

$2mn + 3m^4n^2 - 4n$

Chapter 5 Test

CHAPTER Test Prep VIDEO *The Chapter Test Prep Videos with test solutions are available on DVD, in MyMathLab, and on YouTube—search "LialIntroductoryAlg" and click on "Channels."*

For each polynomial, combine like terms when possible and write the polynomial in descending powers of the variable. Give the degree of the simplified polynomial. Decide whether the simplified polynomial is a monomial, *a* binomial, *a* trinomial, *or* none of these.

1. $5x^2 + 8x - 12x^2$

$-7x^2 + 8x$; 2; binomial

2. $13n^3 - n^2 + n^4 + 3n^4 - 9n^2$

$4n^4 + 13n^3 - 10n^2$; 4; trinomial

Perform the indicated operations.

3. $(5t^4 - 3t^2 + 7t + 3) - (t^4 - t^3 + 3t^2 + 8t + 3)$

$4t^4 + t^3 - 6t^2 - t$

4. $(2y^2 - 8y + 8) + (-3y^2 + 2y + 3) - (y^2 + 3y - 6)$

$-2y^2 - 9y + 17$

5. Subtract.

$$\begin{array}{r} 9t^3 - 4t^2 + 2t + 2 \\ 9t^3 + 8t^2 - 3t - 6 \\ \hline \end{array}$$

$-12t^2 + 5t + 8$

6. $(-2)^3(-2)^2$

-32

7. $\left(\frac{6}{m^2}\right)^3 \quad (m \neq 0)$

$\frac{216}{m^6}$

8. $3x^2(-9x^3 + 6x^2 - 2x + 1)$

$-27x^5 + 18x^4 - 6x^3 + 3x^2$

9. $(2r - 3)(r^2 + 2r - 5)$

$2r^3 + r^2 - 16r + 15$

10. $(t - 8)(t + 3)$

$t^2 - 5t - 24$

11. $(4x + 3y)(2x - y)$

$8x^2 + 2xy - 3y^2$

12. $(5x - 2y)^2$

$25x^2 - 20xy + 4y^2$

13. $(10v + 3w)(10v - 3w)$

$100v^2 - 9w^2$

14. $(x + 1)^3$

$x^3 + 3x^2 + 3x + 1$

15. What expression represents, in appropriate units, the perimeter of this square? The area?

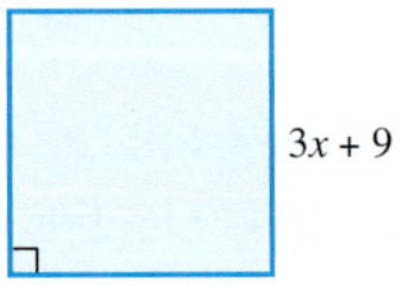

$12x + 36$; $9x^2 + 54x + 81$

16. Determine whether each expression represents a number that is *positive, negative,* or *zero.*

(a) 3^{-4} positive

(b) $(-3)^4$ positive

(c) -3^4 negative

(d) 3^0 positive

(e) $(-3)^0 - 3^0$ zero

(f) $(-3)^{-3}$ negative

Evaluate each expression.

17. 5^{-4}

$\frac{1}{625}$

18. $(-3)^0 + 4^0$

2

19. $4^{-1} + 3^{-1}$

$\frac{7}{12}$

Perform the indicated operations. In Exercises 20 and 21, write each answer using only positive exponents. Assume that variables represent nonzero numbers.

20. $\dfrac{8^{-1} \cdot 8^4}{8^{-2}}$

8^5

21. $\dfrac{(x^{-3})^{-2}(x^{-1}y)^2}{(xy^{-2})^2}$

x^2y^6

22. $\dfrac{8y^3 - 6y^2 + 4y + 10}{2y}$

$4y^2 - 3y + 2 + \frac{5}{y}$

23. $(-9x^2y^3 + 6x^4y^3 + 12xy^3) \div (3xy)$

$-3xy^2 + 2x^3y^2 + 4y^2$

24. $\dfrac{2x^2 + x - 36}{x - 4}$

$2x + 9$

25. $(3x^3 - x + 4) \div (x - 2)$

$3x^2 + 6x + 11 + \frac{26}{x - 2}$

26. Write each number in scientific notation.

(a) 344,000,000,000

3.44×10^{11}

(b) 0.00000557

5.57×10^{-6}

27. Write each number without exponents.

(a) 2.96×10^7

29,600,000

(b) 6.07×10^{-8}

0.0000000607

28. A satellite galaxy of our own Milky Way, known as the Large Magellanic Cloud, is ***1000*** light-years across. A light-year is equal to ***5,890,000,000,000*** mi. (*Source:* "Images of Brightest Nebula Unveiled," *USA Today*.)

(a) Write the two boldface italic numbers in scientific notation.

1×10^3; 5.89×10^{12}

(b) How many miles across is the Large Magellanic Cloud?

5.89×10^{15} mi

Chapters R–5 Cumulative Review Exercises

Perform each operation.

1. $\frac{2}{3}+\frac{1}{8}$

$\frac{19}{24}$

2. $\frac{7}{4}-\frac{9}{5}$

$-\frac{1}{20}$

3. $8.32 - 4.6$

3.72

4. 7.21×8.6

62.006

5. A retailer has \$34,000 invested in her business. She finds that last year she earned 5.4% on this investment. How much did she earn?

\$1836

Find the value of each expression for $x = -2$ and $y = 4$.

6. $\frac{4x-2y}{x+y}$

-8

7. $x^3 - 4xy$

24

Perform the indicated operations.

8. $\frac{(-13+15)-(3+2)}{6-12}$

$\frac{1}{2}$

9. $-7-3[2+(5-8)]$

-4

Decide what property justifies each statement.

10. $(9+2)+3=9+(2+3)$

associative property

11. $-7+7=0$

inverse property

12. $6(4+2)=6(4)+6(2)$

distributive property

Solve each equation.

13. $2x-7x+8x=30$

$\{10\}$

14. $2-3(t-5)=4+t$

$\left\{\frac{13}{4}\right\}$

15. $2(5h+1)=10h+4$

$\emptyset$

16. $d = rt$ for r

$r=\frac{d}{t}$

17. $\frac{x}{5}=\frac{x-2}{7}$

$\{-5\}$

18. $\frac{1}{3}p-\frac{1}{6}p=-2$

$\{-12\}$

19. $0.05x+0.15(50-x)=5.50$

$\{20\}$

20. $4-(3x+12)=(2x-9)-(5x-1)$

{all real numbers}

Solve each problem.

21. A 1-oz mouse takes about 16 times as many breaths as does a 3-ton elephant. (*Source: Dinosaurs, Spitfires, and Sea Dragons*, McGowan, C., Harvard University Press.) If the two animals take a combined total of 170 breaths per minute, how many breaths does each take during that time period?

mouse: 160; elephant: 10

22. If a number is subtracted from 8 and this difference is tripled, the result is three times the number. Find this number to learn how many times a dolphin rests during a 24-hr period.

4

Solve each inequality. Write the solution set in interval notation.

23. $-8x \leq -80$

$[10, \infty)$

24. $-2(x + 4) > 3x + 6$

$\left(-\infty, -\frac{14}{5}\right)$

25. $-3 \leq 2x + 5 < 9$

$[-4, 2)$

Given $2x - 3y = -6$, *find the following.*

26. The intercepts of the graph

$(0, 2)$ and $(-3, 0)$

27. The graph of the equation

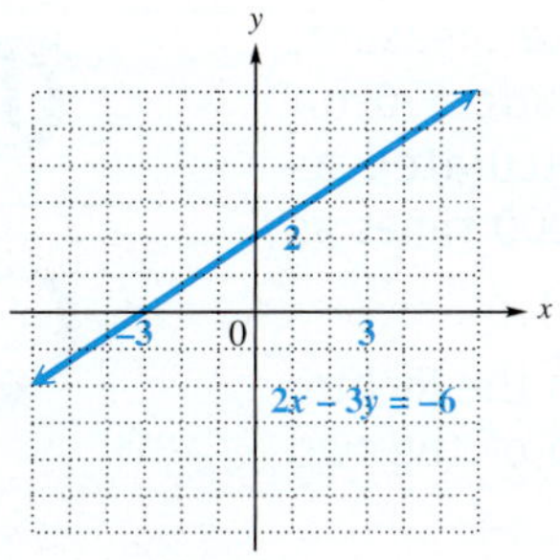

28. The slope of the line

$\frac{2}{3}$

Solve the system using the method indicated.

29. $y = 2x + 5$
$x + y = -4$ (Substitution)

$\{(-3, -1)\}$

30. $3x + 2y = 2$
$2x + 3y = -7$ (Elimination)

$\{(4, -5)\}$

Evaluate each expression.

31. $4^{-1} + 3^0$

$\frac{5}{4}$, or $1\frac{1}{4}$

32. $2^{-4} \cdot 2^5$

2

33. $\frac{8^{-5} \cdot 8^7}{8^2}$

1

34. Write $\frac{(a^{-3}b^2)^2}{(2a^{-4}b^{-3})^{-1}}$ with positive exponents only.

$\frac{2b}{a^{10}}$

35. Write 34,500 in scientific notation.

3.45×10^4

Perform the indicated operations.

36. $(7x^3 - 12x^2 - 3x + 8) + (6x^2 + 4) - (-4x^3 + 8x^2 - 2x - 2)$ $11x^3 - 14x^2 - x + 14$

37. $6x^5(3x^2 - 9x + 10)$

$18x^7 - 54x^6 + 60x^5$

38. $(7x + 4)(9x + 3)$

$63x^2 + 57x + 12$

39. $(5x + 8)^2$

$25x^2 + 80x + 64$

40. $\frac{y^3 - 3y^2 + 8y - 6}{y - 1}$

$y^2 - 2y + 6$

Math in the Media

FLOODS, HURRICANES, AND EARTHQUAKES, OH MY!

Charles F. Richter devised a scale in 1935 to compare the intensities, or relative powers, of earthquakes. The **intensity** of an earthquake is measured relative to the intensity of a standard **zero-level** earthquake of intensity I_0. The relationship is equivalent to $I = I_0 \times 10^R$, where R is the **Richter scale** measure. For example, if an earthquake has magnitude 5.0 on the Richter scale, then its intensity is calculated as $I = I_0 \times 10^{5.0} = I_0 \times 100{,}000$, which is 100,000 times as intense as a zero-level earthquake.

To compare an earthquake that measures 8.0 on the Richter scale to one that measures 5.0, find the ratio of the two intensities.

$$\frac{\text{intensity } 8.0}{\text{intensity } 5.0} = \frac{I_0 \times 10^{8.0}}{I_0 \times 10^{5.0}} = \frac{10^8}{10^5} = 10^{8-5} = 10^3 = 1000$$

Therefore, an earthquake that measures 8.0 is 1000 times as intense as one that measures 5.0.

The Gazette

Quake rattles Iowans

Chances of 'big one' happening here remote UNI professor says

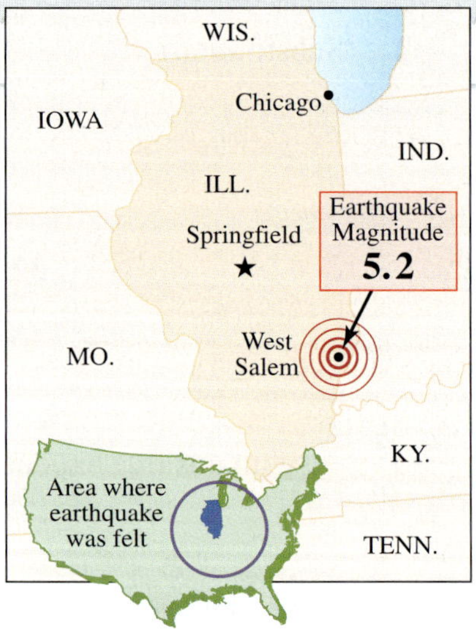

Source: WSRI; USGS.

Year	Earthquake	Richter Scale Measurement
2011	Northeast Japan	9.0
2005	Northern Sumatra, Indonesia	8.6
2004	West coast of Northern Sumatra	9.1
2003	Southeastern Iran	6.6
1998	Balleny Islands region	8.1
1906	San Francisco, CA	7.7

Source: U.S. Geological Survey.

1. Compare the intensity of the 2004 west coast of northern Sumatra earthquake to that of the 1998 Balleny Islands region earthquake.

   The 2004 quake was 10 times as intense as the 1998 quake.

2. Compare the intensity of the 2005 northern Sumatra, Indonesia earthquake to that of the 2003 southeastern Iran earthquake.

   The 2005 quake was 100 times as intense as the 2003 quake.

3. Compare the intensity of the 2011 northeast Japan earthquake to that of the 1906 San Francisco earthquake, the most powerful to strike the United States. (*Hint*: Use the exponential key of a scientific calculator to compute the required power of 10.)

   The 2011 quake was about 20 times as intense as the 1906 quake.

4. Suppose an earthquake measures a value of x on the Richter scale. How would the intensity of a second earthquake compare if its Richter scale measure is $x + 4.0$? How would it compare if its Richter scale measure is $x - 1.0$?

   $10^4 = 10{,}000$ times as intense; $10^{-1} = \frac{1}{10}$ times as intense.

6 Factoring and Applications

Formulas associated with the mathematicians Pythagoras (c. 380–300 B.C.) and Galileo (1564–1642) are used in applications that involve *factoring* polynomials, the subject of this chapter.

6.1 Factors; The Greatest Common Factor

OBJECTIVES

1. Find the greatest common factor of a list of numbers.
2. Find the greatest common factor of a list of variable terms.
3. Factor out the greatest common factor.
4. Factor by grouping.

Teaching Tip
The greatest common factor is often confused with the least common multiple (as in common denominator). Distinguish clearly between these two concepts.

To **factor** a number means to write it as the product of two or more numbers. The product is the **factored form** of the number. Consider an example.

$$12 = \underbrace{\overbrace{6 \cdot 2}^{\text{Factors}}}_{\text{Factored form}}$$

Factoring is a process that "undoes" multiplying. We multiply $6 \cdot 2$ to get 12, but we factor 12 by writing it as $6 \cdot 2$.

OBJECTIVE 1 Find the greatest common factor of a list of numbers. An integer that is a factor of two or more integers is a **common factor** of those integers. For example, 6 is a common factor of 18 and 24 because 6 is a factor of both 18 and 24. Other common factors of 18 and 24 are 1, 2, and 3.

The **greatest common factor (GCF)** of a list of integers is the largest common factor of those integers. This means 6 is the greatest common factor of 18 and 24, since it is the largest of their common factors.

Note

Factors of a number are also ***divisors*** of the number. The ***greatest common factor*** is the same as the ***greatest common divisor.*** Here are some divisibility rules for deciding what numbers divide into a given number.

A Whole Number Divisible by	Must Have the Following Property
2	Ends in 0, 2, 4, 6, or 8
3	Sum of digits divisible by 3
4	Last two digits form a number divisible by 4
5	Ends in 0 or 5
6	Divisible by both 2 and 3
8	Last three digits form a number divisible by 8
9	Sum of digits divisible by 9
10	Ends in 0

EXAMPLE 1 Finding the Greatest Common Factor for Numbers

Find the greatest common factor for each list of numbers.

(a) 30, 45

$$30 = 2 \cdot 3 \cdot 5$$
$$45 = 3 \cdot 3 \cdot 5$$
Write the prime factored form of each number.

Use each prime the* least *number of times it appears in* all *the factored forms. There is no 2 in the prime factored form of 45, so there will be no 2 in the greatest common factor. The least number of times 3 appears in all the factored forms is 1. The least number of times 5 appears is also 1.

$$\text{GCF} = 3^1 \cdot 5^1 = 15 \qquad 3^1 = 3 \text{ and } 5^1 = 5.$$

Continued on Next Page

Teaching Tip
When writing numbers in prime factored form, it may help some students to line up like factors vertically.

$$\begin{aligned} 30 &= 2 \cdot 3 \cdot 5 \\ 45 &= \quad\;\; 3 \cdot 5 \cdot 3 \\ \text{GCF} &= \quad\;\; 3 \cdot 5 \quad = 15 \end{aligned}$$

(b) 72, 120, 432

$$72 = 2 \cdot 2 \cdot 2 \cdot 3 \cdot 3$$
$$120 = 2 \cdot 2 \cdot 2 \cdot 3 \cdot 5$$
$$432 = 2 \cdot 2 \cdot 2 \cdot 2 \cdot 3 \cdot 3 \cdot 3$$

Write the prime factored form of each number.

The least number of times 2 appears in all the factored forms is 3, and the least number of times 3 appears is 1. There is no 5 in the prime factored form of either 72 or 432.

$$\text{GCF} = 2^3 \cdot 3^1 = 24$$ $2^3 = 8$ and $3^1 = 3$.

(c) 10, 11, 14

$$10 = 2 \cdot 5$$
$$11 = 11$$
$$14 = 2 \cdot 7$$

Write the prime factored form of each number.

There are no primes common to all three numbers, so the GCF is 1.

Work Problem 1 at the Side. ▶

OBJECTIVE 2 Find the greatest common factor of a list of variable terms. The terms x^4, x^5, x^6, and x^7 have x^4 as the greatest common factor because the least exponent on the variable x in the factored forms is 4.

$$x^4 = 1 \cdot x^4, \quad x^5 = x \cdot x^4, \quad x^6 = x^2 \cdot x^4, \quad x^7 = x^3 \cdot x^4$$
$$\text{GCF} = x^4$$

Note

The exponent on a variable in the GCF is the* least *exponent that appears on that variable in* all *the terms.

EXAMPLE 2 Finding the Greatest Common Factor for Variable Terms

Find the greatest common factor for each list of terms.

(a) $21m^7$, $18m^6$, $45m^8$, $24m^5$

$$21m^7 = 3 \cdot 7 \cdot m^7$$
$$18m^6 = 2 \cdot 3 \cdot 3 \cdot m^6$$
$$45m^8 = 3 \cdot 3 \cdot 5 \cdot m^8$$
$$24m^5 = 2 \cdot 2 \cdot 2 \cdot 3 \cdot m^5$$

Here, 3 is the greatest common factor of the coefficients 21, 18, 45, and 24. The least exponent on m is 5.

$$\text{GCF} = 3m^5$$

(b) x^4y^2, x^7y^5, x^3y^7, y^{15}

$$x^4y^2 = x^4 \cdot y^2$$
$$x^7y^5 = x^7 \cdot y^5$$
$$x^3y^7 = x^3 \cdot y^7$$
$$y^{15} = y^{15}$$

There is no x in the last term, y^{15}, so x will not appear in the greatest common factor. There is a y in each term, however, and 2 is the least exponent on y.

$$\text{GCF} = y^2$$

---

1 Find the greatest common factor for each list of numbers.

GS **(a)** 30, 20, 15

$30 = 2 \cdot 3 \cdot 5$

$20 = 2 \cdot$ ____ $\cdot$ ____

$15 = 3 \cdot$ ____

GCF = ____

(b) 42, 28, 35

(c) 12, 18, 26, 32

(d) 10, 15, 21

Answers

1. **(a)** 2; 5; 5; 5 **(b)** 7 **(c)** 2 **(d)** 1

2 Find the greatest common factor for each list of terms.

GS **(a)** $6m^4, 9m^2, 12m^5$

$6m^4 = 2 \cdot \_\_\_\_ \cdot m^4$

$9m^2 = 3 \cdot \_\_\_\_ \cdot \_\_\_\_$

$12m^5 = 2 \cdot 2 \cdot \_\_\_\_ \cdot \_\_\_\_$

GCF = ____

(b) $12p^5, 18q^4$

(c) y^4z^2, y^6z^8, z^9

(d) $12p^{11}, 17q^5$

Teaching Tip

Point out that factoring out the GCF and using the distributive property to multiply are inverse operations.

Answers

2. **(a)** 3; 3; m^2; 3; m^5; $3m^2$ **(b)** 6 **(c)** z^2 **(d)** 1

Finding the Greatest Common Factor (GCF)

Step 1 **Factor.** Write each number in prime factored form.

Step 2 **List common factors.** List each prime number or each variable that is a factor of every term in the list. (If a prime does not appear in one of the prime factored forms, it *cannot* appear in the greatest common factor.)

Step 3 **Choose least exponents.** Use as exponents on the common prime factors the *least* exponents from the prime factored forms.

Step 4 **Multiply** the primes from Step 3. If there are no primes left after Step 3, the greatest common factor is 1.

◀ Work Problem 2 at the Side.

OBJECTIVE 3 Factor out the greatest common factor. Writing a polynomial (a sum) in factored form as a product is called **factoring** the polynomial. For example, the polynomial

$$3m + 12$$

has two terms, $3m$ and 12. The greatest common factor of these two terms is 3. We can write $3m + 12$ so that each term is a product with 3 as one factor.

$$3m + 12$$

$$= 3 \cdot m + 3 \cdot 4 \quad \text{GCF} = 3$$

$$= 3(m + 4) \quad \text{Distributive property}$$

The factored form of $3m + 12$ is $3(m + 4)$. This process is called **factoring out the greatest common factor.**

CAUTION

The polynomial $3m + 12$ is *not* in factored form when written as

$3 \cdot m + 3 \cdot 4.$ Not in factored form

The terms are factored, but the polynomial is not. The factored form of $3m + 12$ is the *product*

$3(m + 4).$ In factored form

EXAMPLE 3 Factoring Out the Greatest Common Factor

Write in factored form by factoring out the greatest common factor.

(a) $5y^2 + 10y$

$= 5y(y) + 5y(2)$ GCF $= 5y$

$= 5y(y + 2)$ Distributive property

CHECK Multiply the factored form.

$5y(y + 2)$

$= 5y(y) + 5y(2)$ Distributive property

$= 5y^2 + 10y$ ✓ Original polynomial

Continued on Next Page

(b) $20m^5 + 10m^4 - 15m^3$

$= \mathbf{5m^3}(4m^2) + \mathbf{5m^3}(2m) - \mathbf{5m^3}(3)$ GCF $= 5m^3$

$= \mathbf{5m^3}(4m^2 + 2m - 3)$ Factor out $5m^3$.

CHECK $5m^3(4m^2 + 2m - 3)$

$= 5m^3(4m^2) + 5m^3(2m) + 5m^3(-3)$ Distributive property

$= 20m^5 + 10m^4 - 15m^3$ ✓ Original polynomial

(c) $x^5 + x^3$

$= \mathbf{x^3}(x^2) + \mathbf{x^3}(\mathbf{1})$ GCF $= x^3$

$= \mathbf{x^3}(x^2 + \mathbf{1})$ Don't forget the 1.

Check mentally by distributing x^3 over each term inside the parentheses.

(d) $20m^7p^2 - 36m^3p^4$

$= 4m^3p^2(5m^4) - 4m^3p^2(9p^2)$ GCF $= 4m^3p^2$

$= 4m^3p^2(5m^4 - 9p^2)$ Factor out $4m^3p^2$.

Check mentally by distributing $4m^3p^2$ over each term inside the parentheses.

CAUTION

Be sure to include the 1 in a problem like **Example 3(c)**. ***Check that the factored form can be multiplied out to give the original polynomial.***

Work Problem **3** at the Side. ▶

EXAMPLE 4 Factoring Out a Negative Common Factor

Write $-8x^4 + 16x^3 - 4x^2$ in factored form.

We can factor out either $4x^2$ or $-4x^2$ here. We usually factor out $-4x^2$ so that the coefficient of the first term in the trinomial factor will be positive.

$-8x^4 + 16x^3 - 4x^2$ Be careful with signs.

$= \mathbf{-4x^2}(2x^2) \mathbf{- 4x^2}(-4x) \mathbf{- 4x^2}(\mathbf{1})$ $-4x^2$ is a common factor.

$= \mathbf{-4x^2}(2x^2 - 4x + \mathbf{1})$ Factor out $-4x^2$.

CHECK $-4x^2(2x^2 - 4x + 1)$

$= -4x^2(2x^2) - 4x^2(-4x) - 4x^2(1)$ Distributive property

$= -8x^4 + 16x^3 - 4x^2$ ✓ Original polynomial

Work Problem **4** at the Side. ▶

Note

Whenever we factor a polynomial in which the coefficient of the first term is negative, we will factor out the negative common factor, even if it is just -1. However, it would also be correct to factor out $4x^2$ in **Example 4** to obtain

$$4x^2(-2x^2 + 4x - 1).$$

3 Write in factored form by factoring out the greatest common factor.

GS **(a)** $4x^2 + 6x$

$= \_\_\_(\_\_\_) + 2x(\_\_\_)$

$= \_\_\_(\_\_\_ + \_\_\_)$

(b) $10y^5 - 8y^4 + 6y^2$

GS **(c)** $m^7 + m^9$

$= \_\_\_(\_\_\_) + \_\_\_(m^2)$

$= \_\_\_(\_\_\_ + \_\_\_)$

(d) $15x^3 - 10x^2 + 5x$

(e) $8p^5q^2 + 16p^6q^3 - 12p^4q^7$

4 Write

$$-14a^3b^2 - 21a^2b^3 + 7ab$$

in factored form by factoring out a negative common factor.

Answers

3. **(a)** $2x$; $2x$; 3; $2x$; $2x$; 3
(b) $2y^2(5y^3 - 4y^2 + 3)$
(c) m^7; 1; m^7; m^7; 1; m^2
(d) $5x(3x^2 - 2x + 1)$
(e) $4p^4q^2(2p + 4p^2q - 3q^5)$

4. $-7ab(2a^2b + 3ab^2 - 1)$

5 Write in factored form by factoring out the greatest common factor.

GS **(a)** $r(t-4)+5(t-4)$
$=(t-4)(\_\_\_\_\_)$

(b) $x(x+2)+7(x+2)$

(c) $y^2(y+2)-3(y+2)$

(d) $x(x-1)-5(x-1)$

Teaching Tip
You will need to provide several examples of factoring by grouping to help students understand this concept.

Answers
5. **(a)** $r+5$ **(b)** $(x+2)(x+7)$
(c) $(y+2)(y^2-3)$ **(d)** $(x-1)(x-5)$

EXAMPLE 5 Factoring Out a Common Binomial Factor

Write in factored form by factoring out the greatest common factor.

Same

(a) $a(a+3)+4(a+3)$ — The binomial $a+3$ is the greatest common factor.
$=(a+3)(a+4)$ — Factor out $a+3$.

(b) $x^2(x+1)-5(x+1)$
$=(x+1)(x^2-5)$ — Factor out $x+1$.

◀ Work Problem **5** at the Side.

Note

In factored forms like those in **Example 5,** the order of the factors does not matter because of the commutative property of multiplication.

$(a+3)(a+4)$ can also be written $(a+4)(a+3)$.

OBJECTIVE 4 Factor by grouping. ***When a polynomial has four terms, common factors can sometimes be used to factor by grouping.***

EXAMPLE 6 Factoring by Grouping

Factor by grouping.

(a) $2x+6+ax+3a$

Group the first two terms and the last two terms, since the first two terms have a common factor of 2 and the last two terms have a common factor of a.

$2x+6+ax+3a$
$=(2x+6)+(ax+3a)$ — Group the terms.
$=2(x+3)+a(x+3)$ — Factor each group.

The expression is still not in factored form because it is the ***sum*** of two terms. Now, however, $x+3$ is a common factor and can be factored out.

$=2(x+3)+a(x+3)$ — $x+3$ is a common factor.
$=(x+3)(2+a)$ — Factor out $x+3$.

($(2+a)(x+3)$ is also correct.)

The final result is in factored form because it is a ***product.***

CHECK $(x+3)(2+a)$
$=x(2)+x(a)+3(2)+3(a)$ — Multiply using the FOIL method (**Section 5.3**).
$=2x+ax+6+3a$ — Simplify.
$=2x+6+ax+3a$ ✓ — Rearrange terms to obtain the original polynomial.

Continued on Next Page

(b) $6ax + 24x + a + 4$

$= (6ax + 24x) + (a + 4)$ Group the terms.

$= 6x(a + 4) + 1(a + 4)$ Factor each group.

Remember the 1.

$= (a + 4)(6x + 1)$ Factor out $a + 4$.

CHECK $(a + 4)(6x + 1)$

$= 6ax + a + 24x + 4$ FOIL method

$= 6ax + 24x + a + 4$ ✓ Rearrange terms to obtain the original polynomial.

(c) $2x^2 - 10x + 3xy - 15y$

$= (2x^2 - 10x) + (3xy - 15y)$ Group the terms.

$= 2x(x - 5) + 3y(x - 5)$ Factor each group.

$= (x - 5)(2x + 3y)$ Factor out the common factor, $x - 5$.

CHECK $(x - 5)(2x + 3y)$

$= 2x^2 + 3xy - 10x - 15y$ FOIL method

$= 2x^2 - 10x + 3xy - 15y$ ✓ Original polynomial

(d) $t^3 + 2t^2 - 3t - 6$

Write a + sign between the groups.

$= (t^3 + 2t^2) + (-3t - 6)$ Group the terms.

$= t^2(t + 2) - 3(t + 2)$ Factor out -3 so there is a common factor, $t + 2$; $-3(t + 2) = -3t - 6$.

Be careful with signs.

$= (t + 2)(t^2 - 3)$ Factor out $t + 2$.

Check by multiplying using the FOIL method.

CAUTION

Be careful with signs when grouping in a problem like **Example 6(d).** It is wise to check the factoring in the second step, as shown in the example side comment, before continuing.

Work Problem **6** at the Side. ▶

Factoring a Polynomial with Four Terms by Grouping

Step 1 **Group terms.** Collect the terms into two groups so that each group has a common factor.

Step 2 **Factor within groups.** Factor out the greatest common factor from each group.

Step 3 **Factor the entire polynomial.** Factor a common binomial factor from the results of Step 2.

Step 4 **If necessary, rearrange terms.** If Step 2 does not result in a common binomial factor, try a different grouping.

6 Factor by grouping.

GS **(a)** $pq + 5q + 2p + 10$

$= (\_\_ + 5q) + (\_\_ + 10)$

$= \_\_(p + 5) + \_\_(p + 5)$

$= (\_\_\_\_)(\_\_\_\_)$

(b) $2xy + 3y + 2x + 3$

(c) $2a^2 - 4a + 3ab - 6b$

(d) $x^3 + 3x^2 - 5x - 15$

Answers

6. **(a)** pq; $2p$; q; 2; $p + 5$; $q + 2$
(b) $(2x + 3)(y + 1)$
(c) $(a - 2)(2a + 3b)$
(d) $(x + 3)(x^2 - 5)$

7 Factor by grouping.

(a) $6y^2 - 20w + 15y - 8yw$

(b) $9mn - 4 + 12m - 3n$

(c) $12p^2 - 28q - 16pq + 21p$

EXAMPLE 7 **Rearranging Terms before Factoring by Grouping**

Factor by grouping.

(a) $10x^2 - 12y + 15x - 8xy$

Factoring out the common factor 2 from the first two terms and the common factor x from the last two terms gives the following.

$$10x^2 - 12y + 15x - 8xy$$
$$= 2(5x^2 - 6y) + x(15 - 8y)$$

This does not lead to a common factor, so we try rearranging the terms. There is usually more than one way to do this. We try the following.

$$10x^2 - 12y + 15x - 8xy$$

$= 10x^2 - 8xy - 12y + 15x$ Commutative property

$= (10x^2 - 8xy) + (-12y + 15x)$ Group the terms.

$= 2x(5x - 4y) + 3(-4y + 5x)$ Factor each group.

$= 2x(5x - 4y) + 3(5x - 4y)$ Rewrite $-4y + 5x$.

$= (5x - 4y)(2x + 3)$ Factor out $5x - 4y$.

CHECK $(5x - 4y)(2x + 3)$

$= 10x^2 + 15x - 8xy - 12y$ FOIL method

$= 10x^2 - 12y + 15x - 8xy$ ✓ Original polynomial

(b) $2xy + 12 - 3y - 8x$

We must rearrange the terms to get two groups that each have a common factor. Trial and error suggests the following grouping.

$$2xy + 12 - 3y - 8x$$

Write a + sign between the two groups.

$= (2xy - 3y) + (-8x + 12)$ Group the terms.

$= y(2x - 3) - 4(2x - 3)$ Factor each group; $-4(2x - 3) = -8x + 12$

Be careful with signs.

$= (2x - 3)(y - 4)$ Factor out $2x - 3$.

Since the quantities in parentheses in the second step must be the same, we factored out -4 rather than 4.

CHECK $(2x - 3)(y - 4)$

$= 2xy - 8x - 3y + 12$ FOIL method

$= 2xy + 12 - 3y - 8x$ ✓ Original polynomial

CAUTION

Use negative signs carefully when grouping, as in **Example 7(b)**, or a sign error will occur. ***Always check by multiplying.***

◀ **Work Problem 7 at the Side.**

Answers

7. **(a)** $(2y + 5)(3y - 4w)$
(b) $(3m - 1)(3n + 4)$
(c) $(3p - 4q)(4p + 7)$

6.1 Exercises

FOR EXTRA HELP

Download the MyDashBoard App

MyMathLab®

CONCEPT CHECK *Complete each statement.*

1. To factor a number or quantity means to write it as a <u>product</u>. Factoring is the opposite, or inverse, process of <u>multiplying</u>.

2. An integer or variable expression that is a factor of two or more terms is a <u>common</u> <u>factor</u>. For example, 12 (<u>*is*</u> / *is not*) a common factor of both 36 and 72 since it <u>divides</u> evenly into both integers.

Find the greatest common factor for each list of numbers. ***See Example 1.***

3. 12, 16 4

4. 18, 24 6

5. 40, 20, 4 4

6. 50, 30, 5 5

7. 18, 24, 36, 48 6

8. 15, 30, 45, 75 15

9. 4, 9, 12 1

10. 9, 16, 24 1

Find the greatest common factor for each list of terms. ***See Example 2.***

11. $16y$, 24 8

12. $18w$, 27 9

13. $30x^3$, $40x^6$, $50x^7$ $10x^3$

14. $60z^4$, $70z^8$, $90z^9$ $10z^4$

15. x^4y^3, xy^2 xy^2

16. a^4b^5, a^3b a^3b

17. $42ab^3$, $36a$, $90b$, $48ab$ 6

18. $45c^3d$, $75c$, $90d$, $105cd$ 15

19. $12m^3n^2$, $18m^5n^4$, $36m^8n^3$ $6m^3n^2$

20. $25p^5r^7$, $30p^7r^8$, $50p^5r^3$ $5p^5r^3$

CONCEPT CHECK *An expression is factored when it is written as a product, not a sum. Determine whether each expression is* factored *or* not factored.

21. $2k^2(5k)$ factored

22. $2k^2(5k + 1)$ factored

23. $2k^2 + (5k + 1)$ not factored

24. $(2k^2 + 5k) + 1$ not factored

25. **CONCEPT CHECK** A student factored $18x^3y^2 + 9xy$ as $9xy(2x^2y)$. **What Went Wrong?** Factor correctly.

$18x^3y^2 + 9xy = 9xy(2x^2y + 1)$; If a polynomial has two terms, the product of the factors must have two terms. $9xy(2x^2y) = 18x^3y^2$ is just one term.

26. How can we check an answer when we factor a polynomial?

Verify that we have factored completely. Then multiply the factors. The product should be the original polynomial.

Complete each factoring by writing each polynomial as the product of two factors. ***See Example 3.***

27. $9m^4 = 3m^2(\underline{3m^2})$

28. $12p^5 = 6p^3(\underline{2p^2})$

29. $-8z^9 = -4z^5(\underline{2z^4})$

30. $-15k^{11} = -5k^8(\underline{3k^3})$

31. $6m^4n^5 = 3m^3n(\underline{2mn^4})$

32. $27a^3b^2 = 9a^2b(\underline{3ab})$

33. $12y + 24 = 12(\underline{y + 2})$

34. $18p + 36 = 18(\underline{p + 2})$

35. $10a^2 - 20a = 10a(\underline{a - 2})$

36. $15x^2 - 30x = 15x(\underline{x - 2})$

37. $8x^2y + 12x^3y^2 = 4x^2y(\underline{2 + 3xy})$

38. $18s^3t^2 + 10st = 2st(\underline{9s^2t + 5})$

Write in factored form by factoring out the greatest common factor (or a negative common factor if the coefficient of the term of greatest degree is negative). ***See Examples 3–5.***

39. $x^2 - 4x$ — $x(x-4)$

40. $m^2 - 7m$ — $m(m-7)$

41. $6t^2 + 15t$ — $3t(2t+5)$

42. $8x^2 + 6x$ — $2x(4x+3)$

43. $m^3 - m^2$ — $m^2(m-1)$

44. $p^3 - p^2$ — $p^2(p-1)$

45. $-12x^3 - 6x^2$ — $-6x^2(2x+1)$

46. $-21b^3 + 7b^2$ — $-7b^2(3b-1)$

47. $65y^{10} + 35y^6$ — $5y^6(13y^4+7)$

48. $100a^5 + 16a^3$ — $4a^3(25a^2+4)$

49. $11w^3 - 100$ — no common factor (except 1)

50. $13z^5 - 80$ — no common factor (except 1)

51. $8mn^3 + 24m^2n^3$ — $8mn^3(1+3m)$

52. $19p^2y + 38p^2y^3$ — $19p^2y(1+2y^2)$

53. $-4x^3 + 10x^2 - 6x$ — $-2x(2x^2-5x+3)$

54. $-9z^3 + 6z^2 - 12z$ — $-3z(3z^2-2z+4)$

55. $13y^8 + 26y^4 - 39y^2$ — $13y^2(y^6+2y^2-3)$

56. $5x^5 + 25x^4 - 20x^3$ — $5x^3(x^2+5x-4)$

57. $45q^4p^5 + 36qp^6 + 81q^2p^3$ — $9qp^3(5q^3p^2+4p^3+9q)$

58. $125a^3z^5 + 60a^4z^4 - 85a^5z^2$ — $5a^3z^2(25z^3+12az^2-17a^2)$

59. $c(x+2) + d(x+2)$ — $(x+2)(c+d)$

60. $r(5-x) + t(5-x)$ — $(5-x)(r+t)$

61. $a^2(2a+b) - b(2a+b)$ — $(2a+b)(a^2-b)$

62. $3x(x^2+5) - y(x^2+5)$ — $(x^2+5)(3x-y)$

63. $q(p+4) - 1(p+4)$ — $(p+4)(q-1)$

64. $y^2(x-4) + 1(x-4)$ — $(x-4)(y^2+1)$

CONCEPT CHECK *Students often have difficulty when factoring by grouping because they are not able to tell when the polynomial is completely factored. For example,*

$$5y(2x-3) + 8t(2x-3) \quad \text{Not in factored form}$$

is not in factored form, because it is the ***sum*** *of two terms:* $5y(2x-3)$ *and* $8t(2x-3)$. *However, because* $2x-3$ *is a common factor of these two terms, the expression can now be factored.*

$$(2x-3)(5y+8t) \quad \text{In factored form}$$

The factored form is a ***product*** *of the two factors* $2x-3$ *and* $5y+8t$.

Determine whether each expression is in factored form *or is* not in factored form. *If it is not in factored form, factor it if possible.*

65. $8(7t+4) + x(7t+4)$ — not in factored form; $(7t+4)(8+x)$

66. $3r(5x-1) + 7(5x-1)$ — not in factored form; $(5x-1)(3r+7)$

67. $(8+x)(7t+4)$ — in factored form

68. $(3r+7)(5x-1)$ — in factored form

69. $18x^2(y+4) + 7(y-4)$ — not in factored form

70. $12k^3(s-3) + 7(s+3)$ — not in factored form

Factor by grouping. ***See Examples 6 and 7.***

71. $5m + mn + 20 + 4n$
$(5 + n)(m + 4)$

72. $ts + 5t + 2s + 10$
$(s + 5)(t + 2)$

73. $6xy - 21x + 8y - 28$
$(2y - 7)(3x + 4)$

74. $2mn - 8n + 3m - 12$
$(m - 4)(2n + 3)$

75. $3xy + 9x + y + 3$
$(y + 3)(3x + 1)$

76. $6n + 4mn + 3 + 2m$
$(3 + 2m)(2n + 1)$

77. $7z^2 + 14z - az - 2a$
$(z + 2)(7z - a)$

78. $2b^2 + 3b - 8ab - 12a$
$(2b + 3)(b - 4a)$

79. $18r^2 + 12ry - 3xr - 2xy$
$(3r + 2y)(6r - x)$

80. $5m^2 + 15mp - 2mr - 6pr$
$(m + 3p)(5m - 2r)$

81. $w^3 + w^2 + 9w + 9$
$(w + 1)(w^2 + 9)$

82. $y^3 + y^2 + 6y + 6$
$(y + 1)(y^2 + 6)$

83. $3a^3 + 6a^2 - 2a - 4$
$(a + 2)(3a^2 - 2)$

84. $10x^3 + 15x^2 - 8x - 12$
$(2x + 3)(5x^2 - 4)$

85. $16m^3 - 4m^2p^2 - 4mp + p^3$
$(4m - p^2)(4m^2 - p)$

86. $10t^3 - 2t^2s^2 - 5ts + s^3$
$(5t - s^2)(2t^2 - s)$

87. $y^2 + 3x + 3y + xy$
$(y + 3)(y + x)$

88. $m^2 + 14p + 7m + 2mp$
$(m + 7)(m + 2p)$

89. $2z^2 + 6w - 4z - 3wz$
$(z - 2)(2z - 3w)$

90. $2a^2 + 20b - 8a - 5ab$
$(a - 4)(2a - 5b)$

91. $5m - 6p - 2mp + 15$
$(5 - 2p)(m + 3)$

92. $7y - 9x - 3xy + 21$
$(7 - 3x)(y + 3)$

93. $18r^2 - 2ty + 12ry - 3rt$
$(3r + 2y)(6r - t)$

94. $12a^2 - 4bc + 16ac - 3ab$
$(3a + 4c)(4a - b)$

Relating Concepts (Exercises 95–98) For Individual or Group Work

In many cases, the choice of which pairs of terms to group when factoring by grouping can be made in different ways. To see this for ***Example 7(b), work Exercises 95–98 in order.***

95. Start with the polynomial from **Example 7(b),** $2xy + 12 - 3y - 8x$, and rearrange the terms as follows: $2xy - 8x - 3y + 12$. What property from **Section 1.7** allows this?
commutative property

96. Group the first two terms and the last two terms of the rearranged polynomial in **Exercise 95.** Then factor each group.
$2x(y - 4) - 3(y - 4)$

97. Is your result from **Exercise 96** in factored form? Explain your answer.
No, because it is not a product. It is the difference between $2x(y - 4)$ and $3(y - 4)$.

98. If your answer to **Exercise 97** is *no*, factor the polynomial. Is the result the same as the one shown for **Example 7(b)**?
$(y - 4)(2x - 3)$, or $(2x - 3)(y - 4)$; yes

6.2 Factoring Trinomials

OBJECTIVES

1. Factor trinomials with a coefficient of 1 for the second-degree term.
2. Factor such trinomials after factoring out the greatest common factor.

Using the FOIL method, we can find the product of the binomials $k - 3$ and $k + 1$.

$$(k-3)(k+1) = k^2 - 2k - 3 \qquad \text{Multiplying}$$

Suppose instead that we are given the polynomial $k^2 - 2k - 3$ and want to rewrite it as the product $(k-3)(k+1)$.

$$k^2 - 2k - 3 = (k-3)(k+1) \qquad \text{Factoring}$$

Recall from **Section 6.1** that this process is called *factoring* the polynomial. Factoring reverses or, "undoes," multiplying.

OBJECTIVE 1 Factor trinomials with a coefficient of 1 for the second-degree term. When factoring polynomials with integer coefficients, we use only integers in the factors. For example, we can factor $x^2 + 5x + 6$ by finding integers m and n such that

$$x^2 + 5x + 6 \quad \text{is written as} \quad (x + m)(x + n).$$

To find these integers m and n, we multiply the two binomials on the right.

$$\begin{aligned} &(x+m)(x+n) \\ &= x^2 + nx + mx + mn && \text{FOIL method} \\ &= x^2 + (n+m)x + mn && \text{Distributive property} \end{aligned}$$

Comparing this result with $x^2 + 5x + 6$ shows that we must find integers m and n having a sum of 5 and a product of 6.

Product of m and n is 6.

$$x^2 + 5x + 6 = x^2 + (n + m)x + mn$$

Sum of m and n is 5.

Because many pairs of integers have a sum of 5, it is best to begin by listing those pairs of integers whose product is 6. Both 5 and 6 are positive, so we consider only pairs in which both integers are positive.

◀ **Work Problem 1 at the Side.**

From **Margin Problem 1,** we see that the numbers 6 and 1 and the numbers 3 and 2 both have a product of 6, but only the pair 3 and 2 has a sum of 5. So 3 and 2 are the required integers.

$$x^2 + 5x + 6 \quad \text{is factored as} \quad (x + 3)(x + 2).$$

Check by multiplying the binomials using the FOIL method. ***Make sure that the sum of the outer and inner products produces the correct middle term.***

CHECK $\quad (x+3)(x+2) = x^2 + 5x + 6$ ✓ Correct

$3x$
$2x$
$5x$ Add.

This method of factoring can be used only for trinomials that have 1 as the coefficient of the second-degree (squared) term.

1 **(a)** Complete the table to find pairs of positive integers whose product is 6. Then find the sum of each pair.

Factors of 6	Sums of Factors
6, ___	6 + ___ = ___
___, 2	___ + 2 = ___

(b) Which pair of factors from the table in part (a) has a sum of 5?

Answers

1. **(a)** 1; 1; 7; 3; 3; 5 **(b)** 3, 2

EXAMPLE 1 **Factoring a Trinomial (All Positive Terms)**

Factor $m^2 + 9m + 14$.

Look for two integers whose product is 14 and whose sum is 9. List the pairs of integers whose products are 14, and examine the sums. Only positive integers are needed since all signs in $m^2 + 9m + 14$ are positive.

Factors of 14	Sums of Factors
14, 1	$14 + 1 = 15$
7, 2	$7 + 2 = \mathbf{9}$

Sum is 9.

From the table, 7 and 2 are the required integers, since $7 \cdot 2 = 14$ and $7 + 2 = 9$.

$$m^2 + 9m + 14 \quad \text{factors as} \quad (m + 7)(m + 2).$$

$(m + 2)(m + 7)$ is also correct.

CHECK $(m + 7)(m + 2)$

$= m^2 + 2m + 7m + 14$ FOIL method

$= m^2 + 9m + 14$ ✓ Original polynomial

Work Problem 2 at the Side. ▶

Note

Remember that because of the commutative property of multiplication, the order of the factors does not matter. ***Always check by multiplying.***

EXAMPLE 2 **Factoring a Trinomial (Negative Middle Term)**

Factor $x^2 - 9x + 20$.

Find two integers whose product is 20 and whose sum is -9. Since the numbers we are looking for have a *positive product* and a *negative sum,* we consider only pairs of negative integers.

Factors of 20	Sums of Factors
$-20, -1$	$-20 + (-1) = -21$
$-10, -2$	$-10 + (-2) = -12$
$\mathbf{-5, -4}$	$-5 + (-4) = \mathbf{-9}$

Sum is -9.

The required integers are -5 and -4.

$$x^2 - 9x + 20 \quad \text{factors as} \quad (x - 5)(x - 4).$$

The order of the factors does not matter.

CHECK $(x - 5)(x - 4)$

$= x^2 - 4x - 5x + 20$ FOIL method

$= x^2 - 9x + 20$ ✓ Original polynomial

Work Problem 3 at the Side. ▶

2 Factor each trinomial.

GS **(a)** $y^2 + 12y + 20$

Find two integers whose product is ____ and whose sum is ____. Complete the table.

Factors of 20	Sums of Factors
20, 1	$20 + 1 = 21$
10, ___	10 + ___ = ___
5, ___	5 + ___ = ___

Which pair of factors has the required sum? _______

Now factor the trinomial.

(b) $x^2 + 9x + 18$

3 Factor each trinomial.

GS **(a)** $t^2 - 12t + 32$

Find two integers whose product is ____ and whose sum is ____. Complete the table.

Factors of 32	Sums of Factors
$-32, -1$	$-32 + (-1) = -33$
-16, ___	-16 + (___) = ___
-8, ___	-8 + (___) = ___

Which pair of factors has the required sum? _______

Now factor the trinomial.

(b) $y^2 - 10y + 24$

Answers

2. **(a)** 20; 12; 2; 2; 12; 4; 4; 9; 10 and 2; $(y + 10)(y + 2)$
(b) $(x + 3)(x + 6)$

3. **(a)** 32; -12; -2; -2; -18; -4; -4; -12; -8 and -4; $(t - 8)(t - 4)$
(b) $(y - 6)(y - 4)$

4 Factor each trinomial.

(a) $z^2 + z - 30$

Factors of -30	Sums of Factors

(b) $x^2 + x - 42$

5 Factor each trinomial.

(a) $a^2 - 9a - 22$

Factors of -22	Sums of Factors

(b) $r^2 - 6r - 16$

Answers

4. **(a)** $(z + 6)(z - 5)$ **(b)** $(x + 7)(x - 6)$
5. **(a)** $(a - 11)(a + 2)$ **(b)** $(r - 8)(r + 2)$

EXAMPLE 3 Factoring a Trinomial (Negative Last (Constant) Term)

Factor $x^2 + x - 6$.

We must find two integers whose product is -6 and whose sum is 1 (since the coefficient of x, or $1x$, is 1). To get a *negative product,* the pairs of integers must have different signs.

Once we find the required pair, we can stop listing factors.

Factors of -6	Sums of Factors	
$6, -1$	$6 + (-1) = 5$	
$-6, 1$	$-6 + 1 = -5$	
$3, -2$	$3 + (-2) = 1$	Sum is 1.

The required integers are 3 and -2.

$$x^2 + x - 6 \quad \text{factors as} \quad (x + 3)(x - 2).$$

CHECK $(x + 3)(x - 2)$

$= x^2 - 2x + 3x - 6$ FOIL method

$= x^2 + x - 6$ ✓ Original polynomial

◀ Work Problem **4** at the Side.

EXAMPLE 4 Factoring a Trinomial (Two Negative Terms)

Factor $p^2 - 2p - 15$.

Find two integers whose product is -15 and whose sum is -2. Because the constant term, -15, is negative, we need pairs of integers with different signs.

Factors of -15	Sums of Factors	
$15, -1$	$15 + (-1) = 14$	
$-15, 1$	$-15 + 1 = -14$	
$5, -3$	$5 + (-3) = 2$	
$-5, 3$	$-5 + 3 = -2$	Sum is -2.

The required integers are -5 and 3.

$$p^2 - 2p - 15 \quad \text{factors as} \quad (p - 5)(p + 3).$$

CHECK Multiply $(p - 5)(p + 3)$ to obtain $p^2 - 2p - 15$. ✓

Note

In **Examples 1–4,** we listed factors in descending order (disregarding their signs) when we were looking for the required pair of integers. This helps avoid skipping the correct combination.

◀ Work Problem **5** at the Side.

Some trinomials cannot be factored using only integers. Such trinomials are **prime polynomials.**

EXAMPLE 5 **Deciding Whether Polynomials Are Prime**

Factor each trinomial.

(a) $x^2 - 5x + 12$

As in **Example 2,** both factors must be negative to give a positive product, 12, and a negative sum, −5. List pairs of negative integers whose product is 12, and examine the sums.

Factors of 12	Sums of Factors
$-12, -1$	$-12 + (-1) = -13$
$-6, -2$	$-6 + (-2) = -8$
$-4, -3$	$-4 + (-3) = -7$

No sum is −5.

None of the pairs of integers has a sum of −5. Therefore, the trinomial $x^2 - 5x + 12$ *cannot be factored using only integers.* It is a *prime polynomial.*

(b) $k^2 - 8k + 11$

There is no pair of integers whose product is 11 and whose sum is −8, so $k^2 - 8k + 11$ is a prime polynomial.

Work Problem 6 at the Side. ▶

Factoring $x^2 + bx + c$

Find two integers whose product is c and whose sum is b.

1. Both integers must be positive if b and c are positive. (See **Example 1.**)
2. Both integers must be negative if c is positive and b is negative. (See **Example 2.**)
3. One integer must be positive and one must be negative if c is negative. (See **Examples 3 and 4.**)

EXAMPLE 6 **Factoring a Trinomial with Two Variables**

Factor $z^2 - 2bz - 3b^2$.

Here, the coefficient of z in the middle term is $-2b$, so we need to find two expressions whose product is $-3b^2$ and whose sum is $-2b$.

Factors of $-3b^2$	Sums of Factors
$3b, -b$	$3b + (-b) = 2b$
$-3b, b$	$-3b + b = -2b$

Sum is $-2b$.

$$z^2 - 2bz - 3b^2 \quad \text{factors as} \quad (z - 3b)(z + b).$$

CHECK $(z - 3b)(z + b)$

$= z^2 + zb - 3bz - 3b^2$ FOIL method

$= z^2 + 1bz - 3bz - 3b^2$ Identity and commutative properties

$= z^2 - 2bz - 3b^2$ ✓ Combine like terms.

Work Problem 7 at the Side. ▶

6 Factor each trinomial, if possible.

(a) $x^2 + 5x + 8$

(b) $r^2 - 3r - 4$

(c) $m^2 - 2m + 5$

7 Factor each trinomial.

GS **(a)** $b^2 - 3ab - 4a^2$

We need two expressions whose product is ____ and whose sum is ____.

Complete the table.

Factors of $-4a^2$	Sums of Factors
$4a$, ___	$4a + ($___$) =$ ___
$-4a$, ___	$-4a +$ ___ $=$ ___
$2a, -2a$	$2a + (-2a) = 0$

Which pair of factors has the required sum? ________

Now factor the trinomial.

(b) $r^2 - 6rs + 8s^2$

Answers

6. **(a)** prime **(b)** $(r - 4)(r + 1)$
(c) prime

7. **(a)** $-4a^2$; $-3a$; $-a$; $-a$; $3a$; a; a; $-3a$; $-4a$ and a; $(b - 4a)(b + a)$
(b) $(r - 4s)(r - 2s)$

8 Factor each trinomial completely.

GS **(a)** $2p^3 + 6p^2 - 8p$

$= \_\_\_(\_\_\_ + 3p - \_\_\_)$

$= \_\_\_(\_\_\_\_\_)(\_\_\_\_\_)$

(b) $3y^4 - 27y^3 + 60y^2$

(c) $-3x^4 + 15x^3 - 18x^2$

Teaching Tip

Summarize the signs of the binomials when factoring a trinomial whose leading coefficient is positive.

1. **If the last term of the trinomial is positive, both binomials will have the same "middle" sign as the second term.**
2. **If the last term of the trinomial is negative, the binomials will have one plus and one minus "middle" sign.**

Answers

8. (a) $2p$; p^2; 4; $2p$; $p + 4$; $p - 1$
(b) $3y^2(y - 5)(y - 4)$
(c) $-3x^2(x - 3)(x - 2)$

OBJECTIVE 2 Factor such trinomials after factoring out the greatest common factor. If the terms of a trinomial have a common factor, first factor it out. Then factor the remaining trinomial as in **Examples 1–6.**

EXAMPLE 7 Factoring a Trinomial with a Common Factor

Factor $4x^5 - 28x^4 + 40x^3$.

There is no second-degree term. Look for a common factor.

$$4x^5 - 28x^4 + 40x^3$$

$$= 4x^3(x^2 - 7x + 10) \quad \text{Factor out the greatest common factor, } 4x^3.$$

Now factor $x^2 - 7x + 10$. The integers -5 and -2 have a product of 10 and a sum of -7.

Include $4x^3$.

$$= 4x^3(x - 5)(x - 2) \quad \text{Completely factored form}$$

CHECK $4x^3(x - 5)(x - 2)$

$= 4x^3(x^2 - 2x - 5x + 10)$ FOIL method

$= 4x^3(x^2 - 7x + 10)$ Combine like terms.

$= 4x^5 - 28x^4 + 40x^3$ ✓ Distributive property

◀ Work Problem **8** at the Side.

CAUTION

When factoring, always look for a common factor first. Remember to include the common factor as part of the answer. Check by multiplying out the completely factored form.

6.2 Exercises

FOR EXTRA HELP

CONCEPT CHECK *Answer each question.*

1. When factoring a trinomial in x as $(x + a)(x + b)$, what must be true of a and b, if the coefficient of the constant term of the trinomial is negative?

a and b must have different signs.

2. In **Exercise 1,** what must be true of a and b if the coefficient of the constant term is positive?

a and b must have the same sign.

3. Which one of the following is the correct factored form of $x^2 - 12x + 32$? **C**

A. $(x - 8)(x + 4)$ **B.** $(x + 8)(x - 4)$

C. $(x - 8)(x - 4)$ **D.** $(x + 8)(x + 4)$

4. What would be the first step in factoring

$$2x^3 + 8x^2 - 10x?$$

(See **Example 7.**)

Factor out the greatest common factor, $2x$.

5. What polynomial can be factored as

$$(a + 9)(a + 4)?$$

$a^2 + 13a + 36$

6. What polynomial can be factored as

$$(y - 7)(y + 3)?$$

$y^2 - 4y - 21$

7. What is meant by a *prime polynomial?*

A prime polynomial is one that cannot be factored using only integers in the factors.

8. How can we check our work when factoring a trinomial? Does the check ensure that the trinomial is *completely* factored?

To check a factored form, we multiply the factors. We should get the trinomial we started with. If we forget to factor out any common factors, the check will *not* indicate it.

In Exercises 9–12, list all pairs of integers with the given product. Then find the pair whose sum is given. ***See the tables in Examples 1–4.***

9. Product: 12; Sum: 7

1 and 12, −1 and −12, 2 and 6, −2 and −6, 3 and 4, −3 and −4; The pair with a sum of 7 is 3 and 4.

10. Product: 18; Sum: 9

1 and 18, −1 and −18, 2 and 9, −2 and −9, 3 and 6, −3 and −6; The pair with a sum of 9 is 3 and 6.

11. Product: −24; Sum: −5

1 and −24, −1 and 24, 2 and −12, −2 and 12, 3 and −8, −3 and 8, 4 and −6, −4 and 6; The pair with a sum of −5 is 3 and −8.

12. Product: −36; Sum: −16

1 and −36, −1 and 36, 2 and −18, −2 and 18, 3 and −12, −3 and 12, 4 and −9, −4 and 9, 6 and −6; The pair with a sum of −16 is 2 and −18.

GS *Complete each factoring.* ***See Examples 1–4.***

13. $p^2 + 11p + 30$
$= (p + 5)(\underline{p + 6})$

14. $x^2 + 10x + 21$
$= (x + 7)(\underline{x + 3})$

15. $x^2 + 15x + 44$
$= (x + 4)(\underline{x + 11})$

16. $r^2 + 15r + 56$
$= (r + 7)(\underline{r + 8})$

17. $x^2 - 9x + 8$
$= (x - 1)(\underline{x - 8})$

18. $t^2 - 14t + 24$
$= (t - 2)(\underline{t - 12})$

19. $y^2 - 2y - 15$
$= (y + 3)(\underline{y - 5})$

20. $t^2 - t - 42$
$= (t + 6)(\underline{t - 7})$

21. $x^2 + 9x - 22$
$= (x - 2)(\underline{x + 11})$

22. $x^2 + 6x - 27$
$= (x - 3)(\underline{x + 9})$

23. $y^2 - 7y - 18$
$= (y + 2)(\underline{y - 9})$

24. $y^2 - 2y - 24$
$= (y + 4)(\underline{y - 6})$

Factor completely. If a polynomial cannot be factored, write prime. **See Examples 1–5.**

25. $y^2 + 9y + 8$
$(y + 8)(y + 1)$

26. $a^2 + 9a + 20$
$(a + 4)(a + 5)$

27. $b^2 + 8b + 15$
$(b + 3)(b + 5)$

28. $x^2 + 6x + 8$
$(x + 4)(x + 2)$

29. $m^2 + m - 20$
$(m + 5)(m - 4)$

30. $p^2 + 4p - 5$
$(p + 5)(p - 1)$

31. $x^2 + 3x - 40$
$(x + 8)(x - 5)$

32. $d^2 + 4d - 45$
$(d + 9)(d - 5)$

33. $x^2 + 4x + 5$
prime

34. $t^2 + 11t + 12$
prime

35. $y^2 - 8y + 15$
$(y - 5)(y - 3)$

36. $y^2 - 6y + 8$
$(y - 4)(y - 2)$

37. $z^2 - 15z + 56$
$(z - 8)(z - 7)$

38. $x^2 - 13x + 36$
$(x - 9)(x - 4)$

39. $r^2 - r - 30$
$(r - 6)(r + 5)$

40. $q^2 - q - 42$
$(q - 7)(q + 6)$

41. $a^2 - 8a - 48$
$(a - 12)(a + 4)$

42. $m^2 - 10m - 24$
$(m - 12)(m + 2)$

Factor completely. **See Example 6.**

43. $r^2 + 3ra + 2a^2$
$(r + 2a)(r + a)$

44. $x^2 + 5xa + 4a^2$
$(x + 4a)(x + a)$

45. $x^2 + 4xy + 3y^2$
$(x + y)(x + 3y)$

46. $p^2 + 9pq + 8q^2$
$(p + q)(p + 8q)$

47. $t^2 - tz - 6z^2$
$(t + 2z)(t - 3z)$

48. $a^2 - ab - 12b^2$
$(a + 3b)(a - 4b)$

49. $v^2 - 11vw + 30w^2$
$(v - 5w)(v - 6w)$

50. $v^2 - 11vx + 24x^2$
$(v - 8x)(v - 3x)$

51. $a^2 + 2ab - 15b^2$
$(a + 5b)(a - 3b)$

52. $m^2 + 4mn - 12n^2$
$(m + 6n)(m - 2n)$

Factor completely. See **Example 7.**

53. $4x^2 + 12x - 40$
$4(x + 5)(x - 2)$

54. $5y^2 - 5y - 30$
$5(y + 2)(y - 3)$

55. $2t^3 + 8t^2 + 6t$
$2t(t + 1)(t + 3)$

56. $3t^3 + 27t^2 + 24t$
$3t(t + 8)(t + 1)$

57. $-2x^6 - 8x^5 + 42x^4$
$-2x^4(x - 3)(x + 7)$

58. $-4y^5 - 12y^4 + 40y^3$
$-4y^3(y - 2)(y + 5)$

59. $a^5 + 3a^4b - 4a^3b^2$
$a^3(a + 4b)(a - b)$

60. $z^{10} - 4z^9y - 21z^8y^2$
$z^8(z - 7y)(z + 3y)$

61. $5m^5 + 25m^4 - 40m^2$
$5m^2(m^3 + 5m^2 - 8)$

62. $12k^5 - 6k^3 + 10k^2$
$2k^2(6k^3 - 3k + 5)$

63. $m^3n - 10m^2n^2 + 24mn^3$
$mn(m - 6n)(m - 4n)$

64. $y^3z + 3y^2z^2 - 54yz^3$
$yz(y - 6z)(y + 9z)$

6.3 Factoring Trinomials by Grouping

OBJECTIVE 1 Factor trinomials by grouping when the coefficient of the second-degree term is not 1. We now extend our work to factor a trinomial such as $2x^2 + 7x + 6$, where the coefficient of x^2 is *not* 1.

EXAMPLE 1 Factoring a Trinomial by Grouping (Coefficient of the Second-Degree Term Not 1)

Factor $2x^2 + 7x + 6$.

To factor this trinomial, we look for two positive integers whose product is $2 \cdot 6 = 12$ and whose sum is 7.

$$2x^2 + 7x + 6$$

Sum is 7. (7x)

Product is $2 \cdot 6 = 12$. (2 and 6)

The required integers are 3 and 4, since $3 \cdot 4 = 12$ and $3 + 4 = 7$. We use these integers to write the middle term $7x$ as $3x + 4x$.

$$\begin{aligned}
&2x^2 + 7x + 6 \\
&= 2x^2 + \underbrace{3x + 4x}_{7x} + 6 \\
&= (2x^2 + 3x) + (4x + 6) && \text{Group the terms.} \\
&= x(2x + 3) + 2(2x + 3) && \text{Factor each group.} \\
&= (2x + 3)(x + 2) && \text{Factor out } 2x + 3.
\end{aligned}$$

Must be the same (the two factors $2x + 3$)

CHECK Multiply $(2x + 3)(x + 2)$ to obtain $2x^2 + 7x + 6$. ✓

Work Problem 1 at the Side. ▶

EXAMPLE 2 Factoring Trinomials by Grouping

Factor each trinomial.

(a) $6r^2 + r - 1$

We must find two integers with a product of $6(-1) = -6$ and a sum of 1 (since the coefficient of r, or $1r$, is 1). The integers are -2 and 3. We write the middle term r as $-2r + 3r$.

$$\begin{aligned}
&6r^2 + r - 1 \\
&= 6r^2 - 2r + 3r - 1 && r = -2r + 3r \\
&= (6r^2 - 2r) + (3r - 1) && \text{Group the terms.} \\
&= 2r(3r - 1) + 1(3r - 1) && \text{The binomials must be the same.} \\
&= (3r - 1)(2r + 1) && \text{Factor out } 3r - 1.
\end{aligned}$$

Remember the 1.

CHECK Multiply $(3r - 1)(2r + 1)$ to obtain $6r^2 + r - 1$. ✓

Continued on Next Page

OBJECTIVE

1. Factor trinomials by grouping when the coefficient of the second-degree term is not 1.

1 **(a)** In **Example 1,** we factored

$$2x^2 + 7x + 6$$

by writing $7x$ as $3x + 4x$. This trinomial can also be factored by writing $7x$ as $4x + 3x$.

Complete the following.

$$\begin{aligned}
&2x^2 + \mathbf{7x} + 6 \\
&= 2x^2 + \mathbf{4x} + \mathbf{3x} + 6 \\
&= (2x^2 + ____) + (3x + ____) \\
&= 2x(x + ____) + 3(x + ____) \\
&= (______)(2x + 3)
\end{aligned}$$

(b) Is the answer in part (a) the same as in **Example 1?** (Remember that the order of the factors does not matter.)

(c) Factor $2z^2 + 5z + 3$ by grouping.

Answers

1. **(a)** $4x$; 6; 2; 2; $x + 2$ **(b)** yes
(c) $(2z + 3)(z + 1)$

2 Factor each trinomial by grouping.

(a) $2m^2 + 7m + 3$

(b) $5p^2 - 2p - 3$

(c) $15k^2 - km - 2m^2$

3 Factor the trinomial completely.

GS $4x^2 - 2x - 30$

$= \_\_\_(2x^2 - \_\_\_ - 15)$

$= \_\_\_(2x^2 - \_\_\_ + \_\_\_ - 15)$

$= 2[(2x^2 - \_\_\_) + (5x - 15)]$

$= 2[\_\_\_(\_\_\_) + 5(x - 3)]$

$= \_\_\_\_\_\_$

4 Factor each trinomial completely.

(a) $18p^4 + 63p^3 + 27p^2$

(b) $6a^2 + 3ab - 18b^2$

(b) $12z^2 - 5z - 2$

Look for two integers whose product is $12(-2) = -24$ and whose sum is -5. The required integers are 3 and -8.

$$12z^2 - 5z - 2$$

$$= 12z^2 + 3z - 8z - 2 \qquad -5z = 3z - 8z$$

$$= (12z^2 + 3z) + (-8z - 2) \qquad \text{Group the terms.}$$

$$= 3z(4z + 1) - 2(4z + 1) \qquad \text{Factor each group.}$$

Be careful with signs.

$$= (4z + 1)(3z - 2) \qquad \text{Factor out } 4z + 1.$$

CHECK Multiply $(4z + 1)(3z - 2)$ to obtain $12z^2 - 5z - 2$. ✓

(c) $10m^2 + mn - 3n^2$

Two integers whose product is $10(-3) = -30$ and whose sum is 1 are -5 and 6. Rewrite the trinomial with four terms.

$$10m^2 + mn - 3n^2$$

$$= 10m^2 - 5mn + 6mn - 3n^2 \qquad mn = -5mn + 6mn$$

$$= 5m(2m - n) + 3n(2m - n) \qquad \text{Group the terms. Factor each group.}$$

$$= (2m - n)(5m + 3n) \qquad \text{Factor out } 2m - n.$$

CHECK Multiply $(2m - n)(5m + 3n)$ to obtain $10m^2 + mn - 3n^2$. ✓

◀ **Work Problem 2 at the Side.**

EXAMPLE 3 Factoring a Trinomial with a Common Factor by Grouping

Factor $28x^5 - 58x^4 - 30x^3$.

$$28x^5 - 58x^4 - 30x^3$$

$$= 2x^3(14x^2 - 29x - 15) \qquad \text{Factor out the greatest common factor, } 2x^3.$$

To factor $14x^2 - 29x - 15$, find two integers whose product is $14(-15) = -210$ and whose sum is -29. Factoring 210 into prime factors helps find these integers.

$$210 = 2 \cdot 3 \cdot 5 \cdot 7$$

Combine the prime factors in pairs in different ways, using one positive factor and one negative factor to get -210. The factors 6 and -35 have the correct sum, -29.

$$28x^5 - 58x^4 - 30x^3$$

Remember the common factor.

$$= 2x^3(14x^2 - 29x - 15) \qquad \text{From above}$$

$$= 2x^3(14x^2 + 6x - 35x - 15) \qquad -29x = 6x - 35x$$

$$= 2x^3[(14x^2 + 6x) + (-35x - 15)] \qquad \text{Group the terms.}$$

$$= 2x^3[2x(7x + 3) - 5(7x + 3)] \qquad \text{Factor each group.}$$

$$= 2x^3[(7x + 3)(2x - 5)] \qquad \text{Factor out } 7x + 3.$$

$$= 2x^3(7x + 3)(2x - 5)$$

Check by multiplying.

◀ **Work Problems 3 and 4 at the Side.**

Answers

2. (a) $(2m + 1)(m + 3)$
   (b) $(5p + 3)(p - 1)$
   (c) $(5k - 2m)(3k + m)$
3. 2; x; 2; $6x$; $5x$; $6x$; $2x$; $x - 3$; $2(x - 3)(2x + 5)$
4. (a) $9p^2(2p + 1)(p + 3)$
   (b) $3(2a - 3b)(a + 2b)$

6.3 Exercises

FOR EXTRA HELP

Download the MyDashBoard App

MyMathLab®

1. CONCEPT CHECK Which pair of integers would be used to rewrite the middle term when factoring $12y^2 + 5y - 2$ by grouping? B

A. $-8, 3$ **B.** $8, -3$ **C.** $-6, 4$ **D.** $6, -4$

2. CONCEPT CHECK Which pair of integers would be used to rewrite the middle term when factoring $20b^2 - 13b + 2$ by grouping? D

A. $10, 3$ **B.** $-10, -3$ **C.** $8, 5$ **D.** $-8, -5$

The middle term of each trinomial has been rewritten. Now factor by grouping. ***See Examples 1 and 2.***

3. $m^2 + 8m + 12$
$= m^2 + 6m + 2m + 12$
$(m + 6)(m + 2)$

4. $x^2 + 9x + 14$
$= x^2 + 7x + 2x + 14$
$(x + 7)(x + 2)$

5. $a^2 + 3a - 10$
$= a^2 + 5a - 2a - 10$
$(a + 5)(a - 2)$

6. $y^2 - 2y - 24$
$= y^2 + 4y - 6y - 24$
$(y + 4)(y - 6)$

7. $10t^2 + 9t + 2$
$= 10t^2 + 5t + 4t + 2$
$(2t + 1)(5t + 2)$

8. $6x^2 + 13x + 6$
$= 6x^2 + 9x + 4x + 6$
$(2x + 3)(3x + 2)$

9. $15z^2 - 19z + 6$
$= 15z^2 - 10z - 9z + 6$
$(3z - 2)(5z - 3)$

10. $12p^2 - 17p + 6$
$= 12p^2 - 9p - 8p + 6$
$(4p - 3)(3p - 2)$

11. $8s^2 + 2st - 3t^2$
$= 8s^2 - 4st + 6st - 3t^2$
$(2s - t)(4s + 3t)$

12. $3x^2 - xy - 14y^2$
$= 3x^2 - 7xy + 6xy - 14y^2$
$(3x - 7y)(x + 2y)$

13. $15a^2 + 22ab + 8b^2$
$= 15a^2 + 10ab + 12ab + 8b^2$
$(3a + 2b)(5a + 4b)$

14. $25m^2 + 25mn + 6n^2$
$= 25m^2 + 15mn + 10mn + 6n^2$
$(5m + 3n)(5m + 2n)$

Complete the steps to factor each trinomial by grouping. ***See Examples 1 and 2.***

15. $2m^2 + 11m + 12$

(a) Find two integers whose product is $\underline{2} \cdot \underline{12} = \underline{24}$ and whose sum is $\underline{11}$.

(b) The required integers are $\underline{3}$ and $\underline{8}$. (Order is irrelevant.)

(c) Write the middle term $11m$ as $\underline{3m} + \underline{8m}$.

(d) Rewrite the given trinomial using four terms.
$2m^2 + 3m + 8m + 12$

(e) Factor the polynomial in part (d) by grouping.
$(2m + 3)(m + 4)$

(f) Check by multiplying.
$(2m + 3)(m + 4)$
$= 2m^2 + 8m + 3m + 12$
$= 2m^2 + 11m + 12$

16. $6y^2 - 19y + 10$

(a) Find two integers whose product is $\underline{6} \cdot \underline{10} = \underline{60}$ and whose sum is $\underline{-19}$.

(b) The required integers are $\underline{-4}$ and $\underline{-15}$. (Order is irrelevant.)

(c) Write the middle term $-19y$ as $\underline{-4y} + \underline{(-15y)}$.

(d) Rewrite the given trinomial using four terms.
$6y^2 - 4y - 15y + 10$

(e) Factor the polynomial in part (d) by grouping.
$(3y - 2)(2y - 5)$

(f) Check by multiplying.
$(3y - 2)(2y - 5)$
$= 6y^2 - 15y - 4y + 10$
$= 6y^2 - 19y + 10$

Factor each trinomial by grouping. ***See Examples 1–3.*** **The order of the factors is irrelevant in Exercises 17–40.**

17. $2x^2 + 7x + 3$
$(2x + 1)(x + 3)$

18. $3y^2 + 13y + 4$
$(3y + 1)(y + 4)$

19. $4r^2 + r - 3$
$(4r - 3)(r + 1)$

20. $4r^2 + 3r - 10$
$(4r - 5)(r + 2)$

21. $8m^2 - 10m - 3$
$(4m + 1)(2m - 3)$

22. $20x^2 - 28x - 3$
$(10x + 1)(2x - 3)$

23. $21m^2 + 13m + 2$
$(3m + 1)(7m + 2)$

24. $38x^2 + 23x + 2$
$(2x + 1)(19x + 2)$

25. $6b^2 + 7b + 2$
$(2b + 1)(3b + 2)$

26. $6w^2 + 19w + 10$
$(2w + 5)(3w + 2)$

27. $12y^2 - 13y + 3$
$(4y - 3)(3y - 1)$

28. $15a^2 - 16a + 4$
$(3a - 2)(5a - 2)$

29. $16 + 16x + 3x^2$
$(4 + x)(4 + 3x)$, or $(x + 4)(3x + 4)$

30. $18 + 65x + 7x^2$
$(9 + x)(2 + 7x)$, or $(x + 9)(7x + 2)$

31. $24x^2 - 42x + 9$
$3(4x - 1)(2x - 3)$

32. $48b^2 - 74b - 10$
$2(8b + 1)(3b - 5)$

33. $2m^3 + 2m^2 - 40m$
$2m(m - 4)(m + 5)$

34. $3x^3 + 12x^2 - 36x$
$3x(x - 2)(x + 6)$

35. $-32z^5 + 20z^4 + 12z^3$
$-4z^3(z - 1)(8z + 3)$

36. $-18x^5 - 15x^4 + 75x^3$
$-3x^3(2x + 5)(3x - 5)$

37. $12p^2 + 7pq - 12q^2$
$(3p + 4q)(4p - 3q)$

38. $6m^2 - 5mn - 6n^2$
$(3m + 2n)(2m - 3n)$

39. $6a^2 - 7ab - 5b^2$
$(3a - 5b)(2a + b)$

40. $25g^2 - 5gh - 2h^2$
$(5g + h)(5g - 2h)$

41. CONCEPT CHECK On a quiz, a student factored $16x^2 - 24x + 5$ by grouping as follows.

$$16x^2 - 24x + 5$$
$$= 16x^2 - 4x - 20x + 5$$
$$= 4x(4x - 1) - 5(4x - 1) \quad \text{His answer}$$

He thought his answer was correct since it checked by multiplying. **What Went Wrong?** What is the correct factored form?

The student stopped too soon. He needs to factor out the common factor $4x - 1$ to get $(4x - 1)(4x - 5)$ as the correct answer.

42. CONCEPT CHECK On the same quiz, another student factored $3k^3 - 12k^2 - 15k$ by first factoring out the common factor $3k$ to get $3k(k^2 - 4k - 5)$. Then she wrote the following.

$$k^2 - 4k - 5$$
$$= k^2 - 5k + k - 5$$
$$= k(k - 5) + 1(k - 5)$$
$$= (k - 5)(k + 1) \quad \text{Her answer}$$

What Went Wrong? What is the correct factored form?

The student forgot to include the common factor $3k$ in her answer. The correct answer is $3k(k - 5)(k + 1)$.

6.4 Factoring Trinomials by Using the FOIL Method

OBJECTIVE

1 **Factor trinomials by using the FOIL method.**

OBJECTIVE 1 **Factor trinomials by using the FOIL method.** This section shows an alternative method of factoring trinomials that uses trial and error.

EXAMPLE 1 **Factoring a Trinomial Using FOIL (Coefficient of the Second-Degree Term Not 1)**

Factor $2x^2 + 7x + 6$ (the trinomial in **Section 6.3, Example 1**).

We want to write $2x^2 + 7x + 6$ as the product of two binomials.

$$2x^2 + 7x + 6$$
$$= (\_\_\_\_)(\_\_\_\_)$$

We use the FOIL method in reverse.

The product of the two first terms of the binomials is $2x^2$. The possible factors of $2x^2$ are $2x$ and x, or $-2x$ and $-x$. Since all terms of the trinomial are positive, we consider only positive factors. Thus, we have the following.

$$2x^2 + 7x + 6$$
$$= (2x\_\_\_\_)(x\_\_\_\_)$$

The product of the two last terms of the binomials is 6. It can be factored as $1 \cdot 6$, $6 \cdot 1$, $2 \cdot 3$, or $3 \cdot 2$. We want the pair that gives the correct middle term, $7x$. Begin by trying 1 and 6 in $(2x\_\_\_\_)(x\_\_\_\_)$.

$(2x + 1)(x + 6)$ Incorrect

x

$12x$

$13x$ Add.

Now try the pair 6 and 1 in $(2x\_\_\_\_)(x\_\_\_\_)$.

$(2x + 6)(x + 1)$ Incorrect

$6x$

$2x$

$8x$ Add.

Since $2x + 6 = 2(x + 3)$, the terms of the binomial $2x + 6$ have a factor of 2, while the terms of $2x^2 + 7x + 6$ have no common factor other than 1. The product $(2x + 6)(x + 1)$ cannot be correct.

If the terms of the original polynomial have greatest common factor 1, then each of its factors will also have terms with GCF 1.

We try the pair 2 and 3 in $(2x\_\_\_\_)(x\_\_\_\_)$. Because of the common factor 2 in the terms of $2x + 2$, the product $(2x + 2)(x + 3)$ will not work. Finally, we try the pair 3 and 2 in $(2x\_\_\_\_)(x\_\_\_\_)$.

$(2x + 3)(x + 2) = 2x^2 + 7x + 6$ Correct

$3x$

$4x$

$7x$ Add.

Thus, $2x^2 + 7x + 6$ factors as $(2x + 3)(x + 2)$.

CHECK Multiply $(2x + 3)(x + 2)$ to obtain $2x^2 + 7x + 6$. ✓

Work Problem 1 at the Side. ▶

1 Factor each trinomial.

GS **(a)** $2p^2 + 9p + 9$

The possible factors of the first term $2p^2$ are ____ and p.

The possible factors of the last term 9 are 9 and ____, or ____ and 3.

Try various combinations to find the pair of factors that gives the correct middle term, ____.

Possible Pairs of Factors	Middle Term	Correct?
$(2p + 9)(p + 1)$	$11p$	(*Yes* / *No*)
$(2p + 1)(p + 9)$	____	(*Yes* / *No*)
$(2p + 3)(\_\_\_\_)$	____	(*Yes* / *No*)

$2p^2 + 9p + 9$ factors as ______.

(b) $8y^2 + 22y + 5$

Teaching Tip

Stress the concept involving common factors discussed in Example 1 with students. Understanding and using it will save lots of trial and error work.

Answers

1. (a) $2p$; 1; 3; $9p$; No; $19p$; No; $p + 3$; $9p$; Yes; $(2p + 3)(p + 3)$
(b) $(4y + 1)(2y + 5)$

2 Factor each trinomial.

(a) $6p^2 + 19p + 10$

(b) $8x^2 + 14x + 3$

3 Factor each trinomial.

(a) $4y^2 - 11y + 6$

(b) $9x^2 - 21x + 10$

Teaching Tip

See the Note at the top of the next page. Stress the importance of this concept with students.

Answers

2. (a) $(3p + 2)(2p + 5)$
(b) $(4x + 1)(2x + 3)$

3. (a) $(4y - 3)(y - 2)$
(b) $(3x - 5)(3x - 2)$

EXAMPLE 2 Factoring a Trinomial Using FOIL (All Positive Terms)

Factor $8p^2 + 14p + 5$.

The number 8 has several possible pairs of factors, but 5 has only 1 and 5 or -1 and -5, so we begin by considering the factors of 5. We ignore the negative factors, since all coefficients in the trinomial are positive. If $8p^2 + 14p + 5$ can be factored, the factors will have this form.

$$(\_\_\_\_ + 5)(\_\_\_\_ + 1)$$

The possible pairs of factors of $8p^2$ are $8p$ and p, or $4p$ and $2p$. We try various combinations, checking to see if the middle term is $14p$.

$(8p + 5)(p + 1)$ Incorrect
$5p$
$8p$
$13p$ Add.

$(p + 5)(8p + 1)$ Incorrect
$40p$
p
$41p$ Add.

$(4p + 5)(2p + 1)$ Correct
$10p$
$4p$
$14p$ Add.

This last combination produces $14p$, the correct middle term.

$$8p^2 + 14p + 5 \quad \text{factors as} \quad (4p + 5)(2p + 1).$$

CHECK Multiply $(4p + 5)(2p + 1)$ to obtain $8p^2 + 14p + 5$. ✓

◀ **Work Problem 2 at the Side.**

EXAMPLE 3 Factoring a Trinomial Using FOIL (Negative Middle Term)

Factor $6x^2 - 11x + 3$.

Since 3 has only 1 and 3 or -1 and -3 as factors, it is better here to begin by factoring 3. The last term of the trinomial $6x^2 - 11x + 3$ is positive and the middle term has a negative coefficient, so we consider only negative factors. We need two negative factors because the *product* of two negative factors is positive and their *sum* is negative, as required.

We try -3 and -1 as factors of 3.

$$(\_\_\_\_ - 3)(\_\_\_\_ - 1)$$

The factors of $6x^2$ may be either $6x$ and x, or $2x$ and $3x$.

$(6x - 3)(x - 1)$ Incorrect
$-3x$
$-6x$
$-9x$ Add.

$(2x - 3)(3x - 1)$ Correct
$-9x$
$-2x$
$-11x$ Add.

The factors $2x$ and $3x$ produce $-11x$, the correct middle term.

$$6x^2 - 11x + 3 \quad \text{factors as} \quad (2x - 3)(3x - 1).$$

Check by multiplying.

◀ **Work Problem 3 at the Side.**

Note

Also in **Example 3,** our initial attempt to factor $6x^2 - 11x + 3$ as $(6x - 3)(x - 1)$ *cannot* be correct since the terms of $6x - 3$ have a common factor of 3, while those of the original polynomial do not.

EXAMPLE 4 **Factoring a Trinomial Using FOIL (Negative Constant Term)**

Factor $8x^2 + 6x - 9$.

The integer 8 has several possible pairs of factors, as does -9. Since the constant term is negative, one positive factor and one negative factor of -9 are needed. Since the coefficient of the middle term is relatively small, it is wise to avoid large factors, such as 8 or 9. We try $4x$ and $2x$ as factors of $8x^2$, and 3 and -3 as factors of -9, and check the middle term.

$(4x + 3)(2x - 3)$ Incorrect

$6x$
$-12x$
$-6x$ Add.

Interchange 3 and -3, since only the sign of the middle term is incorrect.

$(4x - 3)(2x + 3)$ Correct

$-6x$
$12x$
$6x$ Add.

The combination on the right produces $6x$, the correct middle term.

$$8x^2 + 6x - 9 \quad \text{factors as} \quad (4x - 3)(2x + 3).$$

Check by multiplying.

Work Problem **4** at the Side. ▶

EXAMPLE 5 **Factoring a Trinomial with Two Variables**

Factor $12a^2 - ab - 20b^2$.

There are several pairs of factors of $12a^2$, including

$$12a \text{ and } a, \quad 6a \text{ and } 2a, \quad \text{and} \quad 4a \text{ and } 3a.$$

There are also many possible pairs of factors of $-20b^2$, including

$$20b \text{ and } -b, \quad -20b \text{ and } b, \quad 10b \text{ and } -2b,$$
$$-10b \text{ and } 2b, \quad 4b \text{ and } -5b, \quad \text{and} \quad -4b \text{ and } 5b.$$

Once again, since the coefficient of the middle term is relatively small, avoid the larger factors. Try the factors $6a$ and $2a$, and $4b$ and $-5b$.

$$(6a + 4b)(2a - 5b) \quad \text{Incorrect}$$

This cannot be correct, since the terms of $6a + 4b$ have 2 as a common factor, while the terms of the given trinomial do not. Try $3a$ and $4a$ with $4b$ and $-5b$.

$$(3a + 4b)(4a - 5b)$$
$$= 12a^2 + ab - 20b^2 \quad \text{Incorrect}$$

Here the middle term is ab, rather than $-ab$. Interchange the signs of the last two terms in the factors.

$$(3a - 4b)(4a + 5b)$$
$$= 12a^2 - ab - 20b^2 \quad \text{Correct}$$

Thus, $12a^2 - ab - 20b^2$ factors as $(3a - 4b)(4a + 5b)$.

Work Problem **5** at the Side. ▶

4 Factor each trinomial, if possible.

GS **(a)** $6x^2 + 5x - 4$

$= (3x + \_\_\_)(\_\_\_ - \_\_\_)$

(b) $6m^2 - 11m - 10$

(c) $4x^2 - 3x - 7$

(d) $3y^2 + 8y - 6$

5 Factor each trinomial.

GS **(a)** $2x^2 - 5xy - 3y^2$

$= (2x + \_\_\_)(\_\_\_ - \_\_\_)$

(b) $8a^2 + 2ab - 3b^2$

Answers

4. **(a)** 4; $2x$; 1
(b) $(2m - 5)(3m + 2)$
(c) $(4x - 7)(x + 1)$
(d) prime

5. **(a)** y; x; $3y$
(b) $(4a + 3b)(2a - b)$

6 Factor each trinomial.

(a) $36z^3 - 6z^2 - 72z$

$= \_\_\_(\_\_\_ - \_\_\_ - 12)$

$= \_\_\_(\_\_\_)(\_\_\_)$

(b) $10x^3 + 45x^2 - 90x$

(c) $-24x^3 + 32x^2 + 6x$

Answers

6. (a) $6z$; $6z^2$; z; $6z$; $3z + 4$; $2z - 3$
(b) $5x(2x - 3)(x + 6)$
(c) $-2x(6x + 1)(2x - 3)$

EXAMPLE 6 Factoring Trinomials with Common Factors

Factor each trinomial.

(a) $15y^3 + 55y^2 + 30y$

$$15y^3 + 55y^2 + 30y$$
$$= 5y(3y^2 + 11y + 6) \quad \text{Factor out the greatest common factor, } 5y.$$

Now factor $3y^2 + 11y + 6$. Try $3y$ and y as factors of $3y^2$ and 2 and 3 as factors of 6.

$$(3y + 2)(y + 3)$$
$$= 3y^2 + 11y + 6 \quad \text{Correct}$$

This leads to the completely factored form.

$$15y^3 + 55y^2 + 30y$$
$$= 5y(3y + 2)(y + 3)$$

Remember the common factor.

CHECK $5y(3y + 2)(y + 3)$

$$= 5y(3y^2 + 9y + 2y + 6) \quad \text{FOIL method}$$
$$= 5y(3y^2 + 11y + 6) \quad \text{Combine like terms.}$$
$$= 15y^3 + 55y^2 + 30y \checkmark \quad \text{Distributive property}$$

(b) $-24a^3 - 42a^2 + 45a$

The common factor could be $3a$ or $-3a$. If we factor out $-3a$, the first term of the trinomial will be positive, which makes it easier to factor the remaining trinomial.

$$-24a^3 - 42a^2 + 45a$$
$$= -3a(8a^2 + 14a - 15) \quad \text{Factor out } -3a.$$
$$= -3a(4a - 3)(2a + 5) \quad \text{Use trial and error.}$$

CHECK $-3a(4a - 3)(2a + 5)$

$$= -3a(8a^2 + 14a - 15) \quad \text{FOIL method; Combine like terms.}$$
$$= -24a^3 - 42a^2 + 45a \checkmark \quad \text{Distributive property}$$

CAUTION

Remember to include the common factor in the final factored form.

◀ **Work Problem 6 at the Side.**

6.4 Exercises

FOR EXTRA HELP

Download the MyDashBoard App

MyMathLab®

CONCEPT CHECK *Decide which is the correct factored form of the given polynomial.*

1. $2x^2 - x - 1$
A. $(2x - 1)(x + 1)$ **B.** $(2x + 1)(x - 1)$ B

2. $3a^2 - 5a - 2$
A. $(3a + 1)(a - 2)$ **B.** $(3a - 1)(a + 2)$ A

3. $4y^2 + 17y - 15$
A. $(y + 5)(4y - 3)$ **B.** $(2y - 5)(2y + 3)$ A

4. $12c^2 - 7c - 12$
A. $(6c - 2)(2c + 6)$ **B.** $(4c + 3)(3c - 4)$ B

5. $4k^2 + 13mk + 3m^2$
A. $(4k + m)(k + 3m)$ **B.** $(4k + 3m)(k + m)$ A

6. $2x^2 + 11x + 12$
A. $(2x + 3)(x + 4)$ **B.** $(2x + 4)(x + 3)$ A

Complete each factoring. ***See Examples 1–6.***

7. $6a^2 + 7ab - 20b^2$
$= (3a - 4b)(\underline{2a + 5b})$

8. $9m^2 - 3mn - 2n^2$
$= (3m + n)(\underline{3m - 2n})$

9. $2x^2 + 6x - 8$
$= 2(\underline{x^2 + 3x - 4})$
$= 2(\underline{x + 4})(\underline{x - 1})$

10. $3x^2 - 9x - 30$
$= 3(\underline{x^2 - 3x - 10})$
$= 3(\underline{x - 5})(\underline{x + 2})$

In the last line of Exercises 9–12, the order of the factors is irrelevant.

11. $4z^3 - 10z^2 - 6z$
$= 2z(\underline{2z^2 - 5z - 3})$
$= 2z(\underline{2z + 1})(\underline{z - 3})$

12. $15r^3 - 39r^2 - 18r$
$= 3r(\underline{5r^2 - 13r - 6})$
$= 3r(\underline{5r + 2})(\underline{r - 3})$

13. CONCEPT CHECK A student factoring the trinomial

$$12x^2 + 7x - 12$$

wrote $(4x + 4)$ as one binomial factor. **What Went Wrong?** Factor the trinomial correctly.

$(4x + 4)$ cannot be a factor because its terms have a common factor of 4, but those of the polynomial do not. The correct factored form is $(4x - 3)(3x + 4)$.

14. CONCEPT CHECK Another student completely factored the trinomial

$$4x^2 + 10x - 6$$

as $(4x - 2)(x + 3)$. **What Went Wrong?** Factor the trinomial completely.

The student forgot to factor out the common factor 2 from the terms of the trinomial. The completely factored form is $2(2x - 1)(x + 3)$.

Factor each trinomial completely. ***See Examples 1–6.*** **The order of the factors is irrelevant in Exercises 15–50.**

15. $3a^2 + 10a + 7$
$(3a + 7)(a + 1)$

16. $7r^2 + 8r + 1$
$(7r + 1)(r + 1)$

17. $2y^2 + 7y + 6$
$(2y + 3)(y + 2)$

18. $5z^2 + 12z + 4$
$(5z + 2)(z + 2)$

19. $15m^2 + m - 2$
$(3m - 1)(5m + 2)$

20. $6x^2 + x - 1$
$(2x + 1)(3x - 1)$

21. $12s^2 + 11s - 5$
$(3s - 1)(4s + 5)$

22. $20x^2 + 11x - 3$
$(4x + 3)(5x - 1)$

23. $10m^2 - 23m + 12$
$(5m - 4)(2m - 3)$

24. $6x^2 - 17x + 12$
$(3x - 4)(2x - 3)$

25. $8w^2 - 14w + 3$
$(4w - 1)(2w - 3)$

26. $9p^2 - 18p + 8$
$(3p - 2)(3p - 4)$

27. $20y^2 - 39y - 11$
$(4y + 1)(5y - 11)$

28. $10x^2 - 11x - 6$
$(2x - 3)(5x + 2)$

29. $3x^2 - 15x + 16$
prime

30. $2t^2 + 13t - 18$
prime

31. $20x^2 + 22x + 6$
$2(5x + 3)(2x + 1)$

32. $36y^2 + 81y + 45$
$9(4y + 5)(y + 1)$

33. $-40m^2q - mq + 6q$
$-q(5m + 2)(8m - 3)$

34. $-15a^2b - 22ab - 8b$
$-b(3a + 2)(5a + 4)$

35. $15n^4 - 39n^3 + 18n^2$
$3n^2(5n - 3)(n - 2)$

36. $24a^4 + 10a^3 - 4a^2$
$2a^2(3a + 2)(4a - 1)$

37. $-15x^2y^2 + 7xy^2 + 4y^2$
$-y^2(5x - 4)(3x + 1)$

38. $-14a^2b^3 - 15ab^3 + 9b^3$
$-b^3(7a - 3)(2a + 3)$

39. $5a^2 - 7ab - 6b^2$
$(5a + 3b)(a - 2b)$

40. $6x^2 - 5xy - y^2$
$(6x + y)(x - y)$

41. $12s^2 + 11st - 5t^2$
$(4s + 5t)(3s - t)$

42. $25a^2 + 25ab + 6b^2$
$(5a + 2b)(5a + 3b)$

43. $6m^6n + 7m^5n^2 + 2m^4n^3$
$m^4n(3m + 2n)(2m + n)$

44. $12k^3q^4 - 4k^2q^5 - kq^6$
$kq^4(6k + q)(2k - q)$

If a trinomial has a negative coefficient for the second-degree term, such as $-2x^2 + 11x - 12$, *it may be easier to factor by first factoring out* **−1**.

$$-2x^2 + 11x - 12$$
$$= -1(2x^2 - 11x + 12) \quad \text{Factor out } -1.$$
$$= -1(2x - 3)(x - 4) \quad \text{Factor the trinomial.}$$

Use this method to factor each trinomial in Exercises 45–50.

45. $-x^2 - 4x + 21$
$-1(x + 7)(x - 3)$

46. $-x^2 + x + 72$
$-1(x + 8)(x - 9)$

47. $-3x^2 - x + 4$
$-1(3x + 4)(x - 1)$

48. $-5x^2 + 2x + 16$
$-1(5x + 8)(x - 2)$

49. $-2a^2 - 5ab - 2b^2$
$-1(a + 2b)(2a + b)$

50. $-3p^2 + 13pq - 4q^2$
$-1(p - 4q)(3p - q)$

Relating Concepts (Exercises 51–56) For Individual or Group Work

Often there are several different equivalent forms of an answer that are all correct. ***Work Exercises 51–56 in order,*** *to see this for factoring problems.*

51. Factor the integer 35 as the product of two prime numbers. $5 \cdot 7$

52. Factor the integer 35 as the product of the negatives of two prime numbers. $(-5)(-7)$

53. Verify that $6x^2 - 11x + 4$ factors as $(3x - 4)(2x - 1)$.
The product of $3x - 4$ and $2x - 1$ is $6x^2 - 11x + 4$.

54. Verify that $6x^2 - 11x + 4$ factors as $(4 - 3x)(1 - 2x)$.
The product of $4 - 3x$ and $1 - 2x$ is $6x^2 - 11x + 4$.

55. Compare the two valid factored forms in **Exercises 53 and 54.** How do the factors in each case compare?
The factors in Exercise 53 are the opposites of the factors in Exercise 54.

56. Suppose you know that the correct factored form of a particular trinomial is $(7t - 3)(2t - 5)$. From **Exercises 51–55,** what is another valid factored form?
$(3 - 7t)(5 - 2t)$

6.5 Special Factoring Techniques

OBJECTIVES

1. Factor a difference of squares.
2. Factor a perfect square trinomial.

By reversing the rules for multiplication of binomials from **Section 5.4,** we obtain rules for factoring polynomials in certain forms.

OBJECTIVE 1 Factor a difference of squares. The formula for the product of the sum and difference of the same two terms is

$$(a+b)(a-b)=a^2-b^2.$$

Reversing this rule leads to the following special factoring rule.

Factoring a Difference of Squares

$$a^2-b^2=(a+b)(a-b)$$

For example,

$$m^2-16$$
$$=m^2-4^2$$
$$=(m+4)(m-4).$$

The following must be true for a binomial to be a difference of squares.

1. Both terms of the binomial must be squares, such as

$$x^2,\quad 9y^2=(3y)^2,\quad 25=5^2,\quad 1=1^2,\quad m^4=(m^2)^2.$$

2. The terms of the binomial must have different signs (one positive and one negative).

EXAMPLE 1 Factoring Differences of Squares

Factor each binomial, if possible.

$$a^2-b^2=(a+b)(a-b)$$

(a) $x^2-49=x^2-7^2=(x+7)(x-7)$

(b) $y^2-m^2=(y+m)(y-m)$

(c) x^2-8

Because 8 is not the square of an integer, this binomial is not a difference of squares. It is a prime polynomial.

(d) p^2+16

Since p^2+16 is a *sum* of squares, it is not equal to $(p+4)(p-4)$. Also, we use the FOIL method and try the following.

$(p-4)(p-4)$
$=p^2-8p+16,$ not $p^2+16.$

$(p+4)(p+4)$
$=p^2+8p+16,$ not $p^2+16.$

Thus, p^2+16 is a prime polynomial.

Work Problem 1 at the Side. ▶

CAUTION

As Example 1(d) suggests, after any common factor is removed, a sum of squares cannot be factored.

1 Factor each binomial, if possible.

GS **(a)** x^2-81
$=(x+\_\_\_\_)(x-\_\_\_\_)$

(b) p^2-100

Teaching Tip
To help students recognize perfect squares, you may wish to have them work Exercise 1 in the exercise set for this section.

(c) t^2-s^2

(d) y^2-10

(e) x^2+36

(f) x^2+y^2

Answers

1. **(a)** 9; 9
(b) $(p+10)(p-10)$
(c) $(t+s)(t-s)$
(d) prime **(e)** prime **(f)** prime

2 Factor each difference of squares.

GS **(a)** $9m^2 - 49$

$= (\_\_\_)^2 - \_\_\_^2$

$= (\_\_\_\_)(\_\_\_\_)$

(b) $64a^2 - 25$

(c) $25a^2 - 64b^2$

3 Factor completely.

GS **(a)** $50r^2 - 32$

$= \_\_\_(\_\_\_\_)$

$= \_\_\_[(\_\_\_)^2 - \_\_\_^2]$

$= \_\_\_(\_\_\_\_)(\_\_\_\_)$

(b) $27y^2 - 75$

(c) $k^4 - 49$

(d) $81r^4 - 16$

Answers

2. **(a)** $3m$; 7; $3m + 7$; $3m - 7$
(b) $(8a + 5)(8a - 5)$
(c) $(5a + 8b)(5a - 8b)$

3. **(a)** 2; $25r^2 - 16$; 2; $5r$; 4; 2; $5r + 4$; $5r - 4$
(b) $3(3y + 5)(3y - 5)$
(c) $(k^2 + 7)(k^2 - 7)$
(d) $(9r^2 + 4)(3r + 2)(3r - 2)$

EXAMPLE 2 **Factoring Differences of Squares**

Factor each difference of squares.

$$a^2 - b^2 = (a + b)(a - b)$$

(a) $25m^2 - 16 = (5m)^2 - 4^2 = (5m + 4)(5m - 4)$

(b) $49z^2 - 64$

$= (7z)^2 - 8^2$ Write each term as a square.

$= (7z + 8)(7z - 8)$ Factor the difference of squares.

(c) $9x^2 - 4z^2$

$= (3x)^2 - (2z)^2$ Write each term as a square.

$= (3x + 2z)(3x - 2z)$ Factor the difference of squares.

◀ Work Problem **2** at the Side.

Note

Always check a factored form by multiplying.

EXAMPLE 3 **Factoring More Complex Differences of Squares**

Factor completely.

(a) $81y^2 - 36$

$= 9(9y^2 - 4)$ Factor out the GCF, 9.

$= 9[(3y)^2 - 2^2]$ Write each term as a square.

$= 9(3y + 2)(3y - 2)$ Factor the difference of squares.

(b) $p^4 - 36$

$= (p^2)^2 - 6^2$ Write each term as a square.

$= (p^2 + 6)(p^2 - 6)$ Factor the difference of squares.

Neither binomial can be factored further.

(c) $m^4 - 16$

$= (m^2)^2 - 4^2$ Write each term as a square.

$= (m^2 + 4)(m^2 - 4)$ Factor the difference of squares.

$= (m^2 + 4)(m + 2)(m - 2)$ Factor the difference of squares again.

Don't stop here.

CAUTION

Factor again when any of the factors is a difference of squares, as in **Example 3(c).** Check by multiplying.

◀ Work Problem **3** at the Side.

OBJECTIVE 2 Factor a perfect square trinomial. The expressions 144, $4x^2$, and $81m^6$ are **perfect squares** because

$$144 = 12^2, \quad 4x^2 = (2x)^2, \quad \text{and} \quad 81m^6 = (9m^3)^2.$$

A **perfect square trinomial** is a trinomial that is the square of a binomial. For example, $x^2 + 8x + 16$ is a perfect square trinomial because it is the square of the binomial $x + 4$.

$$x^2 + 8x + 16$$
$$= (x + 4)(x + 4)$$
$$= (x + 4)^2$$

On the one hand, a necessary condition for a trinomial to be a perfect square is that *two of its terms must be perfect squares*. For this reason,

$$16x^2 + 4x + 15$$

is *not* a perfect square trinomial because only the term $16x^2$ is a perfect square.

On the other hand, even if two of the terms are perfect squares, the trinomial may *not* be a perfect square trinomial. For example,

$$x^2 + 6x + 36$$

has two perfect square terms, but it is *not* a perfect square trinomial.

We can multiply to see that the square of a binomial gives one of the following perfect square trinomials.

Factoring Perfect Square Trinomials

$$a^2 + 2ab + b^2 = (a + b)^2$$
$$a^2 - 2ab + b^2 = (a - b)^2$$

The middle term of a perfect square trinomial is always twice the product of the two terms in the squared binomial. (See Section 5.4.) Use this to check any attempt to factor a trinomial that appears to be a perfect square.

EXAMPLE 4 Factoring a Perfect Square Trinomial

Factor $x^2 + 10x + 25$.

The x^2-term is a perfect square, and so is 25. Try to factor the trinomial

$$x^2 + 10x + 25 \quad \text{as} \quad (x + 5)^2.$$

To check, take twice the product of the two terms in the squared binomial.

$$2 \cdot x \cdot 5 = 10x \leftarrow \text{Middle term of } x^2 + 10x + 25$$

Twice; First term of binomial (x); Last term of binomial (5)

Since $10x$ is the middle term of the given trinomial, the trinomial is a perfect square.

$$x^2 + 10x + 25 \quad \text{factors as} \quad (x + 5)^2.$$

Work Problem 4 at the Side. ▶

4 Factor each trinomial.

GS **(a)** $p^2 + 14p + 49$

The term p^2 is a perfect square. Since $49 = \_\_\_\_^2$, 49 (*is* / *is not*) a perfect square. Try to factor the trinomial

$$p^2 + 14p + 49$$

as $(p + \_\_\_\_)^2$.

Check by taking twice the _____ of the two terms of the squared binomial.

$$2 \cdot p \cdot \_\_\_\_ = \_\_\_\_$$

The result (*is* / *is not*) the middle term of the given trinomial. The trinomial is a perfect square and factors as _____.

(b) $m^2 + 8m + 16$

(c) $x^2 + 2x + 1$

Answers

4. (a) 7; is; 7; product; 7; $14p$; is; $(p + 7)^2$
(b) $(m + 4)^2$ **(c)** $(x + 1)^2$

5 Factor each trinomial.

(a) $p^2 - 18p + 81$

(b) $16a^2 + 56a + 49$

(c) $121p^2 + 110p + 100$

(d) $64x^2 - 48x + 9$

(e) $27y^3 + 72y^2 + 48y$

Teaching Tip

Remind students to be sure to include the common factor in their answer when factoring a polynomial such as the one in Example 5(d).

EXAMPLE 5 Factoring Perfect Square Trinomials

Factor each trinomial.

(a) $x^2 - 22x + 121$

The first and last terms are perfect squares ($121 = 11^2$ or $(-11)^2$). Check to see whether the middle term of $x^2 - 22x + 121$ is twice the product of the first and last terms of the binomial $x - 11$.

$$2 \cdot x \cdot (-11) = -22x \leftarrow \text{Middle term of } x^2 - 22x + 121$$

Twice ↑ ↑ First term ↑ Last term

Thus, $x^2 - 22x + 121$ is a perfect square trinomial.

$$x^2 - 22x + 121 \quad \text{factors as} \quad (x - 11)^2.$$

Same sign

Notice that the sign of the second term in the squared binomial is the same as the sign of the middle term in the trinomial.

(b) $9m^2 - 24m + 16 = (3m)^2 + 2(3m)(-4) + (-4)^2 = (3m - 4)^2$

Twice ↑ ↑ First term ↑ Last Term

(c) $25y^2 + 20y + 16$

The first and last terms are perfect squares.

$$25y^2 = (5y)^2 \quad \text{and} \quad 16 = 4^2$$

Twice the product of the first and last terms of the binomial $5y + 4$ is

$$2 \cdot 5y \cdot 4 = 40y,$$

which is *not* the middle term of

$$25y^2 + 20y + 16.$$

This trinomial is not a perfect square. In fact, the trinomial cannot be factored even with the methods of the previous sections. It is a prime polynomial.

(d) $12z^3 + 60z^2 + 75z$

$= 3z(4z^2 + 20z + 25)$ — Factor out the common factor, $3z$.

$= 3z[(2z)^2 + 2(2z)(5) + 5^2]$ — $4z^2 + 20z + 25$ is a perfect square trinomial.

$= 3z(2z + 5)^2$ — Factor.

Note

1. The sign of the second term in the squared binomial is always the same as the sign of the middle term in the trinomial.
2. The first and last terms of a perfect square trinomial must be *positive*, because they are squares. For example, the polynomial $x^2 - 2x - 1$ cannot be a perfect square because the last term is negative.
3. Perfect square trinomials can also be factored using grouping or the FOIL method. Using the method of this section is often easier.

◀ Work Problem 5 at the Side.

Answers

5. **(a)** $(p - 9)^2$ **(b)** $(4a + 7)^2$ **(c)** prime **(d)** $(8x - 3)^2$ **(e)** $3y(3y + 4)^2$

6.5 Exercises

FOR EXTRA HELP

Download the MyDashBoard App

MyMathLab®

CONCEPT CHECK *Work each problem.*

1. To help factor a difference of squares, complete the following list of squares.

$1^2 =$ **1**	$2^2 =$ **4**	$3^2 =$ **9**	$4^2 =$ **16**	$5^2 =$ **25**
$6^2 =$ **36**	$7^2 =$ **49**	$8^2 =$ **64**	$9^2 =$ **81**	$10^2 =$ **100**
$11^2 =$ **121**	$12^2 =$ **144**	$13^2 =$ **169**	$14^2 =$ **196**	$15^2 =$ **225**
$16^2 =$ **256**	$17^2 =$ **289**	$18^2 =$ **324**	$19^2 =$ **361**	$20^2 =$ **400**

2. To use the factoring techniques described in this section, it is helpful to recognize fourth powers of integers. Complete the following list of fourth powers.

$1^4 =$ **1** $\quad 2^4 =$ **16** $\quad 3^4 =$ **81** $\quad 4^4 =$ **256** $\quad 5^4 =$ **625**

3. Which of the following are differences of squares? **A, D**

A. $x^2 - 4$ **B.** $y^2 + 9$ **C.** $2a^2 - 25$ **D.** $9m^2 - 1$

4. On a quiz, a student indicated *prime* when asked to factor $4x^2 + 16$, since she said that a sum of squares cannot be factored. **What Went Wrong?**

The binomial $4x^2 + 16$ can be factored as $4(x^2 + 4)$. *After* any common factor is removed, a sum of squares (like $x^2 + 4$ here) *cannot* be factored.

Factor each binomial completely. ***See Examples 1–3.***

5. $y^2 - 25$
$(y + 5)(y - 5)$

6. $t^2 - 16$
$(t + 4)(t - 4)$

7. $x^2 - 144$
$(x + 12)(x - 12)$

8. $x^2 - 400$
$(x + 20)(x - 20)$

9. $m^2 - 12$
prime

10. $k^2 - 18$
prime

11. $m^2 + 64$
prime

12. $k^2 + 49$
prime

13. $9r^2 - 4$
$(3r + 2)(3r - 2)$

14. $4x^2 - 9$
$(2x + 3)(2x - 3)$

15. $36x^2 - 16$
$4(3x + 2)(3x - 2)$

16. $32a^2 - 8$
$8(2a + 1)(2a - 1)$

17. $196p^2 - 225$
$(14p + 15)(14p - 15)$

18. $361q^2 - 400$
$(19q + 20)(19q - 20)$

19. $16r^2 - 25a^2$
$(4r + 5a)(4r - 5a)$

20. $49m^2 - 100p^2$
$(7m + 10p)(7m - 10p)$

21. $100x^2 + 49$
prime

22. $81w^2 + 16$
prime

23. $p^4 - 49$
$(p^2 + 7)(p^2 - 7)$

24. $r^4 - 25$
$(r^2 + 5)(r^2 - 5)$

25. $x^4 - 1$
$(x^2 + 1)(x + 1)(x - 1)$

26. $y^4 - 10{,}000$
$(y^2 + 100)(y + 10)(y - 10)$

27. $p^4 - 256$
$(p^2 + 16)(p + 4)(p - 4)$

28. $x^4 - 625$
$(x^2 + 25)(x + 5)(x - 5)$

29. **CONCEPT CHECK** Which of the following are perfect square trinomials? **B, C**

A. $y^2 - 13y + 36$ **B.** $x^2 + 6x + 9$ **C.** $4z^2 - 4z + 1$ **D.** $16m^2 + 10m + 1$

30. In the polynomial $9y^2 + 14y + 25$, the first and last terms are perfect squares. Can the polynomial be factored? If it can, factor it. If it cannot, explain why it is not a perfect square trinomial.

No, it is not a perfect square since the middle term would have to be 30y.

Factor each trinomial completely. It may be necessary to factor out the greatest common factor first. ***See Examples 4 and 5.***

31. $w^2 + 2w + 1$ $(w + 1)^2$

32. $p^2 + 4p + 4$ $(p + 2)^2$

33. $x^2 - 8x + 16$ $(x - 4)^2$

34. $x^2 - 10x + 25$ $(x - 5)^2$

35. $x^2 - 10x + 100$ prime

36. $x^2 - 18x + 36$ prime

37. $2x^2 + 24x + 72$ $2(x + 6)^2$

38. $3y^2 + 48y + 192$ $3(y + 8)^2$

39. $4x^2 + 12x + 9$ $(2x + 3)^2$

40. $25x^2 + 10x + 1$ $(5x + 1)^2$

41. $16x^2 - 40x + 25$ $(4x - 5)^2$

42. $36y^2 - 60y + 25$ $(6y - 5)^2$

43. $49x^2 - 28xy + 4y^2$ $(7x - 2y)^2$

44. $4z^2 - 12zw + 9w^2$ $(2z - 3w)^2$

45. $64x^2 + 48xy + 9y^2$ $(8x + 3y)^2$

46. $9t^2 + 24tr + 16r^2$ $(3t + 4r)^2$

47. $-50h^3 + 40h^2y - 8hy^2$ $-2h(5h - 2y)^2$

48. $-18x^3 - 48x^2y - 32xy^2$ $-2x(3x + 4y)^2$

Although we usually factor polynomials using integers, we can apply the same concepts to factoring using fractions and decimals.

$$z^2 - \frac{9}{16}$$

$$= z^2 - \left(\frac{3}{4}\right)^2 \qquad \frac{9}{16} = \left(\frac{3}{4}\right)^2$$

$$= \left(z + \frac{3}{4}\right)\left(z - \frac{3}{4}\right) \qquad \text{Factor the difference of squares.}$$

Factor each binomial or trinomial.

49. $p^2 - \frac{1}{9}$ $\left(p + \frac{1}{3}\right)\left(p - \frac{1}{3}\right)$

50. $q^2 - \frac{1}{4}$ $\left(q + \frac{1}{2}\right)\left(q - \frac{1}{2}\right)$

51. $4m^2 - \frac{9}{25}$ $\left(2m + \frac{3}{5}\right)\left(2m - \frac{3}{5}\right)$

52. $100b^2 - \frac{4}{49}$ $\left(10b + \frac{2}{7}\right)\left(10b - \frac{2}{7}\right)$

53. $x^2 - 0.64$ $(x + 0.8)(x - 0.8)$

54. $y^2 - 0.36$ $(y + 0.6)(y - 0.6)$

55. $t^2 + t + \frac{1}{4}$ $\left(t + \frac{1}{2}\right)^2$

56. $m^2 + \frac{2}{3}m + \frac{1}{9}$ $\left(m + \frac{1}{3}\right)^2$

57. $x^2 - 1.0x + 0.25$ $(x - 0.5)^2$

58. $y^2 - 1.4y + 0.49$ $(y - 0.7)^2$

Summary Exercises *Recognizing and Applying Factoring Strategies*

Summary Exercises provide mixed practice to better prepare students for tests.

When factoring, ask these questions to decide on a suitable factoring technique.

Factoring a Polynomial

Question 1 **Is there a common factor other than 1?** If so, factor it out.

Question 2 **How many terms are in the polynomial?**

Two terms: Check to see whether it is a difference of squares. If so, factor as in **Section 6.5.**

Three terms: Is it a perfect square trinomial? In this case, factor as in **Section 6.5.**

If the trinomial is not a perfect square, check to see whether the coefficient of the second-degree term is 1. If so, use the method of **Section 6.2.**

If the coefficient of the second-degree term is not 1, use the general factoring methods of **Sections 6.3 and 6.4.**

Four terms: Try to factor the polynomial by grouping, as in **Section 6.1.**

Question 3 **Can any factors be factored further?** If so, factor them.

CAUTION

Be careful when checking your answer to a factoring problem.

1. *Check* that the product of all the factors does indeed yield the original polynomial.
2. *Check* that the original polynomial has been factored *completely*. (See Question 3 above.)

Checking by multiplication alone will not ensure that you have factored the original polynomial completely.

Suppose we factored the trinomial

$$6x^2 + 24xy + 24y^2 \quad \text{as} \quad (2x + 4y)(3x + 6y).$$

When we check by multiplying, we might conclude that our answer is correct.

$$(2x + 4y)(3x + 6y)$$
$$= 6x^2 + 24xy + 24y^2 \quad \text{FOIL method}$$

However, we would not have factored the given polynomial *completely* because the terms of the binomial $2x + 4y$ have a common factor of 2 and the terms of the binomial $3x + 6y$ have a common factor of 3. Factor out the GCF, 6 (which is $2 \cdot 3$), at the beginning, as follows.

$$6x^2 + 24xy + 24y^2$$
$$= 6(x^2 + 4xy + 4y^2) \quad \text{Factor out the GCF, 6.}$$
$$= 6(x + 2y)^2 \quad \text{Factor the perfect square trinomial.}$$

Remember to factor out the GCF first.

Teaching Tip

Consider working a few additional classroom examples.

1. $24m^2 - 42my + 9y^2$
2. $32x^2 + 80x + 50$
3. $45x^2 - 20y^2$
4. $3t^2 - t + 6$

Answers:

1. $3(4m - y)(2m - 3y)$
2. $2(4x + 5)^2$
3. $5(3x + 2y)(3x - 2y)$
4. prime

EXAMPLE Applying Factoring Strategies

Factor $12x^2 + 26xy + 12y^2$ completely.

Question 1 **Is there a common factor other than 1?** Yes, 2 is a common factor, so factor it out.

$$12x^2 + 26xy + 12y^2$$
$$= 2(6x^2 + 13xy + 6y^2)$$

Question 2 **How many terms are in the polynomial?** The polynomial $6x^2 + 13xy + 6y^2$ has three terms, but it is not a perfect square. To factor the trinomial by grouping, as in **Section 6.3,** begin by finding two integers with a product of $6 \cdot 6$, or 36, and a sum of 13. These integers are 4 and 9.

$12x^2 + 26xy + 12y^2$

$= 2(6x^2 + 13xy + 6y^2)$ — Factor out the GCF, 2.

$= 2(6x^2 + 4xy + 9xy + 6y^2)$ — $4 \cdot 9 = 36$; $4 + 9 = 13$

$= 2[(6x^2 + 4xy) + (9xy + 6y^2)]$ — Group the terms.

$= 2[2x(3x + 2y) + 3y(3x + 2y)]$ — Factor each group.

$= 2(3x + 2y)(2x + 3y)$ — Factor out the common factor, $3x + 2y$.

The trinomial $6x^2 + 13xy + 6y^2$ could also be factored by trial and error, using the FOIL method in reverse, as in **Section 6.4.**

Question 3 **Can any factors be factored further?** No. The original polynomial has been factored completely.

Teaching Tip

Work **Exercises 1–10** with students, using the questions in the box on the preceding page to help them decide on a suitable factoring technique for each exercise.

CONCEPT CHECK *Match each polynomial in Column I with the best choice for factoring it in Column II. The choices in Column II may be used once, more than once, or not at all.*

I

1. $12x^2 + 20x + 8$ F
2. $x^2 - 17x + 72$ G
3. $-16m^2n + 24mn - 40mn^2$ A
4. $64a^2 - 121b^2$ B
5. $36p^2 - 60pq + 25q^2$ D
6. $z^2 - 4z + 6$ H
7. $625 - r^4$ C
8. $x^6 + 4x^4 - 3x^2 - 12$ E
9. $4w^2 + 49$ H
10. $144 - 24z + z^2$ D

II

A. Factor out the GCF. No further factoring is possible.

B. Factor a difference of squares once.

C. Factor a difference of squares twice.

D. Factor a perfect square trinomial.

E. Factor by grouping.

F. Factor out the GCF. Then factor a trinomial by grouping or trial and error.

G. Factor into two binomials by finding two integers whose product is the constant in the trinomial and whose sum is the coefficient of the middle term.

H. The polynomial is prime.

Factor each polynomial completely. Check by multiplying.

11. $32m^9 + 16m^5 + 24m^3$
$8m^3(4m^6 + 2m^2 + 3)$

12. $2m^2 - 10m - 48$
$2(m + 3)(m - 8)$

13. $14k^3 + 7k^2 - 70k$
$7k(2k + 5)(k - 2)$

14. $9z^2 + 64$
prime

15. $6z^2 + 31z + 5$
$(6z + 1)(z + 5)$

16. $m^2 - 3mn - 4n^2$
$(m + n)(m - 4n)$

17. $49z^2 - 16y^2$
$(7z + 4y)(7z - 4y)$

18. $100n^2r^2 + 30nr^3 - 50n^2r$
$10nr(10nr + 3r^2 - 5n)$

19. $16x^2 + 20x$
$4x(4x + 5)$

20. $20 + 5m + 12n + 3mn$
$(4 + m)(5 + 3n)$

21. $10y^2 - 7yz - 6z^2$
$(5y - 6z)(2y + z)$

22. $y^4 - 81$
$(y^2 + 9)(y + 3)(y - 3)$

23. $m^2 + 2m - 15$
$(m - 3)(m + 5)$

24. $6y^2 - 5y - 4$
$(2y + 1)(3y - 4)$

25. $32z^3 + 56z^2 - 16z$
$8z(4z - 1)(z + 2)$

26. $p^2 - 24p + 144$
$(p - 12)^2$

27. $z^2 - 12z + 36$
$(z - 6)^2$

28. $9m^2 - 64$
$(3m + 8)(3m - 8)$

29. $y^2 - 4yk - 12k^2$
$(y - 6k)(y + 2k)$

30. $16z^2 - 8z + 1$
$(4z - 1)^2$

31. $6y^2 - 6y - 12$
$6(y - 2)(y + 1)$

32. $x^2 + 2x + 16$
prime

33. $p^2 - 17p + 66$
$(p - 6)(p - 11)$

34. $a^2 + 17a + 72$
$(a + 8)(a + 9)$

35. $k^2 + 100$
prime

36. $108m^2 - 36m + 3$
$3(6m - 1)^2$

37. $z^2 - 3za - 10a^2$
$(z + 2a)(z - 5a)$

38. $2a^3 + a^2 - 14a - 7$
$(2a + 1)(a^2 - 7)$

39. $4k^2 - 12k + 9$
$(2k - 3)^2$

40. $a^2 - 3ab - 28b^2$
$(a - 7b)(a + 4b)$

41. $16r^2 + 24rm + 9m^2$
$(4r + 3m)^2$

42. $3k^2 + 4k - 4$
$(3k - 2)(k + 2)$

43. $n^2 - 12n - 35$
prime

44. $a^4 - 625$
$(a^2 + 25)(a + 5)(a - 5)$

45. $16k^2 - 48k + 36$
$4(2k - 3)^2$

46. $8k^2 - 10k - 3$
$(4k + 1)(2k - 3)$

47. $36y^6 - 42y^5 - 120y^4$
$6y^4(3y + 4)(2y - 5)$

48. $5z^3 - 45z^2 + 70z$
$5z(z - 2)(z - 7)$

49. $8p^2 + 23p - 3$
$(8p - 1)(p + 3)$

50. $8k^2 - 2kh - 3h^2$
$(4k - 3h)(2k + h)$

51. $54m^2 - 24z^2$
$6(3m + 2z)(3m - 2z)$

52. $4k^2 - 20kz + 25z^2$
$(2k - 5z)^2$

53. $6a^2 + 10a - 4$
$2(3a - 1)(a + 2)$

54. $15h^2 + 11hg - 14g^2$
$(3h - 2g)(5h + 7g)$

55. $28a^2 - 63b^2$
$7(2a + 3b)(2a - 3b)$

56. $10z^2 - 7z - 6$
$(5z - 6)(2z + 1)$

57. $125m^4 - 400m^3n + 195m^2n^2$
$5m^2(5m - 13n)(5m - 3n)$

58. $9y^2 + 12y - 5$
$(3y - 1)(3y + 5)$

59. $9u^2 + 66uv + 121v^2$
$(3u + 11v)^2$

60. $36x^2 + 32x + 9$
prime

61. $27p^{10} - 45p^9 - 252p^8$
$9p^8(3p + 7)(p - 4)$

62. $10m^2 + 25m - 60$
$5(2m - 3)(m + 4)$

63. $4 - 2q - 6p + 3pq$
$(2 - q)(2 - 3p)$

64. $k^2 - 121$
$(k + 11)(k - 11)$

65. $64p^2 - 100m^2$
$4(4p + 5m)(4p - 5m)$

66. $m^3 + 4m^2 - 6m - 24$
$(m + 4)(m^2 - 6)$

67. $100a^2 - 81y^2$
$(10a + 9y)(10a - 9y)$

68. $8a^2 + 23ab - 3b^2$
$(8a - b)(a + 3b)$

69. $a^2 + 8a + 16$
$(a + 4)^2$

70. $4y^2 - 25$
$(2y + 5)(2y - 5)$

71. $2x^2 + 5x + 6$
prime

72. $-3x^3 + 12xy^2$
$-3x(x + 2y)(x - 2y)$

73. $25a^2 - 70ab + 49b^2$
$(5a - 7b)^2$

74. $8t^4 - 8$
$8(t^2 + 1)(t + 1)(t - 1)$

75. $-4x^2 + 24xy - 36y^2$
$-4(x - 3y)^2$

76. $100a^2 - 25b^2$
$25(2a + b)(2a - b)$

77. $-2x^2 + 26x - 72$
$-2(x - 9)(x - 4)$

78. $2m^2 - 15n - 5mn + 6m$
$(m + 3)(2m - 5n)$

79. $12x^2 + 22x - 20$
$2(2x + 5)(3x - 2)$

80. $y^6 + 5y^4 - 3y^2 - 15$
$(y^2 + 5)(y^4 - 3)$

81. $y^2 - 64$
$(y + 8)(y - 8)$

82. $12p^3 - 54p^2 - 30p$
$6p(2p + 1)(p - 5)$

6.6 Solving Quadratic Equations by Factoring

OBJECTIVES

1. Solve quadratic equations by factoring.
2. Solve other equations by factoring.

Galileo Galilei developed theories to explain physical phenomena. According to legend, Galileo dropped objects of different weights from the Leaning Tower of Pisa to disprove the belief that heavier objects fall faster than lighter objects. He developed a formula for freely falling objects described by

$$d = 16t^2,$$

where d is the distance in feet that an object falls (disregarding air resistance) in t seconds, regardless of weight.

The equation $d = 16t^2$ is a *quadratic equation*. A quadratic equation contains a second-degree (squared) term and no terms of greater degree.

Galileo Galilei (1564–1642)

Quadratic Equation

A **quadratic equation** is an equation that can be written in the form

$$ax^2 + bx + c = 0,$$

where a, b, and c are real numbers, with $a \neq 0$. The given form is called **standard form.**

$x^2 + 5x + 6 = 0, \quad 2t^2 - 5t = 3, \quad y^2 = 4$ Quadratic equations

In these examples, only $x^2 + 5x + 6 = 0$ is in standard form.

Work Problems 1 and 2 at the Side. ▶

We have factored many quadratic *expressions* of the form $ax^2 + bx + c$. We use factored quadratic expressions to solve quadratic *equations*.

OBJECTIVE 1 Solve quadratic equations by factoring. We do this using the **zero-factor property.**

Zero-Factor Property

If a and b are real numbers and $ab = 0$, then $a = 0$ or $b = 0$.

In words, if the product of two numbers is 0, then at least one of the numbers must be 0. One number *must* be 0, but both *may* be 0.

EXAMPLE 1 Using the Zero-Factor Property

Solve each equation.

(a) $(x + 3)(2x - 1) = 0$

The product $(x + 3)(2x - 1)$ is equal to 0. By the zero-factor property, the only way that the product of these two factors can be 0 is if at least one of the factors equals 0. Therefore, either $x + 3 = 0$ or $2x - 1 = 0$.

$\mathbf{x + 3 = 0}$ or $\mathbf{2x - 1 = 0}$ Zero-factor property

$x = -3$ or $2x = 1$ Solve each equation.

$x = \frac{1}{2}$ Divide each side by 2.

Continued on Next Page

1 Which of the following equations are quadratic equations?

A. $y^2 - 4y - 5 = 0$

B. $x^3 - x^2 + 16 = 0$

C. $2z^2 + 7z = -3$

D. $x + 2y = -4$

2 Write each quadratic equation in standard form.

(a) $x^2 - 3x = 4$

(b) $y^2 = 9y - 8$

Answers

1. A, C
2. (a) $x^2 - 3x - 4 = 0$
   (b) $y^2 - 9y + 8 = 0$

3 Solve each equation. Check your solutions.

GS **(a)** $(x-5)(x+2)=0$

To solve, use the zero-factor property.

$\_\_\_\_ = 0$ or $\_\_\_\_ = 0$

Solve these two equations, obtaining

$x = \_\_\_\_$ or $x = \_\_\_\_$.

CHECK Let $x = 5$.

$$(x-5)(x+2)=0$$
$$(\_\_\_\_-5)(\_\_\_\_+2) \stackrel{?}{=} 0$$
$$\_\_\_\_(7)=0$$

(*True* / *False*)

To check the other solution, substitute ____ for x in the original equation. A (*true* / *false*) statement results.

The solution set is $\{\_\_\_, \_\_\_\}$.

(b) $(3x-2)(x+6)=0$

(c) $z(2z+5)=0$

Teaching Tip

To introduce the zero-factor property, tell students that you are thinking of two numbers with product 0. Ask them what they can tell you about the two numbers.

Answers

3. **(a)** $x-5$; $x+2$; 5; -2; 5; 5; 0; True; -2; true; 5; -2

(b) $\left\{-6, \frac{2}{3}\right\}$ **(c)** $\left\{-\frac{5}{2}, 0\right\}$

The original equation, $(x+3)(2x-1)=0$, has two solutions, -3 and $\frac{1}{2}$. *Check* these solutions by substituting -3 for x in this equation. ***Then start over*** and substitute $\frac{1}{2}$ for x.

CHECK Let $x = -3$.

$$(x+3)(2x-1)=0$$
$$(-3+3)[2(-3)-1] \stackrel{?}{=} 0$$
$$0(-7) \stackrel{?}{=} 0$$
$$0 = 0 \ \checkmark \text{ True}$$

Let $x = \frac{1}{2}$.

$$(x+3)(2x-1)=0$$
$$\left(\frac{1}{2}+3\right)\left(2\cdot\frac{1}{2}-1\right) \stackrel{?}{=} 0$$
$$\frac{7}{2}(1-1) \stackrel{?}{=} 0$$
$$0 = 0 \ \checkmark \text{ True}$$

Both -3 and $\frac{1}{2}$ result in true equations, so the solution set is $\left\{-3, \frac{1}{2}\right\}$.

(b)
$$y(3y-4)=0$$

$y=0$ or $3y-4=0$ — Zero-factor property

$3y = 4$ — Add 4.

$y = \frac{4}{3}$ — Divide by 3.

Don't forget that 0 is a solution.

Check these solutions by substituting each one in the original equation. The solution set is $\left\{0, \frac{4}{3}\right\}$.

◀ Work Problem **3** at the Side.

Note

The word *or* as used in **Example 1** means "one or the other or both."

If the polynomial in an equation is not already factored, first make sure that the equation is in standard form. Then factor and solve.

EXAMPLE 2 Solving Quadratic Equations

Solve each equation.

(a) $x^2-5x=-6$

First, write the equation in standard form by adding 6 to each side.

Don't factor x out at this step.

$$x^2-5x=-6$$

$x^2-5x+6=0$ — Add 6.

Now factor x^2-5x+6. Find two numbers whose product is 6 and whose sum is -5. These two numbers are -2 and -3, so we factor as follows.

$(x-2)(x-3)=0$ — Factor.

$x-2=0$ or $x-3=0$ — Zero-factor property

$x=2$ or $x=3$ — Solve each equation.

Continued on Next Page

CHECK Let $x = 2$.

$$x^2 - 5x = -6$$
$$2^2 - 5(2) \stackrel{?}{=} -6$$
$$4 - 10 \stackrel{?}{=} -6$$
$$-6 = -6 \checkmark \text{ True}$$

Let $x = 3$.

$$x^2 - 5x = -6$$
$$3^2 - 5(3) \stackrel{?}{=} -6$$
$$9 - 15 \stackrel{?}{=} -6$$
$$-6 = -6 \checkmark \text{ True}$$

Both solutions check, so the solution set is $\{2, 3\}$.

(b) $y^2 = y + 20$ — Write this equation in standard form.

Standard form ⟶ $y^2 - y - 20 = 0$	Subtract y and 20.
$(y-5)(y+4) = 0$	Factor.
$y - 5 = 0$ or $y + 4 = 0$	Zero-factor property
$y = 5$ or $y = -4$	Solve each equation.

Check by substituting in the original equation. The solution set is $\{-4, 5\}$.

Work Problem 4 at the Side. ▶

4 Solve each equation. Check your solutions.

(a) $m^2 - 3m - 10 = 0$

(b) $r^2 + 2r = 8$

Teaching Tip

Distinguish between solving linear and quadratic equations.

Solving a Quadratic Equation by Factoring

Step 1 **Write the equation in standard form,** that is, with all terms on one side of the equality symbol in descending powers of the variable and 0 on the other side.

Step 2 **Factor** completely.

Step 3 **Use the zero-factor property** to set each factor with a variable equal to 0.

Step 4 **Solve** the resulting equations.

Step 5 **Check** each solution in the original equation.

5 Solve each equation. Check your solutions.

GS **(a)** $10a^2 - 5a - 15 = 0$

$\_\_\_\_(\_\_\_\_ - a - 3) = 0$

Divide each side by ____.

$2a^2 - a - 3 = 0$

$(2a - 3)(\_\_\_\_\_\_) = 0$

$2a - 3 = \_\_\_\_$ or $\_\_\_\_\_\_ = 0$

$a = \_\_\_\_$ or $a = \_\_\_\_$

Check by substituting the solutions in the original equation.

The solution set is ______.

(b) $4x^2 - 2x = 42$

EXAMPLE 3 Solving a Quadratic Equation (Common Factor)

Solve $4p^2 + 40 = 26p$.

$4p^2 + 40 = 26p$	
$4p^2 - 26p + 40 = 0$	Standard form
$2(2p^2 - 13p + 20) = 0$	Factor out 2.
$2p^2 - 13p + 20 = 0$	Divide each side by 2.
$(2p-5)(p-4) = 0$	Factor.
$2p - 5 = 0$ or $p - 4 = 0$	Zero-factor property
$p = \frac{5}{2}$ or $p = 4$	Solve each equation.

This 2 is *not* a solution of the equation.

Check that the solution set is $\{\frac{5}{2}, 4\}$ by substituting in the original equation.

Work Problem 5 at the Side. ▶

CAUTION

A common error is to include the common factor 2 as a solution in **Example 3.** ***Only factors containing variables lead to solutions.***

Answers

4. **(a)** $\{-2, 5\}$ **(b)** $\{-4, 2\}$

5. **(a)** 5; $2a^2$; 5; $a + 1$; 0; $a + 1$; $\frac{3}{2}$; -1; $\left\{-1, \frac{3}{2}\right\}$, or $\left\{\frac{3}{2}, -1\right\}$

(b) $\left\{-3, \frac{7}{2}\right\}$

6 Solve each equation. Check your solutions.

(a) $49m^2 - 9 = 0$

(b) $m^2 = 3m$

(c) $p(4p + 7) = 2$

EXAMPLE 4 Solving Quadratic Equations

Solve each equation.

(a)

$$16m^2 - 25 = 0$$

$$(4m + 5)(4m - 5) = 0 \quad \text{Factor the difference of squares. (Section 6.5)}$$

$$4m + 5 = 0 \quad \text{or} \quad 4m - 5 = 0 \quad \text{Zero-factor property}$$

$$4m = -5 \quad \text{or} \quad 4m = 5 \quad \text{Solve each equation.}$$

$$m = -\frac{5}{4} \quad \text{or} \quad m = \frac{5}{4} \quad \text{Divide by 4.}$$

Check the solutions $-\frac{5}{4}$ and $\frac{5}{4}$ in the original equation. The solution set is $\left\{-\frac{5}{4}, \frac{5}{4}\right\}$.

(b)

$$y^2 = 2y$$

$$y^2 - 2y = 0 \quad \text{Standard form}$$

Don't forget to set the variable factor y equal to 0.

$$y(y - 2) = 0 \quad \text{Factor.}$$

$$y = 0 \quad \text{or} \quad y - 2 = 0 \quad \text{Zero-factor property}$$

$$y = 2 \quad \text{Solve.}$$

The solution set is $\{0, 2\}$.

(c)

$$k(2k + 5) = 3$$

To be in standard form, 0 must be on one side.

$$2k^2 + 5k = 3 \quad \text{Multiply.}$$

$$\text{Standard form} \longrightarrow 2k^2 + 5k - 3 = 0 \quad \text{Subtract 3.}$$

$$(2k - 1)(k + 3) = 0 \quad \text{Factor.}$$

$$2k - 1 = 0 \quad \text{or} \quad k + 3 = 0 \quad \text{Zero-factor property}$$

$$2k = 1 \quad \text{or} \quad k = -3 \quad \text{Solve each equation.}$$

$$k = \frac{1}{2}$$

The solution set is $\left\{-3, \frac{1}{2}\right\}$.

CAUTION

In **Example 4(b),** it is tempting to begin by dividing both sides of

$$y^2 = 2y$$

by y to get $y = 2$. We do not get the other solution, 0, if we divide by a variable. (We *may* divide each side of an equation by a *nonzero* real number, however. In **Example 3** we divided each side by 2.)

In **Example 4(c),** we cannot use the zero-factor property to solve

$$k(2k + 5) = 3$$

in its given form because of the 3 on the right side of the equation. ***The zero-factor property applies only to a product that equals 0.***

Work Problem 6 at the Side.

Answers

6. (a) $\left\{-\frac{3}{7}, \frac{3}{7}\right\}$ **(b)** $\{0, 3\}$ **(c)** $\left\{-2, \frac{1}{4}\right\}$

EXAMPLE 5 Solving Quadratic Equations (Double Solutions)

Solve each equation.

(a) $z^2 - 22z + 121 = 0$

$(z - 11)^2 = 0$ Factor the perfect square trinomial.

$(z - 11)(z - 11) = 0$ $a^2 = a \cdot a$

$z - 11 = 0$ or $z - 11 = 0$ Zero-factor property

Because the two factors are identical, they both lead to the same solution, called a **double solution.**

$z = 11$ Add 11.

CHECK $z^2 - 22z + 121 = 0$ Original equation

$11^2 - 22(11) + 121 \stackrel{?}{=} 0$ Let $z = 11$.

$121 - 242 + 121 \stackrel{?}{=} 0$ Simplify.

$0 = 0$ ✓ True

The solution set is $\{11\}$.

(b) $9t^2 - 30t = -25$

$9t^2 - 30t + 25 = 0$ Standard form

$(3t - 5)^2 = 0$ Factor the perfect square trinomial.

$3t - 5 = 0$ or $3t - 5 = 0$ Zero-factor property

$3t = 5$ Add 5.

$t = \frac{5}{3}$ $\frac{5}{3}$ is a double solution.

CHECK $9t^2 - 30t = -25$ Original equation

$9\left(\frac{5}{3}\right)^2 - 30\left(\frac{5}{3}\right) \stackrel{?}{=} -25$ Let $t = \frac{5}{3}$.

$9\left(\frac{25}{9}\right) - 30\left(\frac{5}{3}\right) \stackrel{?}{=} -25$ Apply the exponent.

$25 - 50 \stackrel{?}{=} -25$ Multiply.

$-25 = -25$ ✓ True

The solution set is $\left\{\frac{5}{3}\right\}$.

Work Problem 7 at the Side. ▶

CAUTION

Each equation in **Example 5** has only *one* distinct solution. ***We will write a double solution only once in a solution set.***

7 Solve each equation. Check your solutions.

GS **(a)** $x^2 + 16x = -64$

$x^2 + 16x + \_\_\_\_ = 0$

$(\_\_\_\_)^2 = 0$

$\_\_\_\_ = 0$ or $x + 8 = \_\_\_\_$

Solve to obtain $x = \_\_\_\_$, so -8 is a ____ solution. The solution set is ____.

(b) $4x^2 - 4x + 1 = 0$

(c) $4z^2 + 20z = -25$

Answers

7. **(a)** 64; $x + 8$; $x + 8$; 0; -8; double; $\{-8\}$

(b) $\left\{\frac{1}{2}\right\}$ **(c)** $\left\{-\frac{5}{2}\right\}$

8 Solve each equation. Check your solutions.

(a) $r^3 - 16r = 0$

(b) $x^3 - 3x^2 - 18x = 0$

9 Solve each equation. Check your solutions.

(a) $(m + 3)(m^2 - 11m + 10) = 0$

(b) $(2x + 5)(4x^2 - 9) = 0$

Answers

8. (a) $\{-4, 0, 4\}$ **(b)** $\{-3, 0, 6\}$

9. (a) $\{-3, 1, 10\}$ **(b)** $\left\{-\frac{5}{2}, -\frac{3}{2}, \frac{3}{2}\right\}$

Note

Not all quadratic equations can be solved by factoring. A more general method for solving such equations is given in **Chapter 9.**

OBJECTIVE 2 Solve other equations by factoring. We can extend the zero-factor property to solve equations that involve more than two factors with variables. (These equations will have at least one term greater than second degree. They are *not* quadratic equations.)

EXAMPLE 6 Solving an Equation with More Than Two Variable Factors

Solve $6z^3 - 6z = 0$.

$$6z^3 - 6z = 0$$

$$6z(z^2 - 1) = 0 \quad \text{Factor out } 6z.$$

$$6z(z + 1)(z - 1) = 0 \quad \text{Factor } z^2 - 1.$$

By an extension of the zero-factor property, this product can equal 0 only if at least one of the factors equals 0. Write and solve three equations, one for each factor with a variable.

$$6z = 0 \quad \text{or} \quad z + 1 = 0 \quad \text{or} \quad z - 1 = 0 \quad \text{Zero-factor property}$$

$$z = 0 \quad \text{or} \quad z = -1 \quad \text{or} \quad z = 1 \quad \text{Solve each equation.}$$

Check by substituting, in turn, 0, -1, and 1 in the original equation. The solution set is $\{-1, 0, 1\}$.

◀ **Work Problem 8 at the Side.**

EXAMPLE 7 Solving an Equation with a Quadratic Factor

Solve $(2x - 1)(x^2 - 9x + 20) = 0$.

$$(2x - 1)(x^2 - 9x + 20) = 0$$

$$(2x - 1)(x - 5)(x - 4) = 0 \quad \text{Factor } x^2 - 9x + 20.$$

$$2x - 1 = 0 \quad \text{or} \quad x - 5 = 0 \quad \text{or} \quad x - 4 = 0 \quad \text{Zero-factor property}$$

$$x = \frac{1}{2} \quad \text{or} \quad x = 5 \quad \text{or} \quad x = 4 \quad \text{Solve each equation.}$$

Check to verify that the solution set is $\left\{\frac{1}{2}, 4, 5\right\}$.

◀ **Work Problem 9 at the Side.**

CAUTION

In **Example 7,** it would be unproductive to begin by multiplying the two factors together. The zero-factor property requires the *product* of two or more factors to equal 0. ***Always consider first whether an equation is given in the appropriate form to apply the zero-factor property.***

6.6 Exercises

FOR EXTRA HELP

Download the MyDashBoard App

MyMathLab®

CONCEPT CHECK *In Exercises 1–4, fill in each blank with the correct response.*

1. A quadratic equation in x is an equation that can be put into the form **$ax^2 + bx + c$** $= 0$.

2. The form $ax^2 + bx + c = 0$ is called **standard** form.

3. If a quadratic equation is in standard form, to solve the equation we should begin by attempting to **factor** the polynomial.

4. If the product of two numbers is 0, then at least one of the numbers is **0**. This is the **zero**-**factor** property.

CONCEPT CHECK *Work each problem.*

5. Identify each equation as *linear* or *quadratic*.

(a) $2x - 5 = 6$ **(b)** $x^2 - 5 = -4$

(c) $x^2 + 2x - 1 = 2x^2$ **(d)** $5^2x + 2 = 0$

(a) linear (b) quadratic (c) quadratic (d) linear

6. The number 9 is a *double solution* of the equation

$$(x - 9)^2 = 0.$$

Why is this so?

Because $(x - 9)^2 = (x - 9)(x - 9)$, applying the zero-factor property leads to two solutions of 9. Thus, 9 is a double solution.

7. Look at this "solution." **What Went Wrong?**

$$2x(3x - 4) = 0$$

$$x = 2 \quad \text{or} \quad x = 0 \quad \text{or} \quad 3x - 4 = 0$$

$$x = \frac{4}{3}$$

The solution set is $\{2, 0, \frac{4}{3}\}$.

Set each *variable* factor equal to 0, to get $2x = 0$ or $3x - 4 = 0$. The solution set is $\{0, \frac{4}{3}\}$.

8. Look at this "solution." **What Went Wrong?**

$$x(7x - 1) = 0$$

$$7x - 1 = 0 \quad \text{Zero-factor property}$$

$$x = \frac{1}{7}$$

The solution set is $\{\frac{1}{7}\}$.

The variable x is another factor to set equal to 0, so the solution set is $\{0, \frac{1}{7}\}$.

Solve each equation, and check your solutions. ***See Example 1.***

9. $(x + 5)(x - 2) = 0$ **$\{-5, 2\}$**

10. $(x - 1)(x + 8) = 0$ **$\{-8, 1\}$**

11. $(2m - 7)(m - 3) = 0$ **$\left\{3, \frac{7}{2}\right\}$**

12. $(6k + 5)(k + 4) = 0$ **$\left\{-4, -\frac{5}{6}\right\}$**

13. $(2x + 1)(6x - 1) = 0$ **$\left\{-\frac{1}{2}, \frac{1}{6}\right\}$**

14. $(3x + 2)(10x - 1) = 0$ **$\left\{-\frac{2}{3}, \frac{1}{10}\right\}$**

15. $t(6t + 5) = 0$ **$\left\{-\frac{5}{6}, 0\right\}$**

16. $w(4w + 1) = 0$ **$\left\{-\frac{1}{4}, 0\right\}$**

17. $2x(3x - 4) = 0$ **$\left\{0, \frac{4}{3}\right\}$**

18. $6x(4x + 9) = 0$ **$\left\{-\frac{9}{4}, 0\right\}$**

19. $(x - 9)(x - 9) = 0$ **$\{9\}$**

20. $(2x + 1)(2x + 1) = 0$ **$\left\{-\frac{1}{2}\right\}$**

Solve each equation, and check your solutions. ***See Examples 2–7.***

21. $y^2 + 3y + 2 = 0$
$\{-2, -1\}$

22. $p^2 + 8p + 7 = 0$
$\{-7, -1\}$

23. $y^2 - 3y + 2 = 0$
$\{1, 2\}$

24. $r^2 - 4r + 3 = 0$
$\{1, 3\}$

25. $x^2 = 24 - 5x$
$\{-8, 3\}$

26. $t^2 = 2t + 15$
$\{-3, 5\}$

27. $x^2 = 3 + 2x$
$\{-1, 3\}$

28. $m^2 = 4 + 3m$
$\{-1, 4\}$

29. $z^2 + 3z = -2$
$\{-2, -1\}$

30. $p^2 - 2p = 3$
$\{-1, 3\}$

31. $m^2 + 8m + 16 = 0$
$\{-4\}$

32. $b^2 - 6b + 9 = 0$
$\{3\}$

33. $3x^2 + 5x - 2 = 0$
$\left\{-2, \frac{1}{3}\right\}$

34. $6r^2 - r - 2 = 0$
$\left\{-\frac{1}{2}, \frac{2}{3}\right\}$

35. $12p^2 = 8 - 10p$
$\left\{-\frac{4}{3}, \frac{1}{2}\right\}$

36. $18x^2 = 12 + 15x$
$\left\{-\frac{1}{2}, \frac{4}{3}\right\}$

37. $9s^2 + 12s = -4$
$\left\{-\frac{2}{3}\right\}$

38. $36x^2 + 60x = -25$
$\left\{-\frac{5}{6}\right\}$

39. $y^2 - 9 = 0$
$\{-3, 3\}$

40. $m^2 - 100 = 0$
$\{-10, 10\}$

41. $16k^2 - 49 = 0$
$\left\{-\frac{7}{4}, \frac{7}{4}\right\}$

42. $4w^2 - 9 = 0$
$\left\{-\frac{3}{2}, \frac{3}{2}\right\}$

43. $n^2 = 121$
$\{-11, 11\}$

44. $x^2 = 400$
$\{-20, 20\}$

45. $x^2 = 7x$
$\{0, 7\}$

46. $t^2 = 9t$
$\{0, 9\}$

47. $6r^2 = 3r$
$\left\{0, \frac{1}{2}\right\}$

48. $10y^2 = -5y$
$\left\{-\frac{1}{2}, 0\right\}$

49. $g(g - 7) = -10$
$\{2, 5\}$

50. $r(r - 5) = -6$
$\{2, 3\}$

51. $z(2z+7)=4$
$\left\{-4, \frac{1}{2}\right\}$

52. $b(2b+3)=9$
$\left\{-3, \frac{3}{2}\right\}$

53. $2(y^2-66)=-13y$
$\left\{-12, \frac{11}{2}\right\}$

54. $3(t^2+4)=20t$
$\left\{\frac{2}{3}, 6\right\}$

55. $5x^3-20x=0$
$\{-2, 0, 2\}$

56. $3x^3-48x=0$
$\{-4, 0, 4\}$

57. $9y^3-49y=0$
$\left\{-\frac{7}{3}, 0, \frac{7}{3}\right\}$

58. $16r^3-9r=0$
$\left\{-\frac{3}{4}, 0, \frac{3}{4}\right\}$

59. $(2r+5)(3r^2-16r+5)=0$
$\left\{-\frac{5}{2}, \frac{1}{3}, 5\right\}$

60. $(3m+4)(6m^2+m-2)=0$
$\left\{-\frac{4}{3}, -\frac{2}{3}, \frac{1}{2}\right\}$

61. $(2x+7)(x^2+2x-3)=0$
$\left\{-\frac{7}{2}, -3, 1\right\}$

62. $(x+1)(6x^2+x-12)=0$
$\left\{-\frac{3}{2}, -1, \frac{4}{3}\right\}$

63. $x^3+x^2-20x=0$
$\{-5, 0, 4\}$

64. $y^3-6y^2+8y=0$
$\{0, 2, 4\}$

65. $r^4=2r^3+15r^2$
$\{-3, 0, 5\}$

66. $x^3=3x+2x^2$
$\{-1, 0, 3\}$

67. $3x(x+1)=(2x+3)(x+1)$
$\{-1, 3\}$

68. $2x(x+3)=(3x+1)(x+3)$
$\{-3, -1\}$

69. Galileo's formula describing the motion of freely falling objects is

$$d=16t^2.$$

The distance d in feet an object falls depends on the time t elapsed, in seconds. (This is an example of an important mathematical concept, a **function.**)

(a) Use Galileo's formula and complete the following table. (*Hint:* Substitute each given value into the formula and solve for the unknown value.)

t **in seconds**	0	1	2	3	4	6
d **in feet**	0	16	64	144	256	576

(b) When $t=0$, $d=0$. Explain this in the context of the problem.

No time has elapsed, so the object hasn't fallen (been released) yet.

70. In **Exercise 69,** when you substituted 256 for d and solved for t, you should have found two solutions: 4 and -4. Why doesn't -4 make sense as an answer?

Time cannot be negative.

6.7 Applications of Quadratic Equations

OBJECTIVES

1. Solve problems involving geometric figures.
2. Solve problems involving consecutive integers.
3. Solve problems by applying the Pythagorean theorem.
4. Solve problems by using given quadratic models.

We can use factoring to solve quadratic equations that arise in applications by following the same six problem-solving steps given in **Section 2.4.**

Solving an Applied Problem

Step 1 **Read** the problem carefully. What information is given? What is to be found?

Step 2 **Assign a variable** to represent the unknown value. Use a sketch, diagram, or table, as needed. Express any other unknown values in terms of the variable.

Step 3 **Write an equation** using the variable expression(s).

Step 4 **Solve** the equation.

Step 5 **State the answer.** Label it appropriately. Does it seem reasonable?

Step 6 **Check** the answer in the words of the *original* problem.

OBJECTIVE 1 **Solve problems involving geometric figures.** Refer to the formulas inside the back cover of the text, if necessary.

EXAMPLE 1 Solving an Area Problem

The Monroes want to plant a rectangular garden in their yard. The width of the garden will be 4 ft less than its length, and they want it to have an area of 96 ft^2. (ft^2 means square feet.) Find the length and width of the garden.

Step 1 **Read** the problem carefully. We need to find the dimensions of a garden with area 96 ft^2.

Step 2 **Assign a variable.**

Let x = the length of the garden.

Then $x - 4$ = the width. (The width is 4 ft less than the length.)

See **Figure 1.**

Figure 1

Step 3 **Write an equation.** The area of a rectangle is given by

$$\text{Area} = \text{Length} \times \text{Width}. \quad \text{Area formula}$$

Substitute 96 for area, x for length, and $x - 4$ for width.

$$A = LW$$

$$96 = x(x - 4) \quad \text{Let } A = 96,\ L = x,\ W = x - 4.$$

Continued on Next Page

Step 4 **Solve.**

$$96 = x(x-4) \quad \text{Equation from Step 3}$$
$$96 = x^2 - 4x \quad \text{Distributive property}$$
$$x^2 - 4x - 96 = 0 \quad \text{Standard form}$$
$$(x-12)(x+8) = 0 \quad \text{Factor.}$$
$$x - 12 = 0 \quad \text{or} \quad x + 8 = 0 \quad \text{Zero-factor property}$$
$$x = 12 \quad \text{or} \quad x = -8 \quad \text{Solve each equation.}$$

Step 5 **State the answer.** The solutions are 12 and -8. A rectangle cannot have a side of negative length, so we discard -8. The length of the garden will be 12 ft. The width will be $12 - 4 = 8$ ft.

Step 6 **Check.** The width of the garden is 4 ft less than the length, and the area is $12 \cdot 8 = 96 \text{ ft}^2$, as required.

Work Problem 1 at the Side. ▶

Problem-Solving Hint

When solving applied problems, always check solutions against physical facts and discard any answers that are not appropriate.

OBJECTIVE 2 **Solve problems involving consecutive integers.** Recall from **Section 2.4** that **consecutive integers** are integers that are next to each other on a number line, such as 1 and 2, or -11 and -10. See **Figure 2.**

Consecutive odd integers are *odd* integers that are next to each other, such as 1 and 3, or -13 and -11. **Consecutive even integers** are defined similarly, so 2 and 4 are consecutive even integers, as are -10 and -8. See **Figure 3.** (We list consecutive integers in increasing order from left to right.)

Problem-Solving Hint

If x represents the lesser integer, then the following apply.

For two consecutive integers, use	$x, \; x+1$.
For three consecutive integers, use	$x, \; x+1, \; x+2$.
For two consecutive even or odd integers, use	$x, \; x+2$.
For three consecutive even or odd integers, use	$x, \; x+2, \; x+4$.

EXAMPLE 2 **Solving a Consecutive Integer Problem**

The product of the numbers on two consecutive post-office boxes is 210. Find the box numbers.

Step 1 **Read** the problem. Note that the boxes are numbered consecutively.

Step 2 **Assign a variable.** See **Figure 4.**

Let x = the first box number.

Then $x + 1$ = the next consecutive box number.

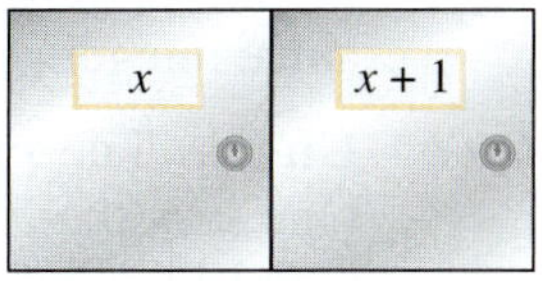

Figure 4

Continued on Next Page

---

1 Solve each problem.

(a) The length of a rectangular room is 2 m more than the width. The area of the floor is 48 m^2. Find the length and width of the room.

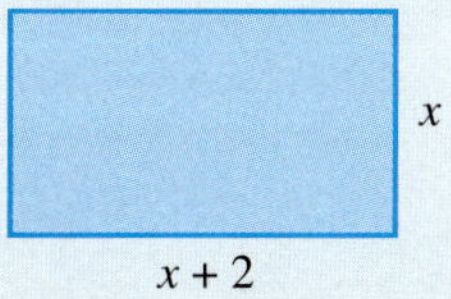

Give the equation using x as the variable, and give the answer.

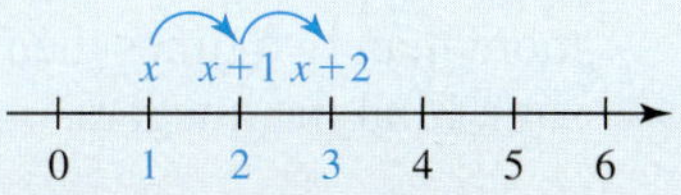

Consecutive integers

Figure 2

Consecutive even integers

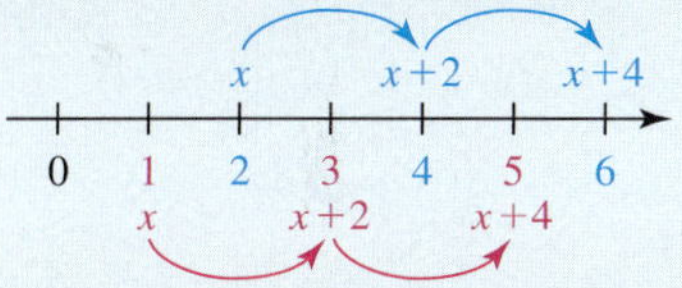

Consecutive odd integers

Figure 3

(b) The length of each side of a square is increased by 4 in. The sum of the areas of the original square and the larger square is 106 in^2. What is the length of a side of the original square?

Answers

1. (a) $48 = (x+2)x$; length: 8 m; width: 6 m
(b) 5 in.

2 Solve the problem.

The product of the numbers on two consecutive lockers at a health club is 132. Find the locker numbers.

3 Solve each problem. Give the equation using x to represent the least integer, and give the answer.

(a) The product of two consecutive even integers is 4 more than two times their sum. Find the integers.

(b) Find three consecutive odd integers such that the product of the least and greatest is 16 more than the middle integer.

Answers

2. 11 and 12

3. (a) $x(x+2)=4+2(x+x+2)$; 4 and 6 or -2 and 0

(b) $x(x+4)=16+(x+2)$; 3, 5, 7

Step 3 **Write an equation.** The product of the box numbers is 210.

$$x(x+1)=210$$

Step 4 **Solve.**

$x^2+x=210$	Distributive property
$x^2+x-210=0$	Standard form
$(x+15)(x-14)=0$	Factor.
$x+15=0$ or $x-14=0$	Zero-factor property
$x=-15$ or $x=14$	Solve each equation.

Step 5 **State the answer.** The solutions are -15 and 14. We discard -15 since a box number cannot be negative. If $x=14$, then $x+1=15$, so the boxes have numbers 14 and 15.

Step 6 **Check.** The numbers 14 and 15 are consecutive and their product is $14 \cdot 15 = 210$, as required.

◀ **Work Problem 2 at the Side.**

EXAMPLE 3 Solving a Consecutive Integer Problem

The product of two consecutive odd integers is 1 less than five times their sum. Find the integers.

Step 1 **Read** carefully. This problem is a little more complicated.

Step 2 **Assign a variable.** We must find two consecutive *odd* integers.

Let $x =$ the lesser integer.

Then $x+2 =$ the next greater odd integer.

Step 3 **Write an equation.**

The product	is	five times the sum	less 1.
↓	↓	↓	↓
$x(x+2)$	$=$	$5(x+x+2)$	$-\ 1$

Step 4 **Solve.**

$x^2+2x=5x+5x+10-1$	Distributive property
$x^2+2x=10x+9$	Combine like terms.
$x^2-8x-9=0$	Standard form
$(x-9)(x+1)=0$	Factor.
$x-9=0$ or $x+1=0$	Zero-factor property
$x=9$ or $x=-1$	Solve each equation.

Step 5 **State the answer.** We need to find two consecutive odd integers.

If $x=9$ is the lesser, then $x+2=9+2=11$ is the greater.

If $x=-1$ is the lesser, then $x+2=-1+2=1$ is the greater.

Do not discard the solution -1 here. There are two sets of answers since integers can be positive *or* negative.

Step 6 **Check.** The product of the first pair of integers is $9 \cdot 11 = 99$. One less than five times their sum is $5(9+11)-1=99$. Thus, 9 and 11 satisfy the problem. Repeat the check with -1 and 1.

◀ **Work Problem 3 at the Side.**

CAUTION

Do *not* use x, $x + 1$, $x + 3$ to represent consecutive odd integers. To see why, let $x = 3$. Then $x + 1 = 3 + 1$ or 4, and $x + 3 = 3 + 3$ or 6. The numbers 3, 4, and 6 are not consecutive odd integers.

Pythagoras (c. 380–300 B.C.)

OBJECTIVE 3 Solve problems by applying the Pythagorean theorem.

The next example requires the Pythagorean theorem from geometry.

Pythagorean Theorem

If a right triangle (a triangle with a 90° angle) has longest side of length c and two other sides of lengths a and b, then

$$a^2 + b^2 = c^2.$$

The longest side, the **hypotenuse,** is opposite the right angle. The two shorter sides are the **legs** of the triangle.

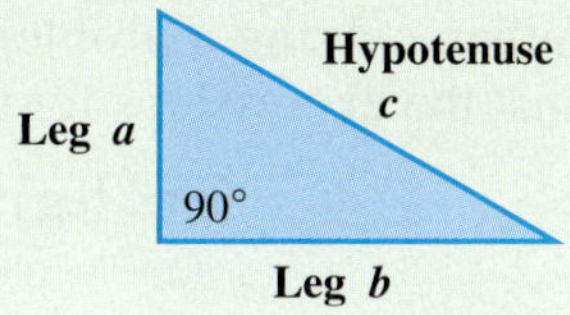

Teaching Tip

Stress that the Pythagorean theorem cannot be applied to triangles that are not right triangles.

EXAMPLE 4 Applying the Pythagorean Theorem

Amy and Kevin leave their office, with Amy traveling north and Kevin traveling east. When Kevin is 1 mi farther than Amy from the office, the distance between them is 2 mi more than Amy's distance from the office. Find their distances from the office and the distance between them.

Step 1 **Read** the problem again. We must find three distances.

Step 2 **Assign a variable.**

Let x = Amy's distance from the office.

Then $x + 1$ = Kevin's distance from the office,

and $x + 2$ = the distance between them.

Place these expressions on a right triangle, as in **Figure 5.**

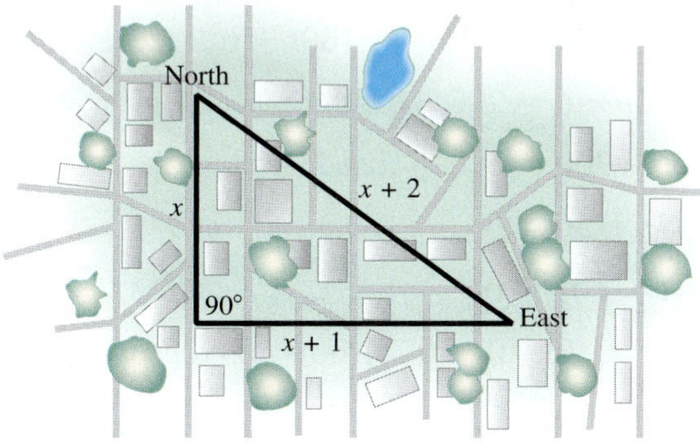

Figure 5

Step 3 **Write an equation.** Substitute into the Pythagorean theorem.

$$a^2 + b^2 = c^2$$

$$x^2 + (x + 1)^2 = (x + 2)^2$$ Be careful to substitute properly.

Continued on Next Page

4 Solve the problem.

The hypotenuse of a right triangle is 3 in. longer than the longer leg. The shorter leg is 3 in. shorter than the longer leg. Find the lengths of the sides of the triangle.

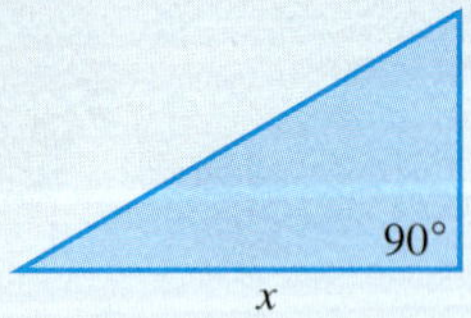

5 Solve each problem.

(a) Refer to **Example 5.** How long will it take for the ball to reach a height of 50 ft?

(b) The number y of impulses fired after a nerve has been stimulated is modeled by

$$y = -x^2 + 2x + 60,$$

where x is in milliseconds (ms) after the stimulation. When will 45 impulses occur? Do we get two solutions? Why is only one answer acceptable?

Answers

4. 9 in., 12 in., 15 in.

5. (a) $\frac{1}{4}$ sec and 11 sec

(b) After 5 ms; There are two solutions, -3 and 5; Only one answer makes sense here because a negative answer is not appropriate.

Step 4 **Solve.**

$$x^2 + x^2 + 2x + 1 = x^2 + 4x + 4$$

$$x^2 - 2x - 3 = 0 \quad \text{Standard form}$$

$$(x - 3)(x + 1) = 0 \quad \text{Factor.}$$

$$x - 3 = 0 \quad \text{or} \quad x + 1 = 0 \quad \text{Zero-factor property}$$

$$x = 3 \quad \text{or} \quad x = -1 \quad \text{Solve each equation.}$$

Step 5 **State the answer.** Since -1 cannot represent a distance, 3 is the only possible answer. Amy's distance is 3 mi, Kevin's distance is $3 + 1 = 4$ mi, and the distance between them is $3 + 2 = 5$ mi.

Step 6 **Check.** Since $3^2 + 4^2 = 5^2$ is true, the answer is correct.

◀ **Work Problem 4 at the Side.**

Problem-Solving Hint

When solving a problem involving the Pythagorean theorem, be sure that the expressions for the sides of the triangle are properly placed.

$$\textbf{(one leg)}^2 + \textbf{(other leg)}^2 = \textbf{hypotenuse}^2$$

OBJECTIVE 4 **Solve problems by using given quadratic models.** In **Examples 1–4,** we wrote quadratic equations to model, or mathematically describe, various situations and then solved the equations. Now we are given the quadratic models and must use them to determine data.

EXAMPLE 5 **Finding the Height of a Ball**

A tennis player can hit a ball 180 ft per sec (123 mph). If she hits a ball directly upward, the height h of the ball in feet at time t in seconds is modeled by the quadratic equation

$$h = -16t^2 + 180t + 6.$$

How long will it take for the ball to reach a height of 206 ft?

A height of 206 ft means $h = 206$, so we substitute 206 for h in the equation and then solve for t.

$$h = -16t^2 + 180t + 6$$

$$206 = -16t^2 + 180t + 6 \quad \text{Let } h = 206.$$

$$-16t^2 + 180t + 6 = 206 \quad \text{Interchange sides.}$$

$$-16t^2 + 180t - 200 = 0 \quad \text{Standard form}$$

$$4t^2 - 45t + 50 = 0 \quad \text{Divide by } -4.$$

$$(4t - 5)(t - 10) = 0 \quad \text{Factor.}$$

$$4t - 5 = 0 \quad \text{or} \quad t - 10 = 0 \quad \text{Zero-factor property}$$

$$t = \frac{5}{4} \quad \text{or} \quad t = 10 \quad \text{Solve each equation.}$$

Since we found two acceptable answers, the ball will be 206 ft above the ground twice (once on its way up and once on its way down)—at $\frac{5}{4}$ sec and at 10 sec after it is hit. See **Figure 6.**

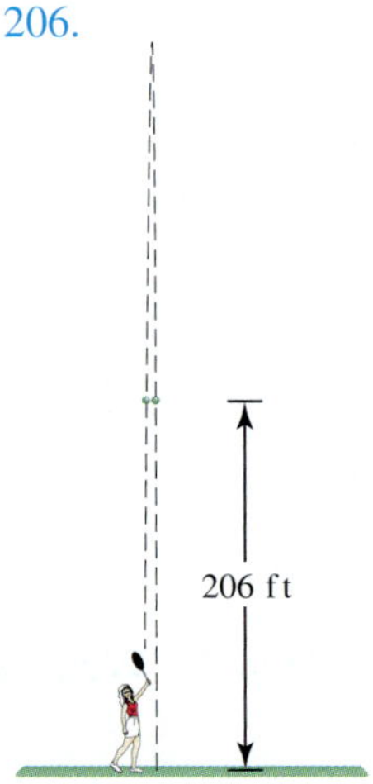

Figure 6

◀ **Work Problem 5 at the Side.**

EXAMPLE 6 **Modeling the Foreign-Born Population of the United States**

The foreign-born population of the United States over the years 1930–2010 can be modeled by the quadratic equation

$$y = 0.009665x^2 - 0.4942x + 15.12,$$

where $x = 0$ represents 1930, $x = 10$ represents 1940, and so on, and y is the number of people in millions. (*Source:* U.S. Census Bureau.)

(a) Use the model to find the foreign-born population in 1980 to the nearest tenth of a million.

Since $x = 0$ represents 1930, $x = 50$ represents 1980. Substitute 50 for x in the equation.

$y = 0.009665(\mathbf{50})^2 - 0.4942(\mathbf{50}) + 15.12$ Let $x = 50$.

$y = 14.6$ Round to the nearest tenth.

In 1980, the foreign-born population of the United States was about 14.6 million.

(b) Repeat part (a) for 2010.

$y = 0.009665(\mathbf{80})^2 - 0.4942(\mathbf{80}) + 15.12$ For 2010, let $x = 80$.

$y = 37.4$ Round to the nearest tenth.

In 2010, the foreign-born population of the United States was about 37.4 million.

(c) The model used in parts (a) and (b) was developed using the data in the table below. How do the results in parts (a) and (b) compare to the actual data from the table?

Year	Foreign-Born Population (millions)
1930	14.2
1940	11.6
1950	10.3
1960	9.7
1970	9.6
1980	14.1
1990	19.8
2000	28.4
2010	37.6

From the table, the actual value for 1980 is 14.1 million. By comparison, our answer in part (a), 14.6 million, is slightly high. For 2010, the actual value is 37.6 million, so our answer of 37.4 million in part (b) is slightly low, but a good estimate.

Work Problem 6 at the Side. ▶

6 Use the model in **Example 6** to find the foreign-born population of the United States in 1990. Give your answer to the nearest tenth of a million. How does it compare to the actual value from the table?

Answer

6. 20.3 million; The actual value is 19.8 million, so our answer using the model is somewhat high.

6.7 Exercises

FOR EXTRA HELP

Download the MyDashBoard App

MyMathLab®

1. **CONCEPT CHECK** To review the six problem-solving steps first introduced in **Section 2.4,** complete each statement.

Step 1: __Read__ the problem carefully.

Step 2: Assign a __variable__ to represent the unknown value.

Step 3: Write a(n) __equation__ using the variable expression(s).

Step 4: __Solve__ the equation.

Step 5: State the __answer__.

Step 6: __Check__ the answer in the words of the __original__ problem.

2. **CONCEPT CHECK** A student solves an applied problem and gets 6 or -3 for the length of the side of a square. Which of these answers is reasonable? Why?

Only 6 is reasonable since a square cannot have a side of negative length.

GS *In Exercises 3–6, a figure and a corresponding geometric formula are given. Using x as the variable, complete Steps 3–6 for each problem. (Refer to the steps in* ***Exercise 1*** *as needed.)*

3.

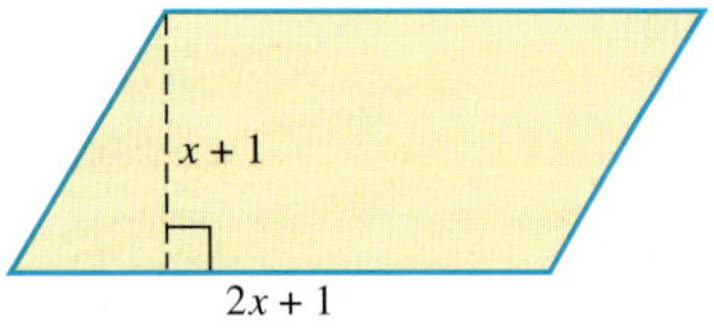

Area of a parallelogram: $A = bh$

The area of this parallelogram is 45 sq. units. Find its base and height.

Step 3: $45 =$ __$(2x + 1)(x + 1)$__

Step 4: $x =$ __4__ or $x =$ __$-\frac{11}{2}$__

Step 5: base: __9__ units; height: __5__ units

Step 6: __$9 \cdot 5$__ $= 45$

4.

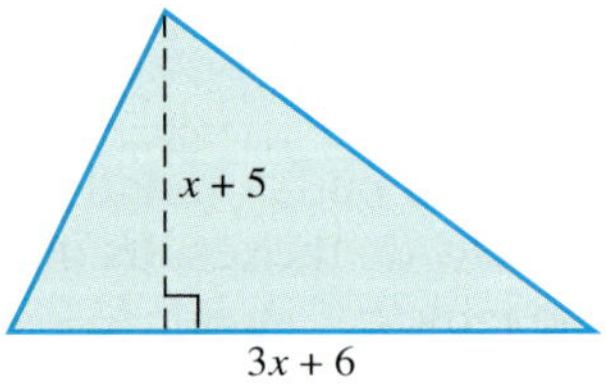

Area of a triangle: $A = \frac{1}{2}bh$

The area of this triangle is 60 sq. units. Find its base and height.

Step 3: $60 =$ __$\frac{1}{2}(3x + 6)(x + 5)$__

Step 4: $x =$ __3__ or $x =$ __-10__

Step 5: base: __15__ units; height: __8__ units

Step 6: __$\frac{1}{2}(15)(8)$__ $= 60$

5.

Volume of a rectangular jewelry box: $V = LWH$

The volume of this box is 192 cu. units. Find its length and width.

Step 3: __192__ $=$ __$4x$__ $(x + 2)$

Step 4: $x =$ __6__ or $x =$ __-8__

Step 5: length: __8__ units; width: __6__ units

Step 6: __$8 \cdot 6$__ $\cdot\, 4 =$ __192__

6.

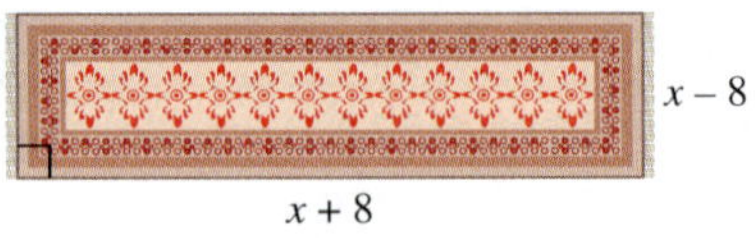

Area of a rectangular rug: $A = LW$

The area of this rug is 80 sq. units. Find its length and width.

Step 3: __80__ $= (x + 8)$ __$(x - 8)$__

Step 4: $x =$ __12__ or $x =$ __-12__

Step 5: length: __20__ units; width: __4__ units

Step 6: __$20 \cdot 4$__ $= 80$

Solve each problem. Check answers to be sure they are reasonable. Refer to the formulas inside the back cover. ***See Example 1.***

7. The length of a standard jewel case is 2 cm more than its width. The area of the rectangular top of the case is 168 cm^2. Find the length and width of the jewel case.

length: 14 cm; width: 12 cm

8. A standard DVD case is 6 cm longer than it is wide. The area of the rectangular top of the case is 247 cm^2. Find the length and width of the case.

length: 19 cm; width: 13 cm

9. The area of a triangle is 30 $in.^2$. The base of the triangle measures 2 in. more than twice the height of the triangle. Find the measures of the base and the height.

base: 12 in.; height: 5 in.

10. A certain triangle has its base equal in measure to its height. The area of the triangle is 72 m^2. Find the equal base and height measure.

12 m

11. The dimensions of a flat-panel monitor are such that its length is 3 in. more than its width. If the length were doubled and if the width were decreased by 1 in., the area would be increased by 150 $in.^2$. What are the length and width of the flat panel?

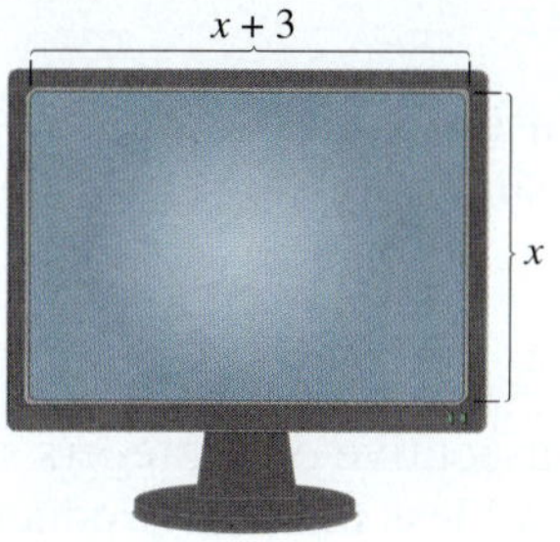

length: 15 in.; width: 12 in.

12. A computer keyboard is 11 in. longer than it is wide. If the length were doubled and if 2 in. were added to the width, the area would be increased by 198 $in.^2$. What are the length and width of the keyboard?

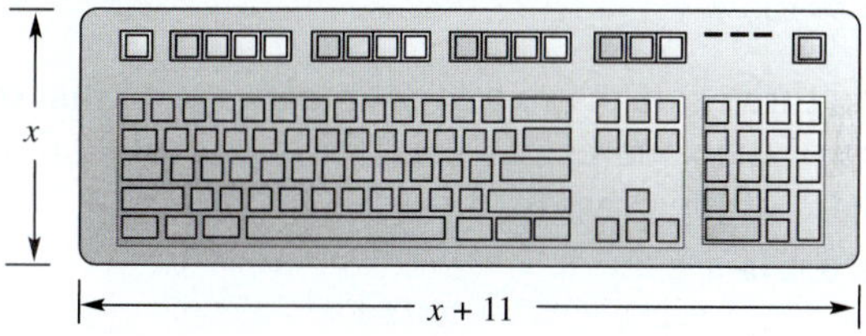

length: 18 in.; width: 7 in.

13. A 10-gal aquarium is 3 in. higher than it is wide. Its length is 21 in., and its volume is 2730 $in.^3$. What are the height and width of the aquarium?

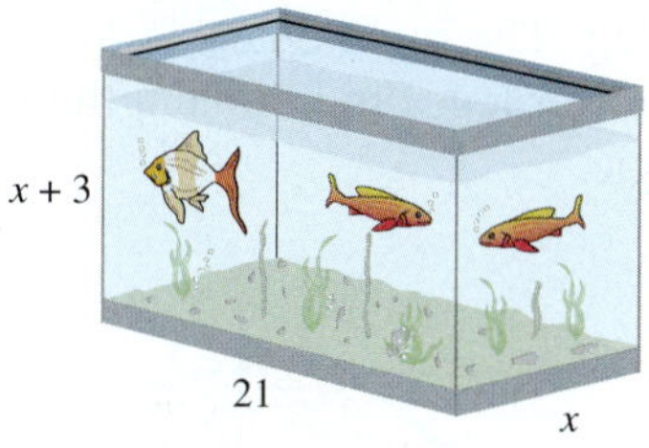

height: 13 in.; width: 10 in.

14. A toolbox is 2 ft high, and its width is 3 ft less than its length. If its volume is 80 ft^3, find the length and width of the box.

length: 8 ft; width: 5 ft

15. A square mirror has sides measuring 2 ft less than the sides of a square painting. If the difference between their areas is 32 ft^2, find the lengths of the sides of the mirror and the painting.

mirror: 7 ft; painting: 9 ft

16. The sides of one square have length 3 m more than the sides of a second square. If the area of the larger square is subtracted from 4 times the area of the smaller square, the result is 36 m^2. What are the lengths of the sides of each square?

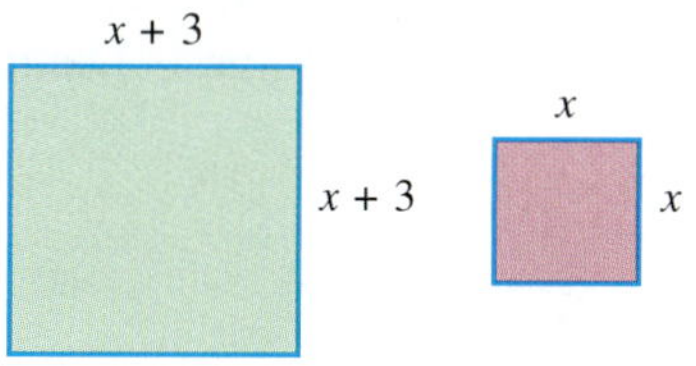

larger square: 8 m; smaller square: 5 m

Solve each problem about consecutive integers. **See Examples 2 and 3.**

17. The product of the numbers on two consecutive volumes of research data is 420. Find the volume numbers.

20, 21

18. The product of the page numbers on two facing pages of a book is 600. Find the page numbers.

24, 25

19. The product of two consecutive integers is 11 more than their sum. Find the integers.

−3, −2 or 4, 5

20. The product of two consecutive integers is 4 less than four times their sum. Find the integers.

0, 1 or 7, 8

21. Find two consecutive odd integers such that their product is 15 more than three times their sum.

−3, −1 or 7, 9

22. Find two consecutive odd integers such that five times their sum is 23 less than their product.

−3, −1 or 11, 13

23. Find three consecutive even integers such that the sum of the squares of the lesser two is equal to the square of the greatest.

−2, 0, 2 or 6, 8, 10

24. Find three consecutive even integers such that the square of the sum of the lesser two is equal to twice the greatest.

−2, 0, 2

25. Find three consecutive odd integers such that 3 times the sum of all three is 18 more than the product of the first and second integers.

7, 9, 11

26. Find three consecutive odd integers such that the sum of all three is 42 less than the product of the second and third integers.

5, 7, 9

Use the Pythagorean theorem to solve each problem. **See Example 4.**

27. The hypotenuse of a right triangle is 1 cm longer than the longer leg. The shorter leg is 7 cm shorter than the longer leg. Find the length of the longer leg of the triangle.

12 cm

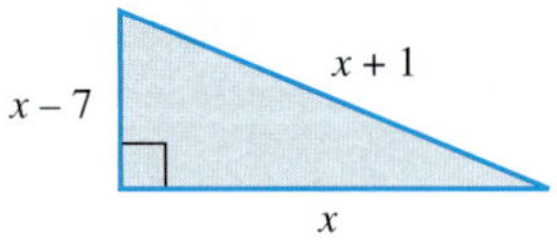

28. The longer leg of a right triangle is 1 m longer than the shorter leg. The hypotenuse is 1 m shorter than twice the shorter leg. Find the length of the shorter leg of the triangle.

3 m

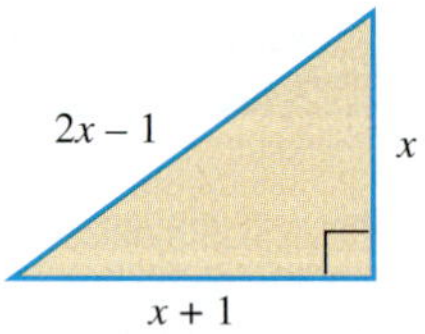

29. Terri works due north of home. Her husband Denny works due east. They leave for work at the same time. By the time Terri is 5 mi from home, the distance between them is 1 mi more than Denny's distance from home. How far from home is Denny?

12 mi

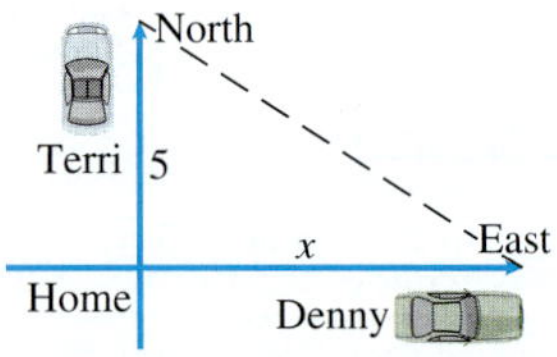

30. Two cars left an intersection at the same time. One traveled north. The other traveled 14 mi farther, but to the east. How far apart were they then, if the distance between them was 4 mi more than the distance traveled east?

34 mi

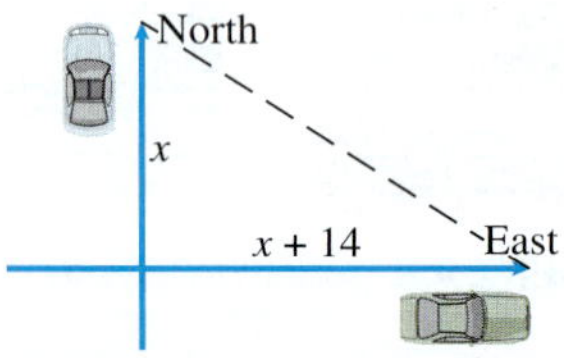

31. A ladder is leaning against a building. The distance from the bottom of the ladder to the building is 4 ft less than the length of the ladder. How high up the side of the building is the top of the ladder if that distance is 2 ft less than the length of the ladder?

8 ft

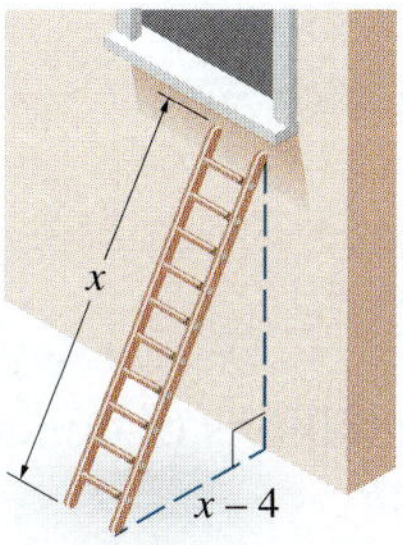

32. A lot has the shape of a right triangle with one leg 2 m longer than the other. The hypotenuse is 2 m less than twice the length of the shorter leg. Find the length of the shorter leg.

6 m

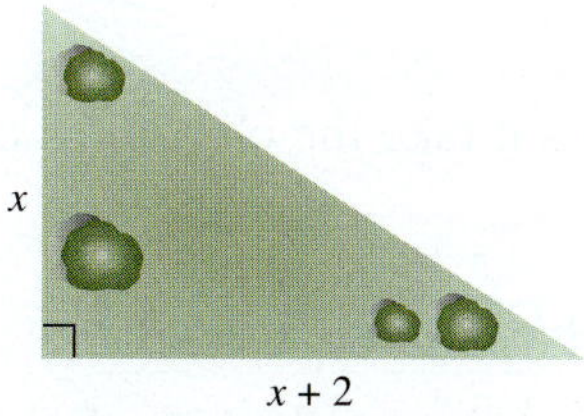

Solve each problem. ***See Example 5.***

33. An object projected from a height of 48 ft with an initial velocity of 32 ft per sec after t seconds has height

$$h = -16t^2 + 32t + 48.$$

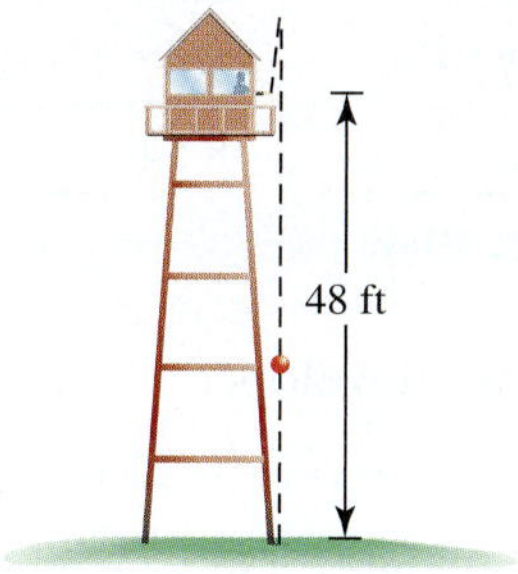

(a) After how many seconds is the height 64 ft? (*Hint:* Let $h = 64$ and solve.)

1 sec

(b) After how many seconds is the height 60 ft?

$\frac{1}{2}$ sec and $1\frac{1}{2}$ sec

(c) After how many seconds does the object hit the ground? (*Hint:* When the object hits the ground, $h = 0$.)

3 sec

(d) The quadratic equation from part (c) has two solutions, yet only one of them is appropriate for answering the question. Why is this so?

The negative solution, −1, does not make sense since t represents time, which cannot be negative.

34. If an object is projected upward from ground level with an initial velocity of 64 ft per sec, its height h in feet t seconds later is

$$h = -16t^2 + 64t.$$

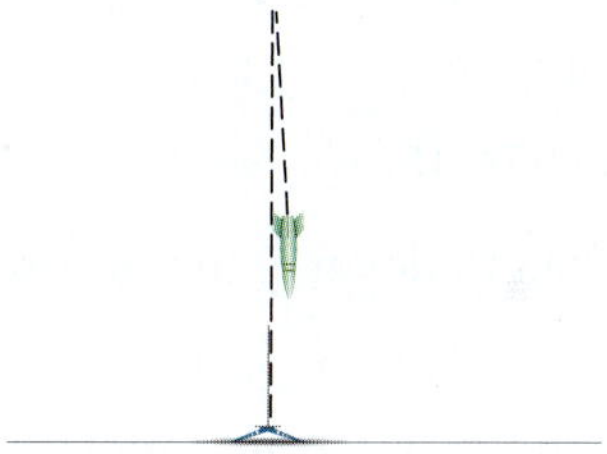

(a) After how many seconds is the height 48 ft? (*Hint:* Let $h = 48$ and solve.)

1 sec and 3 sec

(b) The object reaches its maximum height 2 sec after it is projected. What is this maximum height?

64 ft

(c) After how many seconds does the object hit the ground? (*Hint:* When the object hits the ground, $h = 0$.)

4 sec

(d) The quadratic equation from part (c) has two solutions, yet only one of them is appropriate for answering the question. Why is this so?

The solution 0 represents the time at which it was projected.

If an object is projected upward with an initial velocity of 128 ft per sec, its height h in feet after t seconds is

$$h = 128t - 16t^2.$$

Find the height of the object after each period of time. ***See Example 5.***

35. 1 sec

112 ft

36. 2 sec

192 ft

37. 4 sec

256 ft

38. How long does it take the object just described to return to the ground?

8 sec

Solve each problem. ***See Example 6.***

39. The table shows the number of cellular phone subscribers (in millions) in the United States.

Year	Subscribers (in millions)
1994	24
1996	44
1998	69
2000	109
2002	141
2004	182
2006	233
2008	286
2010	303

Source: CTIA-The Wireless Association.

We used the data to develop the quadratic equation

$$y = 0.347x^2 + 13.14x + 17.98,$$

which models the number of cellular phone subscribers y (in millions) in the year x, where $x = 0$ represents 1994, $x = 2$ represents 1996, and so on.

(a) Use the model to find the number of subscribers in 2000 to the nearest million. How does the result compare to the actual data in the table?

109 million; The result using the model is the same as the actual number from the table for 2000.

(b) What value of x corresponds to 2008?

14

(c) Use the model to find the number of subscribers in 2008 to the nearest million. How does the result compare to the actual data in the table?

270 million; The result is less than 286 million, the actual number for 2008.

(d) Assuming that the trend in the data continues, use the quadratic equation to estimate the number of subscribers in 2013 to the nearest million.

393 million

40. Annual revenue in billions of dollars for eBay is shown in the table.

Year	Annual Revenue (in billions of dollars)
2001	0.749
2002	1.21
2003	2.17
2004	3.27
2005	4.55
2006	5.97
2007	7.67
2008	8.54

Source: eBay.

Using the data, we developed the quadratic equation

$$y = 0.06x^2 + 0.75x + 0.55$$

to model eBay revenues y in year x, where $x = 0$ represents 2001, $x = 1$ represents 2002, and so on.

(a) Use the model to find annual revenue for eBay in 2005 and 2008. How do the results compare with the actual data in the table?

2005: \$4.51 billion; 2008: \$8.74 billion;

The result for 2005 is close to the actual revenue from the table. The result for 2008 is a bit higher than the actual revenue from the table.

(b) Use the model to estimate annual revenue for eBay in 2009, to the nearest tenth.

\$10.4 billion

(c) Actual revenue in 2009 was \$8.73 billion. How does the result from part (b) compare with the actual revenue in 2009?

The result from part (b) is quite a bit higher than the actual revenue.

(d) Should the quadratic equation be used to estimate eBay revenue for years after 2008? Explain.

No, data for later years may not follow the same pattern.

Chapter 6 Summary

Key Terms

6.1

factor For integers a and b, if $a \cdot b = c$, then a and b are factors of c.

factored form An expression is in factored form when it is written as a product.

greatest common factor (GCF) The greatest common factor is the largest quantity that is a factor of each of a group of quantities.

factoring The process of writing a polynomial as a product is called factoring.

6.2

prime polynomial A prime polynomial is a polynomial that cannot be factored using only integers.

6.5

perfect square trinomial A perfect square trinomial is a trinomial that can be factored as the square of a binomial.

6.6

quadratic equation A quadratic equation is an equation that can be written in the form $ax^2 + bx + c = 0$, with $a \neq 0$.

standard form The form $ax^2 + bx + c = 0$ is the standard form of a quadratic equation.

6.7

hypotenuse The longest side of a right triangle, opposite the right angle, is the hypotenuse.

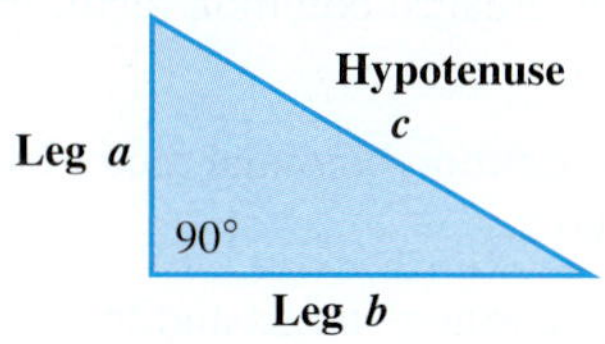

legs The two shorter sides of a right triangle are the legs.

Test Your Word Power

See how well you have learned the vocabulary in this chapter.

1. **Factoring** is
   - **A.** a method of multiplying polynomials
   - **B.** the process of writing a polynomial as a product
   - **C.** the answer in a multiplication problem
   - **D.** a way to add the terms of a polynomial.

2. A polynomial is in **factored form** when
   - **A.** it is prime
   - **B.** it is written as a sum
   - **C.** the second-degree term has a coefficient of 1
   - **D.** it is written as a product.

3. The **greatest common factor** of a polynomial is
   - **A.** the least integer that divides evenly into all its terms
   - **B.** the least expression that is a factor of all its terms
   - **C.** the greatest expression that is a factor of all its terms
   - **D.** the variable that is common to all its terms.

4. A **perfect square trinomial** is a trinomial
   - **A.** that can be factored as the square of a binomial
   - **B.** that cannot be factored
   - **C.** that is multiplied by a binomial
   - **D.** where all terms are perfect squares.

5. A **quadratic equation** is an equation that can be written in the form
   - **A.** $y = mx + b$
   - **B.** $ax^2 + bx + c = 0 \ (a \neq 0)$
   - **C.** $Ax + By = C$
   - **D.** $x = k$.

6. A **hypotenuse** is
   - **A.** either of the two shorter sides of a triangle
   - **B.** the shortest side of a right triangle
   - **C.** the side opposite the right angle in a right triangle
   - **D.** the longest side in any triangle.

Answers to Test Your Word Power

1. B; *Example:* $x^2 - 5x - 14$ factors as $(x - 7)(x + 2)$.
2. D; *Example:* The factored form of $x^2 - 5x - 14$ is $(x - 7)(x + 2)$.
3. C; *Example:* The greatest common factor of $8x^2$, $22xy$, and $16x^3y^2$ is $2x$.
4. A; *Example:* $a^2 + 2a + 1$ is a perfect square trinomial. Its factored form is $(a + 1)^2$.
5. B; *Examples:* $y^2 - 3y + 2 = 0$, $x^2 - 9 = 0$, $2m^2 = 6m + 8$
6. C; *Example:* See the triangle included in the Key Terms above.

Quick Review

6.1 Factors; The Greatest Common Factor

Concepts

Finding the Greatest Common Factor (GCF)

Step 1 Write each number in prime factored form.

Step 2 List each prime number or each variable that is a factor of every term in the list.

Step 3 Use as exponents on the common prime factors the *least* exponents from the prime factored forms.

Step 4 Multiply the primes from Step 3.

Examples

Find the greatest common factor of $4x^2y$, $6x^2y^3$, and $2xy^2$.

$$4x^2y = 2 \cdot 2 \cdot x^2 \cdot y$$
$$6x^2y^3 = 2 \cdot 3 \cdot x^2 \cdot y^3$$
$$2xy^2 = 2 \cdot x \cdot y^2$$

The greatest common factor is $2xy$.

Concepts

Factoring by Grouping

Step 1 Group the terms.

Step 2 Factor out the greatest common factor from each group.

Step 3 Factor out a common binomial factor from the results of Step 2.

Step 4 If necessary, rearrange terms and try a different grouping.

Examples

Factor by grouping.

$$2a^2 + 2ab + a + b$$
$$= (2a^2 + 2ab) + (a + b) \quad \text{Group the terms.}$$
$$= 2a(a + b) + 1(a + b) \quad \text{Factor each group.}$$
$$= (a + b)(2a + 1) \quad \text{Factor out } a + b.$$

6.2 Factoring Trinomials

Concepts

To factor $x^2 + bx + c$, find m and n such that $mn = c$ and $m + n = b$.

$mn = c$ ↓

$x^2 + bx + c$

↑ $m + n = b$

Then $x^2 + bx + c$ factors as $(x + m)(x + n)$.

Check by multiplying.

Examples

Factor $x^2 + 6x + 8$.

$mn = 8$ ↓

$x^2 + 6x + 8$ Here, $m = 2$ and $n = 4$.

↑ $m + n = 6$

$x^2 + 6x + 8$ factors as $(x + 2)(x + 4)$.

CHECK $(x + 2)(x + 4)$

$$= x^2 + 4x + 2x + 8 \quad \text{FOIL method}$$
$$= x^2 + 6x + 8 \checkmark \quad \text{Combine like terms.}$$

6.3 Factoring Trinomials by Grouping

Concepts

To factor $ax^2 + bx + c$, find m and n such that $mn = ac$ and $m + n = b$.

$m + n = b$ ↓

$ax^2 + bx + c$

↑ $mn = ac$ ↑

Then factor $ax^2 + mx + nx + b$ by grouping.

Examples

Factor $3x^2 + 14x - 5$.

↑ -15 ↑

Find two integers with a product of $3(-5) = -15$ and a sum of 14. The integers are -1 and 15.

$$3x^2 + 14x - 5$$
$$= 3x^2 - x + 15x - 5 \quad 14x = -x + 15x$$
$$= (3x^2 - x) + (15x - 5) \quad \text{Group the terms.}$$
$$= x(3x - 1) + 5(3x - 1) \quad \text{Factor each group.}$$
$$= (3x - 1)(x + 5) \quad \text{Factor out } 3x - 1.$$

Concepts	Examples

6.4 Factoring Trinomials by Using the FOIL Method

To factor $ax^2 + bx + c$ by trial and error, use the FOIL method in reverse.

Factor $3x^2 + 14x - 5$.

Since the only positive factors of 3 are 3 and 1, and -5 has possible factors of 1 and -5, or -1 and 5, the only possible factored forms for this trinomial follow.

$(3x + 1)(x - 5)$	Incorrect
$(3x - 5)(x + 1)$	Incorrect
$(3x - 1)(x + 5)$	Correct
$(3x + 5)(x - 1)$	Incorrect

We see by multiplying that the third factorization is correct.

6.5 Special Factoring Techniques

Difference of Squares

$$a^2 - b^2 = (a + b)(a - b)$$

Perfect Square Trinomials

$$a^2 + 2ab + b^2 = (a + b)^2$$

$$a^2 - 2ab + b^2 = (a - b)^2$$

Factor.

$$4x^2 - 9$$
$$= (2x + 3)(2x - 3)$$

$$9x^2 + 6x + 1 = (3x + 1)^2$$

$$4x^2 - 20x + 25 = (2x - 5)^2$$

6.6 Solving Quadratic Equations by Factoring

Zero-Factor Property

If a and b are real numbers and $ab = 0$, then $a = 0$ or $b = 0$.

If $(x - 2)(x + 3) = 0$, then $x - 2 = 0$ or $x + 3 = 0$.

Solving a Quadratic Equation by Factoring

Step 1 Write the equation in standard form.

Step 2 Factor.

Step 3 Use the zero-factor property.

Step 4 Solve the resulting equations.

Step 5 Check.

Solve $2x^2 = 7x + 15$.

$2x^2 - 7x - 15 = 0$	Standard form
$(2x + 3)(x - 5) = 0$	Factor.
$2x + 3 = 0$ or $x - 5 = 0$	Zero-factor property
$2x = -3$ or $x = 5$	Solve each equation.
$x = -\frac{3}{2}$	

The solutions $-\frac{3}{2}$ and 5 satisfy the original equation.

The solution set is $\{-\frac{3}{2}, 5\}$.

6.7 Applications of Quadratic Equations

Pythagorean Theorem

In a right triangle, the square of the hypotenuse equals the sum of the squares of the legs.

$$a^2 + b^2 = c^2$$

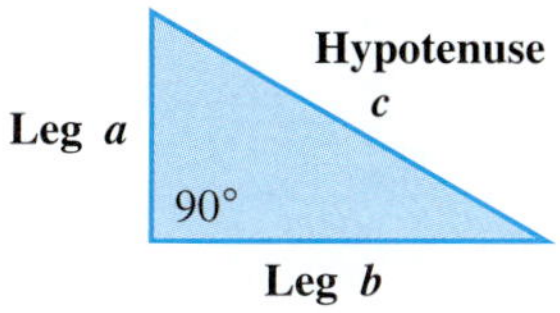

In a right triangle, one leg measures 2 ft longer than the other. The hypotenuse measures 4 ft longer than the shorter leg. Find the lengths of the three sides of the triangle.

Let x = the length of the shorter leg. Then

$$x^2 + (x + 2)^2 = (x + 4)^2.$$

Solve this equation to get $x = 6$ or $x = -2$. Discard -2 as a solution. Check that the sides measure

6 ft, $6 + 2 = 8$ ft, and $6 + 4 = 10$ ft.

Chapter 6 Review Exercises

6.1 *Factor out the greatest common factor or factor by grouping.*

1. $15t + 45$
$15(t + 3)$

2. $60z^3 + 30z$
$30z(2z^2 + 1)$

3. $44x^3 + 55x^2$
$11x^2(4x + 5)$

4. $100m^2n^3 - 50m^3n^4 + 150m^2n^2$
$50m^2n^2(2n - mn^2 + 3)$

5. $2xy - 8y + 3x - 12$
$(x - 4)(2y + 3)$

6. $6y^2 + 9y + 4xy + 6x$
$(2y + 3)(3y + 2x)$

6.2 *Factor completely.*

7. $x^2 + 10x + 21$
$(x + 3)(x + 7)$

8. $y^2 - 13y + 40$
$(y - 5)(y - 8)$

9. $q^2 + 6q - 27$
$(q + 9)(q - 3)$

10. $r^2 - r - 56$
$(r - 8)(r + 7)$

11. $x^2 + x + 1$
prime

12. $3x^2 + 6x + 6$
$3(x^2 + 2x + 2)$

13. $r^2 - 4rs - 96s^2$
$(r + 8s)(r - 12s)$

14. $p^2 + 2pq - 120q^2$
$(p + 12q)(p - 10q)$

15. $-8p^3 + 24p^2 + 80p$
$-8p(p + 2)(p - 5)$

16. $3x^4 + 30x^3 + 48x^2$
$3x^2(x + 2)(x + 8)$

17. $m^2 - 3mn - 18n^2$
$(m + 3n)(m - 6n)$

18. $y^2 - 8yz + 15z^2$
$(y - 3z)(y - 5z)$

19. $p^7 - p^6q - 2p^5q^2$
$p^5(p - 2q)(p + q)$

20. $-3r^5 + 6r^4s + 45r^3s^2$
$-3r^3(r + 3s)(r - 5s)$

6.3–6.4

21. CONCEPT CHECK To begin factoring $6r^2 - 5r - 6$, what are the possible first terms of the two binomial factors, if we consider only positive integer coefficients?
r and $6r$, $2r$ and $3r$

22. CONCEPT CHECK What is the first step you would use to factor $2z^3 + 9z^2 - 5z$?
Factor out z.

Factor completely.

23. $2k^2 - 5k + 2$
$(2k - 1)(k - 2)$

24. $3r^2 + 11r - 4$
$(3r - 1)(r + 4)$

25. $6r^2 - 5r - 6$
$(3r + 2)(2r - 3)$

26. $10z^2 - 3z - 1$
$(5z + 1)(2z - 1)$

27. $5t^2 - 11t + 12$
prime

28. $24x^5 - 20x^4 + 4x^3$
$4x^3(3x - 1)(2x - 1)$

29. $-6x^2 + 3x + 30$
$-3(x + 2)(2x - 5)$

30. $10r^3s + 17r^2s^2 + 6rs^3$
$rs(5r + 6s)(2r + s)$

31. $-30y^3 - 5y^2 + 10y$
$-5y(3y + 2)(2y - 1)$

32. $4z^2 - 5z + 7$
prime

33. $-3m^3n + 19m^2n + 40mn$
$-mn(3m + 5)(m - 8)$

34. $14a^2 - 27ab - 20b^2$
$(2a - 5b)(7a + 4b)$

6.5

35. CONCEPT CHECK Which one of the following is a difference of squares?

A. $32x^2 - 1$ **B.** $4x^2y^2 - 25z^2$

C. $x^2 + 36$ **D.** $25y^3 - 1$

B

36. CONCEPT CHECK Which one of the following is a perfect square trinomial?

A. $x^2 + x + 1$ **B.** $y^2 - 4y + 9$

C. $4x^2 + 10x + 25$ **D.** $x^2 - 20x + 100$

D

Factor completely.

37. $n^2 - 64$
$(n + 8)(n - 8)$

38. $25b^2 - 121$
$(5b + 11)(5b - 11)$

39. $49y^2 - 25w^2$
$(7y + 5w)(7y - 5w)$

40. $144p^2 - 36q^2$
$36(2p + q)(2p - q)$

41. $x^2 + 100$
prime

42. $z^2 + 10z + 25$
$(z + 5)^2$

43. $r^2 - 12r + 36$
$(r - 6)^2$

44. $9t^2 - 42t + 49$
$(3t - 7)^2$

45. $16m^2 + 40mn + 25n^2$
$(4m + 5n)^2$

46. $54x^3 - 72x^2 + 24x$
$6x(3x - 2)^2$

6.6

Solve each equation, and check the solutions.

47. $(4t + 3)(t - 1) = 0$
$\left\{-\frac{3}{4}, 1\right\}$

48. $(x + 7)(x - 4)(x + 3) = 0$
$\{-7, -3, 4\}$

49. $x(2x - 5) = 0$
$\left\{0, \frac{5}{2}\right\}$

50. $z^2 + 4z + 3 = 0$
$\{-3, -1\}$

51. $m^2 - 5m + 4 = 0$
$\{1, 4\}$

52. $x^2 = -15 + 8x$
$\{3, 5\}$

53. $3z^2 - 11z - 20 = 0$
$\left\{-\frac{4}{3}, 5\right\}$

54. $81t^2 - 64 = 0$
$\left\{-\frac{8}{9}, \frac{8}{9}\right\}$

55. $y^2 = 8y$
$\{0, 8\}$

56. $n(n - 5) = 6$
$\{-1, 6\}$

57. $t^2 - 14t + 49 = 0$
$\{7\}$

58. $t^2 = 12(t - 3)$
$\{6\}$

59. $(5z + 2)(z^2 + 3z + 2) = 0$
$\left\{-\frac{2}{5}, -2, -1\right\}$

60. $x^2 = 9$
$\{-3, 3\}$

61. $64x^3 - 9x = 0$
$\left\{-\frac{3}{8}, 0, \frac{3}{8}\right\}$

62. $(2r + 1)(12r^2 + 5r - 3) = 0$
$\left\{-\frac{3}{4}, -\frac{1}{2}, \frac{1}{3}\right\}$

63. $25w^2 - 90w + 81 = 0$
$\left\{\frac{9}{5}\right\}$

64. $r(r - 7) = 30$
$\{-3, 10\}$

6.7

Solve each problem.

65. The length of a rug is 6 ft more than the width. The area is 40 ft^2. Find the length and width of the rug.
length: 10 ft; width: 4 ft

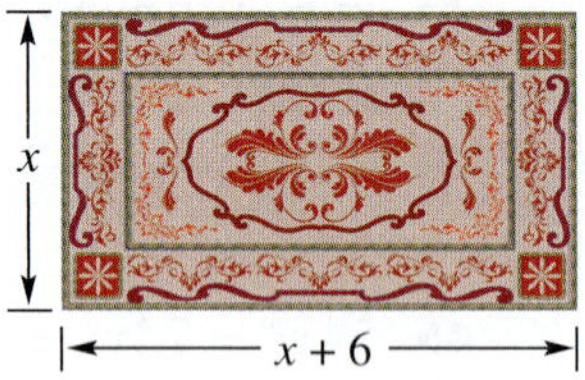

66. The surface area S of a box is given by

$$S = 2WH + 2WL + 2LH.$$

A treasure chest from a sunken galleon has dimensions (in feet) as shown in the figure. Its surface area is 650 ft^2. Find its width. **5 ft**

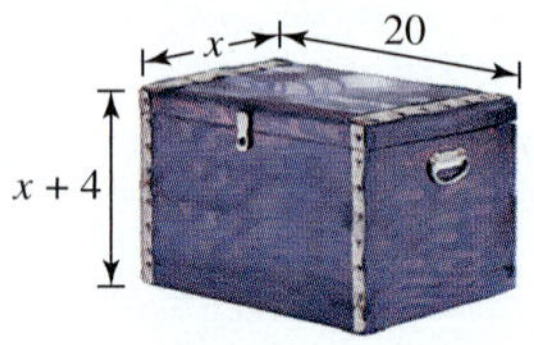

67. The product of two consecutive integers is 29 more than their sum. What are the integers?
6, 7 or −5, −4

68. The product of the lesser two of three consecutive integers is equal to 23 plus the greatest. Find the integers.
−5, −4, −3 or 5, 6, 7

69. Two cars left an intersection at the same time. One traveled west, and the other traveled 14 mi less, but to the south. How far apart were they then, if the distance between them was 16 mi more than the distance traveled south? **26 mi**

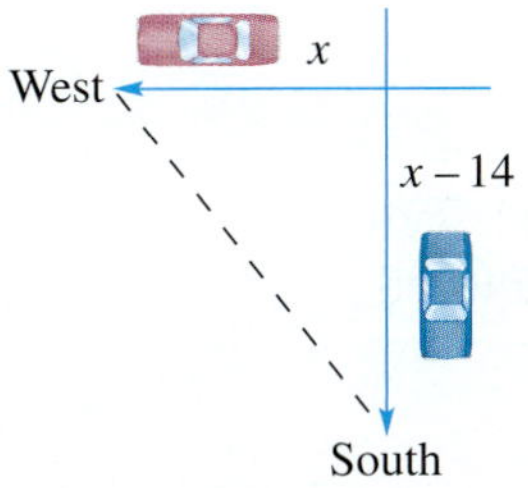

70. The triangular sail of a schooner has an area of 30 m^2. The height of the sail is 4 m more than the base. Find the length of the base of the sail. **6 m**

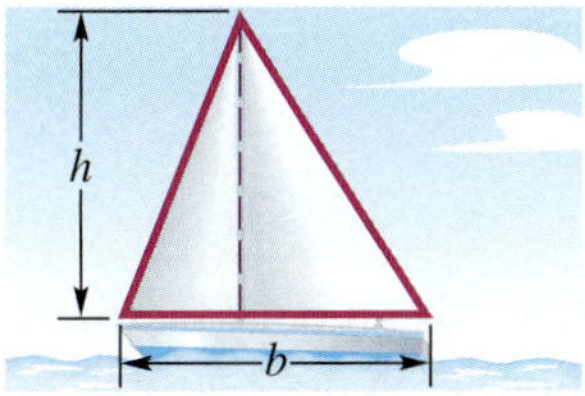

71. The floor plan for a house is a rectangle with length 7 m more than its width. The area is 170 m^2. Find the width and length of the house.

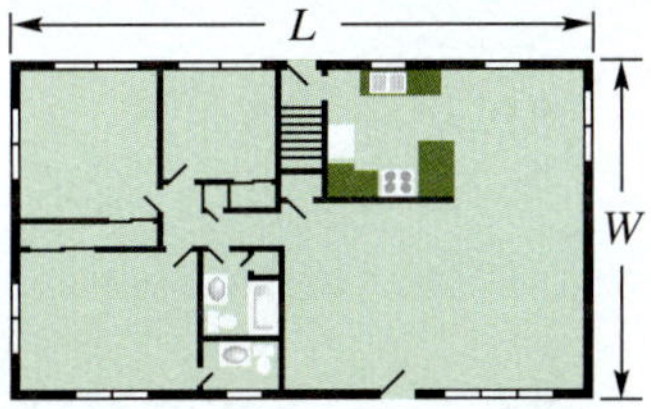

width: 10 m; length: 17 m

72. If an object is dropped, the distance d in feet it falls in t seconds (disregarding air resistance) is given by the quadratic equation

$$d = 16t^2.$$

Find the distance an object would fall in each of the following times.

(a) 4 sec **256 ft** **(b)** 8 sec **1024 ft**

The numbers of alternative-fueled vehicles, in thousands, in use in the United States for selected years over the period 2001–2009 are given in the table.

Year	Number (in thousands)
2001	425
2003	534
2005	592
2007	696
2009	826

Source: Energy Information Administration.

To model the number of vehicles y in year x, we developed the quadratic equation

$$y = 1.57x^2 + 32.5x + 400.$$

Here we used $x = 1$ for 2001, $x = 2$ for 2002, and so on.

73. Use the model to find the number of alternative-fueled vehicles in 2005, to the nearest thousand. How does the result compare with the actual data in the table?

602 thousand; The result is a little higher than the 592 thousand given in the table.

74. What estimate for 2010 is given by the model? Why might the estimate for 2010 be unreliable?

882 thousand; The estimate may be unreliable because the conditions that prevailed in the years 2001–2009 may have changed, causing either a greater increase or a greater decrease predicted by the model for the number of such vehicles.

Mixed Review Exercises

75. CONCEPT CHECK Which of the following is *not* factored completely? **D**

A. $3(7t)$ **B.** $3x(7t + 4)$ **C.** $(3 + x)(7t + 4)$ **D.** $3(7t + 4) + x(7t + 4)$

76. CONCEPT CHECK A student factored $6x^2 + 16x - 32$ as $(2x + 8)(3x - 4)$. Tell why the polynomial is not factored completely, and give the completely factored form.

The terms of $2x + 8$ have a common factor of 2. The completely factored form is $2(x + 4)(3x - 4)$.

Factor completely.

77. $15m^2 + 20mp - 12m - 16p$
$(3m + 4p)(5m - 4)$

78. $24ab^3c^2 - 56a^2bc^3 + 72a^2b^2c$
$8abc(3b^2c - 7ac^2 + 9ab)$

79. $z^2 - 11zx + 10x^2$
$(z - x)(z - 10x)$

80. $3k^2 + 11k + 10$
$(3k + 5)(k + 2)$

81. $y^4 - 625$
$(y^2 + 25)(y + 5)(y - 5)$

82. $6m^3 - 21m^2 - 45m$
$3m(2m + 3)(m - 5)$

83. $25a^2 + 15ab + 9b^2$
prime

84. $2a^5 - 8a^4 - 24a^3$
$2a^3(a + 2)(a - 6)$

85. $-12r^2 - 8rq + 15q^2$
$-1(2r + 3q)(6r - 5q)$

86. $100a^2 - 9$
$(10a + 3)(10a - 3)$

87. $49t^2 + 56t + 16$
$(7t + 4)^2$

Solve each equation, and check the solutions.

88. $t(t - 7) = 0$
$\{0, 7\}$

89. $x^2 + 3x = 10$
$\{-5, 2\}$

90. $25x^2 + 20x + 4 = 0$
$\left\{-\frac{2}{5}\right\}$

Solve each problem.

91. A lot is in the shape of a right triangle. The hypotenuse is 3 m longer than the longer leg. The longer leg is 6 m longer than twice the length of the shorter leg. Find the lengths of the sides of the lot.

15 m, 36 m, 39 m

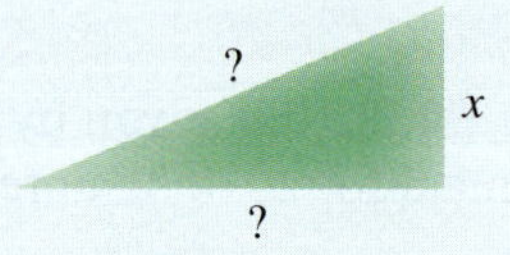

92. A pyramid has a rectangular base with a length that is 2 m more than the width. The height of the pyramid is 6 m, and its volume is 48 m^3. Find the length and width of the base.

length: 6 m; width: 4 m

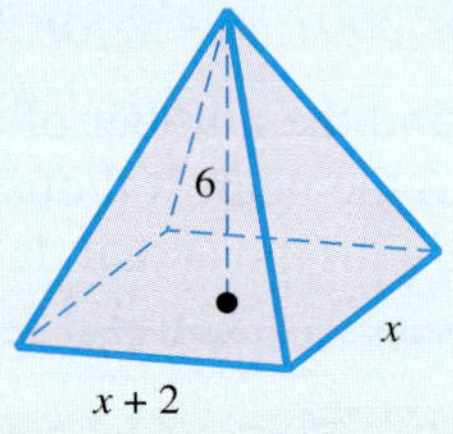

Chapter 6 Test

Chapter Test Prep Video

The Chapter Test Prep Videos with test solutions are available on DVD, in MyMathLab, and on YouTube—search "LialIntroductoryAlg" and click on "Channels."

1. Which one of the following is the correct, completely factored form of $2x^2 - 2x - 24$? D

A. $(2x + 6)(x - 4)$ **B.** $(x + 3)(2x - 8)$

C. $2(x + 4)(x - 3)$ **D.** $2(x + 3)(x - 4)$

Factor each polynomial completely. If the polynomial is prime, *say so.*

2. $12x^2 - 30x$
$6x(2x - 5)$

3. $2m^3n^2 + 3m^3n - 5m^2n^2$
$m^2n(2mn + 3m - 5n)$

4. $2ax - 2bx + ay - by$
$(2x + y)(a - b)$

5. $x^2 - 9x + 14$
$(x - 7)(x - 2)$

6. $2x^2 + x - 3$
$(2x + 3)(x - 1)$

7. $6x^2 - 19x - 7$
$(3x + 1)(2x - 7)$

8. $3x^2 - 12x - 15$
$3(x + 1)(x - 5)$

9. $10z^2 - 17z + 3$
$(5z - 1)(2z - 3)$

10. $t^2 + 6t + 10$
prime

11. $x^2 + 36$
prime

12. $y^2 - 49$
$(y + 7)(y - 7)$

13. $81a^2 - 121b^2$
$(9a + 11b)(9a - 11b)$

14. $x^2 + 16x + 64$
$(x + 8)^2$

15. $4x^2 - 28xy + 49y^2$
$(2x - 7y)^2$

16. $-2x^2 - 4x - 2$
$-2(x + 1)^2$

17. $6t^4 + 3t^3 - 108t^2$
$3t^2(2t + 9)(t - 4)$

18. $4r^2 + 10rt + 25t^2$
prime

19. $4t^3 + 32t^2 + 64t$
$4t(t + 4)^2$

20. $x^4 - 81$
$(x^2 + 9)(x + 3)(x - 3)$

Solve each equation.

21. $(x + 3)(x - 9) = 0$

$\{-3, 9\}$

22. $2r^2 - 13r + 6 = 0$

$\left\{\frac{1}{2}, 6\right\}$

23. $25x^2 - 4 = 0$

$\left\{-\frac{2}{5}, \frac{2}{5}\right\}$

24. $x(x - 20) = -100$

$\{10\}$

25. $t^2 = 3t$

$\{0, 3\}$

26. $(s + 8)(6s^2 + 13s - 5) = 0$

$\left\{-8, -\frac{5}{2}, \frac{1}{3}\right\}$

Solve each problem.

27. The length of a rectangular flower bed is 3 ft less than twice its width. The area of the bed is 54 ft^2. Find the dimensions of the flower bed.

6 ft by 9 ft

28. Find two consecutive integers such that the square of the sum of the two integers is 11 more than the lesser integer.

$-2, -1$

29. A carpenter needs to cut a brace to support a wall stud, as shown in the figure. The brace should be 7 ft less than three times the length of the stud. If the brace will be anchored on the floor 15 ft away from the stud, how long should the brace be?

17 ft

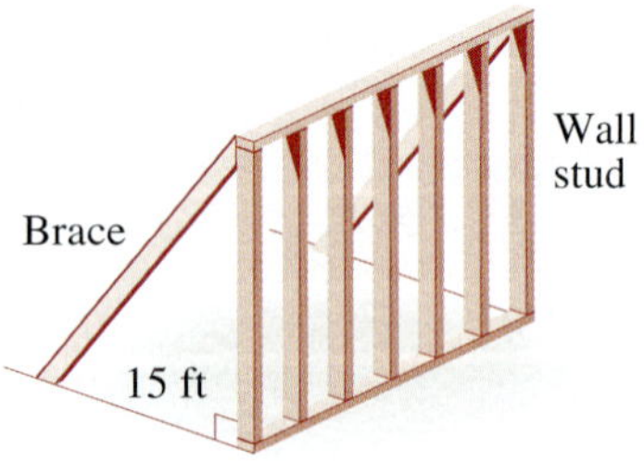

30. The number of pieces of first-class mail sent through the U.S. Postal Service (USPS) from 2001 through 2010 can be approximated by the quadratic equation

$$y = -0.3470x^2 + 1.374x + 100.7,$$

where $x = 1$ represents 2001, $x = 2$ represents 2002, and so on, and y is the number of pieces of mail in billions. Use the model to estimate the number of pieces of first-class mail sent through the USPS in 2009. Round your answer to the nearest billion. (*Source:* U.S. Postal Service.)

85 billion pieces

Chapters R–6 Cumulative Review Exercises

Solve each equation.

1. $3x + 2(x - 4) = 4(x - 2)$
$\{0\}$

2. $0.3x + 0.9x = 0.06$
$\{0.05\}$

3. $\frac{2}{3}n - \frac{1}{2}(n - 4) = 3$
$\{6\}$

4. Solve for t: $A = P + Prt$
$t = \frac{A - P}{Pr}$

5. From a list of "technology-related items," adults were recently surveyed as to those items they couldn't live without. Complete the results shown in the table if 500 adults were surveyed.

Item	Percent That Couldn't Live Without	Number That Couldn't Live Without
Personal computer	46%	230
Cell phone	41%	205
High-speed Internet	38%	190
MP3 player	12%	60

(Other items included digital cable, HDTV, and electronic gaming console.)
Source: Ipsos for AP.

Solve each problem.

6. At the 2010 Winter Olympics in Vancouver, Canada, the top medal winner was the United States, which won a total of 37 medals. The United States won 2 more silver medals than bronze and 4 fewer gold medals than bronze. Find the number of each type of medal won. (*Source: World Almanac and Book of Facts.*)
gold: 9; silver: 15; bronze: 13

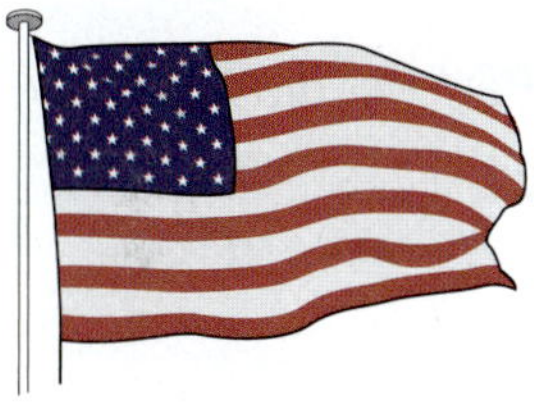

7. In 2010, American women working full time earned, on average, 81.2 cents for every dollar earned by men working full time. The median weekly earnings for full-time male workers were \$824. To the nearest dollar, what were the median weekly earnings for full-time female workers? (*Source*: U.S. Bureau of Labor Statistics.)
\$669

8. Find the measures of the marked angles.
$110°$ and $70°$

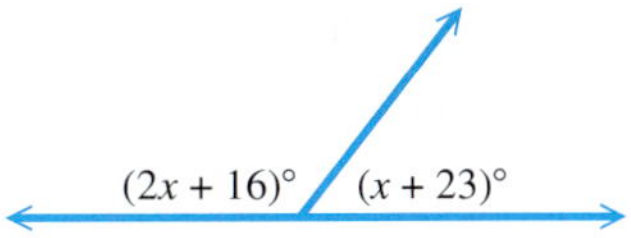

9. Fill in each blank with *positive* or *negative*. The point with coordinates (a, b) is in

(a) quadrant II if a is __negative__ and b is __positive__.

(b) quadrant III if a is __negative__ and b is __negative__.

Consider the equation $y = 4x + 3$. Find the following.

10. The x- and y-intercepts
$\left(-\frac{3}{4}, 0\right), (0, 3)$

11. The slope
4

12. The graph

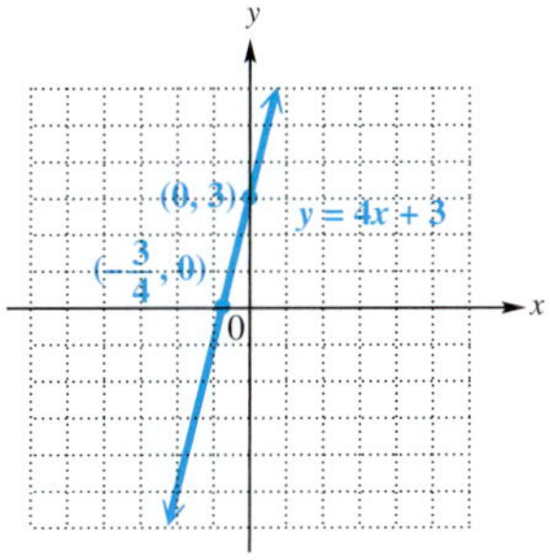

13. The points on the graph show the total retail sales of prescription drugs in the United States in the years 2004–2010, along with a graph of a linear equation that models the data.

(a) Use the ordered pairs shown on the graph to find the slope of the line to the nearest whole number. Interpret the slope.

8; A slope of (approximately) 8 means that the retail sales of prescription drugs increased by about \$8 billion per year.

(b) Use the graph to estimate sales in the year 2008. Write your answer as an ordered pair of the form (year, sales in billions of dollars).

(2008, 250)

Retail Prescription Drug Sales

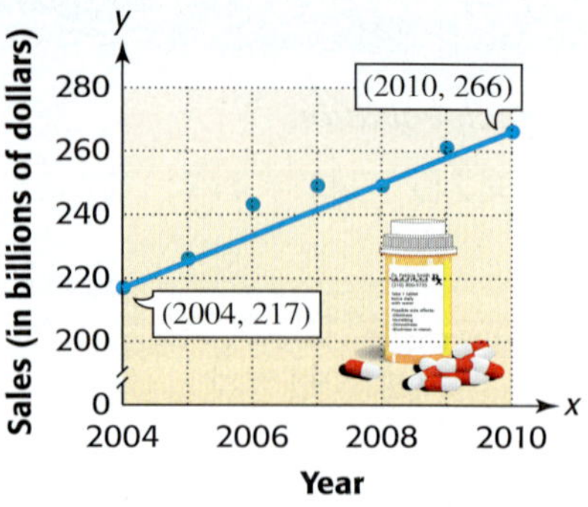

Source: National Association of Chain Drug Stores.

Solve each system of equations.

14. $4x - y = -6$ $\{(-1, 2)\}$

$2x + 3y = 4$

15. $5x + 3y = 10$ $\emptyset$

$2x + \frac{6}{5}y = 5$

Evaluate each expression.

16. $2^{-3} \cdot 2^5$ 4

17. $\left(\frac{3}{4}\right)^{-2}$ $\frac{16}{9}$

18. $\frac{6^5 \cdot 6^{-2}}{6^3}$ 1

19. $\left(\frac{4^{-3} \cdot 4^4}{4^5}\right)^{-1}$ 256

Simplify each expression and write the answer using only positive exponents. Assume no denominators are 0.

20. $\frac{(p^2)^3p^{-4}}{(p^{-3})^{-1}p}$ $\frac{1}{p^2}$

21. $\frac{(m^{-2})^3m}{m^5m^{-4}}$ $\frac{1}{m^6}$

Perform the indicated operations.

22. $(2k^2 + 4k) - (5k^2 - 2) - (k^2 + 8k - 6)$

$-4k^2 - 4k + 8$

23. $(9x + 6)(5x - 3)$

$45x^2 + 3x - 18$

24. $(3p + 2)^2$

$9p^2 + 12p + 4$

25. $\frac{8x^4 + 12x^3 - 6x^2 + 20x}{2x}$

$4x^3 + 6x^2 - 3x + 10$

Factor completely.

26. $2a^2 + 7a - 4$

$(2a - 1)(a + 4)$

27. $10m^2 + 19m + 6$

$(2m + 3)(5m + 2)$

28. $8t^2 + 10tv + 3v^2$

$(4t + 3v)(2t + v)$

29. $4p^2 - 12p + 9$

$(2p - 3)^2$

30. $25r^2 - 81t^2$

$(5r + 9t)(5r - 9t)$

31. $2pq + 6p^3q + 8p^2q$

$2pq(3p + 1)(p + 1)$

Solve each equation.

32. $6m^2 + m - 2 = 0$

$\left\{-\frac{2}{3}, \frac{1}{2}\right\}$

33. $8x^2 = 64x$

$\{0, 8\}$

34. $49x^2 - 56x + 16 = 0$

$\left\{\frac{4}{7}\right\}$

35. The length of the hypotenuse of a right triangle is 3 m more than twice the length of the shorter leg. The longer leg is 7 m longer than the shorter leg. Find the lengths of the sides.

5 m, 12 m, 13 m

7 Rational Expressions and Applications

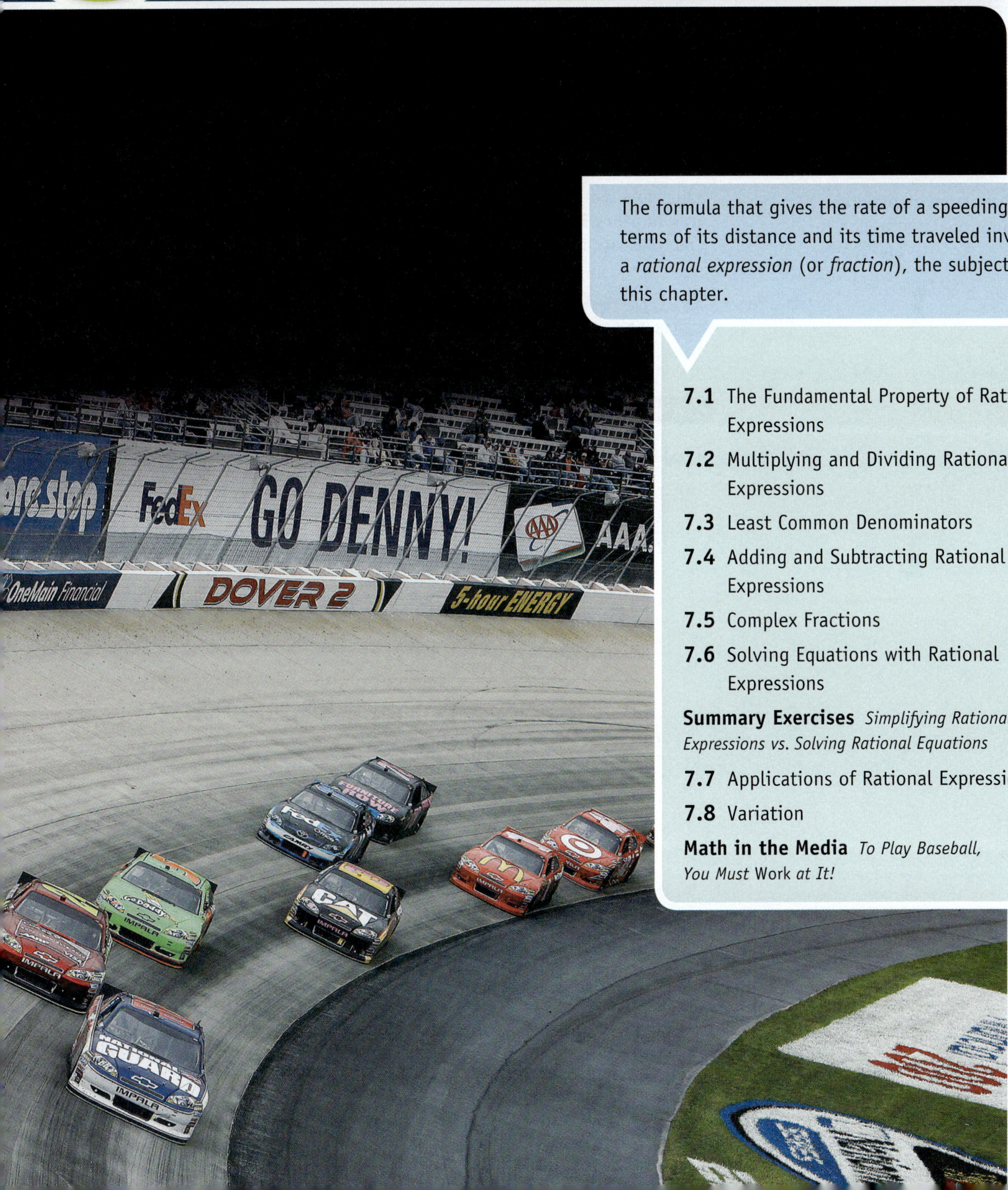

The formula that gives the rate of a speeding car in terms of its distance and its time traveled involves a *rational expression* (or *fraction*), the subject of this chapter.

7.1 The Fundamental Property of Rational Expressions

OBJECTIVES

1. Find the numerical value of a rational expression.
2. Find the values of the variable for which a rational expression is undefined.
3. Write rational expressions in lowest terms.
4. Recognize equivalent forms of rational expressions.

The quotient of two integers (with denominator not 0), such as $\frac{2}{3}$ or $-\frac{3}{4}$, is a *rational number*. In the same way, the quotient of two polynomials with denominator not equal to 0 is a *rational expression.*

Rational Expression

A **rational expression** is an expression of the form $\frac{P}{Q}$, where P and Q are polynomials, with $Q \neq 0$.

$$\frac{-6x}{x^3+8}, \quad \frac{9x}{y+3}, \quad \text{and} \quad \frac{2m^3}{8} \qquad \text{Rational expressions}$$

Our work with rational expressions will require much of what we learned in **Chapters 5 and 6** on polynomials and factoring, as well as the rules for fractions from **Chapter R.**

OBJECTIVE 1 Find the numerical value of a rational expression. Remember that to *evaluate* an expression means to find its *value*. We use substitution to evaluate a rational expression for a given value of the variable.

EXAMPLE 1 Evaluating Rational Expressions

Find the numerical value of $\frac{3x+6}{2x-4}$ for each value of x.

(a) $x = 1$

$$\frac{3x+6}{2x-4}$$

$$= \frac{3(1)+6}{2(1)-4} \qquad \text{Let } x = 1.$$

$$= \frac{9}{-2}, \text{ or } -\frac{9}{2} \qquad \frac{a}{-b} = -\frac{a}{b}$$

(b) $x = 0$

$$\frac{3x+6}{2x-4}$$

$$= \frac{3(0)+6}{2(0)-4} \qquad \text{Let } x = 0.$$

$$= \frac{6}{-4}, \text{ or } -\frac{3}{2} \qquad \text{Lowest terms}$$

(c) $x = 2$

$$\frac{3x+6}{2x-4}$$

$$= \frac{3(2)+6}{2(2)-4} \qquad \text{Let } x = 2.$$

$$= \frac{12}{0}$$

The expression is undefined for $x = 2$.

(d) $x = -2$

$$\frac{3x+6}{2x-4}$$

$$= \frac{3(-2)+6}{2(-2)-4} \qquad \text{Let } x = -2.$$

$$= \frac{0}{-8}, \text{ or } 0 \qquad \frac{0}{b} = 0$$

◀ Work Problem 1 at the Side.

Note

The numerator of a rational expression may be any real number. If the numerator equals 0 and the denominator does not equal 0, then the rational expression equals 0. **See Example 1(d).**

---

1 Find the numerical value of each rational expression for $x = -3$, $x = 0$, and $x = 3$.

GS **(a)** $\frac{x}{2x+1}$

For $x = -3$:

$$\frac{x}{2x+1}$$

$$= \frac{\_\_\_}{2(\_\_\_)+1}$$

$$= \frac{\_\_\_}{\_\_\_+1}$$

$$= \_\_\_$$

(b) $\frac{2x+6}{x-3}$

Teaching Tip

Good factoring skill is required for this and subsequent sections. Review with students the factoring techniques introduced in Chapter 6.

Answers

1. (a) For $x = -3$: -3; -3; -3; -6; $\frac{3}{5}$; for $x = 0$, final answer 0; for $x = 3$, final answer $\frac{3}{7}$

(b) 0; -2; undefined

OBJECTIVE 2 Find the values of the variable for which a rational expression is undefined. In the definition of a rational expression $\frac{P}{Q}$, Q cannot equal 0. ***The denominator of a rational expression cannot equal 0 because division by 0 is undefined.***

For instance, in the rational expression

$$\frac{8x^2}{x-3}, \leftarrow \text{Denominator cannot equal 0.}$$

the variable x can take on any value except 3. When x is 3, the denominator becomes $3 - 3 = 0$, making the expression undefined. Thus, x cannot equal 3. We indicate this restriction by writing $x \neq 3$.

Determining When a Rational Expression Is Undefined

Step 1 Set the denominator of the rational expression equal to 0.

Step 2 Solve this equation.

Step 3 The solutions of the equation are the values that make the rational expression undefined. The variable *cannot* equal these values.

EXAMPLE 2 Finding Values That Make Rational Expressions Undefined

Find any values of the variable for which each rational expression is undefined.

(a) $\frac{x+5}{3x+2}$ We must find any value of x that makes the *denominator* equal to 0, since division by 0 is undefined.

Step 1 Set the denominator equal to 0.

$$3x + 2 = 0$$

Step 2 Solve.

$3x = -2$ Subtract 2.

$x = -\frac{2}{3}$ Divide by 3.

Step 3 The given expression is undefined for $-\frac{2}{3}$, so $x \neq -\frac{2}{3}$.

(b) $\frac{9m^2}{m^2 - 5m + 6}$

$m^2 - 5m + 6 = 0$ Set the denominator equal to 0.

$(m-2)(m-3) = 0$ Factor.

$m - 2 = 0$ or $m - 3 = 0$ Zero-factor property

$m = 2$ or $m = 3$ Solve for m.

The given expression is undefined for 2 and 3, so $m \neq 2, m \neq 3$.

(c) $\frac{2r}{r^2+1}$ This denominator will not equal 0 for any value of r, because r^2 is always greater than or equal to 0, and adding 1 makes the sum greater than or equal to 1. There are no values for which this expression is undefined.

Work Problem 2 at the Side. ▶

2 Find any values of the variable for which each rational expression is undefined. Write answers with the symbol $\neq$.

GS **(a)** $\frac{x+2}{x-5}$

Step 1 ______ $= 0$

Step 2 $x =$ ____

Step 3 The given expression is undefined for ____. Thus, $x\ (= / \neq)\ 5$.

(b) $\frac{3r}{r^2 + 6r + 8}$

(c) $\frac{-5m}{m^2 + 4}$

Teaching Tip

Students will occasionally list 0 as a value that makes an expression like that in Example 2(b) or 2(c) undefined. Remind them that a 0 numerator is allowed, since the fraction will then be equal to 0.

Answers

2. (a) $x - 5$; 5; 5; $\neq$ **(b)** $r \neq -4, r \neq -2$
(c) never undefined

3 Write each expression in lowest terms.

(a) $\frac{15}{40}$

(b) $\frac{5x^4}{15x^2}$

(c) $\frac{6p^3}{2p^2}$

Teaching Tip
Review the arithmetic operations with common fractions from Section R.1. Note for students that these same operations will be applied to algebraic fractions in this chapter.

Answers

3. **(a)** $\frac{3}{8}$ **(b)** $\frac{x^2}{3}$ **(c)** $3p$

OBJECTIVE 3 Write rational expressions in lowest terms. A fraction such as $\frac{2}{3}$ is said to be in *lowest terms.*

Lowest Terms

A rational expression $\frac{P}{Q}$ $(Q \neq 0)$ is in **lowest terms** if the greatest common factor of its numerator and denominator is 1.

We use the **fundamental property of rational expressions** to write a rational expression in lowest terms.

Fundamental Property of Rational Expressions

If $\frac{P}{Q}$ (where $Q \neq 0$) is a rational expression and if K represents any polynomial (where $K \neq 0$), then the following is true.

$$\frac{PK}{QK} = \frac{P}{Q}$$

This property is based on the identity property for multiplication.

$$\frac{PK}{QK} = \frac{P}{Q} \cdot \frac{K}{K} = \frac{P}{Q} \cdot 1 = \frac{P}{Q}$$

EXAMPLE 3 Writing in Lowest Terms

Write each expression in lowest terms.

(a) $\frac{30}{72}$

Begin by factoring.

$$= \frac{2 \cdot 3 \cdot 5}{2 \cdot 2 \cdot 2 \cdot 3 \cdot 3}$$

(b) $\frac{14k^2}{2k^3}$

Write k^2 as $k \cdot k$ and k^3 as $k \cdot k \cdot k$.

$$= \frac{2 \cdot 7 \cdot k \cdot k}{2 \cdot k \cdot k \cdot k}$$

Group any factors common to the numerator and denominator.

$$= \frac{5 \cdot (2 \cdot 3)}{2 \cdot 2 \cdot 3 \cdot (2 \cdot 3)} \qquad = \frac{7(2 \cdot k \cdot k)}{k(2 \cdot k \cdot k)}$$

Use the fundamental property.

$$= \frac{5}{2 \cdot 2 \cdot 3}, \text{ or } \frac{5}{12} \qquad = \frac{7}{k}$$

◀ Work Problem **3** at the Side.

Writing a Rational Expression in Lowest Terms

Step 1 **Factor** the numerator and denominator completely.

Step 2 **Use the fundamental property** to divide out any common factors.

EXAMPLE 4 Writing in Lowest Terms

Write each rational expression in lowest terms.

(a) $\dfrac{3x-12}{5x-20}$

$= \dfrac{3(x-4)}{5(x-4)}$ Factor. (Step 1)

$x \neq 4$, since the denominator is 0 for this value.

$= \dfrac{3}{5}$ Fundamental property (Step 2)

The given expression is equal to $\frac{3}{5}$ for all values of x, where $x \neq 4$ (since the denominator of the original rational expression is 0 when x is 4).

(b) $\dfrac{2y^2-8}{2y+4}$

$= \dfrac{2(y^2-4)}{2(y+2)}$ Factor. (Step 1)

$y \neq -2$, since the denominator is 0 for this value.

$= \dfrac{2(y+2)(y-2)}{2(y+2)}$ Factor the numerator completely.

$= y-2$ Fundamental property (Step 2)

(c) $\dfrac{m^2+2m-8}{2m^2-m-6}$

$= \dfrac{(m+4)(m-2)}{(2m+3)(m-2)}$ Factor. (Step 1)

$m \neq -\frac{3}{2}$, $m \neq 2$

$= \dfrac{m+4}{2m+3}$ Fundamental property (Step 2)

From now on, we write statements of equality of rational expressions with the understanding that they apply only to those real numbers that make neither denominator equal to 0.

CAUTION

Rational expressions cannot be written in lowest terms until after the numerator and denominator have been factored. Only common factors can be divided out, not common terms.

$\dfrac{6x+9}{4x+6} = \dfrac{3(2x+3)}{2(2x+3)} = \dfrac{3}{2}$ (Divide out the common factor.)

$\dfrac{6+x}{4x}$ ← Numerator cannot be factored. (Already in lowest terms)

Work Problem 4 at the Side. ▶

---

4 Write each rational expression in lowest terms.

GS **(a)** $\dfrac{4y+2}{6y+3}$

$= \dfrac{2(\_\_\_\_)}{3(\_\_\_\_)}$

$= \_\_\_\_$

(b) $\dfrac{8p+8q}{5p+5q}$

(c) $\dfrac{x^2+4x+4}{4x+8}$

(d) $\dfrac{a^2-b^2}{a^2+2ab+b^2}$

Teaching Tip

Constantly remind students that ***rational expressions cannot be written in lowest terms until after the numerator and denominator have been factored.***

Answers

4. **(a)** $2y+1$; $2y+1$; $\dfrac{2}{3}$ **(b)** $\dfrac{8}{5}$ **(c)** $\dfrac{x+2}{4}$ **(d)** $\dfrac{a-b}{a+b}$

5 Write $\frac{x-y}{y-x}$ from **Example 5** in lowest terms by factoring -1 from the numerator. How does the result compare to the result in **Example 5?**

EXAMPLE 5 **Writing in Lowest Terms (Factors Are Opposites)**

Write $\frac{x-y}{y-x}$ in lowest terms.

To get a common factor, the denominator $y - x$ can be factored as follows.

$$y - x$$

We are factoring out -1, **NOT** multiplying by it.

$$= -1(-y + x) \quad \text{Factor out } -1.$$

$$= -1(x - y) \quad \text{Commutative property}$$

With this result in mind, we simplify as follows.

$$\frac{x-y}{y-x}$$

$$= \frac{1(x-y)}{-1(x-y)} \quad y - x = -1(x - y) \text{ from above.}$$

$$= \frac{1}{-1}, \text{ or } -1 \quad \text{Fundamental property}$$

◀ **Work Problem 5 at the Side.**

Note

The numerator *or* the denominator could have been factored in the first step in **Example 5.** Factor -1 from the numerator, and confirm that the result is the same.

CAUTION

Although x and y appear in both the numerator and denominator in **Example 5,** we cannot use the fundamental property right away because they are *terms,* not *factors.* ***Terms are added, while factors are multiplied.***

In **Example 5,** notice that $y - x$ is the **opposite** (or **additive inverse**) of $x - y$. A general rule for this situation follows.

Quotient of Opposites

If the numerator and the denominator of a rational expression are opposites, such as in $\frac{x-y}{y-x}$, then the rational expression is equal to -1.

Based on this result, the following are true.

Numerator and denominator are opposites.

$$\frac{q-7}{7-q} = -1 \quad \text{and} \quad \frac{-5a+2b}{5a-2b} = -1$$

However, the following expression cannot be simplified further.

$$\frac{x-2}{x+2}$$

Numerator and denominator are *not* opposites.

Answer

5. The result is -1, the same as in **Example 5.**

EXAMPLE 6 **Writing in Lowest Terms (Factors Are Opposites)**

Write each rational expression in lowest terms.

(a) $\frac{2 - m}{m - 2}$

Since $2 - m$ and $m - 2$ (or $-2 + m$) are opposites, this expression equals -1.

(b) $\frac{4x^2 - 9}{6 - 4x}$

$= \frac{(2x + 3)(2x - 3)}{2(3 - 2x)}$ Factor the numerator and denominator.

$= \frac{(2x + 3)(2x - 3)}{2(-1)(2x - 3)}$ Write $3 - 2x$ as $-1(2x - 3)$.

$= \frac{2x + 3}{2(-1)}$ Fundamental property

$= \frac{2x + 3}{-2}$, or $-\frac{2x + 3}{2}$ $\quad \frac{a}{-b} = -\frac{a}{b}$

(c) $\frac{3 + r}{3 - r}$ $\quad 3 - r$ is *not* the opposite of $3 + r$.

The rational expression is already in lowest terms.

Work Problem 6 at the Side. ▶

OBJECTIVE 4 Recognize equivalent forms of rational expressions. The common fraction $-\frac{5}{6}$ can also be written as $\frac{-5}{6}$ and as $\frac{5}{-6}$.

Consider the final rational expression from **Example 6(b).**

$$-\frac{2x + 3}{2}$$

The $-$ sign representing the factor -1 is in front of the expression, aligned with the fraction bar. The factor -1 may instead be placed in the numerator or in the denominator. Some other equivalent forms of this rational expression are

Use parentheses.

$$\frac{-(2x + 3)}{2} \quad \text{and} \quad \frac{2x + 3}{-2}.$$

By the distributive property,

$$\frac{-(2x + 3)}{2} \quad \text{can also be written} \quad \frac{-2x - 3}{2}.$$

Multiply *each* term in the parentheses by -1.

CAUTION

$\frac{-2x + 3}{2}$ is *not* an equivalent form of $\frac{-(2x + 3)}{2}$. The sign preceding 3 in the numerator of $\frac{-2x + 3}{2}$ should be $-$ rather than $+$. ***Be careful to apply the distributive property correctly.***

6 Write each rational expression in lowest terms.

(a) $\frac{5 - y}{y - 5}$

(b) $\frac{m - n}{n - m}$

GS **(c)** $\frac{25x^2 - 16}{12 - 15x}$

$= \frac{(5x + 4)(\_\_\_\_)}{3(\_\_\_\_)}$

$= \frac{(5x + 4)(5x - 4)}{3(\_\_\_\_)(5x - 4)}$

$= \_\_\_\_$

(d) $\frac{9 - k}{9 + k}$

Answers

6. **(a)** -1 **(b)** -1

(c) $5x - 4$; $4 - 5x$; -1; $\frac{5x + 4}{-3}$, or $-\frac{5x + 4}{3}$

(d) already in lowest terms

7 Decide whether each rational expression is equivalent to

$$-\frac{2x-6}{x+3}.$$

(a) $\frac{-(2x-6)}{x+3}$

(b) $\frac{-2x+6}{x+3}$

(c) $\frac{-2x-6}{x+3}$

(d) $\frac{2x-6}{-(x+3)}$

(e) $\frac{2x-6}{-x-3}$

(f) $\frac{2x-6}{x-3}$

Answers

7. **(a)** equivalent **(b)** equivalent **(c)** not equivalent **(d)** equivalent **(e)** equivalent **(f)** not equivalent

EXAMPLE 7 Writing Equivalent Forms of a Rational Expression

Write four equivalent forms of the following rational expression.

$$-\frac{3x+2}{x-6}$$

If we apply the negative sign to the numerator, we obtain these equivalent forms.

$$① \rightarrow \frac{-(3x+2)}{x-6}, \quad \text{and, by the distributive property,} \quad \frac{-3x-2}{x-6} \leftarrow ②$$

If we apply the negative sign to the denominator of the given rational expression, we obtain two more forms.

$$③ \rightarrow \frac{3x+2}{-(x-6)} \quad \text{and, by distributing again,} \quad \frac{3x+2}{-x+6} \leftarrow ④$$

CAUTION

Recall that $-\frac{5}{6} \neq \frac{-5}{-6}$. Thus, in **Example 7,** it would be incorrect to distribute the negative sign in $-\frac{3x+2}{x-6}$ to *both* the numerator *and* the denominator. (Doing this would actually lead to the *opposite* of the original expression.)

◀ Work Problem **7** at the Side.

7.1 Exercises

FOR EXTRA HELP

Download the MyDashBoard App

MyMathLab®

1. **CONCEPT CHECK** Fill in each blank with the correct response.

 (a) The rational expression $\frac{x+5}{x-3}$ is undefined when $x =$ __**3**__, and is equal to 0 when $x =$ __**−5**__.

 (b) The rational expression $\frac{p-q}{q-p}$ is undefined when $p =$ __**q**__, and in all other cases is equal to __**−1**__.

2. **CONCEPT CHECK** Make the correct choice to complete each statement.

 (a) $\frac{4-r^2}{4+r^2}$ (*is* / *is not*) equal to −1. **is not**

 (b) $\frac{5+2x}{3-x}$ and $\frac{-5-2x}{x-3}$ (*are* / *are not*) equivalent rational expressions. **are**

3. Define *rational expression* in your own words, and give an example.

 A rational expression is a quotient of polynomials, such as $\frac{x+3}{x^2-4}$.

4. Why can't the denominator of a rational expression equal 0?

 Division by 0 is undefined, so if the denominator of a rational expression equals 0, the expression is undefined.

Find the numerical value of each rational expression for **(a)** $x = 2$ *and* **(b)** $x = -3$. ***See Example 1.***

5. $\frac{5x-2}{4x}$ **(a) 1 (b) $\frac{17}{12}$**

6. $\frac{3x+1}{5x}$ **(a) $\frac{7}{10}$ (b) $\frac{8}{15}$**

7. $\frac{x^2-4}{2x+1}$ **(a) 0 (b) −1**

8. $\frac{2x^2-4x}{3x-1}$ **(a) 0 (b) −3**

9. $\frac{(-3x)^2}{4x+12}$ **(a) $\frac{9}{5}$ (b) undefined**

10. $\frac{(-2x)^3}{3x+9}$ **(a) $-\frac{64}{15}$ (b) undefined**

11. $\frac{5x+2}{2x^2+11x+12}$ **(a) $\frac{2}{7}$ (b) $\frac{13}{3}$**

12. $\frac{7-3x}{3x^2-7x+2}$ **(a) undefined (b) $\frac{8}{25}$**

Find any value(s) of the variable for which each rational expression is undefined. Write answers with the symbol ≠. ***See Example 2.***

13. $\frac{2}{5y}$ **$y \neq 0$**

14. $\frac{7}{3z}$ **$z \neq 0$**

15. $\frac{x+1}{x+6}$ **$x \neq -6$**

16. $\frac{m+2}{m+5}$ **$m \neq -5$**

17. $\frac{4x^2}{3x-5}$ **$x \neq \frac{5}{3}$**

18. $\frac{2x^3}{3x-4}$ **$x \neq \frac{4}{3}$**

19. $\frac{m+2}{m^2+m-6}$ **$m \neq -3, m \neq 2$**

20. $\frac{r-5}{r^2-5r+4}$ **$r \neq 1, r \neq 4$**

21. $\frac{x^2-3x}{4}$ **never undefined**

22. $\frac{x^2-4x}{6}$ **never undefined**

23. $\frac{3x}{x^2+2}$ **never undefined**

24. $\frac{4q}{q^2+9}$ **never undefined**

Write each rational expression in lowest terms. ***See Examples 3 and 4.***

25. $\frac{18r^3}{6r}$ **$3r^2$**

26. $\frac{27p^2}{3p}$ **$9p$**

27. $\frac{4(y-2)}{10(y-2)}$ **$\frac{2}{5}$**

28. $\frac{15(m-1)}{9(m-1)}$ **$\frac{5}{3}$**

29. $\frac{(x+1)(x-1)}{(x+1)^2}$ $\frac{x-1}{x+1}$

30. $\frac{(t+5)(t-3)}{(t-1)(t+5)}$ $\frac{t-3}{t-1}$

31. $\frac{7m+14}{5m+10}$ $\frac{7}{5}$

32. $\frac{8z-24}{4z-12}$ 2

33. $\frac{m^2-n^2}{m+n}$ $m-n$

34. $\frac{a^2-b^2}{a-b}$ $a+b$

35. $\frac{12m^2-3}{8m-4}$ $\frac{3(2m+1)}{4}$

36. $\frac{20p^2-45}{6p-9}$ $\frac{5(2p+3)}{3}$

37. $\frac{3m^2-3m}{5m-5}$ $\frac{3m}{5}$

38. $\frac{6t^2-6t}{2t-2}$ $3t$

39. $\frac{9r^2-4s^2}{9r+6s}$ $\frac{3r-2s}{3}$

40. $\frac{16x^2-9y^2}{12x-9y}$ $\frac{4x+3y}{3}$

41. $\frac{2x^2-3x-5}{2x^2-7x+5}$ $\frac{x+1}{x-1}$

42. $\frac{3x^2+8x+4}{3x^2-4x-4}$ $\frac{x+2}{x-2}$

43. $\frac{zw+4z-3w-12}{zw+4z+5w+20}$ $\frac{z-3}{z+5}$

44. $\frac{km+4k+4m+16}{km+4k+5m+20}$ $\frac{k+4}{k+5}$

45. $\frac{ac-ad+bc-bd}{ac-ad-bc+bd}$ $\frac{a+b}{a-b}$

46. $\frac{rt-ru-st+su}{rt-ru+st-su}$ $\frac{r-s}{r+s}$

Write each rational expression in lowest terms. **See Examples 5 and 6.**

47. $\frac{6-t}{t-6}$ -1

48. $\frac{2-k}{k-2}$ -1

49. $\frac{m^2-1}{1-m}$ $-(m+1)$

50. $\frac{a^2-b^2}{b-a}$ $-(a+b)$

51. $\frac{q^2-4q}{4q-q^2}$ -1

52. $\frac{z^2-5z}{5z-z^2}$ -1

53. $\frac{p+6}{p-6}$ It is already in lowest terms.

54. $\frac{5-x}{5+x}$ It is already in lowest terms.

Write four equivalent forms for each rational expression. **See Example 7.**
Answers may vary in Exercises 55–60.

55. $-\frac{x+4}{x-3}$

$\frac{-(x+4)}{x-3}, \frac{-x-4}{x-3}, \frac{x+4}{-(x-3)}, \frac{x+4}{-x+3}$

56. $-\frac{x+6}{x-1}$

$\frac{-(x+6)}{x-1}, \frac{-x-6}{x-1}, \frac{x+6}{-(x-1)}, \frac{x+6}{-x+1}$

57. $-\frac{2x-3}{x+3}$

$\frac{-(2x-3)}{x+3}, \frac{-2x+3}{x+3}, \frac{2x-3}{-(x+3)}, \frac{2x-3}{-x-3}$

58. $-\frac{5x-6}{x+4}$

$\frac{-(5x-6)}{x+4}, \frac{-5x+6}{x+4}, \frac{5x-6}{-(x+4)}, \frac{5x-6}{-x-4}$

59. $-\frac{3x-1}{5x-6}$

$\frac{-(3x-1)}{5x-6}, \frac{-3x+1}{5x-6}, \frac{3x-1}{-(5x-6)}, \frac{3x-1}{-5x+6}$

60. $-\frac{2x-9}{7x-1}$

$\frac{-(2x-9)}{7x-1}, \frac{-2x+9}{7x-1}, \frac{2x-9}{-(7x-1)}, \frac{2x-9}{-7x+1}$

61. The area of the rectangle is represented by x^4+10x^2+21. What is the width? (*Hint:* Use $W = \frac{A}{L}$.)

x^2+3

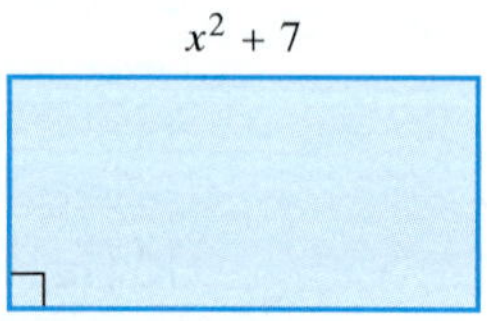

62. The volume of the box is represented by

$$(x^2+8x+15)(x+4).$$

Find the polynomial that represents the area of the bottom of the box.

$x^2+7x+12$

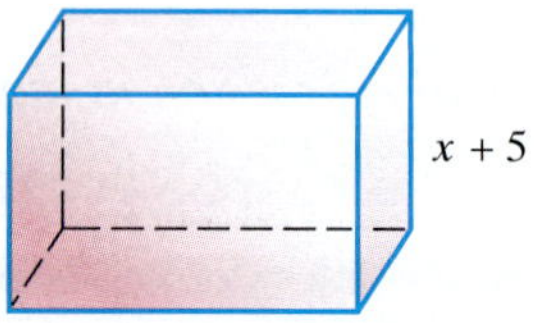

7.2 Multiplying and Dividing Rational Expressions

OBJECTIVES

1. Multiply rational expressions.
2. Find reciprocals.
3. Divide rational expressions.

OBJECTIVE 1 Multiply rational expressions. The product of two fractions is found by multiplying the numerators and multiplying the denominators. Rational expressions are multiplied in the same way.

Multiplying Rational Expressions

The product of the rational expressions $\frac{P}{Q}$ and $\frac{R}{S}$ is defined as follows.

$$\frac{P}{Q} \cdot \frac{R}{S} = \frac{PR}{QS}$$

In words: To multiply rational expressions, multiply the numerators and multiply the denominators.

EXAMPLE 1 Multiplying Rational Expressions

Multiply. Write each answer in lowest terms.

(a) $\frac{3}{10} \cdot \frac{5}{9}$ **(b)** $\frac{6}{x} \cdot \frac{x^2}{12}$

Indicate the product of the numerators and the product of the denominators.

(a) $= \frac{3 \cdot 5}{10 \cdot 9}$ (b) $= \frac{6 \cdot x^2}{x \cdot 12}$

Leave the products in factored form. Factor the numerator and denominator to further identify any common factors. Then use the fundamental property to write each product in lowest terms.

(a) $= \frac{3 \cdot 5}{2 \cdot 5 \cdot 3 \cdot 3}$

$= \frac{1}{6}$ Remember to write 1 in the numerator.

(b) $= \frac{6 \cdot x \cdot x}{2 \cdot 6 \cdot x}$

$= \frac{x}{2}$

Work Problem 1 at the Side. ▶

EXAMPLE 2 Multiplying Rational Expressions

Multiply. Write the answer in lowest terms.

$$\frac{x+y}{2x} \cdot \frac{x^2}{(x+y)^2}$$

$= \frac{(x+y)x^2}{2x(x+y)^2}$ Multiply numerators. Multiply denominators. (Use parentheses here around $x + y$.)

$= \frac{(x+y)x \cdot x}{2x(x+y)(x+y)}$ Factor. Identify the common factors.

$= \frac{x}{2(x+y)}$ $\frac{(x+y)x}{x(x+y)} = 1$; Write in lowest terms.

Work Problem 2 at the Side. ▶

---

1 Multiply. Write each answer in lowest terms.

(a) $\frac{2}{7} \cdot \frac{5}{10}$

(b) $\frac{3m^2}{2} \cdot \frac{10}{m}$

(c) $\frac{8p^2q}{3} \cdot \frac{9}{q^2p}$

2 Multiply. Write each answer in lowest terms.

(a) $\frac{a+b}{5} \cdot \frac{30}{2(a+b)}$

(b) $\frac{3(p-q)}{q^2} \cdot \frac{q}{2(p-q)^2}$

Answers

1. (a) $\frac{1}{7}$ **(b)** $15m$ **(c)** $\frac{24p}{q}$

2. (a) 3 **(b)** $\frac{3}{2q(p-q)}$

3 Multiply. Write each answer in lowest terms.

(a) $\dfrac{x^2+7x+10}{3x+6} \cdot \dfrac{6x-6}{x^2+2x-15}$

Teaching Tip

Repeat the statement "Factor first, divide out second."

(b) $\dfrac{m^2+4m-5}{m+5} \cdot \dfrac{m^2+8m+15}{m-1}$

4 Find the reciprocal of each rational expression.

(a) $\dfrac{5}{8}$

(b) $\dfrac{6b^5}{3r^2b}$

(c) $\dfrac{t^2-4t}{t^2+2t-3}$

Answers

3. (a) $\dfrac{2(x-1)}{x-3}$ (b) $(m+5)(m+3)$

4. (a) $\dfrac{8}{5}$ (b) $\dfrac{3r^2b}{6b^5}$ (c) $\dfrac{t^2+2t-3}{t^2-4t}$

EXAMPLE 3 Multiplying Rational Expressions

Multiply. Write the answer in lowest terms.

$$\frac{x^2+3x}{x^2-3x-4} \cdot \frac{x^2-5x+4}{x^2+2x-3}$$

$$= \frac{(x^2+3x)(x^2-5x+4)}{(x^2-3x-4)(x^2+2x-3)} \quad \text{Definition of multiplication}$$

$$= \frac{x(x+3)(x-4)(x-1)}{(x-4)(x+1)(x+3)(x-1)} \quad \text{Factor.}$$

$$= \frac{x}{x+1} \quad \text{Divide out the common factors.}$$

The quotients $\frac{x+3}{x+3}$, $\frac{x-4}{x-4}$, and $\frac{x-1}{x-1}$ are all equal to 1, justifying the final product $\frac{x}{x+1}$.

◀ Work Problem **3** at the Side.

OBJECTIVE 2 Find reciprocals. If the product of two rational expressions is 1, the rational expressions are **reciprocals** (or **multiplicative inverses**) of each other. The reciprocal of a rational expression is found by interchanging the numerator and the denominator.

$$\frac{2x-1}{x-5} \text{ has reciprocal } \frac{x-5}{2x-1}.$$

EXAMPLE 4 Finding Reciprocals of Rational Expressions

Find the reciprocal of each rational expression.

(a) $\dfrac{4p^3}{9q}$ has reciprocal $\dfrac{9q}{4p^3}$. Interchange the numerator and denominator.

(b) $\dfrac{k^2-9}{k^2-k-20}$ has reciprocal $\dfrac{k^2-k-20}{k^2-9}$.

◀ Work Problem **4** at the Side.

OBJECTIVE 3 Divide rational expressions. Suppose we have $\frac{7}{8}$ gal of milk and want to find how many quarts we have. Since 1 qt is $\frac{1}{4}$ gal, we ask, "How many $\frac{1}{4}$s are there in $\frac{7}{8}$?" This would be interpreted as follows.

$$\frac{7}{8} \div \frac{1}{4}, \quad \text{or} \quad \frac{\frac{7}{8}}{\frac{1}{4}} \leftarrow \text{The fraction bar means division.}$$

The fundamental property of rational expressions can be applied to rational number values of P, Q, and K.

$$\frac{P}{Q} = \frac{P \cdot K}{Q \cdot K} = \frac{\frac{7}{8} \cdot 4}{\frac{1}{4} \cdot 4} = \frac{\frac{7}{8} \cdot 4}{1} = \frac{7}{8} \cdot \frac{4}{1}$$

Let $P = \frac{7}{8}$, $Q = \frac{1}{4}$, and $K = 4$. (K is the reciprocal of Q.)

So, to divide $\frac{7}{8}$ by $\frac{1}{4}$, we multiply $\frac{7}{8}$ by the reciprocal of $\frac{1}{4}$, namely $\frac{4}{1}$, or 4. Since $\frac{7}{8}(4) = \frac{7}{2}$, there are $\frac{7}{2}$ qt, or $3\frac{1}{2}$ qt, in $\frac{7}{8}$ gal.

Dividing Rational Expressions

If $\frac{P}{Q}$ and $\frac{R}{S}$ are any two rational expressions, with $\frac{R}{S} \neq 0$, then their quotient is defined as follows.

$$\frac{P}{Q} \div \frac{R}{S} = \frac{P}{Q} \cdot \frac{S}{R} = \frac{PS}{QR}$$

In words: To divide one rational expression by another rational expression, multiply the first rational expression (dividend) by the reciprocal of the second rational expression (divisor).

EXAMPLE 5 Dividing Rational Expressions

Divide. Write each answer in lowest terms.

(a) $\frac{5}{8} \div \frac{7}{16}$

(b) $\frac{y}{y+3} \div \frac{4y}{y+5}$

Multiply the first expression by the reciprocal of the second.

(a)

$= \frac{5}{8} \cdot \frac{16}{7}$ ← Reciprocal of $\frac{7}{16}$

$= \frac{5 \cdot 16}{8 \cdot 7}$ Multiply.

$= \frac{5 \cdot 8 \cdot 2}{8 \cdot 7}$ Factor 16.

$= \frac{10}{7}$ Lowest terms

(b)

$= \frac{y}{y+3} \cdot \frac{y+5}{4y}$ ← Reciprocal of $\frac{4y}{y+5}$

$= \frac{y(y+5)}{(y+3)(4y)}$ Multiply.

$= \frac{y+5}{4(y+3)}$ Lowest terms

Work Problem 5 at the Side. ▶

EXAMPLE 6 Dividing Rational Expressions

Divide. Write the answer in lowest terms.

$$\frac{(3m)^2}{(2p)^3} \div \frac{6m^3}{16p^2}$$

$= \frac{(3m)^2}{(2p)^3} \cdot \frac{16p^2}{6m^3}$ Multiply by the reciprocal.

$= \frac{9m^2}{8p^3} \cdot \frac{16p^2}{6m^3}$ Power rule for exponents

($(3m)^2 = 3^2m^2$; $(2p)^3 = 2^3p^3$)

$= \frac{9 \cdot 16m^2p^2}{8 \cdot 6p^3m^3}$ Multiply numerators. Multiply denominators.

$= \frac{3 \cdot 3 \cdot 8 \cdot 2 \cdot m^2 \cdot p^2}{8 \cdot 3 \cdot 2 \cdot p^2 \cdot p \cdot m^2 \cdot m}$ Factor.

$= \frac{3}{pm}$, or $\frac{3}{mp}$ Lowest terms

Work Problem 6 at the Side. ▶

5 Divide. Write each answer in lowest terms.

(a) $\frac{3}{4} \div \frac{5}{16}$

(b) $\frac{r}{r-1} \div \frac{3r}{r+4}$

GS **(c)** $\frac{6x-4}{3} \div \frac{15x-10}{9}$

$= \frac{6x-4}{3} \cdot \frac{\_\_\_\_}{\_\_\_\_}$

$= \frac{2(\_\_\_\_)}{3} \cdot \frac{\_\_\_\_}{5(\_\_\_\_)}$

$= \frac{2(3x-2) \cdot 3 \cdot 3}{\_\_\_\_ \cdot 5(3x-2)}$

$= \_\_\_\_$

6 Divide. Write each answer in lowest terms.

(a) $\frac{5a^2b}{2} \div \frac{10ab^2}{8}$

(b) $\frac{(3t)^2}{w} \div \frac{3t^2}{5w^4}$

Answers

5. **(a)** $\frac{12}{5}$ **(b)** $\frac{r+4}{3(r-1)}$

(c) 9; $15x - 10$; $3x - 2$; 9; $3x - 2$; 3; $\frac{6}{5}$

6. **(a)** $\frac{2a}{b}$ **(b)** $15w^3$

7 Divide. Write each answer in lowest terms.

GS **(a)** $\dfrac{y^2+4y+3}{y+3} \div \dfrac{y^2-4y-5}{y-3}$

Rewrite as multiplication, and complete the problem.

$\_\_\_\_\_\_ \cdot \_\_\_\_\_\_$

(b) $\dfrac{4x(x+3)}{2x+1} \div \dfrac{-x^2(x+3)}{4x^2-1}$

8 Divide. Write each answer in lowest terms.

(a) $\dfrac{ab-a^2}{a^2-1} \div \dfrac{a-b}{a-1}$

Teaching Tip

Remind students to write -1 when dividing out factors that are opposites of each other. Emphasize that it may be written in the numerator or in the denominator, but *not in both*.

(b) $\dfrac{x^2-9}{2x+6} \div \dfrac{9-x^2}{4x-12}$

Answers

7. (a) $\dfrac{y^2+4y+3}{y+3} \cdot \dfrac{y-3}{y^2-4y-5}$; $\dfrac{y-3}{y-5}$

(b) $-\dfrac{4(2x-1)}{x}$

8. (a) $\dfrac{-a}{a+1}$ **(b)** $\dfrac{-2x+6}{3+x}$

EXAMPLE 7 Dividing Rational Expressions

Divide. Write the answer in lowest terms.

$$\frac{x^2-4}{(x+3)(x-2)} \div \frac{(x+2)(x+3)}{-2x}$$

$= \dfrac{x^2-4}{(x+3)(x-2)} \cdot \dfrac{-2x}{(x+2)(x+3)}$ Multiply by the reciprocal.

$= \dfrac{-2x(x^2-4)}{(x+3)(x-2)(x+2)(x+3)}$ Multiply numerators. Multiply denominators.

$= \dfrac{-2x(x+2)(x-2)}{(x+3)(x-2)(x+2)(x+3)}$ Factor the numerator.

$= \dfrac{-2x}{(x+3)^2}$, or $-\dfrac{2x}{(x+3)^2}$ Lowest terms; $\frac{-a}{b} = -\frac{a}{b}$

◀ **Work Problem 7 at the Side.**

EXAMPLE 8 Dividing Rational Expressions (Factors Are Opposites)

Divide. Write the answer in lowest terms.

$$\frac{m^2-4}{m^2-1} \div \frac{2m^2+4m}{1-m}$$

$= \dfrac{m^2-4}{m^2-1} \cdot \dfrac{1-m}{2m^2+4m}$ Multiply by the reciprocal.

$= \dfrac{(m^2-4)(1-m)}{(m^2-1)(2m^2+4m)}$ Multiply numerators. Multiply denominators.

$= \dfrac{(m+2)(m-2)(1-m)}{(m+1)(m-1)(2m)(m+2)}$ Factor. $1-m$ and $m-1$ are opposites.

$= \dfrac{-1(m-2)}{2m(m+1)}$ From **Section 7.1**, $\frac{1-m}{m-1} = -1$.

$= \dfrac{-m+2}{2m(m+1)}$, or $\dfrac{2-m}{2m(m+1)}$ Distribute -1 in the numerator. Rewrite $-m+2$ as $2-m$.

◀ **Work Problem 8 at the Side.**

In summary, follow these steps to multiply or divide rational expressions.

Multiplying or Dividing Rational Expressions

Step 1 **Note the operation.** If the operation is division, use the definition of division to rewrite as multiplication.

Step 2 **Multiply** numerators and multiply denominators.

Step 3 **Factor** all numerators and denominators completely.

Step 4 **Write in lowest terms** using the fundamental property.

Steps 2 and 3 may be interchanged based on personal preference.

7.2 Exercises

FOR EXTRA HELP

Download the MyDashBoard App

MyMathLab®

1. **CONCEPT CHECK** Match each multiplication problem in Column I with the correct product in Column II.

I		II
(a) $\frac{5x^3}{10x^4} \cdot \frac{10x^7}{2x}$	B	A. $\frac{2}{5x^5}$
(b) $\frac{10x^4}{5x^3} \cdot \frac{10x^7}{2x}$	D	B. $\frac{5x^5}{2}$
(c) $\frac{5x^3}{10x^4} \cdot \frac{2x}{10x^7}$	C	C. $\frac{1}{10x^7}$
(d) $\frac{10x^4}{5x^3} \cdot \frac{2x}{10x^7}$	A	D. $10x^7$

2. **CONCEPT CHECK** Match each division problem in Column I with the correct quotient in Column II.

I		II
(a) $\frac{5x^3}{10x^4} \div \frac{10x^7}{2x}$	D	A. $\frac{5x^5}{2}$
(b) $\frac{10x^4}{5x^3} \div \frac{10x^7}{2x}$	C	B. $10x^7$
(c) $\frac{5x^3}{10x^4} \div \frac{2x}{10x^7}$	A	C. $\frac{2}{5x^5}$
(d) $\frac{10x^4}{5x^3} \div \frac{2x}{10x^7}$	B	D. $\frac{1}{10x^7}$

Multiply. Write each answer in lowest terms. ***See Examples 1 and 2.***

3. $\frac{10m^2}{7} \cdot \frac{14}{15m}$ $\frac{4m}{3}$

4. $\frac{36z^3}{6} \cdot \frac{28}{z^2}$ $168z$

5. $\frac{16y^4}{18y^5} \cdot \frac{15y^5}{y^2}$ $\frac{40y^2}{3}$

6. $\frac{20x^5}{-2x^2} \cdot \frac{8x^4}{35x^3}$ $-\frac{16x^4}{7}$

7. $\frac{2(c+d)}{3} \cdot \frac{18}{6(c+d)^2}$ $\frac{2}{c+d}$

8. $\frac{4(y-2)}{x} \cdot \frac{3x}{6(y-2)^2}$ $\frac{2}{y-2}$

9. $\frac{(x-y)^2}{2} \cdot \frac{24}{3(x-y)}$ $4(x-y)$

10. $\frac{(a+b)^2}{5} \cdot \frac{30}{2(a+b)}$ $3(a+b)$

Find the reciprocal of each rational expression. ***See Example 4.***

11. $\frac{3p^3}{16q}$ $\frac{16q}{3p^3}$

12. $\frac{6x^4}{9y^2}$ $\frac{9y^2}{6x^4}$

13. $\frac{r^2+rp}{7}$ $\frac{7}{r^2+rp}$

14. $\frac{16}{9a^2+36a}$ $\frac{9a^2+36a}{16}$

15. $\frac{z^2+7z+12}{z^2-9}$ $\frac{z^2-9}{z^2+7z+12}$

16. $\frac{p^2-4p+3}{p^2-3p}$ $\frac{p^2-3p}{p^2-4p+3}$

Divide. Write each answer in lowest terms. ***See Examples 5 and 6.***

17. $\frac{9z^4}{3z^5} \div \frac{3z^2}{5z^3}$ 5

18. $\frac{35q^8}{9q^5} \div \frac{25q^6}{10q^5}$ $\frac{14q^2}{9}$

19. $\frac{4t^4}{2t^5} \div \frac{(2t)^3}{-6}$ $-\frac{3}{2t^4}$

20. $\frac{-12a^6}{3a^2} \div \frac{(2a)^3}{27a}$ $-\frac{27a^2}{2}$

21. $\frac{3}{2y-6} \div \frac{6}{y-3}$ $\frac{1}{4}$

22. $\frac{4m+16}{10} \div \frac{3m+12}{18}$ $\frac{12}{5}$

23. $\frac{(x-3)^2}{6x} \div \frac{x-3}{x^2}$ $\frac{x(x-3)}{6}$

24. $\frac{2a}{a+4} \div \frac{a^2}{(a+4)^2}$ $\frac{2(a+4)}{a}$

Multiply or divide. Write each answer in lowest terms. ***See Examples 3, 7, and 8.***

25. GS $\frac{5x-15}{3x+9}\cdot\frac{4x+12}{6x-18}$

$= \frac{5(\underline{x-3})}{3(\underline{x+3})}\cdot\frac{4(\underline{x+3})}{6(\underline{x-3})}$

$= \underline{\frac{10}{9}}$

26. GS $\frac{8r+16}{24r-24}\cdot\frac{6r-6}{3r+6}$

$= \frac{8(\underline{r+2})}{24(\underline{r-1})}\cdot\frac{6(\underline{r-1})}{3(\underline{r+2})}$

$= \underline{\frac{2}{3}}$

27. $\frac{2-t}{8}\div\frac{t-2}{6}$ $\quad -\frac{3}{4}$

28. $\frac{4}{m-2}\div\frac{16}{2-m}$ $\quad -\frac{1}{4}$

29. $\frac{5-4x}{5+4x}\cdot\frac{4x+5}{4x-5}$ $\quad -1$

30. $\frac{5-x}{5+x}\cdot\frac{x+5}{x-5}$ $\quad -1$

31. $\frac{6(m-2)^2}{5(m+4)^2}\cdot\frac{15(m+4)}{2(2-m)}$

$\frac{9(m-2)}{-(m+4)}$, or $\frac{-9(m-2)}{m+4}$

32. $\frac{7(q-1)}{3(q+1)^2}\cdot\frac{6(q+1)}{3(1-q)^2}$

$\frac{-14}{3(q+1)(1-q)}$, or $\frac{14}{3(q+1)(q-1)}$

33. $\frac{p^2+4p-5}{p^2+7p+10}\div\frac{p-1}{p+4}$

$\frac{p+4}{p+2}$

34. $\frac{z^2-3z+2}{z^2+4z+3}\div\frac{z-1}{z+1}$

$\frac{z-2}{z+3}$

35. $\frac{2k^2-k-1}{2k^2+5k+3}\div\frac{4k^2-1}{2k^2+k-3}$

$\frac{(k-1)^2}{(k+1)(2k-1)}$

36. $\frac{2m^2-5m-12}{m^2+m-20}\div\frac{4m^2-9}{m^2+4m-5}$

$\frac{m-1}{2m-3}$

37. $\frac{2k^2+3k-2}{6k^2-7k+2}\cdot\frac{4k^2-5k+1}{k^2+k-2}$

$\frac{4k-1}{3k-2}$

38. $\frac{2m^2-5m-12}{m^2-10m+24}\div\frac{4m^2-9}{m^2-9m+18}$

$\frac{m-3}{2m-3}$

39. $\frac{m^2+2mp-3p^2}{m^2-3mp+2p^2}\div\frac{m^2+4mp+3p^2}{m^2+2mp-8p^2}$

$\frac{m+4p}{m+p}$

40. $\frac{r^2+rs-12s^2}{r^2-rs-20s^2}\div\frac{r^2-2rs-3s^2}{r^2+rs-30s^2}$

$\frac{r+6s}{r+s}$

41. $\left(\frac{x^2+10x+25}{x^2+10x}\cdot\frac{10x}{x^2+15x+50}\right)\div\frac{x+5}{x+10}$

$\frac{10}{x+10}$

42. $\left(\frac{m^2-12m+32}{8m}\cdot\frac{m^2-8m}{m^2-8m+16}\right)\div\frac{m-8}{m-4}$

$\frac{m-8}{8}$

Solve each problem.

43. If the rational expression $\frac{5x^2y^3}{2pq}$ represents the area of a rectangle and $\frac{2xy}{p}$ represents the length, what rational expression represents the width?

$\frac{5xy^2}{4q}$

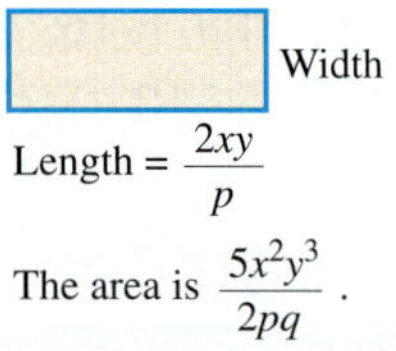

44. If the rational expression $\frac{12a^3b^4}{5cd}$ represents the area of a rectangle and $\frac{9ab^2}{c}$ represents the width, what rational expression represents the length?

$\frac{4a^2b^2}{15d}$

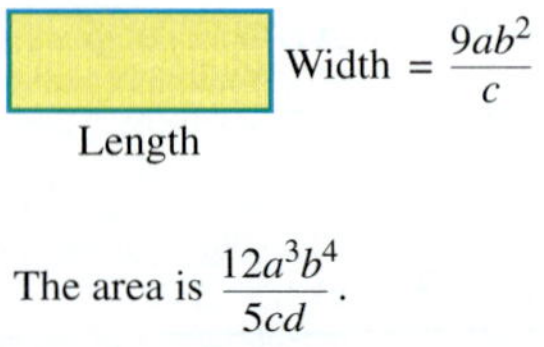

7.3 Least Common Denominators

OBJECTIVES

1. Find the least common denominator for a list of fractions.
2. Write equivalent rational expressions.

OBJECTIVE 1 Find the least common denominator for a list of fractions. Adding or subtracting rational expressions often requires a **least common denominator (LCD).** The LCD is the simplest expression that is divisible by all of the denominators in all of the expressions. For example, the fractions

$$\frac{2}{9} \text{ and } \frac{5}{12} \text{ have LCD 36,}$$

because 36 is the least positive number divisible by both 9 and 12.

We can often find least common denominators by inspection. In other cases, we find the LCD by a procedure similar to that used in **Section 6.1.**

Finding the Least Common Denominator (LCD)

Step 1 **Factor** each denominator into prime factors.

Step 2 **List each different denominator factor** the *greatest* number of times it appears in any of the denominators.

Step 3 **Multiply** the denominator factors from Step 2 to get the LCD.

When each denominator is factored into prime factors, every prime factor must be a factor of the least common denominator.

EXAMPLE 1 Finding Least Common Denominators

Find the LCD for each pair of fractions.

(a) $\frac{1}{24}, \frac{7}{15}$ **(b)** $\frac{1}{8x}, \frac{3}{10x}$

Step 1 Write each denominator in factored form with numerical coefficients in prime factored form.

(a)	(b)
$24 = 2 \cdot 2 \cdot 2 \cdot 3 = 2^3 \cdot 3$	$8x = 2 \cdot 2 \cdot 2 \cdot x = 2^3 \cdot x$
$15 = 3 \cdot 5$	$10x = 2 \cdot 5 \cdot x$

Step 2 We find the LCD by taking each different factor the *greatest* number of times it appears as a factor in any of the denominators.

(a) The factor 2 appears three times in one product and not at all in the other, so the greatest number of times 2 appears is three. The greatest number of times both 3 and 5 appear is one.

(b) Here, 2 appears three times in one product and once in the other, so the greatest number of times 2 appears is three. The greatest number of times 5 appears is one. The greatest number of times x appears is one.

Step 3

(a)
$$\text{LCD} = 2 \cdot 2 \cdot 2 \cdot 3 \cdot 5 = 2^3 \cdot 3 \cdot 5 = 120$$

(b)
$$\text{LCD} = 2 \cdot 2 \cdot 2 \cdot 5 \cdot x = 2^3 \cdot 5 \cdot x = 40x$$

Work Problem 1 at the Side. ▶

1 Find the LCD for each pair of fractions.

GS **(a)** $\frac{7}{10}, \frac{1}{25}$

Step 1 $10 = 2 \cdot$ ____

$25 =$ ____ $\cdot$ ____

Complete Step 2 and Step 3.

The LCD is ____.

(b) $\frac{7}{20p}, \frac{11}{30p}$

(c) $\frac{4}{5x}, \frac{12}{10x}$

Answers

1. **(a)** 5; 5; 5; 50 **(b)** $60p$ **(c)** $10x$

2 Find the LCD.

(a) $\frac{4}{16m^3n}, \frac{5}{9m^5}$

(b) $\frac{3}{25a^2}, \frac{2}{10a^3b}$

3 Find the LCD for the fractions in each list.

(a) $\frac{7}{3a}, \frac{11}{a^2 - 4a}$

Teaching Tip

In Example 3(a), students have a tendency to list the least common denominator for *m* and *m* − 3 as only *m* − 3.

(b) $\frac{2m}{m^2 - 3m + 2}, \frac{5m - 3}{m^2 + 3m - 10}, \frac{4m + 7}{m^2 + 4m - 5}$

(c) $\frac{6}{x - 4}, \frac{3x - 1}{4 - x}$

Answers

2. **(a)** $144m^5n$ **(b)** $50a^3b$

3. **(a)** $3a(a - 4)$
(b) $(m - 1)(m - 2)(m + 5)$
(c) either $x - 4$ or $4 - x$

EXAMPLE 2 Finding the LCD

Find the LCD for $\frac{5}{6r^2}$ and $\frac{3}{4r^3}$.

Step 1 Factor each denominator.

$$6r^2 = 2 \cdot 3 \cdot r^2$$
$$4r^3 = 2^2 \cdot r^3$$

Step 2 The greatest number of times 2 appears is two, the greatest number of times 3 appears is one, and the greatest number of times *r* appears is three.

Step 3 $\text{LCD} = 2^2 \cdot 3 \cdot r^3 = 12r^3$

◀ **Work Problem 2 at the Side.**

CAUTION

When finding the LCD, use each factor the *greatest* number of times it appears in any *single* denominator, not the *total* number of times it appears. For instance, the greatest number of times *r* appears as a factor in one denominator in **Example 2** is 3, *not* 5.

EXAMPLE 3 Finding LCDs

Find the LCD for the fractions in each list.

(a) $\frac{6}{5m}, \frac{4}{m^2 - 3m}$

$$5m = 5 \cdot m$$
$$m^2 - 3m = m(m - 3)$$
Factor each denominator.

Use each different factor the greatest number of times it appears.

$$\text{LCD} = 5 \cdot m \cdot (m - 3) = 5m(m - 3)$$

Be sure to include *m* as a factor in the LCD.

Because *m* is not a *factor* of *m* − 3, both factors, *m* and *m* − 3, must appear in the LCD.

(b) $\frac{1}{r^2 - 4r - 5}, \frac{3}{r^2 - r - 20}, \frac{1}{r^2 - 10r + 25}$

$$r^2 - 4r - 5 = (r - 5)(r + 1)$$
$$r^2 - r - 20 = (r - 5)(r + 4)$$
$$r^2 - 10r + 25 = (r - 5)^2$$
Factor each denominator.

Use each different factor the greatest number of times it appears as a factor.

$$\text{LCD} = (r + 1)(r + 4)(r - 5)^2$$

Be sure to include the exponent 2 on the factor *r* − 5.

(c) $\frac{1}{q - 5}, \frac{3}{5 - q}$

The expressions $q - 5$ and $5 - q$ are opposites of each other because

$$-(q - 5) = -q + 5 = 5 - q.$$

Therefore, either $q - 5$ or $5 - q$ can be used as the LCD.

◀ **Work Problem 3 at the Side.**

OBJECTIVE 2 Write equivalent rational expressions. Once the LCD has been found, the next step in preparing to add or subtract two rational expressions is to use the fundamental property to write equivalent rational expressions.

Writing a Rational Expression with a Specified Denominator

Step 1 **Factor** both denominators.

Step 2 **Decide what factor(s) the denominator must be multiplied by** in order to equal the specified denominator.

Step 3 **Multiply** the rational expression by that factor divided by itself. (That is, multiply by 1.)

EXAMPLE 4 Writing Equivalent Rational Expressions

Write each rational expression as an equivalent expression with the indicated denominator.

(a) $\frac{3}{8} = \frac{?}{40}$ **(b)** $\frac{9k}{25} = \frac{?}{50k}$

Step 1 For each example, first factor the denominator on the right. Then compare the denominator on the left with the one on the right to decide what factors are missing.

$$\frac{3}{8} = \frac{?}{5 \cdot 8} \qquad \frac{9k}{25} = \frac{?}{25 \cdot 2k}$$

Step 2 A factor of 5 is missing. | Factors of 2 and k are missing.

Step 3 Multiply $\frac{3}{8}$ by $\frac{5}{5}$. | Multiply $\frac{9k}{25}$ by $\frac{2k}{2k}$.

$$\frac{3}{8} = \frac{3}{8} \cdot \frac{5}{5} = \frac{15}{40} \qquad \frac{9k}{25} = \frac{9k}{25} \cdot \frac{2k}{2k} = \frac{18k^2}{50k}$$

$\frac{5}{5} = 1$ | $\frac{2k}{2k} = 1$

Work Problem 4 at the Side. ▶

EXAMPLE 5 Writing Equivalent Rational Expressions

Write each rational expression as an equivalent expression with the indicated denominator.

(a) $\frac{8}{3x+1} = \frac{?}{12x+4}$

$\frac{8}{3x+1} = \frac{?}{4(3x+1)}$ Factor the denominator on the right.

The missing factor is 4, so multiply the fraction on the left by $\frac{4}{4}$.

$\frac{8}{3x+1} \cdot \frac{4}{4} = \frac{32}{12x+4}$ Fundamental property

Continued on Next Page

4 Write each rational expression as an equivalent expression with the indicated denominator.

(a) $\frac{3}{4} = \frac{?}{36}$

Teaching Tip

Multiplying by $\frac{5}{5}$ or $\frac{2k}{2k}$ in Example 4 changes only the form of the fraction, not its value. Stress this application of the identity property for multiplication.

GS **(b)** $\frac{7k}{5} = \frac{?}{30p}$

Step 1 Factor the denominator on the right.

$\frac{7k}{5} = \frac{?}{5 \cdot \_\_\_\_}$

Step 2 Factors of ____ and ____ are missing.

Step 3

$\frac{7k}{5} \cdot \frac{\_\_\_\_}{\_\_\_\_} = \frac{\_\_\_\_}{30p}$

Answers

4. (a) $\frac{27}{36}$

(b) $6p$; 6; p; $6p$; $6p$; $42kp$

5 Write each rational expression as an equivalent expression with the indicated denominator.

(a) $\dfrac{9}{2a+5} = \dfrac{?}{6a+15}$

(b) $\dfrac{5k+1}{k^2+2k} = \dfrac{?}{k^3+k^2-2k}$

(b) $\dfrac{12p}{p^2+8p} = \dfrac{?}{p^3+4p^2-32p}$

Factor the denominator in each rational expression.

$$\frac{12p}{p(p+8)} = \frac{?}{p(p+8)(p-4)}$$

p^3+4p^2-32p
$= p(p^2+4p-32)$
$= p(p+8)(p-4)$

The factor $p-4$ is missing, so multiply $\frac{12p}{p(p+8)}$ by $\frac{p-4}{p-4}$.

$$= \frac{12p}{p(p+8)} \cdot \frac{p-4}{p-4} \quad \text{Fundamental property}$$

$$= \frac{12p(p-4)}{p(p+8)(p-4)} \quad \text{Multiply numerators. Multiply denominators.}$$

$$= \frac{12p^2-48p}{p^3+4p^2-32p} \quad \text{Multiply the factors.}$$

Note

In the next section we add and subtract rational expressions, which sometimes requires the steps illustrated in **Examples 4 and 5.** While it may be beneficial to leave the equivalent expression in factored form, we multiplied out the factors in the numerator and the denominator in **Example 5(b),**

$$\frac{12p(p-4)}{p(p+8)(p-4)},$$

in order to give the answer,

$$\frac{12p^2-48p}{p^3+4p^2-32p},$$

in the same form as the original problem.

◀ **Work Problem 5 at the Side.**

Answers

5. (a) $\dfrac{27}{6a+15}$

(b) $\dfrac{(5k+1)(k-1)}{k^3+k^2-2k}$, or $\dfrac{5k^2-4k-1}{k^3+k^2-2k}$

7.3 Exercises

FOR EXTRA HELP

Download the MyDashBoard App

MyMathLab®

CONCEPT CHECK *Choose the correct response in Exercises 1–4.*

1. Suppose that the greatest common factor of a and b is 1. Then the least common denominator for $\frac{1}{a}$ and $\frac{1}{b}$ is

A. a **B.** b **C.** ab **D.** 1. C

2. If a is a factor of b, then the least common denominator for $\frac{1}{a}$ and $\frac{1}{b}$ is

A. a **B.** b **C.** ab **D.** 1. B

3. The least common denominator for $\frac{11}{20}$ and $\frac{1}{2}$ is

A. 40 **B.** 2 **C.** 20 **D.** none of these. C

4. To write the rational expression $\frac{1}{(x-4)^2(y-3)}$ with denominator $(x-4)^3(y-3)^2$, we must multiply both the numerator and the denominator by

A. $(x-4)(y-3)$ **B.** $(x-4)^2$

C. $x-4$ **D.** $(x-4)^2(y-3)$. A

Find the LCD for the fractions in each list. ***See Examples 1 and 2.***

5. $\frac{2}{15}, \frac{3}{10}, \frac{7}{30}$ 30

6. $\frac{5}{24}, \frac{7}{12}, \frac{9}{28}$ 168

7. $\frac{3}{x^4}, \frac{5}{x^7}$ x^7

8. $\frac{2}{y^5}, \frac{3}{y^6}$ y^6

9. $\frac{5}{36q}, \frac{17}{24q}$ $72q$

10. $\frac{4}{30p}, \frac{9}{50p}$ $150p$

11. $\frac{6}{21r^3}, \frac{8}{12r^5}$ $84r^5$

12. $\frac{9}{35t^2}, \frac{5}{49t^6}$ $245t^6$

13. $\frac{5}{12x^2y}, \frac{7}{24x^3y^2}, \frac{-11}{6xy^4}$ $24x^3y^4$

14. $\frac{3}{10a^4b}, \frac{7}{15ab}, \frac{-7}{30ab^7}$ $30a^4b^7$

15. $\frac{1}{x^9y^{14}z^{21}}, \frac{3}{x^8y^{14}z^{22}}, \frac{5}{x^{10}y^{13}z^{23}}$ $x^{10}y^{14}z^{23}$

16. $\frac{2}{t^{15}s^{12}u^9}, \frac{4}{t^{14}s^{12}u^8}, \frac{9}{t^{13}s^{11}u^{10}}$ $t^{15}s^{12}u^{10}$

Find the LCD for the fractions in each list. ***See Examples 1–3.***

17. $\frac{9}{28m^2}, \frac{3}{12m-20}$ $28m^2(3m-5)$

18. $\frac{15}{27a^3}, \frac{8}{9a-45}$ $27a^3(a-5)$

19. $\frac{7}{5b-10}, \frac{11}{6b-12}$ $30(b-2)$

20. $\frac{3}{7x^2+21x}, \frac{1}{5x^2+15x}$ $35x(x+3)$

21. $\frac{5}{c-d}, \frac{8}{d-c}$ $c-d$ or $d-c$

22. $\frac{4}{y-x}, \frac{7}{x-y}$ $y-x$ or $x-y$

23. $\dfrac{3}{k^2+5k}, \dfrac{2}{k^2+3k-10}$
$k(k+5)(k-2)$

24. $\dfrac{1}{z^2-4z}, \dfrac{4}{z^2-3z-4}$
$z(z-4)(z+1)$

25. $\dfrac{6}{a^2+6a}, \dfrac{-5}{a^2+3a-18}$
$a(a+6)(a-3)$

26. $\dfrac{8}{y^2-5y}, \dfrac{-5}{y^2-2y-15}$
$y(y-5)(y+3)$

27. $\dfrac{5}{p^2+8p+15}, \dfrac{3}{p^2-3p-18}, \dfrac{2}{p^2-p-30}$
$(p+3)(p+5)(p-6)$

28. $\dfrac{10}{y^2-10y+21}, \dfrac{2}{y^2-2y-3}, \dfrac{5}{y^2-6y-7}$
$(y-7)(y-3)(y+1)$

Write each rational expression as an equivalent expression with the indicated denominator. ***See Examples 4 and 5.***

29. $\dfrac{4}{11}=\dfrac{?}{55}$
$\dfrac{20}{55}$

30. $\dfrac{6}{7}=\dfrac{?}{42}$
$\dfrac{36}{42}$

31. $\dfrac{-5}{k}=\dfrac{?}{9k}$
$\dfrac{-45}{9k}$

32. $\dfrac{-3}{q}=\dfrac{?}{6q}$
$\dfrac{-18}{6q}$

33. $\dfrac{13}{40y}=\dfrac{?}{80y^3}$
$\dfrac{26y^2}{80y^3}$

34. $\dfrac{5}{27p}=\dfrac{?}{108p^4}$
$\dfrac{20p^3}{108p^4}$

35. $\dfrac{5t^2}{6r}=\dfrac{?}{42r^4}$
$\dfrac{35t^2r^3}{42r^4}$

36. $\dfrac{8y^2}{3x}=\dfrac{?}{30x^3}$
$\dfrac{80x^2y^2}{30x^3}$

37. $\dfrac{5}{2(m+3)}=\dfrac{?}{8(m+3)}$
$\dfrac{20}{8(m+3)}$

38. $\dfrac{7}{4(y-1)}=\dfrac{?}{16(y-1)}$
$\dfrac{28}{16(y-1)}$

39. $\dfrac{-4t}{3t-6}=\dfrac{?}{12-6t}$
$\dfrac{8t}{12-6t}$

40. $\dfrac{-7k}{5k-20}=\dfrac{?}{40-10k}$
$\dfrac{14k}{40-10k}$

41. $\dfrac{14}{z^2-3z}=\dfrac{?}{z(z-3)(z-2)}$
$\dfrac{14(z-2)}{z(z-3)(z-2)}$

42. $\dfrac{12}{x(x+4)}=\dfrac{?}{x(x+4)(x-9)}$
$\dfrac{12(x-9)}{x(x+4)(x-9)}$

43. $\dfrac{2(b-1)}{b^2+b}=\dfrac{?}{b^3+3b^2+2b}$
$\dfrac{2(b-1)(b+2)}{b^3+3b^2+2b}$

44. $\dfrac{3(c+2)}{c(c-1)}=\dfrac{?}{c^3-5c^2+4c}$
$\dfrac{3(c+2)(c-4)}{c^3-5c^2+4c}$

7.4 Adding and Subtracting Rational Expressions

OBJECTIVES

1. Add rational expressions having the same denominator.
2. Add rational expressions having different denominators.
3. Subtract rational expressions.

OBJECTIVE 1 Add rational expressions having the same denominator. We find the sum of two such rational expressions with the same procedure that we used in **Section R.1** for adding two fractions having the same denominator.

Adding Rational Expressions (Same Denominator)

The rational expressions $\frac{P}{Q}$ and $\frac{R}{Q}$ (where $Q \neq 0$) are added as follows.

$$\frac{P}{Q} + \frac{R}{Q} = \frac{P + R}{Q}$$

In words: To add rational expressions with the same denominator, add the numerators and keep the same denominator.

EXAMPLE 1 Adding Rational Expressions (Same Denominator)

Add. Write each answer in lowest terms.

(a) $\frac{4}{9} + \frac{2}{9}$

(b) $\frac{3x}{x+1} + \frac{3}{x+1}$

The denominators are the same, so the sum is found by adding the two numerators and keeping the same (common) denominator.

(a)

$= \frac{4+2}{9}$ Add.

$= \frac{6}{9}$

$= \frac{2 \cdot 3}{3 \cdot 3}$ Factor.

$= \frac{2}{3}$ Lowest terms

(b)

$= \frac{3x+3}{x+1}$ Add.

$= \frac{3(x+1)}{x+1}$ Factor.

$= 3$ Lowest terms

Work Problem 1 at the Side. ▶

1 Add. Write each answer in lowest terms.

(a) $\frac{7}{15} + \frac{3}{15}$

(b) $\frac{3}{y+4} + \frac{2}{y+4}$

(c) $\frac{x}{x+y} + \frac{1}{x+y}$

(d) $\frac{a}{a+b} + \frac{b}{a+b}$

(e) $\frac{x^2}{x+1} + \frac{x}{x+1}$

OBJECTIVE 2 Add rational expressions having different denominators. As in **Section R.1,** we use the following steps to add two rational expressions having different denominators.

Adding Rational Expressions (Different Denominators)

Step 1 **Find the least common denominator (LCD).**

Step 2 **Write each rational expression** as an equivalent rational expression with the LCD as the denominator.

Step 3 **Add** the numerators to get the numerator of the sum. The LCD is the denominator of the sum.

Step 4 **Write in lowest terms** using the fundamental property.

Answers

1. (a) $\frac{2}{3}$ **(b)** $\frac{5}{y+4}$ **(c)** $\frac{x+1}{x+y}$ **(d)** 1 **(e)** x

2 Add. Write each answer in lowest terms.

(a) $\frac{1}{10} + \frac{1}{15}$

Teaching Tip

Adding fractions with the same denominator is an example of adding like quantities. In the case of Example 2(a), five *sixtieths* plus twenty-eight *sixtieths* equals thirty-three *sixtieths*.

GS **(b)** $\frac{6}{5x} + \frac{9}{2x}$

$= \frac{6(\_\_\_)}{5x(\_\_\_)} + \frac{9(\_\_\_)}{2x(\_\_\_)}$

$= \frac{12 + 45}{\_\_\_}$

$= \_\_\_$

(c) $\frac{m}{3n} + \frac{2}{7n}$

Answers

2. (a) $\frac{1}{6}$ **(b)** 2; 2; 5; 5; $10x$; $\frac{57}{10x}$

(c) $\frac{7m + 6}{21n}$

EXAMPLE 2 Adding Rational Expressions (Different Denominators)

Add. Write each answer in lowest terms.

(a) $\frac{1}{12} + \frac{7}{15}$ **(b)** $\frac{2}{3y} + \frac{1}{4y}$

Step 1 First find the LCD, using the methods of the previous section.

(a)	(b)
$12 = 2 \cdot 2 \cdot 3 = 2^2 \cdot 3$	$3y = 3 \cdot y$
$15 = 3 \cdot 5$	$4y = 2 \cdot 2 \cdot y = 2^2 \cdot y$
$\text{LCD} = 2^2 \cdot 3 \cdot 5 = 60$	$\text{LCD} = 2^2 \cdot 3 \cdot y = 12y$

Step 2 Now write each rational expression as an equivalent expression with the LCD (either 60 or $12y$) as the denominator.

(a)	(b)
$\frac{1}{12} + \frac{7}{15}$ The LCD is 60.	$\frac{2}{3y} + \frac{1}{4y}$ The LCD is $12y$.
$= \frac{1(5)}{12(5)} + \frac{7(4)}{15(4)}$	$= \frac{2(4)}{3y(4)} + \frac{1(3)}{4y(3)}$
$= \frac{5}{60} + \frac{28}{60}$	$= \frac{8}{12y} + \frac{3}{12y}$

Step 3 Add the numerators. The LCD is the denominator.

Step 4 Write in lowest terms if necessary.

(a)	(b)
$= \frac{5 + 28}{60}$	$= \frac{8 + 3}{12y}$
$= \frac{33}{60}$, or $\frac{11}{20}$	$= \frac{11}{12y}$

◀ Work Problem **2** at the Side.

EXAMPLE 3 Adding Rational Expressions

Add. Write the answer in lowest terms.

$$\frac{2x}{x^2 - 1} + \frac{-1}{x + 1}$$

Step 1 Since the denominators are different, find the LCD.

$x^2 - 1 = (x + 1)(x - 1)$
$x + 1$ is prime.
} The LCD is $(x + 1)(x - 1)$.

Step 2 Write each rational expression as an equivalent expression with the LCD as the denominator.

$\frac{2x}{x^2 - 1} + \frac{-1}{x + 1}$ The LCD is $(x + 1)(x - 1)$.

$= \frac{2x}{(x + 1)(x - 1)} + \frac{-1(x - 1)}{(x + 1)(x - 1)}$ Multiply the second fraction by $\frac{x - 1}{x - 1}$.

$= \frac{2x}{(x + 1)(x - 1)} + \frac{-x + 1}{(x + 1)(x - 1)}$ Distributive property

Continued on Next Page

Step 3 $= \dfrac{2x - x + 1}{(x+1)(x-1)}$ Add numerators. Keep the same denominator.

$= \dfrac{x+1}{(x+1)(x-1)}$ Combine like terms in the numerator.

Step 4 $= \dfrac{\mathbf{1}(\mathbf{x+1})}{(\mathbf{x+1})(x-1)}$ Identity property for multiplication

Remember to write 1 in the numerator.

$= \dfrac{1}{x-1}$ Divide out the common factors.

Work Problem **3** at the Side. ▶

EXAMPLE 4 **Adding Rational Expressions**

Add. Write the answer in lowest terms.

$$\frac{2x}{x^2+5x+6} + \frac{x+1}{x^2+2x-3}$$

$= \dfrac{2x}{(x+2)(x+3)} + \dfrac{x+1}{(x+3)(x-1)}$ Factor the denominators.

$= \dfrac{2x(\mathbf{x-1})}{(x+2)(x+3)(\mathbf{x-1})} + \dfrac{(x+1)(\mathbf{x+2})}{(\mathbf{x+2})(x+3)(x-1)}$ The LCD is $(x+2)(x+3)(x-1)$.

$= \dfrac{2x(x-1)+(x+1)(x+2)}{(x+2)(x+3)(x-1)}$ Add numerators. Keep the same denominator.

$= \dfrac{2x^2-2x+x^2+3x+2}{(x+2)(x+3)(x-1)}$ Multiply.

$= \dfrac{3x^2+x+2}{(x+2)(x+3)(x-1)}$ Combine like terms.

The numerator cannot be factored here, so the expression is in lowest terms.

Work Problem **4** at the Side. ▶

EXAMPLE 5 **Adding Rational Expressions (Denominators Are Opposites)**

Add. Write the answer in lowest terms.

$\dfrac{y}{\mathbf{y-2}} + \dfrac{8}{\mathbf{2-y}}$ The denominators are opposites.

$= \dfrac{y}{y-2} + \dfrac{8(\mathbf{-1})}{(2-y)(\mathbf{-1})}$ Multiply $\frac{8}{2-y}$ by $\frac{-1}{-1}$ to get a common denominator.

$= \dfrac{y}{y-2} + \dfrac{-8}{\mathbf{-2+y}}$ Multiply. Apply the distributive property in the denominator.

$= \dfrac{y}{y-2} + \dfrac{-8}{\mathbf{y-2}}$ Rewrite $-2+y$ as $y-2$.

$= \dfrac{y-8}{y-2}$ Add numerators. Keep the same denominator.

Work Problem **5** at the Side. ▶

3 Add. Write each answer in lowest terms.

(a) $\dfrac{2p}{3p+3} + \dfrac{5p}{2p+2}$

(b) $\dfrac{4}{y^2-1} + \dfrac{6}{y+1}$

(c) $\dfrac{-2}{p+1} + \dfrac{4p}{p^2-1}$

4 Add. Write each answer in lowest terms.

(a) $\dfrac{2k}{k^2-5k+4} + \dfrac{3}{k^2-1}$

(b) $\dfrac{4m}{m^2+3m+2} + \dfrac{2m-1}{m^2+6m+5}$

5 Add. Write the answer in lowest terms.

$$\frac{m}{2m-3n} + \frac{n}{3n-2m}$$

Answers

3. **(a)** $\dfrac{19p}{6(p+1)}$ **(b)** $\dfrac{2(3y-1)}{(y+1)(y-1)}$

(c) $\dfrac{2}{p-1}$

4. **(a)** $\dfrac{(2k-3)(k+4)}{(k-4)(k-1)(k+1)}$

(b) $\dfrac{6m^2+23m-2}{(m+2)(m+1)(m+5)}$

5. $\dfrac{m-n}{2m-3n}$, or $\dfrac{n-m}{3n-2m}$

6 Subtract. Write each answer in lowest terms.

(a) $\frac{3}{m^2} - \frac{2}{m^2}$

GS **(b)** $\frac{x}{2x+3} - \frac{3x+4}{2x+3}$

$= \frac{x - (\_\_\_\_)}{2x+3}$

$= \frac{x - \_\_\_\_ - \_\_\_\_}{2x+3}$

$= \frac{\_\_\_\_}{2x+3}$

If the numerator is factored, the answer can be written as ______.

(c) $\frac{5t}{t-1} - \frac{5+t}{t-1}$

Answers

6. (a) $\frac{1}{m^2}$

(b) $3x + 4$; $3x$; 4; $-2x - 4$; $\frac{-2(x+2)}{2x+3}$

(c) $\frac{4t-5}{t-1}$

Note

If we had chosen to use $2 - y$ as the common denominator, the final answer would be in the form $\frac{8-y}{2-y}$, which is equivalent to $\frac{y-8}{y-2}$.

$\frac{y}{y-2} + \frac{8}{2-y}$ — See **Example 5.**

$= \frac{y(-1)}{(y-2)(-1)} + \frac{8}{2-y}$ — Multiply $\frac{y}{y-2}$ by $\frac{-1}{-1}$.

$= \frac{-y}{2-y} + \frac{8}{2-y}$ — In the first denominator, $(y-2)(-1) = -y + 2 = 2 - y$.

$= \frac{-y+8}{2-y}$, or $\frac{8-y}{2-y}$ — Add numerators. Keep the same denominator.

OBJECTIVE 3 Subtract rational expressions. To subtract rational expressions having the same denominator, use the following rule.

Subtracting Rational Expressions (Same Denominator)

The rational expressions $\frac{P}{Q}$ and $\frac{R}{Q}$ (where $Q \neq 0$) are subtracted as follows.

$$\frac{P}{Q} - \frac{R}{Q} = \frac{P - R}{Q}$$

In words: To subtract rational expressions with the same denominator, subtract the numerators and keep the same denominator.

EXAMPLE 6 Subtracting Rational Expressions (Same Denominator)

Subtract. Write the answer in lowest terms.

$\frac{2m}{m-1} - \frac{m+3}{m-1}$

Use parentheses around the numerator of the subtrahend.

$= \frac{2m - (m+3)}{m-1}$ — Subtract numerators. Keep the same denominator.

Be careful with signs.

$= \frac{2m - m - 3}{m-1}$ — Distributive property

$= \frac{m-3}{m-1}$ — Combine like terms.

CAUTION

Sign errors often occur in subtraction problems like the one in **Example 6.** The numerator of the fraction being subtracted must be treated as a single quantity. ***Be sure to use parentheses after the subtraction symbol.***

◀ **Work Problem 6 at the Side.**

EXAMPLE 7 **Subtracting Rational Expressions (Different Denominators)**

Subtract. Write the answer in lowest terms.

$$\frac{9}{x-2}-\frac{3}{x}$$ The LCD is $x(x-2)$.

$$=\frac{9x}{x(x-2)}-\frac{3(x-2)}{x(x-2)}$$ Write each expression with the LCD.

$$=\frac{9x-3(x-2)}{x(x-2)}$$ Subtract numerators. Keep the same denominator.

Be careful here.

$$=\frac{9x-3x+6}{x(x-2)}$$ Distributive property

$$=\frac{6x+6}{x(x-2)},\quad \text{or}\quad \frac{6(x+1)}{x(x-2)}$$ Combine like terms. Factor the numerator.

Note

We factored the final numerator in **Example 7** to get $\frac{6(x+1)}{x(x-2)}$. The fundamental property does not apply, however, since there are no common factors to divide out. The answer is in lowest terms.

Work Problem 7 at the Side. ▶

EXAMPLE 8 **Subtracting Rational Expressions (Denominators Are Opposites)**

Subtract. Write the answer in lowest terms.

$$\frac{3x}{x-5}-\frac{2x-25}{5-x}$$ The denominators are opposites. We choose $x-5$ as the common denominator.

$$=\frac{3x}{x-5}-\frac{(2x-25)(-1)}{(5-x)(-1)}$$ Multiply $\frac{2x-25}{5-x}$ by $\frac{-1}{-1}$ to get a common denominator.

$$=\frac{3x}{x-5}-\frac{-2x+25}{x-5}$$ $(2x-25)(-1)=-2x+25=25-2x$; $(5-x)(-1)=-5+x=x-5$

Use parentheses.

$$=\frac{3x-(-2x+25)}{x-5}$$ Subtract numerators.

Be careful with signs.

$$=\frac{3x+2x-25}{x-5}$$ Distributive property

$$=\frac{5x-25}{x-5}$$ Combine like terms.

$$=\frac{5(x-5)}{x-5}$$ Factor.

$$=5$$ Divide out the common factor.

Work Problem 8 at the Side. ▶

---

7 Subtract. Write each answer in lowest terms.

(a) $\frac{1}{k+4}-\frac{2}{k}$

Teaching Tip

Stress the importance of being careful with signs. A common error would have the numerator of the fraction in Example 7 as $9x-3x-6$.

(b) $\frac{6}{a+2}-\frac{1}{a-3}$

8 Subtract. Write each answer in lowest terms.

(a) $\frac{5}{x-1}-\frac{3x}{1-x}$

Teaching Tip

Point out in Example 8 that if we had multiplied $\frac{3x}{x-5}$ by $\frac{-1}{-1}$ and used $5-x$ as the common denominator, we would have found the same answer.

(b) $\frac{2y}{y-2}-\frac{1+y}{2-y}$

Answers

7. (a) $\frac{-k-8}{k(k+4)}$ **(b)** $\frac{5(a-4)}{(a+2)(a-3)}$

8. (a) $\frac{5+3x}{x-1}$, or $\frac{-5-3x}{1-x}$

(b) $\frac{3y+1}{y-2}$, or $\frac{-3y-1}{2-y}$

9 Subtract. Write each answer in lowest terms.

(a) $\dfrac{4y}{y^2-1} - \dfrac{5}{y^2+2y+1}$

GS **(b)** $\dfrac{3r}{r-5} - \dfrac{4}{r^2-10r+25}$

The LCD is ________.

Complete the subtraction.

10 Subtract. Write each answer in lowest terms.

(a) $\dfrac{2}{p^2-5p+4} - \dfrac{3}{p^2-1}$

(b)

$\dfrac{q}{2q^2+5q-3} - \dfrac{3q+4}{3q^2+10q+3}$

Answers

9. (a) $\dfrac{4y^2-y+5}{(y+1)^2(y-1)}$

(b) $(r-5)^2$; $\dfrac{3r-19}{(r-5)^2}$

10. (a) $\dfrac{14-p}{(p-4)(p-1)(p+1)}$

(b) $\dfrac{-3q^2-4q+4}{(2q-1)(q+3)(3q+1)}$

EXAMPLE 9 Subtracting Rational Expressions

Subtract. Write the answer in lowest terms.

$$\frac{6x}{x^2-2x+1} - \frac{1}{x^2-1}$$

$$= \frac{6x}{(x-1)^2} - \frac{1}{(x-1)(x+1)} \qquad \text{Factor the denominators.}$$

From the factored denominators, we identify the LCD, $(x-1)^2(x+1)$. ***We use the factor x − 1 twice*** because it appears twice in the first denominator.

$$= \frac{6x(x+1)}{(x-1)^2(x+1)} - \frac{1(x-1)}{(x-1)(x-1)(x+1)} \qquad \text{Fundamental property}$$

$$= \frac{6x(x+1) - 1(x-1)}{(x-1)^2(x+1)} \qquad \text{Subtract numerators.}$$

$$= \frac{6x^2+6x-x+1}{(x-1)^2(x+1)} \qquad \text{Distributive property}$$

$$= \frac{6x^2+5x+1}{(x-1)^2(x+1)}, \quad \text{or} \quad \frac{(2x+1)(3x+1)}{(x-1)^2(x+1)} \qquad \text{Combine like terms. Factor the numerator.}$$

◀ **Work Problem 9 at the Side.**

EXAMPLE 10 Subtracting Rational Expressions

Subtract. Write the answer in lowest terms.

$$\frac{q}{q^2-4q-5} - \frac{3}{2q^2-13q+15}$$

$$= \frac{q}{(q+1)(q-5)} - \frac{3}{(q-5)(2q-3)} \qquad \text{Factor the denominators. The LCD is } (q+1)(q-5)(2q-3).$$

$$= \frac{q(2q-3)}{(q+1)(q-5)(2q-3)} - \frac{3(q+1)}{(q+1)(q-5)(2q-3)} \qquad \text{Fundamental property}$$

$$= \frac{q(2q-3) - 3(q+1)}{(q+1)(q-5)(2q-3)} \qquad \text{Subtract numerators.}$$

$$= \frac{2q^2-3q-3q-3}{(q+1)(q-5)(2q-3)} \qquad \text{Distributive property}$$

$$= \frac{2q^2-6q-3}{(q+1)(q-5)(2q-3)} \qquad \text{Combine like terms.}$$

The numerator cannot be factored, so the final answer is in lowest terms.

◀ **Work Problem 10 at the Side.**

7.4 Exercises

FOR EXTRA HELP

Download the MyDashBoard App

MyMathLab®

CONCEPT CHECK *Match each expression in Column I with the correct sum or difference in Column II.*

I

1. $\frac{x}{x+6}+\frac{6}{x+6}$ E

2. $\frac{2x}{x-6}-\frac{12}{x-6}$ A

3. $\frac{6}{x-6}-\frac{x}{x-6}$ C

4. $\frac{6}{x+6}-\frac{x}{x+6}$ H

5. $\frac{x}{x+6}-\frac{6}{x+6}$ B

6. $\frac{1}{x}+\frac{1}{6}$ D

7. $\frac{1}{6}-\frac{1}{x}$ G

8. $\frac{1}{6x}-\frac{1}{6x}$ F

II

A. 2

B. $\frac{x-6}{x+6}$

C. -1

D. $\frac{6+x}{6x}$

E. 1

F. 0

G. $\frac{x-6}{6x}$

H. $\frac{6-x}{x+6}$

Add or subtract. Write each answer in lowest terms. ***See Examples 1 and 6.***

9. $\frac{4}{m}+\frac{7}{m}$ $\frac{11}{m}$

10. $\frac{5}{p}+\frac{11}{p}$ $\frac{16}{p}$

11. $\frac{a+b}{2}-\frac{a-b}{2}$ b

12. $\frac{x-y}{2}-\frac{x+y}{2}$ $-y$

13. $\frac{5}{y+4}-\frac{1}{y+4}$ $\frac{4}{y+4}$

14. $\frac{4}{y+3}-\frac{1}{y+3}$ $\frac{3}{y+3}$

15. $\frac{5m}{m+1}-\frac{1+4m}{m+1}$ $\frac{m-1}{m+1}$

16. $\frac{4x}{x+2}-\frac{2+3x}{x+2}$ $\frac{x-2}{x+2}$

17. $\frac{x^2}{x+5}+\frac{5x}{x+5}$ x

18. $\frac{t^2}{t-3}+\frac{-3t}{t-3}$ t

19. GS $\frac{y^2-3y}{y+3}+\frac{-18}{y+3}$

$$= \frac{y^2-3y-18}{y+3}$$

$$= \frac{(y+\underline{3})(y-\underline{6})}{y+3}$$

$$= \underline{y-6}$$

20. GS $\frac{r^2-8r}{r-5}+\frac{15}{r-5}$

$$= \frac{r^2-8r+15}{r-5}$$

$$= \frac{(r-\underline{3})(r-5)}{r-5}$$

$$= \underline{r-3}$$

21. Explain with an example how to add or subtract rational expressions with the same denominator.

Combine the numerators and keep the same denominator. For example, $\frac{3x+2}{x-6}+\frac{-2x-8}{x-6}=\frac{x-6}{x-6}$. Then write in lowest terms: $\frac{x-6}{x-6}=1$.

22. Explain with an example how to add or subtract rational expressions with different denominators.

Find the LCD, write each fraction as an equivalent one with this LCD, and then proceed as in **Exercise 21.** For example, $\frac{3}{x}+\frac{2}{y}=\frac{3y}{xy}+\frac{2x}{xy}=\frac{3y+2x}{xy}$.

Add or subtract. Write each answer in lowest terms. ***See Examples 2, 3, 4, and 7.***

23. $\frac{z}{5}+\frac{1}{3}$ $\frac{3z+5}{15}$

24. $\frac{p}{8}+\frac{3}{5}$ $\frac{5p+24}{40}$

25. $\frac{5}{7}-\frac{r}{2}$ $\frac{10-7r}{14}$

26. $\frac{10}{9}-\frac{z}{3}$ $\frac{10-3z}{9}$

27. $-\frac{3}{4}-\frac{1}{2x}$ $\frac{-3x-2}{4x}$

28. $-\frac{5}{8}-\frac{3}{2a}$ $\frac{-5a-12}{8a}$

29. $\frac{3}{5x}+\frac{9}{4x}$ $\frac{57}{20x}$

30. $\frac{3}{2x}+\frac{4}{7x}$ $\frac{29}{14x}$

31. $\frac{x+1}{6}+\frac{3x+3}{9}$ $\frac{x+1}{2}$

32. $\frac{2x-6}{4}+\frac{x+5}{6}$ $\frac{2x-2}{3}$

33. $\frac{x+3}{3x}+\frac{2x+2}{4x}$ $\frac{5x+9}{6x}$

34. $\frac{x+2}{5x}+\frac{6x+3}{3x}$ $\frac{11x+7}{5x}$

35. $\frac{2}{x+3}+\frac{1}{x}$

The LCD is $x(x+3)$.

Complete the addition.

$\frac{3x+3}{x(x+3)}$

36. $\frac{3}{x-4}+\frac{2}{x}$

The LCD is $x(x-4)$.

Complete the addition.

$\frac{5x-8}{x(x-4)}$

37. $\frac{1}{k+5}-\frac{2}{k}$ $\frac{-k-10}{k(k+5)}$

38. $\frac{3}{m+1}-\frac{4}{m}$ $\frac{-m-4}{m(m+1)}$

39. $\frac{x}{x-2}+\frac{-8}{x^2-4}$ $\frac{x+4}{x+2}$

40. $\frac{2x}{x-1}+\frac{-4}{x^2-1}$ $\frac{2(x+2)}{x+1}$

41. $\frac{x}{x-2}+\frac{4}{x+2}$ $\frac{x^2+6x-8}{(x-2)(x+2)}$

42. $\frac{2x}{x-1}+\frac{3}{x+1}$ $\frac{2x^2+5x-3}{(x-1)(x+1)}$

43. $\frac{t}{t+2}+\frac{5-t}{t}-\frac{4}{t^2+2t}$ $\frac{3}{t}$

44. $\frac{2p}{p-3}+\frac{2+p}{p}-\frac{-6}{p^2-3p}$ $\frac{3p-1}{p-3}$

45. CONCEPT CHECK What are the two possible LCDs that could be used for the following sum?

$$\frac{10}{m-2} + \frac{5}{2-m}$$

$m - 2$ or $2 - m$

46. CONCEPT CHECK If one form of the correct answer to a sum or difference of rational expressions is $\frac{4}{k-3}$, what would be an alternative form of the answer if the denominator is $3 - k$?

$\frac{-4}{3-k}$, or $-\frac{4}{3-k}$

Add or subtract. Write each answer in lowest terms. ***See Examples 5 and 8.***

47. $\frac{4}{x-5} + \frac{6}{5-x}$

$\frac{-2}{x-5}$, or $\frac{2}{5-x}$

48. $\frac{10}{m-2} + \frac{5}{2-m}$

$\frac{5}{m-2}$, or $\frac{-5}{2-m}$

49. $\frac{-1}{1-y} + \frac{3-4y}{y-1}$

-4

50. $\frac{-4}{p-3} - \frac{p+1}{3-p}$

1

51. $\frac{2}{x-y^2} + \frac{7}{y^2-x}$

$\frac{-5}{x-y^2}$, or $\frac{5}{y^2-x}$

52. $\frac{-8}{p-q^2} + \frac{3}{q^2-p}$

$\frac{-11}{p-q^2}$, or $\frac{11}{q^2-p}$

53. $\frac{x}{5x-3y} - \frac{y}{3y-5x}$

$\frac{x+y}{5x-3y}$, or $\frac{-x-y}{3y-5x}$

54. $\frac{t}{8t-9s} - \frac{s}{9s-8t}$

$\frac{t+s}{8t-9s}$, or $\frac{-t-s}{9s-8t}$

55. $\frac{3}{4p-5} + \frac{9}{5-4p}$

$\frac{-6}{4p-5}$, or $\frac{6}{5-4p}$

56. $\frac{8}{3-7y} - \frac{2}{7y-3}$

$\frac{10}{3-7y}$, or $\frac{-10}{7y-3}$

In each subtraction problem, the rational expression that follows the subtraction sign has a numerator with more than one term. ***Be careful with signs*** *and find each difference.* ***See Examples 6–10.***

57. $\frac{2m}{m-n} - \frac{5m+n}{2m-2n}$

$\frac{-(m+n)}{2(m-n)}$

58. $\frac{5p}{p-q} - \frac{3p+1}{4p-4q}$

$\frac{17p-1}{4(p-q)}$

59. $\dfrac{5}{x^2-9}-\dfrac{x+2}{x^2+4x+3}$ $\dfrac{-x^2+6x+11}{(x+3)(x-3)(x+1)}$

60. $\dfrac{1}{a^2-1}-\dfrac{a-1}{a^2+3a-4}$ $\dfrac{-a^2+a+5}{(a+1)(a-1)(a+4)}$

61. $\dfrac{2q+1}{3q^2+10q-8}-\dfrac{3q+5}{2q^2+5q-12}$ $\dfrac{-5q^2-13q+7}{(3q-2)(q+4)(2q-3)}$

62. $\dfrac{4y-1}{2y^2+5y-3}-\dfrac{y+3}{6y^2+y-2}$ $\dfrac{11y^2-y-11}{(2y-1)(y+3)(3y+2)}$

Perform the indicated operations. ***See Examples 1–10.***

63. $\dfrac{4}{r^2-r}+\dfrac{6}{r^2+2r}-\dfrac{1}{r^2+r-2}$ $\dfrac{9r+2}{r(r+2)(r-1)}$

64. $\dfrac{6}{k^2+3k}-\dfrac{1}{k^2-k}+\dfrac{2}{k^2+2k-3}$ $\dfrac{7k-9}{k(k+3)(k-1)}$

65. $\dfrac{x+3y}{x^2+2xy+y^2}+\dfrac{x-y}{x^2+4xy+3y^2}$ $\dfrac{2x^2+6xy+8y^2}{(x+y)(x+y)(x+3y)}$, or $\dfrac{2x^2+6xy+8y^2}{(x+y)^2(x+3y)}$

66. $\dfrac{m}{m^2-1}+\dfrac{m-1}{m^2+2m+1}$ $\dfrac{2m^2-m+1}{(m+1)(m-1)(m+1)}$, or $\dfrac{2m^2-m+1}{(m+1)^2(m-1)}$

67. $\dfrac{r+y}{18r^2+9ry-2y^2}+\dfrac{3r-y}{36r^2-y^2}$ $\dfrac{15r^2+10ry-y^2}{(3r+2y)(6r-y)(6r+y)}$

68. $\dfrac{2x-z}{2x^2+xz-10z^2}-\dfrac{x+z}{x^2-4z^2}$ $\dfrac{-4xz-7z^2}{(x-2z)(x+2z)(2x+5z)}$

69. Refer to the rectangle in the figure.

(a) Find an expression that represents its perimeter. Give the simplified form.

(b) Find an expression that represents its area. Give the simplified form.

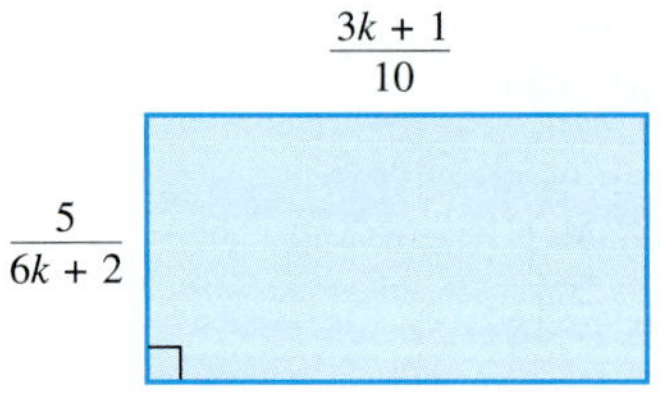

(a) $\dfrac{9k^2+6k+26}{5(3k+1)}$ (b) $\dfrac{1}{4}$

70. Refer to the triangle in the figure.

(a) Find an expression that represents its perimeter. Give the simplified form.

(b) Find an expression that represents its area. Give the simplified form.

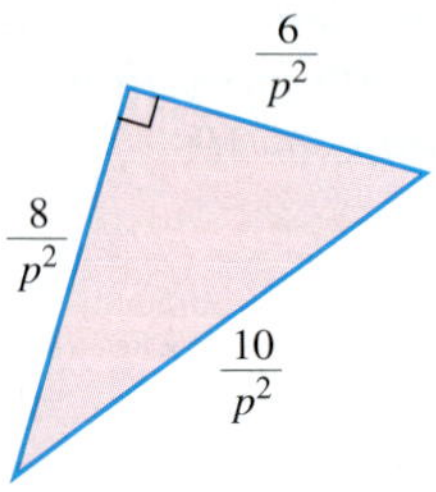

(a) $\dfrac{24}{p^2}$ (b) $\dfrac{24}{p^4}$

7.5 Complex Fractions

OBJECTIVES

1. Simplify a complex fraction by writing it as a division problem (Method 1).
2. Simplify a complex fraction by multiplying numerator and denominator by the least common denominator (Method 2).

The quotient of two mixed numbers in arithmetic, such as $2\frac{1}{2} \div 3\frac{1}{4}$, can be written as a fraction.

$$2\frac{1}{2} \div 3\frac{1}{4}$$

$$= \frac{2\frac{1}{2}}{3\frac{1}{4}}$$

$$= \frac{2 + \frac{1}{2}}{3 + \frac{1}{4}}$$

We do this to illustrate a *complex fraction*.

In algebra, some rational expressions also have fractions in the numerator, or denominator, or both.

Complex Fraction

A rational expression with one or more fractions in the numerator, or denominator, or both, is a **complex fraction.**

$$\frac{2 + \frac{1}{2}}{3 + \frac{1}{4}}, \quad \frac{\frac{3x^2 - 5x}{6x^2}}{2x - \frac{1}{x}}, \quad \text{and} \quad \frac{3 + x}{5 - \frac{2}{x}} \qquad \text{Complex fractions}$$

The parts of a complex fraction are named as follows.

$$\frac{\frac{2}{p} - \frac{1}{q}}{\frac{3}{p} + \frac{5}{q}}$$

← Numerator of complex fraction
← Main fraction bar
← Denominator of complex fraction

Teaching Tip

Encourage students to learn both methods of simplifying complex fractions. Working the same problem side by side with each method may be helpful.

OBJECTIVE 1 Simplify a complex fraction by writing it as a division problem (Method 1). Since the main fraction bar represents division in a complex fraction, one method of simplifying a complex fraction involves division.

Method 1 for Simplifying a Complex Fraction

Step 1 Write both the numerator and denominator as single fractions.

Step 2 Change the complex fraction to a division problem.

Step 3 Perform the indicated division.

1 Simplify each complex fraction using Method 1.

GS **(a)** $\dfrac{\frac{2}{5}+\frac{1}{4}}{\frac{1}{2}+\frac{1}{3}}$

Step 1
Write the numerator as a single fraction. ____

Write the denominator as a single fraction. ____

Step 2
Write the equivalent fraction as a division problem.

____

Step 3
Write the division problem as a multiplication problem.

____

Multiply and write the answer in lowest terms. ____

(b) $\dfrac{6+\frac{1}{x}}{5-\frac{2}{x}}$

(c) $\dfrac{9-\frac{4}{p}}{\frac{2}{p}+1}$

Answers

1. (a) $\frac{13}{20}$; $\frac{5}{6}$; $\frac{13}{20} \div \frac{5}{6}$; $\frac{13}{20} \cdot \frac{6}{5}$; $\frac{39}{50}$
(b) $\frac{6x+1}{5x-2}$ **(c)** $\frac{9p-4}{2+p}$

EXAMPLE 1 Simplifying Complex Fractions (Method 1)

Simplify each complex fraction.

(a) $\dfrac{\frac{2}{3}+\frac{5}{9}}{\frac{1}{4}+\frac{1}{12}}$ **(b)** $\dfrac{6+\frac{3}{x}}{\frac{x}{4}+\frac{1}{8}}$

Step 1 First, write each numerator as a single fraction.

$$\frac{2}{3}+\frac{5}{9} = \frac{2(3)}{3(3)}+\frac{5}{9} = \frac{6}{9}+\frac{5}{9} = \frac{11}{9}$$

$$6+\frac{3}{x} = \frac{6}{1}+\frac{3}{x} = \frac{6x}{x}+\frac{3}{x} = \frac{6x+3}{x}$$

Do the same thing with each denominator.

$$\frac{1}{4}+\frac{1}{12} = \frac{1(3)}{4(3)}+\frac{1}{12} = \frac{3}{12}+\frac{1}{12} = \frac{4}{12}$$

$$\frac{x}{4}+\frac{1}{8} = \frac{x(2)}{4(2)}+\frac{1}{8} = \frac{2x}{8}+\frac{1}{8} = \frac{2x+1}{8}$$

Step 2 Write the equivalent complex fraction as a division problem.

$$\frac{\frac{11}{9}}{\frac{4}{12}} = \frac{11}{9} \div \frac{4}{12}$$

$$\frac{\frac{6x+3}{x}}{\frac{2x+1}{8}} = \frac{6x+3}{x} \div \frac{2x+1}{8}$$

Step 3 Now use the definition of division and multiply by the reciprocal. Then write in lowest terms using the fundamental property.

$$= \frac{11}{9} \cdot \frac{12}{4} = \frac{11 \cdot 3 \cdot 4}{3 \cdot 3 \cdot 4} = \frac{11}{3}$$

$$= \frac{6x+3}{x} \cdot \frac{8}{2x+1} = \frac{3(2x+1)}{x} \cdot \frac{8}{2x+1} = \frac{24}{x}$$

◀ Work Problem 1 at the Side.

EXAMPLE 2 Simplifying a Complex Fraction (Method 1)

Simplify the complex fraction.

$$\frac{\frac{xp}{q^3}}{\frac{p^2}{qx^2}}$$

The numerator and denominator are single fractions, so use the definition of division and then the fundamental property.

$$\frac{xp}{q^3} \div \frac{p^2}{qx^2}$$
$$= \frac{xp}{q^3} \cdot \frac{qx^2}{p^2}$$
$$= \frac{x^3}{q^2p}$$

Work Problem 2 at the Side. ▶

EXAMPLE 3 Simplifying a Complex Fraction (Method 1)

Simplify the complex fraction.

$$\frac{\frac{3}{x+2} - 4}{\frac{2}{x+2} + 1}$$

$$= \frac{\frac{3}{x+2} - \frac{4(x+2)}{x+2}}{\frac{2}{x+2} + \frac{1(x+2)}{x+2}}$$ Write both second terms with a denominator of $x + 2$.

$$= \frac{\frac{3 - 4(x+2)}{x+2}}{\frac{2 + 1(x+2)}{x+2}}$$ Subtract in the numerator. Add in the denominator.

$$= \frac{\frac{3 - 4x - 8}{x+2}}{\frac{2 + x + 2}{x+2}}$$ Distributive property (Be careful with signs.)

$$= \frac{\frac{-5 - 4x}{x+2}}{\frac{4 + x}{x+2}}$$ Combine like terms.

$$= \frac{-5 - 4x}{x+2} \cdot \frac{x+2}{4+x}$$ Multiply by the reciprocal.

$$= \frac{-5 - 4x}{4 + x}$$ Divide out the common factor.

Work Problem 3 at the Side. ▶

OBJECTIVE 2 Simplify a complex fraction by multiplying numerator and denominator by the least common denominator (Method 2). If we multiply both the numerator and the denominator of a complex fraction by the LCD of all the fractions within the complex fraction, the result will no longer be complex. This is Method 2.

2 Simplify each complex fraction using Method 1.

(a) $$\frac{\frac{rs^2}{t}}{\frac{r^2s}{t^2}}$$

(b) $$\frac{\frac{m^2n^3}{p}}{\frac{m^4n}{p^2}}$$

3 Simplify using Method 1.

$$\frac{\frac{2}{x-1} + \frac{1}{x+1}}{\frac{3}{x-1} - \frac{4}{x+1}}$$

Answers

2. (a) $\frac{st}{r}$ **(b)** $\frac{n^2p}{m^2}$

3. $\frac{3x+1}{-x+7}$

Method 2 for Simplifying a Complex Fraction

Step 1 Find the LCD of all fractions within the complex fraction.

Step 2 Multiply both the numerator and the denominator of the complex fraction by this LCD using the distributive property as necessary. Write in lowest terms.

EXAMPLE 4 Simplifying Complex Fractions (Method 2)

Simplify each complex fraction.

(a) $\dfrac{\frac{2}{3}+\frac{5}{9}}{\frac{1}{4}+\frac{1}{12}}$ (In **Example 1,** we simplified these same fractions using Method 1.)

(b) $\dfrac{6+\frac{3}{x}}{\frac{x}{4}+\frac{1}{8}}$

Step 1 Find the LCD for all denominators in the complex fraction.

The LCD for 3, 9, 4, and 12 is 36. | The LCD for x, 4, and 8 is $8x$.

Step 2 Multiply the numerator and denominator of the complex fraction by the LCD.

(a)

$$\frac{\frac{2}{3}+\frac{5}{9}}{\frac{1}{4}+\frac{1}{12}}$$

$$=\frac{36\left(\frac{2}{3}+\frac{5}{9}\right)}{36\left(\frac{1}{4}+\frac{1}{12}\right)}$$

Multiply *each* term by 36.

$$=\frac{36\left(\frac{2}{3}\right)+36\left(\frac{5}{9}\right)}{36\left(\frac{1}{4}\right)+36\left(\frac{1}{12}\right)}$$

$$=\frac{24+20}{9+3}$$

$$=\frac{44}{12}$$

$$=\frac{4\cdot 11}{4\cdot 3}$$

$$=\frac{11}{3}$$

(b)

$$\frac{6+\frac{3}{x}}{\frac{x}{4}+\frac{1}{8}}$$

$$=\frac{8x\left(6+\frac{3}{x}\right)}{8x\left(\frac{x}{4}+\frac{1}{8}\right)}$$

Multiply *each* term by $8x$.

$$=\frac{8x(6)+8x\left(\frac{3}{x}\right)}{8x\left(\frac{x}{4}\right)+8x\left(\frac{1}{8}\right)}$$

$$=\frac{48x+24}{2x^2+x}$$

$$=\frac{24(2x+1)}{x(2x+1)}$$

$$=\frac{24}{x}$$

◀ Work Problem **4** at the Side.

4 Simplify each complex fraction using Method 2.

GS **(a)** $\dfrac{\frac{2}{3}-\frac{1}{4}}{\frac{4}{9}+\frac{1}{2}}$

Step 1
The LCD for ____, ____, ____, and ____ is ____.

Refer to Step 2 in **Example 4,** and simplify the complex fraction.

(b) $\dfrac{2-\frac{6}{a}}{3+\frac{4}{a}}$

(c) $\dfrac{\frac{p}{5-p}}{\frac{4p}{2p+1}}$

Teaching Tip

Although we are finding the LCD, we do not use it as such. In Example 4, multiplying the complex fraction by $\frac{36}{36}$ or by $\frac{8x}{8x}$ is again multiplying by 1.

Answers

4. **(a)** 3; 4; 9; 2; 36; $\frac{15}{34}$
(b) $\frac{2a-6}{3a+4}$ **(c)** $\frac{2p+1}{4(5-p)}$

EXAMPLE 5 Simplifying a Complex Fraction (Method 2)

Simplify the complex fraction.

$$\frac{\frac{3}{5m}-\frac{2}{m^2}}{\frac{9}{2m}+\frac{3}{4m^2}}$$ The LCD for $5m$, m^2, $2m$, and $4m^2$ is $20m^2$.

$$=\frac{20m^2\left(\frac{3}{5m}-\frac{2}{m^2}\right)}{20m^2\left(\frac{9}{2m}+\frac{3}{4m^2}\right)}$$ Multiply numerator and denominator by $20m^2$.

$$=\frac{20m^2\left(\frac{3}{5m}\right)-20m^2\left(\frac{2}{m^2}\right)}{20m^2\left(\frac{9}{2m}\right)+20m^2\left(\frac{3}{4m^2}\right)}$$ Distributive property

$$=\frac{12m-40}{90m+15}, \quad \text{or} \quad \frac{4(3m-10)}{5(18m+3)}$$ Multiply and factor.

Work Problem **5** at the Side. ▶

Some students use Method 1 for problems like **Example 2,** which is the quotient of two fractions, and Method 2 for problems like **Examples 1, 3, 4, and 5,** which have sums or differences in the numerators or denominators.

EXAMPLE 6 Simplifying Complex Fractions

Simplify each complex fraction. Use either method.

(a) $$\frac{\frac{1}{y}+\frac{2}{y+2}}{\frac{4}{y}-\frac{3}{y+2}}$$ There are sums and differences in the numerator and denominator. Use Method 2.

$$=\frac{\left(\frac{1}{y}+\frac{2}{y+2}\right)y(y+2)}{\left(\frac{4}{y}-\frac{3}{y+2}\right)y(y+2)}$$ Multiply numerator and denominator by the LCD, $y(y+2)$.

$$=\frac{\left(\frac{1}{y}\right)y(y+2)+\left(\frac{2}{y+2}\right)y(y+2)}{\left(\frac{4}{y}\right)y(y+2)-\left(\frac{3}{y+2}\right)y(y+2)}$$ Distributive property

$$=\frac{1(y+2)+2y}{4(y+2)-3y}$$ Multiply and simplify.

$$=\frac{y+2+2y}{4y+8-3y}$$ Distributive property

$$=\frac{3y+2}{y+8}$$ Combine like terms.

Continued on Next Page

5 Simplify using Method 2.

GS

$$\frac{\frac{2}{5x}-\frac{3}{x^2}}{\frac{7}{4x}+\frac{1}{2x^2}}$$

The LCD for ____, ____, ____, and ____ is ____.

Simplify the complex fraction.

Teaching Tip

Stress that the same answer is obtained regardless of the method used.

Answer

5. $5x$; x^2; $4x$; $2x^2$; $20x^2$;
$\frac{8x-60}{35x+10}$, or $\frac{4(2x-15)}{5(7x+2)}$

6 Simplify each complex fraction. Use either method.

(a) $$\frac{\frac{1}{x}+\frac{2}{x-1}}{\frac{2}{x}-\frac{4}{x-1}}$$

(b) $$\frac{1-\frac{2}{x}-\frac{15}{x^2}}{1+\frac{5}{x}+\frac{6}{x^2}}$$

(c) $$\frac{\frac{2x+3}{x-4}}{\frac{4x^2-9}{x^2-16}}$$

Answers

6. (a) $\frac{3x-1}{-2x-2}$ **(b)** $\frac{x-5}{x+2}$ **(c)** $\frac{x+4}{2x-3}$

(b) $$\frac{1-\frac{2}{x}-\frac{3}{x^2}}{1-\frac{5}{x}+\frac{6}{x^2}}$$ There are sums and differences in the numerator and denominator. Use Method 2.

$$=\frac{\left(1-\frac{2}{x}-\frac{3}{x^2}\right)x^2}{\left(1-\frac{5}{x}+\frac{6}{x^2}\right)x^2}$$ Multiply numerator and denominator by the LCD, x^2.

$$=\frac{x^2-2x-3}{x^2-5x+6}$$ Distributive property

$$=\frac{(x-3)(x+1)}{(x-3)(x-2)}$$ Factor.

$$=\frac{x+1}{x-2}$$ Divide out the common factor.

(c) $$\frac{\frac{x+2}{x-3}}{\frac{x^2-4}{x^2-9}}$$ This is a quotient of two rational expressions. Use Method 1.

$$=\frac{x+2}{x-3}\div\frac{x^2-4}{x^2-9}$$ Write as a division problem.

$$=\frac{x+2}{x-3}\cdot\frac{x^2-9}{x^2-4}$$ Multiply by the reciprocal.

$$=\frac{(x+2)(x+3)(x-3)}{(x-3)(x+2)(x-2)}$$ Multiply and then factor.

$$=\frac{x+3}{x-2}$$ Divide out the common factors.

◀ **Work Problem 6 at the Side.**

7.5 Exercises

FOR EXTRA HELP

CONCEPT CHECK *In Exercises 1 and 2, answer each question.*

1. In a fraction, what operation does the fraction bar represent?
division

2. What property of real numbers justifies Method 2 of simplifying complex fractions?
identity property for multiplication

3. Consider the complex fraction $\dfrac{\frac{1}{2}-\frac{1}{3}}{\frac{5}{6}-\frac{1}{12}}$. Answer each part, outlining Method 1 for simplifying this complex fraction.

(a) To combine the terms in the numerator, we must find the LCD of $\frac{1}{2}$ and $\frac{1}{3}$. What is this LCD? $\underline{\ 6\ }$
Determine the simplified form of the numerator of the complex fraction. $\underline{\ \frac{1}{6}\ }$

(b) To combine the terms in the denominator, we must find the LCD of $\frac{5}{6}$ and $\frac{1}{12}$. What is this LCD? $\underline{\ 12\ }$
Determine the simplified form of the denominator of the complex fraction. $\underline{\ \frac{3}{4}\ }$

(c) Now use the results from parts (a) and (b) to write the complex fraction as a division problem using the symbol $\div$. $\underline{\ \frac{1}{6} \div \frac{3}{4}\ }$

(d) Perform the operation from part (c) to obtain the final simplification. $\underline{\ \frac{2}{9}\ }$

4. Consider the same complex fraction given in **Exercise 1:** $\dfrac{\frac{1}{2}-\frac{1}{3}}{\frac{5}{6}-\frac{1}{12}}$. Answer each part, outlining Method 2 for simplifying this complex fraction.

(a) We must determine the LCD of all the fractions within the complex fraction. What is this LCD? $\underline{\ 12\ }$

(b) Multiply every term in the complex fraction by the LCD found in part (a), but at this time do not combine the terms in the numerator and the denominator. $\underline{\ \frac{6-4}{10-1}\ }$

(c) Now combine the terms from part (b) to obtain the simplified form of the complex fraction. $\underline{\ \frac{2}{9}\ }$

Simplify each complex fraction. Use either method. ***See Examples 1–6.***

5. $\dfrac{-\frac{4}{3}}{\frac{2}{9}}$ $\quad -6$

6. $\dfrac{-\frac{5}{6}}{\frac{5}{4}}$ $\quad -\frac{2}{3}$

7. $\dfrac{\frac{p}{q^2}}{\frac{p^2}{q}}$ $\quad \frac{1}{pq}$

8. $\dfrac{\frac{a}{x}}{\frac{a^2}{2x}}$ $\quad \frac{2}{a}$

9. $\dfrac{\frac{x}{y^2}}{\frac{x^2}{y}}$ $\quad \frac{1}{xy}$

10. $\dfrac{\frac{p^4}{r}}{\frac{p^2}{r^2}}$ $\quad p^2r$

11. $\dfrac{\frac{4a^4b^3}{3a}}{\frac{2ab^4}{b^2}}$ $\quad \frac{2a^2b}{3}$

12. $\dfrac{\frac{2r^4t^2}{3t}}{\frac{5r^2t^5}{3r}}$ $\quad \frac{2r^3}{5t^4}$

13. $\dfrac{\frac{m+2}{3}}{\frac{m-4}{m}}$ $\frac{m(m+2)}{3(m-4)}$

14. $\dfrac{\frac{q-5}{q}}{\frac{q+5}{3}}$ $\frac{3(q-5)}{q(q+5)}$

15. $\dfrac{\frac{2}{x}-3}{\frac{2-3x}{2}}$ $\frac{2}{x}$

16. $\dfrac{6+\frac{2}{r}}{\frac{3r+1}{4}}$ $\frac{8}{r}$

17. $\dfrac{\frac{1}{x}+x}{\frac{x^2+1}{8}}$ $\frac{8}{x}$

18. $\dfrac{\frac{3}{m}-m}{\frac{3-m^2}{4}}$ $\frac{4}{m}$

19. $\dfrac{a-\frac{5}{a}}{a+\frac{1}{a}}$ $\frac{a^2-5}{a^2+1}$

20. $\dfrac{q+\frac{1}{q}}{q+\frac{4}{q}}$ $\frac{q^2+1}{q^2+4}$

21. $\dfrac{\frac{1}{2}+\frac{1}{p}}{\frac{2}{3}+\frac{1}{p}}$ $\frac{3(p+2)}{2(2p+3)}$

22. $\dfrac{\frac{3}{4}-\frac{1}{r}}{\frac{1}{5}+\frac{1}{r}}$ $\frac{5(3r-4)}{4(r+5)}$

23. $\dfrac{\frac{2}{p^2}-\frac{3}{5p}}{\frac{4}{p}+\frac{1}{4p}}$ $\frac{40-12p}{85p}$

24. $\dfrac{\frac{2}{m^2}-\frac{3}{m}}{\frac{2}{5m^2}+\frac{1}{3m}}$ $\frac{30-45m}{6+5m}$

25. $\dfrac{\frac{t}{t+2}}{\frac{4}{t^2-4}}$ $\frac{t(t-2)}{4}$

26. $\dfrac{\frac{m}{m+1}}{\frac{3}{m^2-1}}$ $\frac{m(m-1)}{3}$

27. $\dfrac{\frac{1}{k+1}-1}{\frac{1}{k+1}+1}$ $\frac{-k}{2+k}$

28. $\dfrac{\frac{2}{p-1}+2}{\frac{3}{p-1}-2}$ $\frac{2p}{5-2p}$

29. $\dfrac{2+\frac{1}{x}-\frac{28}{x^2}}{3+\frac{13}{x}+\frac{4}{x^2}}$ $\frac{2x-7}{3x+1}$

30. $\dfrac{4-\frac{11}{x}-\frac{3}{x^2}}{2-\frac{1}{x}-\frac{15}{x^2}}$ $\frac{4x+1}{2x+5}$

31. $\dfrac{\frac{1}{m-1}+\frac{2}{m+2}}{\frac{2}{m+2}-\frac{1}{m-3}}$ $\frac{3m(m-3)}{(m-1)(m-8)}$

32. $\dfrac{\frac{5}{r+3}-\frac{1}{r-1}}{\frac{2}{r+2}+\frac{3}{r+3}}$ $\frac{4(r-2)(r+2)}{(5r+12)(r-1)}$

33. $2-\dfrac{2}{2+\frac{2}{2+2}}$ $\frac{6}{5}$

34. $3-\dfrac{2}{4+\frac{2}{4-2}}$ $\frac{13}{5}$

7.6 Solving Equations with Rational Expressions

OBJECTIVES

1. Distinguish between operations with rational expressions and equations with terms that are rational expressions.
2. Solve equations with rational expressions.
3. Solve a formula for a specified variable.

OBJECTIVE 1 **Distinguish between operations with rational expressions and equations with terms that are rational expressions.** Before solving equations with rational expressions, we emphasize the difference between sums and differences of terms with rational coefficients, or rational *expressions*, and *equations* with terms that are rational expressions.

Sums and differences are expressions to simplify. Equations are solved.

EXAMPLE 1 Distinguishing between Expressions and Equations

Identify each of the following as an *expression* or an *equation*. Then simplify the expression or solve the equation.

(a) $\frac{3}{4}x - \frac{2}{3}x$ — This is a difference of two terms. It represents an *expression* to simplify since there is no equality symbol.

$= \frac{3 \cdot 3}{3 \cdot 4}x - \frac{4 \cdot 2}{4 \cdot 3}x$ — The LCD is 12. Write each coefficient with this LCD.

$= \frac{9}{12}x - \frac{8}{12}x$ — Multiply.

$= \frac{1}{12}x$ — Combine like terms, using the distributive property: $\frac{9}{12}x - \frac{8}{12}x = \left(\frac{9}{12} - \frac{8}{12}\right)x$.

(b) $\frac{3}{4}x - \frac{2}{3}x = \frac{1}{2}$ — Because there is an equality symbol, this is an *equation* to be solved.

$12\left(\frac{3}{4}x - \frac{2}{3}x\right) = 12\left(\frac{1}{2}\right)$ — Use the multiplication property of equality to clear fractions. Multiply by 12, the LCD.

$12\left(\frac{3}{4}x\right) - 12\left(\frac{2}{3}x\right) = 12\left(\frac{1}{2}\right)$ — Distributive property

Multiply each term by 12.

$9x - 8x = 6$ — Multiply.

$x = 6$ — Combine like terms.

CHECK $\frac{3}{4}x - \frac{2}{3}x = \frac{1}{2}$ — Original equation

$\frac{3}{4}(6) - \frac{2}{3}(6) \stackrel{?}{=} \frac{1}{2}$ — Let $x = 6$.

$\frac{9}{2} - 4 \stackrel{?}{=} \frac{1}{2}$ — Multiply.

$\frac{1}{2} = \frac{1}{2}$ ✓ — True

Since a true statement results, $\{6\}$ is the solution set of the equation.

Work Problem 1 at the Side. ▶

1 Identify each as an *expression* or an *equation*. Then simplify the expression or solve the equation.

(a) $\frac{2x}{3} - \frac{4x}{9}$

(b) $\frac{2x}{3} - \frac{4x}{9} = 2$

Teaching Tip

Again, review the distinction between expressions and equations. Refer back to the discussion is **Section 1.2, Objective 5.**

Teaching Tip

Spend time in class explaining how the least common denominator is used in different ways when adding and subtracting fractions as opposed to clearing fractions in an equation.

Answers

1. **(a)** expression; $\frac{2x}{9}$ **(b)** equation; $\{9\}$

2 Solve each equation, and check your solutions.

(a) $\frac{x}{5} + 3 = \frac{3}{5}$

(b) $\frac{x}{2} - \frac{x}{3} = \frac{5}{6}$

The ideas of **Example 1** can be summarized as follows.

Uses of the LCD

When adding or subtracting rational expressions, keep the LCD throughout the simplification. (See **Example 1(a).**)

When solving an equation with terms that are rational expression, multiply each side by the LCD so that denominators are eliminated. (See **Example 1(b).**)

OBJECTIVE 2 Solve equations with rational expressions. When an equation involves fractions as in **Example 1(b),** we use the multiplication property of equality to clear it of fractions. When we choose the LCD of all the denominators as the multiplier, the resulting equation contains no fractions.

EXAMPLE 2 Solving an Equation with Rational Expressions

Solve, and check the solution.

$$\frac{x}{3} + \frac{x}{4} = 10 + x$$

$$12\left(\frac{x}{3} + \frac{x}{4}\right) = 12(10 + x) \quad \text{Multiply by the LCD, 12, to clear fractions.}$$

$$12\left(\frac{x}{3}\right) + 12\left(\frac{x}{4}\right) = 12(10) + 12x \quad \text{Distributive property}$$

$$4x + 3x = 120 + 12x \quad \text{Multiply.}$$

$$7x = 120 + 12x \quad \text{Combine like terms.}$$

$$-5x = 120 \quad \text{Subtract } 12x.$$

$$x = -24 \quad \text{Divide by } -5.$$

CHECK

$$\frac{x}{3} + \frac{x}{4} = 10 + x \quad \text{Original equation}$$

$$\frac{-24}{3} + \frac{-24}{4} \stackrel{?}{=} 10 - 24 \quad \text{Let } x = -24.$$

$$-8 + (-6) \stackrel{?}{=} -14 \quad \text{Divide. Subtract.}$$

$$-14 = -14 \checkmark \quad \text{True}$$

The solution set is $\{-24\}$.

◀ Work Problem 2 at the Side.

CAUTION

In **Example 2,** we used the multiplication property of equality to multiply each side of an *equation* by the LCD. In **Section 7.5,** we used the fundamental property to multiply a *fraction* by another fraction that had the LCD as both its numerator and denominator. ***Be careful not to confuse these procedures.***

Answers

2. (a) $\{-12\}$ (b) $\{5\}$

EXAMPLE 3 Solving an Equation with Rational Expressions

Solve, and check the solution.

$$\frac{p}{2} - \frac{p-1}{3} = 1$$

$$6\left(\frac{p}{2} - \frac{p-1}{3}\right) = 6 \cdot 1 \quad \text{Multiply by the LCD, 6.}$$

$$6\left(\frac{p}{2}\right) - 6\left(\frac{p-1}{3}\right) = 6 \quad \text{Distributive property}$$

$$3p - 2(p-1) = 6$$

Use parentheses around $p - 1$ to avoid errors.

$$3p - 2(p) - 2(-1) = 6 \quad \text{Distributive property}$$

Be careful with signs.

$$3p - 2p + 2 = 6 \quad \text{Multiply.}$$

$$p = 4 \quad \text{Combine like terms. Subtract 2.}$$

Check that $\{4\}$ is the solution set by replacing p with 4 in the original equation.

Work Problem 3 at the Side. ▶

Recall that division by 0 is undefined. ***Therefore, when solving an equation with rational expressions that have variables in the denominator, the solution cannot be a number that makes the denominator equal 0.***

A value of the variable that appears to be a solution after both sides of an equation with rational expressions are multiplied by a variable expression is a **proposed solution.** ***All proposed solutions must be checked in the original equation.***

EXAMPLE 4 Solving an Equation with Rational Expressions

Solve, and check the proposed solution.

$$\frac{x}{x-2} = \frac{2}{x-2} + 2$$

x cannot equal 2, since 2 causes both denominators to equal 0.

$$(x-2)\left(\frac{x}{x-2}\right) = (x-2)\left(\frac{2}{x-2} + 2\right) \quad \text{Multiply each side by the LCD, } x-2.$$

$$(x-2)\left(\frac{x}{x-2}\right) = (x-2)\left(\frac{2}{x-2}\right) + (x-2)(2) \quad \text{Distributive property}$$

$$x = 2 + 2x - 4 \quad \text{Simplify.}$$

$$x = -2 + 2x \quad \text{Combine like terms.}$$

$$-x = -2 \quad \text{Subtract } 2x.$$

Proposed solution → $x = 2$ Divide by -1.

CHECK

$$\frac{x}{x-2} = \frac{2}{x-2} + 2 \quad \text{Original equation}$$

$$\frac{2}{2-2} \stackrel{?}{=} \frac{2}{2-2} + 2 \quad \text{Let } x = 2.$$

$$\frac{2}{0} \stackrel{?}{=} \frac{2}{0} + 2 \quad \text{Subtract in the denominators.}$$

Division by 0 is undefined.

Thus, the proposed solution 2 must be rejected, and the solution set is $\emptyset$.

Work Problem 4 at the Side. ▶

3 Solve each equation, and check the solutions.

GS **(a)** $\frac{k}{6} - \frac{k+1}{4} = -\frac{1}{2}$

Multiply by the LCD, ____.

$$\_\_\_\left(\frac{k}{6} - \frac{k+1}{4}\right) = \_\_\_\left(-\frac{1}{2}\right)$$

Complete the solution.

(b) $\frac{2m-3}{5} - \frac{m}{3} = -\frac{6}{5}$

4 Solve the equation, and check the proposed solution.

$$1 - \frac{2}{x+1} = \frac{2x}{x+1}$$

Answers

3. **(a)** 12; 12; 12; $\{3\}$ **(b)** $\{-9\}$

4. $\emptyset$ (When the equation is solved, -1 is a proposed solution. However, because $x = -1$ leads to a 0 denominator in the original equation, there is no solution.)

5 Solve each equation, and check the proposed solutions.

(a) $\frac{4}{x^2 - 3x} = \frac{1}{x^2 - 9}$

(b) $\frac{2}{p^2 - 2p} = \frac{3}{p^2 - p}$

Step 1

Factor the denominators.

$p^2 - 2p = p(\_\_\_\_\_\_)$

$p^2 - \ \ p = p(\_\_\_\_\_\_)$

The LCD is ________.

The numbers ____, ____, and ____ cannot be solutions.

Complete the steps to solve the equation.

Teaching Tip

This section is the first time that most students have encountered equations in which, even though their algebra is correct, the apparent solution is not really a solution. Stress that the check is an essential part of the solution process.

A proposed solution that is not an actual solution of the original equation, such as 2 in **Example 4,** is an **extraneous solution,** or **extraneous value.** Some students like to determine which numbers cannot be solutions *before* solving the equation, as we did at the beginning of **Example 4.**

Solving an Equation with Rational Expressions

Step 1 **Multiply each side of the equation by the LCD.** (This clears the equation of fractions.) Be sure to distribute to *every* term on *both* sides of the equation.

Step 2 **Solve** the resulting equation for proposed solutions.

Step 3 **Check** each proposed solution by substituting it in the original equation. Reject any that cause a denominator to equal 0.

EXAMPLE 5 **Solving an Equation with Rational Expressions**

Solve, and check the proposed solution.

$$\frac{2}{x^2 - x} = \frac{1}{x^2 - 1}$$

Step 1 $$\frac{2}{x(x-1)} = \frac{1}{(x+1)(x-1)}$$ Factor the denominators to find the LCD, $x(x+1)(x-1)$.

Notice that 0, −1, and 1 cannot be solutions of this equation. Otherwise, a denominator will equal 0.

$$x(x+1)(x-1)\frac{2}{x(x-1)} = x(x+1)(x-1)\frac{1}{(x+1)(x-1)}$$ Multiply by the LCD.

Step 2

$2(x+1) = x$ Divide out the common factors.

$2x + 2 = x$ Distributive property

$x + 2 = 0$ Subtract x.

Proposed solution → $x = -2$ Subtract 2.

Step 3 The proposed solution is −2, which does not make any denominator equal 0.

CHECK

$\frac{2}{x^2 - x} = \frac{1}{x^2 - 1}$ Original equation

$\frac{2}{(-2)^2 - (-2)} \stackrel{?}{=} \frac{1}{(-2)^2 - 1}$ Let $x = -2$.

$\frac{2}{4+2} \stackrel{?}{=} \frac{1}{4-1}$ Apply the exponents; Definition of subtraction

$\frac{1}{3} = \frac{1}{3}$ ✓ True

The solution set is $\{-2\}$.

◄ Work Problem **5** at the Side.

Answers

5. (a) $\{-4\}$
(b) $p-2$; $p-1$; $p(p-2)(p-1)$; 0; 1; 2 (Order of the numbers 0, 1, and 2 does not matter.); $\{4\}$

EXAMPLE 6 Solving an Equation with Rational Expressions

Solve, and check the proposed solution.

$$\frac{2m}{(m^2-4)}+\frac{1}{m-2}=\frac{2}{m+2}$$

$$\frac{2m}{(m+2)(m-2)}+\frac{1}{m-2}=\frac{2}{m+2}$$ Factor the denominator on the left to find the LCD, $(m+2)(m-2)$.

Notice that -2 and 2 cannot be solutions of the equation.

$$(m+2)(m-2)\left(\frac{2m}{(m+2)(m-2)}+\frac{1}{m-2}\right)$$

$$=(m+2)(m-2)\frac{2}{m+2}$$ Multiply by the LCD.

$$(m+2)(m-2)\frac{2m}{(m+2)(m-2)}+(m+2)(m-2)\frac{1}{m-2}$$

$$=(m+2)(m-2)\frac{2}{m+2}$$ Distributive property

$$2m+m+2=2(m-2)$$ Divide out the common factors.

$$3m+2=2m-4$$ Combine like terms; distributive property

$$m=-6$$ Subtract $2m$. Subtract 2.

A check will verify that $\{-6\}$ is the solution set.

Work Problem **6** at the Side. ▶

6 Solve each equation, and check the proposed solutions.

(a) $\frac{2p}{p^2-1}=\frac{2}{p+1}-\frac{1}{p-1}$

(b) $\frac{8r}{4r^2-1}=\frac{3}{2r+1}+\frac{3}{2r-1}$

EXAMPLE 7 Solving an Equation with Rational Expressions

Solve, and check the proposed solution(s).

$$\frac{1}{x-1}+\frac{1}{2}=\frac{2}{x^2-1}$$

$x \neq 1, -1$ or a denominator is 0.

$$\frac{1}{x-1}+\frac{1}{2}=\frac{2}{(x+1)(x-1)}$$

Factor the denominator on the right. The LCD is $2(x+1)(x-1)$.

$$2(x+1)(x-1)\left(\frac{1}{x-1}+\frac{1}{2}\right)=2(x+1)(x-1)\frac{2}{(x+1)(x-1)}$$

Multiply by the LCD.

$$2(x+1)(x-1)\frac{1}{x-1}+2(x+1)(x-1)\frac{1}{2}$$

$$=2(x+1)(x-1)\frac{2}{(x+1)(x-1)}$$

Distributive property

$$2(x+1)+(x+1)(x-1)=2(2)$$ Divide out the common factors.

$$2x+2+x^2-1=4$$ Distributive property; multiply.

Write in standard form.

$$x^2+2x-3=0$$ Subtract 4. Combine like terms.

Continued on Next Page

Answers

6. (a) $\{-3\}$ (b) $\{0\}$

7 Solve the equation, and check the proposed solution(s).

$$\frac{2}{3x+1}-\frac{1}{x}=\frac{-6x}{3x+1}$$

8 Solve each equation, and check the proposed solution(s).

(a) $\frac{1}{x-2}+\frac{1}{5}=\frac{2}{5(x^2-4)}$

(b) $\frac{6}{5a+10}-\frac{1}{a-5}=\frac{4}{a^2-3a-10}$

Answers

7. $\left\{\frac{1}{2}\right\}$

8. (a) $\{-4, -1\}$ **(b)** $\{60\}$

$$(x+3)(x-1)=0 \quad \text{Factor } x^2+2x-3.$$

$$x+3=0 \quad \text{or} \quad x-1=0 \quad \text{Zero-factor property}$$

$$x=-3 \quad \text{or} \quad x=1 \longleftarrow \text{Proposed solutions}$$

Since 1 makes an original denominator equal 0, the proposed solution 1 is an extraneous value. Check that -3 is a solution.

CHECK

$$\frac{1}{x-1}+\frac{1}{2}=\frac{2}{x^2-1} \quad \text{Original equation}$$

$$\frac{1}{-3-1}+\frac{1}{2}\stackrel{?}{=}\frac{2}{(-3)^2-1} \quad \text{Let } x=-3.$$

$$\frac{1}{-4}+\frac{1}{2}\stackrel{?}{=}\frac{2}{9-1} \quad \text{Subtract. Apply the exponent.}$$

$$\frac{1}{4}=\frac{1}{4} \checkmark \quad \text{True}$$

The check shows that $\{-3\}$ is the solution set.

◀ Work Problem **7** at the Side.

EXAMPLE 8 Solving an Equation with Rational Expressions

Solve, and check the proposed solution(s).

$$\frac{1}{k^2+4k+3}+\frac{1}{2k+2}=\frac{3}{4k+12}$$

$$\frac{1}{(k+1)(k+3)}+\frac{1}{2(k+1)}=\frac{3}{4(k+3)} \quad \text{Factor each denominator. The LCD is } 4(k+1)(k+3).$$

$k \neq -1, -3$

$$4(k+1)(k+3)\left(\frac{1}{(k+1)(k+3)}+\frac{1}{2(k+1)}\right)$$

$$=4(k+1)(k+3)\frac{3}{4(k+3)} \quad \text{Multiply by the LCD.}$$

$$4(k+1)(k+3)\frac{1}{(k+1)(k+3)}+2\cdot 2(k+1)(k+3)\frac{1}{2(k+1)}$$

$$=4(k+1)(k+3)\frac{3}{4(k+3)} \quad \text{Distributive property}$$

Do *not* add 4 + 2 here.

$$4+2(k+3)=3(k+1) \quad \text{Divide out the common factors.}$$

$$4+2k+6=3k+3 \quad \text{Distributive property}$$

$$2k+10=3k+3 \quad \text{Combine like terms.}$$

$$10=k+3 \quad \text{Subtract } 2k.$$

$$7=k \quad \text{Subtract 3.}$$

The proposed solution, 7, does not make an original denominator equal 0. A check shows that the algebra is correct, so $\{7\}$ is the solution set.

◀ Work Problem **8** at the Side.

OBJECTIVE 3 Solve a formula for a specified variable. When solving a formula for a specified variable, ***remember to treat the variable for which you are solving as if it were the only variable, and all others as if they were constants.***

EXAMPLE 9 Solving for Specified Variables

Solve each formula for the specified variable.

(a) $a = \frac{v - w}{t}$ for v

Our goal is to isolate v.

$$a = \frac{v - w}{t}$$

$$at = v - w \qquad \text{Multiply by } t.$$

$$at + w = v, \quad \text{or} \quad v = at + w \qquad \text{Add } w.$$

CHECK Substitute $at + w$ for v in the original equation.

$$a = \frac{v - w}{t} \qquad \text{Original equation}$$

$$a = \frac{at + w - w}{t} \qquad \text{Let } v = at + w.$$

$$a = \frac{at}{t} \qquad \text{Combine like terms.}$$

$$a = a \ ✓ \qquad \text{True}$$

(b) $F = \frac{k}{d - D}$ for d

We must isolate d.

$$F = \frac{k}{d - D}$$

$$F(d - D) = \frac{k}{d - D}(d - D) \qquad \text{Multiply by } d - D \text{ to clear the fraction.}$$

$$F(d - D) = k \qquad \text{Simplify.}$$

$$Fd - FD = k \qquad \text{Distributive property}$$

$$Fd = k + FD \qquad \text{Add } FD.$$

$$d = \frac{k + FD}{F} \qquad \text{Divide by } F.$$

We can write an equivalent form of this answer as follows.

$$d = \frac{k + FD}{F} \qquad \text{Answer from above}$$

$$d = \frac{k}{F} + \frac{FD}{F} \qquad \text{Definition of addition of fractions: } \frac{a+b}{c} = \frac{a}{c} + \frac{b}{c}$$

$$d = \frac{k}{F} + D \qquad \text{Divide out the common factor from } \frac{FD}{F}.$$

Either form of the answer is correct.

Work Problem 9 at the Side. ▶

9 Solve each formula for the specified variable.

GS **(a)** $r = \frac{A - p}{pt}$ for A

The goal is to isolate ____.

Multiply by ____ to obtain the equation ________.

Add ____ to obtain the equation

________ $= A$, or ________.

(b) $p = \frac{x - y}{z}$ for y

(c) $z = \frac{x}{x + y}$ for y

Teaching Tip

Remind students that solving for a specified variable uses the same steps in the same sequence as solving an equation in one variable. As in Section 2.5, encourage students to circle or underline the variable they are solving for.

Answers

9. **(a)** A; pt; $rpt = A - p$; p; $p + prt$; $A = p + prt$
(b) $y = x - pz$
(c) $y = \frac{x - zx}{z}$, or $y = \frac{x}{z} - x$

10 Solve $\frac{2}{x} = \frac{1}{y} + \frac{1}{z}$ for the specified variable.

(a) for z

(b) for y

Answers

10. (a) $z = \frac{xy}{2y - x}$ (b) $y = \frac{xz}{2z - x}$

EXAMPLE 10 Solving for a Specified Variable

Solve the formula $\frac{1}{a} = \frac{1}{b} + \frac{1}{c}$ for c.

$$\frac{1}{a} = \frac{1}{b} + \frac{1}{c}$$ Goal: Isolate c, the specified variable.

$$abc\left(\frac{1}{a}\right) = abc\left(\frac{1}{b} + \frac{1}{c}\right)$$ Multiply by the LCD, abc.

$$abc\left(\frac{1}{a}\right) = abc\left(\frac{1}{b}\right) + abc\left(\frac{1}{c}\right)$$ Distributive property

$$bc = ac + ab$$ Simplify.

$$bc - ac = ab$$ Subtract ac to get both terms with c on the same side.

Pay careful attention here.

$$c(b - a) = ab$$ Factor out c.

$$c = \frac{ab}{b - a}$$ Divide by $b - a$.

CAUTION

Students often have trouble in the step that involves factoring out the variable for which they are solving. In **Example 10,** we needed to get both terms with c on the same side of the equation. This allowed us to factor out c on the left, and then isolate it by dividing each side by $b - a$.

When solving an equation for a specified variable, be sure that the specified variable appears alone on only one side of the equality symbol in the final equation.

◀ Work Problem 10 at the Side.

7.6 Exercises

FOR EXTRA HELP

Download the MyDashBoard App

MyMathLab®

CONCEPT CHECK *Fill in each blank with the correct response.*

1. A value of the variable that appears to be a solution after both sides of an equation with rational expressions are multiplied by a variable expression is a(n) __proposed__ solution. It must be checked in the __original__ equation to determine whether it is an actual solution.

2. A proposed solution that is not an actual solution of an original equation is a(n) __extraneous__ solution, or __extraneous__ value.

Identify each as an expression or an equation. Then simplify the expression or solve the equation. ***See Example 1.***

3. $\frac{7}{8}x + \frac{1}{5}x$ expression; $\frac{43}{40}x$

4. $\frac{4}{7}x + \frac{3}{5}x$ expression; $\frac{41}{35}x$

5. $\frac{7}{8}x + \frac{1}{5}x = 1$ equation; $\left\{\frac{40}{43}\right\}$

6. $\frac{4}{7}x + \frac{3}{5}x = 1$ equation; $\left\{\frac{35}{41}\right\}$

7. $\frac{3}{5}y - \frac{7}{10}y$ expression; $-\frac{1}{10}y$

8. $\frac{3}{5}y - \frac{7}{10}y = 1$ equation; $\{-10\}$

Solve each equation, and check your solutions. ***See Examples 2 and 3.***

9. $\frac{2}{3}x + \frac{1}{2}x = -7$ $\{-6\}$

10. $\frac{1}{4}x - \frac{1}{3}x = 1$ $\{-12\}$

11. $\frac{3x}{5} - 6 = x$ $\{-15\}$

12. $\frac{5t}{4} + t = 9$ $\{4\}$

13. $\frac{4m}{7} + m = 11$ $\{7\}$

14. $a - \frac{3a}{2} = 1$ $\{-2\}$

15. $\frac{z-1}{4} = \frac{z+3}{3}$ $\{-15\}$

16. $\frac{r-5}{2} = \frac{r+2}{3}$ $\{19\}$

17. $\frac{3p+6}{8} = \frac{3p-3}{16}$ $\{-5\}$

18. $\frac{2z+1}{5} = \frac{7z+5}{15}$ $\{-2\}$

19. $\frac{2x+3}{-6} = \frac{3}{2}$ $\{-6\}$

20. $\frac{4x+3}{6} = \frac{5}{2}$ $\{3\}$

21. $\frac{q+2}{3}+\frac{q-5}{5}=\frac{7}{3}$ $\{5\}$

22. $\frac{b+7}{8}-\frac{b-2}{3}=\frac{4}{3}$ $\{1\}$

23. $\frac{t}{6}+\frac{4}{3}=\frac{t-2}{3}$ $\{12\}$

24. $\frac{x}{2}=\frac{5}{4}+\frac{x-1}{4}$ $\{4\}$

25. $\frac{3m}{5}-\frac{3m-2}{4}=\frac{1}{5}$ $\{2\}$

26. $\frac{8p}{5}=\frac{3p-4}{2}+\frac{5}{2}$ $\{5\}$

When solving an equation with variables in denominators, we must determine the values that cause these denominators to equal 0, so that we can reject these values if they appear as proposed solutions. Find all values for which at least one denominator is equal to 0. Write answers using the symbol $\neq$. Do not solve. ***See Examples 4–8.***

27. $\frac{3}{x+2}-\frac{5}{x}=1$
$x \neq -2, x \neq 0$

28. $\frac{7}{x}+\frac{9}{x-4}=5$
$x \neq 0, x \neq 4$

29. $\frac{-1}{(x+3)(x-4)}=\frac{1}{2x+1}$
$x \neq -3, x \neq 4, x \neq -\frac{1}{2}$

30. $\frac{8}{(x-7)(x+3)}=\frac{7}{3x-10}$
$x \neq 7, x \neq -3, x \neq \frac{10}{3}$

31. $\frac{4}{x^2+8x-9}+\frac{1}{x^2-4}=0$
$x \neq -9, x \neq 1, x \neq -2, x \neq 2$

32. $\frac{-3}{x^2+9x-10}-\frac{12}{x^2-49}=0$
$x \neq -10, x \neq 1, x \neq -7, x \neq 7$

Solve each equation, and check your solutions. ***See Examples 4–8.***

33. $\frac{5-2x}{x}=\frac{1}{4}$
$\left\{\frac{20}{9}\right\}$

34. $\frac{2x+3}{x}=\frac{3}{2}$
$\{-6\}$

35. $\frac{k}{k-4}-5=\frac{4}{k-4}$
$\emptyset$

36. $\frac{-5}{a+5}=\frac{a}{a+5}+2$
$\emptyset$

37. $\frac{3}{x-1}+\frac{2}{4x-4}=\frac{7}{4}$
The LCD is $\underline{4(x-1)}$. Complete the steps to solve the equation.
$\{3\}$

38. $\frac{2}{p+3}+\frac{3}{8}=\frac{5}{4p+12}$
The LCD is $\underline{8(p+3)}$. Complete the steps to solve the equation.
$\{-5\}$

39. $\frac{x}{3x+3}=\frac{2x-3}{x+1}-\frac{2x}{3x+3}$
$\{3\}$

40. $\frac{2k+3}{k+1}-\frac{3k}{2k+2}=\frac{-2k}{2k+2}$
$\{-2\}$

41. $\frac{2}{m}=\frac{m}{5m+12}$
$\{-2, 12\}$

42. $\frac{x}{4-x}=\frac{2}{x}$
$\{-4, 2\}$

43. $\frac{5x}{14x+3}=\frac{1}{x}$
$\left\{-\frac{1}{5}, 3\right\}$

44. $\frac{m}{8m+3}=\frac{1}{3m}$
$\left\{-\frac{1}{3}, 3\right\}$

45. $\frac{2}{z-1} - \frac{5}{4} = \frac{-1}{z+1}$

$\left\{-\frac{3}{5}, 3\right\}$

46. $\frac{5}{p-2} = 7 - \frac{10}{p+2}$

$\left\{-\frac{6}{7}, 3\right\}$

47. $\frac{4}{x^2 - 3x} = \frac{1}{x^2 - 9}$

$\{-4\}$

48. $\frac{2}{t^2 - 4} = \frac{3}{t^2 - 2t}$

$\{-6\}$

49. $\frac{-2}{z+5} + \frac{3}{z-5} = \frac{20}{z^2 - 25}$

$\emptyset$

50. $\frac{3}{r+3} - \frac{2}{r-3} = \frac{-12}{r^2 - 9}$

$\emptyset$

51. $\frac{1}{x+4} + \frac{x}{x-4} = \frac{-8}{x^2 - 16}$

$\{-1\}$

52. $\frac{x}{x-3} + \frac{4}{x+3} = \frac{18}{x^2 - 9}$

$\{-10\}$

53. $\frac{4}{3x+6} - \frac{3}{x+3} = \frac{8}{x^2 + 5x + 6}$

$\{-6\}$

54. $\frac{-13}{t^2 + 6t + 8} + \frac{4}{t+2} = \frac{3}{2t+8}$

$\{0\}$

55. $\frac{3x}{x^2 + 5x + 6} = \frac{5x}{x^2 + 2x - 3} - \frac{2}{x^2 + x - 2}$

$\left\{-6, \frac{1}{2}\right\}$

56. $\frac{x+4}{x^2 - 3x + 2} - \frac{5}{x^2 - 4x + 3} = \frac{x-4}{x^2 - 5x + 6}$

$\{6\}$

In Exercises 57 and 58, answer each question.

57. If you are solving a formula for the letter k, and your steps lead to the equation $kr - mr = km$, what would be your next step?

Transform the equation so that the terms with k are on one side and the remaining term is on the other.

58. If you are solving a formula for the letter k, and your steps lead to the equation $kr - km = mr$, what would be your next step?

Factor out k on the left side.

Solve each formula for the specified variable. ***See Example 9.***

59. $m = \frac{kF}{a}$ for F

$F = \frac{ma}{k}$

60. $I = \frac{kE}{R}$ for E

$E = \frac{IR}{k}$

61. $m = \frac{kF}{a}$ for a

$a = \frac{kF}{m}$

62. $I = \frac{kE}{R}$ for R

$R = \frac{kE}{I}$

63. $m = \frac{y - b}{x}$ for y

$y = mx + b$

64. $y = \frac{C - Ax}{B}$ for C

$C = Ax + By$

65. $I = \frac{E}{R + r}$ for R

$R = \frac{E - Ir}{I}$, or $R = \frac{E}{I} - r$

66. $I = \frac{E}{R + r}$ for r

$r = \frac{E - IR}{I}$, or $r = \frac{E}{I} - R$

67. $h = \frac{2A}{B + b}$ for b

$b = \frac{2A - hB}{h}$, or $b = \frac{2A}{h} - B$

68. $h = \frac{2A}{B + b}$ for B

$B = \frac{2A - hb}{h}$, or $B = \frac{2A}{h} - b$

69. $d = \frac{2S}{n(a + L)}$ for a

$a = \frac{2S - dnL}{dn}$, or $a = \frac{2S}{dn} - L$

70. $d = \frac{2S}{n(a + L)}$ for L

$L = \frac{2S - dna}{dn}$, or $L = \frac{2S}{dn} - a$

Solve each equation for the specified variable. ***See Example 10.***

71. $\frac{2}{r} + \frac{3}{s} + \frac{1}{t} = 1$ for t

$t = \frac{rs}{rs - 2s - 3r}$, or $t = \frac{-rs}{-rs + 2s + 3r}$

72. $\frac{5}{p} + \frac{2}{q} + \frac{3}{r} = 1$ for r

$r = \frac{3pq}{pq - 5q - 2p}$, or $r = \frac{-3pq}{-pq + 5q + 2p}$

73. $\frac{1}{a} - \frac{1}{b} - \frac{1}{c} = 2$ for c

$c = \frac{ab}{b - a - 2ab}$, or $c = \frac{-ab}{-b + a + 2ab}$

74. $\frac{-1}{x} + \frac{1}{y} + \frac{1}{z} = 4$ for y

$y = \frac{xz}{z - x + 4xz}$, or $y = \frac{-xz}{-z + x - 4xz}$

75. $9x + \frac{3}{z} = \frac{5}{y}$ for z

$z = \frac{3y}{5 - 9xy}$, or $z = \frac{-3y}{9xy - 5}$

76. $-3t - \frac{4}{p} = \frac{6}{s}$ for p

$p = \frac{4s}{-3st - 6}$, or $p = \frac{-4s}{3st + 6}$

Summary Exercises *Simplifying Rational Expressions vs. Solving Rational Equations*

Summary Exercises provide mixed practice to better prepare students for tests.

Students often confuse *simplifying expressions* with *solving equations.* We review the four operations to simplify the rational expressions $\frac{1}{x}$ and $\frac{1}{x-2}$ as follows.

Add:

$$\frac{1}{x} + \frac{1}{x-2}$$

$$= \frac{1(x-2)}{x(x-2)} + \frac{x(1)}{x(x-2)}$$ Write with a common denominator.

$$= \frac{x-2+x}{x(x-2)}$$ Add numerators. Keep the same denominator.

$$= \frac{2x-2}{x(x-2)}$$ Combine like terms.

Subtract:

$$\frac{1}{x} - \frac{1}{x-2}$$

$$= \frac{1(x-2)}{x(x-2)} - \frac{x(1)}{x(x-2)}$$ Write with a common denominator.

$$= \frac{x-2-x}{x(x-2)}$$ Subtract numerators. Keep the same denominator.

$$= \frac{-2}{x(x-2)}$$ Combine like terms.

Multiply:

$$\frac{1}{x} \cdot \frac{1}{x-2}$$

$$= \frac{1}{x(x-2)}$$ Multiply numerators and multiply denominators.

Divide:

$$\frac{1}{x} \div \frac{1}{x-2}$$

$$= \frac{1}{x} \cdot \frac{x-2}{1}$$ Multiply by the reciprocal of the divisor.

$$= \frac{x-2}{x}$$ Multiply numerators and multiply denominators.

By contrast, consider the following *equation.*

$$\frac{1}{x} + \frac{1}{x-2} = \frac{3}{4}$$ $x \neq 0$ and $x \neq 2$ since a denominator is 0 for these values.

$$4x(x-2)\left(\frac{1}{x} + \frac{1}{x-2}\right) = 4x(x-2)\frac{3}{4}$$ Multiply each side by the LCD, $4x(x-2)$, to clear fractions.

$$4x(x-2)\frac{1}{x} + 4x(x-2)\frac{1}{x-2} = 4x(x-2)\frac{3}{4}$$ Distributive property

$$4(x-2) + 4x = 3x(x-2)$$ Divide out the common factors.

$$4x - 8 + 4x = 3x^2 - 6x$$ Distributive property

$$3x^2 - 14x + 8 = 0$$ Standard form

$$(3x-2)(x-4) = 0$$ Factor.

$$3x - 2 = 0 \quad \text{or} \quad x - 4 = 0$$ Zero-factor property

Proposed solutions $\rightarrow$ $x = \frac{2}{3}$ or $x = 4$ Solve for x.

The proposed solutions are $\frac{2}{3}$ and 4. Neither makes a denominator equal 0. Check by substituting each proposed solution into the original equation to confirm that the solution set is $\{\frac{2}{3}, 4\}$.

Points to Remember when Working with Rational Expressions and Equations

1. When simplifying rational expressions, the fundamental property is applied only after numerators and denominators have been *factored.*
2. When adding and subtracting rational expressions, the common denominator must be kept throughout the problem and in the final result.
3. When simplifying rational expressions, always check to see if the answer is in lowest terms. If it is not, use the fundamental property.
4. When solving equations with rational expressions, the common denominator is used to clear the equation of fractions. Multiply each side by the common denominator. (Notice how this use differs from that of the common denominator in Point 2.)
5. When solving equations with rational expressions, reject any proposed solution that causes an original denominator to equal 0.

For each exercise, indicate "expression" *if an expression is to be simplified or* "equation" *if an equation is to be solved. Then simplify the expression or solve the equation.*

1. $\frac{4}{p}+\frac{6}{p}$

expression; $\frac{10}{p}$

2. $\frac{x^3y^2}{x^2y^4}\cdot\frac{y^5}{x^4}$

expression; $\frac{y^3}{x^3}$

3. $\frac{1}{x^2+x-2}\div\frac{4x^2}{2x-2}$

expression; $\frac{1}{2x^2(x+2)}$

4. $\frac{8}{m-5}=2$

equation; $\{9\}$

5. $\frac{2x^2+x-6}{2x^2-9x+9}\cdot\frac{x^2-2x-3}{x^2-1}$

expression; $\frac{x+2}{x-1}$

6. $\frac{2}{k^2-4k}+\frac{3}{k^2-16}$

expression; $\frac{5k+8}{k(k-4)(k+4)}$

7. $\frac{x-4}{5}=\frac{x+3}{6}$

equation; $\{39\}$

8. $\frac{3t^2-t}{6t^2+15t}\div\frac{6t^2+t-1}{2t^2-5t-25}$

expression; $\frac{t-5}{3(2t+1)}$

9. $\frac{4}{p+2}+\frac{1}{3p+6}$

expression; $\frac{13}{3(p+2)}$

10. $\frac{1}{x}+\frac{1}{x-3}=-\frac{5}{4}$

equation; $\left\{-1, \frac{12}{5}\right\}$

11. $\frac{3}{t-1}+\frac{1}{t}=\frac{7}{2}$

equation; $\left\{\frac{1}{7}, 2\right\}$

12. $\frac{6}{x}-\frac{2}{3x}$

expression; $\frac{16}{3x}$

13. $\frac{5}{4z}-\frac{2}{3z}$

expression; $\frac{7}{12z}$

14. $\frac{k+2}{3}=\frac{2k-1}{5}$

equation; $\{13\}$

15. $\frac{1}{m^2+5m+6}+\frac{2}{m^2+4m+3}$

expression; $\frac{3m+5}{(m+2)(m+3)(m+1)}$

16. $\frac{2k^2-3k}{20k^2-5k}\div\frac{2k^2-5k+3}{4k^2+11k-3}$

expression; $\frac{k+3}{5(k-1)}$

17. $\frac{2}{x+1}+\frac{5}{x-1}=\frac{10}{x^2-1}$

equation; $\emptyset$

18. $\frac{x}{x-2}+\frac{3}{x+2}=\frac{8}{x^2-4}$

equation; $\{-7\}$

7.7 Applications of Rational Expressions

For applications that lead to rational equations, the six-step problem-solving method of **Section 2.4** still applies.

OBJECTIVES

1. Solve problems about numbers.
2. Solve problems about distance, rate, and time.
3. Solve problems about work.

OBJECTIVE 1 Solve problems about numbers.

EXAMPLE 1 Solving a Problem about an Unknown Number

If the same number is added to both the numerator and the denominator of the fraction $\frac{2}{5}$, the result is equivalent to $\frac{2}{3}$. Find the number.

Step 1 **Read** the problem carefully. We are trying to find a number.

Step 2 **Assign a variable.**

Let $x =$ the number added to the numerator and the denominator.

Step 3 **Write an equation.** The fraction

$$\frac{2+x}{5+x}$$

represents the result of adding the same number x to both the numerator and the denominator. This result is equivalent to $\frac{2}{3}$, so we write the following equation.

$$\frac{2+x}{5+x}=\frac{2}{3}$$

Step 4 **Solve.**

$$3(5+x)\frac{2+x}{5+x}=3(5+x)\frac{2}{3} \quad \text{Multiply by the LCD, } 3(5+x).$$

$$3(2+x)=2(5+x) \quad \text{Divide out the common factors.}$$

$$6+3x=10+2x \quad \text{Distributive property}$$

$$x=4 \quad \text{Subtract } 2x. \text{ Subtract 6.}$$

Step 5 **State the answer.** The number is 4.

Step 6 **Check** the solution in the words of the original problem. If 4 is added to both the numerator and the denominator of $\frac{2}{5}$, the result is $\frac{2+4}{5+4}=\frac{6}{9}$, or $\frac{2}{3}$, as required.

Work Problem 1 at the Side. ▶

1 Solve each problem.

GS **(a)** A certain number is added to the numerator and subtracted from the denominator of $\frac{5}{8}$. The new fraction equals the reciprocal of $\frac{5}{8}$. Find the number.

Step 1
We are trying to find a(n) ________.

Step 2
Let $x =$ the number ________ to the numerator and ________ from the denominator.

Step 3
The expression ________ represents the new fraction.

The reciprocal of $\frac{5}{8}$ is ____. Write an equation.

____

Complete Steps 4–6 to solve the problem. Give the answer.

(b) The denominator of a fraction is 1 more than the numerator. If 6 is added to the numerator and subtracted from the denominator, the result is $\frac{15}{4}$. Find the original fraction.

OBJECTIVE 2 Solve problems about distance, rate, and time.

If an automobile travels at an average rate of 65 mph for 2 hr, then it travels

$$65 \times 2 = 130 \text{ mi.} \quad rt=d, \text{ or } d=rt \text{ (Relationship between distance, rate, and time from Section 4.4)}$$

By solving, in turn, for r and t in the formula $d = rt$, we obtain two other equivalent forms of the formula. The three forms are given below.

Distance, Rate, and Time Relationship

$$d=rt \qquad r=\frac{d}{t} \qquad t=\frac{d}{r}$$

Answers

1. (a) number; added; subtracted; $\frac{5+x}{8-x}$; $\frac{8}{5}$; $\frac{5+x}{8-x}=\frac{8}{5}$; The number is 3.
(b) $\frac{9}{10}$

2 Solve each problem.

(a) A new world record in the men's 100-m dash was set by Usain Bolt of Jamaica in 2009. He ran it in 9.58 sec. What was his rate in meters per second, to the nearest hundredth? (*Source: World Almanac and Book of Facts.*)

(b) A new world record for the women's 3000-m steeplechase was set by Gulnara Samitova of Russia in 2008. Her rate was 5.568 m per sec. To the nearest second, what was her time? (*Source: World Almanac and Book of Facts.*)

(c) A small plane flew from Chicago to St. Louis averaging 145 mph. The trip took 2 hr. What is the distance between Chicago and St. Louis?

Answers

2. (a) 10.44 m per sec
(b) 539 sec, or 8 min, 59 sec
(c) 290 mi

EXAMPLE 2 **Finding Distance, Rate, or Time**

Solve each problem using a form of the distance formula.

(a) The speed (rate) of sound is 1088 ft per sec at sea level at 32°F. Find the distance sound travels in 5 sec under these conditions.
We must find distance, given rate and time, using $d = rt$ (or $rt = d$).

$$\underset{\text{Rate}}{\underset{\uparrow}{1088}} \times \underset{\text{Time}}{\underset{\uparrow}{5}} = \underset{\text{Distance}}{\underset{\uparrow}{5440 \text{ ft}}}$$

(b) The winner of the first Indianapolis 500 race (in 1911) was Ray Harroun, driving a Marmon Wasp at an average rate of 74.60 mph. (*Source: World Almanac and Book of Facts.*) How long did it take him to complete the 500 mi?
We must find time, given rate and distance, using $t = \frac{d}{r}$ (or $\frac{d}{r} = t$).

$$\frac{500 \leftarrow \text{Distance}}{74.60 \leftarrow \text{Rate}} = 6.70 \text{ hr (rounded)} \leftarrow \text{Time}$$

To convert 0.70 hr to minutes, we multiply by 60 to get $0.70(60) = 42$. It took Harroun about 6 hr, 42 min, to complete the race.

(c) At the 2008 Olympic Games, Australian swimmer Leisel Jones set an Olympic record of 65.17 sec in the women's 100-m breaststroke swimming event. (*Source: World Almanac and Book of Facts.*) Find her rate.
We must find rate, given distance and time, using $r = \frac{d}{t}$ (or $\frac{d}{t} = r$).

$$\frac{100 \leftarrow \text{Distance}}{65.17 \leftarrow \text{Time}} = 1.53 \text{ m per sec (rounded)} \leftarrow \text{Rate}$$

◀ **Work Problem 2 at the Side.**

Problem-Solving Hint

Many applied problems use the formulas just discussed. The next two examples show how to solve typical applications of the formula $d = rt$.

A helpful strategy for solving such problems is to ***first make a sketch*** showing what is happening in the problem. ***Then make a table*** using the information given, along with the unknown quantities. The table will help organize the information, and the sketch will help set up the equation.

EXAMPLE 3 **Solving a Problem about Distance, Rate, and Time**

Two cars leave Iowa City, Iowa, at the same time and travel east on Interstate 80. One travels at a constant rate of 55 mph. The other travels at a constant rate of 63 mph. In how many hours will the distance between them be 24 mi?

Step 1 **Read** the problem. We must find the time it will take for the distance between the cars to be 24 mi.

Continued on Next Page

Step 2 **Assign a variable.** We are looking for time.

Let t = the number of hours until the distance between the cars is 24 mi.

The sketch in **Figure 1** shows what is happening in the problem.

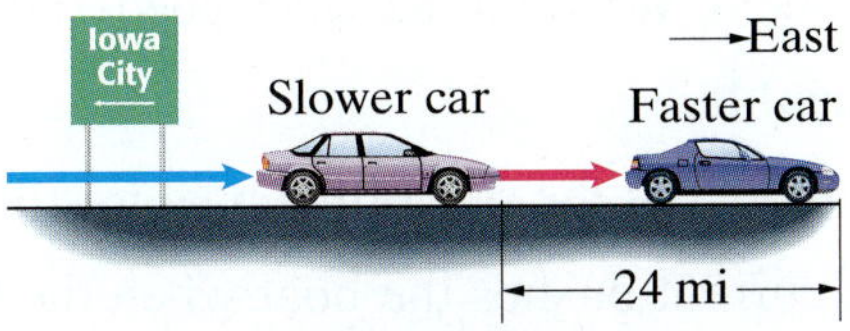

Figure 1

To construct a table, we fill in the information given in the problem, using t for the time traveled by each car. We multiply rate by time to get the expressions for distances traveled.

	Rate	Time	Distance
Faster Car	63	t	$63t$
Slower Car	55	t	$55t$

The quantities $63t$ and $55t$ represent the two distances. The *difference* between the larger distance and the smaller distance is 24 mi.

Step 3 **Write an equation.**

$$63t - 55t = 24$$

Step 4 **Solve.**

$$8t = 24 \quad \text{Combine like terms.}$$

$$t = 3 \quad \text{Divide by 8.}$$

Step 5 **State the answer.** It will take the cars 3 hr to be 24 mi apart.

Step 6 **Check.** After 3 hr, the faster car will have traveled

$$63 \times 3 = 189 \text{ mi}$$

and the slower car will have traveled

$$55 \times 3 = 165 \text{ mi.}$$

The difference is

$$189 - 165 = 24, \quad \text{as required.}$$

Problem-Solving Hint

In problems involving distance, rate, and time like the one in **Example 3,** once we have filled in two pieces of information in each row of the table, we can automatically fill in the third piece of information, using the appropriate form of the distance formula. Then we set up the equation based on our sketch and the information in the table.

Work Problem 3 at the Side. ▶

3 Solve each problem.

GS **(a)** From a point on a straight road, Lupe and Maria ride bicycles in opposite directions. Lupe rides 10 mph and Maria rides 12 mph. In how many hours will they be 55 mi apart?

Steps 1 and 2
Let x = the number of ________ until the distance between Lupe and Maria is ________.

Step 3
Complete the table.

	Rate	× Time	= Distance
Maria	10	t	_____
Lupe	_____	t	_____

Because Lupe and Maria are traveling in opposite directions we must ________ the distances they travel to find the distance between them. Write an equation.

____

Complete Steps 4–6 to solve the problem. Give the answer.

(b) At a given hour, two steamboats leave a city in the same direction on a straight canal. One travels at 18 mph, and the other travels at 25 mph. In how many hours will the boats be 35 mi apart?

Answers

3. **(a)** hours; 55 mi

	Rate	× Time	= Distance
Maria	10	t	$10t$
Lupe	12	t	$12t$

add; $10t + 12t = 55$; $2\frac{1}{2}$ hr

(b) 5 hr

EXAMPLE 4 Solving a Problem about Distance, Rate, and Time

The Tickfaw River has a current of 3 mph. A motorboat takes as long to go 12 mi downstream as to go 8 mi upstream. What is the rate of the boat in still water?

Step 1 **Read** the problem. We want the rate (speed) of the boat in still water.

Step 2 **Assign a variable.**

Let $x =$ the rate of the boat in still water.

Because the current pushes the boat when the boat is going downstream, the rate of the boat downstream will be the *sum* of the rate of the boat and the rate of the current, $(x + 3)$ mph.

Because the current slows down the boat when the boat is going upstream, the boat's rate upstream is given by the *difference* between the rate of the boat in still water and the rate of the current, $(x - 3)$ mph. See **Figure 2.**

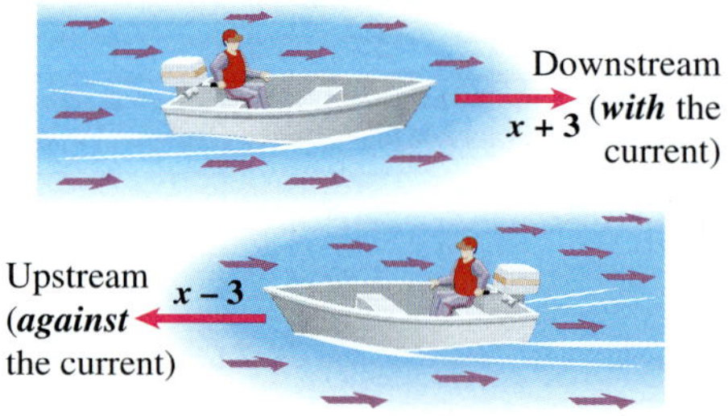

Figure 2

This information is summarized in the following table.

	d	*r*	*t*
Downstream	12	$x + 3$	
Upstream	8	$x - 3$	

Fill in the times by using the formula $t = \frac{d}{r}$.

The time downstream is the distance divided by the rate.

$$t = \frac{d}{r} = \frac{12}{x + 3} \quad \text{Time downstream}$$

The time upstream is also the distance divided by the rate.

$$t = \frac{d}{r} = \frac{8}{x - 3} \quad \text{Time upstream}$$

	d	*r*	*t*
Downstream	12	$x + 3$	$\frac{12}{x + 3}$
Upstream	8	$x - 3$	$\frac{8}{x - 3}$

Times are equal.

Step 3 **Write an equation.**

$$\frac{12}{x + 3} = \frac{8}{x - 3}$$

The time downstream equals the time upstream, so the two times from the table must be equal.

Continued on Next Page

Step 4 **Solve.** $\frac{12}{x+3} = \frac{8}{x-3}$ Equation from Step 3

$$(x+3)(x-3)\frac{12}{x+3} = (x+3)(x-3)\frac{8}{x-3}$$ Multiply by the LCD, $(x+3)(x-3)$.

$$12(x-3) = 8(x+3)$$ Divide out the common factors.

$$12x - 36 = 8x + 24$$ Distributive property

$$4x = 60$$ Subtract $8x$. Add 36.

$$x = 15$$ Divide by 4.

Step 5 **State the answer.** The rate of the boat in still water is 15 mph.

Step 6 **Check.** First find the rate of the boat downstream, which is $15 + 3 = 18$ mph. Divide 12 mi by 18 mph to find the time.

$$t = \frac{d}{r} = \frac{12}{18} = \frac{2}{3} \text{ hr}$$

The rate of the boat upstream is $15 - 3 = 12$ mph. Divide 8 mi by 12 mph to find the time.

$$t = \frac{d}{r} = \frac{8}{12} = \frac{2}{3} \text{ hr}$$

The time upstream equals the time downstream, as required.

Work Problem 4 at the Side. ▶

OBJECTIVE 3 Solve problems about work. Suppose that you can mow your lawn in 4 hr. Then after 1 hr, you will have mowed $\frac{1}{4}$ of the lawn. After 2 hr, you will have mowed $\frac{2}{4}$, or $\frac{1}{2}$, of the lawn, and so on. This idea is generalized as follows.

Rate of Work

If a job can be completed in t units of time, then the rate of work is

$$\frac{1}{t} \text{ job per unit of time.}$$

Problem-Solving Hint

Recall that the formula $d = rt$ says that distance traveled is equal to rate of travel multiplied by time traveled. Similarly, the fractional part of a job accomplished is equal to the rate of work multiplied by the time worked.

In the lawn mowing example, after 3 hr, the fractional part of the job done is found as follows.

$$\underbrace{\frac{1}{4}}_{\text{Rate of work}} \cdot \underbrace{3}_{\text{Time worked}} = \underbrace{\frac{3}{4}}_{\text{Fractional part of job done}}$$

After 4 hr, $\frac{1}{4}(4) = 1$ whole job has been done.

4 Solve each problem.

GS **(a)** A boat can go 10 mi against the current in the same time it can go 30 mi with the current. The current is flowing at 4 mph. Find the rate of the boat with no current.

Steps 1 and 2
Let x = the ________ of the boat with no current.

Step 3
Complete the table.

	d	r	t
Against the Current	10	____	____
With the Current	30	____	____

How are the times going with the current and against the current related?

____

Write an equation.

____

Complete Steps 4–6 to solve the problem. Give the answer.

(b) An airplane, maintaining a constant airspeed, takes as long to go 450 mi with the wind as it does to go 375 mi against the wind. If the wind is blowing at 15 mph, what is the rate of the plane?

Answers

4. (a) rate

	d	r	t
Against the Current	10	$x-4$	$\frac{10}{x-4}$
With the Current	30	$x+4$	$\frac{30}{x+4}$

They are the same; $\frac{10}{x-4} = \frac{30}{x+4}$; 8 mph

(b) 165 mph

Teaching Tip

Be sure that students know that the variable is in the numerators for an example like this. Remind them that the following is true.

$$\frac{1}{8}x = \frac{1}{8} \cdot \frac{x}{1} = \frac{x}{8}$$

EXAMPLE 5 Solving a Problem about Work Rates

With spraying equipment, Mateo can paint the woodwork in a small house in 8 hr. His assistant, Chet, needs 14 hr to complete the same job painting by hand. If both Mateo and Chet work together, how long will it take them to paint the woodwork?

Step 1 **Read** the problem again. We are looking for time working together.

Step 2 **Assign a variable.**

Let $x =$ the number of hours it will take for Mateo and Chet to paint the woodwork, working together.

Begin by making a table. Based on the previous discussion, Mateo's rate alone is $\frac{1}{8}$ job per hour, and Chet's rate is $\frac{1}{14}$ job per hour.

	Rate	Time Working Together	Fractional Part of the Job Done When Working Together
Mateo	$\frac{1}{8}$	x	$\frac{1}{8}x$
Chet	$\frac{1}{14}$	x	$\frac{1}{14}x$

Sum is 1 whole job.

Step 3 **Write an equation.**

$$\underbrace{\text{Fractional part done by Mateo}} + \underbrace{\text{Fractional part done by Chet}} = \underbrace{1 \text{ whole job}}$$

$$\frac{1}{8}x + \frac{1}{14}x = 1$$

Together, Mateo and Chet complete 1 whole job. Add the fractional parts and set the sum equal to 1.

Step 4 **Solve.**

$$56\left(\frac{1}{8}x + \frac{1}{14}x\right) = 56(1) \quad \text{Multiply by the LCD, 56.}$$

$$56\left(\frac{1}{8}x\right) + 56\left(\frac{1}{14}x\right) = 56(1) \quad \text{Distributive property}$$

$$7x + 4x = 56 \quad \text{Multiply.}$$

$$11x = 56 \quad \text{Combine like terms.}$$

$$x = \frac{56}{11} \quad \text{Divide by 11.}$$

Step 5 **State the answer.** Working together, Mateo and Chet can paint the woodwork in $\frac{56}{11}$ hr, or $5\frac{1}{11}$ hr.

Step 6 **Check.** Substitute $\frac{56}{11}$ for x in the equation from Step 3.

$$\frac{1}{8}x + \frac{1}{14}x = 1 \quad \text{Equation from Step 3}$$

$$\frac{1}{8}\left(\frac{56}{11}\right) + \frac{1}{14}\left(\frac{56}{11}\right) \stackrel{?}{=} 1 \quad \text{Let } x = \tfrac{56}{11}.$$

$$\frac{7}{11} + \frac{4}{11} = 1 \quad ✓ \text{ True}$$

Our answer, $\frac{56}{11}$ hr, or $5\frac{1}{11}$ hr, is correct.

Problem-Solving Hint

A common error students make when solving a work problem like that in **Example 5** is to add the two times.

$$8 \text{ hr} + 14 \text{ hr} = \mathbf{22 \text{ hr}} \leftarrow \text{Incorrect answer}$$

The answer 22 hr is unreasonable, because the slower worker (Chet) can do the job *alone* in 14 hr. The correct answer *must* be less than 14 hr.

Keep in mind that the correct time for the answer must also be less than that of the *faster* worker (Mateo, at 8 hr) because he is getting help. Based on this reasoning, does our answer of $5\frac{1}{11}$ hr in **Example 5** make sense?

Work Problem 5 at the Side. ▶

An alternative approach is to consider the part of the job that can be done in 1 hr. For instance, in **Example 5** Mateo can do the entire job in 8 hr, and Chet can do it in 14 hr. Thus, their work rates, as we saw in **Example 5,** are $\frac{1}{8}$ and $\frac{1}{14}$, respectively. Since it takes them x hours to complete the job when working together, in 1 hr they can paint $\frac{1}{x}$ of the woodwork.

The amount painted by Mateo in 1 hr plus the amount painted by Chet in 1 hr must equal the amount they can do together. This leads to the following alternative equation.

Amount by Chet ↓

$$\text{Amount by Mateo} \rightarrow \frac{1}{8} + \frac{1}{14} = \frac{1}{x} \leftarrow \text{Amount together}$$

Compare this with the equation in Step 3 of **Example 5.** Multiplying each side by $56x$ leads to the same equation found in the third line of Step 4 in the example.

$$7x + 4x = 56$$

The same solution results.

5 Solve each problem.

GS **(a)** Michael can paint a room, working alone, in 8 hr. Lindsay can paint the same room, working alone, in 6 hr. How long will it take them if they work together?

Steps 1 and 2

Let $x =$ the number of ________ it will take for Michael and Lindsay to paint the room, working ________.

Step 3

Complete the table.

	Rate	Time Working Together	Fractional Part of the Job Done When Working Together
Michael	____	x	____
Lindsay	____	x	____

Together, Michael and Lindsay complete ____ whole job(s). Write an equation.

____

Complete Steps 4–6 to solve the problem. Give the answer.

(b) Roberto can detail his Camaro in 2 hr working alone. His brother Marco can do the job in 3 hr working alone. How long would it take them if they worked together?

Answers

5. (a) hours; together

	Rate	Time Working Together	Fractional Part of the Job Done When Working Together
Michael	$\frac{1}{8}$	x	$\frac{1}{8}x$
Lindsay	$\frac{1}{6}$	x	$\frac{1}{6}x$

1; $\frac{1}{8}x + \frac{1}{6}x = 1$; $\frac{24}{7}$ hr, or $3\frac{3}{7}$ hr

(b) $\frac{6}{5}$ hr, or $1\frac{1}{5}$ hr

7.7 Exercises

FOR EXTRA HELP

Download the MyDashBoard App

MyMathLab®

1. **CONCEPT CHECK** If a migrating hawk travels m mph in still air, what is its rate when it flies into a steady headwind of 5 mph? What is its rate with a tailwind of 5 mph?

   into a headwind: $(m - 5)$ mph;
   with a tailwind: $(m + 5)$ mph

2. **CONCEPT CHECK** Suppose Stephanie walks D miles at R mph in the same time that Wally walks d miles at r mph. Give an equation relating D, R, d, and r.

   $\frac{D}{R} = \frac{d}{r}$

CONCEPT CHECK *Answer each question.*

3. If it takes Elayn 10 hr to do a job, what is her rate?

   $\frac{1}{10}$ job per hr

4. If it takes Clay 12 hr to do a job, how much of the job does he do in 8 hr?

   $\frac{2}{3}$ of the job

GS *Use Steps 2 and 3 of the six-step problem solving method to set up an equation to use to solve each problem. (Remember that Step 1 is to read the problem carefully.) Do not actually solve the equation.* ***See Example 1.***

5. The numerator of the fraction $\frac{5}{6}$ is increased by an amount so that the value of the resulting fraction is equivalent to $\frac{13}{3}$. By what amount was the numerator increased?

   (a) Let x = **the amount**. (*Step 2*)

   (b) Write an expression for "the numerator of the fraction $\frac{5}{6}$ is increased by an amount." **$5 + x$**

   (c) Set up an equation to solve the problem. (*Step 3*) $\frac{5+x}{6} = \frac{13}{3}$

6. If the same number is added to the numerator and subtracted from the denominator of $\frac{23}{12}$, the resulting fraction is equivalent to $\frac{3}{2}$. What is the number?

   (a) Let x = **the number**. (*Step 2*)

   (b) Write an expression for "a number is added to the numerator of $\frac{23}{12}$." **$23 + x$** Then write an expression for "the same number is subtracted from the denominator of $\frac{23}{12}$." **$12 - x$**

   (c) Set up an equation to solve the problem. (*Step 3*) $\frac{23+x}{12-x} = \frac{3}{2}$

In each problem, state what x represents, write an equation, and answer the question. ***See Example 1.***

7. In a certain fraction, the denominator is 4 less than the numerator. If 3 is added to both the numerator and the denominator, the resulting fraction is equivalent to $\frac{3}{2}$. What was the original fraction?

   x represents the original numerator; $\frac{x+3}{(x-4)+3} = \frac{3}{2}$; $\frac{9}{5}$

8. In a certain fraction, the denominator is 6 more than the numerator. If 3 is added to both the numerator and the denominator, the resulting fraction is equivalent to $\frac{5}{7}$. What was the original fraction (*not* written in lowest terms)?

   x represents the original numerator; $\frac{x+3}{(x+6)+3} = \frac{5}{7}$; $\frac{12}{18}$

9. The denominator of a certain fraction is three times the numerator. If 2 is added to the numerator and subtracted from the denominator, the resulting fraction is equivalent to 1. What was the original fraction (*not* written in lowest terms)?

   x represents the original numerator; $\frac{x+2}{3x-2} = 1$; $\frac{2}{6}$

10. The numerator of a certain fraction is four times the denominator. If 6 is added to both the numerator and the denominator, the resulting fraction is equivalent to 2. What was the original fraction (*not* written in lowest terms)?

    x represents the original denominator; $\frac{4x+6}{x+6} = 2$; $\frac{12}{3}$

11. One-sixth of a number is 5 more than the same number. What is the number?

    x represents the number; $\frac{1}{6}x = x + 5$; -6

12. One-third of a number is 2 more than one-sixth of the same number. What is the number?

    x represents the number; $\frac{1}{3}x = \frac{1}{6}x + 2$; 12

13. A quantity, its $\frac{3}{4}$, its $\frac{1}{2}$, and its $\frac{1}{3}$, added together, become 93. What is the quantity? (*Source: Rhind Mathematical Papyrus.*)

x represents the quantity; $x + \frac{3}{4}x + \frac{1}{2}x + \frac{1}{3}x = 93$; 36

14. A quantity, its $\frac{2}{3}$, its $\frac{1}{2}$, and its $\frac{1}{7}$, added together, become 33. What is the quantity? (*Source: Rhind Mathematical Papyrus.*)

x represents the quantity; $x + \frac{2}{3}x + \frac{1}{2}x + \frac{1}{7}x = 33$; $\frac{1386}{97}$

Solve each problem. ***See Example 2.***

15. In 2007, British explorer and endurance swimmer Lewis Gordon Pugh became the first person to swim at the North Pole. He swam 0.6 mi at 0.0319 mi per min in waters created by melted sea ice. What was his time (to three decimal places)? (*Source: The Gazette.*)

18.809 min

16. In the 2008 Summer Olympics, Britta Steffen of Germany won the women's 100-m freestyle swimming event. Her rate was 1.8825 m per sec. What was her time (to two decimal places)? (*Source: World Almanac and Book of Facts.*)

53.12 sec

17. The winner of the 2012 Daytona 500 (mile) race was Matt Kenseth, who drove his Ford to victory with a rate of 140.256 mph. What was his time (to the nearest thousandth of an hour)? (*Source: World Almanac and Book of Facts.*)

3.565 hr

18. In 2011, the late Dan Wheldon drove his Dellara-Honda Racing car to victory in the Indianapolis 500 (mile) race. His rate was 170.265 mph. What was his time (to the nearest thousandth of an hour)? (*Source: World Almanac and Book of Facts.*)

2.937 hr

19. Tirunesh Dibaba of Ethiopia won the women's 5000-m race in the 2008 Olympics with a time of 15.911 min. What was her rate (to three decimal places)? (*Source: World Almanac and Book of Facts.*)

314.248 m per min

20. The winner of the women's 1500-m run in the 2008 Olympics was Nancy Jebet Langat of Kenya with a time of 4.004 min. What was her rate (to three decimal places)? (*Source: World Almanac and Book of Facts.*)

374.625 m per min

Complete the table and write an equation to use to solve each problem. Do not actually solve the equation. ***See Examples 3 and 4.***

21. Luvenia can row 4 mph in still water. She takes as long to row 8 mi upstream as 24 mi downstream. How fast is the current?

Let x = the rate of the current.

	d	r	t
Upstream	8	$4 - x$	$\frac{8}{4-x}$
Downstream	24	$4 + x$	$\frac{24}{4+x}$

$$\frac{8}{4-x} = \frac{24}{4+x}$$

22. Julio flew his airplane 500 mi against the wind in the same time it took him to fly it 600 mi with the wind. If the rate of the wind was 10 mph, what was the average rate of his plane in still air?

Let x = the rate of the plane in still air.

	d	r	t
Against the Wind	500	$x - 10$	$\frac{500}{x-10}$
With the Wind	600	$x + 10$	$\frac{600}{x+10}$

$$\frac{500}{x-10} = \frac{600}{x+10}$$

Solve each problem. ***See Examples 3 and 4.***

23. A boat can go 20 mi against a current in the same time that it can go 60 mi with the current. The rate of the current is 4 mph. Find the rate of the boat in still water.
8 mph

24. A plane flies 350 mi with the wind in the same time that it can fly 310 mi against the wind. The plane has a still-air rate of 165 mph. Find the wind speed.
10 mph

25. The sanderling is a small shorebird about 6.5 in. long, with a thin, dark bill and a wide, white wing stripe. If a sanderling can fly 30 mi with the wind in the same time it can fly 18 mi against the wind when the wind speed is 8 mph, what is the rate of the bird in still air? (*Source:* U.S. Geological Survey.)

32 mph

26. Airplanes usually fly faster from west to east than from east to west because the prevailing winds go from west to east. The air distance between Chicago and London is about 4000 mi, while the air distance between New York and London is about 3500 mi. If a jet can fly eastbound from Chicago to London in the same time it can fly westbound from London to New York in a 35-mph wind, what is the rate of the plane in still air? (*Source:* www.geobytes.com)

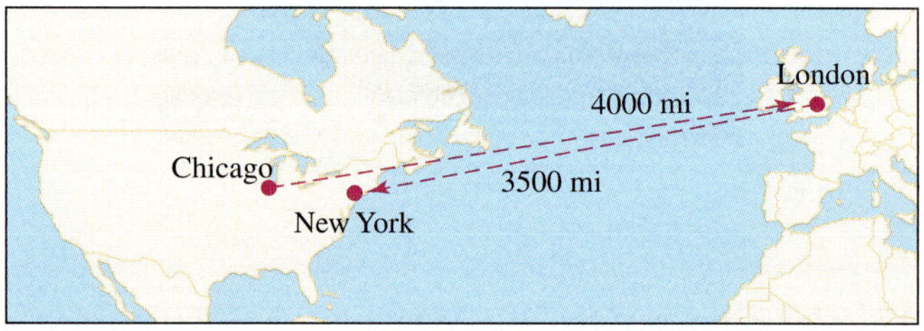

525 mph

27. Perian Herring's boat goes 12 mph. Find the rate of the current of the river if she can go 6 mi upstream in the same amount of time she can go 10 mi downstream.
3 mph

28. Bridget Jarvis can travel 8 mi upstream in the same time it takes her to go 12 mi downstream. Her boat goes 15 mph in still water. What is the rate of the current?
3 mph

Complete the table and write an equation to use to solve each problem. Do not actually solve the equation. ***See Example 5.***

29. Eric Jensen can tune up his Chevy in 2 hr working alone. His son Oscar can do the job in 3 hr working alone. How long would it take them if they worked together?

Let $x =$ the time working together.

	Rate	Time Working Together	Fractional Part of the Job Done When Working Together
Eric	$\frac{1}{2}$	x	$\frac{1}{2}x$
Oscar	$\frac{1}{3}$	x	$\frac{1}{3}x$

$\frac{1}{2}x + \frac{1}{3}x = 1$

30. Working alone, Kyle can paint a room in 8 hr. Julianne can paint the same room working alone in 6 hr. How long will it take them if they work together?

Let $x =$ the time working together.

	Rate	Time Working Together	Fractional Part of the Job Done When Working Together
Kyle	$\frac{1}{8}$	x	$\frac{1}{8}x$
Julianne	$\frac{1}{6}$	x	$\frac{1}{6}x$

$\frac{1}{8}x + \frac{1}{6}x = 1$

Solve each problem. ***See Example 5.***

31. A high school mathematics teacher gave a geometry test. Working alone, it would take her 4 hr to grade the tests. Her student teacher would take 6 hr to grade the same tests. How long would it take them to grade these tests if they work together?

$2\frac{2}{5}$ hr

32. Geraldo and Luisa Hernandez operate a small laundry. Luisa, working alone, can clean a day's laundry in 9 hr. Geraldo can clean a day's laundry in 8 hr. How long would it take them if they work together?

$4\frac{4}{17}$ hr

33. Todd's copier can do a printing job in 7 hr. Scott's copier can do the same job in 12 hr. How long would it take to do the job using both copiers?

$4\frac{8}{19}$ hr

34. A pump can pump the water out of a flooded basement in 10 hr. A smaller pump takes 12 hr. How long would it take to pump the water from the basement using both pumps?

$5\frac{5}{11}$ hr

35. Hilda can paint a room in 6 hr. Working together with Brenda, they can paint the room in $3\frac{3}{4}$ hr. How long would it take Brenda to paint the room by herself?

10 hr

36. Grant can completely mess up his room in 15 min. If his cousin Wade helps him, they can completely mess up the room in $8\frac{4}{7}$ min. How long would it take Wade to mess up the room by himself?

20 min

37. An inlet pipe can fill a swimming pool in 9 hr, and an outlet pipe can empty the pool in 12 hr. Through an error, both pipes are left open. How long will it take to fill the pool?

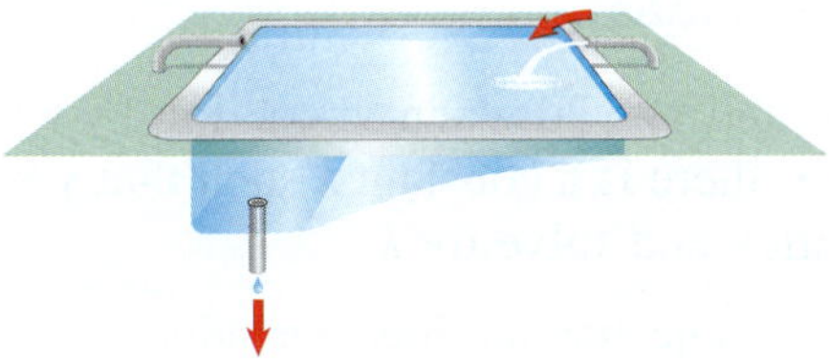

36 hr

38. One pipe can fill a swimming pool in 6 hr, and another pipe can do it in 9 hr. How long will it take the two pipes working together to fill the pool $\frac{3}{4}$ full?

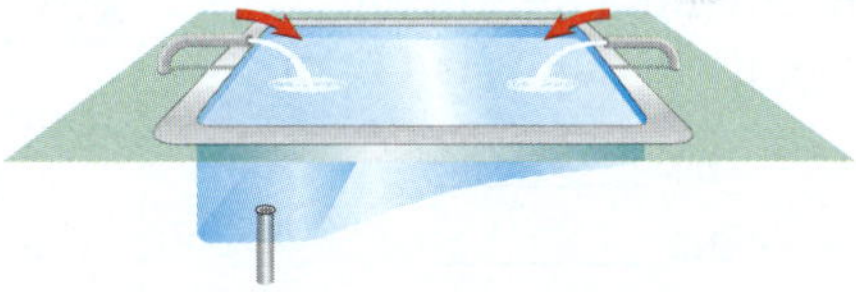

$2\frac{7}{10}$ hr

Solve each problem. (*Source:* Mary Jo Boyer, *Math for Nurses,* Wolter Kluwer.)

39. Nurses use Young's Rule to calculate the pediatric (child's) dose P of a medication, given a child's age c in years and a normal adult dose a. This rule applies to children ages 1 to 12 yr old.

$$P = \frac{c}{c + 12} \cdot a$$

The normal adult dose for milk of magnesia is 30 mL. Use Young's Rule to calculate the correct dose to give a 6-yr-old boy.

10 mL

40. Nurses use Clark's Rule to calculate the pediatric (child's) dose P of a medication, given a child's weight w in pounds and a normal adult dose a. This rule applies to children ages 2 yr or older.

$$P = \frac{w}{150} \cdot a$$

The normal adult dose for ibuprofen is 200 mg. Use Clark's Rule to calculate the correct dose to give a 4-yr-old girl who weighs 30 lb.

40 mg

7.8 Variation

OBJECTIVES

1. Solve direct variation problems.
2. Solve inverse variation problems.

OBJECTIVE 1 Solve direct variation problems. Suppose that gasoline costs \$4.50 per gal. Then 1 gal costs \$4.50, 2 gal cost 2(\$4.50) = \$9.00, 3 gal cost 3(\$4.50) = \$13.50, and so on. Each time, the total cost is obtained by multiplying the number of gallons by the price per gallon. In general, if k equals the price per gallon and x equals the number of gallons, then the total cost y is equal to kx.

As the *number of gallons* **increases,** the *total cost* **increases**.

The following is also true.

As the *number of gallons* **decreases,** the *total cost* **decreases.**

The preceding discussion is an example of *variation*. ***Two variables vary directly if one is a constant multiple of the other.***

Direct Variation

y varies directly as x if there is a constant k such that the following is true.

$$y = kx$$

Also, y is said to be *proportional to* x. The constant k in the equation for direct variation is a numerical value, such as 4.50 in the gasoline price discussion. This value is the **constant of variation.**

EXAMPLE 1 Using Direct Variation

Suppose y varies directly as x, and $y = 20$ when $x = 4$. Find y when $x = 9$.

Since y varies directly as x, there is a constant k such that $y = kx$. We let $y = 20$ and $x = 4$ in this equation and solve for k.

$y = kx$ Equation for direct variation

$20 = k \cdot 4$ Substitute the given values.

$k = 5$ ← Constant of variation

Since $y = kx$ and $k = 5$, we have the following.

$y = 5x$ Let $k = 5$.

Now we can find the value of y when $x = 9$.

$y = 5x = 5 \cdot 9 = 45$ Let $x = 9$.

Thus, $y = 45$ when $x = 9$.

◀ Work Problem 1 at the Side.

OBJECTIVE 2 Solve inverse variation problems. In direct variation, where $k > 0$, as x increases, y increases. Similarly, as x decreases, y decreases. Another type of variation is *inverse variation.* With inverse variation, where $k > 0$,

As one variable ***increases,*** the other variable ***decreases.***

---

1 Solve each problem.

(a) If z varies directly as t, and $z = 11$ when $t = 4$, find z when $t = 32$.

GS **(b)** The circumference of a circle varies directly as the radius. A circle with a radius of 7 cm has a circumference of 43.96 cm. Find the circumference if the radius is 11 cm.

Let C = the ________

and r = the ________.

$C = k \cdot$ ____

$43.96 = k \cdot$ ____

$k =$ ____

The variation equation for this problem is $C =$ ____ r.

Now find the value of C when $r =$ ____.

If the radius is 11 cm, the circumference is ____.

Answers

1. (a) 88
(b) circumference; radius; r; 7; 6.28; 6.28; 11; 69.08 cm

For example, in a closed space, volume decreases as pressure increases, as illustrated by a trash compactor. See **Figure 3.** As the compactor presses down, the pressure on the trash increases. In turn, the trash occupies a smaller space.

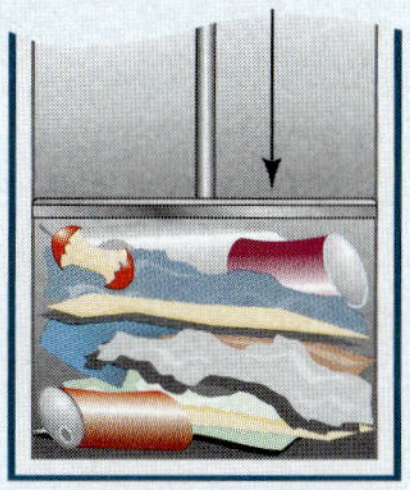

Figure 3

Inverse Variation

y* varies inversely as *x if there exists a constant k such that the following is true.

$$y = \frac{k}{x}$$

EXAMPLE 2 Using Inverse Variation

Suppose y varies inversely as x, and $y = 3$ when $x = 8$. Find y when $x = 6$.

Since y varies inversely as x, there is a constant k such that $y = \frac{k}{x}$. We know that $y = 3$ when $x = 8$, so we can find k.

$$y = \frac{k}{x} \quad \text{Equation for inverse variation}$$

$$3 = \frac{k}{8} \quad \text{Substitute the given values.}$$

$$k = 24 \longleftarrow \text{Constant of variation}$$

Since $y = \frac{24}{x}$, we let $x = 6$ and solve for y.

$$y = \frac{24}{x} = \frac{24}{6} = 4 \quad \text{Let } x = 6.$$

Therefore, when $x = 6$, $y = 4$.

Work Problem 2 at the Side. ▶

EXAMPLE 3 Using Inverse Variation

In the manufacturing of a certain medical syringe, the cost of producing the syringe varies inversely as the number produced. If 10,000 syringes are produced, the cost is \$2 per unit. Find the cost per unit to produce 25,000 syringes.

Let x = the number of syringes produced

and c = the cost per unit.

Since c varies inversely as x, there is a constant k such that $c = \frac{k}{x}$.

$$c = \frac{k}{x} \quad \text{Equation for inverse variation}$$

$$2 = \frac{k}{10{,}000} \quad \text{Substitute the given values.}$$

$$k = 20{,}000 \quad \text{Multiply by 10,000. Rewrite.}$$

Now use $c = \frac{k}{x}$ to find the value of c when $x = 25{,}000$.

$$c = \frac{20{,}000}{25{,}000} = 0.80 \quad \text{Let } k = 20{,}000 \text{ and } x = 25{,}000.$$

The cost per syringe to make 25,000 syringes is \$0.80.

Work Problem 3 at the Side. ▶

2 Solve the problem.

Suppose z varies inversely as t, and $z = 8$ when $t = 2$. Find z when $t = 32$.

Teaching Tip

Some simple examples of variation include

Direct variation: **The harder one pushes on a car's gas pedal, the faster the car goes.**

Inverse variation: **The harder one pushes on a car's brake pedal, the slower the car goes.**

3 Solve the problem.

If the cost of producing pairs of rubber gloves varies inversely as the number of pairs produced, and 5000 pairs can be produced for \$0.50 per pair, how much will it cost per pair to produce 10,000 pairs?

Answers

2. $\frac{1}{2}$

3. \$0.25

7.8 Exercises

FOR EXTRA HELP

Download the MyDashBoard App

MyMathLab®

CONCEPT CHECK *Use personal experience or intuition to determine whether the situation suggests direct or inverse variation.**

1. The rate and the distance traveled by a pickup truck in 3 hr

direct

2. The number of different lottery tickets you buy and your probability of winning that lottery

direct

3. The number of days from now until December 25 and the magnitude of the frenzy of Christmas shopping

inverse

4. Your age and the probability that you believe in Santa Claus

inverse

5. The amount of gasoline that you pump and the amount of empty space left in your tank

inverse

6. The surface area of a balloon and its diameter

direct

7. The amount of pressure put on the accelerator of a car and the rate of the car

direct

8. The number of days until the end of the baseball season and the number of home runs that Jose Bautista has hit during the season

inverse

9. **CONCEPT CHECK** Make the correct choice.

(a) If the constant of variation is positive and y varies directly as x, then as x increases, y (*increases* / *decreases*).

(b) If the constant of variation is positive and y varies inversely as x, then as x increases, y (*increases* / *decreases*).

10. Bill Veeck was the owner of several major league baseball teams in the 1950s and 1960s. He was known to often sit in the stands and enjoy games with his paying customers. Here is a quote attributed to him:

> *"I have discovered in 20 years of moving around a ballpark, that the knowledge of the game is usually in inverse proportion to the price of the seats."*

Explain in your own words the meaning of this statement. (To prove his point, Veeck once allowed the fans to vote on managerial decisions.)

The customers in the lower-priced seats know more about the game than those in the higher-priced seats.

CONCEPT CHECK *Determine whether each equation represents* direct *or* inverse *variation.*

11. $y = \frac{3}{x}$ inverse

12. $y = \frac{5}{x}$ inverse

13. $y = 50x$ direct

14. $y = 200x$ direct

*The authors thank Linda Kodama for suggesting these exercises.

Solve each problem involving direct variation. ***See Example 1.***

15. If z varies directly as x, and $z = 30$ when $x = 8$, find z when $x = 4$.

15

16. If y varies directly as x, and $x = 27$ when $y = 6$, find x when $y = 2$.

9

17. If d varies directly as r, and $d = 200$ when $r = 40$, find d when $r = 60$.

300

18. If d varies directly as t, and $d = 150$ when $t = 3$, find d when $t = 5$.

250

Solve each problem involving inverse variation. ***See Example 2.***

19. If z varies inversely as x, and $z = 50$ when $x = 2$, find z when $x = 25$.

4

20. If x varies inversely as y, and $x = 3$ when $y = 8$, find y when $x = 4$.

6

21. If m varies inversely as r, and $m = 12$ when $r = 8$, find m when $r = 16$.

6

22. If p varies inversely as q, and $p = 7$ when $q = 6$, find p when $q = 2$.

21

Solve each variation problem. ***See Examples 1–3.***

23. For a given base, the area of a triangle varies directly as its height. Find the area of a triangle with a height of 6 in., if the area is 10 in.^{2} when the height is 4 in.

15 in.^{2}

24. The interest on an investment varies directly as the rate of interest. If the interest is \$51 when the interest rate is 4%, find the interest when the rate is 3.2%.

\$40.80

25. Hooke's law for an elastic spring states that the distance a spring stretches varies directly with the force applied. If a force of 75 lb stretches a certain spring 16 in., how much will a force of 200 lb stretch the spring?

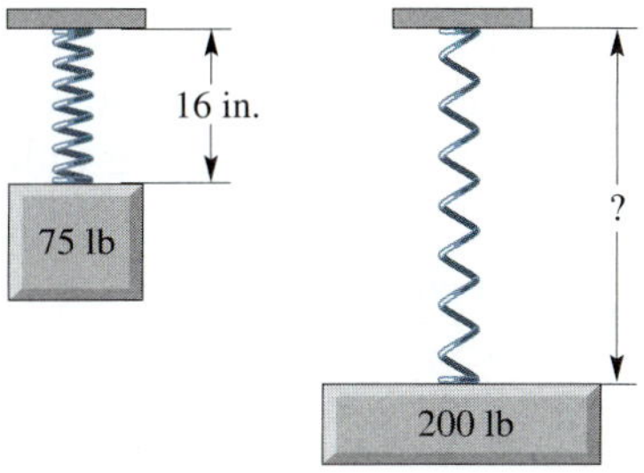

$42\frac{2}{3}$ in.

26. The pressure exerted by water at a given point varies directly with the depth of the point beneath the surface of the water. Water exerts 4.34 lb per in.^{2} for every 10 ft traveled below the water's surface. What is the pressure exerted on a scuba diver at 20 ft?

8.68 lb per in.^{2}

27. For a constant area, the length of a rectangle varies inversely as the width. The length of a rectangle is 27 ft when the width is 10 ft. Find the width of a rectangle with the same area if the length is 18 ft.

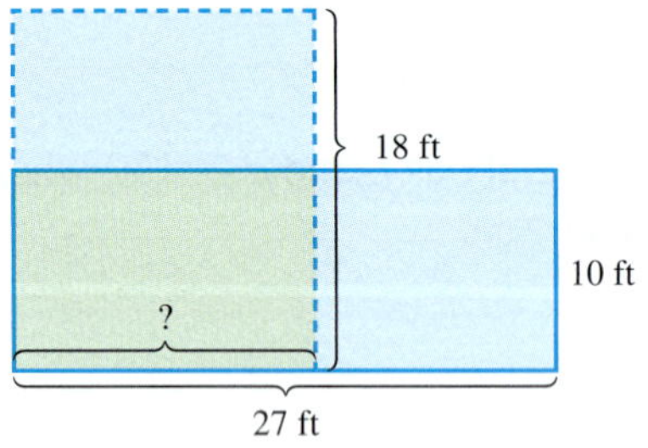

15 ft

28. Over a specified distance, rate varies inversely with time. If a Dodge Viper on a test track goes a certain distance in one-half minute at 160 mph, what rate is needed to go the same distance in three-fourths minute?

$106\frac{2}{3}$ mph

29. If the temperature is constant, the pressure of a gas in a container varies inversely as the volume of the container. If the pressure is 10 lb per ft^2 in a container with volume 3 ft^3, what is the pressure in a container with volume 1.5 ft^3?

20 lb per ft^2

30. The current in a simple electrical circuit varies inversely as the resistance. If the current is 20 amps when the resistance is 5 ohms, find the current when the resistance is 8 ohms.

$12\frac{1}{2}$ amps

31. In the inversion of raw sugar, the rate of change of the amount of raw sugar varies directly as the amount of raw sugar remaining. The rate is 200 kg per hr when there are 800 kg left. What is the rate of change per hour when only 100 kg are left?

25 kg per hr

32. The force required to compress a spring varies directly as the change in the length of the spring. If a force of 12 lb is required to compress a certain spring 3 in., how much force is required to compress the spring 5 in.?

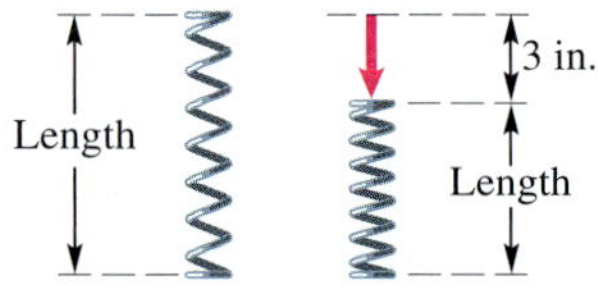

20 lb

33. In rectangles of constant area, length and width vary inversely. When the length is 24, the width is 2. What is the width when the length is 12?

4

34. The speed of a pulley varies inversely as its diameter. One kind of pulley, with diameter 6 in., turns at 150 revolutions per minute. Find the number of revolutions per minute for a similar pulley with diameter 10 in.

90 revolutions per minute

Chapter 7 Summary

Key Terms

7.1

rational expression The quotient of two polynomials with denominator not 0 is a rational expression.

lowest terms A rational expression is written in lowest terms if the greatest common factor of its numerator and denominator is 1.

7.3

least common denominator (LCD) The simplest expression that is divisible by all denominators is the least common denominator.

7.5

complex fraction A rational expression with one or more fractions in the numerator, denominator, or both, is a complex fraction.

7.6

proposed solution A value of the variable that appears to be a solution after both sides of an equation with rational expressions are multiplied by a variable expression is a proposed solution.

extraneous solution (extraneous value) A proposed solution that is not an actual solution of a given equation is an extraneous solution, or extraneous value.

7.8

direct variation y varies directly as x if there is a constant k such that $y = kx$.

constant of variation In the equation $y = kx$ or $y = \frac{k}{x}$, the number k is the constant of variation.

inverse variation y varies inversely as x if there is a constant k such that $y = \frac{k}{x}$.

Test Your Word Power

See how well you have learned the vocabulary in this chapter.

1. A **rational expression** is
   - **A.** an algebraic expression made up of a term or the sum of a finite number of terms with real coefficients and whole number exponents
   - **B.** a polynomial equation of degree 2
   - **C.** an expression with one or more fractions in the numerator, denominator, or both
   - **D.** the quotient of two polynomials with denominator not 0.

2. A **complex fraction** is
   - **A.** an algebraic expression made up of a term or the sum of a finite number of terms with real coefficients and whole number exponents
   - **B.** a polynomial equation of degree 2
   - **C.** a rational expression with one or more fractions in the numerator, denominator, or both
   - **D.** the quotient of two polynomials with denominator not 0.

3. If two positive quantities x and y are in **direct variation,** and the constant of variation is positive, then
   - **A.** as x increases, y decreases
   - **B.** as x increases, y increases
   - **C.** as x increases, y remains constant
   - **D.** as x decreases, y remains constant.

4. If two positive quantities x and y are in **inverse variation,** and the constant of variation is positive, then
   - **A.** as x increases, y decreases
   - **B.** as x increases, y increases
   - **C.** as x increases, y remains constant
   - **D.** as x decreases, y remains constant.

Answers to Test Your Word Power

1. D; *Examples:* $-\frac{3}{4y}, \frac{5x^3}{x+2}, \frac{a+3}{a^2-4a-5}$

2. C; *Examples:* $\frac{\frac{2}{3}}{\frac{4}{7}}, \frac{x-\frac{1}{y}}{x+\frac{1}{y}}, \frac{\frac{2}{a+1}}{a^2-1}$

3. B; *Example:* The equation $y = 3x$ represents direct variation. When $x = 2$, $y = 6$. If x increases to 3, then y increases to $3(3) = 9$.

4. A; *Example:* The equation $y = \frac{3}{x}$ represents inverse variation. When $x = 1$, $y = 3$. If x increases to 2, then y decreases to $\frac{3}{2}$, or $1\frac{1}{2}$.

Quick Review

Concepts	Examples
7.1 The Fundamental Property of Rational Expressions To find the value(s) for which a rational expression is undefined, set the denominator equal to 0 and solve the equation.	Find the values for which $\frac{x-4}{x^2-16}$ is undefined. $x^2 - 16 = 0$ $(x-4)(x+4) = 0$ Factor. $x-4=0$ or $x+4=0$ Zero-factor property $x=4$ or $x=-4$ Solve for x. The rational expression is undefined for 4 and -4, so $x \neq 4$ and $x \neq -4$.
Writing a Rational Expression in Lowest Terms *Step 1* Factor the numerator and denominator. *Step 2* Use the fundamental property to divide out common factors from the numerator and denominator.	Write the expression in lowest terms. $\frac{x^2-1}{(x-1)^2}$ $= \frac{(x-1)(x+1)}{(x-1)(x-1)}$ Factor. $= \frac{x+1}{x-1}$ Lowest terms
There are often several different equivalent forms of a rational expression.	Give four equivalent forms of $-\frac{x-1}{x+2}$. ① $\frac{-(x-1)}{x+2}$, or ② $\frac{-x+1}{x+2}$ Distribute the negative sign in the numerator. ③ $\frac{x-1}{-(x+2)}$, or ④ $\frac{x-1}{-x-2}$ Distribute the negative sign in the denominator.
7.2 Multiplying and Dividing Rational Expressions **Multiplying or Dividing Rational Expressions** *Step 1* Note the operation. If the operation is division, use the definition of division to rewrite as multiplication. *Step 2* Multiply numerators and multiply denominators. *Step 3* Factor numerators and denominators completely. *Step 4* Write in lowest terms, using the fundamental property. ***Steps 2 and 3 may be interchanged based on personal preference.***	Multiply. $\frac{3x+9}{x-5} \cdot \frac{x^2-3x-10}{x^2-9}$ $= \frac{(3x+9)(x^2-3x-10)}{(x-5)(x^2-9)}$ Multiply numerators and denominators. $= \frac{3(x+3)(x-5)(x+2)}{(x-5)(x+3)(x-3)}$ Factor. $= \frac{3(x+2)}{x-3}$ Lowest terms Divide. $\frac{2x+1}{x+5} \div \frac{6x^2-x-2}{x^2-25}$ $= \frac{(2x+1)(x^2-25)}{(x+5)(6x^2-x-2)}$ Multiply by the reciprocal of the divisor. $= \frac{(2x+1)(x+5)(x-5)}{(x+5)(2x+1)(3x-2)}$ Factor. $= \frac{x-5}{3x-2}$ Lowest terms

Concepts	Examples
7.3 Least Common Denominators	
Finding the LCD	Find the LCD for $\frac{3}{k^2 - 8k + 16}$ and $\frac{1}{4k^2 - 16k}$.
Step 1 Factor each denominator into prime factors.	$k^2 - 8k + 16 = (k-4)^2$; $4k^2 - 16k = 4k(k-4)$ — Factor each denominator.
Step 2 List each different factor the greatest number of times it appears.	LCD $= (k-4)^2 \cdot 4 \cdot k$
Step 3 Multiply the factors from Step 2 to get the LCD.	$= 4k(k-4)^2$
Writing Equivalent Rational Expressions	Find the numerator. $\frac{5}{2z^2 - 6z} = \frac{?}{4z^3 - 12z^2}$
Step 1 Factor both denominators.	$\frac{5}{2z(z-3)} = \frac{?}{4z^2(z-3)}$
Step 2 Decide what factors the denominator must be multiplied by to equal the specified denominator.	$2z(z-3)$ must be multiplied by $2z$.
Step 3 Multiply the rational expression by that factor divided by itself. (That is, multiply by 1.)	$\frac{5}{2z(z-3)} \cdot \frac{2z}{2z}$ $= \frac{10z}{4z^2(z-3)}$, or $\frac{10z}{4z^3 - 12z^2}$
7.4 Adding and Subtracting Rational Expressions	Add. $\frac{2}{3m+6} + \frac{m}{m^2-4}$
Adding Rational Expressions	
Step 1 Find the LCD.	$3m + 6 = 3(m+2)$; $m^2 - 4 = (m+2)(m-2)$ — The LCD is $3(m+2)(m-2)$.
Step 2 Rewrite each rational expression with the LCD as denominator.	$= \frac{2(m-2)}{3(m+2)(m-2)} + \frac{3m}{3(m+2)(m-2)}$ — Write with the LCD.
Step 3 Add the numerators to get the numerator of the sum. The LCD is the denominator of the sum.	$= \frac{2m - 4 + 3m}{3(m+2)(m-2)}$ — Add numerators. Keep the same denominator.
Step 4 Write in lowest terms.	$= \frac{5m - 4}{3(m+2)(m-2)}$ — Combine like terms.
Subtracting Rational Expressions	Subtract. $\frac{6}{k+4} - \frac{2}{k}$ — The LCD is $k(k+4)$.
Follow the same steps as for addition, but subtract in Step 3.	$= \frac{6k}{(k+4)k} - \frac{2(k+4)}{k(k+4)}$ — Write with the LCD.
	$= \frac{6k - 2(k+4)}{k(k+4)}$ — Subtract numerators. Keep the same denominator.
	$= \frac{6k - 2k - 8}{k(k+4)}$ — Distributive property
	$= \frac{4k - 8}{k(k+4)}$ — Combine like terms.

Concepts	Examples

7.5 Complex Fractions

Simplifying Complex Fractions

Method 1 Simplify the numerator and denominator separately. Then divide the simplified numerator by the simplified denominator.

Method 2 Multiply the numerator and denominator of the complex fraction by the LCD of all the fractions in the numerator and denominator of the complex fraction. Write in lowest terms.

Simplify.

Method 1

$$\frac{\frac{1}{a}-a}{1-a} = \frac{\frac{1}{a}-\frac{a^2}{a}}{1-a} = \frac{\frac{1-a^2}{a}}{1-a} = \frac{1-a^2}{a} \div (1-a) = \frac{1-a^2}{a}\cdot\frac{1}{1-a} = \frac{(1-a)(1+a)}{a(1-a)} = \frac{1+a}{a}$$

Method 2

$$\frac{\frac{1}{a}-a}{1-a} = \frac{\left(\frac{1}{a}-a\right)a}{(1-a)a} = \frac{\frac{a}{a}-a^2}{(1-a)a} = \frac{1-a^2}{(1-a)a} = \frac{(1+a)(1-a)}{(1-a)a} = \frac{1+a}{a}$$

7.6 Solving Equations with Rational Expressions

Solving Equations with Rational Expressions

Step 1 Multiply each side of the equation by the LCD. (This clears the equation of fractions.)

Solve $\frac{x}{x-3}+\frac{4}{x+3}=\frac{18}{x^2-9}$.

$$\frac{x}{x-3}+\frac{4}{x+3}=\frac{18}{(x-3)(x+3)} \quad \text{Factor.}$$

The LCD is $(x-3)(x+3)$. Note that 3 and -3 cannot be solutions, as they cause a denominator to equal 0.

$$(x-3)(x+3)\left(\frac{x}{x-3}+\frac{4}{x+3}\right) = (x-3)(x+3)\frac{18}{(x-3)(x+3)}$$

Multiply by the LCD.

Step 2 Solve the resulting equation.

$$x(x+3)+4(x-3)=18 \quad \text{Distributive property}$$

$$x^2+3x+4x-12=18 \quad \text{Distributive property}$$

$$x^2+7x-30=0 \quad \text{Standard form}$$

$$(x-3)(x+10)=0 \quad \text{Factor.}$$

$$x-3=0 \quad \text{or} \quad x+10=0 \quad \text{Zero-factor property}$$

Reject → $x=3$ or $x=-10$ Solve for x.

Step 3 Check each proposed solution by substituting it in the original equation. Reject any value that causes an original denominator to equal 0.

Since 3 causes denominators to equal 0, the only solution is -10. Thus, $\{-10\}$ is the solution set.

Concepts	Examples

7.7 Applications of Rational Expressions

Solving Problems about Distance, Rate, and Time

Use the formulas relating d, r, and t.

$$d = rt, \quad r = \frac{d}{t}, \quad t = \frac{d}{r}$$

Solving Problems about Work

Step 1 **Read** the problem carefully.

Step 2 **Assign a variable.** State what the variable represents. Put the information from the problem in a table. If a job is done in t units of time, then the rate is $\frac{1}{t}$.

Step 3 **Write an equation.** The sum of the fractional parts should equal 1 (whole job).

Step 4 **Solve** the equation.

Step 5 **State the answer.**

Step 6 **Check.**

It takes the regular mail carrier 6 hr to cover her route. A substitute takes 8 hr to cover the same route. How long would it take them to cover the route together?

Let x = the number of hours to cover the route together.

The rate of the regular carrier is $\frac{1}{6}$ job per hour, and the rate of the substitute is $\frac{1}{8}$ job per hour. Multiply rate by time to get the fractional part of the job done.

	Rate	Time	Part of the Job Done
Regular	$\frac{1}{6}$	x	$\frac{1}{6}x$
Substitute	$\frac{1}{8}$	x	$\frac{1}{8}x$

$\frac{1}{6}x + \frac{1}{8}x = 1$ The parts add to 1 whole job.

$24\left(\frac{1}{6}x + \frac{1}{8}x\right) = 24(1)$ The LCD is 24.

$4x + 3x = 24$ Distributive property

$7x = 24$ Combine like terms.

$x = \frac{24}{7}$ Divide by 7.

It would take them $\frac{24}{7}$ hr, or $3\frac{3}{7}$ hr, to cover the route together. The solution checks because $\frac{1}{6}\left(\frac{24}{7}\right) + \frac{1}{8}\left(\frac{24}{7}\right) = 1$.

7.8 Variation

Solving Variation Problems

Step 1 Write the variation equation. Use

$y = kx$ Direct variation

or $y = \frac{k}{x}$. Inverse variation

Step 2 Find k by substituting the given values of x and y into the equation.

Step 3 Write the equation with the value of k from Step 2 and the given value of x or y. Solve for the remaining variable.

If y varies inversely as x, and $y = 4$ when $x = 9$, find y when $x = 6$.

$y = \frac{k}{x}$ Equation for inverse variation

$4 = \frac{k}{9}$ Substitute given values.

$k = 36$ Solve for k.

$y = \frac{36}{x}$ Let $k = 36$.

$y = \frac{36}{6}$, or 6 Let $x = 6$.

Chapter 7 Review Exercises

7.1 *Find the value(s) of the variable for which each rational expression is undefined. Write answers with the symbol $\neq$.*

1. $\dfrac{4}{x-3}$ $x \neq 3$

2. $\dfrac{x+3}{2x}$ $x \neq 0$

3. $\dfrac{m-2}{m^2-2m-3}$ $m \neq -1, m \neq 3$

4. $\dfrac{2k+1}{3k^2+17k+10}$ $k \neq -5, k \neq -\frac{2}{3}$

Find the numerical value of each rational expression for **(a)** $x = -2$ *and* **(b)** $x = 4$.

5. $\dfrac{x^2}{x-5}$ (a) $-\frac{4}{7}$ (b) -16

6. $\dfrac{4x-3}{5x+2}$ (a) $\frac{11}{8}$ (b) $\frac{13}{22}$

7. $\dfrac{3x}{x^2-4}$ (a) undefined (b) 1

8. $\dfrac{x-1}{x+2}$ (a) undefined (b) $\frac{1}{2}$

Write each rational expression in lowest terms.

9. $\dfrac{5a^3b^3}{15a^4b^2}$ $\frac{b}{3a}$

10. $\dfrac{m-4}{4-m}$ -1

11. $\dfrac{4x^2-9}{6-4x}$ $\frac{-(2x+3)}{2}$

12. $\dfrac{4p^2+8pq-5q^2}{10p^2-3pq-q^2}$ $\frac{2p+5q}{5p+q}$

Write four equivalent expressions for each fraction. Answers may vary in Exercises 13 and 14.

13. $-\dfrac{4x-9}{2x+3}$ $\frac{-(4x-9)}{2x+3}, \frac{-4x+9}{2x+3}, \frac{4x-9}{-(2x+3)}, \frac{4x-9}{-2x-3}$

14. $-\dfrac{8-3x}{3-6x}$ $\frac{-(8-3x)}{3-6x}, \frac{-8+3x}{3-6x}, \frac{8-3x}{-(3-6x)}, \frac{8-3x}{-3+6x}$

7.2 *Multiply or divide. Write each answer in lowest terms.*

15. $\dfrac{8x^2}{12x^5} \cdot \dfrac{6x^4}{2x}$ 2

16. $\dfrac{9m^2}{(3m)^4} \div \dfrac{6m^5}{36m}$ $\frac{2}{3m^6}$

17. $\dfrac{x-3}{4} \cdot \dfrac{5}{2x-6}$ $\frac{5}{8}$

18. $\dfrac{2r+3}{r-4} \cdot \dfrac{r^2-16}{6r+9}$ $\frac{r+4}{3}$

19. $\dfrac{3q+3}{5-6q} \div \dfrac{4q+4}{2(5-6q)}$ $\frac{3}{2}$

20. $\dfrac{y^2-6y+8}{y^2+3y-18} \div \dfrac{y-4}{y+6}$ $\frac{y-2}{y-3}$

21. $\dfrac{2p^2+13p+20}{p^2+p-12} \cdot \dfrac{p^2+2p-15}{2p^2+7p+5}$ $\frac{p+5}{p+1}$

22. $\dfrac{3z^2+5z-2}{9z^2-1} \cdot \dfrac{9z^2+6z+1}{z^2+5z+6}$ $\frac{3z+1}{z+3}$

7.3 *Find the least common denominator for the fractions in each list.*

23. $\frac{1}{8}, \frac{5}{12}, \frac{7}{32}$

96

24. $\frac{4}{9y}, \frac{7}{12y^2}, \frac{5}{27y^4}$

$108y^4$

25. $\frac{1}{m^2 + 2m}, \frac{4}{m^2 + 7m + 10}$

$m(m + 2)(m + 5)$

26. $\frac{3}{x^2 + 4x + 3}, \frac{5}{x^2 + 5x + 4}, \frac{2}{x^2 + 7x + 12}$

$(x + 3)(x + 1)(x + 4)$

Write each rational expression as an equivalent expression with the indicated denominator.

27. $\frac{5}{8} = \frac{?}{56}$

$\frac{35}{56}$

28. $\frac{10}{k} = \frac{?}{4k}$

$\frac{40}{4k}$

29. $\frac{3}{2a^3} = \frac{?}{10a^4}$

$\frac{15a}{10a^4}$

30. $\frac{9}{x - 3} = \frac{?}{18 - 6x}$

$\frac{-54}{18 - 6x}$

31. $\frac{-3y}{2y - 10} = \frac{?}{50 - 10y}$

$\frac{15y}{50 - 10y}$

32. $\frac{4b}{b^2 + 2b - 3} = \frac{?}{(b + 3)(b - 1)(b + 2)}$

$\frac{4b(b + 2)}{(b + 3)(b - 1)(b + 2)}$

7.4 *Add or subtract. Write each answer in lowest terms.*

33. $\frac{10}{x} + \frac{5}{x}$

$\frac{15}{x}$

34. $\frac{6}{3p} - \frac{12}{3p}$

$-\frac{2}{p}$

35. $\frac{9}{k} - \frac{5}{k - 5}$

$\frac{4k - 45}{k(k - 5)}$

36. $\frac{4}{y} + \frac{7}{7 + y}$

$\frac{28 + 11y}{y(7 + y)}$

37. $\frac{m}{3} - \frac{2 + 5m}{6}$

$\frac{-2 - 3m}{6}$

38. $\frac{12}{x^2} - \frac{3}{4x}$

$\frac{3(16 - x)}{4x^2}$

39. $\frac{5}{a - 2b} + \frac{2}{a + 2b}$

$\frac{7a + 6b}{(a - 2b)(a + 2b)}$

40. $\frac{4}{k^2 - 9} - \frac{k + 3}{3k - 9}$

$\frac{-k^2 - 6k + 3}{3(k + 3)(k - 3)}$

41. $\frac{8}{z^2 + 6z} - \frac{3}{z^2 + 4z - 12}$

$\frac{5z - 16}{z(z + 6)(z - 2)}$

42. $\frac{11}{2p - p^2} - \frac{2}{p^2 - 5p + 6}$

$\frac{-13p + 33}{p(p - 2)(p - 3)}$

7.5 *Simplify each complex fraction.*

43. $\frac{\frac{a^4}{b^2}}{\frac{a^3}{b}}$

$\frac{a}{b}$

44. $\frac{\frac{y - 3}{y}}{\frac{y + 3}{4y}}$

$\frac{4(y - 3)}{y + 3}$

45. $\frac{\frac{3m + 2}{m}}{\frac{2m - 5}{6m}}$

$\frac{6(3m + 2)}{2m - 5}$

46. $\frac{\frac{1}{p} - \frac{1}{q}}{\frac{1}{q - p}}$

$\frac{(q - p)^2}{pq}$

47. $\frac{x + \frac{1}{w}}{x - \frac{1}{w}}$

$\frac{xw + 1}{xw - 1}$

48. $\frac{\frac{1}{r + t} - 1}{\frac{1}{r + t} + 1}$

$\frac{1 - r - t}{1 + r + t}$

7.6 *Solve each equation. Check the solutions.*

49. $\frac{k}{5} - \frac{2}{3} = \frac{1}{2}$

$\left\{\frac{35}{6}\right\}$

50. $\frac{4-z}{z} + \frac{3}{2} = \frac{-4}{z}$

$\{-16\}$

51. $\frac{x}{2} - \frac{x-3}{7} = -1$

$\{-4\}$

52. $\frac{3y-1}{y-2} = \frac{5}{y-2} + 1$

$\emptyset$

53. $\frac{3}{m-2} + \frac{1}{m-1} = \frac{7}{m^2 - 3m + 2}$

$\{3\}$

Solve for the specified variable.

54. $m = \frac{Ry}{t}$ for t

$t = \frac{Ry}{m}$

55. $x = \frac{3y-5}{4}$ for y

$y = \frac{4x+5}{3}$

56. $\frac{1}{r} - \frac{1}{s} = \frac{1}{t}$ for t

$t = \frac{rs}{s-r}$

7.7 *Solve each problem.*

57. In a certain fraction, the denominator is 5 less than the numerator. If 5 is added to both the numerator and the denominator, the resulting fraction is equivalent to $\frac{5}{4}$. Find the original fraction (*not* written in lowest terms).

$\frac{20}{15}$

58. The denominator of a certain fraction is six times the numerator. If 3 is added to the numerator and subtracted from the denominator, the resulting fraction is equivalent to $\frac{2}{5}$. Find the original fraction (*not* written in lowest terms).

$\frac{3}{18}$

59. On June 25, 2011, Marco Andretti won the fifth Iowa Corn Indy 250. He drove a Dallara-Honda the 218.75 mi distance with an average rate of 118.671 mph. What was his time (to the nearest thousandth of an hour)? (*Source:* www.indycar.com)

1.843 hr

60. In the 2010 Winter Olympics in Vancouver, Canada, Sven Kramer of the Netherlands won the men's 5000-m speed skating event in 6.243 min. What was his rate (to three decimal places)? (*Source: World Almanac and Book of Facts.*)

800.897 m per min

61. Zachary and Samuel are brothers who share a bedroom. By himself, Zachary can completely mess up their room in 20 min, while it would take Samuel only 12 min to do the same thing. How long would it take them to mess up the room together?

$7\frac{1}{2}$ min

62. A man can plant his garden in 5 hr, working alone. His daughter can do the same job in 8 hr. How long would it take them if they worked together?

$3\frac{1}{13}$ hr

7.8 *Solve each problem.*

63. If y varies directly as x, and $x = 12$ when $y = 5$, find x when $y = 3$.

$\frac{36}{5}$

64. If a parallelogram has a fixed area, the height varies inversely as the base. A parallelogram has a height of 8 cm and a base of 12 cm. Find the height if the base is changed to 24 cm. 4 cm

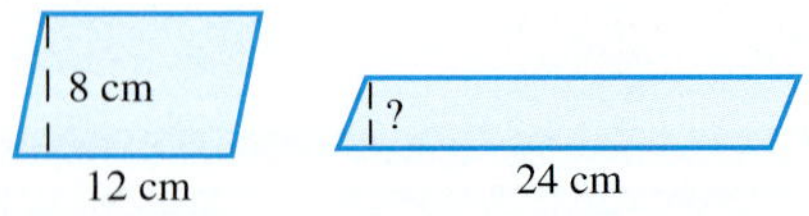

Mixed Review Exercises

Perform the indicated operations. Write each answer in lowest terms.

65. $\frac{4}{m-1} - \frac{3}{m+1}$

$\frac{m+7}{(m-1)(m+1)}$

66. $\frac{8p^5}{5} \div \frac{2p^3}{10}$

$8p^2$

67. $\frac{r-3}{8} \div \frac{3r-9}{4}$

$\frac{1}{6}$

68. $\frac{\frac{5}{x}-1}{\frac{5-x}{3x}}$

3

69. $\frac{4}{z^2-2z+1} - \frac{3}{z^2-1}$

$\frac{z+7}{(z+1)(z-1)^2}$

70. $\frac{2x^2+5x-12}{4x^2-9} \cdot \frac{x^2-3x-28}{x^2+8x+16}$

$\frac{x-7}{2x+3}$

Solve each equation, and check the solutions.

71. $\frac{5t}{6} = \frac{2t-1}{3} + 1$

$\{4\}$

72. $\frac{2}{z} - \frac{z}{z+3} = \frac{1}{z+3}$

$\{-2, 3\}$

73. $\frac{2x}{x^2-16} - \frac{2}{x-4} = \frac{4}{x+4}$

$\{2\}$

74. Solve $a = \frac{v-w}{t}$ for w.

$w = v - at$

Solve each problem.

75. If the same number is added to both the numerator and denominator of the fraction $\frac{4}{11}$, the result is equivalent to $\frac{1}{2}$. Find the number.

3

76. Working together, Seema and Satish can clean their house in 3 hr. Working alone, it takes Seema 5 hr to clean the house. How long would it take Satish to clean the house by himself?

$7\frac{1}{2}$ hr

77. Anne Kelly flew her plane 400 km with the wind in the same time it took her to go 200 km against the wind. The wind speed is 50 km per hr. Find the rate of the plane in still air. Use x as the variable.

150 km per hr

	d	r	t
With the Wind	400	$x + 50$	$\frac{400}{x+50}$
Against the Wind	200	$x - 50$	$\frac{200}{x-50}$

78. At a given hour, two steamboats leave a city in the same direction on a straight canal. One travels at 18 mph, and the other travels at 25 mph. In how many hours will the boats be 70 mi apart?

10 hr

79. The current in a simple electrical circuit varies inversely as the resistance. If the current is 80 amps when the resistance is 10 ohms, find the current if the resistance is 16 ohms.

50 amps

80. **CONCEPT CHECK** If a rectangle of fixed area has its length increased, then its width decreases. Is this an example of direct or inverse variation?

inverse

Chapter 7 Test

CHAPTER Test Prep VIDEO

The Chapter Test Prep Videos with test solutions are available on DVD, in MyMathLab, and on YouTube — search "LialIntroductoryAlg" and click on "Channels."

1. Find any values for which $\frac{3x-1}{x^2-2x-8}$ is undefined. Write your answer with the symbol $\neq$.

$x \neq -2, x \neq 4$

2. Find the numerical value of $\frac{6r+1}{2r^2-3r-20}$ for each value of r.

(a) $r=-2$ $\quad \frac{11}{6}$

(b) $r=4$ $\quad$ undefined

3. Write four rational expressions equivalent to $-\frac{6x-5}{2x+3}$.

(Answers may vary.) $\frac{-(6x-5)}{2x+3}, \frac{-6x+5}{2x+3}, \frac{6x-5}{-(2x+3)}, \frac{6x-5}{-2x-3}$

Write each rational expression in lowest terms.

4. $\frac{-15x^6y^4}{5x^4y}$

$-3x^2y^3$

5. $\frac{6a^2+a-2}{2a^2-3a+1}$

$\frac{3a+2}{a-1}$

Multiply or divide. Write each answer in lowest terms.

6. $\frac{5(d-2)}{9} \div \frac{3(d-2)}{5}$

$\frac{25}{27}$

7. $\frac{6k^2-k-2}{8k^2+10k+3} \cdot \frac{4k^2+7k+3}{3k^2+5k+2}$

$\frac{3k-2}{3k+2}$

8. $\frac{4a^2+9a+2}{3a^2+11a+10} \div \frac{4a^2+17a+4}{3a^2+2a-5}$

$\frac{a-1}{a+4}$

9. $\frac{x^2-10x+25}{9-6x+x^2} \cdot \frac{x-3}{5-x}$

$\frac{x-5}{3-x}$

Find the least common denominator for each list of fractions.

10. $\frac{-3}{10p^2}, \frac{21}{25p^3}, \frac{-7}{30p^5}$

$150p^5$

11. $\frac{r+1}{2r^2+7r+6}, \frac{-2r+1}{2r^2-7r-15}$

$(2r+3)(r+2)(r-5)$

Write each rational expression as an equivalent expression with the indicated denominator.

12. $\frac{15}{4p} = \frac{?}{64p^3}$

$\frac{240p^2}{64p^3}$

13. $\frac{3}{6m-12} = \frac{?}{42m-84}$

$\frac{21}{42m-84}$

Add or subtract. Write each answer in lowest terms.

14. $\frac{4x+2}{x+5} + \frac{-2x+8}{x+5}$

2

15. $\frac{-4}{y+2} + \frac{6}{5y+10}$

$\frac{-14}{5(y+2)}$

16. $\dfrac{x+1}{3-x} - \dfrac{x^2}{x-3}$

$\dfrac{x^2+x+1}{3-x}$, or $\dfrac{-x^2-x-1}{x-3}$

17. $\dfrac{3}{2m^2-9m-5} - \dfrac{m+1}{2m^2-m-1}$

$\dfrac{-m^2+7m+2}{(2m+1)(m-5)(m-1)}$

Simplify each complex fraction.

18. $\dfrac{\frac{2p}{k^2}}{\frac{3p^2}{k^3}}$

$\dfrac{2k}{3p}$

19. $\dfrac{\frac{1}{x+3} - 1}{1 + \frac{1}{x+3}}$

$\dfrac{-2-x}{4+x}$

Solve each equation.

20. $\dfrac{3x}{x+1} = \dfrac{3}{2x}$

$\left\{-\dfrac{1}{2}, 1\right\}$

21. $\dfrac{2}{x-1} - \dfrac{2}{3} = \dfrac{-1}{x+1}$

$\left\{-\dfrac{1}{2}, 5\right\}$

22. $\dfrac{2x}{x-3} + \dfrac{1}{x+3} = \dfrac{-6}{x^2-9}$

$\left\{-\dfrac{1}{2}\right\}$

23. Solve the formula $F = \dfrac{k}{d-D}$ for D.

$D = \dfrac{dF-k}{F}$, or $D = d - \dfrac{k}{F}$

Solve each problem.

24. If the same number is added to the numerator and subtracted from the denominator of $\frac{5}{6}$, the resulting fraction is equivalent to $\frac{1}{10}$. What is the number?

-4

25. A boat travels 7 mph in still water. It takes as long to go 20 mi upstream as 50 mi downstream. Find the rate of the current.

3 mph

26. A man can paint a room in his house, working alone, in 5 hr. His wife can do the job in 4 hr. How long will it take them to paint the room if they work together?

$2\frac{2}{9}$ hr

27. Under certain conditions, the length of time that it takes for fruit to ripen during the growing season varies inversely as the average maximum temperature during the season. If it takes 25 days for fruit to ripen with an average maximum temperature of 80°F, find the number of days it would take at 75°F. Round your answer to the nearest whole number.

27 days

28. If x varies directly as y, and $x = 12$ when $y = 4$, find x when $y = 9$.

27

Chapters R–7 Cumulative Review Exercises

1. Evaluate $3 + 4\left(\frac{1}{2} - \frac{3}{4}\right)$. 2

Solve. Graph the solution set in Exercises 5 and 6.

2. $3(2t - 5) = 2 + 5t$ $\{17\}$

3. $A = \frac{1}{2}bh$ for b $b = \frac{2A}{h}$

4. $\frac{2 + m}{2 - m} = \frac{3}{4}$ $\left\{-\frac{2}{7}\right\}$

5. $5x \leq 6x + 8$ $[-8, \infty)$

6. $5m - 9 > 2m + 3$ $(4, \infty)$

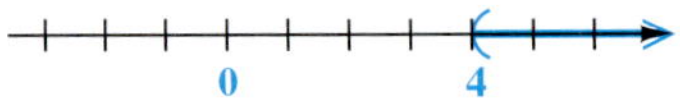

Sketch each graph.

7. $y = -3x + 2$

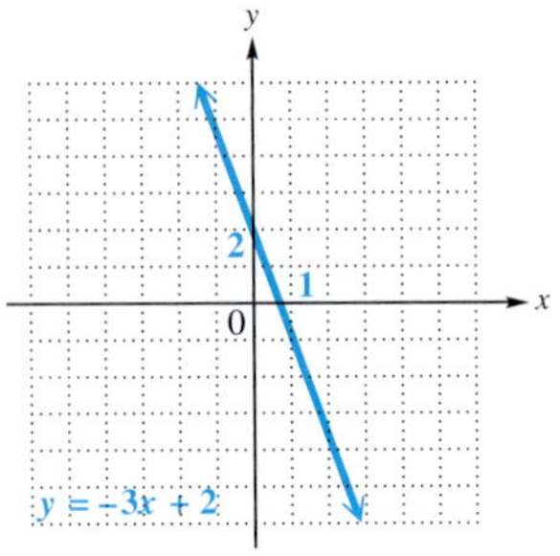

8. $y \geq 2x + 3$

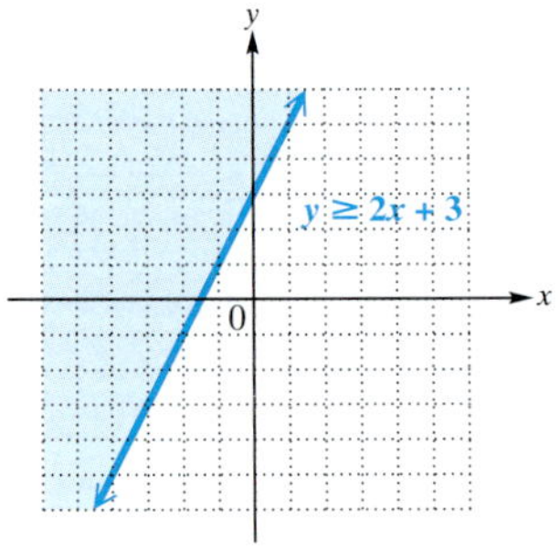

9. Solve by elimination. $\{(-1, 2)\}$

$$3x + 4y = 5$$
$$6x + 7y = 8$$

10. Solve by substitution. $\{(1, -2)\}$

$$y = -3x + 1$$
$$x + 2y = -3$$

Simplify each expression.

11. $\frac{(2x^3)^{-1} \cdot x}{2^3x^5}$ $\frac{1}{16x^7}$

12. $\frac{(m^{-2})^3m}{m^5m^{-4}}$ $\frac{1}{m^6}$

13. $\frac{2p^3q^4}{8p^5q^3}$ $\frac{q}{4p^2}$

Perform the indicated operations.

14. $(2k^2 + 3k) - (k^2 + k - 1)$ $k^2 + 2k + 1$

15. $8x^2y^2(9x^4y^5)$ $72x^6y^7$

16. $(2a - b)^2$ $4a^2 - 4ab + b^2$

17. $(y^2 + 3y + 5)(3y - 1)$ $3y^3 + 8y^2 + 12y - 5$

18. $\frac{12p^3 + 2p^2 - 12p + 4}{2p - 2}$ $6p^2 + 7p + 1 + \frac{3}{p - 1}$

Factor completely.

19. $8t^2 + 10tv + 3v^2$

$(4t + 3v)(2t + v)$

20. $8r^2 - 9rs + 12s^2$

prime

21. $16x^4 - 1$

$(4x^2 + 1)(2x + 1)(2x - 1)$

Solve each equation.

22. $r^2 = 2r + 15$

$\{-3, 5\}$

23. $(r - 5)(2r + 1)(3r - 2) = 0$

$\left\{-\frac{1}{2}, \frac{2}{3}, 5\right\}$

Solve each problem.

24. One number is 4 more than another. The product of the numbers is 2 less than the smaller number. Find the smaller number. -2 or -1

25. The length of a rectangle is 2 m less than twice the width. The area is 60 m^2. Find the width of the rectangle. 6 m

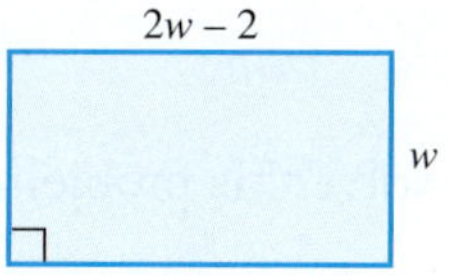

26. For what value(s) of t is $\frac{2 + t}{t^2 - 4}$ undefined? Write your answer with the symbol $\neq$.

$t \neq -2, t \neq 2$

27. One of the following is equal to 1 for *all* real numbers. Which one is it? A

A. $\dfrac{k^2 + 2}{k^2 + 2}$ **B.** $\dfrac{4 - m}{4 - m}$ **C.** $\dfrac{2x + 9}{2x + 9}$ **D.** $\dfrac{x^2 - 1}{x^2 - 1}$

28. Which one of the following rational expressions is *not* equivalent to $\dfrac{4 - 3x}{7}$? D

A. $-\dfrac{-4 + 3x}{7}$ **B.** $-\dfrac{4 - 3x}{-7}$ **C.** $\dfrac{-4 + 3x}{-7}$ **D.** $\dfrac{-(3x + 4)}{7}$

Perform each operation, and write the answer in lowest terms.

29. $\dfrac{5}{q} - \dfrac{1}{q}$ $\quad \dfrac{4}{q}$

30. $\dfrac{3}{7} + \dfrac{4}{r}$ $\quad \dfrac{3r + 28}{7r}$

31. $\dfrac{4}{5q - 20} - \dfrac{1}{3q - 12}$ $\quad \dfrac{7}{15(q - 4)}$

32. $\dfrac{2}{k^2 + k} - \dfrac{3}{k^2 - k}$

$\dfrac{-k - 5}{k(k + 1)(k - 1)}$

33. $\dfrac{7z^2 + 49z + 70}{16z^2 + 72z - 40} \div \dfrac{3z + 6}{4z^2 - 1}$

$\dfrac{7(2z + 1)}{24}$

34. $\dfrac{\dfrac{4}{a} + \dfrac{5}{2a}}{\dfrac{7}{6a} - \dfrac{1}{5a}}$ $\quad \dfrac{195}{29}$

Solve each equation, and check your solutions.

35. $\dfrac{r + 2}{5} = \dfrac{r - 3}{3}$ $\quad \left\{\dfrac{21}{2}\right\}$

36. $\dfrac{1}{x} = \dfrac{1}{x + 1} + \dfrac{1}{2}$ $\quad \{-2, 1\}$

Solve each problem.

37. On a business trip, Carrie Ann traveled to her destination at an average rate of 60 mph. Coming home, her average rate was 50 mph, and the trip took $\frac{1}{2}$ hr longer. How far did she travel each way?

150 mi

38. Len can weed the yard in 3 hr. Bruno can weed the yard in 2 hr. How long would it take them if they worked together?

$1\frac{1}{5}$ hr

Math in the Media

TO PLAY BASEBALL, YOU MUST *WORK* AT IT!

In the 1994 movie *Little Big League,* the young Billy Heywood inherits the Minnesota Twins baseball team and becomes its manager. He leads the team to the Division Championship and then to the playoffs. But before the final playoff game, the biggest game of the year, he can't keep his mind on his job because a homework problem is giving him trouble.

> *If Joe can paint a house in 3 hours, and Sam can paint the same house in 5 hours, how long does it take for them to do it together?*

With the help of one of his players, he is able to solve the problem, and the team goes on to victory.

1. Use the method described in **Example 5** of **Section 7.7** to solve this problem.

$\frac{15}{8}$ **hr, or** $1\frac{7}{8}$ **hr**

2. Before the player was able to solve the problem correctly, Billy got "help" from some of the other players. The incorrect answers they gave him are included in the following.

(a) 15 hr **(b)** 8 hr **(c)** 4 hr

Explain the faulty reasoning behind each of these incorrect answers.

(a) The player multiplied: $5 \times 3 = 15$.
(b) The player added: $5 + 3 = 8$.
(c) The player divided the part (b) incorrect answer by 2, because there were two painters: $\frac{8}{2} = 4$.

3. The player who gave Billy the correct answer solved the problem as follows:

> *Using the simple formula a times b over a plus b, we get our answer of one and seven-eighths.*

Show that if it takes one person a hours to complete one job and another b hours to complete the same job, then the expression stated by the player,

$$\frac{a \cdot b}{a + b}$$

actually does give the number of hours it would take them to do the job together. (*Hint:* Refer to **Example 5** and use a and b rather than 8 and 14. Then solve the resulting formula for x.)

$$\frac{1}{a}x + \frac{1}{b}x = 1$$

$$bx + ax = ab$$

$$x(b + a) = ab$$

$$x = \frac{ab}{b + a}, \quad \text{or} \quad x = \frac{a \cdot b}{a + b}$$

8 Roots and Radicals

Many formulas in mathematics and science, such as that for finding the time it takes for a pendulum to swing from one extreme to the other and back again, include *square roots,* the topic of this chapter.

8.1 Evaluating Roots

OBJECTIVES

1. Find square roots.
2. Decide whether a given root is rational, irrational, or not a real number.
3. Find decimal approximations for irrational square roots.
4. Use the Pythagorean theorem.
5. Find cube, fourth, and other roots.

OBJECTIVE 1 Find square roots. Recall that *squaring* a number means multiplying the number by itself.

7^2 means $7 \cdot 7$, which equals 49. The square of 7 is 49.

The opposite (inverse) of squaring a number is taking its *square root.* This is equivalent to asking

"What number when multiplied by itself equals 49?"

For the example above, one answer is 7, since $7 \cdot 7 = 49$.

This discussion can be generalized.

Square Root

A number b is a **square root** of a if $b^2 = a$.

EXAMPLE 1 Finding All Square Roots of a Number

Find all square roots of 49.

We ask, "What number when multiplied by itself equals 49?" As mentioned above, one square root is 7. Another square root of 49 is −7, because

$$(-7)(-7) = 49.$$

Thus, the number 49 has *two* square roots: 7 and −7. One square root is positive, and one is negative.

◀ Work Problem 1 at the Side.

The **positive** or **principal square root** of a number is written with the symbol $\sqrt{}$. For example, the positive square root of 121 is 11.

$$\sqrt{121} = 11 \qquad 11^2 = 121$$

The symbol $-\sqrt{}$ is used for the **negative square root** of a number. For example, the negative square root of 121 is −11.

$$-\sqrt{121} = -11 \qquad (-11)^2 = 121$$

The **radical symbol** $\sqrt{}$ always represents the positive square root (except that $\sqrt{0} = 0$). The number inside the radical symbol is the **radicand,** and the entire expression—radical symbol and radicand—is a **radical.**

Radical symbol → $\sqrt{a}$ ← Radicand

Radical

An algebraic expression containing a radical is a **radical expression.**

The radical symbol $\sqrt{}$ has been used since sixteenth-century Germany and was probably derived from the letter *R*. The radical symbol at the right comes from the Latin word for root, *radix*. It was first used by Leonardo of Pisa (Fibonacci) in 1220.

Early radical symbol

1 Find all square roots of the given number.

GS (a) 100
Ask "What number when multiplied by itself equals 100?" There are two answers.

(b) 25 (c) 36 (d) $\frac{25}{36}$

Leonardo of Pisa (c. 1170–1250)

Answers

1. (a) 10, −10 (b) 5, −5 (c) 6, −6 (d) $\frac{5}{6}, -\frac{5}{6}$

We summarize our discussion of square roots as follows.

Square Roots of *a*

Let a be a positive real number.

$\sqrt{a}$ is the positive or principal square root of a.

$-\sqrt{a}$ is the negative square root of a.

For nonnegative a, the following hold.

$$\sqrt{a} \cdot \sqrt{a} = (\sqrt{a})^2 = a \quad \text{and} \quad -\sqrt{a} \cdot (-\sqrt{a}) = (-\sqrt{a})^2 = a$$

Also, $\sqrt{0} = 0$.

Calculator Tip

Most calculators have a square root key, usually labeled $\sqrt{x}$, for finding the square root of a number. On some models, the square root key must be used in conjunction with the key marked INV or 2nd.

EXAMPLE 2 Finding Square Roots

Find each square root.

(a) $\sqrt{144}$

The radical $\sqrt{144}$ represents the positive or principal square root of 144. Think of a positive number whose square is 144.

$$12^2 = 144, \quad \text{so} \quad \sqrt{144} = 12.$$

(b) $-\sqrt{1024}$

This symbol represents the negative square root of 1024. A calculator with a square root key can be used to find $\sqrt{1024} = 32$. Then, $-\sqrt{1024} = -32$.

(c) $\sqrt{\frac{4}{9}} = \frac{2}{3}$ **(d)** $-\sqrt{\frac{16}{49}} = -\frac{4}{7}$ **(e)** $\sqrt{0.81} = 0.9$

Work Problem 2 at the Side. ▶

As noted above, when the square root of a positive real number is squared, the result is that positive real number. (Also, $(\sqrt{0})^2 = 0$.)

EXAMPLE 3 Squaring Radical Expressions

Find the *square* of each radical expression.

(a) $\sqrt{13}$ The square of $\sqrt{13}$ is $(\sqrt{13})^2 = 13$. Definition of square root

(b) $-\sqrt{29}$

$(-\sqrt{29})^2 = 29$

The square of a *negative* number is positive.

(c) $\sqrt{p^2+1}$

$(\sqrt{p^2+1})^2 = p^2 + 1$

Work Problem 3 at the Side. ▶

2 Find each square root.

GS **(a)** $\sqrt{16}$

$4^2 =$ ____, so

$\sqrt{16} =$ ____.

GS **(b)** $-\sqrt{169}$

$13^2 =$ ____, so

$-\sqrt{169} =$ ____.

(c) $-\sqrt{225}$ **(d)** $\sqrt{729}$

(e) $-\sqrt{\frac{36}{25}}$ **(f)** $\sqrt{0.49}$

Teaching Tip

Mention that the statement $\sqrt{9} = \pm 3$ is incorrect. It says, in part, that a positive number equals a negative number.

3 Find the *square* of each radical expression.

GS **(a)** $\sqrt{41}$

$(\sqrt{41})^2 =$ ____

GS **(b)** $-\sqrt{39}$

$(-\sqrt{39})^2 =$ ____

(c) $\sqrt{120}$

(d) $\sqrt{2x^2+3}$

Answers

2. **(a)** 16; 4 **(b)** 169; −13 **(c)** −15 **(d)** 27 **(e)** $-\frac{6}{5}$ **(f)** 0.7

3. **(a)** 41 **(b)** 39 **(c)** 120 **(d)** $2x^2 + 3$

OBJECTIVE 2 Decide whether a given root is rational, irrational, or not a real number. Numbers with rational square roots are **perfect squares.**

$$\text{Perfect squares}\left\{\begin{array}{l} 25 \\ 144 \quad \text{are perfect squares since} \\ \dfrac{4}{9} \end{array}\right. \begin{array}{r} \sqrt{25}=5 \\ \sqrt{144}=12 \\ \sqrt{\dfrac{4}{9}}=\dfrac{2}{3} \end{array}\left.\right\}\text{Rational square roots}$$

A number that is not a perfect square has a square root that is not a rational number. For example, $\sqrt{5}$ is not a rational number because it cannot be written as the ratio of two integers. Its decimal equivalent (or approximation) neither terminates nor repeats. However, $\sqrt{5}$ is a real number and corresponds to a point on the number line.

A real number that is not rational is an **irrational number.** The number $\sqrt{5}$ is irrational. ***Many square roots of integers are irrational.***

> If a is a *positive* real number that is *not* a perfect square, then
>
> $\sqrt{a}$ is irrational.

Not every number has a real number square root. For example, there is no real number that can be squared to obtain -36. (The square of a real number can never be negative.) Because of this, $\mathbf{\sqrt{-36}}$ ***is not a real number.***

> If a is a ***negative*** real number, then $\sqrt{a}$ is *not* a real number.

CAUTION

Do not confuse $\sqrt{-36}$ and $-\sqrt{36}$. $\sqrt{-36}$ is not a real number since there is no real number that can be squared to obtain -36. However, $-\sqrt{36}$ is the negative square root of 36, which is -6.

EXAMPLE 4 Identifying Types of Square Roots

Tell whether each square root is *rational, irrational,* or *not a real number.*

(a) $\sqrt{17}$ Because 17 is not a perfect square, $\sqrt{17}$ is irrational.

(b) $\sqrt{64}$ 64 is a perfect square, 8^2, so $\sqrt{64}=8$ is a rational number.

(c) $\sqrt{-25}$ There is no real number whose square is -25. Therefore, $\sqrt{-25}$ is not a real number.

◀ Work Problem 4 at the Side.

Note

Not all irrational numbers are square roots of integers. For example, π (approximately 3.14159) is an irrational number that is not a square root of any integer.

---

4 Tell whether each square root is *rational, irrational,* or *not a real number.*

(a) $\sqrt{9}$

Teaching Tip

Remind students that the real numbers consist of the rationals and the irrationals. To this point, we have worked only with rational numbers. This chapter deals with irrational numbers as well.

Reals < Rationals / Irrationals

(b) $\sqrt{7}$

Teaching Tip

You may want to review the diagram of the real number system given in Section 1.3 with students.

(c) $\sqrt{\dfrac{9}{16}}$

(d) $\sqrt{72}$

(e) $\sqrt{-43}$

Answers

4. (a) rational **(b)** irrational **(c)** rational **(d)** irrational **(e)** not a real number

OBJECTIVE 3 Find decimal approximations for irrational square roots. Even if a number is irrational, a decimal that approximates the number can be found using a calculator.

EXAMPLE 5 Approximating Irrational Square Roots

Find a decimal approximation for each square root. Round answers to the nearest thousandth.

(a) $\sqrt{11}$

Using the square root key of a calculator gives $3.31662479 \approx 3.317$, where $\approx$ means **"is approximately equal to."**

(b) $\sqrt{39} \approx 6.245$ Use a calculator. **(c)** $-\sqrt{740} \approx -27.203$

Work Problem 5 at the Side. ▶

OBJECTIVE 4 Use the Pythagorean theorem. Many applications of square roots use the Pythagorean theorem. Recall from **Section 6.7** and **Figure 1** that if c is the length of the hypotenuse of a right triangle, and a and b are the lengths of the two legs, then

$$a^2 + b^2 = c^2.$$

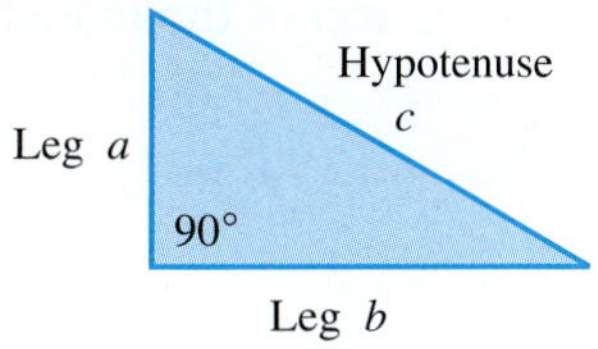

Figure 1

EXAMPLE 6 Using the Pythagorean Theorem

Find the length of the unknown side of each right triangle with sides a, b, and c, where c is the hypotenuse.

(a) $a = 3, b = 4$

$$a^2 + b^2 = c^2 \quad \text{Use the Pythagorean theorem.}$$

$$3^2 + 4^2 = c^2 \quad \text{Let } a = 3 \text{ and } b = 4.$$

$$9 + 16 = c^2 \quad \text{Square.}$$

$$25 = c^2 \quad \text{Add.}$$

Since the length of a side of a triangle must be a positive number, find the positive square root of 25 to get c.

$$c = \sqrt{25} = 5$$

(b) $b = 5, c = 9$

$$a^2 + b^2 = c^2 \quad \text{Use the Pythagorean theorem.}$$

$$a^2 + 5^2 = 9^2 \quad \text{Let } b = 5 \text{ and } c = 9.$$

Solve for a^2.

$$a^2 + 25 = 81 \quad \text{Square.}$$

$$a^2 = 56 \quad \text{Subtract 25.}$$

Use a calculator to find the positive square root of 56 to approximate a.

$$a = \sqrt{56} \approx 7.483$$

Work Problem 6 at the Side. ▶

5 Find a decimal approximation for each square root. Round answers to the nearest thousandth.

(a) $\sqrt{28}$ **(b)** $\sqrt{63}$

(c) $-\sqrt{190}$ **(d)** $\sqrt{1000}$

6 Find the length of the unknown side of each right triangle with sides a, b, and c, where c is the hypotenuse. Give any decimal approximations to the nearest thousandth.

GS **(a)** $a = 7, b = 24$

$$\_\_\_\_^2 + \_\_\_\_^2 = \_\_\_\_^2$$

$$49 + \_\_\_\_ = c^2$$

$$\_\_\_\_ = c^2$$

The positive square root of 625 is ____, so $c =$ ____.

(b) $b = 13, c = 15$

(c)

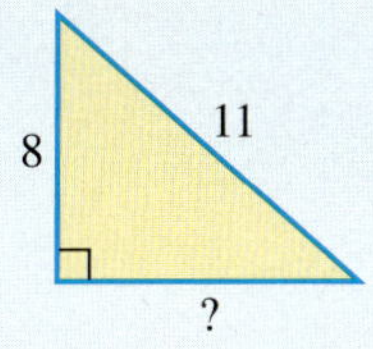

Answers

5. (a) 5.292 **(b)** 7.937 **(c)** −13.784 **(d)** 31.623

6. (a) 7; 24; c; 576; 625; 25; 25 **(b)** $\sqrt{56} \approx 7.483$ **(c)** $\sqrt{57} \approx 7.550$

7 A rectangle has dimensions 5 ft by 12 ft. Find the length of its diagonal.

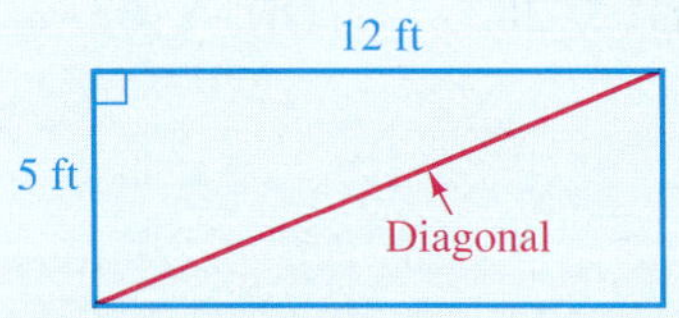

(Note that the diagonal divides the rectangle into two right triangles with itself as the hypotenuse.)

Answer

7. 13 ft

CAUTION

Be careful not to make the common mistake of thinking that $\sqrt{a^2 + b^2}$ equals $a + b$. Consider the following.

$$\sqrt{9 + 16} = \sqrt{25} = 5, \quad \text{but} \quad \sqrt{9} + \sqrt{16} = 3 + 4 = 7.$$

In general, $\sqrt{a^2 + b^2} \neq a + b$.

EXAMPLE 7 Using the Pythagorean Theorem to Solve an Application

A ladder 10 ft long leans against a wall. The foot of the ladder is 6 ft from the base of the wall. How high up the wall does the top of the ladder rest?

Step 1 **Read** the problem again.

Step 2 **Assign a variable.** As shown in **Figure 2,** a right triangle is formed with the ladder as the hypotenuse. Let a represent the height of the top of the ladder when measured straight down to the ground.

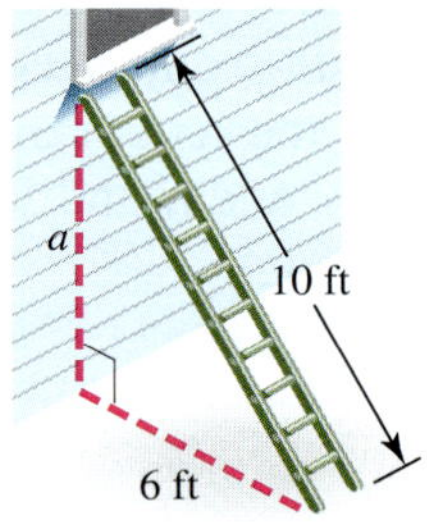

Recall that the symbol ⅂ indicates a 90° or right angle.

Figure 2

Step 3 **Write an equation** using the Pythagorean theorem.

$$a^2 + b^2 = c^2 \qquad \text{Substitute carefully.}$$

$$a^2 + 6^2 = 10^2 \qquad \text{Let } b = 6 \text{ and } c = 10.$$

Step 4 **Solve.**

$$a^2 + 36 = 100 \qquad \text{Square.}$$

$$a^2 = 64 \qquad \text{Subtract 36.}$$

$$a = \sqrt{64} \qquad \text{Solve for } a.$$

$$a = 8 \qquad \sqrt{64} = 8$$

Choose the positive square root of 64 since a represents a length.

Step 5 **State the answer.** The top of the ladder rests 8 ft up the wall.

Step 6 **Check.** From **Figure 2,** we have the following.

$$8^2 + 6^2 \stackrel{?}{=} 10^2 \qquad a^2 + b^2 = c^2$$

$$64 + 36 = 100 \ \checkmark \qquad \text{True}$$

The check confirms that the top of the ladder rests 8 ft up the wall.

◀ Work Problem **7** at the Side.

OBJECTIVE 5 **Find cube, fourth, and other roots.** Finding the square root of a number is the inverse (opposite) of squaring a number. There are inverses to finding the cube of a number and to finding the fourth or greater power of a number. These inverses are, respectively, the **cube root** $\sqrt[3]{a}$, and the **fourth root** $\sqrt[4]{a}$. Similar symbols are used for other roots.

> **$\sqrt[n]{a}$**
>
> The nth root of a, written $\sqrt[n]{a}$, is a number whose nth power equals a. That is,
>
> $$\sqrt[n]{a} = b \quad \text{means} \quad b^n = a.$$

In $\sqrt[n]{a}$, the number n is the **index,** or **order,** of the radical.

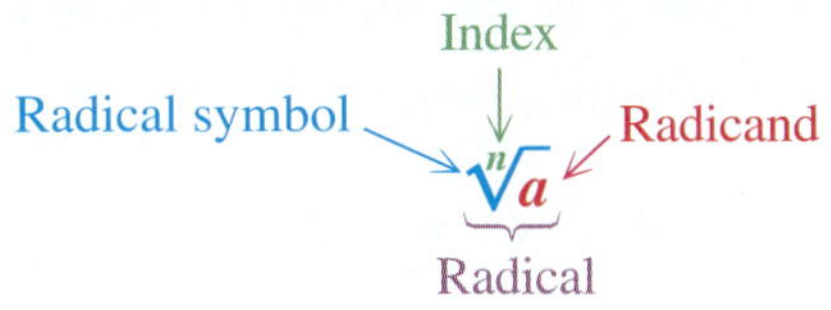

We could write $\sqrt[2]{a}$ instead of $\sqrt{a}$, but the simpler symbol $\sqrt{a}$ is customary since the square root is the most commonly used root.

> **Calculator Tip**
>
> A calculator that has a key marked $\sqrt[x]{y}$, x^y, or y^x (again perhaps in conjunction with the INV or 2nd key) can be used to find other roots.

When working with cube roots or fourth roots, it is helpful to memorize the first few **perfect cubes** ($1^3 = 1$, $2^3 = 8$, $3^3 = 27$, and so on) and the first few **perfect fourth powers** ($1^4 = 1$, $2^4 = 16$, $3^4 = 81$, and so on).

Work Problem 8 at the Side. ▶

EXAMPLE 8 **Finding Cube Roots**

Find each cube root.

(a) $\sqrt[3]{8}$

What number can be cubed to give 8? Because $2^3 = 8$, $\sqrt[3]{8} = 2$.

(b) $\sqrt[3]{-8} = -2$, because $(-2)^3 = -8$.

(c) $\sqrt[3]{216} = 6$, because $6^3 = 216$.

Work Problem 9 at the Side. ▶

Notice in **Example 8(b)** that we can find the cube root of a negative number. (Contrast this with the square root of a negative number, which is not real.) In fact, the cube root of a positive number is positive, and the cube root of a negative number is negative. ***There is only one real number cube root for each real number.***

When a radical has an ***even index*** (square root, fourth root, and so on), ***the radicand must be nonnegative*** to yield a real number root. Also, for $a > 0$,

$$\sqrt{a},\ \sqrt[4]{a},\ \sqrt[6]{a}, \text{ and so on are positive (principal) roots.}$$

$$-\sqrt{a},\ -\sqrt[4]{a},\ -\sqrt[6]{a}, \text{ and so on are negative roots.}$$

8 Complete the following list of perfect cubes and perfect fourth powers.

Perfect Cubes	Perfect Fourth Powers
$1^3 = 1$	$1^4 = 1$
$2^3 = 8$	$2^4 = 16$
$3^3 = 27$	$3^4 = 81$
$4^3 =$ ____	$4^4 =$ ____
$5^3 =$ ____	$5^4 =$ ____
$6^3 =$ ____	$6^4 =$ ____
$7^3 =$ ____	$7^4 =$ ____
$8^3 =$ ____	$8^4 =$ ____
$9^3 =$ ____	$9^4 =$ ____
$10^3 =$ ____	$10^4 =$ ____

9 Find each cube root.

GS **(a)** $\sqrt[3]{27}$

____$^3 = 27$, so

$\sqrt[3]{27} =$ ____.

(b) $\sqrt[3]{1}$

(c) $\sqrt[3]{-125}$

Teaching Tip

Encourage students to keep the lists of perfect cubes and perfect fourth powers handy for reference when simplifying cube roots and fourth roots.

Answers

8. Perfect cubes: 64; 125; 216; 343; 512; 729; 1000
Perfect fourth powers: 256; 625; 1296; 2401; 4096; 6561; 10,000

9. **(a)** 3; 3 **(b)** 1 **(c)** −5

10 Find each root.

(a) $\sqrt[4]{81}$

(b) $\sqrt[4]{-81}$

(c) $-\sqrt[4]{81}$

(d) $\sqrt[5]{243}$

(e) $\sqrt[5]{-243}$

Answers

10. (a) 3 **(b)** not a real number **(c)** -3 **(d)** 3 **(e)** -3

EXAMPLE 9 Finding Other Roots

Find each root.

(a) $\sqrt[4]{16} = 2$, because 2 is positive and $2^4 = 16$.

(b) $-\sqrt[4]{16}$

From part (a), $\sqrt[4]{16} = 2$, so the negative root is $-\sqrt[4]{16} = -2$.

(c) $\sqrt[4]{-16}$

For a real number fourth root, the radicand must be nonnegative. There is no real number that equals $\sqrt[4]{-16}$.

(d) $-\sqrt[5]{32}$

First find $\sqrt[5]{32}$. Because 2 is the number whose fifth power is 32, $\sqrt[5]{32} = 2$. Since $\sqrt[5]{32} = 2$, it follows that

$$-\sqrt[5]{32} = -2.$$

(e) $\sqrt[5]{-32} = -2$, because $(-2)^5 = -32$.

◀ **Work Problem 10 at the Side.**

8.1 Exercises

FOR EXTRA HELP

Download the MyDashBoard App

MyMathLab®

CONCEPT CHECK *Decide whether each statement is* true *or* false. *If false, tell why.*

1. Every positive number has two real square roots.
true

2. A negative number has negative square roots.
false; A negative number has no real square roots.

3. Every nonnegative number has two real square roots.
false; Zero has only one square root.

4. The positive square root of a positive number is its principal square root.
true

5. The cube root of every real number has the same sign as the number itself.
true

6. Every positive number has three real cube roots.
false; A positive number has just one real cube root.

Find all square roots of each number. ***See Example 1.***

7. 9
$-3, 3$

8. 16
$-4, 4$

9. 64
$-8, 8$

10. 100
$-10, 10$

11. 169
$-13, 13$

12. 225
$-15, 15$

13. $\frac{25}{196}$
$-\frac{5}{14}, \frac{5}{14}$

14. $\frac{81}{400}$
$-\frac{9}{20}, \frac{9}{20}$

15. 900
$-30, 30$

16. 1600
$-40, 40$

CONCEPT CHECK *What must be true about the value of the variable a for each statement in Exercises 17–20 to be true?*

17. $\sqrt{a}$ represents a positive number.
a must be positive.

18. $-\sqrt{a}$ represents a negative number.
a must be positive.

19. $\sqrt{a}$ is not a real number.
a must be negative.

20. $-\sqrt{a}$ is not a real number.
a must be negative.

Find each square root. ***See Examples 2 and 4(c).***

21. $\sqrt{64}$
GS $8^2 =$ __64__, so
$\sqrt{64} =$ __8__.

22. $\sqrt{9}$
GS $3^2 =$ __9__, so
$\sqrt{9} =$ __3__.

23. $\sqrt{1}$
1

24. $\sqrt{4}$
2

25. $\sqrt{49}$
7

26. $\sqrt{81}$
9

27. $-\sqrt{256}$
-16

28. $-\sqrt{196}$
-14

29. $-\sqrt{\frac{144}{121}}$
$-\frac{12}{11}$

30. $-\sqrt{\frac{49}{36}}$
$-\frac{7}{6}$

31. $\sqrt{0.64}$
0.8

32. $\sqrt{0.16}$
0.4

33. $\sqrt{-121}$
not a real number

34. $\sqrt{-64}$
not a real number

35. $-\sqrt{-49}$
not a real number

36. $-\sqrt{-100}$
not a real number

Find the square of each radical expression. ***See Example 3.***

37. $\sqrt{100}$ GS $\left(\sqrt{100}\right)^2 =$ __100__

38. $\sqrt{36}$ GS $\left(\sqrt{36}\right)^2 =$ __36__

39. $-\sqrt{19}$ 19

40. $-\sqrt{99}$ 99

41. $\sqrt{\frac{2}{3}}$ $\frac{2}{3}$

42. $\sqrt{\frac{5}{7}}$ $\frac{5}{7}$

43. $\sqrt{3x^2+4}$ $3x^2+4$

44. $\sqrt{9y^2+3}$ $9y^2+3$

CONCEPT CHECK *Without using a calculator, determine between which two consecutive integers each square root lies. For example,*

$$\sqrt{75} \text{ is between 8 and 9, because } \sqrt{64}=8,\ \sqrt{81}=9, \text{ and } 64<75<81.$$

45. $\sqrt{94}$ 9 and 10

46. $\sqrt{43}$ 6 and 7

47. $\sqrt{51}$ 7 and 8

48. $\sqrt{30}$ 5 and 6

49. $\sqrt{23.2}$ 4 and 5

50. $\sqrt{10.3}$ 3 and 4

Determine whether each number is rational, irrational, *or* not a real number. *If a number is rational, give its exact value. If a number is irrational, give a decimal approximation to the nearest thousandth. Use a calculator as necessary.* ***See Examples 4 and 5.***

51. $\sqrt{25}$ rational; 5

52. $\sqrt{169}$ rational; 13

53. $\sqrt{29}$ irrational; 5.385

54. $\sqrt{33}$ irrational; 5.745

55. $-\sqrt{64}$ rational; -8

56. $-\sqrt{81}$ rational; -9

57. $-\sqrt{300}$ irrational; -17.321

58. $-\sqrt{500}$ irrational; -22.361

59. $\sqrt{-29}$ not a real number

60. $\sqrt{-47}$ not a real number

61. $\sqrt{1200}$ irrational; 34.641

62. $\sqrt{1500}$ irrational; 38.730

Work Exercises 63 and 64 without using a calculator.

63. Choose the best estimate for the length and width (in meters) of this rectangle. C

A. 11 by 6 **B.** 11 by 7 **C.** 10 by 7 **D.** 10 by 6

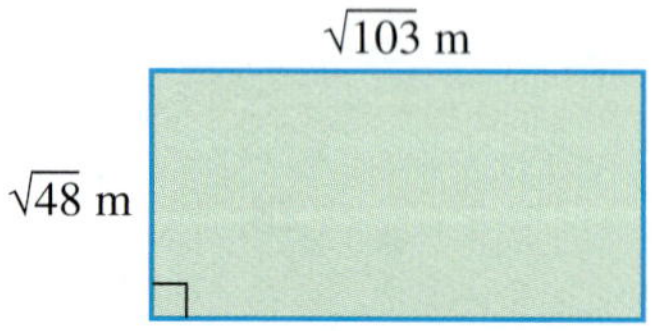

64. Choose the best estimate for the base and height (in feet) of this triangle. A

A. $b=8, h=5$ **B.** $b=8, h=4$

C. $b=9, h=5$ **D.** $b=9, h=4$

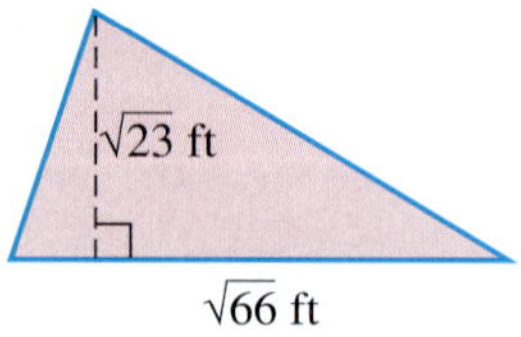

Find the length of the unknown side of each right triangle with sides a, b, and c, where c is the hypotenuse. Give any decimal approximations to the nearest thousandth. ***See*** **Figure 1** ***and Example 6.***

65. $a=8, b=15$ $c=17$

66. $a=24, b=10$ $c=26$

67. $a=6, c=10$ $b=8$

68. $a=5, c=13$ $b=12$

69. $a=11, b=4$ $c \approx 11.705$

70. $a=13, b=9$ $c \approx 15.811$

Solve each problem. Give any decimal approximations to the nearest tenth.
See Example 7.

71. The diagonal of a rectangle measures 25 cm. The width of the rectangle is 7 cm. Find the length of the rectangle. **24 cm**

72. The length of a rectangle is 40 m, and the width is 9 m. Find the measure of the diagonal of the rectangle. **41 m**

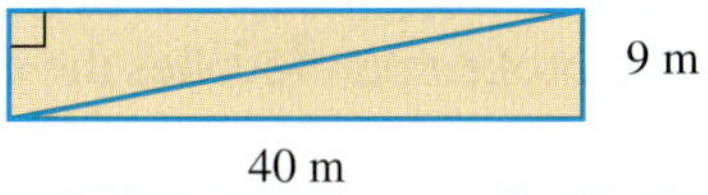

73. Tyler is flying a kite on 100 ft of string. How high is it above his hand (vertically) if the horizontal distance between Tyler and the kite is 60 ft?
80 ft

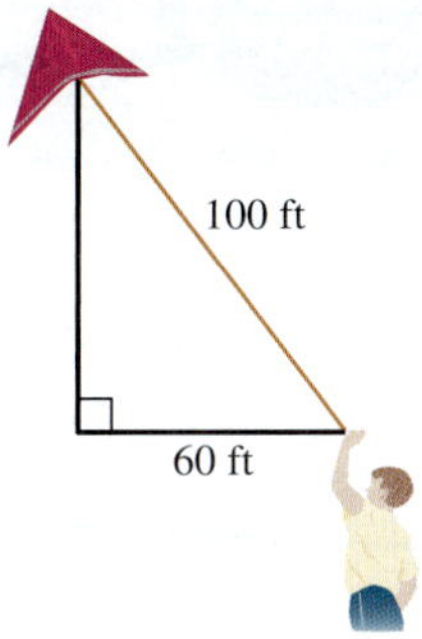

74. A guy wire is attached to the mast of a transmitting antenna. It is attached 96 ft above ground level. If the wire is staked to the ground 72 ft from the base of the mast, how long is the wire?
120 ft

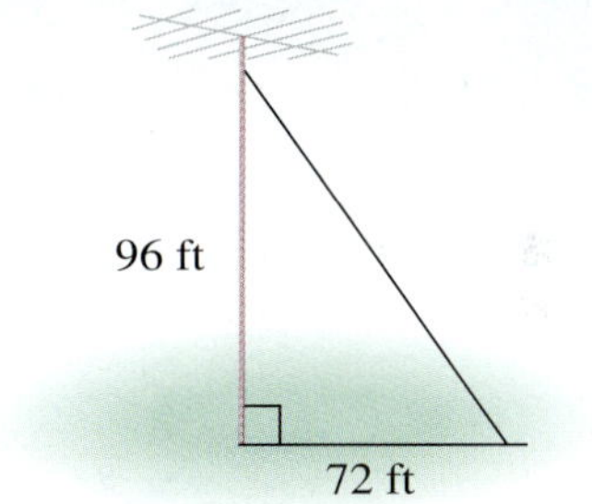

75. A surveyor measured the distances shown in the figure. Find the distance across the lake between points *R* and *S*. **195 ft**

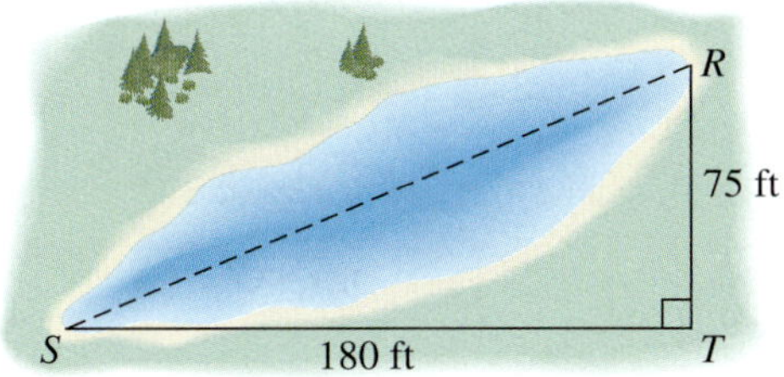

76. A boat is being pulled toward a dock with a rope attached at water level. When the boat is 24 ft from the dock, 30 ft of rope is extended. What is the height of the dock above the water? **18 ft**

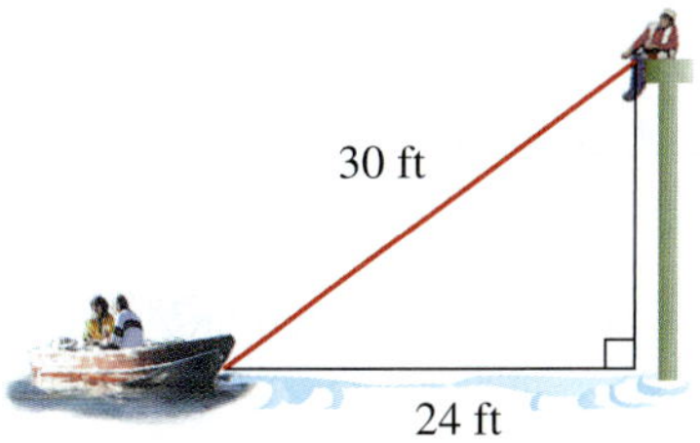

77. A surveyor wants to find the height of a building. At a point 110.0 ft from the base of the building he sights to the top of the building and finds the distance to be 193.0 ft. How high is the building?
158.6 ft

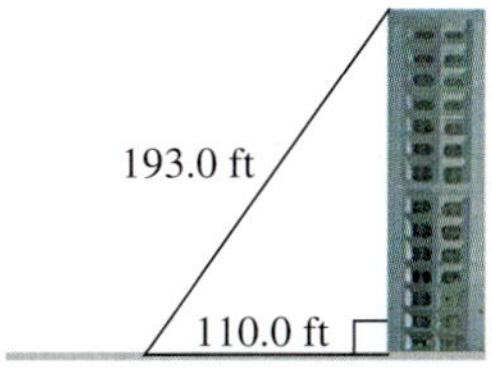

78. Two towns are separated by a dense forest. To go from Town B to Town A, it is necessary to travel due west for 19.0 mi, then turn due north and travel for 14.0 mi. How far apart are the towns?
23.6 mi

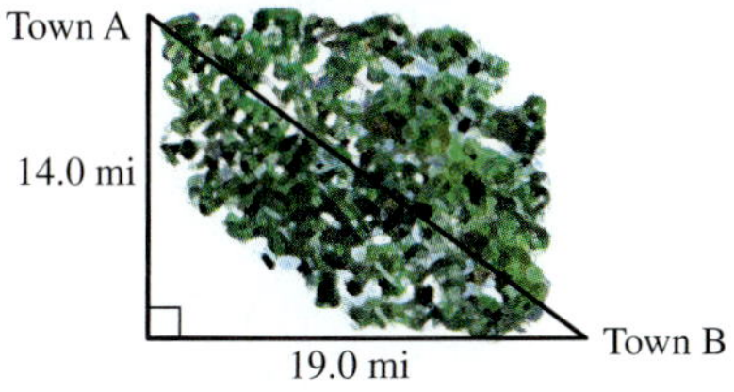

79. Following Hurricane Katrina, thousands of pine trees in southeastern Louisiana formed right triangles as shown in the photo. Suppose that, for a small such tree, the vertical distance from the base of the broken tree to the point of the break is 4.5 ft. The length of the broken part is 12.0 ft. How far along the ground is it from the base of the tree to the point where the broken part touches the ground?
11.1 ft

80. One of the authors of this text purchased a Samsung UN46C6300 LED HDTV. A television set is "sized" according to the diagonal measurement of the viewing screen. The author purchased a 46-in. TV, so the TV measures 46 in. from one corner of the viewing screen diagonally to the other corner. The viewing screen is 40 in. wide. Find the height of the viewing screen. 22.7 in.

81. What is the value of x (to the nearest thousandth) in the figure? 9.434

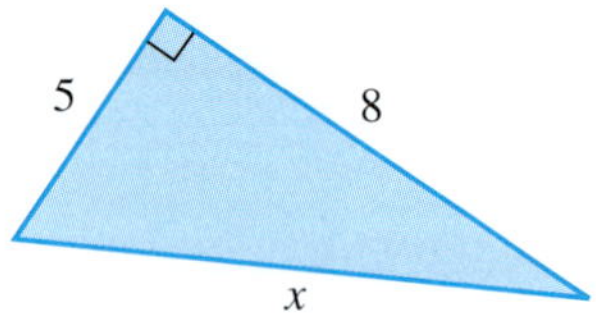

82. What is the value of y (to the nearest thousandth) in the figure? 9.747

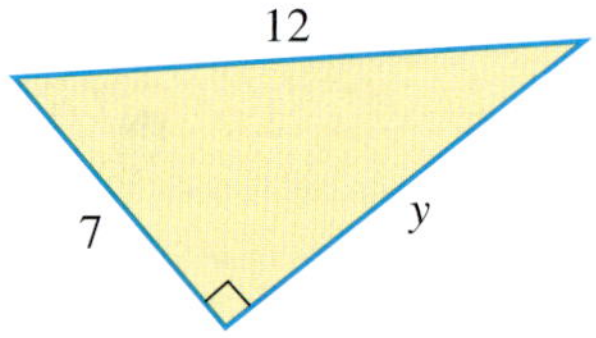

Find each root. ***See Examples 8 and 9.***

83. $\sqrt[3]{64}$
GS $\underline{\ 4\ }^3 = 64$, so
$\sqrt[3]{64} = \underline{\ 4\ }$.

84. $\sqrt[3]{343}$
GS $\underline{\ 7\ }^3 = 343$, so
$\sqrt[3]{343} = \underline{\ 7\ }$.

85. $\sqrt[3]{125}$ 5

86. $\sqrt[3]{729}$ 9

87. $\sqrt[3]{512}$ 8

88. $\sqrt[3]{1000}$ 10

89. $\sqrt[3]{-27}$ −3

90. $\sqrt[3]{-64}$ −4

91. $\sqrt[3]{-216}$ −6

92. $\sqrt[3]{-343}$ −7

93. $-\sqrt[3]{-8}$ 2

94. $-\sqrt[3]{-216}$ 6

95. $\sqrt[4]{256}$ 4

96. $\sqrt[4]{625}$ 5

97. $\sqrt[4]{1296}$ 6

98. $\sqrt[4]{10{,}000}$ 10

99. $\sqrt[4]{-1}$ not a real number

100. $\sqrt[4]{-625}$ not a real number

101. $-\sqrt[4]{625}$ −5

102. $-\sqrt[4]{256}$ −4

103. $\sqrt[5]{-1024}$ −4

104. $\sqrt[5]{-100{,}000}$ −10

8.2 Multiplying, Dividing, and Simplifying Radicals

OBJECTIVES

1. Multiply square root radicals.
2. Simplify radicals by using the product rule.
3. Simplify radicals by using the quotient rule.
4. Simplify radicals involving variables.
5. Simplify other roots.

OBJECTIVE 1 Multiply square root radicals. Consider the following.

$$\sqrt{4} \cdot \sqrt{9} = 2 \cdot 3 = 6 \quad \text{and} \quad \sqrt{4 \cdot 9} = \sqrt{36} = 6$$

This shows that $\sqrt{4} \cdot \sqrt{9} = \sqrt{4 \cdot 9}$.

The result here is a particular case of the **product rule for radicals.**

Product Rule for Radicals

If a and b are nonnegative real numbers, then the following hold.

$$\sqrt{a} \cdot \sqrt{b} = \sqrt{a \cdot b} \quad \text{and} \quad \sqrt{a \cdot b} = \sqrt{a} \cdot \sqrt{b}$$

In words, the product of two square roots is the square root of the product. The square root of a product is the product of the two square roots.

EXAMPLE 1 Using the Product Rule to Multiply Radicals

Use the product rule for radicals to find each product.

(a) $\sqrt{2} \cdot \sqrt{3}$
$= \sqrt{2 \cdot 3}$
$= \sqrt{6}$

(b) $\sqrt{7} \cdot \sqrt{5}$
$= \sqrt{35}$

(c) $\sqrt{11} \cdot \sqrt{a}$ $(a \geq 0)$
$= \sqrt{11a}$

Work Problem 1 at the Side. ▶

OBJECTIVE 2 Simplify radicals by using the product rule. ***A square root radical is simplified when no perfect square factor other than 1 remains under the radical symbol.*** This is accomplished by using the product rule.

EXAMPLE 2 Using the Product Rule to Simplify Radicals

Simplify each radical.

(a) $\sqrt{20}$ (20 has a perfect square factor of 4.)

$= \sqrt{4 \cdot 5}$ Factor; 4 is a perfect square.

$= \sqrt{4} \cdot \sqrt{5}$ Product rule

$= 2\sqrt{5}$ $\sqrt{4} = 2$

Thus, $\sqrt{20} = 2\sqrt{5}$. Because 5 has no perfect square factor (other than 1), $2\sqrt{5}$ is the **simplified form** of $\sqrt{20}$. Note that $2\sqrt{5}$ represents a product whose factors are 2 and $\sqrt{5}$.

We could factor 20 into prime factors and look for pairs of like factors.

$\sqrt{20}$

$= \sqrt{2 \cdot 2 \cdot 5}$ Each pair of like factors produces one factor outside the radical.

$= 2\sqrt{5}$

Continued on Next Page

1. Use the product rule for radicals to find each product.

GS **(a)** $\sqrt{6} \cdot \sqrt{11}$
$= \sqrt{\_\_\_ \cdot \_\_\_}$
$= \sqrt{\_\_\_}$

(b) $\sqrt{2} \cdot \sqrt{5}$

(c) $\sqrt{10} \cdot \sqrt{r}$ $(r \geq 0)$

Answers

1. (a) 6; 11; 66 **(b)** $\sqrt{10}$ **(c)** $\sqrt{10r}$

2 Simplify each radical.

GS **(a)** $\sqrt{8}$

$= \sqrt{\_\_\_\_ \cdot 2}$

$= \sqrt{\_\_\_\_} \cdot \sqrt{2}$

$= \_\_\_\_$

(b) $\sqrt{27}$

Teaching Tip

In Example 2(c), again stress the use of the greatest perfect square factor in 300. Although

$300 = 4 \cdot 75$ or $300 = 25 \cdot 12$

could also be used, they would require additional steps.

(c) $\sqrt{50}$

(d) $\sqrt{60}$

(e) $\sqrt{30}$

Answers

2. **(a)** 4; 4; $2\sqrt{2}$ **(b)** $3\sqrt{3}$ **(c)** $5\sqrt{2}$ **(d)** $2\sqrt{15}$ **(e)** cannot be simplified

(b) $\sqrt{72}$ — Look for the *greatest* perfect square factor of 72.

$= \sqrt{36 \cdot 2}$ Factor; 36 is a perfect square.

$= \sqrt{36} \cdot \sqrt{2}$ Product rule

$= 6\sqrt{2}$ $\sqrt{36} = 6$

We could factor 72 into prime factors and look for pairs of like factors.

$\sqrt{72}$

$= \sqrt{2 \cdot 2 \cdot 2 \cdot 3 \cdot 3}$ Factor into primes.

$= 2 \cdot 3 \cdot \sqrt{2}$ $\sqrt{2 \cdot 2} = 2$; $\sqrt{3 \cdot 3} = 3$

$= 6\sqrt{2}$ The result is the same.

(c) $\sqrt{300}$

$= \sqrt{100 \cdot 3}$ 100 is a perfect square.

$= \sqrt{100} \cdot \sqrt{3}$ Product rule

$= 10\sqrt{3}$ $\sqrt{100} = 10$

(d) $\sqrt{15}$ Since 15 has no perfect square factors (except 1), $\sqrt{15}$ cannot be simplified.

◀ Work Problem 2 at the Side.

EXAMPLE 3 Multiplying and Simplifying Radicals

Find each product and simplify.

(a) $\sqrt{9} \cdot \sqrt{75}$

$= 3\sqrt{75}$ $\sqrt{9} = 3$

$= 3\sqrt{25 \cdot 3}$ Factor; 25 is a perfect square.

$= 3\sqrt{25} \cdot \sqrt{3}$ Product rule

$= 3 \cdot 5 \cdot \sqrt{3}$ $\sqrt{25} = 5$

$= 15\sqrt{3}$ Multiply.

We could have used the product rule to get $\sqrt{9} \cdot \sqrt{75} = \sqrt{675}$, and then simplified. The method above uses smaller numbers.

(b) $\sqrt{8} \cdot \sqrt{12}$

$= \sqrt{8 \cdot 12}$ Product rule

$= \sqrt{4 \cdot 2 \cdot 4 \cdot 3}$ Factor; 4 is a perfect square.

$= \sqrt{4} \cdot \sqrt{4} \cdot \sqrt{2 \cdot 3}$ Commutative property; product rule

$= 2 \cdot 2 \cdot \sqrt{6}$ $\sqrt{4} = 2$

$= 4\sqrt{6}$ Multiply.

Continued on Next Page

(c) $2\sqrt{3} \cdot 3\sqrt{6}$

$= 2 \cdot 3 \cdot \sqrt{3 \cdot 6}$ Commutative property; product rule

$= 6\sqrt{18}$ Multiply.

$= 6\sqrt{9 \cdot 2}$ Factor; 9 is a perfect square.

$= 6\sqrt{9} \cdot \sqrt{2}$ Product rule

$= 6 \cdot 3 \cdot \sqrt{2}$ $\sqrt{9} = 3$

$= 18\sqrt{2}$ Multiply.

Work Problem 3 at the Side. ▶

Note

There is often more than one way to find a product.

$\sqrt{8} \cdot \sqrt{12}$ **Example 3(b)**

$= \sqrt{4 \cdot 2} \cdot \sqrt{4 \cdot 3}$ Factor.

$= 2\sqrt{2} \cdot 2\sqrt{3}$ $\sqrt{4} = 2$

$= 2 \cdot 2 \cdot \sqrt{2} \cdot \sqrt{3}$ Commutative property

Same result → $= 4\sqrt{6}$ Multiply; product rule

OBJECTIVE 3 Simplify radicals by using the quotient rule. The **quotient rule for radicals** is very similar to the product rule.

Quotient Rule for Radicals

If a and b are nonnegative real numbers and $b \neq 0$, then the following hold.

$$\sqrt{\frac{a}{b}} = \frac{\sqrt{a}}{\sqrt{b}} \quad \text{and} \quad \frac{\sqrt{a}}{\sqrt{b}} = \sqrt{\frac{a}{b}}$$

In words, the square root of a quotient is the quotient of the two square roots. The quotient of two square roots is the square root of the quotient.

EXAMPLE 4 Using the Quotient Rule to Simplify Radicals

Use the quotient rule to simplify each radical.

(a) $\sqrt{\frac{25}{9}} = \frac{\sqrt{25}}{\sqrt{9}} = \frac{5}{3}$

(b) $\frac{\sqrt{288}}{\sqrt{2}} = \sqrt{\frac{288}{2}} = \sqrt{144} = 12$

(c) $\sqrt{\frac{3}{4}} = \frac{\sqrt{3}}{\sqrt{4}} = \frac{\sqrt{3}}{2}$

Work Problem 4 at the Side. ▶

3 Find each product and simplify.

GS **(a)** $\sqrt{3} \cdot \sqrt{15}$

$= \sqrt{3 \cdot \_\_\_\_}$

$= \sqrt{\_\_\_\_}$

$= \sqrt{\_\_\_\_ \cdot 5}$

$= \sqrt{\_\_\_\_} \cdot \sqrt{5}$

$= \_\_\_\_$

(b) $\sqrt{10} \cdot \sqrt{50}$

(c) $\sqrt{12} \cdot \sqrt{2}$

(d) $3\sqrt{5} \cdot 4\sqrt{10}$

4 Use the quotient rule to simplify each radical.

GS **(a)** $\sqrt{\frac{81}{16}}$

$= \frac{\sqrt{\_\_\_\_}}{\sqrt{\_\_\_\_}}$

$= \_\_\_\_$

(b) $\frac{\sqrt{192}}{\sqrt{3}}$

(c) $\sqrt{\frac{10}{49}}$

Answers

3. **(a)** 15; 45; 9; 9; $3\sqrt{5}$ **(b)** $10\sqrt{5}$ **(c)** $2\sqrt{6}$ **(d)** $60\sqrt{2}$

4. **(a)** 81; 16; $\frac{9}{4}$ **(b)** 8 **(c)** $\frac{\sqrt{10}}{7}$

5 Simplify $\dfrac{8\sqrt{50}}{4\sqrt{5}}$.

6 Simplify.

(a) $\sqrt{\dfrac{3}{8}} \cdot \sqrt{\dfrac{7}{2}}$

GS (b) $\sqrt{\dfrac{5}{6}} \cdot \sqrt{120}$

$= \sqrt{\dfrac{5}{6} \cdot \underline{\quad}}$

$= \sqrt{\underline{\quad}}$

$= \underline{\quad}$

Answers

5. $2\sqrt{10}$

6. (a) $\dfrac{\sqrt{21}}{4}$ (b) 120; 100; 10

EXAMPLE 5 Using the Quotient Rule to Divide Radicals

Simplify.

$$\frac{27\sqrt{15}}{9\sqrt{3}}$$

$$= \frac{27}{9} \cdot \frac{\sqrt{15}}{\sqrt{3}} \quad \text{Multiplication of fractions}$$

$$= \frac{27}{9} \cdot \sqrt{\frac{15}{3}} \quad \text{Quotient rule}$$

$$= 3\sqrt{5} \quad \text{Divide.}$$

◀ **Work Problem 5 at the Side.**

EXAMPLE 6 Using Both the Product and Quotient Rules

Simplify.

$$\sqrt{\frac{3}{5}} \cdot \sqrt{\frac{1}{5}}$$

$$= \sqrt{\frac{3}{5} \cdot \frac{1}{5}} \quad \text{Product rule}$$

$$= \sqrt{\frac{3}{25}} \quad \text{Multiply fractions.}$$

$$= \frac{\sqrt{3}}{\sqrt{25}}, \text{ or } \frac{\sqrt{3}}{5} \quad \text{Quotient rule; } \sqrt{25} = 5$$

◀ **Work Problem 6 at the Side.**

OBJECTIVE 4 **Simplify radicals involving variables.** Consider a radical with variable radicand, such as $\sqrt{x^2}$.

If x represents a nonnegative number, then $\sqrt{x^2} = x$.

If x represents a negative number, then $\sqrt{x^2} = -x$, the *opposite* of x (which is positive).

For example, $\sqrt{5^2} = 5$, but $\sqrt{(-5)^2} = \sqrt{25} = 5$, the *opposite* of -5.

This means that the square root of a squared number is always nonnegative. We can use absolute value to express this.

$\sqrt{a^2}$

For any real number a, the following holds.

$$\sqrt{a^2} = |a|$$

The product and quotient rules apply when variables appear under radical symbols, as long as the variables represent only *nonnegative* real numbers. ***To avoid negative radicands, we assume variables under radical symbols are nonnegative in this text unless otherwise specified.*** In such cases, absolute value bars are not necessary, since for all $x \geq 0$, $|x| = x$.

EXAMPLE 7 **Simplifying Radicals Involving Variables**

Simplify each radical. Assume that all variables represent nonnegative real numbers.

(a) $\sqrt{x^4} = x^2$, since $(x^2)^2 = x^4$.

(b) $\sqrt{25m^6}$

$= \sqrt{25} \cdot \sqrt{m^6}$ Product rule

$= 5m^3$ $(m^3)^2 = m^6$

(c) $\sqrt{8p^{10}}$

$= \sqrt{4 \cdot 2 \cdot p^{10}}$ Factor; 4 is a perfect square.

$= \sqrt{4} \cdot \sqrt{2} \cdot \sqrt{p^{10}}$ Product rule

$= 2 \cdot \sqrt{2} \cdot p^5$ $(p^5)^2 = p^{10}$

$= 2p^5\sqrt{2}$ Commutative property

(d) $\sqrt{r^9}$

$= \sqrt{r^8 \cdot r}$

$= \sqrt{r^8} \cdot \sqrt{r}$ Product rule

$= r^4\sqrt{r}$ $(r^4)^2 = r^8$

(e) $\sqrt{\frac{5}{x^2}}$

$= \frac{\sqrt{5}}{\sqrt{x^2}}$ Quotient rule

$= \frac{\sqrt{5}}{x}$ $(x \neq 0)$

Note

A quick way to find the square root of a variable raised to an even power is to divide the exponent by the index, 2.

Examples: $\sqrt{x^6} = x^3$ and $\sqrt{x^{10}} = x^5$

$6 \div 2 = 3$ $\quad$ $10 \div 2 = 5$

Work Problem 7 at the Side. ▶

OBJECTIVE 5 **Simplify other roots.** The product and quotient rules for radicals also apply to other roots. To simplify cube roots, look for factors that are perfect cubes, as in **Section 8.1.** Recall that a perfect cube is a number with a rational cube root. For example, $\sqrt[3]{64} = 4$, and because 4 is a rational number, 64 is a perfect cube.

Other roots are handled in a similar manner.

Properties of Radicals

For all real numbers a and b where the indicated roots exist, the following hold.

$$\sqrt[n]{a} \cdot \sqrt[n]{b} = \sqrt[n]{ab} \quad \text{and} \quad \frac{\sqrt[n]{a}}{\sqrt[n]{b}} = \sqrt[n]{\frac{a}{b}} \quad (b \neq 0)$$

7 Simplify each radical. Assume that all variables represent nonnegative real numbers.

GS **(a)** $\sqrt{x^8}$

$(\_\_\_)^2 = x^8$, so

$\sqrt{x^8} = \_\_\_$.

GS **(b)** $\sqrt{36y^6}$

$= \sqrt{36} \cdot \sqrt{\_\_\_}$

$= \_\_\_$

(c) $\sqrt{100p^{12}}$

(d) $\sqrt{12z^2}$

(e) $\sqrt{a^5}$

(f) $\sqrt{\frac{10}{n^4}}$ $(n \neq 0)$

Answers

7. (a) x^4; x^4 **(b)** y^6; $6y^3$ **(c)** $10p^6$ **(d)** $2z\sqrt{3}$ **(e)** $a^2\sqrt{a}$ **(f)** $\frac{\sqrt{10}}{n^2}$

8 Simplify each radical.

GS **(a)** $\sqrt[3]{108}$

$= \sqrt[3]{\_\_\_\_ \cdot 4}$

$= \sqrt[3]{\_\_\_\_} \cdot \sqrt[3]{4}$

$= \_\_\_\_$

(b) $\sqrt[3]{250}$

(c) $\sqrt[4]{160}$

(d) $\sqrt[4]{\dfrac{16}{625}}$

9 Simplify each radical.

(a) $\sqrt[3]{z^9}$

(b) $\sqrt[3]{8x^6}$

(c) $\sqrt[3]{54t^5}$

(d) $\sqrt[3]{\dfrac{a^{15}}{64}}$

Teaching Tip

Students often forget to write the root index when the radical represents a root other than the square root. To show them the importance of the index, compare

$$\sqrt{64} = 8 \quad \text{and} \quad \sqrt[3]{64} = 4.$$

Answers

8. (a) 27; 27; $3\sqrt[3]{4}$ **(b)** $5\sqrt[3]{2}$ **(c)** $2\sqrt[4]{10}$ **(d)** $\frac{2}{5}$

9. (a) z^3 **(b)** $2x^2$ **(c)** $3t\sqrt[3]{2t^2}$ **(d)** $\frac{a^5}{4}$

EXAMPLE 8 Simplifying Other Roots

Simplify each radical.

(a) $\sqrt[3]{32}$

$= \sqrt[3]{8 \cdot 4}$ Factor; 8 is a perfect cube.

$= \sqrt[3]{8} \cdot \sqrt[3]{4}$ Product rule

$= 2\sqrt[3]{4}$ Take the cube root.

Remember to write the root index 3 in each radical.

(b) $\sqrt[4]{32}$

$= \sqrt[4]{16 \cdot 2}$ Factor; 16 is a perfect fourth power.

$= \sqrt[4]{16} \cdot \sqrt[4]{2}$ Product rule

$= 2\sqrt[4]{2}$ Take the fourth root.

Remember to write the root index 4 in each radical.

(c) $\sqrt[3]{\dfrac{27}{125}}$

$= \dfrac{\sqrt[3]{27}}{\sqrt[3]{125}}$ Quotient rule

$= \dfrac{3}{5}$ Take cube roots.

◀ **Work Problem 8 at the Side.**

Other roots of radicals involving variables can also be simplified. To simplify cube roots with variables, use the fact that for *any* real number a,

$$\sqrt[3]{a^3} = a.$$

EXAMPLE 9 Simplifying Cube Roots Involving Variables

Simplify each radical.

(a) $\sqrt[3]{m^6}$

$= m^2$ $(m^2)^3 = m^6$

(b) $\sqrt[3]{27x^{12}}$

$= \sqrt[3]{27} \cdot \sqrt[3]{x^{12}}$ Product rule

$= 3x^4$ $3^3 = 27$; $(x^4)^3 = x^{12}$

(c) $\sqrt[3]{32a^4}$

$= \sqrt[3]{8a^3 \cdot 4a}$ Factor; 8 is a perfect cube.

$= \sqrt[3]{8a^3} \cdot \sqrt[3]{4a}$ Product rule

$= 2a\sqrt[3]{4a}$ $(2a)^3 = 8a^3$

(d) $\sqrt[3]{\dfrac{y^3}{125}}$

$= \dfrac{\sqrt[3]{y^3}}{\sqrt[3]{125}}$ Quotient rule

$= \dfrac{y}{5}$ Take cube roots.

◀ **Work Problem 9 at the Side.**

8.2 Exercises

FOR EXTRA HELP

Download the MyDashBoard App

MyMathLab®

CONCEPT CHECK *Decide whether each statement is* true *or* false. *If false, show why.*

1. $\sqrt{(-6)^2} = -6$
false; $\sqrt{(-6)^2} = \sqrt{36} = 6$

2. $\sqrt[3]{(-6)^3} = -6$
true

3. The radical $2\sqrt{7}$ represents a sum.
false; $2\sqrt{7}$ represents the product of 2 and $\sqrt{7}$.

4. In the radical $3\sqrt{11}$, the numbers 3 and $\sqrt{11}$ are factors. true

Use the product rule for radicals to find each product. ***See Example 1.***

5. $\sqrt{3} \cdot \sqrt{5}$
$\sqrt{15}$

6. $\sqrt{3} \cdot \sqrt{7}$
$\sqrt{21}$

7. $\sqrt{2} \cdot \sqrt{11}$
$\sqrt{22}$

8. $\sqrt{2} \cdot \sqrt{15}$
$\sqrt{30}$

9. $\sqrt{6} \cdot \sqrt{7}$
$\sqrt{42}$

10. $\sqrt{5} \cdot \sqrt{6}$
$\sqrt{30}$

11. $\sqrt{13} \cdot \sqrt{r}$ $(r \geq 0)$
$\sqrt{13r}$

12. $\sqrt{19} \cdot \sqrt{k}$ $(k \geq 0)$
$\sqrt{19k}$

13. CONCEPT CHECK Which one of the following radicals is simplified? A

A. $\sqrt{47}$ **B.** $\sqrt{45}$ **C.** $\sqrt{48}$ **D.** $\sqrt{44}$

14. If p is a prime number, is $\sqrt{p}$ in simplified form? Explain.
Yes; because a prime number cannot be factored (except as the product of itself and 1), it cannot have any perfect square factors (except 1).

Simplify each radical. ***See Example 2.***

15. $\sqrt{45}$
$3\sqrt{5}$

16. $\sqrt{63}$
$3\sqrt{7}$

17. $\sqrt{24}$
$2\sqrt{6}$

18. $\sqrt{44}$
$2\sqrt{11}$

19. $\sqrt{90}$
$3\sqrt{10}$

20. $\sqrt{56}$
$2\sqrt{14}$

21. $\sqrt{75}$
$5\sqrt{3}$

22. $\sqrt{18}$
$3\sqrt{2}$

23. $\sqrt{125}$
$5\sqrt{5}$

24. $\sqrt{80}$
$4\sqrt{5}$

25. $\sqrt{145}$
cannot be simplified

26. $\sqrt{110}$
cannot be simplified

27. $\sqrt{160}$
$4\sqrt{10}$

28. $\sqrt{128}$
$8\sqrt{2}$

29. $-\sqrt{700}$
$-10\sqrt{7}$

30. $-\sqrt{600}$
$-10\sqrt{6}$

31. $3\sqrt{52}$
$6\sqrt{13}$

32. $9\sqrt{8}$
$18\sqrt{2}$

33. $5\sqrt{50}$
$25\sqrt{2}$

34. $6\sqrt{40}$
$12\sqrt{10}$

Use the Pythagorean theorem to find the length of the unknown side of each right triangle. Express answers as simplified radicals. **See Section 8.1.**

35.
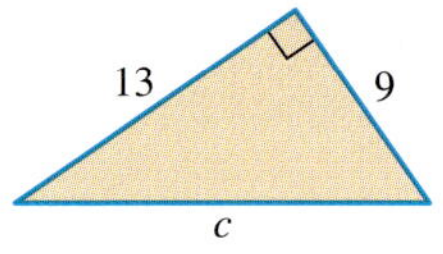

$5\sqrt{10}$

36.
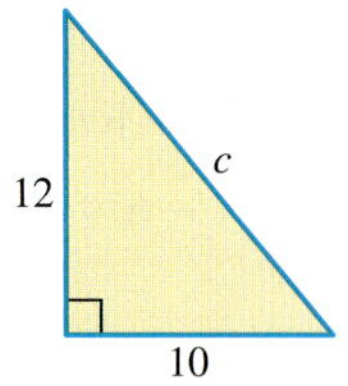

$2\sqrt{61}$

37.
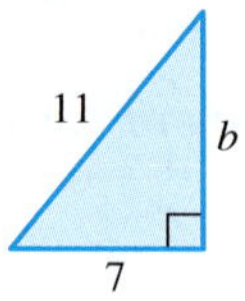

$6\sqrt{2}$

38.
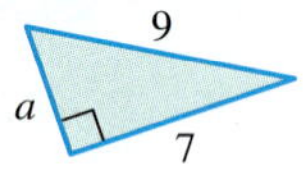

$4\sqrt{2}$

Find each product and simplify. **See Example 3.**

39. $\sqrt{3} \cdot \sqrt{18}$ $3\sqrt{6}$

40. $\sqrt{3} \cdot \sqrt{21}$ $3\sqrt{7}$

41. $\sqrt{9} \cdot \sqrt{32}$ $12\sqrt{2}$

42. $\sqrt{9} \cdot \sqrt{50}$ $15\sqrt{2}$

43. $\sqrt{12} \cdot \sqrt{48}$ 24

44. $\sqrt{50} \cdot \sqrt{72}$ 60

45. $\sqrt{12} \cdot \sqrt{30}$ $6\sqrt{10}$

46. $\sqrt{30} \cdot \sqrt{24}$ $12\sqrt{5}$

47. $2\sqrt{10} \cdot 3\sqrt{2}$ $12\sqrt{5}$

48. $5\sqrt{6} \cdot 2\sqrt{10}$ $20\sqrt{15}$

49. $5\sqrt{3} \cdot 2\sqrt{15}$ $30\sqrt{5}$

50. $4\sqrt{6} \cdot 3\sqrt{2}$ $24\sqrt{3}$

51. Simplify the product $\sqrt{8} \cdot \sqrt{32}$ in two ways.

Method 1: Multiply 8 by 32 and simplify the square root of this product.

Method 2: Simplify $\sqrt{8}$, simplify $\sqrt{32}$, and then multiply.

How do the answers compare? Make a conjecture (an educated guess) about using these methods when simplifying a product such as this.

Method 1: $\sqrt{8} \cdot \sqrt{32} = \sqrt{8 \cdot 32} = \sqrt{256} = 16$

Method 2: $\sqrt{8} = 2\sqrt{2}$ and $\sqrt{32} = 4\sqrt{2}$, so
$\sqrt{8} \cdot \sqrt{32} = 2\sqrt{2} \cdot 4\sqrt{2} = 8 \cdot 2 = 16$

The same answer results. Either method can be used to obtain the correct answer.

52. Simplify the radical $\sqrt{288}$ in two ways.

Method 1: Factor 288 as $144 \cdot 2$ and then simplify.

Method 2: Factor 288 as $48 \cdot 6$ and then simplify.

How do the answers compare? Make a conjecture concerning the quickest way to simplify such a radical.

Method 1: $\sqrt{288} = \sqrt{144 \cdot 2} = \sqrt{144} \cdot \sqrt{2} = 12\sqrt{2}$

Method 2: $\sqrt{288} = \sqrt{48 \cdot 6} = \sqrt{48} \cdot \sqrt{6} = \sqrt{16 \cdot 3} \cdot \sqrt{6} =$
$4\sqrt{3} \cdot \sqrt{6} = 4\sqrt{18} = 4\sqrt{9 \cdot 2} = 4 \cdot 3\sqrt{2} = 12\sqrt{2}$

The same answer results. Method 1 yields the answer more quickly using the greatest perfect square factor first.

Simplify each radical expression. **See Examples 4–6.**

53. $\sqrt{\frac{16}{225}}$ $\frac{4}{15}$

54. $\sqrt{\frac{9}{100}}$ $\frac{3}{10}$

55. $\sqrt{\frac{7}{16}}$ $\frac{\sqrt{7}}{4}$

56. $\sqrt{\frac{13}{25}}$ $\frac{\sqrt{13}}{5}$

57. $\sqrt{\frac{4}{50}}$ $\frac{\sqrt{2}}{5}$

58. $\sqrt{\frac{14}{72}}$ $\frac{\sqrt{7}}{6}$

59. $\frac{\sqrt{75}}{\sqrt{3}}$ 5

60. $\frac{\sqrt{200}}{\sqrt{2}}$ 10

61. $\sqrt{\frac{5}{2}} \cdot \sqrt{\frac{125}{8}}$ $\frac{25}{4}$

62. $\sqrt{\frac{8}{3}} \cdot \sqrt{\frac{512}{27}}$ $\frac{64}{9}$

63. $\frac{30\sqrt{10}}{5\sqrt{2}}$ $6\sqrt{5}$

64. $\frac{50\sqrt{20}}{2\sqrt{10}}$ $25\sqrt{2}$

Simplify each radical. Assume that all variables represent nonnegative real numbers. ***See Example 7.***

65. $\sqrt{m^2}$ — m

66. $\sqrt{k^2}$ — k

67. $\sqrt{y^4}$ — y^2

68. $\sqrt{s^4}$ — s^2

69. $\sqrt{36z^2}$ — $6z$

70. $\sqrt{49n^2}$ — $7n$

71. $\sqrt{400x^6}$ — $20x^3$

72. $\sqrt{900y^8}$ — $30y^4$

73. $\sqrt{18x^8}$ — $3x^4\sqrt{2}$

74. $\sqrt{20r^{10}}$ — $2r^5\sqrt{5}$

75. $\sqrt{45c^{14}}$ — $3c^7\sqrt{5}$

76. $\sqrt{50d^{20}}$ — $5d^{10}\sqrt{2}$

77. $\sqrt{z^5}$ — $z^2\sqrt{z}$

78. $\sqrt{y^3}$ — $y\sqrt{y}$

79. $\sqrt{a^{13}}$ — $a^6\sqrt{a}$

80. $\sqrt{p^{17}}$ — $p^8\sqrt{p}$

81. $\sqrt{64x^7}$ — $8x^3\sqrt{x}$

82. $\sqrt{25t^{11}}$ — $5t^5\sqrt{t}$

83. $\sqrt{x^6y^{12}}$ — x^3y^6

84. $\sqrt{a^8b^{10}}$ — a^4b^5

85. $\sqrt{81m^4n^2}$ — $9m^2n$

86. $\sqrt{100c^4d^6}$ — $10c^2d^3$

87. $\sqrt{\dfrac{7}{x^{10}}}$ $(x \neq 0)$ — $\dfrac{\sqrt{7}}{x^5}$

88. $\sqrt{\dfrac{14}{z^{12}}}$ $(z \neq 0)$ — $\dfrac{\sqrt{14}}{z^6}$

89. $\sqrt{\dfrac{y^4}{100}}$ — $\dfrac{y^2}{10}$

90. $\sqrt{\dfrac{w^8}{144}}$ — $\dfrac{w^4}{12}$

91. $\sqrt{\dfrac{x^6}{y^8}}$ $(y \neq 0)$ — $\dfrac{x^3}{y^4}$

92. $\sqrt{\dfrac{a^4}{b^6}}$ $(b \neq 0)$ — $\dfrac{a^2}{b^3}$

Simplify each radical. ***See Example 8.***

93. $\sqrt[3]{40}$ — $2\sqrt[3]{5}$

94. $\sqrt[3]{48}$ — $2\sqrt[3]{6}$

95. $\sqrt[3]{54}$ — $3\sqrt[3]{2}$

96. $\sqrt[3]{135}$ — $3\sqrt[3]{5}$

97. $\sqrt[3]{128}$ — $4\sqrt[3]{2}$

98. $\sqrt[3]{192}$ — $4\sqrt[3]{3}$

99. $\sqrt[4]{80}$ — $2\sqrt[4]{5}$

100. $\sqrt[4]{243}$ — $3\sqrt[4]{3}$

101. $\sqrt[3]{\dfrac{8}{27}}$ — $\dfrac{2}{3}$

102. $\sqrt[3]{\dfrac{64}{125}}$ — $\dfrac{4}{5}$

103. $\sqrt[3]{-\dfrac{216}{125}}$ — $-\dfrac{6}{5}$

104. $\sqrt[3]{-\dfrac{1}{64}}$ — $-\dfrac{1}{4}$

Simplify each radical. ***See Example 9.***

105. $\sqrt[3]{p^3}$ p

106. $\sqrt[3]{w^3}$ w

107. $\sqrt[3]{x^9}$ x^3

108. $\sqrt[3]{y^{18}}$ y^6

109. $\sqrt[3]{64z^6}$ $4z^2$

110. $\sqrt[3]{125a^{15}}$ $5a^5$

111. $\sqrt[3]{343a^9b^3}$ $7a^3b$

112. $\sqrt[3]{216m^3n^6}$ $6mn^2$

113. $\sqrt[3]{16t^5}$ $2t\sqrt[3]{2t^2}$

114. $\sqrt[3]{24x^4}$ $2x\sqrt[3]{3x}$

115. $\sqrt[3]{\dfrac{m^{12}}{8}}$ $\dfrac{m^4}{2}$

116. $\sqrt[3]{\dfrac{n^9}{27}}$ $\dfrac{n^3}{3}$

The volume V of a cube is found with the formula $V = s^3$, where s is the length of an edge of the cube. Use this information in Exercises 117 and 118.

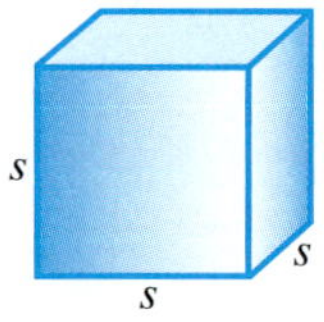

117. A container in the shape of a cube has a volume of 216 cm^3. What is the depth of the container?
6 cm

118. A cube-shaped box must be constructed to contain 128 ft^3. What should the dimensions (height, width, and length) of the box be?
Each dimension is $4\sqrt[3]{2}$ ft.

The volume V of a sphere is found with the formula $V = \frac{4}{3}\pi r^3$, where r is the length of the radius of the sphere. Use this information in Exercises 119 and 120.

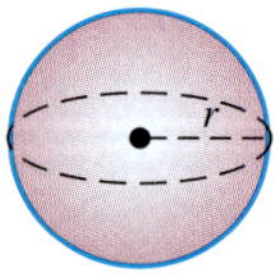

119. A ball in the shape of a sphere has a volume of 288π in.^3. What is the radius of the ball?
6 in.

120. Suppose that the volume of the ball described in **Exercise 119** is multiplied by 8. How is the radius affected?
It is doubled to 12 in.

Work Exercises 121 and 122 without using a calculator.

121. Choose the best estimate for the area (in square inches) of this rectangle. D

A. 45 **B.** 72 **C.** 80 **D.** 90

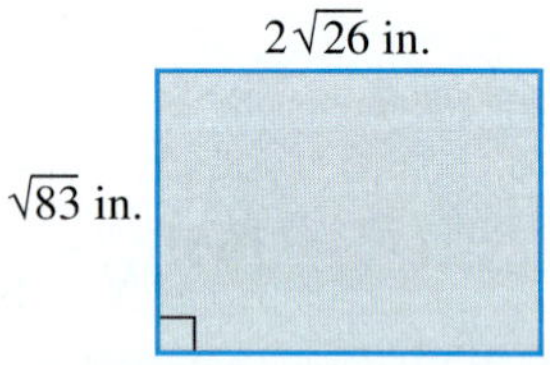

122. Choose the best estimate for the area (in square feet) of the triangle. B

A. 20 **B.** 40 **C.** 60 **D.** 80

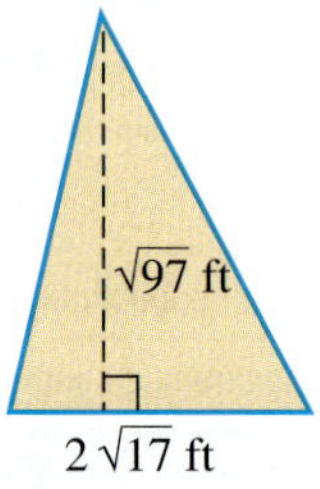

8.3 Adding and Subtracting Radicals

OBJECTIVES

1. Add and subtract radicals.
2. Simplify radical sums and differences.
3. Simplify *more* complicated radical expressions.

OBJECTIVE 1 Add and subtract radicals. We use the distributive property.

$8\sqrt{3} + 6\sqrt{3}$
$= (8+6)\sqrt{3}$ Distributive property
$= 14\sqrt{3}$ Add.

$2\sqrt{11} - 7\sqrt{11}$
$= (2-7)\sqrt{11}$ Distributive property
$= -5\sqrt{11}$ Subtract.

Only **like radicals**—those that are ***multiples of the same root of the same number***—can be combined in this way. Examples of **unlike radicals** are

$2\sqrt{5}$ and $2\sqrt{3}$, Radicands are different.

as well as $2\sqrt{3}$ and $2\sqrt[3]{3}$. Indexes are different.

EXAMPLE 1 Adding and Subtracting Like Radicals

Add or subtract, as indicated.

(a) $3\sqrt{6} + 5\sqrt{6}$ These are like radicals.
$= (3+5)\sqrt{6}$ We are factoring out $\sqrt{6}$ here.
$= 8\sqrt{6}$

(b) $5\sqrt{10} - 7\sqrt{10}$ These are like radicals.
$= (5-7)\sqrt{10}$
$= -2\sqrt{10}$

(c) $\sqrt{7} + 2\sqrt{7}$
$= 1\sqrt{7} + 2\sqrt{7}$
$= (1+2)\sqrt{7}$
$= 3\sqrt{7}$

(d) $\sqrt{5} + \sqrt{5}$
$= 1\sqrt{5} + 1\sqrt{5}$
$= (1+1)\sqrt{5}$
$= 2\sqrt{5}$

(e) $\sqrt{3} + \sqrt{7}$ cannot be combined. They are unlike radicals.

Work Problems 1 and 2 at the Side. ▶

OBJECTIVE 2 Simplify radical sums and differences.

EXAMPLE 2 Simplifying Radicals to Add or Subtract

Add or subtract, as indicated.

(a) $3\sqrt{2} + \sqrt{8}$
$= 3\sqrt{2} + \sqrt{4 \cdot 2}$ Factor.
$= 3\sqrt{2} + \sqrt{4} \cdot \sqrt{2}$ Product rule
$= 3\sqrt{2} + 2\sqrt{2}$ $\sqrt{4} = 2$
$= 5\sqrt{2}$ Add like radicals.

(b) $\sqrt{18} - \sqrt{27}$
$= \sqrt{9 \cdot 2} - \sqrt{9 \cdot 3}$ Factor; 9 is a perfect square.
$= \sqrt{9} \cdot \sqrt{2} - \sqrt{9} \cdot \sqrt{3}$ Product rule
$= 3\sqrt{2} - 3\sqrt{3}$ $\sqrt{9} = 3$

These are unlike radicals. They cannot be combined.

Continued on Next Page

1 Are the radicals in each pair *like* or *unlike*? Tell why.

(a) $5\sqrt{6}$ and $4\sqrt{6}$

(b) $2\sqrt{3}$ and $3\sqrt{2}$

(c) $\sqrt{10}$ and $\sqrt[3]{10}$

(d) $7\sqrt{2x}$ and $8\sqrt{2x}$

(e) $\sqrt{3y}$ and $\sqrt{6y}$

2 Add or subtract, as indicated.

GS **(a)** $8\sqrt{5} + 2\sqrt{5}$
$= (\_\_\_ + \_\_\_)\_\_\_$
$= \_\_\_$

(b) $4\sqrt{3} - 9\sqrt{3}$

(c) $4\sqrt{11} - 3\sqrt{11}$

(d) $\sqrt{15} + \sqrt{15}$

(e) $2\sqrt{7} + 2\sqrt{10}$

Answers

1. **(a)** like; Both are square roots and have the same radicand, 6.
(b) unlike; The radicands are different—one is 3 and one is 2.
(c) unlike; The indexes are different—one is a square root and one is a cube root.
(d) like; Both are square roots and have the same radicand, $2x$.
(e) unlike; The radicands are different—one is $3y$ and one is $6y$.

2. **(a)** 8; 2; $\sqrt{5}$; $10\sqrt{5}$ **(b)** $-5\sqrt{3}$
(c) $\sqrt{11}$ **(d)** $2\sqrt{15}$
(e) These are unlike radicals and cannot be combined.

3 Add or subtract, as indicated.

GS **(a)** $\sqrt{8} + 4\sqrt{2}$

$= \sqrt{\_\_\_ \cdot 2} + 4\sqrt{2}$

$= \sqrt{\_\_\_} \cdot \sqrt{2} + 4\sqrt{2}$

$= \_\_\_ + 4\sqrt{2}$

$= \_\_\_$

(b) $\sqrt{18} - \sqrt{2}$

(c) $\sqrt{27} + \sqrt{12}$

(d) $5\sqrt{200} - 6\sqrt{18}$

4 Simplify each radical expression. Assume that all variables represent nonnegative real numbers.

(a) $\sqrt{7} \cdot \sqrt{21} + 2\sqrt{27}$

(b) $\sqrt{3r} \cdot \sqrt{6} + \sqrt{8r}$

(c) $y\sqrt{72} - \sqrt{18y^2}$

(d) $\sqrt[3]{81x^4} + 5\sqrt[3]{24x^4}$

Teaching Tip

Warn students to avoid common errors when adding radicals. For example,

$$3\sqrt{2} + \sqrt{8} \neq 3\sqrt{10}.$$

Answers

3. **(a)** 4; 4; $2\sqrt{2}$; $6\sqrt{2}$ **(b)** $2\sqrt{2}$
(c) $5\sqrt{3}$ **(d)** $32\sqrt{2}$

4. **(a)** $13\sqrt{3}$ **(b)** $5\sqrt{2r}$ **(c)** $3y\sqrt{2}$
(d) $13x\sqrt[3]{3x}$

(c) $2\sqrt{12} + 3\sqrt{75}$

$= 2(\sqrt{4} \cdot \sqrt{3}) + 3(\sqrt{25} \cdot \sqrt{3})$	Product rule
$= 2(2\sqrt{3}) + 3(5\sqrt{3})$	$\sqrt{4} = 2$; $\sqrt{25} = 5$
$= 4\sqrt{3} + 15\sqrt{3}$	Multiply.
$= 19\sqrt{3}$ (Think: $(4 + 15)\sqrt{3}$)	Add like radicals.

◀ **Work Problem 3 at the Side.**

OBJECTIVE 3 Simplify more complicated radical expressions.

EXAMPLE 3 Simplifying Radical Expressions

Simplify each radical expression. Assume that all variables represent nonnegative real numbers.

(a) $\sqrt{5} \cdot \sqrt{15} + 4\sqrt{3}$ (We simplify this expression by performing the operations.)

$= \sqrt{5 \cdot 15} + 4\sqrt{3}$	Product rule
$= \sqrt{75} + 4\sqrt{3}$	Multiply.
$= \sqrt{25 \cdot 3} + 4\sqrt{3}$	Factor; 25 is a perfect square.
$= \sqrt{25} \cdot \sqrt{3} + 4\sqrt{3}$	Product rule
$= 5\sqrt{3} + 4\sqrt{3}$	$\sqrt{25} = 5$
$= 9\sqrt{3}$	Add like radicals.

(b) $\sqrt{12k} + \sqrt{27k}$

$= \sqrt{4 \cdot 3k} + \sqrt{9 \cdot 3k}$	Factor.
$= \sqrt{4} \cdot \sqrt{3k} + \sqrt{9} \cdot \sqrt{3k}$	Product rule
$= 2\sqrt{3k} + 3\sqrt{3k}$	$\sqrt{4} = 2$; $\sqrt{9} = 3$
$= 5\sqrt{3k}$	Add like radicals.

(c) $3x\sqrt{50} + \sqrt{2x^2}$

$= 3x\sqrt{25 \cdot 2} + \sqrt{x^2 \cdot 2}$	Factor.
$= 3x\sqrt{25} \cdot \sqrt{2} + \sqrt{x^2} \cdot \sqrt{2}$	Product rule
$= 3x \cdot 5\sqrt{2} + x\sqrt{2}$	$\sqrt{25} = 5$; $\sqrt{x^2} = x$
$= 15x\sqrt{2} + x\sqrt{2}$	Multiply.
$= 16x\sqrt{2}$ (Think: $(15x + 1x)\sqrt{2}$)	Add like radicals.

(d) $2\sqrt[3]{32m^3} - \sqrt[3]{108m^3}$

$= 2\sqrt[3]{(8m^3)4} - \sqrt[3]{(27m^3)4}$	Factor.
$= 2(2m)\sqrt[3]{4} - 3m\sqrt[3]{4}$	$\sqrt[3]{8m^3} = 2m$; $\sqrt[3]{27m^3} = 3m$
$= 4m\sqrt[3]{4} - 3m\sqrt[3]{4}$	Multiply.
$= m\sqrt[3]{4}$	Subtract like radicals.

◀ **Work Problem 4 at the Side.**

8.3 Exercises

FOR EXTRA HELP

Download the MyDashBoard App

MyMathLab®

CONCEPT CHECK *Complete each statement.*

1. Like radicals have the same ___radicand___ and the same ___index___, or order. For example, $5\sqrt{2}$ and $-3\sqrt{2}$ are (*like* / *unlike*) radicals, as are $\sqrt{7}$, $-\sqrt{\_7\_}$, and $8\sqrt{\_7\_}$.

2. The radicals $\sqrt[4]{3xy^3}$ and $-6\sqrt[4]{3xy^3}$ are (*like* / *unlike*) radicals because both have the same root index, ___4___, and the same radicand, ___$3xy^3$___.

3. $\sqrt{5} + 5\sqrt{3}$ cannot be simplified because the ___radicands___ are different. They are (*like* / *unlike*) radicals.

4. $4\sqrt[3]{2} + 3\sqrt{2}$ cannot be simplified because the ___indexes or roots___ are different. They are (*like* / *unlike*) radicals.

5. Simplifying the expression $5\sqrt{2} + 6\sqrt{2}$ as $(5 + 6)\sqrt{2}$, or $11\sqrt{2}$, is an application of the ___distributive___ property.

6. The radicals $\sqrt{5x}$ and $4\sqrt{5x}$ are (*like* / *unlike*) radicals. To add them, use the identity property for multiplication to write $\sqrt{5x}$ as ___1___$\sqrt{5x}$. Thus, $\sqrt{5x} + 4\sqrt{5x}$ can be added to obtain ___$5\sqrt{5x}$___.

Simplify and add or subtract wherever possible. ***See Examples 1 and 2.***

7. $2\sqrt{3} + 5\sqrt{3}$
$7\sqrt{3}$

8. $6\sqrt{5} + 8\sqrt{5}$
$14\sqrt{5}$

9. $14\sqrt{7} - 19\sqrt{7}$
$-5\sqrt{7}$

10. $16\sqrt{2} - 18\sqrt{2}$
$-2\sqrt{2}$

11. $\sqrt{17} + 4\sqrt{17}$
$5\sqrt{17}$

12. $5\sqrt{19} + \sqrt{19}$
$6\sqrt{19}$

13. $6\sqrt{7} - \sqrt{7}$
$5\sqrt{7}$

14. $11\sqrt{14} - \sqrt{14}$
$10\sqrt{14}$

15. $\sqrt{6} + \sqrt{6}$
$2\sqrt{6}$

16. $\sqrt{11} + \sqrt{11}$
$2\sqrt{11}$

17. $\sqrt{6} + \sqrt{7}$
These unlike radicals cannot be combined.

18. $\sqrt{14} + \sqrt{17}$
These unlike radicals cannot be combined.

19. $5\sqrt{3} + \sqrt{12}$
$7\sqrt{3}$

20. $3\sqrt{2} + \sqrt{50}$
$8\sqrt{2}$

21. $\sqrt{45} + 4\sqrt{20}$
$11\sqrt{5}$

22. $\sqrt{24} + 6\sqrt{54}$
$20\sqrt{6}$

23. $5\sqrt{72} - 3\sqrt{50}$
$15\sqrt{2}$

24. $6\sqrt{18} - 5\sqrt{32}$
$-2\sqrt{2}$

25. $-5\sqrt{32} + 2\sqrt{98}$
$-6\sqrt{2}$

26. $-4\sqrt{75} + 3\sqrt{12}$
$-14\sqrt{3}$

27. $5\sqrt{7} - 3\sqrt{28} + 6\sqrt{63}$
$17\sqrt{7}$

28. $3\sqrt{11} + 5\sqrt{44} - 8\sqrt{99}$
$-11\sqrt{11}$

29. $2\sqrt{8} - 5\sqrt{32} - 2\sqrt{48}$
$-16\sqrt{2} - 8\sqrt{3}$

30. $5\sqrt{72} - 3\sqrt{48} + 4\sqrt{128}$
$62\sqrt{2} - 12\sqrt{3}$

31. $4\sqrt{50} + 3\sqrt{12} - 5\sqrt{45}$
$20\sqrt{2} + 6\sqrt{3} - 15\sqrt{5}$

32. $6\sqrt{18} + 2\sqrt{48} + 6\sqrt{28}$
$18\sqrt{2} + 8\sqrt{3} + 12\sqrt{7}$

33. $\frac{1}{4}\sqrt{288} + \frac{1}{6}\sqrt{72}$
$4\sqrt{2}$

34. $\frac{2}{3}\sqrt{27} + \frac{3}{4}\sqrt{48}$
$5\sqrt{3}$

Find the perimeter of each figure.

35.

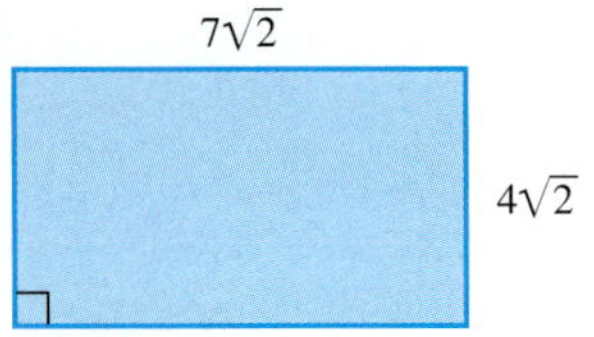

$22\sqrt{2}$

36.

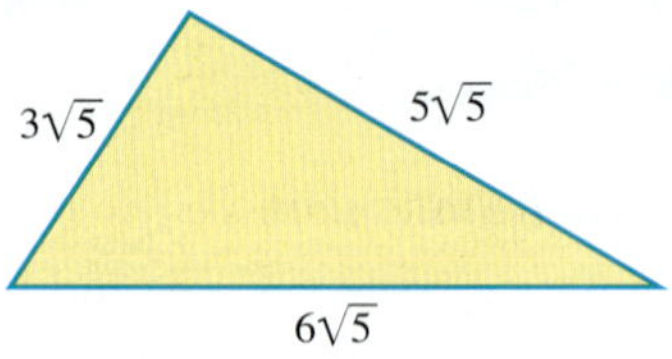

$14\sqrt{5}$

Perform the indicated operations. Assume that all variables represent nonnegative real numbers. **See Example 3.**

37. $\sqrt{6} \cdot \sqrt{2} + 9\sqrt{3}$
$11\sqrt{3}$

38. $4\sqrt{15} \cdot \sqrt{3} + 4\sqrt{5}$
$16\sqrt{5}$

39. $\sqrt{3} \cdot \sqrt{7} + 2\sqrt{21}$
$3\sqrt{21}$

40. $\sqrt{13} \cdot \sqrt{2} + 3\sqrt{26}$
$4\sqrt{26}$

41. $\sqrt{32x} - \sqrt{18x}$
$\sqrt{2x}$

42. $\sqrt{125t} - \sqrt{80t}$
$\sqrt{5t}$

43. $\sqrt{27r} + \sqrt{48r}$
$7\sqrt{3r}$

44. $\sqrt{24x} + \sqrt{54x}$
$5\sqrt{6x}$

45. $\sqrt{9x} + \sqrt{49x} - \sqrt{25x}$
$5\sqrt{x}$

46. $\sqrt{4a} - \sqrt{16a} + \sqrt{100a}$
$8\sqrt{a}$

47. $\sqrt{6x^2} + x\sqrt{24}$
$3x\sqrt{6}$

48. $\sqrt{75x^2} + x\sqrt{108}$
$11x\sqrt{3}$

49. $3\sqrt{8x^2} - 4x\sqrt{2} - x\sqrt{8}$
0

50. $\sqrt{2b^2} + 3b\sqrt{18} - b\sqrt{200}$
0

51. $-8\sqrt{32k} + 6\sqrt{8k}$
$-20\sqrt{2k}$

52. $4\sqrt{12x} + 2\sqrt{27x}$
$14\sqrt{3x}$

53. $2\sqrt{125x^2z} + 8x\sqrt{80z}$
$42x\sqrt{5z}$

54. $\sqrt{48x^2y} + 5x\sqrt{27y}$
$19x\sqrt{3y}$

55. $4\sqrt[3]{16} - 3\sqrt[3]{54}$
$-\sqrt[3]{2}$

56. $5\sqrt[3]{128} + 3\sqrt[3]{250}$
$35\sqrt[3]{2}$

57. $6\sqrt[3]{8p^2} - 2\sqrt[3]{27p^2}$
$6\sqrt[3]{p^2}$

58. $8k\sqrt[3]{54k} + 6\sqrt[3]{16k^4}$
$36k\sqrt[3]{2k}$

59. $5\sqrt[4]{m^3} + 8\sqrt[4]{16m^3}$
$21\sqrt[4]{m^3}$

60. $5\sqrt[4]{m^5} + 3\sqrt[4]{81m^5}$
$14m\sqrt[4]{m}$

8.4 Rationalizing the Denominator

OBJECTIVES

1. Rationalize denominators with square roots.
2. Write radicals in simplified form.
3. Rationalize denominators with cube roots.

OBJECTIVE 1 **Rationalize denominators with square roots.** Although calculators now make it fairly easy to divide by a radical in an expression such as $\frac{1}{\sqrt{2}}$, it is sometimes easier to work with radical expressions if the denominators do not contain any radicals.

For example, the radical in the denominator of $\frac{1}{\sqrt{2}}$ can be eliminated by multiplying the numerator and denominator by $\sqrt{2}$, since $\sqrt{2} \cdot \sqrt{2} = \sqrt{4} = 2$.

$$\frac{1}{\sqrt{2}} = \frac{1 \cdot \sqrt{2}}{\sqrt{2} \cdot \sqrt{2}} = \frac{\sqrt{2}}{2} \qquad \text{Multiply by } \frac{\sqrt{2}}{\sqrt{2}} = 1.$$

This process of changing a denominator from one with a radical to one without a radical is called **rationalizing the denominator.** ***The value of the radical expression is not changed. Only the form is changed, because the expression has been multiplied by 1 in the form of*** $\frac{\sqrt{2}}{\sqrt{2}}$.

EXAMPLE 1 Rationalizing Denominators

Rationalize each denominator.

(a) $\frac{9}{\sqrt{6}}$

$= \frac{9 \cdot \sqrt{6}}{\sqrt{6} \cdot \sqrt{6}}$ Multiply by $\frac{\sqrt{6}}{\sqrt{6}} = 1$.

$= \frac{9\sqrt{6}}{6}$ In the denominator, $\sqrt{6} \cdot \sqrt{6} = \sqrt{36} = 6$.

$= \frac{3\sqrt{6}}{2}$ Write in lowest terms.

The denominator is now a rational number.

(b) $\frac{12}{\sqrt{8}}$ While the denominator could be rationalized by multiplying by $\sqrt{8}$, simplifying the denominator first is more direct.

$\frac{12}{\sqrt{8}}$

$= \frac{12}{2\sqrt{2}}$ $\sqrt{8} = \sqrt{4} \cdot \sqrt{2} = 2\sqrt{2}$

$= \frac{12 \cdot \sqrt{2}}{2\sqrt{2} \cdot \sqrt{2}}$ Multiply by $\frac{\sqrt{2}}{\sqrt{2}} = 1$.

$= \frac{12 \cdot \sqrt{2}}{2 \cdot 2}$ $\sqrt{2} \cdot \sqrt{2} = \sqrt{4} = 2$

$= \frac{12\sqrt{2}}{4}$ Multiply.

$= 3\sqrt{2}$ Write in lowest terms; $\frac{12}{4} = 3$

Work Problem 1 at the Side. ▶

1 Rationalize each denominator.

GS **(a)** $\frac{3}{\sqrt{5}}$

$= \frac{3 \cdot \_\_\_}{\sqrt{5} \cdot \_\_\_}$

$= \_\_\_$

(b) $\frac{-6}{\sqrt{11}}$

(c) $-\frac{\sqrt{7}}{\sqrt{2}}$

(d) $\frac{20}{\sqrt{18}}$

Answers

1. **(a)** $\sqrt{5}$; $\sqrt{5}$; $\frac{3\sqrt{5}}{5}$ **(b)** $\frac{-6\sqrt{11}}{11}$
(c) $-\frac{\sqrt{14}}{2}$ **(d)** $\frac{10\sqrt{2}}{3}$

2 Simplify.

(a) $\sqrt{\dfrac{16}{11}}$

$= \dfrac{\sqrt{16}}{\_\_\_\_}$

$= \dfrac{\sqrt{16} \cdot \_\_\_\_}{\sqrt{11} \cdot \_\_\_\_}$

$= \dfrac{\_\_\_\_\sqrt{11}}{\_\_\_\_}$

(b) $\sqrt{\dfrac{5}{18}}$

(c) $\sqrt{\dfrac{8}{32}}$

Note

In **Example 1(b),** we could also have rationalized the original denominator $\sqrt{8}$ by multiplying by $\sqrt{2}$, since $\sqrt{8} \cdot \sqrt{2} = \sqrt{16} = 4$.

$$\frac{12}{\sqrt{8}} = \frac{12 \cdot \sqrt{2}}{\sqrt{8} \cdot \sqrt{2}} = \frac{12\sqrt{2}}{\sqrt{16}} = \frac{12\sqrt{2}}{4} = 3\sqrt{2}$$

Either approach yields the same correct answer.

OBJECTIVE 2 Write radicals in simplified form. A radical is considered to be in simplified form if the following three conditions are met.

Conditions for Simplified Form of a Radical

1. The radicand contains no factor (except 1) that is a perfect square (when dealing with square roots), a perfect cube (when dealing with cube roots), and so on.
2. The radicand has no fractions.
3. No denominator contains a radical.

EXAMPLE 2 Simplifying a Radical

Simplify.

$\sqrt{\dfrac{27}{5}}$

$= \dfrac{\sqrt{27}}{\sqrt{5}}$ Quotient rule

$= \dfrac{\sqrt{27} \cdot \sqrt{5}}{\sqrt{5} \cdot \sqrt{5}}$ Rationalize the denominator.

$= \dfrac{\sqrt{27} \cdot \sqrt{5}}{5}$ $\sqrt{5} \cdot \sqrt{5} = \sqrt{25} = 5$

$= \dfrac{\sqrt{9 \cdot 3} \cdot \sqrt{5}}{5}$ Factor.

$= \dfrac{\sqrt{9} \cdot \sqrt{3} \cdot \sqrt{5}}{5}$ Product rule

$= \dfrac{3 \cdot \sqrt{3} \cdot \sqrt{5}}{5}$ $\sqrt{9} = 3$

$= \dfrac{3\sqrt{15}}{5}$ Product rule

The three conditions for a simplified radical are met.

◀ Work Problem **2** at the Side.

Answers

2. (a) $\sqrt{11}$; $\sqrt{11}$; $\sqrt{11}$; 4; 11

(b) $\dfrac{\sqrt{10}}{6}$ (c) $\dfrac{1}{2}$

EXAMPLE 3 Simplifying a Product of Radicals

Simplify.

$$\sqrt{\frac{5}{8}} \cdot \sqrt{\frac{1}{6}}$$

$= \sqrt{\frac{5}{8} \cdot \frac{1}{6}}$ — Product rule

$= \sqrt{\frac{5}{48}}$ — Multiply fractions.

$= \frac{\sqrt{5}}{\sqrt{48}}$ — Quotient rule

$= \frac{\sqrt{5}}{\sqrt{16} \cdot \sqrt{3}}$ — Product rule

$= \frac{\sqrt{5}}{4\sqrt{3}}$ — $\sqrt{16} = 4$

$= \frac{\sqrt{5} \cdot \sqrt{3}}{4\sqrt{3} \cdot \sqrt{3}}$ — Rationalize the denominator.

$= \frac{\sqrt{15}}{4 \cdot 3}$ — Product rule; $\sqrt{3} \cdot \sqrt{3} = 3$

$= \frac{\sqrt{15}}{12}$ — Multiply.

Work Problem **3** at the Side. ▶

EXAMPLE 4 Simplifying Quotients Involving Radicals

Simplify. Assume that x and y represent positive real numbers.

(a) $\frac{\sqrt{4x}}{\sqrt{y}}$

$= \frac{\sqrt{4x} \cdot \sqrt{y}}{\sqrt{y} \cdot \sqrt{y}}$ — Rationalize the denominator.

$= \frac{\sqrt{4xy}}{y}$ — Product rule; $\sqrt{y} \cdot \sqrt{y} = y$

$= \frac{\sqrt{4} \cdot \sqrt{xy}}{y}$ — Product rule

$= \frac{2\sqrt{xy}}{y}$ — $\sqrt{4} = 2$

(b) $\sqrt{\frac{2x^2y}{3}}$

$= \frac{\sqrt{2x^2y}}{\sqrt{3}}$ — Quotient rule

$= \frac{\sqrt{2x^2y} \cdot \sqrt{3}}{\sqrt{3} \cdot \sqrt{3}}$ — Rationalize the denominator.

$= \frac{\sqrt{6x^2y}}{3}$ — Product rule; $\sqrt{3} \cdot \sqrt{3} = 3$

$= \frac{\sqrt{x^2} \cdot \sqrt{6y}}{3}$ — Product rule

$= \frac{x\sqrt{6y}}{3}$ — $\sqrt{x^2} = x$, since $x > 0$.

Work Problem **4** at the Side. ▶

3 Simplify.

(a) $\sqrt{\frac{1}{2}} \cdot \sqrt{\frac{5}{6}}$

GS **(b)** $\sqrt{\frac{1}{10}} \cdot \sqrt{20}$

$= \sqrt{\frac{1}{10}} \cdot \sqrt{\frac{20}{\_\_\_\_}}$

$= \sqrt{\frac{\_\_\_\_}{10}}$

$= \_\_\_\_$

(c) $\sqrt{\frac{5}{8}} \cdot \sqrt{\frac{24}{10}}$

4 Simplify. Assume that all variables represent positive real numbers.

(a) $\frac{\sqrt{5p}}{\sqrt{q}}$

(b) $\sqrt{\frac{5r^2t^2}{7}}$

Answers

3. (a) $\frac{\sqrt{15}}{6}$ (b) 1; 20; $\sqrt{2}$ (c) $\frac{\sqrt{6}}{2}$

4. (a) $\frac{\sqrt{5pq}}{q}$ (b) $\frac{rt\sqrt{35}}{7}$

5 Rationalize each denominator.

GS **(a)** $\sqrt[3]{\frac{5}{7}}$

$= \frac{\sqrt[3]{5}}{\sqrt[3]{7}}$

To get a perfect cube in the denominator radicand, multiply the ________ and denominator by

$\sqrt[3]{\_\_\_\_ \cdot \_\_\_\_}$.

Complete the solution.

(b) $\frac{\sqrt[3]{5}}{\sqrt[3]{9}}$

(c) $\frac{\sqrt[3]{4}}{\sqrt[3]{25y}} \quad (y \neq 0)$

Answers

5. **(a)** numerator; 7; 7; $\frac{\sqrt[3]{245}}{7}$ **(b)** $\frac{\sqrt[3]{15}}{3}$

(c) $\frac{\sqrt[3]{20y^2}}{5y}$

OBJECTIVE 3 Rationalize denominators with cube roots. To rationalize a denominator with a cube root, we change the radicand in the denominator to a perfect cube.

EXAMPLE 5 Rationalizing Denominators with Cube Roots

Rationalize each denominator.

(a) $\sqrt[3]{\frac{3}{2}}$

First write the expression as a quotient of radicals. Then multiply the numerator and denominator by the appropriate number of factors of 2 to make the radicand in the denominator a perfect cube. This will eliminate the radical in the denominator. Here, multiply by $\sqrt[3]{2 \cdot 2}$, or $\sqrt[3]{2^2}$.

$$\sqrt[3]{\frac{3}{2}} = \frac{\sqrt[3]{3}}{\sqrt[3]{2}} = \frac{\sqrt[3]{3} \cdot \sqrt[3]{2 \cdot 2}}{\sqrt[3]{2} \cdot \sqrt[3]{2 \cdot 2}} = \frac{\sqrt[3]{3 \cdot 2 \cdot 2}}{\sqrt[3]{2 \cdot 2 \cdot 2}} = \frac{\sqrt[3]{12}}{2}$$

We need 3 factors of 2 in the radicand in the denominator.

$\sqrt[3]{2 \cdot 2 \cdot 2} = \sqrt[3]{2^3} = 2$
Denominator radicand is a perfect cube.

(b) $\frac{\sqrt[3]{3}}{\sqrt[3]{4}}$

Since $\sqrt[3]{4} = \sqrt[3]{2 \cdot 2}$, multiply the numerator and denominator by a sufficient number of factors of 2 to get a perfect cube in the radicand in the denominator.

$$\frac{\sqrt[3]{3}}{\sqrt[3]{4}} = \frac{\sqrt[3]{3} \cdot \sqrt[3]{2}}{\sqrt[3]{2 \cdot 2} \cdot \sqrt[3]{2}} = \frac{\sqrt[3]{6}}{\sqrt[3]{2 \cdot 2 \cdot 2}} = \frac{\sqrt[3]{6}}{2}$$

(c) $\frac{\sqrt[3]{2}}{\sqrt[3]{3x^2}} \quad (x \neq 0)$

Multiply the numerator and denominator by a sufficient number of factors of 3 and of x to get a perfect cube in the radicand in the denominator.

$$\frac{\sqrt[3]{2}}{\sqrt[3]{3x^2}} = \frac{\sqrt[3]{2} \cdot \sqrt[3]{3 \cdot 3 \cdot x}}{\sqrt[3]{3 \cdot x \cdot x} \cdot \sqrt[3]{3 \cdot 3 \cdot x}} = \frac{\sqrt[3]{18x}}{\sqrt[3]{(3x)^3}} = \frac{\sqrt[3]{18x}}{3x}$$

We need 3 factors of 3 and 3 factors of x in the radicand in the denominator.

Denominator radicand is a perfect cube.

CAUTION

A common error in a problem like the one in **Example 5(a)** is to multiply by $\sqrt[3]{2}$ instead of $\sqrt[3]{2^2}$. Doing this would give a denominator of

$$\sqrt[3]{2} \cdot \sqrt[3]{2} = \sqrt[3]{4}.$$

Because 4 is not a perfect cube, the denominator is still not rationalized.

◀ Work Problem 5 at the Side.

8.4 Exercises

FOR EXTRA HELP

Download the MyDashBoard App

MyMathLab®

1. CONCEPT CHECK When we rationalize the denominator in an expression such as $\frac{4}{\sqrt{3}}$, we multiply both the numerator and denominator by $\sqrt{3}$. By what number are we actually multiplying the given expression, and what property of real numbers justifies the fact that our result is equal to the given expression?

We are actually multiplying by 1. The identity property for multiplication justifies our result.

2. In **Example 1(a),** we show algebraically that $\frac{9}{\sqrt{6}}$ is equal to $\frac{3\sqrt{6}}{2}$. Support this result numerically by finding the decimal approximation of $\frac{9}{\sqrt{6}}$ on your calculator, and then finding the decimal approximation of $\frac{3\sqrt{6}}{2}$. What do you notice?

Both approximations are 3.674234614. They are the same.

Rationalize each denominator. ***See Examples 1 and 2.***

3. $\frac{6}{\sqrt{5}}$ **$\frac{6\sqrt{5}}{5}$**

4. $\frac{3}{\sqrt{2}}$ **$\frac{3\sqrt{2}}{2}$**

5. $\frac{5}{\sqrt{5}}$ **$\sqrt{5}$**

6. $\frac{15}{\sqrt{15}}$ **$\sqrt{15}$**

7. $\frac{8}{\sqrt{2}}$ **$4\sqrt{2}$**

8. $\frac{12}{\sqrt{3}}$ **$4\sqrt{3}$**

9. $\frac{-\sqrt{11}}{\sqrt{3}}$ **$\frac{-\sqrt{33}}{3}$**

10. $\frac{-\sqrt{13}}{\sqrt{5}}$ **$\frac{-\sqrt{65}}{5}$**

11. $\frac{7\sqrt{3}}{\sqrt{5}}$ **$\frac{7\sqrt{15}}{5}$**

12. $\frac{4\sqrt{6}}{\sqrt{5}}$ **$\frac{4\sqrt{30}}{5}$**

13. $\frac{24\sqrt{10}}{16\sqrt{3}}$ **$\frac{\sqrt{30}}{2}$**

14. $\frac{18\sqrt{15}}{12\sqrt{2}}$ **$\frac{3\sqrt{30}}{4}$**

15. $\frac{16}{\sqrt{27}}$ **$\frac{16\sqrt{3}}{9}$**

16. $\frac{24}{\sqrt{18}}$ **$4\sqrt{2}$**

17. $\frac{-3}{\sqrt{50}}$ **$\frac{-3\sqrt{2}}{10}$**

18. $\frac{-5}{\sqrt{75}}$ **$\frac{-\sqrt{3}}{3}$**

19. $\frac{63}{\sqrt{45}}$ **$\frac{21\sqrt{5}}{5}$**

20. $\frac{27}{\sqrt{32}}$ **$\frac{27\sqrt{2}}{8}$**

21. $\frac{\sqrt{24}}{\sqrt{8}}$ **$\sqrt{3}$**

22. $\frac{\sqrt{36}}{\sqrt{18}}$ **$\sqrt{2}$**

23. $\sqrt{\frac{1}{2}}$ **$\frac{\sqrt{2}}{2}$**

24. $\sqrt{\frac{1}{3}}$ **$\frac{\sqrt{3}}{3}$**

25. $\sqrt{\frac{13}{5}}$ **$\frac{\sqrt{65}}{5}$**

26. $\sqrt{\frac{17}{11}}$ **$\frac{\sqrt{187}}{11}$**

Simplify each product of radicals. ***See Example 3.***

27. $\sqrt{\frac{7}{13}} \cdot \sqrt{\frac{13}{3}}$ **$\frac{\sqrt{21}}{3}$**

28. $\sqrt{\frac{19}{20}} \cdot \sqrt{\frac{20}{3}}$ **$\frac{\sqrt{57}}{3}$**

29. $\sqrt{\frac{21}{7}} \cdot \sqrt{\frac{21}{8}}$ **$\frac{3\sqrt{14}}{4}$**

30. $\sqrt{\frac{5}{8}} \cdot \sqrt{\frac{5}{6}}$ **$\frac{5\sqrt{3}}{12}$**

31. $\sqrt{\frac{1}{12}} \cdot \sqrt{\frac{1}{3}}$ **$\frac{1}{6}$**

32. $\sqrt{\frac{1}{8}} \cdot \sqrt{\frac{1}{2}}$ **$\frac{1}{4}$**

33. $\sqrt{\frac{2}{9}} \cdot \sqrt{\frac{9}{2}}$ **1**

34. $\sqrt{\frac{4}{3}} \cdot \sqrt{\frac{3}{4}}$ **1**

Simplify each radical. Assume that all variables represent positive real numbers. ***See Example 4.***

35. $\dfrac{\sqrt{7}}{\sqrt{x}}$ $\dfrac{\sqrt{7x}}{x}$

36. $\dfrac{\sqrt{19}}{\sqrt{y}}$ $\dfrac{\sqrt{19y}}{y}$

37. $\dfrac{\sqrt{4x^3}}{\sqrt{y}}$ $\dfrac{2x\sqrt{xy}}{y}$

38. $\dfrac{\sqrt{9t^3}}{\sqrt{s}}$ $\dfrac{3t\sqrt{st}}{s}$

39. $\sqrt{\dfrac{5x^3z}{6}}$ $\dfrac{x\sqrt{30xz}}{6}$

40. $\sqrt{\dfrac{3st^3}{5}}$ $\dfrac{t\sqrt{15st}}{5}$

41. $\sqrt{\dfrac{9a^2r^5}{7t}}$ $\dfrac{3ar^2\sqrt{7rt}}{7t}$

42. $\sqrt{\dfrac{16x^3y^2}{13z}}$ $\dfrac{4xy\sqrt{13xz}}{13z}$

43. CONCEPT CHECK Which one of the following would be an appropriate choice for multiplying the numerator and denominator of $\frac{\sqrt[3]{2}}{\sqrt[3]{5}}$ by in order to rationalize the denominator? B

A. $\sqrt[3]{5}$ **B.** $\sqrt[3]{25}$ **C.** $\sqrt[3]{2}$ **D.** $\sqrt[3]{4}$

44. CONCEPT CHECK In **Example 5(b),** we multiplied the numerator and denominator of $\frac{\sqrt[3]{3}}{\sqrt[3]{4}}$ by $\sqrt[3]{2}$ to rationalize the denominator. Suppose we had chosen to multiply by $\sqrt[3]{16}$ instead. Would we have obtained the correct answer after all simplifications were done? yes

Rationalize each denominator. Assume that variables in the denominator represent nonzero real numbers. ***See Example 5.***

45. $\sqrt[3]{\dfrac{1}{2}}$ $\dfrac{\sqrt[3]{4}}{2}$

46. $\sqrt[3]{\dfrac{1}{4}}$ $\dfrac{\sqrt[3]{2}}{2}$

47. $\sqrt[3]{\dfrac{5}{9}}$ $\dfrac{\sqrt[3]{15}}{3}$

48. $\sqrt[3]{\dfrac{2}{5}}$ $\dfrac{\sqrt[3]{50}}{5}$

49. $\dfrac{\sqrt[3]{4}}{\sqrt[3]{7}}$ $\dfrac{\sqrt[3]{196}}{7}$

50. $\dfrac{\sqrt[3]{5}}{\sqrt[3]{10}}$ $\dfrac{\sqrt[3]{4}}{2}$

51. $\sqrt[3]{\dfrac{3}{4y^2}}$ $\dfrac{\sqrt[3]{6y}}{2y}$

52. $\sqrt[3]{\dfrac{3}{25x^2}}$ $\dfrac{\sqrt[3]{15x}}{5x}$

53. $\dfrac{\sqrt[3]{7m}}{\sqrt[3]{36n}}$ $\dfrac{\sqrt[3]{42mn^2}}{6n}$

54. $\dfrac{\sqrt[3]{11p}}{\sqrt[3]{49q}}$ $\dfrac{\sqrt[3]{77pq^2}}{7q}$

*In Exercises 55 and 56, **(a)** give the answer as a simplified radical and **(b)** use a calculator to give the answer correct to the nearest thousandth.*

55. The period p of a pendulum is the time it takes for it to swing from one extreme to the other and back again. The value of p in seconds is given by

$$p = k \cdot \sqrt{\frac{L}{g}},$$

where L is the length of the pendulum, g is the acceleration due to gravity, and k is a constant. Find the period when $k = 6$, $L = 9$ ft, and $g = 32$ ft per sec per sec.

(a) $\dfrac{9\sqrt{2}}{4}$ sec
(b) 3.182 sec

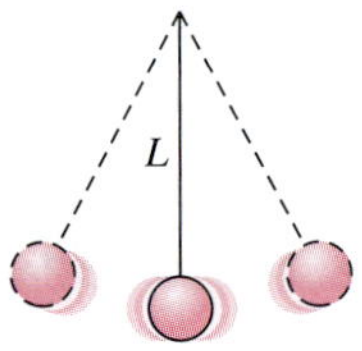

56. The velocity v in kilometers per second of a meteorite approaching Earth is given by

$$v = \frac{k}{\sqrt{d}},$$

where d is its distance from the center of Earth and k is a constant. What is the velocity of a meteorite that is 6000 km away from the center of Earth, if $k = 450$?

(a) $\dfrac{3\sqrt{15}}{2}$ km per sec (b) 5.809 km per sec

8.5 More Simplifying and Operations with Radicals

OBJECTIVES

1. Simplify products of radical expressions.
2. Use conjugates to rationalize denominators of radical expressions.
3. Write radical expressions with quotients in lowest terms.

Apply these guidelines when simplifying radical expressions.

Simplifying Radical Expressions

1. If a radical represents a rational number, use that rational number in place of the radical.

Examples: $\sqrt{49} = 7$, $\sqrt{\frac{169}{9}} = \frac{13}{3}$

2. If a radical expression contains products of radicals, use the product rule for radicals, $\sqrt[n]{x} \cdot \sqrt[n]{y} = \sqrt[n]{xy}$, to get a single radical.

Examples: $\sqrt{3} \cdot \sqrt{2} = \sqrt{6}$, $\sqrt[3]{5} \cdot \sqrt[3]{x} = \sqrt[3]{5x}$

3. If a radicand of a square root radical has a factor that is a perfect square, express the radical as the product of the positive square root of the perfect square and the remaining radicand factor. A similar statement applies to higher roots.

Examples: $\sqrt{20} = \sqrt{4 \cdot 5} = \sqrt{4} \cdot \sqrt{5} = 2\sqrt{5}$

$\sqrt[3]{16} = \sqrt[3]{8 \cdot 2} = \sqrt[3]{8} \cdot \sqrt[3]{2} = 2\sqrt[3]{2}$

4. If a radical expression contains sums or differences of radicals, use the distributive property to combine like radicals.

Examples: $3\sqrt{2} + 4\sqrt{2}$ can be combined to obtain $7\sqrt{2}$.

$3\sqrt{2} + 4\sqrt{3}$ cannot be simplified further.

5. Rationalize any denominator containing a radical.

Examples: $\frac{5}{\sqrt{3}} = \frac{5 \cdot \sqrt{3}}{\sqrt{3} \cdot \sqrt{3}} = \frac{5\sqrt{3}}{3}$

$\sqrt[3]{\frac{1}{4}} = \frac{\sqrt[3]{1}}{\sqrt[3]{4}} = \frac{\sqrt[3]{1} \cdot \sqrt[3]{2}}{\sqrt[3]{4} \cdot \sqrt[3]{2}} = \frac{\sqrt[3]{2}}{\sqrt[3]{8}} = \frac{\sqrt[3]{2}}{2}$

OBJECTIVE 1 Simplify products of radical expressions.

EXAMPLE 1 Multiplying Radical Expressions

Find each product, and simplify.

(a) $\sqrt{5}(\sqrt{8} - \sqrt{32})$ — Simplify inside the parentheses.

$= \sqrt{5}(2\sqrt{2} - 4\sqrt{2})$ — $\sqrt{8} = 2\sqrt{2}$; $\sqrt{32} = 4\sqrt{2}$

$= \sqrt{5}(-2\sqrt{2})$ — Subtract like radicals.

$= -2\sqrt{5 \cdot 2}$ — Product rule

$= -2\sqrt{10}$ — Multiply.

Continued on Next Page

1 Find each product, and simplify.

GS **(a)** $\sqrt{7}(\sqrt{2}+\sqrt{5})$

$= \_\_\_\_ \cdot \sqrt{2} + \sqrt{7} \cdot \_\_\_\_$

$= \_\_\_\_\_\_\_\_$

(b) $\sqrt{2}(\sqrt{8}+\sqrt{20})$

(c) $(\sqrt{2}+5\sqrt{3})(\sqrt{3}-2\sqrt{2})$

(d) $(\sqrt{2}-\sqrt{5})(\sqrt{10}+\sqrt{2})$

2 Find each product. Simplify the answers. In part (c), assume that $x \geq 0$.

GS **(a)** $(\sqrt{5}-3)^2$

$= (\_\_\_\_)^2 - 2(\_\_\_\_)(3) + \_\_\_\_^2$

$= \_\_\_\_ - \_\_\_\_ + 9$

$= \_\_\_\_\_\_\_\_$

(b) $(4\sqrt{2}+5)^2$

(c) $(6+\sqrt{x})^2$

Answers

1. **(a)** $\sqrt{7}$; $\sqrt{5}$; $\sqrt{14}+\sqrt{35}$
(b) $4+2\sqrt{10}$ **(c)** $11-9\sqrt{6}$
(d) $2\sqrt{5}+2-5\sqrt{2}-\sqrt{10}$

2. **(a)** $\sqrt{5}$; $\sqrt{5}$; 3; 5; $6\sqrt{5}$; $14-6\sqrt{5}$
(b) $57+40\sqrt{2}$ **(c)** $36+12\sqrt{x}+x$

(b) $(\sqrt{3}+2\sqrt{5})(\sqrt{3}-4\sqrt{5})$ — Use the FOIL method to multiply.

$= \underbrace{\sqrt{3}(\sqrt{3})}_{\text{First}} + \underbrace{\sqrt{3}(-4\sqrt{5})}_{\text{Outer}} + \underbrace{2\sqrt{5}(\sqrt{3})}_{\text{Inner}} + \underbrace{2\sqrt{5}(-4\sqrt{5})}_{\text{Last}}$

$= 3 - 4\sqrt{15} + 2\sqrt{15} - 8 \cdot 5$ Product rule

$= 3 - 2\sqrt{15} - 40$ Add like radicals. Multiply.

$= -37 - 2\sqrt{15}$ Combine like terms.

This does *not* equal $-39\sqrt{15}$.

(c) $(\sqrt{3}+\sqrt{21})(\sqrt{3}-\sqrt{7})$

$= \sqrt{3}(\sqrt{3}) + \sqrt{3}(-\sqrt{7}) + \sqrt{21}(\sqrt{3}) + \sqrt{21}(-\sqrt{7})$ FOIL method

$= 3 - \sqrt{21} + \sqrt{63} - \sqrt{147}$ Product rule

$= 3 - \sqrt{21} + \sqrt{9}\cdot\sqrt{7} - \sqrt{49}\cdot\sqrt{3}$ Factor; 9 and 49 are perfect squares.

$= 3 - \sqrt{21} + 3\sqrt{7} - 7\sqrt{3}$ $\sqrt{9}=3$; $\sqrt{49}=7$

Since there are no like radicals, no terms can be combined.

◀ Work Problem **1** at the Side.

Example 2 uses the following rules for the square of a binomial from **Section 5.4.**

$$(a+b)^2 = a^2 + 2ab + b^2 \quad \text{and} \quad (a-b)^2 = a^2 - 2ab + b^2$$

EXAMPLE 2 Using Special Products with Radicals

Find each product. In part (c), assume that $x \geq 0$.

(a) $(\sqrt{10}-7)^2$

$= (\sqrt{10})^2 - 2(\sqrt{10})(7) + 7^2$ $(a-b)^2 = a^2 - 2ab + b^2$; Let $a = \sqrt{10}$ and $b = 7$.

$= 10 - 14\sqrt{10} + 49$ $(\sqrt{10})^2 = 10$; $7^2 = 49$

$= 59 - 14\sqrt{10}$ Add 10 and 49.

Do *not* try to combine further here. $59 - 14\sqrt{10} \neq 45\sqrt{10}$

(b) $(2\sqrt{3}+4)^2$

$= (2\sqrt{3})^2 + 2(2\sqrt{3})(4) + 4^2$ $(a+b)^2 = a^2 + 2ab + b^2$; Let $a = 2\sqrt{3}$ and $b = 4$.

$= 12 + 16\sqrt{3} + 16$ $(2\sqrt{3})^2 = 4 \cdot 3 = 12$

$= 28 + 16\sqrt{3}$ Add 12 and 16.

Do *not* try to combine further here. $28 + 16\sqrt{3} \neq 44\sqrt{3}$

(c) $(5-\sqrt{x})^2$

$= 5^2 - 2(5)(\sqrt{x}) + (\sqrt{x})^2$ Square the binomial.

$= 25 - 10\sqrt{x} + x$ Simplify.

◀ Work Problem **2** at the Side.

Example 3 involves the product of the sum and difference of two terms.

$$(a + b)(a - b) = a^2 - b^2$$

EXAMPLE 3 **Using a Special Product with Radicals**

Find each product. In part (b), assume that $x \geq 0$.

(a) $\left(4 + \sqrt{3}\right)\left(4 - \sqrt{3}\right)$ — $(a + b)(a - b) = a^2 - b^2$; Let $a = 4$ and $b = \sqrt{3}$.

$= 4^2 - \left(\sqrt{3}\right)^2$

$= 16 - 3$ — $4^2 = 16$; $\left(\sqrt{3}\right)^2 = 3$

$= 13$ — Subtract.

(b) $\left(\sqrt{x} - \sqrt{6}\right)\left(\sqrt{x} + \sqrt{6}\right)$ — $(a + b)(a - b) = a^2 - b^2$; Let $a = \sqrt{x}$ and $b = \sqrt{6}$.

$= \left(\sqrt{x}\right)^2 - \left(\sqrt{6}\right)^2$

$= x - 6$ — $\left(\sqrt{x}\right)^2 = x$; $\left(\sqrt{6}\right)^2 = 6$

Work Problem 3 at the Side. ▶

In **Example 3,** the pair of expressions $4 + \sqrt{3}$ and $4 - \sqrt{3}$ are **conjugates** of each other, as are $\sqrt{x} - \sqrt{6}$ and $\sqrt{x} + \sqrt{6}$. Recall from **Section 5.4** that the expressions $a + b$ and $a - b$ are conjugates.

OBJECTIVE 2 **Use conjugates to rationalize denominators of radical expressions.** To rationalize the denominator in a quotient such as

$$\frac{2}{4 + \sqrt{3}},$$

we multiply the numerator and denominator by the conjugate of the denominator, here $4 - \sqrt{3}$.

$$\frac{2\left(4 - \sqrt{3}\right)}{\left(4 + \sqrt{3}\right)\left(4 - \sqrt{3}\right)} \quad \text{gives} \quad \frac{2\left(4 - \sqrt{3}\right)}{13}.$$

See Example 3(a).

EXAMPLE 4 **Using Conjugates to Rationalize Denominators**

Simplify by rationalizing each denominator. In part (c), assume that $x \geq 0$.

(a) $\dfrac{5}{3 + \sqrt{5}}$

$= \dfrac{5\left(3 - \sqrt{5}\right)}{\left(3 + \sqrt{5}\right)\left(3 - \sqrt{5}\right)}$ — Multiply the numerator and denominator by the conjugate of the denominator.

$= \dfrac{5\left(3 - \sqrt{5}\right)}{3^2 - \left(\sqrt{5}\right)^2}$ — $(a + b)(a - b) = a^2 - b^2$

$= \dfrac{5\left(3 - \sqrt{5}\right)}{9 - 5}$, or $\dfrac{5\left(3 - \sqrt{5}\right)}{4}$ — $3^2 = 9$ and $\left(\sqrt{5}\right)^2 = 5$; Subtract.

Continued on Next Page

---

3 Find each product. Simplify the answers. In part (d), assume that $x \geq 0$.

GS **(a)** $\left(3 + \sqrt{5}\right)\left(3 - \sqrt{5}\right)$

$= \underline{\quad}^2 - (\underline{\quad})^2$

$= \underline{\quad} - \underline{\quad}$

$= \underline{\quad}$

(b) $\left(\sqrt{3} - 2\right)\left(\sqrt{3} + 2\right)$

(c) $\left(\sqrt{5} + \sqrt{3}\right)\left(\sqrt{5} - \sqrt{3}\right)$

(d) $\left(\sqrt{10} - \sqrt{x}\right)\left(\sqrt{10} + \sqrt{x}\right)$

Answers

3. **(a)** 3; $\sqrt{5}$; 9; 5; 4 **(b)** -1 **(c)** 2 **(d)** $10 - x$

4 Rationalize each denominator. In part (c), assume that $x \geq 0$.

GS **(a)** $\dfrac{5}{4+\sqrt{2}}$

Multiply the numerator and denominator by the ________ of the denominator.

This number is ________.

Complete the solution.

(b) $\dfrac{\sqrt{5}+3}{2-\sqrt{5}}$

(c) $\dfrac{7}{5-\sqrt{x}}$

5 Write each quotient in lowest terms.

GS **(a)** $\dfrac{5\sqrt{3}-15}{10}$

$= \dfrac{\_\_\_(\sqrt{3}-\_\_\_)}{\_\_\_ \cdot 2}$

$= \_\_\_\_\_$

(b) $\dfrac{12+8\sqrt{5}}{16}$

Teaching Tip

Be sure to assign Exercises 55–62 in anticipation of using the quadratic formula.

Answers

4. **(a)** conjugate; $4-\sqrt{2}$; $\dfrac{5(4-\sqrt{2})}{14}$

(b) $-11-5\sqrt{5}$

(c) $\dfrac{7(5+\sqrt{x})}{25-x}$ $(x \neq 25)$

5. **(a)** 5; 3; 5; $\dfrac{\sqrt{3}-3}{2}$ **(b)** $\dfrac{3+2\sqrt{5}}{4}$

(b) $\dfrac{6+\sqrt{2}}{\sqrt{2}-5}$

$= \dfrac{(6+\sqrt{2})(\sqrt{2}+5)}{(\sqrt{2}-5)(\sqrt{2}+5)}$ Multiply the numerator and denominator by the conjugate of the denominator.

$= \dfrac{6\sqrt{2}+30+2+5\sqrt{2}}{2-25}$ FOIL method; $(a+b)(a-b)=a^2-b^2$

$= \dfrac{11\sqrt{2}+32}{-23}$ Combine like terms.

$= \dfrac{-11\sqrt{2}-32}{23}$ $\frac{a}{-b}=\frac{-a}{b}$

(c) $\dfrac{4}{3-\sqrt{x}}$

$= \dfrac{4(3+\sqrt{x})}{(3-\sqrt{x})(3+\sqrt{x})}$ Multiply by $\frac{3+\sqrt{x}}{3+\sqrt{x}}=1$.

$= \dfrac{4(3+\sqrt{x})}{9-x}$ $3^2=9$; $(\sqrt{x})^2=x$

We assume here that $x \neq 9$.

◀ Work Problem **4** at the Side.

OBJECTIVE 3 Write radical expressions with quotients in lowest terms.

EXAMPLE 5 Writing a Radical Quotient in Lowest Terms

Write $\dfrac{3\sqrt{3}+9}{12}$ in lowest terms.

$\dfrac{3\sqrt{3}+9}{12}$ Don't simplify yet.

$= \dfrac{3(\sqrt{3}+3)}{3(4)}$ Factor.

$= 1 \cdot \dfrac{\sqrt{3}+3}{4}$ Divide out the common factor; $\frac{3}{3}=1$

$= \dfrac{\sqrt{3}+3}{4}$ Identity property; lowest terms

◀ Work Problem **5** at the Side.

CAUTION

An expression like the one in **Example 5** can only be simplified by factoring a common factor from the denominator and *each* term of the numerator. ***First factor, and then divide out the common factor.***

Factor $\dfrac{4+8\sqrt{5}}{4}$ as $\dfrac{4(1+2\sqrt{5})}{4}$ to obtain $1+2\sqrt{5}$.

8.5 Exercises

FOR EXTRA HELP

Download the MyDashBoard App

MyMathLab®

CONCEPT CHECK *In Exercises 1–4, perform the operations mentally, and write the answers without doing intermediate steps.*

1. $\sqrt{49} + \sqrt{36}$
13

2. $\sqrt{100} - \sqrt{81}$
1

3. $\sqrt{2} \cdot \sqrt{8}$
4

4. $\sqrt{8} \cdot \sqrt{8}$
8

5. CONCEPT CHECK In **Example 1(b),** a student simplified $-37 - 2\sqrt{15}$ by combining the -37 and the -2 to get $-39\sqrt{15}$, which is incorrect. **What Went Wrong?**

Because multiplication must be performed before addition, it is incorrect to add -37 and -2. $-2\sqrt{15}$ cannot be simplified further. The final answer is $-37 - 2\sqrt{15}$.

6. CONCEPT CHECK Find each product mentally.

(a) $(\sqrt{x} + \sqrt{y})(\sqrt{x} - \sqrt{y})$
$x - y$

(b) $(\sqrt{28} - \sqrt{14})(\sqrt{28} + \sqrt{14})$
14

Simplify each expression. Use the five guidelines given in this section. Assume that all variables represent nonnegative real numbers. ***See Examples 1–3.***

7. $\sqrt{5}(\sqrt{3} - \sqrt{7})$
$\sqrt{15} - \sqrt{35}$

8. $\sqrt{7}(\sqrt{10} + \sqrt{3})$
$\sqrt{70} + \sqrt{21}$

9. $2\sqrt{5}(\sqrt{2} + 3\sqrt{5})$
$2\sqrt{10} + 30$

10. $3\sqrt{7}(2\sqrt{7} + 4\sqrt{5})$
$42 + 12\sqrt{35}$

11. $3\sqrt{14} \cdot \sqrt{2} - \sqrt{28}$
$4\sqrt{7}$

12. $7\sqrt{6} \cdot \sqrt{3} - 2\sqrt{18}$
$15\sqrt{2}$

13. $(2\sqrt{6} + 3)(3\sqrt{6} + 7)$
$57 + 23\sqrt{6}$

14. $(4\sqrt{5} - 2)(2\sqrt{5} - 4)$
$48 - 20\sqrt{5}$

15. $(5\sqrt{7} - 2\sqrt{3})(3\sqrt{7} + 4\sqrt{3})$
$81 + 14\sqrt{21}$

16. $(2\sqrt{10} + 5\sqrt{2})(3\sqrt{10} - 3\sqrt{2})$
$30 + 18\sqrt{5}$

17. $(8 - \sqrt{7})^2$
$71 - 16\sqrt{7}$

18. $(6 - \sqrt{11})^2$
$47 - 12\sqrt{11}$

19. $(2\sqrt{7} + 3)^2$
$37 + 12\sqrt{7}$

20. $(4\sqrt{5} + 5)^2$
$105 + 40\sqrt{5}$

21. $(\sqrt{a} + 1)^2$
$a + 2\sqrt{a} + 1$

22. $(\sqrt{y} + 4)^2$
$y + 8\sqrt{y} + 16$

23. $(5 - \sqrt{2})(5 + \sqrt{2})$
23

24. $(3 - \sqrt{5})(3 + \sqrt{5})$
4

25. $(\sqrt{8}-\sqrt{7})(\sqrt{8}+\sqrt{7})$ 1

26. $(\sqrt{12}-\sqrt{11})(\sqrt{12}+\sqrt{11})$ 1

27. $(\sqrt{y}-\sqrt{10})(\sqrt{y}+\sqrt{10})$ $y-10$

28. $(\sqrt{t}-\sqrt{13})(\sqrt{t}+\sqrt{13})$ $t-13$

29. $(\sqrt{2}+\sqrt{3})(\sqrt{6}-\sqrt{2})$ $2\sqrt{3}-2+3\sqrt{2}-\sqrt{6}$

30. $(\sqrt{3}+\sqrt{5})(\sqrt{15}-\sqrt{5})$ $3\sqrt{5}-\sqrt{15}+5\sqrt{3}-5$

31. $(\sqrt{10}-\sqrt{5})(\sqrt{5}+\sqrt{20})$ $15\sqrt{2}-15$

32. $(\sqrt{6}-\sqrt{3})(\sqrt{3}+\sqrt{18})$ $3\sqrt{2}+6\sqrt{3}-3-3\sqrt{6}$

33. $(\sqrt{5}+\sqrt{30})(\sqrt{6}+\sqrt{3})$ $\sqrt{30}+\sqrt{15}+6\sqrt{5}+3\sqrt{10}$

34. $(\sqrt{10}-\sqrt{20})(\sqrt{2}-\sqrt{5})$ $2\sqrt{5}-5\sqrt{2}-2\sqrt{10}+10$

35. $(\sqrt{5}-\sqrt{10})(\sqrt{x}-\sqrt{2})$ $\sqrt{5x}-\sqrt{10}-\sqrt{10x}+2\sqrt{5}$

36. $(\sqrt{x}+\sqrt{6})(\sqrt{10}+\sqrt{3})$ $\sqrt{10x}+\sqrt{3x}+2\sqrt{15}+3\sqrt{2}$

37. CONCEPT CHECK Determine the expression by which you should multiply the numerator and denominator to rationalize each denominator.

(a) $\dfrac{1}{\sqrt{5}+\sqrt{3}}$ $\sqrt{5}-\sqrt{3}$

(b) $\dfrac{3}{\sqrt{6}-\sqrt{5}}$ $\sqrt{6}+\sqrt{5}$

38. CONCEPT CHECK A student tried to rationalize the denominator of $\frac{2}{4+\sqrt{3}}$ by multiplying by $\frac{4+\sqrt{3}}{4+\sqrt{3}}$. **What Went Wrong?** By what should he multiply?

There would still be a radical in the denominator. He should multiply by $\frac{4-\sqrt{3}}{4-\sqrt{3}}$.

Rationalize each denominator. Write quotients in lowest terms. Assume that all variables represent nonnegative real numbers. ***See Examples 4 and 5.***

39. $\dfrac{1}{3+\sqrt{2}}$ $\dfrac{3-\sqrt{2}}{7}$

40. $\dfrac{1}{4-\sqrt{3}}$ $\dfrac{4+\sqrt{3}}{13}$

41. $\dfrac{14}{2-\sqrt{11}}$ $-4-2\sqrt{11}$

42. $\dfrac{19}{5-\sqrt{6}}$ $5+\sqrt{6}$

43. $\dfrac{\sqrt{2}}{2-\sqrt{2}}$ $1+\sqrt{2}$

44. $\dfrac{\sqrt{7}}{7-\sqrt{7}}$ $\dfrac{\sqrt{7}+1}{6}$

45. $\dfrac{\sqrt{5}}{\sqrt{2}+\sqrt{3}}$ $-\sqrt{10}+\sqrt{15}$

46. $\dfrac{\sqrt{3}}{\sqrt{2}+\sqrt{3}}$ $-\sqrt{6}+3$

47. $\dfrac{\sqrt{5}+2}{2-\sqrt{3}}$ $2\sqrt{5}+\sqrt{15}+4+2\sqrt{3}$

48. $\dfrac{\sqrt{7}+3}{4-\sqrt{5}}$ $\dfrac{4\sqrt{7}+\sqrt{35}+12+3\sqrt{5}}{11}$

49. $\dfrac{6-\sqrt{5}}{\sqrt{2}+2}$ $\dfrac{-6\sqrt{2}+12+\sqrt{10}-2\sqrt{5}}{2}$

50. $\dfrac{3+\sqrt{2}}{\sqrt{2}+1}$ $-1+2\sqrt{2}$

51. $\dfrac{12}{\sqrt{x}+1}$ $\dfrac{12(\sqrt{x}-1)}{x-1}$ $(x \neq 1)$

52. $\dfrac{10}{\sqrt{x}-4}$ $(x \neq 16)$ $\dfrac{10(\sqrt{x}+4)}{x-16}$

53. $\dfrac{3}{7-\sqrt{x}}$ $(x \neq 49)$ $\dfrac{3(7+\sqrt{x})}{49-x}$

54. $\dfrac{1}{6-\sqrt{z}}$ $(z \neq 36)$ $\dfrac{6+\sqrt{z}}{36-z}$

Write each quotient in lowest terms. ***See Example 5.***

55. $\dfrac{6\sqrt{11}-12}{6}$ $\sqrt{11}-2$

56. $\dfrac{12\sqrt{5}-24}{12}$ $\sqrt{5}-2$

57. $\dfrac{2\sqrt{3}+10}{16}$ $\dfrac{\sqrt{3}+5}{8}$

58. $\dfrac{4\sqrt{6}+24}{20}$ $\dfrac{\sqrt{6}+6}{5}$

59. $\dfrac{12-\sqrt{40}}{4}$ $\dfrac{6-\sqrt{10}}{2}$

60. $\dfrac{9-\sqrt{72}}{12}$ $\dfrac{3-2\sqrt{2}}{4}$

61. $\dfrac{16+\sqrt{128}}{24}$ $\dfrac{2+\sqrt{2}}{3}$

62. $\dfrac{25+\sqrt{75}}{10}$ $\dfrac{5+\sqrt{3}}{2}$

Solve each problem.

63. The radius r of the circular top or bottom of a tin can with a surface area S and a height h is given by

$$r = \frac{-h + \sqrt{h^2 + 0.64S}}{2}.$$

(a) What radius should be used to make a can with a height of 12 in. and a surface area of 400 in.^{2}?

4 in.

(b) What radius should be used to make a can with a height of 6 in. and a surface area of 200 in.^{2}? Round the answer to the nearest tenth.

3.4 in.

64. If an investment of P dollars grows to A dollars in 2 yr, the annual rate of return on the investment is given by

$$r = \frac{\sqrt{A} - \sqrt{P}}{\sqrt{P}}.$$

First rationalize the denominator. Then find the annual rate of return r (as a percent) if \$50,000 increases to \$54,080.

$\frac{\sqrt{AP} - P}{P}$; 4%

Relating Concepts (Exercises 65–70) For Individual or Group Work

Work Exercises 65–70 in order, *to see why a common student error is indeed an error.*

65. Use the distributive property to write $6(5 + 3x)$ as a sum.

$30 + 18x$

66. Your answer in **Exercise 65** should be $30 + 18x$. Why can we not combine these two terms to get $48x$?

They are not like terms.

67. Simplify $(2\sqrt{10} + 5\sqrt{2})(3\sqrt{10} - 3\sqrt{2})$. (See **Exercise 16.**)

$30 + 18\sqrt{5}$

68. Your answer in **Exercise 67** should be $30 + 18\sqrt{5}$. Many students will, in error, try to combine these terms to get $48\sqrt{5}$. Why is this wrong?

They are not like radicals.

69. Write the expression similar to $30 + 18x$ that simplifies to $48x$. Then write the expression similar to $30 + 18\sqrt{5}$ that simplifies to $48\sqrt{5}$.

Make the first term $30x$, so that $30x + 18x = 48x$. Make the first term $30\sqrt{5}$, so that $30\sqrt{5} + 18\sqrt{5} = 48\sqrt{5}$.

70. Write a short explanation of the similarities between combining like terms and combining like radicals.

Both like terms and like radicals are combined by adding their numerical coefficients. The variables in like terms are replaced by radicals in like radicals.

Summary Exercises *Applying Operations with Radicals*

Summary Exercises provide mixed practice to better prepare students for tests.

Perform all indicated operations and express each answer in simplest form. Assume that all variables represent positive real numbers.

1. $5\sqrt{10} - 8\sqrt{10}$

$-3\sqrt{10}$

2. $\sqrt{5}(\sqrt{5} - \sqrt{3})$

$5 - \sqrt{15}$

3. $(1 + \sqrt{3})(2 - \sqrt{6})$

$2 - \sqrt{6} + 2\sqrt{3} - 3\sqrt{2}$

4. $\sqrt{98} - \sqrt{72} + \sqrt{50}$

$6\sqrt{2}$

5. $(3\sqrt{5} - 2\sqrt{7})^2$

$73 - 12\sqrt{35}$

6. $\dfrac{3}{\sqrt{6}}$

$\dfrac{\sqrt{6}}{2}$

7. $\sqrt[3]{16t^2} - \sqrt[3]{54t^2} + \sqrt[3]{128t^2}$

$3\sqrt[3]{2t^2}$

8. $\dfrac{8}{\sqrt{7} - \sqrt{5}}$

$4\sqrt{7} + 4\sqrt{5}$

9. $\dfrac{1 + \sqrt{2}}{1 - \sqrt{2}}$

$-3 - 2\sqrt{2}$

10. $(1 + \sqrt[3]{3})(1 - \sqrt[3]{3} + \sqrt[3]{9})$

4

11. $(\sqrt{3} + 6)(\sqrt{3} - 6)$

-33

12. $\dfrac{1}{\sqrt{t} + \sqrt{3}}$

$\dfrac{\sqrt{t} - \sqrt{3}}{t - 3}$ $(t \neq 3)$

13. $\sqrt[3]{8x^3y^5z^6}$

$2xyz^2\sqrt[3]{y^2}$

14. $\dfrac{12}{\sqrt[3]{9}}$

$4\sqrt[3]{3}$

15. $\dfrac{5}{\sqrt{6} - 1}$

$\sqrt{6} + 1$

16. $\sqrt{\dfrac{2}{3x}}$

$\dfrac{\sqrt{6x}}{3x}$

17. $\dfrac{6\sqrt{3}}{5\sqrt{12}}$

$\dfrac{3}{5}$

18. $\dfrac{8\sqrt{50}}{2\sqrt{25}}$

$4\sqrt{2}$

19. $\dfrac{-4}{\sqrt[3]{4}}$

$-2\sqrt[3]{2}$

20. $\dfrac{\sqrt{6} - \sqrt{5}}{\sqrt{6} + \sqrt{5}}$

$11 - 2\sqrt{30}$

21. $\sqrt{75x} - \sqrt{12x}$

$3\sqrt{3x}$

22. $(5 + 3\sqrt{3})^2$

$52 + 30\sqrt{3}$

23. $(\sqrt{7} - \sqrt{6})(\sqrt{7} + \sqrt{6})$

1

24. $\sqrt[3]{\frac{16}{81}}$

$\frac{2\sqrt[3]{18}}{9}$

25. $x\sqrt[4]{x^5} - 3\sqrt[4]{x^9} + x^2\sqrt[4]{x}$

$-x^2\sqrt[4]{x}$

26. $\sqrt{7} + \sqrt{7}$

$2\sqrt{7}$

27. $\sqrt{14} + \sqrt{5}$

cannot be combined

28. $9\sqrt{24} - 2\sqrt{54} + 3\sqrt{20}$

$12\sqrt{6} + 6\sqrt{5}$

29. $\sqrt{\frac{3}{4}} \cdot \sqrt{\frac{1}{5}}$

$\frac{\sqrt{15}}{10}$

30. $\frac{5}{\sqrt{5}}$

$\sqrt{5}$

31. $\sqrt[3]{24} + 6\sqrt[3]{81}$

$20\sqrt[3]{3}$

32. $\frac{8}{4 - \sqrt{x}}$ $(x \neq 16)$

$\frac{8(4 + \sqrt{x})}{16 - x}$

33. $\sqrt[3]{4}(\sqrt[3]{2} - 3)$

$2 - 3\sqrt[3]{4}$

34. $\sqrt{32x} - \sqrt{18x}$

$\sqrt{2x}$

35. $\sqrt{\frac{5}{8}}$

$\frac{\sqrt{10}}{4}$

36. $(7 + \sqrt{x})^2$

$49 + 14\sqrt{x} + x$

A biologist has shown that the number of different plant species S on a Galápagos Island is related to the area of the island, A (in square miles), by the following formula.

$$S = 28.6\sqrt[3]{A}$$

How many plant species (to the nearest whole number) would exist on such an island with the following areas?

37. 8 mi^2

57 species

38. 27,000 mi^2

858 species

8.6 Solving Equations with Radicals

A **radical equation** is an equation with a variable in a radicand.

$$\sqrt{x} = 6, \quad \sqrt{x+1} = 3, \quad \text{and} \quad 3\sqrt{x} = \sqrt{8x+9} \qquad \text{Radical equations}$$

OBJECTIVE 1 Solve radical equations having square root radicals. To solve radical equations, we use a new property, called the ***squaring property of equality.***

Squaring Property of Equality

If each side of a given equation is squared, all solutions of the original equation are *among* the solutions of the squared equation.

CAUTION

Using the squaring property can give a new equation with *more* solutions than the original equation. For example, starting with the equation $x = 4$ and squaring each side gives

$$x^2 = 4^2, \quad \text{or} \quad x^2 = 16.$$

This last equation, $x^2 = 16$, has *two* solutions, 4 or -4, while the original equation, $x = 4$, has only *one* solution, 4.

Because of this possibility, checking is more than just a guard against algebraic errors when solving an equation with radicals. It is an essential part of the solution process. ***All proposed solutions from the squared equation must be checked in the original equation.***

EXAMPLE 1 Using the Squaring Property of Equality

Solve $\sqrt{p+1} = 3$.

$$\sqrt{p+1} = 3$$

$$(\sqrt{p+1})^2 = 3^2 \qquad \text{Use the squaring property to square each side.}$$

$$p + 1 = 9 \qquad (\sqrt{p+1})^2 = p + 1$$

Proposed solution ⟶ $p = 8$ Subtract 1.

CHECK (A check is essential.)

$$\sqrt{p+1} = 3 \qquad \text{Original equation}$$

$$\sqrt{8+1} \stackrel{?}{=} 3 \qquad \text{Let } p = 8.$$

$$\sqrt{9} \stackrel{?}{=} 3 \qquad \text{Add.}$$

$$3 = 3 \checkmark \qquad \text{True}$$

Because this statement is true, $\{8\}$ is the solution set of $\sqrt{p+1} = 3$. In this case the equation obtained by squaring had just one solution, which also satisfied the original equation.

Work Problem 1 at the Side. ▶

OBJECTIVES

1. Solve radical equations having square root radicals.
2. Identify equations with no solutions.
3. Solve equations by squaring a binomial.
4. Solve problems using formulas that involve radicals.

1 Solve each equation. Be sure to check your solutions.

GS **(a)** $\sqrt{k} = 3$

$$(\sqrt{k})^{\_\_} = 3^2$$

$$\_\_\_\_ = 9$$

CHECK $\sqrt{k} = 3$

$$\sqrt{\_\_\_\_} \stackrel{?}{=} 3$$

$$\_\_\_\_ = 3$$

(*True / False*)

The solution set is ____.

(b) $\sqrt{x-2} = 4$

Teaching Tip

Using the squaring property may produce a linear or a quadratic equation. Some review may be needed here as to the appropriate method to use to solve these equations.

(c) $\sqrt{9-t} = 4$

Answers

1. (a) 2; k; 9; 3; True; $\{9\}$ **(b)** $\{18\}$ **(c)** $\{-7\}$

2 Solve each equation.

(a) $\sqrt{3x+9} = 2\sqrt{x}$

Teaching Tip

Emphasize that even though the algebraic work may be done perfectly, the answer produced may not make the original equation true.

(b) $5\sqrt{x} = \sqrt{20x+5}$

3 Solve each equation.

(a) $\sqrt{x} = -4$

GS **(b)** $\sqrt{x} + 6 = 0$

$$\sqrt{x} = \_\_\_\_$$
$$(\sqrt{x})^2 = (\_\_\_\_)^2$$
$$x = \_\_\_\_$$

CHECK $\sqrt{x} + 6 = 0$

$$\sqrt{\_\_\_\_} + 6 \stackrel{?}{=} 0$$
$$\_\_\_\_ = 0$$

(*True* / *False*)

The solution set is ____.

Answers

2. **(a)** $\{9\}$ **(b)** $\{1\}$

3. **(a)** $\emptyset$ **(b)** -6; -6; 36; 36; 12; False; $\emptyset$

EXAMPLE 2 Using the Squaring Property with a Radical on Each Side

Solve $3\sqrt{x} = \sqrt{x+8}$.

$$3\sqrt{x} = \sqrt{x+8}$$

$$(3\sqrt{x})^2 = (\sqrt{x+8})^2 \quad \text{Squaring property}$$

$$3^2(\sqrt{x})^2 = (\sqrt{x+8})^2 \quad (ab)^2 = a^2b^2$$

$$9x = x + 8 \quad (\sqrt{x})^2 = x;\ (\sqrt{x+8})^2 = x+8$$

Be careful here.

$$8x = 8 \quad \text{Subtract } x.$$

Proposed solution ⟶ $x = 1$ Divide by 8.

CHECK

$$3\sqrt{x} = \sqrt{x+8} \quad \text{Original equation}$$

$$3\sqrt{1} \stackrel{?}{=} \sqrt{1+8} \quad \text{Let } x = 1.$$

$$3(1) \stackrel{?}{=} \sqrt{9} \quad \text{Simplify.}$$

$$3 = 3 \ ✓ \quad \text{True}$$

This is *not* the solution.

The solution set is $\{1\}$.

◀ Work Problem **2** at the Side.

OBJECTIVE 2 **Identify equations with no solutions.** Not all radical equations have solutions.

EXAMPLE 3 Using the Squaring Property When One Side Is Negative

Solve $\sqrt{x} = -3$.

$$\sqrt{x} = -3$$

$$(\sqrt{x})^2 = (-3)^2 \quad \text{Squaring property}$$

$$x = 9 \longleftarrow \text{Proposed solution}$$

CHECK

$$\sqrt{x} = -3 \quad \text{Original equation}$$

$$\sqrt{9} \stackrel{?}{=} -3 \quad \text{Let } x = 9.$$

$$3 = -3 \quad \text{False}$$

Because the statement $3 = -3$ is false, the number 9 is *not* a solution of the given equation. It is an **extraneous solution** and must be rejected. In fact, $\sqrt{x} = -3$ has no solution. The solution set is $\emptyset$.

Note

Because $\sqrt{x}$ represents the *principal* or *nonnegative* square root of x in **Example 3,** we might have seen immediately that there is no solution.

◀ Work Problem **3** at the Side.

We use the following steps when solving an equation with radicals.

Solving a Radical Equation

Step 1 **Isolate a radical.** Arrange the terms so that a radical is alone on one side of the equation.

Step 2 **Square each side.**

Step 3 **Combine like terms.**

Step 4 **Repeat Steps 1–3,** if there is still a term with a radical.

Step 5 **Solve the equation.** Find all proposed solutions.

Step 6 **Check all proposed solutions** in the original equation.

EXAMPLE 4 **Using the Squaring Property with a Quadratic Expression**

Solve $x = \sqrt{x^2 + 5x + 10}$.

Step 1 The radical is already isolated on the right side of the equation.

Step 2 Square each side.

$$x^2 = \left(\sqrt{x^2 + 5x + 10}\right)^2 \qquad \text{Squaring property}$$

$$x^2 = x^2 + 5x + 10 \qquad \left(\sqrt{a}\right)^2 = a$$

Step 3

$$0 = 5x + 10 \qquad \text{Subtract } x^2.$$

Step 4 This step is not needed.

Step 5

$$-10 = 5x \qquad \text{Subtract 10.}$$

Proposed solution → $\mathbf{-2 = x}$ Divide by 5.

Step 6 CHECK

$$x = \sqrt{x^2 + 5x + 10} \qquad \text{Original equation}$$

$$-2 \stackrel{?}{=} \sqrt{(-2)^2 + 5(-2) + 10} \qquad \text{Let } x = -2.$$

$$-2 \stackrel{?}{=} \sqrt{4 - 10 + 10} \qquad \text{Multiply.}$$

$$-2 = 2 \qquad \text{False}$$

The principal square root of a quantity *cannot* be negative.

Since substituting -2 for x leads to a false result, the equation has no solution, and the solution set is $\emptyset$.

Work Problem 4 at the Side. ▶

OBJECTIVE 3 **Solve equations by squaring a binomial.** Recall the rules for squaring binomials from **Section 5.4.**

$$(a + b)^2 = a^2 + 2ab + b^2 \quad \text{and} \quad (a - b)^2 = a^2 - 2ab + b^2$$

We apply the second rule in **Example 5** on the next page with $(x - 3)^2$.

$$(x - 3)^2$$
$$= x^2 - 2x(3) + 3^2$$
$$= x^2 - 6x + 9$$

Remember the middle term when squaring.

Work Problem 5 at the Side. ▶

4 Solve $x = \sqrt{x^2 - 4x - 16}$.

5 Square each expression.

GS **(a)** $w - 5$

$(w - 5)^2$

$= \underline{\quad}^2 - \underline{\quad} w (\underline{\quad})$

$+ \underline{\quad}^2$

$= \underline{\qquad}$

(b) $2k - 5$

(c) $3m - 2p$

Teaching Tip

Watch for the following ERROR.

$$(x - 3)^2 = x^2 + 9$$

Answers

4. $\emptyset$

5. (a) w; 2; 5; 5; $w^2 - 10w + 25$
(b) $4k^2 - 20k + 25$
(c) $9m^2 - 12mp + 4p^2$

6 Solve each equation.

GS **(a)** $\sqrt{6w+6} = w+1$

$(\sqrt{6w+6})^{—} = (w+1)^2$

$\_\_\_\_\_ = w^2 + \_\_\_\_\_$

$w^2 - \_\_\_\_\_ = 0$

$(\_\_\_\_\_)(\_\_\_\_\_) = 0$

Complete the solution.

(b) $2u - 1 = \sqrt{10u+9}$

Answers

6. **(a)** 2; $6w+6$; $2w+1$; $4w-5$; $w+1$; $w-5$; $\{-1, 5\}$
(b) $\{4\}$

EXAMPLE 5 **Using the Squaring Property When One Side Has Two Terms**

Solve $\sqrt{2x-3} = x-3$.

$$\sqrt{2x-3} = x-3$$

$$(\sqrt{2x-3})^2 = (x-3)^2 \quad \text{Square each side.}$$

$$2x-3 = x^2 - 6x + 9 \quad (x-y)^2 = x^2 - 2xy + y^2$$

This equation is quadratic because of the presence of the x^2-term. To solve it, as shown in **Section 6.6,** we must write the equation in standard form.

$$\text{Standard form} \rightarrow x^2 - 8x + 12 = 0 \quad \text{Subtract } 2x \text{, add 3, and interchange sides.}$$

$$(x-6)(x-2) = 0 \quad \text{Factor.}$$

$$x-6=0 \quad \text{or} \quad x-2=0 \quad \text{Zero-factor property}$$

$$x = 6 \quad \text{or} \quad x = 2 \leftarrow \text{Proposed solutions}$$

CHECK Let $x = 6$.

$$\sqrt{2x-3} = x-3$$
$$\sqrt{2(6)-3} \stackrel{?}{=} 6-3$$
$$\sqrt{12-3} \stackrel{?}{=} 3$$
$$\sqrt{9} \stackrel{?}{=} 3$$
$$3 = 3 \checkmark \text{ True}$$

Let $x = 2$.

$$\sqrt{2x-3} = x-3$$
$$\sqrt{2(2)-3} \stackrel{?}{=} 2-3$$
$$\sqrt{4-3} \stackrel{?}{=} -1$$
$$\sqrt{1} \stackrel{?}{=} -1$$
$$1 = -1 \quad \text{False}$$

Only 6 is a valid solution. (2 is extraneous.) The solution set is $\{6\}$.

◀ **Work Problem 6 at the Side.**

EXAMPLE 6 **Rewriting an Equation before Using the Squaring Property**

Solve $\sqrt{9x} - 1 = 2x$.

We must apply Step 1 to isolate the radical *before* squaring each side. If we skip Step 1 and begin by squaring, we obtain the following.

$$(\sqrt{9x} - 1)^2 = (2x)^2$$

$$9x - 2\sqrt{9x} + 1 = 4x^2$$

Don't do this.

This equation still contains a radical. Follow the steps below.

$$\sqrt{9x} - 1 = 2x$$

This is a key step.

$$\sqrt{9x} = 2x + 1 \quad \text{Add 1 to isolate the radical. (Step 1)}$$

$$(\sqrt{9x})^2 = (2x+1)^2 \quad \text{Square each side. (Step 2)}$$

No terms contain radicals.

$$9x = 4x^2 + 4x + 1 \quad (a+b)^2 = a^2 + 2ab + b^2$$

$$4x^2 - 5x + 1 = 0 \quad \text{Standard form (Step 3)}$$

$$(4x-1)(x-1) = 0 \quad \text{Factor. (Step 5; Step 4 is not needed.)}$$

$$4x - 1 = 0 \quad \text{or} \quad x - 1 = 0 \quad \text{Zero-factor property}$$

$$x = \frac{1}{4} \quad \text{or} \quad x = 1 \leftarrow \text{Proposed solutions}$$

Continued on Next Page

CHECK Let $x = \frac{1}{4}$. (Step 6)

$$\sqrt{9x} - 1 = 2x$$

$$\sqrt{9\left(\frac{1}{4}\right)} - 1 \stackrel{?}{=} 2\left(\frac{1}{4}\right)$$

$$\frac{3}{2} - 1 \stackrel{?}{=} \frac{1}{2}$$

$$\frac{1}{2} = \frac{1}{2} \checkmark \quad \text{True}$$

Let $x = 1$. (Step 6)

$$\sqrt{9x} - 1 = 2x$$

$$\sqrt{9(1)} - 1 \stackrel{?}{=} 2(1)$$

$$3 - 1 \stackrel{?}{=} 2$$

$$2 = 2 \checkmark \quad \text{True}$$

Both proposed solutions check, so the solution set is $\left\{\frac{1}{4}, 1\right\}$.

CAUTION

When squaring each side of

$$\sqrt{9x} = 2x + 1$$

in **Example 6,** the *entire* binomial $2x + 1$ must be squared to get $4x^2 + 4x + 1$. It is incorrect to square the $2x$ and the 1 separately to get $4x^2 + 1$.

Work Problem 7 at the Side. ▶

Some radical equations require squaring twice, as in the next example.

EXAMPLE 7 Using the Squaring Property Twice

Solve $\sqrt{21 + x} = 3 + \sqrt{x}$.

$$\sqrt{21 + x} = 3 + \sqrt{x}$$

$$\left(\sqrt{21 + x}\right)^2 = \left(3 + \sqrt{x}\right)^2 \quad \text{Square each side.}$$

$$21 + x = 9 + 6\sqrt{x} + x \quad \text{Be careful here.}$$

$$12 = 6\sqrt{x} \quad \text{Subtract 9. Subtract } x.$$

$$2 = \sqrt{x} \quad \text{Divide by 6.}$$

$$2^2 = \left(\sqrt{x}\right)^2 \quad \text{Square each side again.}$$

Proposed solution ⟶ $4 = x$ Apply the exponents.

CHECK

$$\sqrt{21 + x} = 3 + \sqrt{x} \quad \text{Original equation}$$

$$\sqrt{21 + 4} \stackrel{?}{=} 3 + \sqrt{4} \quad \text{Let } x = 4.$$

$$\sqrt{25} \stackrel{?}{=} 3 + 2 \quad \text{Simplify.}$$

$$5 = 5 \checkmark \quad \text{True}$$

The solution set is $\{4\}$.

Work Problem 8 at the Side. ▶

---

7 Solve each equation.

GS **(a)** $\sqrt{x} - 3 = x - 15$

$$\sqrt{x} = x - \underline{\quad}$$

$$\left(\sqrt{x}\right)^{\underline{\quad}} = (x - \underline{\quad})^2$$

$$\underline{\quad} = x^2 - \underline{\qquad}$$

$$\underline{\qquad} = 0$$

Complete the solution.

(b) $\sqrt{z + 5} + 2 = z + 5$

8 Solve each equation.

(a) $\sqrt{p + 1} - \sqrt{p - 4} = 1$

Teaching Tip

Be aware that students often will erroneously try to solve an equation such as

$$\sqrt{2x + 3} + \sqrt{x + 1} = 1$$

by squaring each term. Sometimes they will actually get the correct answer using the wrong method! (Try this one.) Caution students about this common error.

(b) $\sqrt{2x + 1} + \sqrt{x + 4} = 3$

Answers

7. (a) 12; 2; 12; x; $24x + 144$; $x^2 - 25x + 144$; $\{16\}$
(b) $\{-1\}$

8. (a) $\{8\}$ **(b)** $\{0\}$

Heron of Alexandria

First century Greek mathematician

OBJECTIVE 4 Solve problems using formulas that involve radicals.

The most common formula for the area of a triangle is $A = \frac{1}{2}bh$, where b is the length of the base and h is the height. What if the height h is not known? **Heron's formula** allows us to calculate the area of a triangle if we know only the lengths of the sides a, b, and c. To use Heron's formula, we first find the **semiperimeter** s, which is one-half the perimeter.

$$s = \frac{1}{2}(a + b + c) \quad \text{Semiperimeter}$$

The area $\mathscr{A}$ is then given by the following formula.

$$\mathscr{A} = \sqrt{s(s-a)(s-b)(s-c)} \quad \text{Heron's formula}$$

For example, a 3-4-5 right triangle has area

$$\mathscr{A} = \frac{1}{2}(3)(4) = 6 \text{ square units,}$$

calculated using the familiar formula. We can also use Heron's formula, with $s = \frac{1}{2}(3 + 4 + 5) = 6$.

$$\mathscr{A} = \sqrt{6(6-3)(6-4)(6-5)} \quad \text{Let } s = 6,\ a = 3,\ b = 4, \text{ and } c = 5.$$

$$\mathscr{A} = \sqrt{6 \cdot 3 \cdot 2 \cdot 1} \quad \text{Subtract.}$$

$$\mathscr{A} = \sqrt{36} \quad \text{Multiply.}$$

$$\mathscr{A} = 6 \quad \text{Take the positive root.}$$

The area is again 6 square units, as expected.

9 Find the area of a triangle with sides of lengths 7 cm, 15 cm, and 20 cm.

EXAMPLE 8 Using Heron's Formula to Find the Area of a Triangle

The sides of a triangular garden plot have lengths 20 ft, 34 ft, and 42 ft. See **Figure 3**. Find its area.

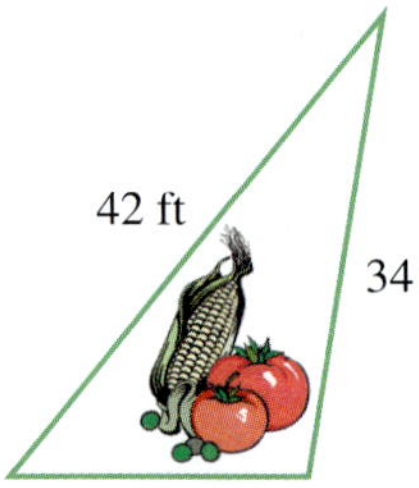

Figure 3

First find the semiperimeter of the triangle.

$$s = \frac{1}{2}(a + b + c)$$

$$s = \frac{1}{2}(20 + 34 + 42) \quad \text{Let } a = 20,\ b = 34, \text{ and } c = 42.$$

$$s = \frac{1}{2}(96), \quad \text{or} \quad 48 \quad \text{Add, and then multiply.}$$

Now use Heron's formula to find the area.

$$\mathscr{A} = \sqrt{s(s-a)(s-b)(s-c)} \quad \text{Heron's formula}$$

$$\mathscr{A} = \sqrt{48(48-20)(48-34)(48-42)} \quad \text{Substitute.}$$

$$\mathscr{A} = \sqrt{48(28)(14)(6)} \quad \text{Subtract.}$$

$$\mathscr{A} = \sqrt{112{,}896} \quad \text{Multiply.}$$

$$\mathscr{A} = 336 \quad \text{Use a calculator.}$$

The area of the garden plot is 336 ft^2.

◀ **Work Problem 9 at the Side.**

Answer

9. 42 cm^2

8.6 Exercises

FOR EXTRA HELP

Download the MyDashBoard App

MyMathLab®

CONCEPT CHECK *Fill in the blanks to complete the following.*

1. To solve an equation involving a radical, such as $\sqrt{2x-1}=5$, use the ___squaring___ property of equality. This property says that if each side of an equation is ___squared___, all solutions of the ___original___ equation are among the solutions of the squared equation.

2. Solving some radical equations involves squaring a binomial using the following rules.

$$(a+b)^2 = \underline{a^2+2ab+b^2}$$

$$(a-b)^2 = \underline{a^2-2ab+b^2}$$

Solve each equation. ***See Examples 1–4.***

3. $\sqrt{x}=7$ $\{49\}$

4. $\sqrt{k}=10$ $\{100\}$

5. $\sqrt{t+2}=3$ $\{7\}$

6. $\sqrt{x+7}=5$ $\{18\}$

7. $\sqrt{r-4}=9$ $\{85\}$

8. $\sqrt{k-12}=3$ $\{21\}$

9. $\sqrt{t}=-5$ $\emptyset$

10. $\sqrt{x}=-8$ $\emptyset$

11. $\sqrt{4-t}=7$ $\{-45\}$

12. $\sqrt{9-s}=5$ $\{-16\}$

13. $\sqrt{2t+3}=0$ $\left\{-\frac{3}{2}\right\}$

14. $\sqrt{5x-4}=0$ $\left\{\frac{4}{5}\right\}$

15. $\sqrt{3x-8}=-2$ $\emptyset$

16. $\sqrt{6x+4}=-3$ $\emptyset$

17. $\sqrt{w}-4=7$ $\{121\}$

18. $\sqrt{t}+3=10$ $\{49\}$

19. $\sqrt{10x-8}=3\sqrt{x}$ $\{8\}$

20. $\sqrt{17t-4}=4\sqrt{t}$ $\{4\}$

21. $5\sqrt{x}=\sqrt{10x+15}$ $\{1\}$

22. $4\sqrt{z}=\sqrt{20z-16}$ $\{4\}$

23. $\sqrt{3x-5}=\sqrt{2x+1}$ $\{6\}$

24. $\sqrt{5x+2}=\sqrt{3x+8}$ $\{3\}$

25. $k=\sqrt{k^2-5k-15}$ $\emptyset$

26. $s=\sqrt{s^2-2s-6}$ $\emptyset$

27. $7x=\sqrt{49x^2+2x-10}$ $\{5\}$

28. $6x=\sqrt{36x^2+5x-5}$ $\{1\}$

29. CONCEPT CHECK Consider the following incorrect "solution." **What Went Wrong?**

$$-\sqrt{x-1} = -4$$
$$-(x-1) = 16 \quad \text{Square each side.}$$
$$-x+1 = 16 \quad \text{Distributive property}$$
$$-x = 15 \quad \text{Subtract 1.}$$
$$x = -15 \quad \text{Multiply by } -1.$$

Solution set: $\{-15\}$

When the left side is squared, the result should be $x - 1$, not $-(x - 1)$. The correct solution set is $\{17\}$.

30. CONCEPT CHECK The first step in solving the equation

$$\sqrt{2x+1} = x - 7$$

is to square each side of the equation. Errors often occur in solving equations such as this one when the right side of the equation is squared incorrectly. What is the square of the right side?

$x^2 - 14x + 49$

Solve each equation. ***See Examples 5 and 6.***

31. $\sqrt{2x+1} = x - 7$ $\{12\}$

32. $\sqrt{3x+3} = x - 5$ $\{11\}$

33. $\sqrt{3k+10} + 5 = 2k$ $\{5\}$

34. $\sqrt{4t+13} + 1 = 2t$ $\{3\}$

35. $\sqrt{5x+1} - 1 = x$ $\{0, 3\}$

36. $\sqrt{x+1} - x = 1$ $\{-1, 0\}$

37. $\sqrt{6t+7} + 3 = t + 5$ $\{-1, 3\}$

38. $\sqrt{10x+24} = x + 4$ $\{-2, 4\}$

39. $x - 4 - \sqrt{2x} = 0$ $\{8\}$

40. $x - 3 - \sqrt{4x} = 0$ $\{9\}$

41. $\sqrt{x} + 6 = 2x$ $\{4\}$

42. $\sqrt{k} + 12 = k$ $\{16\}$

Solve each equation. ***See Example 7.***

43. $\sqrt{x+1} - \sqrt{x-4} = 1$ $\{8\}$

44. $\sqrt{2x+3} + \sqrt{x+1} = 1$ $\{-1\}$

45. $\sqrt{x} = \sqrt{x-5} + 1$ $\{9\}$

46. $\sqrt{2x} = \sqrt{x+7} - 1$ $\{2\}$

47. $\sqrt{3x+4} - \sqrt{2x-4} = 2$ $\{4, 20\}$

48. $\sqrt{1-x} + \sqrt{x+9} = 4$ $\{-8, 0\}$

49. $\sqrt{2x+11} + \sqrt{x+6} = 2$ $\{-5\}$

50. $\sqrt{x+9} + \sqrt{x+16} = 7$ $\{0\}$

Solve each problem.

51. The square root of the sum of a number and 4 is 5. Find the number. **21**

52. The square root of the sum of a number and 7 is 4. Find the number. **9**

53. Three times the square root of 2 equals the square root of the sum of some number and 10. Find the number. **8**

54. The negative square root of a number equals that number decreased by 2. Find the number. **1**

Use Heron's formula to solve each problem. ***See Example 8.***

55. A surveyor has measured the lengths of the sides of a triangular plot of land as 180 ft, 200 ft, and 240 ft. Find the area of the plot. (Round to the nearest whole number.) **17,616 ft²**

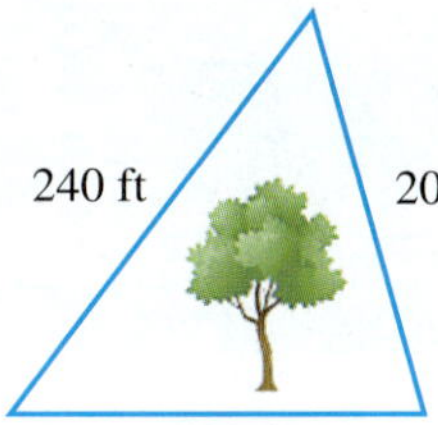

56. A triangular indoor play space in a day care center has sides of length 30 ft, 45 ft, and 65 ft. How much carpeting would be needed to cover this space? (Round to the nearest whole number.) **592 ft²**

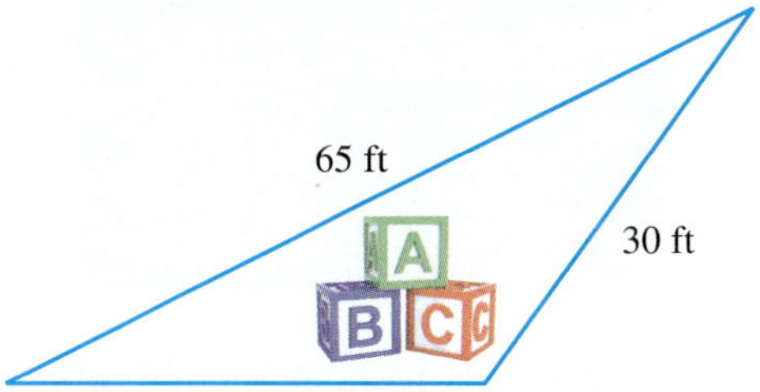

Solve each problem. Give answers to the nearest tenth.

57. To estimate the speed at which a car was traveling at the time of an accident, a police officer drives the car involved in the accident under conditions similar to those during which the accident took place and then skids to a stop. If the car is driven at 30 mph, then the speed at the time of the accident is given by

$$s = 30\sqrt{\frac{a}{p}},$$

where a is the length of the skid marks left at the time of the accident and p is the length of the skid marks in the police test. Find s for the following values of a and p.

(a) $a = 862$ ft; $p = 156$ ft **70.5 mph**

(b) $a = 382$ ft; $p = 96$ ft **59.8 mph**

(c) $a = 84$ ft; $p = 26$ ft **53.9 mph**

58. A formula for calculating the distance, d, one can see from an airplane to the horizon on a clear day is

$$d = 1.22\sqrt{x},$$

where x is the altitude of the plane in feet and d is given in miles.

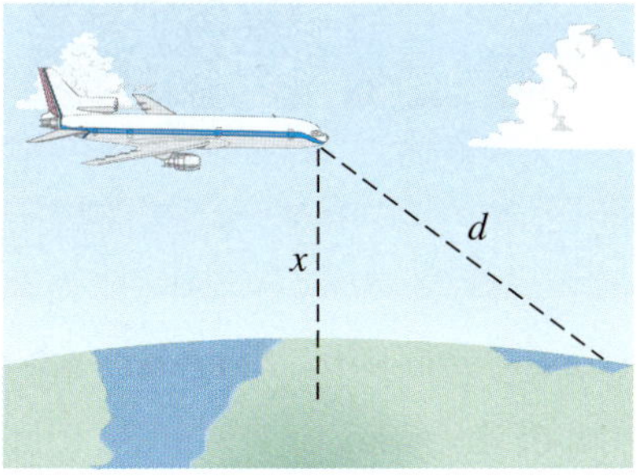

How far can one see to the horizon in a plane flying at the following altitudes?

(a) 15,000 ft **149.4 mi**

(b) 18,000 ft **163.7 mi**

(c) 24,000 ft **189.0 mi**

On a clear day, the maximum distance in kilometers that one can see from a structure is given by the formula

$$\text{sight distance} = 111.7\sqrt{\text{height of structure in kilometers}}.$$

Use this formula and the conversion equations $1 \text{ ft} \approx 0.3048 \text{ m}$ *and* $1 \text{ km} \approx 0.621371 \text{ mi}$ *as necessary to solve each problem. Round your answers to the nearest mile.* (*Source: A Sourcebook of Applications of School Mathematics,* NCTM.)

59. The London Eye is a unique structure that features 32 observation capsules. It is the fourth-tallest structure in London, with a diameter of 135 m. Does the formula justify the claim that on a clear day passengers on the London Eye can see Windsor Castle, 25 mi away? (*Source:* www.londoneye.com) **yes; 26 mi**

60. The Empire State Building in New York City is 1250 ft high. (The antenna reaches to 1454 ft.) The observation deck, located on the 102nd floor, is at a height of 1050 ft. How far could you see on a clear day from the observation deck? (*Source:* www.esbnyc.com) **39 mi**

61. The twin Petronas Towers in Kuala Lumpur, Malaysia, are 1483 ft high (including the spires). How far would one of the builders have been able to see on a clear day from the top of a spire? (*Source*: *World Almanac and Book of Facts.*) **47 mi**

62. The Khufu Pyramid in Giza was built in about 2566 B.C. to a height, at that time, of 481 ft. It is now only about 449 ft high. How far would one of the original builders of the pyramid have been able to see from the top of the pyramid? (*Source*: www.archaeology.com) **27 mi**

Relating Concepts (Exercises 63–68) For Individual or Group Work

Consider the figure, and **work Exercises 63–68 in order.**

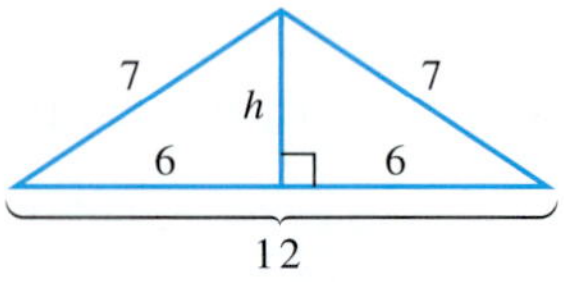

63. The lengths of the sides of the entire triangle are 7, 7, and 12. Find the semiperimeter s. **$s = 13$ units**

64. Now use Heron's formula to find the area of the entire triangle. Write it as a simplified radical. **$6\sqrt{13}$ sq. units**

65. Find the value of h by using the Pythagorean theorem. **$h = \sqrt{13}$ units**

66. Find the area of each of the congruent right triangles forming the entire triangle by using the formula $A = \frac{1}{2}bh$. **$3\sqrt{13}$ sq. units**

67. Double your result from **Exercise 66** to determine the area of the entire triangle. **$6\sqrt{13}$ sq. units**

68. How do your answers in **Exercises 64 and 67** compare? **They are both $6\sqrt{13}$.**

Chapter 8 Summary

Key Terms

8.1

square root A number a is a square root of b if $a^2 = b$.

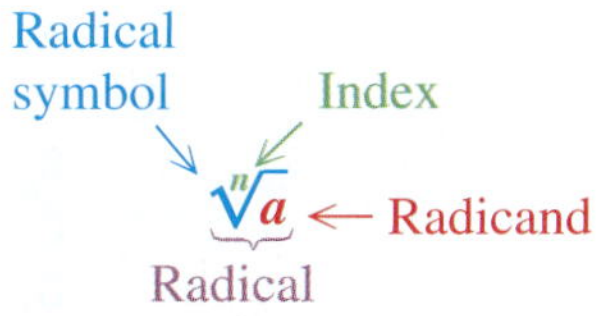

principal square root The positive square root of a number is its principal square root.

radicand The number or expression inside a radical symbol is the radicand.

radical A radical symbol with a radicand is a radical.

radical expression An algebraic expression containing a radical is a radical expression.

perfect square A number with a rational square root is a perfect square.

irrational number A real number that is not rational is an irrational number.

cube root A number a is a cube root of b if $a^3 = b$.

index (order) In a radical of the form $\sqrt[n]{a}$, the number n is the index, or order.

8.2

perfect cube A number with a rational cube root is a perfect cube.

8.3

like radicals Like radicals are multiples of the same root of the same number.

8.4

rationalizing the denominator The process of changing the denominator of a fraction from a radical (irrational number) to an expression not involving a radical is called rationalizing the denominator.

8.5

conjugate The conjugate of $a + b$ is $a - b$.

8.6

radical equation An equation with a variable in a radicand is a radical equation.

extraneous solution A proposed solution that does not satisfy the given equation is an extraneous solution.

New Symbols

$\sqrt{}$	radical symbol
$\approx$	is approximately equal to
$\sqrt[3]{a}$	cube root of a
$\sqrt[n]{a}$	nth root of a

Test Your Word Power

See how well you have learned the vocabulary in this chapter.

1. A **square root** of a number is
   - **A.** the number raised to the second power
   - **B.** the number under a radical symbol
   - **C.** a number that when multiplied by itself gives the original number
   - **D.** the inverse of the number.

2. A **radicand** is
   - **A.** the index of a radical
   - **B.** the number or expression inside the radical symbol
   - **C.** the positive root of a number
   - **D.** the radical symbol.

3. A **radical** is
   - **A.** a symbol that indicates the nth root
   - **B.** an algebraic expression containing a square root
   - **C.** the positive nth root of a number
   - **D.** a radical symbol and the number or expression inside it.

4. The **principal root** of a positive number with even index n is
   - **A.** the positive nth root of the number
   - **B.** the negative nth root of the number
   - **C.** the square root of the number
   - **D.** the cube root of the number.

5. An **irrational number** is
   - **A.** the quotient of two integers, with denominator not 0
   - **B.** a decimal number that neither terminates nor repeats
   - **C.** the principal square root of a number
   - **D.** a nonreal number.

(continued)

Test Your Word Power *(continued)*

6. The **Pythagorean theorem** states that, in a right triangle,
   A. the sum of the measures of the angles is 180°
   B. the sum of the lengths of the two shorter sides equals the length of the longest side
   C. the longest side is opposite the right angle
   D. the square of the length of the longest side equals the sum of the squares of the lengths of the two shorter sides.

7. **Like radicals** are
   A. radicals in simplest form
   B. algebraic expressions containing radicals
   C. multiples of the same root of the same number
   D. radicals with the same index.

8. The **conjugate** of $a + b$ is
   A. $a - b$
   B. $a \cdot b$
   C. $a \div b$
   D. $(a + b)^2$.

9. **Rationalizing the denominator** is the process of
   A. eliminating fractions from a radical expression
   B. changing the denominator of a fraction from a radical expression to an expression not involving a radical
   C. clearing a radical expression of radicals
   D. multiplying radical expressions.

10. An **extraneous solution** is a value
   A. that makes an equation false and must be discarded
   B. that makes an equation true
   C. that makes an expression equal 0
   D. that checks in the original equation.

Answers to Test Your Word Power

1. C; *Examples:* 6 is a square root of 36 since $6^2 = 6 \cdot 6 = 36$. -6 is also a square root of 36.
2. B; *Example:* In $\sqrt{3xy}$, $3xy$ is the radicand.
3. D; *Examples:* $\sqrt{144}$, $\sqrt{4xy^2}$, and $\sqrt{4 + t^2}$
4. A; *Examples:* $\sqrt{36} = 6$, $\sqrt[4]{81} = 3$, and $\sqrt[6]{64} = 2$
5. B; *Examples:* π, $\sqrt{2}$, $-\sqrt{5}$
6. D; *Example:* In a right triangle where $a = 6$, $b = 8$, and $c = 10$, $6^2 + 8^2 = 10^2$.
7. C; *Examples:* $\sqrt{7}$ and $3\sqrt{7}$ are like radicals, as are $2\sqrt[3]{6k}$ and $5\sqrt[3]{6k}$.
8. A; *Example:* The conjugate of $\sqrt{3} + 1$ is $\sqrt{3} - 1$.
9. B; *Example:* To rationalize the denominator of $\frac{5}{\sqrt{3} + 1}$, multiply the numerator and denominator by $\sqrt{3} - 1$ (the conjugate of the denominator) to get $\frac{5(\sqrt{3} - 1)}{2}$.
10. A; *Example:* The proposed solution 2 is extraneous when $\sqrt{5q - 1} + 3 = 0$ is solved using the method of **Section 8.6.**

Quick Review

8.1 Evaluating Roots

Concepts	Examples
Let a be a positive real number.	
$\sqrt{a}$ is the positive or principal square root of a.	$\sqrt{49} = 7$
$-\sqrt{a}$ is the negative square root of a. Also, $\sqrt{0} = 0$.	$-\sqrt{81} = -9$
If a is a negative real number, then $\sqrt{a}$ is not a real number.	$\sqrt{-25}$ is not a real number.
If a is a positive rational number, then $\sqrt{a}$ is rational if a is a perfect square. $\sqrt{a}$ is irrational if a is not a perfect square.	$\sqrt{\frac{4}{9}}$ and $\sqrt{16}$ are rational. $\sqrt{\frac{2}{3}}$ and $\sqrt{21}$ are irrational.
Every real number has exactly one real cube root.	$\sqrt[3]{27} = 3$ $\quad \sqrt[3]{-8} = -2$
Pythagorean Theorem If c is the length of the longest side (hypotenuse) of a right triangle and a and b are the lengths of the shorter sides (legs), then $$a^2 + b^2 = c^2.$$	Find a for the triangle in the figure. $a^2 + 10^2 = 13^2$; $a^2 + 100 = 169$; $a^2 = 69$; $a = \sqrt{69} \approx 8.3$

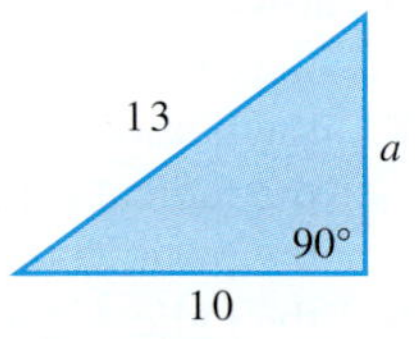

Concepts	Examples
8.2 Multiplying, Dividing, and Simplifying Radicals	
Product Rule for Radicals If a and b are nonnegative real numbers, then the following hold. $\sqrt{a} \cdot \sqrt{b} = \sqrt{ab}$ and $\sqrt{ab} = \sqrt{a} \cdot \sqrt{b}$	$\sqrt{5} \cdot \sqrt{7} = \sqrt{5 \cdot 7} = \sqrt{35}$ $\sqrt{48} = \sqrt{16 \cdot 3} = \sqrt{16} \cdot \sqrt{3} = 4\sqrt{3}$
Quotient Rule for Radicals If a and b are nonnegative real numbers and $b \neq 0$, then the following hold. $\sqrt{\frac{a}{b}} = \frac{\sqrt{a}}{\sqrt{b}}$ and $\frac{\sqrt{a}}{\sqrt{b}} = \sqrt{\frac{a}{b}}$	$\sqrt{\frac{25}{64}} = \frac{\sqrt{25}}{\sqrt{64}} = \frac{5}{8}$ $\quad \frac{\sqrt{8}}{\sqrt{2}} = \sqrt{\frac{8}{2}} = \sqrt{4} = 2$
For all real numbers a and b where the indicated roots exist, the following hold. $\sqrt[n]{a} \cdot \sqrt[n]{b} = \sqrt[n]{ab}$ and $\frac{\sqrt[n]{a}}{\sqrt[n]{b}} = \sqrt[n]{\frac{a}{b}} \quad (b \neq 0)$	$\sqrt[3]{5} \cdot \sqrt[3]{3} = \sqrt[3]{15}$ $\quad \frac{\sqrt[4]{12}}{\sqrt[4]{4}} = \sqrt[4]{\frac{12}{4}} = \sqrt[4]{3}$
8.3 Adding and Subtracting Radicals	
Add and subtract like radicals by using the distributive property. ***Only like radicals can be combined in this way.***	$2\sqrt{5} + 4\sqrt{5} = (2 + 4)\sqrt{5} = 6\sqrt{5}$ $\sqrt{8} - \sqrt{32} = 2\sqrt{2} - 4\sqrt{2} = -2\sqrt{2}$
8.4 Rationalizing the Denominator	
The denominator of a radical expression can be rationalized by multiplying both the numerator and denominator by a number that will eliminate the radical from the denominator.	$\frac{2}{\sqrt{3}} = \frac{2 \cdot \sqrt{3}}{\sqrt{3} \cdot \sqrt{3}} = \frac{2\sqrt{3}}{3}$ $\sqrt[3]{\frac{5}{121}} = \frac{\sqrt[3]{5} \cdot \sqrt[3]{11}}{\sqrt[3]{11^2} \cdot \sqrt[3]{11}} = \frac{\sqrt[3]{55}}{11}$
8.5 More Simplifying and Operations with Radicals	
When appropriate, use the rules for adding and multiplying polynomials to simplify radical expressions.	$\sqrt{6}(\sqrt{5} - \sqrt{7}) = \sqrt{30} - \sqrt{42}$ $(\sqrt{3} + 1)(\sqrt{3} - 2)$ $= 3 - 2\sqrt{3} + \sqrt{3} - 2$ FOIL method $= 1 - \sqrt{3}$ Combine like terms.
The formulas $(a + b)^2 = a^2 + 2ab + b^2,$ $(a - b)^2 = a^2 - 2ab + b^2,$ and $(a + b)(a - b) = a^2 - b^2$ are useful when simplifying radical expressions.	$(\sqrt{13} - \sqrt{2})^2$ $= (\sqrt{13})^2 - 2(\sqrt{13})(\sqrt{2}) + (\sqrt{2})^2$ $= 13 - 2\sqrt{26} + 2$ $= 15 - 2\sqrt{26}$ $(\sqrt{5} + \sqrt{3})(\sqrt{5} - \sqrt{3})$ $= 5 - 3$ $= 2$

(continued)

Concepts | **Examples**

8.5 More Simplifying and Operations with Radicals *(continued)*

Any denominators with radicals should be rationalized.

$$\frac{3}{\sqrt{6}} = \frac{3 \cdot \sqrt{6}}{\sqrt{6} \cdot \sqrt{6}} = \frac{3\sqrt{6}}{6} = \frac{\sqrt{6}}{2}$$

If a radical expression contains two terms in the denominator and at least one of those terms is a square root radical, multiply both the numerator and denominator by the conjugate of the denominator. In general, the expressions $a + b$ and $a - b$ are conjugates.

$$\frac{6}{\sqrt{7} - \sqrt{2}}$$

$$= \frac{6(\sqrt{7} + \sqrt{2})}{(\sqrt{7} - \sqrt{2})(\sqrt{7} + \sqrt{2})}$$ $\sqrt{7} + \sqrt{2}$ is the conjugate of $\sqrt{7} - \sqrt{2}$.

$$= \frac{6(\sqrt{7} + \sqrt{2})}{7 - 2}$$ Multiply.

$$= \frac{6(\sqrt{7} + \sqrt{2})}{5}$$ Subtract.

8.6 Solving Equations with Radicals

Solving a Radical Equation

Step 1 Isolate a radical.

Step 2 Square each side. (By the squaring property of equality, all solutions of the original equation are *among* the solutions of the squared equation.)

Step 3 Combine like terms.

Step 4 If there is still a term with a radical, repeat Steps 1–3.

Step 5 Solve the equation for proposed solutions.

Step 6 Check all proposed solutions from Step 5 in the original equation.

Solve.

$$\sqrt{2x - 3} + x = 3$$

$$\sqrt{2x - 3} = 3 - x$$ Isolate the radical.

$$\left(\sqrt{2x - 3}\right)^2 = (3 - x)^2$$ Square each side.

$$2x - 3 = 9 - 6x + x^2$$ Remember the middle term when squaring $3 - x$.

$$x^2 - 8x + 12 = 0$$ Standard form

$$(x - 2)(x - 6) = 0$$ Factor.

$$x - 2 = 0 \quad \text{or} \quad x - 6 = 0$$ Zero-factor property

$$x = 2 \quad \text{or} \quad x = 6$$ Proposed solutions

A check is essential here. As shown below, 6 is extraneous.

CHECK $\sqrt{2x - 3} + x = 3$ Original equation

$$\sqrt{2(6) - 3} + 6 \stackrel{?}{=} 3$$ Let $x = 6$.

$$\sqrt{9} + 6 \stackrel{?}{=} 3$$

$$9 = 3$$ False

Verify that 2 is the only solution. The solution set is $\{2\}$.

Chapter 8 Review Exercises

8.1 *Find all square roots of each number.*

1. 49 $-7, 7$

2. 81 $-9, 9$

3. 196 $-14, 14$

4. 121 $-11, 11$

5. 256 $-16, 16$

6. 729 $-27, 27$

Find each root.

7. $\sqrt{16}$ 4

8. $-\sqrt{0.36}$ -0.6

9. $\sqrt[3]{-512}$ -8

10. $\sqrt[4]{81}$ 3

11. $\sqrt{-8100}$ not a real number

12. $-\sqrt{4225}$ -65

13. $\sqrt{\frac{144}{169}}$ $\frac{12}{13}$

14. $-\sqrt{\frac{100}{81}}$ $-\frac{10}{9}$

15. Find the value of x. 8

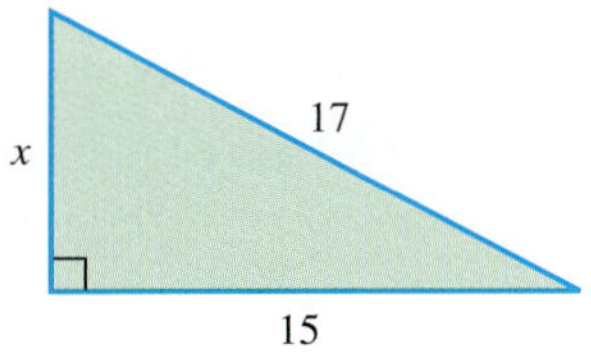

16. An HP f1905 computer monitor has viewing screen dimensions as shown in the figure. Find the diagonal measure of the viewing screen to the nearest tenth. 48.3 cm

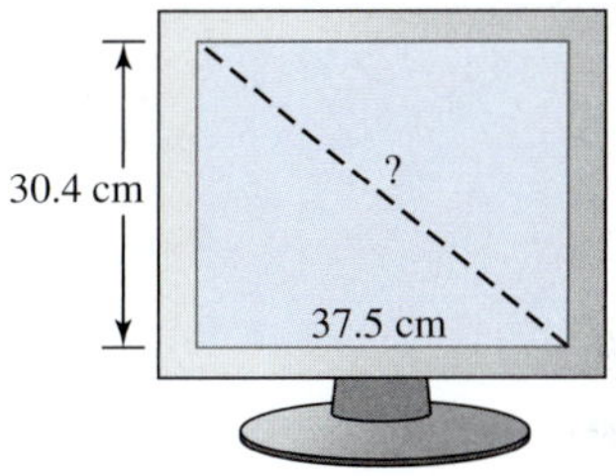

Write rational, irrational, *or* not a real number *for each number. If a number is rational, give its exact value. If a number is irrational, give a decimal approximation for the number. Round approximations to the nearest thousandth.*

17. $\sqrt{73}$ irrational; 8.544

18. $\sqrt{169}$ rational; 13

19. $-\sqrt{625}$ rational; -25

20. $\sqrt{-19}$ not a real number

8.2 *Simplify each expression.*

21. $\sqrt{48}$ $4\sqrt{3}$

22. $-\sqrt{288}$ $-12\sqrt{2}$

23. $\sqrt[3]{16}$ $2\sqrt[3]{2}$

24. $\sqrt[3]{375}$ $5\sqrt[3]{3}$

25. $\sqrt{12} \cdot \sqrt{27}$ 18

26. $\sqrt{32} \cdot \sqrt{48}$ $16\sqrt{6}$

27. $-\sqrt{\frac{121}{400}}$ $-\frac{11}{20}$

28. $\sqrt{\frac{7}{169}}$ $\frac{\sqrt{7}}{13}$

29. $\sqrt{\frac{1}{6}} \cdot \sqrt{\frac{5}{6}}$ $\frac{\sqrt{5}}{6}$

30. $\sqrt{\frac{2}{5}} \cdot \sqrt{\frac{2}{45}}$ $\frac{2}{15}$

31. $\frac{3\sqrt{10}}{\sqrt{5}}$ $3\sqrt{2}$

32. $\frac{8\sqrt{150}}{4\sqrt{75}}$ $2\sqrt{2}$

Simplify each expression. Assume that all variables represent nonnegative real numbers.

33. $\sqrt{r^{18}}$ r^9

34. $\sqrt{x^{10}y^{16}}$ x^5y^8

35. $\sqrt{162x^9}$ $9x^4\sqrt{2x}$

36. $\sqrt{\frac{36}{p^2}}$ $(p \neq 0)$ $\frac{6}{p}$

37. $\sqrt{a^{15}b^{21}}$ $a^7b^{10}\sqrt{ab}$

38. $\sqrt{121x^6y^{10}}$ $11x^3y^5$

39. $\sqrt[3]{y^6}$ y^2

40. $\sqrt[3]{216x^{15}}$ $6x^5$

8.3 *Simplify and combine terms where possible.*

41. $7\sqrt{11} + \sqrt{11}$ $8\sqrt{11}$

42. $3\sqrt{2} + 6\sqrt{2}$ $9\sqrt{2}$

43. $3\sqrt{75} + 2\sqrt{27}$ $21\sqrt{3}$

44. $4\sqrt{12} + \sqrt{48}$ $12\sqrt{3}$

45. $4\sqrt{24} - 3\sqrt{54} + \sqrt{6}$ 0

46. $2\sqrt{7} - 4\sqrt{28} + 3\sqrt{63}$ $3\sqrt{7}$

47. $\frac{2}{5}\sqrt{75} + \frac{3}{4}\sqrt{160}$ $2\sqrt{3} + 3\sqrt{10}$

48. $\frac{1}{3}\sqrt{18} + \frac{1}{4}\sqrt{32}$ $2\sqrt{2}$

49. $\sqrt{15} \cdot \sqrt{2} + 5\sqrt{30}$ $6\sqrt{30}$

Simplify each expression. Assume that all variables represent nonnegative real numbers.

50. $\sqrt{4x} + \sqrt{36x} - \sqrt{9x}$ $5\sqrt{x}$

51. $\sqrt{16p} + 3\sqrt{p} - \sqrt{49p}$ 0

52. $3k\sqrt{8k^2n} + 5k^2\sqrt{2n}$ $11k^2\sqrt{2n}$

8.4 *Perform the indicated operations, and write all answers in simplest form. Rationalize all denominators. Assume that all variables represent nonnegative real numbers.*

53. $\frac{10}{\sqrt{3}}$ $\frac{10\sqrt{3}}{3}$

54. $\frac{8\sqrt{2}}{\sqrt{5}}$ $\frac{8\sqrt{10}}{5}$

55. $\frac{12}{\sqrt{24}}$ $\sqrt{6}$

56. $\sqrt{\frac{2}{5}}$ $\frac{\sqrt{10}}{5}$

57. $\sqrt{\frac{5}{14}} \cdot \sqrt{28}$ $\sqrt{10}$

58. $\sqrt{\frac{2}{7}} \cdot \sqrt{\frac{1}{3}}$ $\frac{\sqrt{42}}{21}$

59. $\sqrt{\frac{r^2}{16x}}$ $(x \neq 0)$ $\frac{r\sqrt{x}}{4x}$

60. $\sqrt[3]{\frac{1}{3}}$ $\frac{\sqrt[3]{9}}{3}$

8.5 *Simplify each expression.*

61. $-\sqrt{3}(\sqrt{5}+\sqrt{27})$
$-\sqrt{15}-9$

62. $3\sqrt{2}(\sqrt{3}+2\sqrt{2})$
$3\sqrt{6}+12$

63. $(2\sqrt{3}-4)(5\sqrt{3}+2)$
$22-16\sqrt{3}$

64. $(\sqrt{7}+2\sqrt{6})(\sqrt{12}-\sqrt{2})$
$2\sqrt{21}-\sqrt{14}+12\sqrt{2}-4\sqrt{3}$

65. $(2\sqrt{3}+5)(2\sqrt{3}-5)$
-13

66. $(\sqrt{x}+2)^2 \quad (x \geq 0)$
$x+4\sqrt{x}+4$

Rationalize each denominator. Assume that all variables represent nonnegative real numbers.

67. $\dfrac{1}{2+\sqrt{5}}$
$-2+\sqrt{5}$

68. $\dfrac{3}{1+\sqrt{x}}$
$\dfrac{3(1-\sqrt{x})}{1-x} \quad (x \neq 1)$

69. $\dfrac{\sqrt{5}-1}{\sqrt{2}+3}$
$\dfrac{-\sqrt{10}+3\sqrt{5}+\sqrt{2}-3}{7}$

Write each quotient in lowest terms.

70. $\dfrac{15+10\sqrt{6}}{15}$
$\dfrac{3+2\sqrt{6}}{3}$

71. $\dfrac{3+9\sqrt{7}}{12}$
$\dfrac{1+3\sqrt{7}}{4}$

72. $\dfrac{6+\sqrt{192}}{2}$
$3+4\sqrt{3}$

8.6 *Solve each equation.*

73. $\sqrt{x}+5=0$
$\emptyset$

74. $\sqrt{k+1}=7$
$\{48\}$

75. $\sqrt{5t+4}=3\sqrt{t}$
$\{1\}$

76. $\sqrt{2p+3}=\sqrt{5p-3}$
$\{2\}$

77. $\sqrt{4x+1}=x-1$
$\{6\}$

78. $\sqrt{13+4t}=t+4$
$\{-3, -1\}$

79. $\sqrt{2-x}+3=x+7$
$\{-2\}$

80. $\sqrt{x}-x+2=0$
$\{4\}$

81. $\sqrt{x+2}-\sqrt{x-3}=1$
$\{7\}$

Mixed Review Exercises

82. CONCEPT CHECK Match each radical in Column I with the equivalent choice in Column II. Choices may be used once, more than once, or not at all.

I		II	
(a) $\sqrt{64}$ B	(b) $-\sqrt{64}$ F	A. 4	B. 8
(c) $\sqrt{-64}$ D	(d) $\sqrt[3]{64}$ A	C. -4	D. Not a real number
(e) $\sqrt[3]{-64}$ C	(f) $-\sqrt[3]{-64}$ A	E. 16	F. -8

Simplify each expression if possible. Assume that all variables represent nonnegative real numbers.

83. $2\sqrt{27} + 3\sqrt{75} - \sqrt{300}$
$11\sqrt{3}$

84. $\dfrac{1}{5 + \sqrt{2}}$
$\dfrac{5 - \sqrt{2}}{23}$

85. $\sqrt{\dfrac{1}{3}} \cdot \sqrt{\dfrac{24}{5}}$
$\dfrac{2\sqrt{10}}{5}$

86. $\sqrt[3]{54a^7b^{10}}$
$3a^2b^3\sqrt[3]{2ab}$

87. $-\sqrt{5}\left(\sqrt{2} + \sqrt{75}\right)$
$-\sqrt{10} - 5\sqrt{15}$

88. $\sqrt{\dfrac{16r^3}{3s}}$ $(s \neq 0)$
$\dfrac{4r\sqrt{3rs}}{3s}$

89. $\dfrac{12 + 6\sqrt{13}}{12}$
$\dfrac{2 + \sqrt{13}}{2}$

90. $\left(\sqrt{5} - \sqrt{2}\right)^2$
$7 - 2\sqrt{10}$

91. $\left(6\sqrt{7} + 2\right)\left(4\sqrt{7} - 1\right)$
$166 + 2\sqrt{7}$

Solve each equation.

92. $\sqrt{x + 2} = x - 4$
$\{7\}$

93. $\sqrt{k} + 3 = 0$
$\emptyset$

94. $\sqrt{1 + 3t} - t = -3$
$\{8\}$

95. The diagram at the right can be used to verify the Pythagorean theorem, as follows.

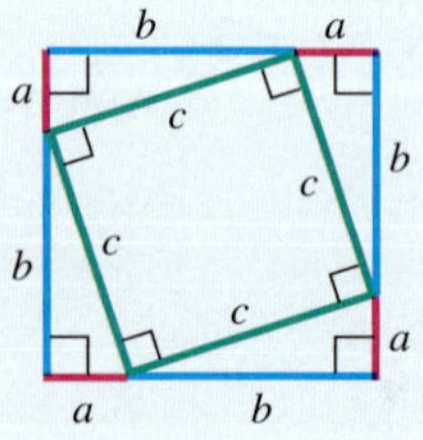

(a) Find the area of the large square. $(a + b)^2$, or $a^2 + 2ab + b^2$

(b) Find the sum of the areas of the smaller square and the four right triangles.
$c^2 + 2ab$

(c) Set the areas equal and simplify as much as possible.
Subtract $2ab$ from each side to get $a^2 + b^2 = c^2$.

Chapter 8 Test

CHAPTER Test Prep VIDEO

The Chapter Test Prep Videos with test solutions are available on DVD, in MyMathLab, and on YouTube—search "LialIntroductoryAlg" and click on "Channels."

1. Find all square roots of 400.
$-20, 20$

2. Consider $\sqrt{142}$.

(a) Determine whether this number is rational or irrational.
irrational

(b) Find a decimal approximation to the nearest thousandth.
11.916

3. If $\sqrt{a}$ is not a real number, then what kind of number must a be?
a must be negative.

Simplify where possible. Assume that all variables represent nonnegative real numbers.

4. $\sqrt[3]{216}$
6

5. $-\sqrt{54}$
$-3\sqrt{6}$

6. $\sqrt{\dfrac{128}{25}}$
$\dfrac{8\sqrt{2}}{5}$

7. $\sqrt[3]{32}$
$2\sqrt[3]{4}$

8. $\dfrac{20\sqrt{18}}{5\sqrt{3}}$
$4\sqrt{6}$

9. $3\sqrt{28} + \sqrt{63}$
$9\sqrt{7}$

10. $(\sqrt{5} + \sqrt{6})^2$
$11 + 2\sqrt{30}$

11. $\sqrt{32x^2y^3}$
$4xy\sqrt{2y}$

12. $(6 - \sqrt{5})(6 + \sqrt{5})$
31

13. $(2 - \sqrt{7})(3\sqrt{2} + 1)$
$6\sqrt{2} + 2 - 3\sqrt{14} - \sqrt{7}$

14. $3\sqrt{27x} - 4\sqrt{48x} + 2\sqrt{3x}$
$-5\sqrt{3x}$

Solve each problem.

15. Find the measure of the unknown leg of this right triangle.

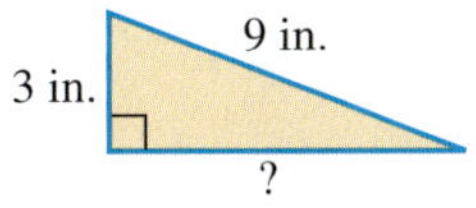

(a) Give its length in simplified radical form.
$6\sqrt{2}$ in.

(b) Round the answer to the nearest thousandth.
8.485 in.

16. In electronics, the impedance Z of an alternating current circuit is given by the following formula.

$$Z = \sqrt{R^2 + X^2}$$

Here, R is the resistance and X is the reactance, both in ohms. Find the value of the impedance Z if $R = 40$ ohms and $X = 30$ ohms. (*Source:* Cooke, N., and J. Orleans, *Mathematics Essential to Electricity and Radio*, McGraw-Hill.) **50 ohms**

Rationalize each denominator.

17. $\dfrac{5\sqrt{2}}{\sqrt{7}}$

$\dfrac{5\sqrt{14}}{7}$

18. $\sqrt{\dfrac{2}{3x}} \quad (x > 0)$

$\dfrac{\sqrt{6x}}{3x}$

19. $\dfrac{-2}{\sqrt[3]{4}}$

$-\sqrt[3]{2}$

20. $\dfrac{-3}{4 - \sqrt{3}}$

$\dfrac{-3(4 + \sqrt{3})}{13}$

21. Write $\dfrac{\sqrt{12} + 3\sqrt{128}}{6}$ in lowest terms.

$\dfrac{\sqrt{3} + 12\sqrt{2}}{3}$

Solve each equation.

22. $\sqrt{p} + 4 = 0$

$\emptyset$

23. $\sqrt{x + 1} = 5 - x$

$\{3\}$

24. $3\sqrt{x} - 2 = x$

$\{1, 4\}$

25. $\sqrt{x + 7} - \sqrt{x} = 1$

$\{9\}$

26. Consider the following incorrect "solution." **What Went Wrong?** Give the correct solution set.

$$\sqrt{2x + 1} + 5 = 0$$

$$\sqrt{2x + 1} = -5 \quad \text{Subtract 5.}$$

$$2x + 1 = 25 \quad \text{Square each side.}$$

$$2x = 24 \quad \text{Subtract 1.}$$

$$x = 12 \quad \text{Divide by 2.}$$

The solution set is $\{12\}$.

12 is not a solution. A check shows that it does not satisfy the original equation. The solution set is $\emptyset$.

Chapters R–8 Cumulative Review Exercises

Simplify each expression.

1. $3(6+7)+6\cdot 4-3^2$
54

2. $\dfrac{3(6+7)+3}{2(4)-1}$
6

3. $|-6|-|-3|$
3

Solve each equation or inequality.

4. $5(k-4)-k=k-11$
$\{3\}$

5. $-\dfrac{3}{4}x \le 12$
$[-16, \infty)$

6. $5z+3-4>2z+9+z$
$(5, \infty)$

7. Trevor Brazile won the ProRodeo All-Around Championship in both 2009 and 2010. He won a total of \$854,700 in these two years, winning \$161,142 more in 2010 than in 2009. How much did he win each year? (*Source: World Almanac and Book of Facts.*)
2009: \$346,779; 2010: \$507,921

Graph.

8. $-4x+5y=-20$

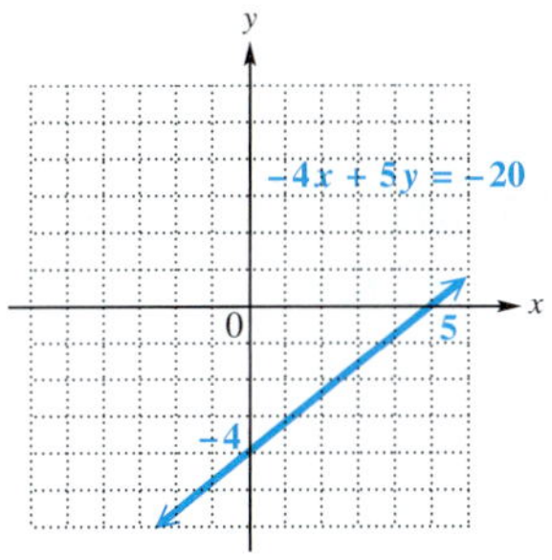

9. $x=2$

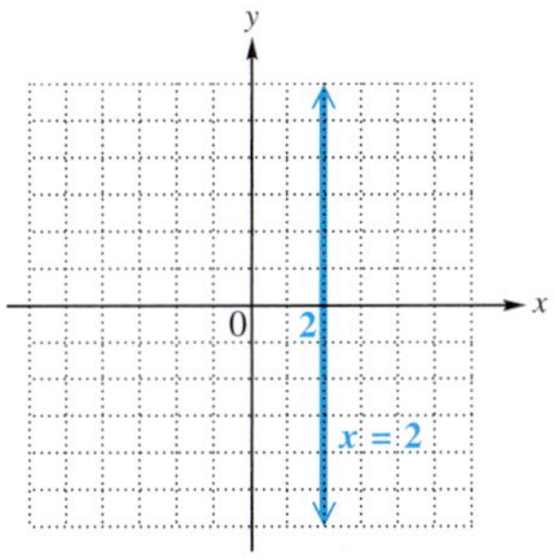

10. $2x-5y>10$

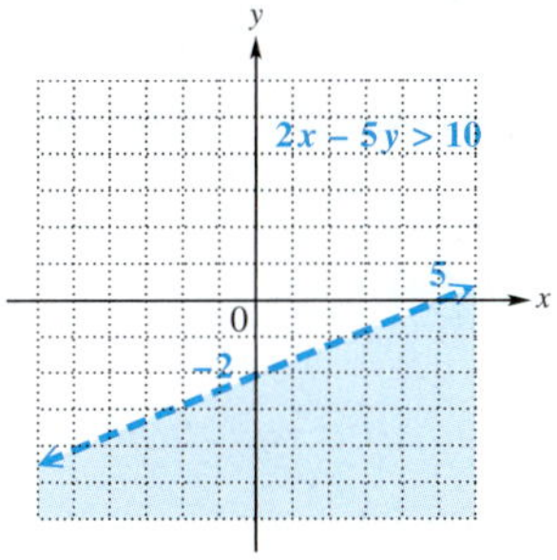

The graph shows a line that models the number of cell phone subscribers in millions in the United States from 2000 to 2010.

11. Use the ordered pairs shown on the graph to find the slope of the line to the nearest tenth. Interpret the slope.
19.3; The number of subscribers increased by an average of 19.3 million per yr.

12. Use the slope from **Exercise 11** and the ordered pair (0, 109.5) to find the equation of the line that models the data, where $x=0$ represents 2000.
$y = 19.3x + 109.5$

13. Use the equation from **Exercise 12** to estimate the number of cell phone subscribers in 2012. Round to the nearest tenth of a million.
341.1 million

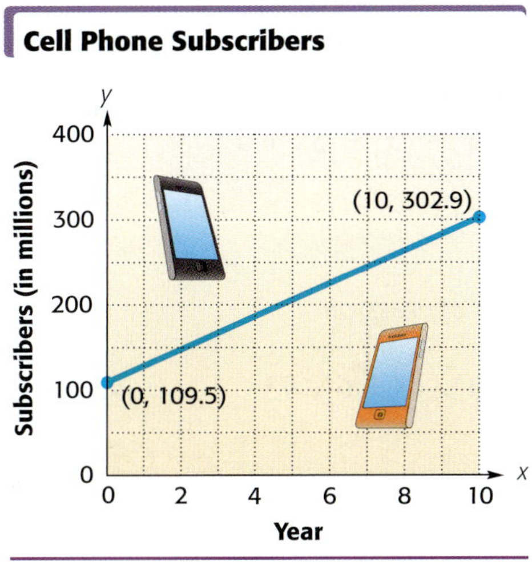

Source: *CITA–The Wireless Association.*

Solve each system of equations.

14. $4x - y = 19$
$3x + 2y = -5$
$\{(3, -7)\}$

15. $2x - y = 6$
$3y = 6x - 18$
$\{(x, y) \mid 2x - y = 6\}$

16. Des Moines and Chicago are 345 mi apart. Two cars start from these cities traveling toward each other. They meet after 3 hr. The car from Chicago has an average rate 7 mph faster than the other car. Find the average rate of each car. (*Source: State Farm Road Atlas.*)

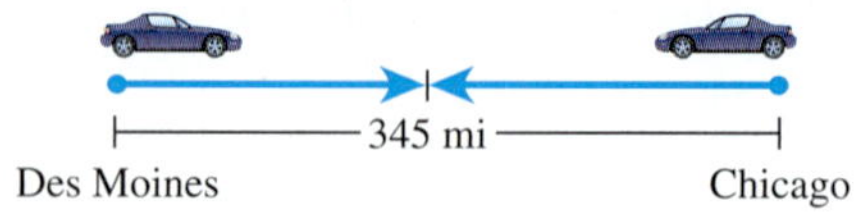

from Chicago: 61 mph; from Des Moines: 54 mph

Simplify and write each expression without negative exponents. Assume that variables represent positive numbers.

17. $(3x^6)(2x^2y)^2$ $12x^{10}y^2$

18. $\left(\dfrac{3^2y^{-2}}{2^{-1}y^3}\right)^{-3}$ $\dfrac{y^{15}}{5832}$

19. Subtract $7x^3 - 8x^2 + 4$ from $10x^3 + 3x^2 - 9$.
$3x^3 + 11x^2 - 13$

20. Divide $\dfrac{8t^3 - 4t^2 - 14t + 15}{2t + 3}$.
$4t^2 - 8t + 5$

21. China is the world's most populous country. In 2011, the population of China was about 1.34×10^9. Write the population without scientific notation. (*Source*: *World Almanac and Book of Facts.*)
1,340,000,000

Factor each polynomial completely.

22. $m^2 + 12m + 32$
$(m + 8)(m + 4)$

23. $25t^4 - 36$
$(5t^2 + 6)(5t^2 - 6)$

24. $12a^2 + 4ab - 5b^2$
$(6a + 5b)(2a - b)$

25. $81z^2 + 72z + 16$
$(9z + 4)^2$

Solve each quadratic equation.

26. $x^2 - 7x = -12$
$\{3, 4\}$

27. $(x + 4)(x - 1) = -6$
$\{-2, -1\}$

28. $z^2 + 144 = 24z$
$\{12\}$

Perform the indicated operations. Express answers in lowest terms.

29. $\dfrac{x^2 - 3x - 4}{x^2 + 3x} \cdot \dfrac{x^2 + 2x - 3}{x^2 - 5x + 4}$
$\dfrac{x + 1}{x}$

30. $\dfrac{t^2 + 4t - 5}{t + 5} \div \dfrac{t - 1}{t^2 + 8t + 15}$
$(t + 5)(t + 3)$

31. $\dfrac{2}{x + 3} - \dfrac{4}{x - 1}$
$\dfrac{-2x - 14}{(x + 3)(x - 1)}$

Simplify each expression.

32. $\sqrt{27} - 2\sqrt{12} + 6\sqrt{75}$
$29\sqrt{3}$

33. $\dfrac{2}{\sqrt{3} + \sqrt{5}}$
$-\sqrt{3} + \sqrt{5}$

34. $(3\sqrt{2} + 1)(4\sqrt{2} - 3)$
$21 - 5\sqrt{2}$

Solve each radical equation.

35. $\sqrt{x + 2} = -8$
$\emptyset$

36. $\sqrt{x} + 2 = x - 10$
$\{16\}$

9 Quadratic Equations

The trajectory of a water spout follows a *parabolic* path, which can be described by a *quadratic* (second-degree) *equation*, the topic of this chapter.

9.1 Solving Quadratic Equations by the Square Root Property

OBJECTIVES

1. Review the zero-factor property.
2. Solve equations of the form $x^2 = k$, where $k > 0$.
3. Solve equations of the form $(ax + b)^2 = k$, where $k > 0$.
4. Use formulas involving squared variables.

1 Solve each equation by using the zero-factor property.

GS **(a)** $x^2 - 2x - 15 = 0$

$(x + \_\_\_\_)(x - \_\_\_\_) = 0$

$x + \_\_\_\_ = 0$ or $x - \_\_\_\_ = 0$

$x = \_\_\_\_$ or $x = \_\_\_\_$

The solution set is $\{\_\_\_\_, \_\_\_\_\}$.

(b) $2x^2 - 3x + 1 = 0$

(c) $x^2 = 400$

Answers

1. (a) 3; 5; 3; 5; −3; 5
(b) $\left\{\frac{1}{2}, 1\right\}$ (c) $\{-20, 20\}$

OBJECTIVE 1 Review the zero-factor property. Recall that a **quadratic equation** is an equation that can be written in the form

$$ax^2 + bx + c = 0, \quad \text{Standard form}$$

for real numbers a, b, and c, with $a \neq 0$. In **Section 6.6** we solved quadratic equations by factoring, using the zero-factor property.

Zero-Factor Property

If a and b are real numbers and if $ab = 0$, then $a = 0$ or $b = 0$.

EXAMPLE 1 Solving Quadratic Equations by the Zero-Factor Property

Solve each equation by using the zero-factor property.

(a)

$$x^2 + 4x + 3 = 0$$

$$(x + 3)(x + 1) = 0 \quad \text{Factor.}$$

$$x + 3 = 0 \quad \text{or} \quad x + 1 = 0 \quad \text{Zero-factor property}$$

$$x = -3 \quad \text{or} \quad x = -1 \quad \text{Solve each equation.}$$

The solution set is $\{-3, -1\}$.

(b)

$$x^2 = 9$$

$$x^2 - 9 = 0 \quad \text{Subtract 9.}$$

$$(x + 3)(x - 3) = 0 \quad \text{Factor.}$$

$$x + 3 = 0 \quad \text{or} \quad x - 3 = 0 \quad \text{Zero-factor property}$$

$$x = -3 \quad \text{or} \quad x = 3 \quad \text{Solve each equation.}$$

The solution set is $\{-3, 3\}$.

◀ **Work Problem 1 at the Side.**

Not all quadratic equations can easily be solved by factoring, so we must develop other methods.

OBJECTIVE 2 Solve equations of the form $x^2 = k$, where $k > 0$. In **Example 1(b),** we might also have solved $x^2 = 9$ by noticing that x must be a number whose square is 9. Thus, $x = \sqrt{9} = 3$ or $x = -\sqrt{9} = -3$. This is generalized as the **square root property.**

Square Root Property

If k is a positive number and if $x^2 = k$, then

$$x = \sqrt{k} \quad \text{or} \quad x = -\sqrt{k}.$$

The solution set is $\{-\sqrt{k}, \sqrt{k}\}$, which can be written $\{\pm\sqrt{k}\}$. (The symbol $\pm$ is read "positive or negative" or "plus or minus.")

EXAMPLE 2 **Solving Quadratic Equations of the Form $x^2 = k$**

Solve each equation. Write radicals in simplified form.

(a) $x^2 = 16$

By the square root property, if $x^2 = 16$, then

$$x = \sqrt{16} = 4 \quad \text{or} \quad x = -\sqrt{16} = -4.$$

Check each solution by substituting it for x in the original equation. The solution set is $\{-4, 4\}$, or $\{\pm 4\}$.

This notation indicates two solutions, one positive and one negative.

(b) $z^2 = 5$

The solutions are $z = \sqrt{5}$ or $z = -\sqrt{5}$, so the solution set is $\{-\sqrt{5}, \sqrt{5}\}$, or $\{\pm\sqrt{5}\}$.

(c)

$$5m^2 - 32 = 8$$

$$5m^2 = 40 \quad \text{Add 32.}$$

$$m^2 = 8 \quad \text{Divide by 5.}$$

Don't stop here. Simplify the radicals.

$$m = \sqrt{8} \quad \text{or} \quad m = -\sqrt{8} \quad \text{Square root property}$$

$$m = 2\sqrt{2} \quad \text{or} \quad m = -2\sqrt{2} \quad \sqrt{8} = \sqrt{4} \cdot \sqrt{2} = 2\sqrt{2}$$

The solution set is $\{-2\sqrt{2}, 2\sqrt{2}\}$, or $\{\pm 2\sqrt{2}\}$.

(d) $x^2 = -4$

Because -4 is a negative number and because the square of a real number cannot be negative, ***there is no real number solution*** of this equation. (In this book, we are concerned with finding only *real number* solutions.) The solution set is $\emptyset$.

Work Problem 2 at the Side. ▶

OBJECTIVE 3 **Solve equations of the form $(ax + b)^2 = k$, where $k > 0$.** In each equation in **Example 2,** the exponent 2 had a single variable as its base. We can extend the square root property to solve equations in which the base is a binomial.

EXAMPLE 3 **Solving Quadratic Equations of the Form $(x + b)^2 = k$**

Solve each equation.

(a) Use $x - 3$ as the base.

$$(x - 3)^2 = 16$$

$$x - 3 = \sqrt{16} \quad \text{or} \quad x - 3 = -\sqrt{16} \quad \text{Square root property}$$

$$x - 3 = 4 \quad \text{or} \quad x - 3 = -4 \quad \sqrt{16} = 4$$

$$x = 7 \quad \text{or} \quad x = -1 \quad \text{Add 3.}$$

CHECK

$(x - 3)^2 = 16$		$(x - 3)^2 = 16$	
$(7 - 3)^2 \stackrel{?}{=} 16$	Let $x = 7$.	$(-1 - 3)^2 \stackrel{?}{=} 16$	Let $x = -1$.
$4^2 \stackrel{?}{=} 16$	Subtract.	$(-4)^2 \stackrel{?}{=} 16$	Subtract.
$16 = 16$ ✓	True	$16 = 16$ ✓	True

The solutions are 7 and -1, and the solution set is $\{-1, 7\}$.

Continued on Next Page

2 Solve each equation. Write radicals in simplified form.

(a) $x^2 = 49$

(b) $x^2 = 11$

(c) $2x^2 + 8 = 32$

(d) $x^2 = -9$

Teaching Tip

Emphasize that when the square root property is used, both the positive and the negative square roots must be given.

Answers

2. (a) $\{-7, 7\}$ **(b)** $\{-\sqrt{11}, \sqrt{11}\}$ **(c)** $\{-2\sqrt{3}, 2\sqrt{3}\}$ **(d)** $\emptyset$

3 Solve each equation.

(a) $(x+2)^2 = 36$

(b) $(x+1)^2 = 6$

$x+1=\sqrt{6}$ or $x+1=-\sqrt{6}$ Square root property

$x=-1+\sqrt{6}$ or $x=-1-\sqrt{6}$ Add -1.

CHECK $(-1+\sqrt{6}+1)^2=(\sqrt{6})^2=6$ ✓ Let $x=-1+\sqrt{6}$.

$(-1-\sqrt{6}+1)^2=(-\sqrt{6})^2=6$ ✓ Let $x=-1-\sqrt{6}$.

The solution set is $\{-1+\sqrt{6}, -1-\sqrt{6}\}$, or $\{-1 \pm \sqrt{6}\}$.

◀ Work Problem **3** at the Side.

GS **(b)** $(x-4)^2 = 3$

$x-4=$ ____ or $x-4=$ ____

$x=$ ____ or $x=$ ____

The solution set is ________.

EXAMPLE 4 Solving a Quadratic Equation of the Form $(ax+b)^2=k$

Solve $(3r-2)^2=27$.

$$(3r-2)^2=27$$

$3r-2=\sqrt{27}$ or $3r-2=-\sqrt{27}$ Square root property

$3r-2=3\sqrt{3}$ or $3r-2=-3\sqrt{3}$ $\sqrt{27}=\sqrt{9}\cdot\sqrt{3}=3\sqrt{3}$

$3r=2+3\sqrt{3}$ or $3r=2-3\sqrt{3}$ Add 2.

$r=\dfrac{2+3\sqrt{3}}{3}$ or $r=\dfrac{2-3\sqrt{3}}{3}$ Divide by 3.

CHECK $(3r-2)^2=27$ Original equation

$\left(3\cdot\dfrac{2+3\sqrt{3}}{3}-2\right)^2 \stackrel{?}{=} 27$ Let $r=\dfrac{2+3\sqrt{3}}{3}$.

$(2+3\sqrt{3}-2)^2 \stackrel{?}{=} 27$ Multiply.

$(3\sqrt{3})^2 \stackrel{?}{=} 27$ Subtract. ($(ab)^2=a^2b^2$)

$27=27$ ✓ True

The check of the other solution is similar. The solution set is

$$\left\{\frac{2+3\sqrt{3}}{3}, \frac{2-3\sqrt{3}}{3}\right\}.$$

◀ Work Problem **4** at the Side.

4 Solve $(2x-5)^2=18$.

Teaching Tip

In **Example 4,** point out the common error of simplifying the expression $\dfrac{2+3\sqrt{3}}{3}$ as $2+\sqrt{3}$. Review the idea that *only common factors may be divided out.* Show why 3 is not a factor in the numerator.

CAUTION

The solutions in **Example 4** are fractions that cannot be simplified. Note that 3 is *not* a common factor in the numerator.

5 Solve each equation.

(a) $(5x+1)^2=7$

(b) $(7x-1)^2=-1$

EXAMPLE 5 Recognizing When There Is No Real Solution

Solve $(x+3)^2=-9$.

Because the square root of -9 is not a real number, there is no real number solution for this equation. The solution set is $\emptyset$.

◀ Work Problem **5** at the Side.

Answers

3. (a) $\{-8, 4\}$
(b) $\sqrt{3}$; $-\sqrt{3}$; $4+\sqrt{3}$; $4-\sqrt{3}$; $\{4+\sqrt{3}, 4-\sqrt{3}\}$

4. $\left\{\dfrac{5+3\sqrt{2}}{2}, \dfrac{5-3\sqrt{2}}{2}\right\}$

5. (a) $\left\{\dfrac{-1+\sqrt{7}}{5}, \dfrac{-1-\sqrt{7}}{5}\right\}$ (b) $\emptyset$

OBJECTIVE 4 Use formulas involving squared variables.

EXAMPLE 6 Finding the Length of a Bass

We can approximate the weight of a bass, in pounds, given its length L and its girth (distance around) g, where both are measured in inches, using the following formula.

$$w = \frac{L^2 g}{1200}$$

Approximate the length of a bass weighing 2.20 lb and having girth 10 in. (*Source: Sacramento Bee.*)

$$w = \frac{L^2 g}{1200} \qquad \text{Given formula}$$

$$2.20 = \frac{L^2 \cdot 10}{1200} \qquad w = 2.20,\ g = 10$$

$$2640 = 10L^2 \qquad \text{Multiply by 1200.}$$

$$L^2 = 264 \qquad \text{Divide by 10. Interchange sides.}$$

$$L = \sqrt{264} \quad \text{or} \quad L = -\sqrt{264} \qquad \text{Square root property}$$

A calculator shows that $\sqrt{264} \approx 16.25$, so the length of the bass is a little more than 16 in. (We discard the negative solution $-\sqrt{264} \approx -16.25$, since L represents length.)

Work Problem 6 at the Side. ▶

6 Use the formula in **Example 6** to approximate the length to the nearest hundredth of a bass weighing 2.80 lb and having girth 11 in.

Answer

6. approximately 17.48 in.

9.1 Exercises

FOR EXTRA HELP

Download the MyDashBoard App

MyMathLab®

CONCEPT CHECK *Match each equation in Column I with the correct description of its solution in Column II.*

I		II
1. $x^2 = 12$	C	**A.** No real number solutions
2. $x^2 = -9$	A	**B.** Two integer solutions
3. $x^2 = \frac{25}{36}$	D	**C.** Two irrational solutions
4. $x^2 = 16$	B	**D.** Two rational solutions that are not integers

5. CONCEPT CHECK When a student was asked to solve $x^2 = 81$, she wrote $\{9\}$ as her answer. Her teacher did not give her full credit. The student argued that because $9^2 = 81$, her answer had to be correct. **What Went Wrong?** Give the correct solution set.

According to the square root property, -9 is also a solution, so her answer was not completely correct. The solution set is $\{-9, 9\}$, or $\{\pm 9\}$.

6. Give an example of an equation that can be solved by the square root property for equations. Then solve it.

The equation $x^2 = 21$ can be solved with the square root property. Because the square of x is 21, x can be either the positive or the negative square root of 21. The solution set is $\{-\sqrt{21}, \sqrt{21}\}$, or $\{\pm\sqrt{21}\}$.

Solve each equation by using the zero-factor property. ***See Example 1.***

7. $x^2 - x - 56 = 0$ $\{-7, 8\}$

8. $x^2 - 2x - 99 = 0$ $\{-9, 11\}$

9. $x^2 = 121$ $\{-11, 11\}$

10. $x^2 = 144$ $\{-12, 12\}$

11. $3x^2 - 13x = 30$ $\left\{-\frac{5}{3}, 6\right\}$

12. $5x^2 - 14x = 3$ $\left\{-\frac{1}{5}, 3\right\}$

Solve each equation by using the square root property. Write all radicals in simplest form. ***See Example 2.***

13. $x^2 = 81$ $\{-9, 9\}$

14. $x^2 = 121$ $\{-11, 11\}$

15. $k^2 = 14$ $\{-\sqrt{14}, \sqrt{14}\}$

16. $m^2 = 22$ $\{-\sqrt{22}, \sqrt{22}\}$

17. $t^2 = 48$ $\{-4\sqrt{3}, 4\sqrt{3}\}$

18. $x^2 = 54$ $\{-3\sqrt{6}, 3\sqrt{6}\}$

19. $x^2 = \frac{25}{4}$ $\left\{-\frac{5}{2}, \frac{5}{2}\right\}$

20. $m^2 = \frac{36}{121}$ $\left\{-\frac{6}{11}, \frac{6}{11}\right\}$

21. $x^2 = -100$ $\emptyset$

22. $x^2 = -64$ $\emptyset$

23. $z^2 = 2.25$ $\{-1.5, 1.5\}$

24. $w^2 = 56.25$ $\{-7.5, 7.5\}$

25. $r^2 - 3 = 0$

$\{-\sqrt{3}, \sqrt{3}\}$

26. $x^2 - 13 = 0$

$\{-\sqrt{13}, \sqrt{13}\}$

27. $7x^2 = 4$

$\left\{-\frac{2\sqrt{7}}{7}, \frac{2\sqrt{7}}{7}\right\}$

28. $3x^2 = 10$

$\left\{-\frac{\sqrt{30}}{3}, \frac{\sqrt{30}}{3}\right\}$

29. $4x^2 - 72 = 0$

$\{-3\sqrt{2}, 3\sqrt{2}\}$

30. $5z^2 - 200 = 0$

$\{-2\sqrt{10}, 2\sqrt{10}\}$

31. $3x^2 - 8 = 64$

$\{-2\sqrt{6}, 2\sqrt{6}\}$

32. $2x^2 + 7 = 61$

$\{-3\sqrt{3}, 3\sqrt{3}\}$

33. $5x^2 + 4 = 8$

$\left\{-\frac{2\sqrt{5}}{5}, \frac{2\sqrt{5}}{5}\right\}$

34. $4x^2 - 3 = 7$

$\left\{-\frac{\sqrt{10}}{2}, \frac{\sqrt{10}}{2}\right\}$

Solve each equation by using the square root property. Express all radicals in simplest form. ***See Examples 3–5.***

35. $(x - 3)^2 = 25$

$\{-2, 8\}$

36. $(x - 7)^2 = 16$

$\{3, 11\}$

37. $(x + 5)^2 = -13$

$\emptyset$

38. $(x + 2)^2 = -17$

$\emptyset$

39. $(x - 8)^2 = 27$

$\{8 + 3\sqrt{3}, 8 - 3\sqrt{3}\}$

40. $(x - 5)^2 = 40$

$\{5 + 2\sqrt{10}, 5 - 2\sqrt{10}\}$

41. $(3x + 2)^2 = 49$

$\left\{-3, \frac{5}{3}\right\}$

42. $(5x + 3)^2 = 36$

$\left\{-\frac{9}{5}, \frac{3}{5}\right\}$

43. $(4x - 3)^2 = 9$

$\left\{0, \frac{3}{2}\right\}$

44. $(7x - 5)^2 = 25$

$\left\{0, \frac{10}{7}\right\}$

45. $(5 - 2x)^2 = 30$

$\left\{\frac{5 + \sqrt{30}}{2}, \frac{5 - \sqrt{30}}{2}\right\}$

46. $(3 - 2x)^2 = 70$

$\left\{\frac{3 + \sqrt{70}}{2}, \frac{3 - \sqrt{70}}{2}\right\}$

47. $(3x + 1)^2 = 18$
$\left\{\frac{-1 + 3\sqrt{2}}{3}, \frac{-1 - 3\sqrt{2}}{3}\right\}$

48. $(5x + 6)^2 = 75$
$\left\{\frac{-6 + 5\sqrt{3}}{5}, \frac{-6 - 5\sqrt{3}}{5}\right\}$

49. $\left(\frac{1}{2}x + 5\right)^2 = 12$
$\{-10 + 4\sqrt{3}, -10 - 4\sqrt{3}\}$

50. $\left(\frac{1}{3}x + 4\right)^2 = 27$
$\{-12 + 9\sqrt{3}, -12 - 9\sqrt{3}\}$

51. $(4x - 1)^2 - 48 = 0$
$\left\{\frac{1 + 4\sqrt{3}}{4}, \frac{1 - 4\sqrt{3}}{4}\right\}$

52. $(2x - 5)^2 - 180 = 0$
$\left\{\frac{5 + 6\sqrt{5}}{2}, \frac{5 - 6\sqrt{5}}{2}\right\}$

Solve each problem. ***See Example 6.***

53. One expert at marksmanship can hold a silver dollar at forehead level, drop it, draw his gun, and shoot the coin as it passes waist level. The distance traveled by a falling object is given by

$$d = 16t^2,$$

where d is the distance (in feet) the object falls in t seconds. If the coin falls about 4 ft, use the formula to estimate the time that elapses between the dropping of the coin and the shot.

about $\frac{1}{2}$ sec

54. The illumination produced by a light source depends on the distance from the source. For a particular light source, this relationship can be expressed as

$$d^2 = \frac{4050}{I},$$

where d is the distance from the source (in feet) and I is the amount of illumination in foot-candles. How far from the source is the illumination equal to 50 foot-candles?

9 ft

55. The area A of a circle with radius r is given by the formula

$$A = \pi r^2.$$

If a circle has area 81π in.^{2}, what is its radius?

9 in.

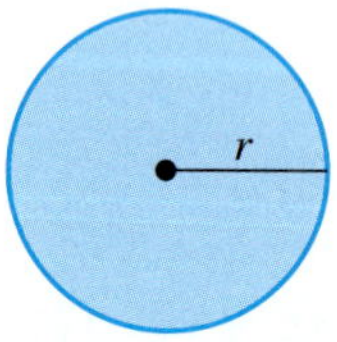

$A = \pi r^2$

56. The surface area S of a sphere with radius r is given by the formula

$$S = 4\pi r^2.$$

If a sphere has surface area 36π ft^2, what is its radius?

3 ft

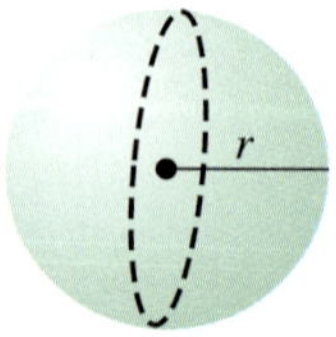

$S = 4\pi r^2$

The amount A that P dollars invested at an annual rate of interest r will grow to in 2 yr is $A = P(1 + r)^2$.

57. At what interest rate will \$100 grow to \$110.25 in 2 yr?
5%

58. At what interest rate will \$200 grow to \$208.08 in 2 yr?
2%

9.2 Solving Quadratic Equations by Completing the Square

OBJECTIVES

1. Solve quadratic equations by completing the square when the coefficient of the second-degree term is 1.
2. Solve quadratic equations by completing the square when the coefficient of the second-degree term is not 1.
3. Simplify the terms of an equation before solving.
4. Solve applied problems that require quadratic equations.

OBJECTIVE 1 **Solve quadratic equations by completing the square when the coefficient of the second-degree term is 1.** The methods we have studied so far are not enough to solve the equation

$$x^2 + 6x + 7 = 0.$$

If we could write the equation in the form $(x + 3)^2$ equals a constant, we could solve it with the square root property discussed in **Section 9.1.** To do that, we need to have a perfect square trinomial on one side of the equation.

Recall from **Section 6.5** that a perfect square trinomial has the form

$$x^2 + \mathbf{2kx} + k^2 \quad \text{or} \quad x^2 - \mathbf{2kx} + k^2,$$

where k represents a positive number.

EXAMPLE 1 Creating Perfect Square Trinomials

Complete each trinomial so that it is a perfect square. Then factor the trinomial.

(a) $x^2 + \mathbf{8x} +$ ______

The perfect square trinomial will have the form $x^2 + \mathbf{2kx} + k^2$. Thus, the middle term, $8x$, must equal $2kx$.

$$8x = 2kx \quad \leftarrow \text{Solve this equation for } k.$$

$$4 = k \quad \text{Divide each side by } 2x.$$

Therefore, $k = 4$ and $k^2 = 4^2 = \mathbf{16}$. The required perfect square trinomial is

$$x^2 + 8x + \mathbf{16}, \quad \text{which factors as} \quad (x + 4)^2.$$

(b) $x^2 - \mathbf{18x} +$ ______

Here the perfect square trinomial will have the form $x^2 - \mathbf{2kx} + k^2$. The middle term, $-18x$, must equal $-2kx$.

$$-18x = -2kx \quad \leftarrow \text{Solve this equation for } k.$$

$$9 = k \quad \text{Divide each side by } -2x.$$

Thus, $k = 9$ and $k^2 = 9^2 = \mathbf{81}$. The required perfect square trinomial is

$$x^2 - 18x + \mathbf{81}, \quad \text{which factors as} \quad (x - 9)^2.$$

Work Problem 1 at the Side. ▶

1 Complete each trinomial so that it is a perfect square. Then factor the trinomial.

(a) $x^2 + 12x +$ ____ factors as ________.

(b) $x^2 - 14x +$ ____ factors as ________.

(c) $x^2 - 2x +$ ____ factors as ________.

EXAMPLE 2 Rewriting an Equation to Use the Square Root Property

Solve $x^2 + 6x + 7 = 0$.

$$x^2 + 6x = -7 \quad \text{Subtract 7 from each side.}$$

To use the square root property, the expression on the left, $x^2 + 6x$, must be written as a perfect square trinomial in the form $x^2 + \mathbf{2kx} + k^2$.

$$x^2 + \mathbf{6x} + \text{\_\_\_\_}$$ (A square must go here.)

Here, $2kx = 6x$, so $k = 3$ and $k^2 = \mathbf{9}$. The required perfect square trinomial is

$$x^2 + 6x + \mathbf{9}, \quad \text{which factors as} \quad (x + 3)^2.$$

Continued on Next Page

Answers

1. (a) 36; $(x + 6)^2$
(b) 49; $(x - 7)^2$
(c) 1; $(x - 1)^2$

2 Solve $x^2 - 4x - 1 = 0$.

3 Solve each equation by completing the square.

GS **(a)** $x^2 + 4x = 1$

Take half the coefficient of x and square it.

$\frac{1}{2} \cdot$ ____ $=$ ____,

and ____^{2} = ____.

Add ____ to each side of the equation.

$x^2 + 4x +$ ____ $= 1 +$ ____

Factor and add.

____

Complete the solution.

(b) $z^2 + 6z - 3 = 0$

Answers

2. $\{2 + \sqrt{5}, 2 - \sqrt{5}\}$

3. **(a)** 4; 2; 2; 4; 4; 4; 4; $(x + 2)^2 = 5$; $\{-2 + \sqrt{5}, -2 - \sqrt{5}\}$

(b) $\{-3 + 2\sqrt{3}, -3 - 2\sqrt{3}\}$

Therefore, if we add 9 to *each* side of $x^2 + 6x = -7$, the equation will have a perfect square trinomial on the left side, as needed.

This is a key step.

$$x^2 + 6x + 9 = -7 + 9 \quad \text{Add 9.}$$

$$(x + 3)^2 = 2 \quad \text{Factor on the left. Add on the right.}$$

Now use the square root property to complete the solution.

$$x + 3 = \sqrt{2} \quad \text{or} \quad x + 3 = -\sqrt{2}$$

$$x = -3 + \sqrt{2} \quad \text{or} \quad x = -3 - \sqrt{2} \quad \text{Add } -3.$$

Check by first substituting $-3 + \sqrt{2}$ and then substituting $-3 - \sqrt{2}$ for x in the original equation. The solution set is $\{-3 + \sqrt{2}, -3 - \sqrt{2}\}$.

◀ **Work Problem 2 at the Side.**

The process of changing the form of the equation in **Example 2** from

$$x^2 + 6x + 7 = 0 \quad \text{to} \quad (x + 3)^2 = 2$$

is called **completing the square.** Completing the square changes only the form of the equation. To see this, multiply out the left side of $(x + 3)^2 = 2$. Then write the equation in standard form to get $x^2 + 6x + 7 = 0$.

Look again at the original equation in **Example 2.**

$$x^2 + 6x + 7 = 0$$

If we take half the coefficient of x, which is 6 here, and square it, we get 9.

$$\frac{1}{2} \cdot 6 = 3 \quad \text{and} \quad 3^2 = 9$$

Coefficient of x (6) — Quantity added to each side (9)

To complete the square in **Example 2,** we added 9 to each side.

EXAMPLE 3 Completing the Square to Solve a Quadratic Equation

Complete the square to solve $x^2 - 8x = 5$.

To complete the square on $x^2 - 8x$, take half the coefficient of x and square it.

$$\frac{1}{2}(-8) = -4 \quad \text{and} \quad (-4)^2 = 16$$

Coefficient of x (−8)

Add the result, 16, to each side of the equation.

$$x^2 - 8x = 5 \quad \text{Given equation}$$

$$x^2 - 8x + 16 = 5 + 16 \quad \text{Add 16.}$$

$$(x - 4)^2 = 21 \quad \text{Factor. Add.}$$

$$x - 4 = \sqrt{21} \quad \text{or} \quad x - 4 = -\sqrt{21} \quad \text{Square root property}$$

$$x = 4 + \sqrt{21} \quad \text{or} \quad x = 4 - \sqrt{21} \quad \text{Add 4.}$$

A check indicates that the solution set is $\{4 + \sqrt{21}, 4 - \sqrt{21}\}$.

◀ **Work Problem 3 at the Side.**

OBJECTIVE 2 Solve quadratic equations by completing the square when the coefficient of the second-degree term is not 1. If a quadratic equation has the form

$$ax^2 + bx + c = 0, \quad \text{where } a \neq 1,$$

then to obtain 1 as the coefficient of x^2, we first divide each side by a.

Solving a Quadratic Equation by Completing the Square

Step 1 **Be sure the second-degree term has coefficient 1.** If the coefficient of the second-degree term is 1, proceed to Step 2. If it is not 1, but some other nonzero number a, divide each side of the equation by a.

Step 2 **Write in correct form.** Make sure that all terms with variables are on one side of the equality symbol and that all constant terms are on the other side.

Step 3 **Complete the square.** Take half the coefficient of the first-degree term, and square it. Add the square to each side of the equation. Factor the variable side, and simplify on the other side.

Step 4 **Solve** the equation by using the square root property.

EXAMPLE 4 Solving a Quadratic Equation by Completing the Square

Solve $4x^2 + 16x = 9$.

Step 1 ***Before completing the square, the coefficient of x^2 must be 1,*** not 4. We get 1 as the coefficient of x^2 here by dividing each side by 4.

The coefficient of x^2 must be 1.

$$x^2 + 4x = \frac{9}{4} \quad \text{Divide by 4.}$$

Step 2 The equation is already in the correct form, with the variable terms on one side and the constant on the other.

Step 3 Complete the square. Take half the coefficient of x and square it.

$$\frac{1}{2}(4) = 2 \quad \text{and} \quad 2^2 = 4$$

$$x^2 + 4x + 4 = \frac{9}{4} + 4 \quad \text{Add 4.}$$

$$(x + 2)^2 = \frac{25}{4} \quad \text{Factor; } \frac{9}{4} + 4 = \frac{9}{4} + \frac{16}{4} = \frac{25}{4}.$$

Step 4 Solve the equation by using the square root property.

$$x + 2 = \sqrt{\frac{25}{4}} \quad \text{or} \quad x + 2 = -\sqrt{\frac{25}{4}} \quad \text{Square root property}$$

$$x + 2 = \frac{5}{2} \quad \text{or} \quad x + 2 = -\frac{5}{2} \quad \text{Take square roots.}$$

$$x = \frac{1}{2} \quad \text{or} \quad x = -\frac{9}{2} \quad \text{Subtract } 2 = \frac{4}{2}.$$

Continued on Next Page

Teaching Tip

Point out that if the solutions are rational numbers, as they are in Examples 4 and 5, the equation can also be solved by factoring. Emphasize that the method of completing the square is a more powerful method than factoring because it allows us to solve *any* quadratic equation.

4 Solve each equation by completing the square.

(a) $9x^2 + 18x = -5$

(b) $4t^2 - 24t + 11 = 0$

5 Solve each equation by completing the square.

GS **(a)** $3x^2 + 5x - 2 = 0$

Step 1
Divide each side by ____.

____

Step 2
Add ____ to each side.

____

Step 3
Take half the coefficient of x and square it. Add ____ to each side to complete the square.

$$x^2 + \frac{5}{3}x + \_\_\_\_ = \frac{2}{3} + \frac{25}{36}$$

Factor and add.

$$(\_\_\_\_\_\_)^2 = \_\_\_\_$$

Continue with Step 4 to complete the solution.

(b) $2x^2 - 4x - 1 = 0$

Answers

4. (a) $\left\{-\frac{1}{3}, -\frac{5}{3}\right\}$ **(b)** $\left\{\frac{11}{2}, \frac{1}{2}\right\}$

5. (a) 3; $x^2 + \frac{5}{3}x - \frac{2}{3} = 0$; $\frac{2}{3}$; $x^2 + \frac{5}{3}x = \frac{2}{3}$; $\frac{25}{36}$; $\frac{25}{36}$; $x + \frac{5}{6}$; $\frac{49}{36}$; $\left\{-2, \frac{1}{3}\right\}$

(b) $\left\{\frac{2+\sqrt{6}}{2}, \frac{2-\sqrt{6}}{2}\right\}$

CHECK Substitute each solution in the original equation.

$$4x^2 + 16x = 9$$
$$4\left(\frac{1}{2}\right)^2 + 16\left(\frac{1}{2}\right) \stackrel{?}{=} 9 \quad \text{Let } x = \tfrac{1}{2}.$$
$$4\left(\frac{1}{4}\right) + 8 \stackrel{?}{=} 9$$
$$1 + 8 \stackrel{?}{=} 9$$
$$9 = 9 \ \checkmark \ \text{True}$$

$$4x^2 + 16x = 9$$
$$4\left(-\frac{9}{2}\right)^2 + 16\left(-\frac{9}{2}\right) \stackrel{?}{=} 9 \quad \text{Let } x = -\tfrac{9}{2}.$$
$$4\left(\frac{81}{4}\right) - 72 \stackrel{?}{=} 9$$
$$81 - 72 \stackrel{?}{=} 9$$
$$9 = 9 \ \checkmark \ \text{True}$$

The two solutions $\frac{1}{2}$ and $-\frac{9}{2}$ check, so the solution set is $\left\{-\frac{9}{2}, \frac{1}{2}\right\}$.

◀ Work Problem **4** at the Side.

EXAMPLE 5 **Solving a Quadratic Equation by Completing the Square**

Solve $2x^2 - 7x - 9 = 0$.

Step 1 Transform so that 1 is the coefficient of the x^2-term.

$$x^2 - \frac{7}{2}x - \frac{9}{2} = 0 \quad \text{Divide by 2.}$$

Step 2 Add $\frac{9}{2}$ to each side to get the variable terms on the left and the constant on the right.

$$x^2 - \frac{7}{2}x = \frac{9}{2} \quad \text{Add } \tfrac{9}{2}.$$

Step 3 To complete the square, take half the coefficient of x and square it.

$$\left[\frac{1}{2}\left(-\frac{7}{2}\right)\right]^2 = \left(-\frac{7}{4}\right)^2 = \frac{49}{16}$$

Add the result, $\frac{49}{16}$, to each side of the equation.

$$x^2 - \frac{7}{2}x + \frac{49}{16} = \frac{9}{2} + \frac{49}{16}$$

Be sure to add $\frac{49}{16}$ to *each* side.

$$\left(x - \frac{7}{4}\right)^2 = \frac{121}{16} \quad \text{Factor; } \tfrac{9}{2} + \tfrac{49}{16} = \tfrac{72}{16} + \tfrac{49}{16} = \tfrac{121}{16}.$$

Step 4 Solve by using the square root property.

$$x - \frac{7}{4} = \sqrt{\frac{121}{16}} \quad \text{or} \quad x - \frac{7}{4} = -\sqrt{\frac{121}{16}} \quad \text{Square root property}$$
$$x = \frac{7}{4} + \frac{11}{4} \quad \text{or} \quad x = \frac{7}{4} - \frac{11}{4} \quad \text{Add } \tfrac{7}{4};\ \sqrt{\tfrac{121}{16}} = \tfrac{11}{4}.$$
$$x = \frac{18}{4} \quad \text{or} \quad x = -\frac{4}{4} \quad \text{Add and subtract.}$$
$$x = \frac{9}{2} \quad \text{or} \quad x = -1 \quad \text{Lowest terms}$$

A check confirms that the solution set is $\left\{-1, \frac{9}{2}\right\}$.

◀ Work Problem **5** at the Side.

EXAMPLE 6 **Solving a Quadratic Equation by Completing the Square**

Solve $4p^2 + 8p + 5 = 0$.

$$4p^2 + 8p + 5 = 0$$

The coefficient of the second-degree term must be 1.

$$p^2 + 2p + \frac{5}{4} = 0 \qquad \text{Divide by 4.}$$

$$p^2 + 2p = -\frac{5}{4} \qquad \text{Subtract } \tfrac{5}{4}.$$

The coefficient of p is 2. Take half of 2, square the result, and add it to each side.

$$p^2 + 2p + 1 = -\frac{5}{4} + 1 \qquad \left[\tfrac{1}{2}(2)\right]^2 = 1^2 = 1;\ \text{Add 1.}$$

$$(p + 1)^2 = -\frac{1}{4} \qquad \text{Factor. Add.}$$

We cannot use the square root property to solve this equation, because the square root of $-\frac{1}{4}$ is not a real number. This equation has no real number solution.* The solution set is $\emptyset$.

Work Problem 6 at the Side. ▶

OBJECTIVE 3 Simplify the terms of an equation before solving.

EXAMPLE 7 **Simplifying before Completing the Square**

Solve $(x + 3)(x - 1) = 2$.

$$(x + 3)(x - 1) = 2 \qquad \text{Given equation}$$

$$x^2 + 2x - 3 = 2 \qquad \text{Multiply using the FOIL method.}$$

$$x^2 + 2x = 5 \qquad \text{Add 3.}$$

$$x^2 + 2x + 1 = 5 + 1 \qquad \text{Add } \left[\tfrac{1}{2}(2)\right]^2 = 1^2 = 1.$$

$$(x + 1)^2 = 6 \qquad \text{Factor on the left. Add on the right.}$$

$$x + 1 = \sqrt{6} \quad \text{or} \quad x + 1 = -\sqrt{6} \qquad \text{Square root property}$$

$$x = -1 + \sqrt{6} \quad \text{or} \quad x = -1 - \sqrt{6} \qquad \text{Subtract 1.}$$

The solution set is $\{-1 + \sqrt{6}, -1 - \sqrt{6}\}$.

Work Problem 7 at the Side. ▶

Note

The solutions $-1 + \sqrt{6}$ and $-1 - \sqrt{6}$ given in **Example 7** are *exact*. In applications, decimal solutions are more appropriate. Using the square root key of a calculator, $\sqrt{6} \approx 2.449$, which results in the following approximate solutions.

$$x \approx 1.449 \quad \text{and} \quad x \approx -3.449$$

*The equation in **Example 6** has no solution over the *real number system*. In the **complex number system,** however, this equation does have solutions. The complex numbers include numbers whose squares are negative. These numbers are discussed in intermediate and college algebra courses.

6 Solve $5x^2 + 3x + 1 = 0$ by completing the square.

7 Solve each equation.

(a) $r(r - 3) = -1$

(b) $(x + 2)(x + 1) = 5$

Answers

6. $\emptyset$

7. (a) $\left\{\frac{3 + \sqrt{5}}{2}, \frac{3 - \sqrt{5}}{2}\right\}$

(b) $\left\{\frac{-3 + \sqrt{21}}{2}, \frac{-3 - \sqrt{21}}{2}\right\}$

8 Solve each problem.

(a) Suppose a ball is projected upward from ground level with an initial velocity of 128 ft per sec. Its altitude (height) s at time t (in seconds) is given by

$$s = -16t^2 + 128t,$$

where s is in feet. At what times will the ball be 48 ft above the ground? Give answers to the nearest tenth.

(b) At what times will the ball in **Example 8** be 28 ft above the ground?

Answers

8. **(a)** 0.4 sec and 7.6 sec
(b) 0.5 sec and 3.5 sec

OBJECTIVE 4 Solve applied problems that require quadratic equations. The next example illustrates an application of quadratic equations from physics.

EXAMPLE 8 Solving a Velocity Problem

If a ball is projected into the air from ground level with an initial velocity of 64 ft per sec, its altitude (height) s in feet in t seconds is given by the formula

$$s = -16t^2 + 64t.$$

At what times will the ball be 48 ft above the ground?

Since s represents the height, we substitute **48** for s in the formula and then solve this equation for time t by completing the square.

$48 = -16t^2 + 64t$	Let $s = 48$.
$-3 = t^2 - 4t$	Divide by -16.
$t^2 - 4t = -3$	Interchange sides.
$t^2 - 4t + 4 = -3 + 4$	Add $\left[\frac{1}{2}(-4)\right]^2 = (-2)^2 = 4$.
$(t-2)^2 = 1$	Factor. Add.
$t - 2 = 1$ or $t - 2 = -1$	Square root property
$t = 3$ or $t = 1$	Add 2.

The ball reaches a height of 48 ft twice, once on the way up and again on the way down. It takes 1 sec to reach 48 ft on the way up, and then after 3 sec, the ball reaches 48 ft again on the way down.

Work Problem 8 at the Side.

9.2 Exercises

FOR EXTRA HELP

Download the MyDashBoard App

MyMathLab®

1. CONCEPT CHECK Which step is an appropriate way to begin solving the quadratic equation

$$2x^2 - 4x = 9$$

by completing the square? D

A. Add 4 to each side of the equation.

B. Factor the left side as $2x(x - 2)$.

C. Factor the left side as $x(2x - 4)$.

D. Divide each side by 2.

2. CONCEPT CHECK In **Example 3** of **Section 6.6,** we solved the quadratic equation

$$4p^2 - 26p + 40 = 0$$

by factoring. If we were to solve by completing the square, would we get the same solutions, $\frac{5}{2}$ and 4? yes

Complete each trinomial so that it is a perfect square. Then factor the trinomial. ***See Example 1.***

3. $x^2 + 10x +$ __25__ $(x + 5)^2$

4. $x^2 + 16x +$ __64__ $(x + 8)^2$

5. $x^2 + 2x +$ __1__ $(x + 1)^2$

6. $m^2 - 2m +$ __1__ $(m - 1)^2$

7. $p^2 - 5p +$ __$\frac{25}{4}$__ $\left(p - \frac{5}{2}\right)^2$

8. $x^2 + 3x +$ __$\frac{9}{4}$__ $\left(x + \frac{3}{2}\right)^2$

Solve each equation by completing the square. ***See Examples 2 and 3.***

9. $x^2 - 4x = -3$ $\{1, 3\}$

10. $x^2 - 2x = 8$ $\{-2, 4\}$

11. $x^2 + 5x + 6 = 0$ $\{-3, -2\}$

12. $x^2 + 6x + 5 = 0$ $\{-5, -1\}$

13. $x^2 + 2x - 5 = 0$ $\{-1 + \sqrt{6}, -1 - \sqrt{6}\}$

14. $x^2 + 4x + 1 = 0$ $\{-2 + \sqrt{3}, -2 - \sqrt{3}\}$

15. $x^2 - 8x = -4$ $\{4 + 2\sqrt{3}, 4 - 2\sqrt{3}\}$

16. $m^2 - 4m = 14$ $\{2 + 3\sqrt{2}, 2 - 3\sqrt{2}\}$

17. $t^2 + 6t + 9 = 0$ $\{-3\}$

18. $k^2 - 8k + 16 = 0$ $\{4\}$

19. $x^2 + x - 1 = 0$ $\left\{\frac{-1 + \sqrt{5}}{2}, \frac{-1 - \sqrt{5}}{2}\right\}$

20. $x^2 + x - 3 = 0$ $\left\{\frac{-1 + \sqrt{13}}{2}, \frac{-1 - \sqrt{13}}{2}\right\}$

Solve each equation by completing the square. ***See Examples 4–7.***

21. $4x^2 + 4x - 3 = 0$ $\left\{-\frac{3}{2}, \frac{1}{2}\right\}$

22. $9x^2 + 3x - 2 = 0$ $\left\{-\frac{2}{3}, \frac{1}{3}\right\}$

23. $2x^2 - 4x = 5$ $\left\{\frac{2+\sqrt{14}}{2}, \frac{2-\sqrt{14}}{2}\right\}$

24. $2x^2 - 6x = 3$ $\left\{\frac{3+\sqrt{15}}{2}, \frac{3-\sqrt{15}}{2}\right\}$

25. $2p^2 - 2p + 3 = 0$ $\emptyset$

26. $3q^2 - 3q + 4 = 0$ $\emptyset$

27. $3k^2 + 7k = 4$ $\left\{\frac{-7+\sqrt{97}}{6}, \frac{-7-\sqrt{97}}{6}\right\}$

28. $2k^2 + 5k = 1$ $\left\{\frac{-5+\sqrt{33}}{4}, \frac{-5-\sqrt{33}}{4}\right\}$

29. $(x + 3)(x - 1) = 5$ $\{-4, 2\}$

30. $(y - 8)(y + 2) = 24$ $\{-4, 10\}$

31. $(r - 3)(r - 5) = 2$ $\{4 + \sqrt{3}, 4 - \sqrt{3}\}$

32. $(k - 1)(k - 7) = 1$ $\{4 + \sqrt{10}, 4 - \sqrt{10}\}$

33. $-x^2 + 2x = -5$ $\{1 + \sqrt{6}, 1 - \sqrt{6}\}$

34. $-r^2 + 3r = -2$ $\left\{\frac{3+\sqrt{17}}{2}, \frac{3-\sqrt{17}}{2}\right\}$

Solve each problem. ***See Example 8.***

35. If an object is projected upward from ground level on Earth with an initial velocity of 96 ft per sec, its altitude (height) s in feet in t seconds is given by the formula $s = -16t^2 + 96t$. At what times will the object be 80 ft above the ground?
1 sec and 5 sec

36. At what times will the object described in **Exercise 35** be 100 ft above the ground? Round your answers to the nearest tenth.
1.3 sec and 4.7 sec

37. If an object is projected upward on the surface of Mars from ground level with an initial velocity of 104 ft per sec, its altitude (height) s in feet in t seconds is given by the formula $s = -13t^2 + 104t$. At what times will the object be 195 ft above the surface?
3 sec and 5 sec

38. After how many seconds will the object in **Exercise 37** return to the surface? (*Hint:* When it returns to the surface, $s = 0$.)
8 sec

39. A farmer has a rectangular cattle pen with perimeter 350 ft and area 7500 ft^2. What are the dimensions of the pen? (*Hint:* Use the figure to set up the equation.)

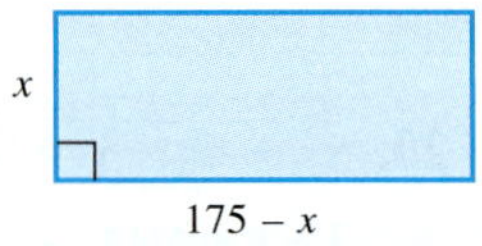

75 ft by 100 ft

40. The base of a triangle measures 1 m more than three times the height of the triangle. Its area is 15 m^2. Find the lengths of the base and the height.

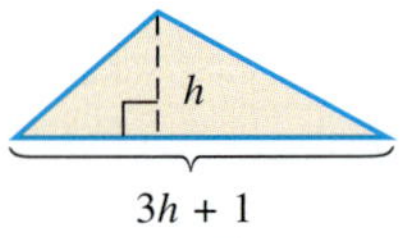

base: 10 m; height: 3 m

9.3 Solving Quadratic Equations by the Quadratic Formula

OBJECTIVES

1. Identify the values of a, b, and c in a quadratic equation.
2. Use the quadratic formula to solve quadratic equations.
3. Solve quadratic equations with only one solution.
4. Solve quadratic equations with fractions.

We can solve any quadratic equation by completing the square, but the method can be tedious. In this section we complete the square on the general quadratic equation

$$ax^2 + bx + c = 0, \quad \text{with} \quad a \neq 0, \quad \text{Standard form}$$

to obtain the *quadratic formula,* which gives the solution(s) of any quadratic equation.

Note

In $ax^2 + bx + c = 0$, there is a restriction that a is not zero. If it were, the equation would be linear, not quadratic.

OBJECTIVE 1 Identify the values of a, b, and c in a quadratic equation. To solve a quadratic equation by this new method, we must first identify the values of a, b, and c in the standard form.

EXAMPLE 1 Identifying Values of a, b, and c in Quadratic Equations

Identify the values of the variables a, b, and c in each quadratic equation $ax^2 + bx + c = 0$.

(a) $2x^2 + 3x - 5 = 0$ (with a, b, c marked above 2, 3, -5) This equation is in standard form.

Here, $a = 2$, $b = 3$, and $c = -5$.

(b) $-x^2 + 2 = 6x$

Subtract $6x$ to write the equation in standard form $ax^2 + bx + c = 0$.

$$-x^2 - 6x + 2 = 0 \quad \text{Subtract } 6x.$$

($-x^2$ means $-1x^2$.)

Here, $a = -1$, $b = -6$, and $c = 2$.

(c) $5x^2 - 12 = 0$

The x-term is missing, so write the equation as follows.

$$5x^2 + 0x - 12 = 0$$

Then, $a = 5$, $b = 0$, and $c = -12$.

(d) $-4x^2 = -x$

In standard form, this equation is written $-4x^2 + x = 0$. The constant term c is missing, so $a = -4$, $b = 1$, and $c = 0$.

(e)

$$(2x - 7)(x + 4) = -23$$

$$2x^2 + x - 28 = -23 \quad \text{Multiply using the FOIL method.}$$

$$2x^2 + x - 5 = 0 \quad \text{Add 23. The equation is now in standard form.}$$

(The equation is not in standard form.)

Identify the required values. We see that $a = 2$, $b = 1$, and $c = -5$.

Work Problem 1 at the Side. ▶

1 Identify the values of the variables a, b, and c in each quadratic equation $ax^2 + bx + c = 0$.

(a) $5x^2 + 2x - 1 = 0$

(b) $3x^2 = x - 2$

(c) $9x^2 - 13 = 0$

(d) $-x^2 + x = 0$

(e) $(3x + 2)(x - 1) = 8$

Answers

1. (a) $a = 5, b = 2, c = -1$
(b) $a = 3, b = -1, c = 2$
(c) $a = 9, b = 0, c = -13$
(d) $a = -1, b = 1, c = 0$
(e) $a = 3, b = -1, c = -10$

Teaching Tip

Explain that in this section we develop a formula for solving quadratic equations by using the method of completing the square for the general equation

$$ax^2 + bx + c = 0.$$

This may be the first time students have seen such a derivation. Explain that mathematical formulas do not just appear magically, but are derived from generalizations such as this one.

OBJECTIVE 2 Use the quadratic formula to solve quadratic equations.

To develop the quadratic formula, we complete the square on $ax^2 + bx + c = 0$ (where $a > 0$). For comparison, we also show the corresponding steps for solving $2x^2 + x - 5 = 0$ (from **Example 1(e)**).

Step 1 Transform so that the coefficient of x^2 is equal to 1.

$2x^2 + x - 5 = 0$	Standard form	$ax^2 + bx + c = 0$	Standard form
$x^2 + \frac{1}{2}x - \frac{5}{2} = 0$	Divide by 2.	$x^2 + \frac{b}{a}x + \frac{c}{a} = 0$	Divide by a.

Step 2 Write so that the variable terms with x are alone on the left side.

$x^2 + \frac{1}{2}x = \frac{5}{2}$	Add $\frac{5}{2}$.	$x^2 + \frac{b}{a}x = -\frac{c}{a}$	Subtract $\frac{c}{a}$.

Step 3 Add the square of half the coefficient of x to each side, factor the left side, and combine terms on the right.

$x^2 + \frac{1}{2}x + \frac{1}{16} = \frac{5}{2} + \frac{1}{16}$	Add $\frac{1}{16}$.	$x^2 + \frac{b}{a}x + \frac{b^2}{4a^2} = -\frac{c}{a} + \frac{b^2}{4a^2}$	Add $\frac{b^2}{4a^2}$.
$\left(x + \frac{1}{4}\right)^2 = \frac{41}{16}$	Factor. Add.	$\left(x + \frac{b}{2a}\right)^2 = \frac{b^2 - 4ac}{4a^2}$	Factor. Add.

Step 4 Use the square root property to complete the solution.

$x + \frac{1}{4} = \pm\sqrt{\frac{41}{16}}$	$x + \frac{b}{2a} = \pm\sqrt{\frac{b^2 - 4ac}{4a^2}}$
$x + \frac{1}{4} = \pm\frac{\sqrt{41}}{4}$	$x + \frac{b}{2a} = \pm\frac{\sqrt{b^2 - 4ac}}{2a}$
$x = -\frac{1}{4} \pm \frac{\sqrt{41}}{4}$	$x = -\frac{b}{2a} \pm \frac{\sqrt{b^2 - 4ac}}{2a}$
$x = \frac{-1 \pm \sqrt{41}}{4}$	$\boldsymbol{x = \frac{-b \pm \sqrt{b^2 - 4ac}}{2a}}$

The final result on the right is the **quadratic formula,** which is also valid for $a < 0$. ***It is a key result that should be memorized. It gives two values: one for the + sign and one for the − sign.***

Teaching Tip

Stress that the quadratic formula should be memorized and that the values of a, b, and c are determined only after the equation is written in standard form.

Also emphasize that $-b \pm \sqrt{b^2 - 4ac}$ is written all over $2a$. See Exercise 1.

Quadratic Formula

The quadratic equation $ax^2 + bx + c = 0$ (where $a \neq 0$) has solutions

$$x = \frac{-b + \sqrt{b^2 - 4ac}}{2a} \quad \text{and} \quad x = \frac{-b - \sqrt{b^2 - 4ac}}{2a},$$

or in compact form, $$x = \frac{-b \pm \sqrt{b^2 - 4ac}}{2a}.$$

CAUTION

In the quadratic formula, the fraction bar extends under $-b$ as well as the radical. ***Be sure to find the values of $-b \pm \sqrt{b^2 - 4ac}$ first, and then divide those results by the value of $2a$.***

EXAMPLE 2 Solving a Quadratic Equation by the Quadratic Formula

Solve $2x^2 - 7x - 9 = 0$.

$$x = \frac{-b \pm \sqrt{b^2 - 4ac}}{2a} \quad \text{Quadratic formula}$$

Be sure to write $-b$ in the numerator.

$$x = \frac{-(-7) \pm \sqrt{(-7)^2 - 4(2)(-9)}}{2(2)} \quad \text{Substitute } a = 2, b = -7, \text{ and } c = -9.$$

$$x = \frac{7 \pm \sqrt{49 + 72}}{4} \quad \text{Simplify.}$$

$$x = \frac{7 \pm \sqrt{121}}{4} \quad \text{Add.}$$

This represents **two** solutions.

$$x = \frac{7 \pm 11}{4} \quad \sqrt{121} = 11$$

Find the two solutions by first using the plus symbol and then using the minus symbol.

$$x = \frac{7 + 11}{4} = \frac{18}{4} = \frac{9}{2} \quad \text{or} \quad x = \frac{7 - 11}{4} = \frac{-4}{4} = -1$$

Check each solution in the original equation. The solution set is $\left\{-1, \frac{9}{2}\right\}$.

Work Problem 2 at the Side. ▶

EXAMPLE 3 Rewriting a Quadratic Equation before Solving

Solve $x^2 = 2x + 1$.

Write the equation in standard form as $x^2 - 2x - 1 = 0$.

$$x = \frac{-b \pm \sqrt{b^2 - 4ac}}{2a} \quad \text{Quadratic formula}$$

Be careful substituting the negative values.

$$x = \frac{-(-2) \pm \sqrt{(-2)^2 - 4(1)(-1)}}{2(1)} \quad \text{Substitute } a = 1, b = -2, \text{ and } c = -1.$$

$$x = \frac{2 \pm \sqrt{4 + 4}}{2} \quad \text{Simplify.}$$

$$x = \frac{2 \pm \sqrt{8}}{2} \quad \text{Add.}$$

$$x = \frac{2 \pm 2\sqrt{2}}{2} \quad \sqrt{8} = \sqrt{4} \cdot \sqrt{2} = 2\sqrt{2}$$

Factor first. Then divide out the common factor.

$$x = \frac{2(1 \pm \sqrt{2})}{2} \quad \text{Factor to write in lowest terms.}$$

$$x = 1 \pm \sqrt{2} \quad \text{Divide out the common factor.}$$

The solution set is $\{1 + \sqrt{2}, 1 - \sqrt{2}\}$.

Work Problem 3 at the Side. ▶

---

2 Solve each equation by using the quadratic formula.

(a) $2x^2 + 3x - 5 = 0$

(b) $6x^2 + x - 1 = 0$

Teaching Tip
Point out that the equation in Example 2 was solved by completing the square in Section 9.2, Example 5.

3 Solve $x^2 + 1 = -8x$.

GS Write the equation in standard form.

____

Substitute $a =$ ____, $b =$ ____, and $c =$ ____ in the quadratic formula and complete the solution.

Answers

2. (a) $\left\{1, -\frac{5}{2}\right\}$ **(b)** $\left\{-\frac{1}{2}, \frac{1}{3}\right\}$

3. $x^2 + 8x + 1 = 0$; 1; 8; 1; $\{-4 + \sqrt{15}, -4 - \sqrt{15}\}$

4 Solve $9x^2 - 12x + 4 = 0$.

5 Solve.

(a) $x^2 - \frac{4}{3}x + \frac{2}{3} = 0$

Teaching Tip

In Example 5, explain that the quadratic formula could be used with the fractions as coefficients, but using integer coefficients is much easier.

GS **(b)** $x^2 - \frac{9}{5}x = \frac{2}{5}$

Multiply by the LCD, ____, to clear fractions.

Simplify and write the resulting equation in standard form.

____

Substitute $a =$ ____, $b =$ ____, and $c =$ ____ in the quadratic formula. Complete the solution.

Answers

4. $\left\{\frac{2}{3}\right\}$

5. **(a)** $\emptyset$

(b) 5; $5x^2 - 9x - 2 = 0$; 5; -9; -2; $\left\{-\frac{1}{5}, 2\right\}$

OBJECTIVE 3 Solve quadratic equations with only one solution.

EXAMPLE 4 Solving a Quadratic Equation with One Solution

Solve $4x^2 + 25 = 20x$.

Write the equation in standard form.

$$4x^2 - 20x + 25 = 0 \qquad \text{Subtract } 20x.$$

Here, $a = 4$, $b = -20$, and $c = 25$. Substitute in the quadratic formula.

$$x = \frac{-(-20) \pm \sqrt{(-20)^2 - 4(4)(25)}}{2(4)} = \frac{20 \pm 0}{8} = \frac{5}{2}$$

In this case, $b^2 - 4ac = 0$, and the trinomial $4x^2 - 20x + 25$ is a perfect square. There is one distinct solution in the solution set $\left\{\frac{5}{2}\right\}$.

◀ Work Problem **4** at the Side.

Note

The single solution of the equation in **Example 4** is a rational number. If all solutions of a quadratic equation are rational, the equation can be solved by factoring as well.

OBJECTIVE 4 Solve quadratic equations with fractions.

EXAMPLE 5 Solving a Quadratic Equation with Fractions

Solve $\frac{1}{10}t^2 = \frac{2}{5}t - \frac{1}{2}$.

$$\frac{1}{10}t^2 = \frac{2}{5}t - \frac{1}{2}$$

$$10\left(\frac{1}{10}t^2\right) = 10\left(\frac{2}{5}t - \frac{1}{2}\right) \qquad \text{Clear fractions. Multiply by the LCD, 10.}$$

$$10\left(\frac{1}{10}t^2\right) = 10\left(\frac{2}{5}t\right) - 10\left(\frac{1}{2}\right) \qquad \text{Distributive property}$$

$$t^2 = 4t - 5 \qquad \text{Multiply.}$$

$$t^2 - 4t + 5 = 0 \qquad \text{Subtract } 4t. \text{ Add } 5.$$

Identify $a = 1$, $b = -4$, and $c = 5$.

$$t = \frac{-(-4) \pm \sqrt{(-4)^2 - 4(1)(5)}}{2(1)} \qquad \text{Substitute in the quadratic formula.}$$

$$t = \frac{4 \pm \sqrt{16 - 20}}{2} \qquad \text{Simplify.}$$

$$t = \frac{4 \pm \sqrt{-4}}{2} \qquad \text{Subtract.}$$

Stop here.

Because $\sqrt{-4}$ does not represent a real number, the solution set is $\emptyset$.

◀ Work Problem **5** at the Side.

9.3 Exercises

FOR EXTRA HELP

Download the MyDashBoard App

MyMathLab®

1. **CONCEPT CHECK** A student writes the quadratic formula as

$$x = -b \pm \frac{\sqrt{b^2 - 4ac}}{2a}.$$

What Went Wrong? Explain the error, and give the correct formula.

$2a$ should be the denominator for $-b$ as well. The correct formula is $x = \frac{-b \pm \sqrt{b^2 - 4ac}}{2a}$.

2. **CONCEPT CHECK** If we apply the quadratic formula and find that the value of $b^2 - 4ac$ is negative, what can we conclude?

There is no real number solution, and thus the solution set is $\emptyset$.

Write each equation in standard form $ax^2 + bx + c = 0$, if necessary. Then identify the values of a, b, and c. Do not actually solve the equation. ***See Example 1.***

3. $4x^2 + 5x - 9 = 0$

$a =$ 4, $b =$ 5, $c =$ −9

4. $8x^2 + 3x - 4 = 0$

$a =$ 8, $b =$ 3, $c =$ −4

5. $3x^2 = 4x + 2$

$a =$ 3, $b =$ −4, $c =$ −2

6. $5x^2 = 3x - 6$

$a =$ 5, $b =$ −3, $c =$ 6

7. $3x^2 = -7x$

$a =$ 3, $b =$ 7, $c =$ 0

8. $9x^2 = 8x$

$a =$ 9, $b =$ −8, $c =$ 0

Use the quadratic formula to solve each equation. Write all radicals in simplified form, and write all answers in lowest terms. ***See Examples 2–4.***

9. $k^2 + 12k - 13 = 0$

$\{-13, 1\}$

10. $r^2 - 8r - 9 = 0$

$\{-1, 9\}$

11. $p^2 - 4p + 4 = 0$

$\{2\}$

12. $9x^2 + 6x + 1 = 0$

$\left\{-\frac{1}{3}\right\}$

13. $2x^2 = 5 + 3x$

$\left\{-1, \frac{5}{2}\right\}$

14. $2z^2 = 30 + 7z$

$\left\{-\frac{5}{2}, 6\right\}$

15. $2x^2 + 12x = -5$

$\left\{\frac{-6 + \sqrt{26}}{2}, \frac{-6 - \sqrt{26}}{2}\right\}$

16. $5m^2 + m = 1$

$\left\{\frac{-1 + \sqrt{21}}{10}, \frac{-1 - \sqrt{21}}{10}\right\}$

17. $6x^2 + 6x = 0$

$\{-1, 0\}$

18. $4n^2 - 12n = 0$

$\{0, 3\}$

19. $-2x^2 = -3x + 2$

$\emptyset$

20. $-x^2 = -5x + 20$

$\emptyset$

21. $3x^2 + 5x + 1 = 0$

$\left\{\frac{-5 + \sqrt{13}}{6}, \frac{-5 - \sqrt{13}}{6}\right\}$

22. $6x^2 - 6x + 1 = 0$

$\left\{\frac{3 + \sqrt{3}}{6}, \frac{3 - \sqrt{3}}{6}\right\}$

23. $7x^2 = 12x$

$\left\{0, \frac{12}{7}\right\}$

24. $9r^2 = 11r$

$\left\{0, \frac{11}{9}\right\}$

25. $x^2 - 24 = 0$

$\{-2\sqrt{6}, 2\sqrt{6}\}$

26. $z^2 - 96 = 0$

$\{-4\sqrt{6}, 4\sqrt{6}\}$

27. $25x^2 - 4 = 0$

$\left\{-\frac{2}{5}, \frac{2}{5}\right\}$

28. $16x^2 - 9 = 0$

$\left\{-\frac{3}{4}, \frac{3}{4}\right\}$

29. $3x^2 - 2x + 5 = 10x + 1$

$\left\{\frac{6 + 2\sqrt{6}}{3}, \frac{6 - 2\sqrt{6}}{3}\right\}$

30. $4x^2 - x + 4 = x + 7$

$\left\{\frac{1 + \sqrt{13}}{4}, \frac{1 - \sqrt{13}}{4}\right\}$

31. $2x^2 + x + 5 = 0$

$\emptyset$

32. $3x^2 + 2x + 8 = 0$

$\emptyset$

Use the quadratic formula to solve each equation. ***See Example 5.***

33. $\frac{3}{2}k^2 - k - \frac{4}{3} = 0$

$\left\{-\frac{2}{3}, \frac{4}{3}\right\}$

34. $\frac{2}{5}x^2 - \frac{3}{5}x - 1 = 0$

$\left\{-1, \frac{5}{2}\right\}$

35. $\frac{1}{2}x^2 + \frac{1}{6}x = 1$

$\left\{\frac{-1 + \sqrt{73}}{6}, \frac{-1 - \sqrt{73}}{6}\right\}$

36. $\frac{2}{3}t^2 - \frac{4}{9}t = \frac{1}{3}$

$\left\{\frac{2 + \sqrt{22}}{6}, \frac{2 - \sqrt{22}}{6}\right\}$

37. $\frac{3}{8}x^2 - x + \frac{17}{24} = 0$

$\emptyset$

38. $\frac{1}{3}x^2 + \frac{8}{9}x + \frac{7}{9} = 0$

$\emptyset$

39. $0.6x - 0.4x^2 = -1$

$\left\{-1, \frac{5}{2}\right\}$

40. $0.5x^2 = x + 0.5$

$\{1 + \sqrt{2}, 1 - \sqrt{2}\}$

41. $0.25x^2 = -1.5x - 1$

$\{-3 + \sqrt{5}, -3 - \sqrt{5}\}$

Solve each problem.

42. An astronaut on the moon throws a baseball upward. The altitude (height) h of the ball, in feet, x seconds after he throws it, is given by the equation

$$h = -2.7x^2 + 30x + 6.5.$$

At what times is the ball 12 ft above the moon's surface? Give answer(s) to the nearest tenth.

0.2 sec and 10.9 sec

43. A frog is sitting on a stump 3 ft above the ground. He hops off the stump and lands on the ground 4 ft away. During his leap, his height h is given by the equation

$$h = -0.5x^2 + 1.25x + 3,$$

where x is the distance in feet from the base of the stump, and h is in feet. How far was the frog from the base of the stump when he was 1.25 ft above the ground?

3.5 ft

44. An old Babylonian problem asks for the length of the side of a square, given that the area of the square minus the length of a side is 870. Find the length of the side. (*Source:* Eves, H., *An Introduction to the History of Mathematics*, Saunders College Publishing.)

30

45. A rule for estimating the number of board feet of lumber that can be cut from a log depends on the diameter of the log. To find the diameter d required to get 9 board feet of lumber, we use the equation

$$\left(\frac{d - 4}{4}\right)^2 = 9.$$

Solve this equation for d. Are both answers reasonable?

−8, 16; Only 16 board feet is a reasonable answer.

Summary Exercises *Applying Methods for Solving Quadratic Equations*

Summary Exercises provide mixed practice to better prepare students for tests.

Four algebraic methods have now been introduced for solving quadratic equations written in the form $ax^2 + bx + c = 0$. The following table shows some advantages and some disadvantages of each method.

Method	Advantages	Disadvantages
1. Factoring	It is usually the fastest method.	Not all equations can be solved by factoring. Some factorable polynomials are difficult to factor.
2. Square root property	It is the simplest method for solving equations of the form $(ax + b)^2 =$ a number.	Few equations are given in this form.
3. Completing the square	It can always be used. (Also, the procedure is useful in other areas of mathematics.)	It requires more steps than other methods.
4. Quadratic formula	It can always be used.	Sign errors are common because of the presence of the expression $\sqrt{b^2 - 4ac}$.

Solve each quadratic equation by the method of your choice.

1. $x^2 = 36$
$\{-6, 6\}$

2. $x^2 + 3x = -1$
$\left\{\frac{-3 + \sqrt{5}}{2}, \frac{-3 - \sqrt{5}}{2}\right\}$

3. $x^2 - \frac{100}{81} = 0$
$\left\{-\frac{10}{9}, \frac{10}{9}\right\}$

4. $81t^2 = 49$
$\left\{-\frac{7}{9}, \frac{7}{9}\right\}$

5. $z^2 - 4z + 3 = 0$
$\{1, 3\}$

6. $w^2 + 3w + 2 = 0$
$\{-2, -1\}$

7. $z(z - 9) = -20$
$\{4, 5\}$

8. $x^2 + 3x - 2 = 0$
$\left\{\frac{-3 + \sqrt{17}}{2}, \frac{-3 - \sqrt{17}}{2}\right\}$

9. $(3k - 2)^2 = 9$
$\left\{-\frac{1}{3}, \frac{5}{3}\right\}$

10. $(2s - 1)^2 = 10$
$\left\{\frac{1 + \sqrt{10}}{2}, \frac{1 - \sqrt{10}}{2}\right\}$

11. $(x + 6)^2 = 121$
$\{-17, 5\}$

12. $(5k + 1)^2 = 36$
$\left\{-\frac{7}{5}, 1\right\}$

13. $(3r - 7)^2 = 24$
$\left\{\frac{7 + 2\sqrt{6}}{3}, \frac{7 - 2\sqrt{6}}{3}\right\}$

14. $(7p - 1)^2 = 32$
$\left\{\frac{1 + 4\sqrt{2}}{7}, \frac{1 - 4\sqrt{2}}{7}\right\}$

15. $(5x - 8)^2 = -6$
$\emptyset$

16. $2t^2 + 1 = t$
$\emptyset$

17. $-2x^2 = -3x - 2$
$\left\{-\frac{1}{2}, 2\right\}$

18. $-2x^2 + x = -1$
$\left\{-\frac{1}{2}, 1\right\}$

19. $8z^2 = 15 + 2z$

$\left\{-\frac{5}{4}, \frac{3}{2}\right\}$

20. $3k^2 = 3 - 8k$

$\left\{-3, \frac{1}{3}\right\}$

21. $0 = -x^2 + 2x + 1$

$\{1 + \sqrt{2}, 1 - \sqrt{2}\}$

22. $3x^2 + 5x = -1$

$\left\{\frac{-5 + \sqrt{13}}{6}, \frac{-5 - \sqrt{13}}{6}\right\}$

23. $5x^2 - 22x = -8$

$\left\{\frac{2}{5}, 4\right\}$

24. $x(x + 6) + 4 = 0$

$\{-3 + \sqrt{5}, -3 - \sqrt{5}\}$

25. $(x + 2)(x + 1) = 10$

$\left\{\frac{-3 + \sqrt{41}}{2}, \frac{-3 - \sqrt{41}}{2}\right\}$

26. $16x^2 + 40x + 25 = 0$

$\left\{-\frac{5}{4}\right\}$

27. $4x^2 = -1 + 5x$

$\left\{\frac{1}{4}, 1\right\}$

28. $2p^2 = 2p + 1$

$\left\{\frac{1 + \sqrt{3}}{2}, \frac{1 - \sqrt{3}}{2}\right\}$

29. $3x(3x + 4) = 7$

$\left\{\frac{-2 + \sqrt{11}}{3}, \frac{-2 - \sqrt{11}}{3}\right\}$

30. $5x - 1 + 4x^2 = 0$

$\left\{\frac{-5 + \sqrt{41}}{8}, \frac{-5 - \sqrt{41}}{8}\right\}$

31. $\frac{x^2}{2} + \frac{7x}{4} + \frac{11}{8} = 0$

$\left\{\frac{-7 + \sqrt{5}}{4}, \frac{-7 - \sqrt{5}}{4}\right\}$

32. $t(15t + 58) = -48$

$\left\{-\frac{8}{3}, -\frac{6}{5}\right\}$

33. $9k^2 = 16(3k + 4)$

$\left\{\frac{8 + 8\sqrt{2}}{3}, \frac{8 - 8\sqrt{2}}{3}\right\}$

34. $\frac{1}{5}x^2 + x + 1 = 0$

$\left\{\frac{-5 + \sqrt{5}}{2}, \frac{-5 - \sqrt{5}}{2}\right\}$

35. $x^2 - x + 3 = 0$

$\emptyset$

36. $4x^2 - 11x + 8 = -2$

$\emptyset$

37. $-3x^2 + 4x = -4$

$\left\{-\frac{2}{3}, 2\right\}$

38. $z^2 - \frac{5}{12}z = \frac{1}{6}$

$\left\{-\frac{1}{4}, \frac{2}{3}\right\}$

39. $5k^2 + 19k = 2k + 12$

$\left\{-4, \frac{3}{5}\right\}$

40. $\frac{1}{2}x^2 - x = \frac{15}{2}$

$\{-3, 5\}$

41. $x^2 - \frac{4}{15} = -\frac{4}{15}x$

$\left\{-\frac{2}{3}, \frac{2}{5}\right\}$

42. $(x + 2)(x - 4) = 16$

$\{-4, 6\}$

9.4 Graphing Quadratic Equations

OBJECTIVES

1. Graph quadratic equations.
2. Find the vertex of a parabola.

OBJECTIVE 1 **Graph quadratic equations.** In this section, we graph quadratic equations in two variables, of the form

$$y = ax^2 + bx + c.$$

The simplest quadratic equation, $y = x^2$ (or $y = 1x^2 + 0x + 0$), can be graphed in much the same way that straight lines were graphed in **Chapter 3,** by finding ordered pairs that satisfy the equation.

EXAMPLE 1 Graphing a Quadratic Equation

Graph $y = x^2$.

Select several values for x. Then find the corresponding y-values. For example, selecting $x = 2$ and substituting in $y = x^2$ gives

$$y = 2^2 = 4,$$

and so the point $(2, 4)$ is on the graph of $y = x^2$. ***(Recall that in an ordered pair such as (2, 4), the x-value comes first and the y-value second.)***

Work Problem 1 at the Side. ▶

If we plot the points from **Margin Problem 1** on a coordinate system and draw a smooth curve through them, we obtain the graph in **Figure 1.** A table of values is shown with the graph.

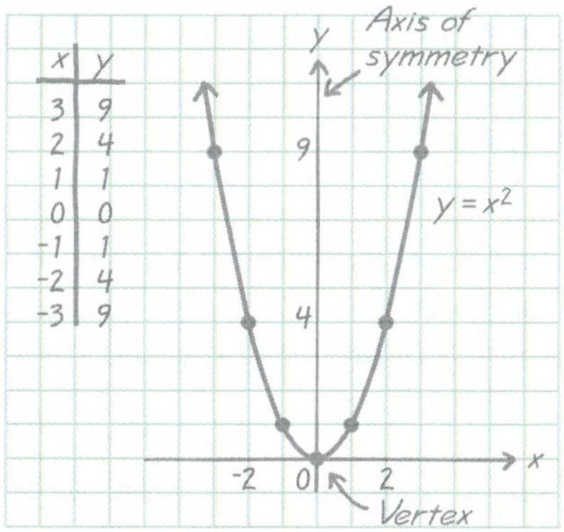

Figure 1

Work Problem 2 at the Side. ▶

The curve in **Figure 1** is a **parabola.** Every equation of the form

$$y = ax^2 + bx + c, \quad \text{with } a \neq 0,$$

has a graph that is a parabola. The point $(0, 0)$, the *lowest* point on this graph, is the **vertex** of the parabola. (If a parabola opens downward, the vertex will be the *highest* point. See **Figure 2** on the next page.)

The vertical line through the vertex of a parabola that opens upward or downward is the **axis of symmetry.** The two halves of the parabola are mirror images of each other across this line. For the parabola shown in **Figure 1,** the axis of symmetry is the y-axis.

1 Complete the table of values for $y = x^2$.

x	y
3	
2	4
1	
0	
−1	
−2	
−3	

2 Graph $y = \frac{1}{2}x^2$ by first completing the table of values.

x	y
−2	
−1	
0	
1	
2	

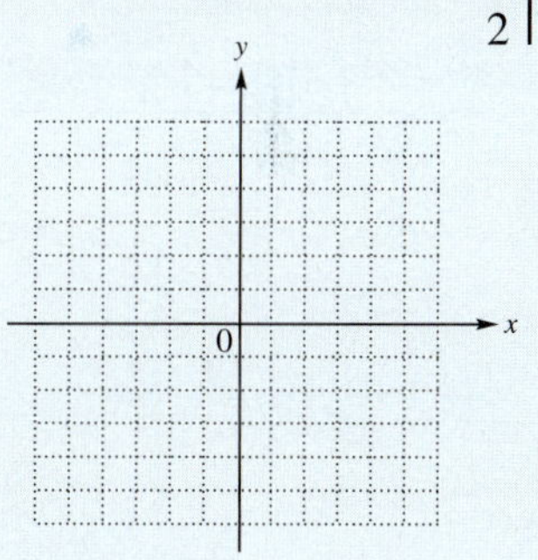

Answers

1. See the table beside the graph in **Figure 1.**

2.

x	y
−2	2
−1	$\frac{1}{2}$
0	0
1	$\frac{1}{2}$
2	2

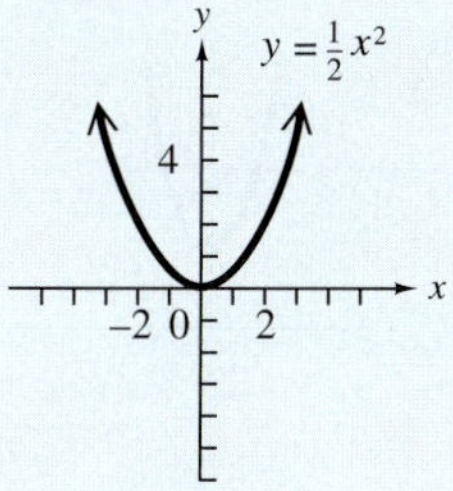

3 Complete each ordered pair for $y = -x^2 + 3$.

$(-2, \_\_)$, $(-1, \_\_)$,

$(1, \_\_)$, $(2, \_\_)$

4 Graph each equation, and identify each vertex.

(a) $y = -x^2 - 3$

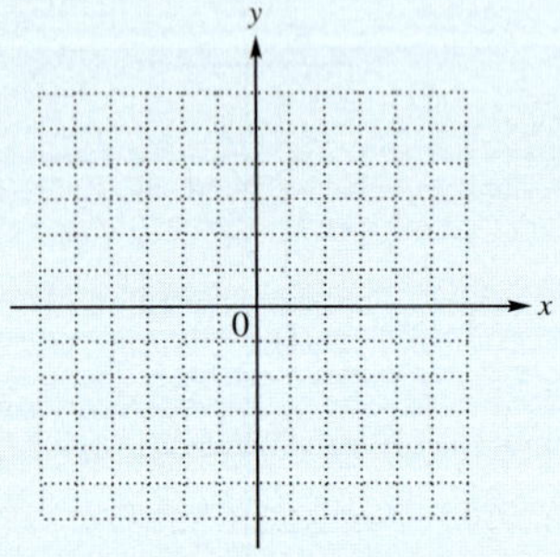

(b) $y = x^2 + 3$

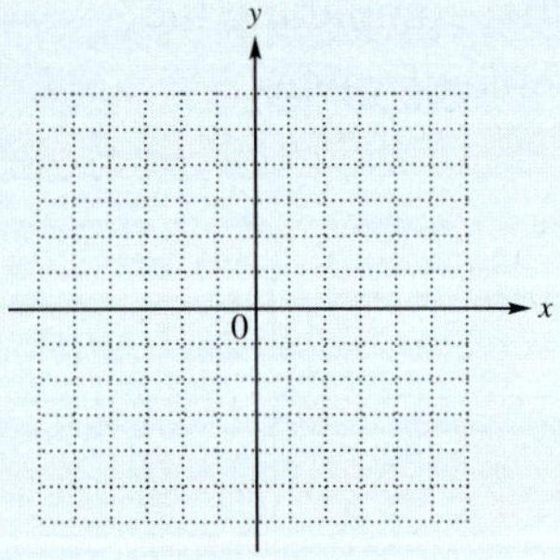

Answers

3. $(-2, -1), (-1, 2), (1, 2), (2, -1)$

4. (a)

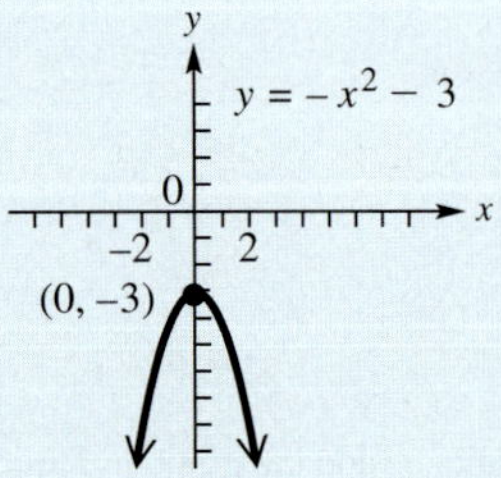

(b)

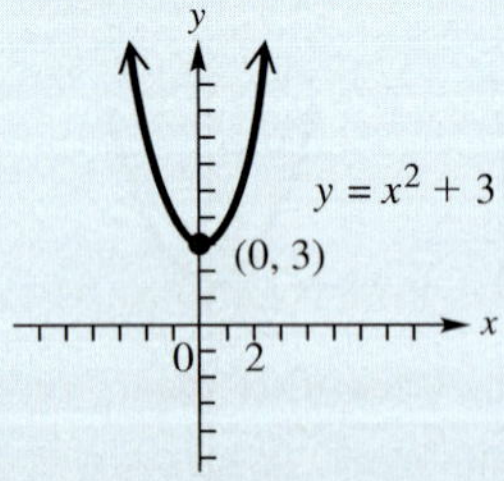

EXAMPLE 2 Graphing a Parabola by Plotting Points

Graph $y = -x^2 + 3$.

Find several ordered pairs. Let $x = 0$ to find the y-intercept.

$$y = -x^2 + 3 = -0^2 + 3 = 3$$

This gives the ordered pair $(0, 3)$.

◀ Work Problem **3** at the Side.

Plot the ordered pair $(0, 3)$ and the ordered pairs from **Margin Problem 3** at the side and join them with a smooth curve as shown in **Figure 2.** The vertex of this parabola is $(0, 3)$. The graph opens downward because x^2 has a negative coefficient, so the vertex is the *highest* point of the graph.

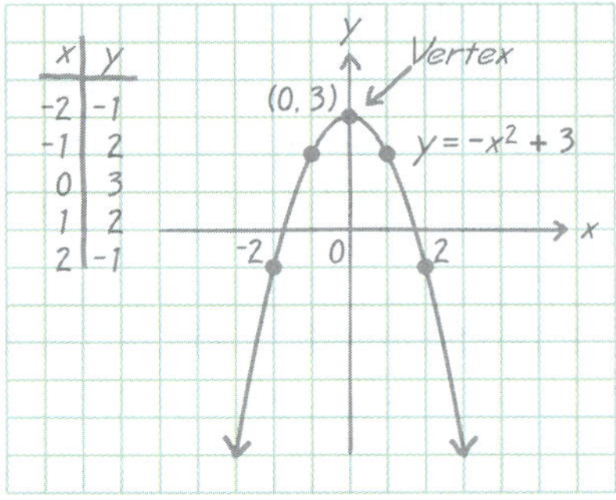

Figure 2

◀ Work Problem **4** at the Side.

OBJECTIVE 2 Find the vertex of a parabola. The vertex is the most important point to locate when graphing a quadratic equation.

EXAMPLE 3 Graphing a Parabola

Graph $y = x^2 - 2x - 3$.

Because of its symmetry, if a parabola has two x-intercepts, the x-value of the vertex is exactly halfway between them. Therefore, we begin by finding the x-intercepts. We let $y = 0$ in the equation, and solve for x.

$$0 = x^2 - 2x - 3 \qquad \text{Let } y = 0.$$

$$0 = (x + 1)(x - 3) \qquad \text{Factor.}$$

$$x + 1 = 0 \quad \text{or} \quad x - 3 = 0 \qquad \text{Zero-factor property}$$

$$x = -1 \quad \text{or} \quad x = 3 \qquad \text{Solve each equation.}$$

There are two x-intercepts, $(-1, 0)$ and $(3, 0)$. Since the x-value of the vertex is halfway between the x-values of the two x-intercepts, it is half their sum.

$$x = \frac{1}{2}(-1 + 3)$$

$$= 1 \leftarrow x\text{-value of the vertex}$$

We find the corresponding y-value by substituting 1 for x in the equation.

$$y = 1^2 - 2(1) - 3 \qquad \text{Let } x = 1.$$

$$= -4 \leftarrow y\text{-value of the vertex}$$

The vertex is $(1, -4)$. The axis of symmetry is the vertical line $x = 1$.

Continued on Next Page

To find the y-intercept, we substitute 0 for x in the equation.

$$y = 0^2 - 2(0) - 3 \quad \text{Let } x = 0.$$
$$= -3 \quad \text{Simplify.}$$

The y-intercept is $(0, -3)$.

We plot the three intercepts and the vertex, and find additional ordered pairs as needed. For example, we let $x = 2$.

$$y = 2^2 - 2(2) - 3 \quad \text{Let } x = 2.$$
$$= -3 \quad \text{Simplify.}$$

This leads to the ordered pair $(2, -3)$. A table that includes the ordered pairs we found is shown with the graph in **Figure 3.**

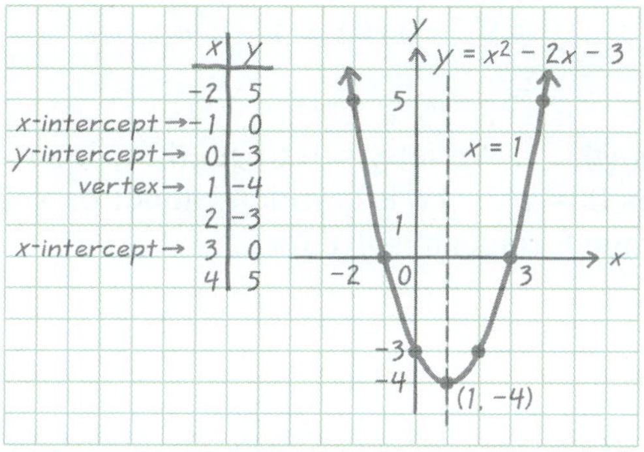

Figure 3

Work Problem 5 at the Side. ▶

We can generalize from **Example 3.** The x-coordinates of the x-intercepts for $y = ax^2 + bx + c$, by the quadratic formula, are

$$x = \frac{-b + \sqrt{b^2 - 4ac}}{2a} \quad \text{and} \quad x = \frac{-b - \sqrt{b^2 - 4ac}}{2a}.$$

Thus, the x-value of the vertex is halfway between the x-values of the x-intercepts.

$$x = \frac{1}{2}\left(\frac{-b + \sqrt{b^2 - 4ac}}{2a} + \frac{-b - \sqrt{b^2 - 4ac}}{2a}\right)$$

$$x = \frac{1}{2}\left(\frac{-b + \sqrt{b^2 - 4ac} - b - \sqrt{b^2 - 4ac}}{2a}\right)$$

$$x = \frac{1}{2}\left(\frac{-2b}{2a}\right) \quad \text{Combine like terms.}$$

$$x = -\frac{b}{2a} \quad \text{Multiply. Write in lowest terms.}$$

For the equation in **Example 3,** $y = x^2 - 2x - 3$, we have $a = 1$ and $b = -2$. Thus, the x-value of the vertex is found as follows.

$$x = -\frac{b}{2a} = -\frac{-2}{2(1)} = 1 \quad \leftarrow \text{This is the same } x\text{-value for the vertex found in } \textbf{Example 3.}$$

Note

The x-value of the vertex is $x = -\frac{b}{2a}$, even if the graph has no x-intercepts.

5 Graph $y = x^2 + 2x - 8$.

GS To find the x-intercepts, solve the equation.

$0 =$ ______________

$0 = (x + 4)(x - \_\_\_\_)$

$x =$ ____ or $x = 2$

The x-intercepts are ____ and $(2, 0)$.

The x-value of the vertex is

$\frac{1}{2}(-4 + \_\_\_\_) = \_\_\_\_.$

The y-value of the vertex is ____.

The vertex is ____.

The y-intercept is ____.

The axis of symmetry is the line $x =$ ____.

Find additional ordered pairs as needed and sketch the graph.

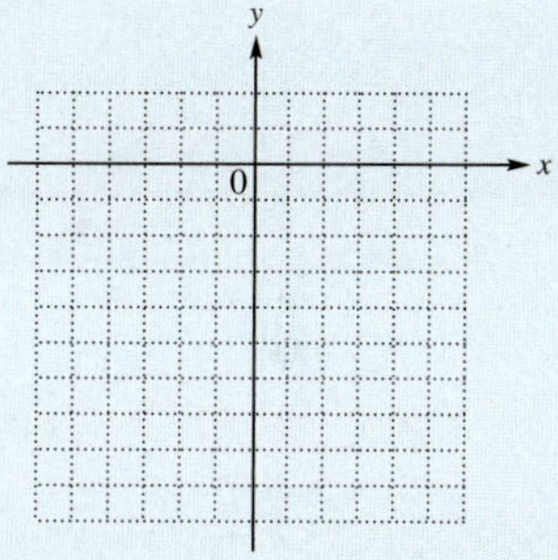

Answers

5. $x^2 + 2x - 8$; 2; -4; $(-4, 0)$; 2; -1; -9; $(-1, -9)$; $(0, -8)$; -1

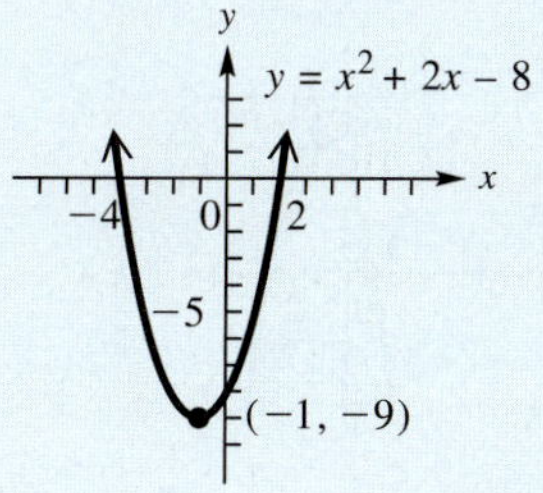

6 Complete each ordered pair for $y = x^2 - 4x + 1$.

$(5, \_\_)$, $(4, \_\_)$, $(-1, \_\_)$

7 Graph each parabola.

(a) $y = x^2 - 3x - 3$

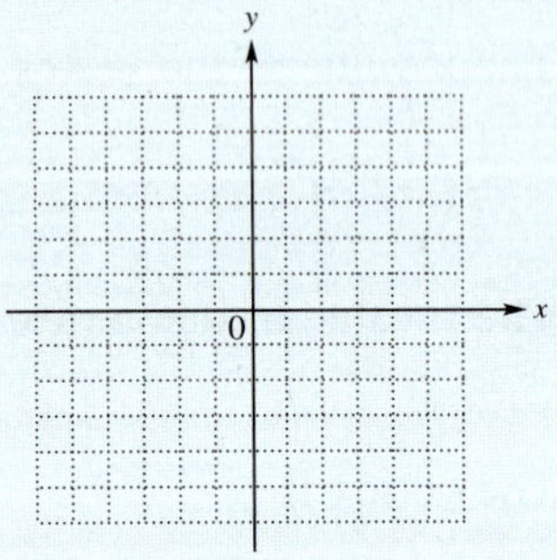

(b) $y = -x^2 + 2x + 4$

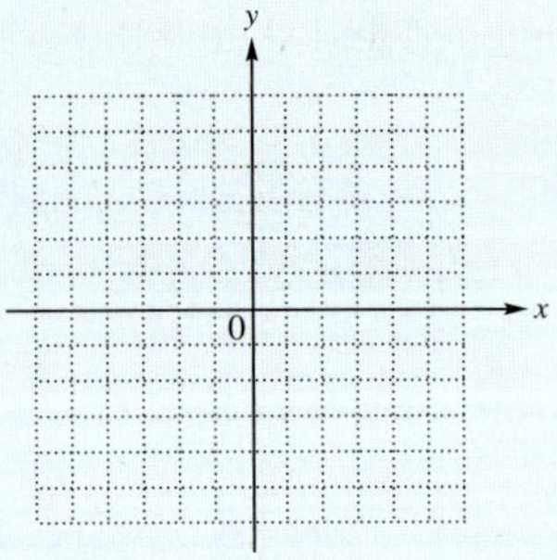

Answers

6. $(5, 6)$, $(4, 1)$, $(-1, 6)$

7. (a)

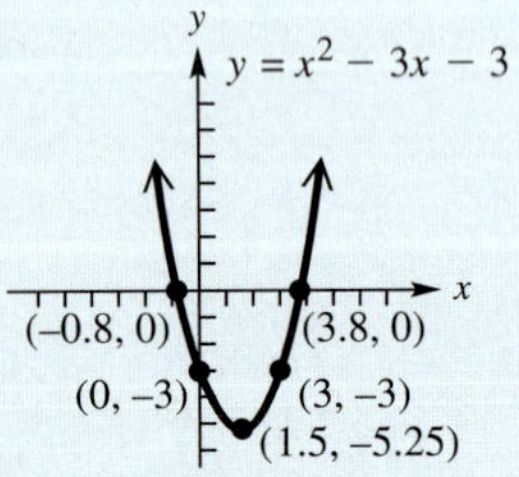

(b)

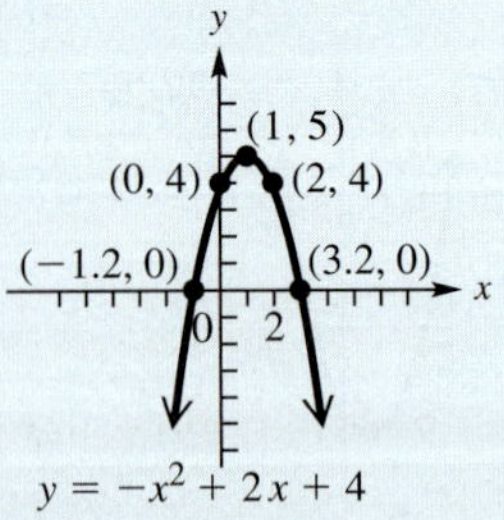

EXAMPLE 4 Graphing a Parabola

Graph $y = x^2 - 4x + 1$.

Here, $a = 1$ and $b = -4$, so we substitute to find the x-value of the vertex.

$$x = -\frac{b}{2a} = -\frac{-4}{2(1)} = 2$$

The y-value of the vertex is found as follows.

$$y = 2^2 - 4(2) + 1 \quad \text{Let } x = 2.$$

$$y = -3 \quad \text{Simplify.}$$

The vertex is $(2, -3)$. The axis is the line $x = 2$.

Now we find the intercepts. Let $x = 0$ in $y = x^2 - 4x + 1$.

$$y = 0^2 - 4(0) + 1 \quad \text{Let } x = 0.$$

$$y = 1$$

The y-intercept is $(0, 1)$. We let $y = 0$ to get the x-intercepts. If $y = 0$, the equation becomes $0 = x^2 - 4x + 1$, which cannot be solved by factoring. We use the quadratic formula to solve for x.

$$x = \frac{-(-4) \pm \sqrt{(-4)^2 - 4(1)(1)}}{2(1)} \quad \text{Let } a = 1, b = -4, c = 1 \text{ in the quadratic formula.}$$

$$x = \frac{4 \pm \sqrt{12}}{2} \quad \text{Simplify.}$$

$$x = \frac{4 \pm 2\sqrt{3}}{2} \quad \sqrt{12} = \sqrt{4} \cdot \sqrt{3} = 2\sqrt{3}$$

Factor first. Then divide out the common factor.

$$x = \frac{2(2 \pm \sqrt{3})}{2} \quad \text{Factor.}$$

$$x = 2 \pm \sqrt{3} \quad \text{Divide out 2.}$$

Using a calculator, we find that the x-intercepts are $(3.7, 0)$ and $(0.3, 0)$, to the nearest tenth.

◀ **Work Problem 6 at the Side.**

We plot the intercepts, vertex, and the points found in **Margin Problem 6,** and join these points with a smooth curve. The graph is shown in **Figure 4.**

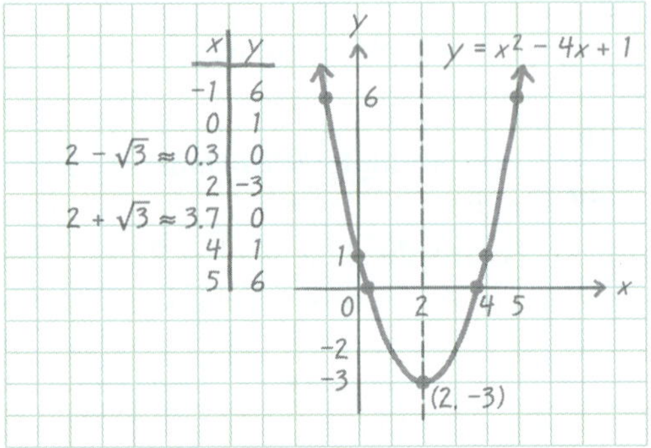

Figure 4

◀ **Work Problem 7 at the Side.**

9.4 Exercises

FOR EXTRA HELP

Download the MyDashBoard App

MyMathLab®

CONCEPT CHECK *Fill in each blank with the correct response.*

1. Every equation of the from $y = ax^2 + bx + c$, with $a \neq 0$, has a graph that is a(n) **parabola**.

2. The **vertex** of a parabola is the lowest point of the graph if the parabola opens upward and the **highest** point of the graph if the parabola opens downward.

3. The vertical line through the vertex of a parabola that opens upward or downward is the **axis of symmetry** of the parabola. The two halves of the parabola are **mirror** images of each other across this line.

4. The graph of an equation of the form $y = ax^2 + bx + c$, with $a \neq 0$, has exactly **one** y-intercept(s), but may have either two, **one**, or **no** x-intercepts.

Graph each equation. Give the coordinates of the vertex in each case. ***See Examples 1–4.***

5. $y = 2x^2$

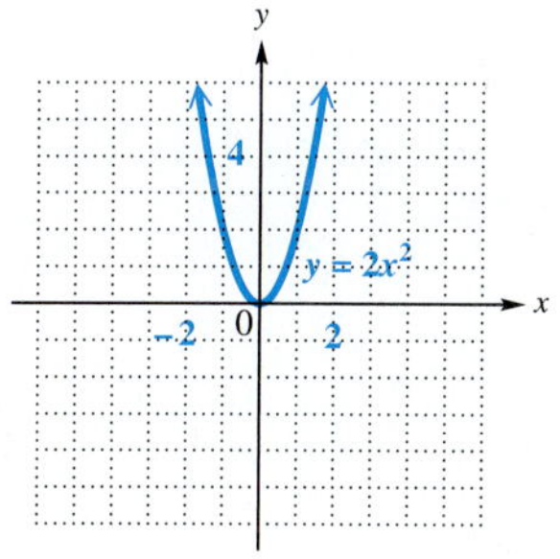

(0, 0)

6. $y = \frac{1}{4}x^2$

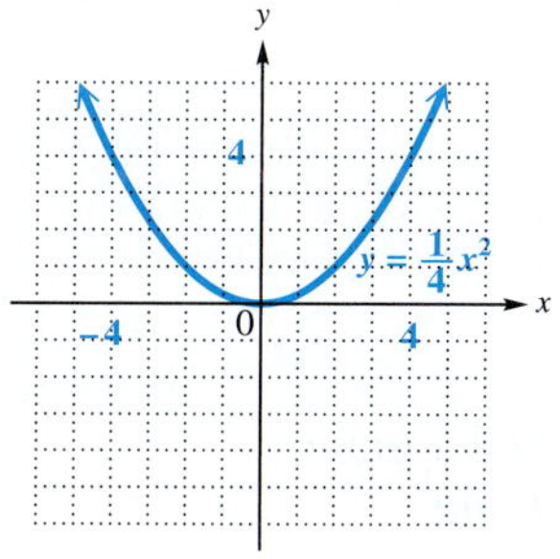

(0, 0)

7. $y = x^2 - 4$

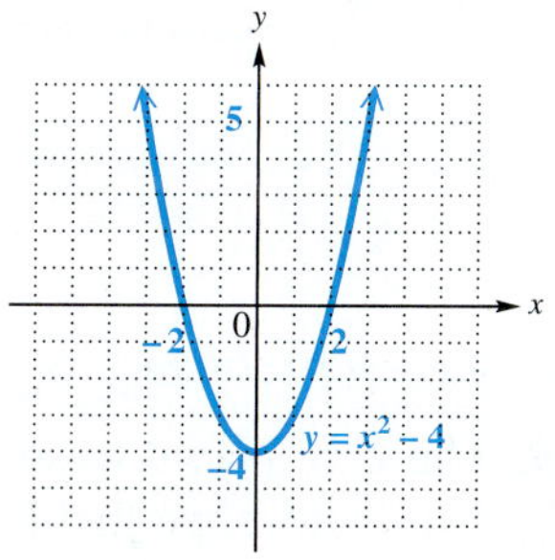

(0, −4)

8. $y = x^2 - 2$

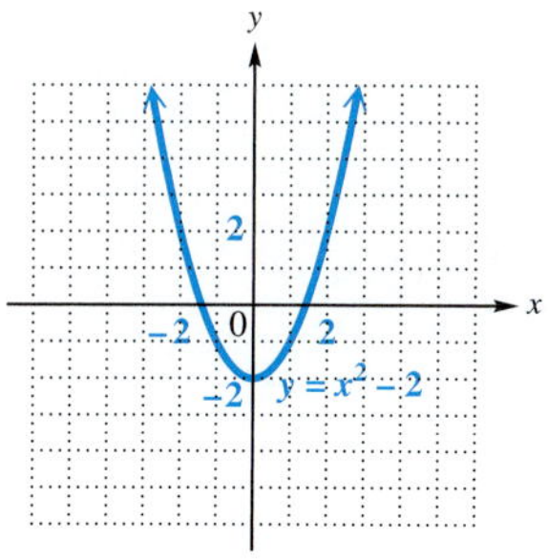

(0, −2)

9. $y = -x^2 + 2$

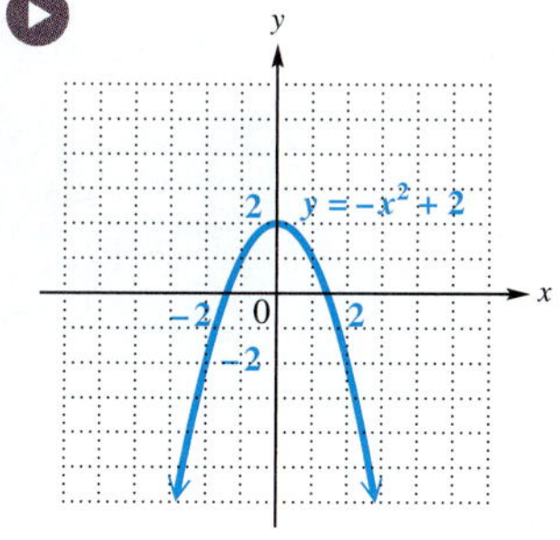

(0, 2)

10. $y = -x^2 + 4$

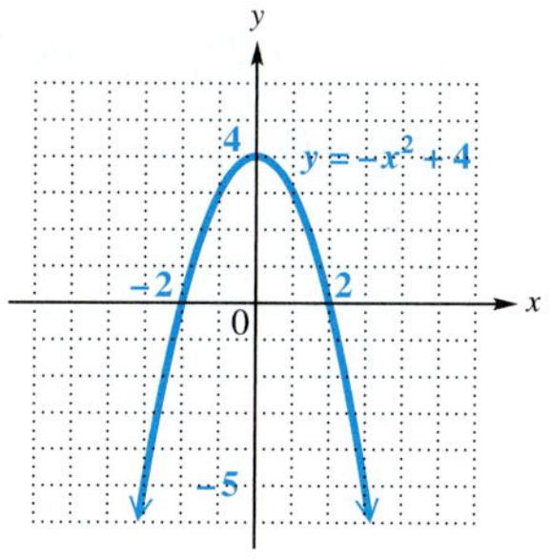

(0, 4)

11. $y = (x + 1)^2$

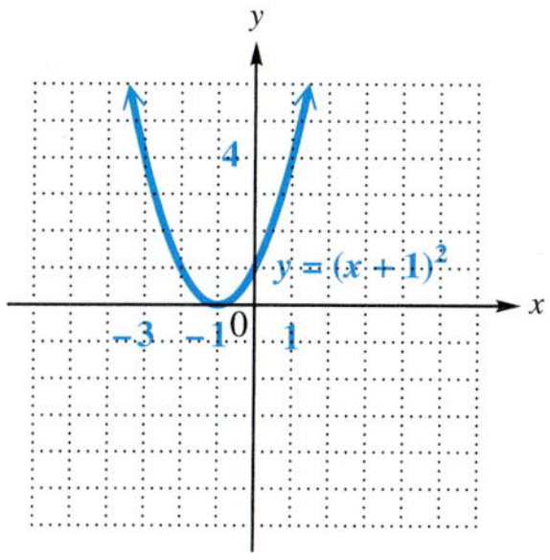

(−1, 0)

12. $y = (x - 2)^2$

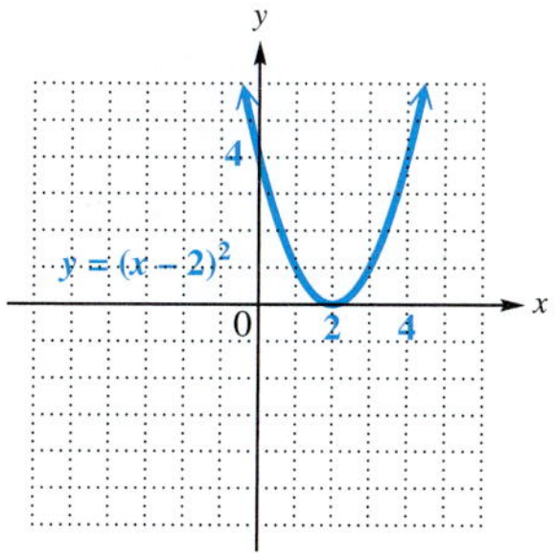

(2, 0)

13. $y = x^2 + 2x + 3$

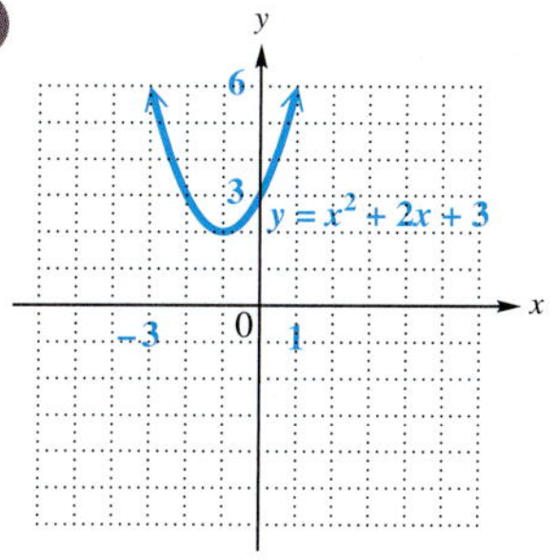

(−1, 2)

14. $y = x^2 - 4x + 3$

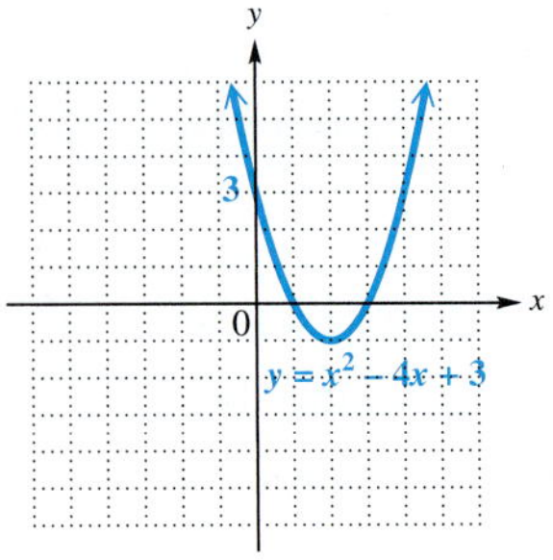

(2, −1)

15. $y = -x^2 + 6x - 5$

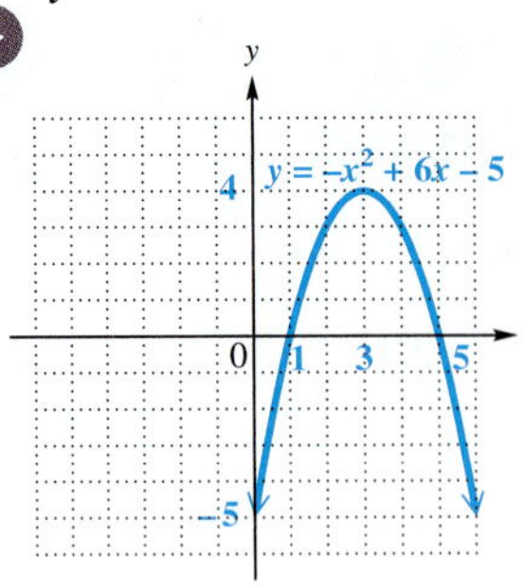

(3, 4)

16. $y = -x^2 - 4x - 3$

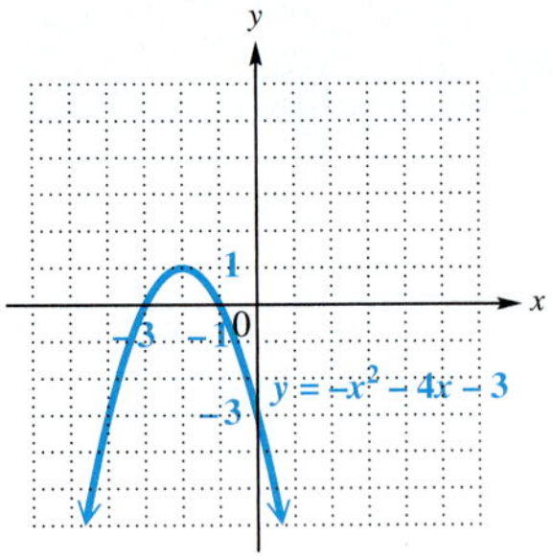

(−2, 1)

17. CONCEPT CHECK Based on your work in **Exercises 5–16**, what seems to be the direction in which the parabola

$$y = ax^2 + bx + c$$

opens if $a > 0$? If $a < 0$?

If $a > 0$, it opens upward, and if $a < 0$, it opens downward.

18. CONCEPT CHECK See **Exercises 12–14.** How many real solutions does a quadratic equation have if its corresponding graph has

(a) no x-intercepts **none**

(b) one x-intercept **one**

(c) two x-intercepts? **two**

Solve each problem.

19. The U.S. Naval Research Laboratory designed a giant radio telescope that had a diameter of 300 ft and maximum depth of 44 ft. The graph on the right below describes a cross section of this telescope. Find the equation of this parabola. (*Source:* Mar, J. and Liebowitz, H., *Structure Technology for Large Radio and Radar Telescope Systems*, The MIT Press.)

$y = \frac{11}{5625}x^2$

20. Suppose the telescope in **Exercise 19** had a diameter of 400 ft and maximum depth of 50 ft. Find the equation of this parabola.

$y = \frac{1}{800}x^2$

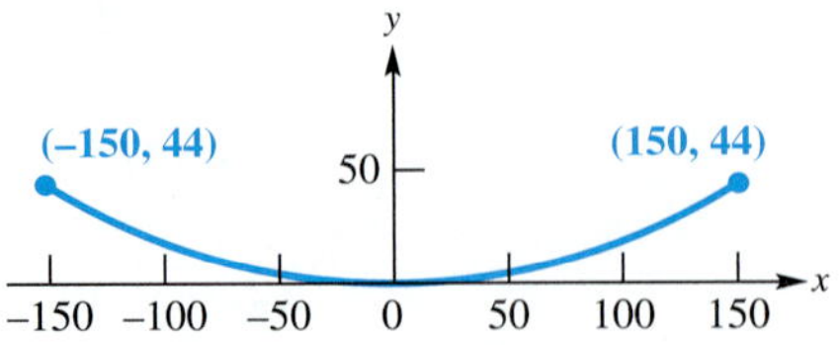

9.5 Introduction to Functions

OBJECTIVES

1. Understand the definition of a relation.
2. Understand the definition of a function.
3. Determine whether an equation represents a function.
4. Use function notation.
5. Apply the function concept in an application.

If gasoline costs \$4.00 per gal and we purchase 1 gal, then we must pay \$4.00(1) = \$4.00. For 2 gal, the cost is \$4.00(2) = \$8.00. For 3 gal, the cost is \$4.00(3) = \$12.00, and so on. Generalizing, if x represents the number of gallons, then the cost y is \$4.00x. The equation

$$y = 4.00x$$

relates the number of gallons, x, to the cost in dollars, y. The set of ordered pairs (x, y) that satisfy this equation forms a *relation.*

OBJECTIVE 1 Understand the definition of a relation. In an ordered pair (x, y), x and y are the **components** of the ordered pair. Any set of ordered pairs is a **relation.** The set of all first components in the ordered pairs of a relation is the **domain** of the relation, and the set of all second components in the ordered pairs is the **range** of the relation.

EXAMPLE 1 Identifying Domains and Ranges of Relations

Identify the domain and range of each relation.

(a) $\{(0, 1), (2, 5), (3, 8), (4, 2)\}$

This relation has

domain $\{0, 2, 3, 4\}$

and range $\{1, 5, 8, 2\}$.

The correspondence between the elements of the domain and the elements of the range is shown in **Figure 5.**

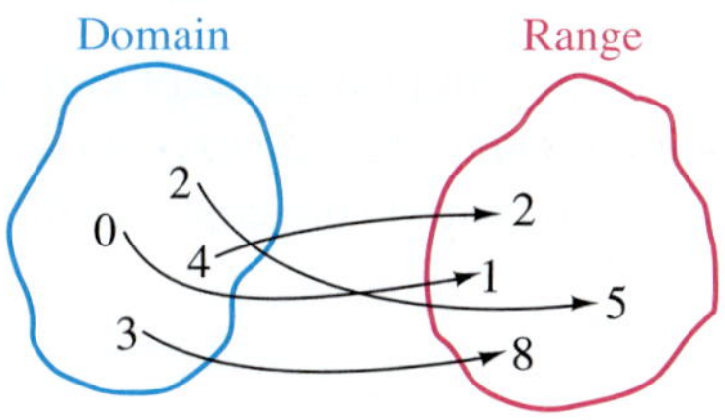

Figure 5

(b) $\{(3, 5), (3, 6), (3, 7), (3, 8)\}$

The relation has domain $\{3\}$ and range $\{5, 6, 7, 8\}$.

Work Problem 1 at the Side. ▶

1 Identify the domain and the range of each relation.

(a) $\{(5, 10), (15, 20), (25, 30), (35, 40)\}$

(b) $\{(1, 4), (2, 4), (3, 4)\}$

OBJECTIVE 2 Understand the definition of a function. We now investigate an important type of relation, called a *function.*

Function

A **function** is a set of ordered pairs in which each distinct first component corresponds to exactly one second component.

The relation in **Example 1(a)** is a function. The relation in **Example 1(b)** is *not* a function because the first component, 3, corresponds to more than one second component. If the components of the ordered pairs in **Example 1(b)** were interchanged, giving the relation

$$\{(5, 3), (6, 3), (7, 3), (8, 3)\},$$

then the relation *would* be a function. ***In that case, each domain element (first component) corresponds to exactly one range element (second component).***

Answers

1. (a) domain: $\{5, 15, 25, 35\}$; range: $\{10, 20, 30, 40\}$
   (b) domain: $\{1, 2, 3\}$; range: $\{4\}$

2 Determine whether each relation is a function.

(a) $\{(-2, 8), (-1, 1), (0, 0), (1, 1), (2, 8)\}$

(b) $\{(5, 2), (5, 1), (5, 0)\}$

(c) $\{(-1, -3), (0, 2), (3, 1), (8, 1)\}$

Answers

2. **(a)** function **(b)** not a function **(c)** function

EXAMPLE 2 **Determining Whether Relations Are Functions**

Determine whether each relation is a function.

(a) $\{(-2, 4), (-1, 1), (0, 0), (1, 1), (2, 4)\}$

Each first component appears once and only once. The relation is a function.

(b) $\{(9, 3), (9, -3), (4, 2)\}$

The first component 9 appears in two ordered pairs and corresponds to two different second components. Therefore, this relation is not a function.

◀ Work Problem **2** at the Side.

Functions often have an infinite number of ordered pairs and may be represented by equations that tell how to find the second components (outputs), given the first components (inputs). Here are some everyday examples of functions.

1. The **cost** y in dollars charged by an express mail company is a function of the **weight** x in pounds determined by the equation $y = 1.5(x - 1) + 9$.
2. In Cedar Rapids, Iowa, sales tax is 7% of the price of an item. The **tax** y on a particular item is a function of the **price** x, because $y = 0.07x$.
3. The **distance** d traveled by a car moving at a constant speed of 45 mph is a function of the **time** t. Thus, $d = 45t$.

The function concept can be illustrated by an input-output "machine," as seen in **Figure 6.** The express mail company equation $y = 1.5(x - 1) + 9$ provides an output (the cost y in dollars) for a given input (the weight x in pounds).

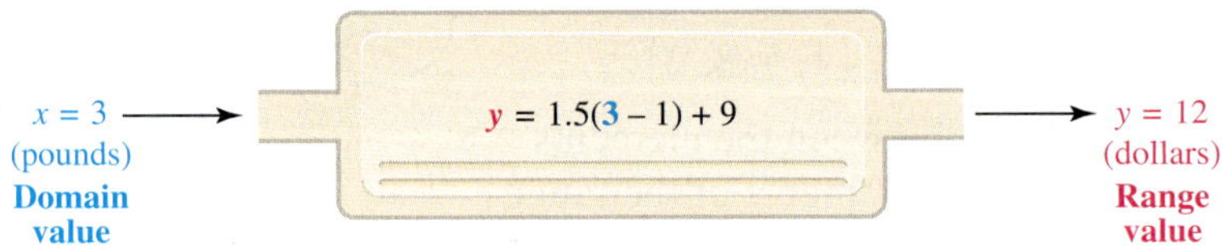

An input-output (function) machine for $y = 1.5(x - 1) + 9$

Figure 6

OBJECTIVE 3 **Determine whether an equation represents a function.** Given the graph of an equation, the definition of a function can be used to decide whether or not the graph represents a function. By definition, each x-value of a function must lead to exactly one y-value.

In **Figure 7(a),** the indicated x-value leads to two y-values, so this graph is not the graph of a function. A vertical line can be drawn that intersects this graph in more than one point.

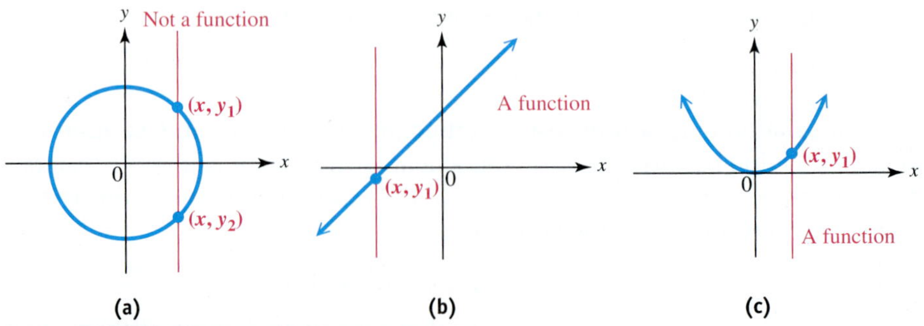

Figure 7

By contrast, in **Figure 7(b)** and **Figure 7(c)** any vertical line will intersect each graph in no more than one point, so these graphs are graphs of functions. This idea leads to the **vertical line test** for a function.

Vertical Line Test

If a vertical line intersects a graph in more than one point, then the graph is not the graph of a function.

As **Figure 7(b)** suggests, any nonvertical line is the graph of a function. ***Thus, any linear equation of the form $y = mx + b$ defines a function.*** (Recall that a vertical line has undefined slope.) Also, any vertical parabola, as in **Figure 7(c)**, is the graph of a function, so ***any quadratic equation of the form $y = ax^2 + bx + c$ (with $a \neq 0$) defines a function.***

EXAMPLE 3 **Deciding Whether Relations Are Functions**

Determine whether each relation represented by a graph or an equation is a function.

(a)

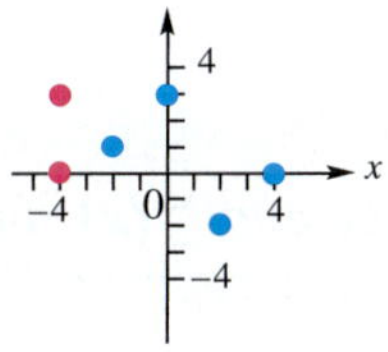

Because there are two ordered pairs with first component −4, as shown in red, this is not the graph of a function.

(b)

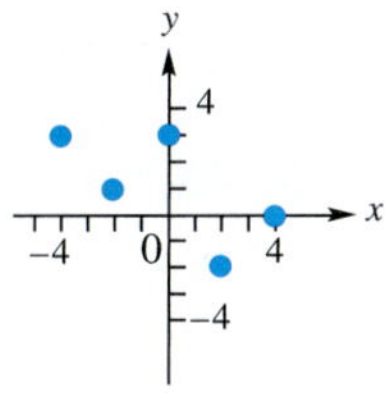

Every first component is paired with one and only one second component, and as a result, no vertical line intersects the graph in more than one point. Therefore, this is the graph of a function.

(c) $y = 2x - 9$

This linear equation is in the form $y = mx + b$. Since the graph of this equation is a line that is not vertical, the equation defines a function.

(d)

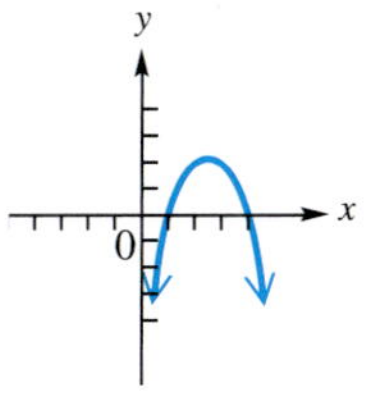

Use the vertical line test. Any vertical line intersects the graph of a vertical parabola just once, so this is the graph of a function.

(e)

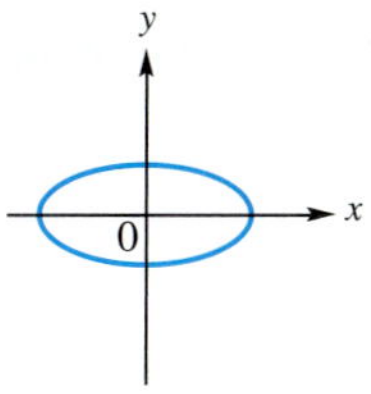

The vertical line test shows that this graph is not the graph of a function—a vertical line could intersect the graph twice.

(f) $x = 4$

The graph of $x = 4$ is a vertical line, so the equation does *not* define a function. (Every ordered pair has x-value 4.)

Work Problem 3 at the Side. ▶

3 Determine whether each relation represented by a graph or an equation is a function.

(a)

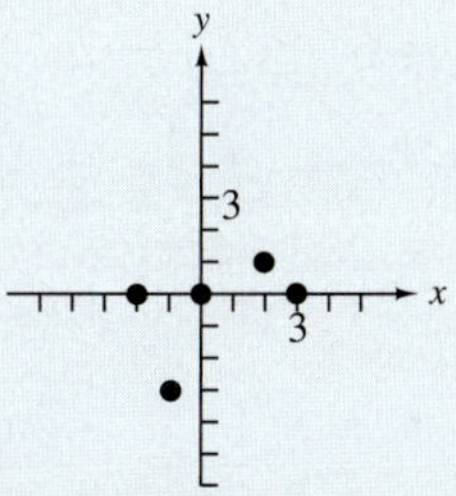

(b)

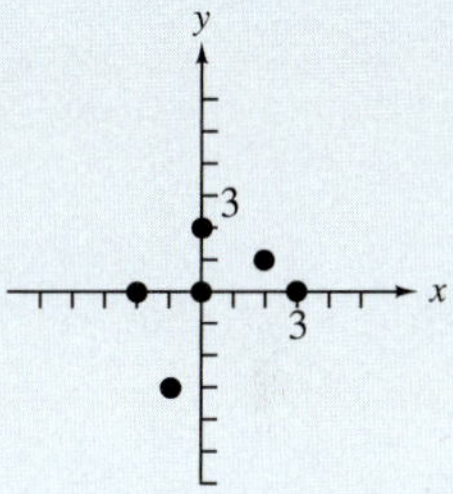

(c)

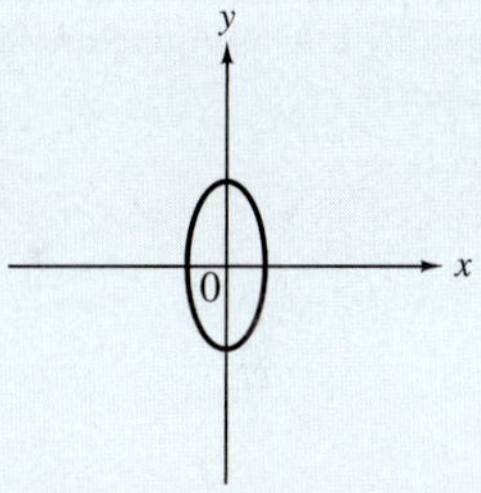

(d)

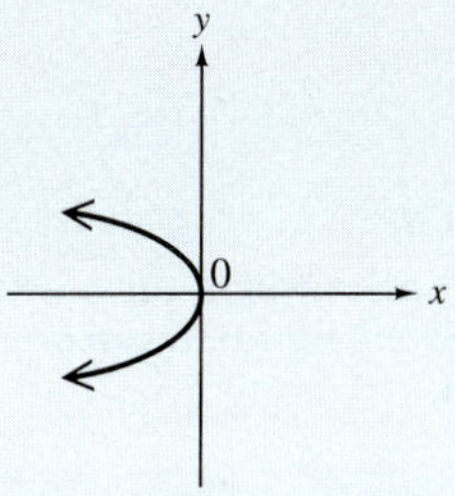

(e) $y = 3$

Answers

3. (a) function **(b)** not a function **(c)** not a function **(d)** not a function **(e)** function

4 For the function $f(x) = 6x - 2$, find each function value.

GS **(a)** $f(-1)$

$= 6(\_\_\_\_) - 2$

$= \_\_\_\_ - 2$

$= \_\_\_\_$

(b) $f(0)$

Teaching Tip

Stress that y and *f(x)* can be used interchangeably in the equation that defines a function.

(c) $f(1)$

Teaching Tip

Function notation will require several examples, as the notation can be confusing for students.

Answers

4. (a) -1; -6; -8
(b) -2 **(c)** 4

OBJECTIVE 4 Use function notation. The letters f, g, and h are commonly used to name functions. For example, the function $y = 3x + 5$ may be written

$$f(x) = 3x + 5,$$

where $f(x)$, which represents the value of f at x, is read "***f* of *x***." The notation $f(x)$ is another way of expressing the range element y for a function f. For the function $f(x) = 3x + 5$, if $x = 7$ then we find $f(7)$ as follows.

$f(7) = 3 \cdot 7 + 5$ Let $x = 7$.

$f(7) = 21 + 5$ Multiply.

$f(7) = 26$ Add.

Read this result, $f(7) = 26$, as "f of 7 equals 26." The notation $f(7)$ represents the value of y when x is 7. The statement $f(7) = 26$ says that the value of y is 26 when x is 7. It also indicates that the point $(7, 26)$ lies on the graph of f.

Similarly, to find $f(-3)$ for $f(x) = 3x + 5$, substitute -3 for x.

Use parentheses to avoid errors.

$f(-3) = 3(-3) + 5$ Let $x = -3$.

$f(-3) = -9 + 5$ Multiply.

$f(-3) = -4$ Add.

CAUTION

The notation $f(x)$ does *not* mean f times x. ***The symbol f(x) means the value of the function f when evaluated for x. It represents the y-value that corresponds to x.***

Function Notation

In the notation $f(x)$,

	f	is the name of the function,
	x	is the domain value,
and	$f(x)$	is the range value y for the domain value x.

EXAMPLE 4 Using Function Notation

For the function $f(x) = x^2 - 3$, find each function value.

(a) $f(4)$

Substitute 4 for x.

$f(x) = x^2 - 3$ Given function

$f(4) = 4^2 - 3$ Let $x = 4$.

4^2 means $4 \cdot 4$. $= 16 - 3$ Apply the exponent.

$f(4) = 13$ Subtract.

(b) $f(0) = 0^2 - 3$ Let $x = 0$.

$f(0) = 0 - 3$

$f(0) = -3$

(c) $f(-3) = (-3)^2 - 3$ Let $x = -3$.

$f(-3) = 9 - 3$ $(-3)^2$ means $-3 \cdot (-3)$.

$f(-3) = 6$

◀ Work Problem 4 at the Side.

OBJECTIVE 5 Apply the function concept in an application. Because a function assigns to each element in its domain exactly one element in its range, the function concept is used in real-data applications in which two quantities are related.

EXAMPLE 5 Applying the Function Concept to Population

Asian-American populations (in millions) are shown in the table.

ASIAN-AMERICAN POPULATION

Year	Population (in millions)
2000	11.2
2004	13.1
2006	14.9
2010	16.7

Source: U.S. Census Bureau.

(a) Use the table to write a set of ordered pairs that defines a function f.

If we choose the years as the domain elements and the populations in millions as the range elements, the information in the table can be written as a set of four ordered pairs. In set notation, the function f is

$$f = \{(2000, 11.2), (2004, 13.1), (2006, 14.9), (2010, 16.7)\}.$$

y-values are in millions.

(b) What is the domain of f? What is the range?

The domain is the set of years, or x-values.

$$\{2000, 2004, 2006, 2010\}$$

The range is the set of populations in millions, or y-values.

$$\{11.2, 13.1, 14.9, 16.7\}$$

(c) Find $f(2000)$ and $f(2006)$.

We refer to the table or the ordered pairs in part (a) to find that

$$f(\mathbf{2000}) = 11.2 \text{ million} \quad \text{and} \quad f(\mathbf{2006}) = 14.9 \text{ million.}$$

(d) For what x-value does $f(x)$ equal 16.7 million? 13.1 million?

We use the table or the ordered pairs found in part (a) to determine

$$f(\mathbf{2010}) = 16.7 \text{ million} \quad \text{and} \quad f(\mathbf{2004}) = 13.1 \text{ million.}$$

Work Problem 5 at the Side. ▶

5 The median age at first marriage for men in the United States for selected years is given in the table.

Year	Age (in years)
2004	27.4
2006	27.5
2008	27.6
2010	28.2

Source: U.S. Census Bureau.

(a) Write a set of ordered pairs that defines a function f for the data.

(b) Give the domain and range of f.

(c) Find $f(2010)$.

(d) For what x-value does $f(x)$ equal 27.5?

Answers

5. (a) $f = \{(2004, 27.4), (2006, 27.5), (2008, 27.6), (2010, 28.2)\}$

(b) domain: $\{2004, 2006, 2008, 2010\}$; range: $\{27.4, 27.5, 27.6, 28.2\}$

(c) 28.2 **(d)** 2006

9.5 Exercises

FOR EXTRA HELP

Download the MyDashBoard App

MyMathLab®

CONCEPT CHECK *Complete the following table for the function* $f(x) = x + 2$.

	x	$x + 2$	$f(x)$	(x, y)
	0	2	2	(0, 2)
1.	1	3	3	(1, 3)
2.	2	4	4	(2, 4)
3.	3	5	5	(3, 5)
4.	4	6	6	(4, 6)

5. CONCEPT CHECK Describe the graph of function f in **Exercises 1–4** if the domain is $\{0, 1, 2, 3, 4\}$.

The graph consists of the five points (0, 2), (1, 3), (2, 4), (3, 5), and (4, 6).

6. CONCEPT CHECK Describe the graph of function f in **Exercises 1–4** if the domain is the set of all real numbers.

The graph is the line through the points (0, 2), (1, 3), (2, 4), (3, 5), and (4, 6).

Determine whether each relation is a function. In Exercises 7–12, give the domain and the range. ***See Examples 1–3.***

7. $\{(-4, 3), (-2, 1), (0, 5), (-2, -8)\}$

not a function; domain: $\{-4, -2, 0\}$; range: $\{3, 1, 5, -8\}$

8. $\{(3, 7), (1, 4), (0, -2), (-1, -1), (-2, 5)\}$

function; domain: $\{3, 1, 0, -1, -2\}$; range: $\{7, 4, -2, -1, 5\}$

9.

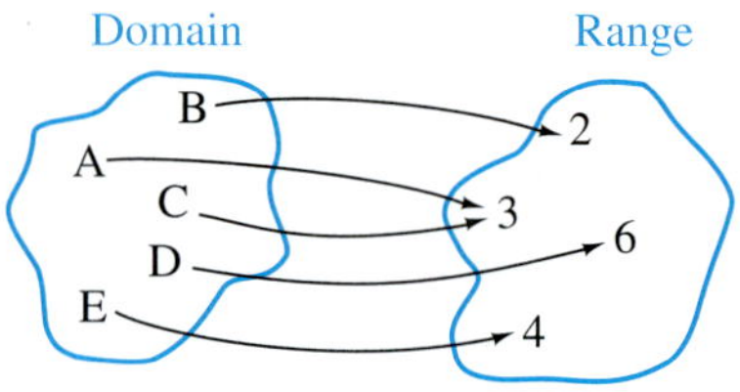

function; domain: $\{A, B, C, D, E\}$; range: $\{2, 3, 6, 4\}$

10.

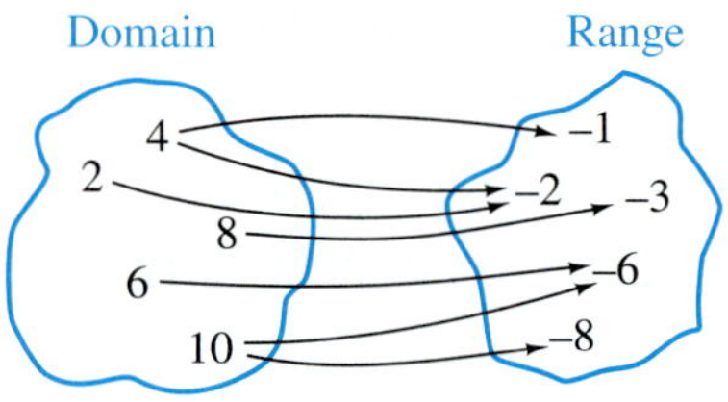

not a function; domain: $\{4, 2, 8, 6, 10\}$; range: $\{-1, -2, -3, -6, -8\}$

11.

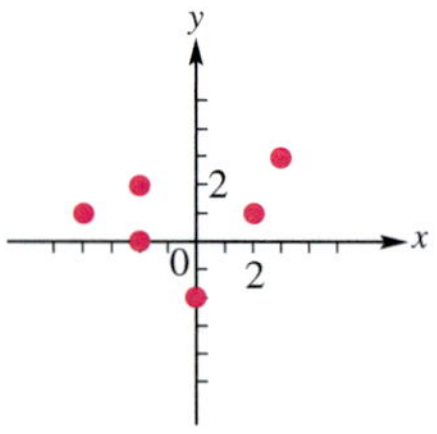

not a function; domain: $\{-4, -2, 0, 2, 3\}$; range: $\{-2, 0, 1, 2, 3\}$

12.

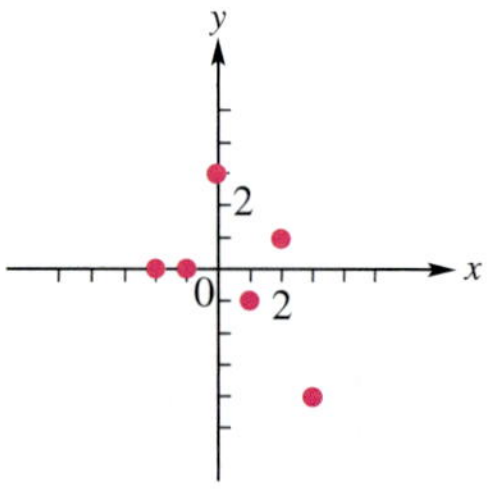

function; domain: $\{-2, -1, 0, 1, 2, 3\}$; range: $\{-4, -1, 0, 1, 3\}$

13.

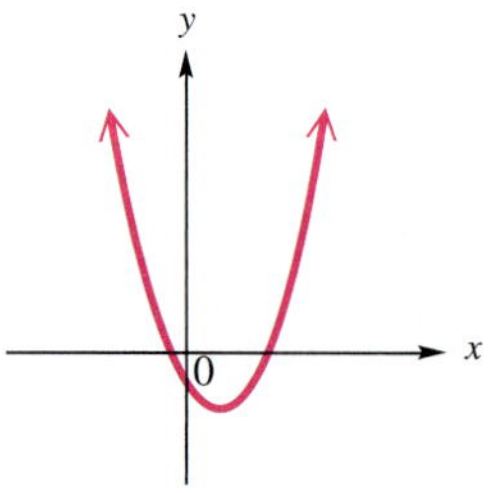

function

14.

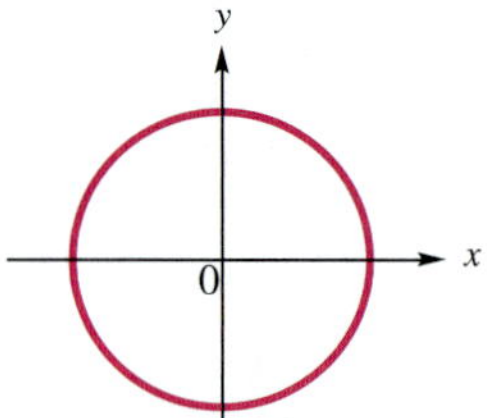

not a function

15.

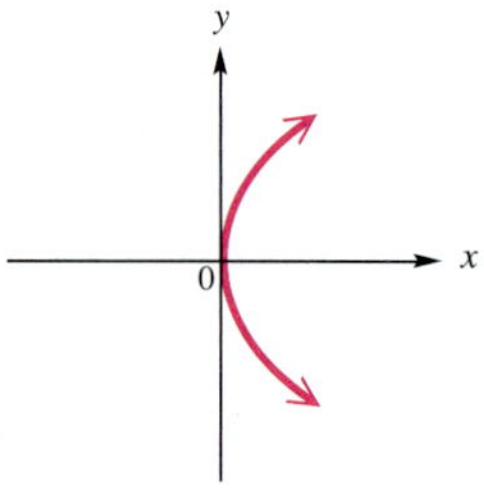

not a function

16.

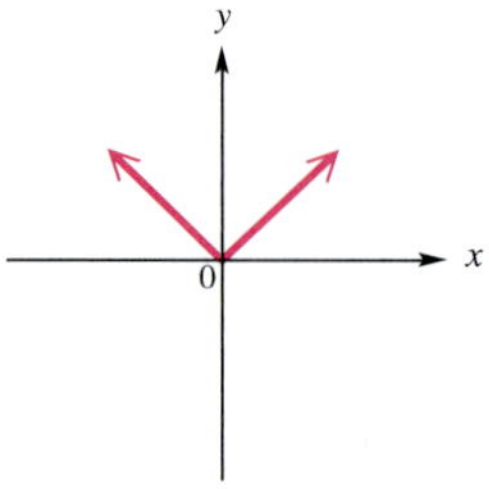

function

Decide whether each equation defines y as a function of x. ***See Example 3.***

17. $y = 5x + 3$
function

18. $y = -7x + 12$
function

19. $y = \sqrt{x}$
function

20. $y = \pm\sqrt{x}$
not a function

21. $x = -7$
not a function

22. $x = 0$
not a function

23. $y = 5$
function

24. $y = -8$
function

For each function f, find ***(a)*** *$f(2)$,* ***(b)*** *$f(0)$, and* ***(c)*** *$f(-3)$.* ***See Example 4.***

25. $f(x) = 4x + 3$
(a) 11 (b) 3 (c) -9

26. $f(x) = -3x + 5$
(a) -1 (b) 5 (c) 14

27. $f(x) = x^2 - x + 2$
(a) 4 (b) 2 (c) 14

28. $f(x) = x^3 + x$
(a) 10 (b) 0 (c) -30

29. $f(x) = |x|$

$|x|$ is read "the absolute value of x."

(a) $f(2) = |\underline{\ 2\ }|$

$f(2) = \underline{\ 2\ }$

(b) Find $f(0)$.
0

(c) Find $f(-3)$.
3

30. $f(x) = |x + 7|$

$|x + 7|$ is read "the absolute value of the quantity x plus $\underline{\ 7\ }$."

(a) $f(2) = |\underline{\ 2\ } + 7|$

$f(2) = |\underline{\ 9\ }|$

$f(2) = \underline{\ 9\ }$

(b) Find $f(0)$.
7

(c) Find $f(-3)$.
4

The graph shows the number of foreign-born residents (in millions) for selected years. Use the information in the graph for Exercises 31–35. ***See Example 5.***

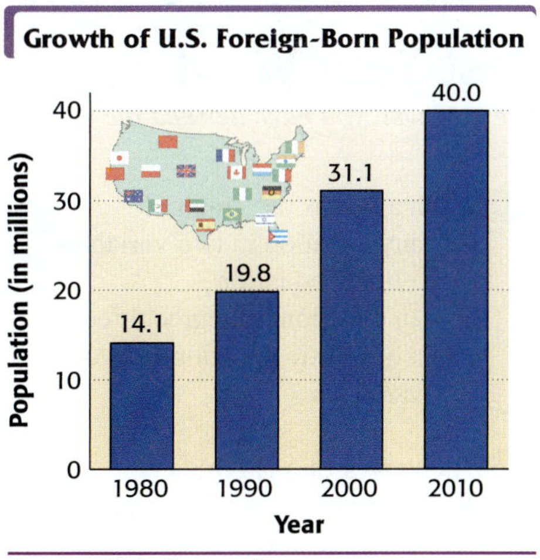

Source: U.S. Census Bureau.

31. Write the information in the graph as a set of ordered pairs. Does this set represent a function?
{(1980, 14.1), (1990, 19.8), (2000, 31.1), (2010, 40.0)}; yes

32. Suppose that g is the name given to this relation. Give the domain and range of g.
domain: {1980, 1990, 2000, 2010}; range: {14.1, 19.8, 31.1, 40.0}

33. Find $g(1980)$ and $g(2000)$.
$g(1980) = 14.1$ (million); $g(2000) = 31.1$ (million)

34. For what value of x does $g(x) = 40.0$ (million)?
2010

35. Suppose $g(2008) = 38.0$ (million). What does this tell you in the context of the application?
For the year 2008, the function gives 38.0 million foreign-born residents in the United States.

Chapter 9 Summary

Key Terms

9.4

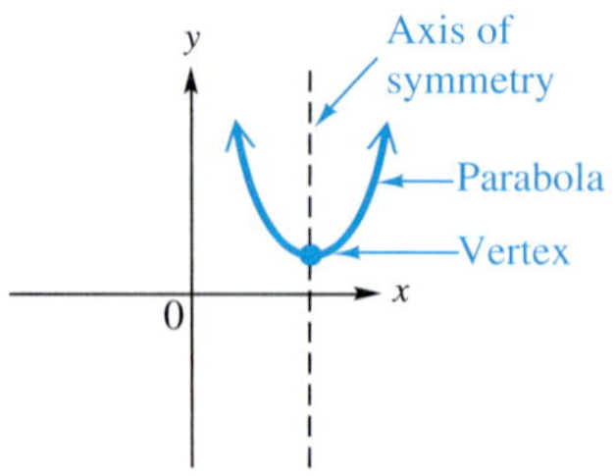

parabola The graph of the quadratic equation $y = ax^2 + bx + c$ is a parabola.

vertex The vertex of a parabola that opens upward or downward is the lowest or highest point on the graph.

axis of symmetry The axis of symmetry of a parabola that opens upward or downward is the vertical line through the vertex.

9.5

components In an ordered pair (x, y), x and y are the components.

relation Any set of ordered pairs is a relation.

domain The set of all first components in the ordered pairs of a relation is the domain of the relation.

range The set of all second components in the ordered pairs of a relation is the range of the relation.

function A function is a set of ordered pairs in which each distinct first component corresponds to exactly one second component.

New Symbols

$\pm$ positive or negative (plus or minus)

$f(x)$ function f of x

Test Your Word Power

See how well you have learned the vocabulary in this chapter.

1. A **parabola** is the graph of
   - **A.** any equation in two variables
   - **B.** a linear equation
   - **C.** an equation of degree three
   - **D.** a quadratic equation in two variables.

2. The **vertex** of a parabola is
   - **A.** the point where the graph intersects the y-axis
   - **B.** the point where the graph intersects the x-axis
   - **C.** the lowest point on a parabola that opens up or the highest point on a parabola that opens down
   - **D.** the origin.

3. A **relation** is
   - **A.** any set of ordered pairs
   - **B.** a set of ordered pairs in which each distinct first component corresponds to exactly one second component
   - **C.** two sets of ordered pairs that are related
   - **D.** a graph of ordered pairs.

4. The **domain** of a relation is
   - **A.** the set of all x- and y-values in the ordered pairs of the relation
   - **B.** the difference between the components in an ordered pair of the relation
   - **C.** the set of all first components in the ordered pairs of the relation
   - **D.** the set of all second components in the ordered pairs of the relation.

5. The **range** of a relation is
   - **A.** the set of all x- and y-values in the ordered pairs of the relation
   - **B.** the difference between the components in an ordered pair of the relation
   - **C.** the set of all first components in the ordered pairs of the relation
   - **D.** the set of all second components in the ordered pairs of the relation.

6. A **function** is
   - **A.** any set of ordered pairs
   - **B.** a set of ordered pairs in which each distinct first component corresponds to exactly one second component
   - **C.** two sets of ordered pairs that are related
   - **D.** a graph of ordered pairs.

Answers to Test Your Word Power

1. D; *Examples:* See **Figures 1–4** in **Section 9.4.**
2. C; *Example:* The graph of $y = (x + 3)^2$ has vertex $(-3, 0)$, which is the lowest point on the graph.
3. A; *Example:* $\{(0, 2), (2, 4), (3, 6), (-1, 3)\}$
4. C; *Example:* The domain in the relation given in Answer 3 is the set of x-values $\{0, 2, 3, -1\}$.
5. D; *Example*: The range of the relation given in Answer 3 is the set of y-values $\{2, 4, 6, 3\}$.
6. B; *Example*: The relation given in Answer 3 is a function since each x-value corresponds to exactly one y-value.

Quick Review

Concepts	Examples
9.1 Solving Quadratic Equations by the Square Root Property **Square Root Property of Equations** If k is positive, and if $x^2 = k$, then $$x = \sqrt{k} \quad \text{or} \quad x = -\sqrt{k}.$$ The solution set $\{-\sqrt{k}, \sqrt{k}\}$ can also be written $\{\pm\sqrt{k}\}$.	Solve $(2x + 1)^2 = 5$. $2x + 1 = \sqrt{5}$ or $2x + 1 = -\sqrt{5}$ $2x = -1 + \sqrt{5}$ or $2x = -1 - \sqrt{5}$ $x = \frac{-1 + \sqrt{5}}{2}$ or $x = \frac{-1 - \sqrt{5}}{2}$ Solution set: $\left\{\frac{-1 + \sqrt{5}}{2}, \frac{-1 - \sqrt{5}}{2}\right\}$
9.2 Solving Quadratic Equations by Completing the Square **Solving a Quadratic Equation by Completing the Square** *Step 1* If the coefficient of the second-degree term is 1, go to Step 2. If it is not 1, divide each side of the equation by this coefficient. *Step 2* Make sure that all variable terms are on one side of the equation and all constant terms are on the other. *Step 3* Take half the coefficient of x, square it, and add the square to each side of the equation. Factor the variable side, and combine terms on the other side. *Step 4* Use the square root property to solve the equation.	Solve $2x^2 + 4x - 1 = 0$. $x^2 + 2x - \frac{1}{2} = 0$ Divide by 2. $x^2 + 2x = \frac{1}{2}$ Add $\frac{1}{2}$. $x^2 + 2x + 1 = \frac{1}{2} + 1$ $\left[\frac{1}{2}(2)\right]^2 = 1^2 = 1$ $(x + 1)^2 = \frac{3}{2}$ Factor. Add. $x + 1 = \sqrt{\frac{3}{2}}$ or $x + 1 = -\sqrt{\frac{3}{2}}$ $x + 1 = \frac{\sqrt{3} \cdot \sqrt{2}}{\sqrt{2} \cdot \sqrt{2}}$ or $x + 1 = -\frac{\sqrt{3} \cdot \sqrt{2}}{\sqrt{2} \cdot \sqrt{2}}$ $x + 1 = \frac{\sqrt{6}}{2}$ or $x + 1 = -\frac{\sqrt{6}}{2}$ $x = -1 + \frac{\sqrt{6}}{2}$ or $x = -1 - \frac{\sqrt{6}}{2}$ $x = \frac{-2}{2} + \frac{\sqrt{6}}{2}$ or $x = \frac{-2}{2} - \frac{\sqrt{6}}{2}$ $x = \frac{-2 + \sqrt{6}}{2}$ or $x = \frac{-2 - \sqrt{6}}{2}$ Solution set: $\left\{\frac{-2 + \sqrt{6}}{2}, \frac{-2 - \sqrt{6}}{2}\right\}$

Concepts	Examples

9.3 Solving Quadratic Equations by the Quadratic Formula

Quadratic Formula

The solutions of $ax^2 + bx + c = 0$ (with $a \neq 0$) are

$$x = \frac{-b \pm \sqrt{b^2 - 4ac}}{2a}.$$

Solve $3x^2 - 4x - 2 = 0$.

$$x = \frac{-(-4) \pm \sqrt{(-4)^2 - 4(3)(-2)}}{2(3)}$$ Let $a = 3$, $b = -4$, $c = -2$.

$$x = \frac{4 \pm \sqrt{40}}{6}$$ Simplify.

$$x = \frac{4 \pm 2\sqrt{10}}{6}$$ $\sqrt{40} = \sqrt{4} \cdot \sqrt{10} = 2\sqrt{10}$

$$x = \frac{2(2 \pm \sqrt{10})}{2(3)}$$ Factor.

$$x = \frac{2 \pm \sqrt{10}}{3}$$ Divide out 2.

Solution set: $\left\{\frac{2 + \sqrt{10}}{3}, \frac{2 - \sqrt{10}}{3}\right\}$

9.4 Graphing Quadratic Equations

To graph $y = ax^2 + bx + c$, follow these steps.

Step 1 Find the vertex. Use $x = -\frac{b}{2a}$ and find y by substituting this value for x in the equation.

Graph $y = 2x^2 - 5x - 3$.

$$x = -\frac{b}{2a} = -\frac{-5}{2(2)} = \frac{5}{4}$$ $a = 2, b = -5$

$$y = 2\left(\frac{5}{4}\right)^2 - 5\left(\frac{5}{4}\right) - 3$$ Let $x = \frac{5}{4}$.

$$y = 2\left(\frac{25}{16}\right) - \frac{25}{4} - 3$$ Apply the exponent. Multiply.

$$y = \frac{25}{8} - \frac{50}{8} - \frac{24}{8}$$ Multiply. Find a common denominator.

$$y = -\frac{49}{8}$$ Subtract.

The vertex is $\left(\frac{5}{4}, -\frac{49}{8}\right)$.

Step 2 Let $x = 0$ to find the y-intercept.

$$y = 2(0)^2 - 5(0) - 3$$ Let $x = 0$.

$$y = -3$$ Simplify.

The y-intercept is $(0, -3)$.

Step 3 Let $y = 0$ and solve to find the x-intercepts (if they exist).

$$0 = 2x^2 - 5x - 3$$ Let $y = 0$.

$$0 = (2x + 1)(x - 3)$$ Factor.

$2x + 1 = 0$ or $x - 3 = 0$ Zero-factor property

$2x = -1$ or $x = 3$ Solve each equation.

$$x = -\frac{1}{2}$$

The x-intercepts are $\left(-\frac{1}{2}, 0\right)$ and $(3, 0)$.

Concepts	Examples

Step 4 Plot the intercepts and the vertex.

Step 5 Find and plot additional ordered pairs near the vertex and intercepts as needed.

x	y
$-\frac{1}{2}$	0
0	-3
1	-6
$\frac{5}{4}$	$-\frac{49}{8}$
2	-5
3	0

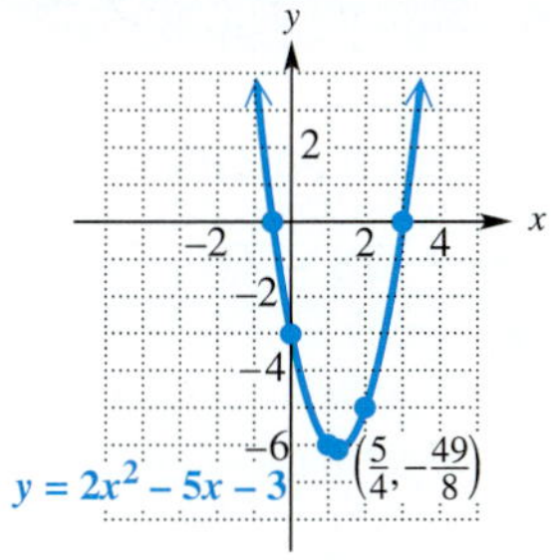

9.5 Introduction to Functions

Vertical Line Test
If a vertical line intersects a graph in more than one point, then the graph is not the graph of a function.

By the vertical line test, the graph shown is not the graph of a function.

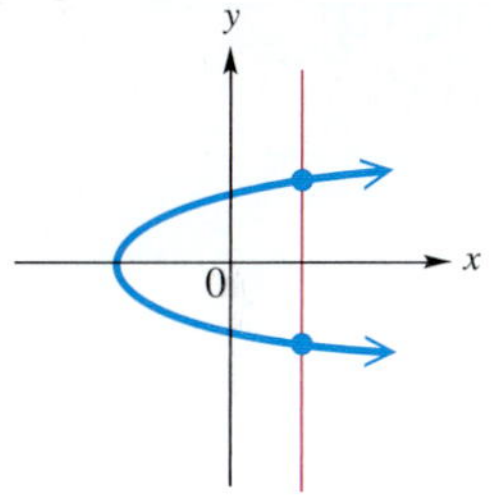

The set of all first components in the ordered pairs of a relation is the **domain** of the relation, and the set of all second components in the ordered pairs is the **range** of the relation.

The function

$$\{(10, 5), (20, 15), (30, 25)\}$$

has domain $\{10, 20, 30\}$ and range $\{5, 15, 25\}$.

To find $f(x)$ for a specific value of x, replace x by that value in the expression for the function f.

If $f(x) = 2x + 7$, then we find $f(3)$ as follows.

$f(x) = 2x + 7$

$f(3) = 2(3) + 7$ Let $x = 3$.

$f(3) = 13$ Multiply. Add.

Chapter 9 Review Exercises

9.1 *Solve each equation by using the zero-factor property.*

1. $x^2 + 3x - 28 = 0$ $\{-7, 4\}$

2. $x^2 + 14x + 45 = 0$ $\{-9, -5\}$

3. $2z^2 + 7z = 15$ $\left\{-5, \frac{3}{2}\right\}$

4. $r^2 - 169 = 0$ $\{-13, 13\}$

Solve each equation by using the square root property. Express all radicals in simplest form.

5. $y^2 = 144$ $\{-12, 12\}$

6. $x^2 = 37$ $\{-\sqrt{37}, \sqrt{37}\}$

7. $m^2 = 128$ $\{-8\sqrt{2}, 8\sqrt{2}\}$

8. $(k + 2)^2 = 25$ $\{-7, 3\}$

9. $(r - 3)^2 = 10$ $\{3 + \sqrt{10}, 3 - \sqrt{10}\}$

10. $(2p + 1)^2 = 14$ $\left\{\frac{-1 + \sqrt{14}}{2}, \frac{-1 - \sqrt{14}}{2}\right\}$

11. $(3k + 2)^2 = -3$ $\emptyset$

12. $(3x + 5)^2 = 0$ $\left\{-\frac{5}{3}\right\}$

9.2 *Solve each equation by completing the square.*

13. $m^2 + 6m + 5 = 0$ $\{-5, -1\}$

14. $p^2 + 4p = 7$ $\{-2 + \sqrt{11}, -2 - \sqrt{11}\}$

15. $-x^2 + 5 = 2x$ $\{-1 + \sqrt{6}, -1 - \sqrt{6}\}$

16. $2x^2 - 3 = -8x$ $\left\{\frac{-4 + \sqrt{22}}{2}, \frac{-4 - \sqrt{22}}{2}\right\}$

17. $4(x^2 + 7x) + 29 = -20$ $\left\{-\frac{7}{2}\right\}$

18. $(4x + 1)(x - 1) = -7$ $\emptyset$

Solve each problem.

19. If an object is projected upward on Earth from a height of 50 ft, with an initial velocity of 32 ft per sec, then its height h after t seconds is given by

$$h = -16t^2 + 32t + 50,$$

where h is in feet. At what time(s) will it be at a height of 30 ft? **2.5 sec**

20. Find the lengths of the three sides of the right triangle shown. **6, 8, 10**

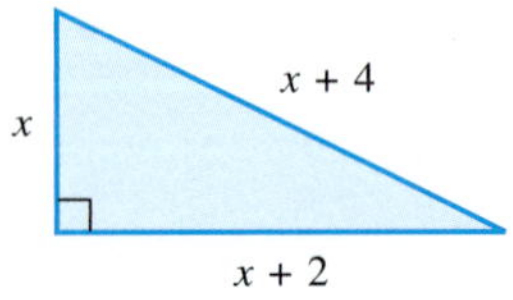

Solve each equation by using the quadratic formula.

21. $m^2 - 5m - 36 = 0$
$\{-4, 9\}$

22. $5r^2 = 14r$
$\left\{0, \frac{14}{5}\right\}$

23. $2x^2 - x + 3 = 0$
$\emptyset$

24. $4w^2 - 12w + 9 = 0$
$\left\{\frac{3}{2}\right\}$

25. $-4x^2 - 2x + 7 = 0$
$\left\{\frac{-1 + \sqrt{29}}{4}, \frac{-1 - \sqrt{29}}{4}\right\}$

26. $2x^2 + 8 = 4x + 11$
$\left\{\frac{2 + \sqrt{10}}{2}, \frac{2 - \sqrt{10}}{2}\right\}$

27. $x(5x - 1) = 1$
$\left\{\frac{1 + \sqrt{21}}{10}, \frac{1 - \sqrt{21}}{10}\right\}$

28. $\frac{1}{4}x^2 = 2 - \frac{3}{4}x$
$\left\{\frac{-3 + \sqrt{41}}{2}, \frac{-3 - \sqrt{41}}{2}\right\}$

29. $\frac{1}{2}x^2 + 3x = 5$
$\{-3 + \sqrt{19}, -3 - \sqrt{19}\}$

30. $0.2x^2 = 0.4x - 0.1$
$\left\{\frac{2 + \sqrt{2}}{2}, \frac{2 - \sqrt{2}}{2}\right\}$

9.4 *Sketch the graph of each equation. Identify each vertex.*

31. $y = -x^2 + 5$

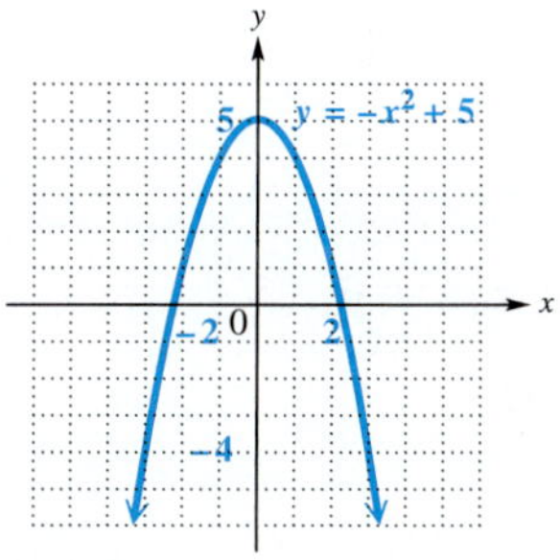

$(0, 5)$

32. $y = (x - 1)^2$

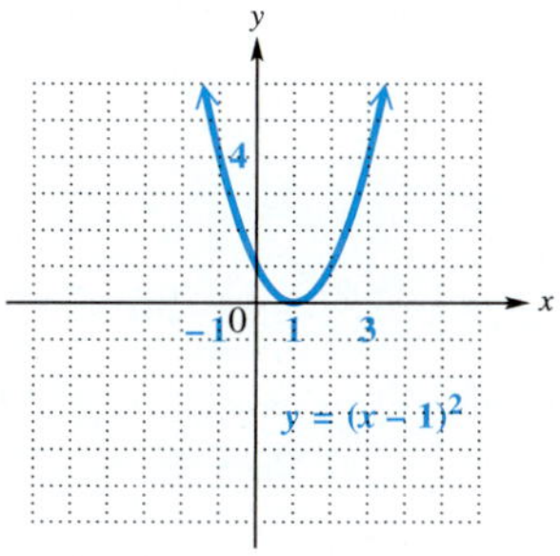

$(1, 0)$

33. $y = -x^2 + 2x + 3$

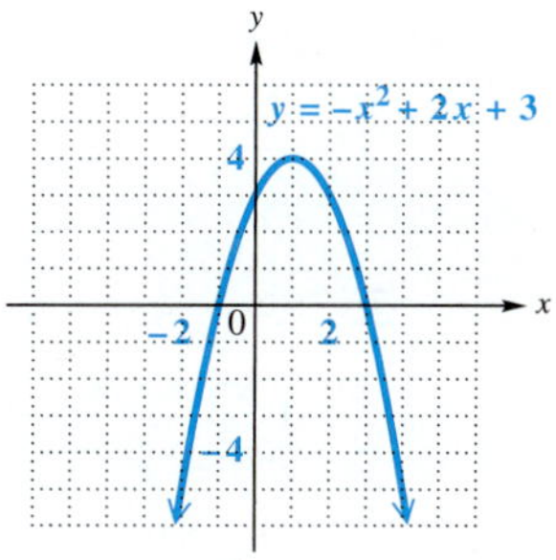

$(1, 4)$

34. $y = x^2 + 4x + 2$

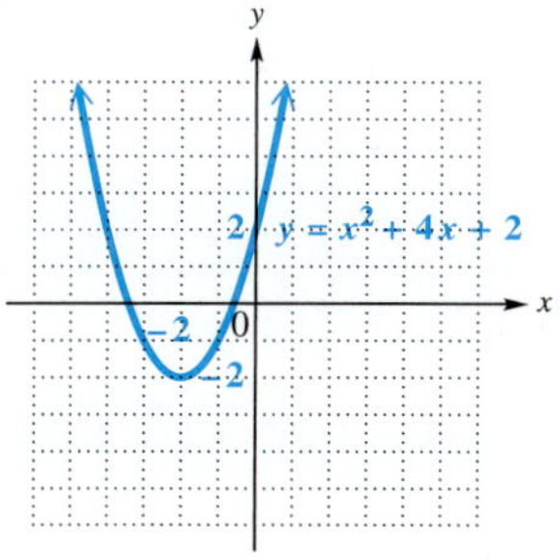

$(-2, -2)$

9.5 *Determine whether each relation is or is not a function. In Exercises 35 and 36, give the domain and the range.*

35. $\{(-2, 4), (0, 8), (2, 5), (2, 3)\}$
not a function; domain: $\{-2, 0, 2\}$;
range: $\{4, 8, 5, 3\}$

36. $\{(8, 3), (7, 4), (6, 5), (5, 6), (4, 7)\}$
function; domain: $\{8, 7, 6, 5, 4\}$;
range: $\{3, 4, 5, 6, 7\}$

37.

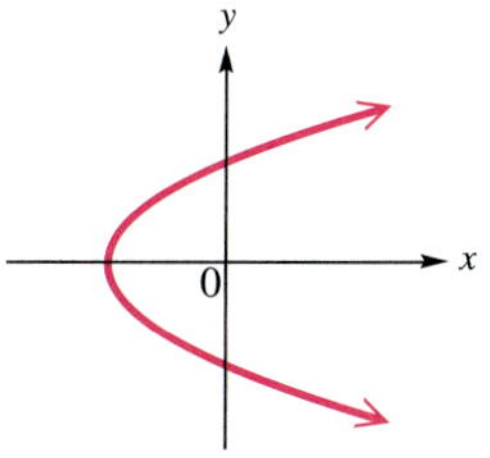

not a function

38.

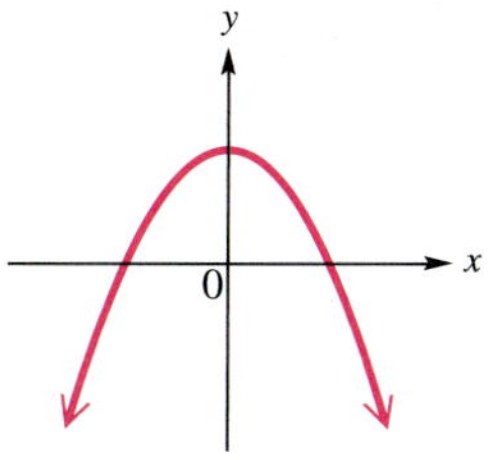

function

39. $2x + 3y = 12$
function

40. $y = x^2$
function

*Find **(a)** $f(2)$ and **(b)** $f(-1)$.*

41. $f(x) = 3x + 2$
(a) 8 (b) -1

42. $f(x) = 2x^2 - 1$
(a) 7 (b) 1

43. $f(x) = |x + 3|$
(a) 5 (b) 2

44. $f(x) = -x^2 + 2x - 3$
(a) -3 (b) -6

Mixed Review Exercises

Solve by any method.

45. $(2t-1)(t+1)=54$

$\left\{-\frac{11}{2}, 5\right\}$

46. $(2p+1)^2=100$

$\left\{-\frac{11}{2}, \frac{9}{2}\right\}$

47. $(k+2)(k-1)=3$

$\left\{\frac{-1+\sqrt{21}}{2}, \frac{-1-\sqrt{21}}{2}\right\}$

48. $6t^2+7t-3=0$

$\left\{-\frac{3}{2}, \frac{1}{3}\right\}$

49. $2x^2+3x+2=x^2-2x$

$\left\{\frac{-5+\sqrt{17}}{2}, \frac{-5-\sqrt{17}}{2}\right\}$

50. $x^2+2x+5=7$

$\{-1+\sqrt{3}, -1-\sqrt{3}\}$

51. $m^2-4m+10=0$

$\emptyset$

52. $k^2-9k+10=0$

$\left\{\frac{9+\sqrt{41}}{2}, \frac{9-\sqrt{41}}{2}\right\}$

53. $(5x+6)^2=0$

$\left\{-\frac{6}{5}\right\}$

54. $\frac{1}{2}r^2=\frac{7}{2}-r$

$\{-1+2\sqrt{2}, -1-2\sqrt{2}\}$

55. $x^2+4x=1$

$\{-2+\sqrt{5}, -2-\sqrt{5}\}$

56. $7x^2-8=5x^2+8$

$\{-2\sqrt{2}, 2\sqrt{2}\}$

57. Consider the quadratic equation $x^2-9=0$.

(a) Solve the equation by factoring. $\{-3, 3\}$

(b) Solve the equation by the square root property. $\{-3, 3\}$

(c) Solve the equation by the quadratic formula. $\{-3, 3\}$

(d) Compare your answers. If a quadratic equation can be solved by both factoring and the quadratic formula, should we always get the same results? Explain.

We will always get the same results, no matter which method of solution is used.

58. CONCEPT CHECK A student writes the quadratic formula as

$$x=-b\pm\sqrt{\frac{b^2-4ac}{2a}}.$$

What Went Wrong? Explain the error, and give the correct formula.

The fraction bar should be under both $-b$ and $\sqrt{b^2-4ac}$. The term $2a$ should not be in the radicand. The correct formula is

$$x=\frac{-b\pm\sqrt{b^2-4ac}}{2a}.$$

Chapter 9 Test

CHAPTER Test Prep VIDEO — The Chapter Test Prep Videos with test solutions are available on DVD, in MyMathLab, and on YouTube—search "LialIntroductoryAlg" and click on "Channels."

Solve by using the square root property.

1. $x^2 = 39$

$\{-\sqrt{39}, \sqrt{39}\}$

2. $(x + 3)^2 = 64$

$\{-11, 5\}$

3. $(4x + 3)^2 = 24$

$\left\{\frac{-3 + 2\sqrt{6}}{4}, \frac{-3 - 2\sqrt{6}}{4}\right\}$

Solve by completing the square.

4. $x^2 - 4x = 6$

$\{2 + \sqrt{10}, 2 - \sqrt{10}\}$

5. $2x^2 + 12x - 3 = 0$

$\left\{\frac{-6 + \sqrt{42}}{2}, \frac{-6 - \sqrt{42}}{2}\right\}$

Solve by the quadratic formula.

6. $2x^2 + 5x - 3 = 0$

$\left\{-3, \frac{1}{2}\right\}$

7. $3w^2 + 2 = 6w$

$\left\{\frac{3 + \sqrt{3}}{3}, \frac{3 - \sqrt{3}}{3}\right\}$

8. $4x^2 + 8x + 11 = 0$

$\emptyset$

9. $t^2 - \frac{5}{3}t + \frac{1}{3} = 0$

$\left\{\frac{5 + \sqrt{13}}{6}, \frac{5 - \sqrt{13}}{6}\right\}$

Solve by the method of your choice.

10. $p^2 - 2p - 1 = 0$

$\{1 + \sqrt{2}, 1 - \sqrt{2}\}$

11. $(2x + 1)^2 = 18$

$\left\{\frac{-1 + 3\sqrt{2}}{2}, \frac{-1 - 3\sqrt{2}}{2}\right\}$

12. $(x - 5)(2x - 1) = 1$

$\left\{\frac{11 + \sqrt{89}}{4}, \frac{11 - \sqrt{89}}{4}\right\}$

13. $t^2 + 25 = 10t$

$\{5\}$

Solve each problem.

14. If an object is projected into the air from ground level with an initial velocity of 64 ft per sec, its altitude (height) s in feet after t seconds is given by the formula

$$s = -16t^2 + 64t.$$

At what time(s) will the object be at a height of 64 ft?

2 sec

15. Find the lengths of the three sides of the right triangle.

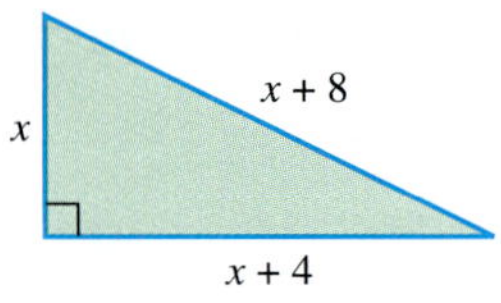

12, 16, 20

Sketch the graph of each equation. Identify each vertex.

16. $y = (x - 3)^2$

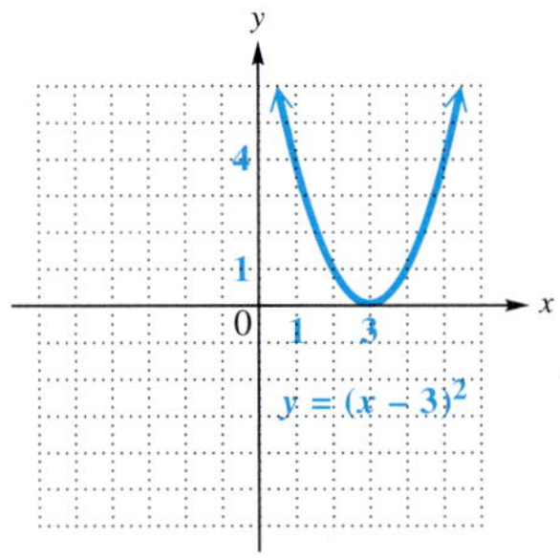

(3, 0)

17. $y = -x^2 - 2x - 4$

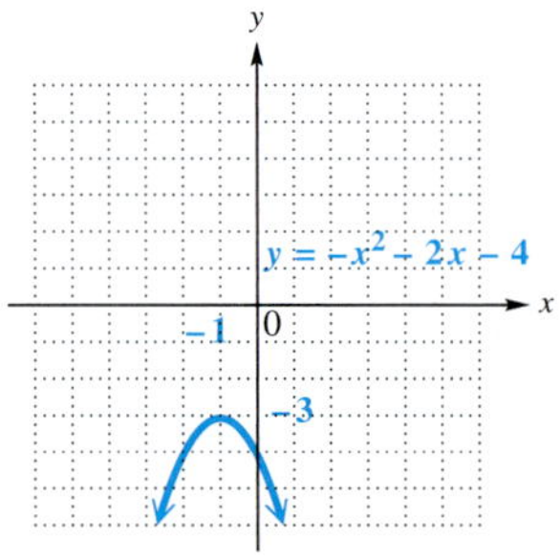

(−1, −3)

18. Determine whether each relation is a function. If it is, give the domain and the range.

(a) $\{(2, 3), (2, 4), (2, 5)\}$

not a function

(b) $\{(0, 2), (1, 2), (2, 2)\}$

function; domain: {0, 1, 2}; range: {2}

19. Use the vertical line test to determine whether the graph is that of a function.

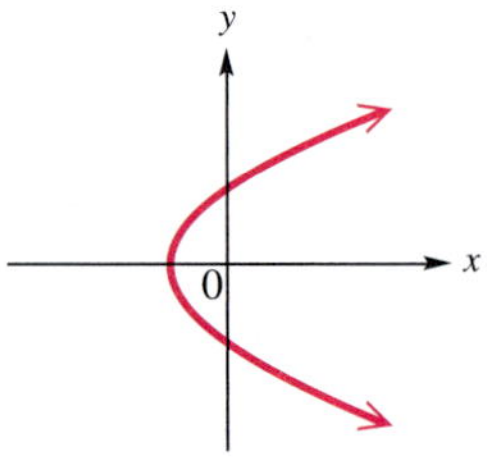

not a function

20. If $f(x) = 3x + 7$, find $f(-2)$.

1

Chapters R–9 Cumulative Review Exercises

Perform the indicated operations, wherever possible.

1. $\dfrac{-4 \cdot 3^2 + 2 \cdot 3}{2 - 4 \cdot 1}$

15

2. $|-3| - |1 - 6|$

-2

3. $5(4m - 2) - (m + 7)$

$19m - 17$

Solve each equation.

4. $6x - 5 = 13$

$\{3\}$

5. $3k - 9k - 8k + 6 = -64$

$\{5\}$

6. $2(m - 1) - 6(3 - m) = -4$

$\{2\}$

7. The perimeter of a basketball court is 288 ft. The width of the court is 44 ft less than the length. What are the dimensions of the court?

width: 50 ft; length: 94 ft

8. Find the measures of the marked angles.

100°; 80°

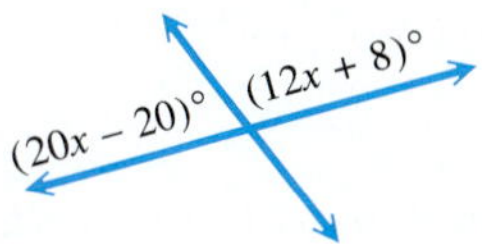

9. Solve the formula $P = 2L + 2W$ for L.

$L = \dfrac{P - 2W}{2}$, or $L = \dfrac{P}{2} - W$

10. Solve the inequality. Graph the solutions.

$-9p + 2(8 - p) - 6 \geq 4p - 50$ $(-\infty, 4]$

Graph the following.

11. $2x + 3y = 6$

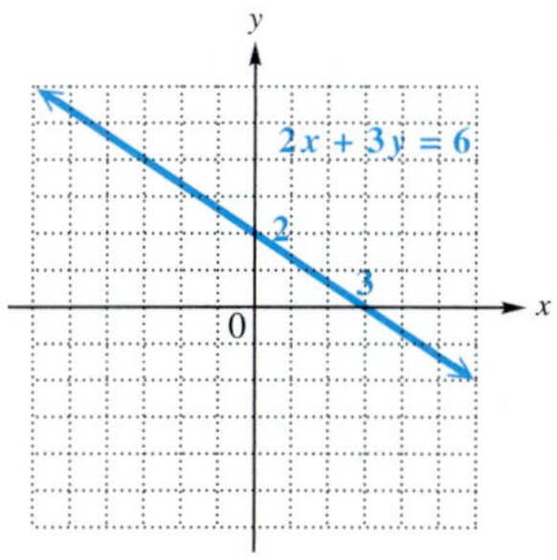

12. $y = 3$

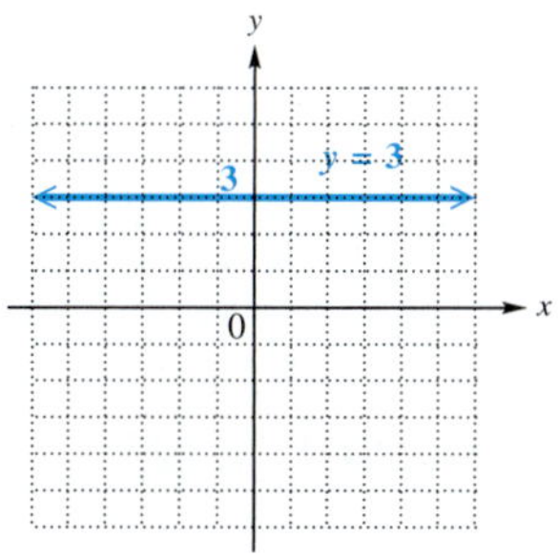

13. $2x - 5y < 10$

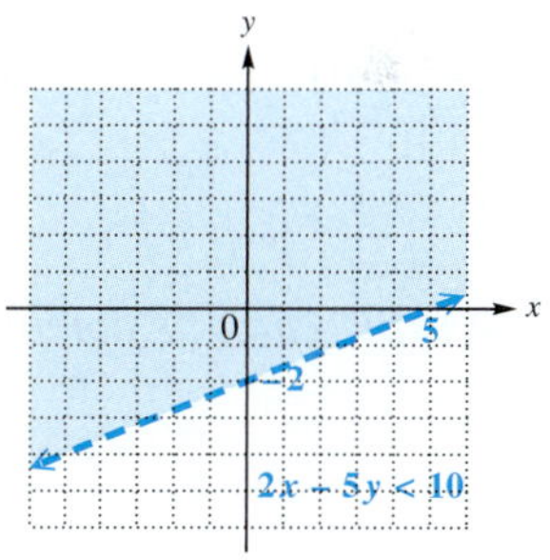

14. Find the slope of the line through $(-1, 4)$ and $(5, 2)$.

$-\dfrac{1}{3}$

15. Write an equation of a line with slope 2 and y-intercept $(0, 3)$. Give it in the form $Ax + By = C$.

$2x - y = -3$

Solve each system of equations.

16. $2x + y = -4$
$-3x + 2y = 13$

$\{(-3, 2)\}$

17. $3x - 5y = 8$
$-6x + 10y = 16$

$\emptyset$

18. Based on January 2012 prices on a Pay-as-You-Go Internet site, we could purchase 3 Motorola Theory cell phones and 2 Motorola i412 cell phones for \$379.95. We could also purchase 2 Theory cell phones and 3 i412 phones for \$369.95. Find the price for a single phone of each model. (*Source:* www.boostmobile.com)

Motorola Theory: \$79.99; Motorola i412: \$69.99

Simplify. Write answers with positive exponents.

19. $(3^2 \cdot x^{-4})^{-1}$ $\frac{x^4}{9}$

20. $\left(\frac{b^{-3}c^4}{b^5c^3}\right)^{-2}$ $\frac{b^{16}}{c^2}$

21. $\left(\frac{5}{3}\right)^{-3}$ $\frac{27}{125}$

Perform the indicated operations.

22. $(5x^5 - 9x^4 + 8x^2) - (9x^2 + 8x^4 - 3x^5)$ $8x^5 - 17x^4 - x^2$

23. $(2x - 5)(x^3 + 3x^2 - 2x - 4)$ $2x^4 + x^3 - 19x^2 + 2x + 20$

Factor as completely as possible.

24. $16x^3 - 48x^2y$ $16x^2(x - 3y)$

25. $2a^2 - 5a - 3$ $(2a + 1)(a - 3)$

26. $25m^2 - 20m + 4$ $(5m - 2)^2$

27. Solve $3x^2 = x + 4$. $\left\{-1, \frac{4}{3}\right\}$

28. The length of a rectangle is 2.5 times its width. The area is 1000 m^2. Find the length. 50 m

Perform the indicated operations. Write all answers in lowest terms.

29. $\frac{2}{a-3} \div \frac{5}{2a-6}$ $\frac{4}{5}$

30. $\frac{1}{k} - \frac{2}{k-1}$ $\frac{-k-1}{k(k-1)}$

31. $\frac{6 + \frac{1}{x}}{3 - \frac{1}{x}}$ $\frac{6x+1}{3x-1}$

Solve each equation.

32. $\frac{1}{x+3} + \frac{1}{x} = \frac{7}{10}$ $\left\{-\frac{15}{7}, 2\right\}$

33. $\sqrt{x+2} = x - 4$ $\{7\}$

34. $2x^2 - 2x = 1$ $\left\{\frac{1+\sqrt{3}}{2}, \frac{1-\sqrt{3}}{2}\right\}$

Simplify each expression as completely as possible.

35. $\frac{6\sqrt{6}}{\sqrt{5}}$ $\frac{6\sqrt{30}}{5}$

36. $3\sqrt{5} - 2\sqrt{20} + \sqrt{125}$ $4\sqrt{5}$

37. $\sqrt[3]{16a^3b^4} - \sqrt[3]{54a^3b^4}$ $-ab\sqrt[3]{2b}$

38. Graph the equation $y = x^2 - 4x$. Identify the vertex. What are the x-intercepts?

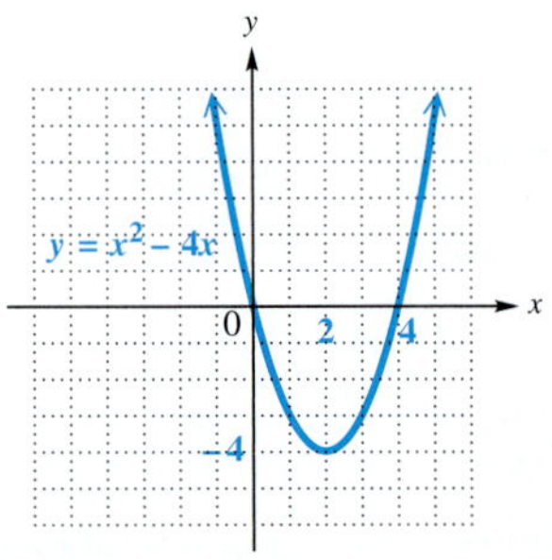

(2, −4); (0, 0) and (4, 0)

39. (a) Is $\{(0, 4), (1, 2), (3, 5)\}$ a function? yes

(b) Give the domain and the range of the relation in part (a). domain: {0, 1, 3}; range: {4, 2, 5}

40. If $f(x) = -2x + 7$, find $f(-2)$. 11

Appendix: Factoring Sums and Differences of Cubes

OBJECTIVES

1. Factor a difference of cubes.
2. Factor a sum of cubes.

OBJECTIVE 1 **Factor a difference of cubes.** The difference of squares was factored in **Section 6.5.** We can also factor a **difference of cubes.**

Factoring a Difference of Cubes

$$a^3 - b^3 = (a - b)(a^2 + ab + b^2)$$

This pattern should be memorized. To see that the pattern is correct, multiply

$$(a - b)(a^2 + ab + b^2),$$

as shown in the margin.

$$\begin{array}{r} a^2 + ab + b^2 \\ a - b \\ \hline -a^2b - ab^2 - b^3 \\ a^3 + a^2b + ab^2 \\ \hline a^3 - b^3 \end{array}$$

Multiply vertically. (Section 5.3)

Add.

Notice the pattern of the terms in the factored form of $a^3 - b^3$.

- $a^3 - b^3$ factors as (a binomial factor) · (a trinomial factor).
- The binomial factor has the difference of the cube roots of the given terms.
- The terms in the trinomial factor are all positive.
- The terms in the binomial factor determine the trinomial factor.

$$a^3 - b^3 = (a - b)(\underbrace{a^2}_{\text{First term squared}} + \underbrace{ab}_{\text{Positive product of the terms}} + \underbrace{b^2}_{\text{Second term squared}})$$

Teaching Tip

To help students recognize perfect cubes, you may wish to have them work Exercise 1 in the exercise set for this appendix.

EXAMPLE 1 **Factoring Differences of Cubes**

Factor each difference of cubes.

(a) $m^3 - 125$ Use the pattern for a difference of cubes.

$$a^3 - b^3 = (a - b)(a^2 + ab + b^2)$$

$$m^3 - 125 = m^3 - 5^3 = (m - 5)(m^2 + 5m + 5^2)$$

$$= (m - 5)(m^2 + 5m + 25)$$

(b) $8p^3 - 27$

$= (2p)^3 - 3^3$ — $8p^3 = (2p)^3$ and $27 = 3^3$.

$= (2p - 3)[(2p)^2 + (2p)3 + 3^2]$ — Let $a = 2p$ and $b = 3$.

$= (2p - 3)(4p^2 + 6p + 9)$ — Apply the exponents. Multiply.

$(2p)^2 = 2^2p^2 = 4p^2$

Continued on Next Page

(c) $4m^3 - 32n^3$

$= 4(m^3 - 8n^3)$ — Factor out the common factor.

$= 4[m^3 - (2n)^3]$ — $8n^3 = (2n)^3$

$= 4(m - 2n)[m^2 + m(2n) + (2n)^2]$ — Let $a = m$ and $b = 2n$.

$= 4(m - 2n)(m^2 + 2mn + 4n^2)$ — Apply the exponents. Multiply.

◀ **Work Problem 1 at the Side.**

CAUTION

A common error in factoring $a^3 - b^3 = (a - b)(a^2 + ab + b^2)$ is to try to factor $a^2 + ab + b^2$. This is usually not possible.

OBJECTIVE 2 **Factor a sum of cubes.** A sum of squares, such as $m^2 + 25$, cannot be factored using real numbers, but a **sum of cubes** can.

Factoring a Sum of Cubes

$$a^3 + b^3 = (a + b)(a^2 - ab + b^2)$$

Observe the positive and negative signs in these factoring patterns.

$$a^3 - b^3 = (a - b)(a^2 + ab + b^2)$$

Difference of cubes (Same sign: $-$ and $-$; Opposite sign: $+$; Positive: ab term sign... b^2 positive)

Positive

Same sign — Opposite sign

The only difference between the patterns is the positive and negative signs.

Positive

$$a^3 + b^3 = (a + b)(a^2 - ab + b^2)$$

Sum of cubes

Same sign — Opposite sign

EXAMPLE 2 **Factoring Sums of Cubes**

Factor each sum of cubes.

(a) $k^3 + 27$

$= k^3 + 3^3$ — $27 = 3^3$

$= (k + 3)(k^2 - 3k + 3^2)$ — Let $a = k$ and $b = 3$.

$= (k + 3)(k^2 - 3k + 9)$ — Apply the exponent.

(b) $8m^3 + 125p^3$

$= (2m)^3 + (5p)^3$ — $8m^3 = (2m)^3$; $125p^3 = (5p)^3$

$= (2m + 5p)[(2m)^2 - (2m)(5p) + (5p)^2]$ — Let $a = 2m$ and $b = 5p$.

$= (2m + 5p)(4m^2 - 10mp + 25p^2)$ — Apply the exponents. Multiply.

◀ **Work Problem 2 at the Side.**

1 Factor each difference of cubes.

GS **(a)** $t^3 - 64$

$= \_\_\_^3 - \_\_\_^3$

$= (\_\_\_ - \_\_\_) \cdot (\_\_\_^2 + \_\_\_ + \_\_\_^2)$

$= \_\_\_\_\_\_$

(b) $2x^3 - 54$

(c) $8k^3 - y^3$

2 Factor each sum of cubes.

GS **(a)** $x^3 + 8$

$= \_\_\_^3 + \_\_\_^3$

$= (\_\_\_ + \_\_\_) \cdot (\_\_\_^2 - \_\_\_ + \_\_\_^2)$

$= \_\_\_\_\_\_$

(b) $64y^3 + 1$

(c) $27m^3 + 343n^3$

Answers

1. **(a)** t; 4; t; 4; t; $4t$; 4; $(t - 4)(t^2 + 4t + 16)$
(b) $2(x - 3)(x^2 + 3x + 9)$
(c) $(2k - y)(4k^2 + 2ky + y^2)$

2. **(a)** x; 2; x; 2; x; $2x$; 2; $(x + 2)(x^2 - 2x + 4)$
(b) $(4y + 1)(16y^2 - 4y + 1)$
(c) $(3m + 7n)(9m^2 - 21mn + 49n^2)$

Appendix Exercises

FOR EXTRA HELP

Download the MyDashBoard App

MyMathLab®

CONCEPT CHECK *Work each problem.*

1. To help factor the sum or difference of cubes, complete the following list of cubes.

$1^3 =$ 1 $\quad 2^3 =$ 8 $\quad 3^3 =$ 27 $\quad 4^3 =$ 64 $\quad 5^3 =$ 125

$6^3 =$ 216 $\quad 7^3 =$ 343 $\quad 8^3 =$ 512 $\quad 9^3 =$ 729 $\quad 10^3 =$ 1000

2. The following powers of x are all perfect cubes: $x^3, x^6, x^9, x^{12}, x^{15}$. Based on this observation, we may make a conjecture that if the power of a variable is divisible by 3 (with 0 remainder), then we have a perfect cube.

3. Which of the following are differences of cubes?

A. $9x^3 - 125$ **B.** $x^3 - 16$ **C.** $x^3 - 1$ **D.** $8x^3 - 27y^3$ C, D

4. Which of the following are sums of cubes?

A. $x^3 + 1$ **B.** $x^3 + 36$ **C.** $12x^3 + 27$ **D.** $64x^3 + 216y^3$ A, D

*Factor. Use your answers in **Exercises 1 and 2** as necessary. **See Examples 1 and 2.***

5. $x^3 + 1$
$(x + 1)(x^2 - x + 1)$

6. $m^3 + 8$
$(m + 2)(m^2 - 2m + 4)$

7. $x^3 - 1$
$(x - 1)(x^2 + x + 1)$

8. $m^3 - 8$
$(m - 2)(m^2 + 2m + 4)$

9. $p^3 + q^3$
$(p + q)(p^2 - pq + q^2)$

10. $w^3 + z^3$
$(w + z)(w^2 - wz + z^2)$

11. $y^3 - 216$
$(y - 6)(y^2 + 6y + 36)$

12. $x^3 - 343$
$(x - 7)(x^2 + 7x + 49)$

13. $k^3 + 1000$
$(k + 10)(k^2 - 10k + 100)$

14. $p^3 + 512$
$(p + 8)(p^2 - 8p + 64)$

15. $27x^3 - 1$
$(3x - 1)(9x^2 + 3x + 1)$

16. $64y^3 - 27$
$(4y - 3)(16y^2 + 12y + 9)$

17. $125x^3 + 8$
$(5x + 2)(25x^2 - 10x + 4)$

18. $216x^3 + 125$
$(6x + 5)(36x^2 - 30x + 25)$

19. $y^3 - 8x^3$
$(y - 2x)(y^2 + 2xy + 4x^2)$

20. $w^3 - 216z^3$
$(w - 6z)(w^2 + 6wz + 36z^2)$

21. $27x^3 - 64y^3$
$(3x - 4y)(9x^2 + 12xy + 16y^2)$

22. $125m^3 - 8n^3$
$(5m - 2n)(25m^2 + 10mn + 4n^2)$

23. $8p^3 + 729q^3$
$(2p + 9q)(4p^2 - 18pq + 81q^2)$

24. $27x^3 + 1000y^3$
$(3x + 10y)(9x^2 - 30xy + 100y^2)$

25. $16t^3 - 2$
$2(2t - 1)(4t^2 + 2t + 1)$

26. $3p^3 - 81$
$3(p - 3)(p^2 + 3p + 9)$

27. $40w^3 + 135$

$5(2w + 3)(4w^2 - 6w + 9)$

28. $32z^3 + 500$

$4(2z + 5)(4z^2 - 10z + 25)$

29. $x^3 + y^6$

$(x + y^2)(x^2 - xy^2 + y^4)$

30. $p^9 + q^3$

$(p^3 + q)(p^6 - p^3q + q^2)$

31. $125k^3 - 8m^9$

$(5k - 2m^3)(25k^2 + 10km^3 + 4m^6)$

32. $125c^6 - 216d^3$

$(5c^2 - 6d)(25c^4 + 30c^2d + 36d^2)$

Relating Concepts (Exercises 33–40) For Individual or Group Work

*A binomial may be both a difference of squares and a difference of cubes. One example of such a binomial is $x^6 - 1$. Using the techniques of **Section 6.5**, one factoring method will give the completely factored form, while the other will not. **Work Exercises 33–40 in order,** to determine the method to use.*

33. Factor $x^6 - 1$ as the difference of two squares.

$(x^3 - 1)(x^3 + 1)$

34. The factored form obtained in **Exercise 33** consists of a difference of cubes multiplied by a sum of cubes. Factor each binomial further.

$(x - 1)(x^2 + x + 1)(x + 1)(x^2 - x + 1)$

35. Now start over and factor $x^6 - 1$ as a difference of cubes.

$(x^2 - 1)(x^4 + x^2 + 1)$

36. The factored form obtained in **Exercise 35** consists of a binomial that is a difference of squares and a trinomial. Factor the binomial further.

$(x - 1)(x + 1)(x^4 + x^2 + 1)$

37. Compare the results in **Exercises 34 and 36.** Which one of these is the completely factored form?

The result in Exercise 34 is completely factored.

38. Verify that the trinomial in the factored form in **Exercise 36** is the product of the two trinomials in the factored form in **Exercise 34.**

Show that $x^4 + x^2 + 1 = (x^2 + x + 1)(x^2 - x + 1)$.

39. Use the results of **Exercises 33–38** to complete the following statement.

In general, if I must choose between factoring first using the method for a difference of squares or the method for a difference of cubes, I should choose the difference of squares method to eventually obtain the completely factored form.

40. Find the *completely* factored form of $x^6 - 729$ using the knowledge gained in **Exercises 33–39.**

$(x - 3)(x^2 + 3x + 9)(x + 3)(x^2 - 3x + 9)$

Answers to Selected Exercises

In this section we provide the answers that we think most students will obtain when they work the exercises using the methods explained in the text. If your answer does not look exactly like the one given here, it is not necessarily wrong. In many cases there are equivalent forms of the answer that are correct. For example, if the answer section shows $\frac{3}{4}$ and your answer is 0.75, you have obtained the correct answer but written it in a different (yet equivalent) form. Unless the directions specify otherwise, 0.75 is just as valid an answer as $\frac{3}{4}$.

In general, if your answer does not agree with the one given in the text, see whether it can be transformed into the other form. If it can, then it is the correct answer. If you still have doubts, talk with your instructor.

CHAPTER R Prealgebra Review

SECTION R.1 (pages 11–14)

1. true **2.** true **3.** false; This is an improper fraction. Its value is 1. **4.** false; The number 1 is neither prime nor composite. **5.** false; The fraction $\frac{17}{51}$ can be simplified to $\frac{1}{3}$. **6.** false; The reciprocal of $\frac{8}{2} = 4$ is $\frac{2}{8} = \frac{1}{4}$. **7.** false; *Product* indicates multiplication, so the product of 8 and 2 is 16. **8.** false; *Difference* indicates subtraction, so the difference between 12 and 2 is 10. **9.** prime **11.** composite **13.** composite **15.** prime **17.** $2 \cdot 3 \cdot 5$ **19.** $2 \cdot 2 \cdot 3 \cdot 3 \cdot 7$ **21.** $2 \cdot 2 \cdot 31$ **23.** 29 **25.** $\frac{1}{2}$ **27.** $\frac{5}{6}$ **29.** $\frac{1}{5}$ **31.** $\frac{6}{5}$ **33.** $1\frac{5}{7}$ **35.** $6\frac{5}{12}$ **37.** $7\frac{6}{11}$ **39.** $\frac{13}{5}$ **41.** $\frac{83}{8}$ **43.** $\frac{51}{5}$ **45.** A **46.** C **47.** $\frac{24}{35}$ **49.** $\frac{6}{25}$ **51.** $\frac{6}{5}$, or $1\frac{1}{5}$ **53.** $\frac{232}{15}$, or $15\frac{7}{15}$ **55.** $\frac{10}{3}$, or $3\frac{1}{3}$ **57.** 12 **59.** $\frac{1}{16}$ **61.** $\frac{84}{47}$, or $1\frac{37}{47}$ **63.** $\frac{2}{3}$ **65.** $\frac{8}{9}$ **67.** $\frac{27}{8}$, or $3\frac{3}{8}$ **69.** $\frac{17}{36}$ **71.** $\frac{11}{12}$ **73.** $\frac{4}{3}$, or $1\frac{1}{3}$ **75.** 6 cups **77.** $618\frac{3}{4}$ ft **79.** $\frac{9}{16}$ in. **81.** $\frac{7}{16}$ in. **83.** $16\frac{5}{8}$ yd **85.** 8 cakes (There will be some sugar left over.) **87.** $\frac{1}{15}$ **89.** about $4\frac{1}{2}$ million, or 4,500,000

SECTION R.2 (pages 23–24)

1. (a) 6 **(b)** 9 **(c)** 1 **(d)** 7 **(e)** 4 **2.** Answers will vary. One example is 5243.0164. **3. (a)** 46.25 **(b)** 46.2 **(c)** 46 **(d)** 50 **4. (a)** 0.889 **(b)** 0.556 **(c)** 0.976 **(d)** 0.864 **5.** C **6.** D **7.** B **8.** C **9.** $\frac{4}{10}$ **11.** $\frac{64}{100}$ **13.** $\frac{138}{1000}$ **15.** $\frac{3805}{1000}$ **17.** 139; 143.094 **19.** 27; 25.61 **21.** 10; 15.33 **23.** 82; 81.716 **25.** 17; 15.211 **27.** 0.006; 0.006 **29.** 90; 116.48 **31.** 8; 7.15 **33.** 2; 2.05 **35.** 6000; 5711.6 **37.** 0.2; 0.162 **39.** 0.125 **41.** 1.25 **43.** $0.\overline{5}$; 0.556 **45.** $0.1\overline{6}$; 0.167 **47.** To convert a decimal to a percent, move the decimal point two places to the right and attach a percent symbol (%). **48.** To convert a percent to a decimal, move the decimal point two places to the left and drop the percent symbol (%). **49.** 0.54 **51.** 1.17 **53.** 0.024 **55.** 0.0625 **57.** 0.008 **59.** 73% **61.** 0.4% **63.** 128% **65.** 30% **67.** 600% **69.** 80% **71.** $18.\overline{18}\%$ **73.** 175% **75.** $216.\overline{6}\%$ **77.** \$15 **79.** 70%

CHAPTER 1 The Real Number System

SECTION 1.1 (pages 33–35)

1. true **2.** false; +, −, ·, and ÷ are operation symbols. **3.** false; Using the rules for order of operations gives $4 + 3(8 - 2) = 4 + 3(6) = 4 + 18 = 22$. **4.** false; $3^3 = 3 \cdot 3 \cdot 3 = 27$ **5.** false; The correct translation is $4 = 16 - 12$. **6.** false; The correct translation is $6 = 10 - 4$. **7.** The 4 would be applied last because we work first inside the parentheses. **8.** exponents **9.** 49 **11.** 144 **13.** 64 **15.** 1000 **17.** 81 **19.** 1024 **21.** $\frac{16}{81}$ **23.** 0.000064 **25.** ②①; 45; 58 **27.** ②①③; 12; 8; 13 **29.** ①②; 32 **31.** ①②; 19 **33.** $\frac{49}{30}$ **35.** 12 **37.** 36.14 **39.** 26 **41.** 4 **43.** 95 **45.** 12 **47.** 14 **49.** $\frac{19}{2}$ **51.** $3 \cdot (6 + 4) \cdot 2 = 60$ **52.** $2 \cdot (8 - 1) \cdot 3 = 42$ **53.** $10 - (7 - 3) = 6$ **54.** $15 - (10 - 2) = 7$ **55.** $(8 + 2)^2 = 100$ **56.** $(4 + 2)^2 = 36$ **57.** false **59.** true **61.** true **63.** false **65.** false **67.** true **69.** $15 = 5 + 10$ **71.** $9 > 5 - 4$ **73.** $16 \neq 19$ **75.** $2 \leq 3$ **77.** Seven is less than nineteen; true **79.** One-third is not equal to three-tenths; true **81.** Eight is greater than or equal to eleven; false **83.** $30 > 5$ **85.** $3 \leq 12$ **87.** $1.3 \leq 2.5$ **89.** $\frac{3}{4} < \frac{4}{5}$ **91. (a)** $14.7 - 40 \cdot 0.13$ **(b)** 9.5 **(c)** 8.075; walking (5 mph) **(d)** $14.7 - 55 \cdot 0.11$; 8.65; 7.3525, swimming **93.** Alaska, Texas, California, Idaho **95.** Alaska, Texas, California, Idaho, Missouri

SECTION 1.2 (pages 41–44)

1. B **2.** C **3.** A **4.** B, C **5.** 11 **6.** 10 **7.** $13 + x$; 16 **8.** expression; equation **9.** The equation would be $5x - 9 = 49$. **10.** $2x^3 = 2 \cdot x \cdot x \cdot x$, while $2x \cdot 2x \cdot 2x = (2x)^3$. **11.** Answers will vary. Two such pairs are $x = 0, y = 6$ and $x = 1, y = 4$. For each pair, the value of $2x + y$ is 6. **12.** The exponent 2 applies only to the base x. The exponential must be evaluated before the product. **13. (a)** 64 **(b)** 144 **15. (a)** $\frac{7}{8}$ **(b)** $\frac{13}{12}$ **17. (a)** 9.569 **(b)** 14.353 **19. (a)** 52 **(b)** 114 **21. (a)** 12 **(b)** 33 **23. (a)** 6 **(b)** $\frac{9}{5}$ **25. (a)** $\frac{4}{3}$ **(b)** $\frac{13}{6}$ **27. (a)** $\frac{2}{7}$ **(b)** $\frac{16}{27}$ **29. (a)** 12 **(b)** 55 **31. (a)** 1 **(b)** $\frac{28}{17}$ **33. (a)** 3.684 **(b)** 8.841 **35.** $12x$ **37.** $x - 2$ **39.** $7 - \frac{1}{3}x$ **41.** $2x - 6$ **43.** $\frac{12}{x + 3}$ **45.** $6(x - 4)$ **47.** An equation cannot be *evaluated*. An equation that contains a variable can be *solved* to find the value or values of the variable that make it a true statement. **48.** An expression cannot be *solved*. It indicates a series of operations to be performed. An equation can be solved. **49.** no **51.** yes **53.** yes **55.** no **57.** yes **59.** yes **61.** $x + 8 = 18$ **63.** $2x + 5 = 5$ **65.** $16 - \frac{3}{4}x = 13$ **67.** $3x = 2x + 8$ **69.** expression **71.** equation **73.** equation **75.** expression **77.** 69 yr **78.** 74 yr **79.** 77 yr **80.** 79 yr

SECTION 1.3 (pages 52–54)

1. 0 **2.** integers **3.** positive **4.** right **5.** quotient; denominator **6.** irrational **7.** 2,632,000 **9.** −7067

11. 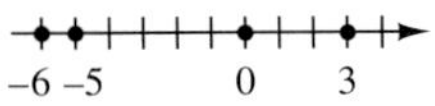

13. 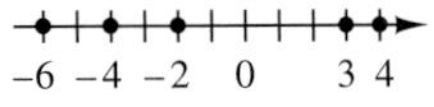

15. 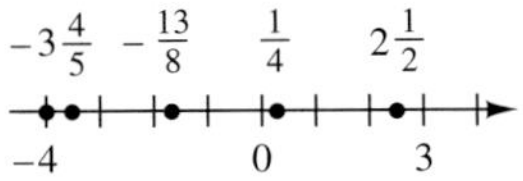

17. **(a)** 3, 7 **(b)** 0, 3, 7 **(c)** −9, 0, 3, 7 **(d)** $-9, -1\frac{1}{4}, -\frac{3}{5}, 0, 3, 5.9, 7$ **(e)** $-\sqrt{7}, \sqrt{5}$ **(f)** All are real numbers. **19.** **(a)** 11 **(b)** 0, 11 **(c)** 0, 11, −6 **(d)** $\frac{7}{9}, -2.\overline{3}, 0, -8\frac{3}{4}, 11, -6$ **(e)** $\sqrt{3}, \pi$ **(f)** All are real numbers.

21. 4 **22.** One example is 2.85. There are others. **23.** 0 **24.** One example is 4. There are others. **25.** One example is $\sqrt{13}$. There are others. **26.** 0 **27.** true **28.** false **29.** true **30.** true **31.** false **32.** true *In Exercises 33–38, answers will vary.* **33.** $\frac{1}{2}, \frac{5}{8}, 1\frac{3}{4}$ **34.** $-1, -\frac{3}{4}, -5$ **35.** $-3\frac{1}{2}, -\frac{2}{3}, \frac{3}{7}$ **36.** $\frac{1}{2}, -\frac{2}{3}, \frac{2}{7}$ **37.** $\sqrt{5}, \pi, -\sqrt{3}$ **38.** $\frac{2}{3}, \frac{5}{6}, \frac{5}{2}$ **39.** −11 **41.** −21 **43.** −100 **45.** $-\frac{2}{3}$ **47.** false **49.** true **51.** **(a)** 2 **(b)** 2 **53.** **(a)** −6 **(b)** 6 **55.** **(a)** $\frac{3}{4}$ **(b)** $\frac{3}{4}$ **57.** **(a)** −4.95 **(b)** 4.95 **59.** **(a)** A **(b)** A **(c)** B **(d)** B **60.** 5; 5; 5; −5 **61.** 7 **63.** −12 **65.** $-\frac{2}{3}$ **67.** 9 **69.** false **71.** true **73.** Energy, 2008 to 2009 **75.** 2009 to 2010

SECTION 1.4 (pages 60–63)

1. negative; −5 **2.** zero (0) **3.** negative; −2

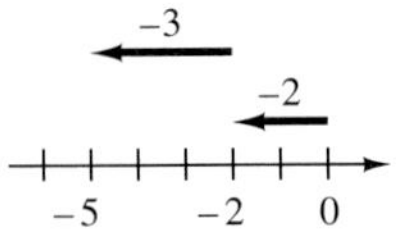

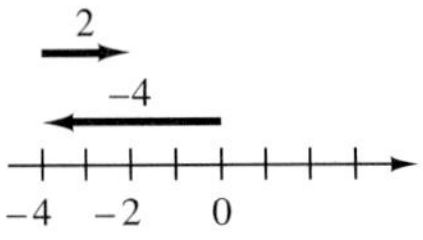

4. −3; 5 **5.** Add −2 and 5. **6.** Add −6 and 2. **7.** Add −1 and −3. **8.** Add −8 and 4. **9.** 2 **11.** −3 **13.** −10 **15.** −13 **17.** −15.9 **19.** −5; 5 **21.** 13 **23.** 0 **25.** −8 **27.** $\frac{3}{10}$ **29.** $\frac{1}{2}$ **31.** $-\frac{3}{4}$ **33.** −1.6 **35.** −8.7 **37.** −11; −14; −25 **39.** $-\frac{1}{4}$, or −0.25 **41.** true **43.** false **45.** true **47.** false **49.** true **51.** false **53.** −5 + 12 + 6; 13 **55.** [−19 + (−4)] + 14; −9 **57.** [−4 + (−10)] + 12; −2 **59.** $\left[\frac{5}{7} + \left(-\frac{9}{7}\right)\right] + \frac{2}{7}; -\frac{2}{7}$ **61.** −\$62 **63.** −184 m **65.** 17 **67.** 37 yd **69.** 120°F **71.** −\$107 **73.** −6

SECTION 1.5 (pages 68–72)

1. −8; −6; 2 **2.** 5; −4 **3.** 7 − 12; 12 − 7 **4.** additive; inverse; opposite **5.** −4 **6.** −22 **7.** −3; −10 **9.** −16 **11.** 4; 11 **13.** 19 **15.** −4 **17.** 5 **19.** 0 **21.** $\frac{3}{4}$ **23.** $-\frac{11}{8}$ **25.** $\frac{15}{8}$ **27.** 13.6 **29.** −11.9 **31.** 8 **33.** −2.8 **35.** −6.3 **37.** −28 **39.** $\frac{37}{12}$ **41.** −42.04 **43.** 4 − (−8); 12 **45.** −2 − 8; −10 **47.** [9 + (−4)] − 7; −2 **49.** [8 − (−5)] − 12; 1 **51.** −69°F **53.** 14,776 ft **55.** −\$80 **57.** −176.9°F **59.** \$1045.55 **61.** 469 B.C. **63.** \$323.83 **65.** 14 ft **67.** 40,776 ft **69.** 4.4% **70.** Americans spent more money than they earned, which means that they had to dip into savings or borrow money. **71.** \$1530 billion **73.** \$2900 **75.** −\$20,900 **77.** 136 ft **79.** negative **80.** positive **81.** positive **82.** negative **83.** positive **84.** negative

SECTION 1.6 (pages 81–84)

1. greater than 0 **2.** less than 0 **3.** less than 0 **4.** less than 0 **5.** greater than 0 **6.** less than 0 **7.** −28 **9.** 30 **11.** 0 **13.** $\frac{5}{6}$ **15.** −2.38 **17.** $\frac{3}{2}$ **19.** −3 **21.** −2 **23.** 16 **25.** 0 **27.** undefined **29.** $\frac{3}{2}$ **31.** C **32.** A **33.** 30; 10; 3 **35.** 7 **37.** 4 **39.** −3 **41.** −1 **43.** $\frac{7}{4}$ **45.** 68 **47.** −228 **49.** 1 **51.** 0 **53.** −6 **55.** undefined **57.** −12 + 4(−7); −40 **59.** −1 − 2(−8)(2); 31 **61.** −3[3 − (−7)]; −30 **63.** $\frac{3}{10}[-2 + (-28)]; -9$ **65.** $\frac{-20}{-8 + (-2)}; 2$ **67.** $\frac{-18 + (-6)}{2(-4)}; 3$ **69.** $\frac{-\frac{2}{3}\left(-\frac{1}{5}\right)}{\frac{1}{7}}; \frac{14}{15}$ **71.** $9x = -36$ **73.** $\frac{x}{4} = -1$ **75.** $x - \frac{9}{11} = 5$ **77.** $\frac{6}{x} = -3$ **79.** 29 **80.** The incorrect answer, 92, was obtained by performing all of the operations in order from left to right rather than following the rules for order of operations. The multiplications and divisions need to be done in order, before the additions and subtractions. **81.** 42 **82.** 5 **83.** $8\frac{2}{5}$ **84.** $8\frac{2}{5}$ **85.** 2 **86.** $-12\frac{1}{2}$

SUMMARY EXERCISES Performing Operations with Real Numbers (pages 85–86)

1. −16 **2.** 4 **3.** 0 **4.** −24 **5.** −17 **6.** 76 **7.** −18 **8.** 90 **9.** 38 **10.** 4 **11.** −5 **12.** 5 **13.** $-\frac{7}{2}$, or $-3\frac{1}{2}$ **14.** 4 **15.** 13 **16.** $\frac{5}{4}$, or $1\frac{1}{4}$ **17.** 9 **18.** $\frac{37}{10}$, or $3\frac{7}{10}$ **19.** 0 **20.** 25 **21.** 14 **22.** 0 **23.** −4 **24.** $\frac{6}{5}$, or $1\frac{1}{5}$ **25.** −1 **26.** $\frac{52}{37}$, or $1\frac{15}{37}$ **27.** $\frac{17}{16}$, or $1\frac{1}{16}$ **28.** $-\frac{2}{3}$ **29.** 3.33 **30.** 1.02 **31.** −13 **32.** 0 **33.** 24 **34.** −7 **35.** 37 **36.** −3 **37.** −1 **38.** $\frac{1}{2}$ **39.** $-\frac{5}{13}$ **40.** 5 **41.** undefined **42.** 0

SECTION 1.7 (pages 93–96)

1. **(a)** B **(b)** F **(c)** C **(d)** I **(e)** B **(f)** D, F **(g)** B **(h)** A **(i)** G **(j)** H **2.** order; grouping **3.** yes **4.** yes **5.** no **6.** no **7.** no **8.** no **9.** (foreign sales) clerk; foreign (sales clerk) **10.** (defective merchandise) counter; defective (merchandise counter) **11.** −15; commutative property **13.** 3; commutative property **15.** 6; associative property **17.** 7; associative property **19.** Subtraction is not associative. **20.** Division is not associative. **21.** row 1: $-5, \frac{1}{5}$; row 2: $10, -\frac{1}{10}$; row 3: $\frac{1}{2}, -2$; row 4: $-\frac{3}{8}, \frac{8}{3}$; row 5: $-x, \frac{1}{x}$ $(x \neq 0)$; row 6: $y, -\frac{1}{y}$ $(y \neq 0)$; opposite; the same **22.** identity property **23.** commutative property **25.** associative property **27.** inverse property

29. inverse property **31.** identity property
33. commutative property **35.** distributive property
37. identity property **39.** distributive property
41. The expression following the first equality symbol should be $-3(4) - 3(-6)$. This simplifies to $-12 + 18$, which equals 6.
42. We must multiply $\frac{3}{4}$ by 1 in the form of a fraction, $\frac{3}{3}$: $\frac{3}{4} \cdot \frac{3}{3} = \frac{9}{12}$.
43. $7 + r$ **45.** s **47.** $-6x + (-6)7$; $-6x - 42$
49. $w + [5 + (-3)]$; $w + 2$ **51.** 6700 **53.** 2 **55.** 0.77 **57.** $4t + 12$
59. $-8r - 24$ **61.** y; -4; $-5y + 20$ **63.** $-16y - 20z$ **65.** $8(z + w)$
67. $5(3 + 17)$; 100 **69.** $7(2v + 5r)$ **71.** $24r + 32s - 40y$
73. $-24x - 9y - 12z$ **75.** $-4t - 5m$ **77.** $5c + 4d$ **79.** $3q - 5r + 8s$

SECTION 1.8 (pages 101–103)

1. B **2.** C **3.** A **4.** B **5.** $15x$ **7.** $13b$ **9.** $4r + 11$
11. $5 + 2x - 6y$ **13.** $-7 + 3p$ **15.** $-\frac{4}{3}y - 10$ **17.** -12 **19.** 5
21. 1 **23.** -1 **25.** 74 **27.** $-\frac{3}{8}$ **29.** $\frac{1}{2}$ **31.** $\frac{2}{5}$ **33.** -1.28
35. like **37.** unlike **39.** like **41.** unlike **43.** x; -3; $2x$; 6; $1 - 2x$
45. $-\frac{1}{3}t - \frac{28}{3}$ **47.** $-4.1r + 4.2$ **49.** $-2y^2 + 3y^3$ **51.** $-19p + 16$
53. $-2x + 4$ **55.** $-\frac{14}{3}x - \frac{22}{3}$ **57.** $-\frac{3}{2}y + 16$ **59.** $-16y + 63$
61. $4r + 15$ **63.** $12k - 5$ **65.** $-2k - 3$ **67.** $4x - 7$
69. $-23.7y - 12.6$ **71.** $(x + 3) + 5x$; $6x + 3$
73. $(13 + 6x) - (-7x)$; $13 + 13x$ **75.** $2(3x + 4) - (-4 + 6x)$; 12
77. Wording may vary. One example is "the difference between 9 times a number and the sum of the number and 2." **79.** The student made a sign error when applying the distributive property. $7x - 2(3 - 2x)$ means $7x - 2(3) - 2(-2x)$, which simplifies to $7x - 6 + 4x$, or $11x - 6$. **80.** The student incorrectly started by adding $3 + 2$. As the first step, 2 must be multiplied by $4x - 5$. Thus, $3 + 2(4x - 5)$ equals $3 + 8x - 10$, which simplifies to $8x - 7$. **81.** $1000 + 5x$ (dollars)
82. $750 + 3y$ (dollars) **83.** $1000 + 5x + 750 + 3y$ (dollars)
84. $1750 + 5x + 3y$ (dollars)

Chapter 1 REVIEW EXERCISES (pages 109–113)

1. 625 **2.** 0.00000081 **3.** 0.009261 **4.** $\frac{125}{8}$ **5.** 27 **6.** 200 **7.** 7
8. 4 **9.** $13 < 17$ **10.** $5 + 2 \neq 10$ **11.** Six is less than fifteen.
12. Answers will vary. One example is $2 + 5 \geq \frac{16}{2}$. **13.** 30 **14.** 60
15. 14 **16.** 13 **17.** $x + 6$ **18.** $8 - x$ **19.** $6x - 9$ **20.** $12 + \frac{3}{5}x$
21. yes **22.** no **23.** $2x - 6 = 10$ **24.** $4x = 8$ **25.** equation
26. expression

27.
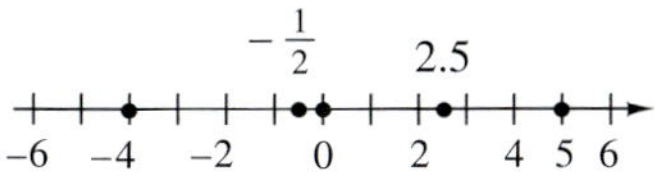

28.
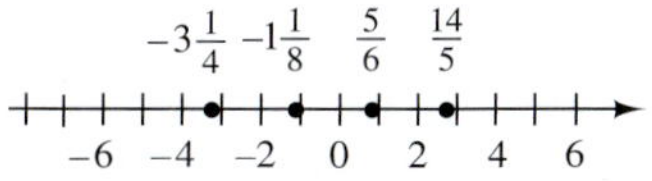

29. -10 **30.** -9 **31.** $-\frac{3}{4}$ **32.** $-|23|$ **33.** true **34.** true **35.** true
36. false **37.** -3 **38.** -19 **39.** -7 **40.** 9 **41.** -6 **42.** -4
43. -17 **44.** $-\frac{29}{36}$ **45.** -10 **46.** -19 **47.** $(-31 + 12) + 19$; 0
48. $[-4 + (-8)] + 13$; 1 **49.** \$26.25 **50.** -10°F **51.** -11
52. -1 **53.** 7 **54.** $-\frac{43}{35}$ **55.** 10.31 **56.** -12 **57.** 2 **58.** 1
59. $-4 - (-6)$; 2 **60.** $[4 + (-8)] - 5$; -9
61. $[18 - (-23)] - 15$; 26 **62.** $19 - (-7 - 12)$; 38 **63.** 38
64. 12,392.69 **65.** -126 thousand **66.** 267 thousand
67. 681 thousand **68.** -857 thousand **69.** 36 **70.** -105
71. $\frac{1}{2}$ **72.** 10.08 **73.** -20 **74.** -10 **75.** -24
76. -35 **77.** 4 **78.** -20 **79.** $-\frac{3}{4}$ **80.** 11.3 **81.** -1
82. undefined **83.** 1 **84.** 0 **85.** -18 **86.** -18 **87.** 125
88. -423 **89.** $-4(5) - 9$; -29 **90.** $\frac{5}{6}[12 + (-6)]$; 5
91. $\frac{12}{8 + (-4)}$; 3 **92.** $\frac{-20(12)}{15 - (-15)}$; -8 **93.** $\frac{x}{x + 5} = -2$
94. $8x - 3 = -7$ **95.** identity property **96.** identity property
97. inverse property **98.** inverse property **99.** distributive property
100. associative property **101.** associative property **102.** commutative property **103.** $(7 + 1)y$; $8y$ **104.** $-12 \cdot 4 - 12(-t)$; $-48 + 12t$
105. $3(2s + 4y)$; $6s + 12y$ **106.** $-1(-4r) + (-1)(5s)$; $4r - 5s$
107. $17p^2$ **108.** $16r^2 + 7r$ **109.** $-19k + 54$ **110.** $5s - 6$
111. $-45t - 23$ **112.** $-45t^2 - 23.4t$ **113.** -6 **114.** $\frac{25}{36}$ **115.** -26
116. $\frac{8}{3}$ **117.** $-\frac{1}{24}$ **118.** $\frac{7}{2}$ **119.** 2 **120.** 77.6 **121.** $-1\frac{1}{2}$
122. 11 **123.** $-\frac{28}{15}$ **124.** 24 **125.** -11 **126.** 16 **127.** -47°F
128. 27 ft

Chapter 1 TEST (pages 114–115)

1. true **2.** false **3.**
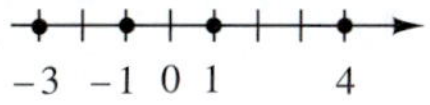

4. $-|-8|$ (or -8)
5. -1.277 **6.** $\frac{-6}{2 + (-8)}$; 1 **7.** negative **8.** 4 **9.** $-2\frac{5}{6}$ **10.** 6
11. 2 **12.** 108 **13.** 11 **14.** $\frac{30}{7}$ **15.** -70 **16.** 3 **17.** 178°F
18. 15 **19.** $-\$1.42$ trillion **20.** D **21.** A **22.** E **23.** B **24.** C
25. $21x$ **26.** $-9x^2 - 6x - 8$ **27.** identity and distributive properties
28. **(a)** -18 **(b)** -18 **(c)** The distributive property tells us that the two methods produce equal results.

CHAPTER 2 Equations, Inequalities, and Applications

SECTION 2.1 (pages 123–124)

1. equality; expression **2.** linear; = **3.** equivalent **4.** added to; subtracted from **5.** C **6.** A, B **7.** **(a)** expression; $x + 15$ **(b)** expression; $m + 7$ **(c)** equation; $\{-1\}$ **(d)** equation; $\{-17\}$
8. Replace the variable(s) in the original equation with the proposed solution. A true statement will result if the proposed solution is correct.
9. $\{12\}$ **11.** $\{-3\}$ **13.** $\{4\}$ **15.** $\{-9\}$ **17.** $\left\{-\frac{3}{4}\right\}$ **19.** $\{6.3\}$
21. $\{-16.9\}$ **23.** $\{-10\}$ **25.** $\{-13\}$ **27.** $\left\{\frac{4}{15}\right\}$ **29.** $\{7\}$

ANSWERS

31. $\{-3\}$ **33.** $\{-4\}$ **35.** $\{3\}$ **37.** $\{-2\}$ **39.** $\{4\}$ **41.** $\{-16\}$ **43.** $\{2\}$ **45.** $\{2\}$ **47.** $\{-4\}$ **49.** $\{4\}$ **51.** $\{0\}$ **53.** $\left\{\frac{7}{15}\right\}$ **55.** $\{7\}$ **57.** $\{-4\}$ **59.** $\{13\}$ **61.** $\{29\}$ **63.** $\{18\}$ **65.** Answers will vary. One example is $x - 6 = -8$. **66.** Answers will vary. One example is $x + \frac{1}{2} = 1$.

SECTION 2.2 (pages 130–131)

1. **(a)** and **(c)**: multiplication property of equality; **(b)** and **(d)**: addition property of equality **2.** C **3.** $\frac{3}{2}$ **4.** $\frac{5}{4}$ **5.** 10 **6.** 100 **7.** $-\frac{2}{9}$ **8.** $-\frac{3}{8}$ **9.** -1 **10.** -1 **11.** 6 **12.** 7 **13.** -4 **14.** -13 **15.** 0.12 **16.** 0.21 **17.** -1 **18.** -1 **19.** B **20.** A **21.** $\{6\}$ **23.** $\left\{\frac{15}{2}\right\}$ **25.** $\{-5\}$ **27.** $\left\{-\frac{18}{5}\right\}$ **29.** $\{12\}$ **31.** 2; 2; 0; $\{0\}$ **33.** $\{-12\}$ **35.** $\{40\}$ **37.** $\{-12.2\}$ **39.** $\{-48\}$ **41.** $\{72\}$ **43.** $\{-35\}$ **45.** $\{14\}$ **47.** $\left\{-\frac{27}{35}\right\}$ **49.** $\{3\}$ **51.** $\{-5\}$ **53.** $\{20\}$ **55.** $\{7\}$ **57.** 1; $\{-12\}$ **59.** $\{0\}$ **61.** $\{-6\}$ **63.** Answers will vary. One example is $\frac{3}{2}x = -6$. **64.** Answers will vary. One example is $100x = 17$. **65.** $-4x = 10$; $-\frac{5}{2}$ **67.** $\frac{x}{-5} = 2$; -10

SECTION 2.3 (pages 141–144)

1. addition; subtract **2.** left; like **3.** distributive; parentheses **4.** multiplication; $\frac{4}{3}$ **5.** fractions; 6 **6.** decimals; 10 **7.** **(a)** identity; B **(b)** conditional; A **(c)** contradiction; C **8.** D **9.** A **10.** D **11.** $\{4\}$ **13.** $\{-5\}$ **15.** $\left\{\frac{5}{2}\right\}$ **17.** $\left\{-\frac{1}{2}\right\}$ **19.** $\{5\}$ **21.** $\{1\}$ **23.** $\left\{-\frac{5}{3}\right\}$ **25.** $\{2\}$ **27.** $\left\{-\frac{5}{3}\right\}$ **29.** $\emptyset$ **31.** $\{0\}$ **33.** $\{-1\}$ **35.** $\emptyset$ **37.** {all real numbers} **39.** $\{0\}$ **41.** 14; $\{7\}$ **43.** $\{12\}$ **45.** $\{11\}$ **47.** $\{0\}$ **49.** $\left\{\frac{3}{25}\right\}$ **51.** 100; $\{60\}$ **53.** $\{4\}$ **55.** $\{5000\}$ **57.** $\left\{-\frac{72}{11}\right\}$ **59.** $\{0\}$ **61.** $\emptyset$ **63.** {all real numbers} **65.** $\{-6\}$ **67.** $\{15\}$ **69.** $12 - q$ **71.** $\frac{9}{z}$ **73.** $x + 29$ **75.** $a + 12$; $a - 2$ **77.** $25r$ **79.** $\frac{t}{5}$ **81.** $3x + 2y$

SUMMARY EXERCISES Applying Methods for Solving Linear Equations (pages 146–147)

1. $\{-5\}$ **2.** $\{4\}$ **3.** $\{-5.1\}$ **4.** $\{25\}$ **5.** $\{-25\}$ **6.** $\{-6\}$ **7.** $\{-3\}$ **8.** $\{-16\}$ **9.** $\{0\}$ **10.** $\left\{-\frac{96}{5}\right\}$ **11.** {all real numbers} **12.** $\{23.7\}$ **13.** $\{7\}$ **14.** $\{0\}$ **15.** $\{5\}$ **16.** $\{1\}$ **17.** $\{-16\}$ **18.** $\emptyset$ **19.** $\{6\}$ **20.** $\{3\}$ **21.** $\emptyset$ **22.** $\left\{\frac{7}{3}\right\}$ **23.** $\{25\}$ **24.** $\{-10.8\}$ **25.** $\{3\}$ **26.** $\{7\}$ **27.** $\{2\}$ **28.** {all real numbers} **29.** $\left\{-\frac{2}{7}\right\}$ **30.** $\{10\}$ **31.** $\left\{\frac{14}{17}\right\}$ **32.** $\left\{-\frac{5}{2}\right\}$ **33.** {all real numbers} **34.** $\{64\}$

SECTION 2.4 (pages 156–161)

1. *Step 1:* Read the problem carefully; *Step 2:* Assign a variable to represent the unknown; *Step 3:* Write an equation; *Step 4:* Solve the equation; *Step 5:* State the answer; *Step 6:* Check the answer. **2.** Some examples are *is, are, was,* and *were*. **3.** D; There cannot be a fractional number of cars. **4.** D; A day cannot have more than 24 hr. **5.** A; Distance cannot be negative. **6.** C; Time cannot be negative. **7.** 1; 16 (or 14); -7 (or -9) **8.** odd; 2; 11 (or 15); even; 2; 10 (or 14) **9.** complementary; supplementary **10.** 90°; 180° **11.** yes, 90°; yes, 45° **12.** $x - 1$; $x - 2$ **13.** $8 \cdot (x + 6) = 104$; 7 **15.** $5x + 2 = 4x + 5$; 3 **17.** $3x - 2 = 5x + 14$; -8 **19.** $3(x - 2) = x + 6$; 6 **21.** $3x + (x + 7) = -11 - 2x$; -3 **23.** *Step 1:* We are asked to find the number of drive-in movie screens in the two states; *Step 2:* the number of screens in Pennsylvania; *Step 3:* x; $x + 3$; *Step 4:* 28; *Step 5:* 28; 28; 31; *Step 6:* 3; screens in New York; 31; 59 **25.** Democrats: 193; Republicans: 242 **27.** Bon Jovi: \$108.2 million; Roger Waters: \$89.5 million **29.** wins: 57; losses: 25 **31.** orange: 97 mg; pineapple: 25 mg **33.** 420 lb **35.** active: 225 mg; inert: 25 mg **37.** 1950 Denver nickel: \$16.00; 1944 Philadelphia nickel: \$12.00 **39.** onions: 81.3 kg; grilled steak: 536.3 kg **41.** American: 18; United: 11; Southwest: 26 **43.** $x + 5$; $x + 9$; shortest piece: 15 in.; middle piece: 20 in.; longest piece: 24 in. **45.** gold: 46; silver: 29; bronze: 29 **47.** 36 million mi **49.** *A* and *B*: 40°; *C*: 100° **51.** 68, 69 **53.** 146, 147 **55.** 10, 12 **57.** 17, 19 **59.** 10, 11 **61.** 18 **63.** 18° **65.** 20° **67.** 39° **69.** 50°

SECTION 2.5 (pages 168–173)

1. The perimeter of a plane geometric figure is the distance around the figure. **2.** The area of a plane geometric figure is the measure of the surface covered or enclosed by the figure. **3.** area **4.** area **5.** perimeter **6.** perimeter **7.** area **8.** area **9.** area **10.** area **11.** $P = 26$ **13.** $A = 64$ **15.** $b = 4$ **17.** $t = 5.6$ **19.** $h = 7$ **21.** $r = 2.6$ **23.** $A = 50.24$ **25.** $V = 150$ **27.** $V = 52$ **29.** $V = 7234.56$ **31.** $I = \$600$ **33.** $p = \$550$ **35.** 0.025; $t = 1.5$ yr **37.** length: 18 in.; width: 9 in. **39.** length: 14 m; width: 4 m **41.** shortest: 5 in.; medium: 7 in.; longest: 8 in. **43.** two equal sides: 7 m; third side: 10 m **45.** perimeter: 5.4 m; area: 1.8 m^2 **47.** 10 ft **49.** about 154,000 ft^2 **51.** 194.48 ft^2; 49.42 ft **53.** 23,800.10 ft^2 **55.** length: 36 in.; maximum volume: 11,664 in.^{3} **57.** 48°, 132° **59.** 55°, 35° **61.** 51°, 51° **63.** 105°, 105° **65.** $t = \frac{d}{r}$ **67.** $H = \frac{V}{LW}$ **69.** $b = P - a - c$ **71.** $r = \frac{C}{2\pi}$ **73.** $r = \frac{I}{pt}$ **75.** $h = \frac{2A}{b}$ **77.** $h = \frac{3V}{\pi r^2}$ **79.** $W = \frac{P - 2L}{2}$ **81.** $m = \frac{y - b}{x}$ **83.** $y = \frac{C - Ax}{B}$ **85.** $r = \frac{M - C}{C}$ **87.** $a = \frac{P - 2b}{2}$

We give one possible answer for Exercises 89–95. There are other correct forms.

89. $y = -6x + 4$ **91.** $y = 5x - 2$ **93.** $y = \frac{3}{5}x - 3$ **95.** $y = \frac{1}{3}x - 4$

SECTION 2.6 (pages 180–185)

1. compare; A, D **2.** ratios; proportion; cross products **3.** **(a)** C **(b)** D **(c)** B **(d)** A **4.** C, E **5.** $\frac{6}{7}$ **7.** $\frac{18}{55}$ **9.** $\frac{5}{16}$ **11.** $\frac{4}{15}$ **13.** $\frac{6}{5}$ **15.** 10 lb; \$0.749 **17.** 64 oz; \$0.047 **19.** 32 oz; \$0.531 **21.** 32 oz; \$0.056 **23.** $\{35\}$ **25.** $\{7\}$ **27.** $\left\{\frac{8}{5}\right\}$ **29.** $\{2\}$ **31.** $\{-1\}$ **33.** $\{5\}$

35. $\left\{-\frac{31}{5}\right\}$ **37.** \$30.00 **39.** \$56.85 **41.** 50,000 fish **43.** 4 ft **45.** $\frac{87}{100}$; 209.67 million drivers **47.** $2\frac{5}{8}$ cups **49.** \$391.53 **51.** 9; $x = 4$ **53.** $x = 8$ **55.** $x = 1.5$; $y = 4.5$ **57.** side of triangle labeled Chair: 18 ft; side of triangle labeled Pole: 12 ft; One proportion is $\frac{x}{12} = \frac{18}{4}$; 54 ft **59. (a)** 2625 mg **(b)** $\frac{125\text{ mg}}{5\text{ mL}} = \frac{2625\text{ mg}}{x\text{ mL}}$ **(c)** 105 mL **61.** \$238 **63.** \$278 **65.** C **66.** A **67.** 109.2 **69.** 700 **71.** 425 **73.** 8% **75.** 120% **77.** \$119.25; \$675.75 **79.** 80% **81.** \$3000 **83.** \$304 **85.** 14,825,000; • ; 153,889,000; 9.6% **87.** 79.3% **89.** 892% **91.** 30 **92. (a)** $5x = 12$ **(b)** $\left\{\frac{12}{5}\right\}$ **93.** $\left\{\frac{12}{5}\right\}$ **94.** Both methods give the same solution set.

SUMMARY EXERCISES Applying Problem-Solving Techniques (pages 186–187)

1. 48 **2.** 80° **3.** 4 **4.** 104°, 104° **5.** 3 **6.** 18, 20 **7.** 140°, 40° **8.** 36 quart cartons **9.** 24.34 in.; 727.28 in.^{2} **10.** 6.2% **11.** Barcelona: \$7.9 million; Real Madrid: \$7.4 million; New York Yankees: \$6.8 million **12.** 12.42 cm **13.** $16\frac{2}{3}$% **14.** 510 calories **15.** 4000 calories **16.** Charl Schwartzel: 274; Adam Scott: 276 **17.** 24 oz; \$0.074 **18.** 14 gold medals **19.** \$49.21 **20.** 71.2%

SECTION 2.7 (pages 197–200)

1. $<$ (or $>$); $>$ (or $<$); $\le$ (or $\ge$); $\ge$ (or $\le$) **2.** false **3.** $(0, \infty)$ **4.** $(-\infty, \infty)$ **5.** $x > -4$ **6.** $x \ge -4$ **7.** $x \le 4$ **8.** $x < 4$ **9.** $-1 < x \le 2$ **10.** $-1 \le x < 2$

11. $(-\infty, 4]$

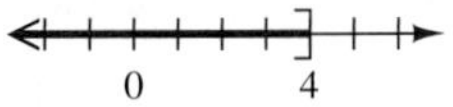

13. $(-3, \infty)$

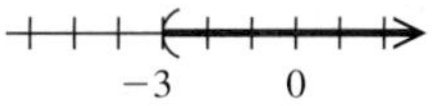

15. $[8, 10]$

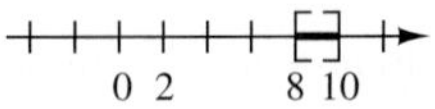

17. $(0, 10]$

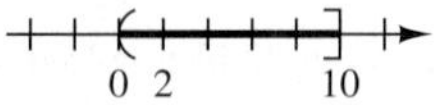

19. $[1, \infty)$

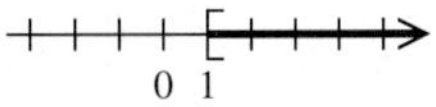

21. $[5, \infty)$

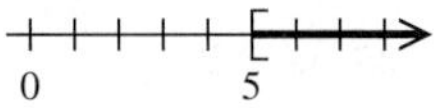

23. $(-\infty, -6)$

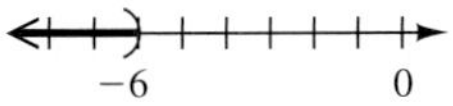

25. It must be reversed when multiplying or dividing by a negative number.

26. Divide by -5 and reverse the direction of the inequality symbol to get $x < -4$.

27. $(-\infty, 6)$

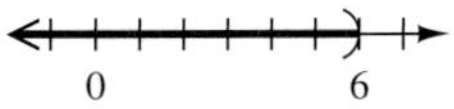

29. $[-10, \infty)$

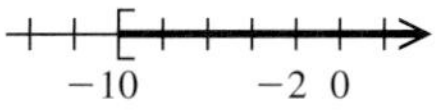

31. $(-\infty, -3)$

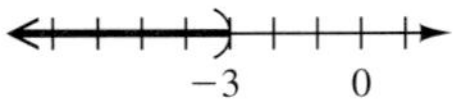

33. $(-\infty, 0]$

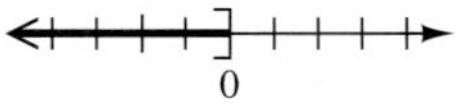

35. $(20, \infty)$

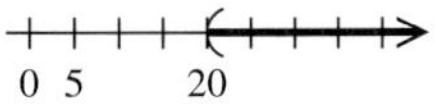

37. $[-3, \infty)$

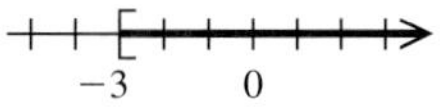

39. $(-\infty, -3]$

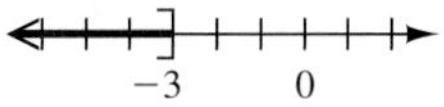

41. $(-1, \infty)$

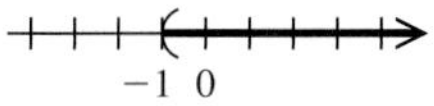

43. $[-5, \infty)$

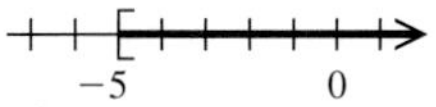

45. $(-\infty, 1)$

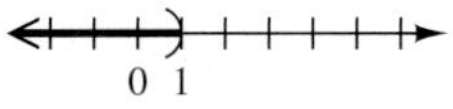

47. $(-\infty, 0]$

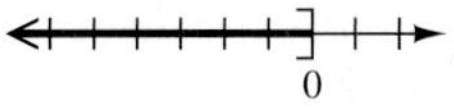

49. $\left(-\frac{1}{2}, \infty\right)$

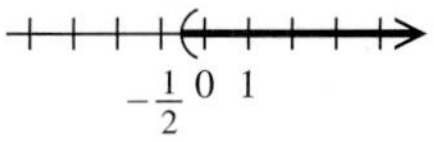

51. $[4, \infty)$

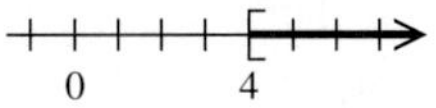

53. $(-\infty, 32)$

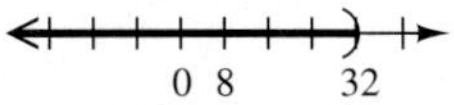

55. $\left[\frac{5}{12}, \infty\right)$

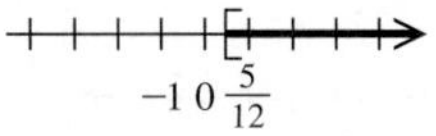

57. $(-21, \infty)$

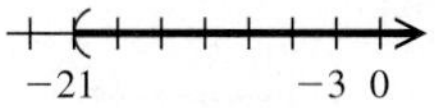

59. $x \ge 16$ **60.** $x < 1$ **61.** $x > 8$ **62.** $x \ge 12$ **63.** $x \le 20$ **64.** $x > 40$ **65.** 88 or more **67.** 80 or more **69.** all numbers greater than 16 **71.** It has never exceeded 40°C. **73.** 32 or greater **75.** 12 min

77. $[-1, 6]$

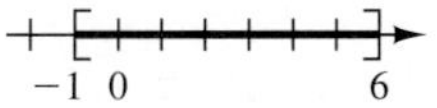

79. $(1, 3)$

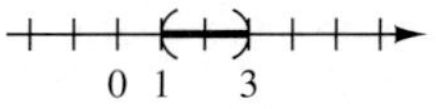

81. $\left(-\frac{11}{6}, -\frac{2}{3}\right]$

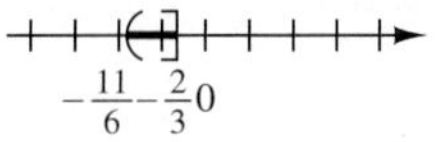

83. $[-26, 6]$

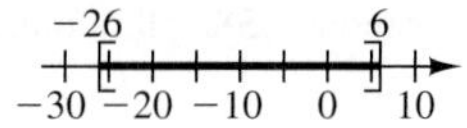

85. $[-3, 6]$

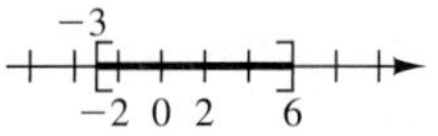

87. $\{4\}$

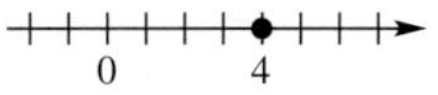

88. (a) $(-\infty, 4)$

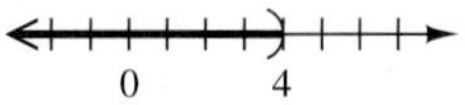

(b) $(4, \infty)$

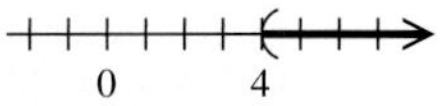

89. It is the set of all real numbers.

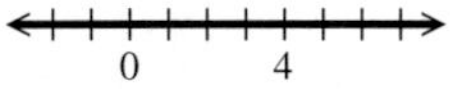

90. The graph would be the set of all real numbers; $(-\infty, \infty)$

Chapter 2 REVIEW EXERCISES (pages 205–208)

1. $\{9\}$ **2.** $\{4\}$ **3.** $\{-6\}$ **4.** $\left\{\frac{3}{2}\right\}$ **5.** $\{20\}$ **6.** $\left\{-\frac{61}{2}\right\}$ **7.** $\{15\}$ **8.** $\{0\}$ **9.** $\emptyset$ **10.** {all real numbers} **11.** $-\frac{7}{2}$ **12.** 20 **13.** Hawaii: 6425 mi^2; Rhode Island: 1212 mi^2 **14.** Seven Falls: 300 ft; Twin Falls: 120 ft **15.** 80° **16.** 11, 13 **17.** $h = 11$ **18.** $A = 28$ **19.** $r = 4.75$ **20.** $V = 3052.08$ **21.** $h = \frac{A}{b}$ **22.** $h = \frac{2A}{b + B}$ **23.** (Other correct forms of the answers are possible.) **(a)** $y = -x + 11$ **(b)** $y = \frac{3}{2}x - 6$ **24.** 135°, 45° **25.** 100°, 100° **26.** perimeter: 326.5 ft; area: 6538.875 ft^2 **27.** diameter: 46.78 ft; radius: 23.39 ft **28.** $\frac{3}{2}$ **29.** $\frac{5}{14}$

ANSWERS

30. $\frac{3}{4}$ **31.** $\left\{\frac{7}{2}\right\}$ **32.** $\left\{-\frac{8}{3}\right\}$ **33.** $\left\{\frac{25}{19}\right\}$ **34.** 40% means $\frac{40}{100}$, or $\frac{2}{5}$. It is the same as the ratio 2 to 5. **35.** $6\frac{2}{3}$ lb **36.** 36 oz **37.** 375 km **38.** 18 oz; \$0.249 **39.** 6 **40.** 175% **41.** $33\frac{1}{3}$% **42.** 2500 **43.** \$27,160 **44.** \$350.46

45. $[-4, \infty)$

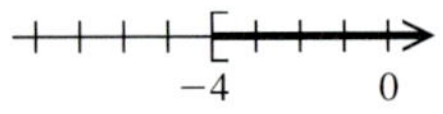

46. $(-\infty, 7)$

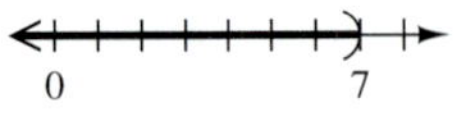

47. $[-5, 6)$

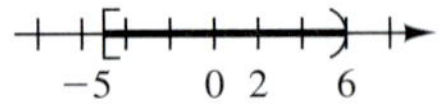

48. $\left[\frac{1}{2}, \infty\right)$

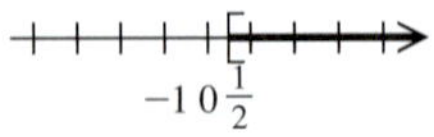

49. $[-3, \infty)$

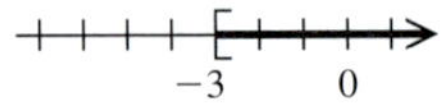

50. $(-\infty, 2)$

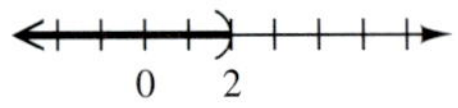

51. $[3, \infty)$

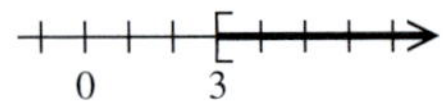

52. $[46, \infty)$

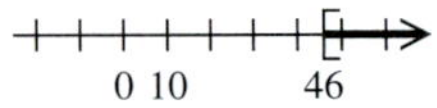

53. $(-\infty, -5)$

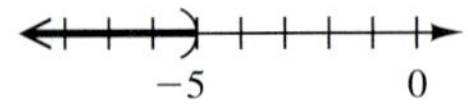

54. $(-\infty, -37)$

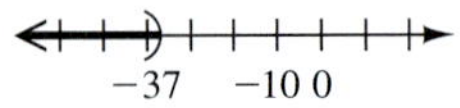

55. $\left[-2, \frac{3}{2}\right)$

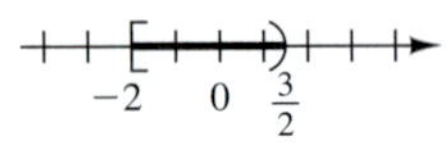

56. $(1, 5]$

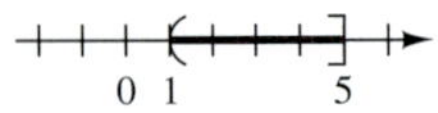

57. 88 or more **58.** all numbers less than or equal to $-\frac{1}{3}$ **59.** $\{7\}$ **60.** $r = \frac{d}{2}$ **61.** $(-\infty, 2)$ **62.** $\{-9\}$ **63.** $\{70\}$ **64.** $\left\{\frac{13}{4}\right\}$ **65.** $\emptyset$ **66.** {all real numbers} **67.** DiGiorno: \$668.7 million; Red Baron: \$268.8 million **68.** 2.0 in. **69.** gold: 14; silver: 7; bronze: 5 **70.** D: 22°; E: 44°; F: 114° **71.** 44 m **72.** 70 ft **73.** 160 oz; \$0.062 **74.** 24°, 66° **75.** 92 or more **76.** \$197.50

Chapter 2 TEST (pages 209–210)

1. $\{6\}$ **2.** $\{-6\}$ **3.** $\left\{\frac{13}{4}\right\}$ **4.** $\{-10.8\}$ **5.** $\{21\}$ **6.** {all real numbers} **7.** $\{30\}$ **8.** $\emptyset$ **9.** wins: 62; losses: 20 **10.** Hawaii: 4021 mi^2; Maui: 728 mi^2; Kauai: 551 mi^2 **11.** 24, 26 **12.** 50° **13. (a)** $W = \frac{P - 2L}{2}$ **(b)** 18 **14.** $y = \frac{5}{4}x - 2$ (Other correct forms of the answer are possible.) **15.** 100°, 80° **16.** 75°, 75° **17.** $\{6\}$ **18.** $\{-29\}$ **19.** 16 oz; \$0.249 **20.** 2300 mi **21.** \$26.82; \$122.18 **22.** 40% **23. (a)** $x < 0$ **(b)** $-2 < x \le 3$

24. $(-\infty, 11)$

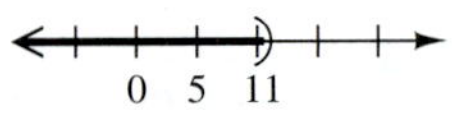

25. $[-3, \infty)$

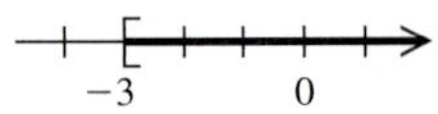

26. $(-\infty, 4]$

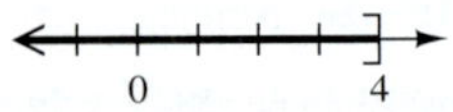

27. $(-2, 6]$

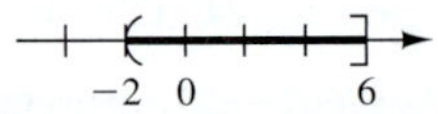

28. 83 or more

Chapters R–2 CUMULATIVE REVIEW EXERCISES (pages 211–212)

1. $\frac{3}{8}$ **2.** $\frac{3}{4}$ **3.** $\frac{31}{20}$ **4.** $\frac{551}{40}$, or $13\frac{31}{40}$ **5.** 6 **6.** $\frac{6}{5}$ **7.** 34.03 **8.** 27.31 **9.** 30.51 **10.** 56.3 **11.** 35 yd **12.** $7\frac{1}{2}$ cups **13.** $3\frac{3}{8}$ in. **14.** \$2599.94 **15.** true **16.** true **17.** 7 **18.** 1 **19.** 13 **20.** −40 **21.** −12 **22.** undefined **23.** −6 **24.** 28 **25.** 1 **26.** 0 **27.** $\frac{73}{18}$ **28.** −64 **29.** −134 **30.** $-\frac{29}{6}$ **31.** distributive property **32.** commutative property **33.** inverse property **34.** identity property **35.** $7p - 14$ **36.** $2k - 11$ **37.** $\{7\}$ **38.** $\{-4\}$ **39.** $\{-1\}$ **40.** $\left\{-\frac{3}{5}\right\}$ **41.** $\{2\}$ **42.** $\{-13\}$ **43.** $\{26\}$ **44.** $\{-12\}$ **45.** $c = P - a - b - B$ **46.** $s = \frac{P}{4}$

47. $(-\infty, 2]$

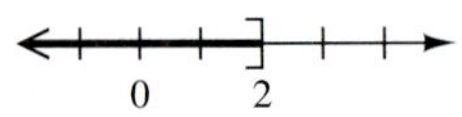

48. $(-\infty, 1)$

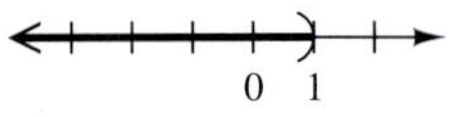

49. \$230.50 **50.** \$7125 **51.** 30 cm **52.** 16 in.

CHAPTER 3 Graphs of Linear Equations and Inequalities in Two Variables

SECTION 3.1 (pages 222–227)

1. does; do not **2.** II **3.** y **4.** 3 **5.** 6 **6.** −4 **7.** No. For two ordered pairs to be equal, their x-values must be equal and their y-values must be equal. Here we have $4 \ne -1$ and $-1 \ne 4$. **8.** Substituting $\frac{1}{3}$ for x in $y = 6x + 2$ gives $y = 6\left(\frac{1}{3}\right) + 2$. Here $6\left(\frac{1}{3}\right) = 2$, so $y = 2 + 2$, or 4. Because $6\left(\frac{1}{7}\right) = \frac{6}{7}$, calculating y for $x = \frac{1}{7}$ requires working with fractions. **9.** negative; negative **10.** negative; positive **11.** positive; negative **12.** positive; positive **13.** 2007, 2008, 2009, 2010 **15.** 2004: about 171 billion lb; 2010: about 193 billion lb **16.** U.S. milk production increased about 22 billion lb during these years, from about 171 billion lb to about 193 billion lb. **17.** 2009 to 2010; about 1.5 million **19.** The number of cars imported each year was decreasing. **21.** yes **23.** yes **25.** no **27.** yes **29.** no **31.** no **33.** 11 **35.** $-\frac{7}{2}$ **37.** −4 **39.** −5 **41.** 4, 6, −6; (0, 4), (6, 0), (−6, 8) **43.** 3, −5, −15; (0, 3), (−5, 0), (−15, −6) **45.** −9, −9, −9; (−9, 6), (−9, 2), (−9, −3) **47.** −6, −6, −6; (8, −6), (4, −6), (−2, −6) **49.** 8, 8, 8; (8, 8), (8, 3), (8, 0) **51.** −2, −2, −2; (9, −2), (2, −2), (0, −2) **53.** (2, 4); I **55.** (−5, 4); II **57.** (3, 0); no quadrant **59.** If $xy < 0$, then either $x < 0$ and $y > 0$ or $x > 0$ and $y < 0$. If $x < 0$ and $y > 0$, then the point lies in quadrant II. If $x > 0$ and $y < 0$, then the point lies in quadrant IV. **60.** If $xy > 0$, then either $x > 0$ and $y > 0$ or $x < 0$ and $y < 0$. If $x > 0$ and $y > 0$, then the point lies in quadrant I. If $x < 0$ and $y < 0$, then the point lies in quadrant III.

61–72.

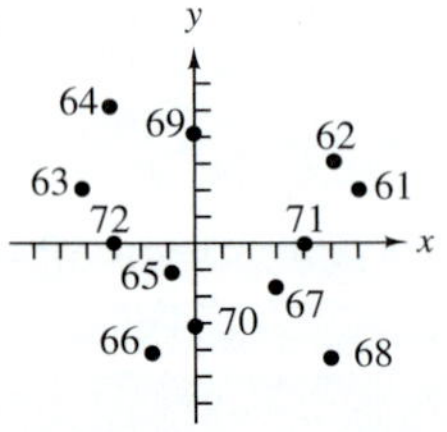

73. −3, 6, −2, 4

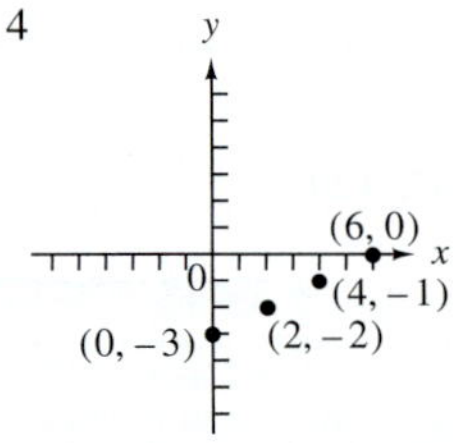

ANSWERS

75. $-3, 4, -6, -\frac{4}{3}$

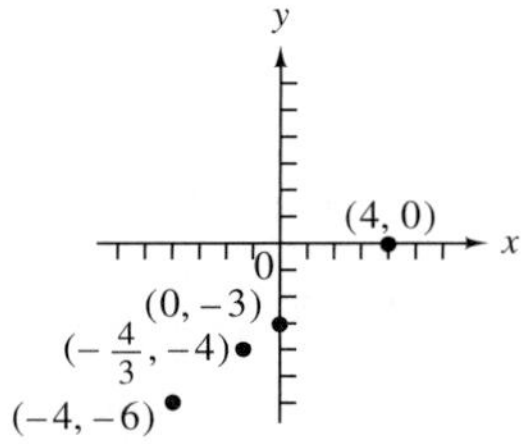

77. $-4, -4, -4, -4$

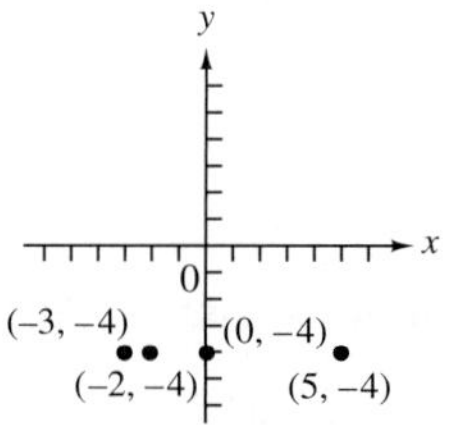

79. The points in each graph appear to lie on a straight line.

80. **(a)** They are the same (all -4). **(b)** They are the same (all 5).

81. **(a)** $(5, 45)$ **(b)** $(6, 50)$ **83.** **(a)** $(2007, 27.1)$, $(2008, 29.3)$, $(2009, 28.3)$, $(2010, 28.0)$, $(2011, 26.9)$ **(b)** $(2000, 32.4)$ means that 32.4 percent of 2-year college students in 2000 received a degree within 3 years.

(c) **2-YEAR COLLEGE STUDENTS COMPLETING A DEGREE WITHIN 3 YEARS**

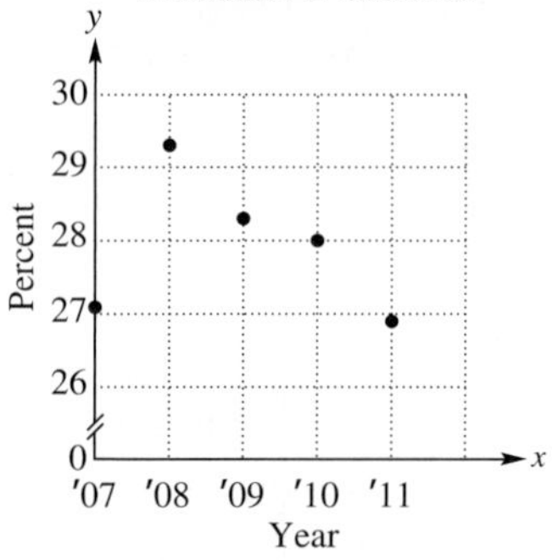

(d) With the exception of the point for 2007, the points lie approximately on a straight line. Rates at which 2-year college students complete a degree within 3 years were generally decreasing.

85. **(a)** 98, 88, 78, 68 **(b)** $(20, 98)$, $(40, 88)$, $(60, 78)$, $(80, 68)$

(c) **TARGET HEART RATE ZONE (Lower Limit)**

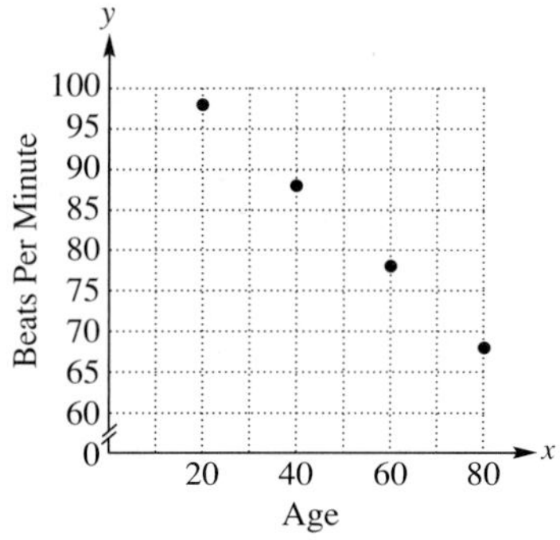

Yes the points lie in a linear pattern.

87. between 98 and 157 beats per minute; between 88 and 141 beats per minute

SECTION 3.2 (pages 236–241)

1. *By*; *C*; 0 **2.** line; solution **3.** **(a)** A **(b)** C **(c)** D **(d)** B **4.** A, C, D **5.** *x*-intercept: $(4, 0)$; *y*-intercept: $(0, -4)$ **6.** *x*-intercept: $(-5, 0)$; *y*-intercept: $(0, 5)$ **7.** *x*-intercept: $(-2, 0)$; *y*-intercept: $(0, -3)$ **8.** *x*-intercept: $(4, 0)$; *y*-intercept: $(0, 3)$

9. 5, 5, 3

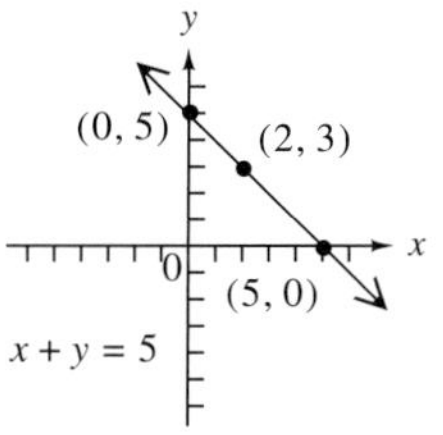

11. 1, 3, −1

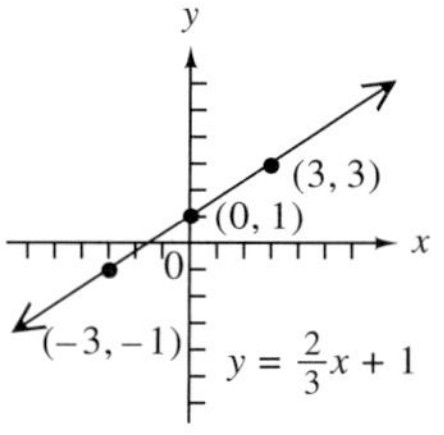

13. −6, −2, −5

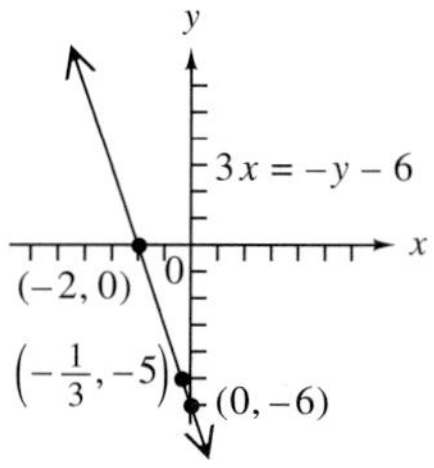

15. 0; $(0, -8)$; 0; $(8, 0)$

17. $(0, -8)$; $(12, 0)$

19. $(0, 0)$; $(0, 0)$

21.

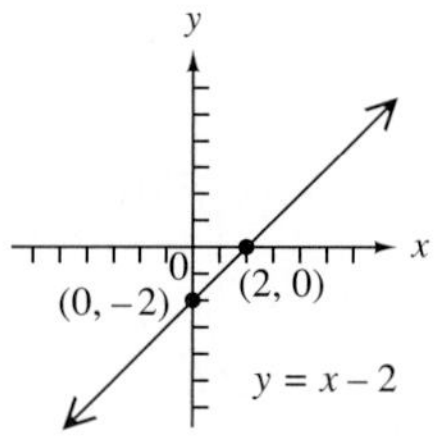

23.

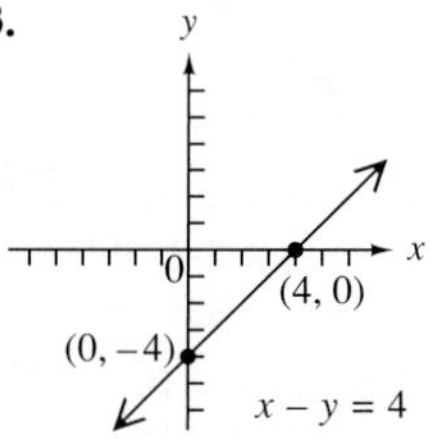

25.

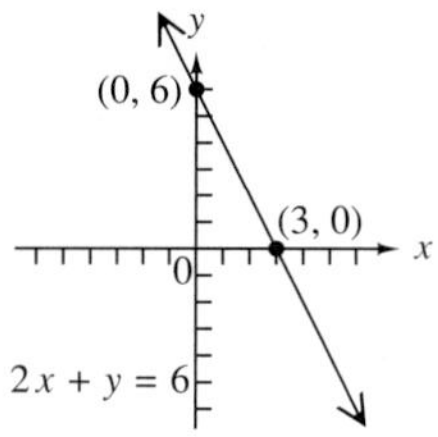

27.

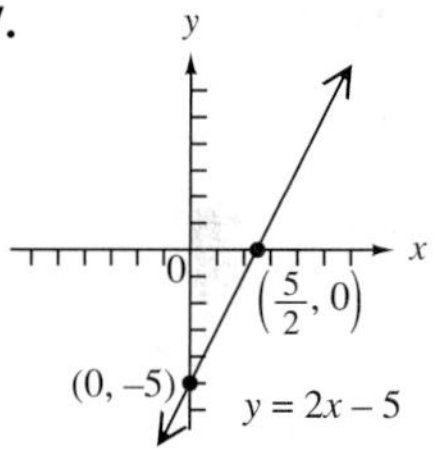

29.

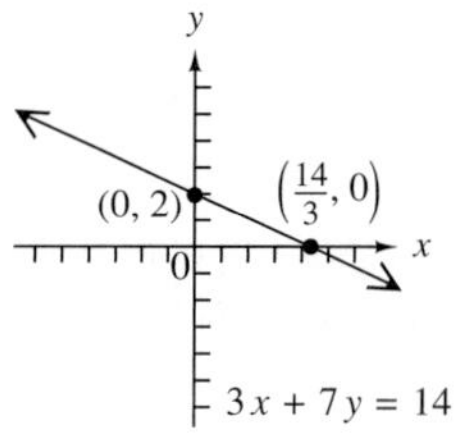

31.

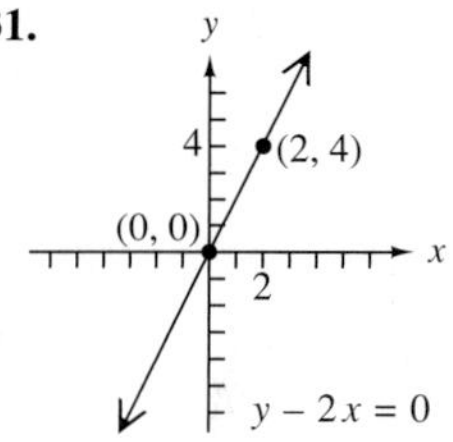

33.

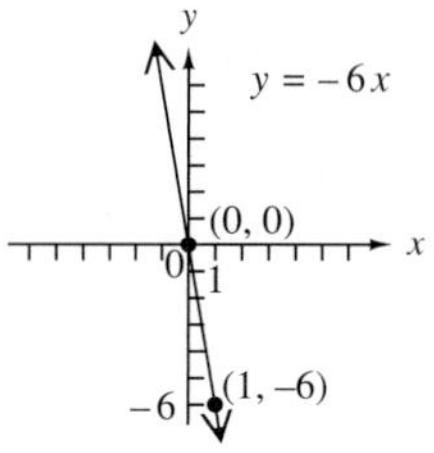

35.

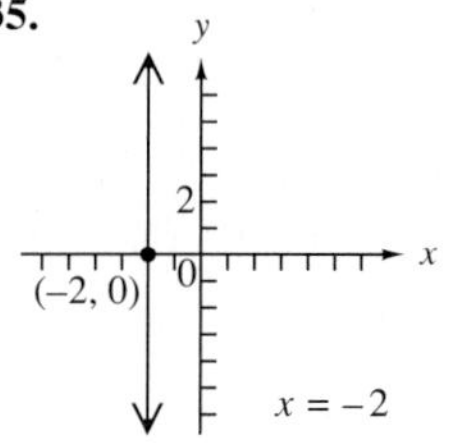

37.

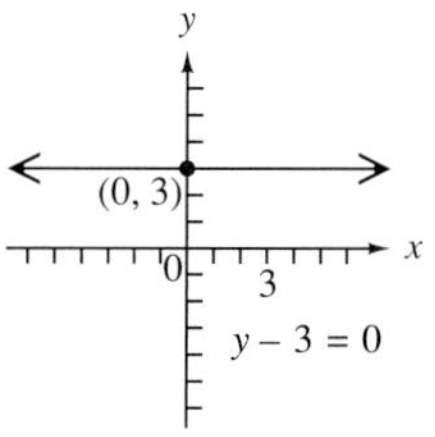

39.

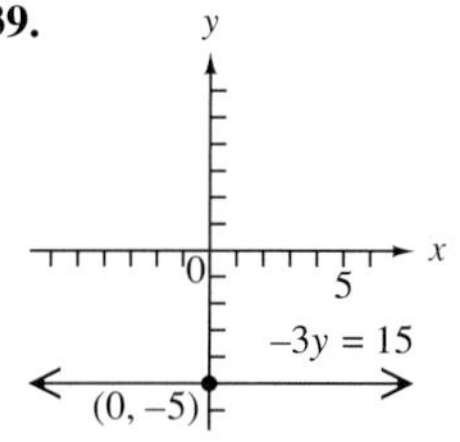

41. **(a)** D **(b)** C **(c)** B **(d)** A **42.** $y = 0$; $x = 0$

In Exercises 43–46, descriptions may vary.

43. The graph is a line with *x*-intercept $(-3, 0)$ and *y*-intercept $(0, 9)$. **44.** The graph is a vertical line with *x*-intercept $(11, 0)$. **45.** The graph is a horizontal line with *y*-intercept $(0, -2)$. **46.** The graph passes through the origin $(0, 0)$ and the points $(2, 1)$ and $(4, 2)$.

47. **(a)** 151.5 cm, 159.3 cm, 174.9 cm **(b)** (20, 151.5), (22, 159.3), (26, 174.9)

(c) HEIGHTS OF WOMEN

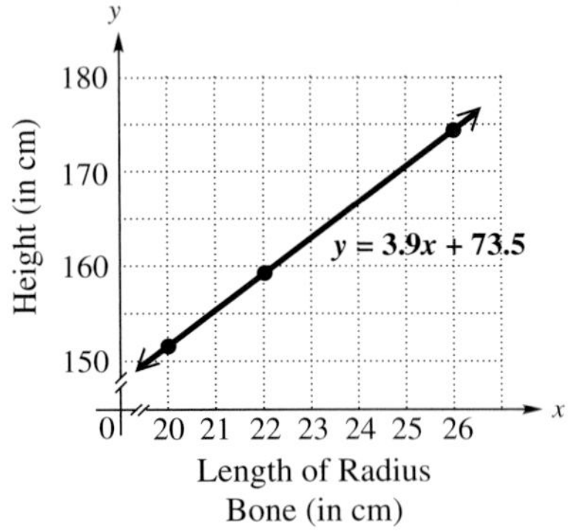

(d) 24 cm; 24 cm

49. **(a)** \$62.50; \$100 **(b)** 200 **(c)** (50, 62.50), (100, 100), (200, 175)

(d) POSTER COSTS

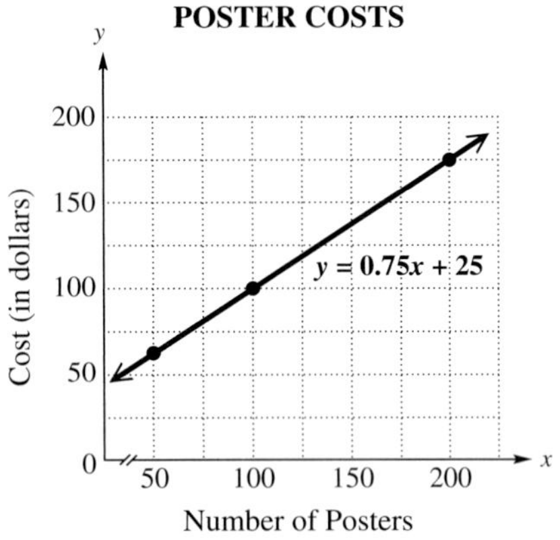

51. **(a)** \$30,000 **(b)** \$15,000 **(c)** \$5000 **(d)** After 5 yr, the SUV has a value of \$5000. **53.** **(a)** 1990: 23.8 lb; 2000: 29.0 lb; 2009: 33.8 lb **(b)** 1990: 25 lb; 2000: 30 lb; 2009: 33 lb **(c)** The values are quite close.

SECTION 3.3 (pages 250–253)

1. steepness; vertical; horizontal **2.** ratio; y; rise; x; run **3.** **(a)** 6 **(b)** 4 **(c)** $\frac{6}{4}$, or $\frac{3}{2}$; slope of the line **(d)** Yes, it doesn't matter which point we start with. The slope would be expressed as the quotient of -6 and -4, which simplifies to $\frac{3}{2}$.

4. **(a)** C **(b)** A **(c)** D **(d)** B **5.** 4 **7.** $-\frac{1}{2}$

9. Answers will vary.

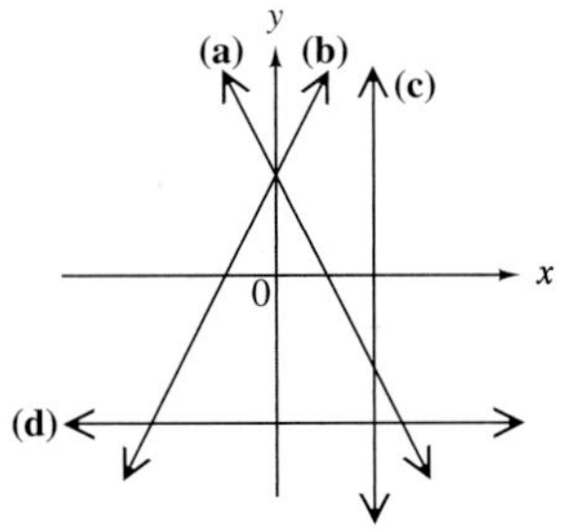

10. **(a)** falls from left to right **(b)** horizontal **(c)** vertical **(d)** rises from left to right **11.** **(a)** negative **(b)** 0 **12.** **(a)** negative **(b)** negative **13.** **(a)** positive **(b)** negative **14.** **(a)** positive **(b)** 0 **15.** **(a)** 0 **(b)** negative **16.** **(a)** 0 **(b)** positive **17.** $\frac{8}{27}$ **18.** $\frac{3}{10}$

19. $-\frac{2}{3}$ **20.** $-\frac{1}{4}$ **21.** Because the student found the difference $3 - 5 = -2$ in the numerator, he or she should have subtracted in the same order in the denominator to get $-1 - 2 = -3$. The correct slope is $\frac{-2}{-3} = \frac{2}{3}$.

22. Slope is defined as $\frac{\text{change in } y}{\text{change in } x}$, but the student found $\frac{\text{change in } x}{\text{change in } y}$. The correct slope is $\frac{-5}{8} = -\frac{5}{8}$. **23.** $\frac{5}{4}$ **25.** $\frac{3}{2}$ **27.** -3 **29.** 0

31. undefined **33.** $\frac{1}{4}$ **35.** $-\frac{1}{2}$ **37.** 5 **39.** $\frac{1}{4}$ **41.** $\frac{3}{2}$ **43.** $-\frac{3}{2}$

45. 0 **47.** undefined **49.** 1 **51.** $\frac{4}{3}$; $\frac{4}{3}$; parallel **53.** $\frac{5}{3}$; $\frac{3}{5}$; neither

55. $\frac{3}{5}$; $-\frac{5}{3}$; perpendicular **57.** 0.42 **58.** positive; increased **59.** 0.42%

60. -0.3 **61.** negative; decreased **62.** 0.3%

SECTION 3.4 (pages 261–266)

1. **(a)** D **(b)** C **(c)** B **(d)** A **2.** A, B, D **3.** **(a)** C **(b)** B **(c)** A **(d)** D **4.** **(a)** the y-axis **(b)** the x-axis **5.** slope: $\frac{5}{2}$; y-intercept: $(0, -4)$ **7.** slope: -1; y-intercept: $(0, 9)$ **9.** slope: $\frac{1}{5}$; y-intercept: $\left(0, -\frac{3}{10}\right)$ **11.** $y = 4x - 3$ **13.** $y = -x - 7$ **15.** $y = 3$

17. $x = 0$ **19.** $y = 3x - 3$ **20.** $y = 2x - 4$ **21.** $y = -x + 3$

22. $y = -x - 2$ **23.** $y = -\frac{1}{2}x + 2$ **24.** $y = \frac{3}{2}x - 3$

25.

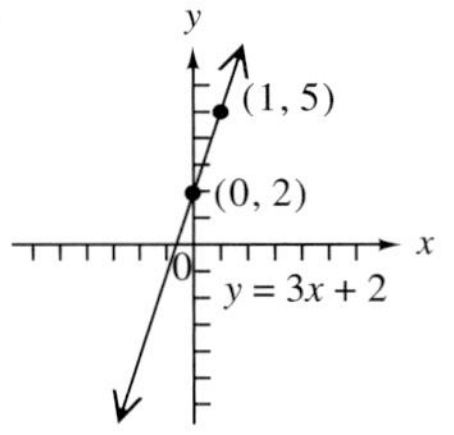

27.

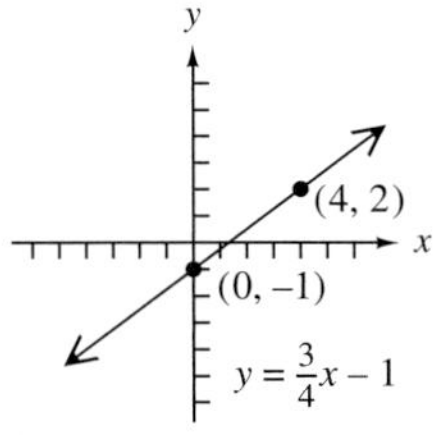

29.

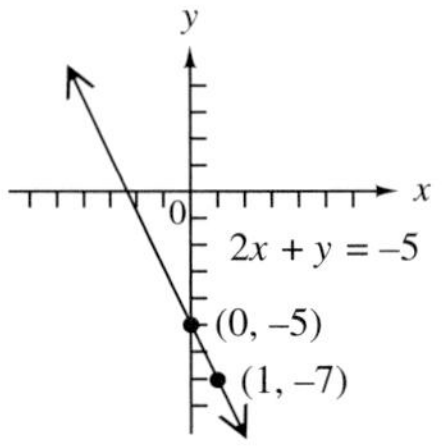

31.

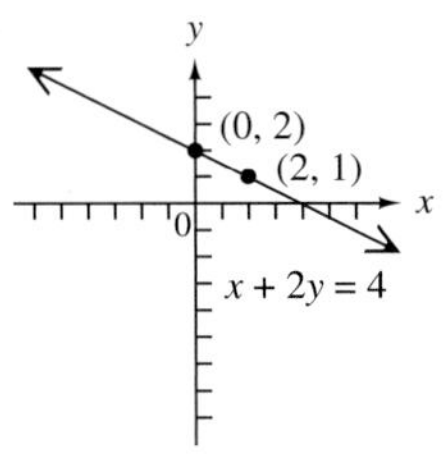

33. $y = \frac{1}{2}x + 4$

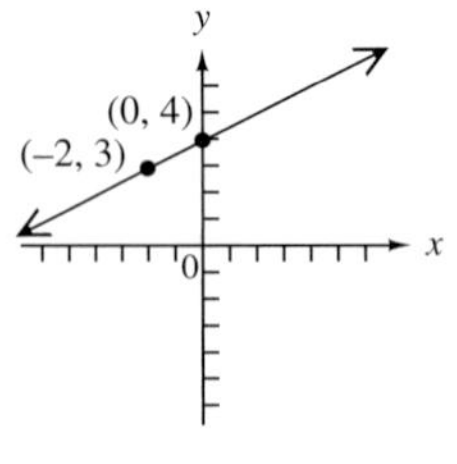

35. $y = -\frac{2}{5}x - \frac{23}{5}$

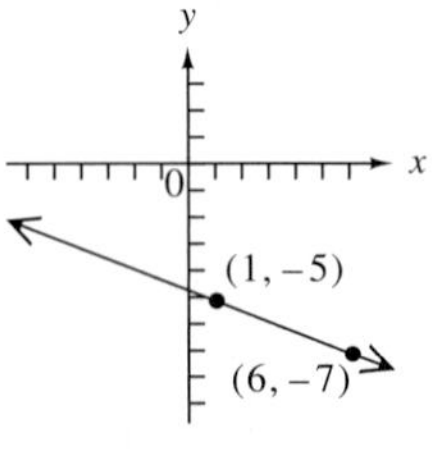

37. $y = 2$

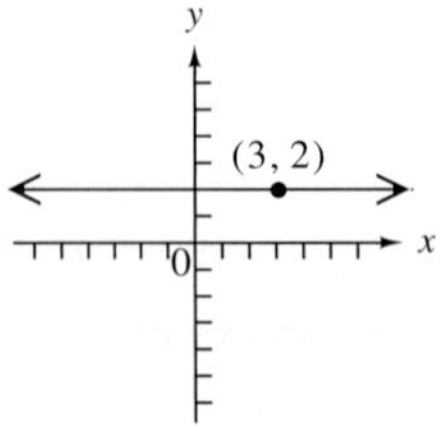

39. $x = 3$ (no slope-intercept form)

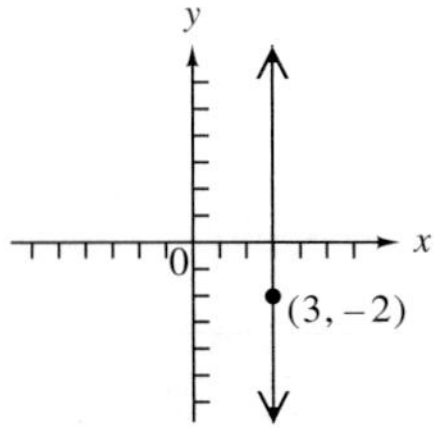

41. $y = \frac{2}{3}x$

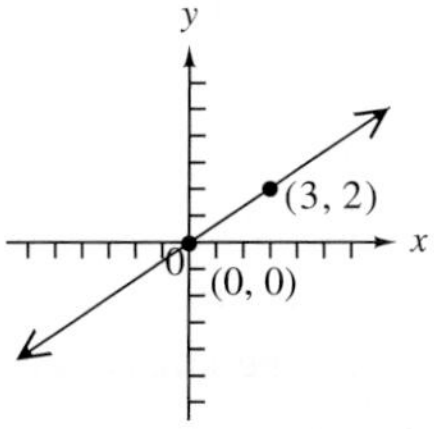

43. $y = 2x - 7$ **45.** $y = -2x - 4$ **47.** $y = \frac{2}{3}x + \frac{19}{3}$ **49.** **(a)** $y = x - 3$ **(b)** $x - y = 3$ **51.** **(a)** $y = -\frac{5}{7}x - \frac{54}{7}$ **(b)** $5x + 7y = -54$ **53.** **(a)** $y = -\frac{2}{3}x - 2$ **(b)** $2x + 3y = -6$ **55.** **(a)** $y = \frac{1}{3}x + \frac{4}{3}$ **(b)** $x - 3y = -4$ **57.** $y = -2x - 3$

59. $y = 4x - 5$ **61.** $y = \frac{3}{4}x - \frac{9}{2}$ **63.** **(a)** \$400 **(b)** \$0.25 **(c)** $y = 0.25x + 400$ **(d)** \$425 **(e)** 1500 **65.** **(a)** (1, 2294), (2, 2372), (3, 2558), (4, 2727), (5, 2963) **(b)** yes

AVERAGE ANNUAL COSTS AT 2-YEAR COLLEGES

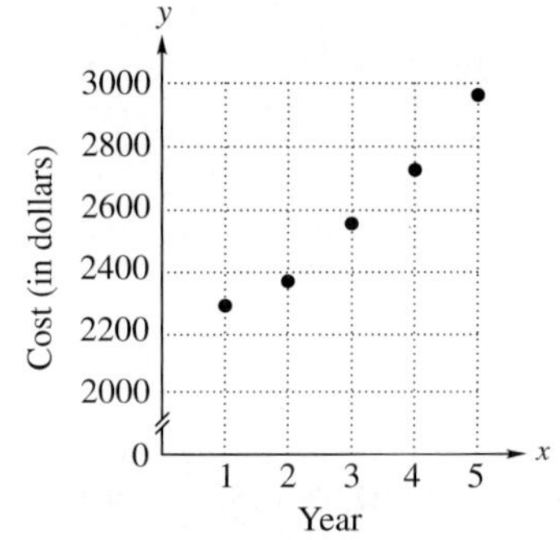

(c) $y = 197x + 1978$ **(d)** \$3160 **67.** $y = 15x - 29{,}978$

69. (0, 32); (100, 212) **70.** $\frac{9}{5}$ **71.** $F - 32 = \frac{9}{5}(C - 0)$

72. $F = \frac{9}{5}C + 32$ **73.** $C = \frac{5}{9}(F - 32)$ **74.** 86° **75.** 10° **76.** −40°

SUMMARY EXERCISES Applying Graphing and Equation-Writing Techniques for Lines (pages 267–268)

1. **(a)** B **(b)** D **(c)** A **(d)** C **2.** A, B

3.

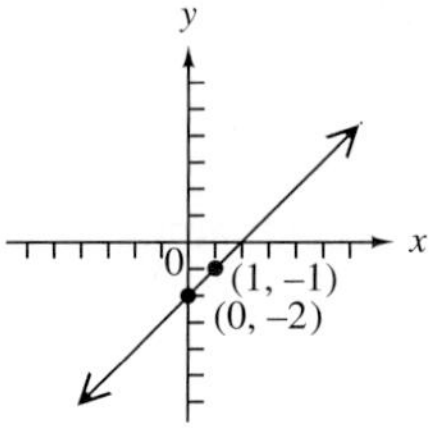

4.

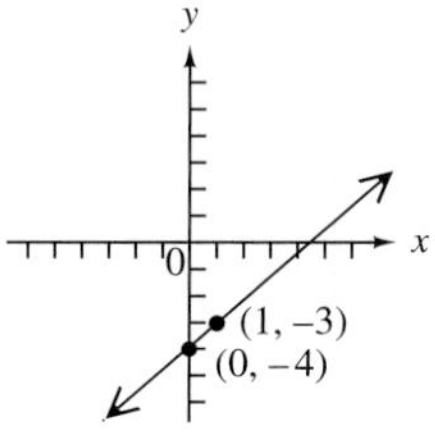

5.

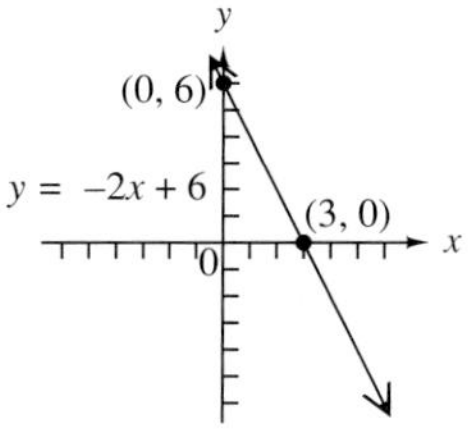

6.

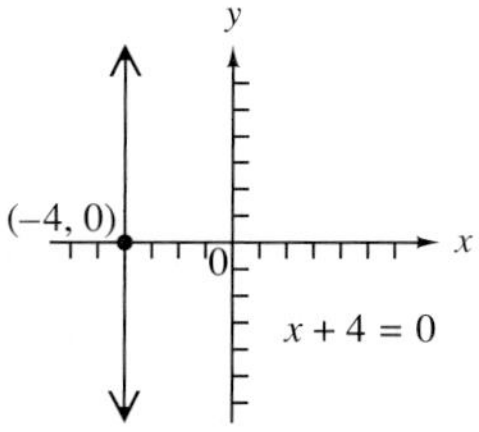

7.

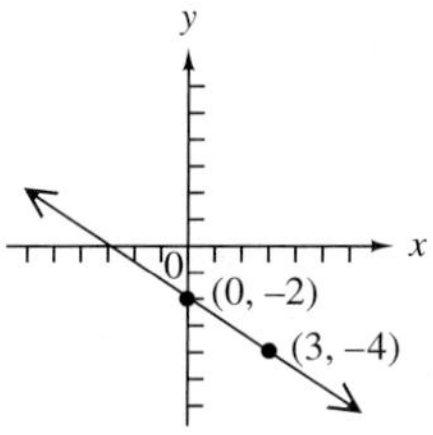

8.

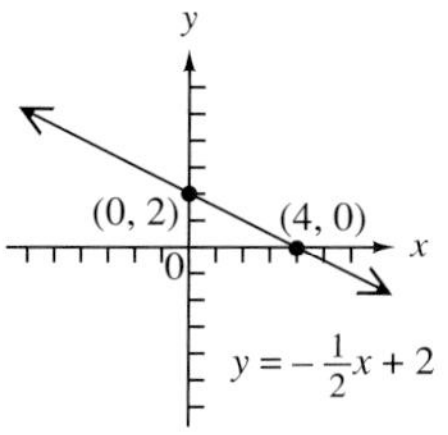

9.

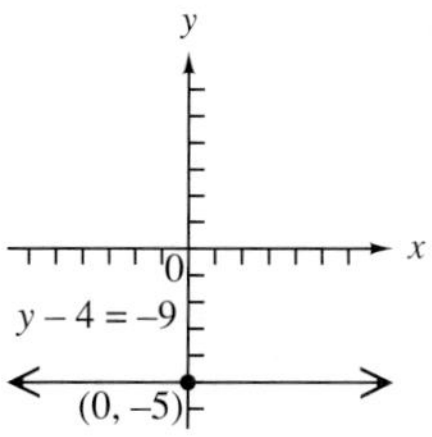

10.

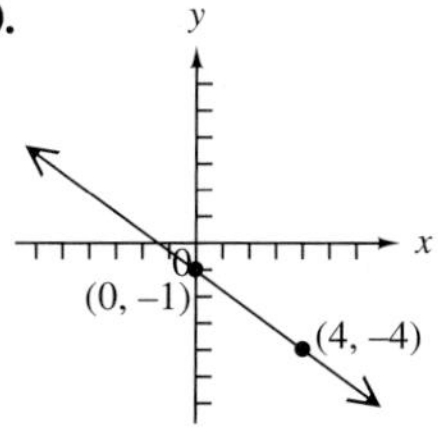

11.

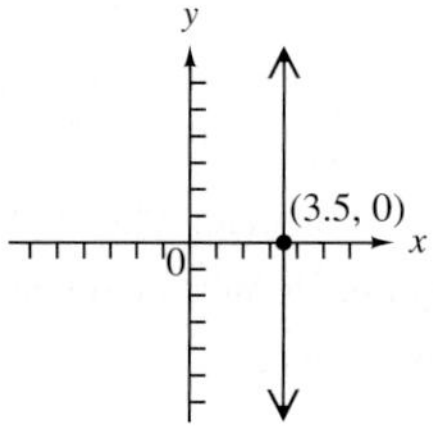

12.

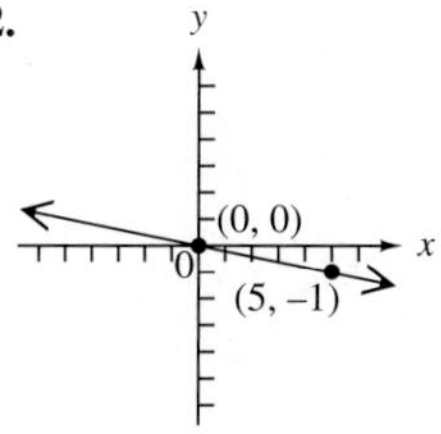

13.

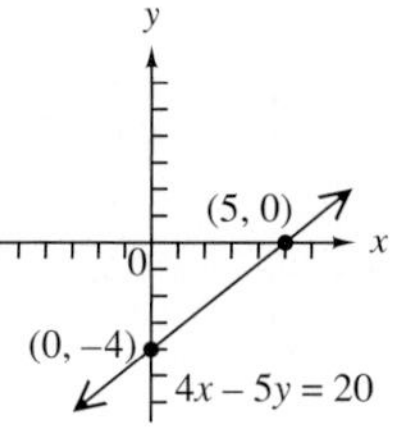

14.

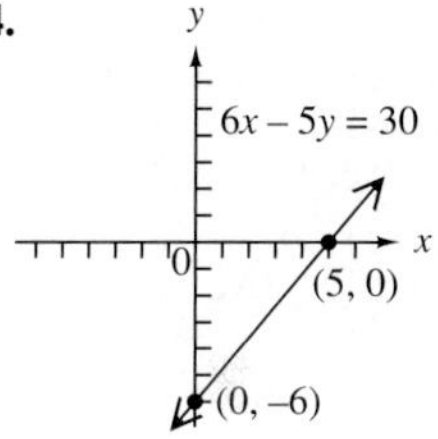

15.

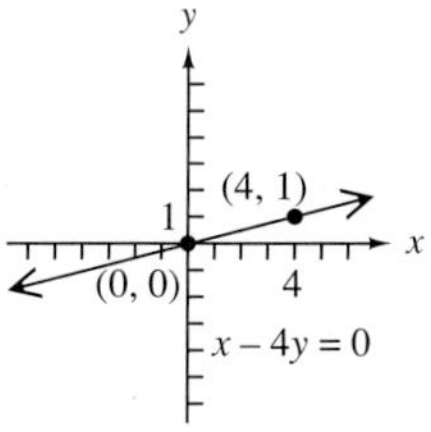

16.

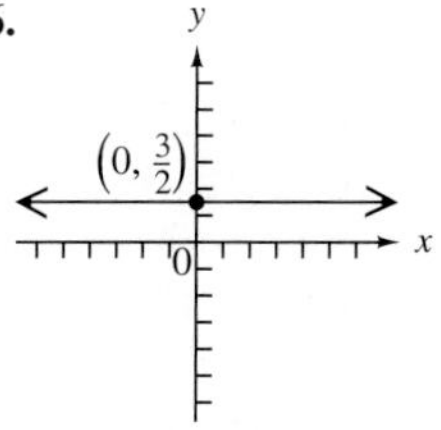

17.

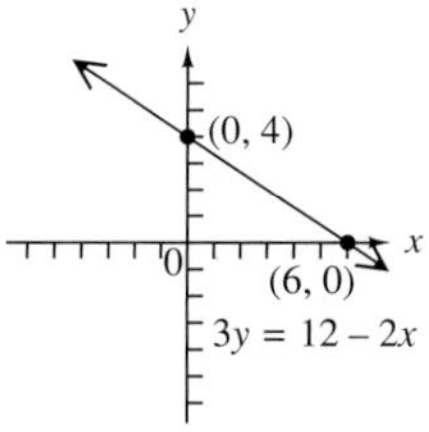

18.

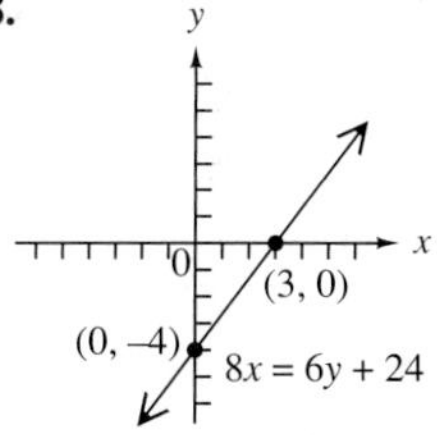

19. $y = -3x - 6$ **20.** $y = -4x - 3$ **21.** $y = \frac{3}{5}x$ **22.** $x = 0$ **23.** $y = 0$ **24.** $y = -2x - 4$ **25.** $y = \frac{5}{3}x + 5$

SECTION 3.5 (pages 273–275)

1. >, > **2.** < **3.** ≤ **4.** ≥ **5.** false; The point (4, 0) lies on the boundary line $3x - 4y = 12$, which is *not* part of the graph because the symbol < does not involve equality. **6.** true **7.** false; Because (0, 0) is on the boundary line $x + 4y = 0$, it cannot be used as a test point. Use a test point *off* the line. **8.** false; Use a solid line for the boundary line because of the equality portion of the ≥ symbol. **9.** true **10.** true

11.

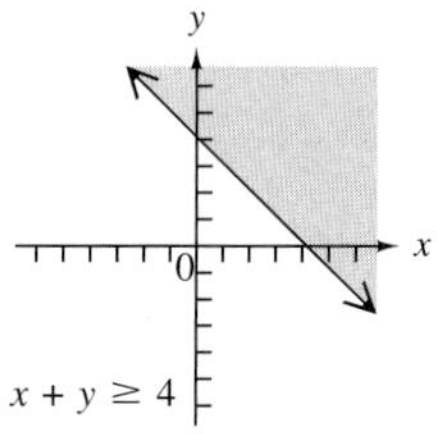

13.

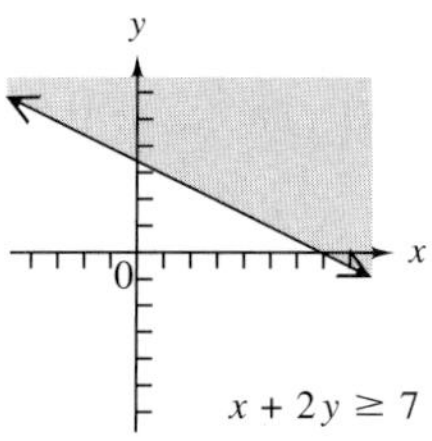

ANSWERS

15.

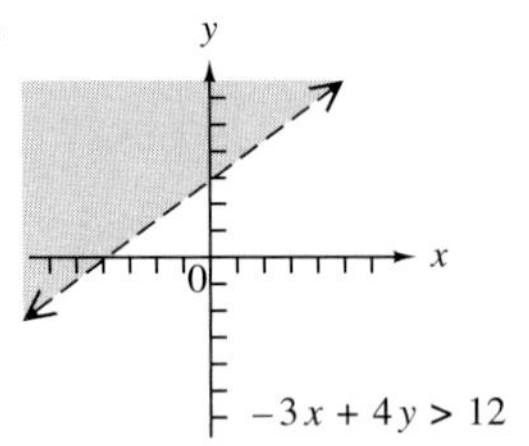

17.

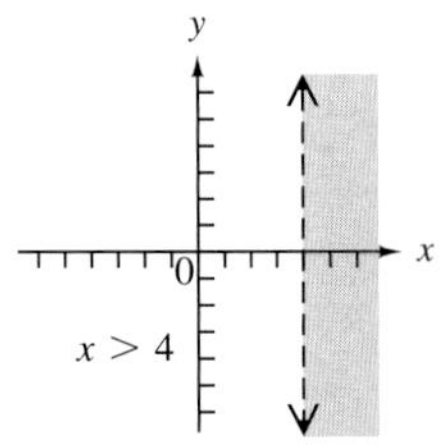

19.

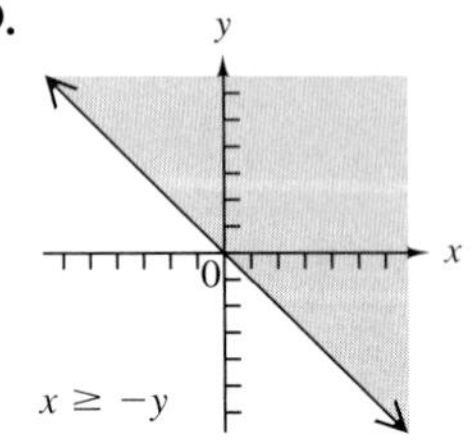

21. Use a dashed line if the symbol is < or >. Use a solid line if the symbol is ≤ or ≥.

22. A test point cannot lie on the boundary line. It must lie on one side of the boundary.

23.

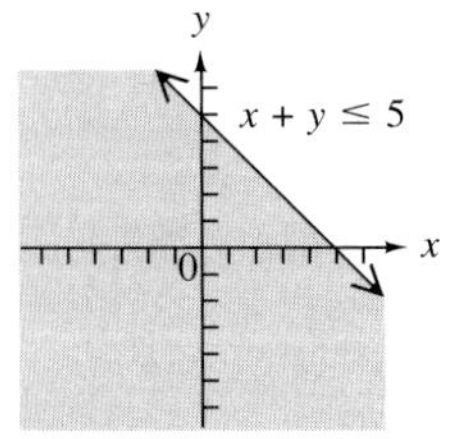

25.

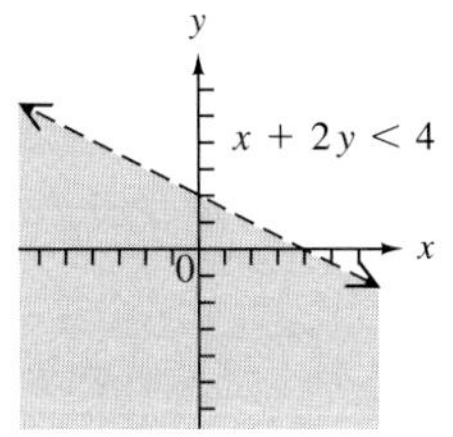

27.

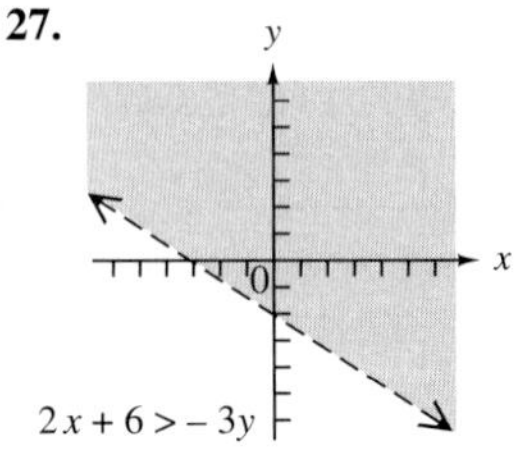

29.

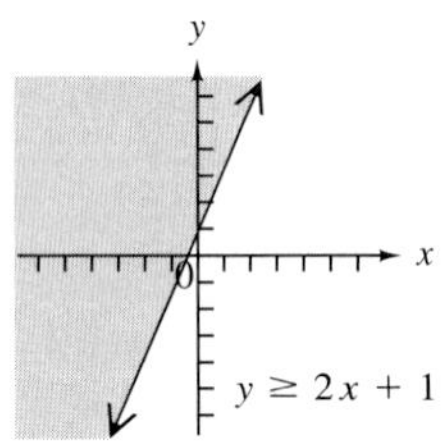

31.

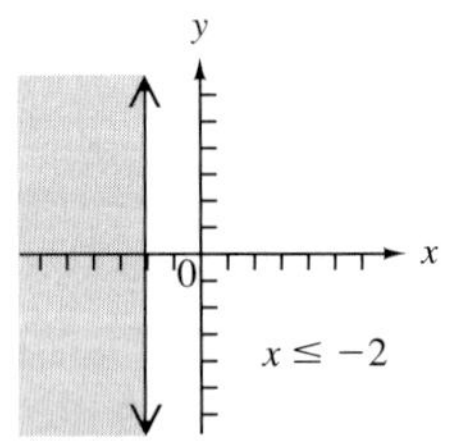

33.

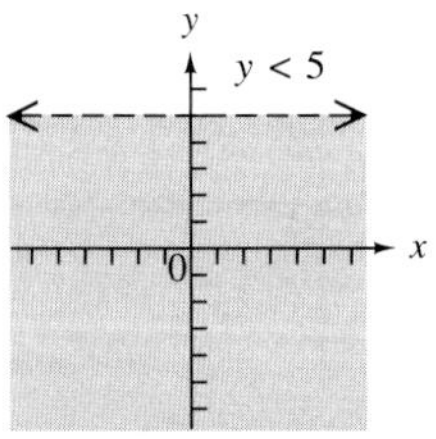

35.

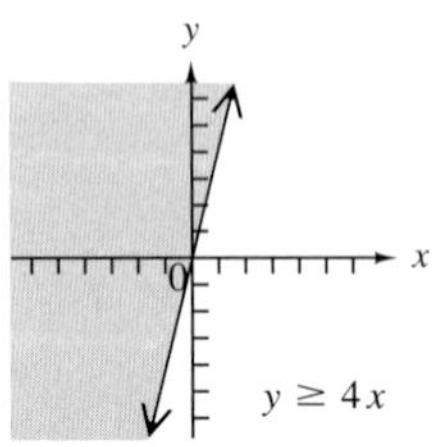

37.

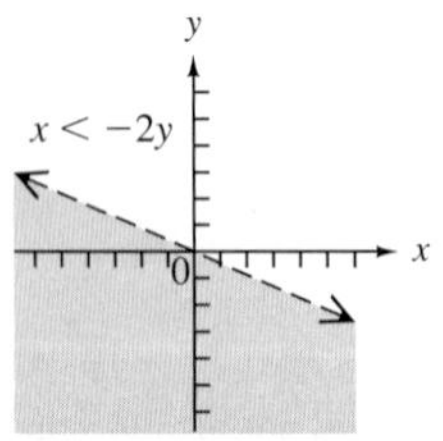

Chapter 3 REVIEW EXERCISES (pages 280–284)

1. (2006, 52.5), (2011, 55.4) **2.** In the year 2000, 52.0% of first-year college students at two-year public institutions returned for a second year. **3.** 2008 to 2009; about 53.7% **4.** 2007 to 2008; 2009 to 2010; about 2% increase each time **5.** -1; 2; 1 **6.** 2; $\frac{3}{2}$; $\frac{14}{3}$ **7.** 0; $\frac{8}{3}$; -9 **8.** 7; 7; 7 **9.** yes **10.** no **11.** yes **12.** no **13.** quadrant I **14.** quadrant II **15.** no quadrant **16.** no quadrant

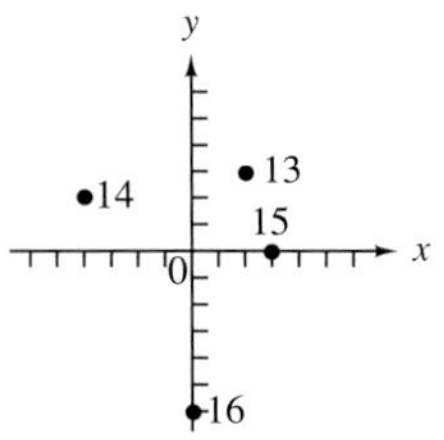

Graph for Exercises 13–16

17. x is positive in quadrants I and IV; y is negative in quadrants III and IV. Thus, if x is positive and y is negative, (x, y) must lie in quadrant IV. **18.** In the ordered pair $(k, 0)$, the y-value is 0, so the point lies on the x-axis. In the ordered pair $(0, k)$, the x-value is 0, so the point lies on the y-axis.

19.

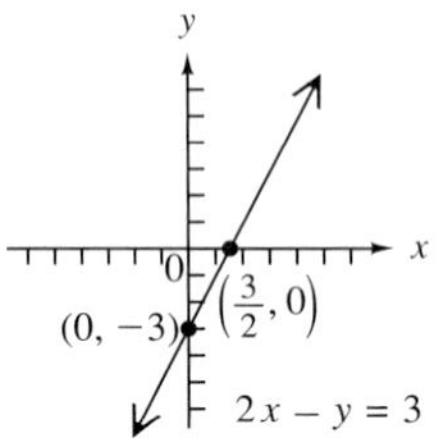

20.

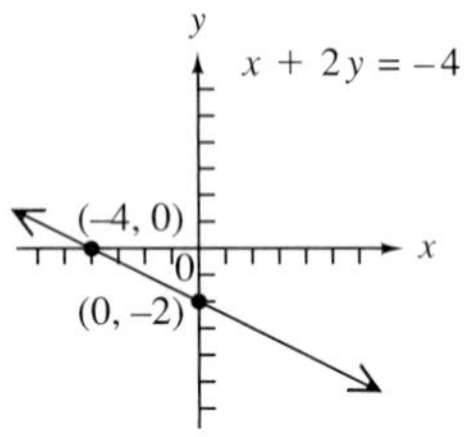

21.

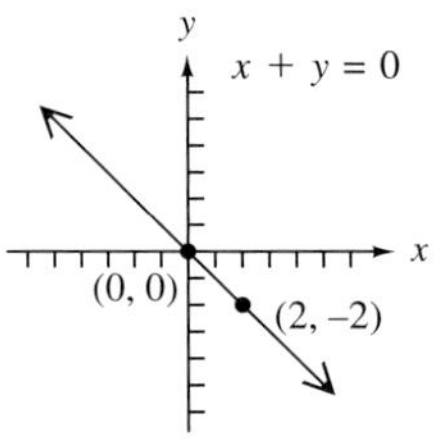

22.

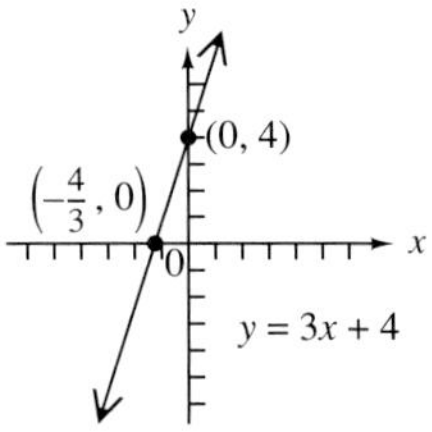

23. $-\frac{1}{2}$ **24.** $-\frac{2}{3}$ **25.** 0 **26.** undefined **27.** 3 **28.** $\frac{2}{3}$ **29.** $\frac{3}{2}$ **30.** $-\frac{1}{3}$ **31.** $\frac{3}{2}$ **32.** undefined **33.** 0 **34.** **(a)** 2 **(b)** $\frac{1}{3}$ **35.** parallel **36.** perpendicular **37.** neither **38.** 0 **39.** $y = -x + \frac{2}{3}$ **40.** $y = -\frac{1}{3}x + 1$ **41.** $y = x - 7$ **42.** $y = \frac{2}{3}x + \frac{14}{3}$ **43.** $y = -\frac{3}{4}x - \frac{1}{4}$ **44.** $y = -\frac{1}{4}x + \frac{3}{2}$ **45.** $y = 1$ **46.** $x = \frac{1}{3}$ **47.** **(a)** $y = -\frac{1}{3}x + 5$ **(b)** slope: $-\frac{1}{3}$; y-intercept: $(0, 5)$

(c)

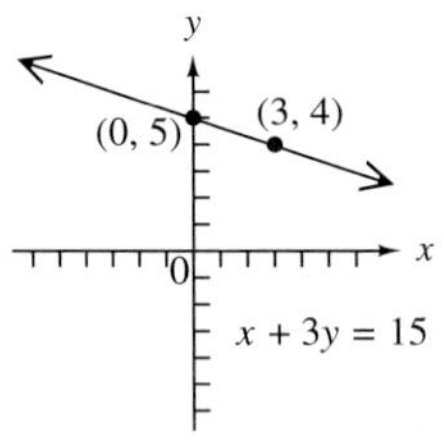

48.

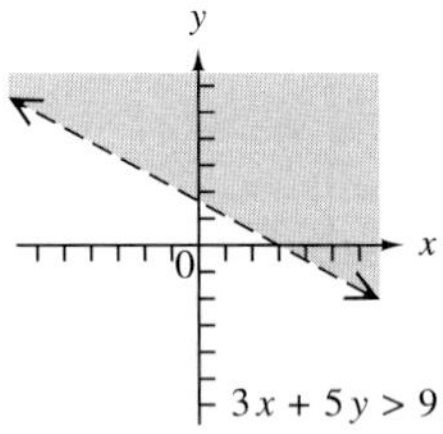

49.

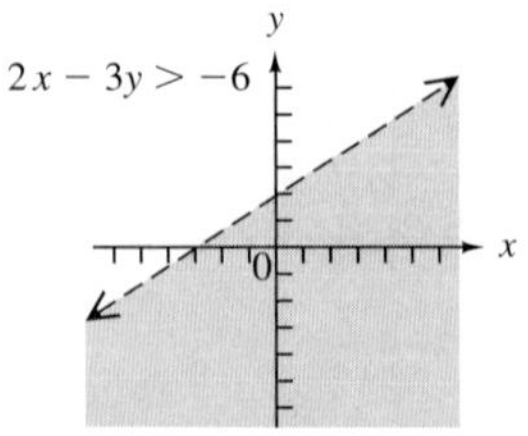

50.

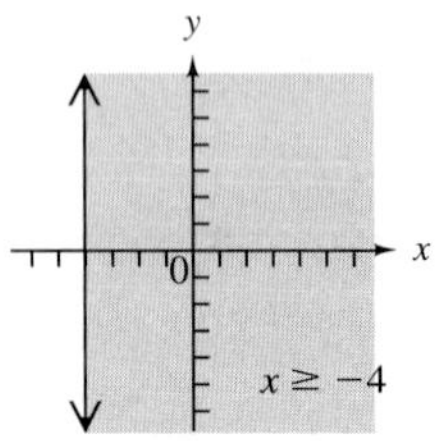

51. A **52.** C, D **53.** A, B, D **54.** D **55.** C **56.** B

57. (0, −5); $\left(-\frac{5}{2}, 0\right)$; −2

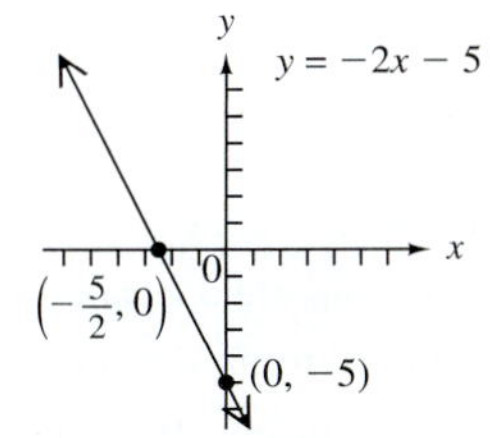

58. (0, 0); (0, 0); $-\frac{1}{3}$

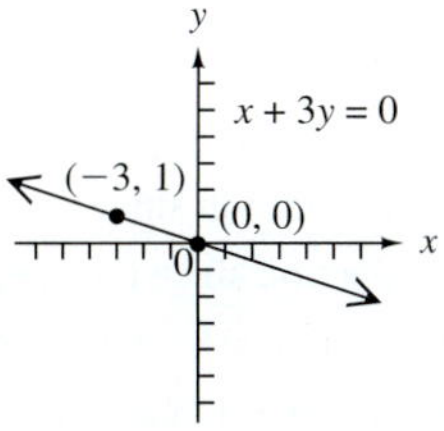

59. (0, 5); no *x*-intercept; 0

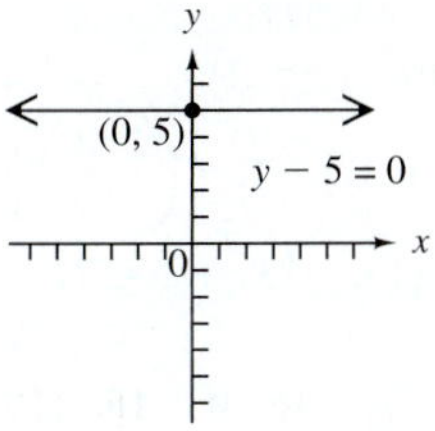

60. no *y*-intercept; (−1, 0); undefined

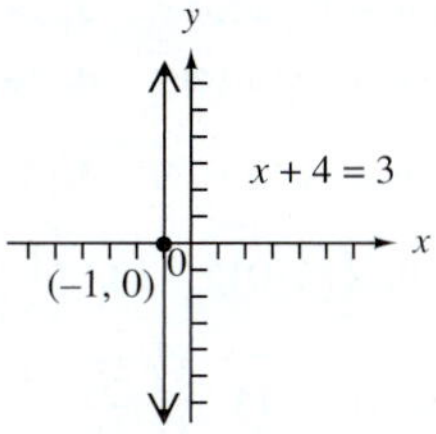

61. (a) $y = -\frac{1}{4}x - \frac{5}{4}$ **(b)** $x + 4y = -5$ **62. (a)** $y = -3x + 30$ **(b)** $3x + y = 30$ **63. (a)** $y = -\frac{4}{7}x - \frac{23}{7}$ **(b)** $4x + 7y = -23$

64.

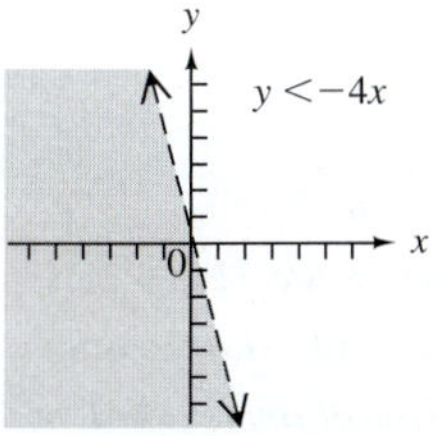

65.

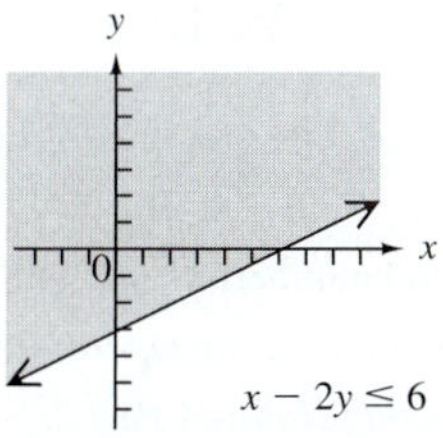

66. 1.7% **67.** Since the graph rises from left to right, the slope is positive. **68.** (2004, 42.3), (2009, 44.0) **69.** $y = 0.34x - 639.06$ **70.** 0.34; yes **71.** 42.6, 43.0, 43.3, 43.7 **72.** 44.3%; No. The equation is based on data only from 2004 through 2009.

Chapter 3 Test (pages 285–286)

1. between 2003 and 2004, 2004 and 2005, and 2005 and 2006 **2.** The unemployment rate was increasing. **3.** 2008: 5.8%; 2009: 9.3%; increase: 3.5%

4. (0, 6); (2, 0)

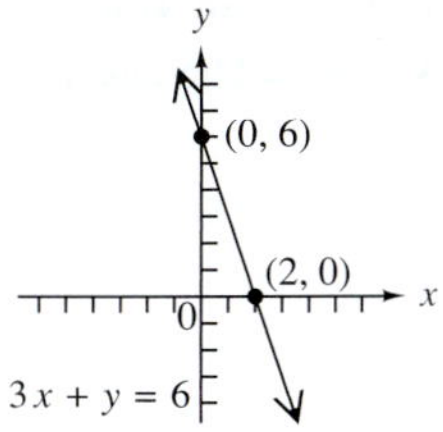

5. (0, 0); (0, 0)

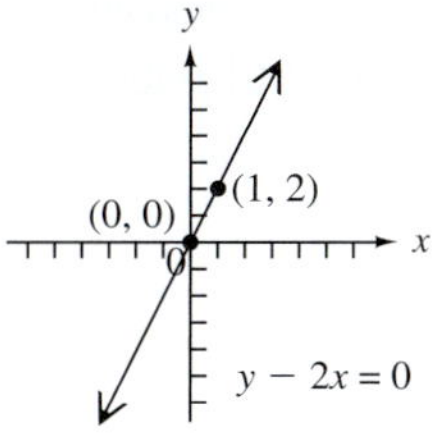

6. none; (−3, 0)

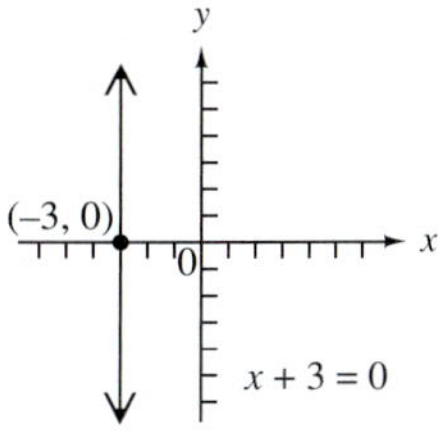

7. (0, 1); none

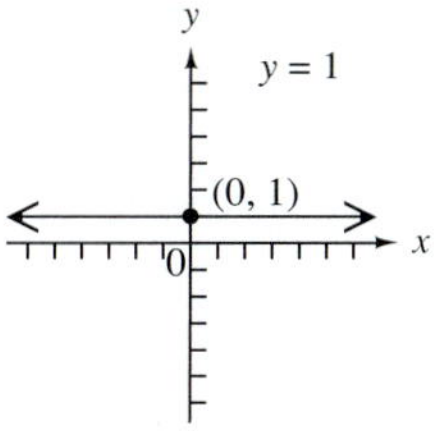

8. (0, −4); (4, 0)

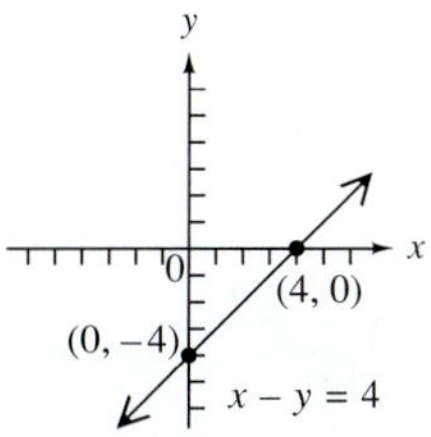

9. $-\frac{8}{3}$ **10.** −2 **11.** undefined **12.** $\frac{5}{2}$ **13.** 0 **14.** $y = 2x + 6$ **15.** $y = \frac{5}{2}x - 4$ **16.** $y = -9x + 12$ **17.** $y = -\frac{3}{2}x + \frac{9}{2}$

18.

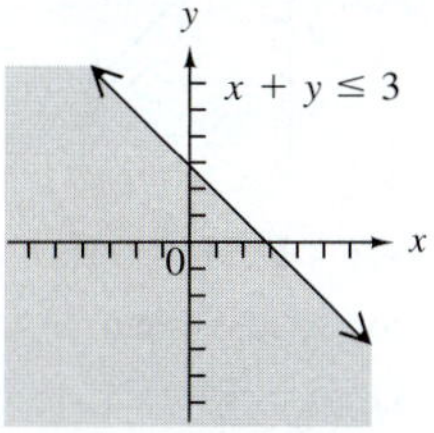

19.

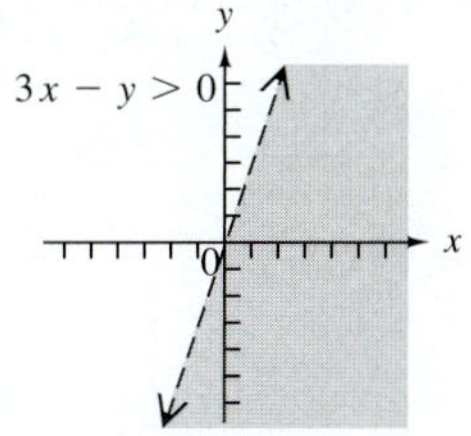

20. The slope is negative since sales are decreasing. **21.** (0, 209), (11, 123); −7.8 **22.** $y = -7.8x + 209$ **23.** 154.4 thousand; The equation gives a number that is lower than the actual sales. **24.** In 2011, worldwide snowmobile sales were 123 thousand.

Chapters R–3 CUMULATIVE REVIEW EXERCISES (pages 287–288)

1. $\frac{301}{40}$, or $7\frac{21}{40}$ **2.** 6 **3.** 7 **4.** $\frac{73}{18}$, or $4\frac{1}{18}$ **5.** true **6.** −43 **7.** distributive property **8.** $-p + 2$ **9.** $h = \frac{3V}{\pi r^2}$ **10.** {−1} **11.** {2} **12.** {−13} **13.** $(-2.6, \infty)$ [number line: −2.6, 0, 1]

14. $(0, \infty)$ [number line: 0] **15.** $(-\infty, -4]$ [number line: −4, 0]

16. high school diploma: \$32,552; bachelor's degree: \$53,976 **17.** 13 mi **18. (a)** 85.63, 77.82, 76.09 **(b)** (20, 85.63), (38, 77.82), (42, 76.09) **(c)** In 1992, the winning time was approximately 80.42 sec. **19. (a)** 19.8 million **(b)** 13.2 million **(c)** 7.8 million **20.** (−4, 0); (0, 3) **21.** $\frac{3}{4}$

22.

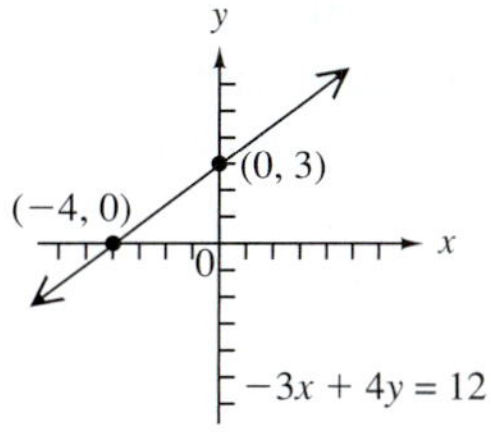

23. 6 **24.** perpendicular **25.** $y = 3x - 11$ **26.** $y = 4$

CHAPTER 4 Systems of Linear Equations and Inequalities

SECTION 4.1 (pages 295–298)

1. system of linear equations; same **2.** ordered pair; true **3.** inconsistent; no; independent **4.** solution; consistent **5.** dependent; consistent; infinitely many **6.** parallel; solution **7.** It is not a solution of the system because it is not a solution of the second equation, $2x + y = 4$.

8. $\{(x, y) \mid 3x - 2y = 4\}$ **9.** B, because the ordered pair must be in quadrant II. **10.** D, because the ordered pair must be on the y-axis, with $y < 0$. **11.** no **13.** yes **15.** yes **17.** no

We show the graphs here only for Exercises 19–23.

19. $\{(4, 2)\}$

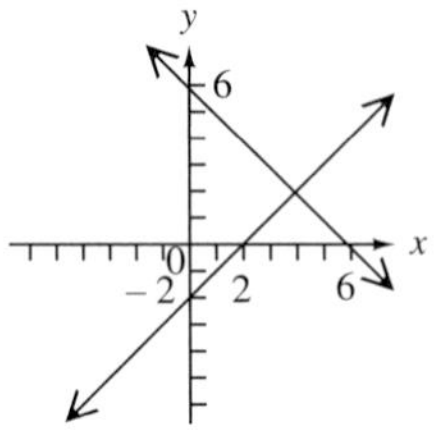

21. $\{(0, 4)\}$

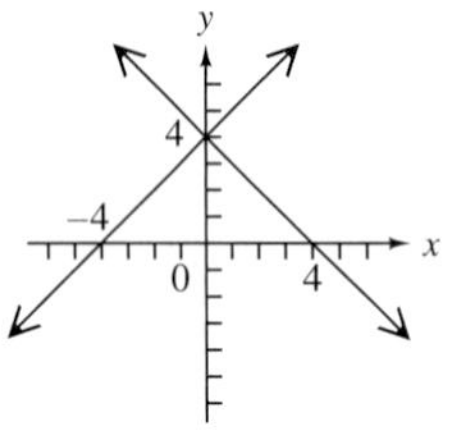

23. $\{(4, -1)\}$

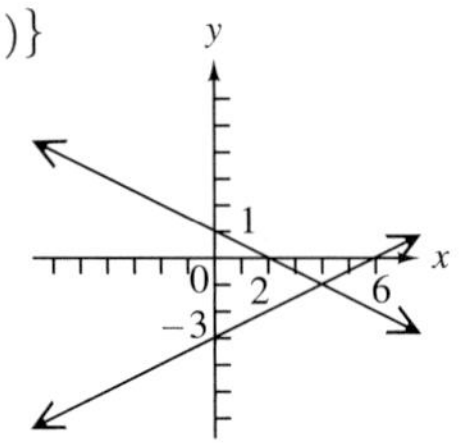

25. $\{(1, 3)\}$ **27.** $\{(0, 2)\}$ **29.** $\emptyset$ (inconsistent system) **31.** $\{(x, y) \mid 5x - 3y = 2\}$ (dependent equations) **33.** $\{(4, -3)\}$ **35.** $\{(x, y) \mid 2x - y = 4\}$ (dependent equations) **37.** $\emptyset$ (inconsistent system) **39. (a)** neither **(b)** intersecting lines **(c)** one solution **41. (a)** dependent **(b)** one line **(c)** infinite number of solutions **43. (a)** inconsistent **(b)** parallel lines **(c)** no solution **45. (a)** neither **(b)** intersecting lines **(c)** one solution **47. (a)** 40 **(b)** 30 **49.** 1980–2000 **51.** 2006; about 600 (million) units **53.** The slope would be negative. Sales of CDs were decreasing during this period. **54.** The slope would be positive. Sales of digital downloads were increasing during this period.

SECTION 4.2 (pages 305–306)

1. The student must find the value of y and write the solution as an ordered pair. The solution set is $\{(3, 0)\}$. **2.** The true result $0 = 0$ means that the system has an infinite number of solutions. The solution set is $\{(x, y) \mid x + y = 4\}$. **3.** A false statement, such as $0 = 3$, occurs. **4.** A true statement, such as $0 = 0$, occurs. **5.** $\{(3, 9)\}$ **7.** $\{(7, 3)\}$ **9.** $\{(-2, 4)\}$ **11.** $\{(-4, 8)\}$ **13.** $\{(3, -2)\}$ **15.** $\{(x, y) \mid 3x - y = 5\}$ **17.** $\emptyset$ **19.** $\{(x, y) \mid 2x - y = -12\}$ **21.** $\left\{\left(\frac{1}{3}, -\frac{1}{2}\right)\right\}$ **23.** $\{(2, -3)\}$ **25.** $\{(2, -4)\}$ **27.** $\{(-4, 2)\}$ **29.** $\{(7, -3)\}$ **31.** $\{(2, 3)\}$ **33.** To find the total cost, multiply the number of bicycles (x) by the cost per bicycle (400 dollars) and add the fixed cost (5000 dollars). Thus $y_1 = 400x + 5000$ gives this total cost (in dollars). **34.** $y_2 = 600x$ **35.** $y_1 = 400x + 5000$, $y_2 = 600x$; solution set: $\{(25, 15{,}000)\}$ **36.** 25; 15,000; 15,000

SECTION 4.3 (pages 311–313)

1. true **2.** false; Multiply by -3. **3.** true **4.** true **5.** $\{(-1, 3)\}$ **7.** $\{(-1, -3)\}$ **9.** $\{(-2, 3)\}$ **11.** $\left\{\left(\frac{1}{2}, 4\right)\right\}$ **13.** $\{(3, -6)\}$ **15.** $\{(7, 4)\}$ **17.** $\{(0, 4)\}$ **19.** $\{(-4, 0)\}$ **21.** $\{(0, 0)\}$ **23.** $\emptyset$ **25.** $\{(x, y) \mid x - 3y = -4\}$ **27.** $\{(0, 7)\}$ **29.** $\{(-6, 5)\}$ **31.** $\left\{\left(-\frac{6}{5}, \frac{4}{5}\right)\right\}$ **33.** $\left\{\left(\frac{1}{8}, -\frac{5}{6}\right)\right\}$ **35.** $\emptyset$ **37.** $\{(x, y) \mid 2x + y = 0\}$ **39.** $\{(11, 15)\}$ **41.** $\left\{\left(13, -\frac{7}{5}\right)\right\}$ **43.** $\{(6, -4)\}$ **45.** $5.66 = 2001a + b$ **46.** $7.89 = 2010a + b$ **47.** $2001a + b = 5.66$, $2010a + b = 7.89$; solution set: $\{(0.248, -490.590)\}$ **48. (a)** $y = 0.248x - 490.590$ **(b)** \$7.39; This is a bit more than the actual figure.

SUMMARY EXERCISES Applying Techniques for Solving Systems of Linear Equations (pages 314–315)

1. (a) Use substitution since the second equation is solved for y. **(b)** Use elimination since the coefficients of the y-terms are opposites. **(c)** Use elimination since the equations are in standard form with no coefficients of 1 or -1. Solving by substitution would involve fractions. **2.** System B is easier to solve by substitution because the second equation is already solved for y. **3. (a)** $\{(1, 4)\}$ **(b)** $\{(1, 4)\}$ **(c)** Answers will vary. **4. (a)** $\{(-5, 2)\}$ **(b)** $\{(-5, 2)\}$ **(c)** Answers will vary. **5.** $\{(2, 6)\}$ **6.** $\{(-3, 2)\}$ **7.** $\left\{\left(\frac{1}{3}, \frac{1}{2}\right)\right\}$ **8.** $\emptyset$ **9.** $\{(3, 0)\}$ **10.** $\left\{\left(\frac{3}{2}, -\frac{3}{2}\right)\right\}$ **11.** $\{(x, y) \mid 3x + y = 7\}$ **12.** $\{(9, 4)\}$ **13.** $\left\{\left(\frac{45}{31}, \frac{4}{31}\right)\right\}$ **14.** $\{(4, -5)\}$ **15.** $\emptyset$ **16.** $\{(0, 0)\}$ **17.** $\left\{\left(\frac{19}{3}, -5\right)\right\}$ **18.** $\left\{\left(\frac{22}{13}, -\frac{23}{13}\right)\right\}$ **19.** $\{(-12, -60)\}$ **20.** $\{(2, -3)\}$ **21.** $\{(24, -12)\}$ **22.** $\{(-2, 1)\}$ **23.** $\{(-35, 13)\}$ **24.** $\{(10, -9)\}$ **25.** $\{(-4, 6)\}$ **26.** $\{(5, 3)\}$

SECTION 4.4 (pages 322–327)

1. D **2.** A **3.** B **4.** B **5.** D **6.** D **7.** C **8.** C **9.** the second number; $x - y = 48$; The two numbers are 73 and 25. **11.** *The Phantom of the Opera:* 9803; *Cats:* 7485 **13.** *Harry Potter and the Deathly Hallows Part 2:* \$381 million; *Transformers: Dark of the Moon:* \$352 million **15.** Terminal Tower: 708 ft; Key Tower: 947 ft **17.** variables; width; Equation (1): $x = 38 + y$; length; twice; Equation (2): $2x + 2y = 188$; length: 66 yd; width: 28 yd **19. (a)** 45 units **(b)** Do not produce—the product will lead to a loss. **21.** table entries: (third column) $1x$ (or x), $10y$; 46 ones; 28 tens **23.** 5 DVDs of *Moneyball;* 2 Blu-ray discs of *The Concert for George* **25.** table entries: (second column) 4%, or 0.04; (third column) $0.04y$; Equation (1): $x = 2y$; Equation (2): $0.05x + 0.04y = 350$; \$2500 at 4%; \$5000 at 5% **27.** U2: \$92; Taylor Swift: \$72 **29.** table entries: (third column) $0.40x$, $0.70y$, $0.50(120)$ (or 60); 80 L of 40% solution; 40 L of 70% solution **31.** table entries: (second column) 3, 4; (third column) $6x$, $3y$, $4(90)$ (or 360); 30 lb at \$6 per lb; 60 lb at \$3 per lb **33.** nuts: 40 lb; raisins: 20 lb **35.** table entries: (third column) 4.5, 4.5; (fourth column) $4.5x$, $4.5y$; Equation (1): $4.5x + 4.5y = 495$; Equation (2): $x = 10 + y$; 60 mph; 50 mph **37.** bicycle: 13.5 mph; car: 49.3 mph **39.** car leaving Cincinnati: 55 mph; car leaving Toledo: 70 mph **41.** Roberto: 17.5 mph; Juana: 12.5 mph **43.** table entries: (third column) 3, 3; (fourth column) 24; boat: 10 mph; current: 2 mph **45.** plane: 470 mph; wind: 30 mph

SECTION 4.5 (pages 331–332)

1. C **2.** A **3.** B **4.** D

5.

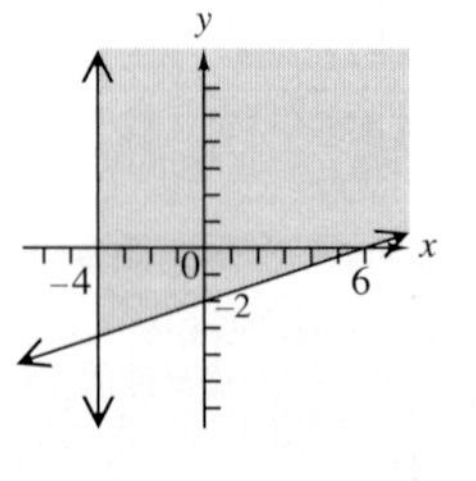

6.

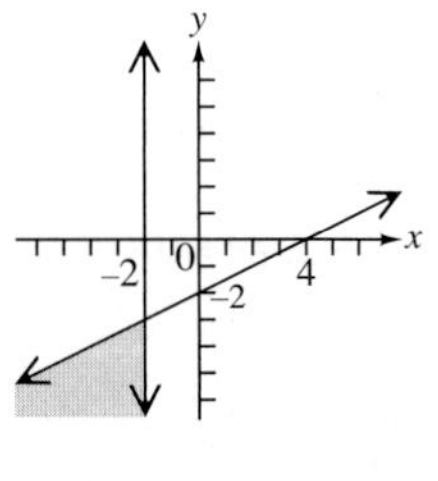

7.

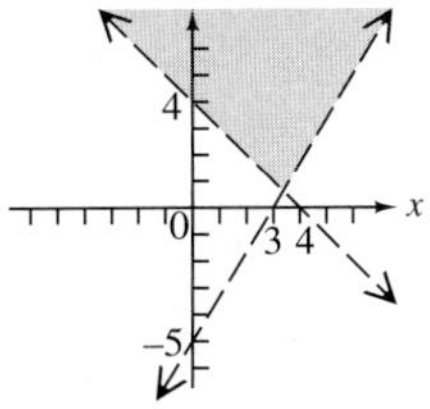

8.

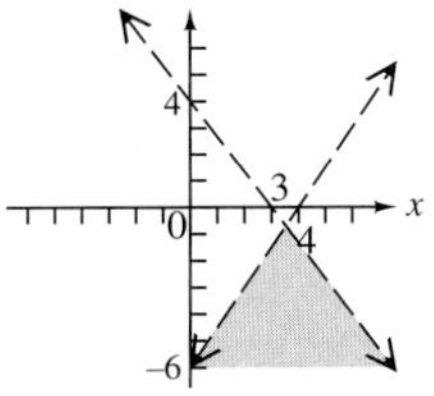

9.

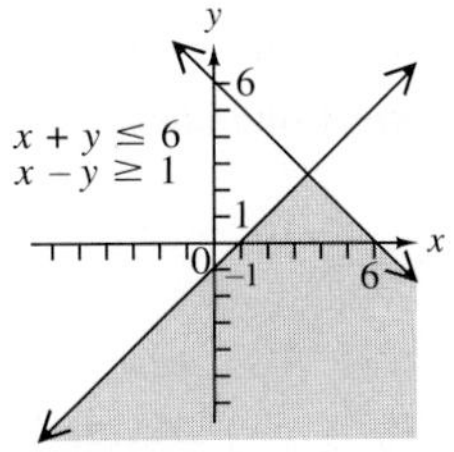

11.

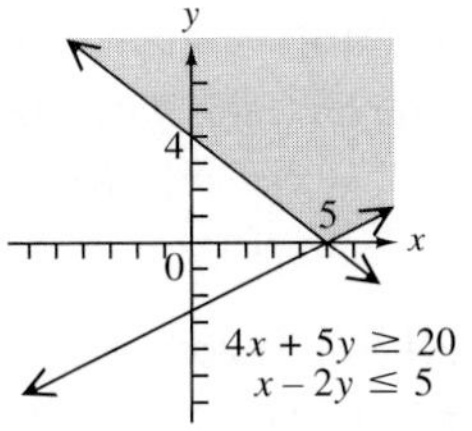

13.

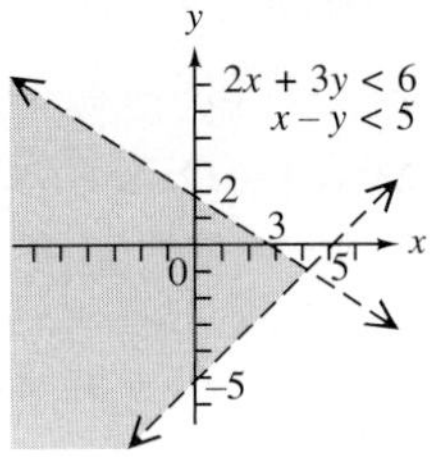

15.

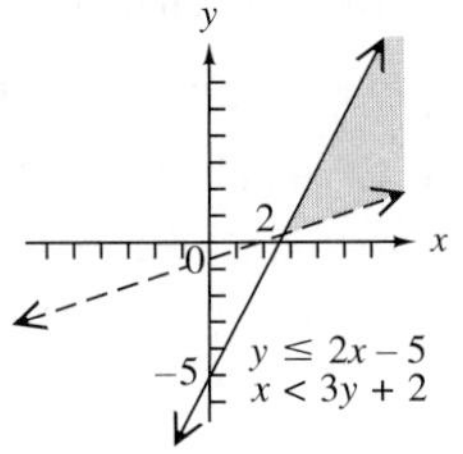

17.

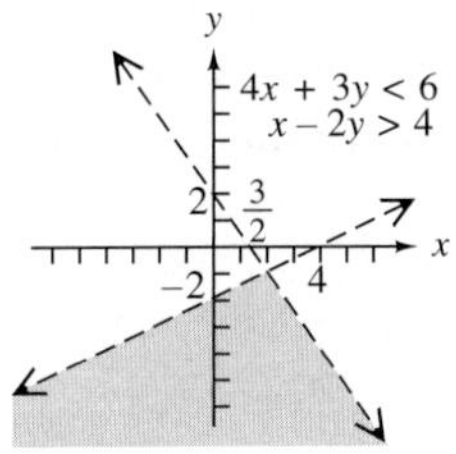

19.

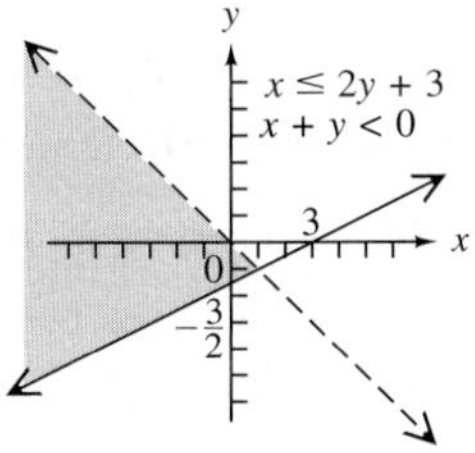

21.

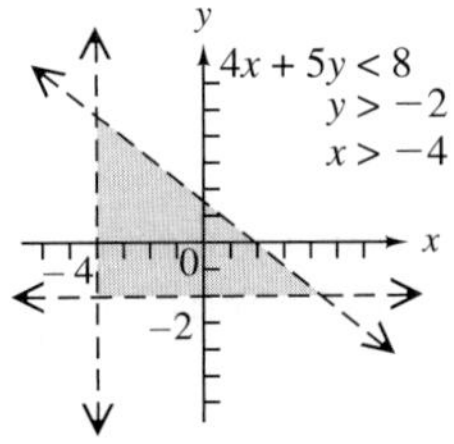

23.

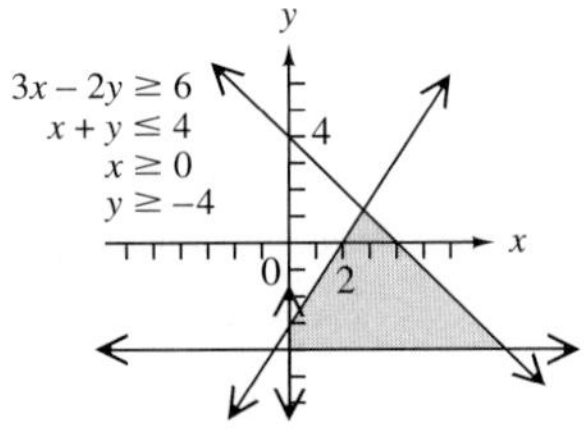

25. There are infinitely many points in any of the shaded regions. There are, however, points that are not in the shaded regions, and thus there are ordered pairs that are not solutions.

Chapter 4 REVIEW EXERCISES (pages 336–338)

1. yes **2.** no

We do not show the graphs for Exercises 3–6.

3. $\{(3, 1)\}$ **4.** $\{(0, -2)\}$ **5.** $\{(x, y) \mid x - 2y = 2\}$ **6.** $\emptyset$

7. $\{(2, 1)\}$ **8.** $\{(3, 5)\}$ **9.** $\{(6, 4)\}$ **10.** $\emptyset$ **11.** $\{(7, 1)\}$

12. $\{(-5, -2)\}$ **13.** $\{(-4, 3)\}$ **14.** $\{(x, y) \mid 3x - 4y = 9\}$

15. $\{(9, 2)\}$ **16.** $\left\{\left(\frac{10}{7}, -\frac{9}{7}\right)\right\}$ **17.** $\{(8, 9)\}$ **18.** $\{(2, 1)\}$

19. $\{(7, -2)\}$ **20.** $\{(-4, 2)\}$ **21.** Pizza Hut: 7542 locations; Domino's: 4929 locations **22.** *Reader's Digest*: 5.7 million; *People:* 3.6 million **23.** table entries: (first column) x; (second column) 0.90; (third column) $0.90y$, 100 (1) (or 100); 25 lb of \$1.30 candy; 75 lb of \$0.90 candy **24.** table entries: (first column) y, 20; (second column) 20; (third column) $10x$; 13 twenties; 7 tens **25.** length: 27 m; width: 18 m **26.** plane: 250 mph; wind: 20 mph **27.** table entries: (second column) 0.04; (third column) $0.03x$, $0.04y$, 650; \$7000 at 3%; \$11,000 at 4% **28.** table entries: (second column) 0.70; (third column) $0.40x$, $0.70y$, 0.50(90) (or 45); 60 L of 40% solution; 30 L of 70% solution **29.** B **30.** B

31.

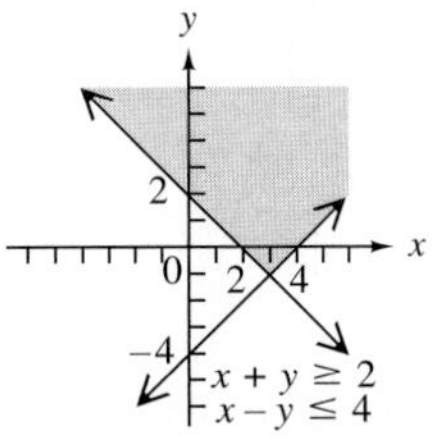

32.

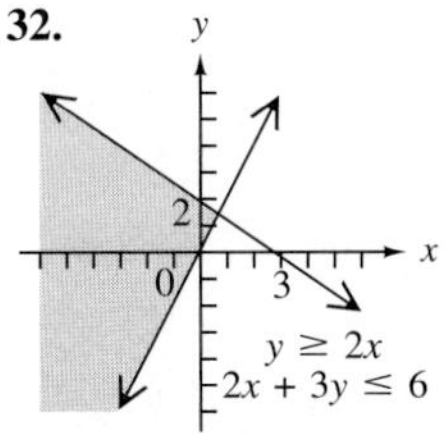

33.

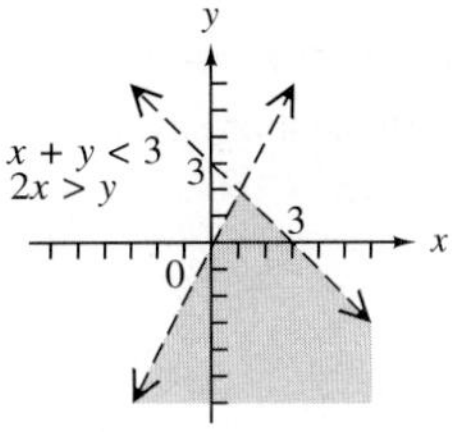

34. $\{(2, 0)\}$ **35.** $\{(-4, 15)\}$

36. $\emptyset$

37.

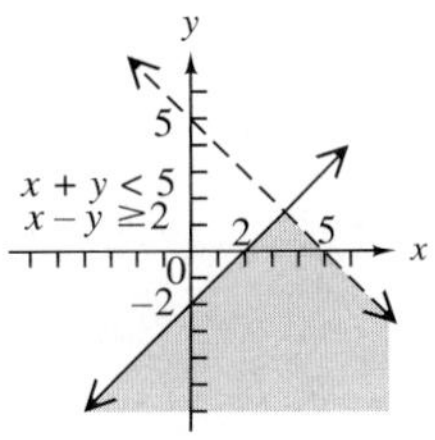

38.

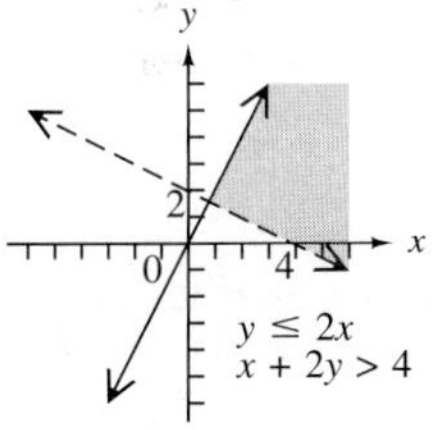

39.

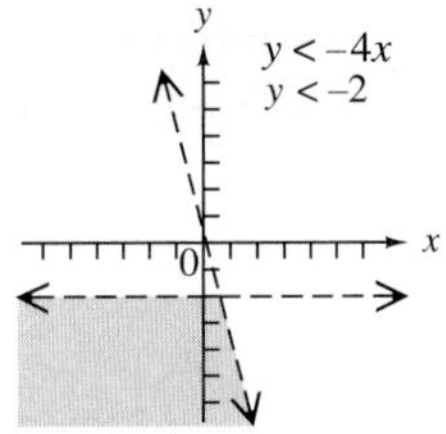

40. 8 in., 8 in., and 13 in.

41. Giants: 21; Patriots: 17

42. **(a)** years 0–6 **(b)** year 6; about \$650

Chapter 4 TEST (pages 339–340)

1. **(a)** no **(b)** no **(c)** yes **2.** $\{(2, -3)\}$ **3.** It has no solution.

4. $\{(1, -6)\}$ **5.** $\{(-35, 35)\}$ **6.** $\{(5, 6)\}$ **7.** $\{(-1, 3)\}$

8. $\{(-1, 3)\}$ **9.** $\{(0, 0)\}$ **10.** $\emptyset$ **11.** $\{(x, y) \mid 3x - y = 6\}$

12. $\{(-15, 6)\}$ **13.** Memphis and Atlanta: 394 mi; Minneapolis and Houston: 1176 mi **14.** Statue of Liberty: 3.8 million; National World War II Memorial: 4.0 million **15.** 20 L of 15% solution; 30 L of 40% solution **16.** slower car: 45 mph; faster car: 60 mph

17.

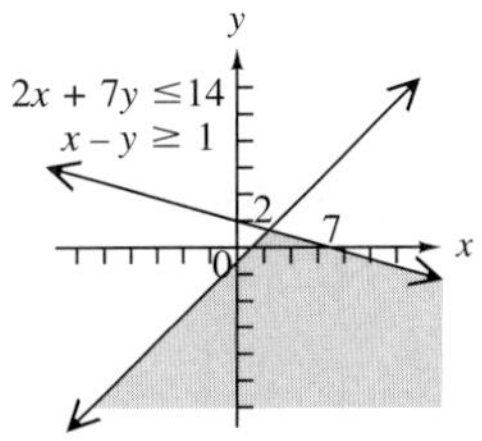

18.

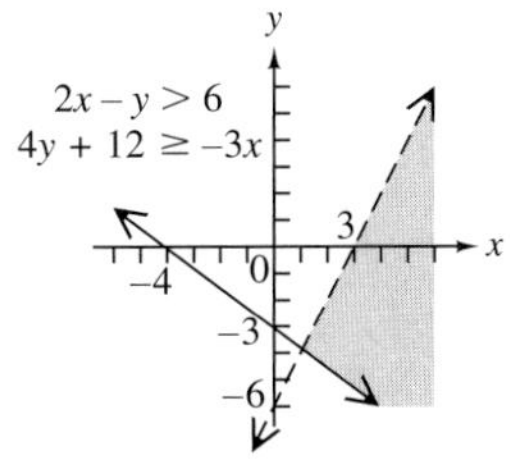

Chapters R–4 CUMULATIVE REVIEW EXERCISES (pages 341–342)

1. −1, 1, −2, 2, −4, 4, −5, 5, −8, 8, −10, 10, −20, 20, −40, 40 **2.** 1

3. commutative property **4.** distributive property **5.** inverse property

ANSWERS

6. 46 **7.** $T = \frac{PV}{k}$ **8.** $\left\{-\frac{13}{11}\right\}$ **9.** $\left\{\frac{9}{11}\right\}$ **10.** $(-18, \infty)$

11. $\left(-\frac{11}{2}, \infty\right)$ **12.** length: 9.50 in.; width: 7.31 in.

13.

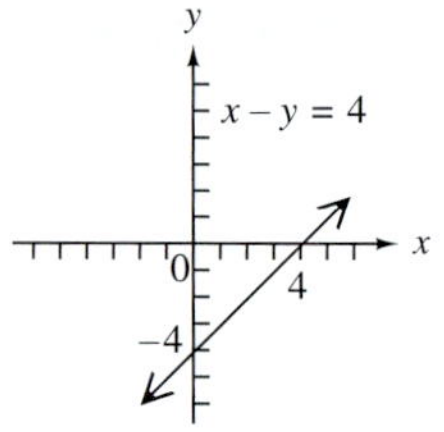

14.

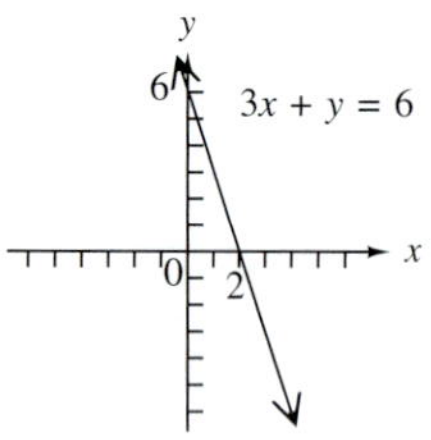

15. $-\frac{4}{3}$ **16.** $-\frac{1}{4}$ **17.** $y = \frac{1}{2}x + 3$ **18.** $y = 2x + 1$

19. **(a)** $x = 9$ **(b)** $y = -1$ **20.** $\{(-1, 6)\}$ **21.** $\{(3, -4)\}$

22. $\emptyset$ **23.** table entries: (third column) 2; (fourth column) 2*y*, 2528; 405 adults, 49 children **24.** 19 in., 19 in., 15 in.

25.

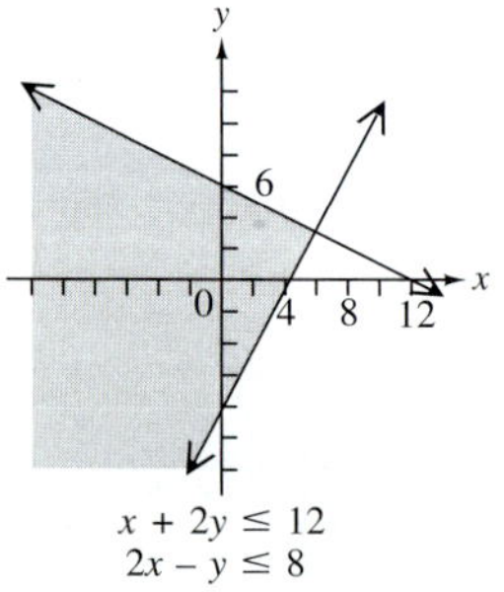

CHAPTER 5 Exponents and Polynomials

SECTION 5.1 (pages 349–352)

1. 7; 5 **2.** two **3.** 8 **4.** is not **5.** 26 **6.** $5x^9$ **7.** 1; 6 **9.** 1; 1 **11.** 1; $\frac{1}{5}$ **13.** 2; −19, −1 **15.** 3; 1, −8, $\frac{2}{3}$ **17.** $2m^5$ **19.** $-r^5$ **21.** $\frac{2}{3}x^4$ **23.** cannot be simplified; $0.2m^5 - 0.5m^2$ **25.** $-5x^5$ **27.** $5p^9 + 4p^7$ **29.** $-2y^2$ **31.** already simplified; 4; binomial **33.** already simplified; $6m^5 + 5m^4 - 7m^3 - 3m^2$; 5; none of these **35.** $x^4 + \frac{1}{3}x^2 - 4$; 4; trinomial **37.** 7; 0; monomial **39.** $1.5x^2 - 0.5x$; 2; binomial **41.** **(a)** −1 **(b)** 5 **43.** **(a)** 19 **(b)** −2 **45.** **(a)** 36 **(b)** −12 **47.** **(a)** −124 **(b)** 5 **49.** $5m^2 + 3m$ **51.** $4x^4 - 4x^2$ **53.** $\frac{7}{6}x^2 - \frac{2}{15}x + \frac{5}{6}$ **55.** $12m^3 - 13m^2 + 6m + 11$ **57.** $2.9x^3 - 3.5x^2 - 1.5x - 9$ **59.** $8r^2 + 5r - 12$ **61.** $5m^2 - 14m + 6$ **63.** $4x^3 + 2x^2 + 5x$ **65.** $-18y^5 + 7y^4 + 5y^3 + 3y^2 + y$ **67.** $-2m^3 + 7m^2 + 8m - 9$ **69.** $-11x^2 - 3x - 3$ **71.** $2x^2 + 8x$ **73.** $8x^2 + 8x + 6$ **75.** $8t^2 + 8t + 13$ **77.** $13a^2b - 7a^2 - b$ **79.** $c^4d - 5c^2d^2 + d^2$ **81.** $12m^3n - 11m^2n^2 - 4mn^2$ **83.** **(a)** $23y + 5t$ **(b)** 25°, 67°, 88° **85.** 5; 175 **86.** 87 ft; (1, 87) **87.** 6; \$27 **88.** 2.5; 130

SECTION 5.2 (pages 359–360)

1. false **2.** true **3.** false **4.** true **5.** 1 **6.** three squared; five cubed; seven to the fourth power **7.** t^7 **9.** $\left(\frac{1}{2}\right)^5$ **11.** $(-8p)^2$ **13.** base: 3; exponent: 5; 243 **15.** base: −3; exponent: 5; −243 **17.** base: −6*x*; exponent: 4 **19.** base: *x*; exponent: 4 **21.** 5^8 **23.** 4^{12} **25.** $(-7)^9$ **27.** t^{24} **29.** $-56r^7$ **31.** $42p^{10}$ **33.** The product rule does not apply. **35.** The product rule does not apply. **37.** 4^6 **39.** t^{20} **41.** $343r^3$ **43.** 5^{12} **45.** -8^{15} **47.** $5^5x^5y^5$ **49.** $8q^3r^3$ **51.** $\frac{1}{8}$ **53.** $\frac{a^3}{b^3}$ **55.** $\frac{9^8}{5^8}$ **57.** $-8x^6y^3$ **59.** $9a^6b^4$ **61.** $\frac{5^5}{2^5}$ **63.** $\frac{9^5}{8^3}$ **65.** $2^{12}x^{12}$ **67.** -6^5p^5 **69.** $6^5x^{10}y^{15}$ **71.** x^{21} **73.** $4w^4x^{26}y^7$ **75.** $-r^{18}s^{17}$ **77.** $\frac{125a^6b^{15}}{c^{18}}$ **79.** $25m^6p^{14}q^5$ **81.** $16x^{10}y^{16}z^{10}$ **83.** Using power rule (a), simplify as follows: $(10^2)^3 = 10^{2 \cdot 3} = 10^6 = 1{,}000{,}000$. **84.** The 4 is used as an *exponent* on 3. It is *not* multiplied by 3. The correct simplification is $\mathbf{3^4} \cdot x^8 \cdot y^{12} = \mathbf{81}x^8y^{12}$. **85.** $30x^7$ **87.** $6p^7$

SECTION 5.3 (pages 365–367)

1. **(a)** B **(b)** D **(c)** A **(d)** C **2.** **(a)** C **(b)** A **(c)** B **(d)** D **3.** distributive **4.** binomials **5.** $15pq^2$ **7.** $-18m^3n^2$ **9.** $9y^{10}$ **11.** $-8x^{10}$ **13.** $-6m^2 - 4m$ **15.** $6p - \frac{9}{2}p^2 + 9p^4$ **17.** $10y^9 + 4y^6 + 6y^5$ **19.** $6y^6 + 4y^4 + 2y^3$ **21.** $28r^5 - 32r^4 + 36r^3$ **23.** $6a^4 - 12a^3b + 15a^2b^2$ **25.** $3m^2$; $2mn$; n^3; $21m^5n^2 + 14m^4n^3 - 7m^3n^5$ **27.** $12x^3 + 26x^2 + 10x + 1$ **29.** $6r^3 + 5r^2 - 12r + 4$ **31.** $20m^4 - m^3 - 8m^2 - 17m - 15$ **33.** $5x^4 - 13x^3 + 20x^2 + 7x + 5$ **35.** $3x^5 + 18x^4 - 2x^3 - 8x^2 + 24x$ **37.** first row: x^2; $4x$; second row: $3x$; 12; Product: $x^2 + 7x + 12$ **39.** first row: $2x^3$; $6x^2$; $4x$; second row: x^2; $3x$; 2; Product: $2x^3 + 7x^2 + 7x + 2$ **41.** $m^2 + 12m + 35$ **43.** $n^2 + n - 6$ **45.** $8r^2 - 10r - 3$ **47.** $9x^2 - 4$ **49.** $9q^2 + 6q + 1$ **51.** $15xy - 40x + 21y - 56$ **53.** $6t^2 + 23st + 20s^2$ **55.** $-0.3t^2 + 0.22t + 0.24$ **57.** $x^2 - \frac{5}{12}x - \frac{1}{6}$ **59.** $\frac{15}{16} - \frac{1}{4}r - 2r^2$ **61.** $2x^3 + x^2 - 15x$ **63.** $6y^5 - 21y^4 - 45y^3$ **65.** $-200r^7 + 32r^3$ **67.** **(a)** $3y^2 + 10y + 7$ **(b)** $8y + 16$ **69.** $30x + 60$ **70.** $30x + 60 = 600$; $\{18\}$ **71.** 10 yd by 60 yd **72.** 140 yd **73.** \$2100 **74.** \$1260

SECTION 5.4 (pages 372–374)

1. square; twice; product; square **2.** positive; negative **3.** difference; squares **4.** conjugates **5.** **(a)** $4x^2$ **(b)** $12x$ **(c)** 9 **(d)** $4x^2 + 12x + 9$ **7.** $p^2 + 4p + 4$ **9.** $z^2 - 10z + 25$ **11.** $x^2 - \frac{3}{2}x + \frac{9}{16}$ **13.** $v^2 + 0.8v + 0.16$ **15.** $16x^2 - 24x + 9$ **17.** $100z^2 + 120z + 36$ **19.** $4p^2 + 20pq + 25q^2$ **21.** $0.64t^2 + 1.12ts + 0.49s^2$ **23.** $25x^2 + 4xy + \frac{4}{25}y^2$ **25.** $9t^3 - 6t^2 + t$ **27.** $48t^3 + 24t^2 + 3t$ **29.** $-16r^2 + 16r - 4$ **31.** **(a)** $49x^2$ **(b)** 0 **(c)** $-9y^2$ **(d)** $49x^2 - 9y^2$; Because 0 is the identity element for addition, it is not necessary to write "+ 0." **33.** $q^2 - 4$ **35.** $r^2 - \frac{9}{16}$ **37.** $s^2 - 6.25$ **39.** $4w^2 - 25$ **41.** $100x^2 - 9y^2$ **43.** $4x^4 - 25$ **45.** $49x^2 - \frac{9}{49}$ **47.** $9p^3 - 49p$ **49.** $m^3 - 15m^2 + 75m - 125$ **51.** $y^3 + 6y^2 + 12y + 8$ **53.** $8a^3 + 12a^2 + 6a + 1$ **55.** $81r^4 - 216r^3t + 216r^2t^2 - 96rt^3 + 16t^4$ **57.** $3x^5 - 27x^4 + 81x^3 - 81x^2$ **59.** $-8x^6y - 32x^5y^2 - 48x^4y^3 - 32x^3y^4 - 8x^2y^5$ **61.** $\frac{1}{2}m^2 - 2n^2$ **63.** $9a^2 - 4$ **65.** $\pi x^2 + 4\pi x + 4\pi$ **67.** $x^3 + 6x^2 + 12x + 8$ **69.** $(a + b)^2$ **70.** a^2 **71.** $2ab$ **72.** b^2 **73.** $a^2 + 2ab + b^2$ **74.** They both represent the area of the entire large square. **75.** 1225 **76.** $30^2 + 2(30)(5) + 5^2$ **77.** 1225 **78.** They are equal.

SECTION 5.5 (pages 382–383)

1. negative **2.** positive **3.** negative **4.** negative **5.** positive **6.** positive **7.** 0 **8.** 0 **9.** (a) B (b) C (c) D (d) B (e) E (f) B **10.** (a) C (b) F (c) F (d) B (e) E (f) B **11.** 1 **13.** 1 **15.** -1 **17.** 0 **19.** 0 **21.** 2 **23.** $\frac{1}{64}$ **25.** 16 **27.** $\frac{49}{36}$ **29.** $\frac{1}{81}$ **31.** $\frac{8}{15}$ **33.** $-\frac{7}{18}$ **35.** $\frac{1}{9}$ **37.** $\frac{1}{6^5}$, or $\frac{1}{7776}$ **39.** 216 **41.** $2r^4$ **43.** $\frac{25}{64}$ **45.** $\frac{p^5}{q^8}$ **47.** r^9 **49.** $\frac{x^5}{6}$ **51.** $3y^2$ **53.** x^3 **55.** $\frac{yz^2}{4x^3}$ **57.** $a+b$ **59.** 343 **61.** $\frac{1}{x^2}$ **63.** $\frac{64x}{9}$ **65.** $\frac{x^2z^4}{y^2}$ **67.** $6x$ **69.** $\frac{1}{m^{10}n^5}$ **71.** $\frac{5}{16x^5}$ **73.** $\frac{36q^2}{m^4p^2}$ **75.** The student attempted to use the quotient rule with unequal bases. The correct way to simplify is $\frac{16^3}{2^2} = \frac{(2^4)^3}{2^2} = \frac{2^{12}}{2^2} = 2^{12-2} = 2^{10} = 1024$. **76.** The student incorrectly assumed that the negative exponent indicated a negative number. The correct way to simplify is $5^{-4} = \frac{1}{5^4} = \frac{1}{625}$.

SUMMARY EXERCISES Applying the Rules for Exponents (pages 384–385)

1. positive **2.** negative **3.** negative **4.** negative **5.** positive **6.** negative **7.** positive **8.** negative **9.** $\frac{6^{12}x^{24}}{5^{12}}$ **10.** $\frac{r^6s^{12}}{729t^6}$ **11.** $100{,}000x^7y^{14}$ **12.** $-128a^{10}b^{15}c^4$ **13.** $\frac{729w^3x^9}{y^{12}}$ **14.** $\frac{x^4y^6}{16}$ **15.** c^{22} **16.** $\frac{1}{k^4t^{12}}$ **17.** $\frac{11}{30}$ **18.** $y^{12}z^3$ **19.** $\frac{x^6}{y^5}$ **20.** 0 **21.** $\frac{1}{z^2}$ **22.** $\frac{9}{r^2s^2t^{10}}$ **23.** $\frac{300x^3}{y^3}$ **24.** $\frac{3}{5x^6}$ **25.** x^8 **26.** $\frac{y^{11}}{x^{11}}$ **27.** $\frac{a^6}{b^4}$ **28.** $6ab$ **29.** $\frac{61}{900}$ **30.** 1 **31.** $\frac{343a^6b^9}{8}$ **32.** 1 **33.** -1 **34.** 0 **35.** $\frac{27y^{18}}{4x^8}$ **36.** $\frac{1}{a^8b^{12}c^{16}}$ **37.** $\frac{x^{15}}{216z^9}$ **38.** $\frac{q}{8p^6r^3}$ **39.** x^6y^6 **40.** 0 **41.** $\frac{343}{x^{15}}$ **42.** $\frac{9}{x^6}$ **43.** $5p^{10}q^9$ **44.** $\frac{7}{24}$ **45.** $\frac{r^{14}t}{2s^2}$ **46.** 1 **47.** $8p^{10}q$ **48.** $\frac{1}{mn^3p^3}$ **49.** -1 **50.** $\frac{3}{40}$

SECTION 5.6 (pages 388–389)

1. $6x^2+8$; 2; $3x^2+4$ **2.** 0 **3.** $3x^2+4$; 2 (These may be reversed.); $6x^2+8$ **4.** is not **5.** To use the method of this section, the divisor must be a monomial. This is true of the first problem, but not the second. **6.** 8; 13; no **7.** 2 **8.** -30 **9.** $2m^2-m$ **11.** $30x^3-10x+5$ **13.** $-4m^3+2m^2-1$ **15.** $4t^4-2t^2+2t$ **17.** $a^4-a+\frac{2}{a}$ **19.** $-2x^3+\frac{2x^2}{3}-x$ **21.** $-9x^2+5x+1$ **23.** $\frac{4x^2}{3}+x-\frac{2}{3x}$ **25.** $27r^4$; $36r^3$; $6r^2$; $3r$; 2; $9r^3-12r^2-2r+1-\frac{2}{3r}$ **27.** $-m^2+3m-\frac{4}{m}$ **29.** $\frac{12}{x}-\frac{6}{x^2}+\frac{14}{x^3}-\frac{10}{x^4}$ **31.** $-4b^2+3ab-\frac{5}{a}$ **33.** $6x-2+\frac{1}{x}$ **35.** $15x^5-35x^4+35x^3$ **36.** $-72y^6+60y^5-24y^4+36y^3-84y^2$

SECTION 5.7 (pages 394–395)

1. The divisor is $2x+5$. The quotient is $2x^3-4x^2+3x+2$. **2.** Stop when the degree of the remainder is less than the degree of the divisor, or when the remainder is 0. **3.** Divide $12m^2$ by $2m$ to get $6m$. **4.** Multiply $6m$ by $2m-3$ to get $12m^2-18m$. **5.** $x+2$ **7.** $2y-5$ **9.** $p-4+\frac{44}{p+6}$ **11.** $r-5$ **13.** $2a-14+\frac{74}{2a+3}$ **15.** $4x^2-7x+3$ **17.** $3y^2-2y+2$ **19.** $2x^2-2x+3+\frac{-1}{x+1}$ **21.** $3k-4+\frac{2}{k^2-2}$ **23.** x^2+1 **25.** x^2+1 **27.** $2p^2-5p+4+\frac{6}{3p^2+1}$ **29.** x^3+6x-7 **31.** $2x^2+\frac{3}{5}x+\frac{1}{5}$ **33.** (x^2+x-3) units **35.** 33 **36.** 33 **37.** They are the same. **38.** The answers should agree.

SECTION 5.8 (pages 400–402)

1. (a) C (b) A (c) B (d) D **2.** (a) A (b) C (c) B (d) D **3.** in scientific notation **4.** in scientific notation **5.** not in scientific notation; 5.6×10^6 **6.** not in scientific notation; 3.4×10^4 **7.** not in scientific notation; 4×10^{-3} **8.** not in scientific notation; 7×10^{-4} **9.** not in scientific notation; 8×10^1 **10.** not in scientific notation; 9×10^2 **11.** A number is written in scientific notation if it is the product of a number whose absolute value is between 1 and 10 (inclusive of 1) and a power of 10. **12.** To multiply by a positive power of 10, move the decimal point to the right as many places as the exponent on 10. With a negative power of 10, move the decimal point to the left as many places as the absolute value of the exponent on 10. **13.** 5.876×10^9 **15.** 8.235×10^4 **17.** 7×10^{-6} **19.** -2.03×10^{-3} **21.** 750,000 **23.** 5,677,000,000,000 **25.** 1,000,000,000,000 **27.** -6.21 **29.** 0.00078 **31.** 0.000000005134 **33.** 6×10^{11}; 600,000,000,000 **35.** 1.5×10^7; 15,000,000 **37.** 8×10^{-3}; 0.008 **39.** 2.4×10^2; 240 **41.** 6.3×10^{-2}; 0.063 **43.** 6.426×10^4; 64,260 **45.** 3×10^{-4}; 0.0003 **47.** 4×10^1; 40 **49.** 1.3×10^{-5}; 0.000013 **51.** 5×10^2; 500 **53.** 2.6×10^{-3}; 0.0026 **55.** 7.205×10^{-6}; 0.000007205 **57.** 0.000002 **59.** 4.2×10^{42} **61.** 1.5×10^{17} mi **63.** \$3209 **65.** \$52,733 **67.** 3.59×10^2 sec, or 359 sec **69.** \$86.40

Chapter 5 REVIEW EXERCISES (pages 406–409)

1. $22m^2$; degree 2; monomial **2.** p^3-p^2+4p+2; degree 3; none of these **3.** already in descending powers; degree 5; none of these **4.** $-8y^5-7y^4+9y$; degree 5; trinomial **5.** $-5a^3+4a^2$ **6.** $2r^3-3r^2+9r$ **7.** $11y^2-10y+9$ **8.** $-13k^4-15k^2-4k-6$ **9.** $10m^3-6m^2-3$ **10.** $-y^2-4y+26$ **11.** $10p^2-3p-11$ **12.** $7r^4-4r^3-1$ **13.** 4^{11} **14.** -5^{11} **15.** $-72x^7$ **16.** $10x^{14}$ **17.** 19^5x^5 **18.** -4^7y^7 **19.** $5p^4t^4$ **20.** $\frac{7^6}{5^6}$ **21.** $27x^6y^9$ **22.** t^{42} **23.** $36x^{16}y^4z^{16}$ **24.** $\frac{8m^9n^3}{p^6}$ **25.** $125x^6$ **26.** The product rule for exponents does not apply here because we want the sum of 7^2 and 7^4, not their product. **27.** $10x^2+70x$ **28.** $-6p^5+15p^4$ **29.** $6r^3+8r^2-17r+6$ **30.** $8y^3+27$ **31.** $5p^5-2p^4-3p^3+25p^2+15p$ **32.** $x^2+3x-18$ **33.** $6k^2-9k-6$ **34.** $12p^2-48pq+21q^2$ **35.** $2m^4+5m^3-16m^2-28m+9$ **36.** $a^2+8a+16$ **37.** $9p^2-12p+4$ **38.** $4r^2+20rs+25s^2$ **39.** $r^3+6r^2+12r+8$ **40.** $8x^3-12x^2+6x-1$ **41.** $4z^2-49$ **42.** $36m^2-25$ **43.** $25a^2-36b^2$ **44.** $4x^4-25$ **45.** three; two **46.** $(a+b)^2=(a+b)(a+b)=a^2+2ab+b^2$. The term $2ab$ is not in a^2+b^2. **47.** 2 **48.** $\frac{1}{32}$ **49.** $\frac{25}{36}$ **50.** $-\frac{3}{16}$ **51.** 36 **52.** x^2 **53.** $\frac{1}{p^{12}}$ **54.** r^4 **55.** 2^8 **56.** $\frac{1}{9^6}$ **57.** 5^8 **58.** $\frac{1}{8^{12}}$ **59.** $\frac{1}{m^2}$ **60.** y^7 **61.** r^{13}

62. $25m^6$ **63.** $\frac{y^{12}}{8}$ **64.** $\frac{1}{a^3b^5}$ **65.** $72r^5$ **66.** $\frac{8n^{10}}{3m^{13}}$ **67.** $\frac{5y^2}{3}$
68. $-2x^2y$ **69.** $-y^3+2y-3$ **70.** $p-3+\frac{5}{2p}$
71. $-x^9+2x^8-4x^3+7x$ **72.** $-2m^2n+mn^2+\frac{6n^3}{5}$ **73.** $2r+7$
74. $4m+3+\frac{5}{3m-5}$ **75.** $2a+1+\frac{-8a+12}{5a^2-3}$
76. $k^2+2k+4+\frac{-2k-12}{2k^2+1}$ **77.** 4.8×10^7 **78.** 2.8988×10^{10}
79. 6.5×10^{-5} **80.** 8.24×10^{-8} **81.** 24,000 **82.** 78,300,000
83. 0.000000897 **84.** 0.00000000000995 **85.** 8×10^2; 800
86. 4×10^6; 4,000,000 **87.** 2.5×10^{-2}; 0.025 **88.** 1×10^{-2}; 0.01
89. 2.796×10^{10} calculations; 1.6776×10^{12} calculations
90. about 3.3 **91.** 0 **92.** $\frac{243}{p^3}$ **93.** $\frac{1}{49}$ **94.** $49-28k+4k^2$
95. y^2+5y+1 **96.** $\frac{1296r^8s^4}{625}$ **97.** $-8m^7-10m^6-6m^5$ **98.** 32
99. $5xy^3-\frac{8y^2}{5}+3x^2y$ **100.** $\frac{r^2}{6}$ **101.** $8x^3+12x^2y+6xy^2+y^3$
102. $\frac{3}{4}$ **103.** a^3-2a^2-7a+2 **104.** $8y^3-9y^2+5$
105. $10r^2+21r-10$ **106.** $144a^2-1$ **107.** $2x^2+x-6$; $6x-2$
108. $20x^4+8x^2$; $25x^8+20x^6+4x^4$ **109.** The friend wrote the second term of the quotient as $-12x$ rather than $-2x$. Here is the correct method: $\frac{6x^2-12x}{6}=\frac{6x^2}{6}-\frac{12x}{6}=x^2-2x$. **110.** $2mn+3m^4n^2-4n$

Chapter 5 TEST (pages 410–411)

1. $-7x^2+8x$; 2; binomial **2.** $4n^4+13n^3-10n^2$; 4; trinomial
3. $4t^4+t^3-6t^2-t$ **4.** $-2y^2-9y+17$ **5.** $-12t^2+5t+8$ **6.** -32
7. $\frac{216}{m^6}$ **8.** $-27x^5+18x^4-6x^3+3x^2$ **9.** $2r^3+r^2-16r+15$
10. $t^2-5t-24$ **11.** $8x^2+2xy-3y^2$ **12.** $25x^2-20xy+4y^2$
13. $100v^2-9w^2$ **14.** x^3+3x^2+3x+1 **15.** $12x+36$; $9x^2+54x+81$
16. **(a)** positive **(b)** positive **(c)** negative **(d)** positive **(e)** zero **(f)** negative **17.** $\frac{1}{625}$ **18.** 2 **19.** $\frac{7}{12}$ **20.** 8^5 **21.** x^2y^6
22. $4y^2-3y+2+\frac{5}{y}$ **23.** $-3xy^2+2x^3y^2+4y^2$ **24.** $2x+9$
25. $3x^2+6x+11+\frac{26}{x-2}$ **26.** **(a)** 3.44×10^{11} **(b)** 5.57×10^{-6}
27. **(a)** 29,600,000 **(b)** 0.0000000607
28. **(a)** 1×10^3; 5.89×10^{12} **(b)** 5.89×10^{15} mi

Chapters R–5 CUMULATIVE REVIEW EXERCISES (pages 412–413)

1. $\frac{19}{24}$ **2.** $-\frac{1}{20}$ **3.** 3.72 **4.** 62.006 **5.** \$1836 **6.** -8 **7.** 24
8. $\frac{1}{2}$ **9.** -4 **10.** associative property **11.** inverse property
12. distributive property **13.** $\{10\}$ **14.** $\left\{\frac{13}{4}\right\}$ **15.** $\emptyset$ **16.** $r=\frac{d}{t}$
17. $\{-5\}$ **18.** $\{-12\}$ **19.** $\{20\}$ **20.** {all real numbers}
21. mouse: 160; elephant: 10 **22.** 4 **23.** $[10,\infty)$ **24.** $\left(-\infty,-\frac{14}{5}\right)$
25. $[-4,2)$ **26.** $(0,2)$ and $(-3,0)$
27.

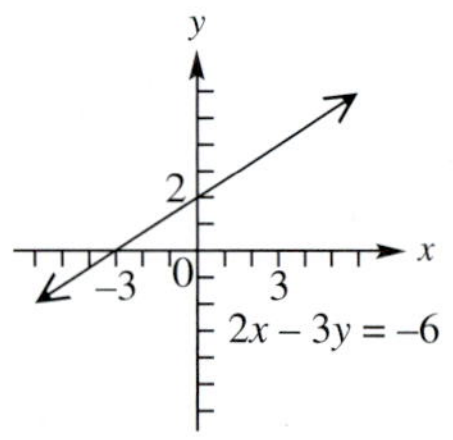

28. $\frac{2}{3}$ **29.** $\{(-3,-1)\}$
30. $\{(4,-5)\}$ **31.** $\frac{5}{4}$, or $1\frac{1}{4}$
32. 2 **33.** 1 **34.** $\frac{2b}{a^{10}}$
35. 3.45×10^4
36. $11x^3-14x^2-x+14$ **37.** $18x^7-54x^6+60x^5$
38. $63x^2+57x+12$ **39.** $25x^2+80x+64$ **40.** y^2-2y+6

CHAPTER 6 Factoring and Applications

SECTION 6.1 (pages 423–425)

1. product; multiplying **2.** common factor; is; divides **3.** 4 **5.** 4
7. 6 **9.** 1 **11.** 8 **13.** $10x^3$ **15.** xy^2 **17.** 6 **19.** $6m^3n^2$
21. factored **22.** factored **23.** not factored **24.** not factored
25. The correct factored form is $18x^3y^2+9xy=9xy(2x^2y+1)$. If a polynomial has two terms, the product of the factors must have two terms. $9xy(2x^2y)=18x^3y^2$ is just one term. **26.** First, we verify that we have factored completely. Then we multiply the factors. The product should be the original polynomial. **27.** $3m^2$ **29.** $2z^4$ **31.** $2mn^4$ **33.** $y+2$
35. $a-2$ **37.** $2+3xy$ **39.** $x(x-4)$ **41.** $3t(2t+5)$ **43.** $m^2(m-1)$
45. $-6x^2(2x+1)$ **47.** $5y^6(13y^4+7)$ **49.** no common factor (except 1)
51. $8mn^3(1+3m)$ **53.** $-2x(2x^2-5x+3)$ **55.** $13y^2(y^6+2y^2-3)$
57. $9qp^3(5q^3p^2+4p^3+9q)$ **59.** $(x+2)(c+d)$ **61.** $(2a+b)(a^2-b)$
63. $(p+4)(q-1)$ **65.** not in factored form; $(7t+4)(8+x)$
66. not in factored form; $(5x-1)(3r+7)$ **67.** in factored form
68. in factored form **69.** not in factored form **70.** not in factored form
71. $(5+n)(m+4)$ **73.** $(2y-7)(3x+4)$ **75.** $(y+3)(3x+1)$
77. $(z+2)(7z-a)$ **79.** $(3r+2y)(6r-x)$ **81.** $(w+1)(w^2+9)$
83. $(a+2)(3a^2-2)$ **85.** $(4m-p^2)(4m^2-p)$ **87.** $(y+3)(y+x)$
89. $(z-2)(2z-3w)$ **91.** $(5-2p)(m+3)$ **93.** $(3r+2y)(6r-t)$
95. commutative property **96.** $2x(y-4)-3(y-4)$ **97.** No, because it is not a product. It is the difference between $2x(y-4)$ and $3(y-4)$.
98. $(y-4)(2x-3)$, or $(2x-3)(y-4)$; yes

SECTION 6.2 (pages 431–432)

1. a and b must have different signs. **2.** a and b must have the same sign.
3. C **4.** Factor out the greatest common factor, $2x$. **5.** $a^2+13a+36$
6. $y^2-4y-21$ **7.** A prime polynomial is one that cannot be factored using only integers in the factors. **8.** To check a factored form, we multiply the factors. We should get the trinomial we started with. If we forget to factor out any common factors, the check will *not* indicate it. **9.** 1 and 12, -1 and -12, 2 and 6, -2 and -6, 3 and 4, -3 and -4; The pair with a sum of 7 is 3 and 4. **11.** 1 and -24, -1 and 24, 2 and -12, -2 and 12, 3 and -8, -3 and 8, 4 and -6, -4 and 6; The pair with a sum of -5 is 3 and -8. **13.** $p+6$ **15.** $x+11$ **17.** $x-8$ **19.** $y-5$ **21.** $x+11$
23. $y-9$ **25.** $(y+8)(y+1)$ **27.** $(b+3)(b+5)$
29. $(m+5)(m-4)$ **31.** $(x+8)(x-5)$ **33.** prime
35. $(y-5)(y-3)$ **37.** $(z-8)(z-7)$ **39.** $(r-6)(r+5)$
41. $(a-12)(a+4)$ **43.** $(r+2a)(r+a)$ **45.** $(x+y)(x+3y)$
47. $(t+2z)(t-3z)$ **49.** $(v-5w)(v-6w)$ **51.** $(a+5b)(a-3b)$

53. $4(x+5)(x-2)$ **55.** $2t(t+1)(t+3)$ **57.** $-2x^4(x-3)(x+7)$
59. $a^3(a+4b)(a-b)$ **61.** $5m^2(m^3+5m^2-8)$
63. $mn(m-6n)(m-4n)$

SECTION 6.3 (pages 435–436)

1. B **2.** D **3.** $(m+6)(m+2)$ **5.** $(a+5)(a-2)$
7. $(2t+1)(5t+2)$ **9.** $(3z-2)(5z-3)$ **11.** $(2s-t)(4s+3t)$
13. $(3a+2b)(5a+4b)$ **15. (a)** 2; 12; 24; 11 **(b)** 3; 8 (Order is irrelevant.) **(c)** $3m$; $8m$ **(d)** $2m^2+3m+8m+12$
(e) $(2m+3)(m+4)$ **(f)** $(2m+3)(m+4)=2m^2+11m+12$
17. $(2x+1)(x+3)$ **19.** $(4r-3)(r+1)$ **21.** $(4m+1)(2m-3)$
23. $(3m+1)(7m+2)$ **25.** $(2b+1)(3b+2)$ **27.** $(4y-3)(3y-1)$
29. $(4+x)(4+3x)$, or $(x+4)(3x+4)$ **31.** $3(4x-1)(2x-3)$
33. $2m(m-4)(m+5)$ **35.** $-4z^3(z-1)(8z+3)$
37. $(3p+4q)(4p-3q)$ **39.** $(3a-5b)(2a+b)$ **41.** The student stopped too soon. He needs to factor out the common factor $4x-1$ to get $(4x-1)(4x-5)$ as the correct answer. **42.** The student forgot to include the common factor $3k$ in her answer. The correct answer is $3k(k-5)(k+1)$.

SECTION 6.4 (pages 441–442)

1. B **2.** A **3.** A **4.** B **5.** A **6.** A **7.** $2a+5b$
9. x^2+3x-4; $x+4$, $x-1$ **11.** $2z^2-5z-3$; $2z+1$, $z-3$
13. $(4x+4)$ cannot be a factor because its terms have a common factor of 4, but those of the polynomial do not. The correct factored form is $(4x-3)(3x+4)$. **14.** The student forgot to factor out the common factor 2 from the terms of the trinomial. The completely factored form is $2(2x-1)(x+3)$. **15.** $(3a+7)(a+1)$ **17.** $(2y+3)(y+2)$
19. $(3m-1)(5m+2)$ **21.** $(3s-1)(4s+5)$ **23.** $(5m-4)(2m-3)$
25. $(4w-1)(2w-3)$ **27.** $(4y+1)(5y-11)$ **29.** prime
31. $2(5x+3)(2x+1)$ **33.** $-q(5m+2)(8m-3)$
35. $3n^2(5n-3)(n-2)$ **37.** $-y^2(5x-4)(3x+1)$
39. $(5a+3b)(a-2b)$ **41.** $(4s+5t)(3s-t)$
43. $m^4n(3m+2n)(2m+n)$ **45.** $-1(x+7)(x-3)$
47. $-1(3x+4)(x-1)$ **49.** $-1(a+2b)(2a+b)$ **51.** $5\cdot 7$
52. $(-5)(-7)$ **53.** The product of $3x-4$ and $2x-1$ is $6x^2-11x+4$. **54.** The product of $4-3x$ and $1-2x$ is $6x^2-11x+4$. **55.** The factors in **Exercise 53** are the opposites of the factors in **Exercise 54.**
56. $(3-7t)(5-2t)$

SECTION 6.5 (pages 447–448)

1. 1; 4; 9; 16; 25; 36; 49; 64; 81; 100; 121; 144; 169; 196; 225; 256; 289; 324; 361; 400 **2.** 1; 16; 81; 256; 625 **3.** A, D **4.** The binomial $4x^2+16$ can be factored as $4(x^2+4)$. *After* any common factor is removed, a sum of squares (like x^2+4 here) *cannot* be factored.
5. $(y+5)(y-5)$ **7.** $(x+12)(x-12)$ **9.** prime **11.** prime
13. $(3r+2)(3r-2)$ **15.** $4(3x+2)(3x-2)$
17. $(14p+15)(14p-15)$ **19.** $(4r+5a)(4r-5a)$ **21.** prime
23. $(p^2+7)(p^2-7)$ **25.** $(x^2+1)(x+1)(x-1)$
27. $(p^2+16)(p+4)(p-4)$ **29.** B, C **30.** No, it is not a perfect square since the middle term would have to be $30y$. **31.** $(w+1)^2$
33. $(x-4)^2$ **35.** prime **37.** $2(x+6)^2$ **39.** $(2x+3)^2$
41. $(4x-5)^2$ **43.** $(7x-2y)^2$ **45.** $(8x+3y)^2$ **47.** $-2h(5h-2y)^2$

49. $\left(p+\frac{1}{3}\right)\left(p-\frac{1}{3}\right)$ **51.** $\left(2m+\frac{3}{5}\right)\left(2m-\frac{3}{5}\right)$

53. $(x+0.8)(x-0.8)$ **55.** $\left(t+\frac{1}{2}\right)^2$ **57.** $(x-0.5)^2$

SUMMARY EXERCISES Recognizing and Applying Factoring Strategies (pages 450–452)

1. F **2.** G **3.** A **4.** B **5.** D **6.** H **7.** C **8.** E **9.** H **10.** D
11. $8m^3(4m^6+2m^2+3)$ **12.** $2(m+3)(m-8)$
13. $7k(2k+5)(k-2)$ **14.** prime **15.** $(6z+1)(z+5)$
16. $(m+n)(m-4n)$ **17.** $(7z+4y)(7z-4y)$
18. $10nr(10nr+3r^2-5n)$ **19.** $4x(4x+5)$ **20.** $(4+m)(5+3n)$
21. $(5y-6z)(2y+z)$ **22.** $(y^2+9)(y+3)(y-3)$
23. $(m-3)(m+5)$ **24.** $(2y+1)(3y-4)$ **25.** $8z(4z-1)(z+2)$
26. $(p-12)^2$ **27.** $(z-6)^2$ **28.** $(3m+8)(3m-8)$
29. $(y-6k)(y+2k)$ **30.** $(4z-1)^2$ **31.** $6(y-2)(y+1)$
32. prime **33.** $(p-6)(p-11)$ **34.** $(a+8)(a+9)$ **35.** prime
36. $3(6m-1)^2$ **37.** $(z+2a)(z-5a)$ **38.** $(2a+1)(a^2-7)$
39. $(2k-3)^2$ **40.** $(a-7b)(a+4b)$ **41.** $(4r+3m)^2$
42. $(3k-2)(k+2)$ **43.** prime **44.** $(a^2+25)(a+5)(a-5)$
45. $4(2k-3)^2$ **46.** $(4k+1)(2k-3)$ **47.** $6y^4(3y+4)(2y-5)$
48. $5z(z-2)(z-7)$ **49.** $(8p-1)(p+3)$ **50.** $(4k-3h)(2k+h)$
51. $6(3m+2z)(3m-2z)$ **52.** $(2k-5z)^2$ **53.** $2(3a-1)(a+2)$
54. $(3h-2g)(5h+7g)$ **55.** $7(2a+3b)(2a-3b)$
56. $(5z-6)(2z+1)$ **57.** $5m^2(5m-13n)(5m-3n)$
58. $(3y-1)(3y+5)$ **59.** $(3u+11v)^2$ **60.** prime
61. $9p^8(3p+7)(p-4)$ **62.** $5(2m-3)(m+4)$
63. $(2-q)(2-3p)$ **64.** $(k+11)(k-11)$
65. $4(4p+5m)(4p-5m)$ **66.** $(m+4)(m^2-6)$
67. $(10a+9y)(10a-9y)$ **68.** $(8a-b)(a+3b)$ **69.** $(a+4)^2$
70. $(2y+5)(2y-5)$ **71.** prime **72.** $-3x(x+2y)(x-2y)$
73. $(5a-7b)^2$ **74.** $8(t^2+1)(t+1)(t-1)$ **75.** $-4(x-3y)^2$
76. $25(2a+b)(2a-b)$ **77.** $-2(x-9)(x-4)$
78. $(m+3)(2m-5n)$ **79.** $2(2x+5)(3x-2)$ **80.** $(y^2+5)(y^4-3)$
81. $(y+8)(y-8)$ **82.** $6p(2p+1)(p-5)$

SECTION 6.6 (pages 459–461)

1. ax^2+bx+c **2.** standard **3.** factor **4.** 0; zero; factor
5. (a) linear **(b)** quadratic **(c)** quadratic **(d)** linear **6.** Because $(x-9)^2=(x-9)(x-9)$, applying the zero-factor property leads to two solutions of 9. Thus, 9 is a double solution. **7.** Set each *variable* factor equal to 0, to get $2x=0$ or $3x-4=0$. The solution set is $\left\{0, \frac{4}{3}\right\}$.
8. The variable x is another factor to set equal to 0, so the solution set is $\left\{0, \frac{1}{7}\right\}$. **9.** $\{-5, 2\}$ **11.** $\left\{3, \frac{7}{2}\right\}$ **13.** $\left\{-\frac{1}{2}, \frac{1}{6}\right\}$ **15.** $\left\{-\frac{5}{6}, 0\right\}$
17. $\left\{0, \frac{4}{3}\right\}$ **19.** $\{9\}$ **21.** $\{-2, -1\}$ **23.** $\{1, 2\}$ **25.** $\{-8, 3\}$
27. $\{-1, 3\}$ **29.** $\{-2, -1\}$ **31.** $\{-4\}$ **33.** $\left\{-2, \frac{1}{3}\right\}$
35. $\left\{-\frac{4}{3}, \frac{1}{2}\right\}$ **37.** $\left\{-\frac{2}{3}\right\}$ **39.** $\{-3, 3\}$ **41.** $\left\{-\frac{7}{4}, \frac{7}{4}\right\}$
43. $\{-11, 11\}$ **45.** $\{0, 7\}$ **47.** $\left\{0, \frac{1}{2}\right\}$ **49.** $\{2, 5\}$ **51.** $\left\{-4, \frac{1}{2}\right\}$
53. $\left\{-12, \frac{11}{2}\right\}$ **55.** $\{-2, 0, 2\}$ **57.** $\left\{-\frac{7}{3}, 0, \frac{7}{3}\right\}$ **59.** $\left\{-\frac{5}{2}, \frac{1}{3}, 5\right\}$

61. $\left\{-\frac{7}{2}, -3, 1\right\}$ **63.** $\{-5, 0, 4\}$ **65.** $\{-3, 0, 5\}$ **67.** $\{-1, 3\}$
69. (a) 64; 144; 4; 6 **(b)** No time has elapsed, so the object hasn't fallen (been released) yet. **70.** Time cannot be negative.

SECTION 6.7 (pages 468–472)

1. Read; variable; equation; Solve; answer; Check; original **2.** Only 6 is reasonable since a square cannot have a side of negative length.
3. *Step 3:* $(2x + 1)(x + 1)$; *Step 4:* 4; $-\frac{11}{2}$; *Step 5:* 9; 5; *Step 6:* $9 \cdot 5$
5. *Step 3:* 192; $4x$; *Step 4:* 6; -8; *Step 5:* 8; 6; *Step 6:* $8 \cdot 6$; 192
7. length: 14 cm; width: 12 cm **9.** base: 12 in.; height: 5 in.
11. length: 15 in.; width: 12 in. **13.** height: 13 in.; width: 10 in.
15. mirror: 7 ft; painting: 9 ft **17.** 20, 21 **19.** $-3, -2$ or 4, 5
21. $-3, -1$ or 7, 9 **23.** $-2, 0, 2$ or 6, 8, 10 **25.** 7, 9, 11 **27.** 12 cm
29. 12 mi **31.** 8 ft **33. (a)** 1 sec **(b)** $\frac{1}{2}$ sec and $1\frac{1}{2}$ sec **(c)** 3 sec
(d) The negative solution, -1, does not make sense since t represents time, which cannot be negative. **35.** 112 ft **37.** 256 ft **39. (a)** 109 million; The result using the model is the same as the actual number from the table for 2000. **(b)** 14 **(c)** 270 million; The result is a little less than 286 million, the actual number for 2008. **(d)** 393 million

Chapter 6 REVIEW EXERCISES (pages 476–480)

1. $15(t + 3)$ **2.** $30z(2z^2 + 1)$ **3.** $11x^2(4x + 5)$
4. $50m^2n^2(2n - mn^2 + 3)$ **5.** $(x - 4)(2y + 3)$ **6.** $(2y + 3)(3y + 2x)$
7. $(x + 3)(x + 7)$ **8.** $(y - 5)(y - 8)$ **9.** $(q + 9)(q - 3)$
10. $(r - 8)(r + 7)$ **11.** prime **12.** $3(x^2 + 2x + 2)$
13. $(r + 8s)(r - 12s)$ **14.** $(p + 12q)(p - 10q)$
15. $-8p(p + 2)(p - 5)$ **16.** $3x^2(x + 2)(x + 8)$
17. $(m + 3n)(m - 6n)$ **18.** $(y - 3z)(y - 5z)$
19. $p^5(p - 2q)(p + q)$ **20.** $-3r^3(r + 3s)(r - 5s)$
21. r and $6r$, $2r$ and $3r$ **22.** Factor out z. **23.** $(2k - 1)(k - 2)$
24. $(3r - 1)(r + 4)$ **25.** $(3r + 2)(2r - 3)$ **26.** $(5z + 1)(2z - 1)$
27. prime **28.** $4x^3(3x - 1)(2x - 1)$ **29.** $-3(x + 2)(2x - 5)$
30. $rs(5r + 6s)(2r + s)$ **31.** $-5y(3y + 2)(2y - 1)$ **32.** prime
33. $-mn(3m + 5)(m - 8)$ **34.** $(2a - 5b)(7a + 4b)$ **35.** B
36. D **37.** $(n + 8)(n - 8)$ **38.** $(5b + 11)(5b - 11)$
39. $(7y + 5w)(7y - 5w)$ **40.** $36(2p + q)(2p - q)$ **41.** prime
42. $(z + 5)^2$ **43.** $(r - 6)^2$ **44.** $(3t - 7)^2$ **45.** $(4m + 5n)^2$
46. $6x(3x - 2)^2$ **47.** $\left\{-\frac{3}{4}, 1\right\}$ **48.** $\{-7, -3, 4\}$ **49.** $\left\{0, \frac{5}{2}\right\}$
50. $\{-3, -1\}$ **51.** $\{1, 4\}$ **52.** $\{3, 5\}$ **53.** $\left\{-\frac{4}{3}, 5\right\}$
54. $\left\{-\frac{8}{9}, \frac{8}{9}\right\}$ **55.** $\{0, 8\}$ **56.** $\{-1, 6\}$ **57.** $\{7\}$ **58.** $\{6\}$
59. $\left\{-\frac{2}{5}, -2, -1\right\}$ **60.** $\{-3, 3\}$ **61.** $\left\{-\frac{3}{8}, 0, \frac{3}{8}\right\}$
62. $\left\{-\frac{3}{4}, -\frac{1}{2}, \frac{1}{3}\right\}$ **63.** $\left\{\frac{9}{5}\right\}$ **64.** $\{-3, 10\}$
65. length: 10 ft; width: 4 ft **66.** 5 ft **67.** 6, 7 or $-5, -4$
68. $-5, -4, -3$ or 5, 6, 7 **69.** 26 mi **70.** 6 m **71.** width: 10 m; length: 17 m **72. (a)** 256 ft **(b)** 1024 ft **73.** 602 thousand; The result is a little higher than the 592 thousand given in the table.
74. 882 thousand; The estimate may be unreliable because the conditions that prevailed in the years 2001–2009 may have changed, causing either a greater increase or a greater decrease predicted by the model for the number of such vehicles. **75.** D **76.** The terms of $2x + 8$ have a common factor of 2. The completely factored form is $2(x + 4)(3x - 4)$. **77.** $(3m + 4p)(5m - 4)$
78. $8abc(3b^2c - 7ac^2 + 9ab)$ **79.** $(z - x)(z - 10x)$
80. $(3k + 5)(k + 2)$ **81.** $(y^2 + 25)(y + 5)(y - 5)$
82. $3m(2m + 3)(m - 5)$ **83.** prime **84.** $2a^3(a + 2)(a - 6)$
85. $-1(2r + 3q)(6r - 5q)$ **86.** $(10a + 3)(10a - 3)$ **87.** $(7t + 4)^2$
88. $\{0, 7\}$ **89.** $\{-5, 2\}$ **90.** $\left\{-\frac{2}{5}\right\}$ **91.** 15 m, 36 m, 39 m
92. length: 6 m; width: 4 m

Chapter 6 TEST (pages 481–482)

1. D **2.** $6x(2x - 5)$ **3.** $m^2n(2mn + 3m - 5n)$ **4.** $(2x + y)(a - b)$
5. $(x - 7)(x - 2)$ **6.** $(2x + 3)(x - 1)$ **7.** $(3x + 1)(2x - 7)$
8. $3(x + 1)(x - 5)$ **9.** $(5z - 1)(2z - 3)$ **10.** prime **11.** prime
12. $(y + 7)(y - 7)$ **13.** $(9a + 11b)(9a - 11b)$ **14.** $(x + 8)^2$
15. $(2x - 7y)^2$ **16.** $-2(x + 1)^2$ **17.** $3t^2(2t + 9)(t - 4)$ **18.** prime
19. $4t(t + 4)^2$ **20.** $(x^2 + 9)(x + 3)(x - 3)$ **21.** $\{-3, 9\}$ **22.** $\left\{\frac{1}{2}, 6\right\}$
23. $\left\{-\frac{2}{5}, \frac{2}{5}\right\}$ **24.** $\{10\}$ **25.** $\{0, 3\}$ **26.** $\left\{-8, -\frac{5}{2}, \frac{1}{3}\right\}$
27. 6 ft by 9 ft **28.** $-2, -1$ **29.** 17 ft **30.** 85 billion pieces

Chapters R–6 CUMULATIVE REVIEW EXERCISES (pages 483–484)

1. $\{0\}$ **2.** $\{0.05\}$ **3.** $\{6\}$ **4.** $t = \frac{A - P}{Pr}$ **5.** second column: 38%, 12%; third column: 230, 205 **6.** gold: 9; silver: 15; bronze: 13 **7.** $669
8. 110° and 70° **9. (a)** negative; positive **(b)** negative; negative
10. $\left(-\frac{3}{4}, 0\right)$, $(0, 3)$ **11.** 4
12.

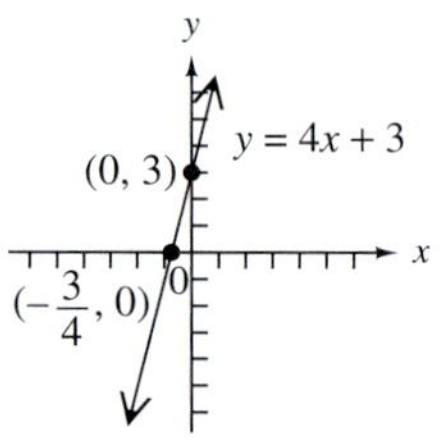

13. (a) 8; A slope of (approximately) 8 means that the retail sales of prescription drugs increased by about $8 billion per year. **(b)** (2008, 250)
14. $\{(-1, 2)\}$ **15.** $\emptyset$ **16.** 4
17. $\frac{16}{9}$ **18.** 1 **19.** 256 **20.** $\frac{1}{p^2}$
21. $\frac{1}{m^6}$ **22.** $-4k^2 - 4k + 8$ **23.** $45x^2 + 3x - 18$ **24.** $9p^2 + 12p + 4$
25. $4x^3 + 6x^2 - 3x + 10$ **26.** $(2a - 1)(a + 4)$
27. $(2m + 3)(5m + 2)$ **28.** $(4t + 3v)(2t + v)$ **29.** $(2p - 3)^2$
30. $(5r + 9t)(5r - 9t)$ **31.** $2pq(3p + 1)(p + 1)$ **32.** $\left\{-\frac{2}{3}, \frac{1}{2}\right\}$
33. $\{0, 8\}$ **34.** $\left\{\frac{4}{7}\right\}$ **35.** 5 m, 12 m, 13 m

CHAPTER 7 Rational Expressions and Applications

Note: In work with rational expressions, several different equivalent forms of the answer often exist. If your answer does not look exactly like the one given here, check to see if you have written an equivalent form.

SECTION 7.1 (pages 493–494)

1. (a) 3; -5 **(b)** q; -1 **2. (a)** is not **(b)** are **3.** A rational expression is a quotient of polynomials, such as $\frac{x + 3}{x^2 - 4}$. **4.** Division by 0 is undefined, so if the denominator of a rational expression equals 0, the expression is undefined. **5. (a)** 1 **(b)** $\frac{17}{12}$ **7. (a)** 0 **(b)** -1 **9. (a)** $\frac{9}{5}$ **(b)** undefined **11. (a)** $\frac{2}{7}$ **(b)** $\frac{13}{3}$ **13.** $y \neq 0$ **15.** $x \neq -6$ **17.** $x \neq \frac{5}{3}$ **19.** $m \neq -3, m \neq 2$ **21.** never undefined **23.** never undefined **25.** $3r^2$ **27.** $\frac{2}{5}$ **29.** $\frac{x-1}{x+1}$ **31.** $\frac{7}{5}$ **33.** $m-n$ **35.** $\frac{3(2m+1)}{4}$ **37.** $\frac{3m}{5}$ **39.** $\frac{3r-2s}{3}$ **41.** $\frac{x+1}{x-1}$ **43.** $\frac{z-3}{z+5}$ **45.** $\frac{a+b}{a-b}$ **47.** -1 **49.** $-(m+1)$ **51.** -1 **53.** It is already in lowest terms.

Answers may vary in Exercises 55–59.

55. $\frac{-(x+4)}{x-3}, \frac{-x-4}{x-3}, \frac{x+4}{-(x-3)}, \frac{x+4}{-x+3}$
57. $\frac{-(2x-3)}{x+3}, \frac{-2x+3}{x+3}, \frac{2x-3}{-(x+3)}, \frac{2x-3}{-x-3}$
59. $\frac{-(3x-1)}{5x-6}, \frac{-3x+1}{5x-6}, \frac{3x-1}{-(5x-6)}, \frac{3x-1}{-5x+6}$ **61.** x^2+3

SECTION 7.2 (pages 499–500)

1. (a) B **(b)** D **(c)** C **(d)** A **2. (a)** D **(b)** C **(c)** A **(d)** B **3.** $\frac{4m}{3}$ **5.** $\frac{40y^2}{3}$ **7.** $\frac{2}{c+d}$ **9.** $4(x-y)$ **11.** $\frac{16q}{3p^3}$ **13.** $\frac{7}{r^2+rp}$ **15.** $\frac{z^2-9}{z^2+7z+12}$ **17.** 5 **19.** $-\frac{3}{2t^4}$ **21.** $\frac{1}{4}$ **23.** $\frac{x(x-3)}{6}$ **25.** $x-3$; $x+3$; $x+3$; $x-3$; $\frac{10}{9}$ **27.** $-\frac{3}{4}$ **29.** -1 **31.** $\frac{9(m-2)}{-(m+4)}$, or $\frac{-9(m-2)}{m+4}$ **33.** $\frac{p+4}{p+2}$ **35.** $\frac{(k-1)^2}{(k+1)(2k-1)}$ **37.** $\frac{4k-1}{3k-2}$ **39.** $\frac{m+4p}{m+p}$ **41.** $\frac{10}{x+10}$ **43.** $\frac{5xy^2}{4q}$

SECTION 7.3 (pages 505–506)

1. C **2.** B **3.** C **4.** A **5.** 30 **7.** x^7 **9.** $72q$ **11.** $84r^5$ **13.** $24x^3y^4$ **15.** $x^{10}y^{14}z^{23}$ **17.** $28m^2(3m-5)$ **19.** $30(b-2)$ **21.** $c-d$ or $d-c$ **23.** $k(k+5)(k-2)$ **25.** $a(a+6)(a-3)$ **27.** $(p+3)(p+5)(p-6)$ **29.** $\frac{20}{55}$ **31.** $\frac{-45}{9k}$ **33.** $\frac{26y^2}{80y^3}$ **35.** $\frac{35t^2r^3}{42r^4}$ **37.** $\frac{20}{8(m+3)}$ **39.** $\frac{8t}{12-6t}$ **41.** $\frac{14(z-2)}{z(z-3)(z-2)}$ **43.** $\frac{2(b-1)(b+2)}{b^3+3b^2+2b}$

SECTION 7.4 (pages 513–516)

1. E **2.** A **3.** C **4.** H **5.** B **6.** D **7.** G **8.** F **9.** $\frac{11}{m}$ **11.** b **13.** $\frac{4}{y+4}$ **15.** $\frac{m-1}{m+1}$ **17.** x **19.** $y^2-3y-18$; 3; 6; $y-6$ **21.** Combine the numerators and keep the same denominator. For example, $\frac{3x+2}{x-6}+\frac{-2x-8}{x-6}=\frac{x-6}{x-6}$. Then write in lowest terms: $\frac{x-6}{x-6}=1$. **22.** Find the LCD, write each fraction as an equivalent one with this LCD, and then proceed as in **Exercise 21.** For example, $\frac{3}{x}+\frac{2}{y}=\frac{3y}{xy}+\frac{2x}{xy}=\frac{3y+2x}{xy}$. **23.** $\frac{3z+5}{15}$ **25.** $\frac{10-7r}{14}$ **27.** $\frac{-3x-2}{4x}$ **29.** $\frac{57}{20x}$ **31.** $\frac{x+1}{2}$ **33.** $\frac{5x+9}{6x}$ **35.** $x(x+3)$; $\frac{3x+3}{x(x+3)}$ **37.** $\frac{-k-10}{k(k+5)}$ **39.** $\frac{x+4}{x+2}$ **41.** $\frac{x^2+6x-8}{(x-2)(x+2)}$ **43.** $\frac{3}{t}$ **45.** $m-2$ or $2-m$ **46.** $\frac{-4}{3-k}$, or $-\frac{4}{3-k}$ **47.** $\frac{-2}{x-5}$, or $\frac{2}{5-x}$ **49.** -4 **51.** $\frac{-5}{x-y^2}$, or $\frac{5}{y^2-x}$ **53.** $\frac{x+y}{5x-3y}$, or $\frac{-x-y}{3y-5x}$ **55.** $\frac{-6}{4p-5}$, or $\frac{6}{5-4p}$ **57.** $\frac{-(m+n)}{2(m-n)}$ **59.** $\frac{-x^2+6x+11}{(x+3)(x-3)(x+1)}$ **61.** $\frac{-5q^2-13q+7}{(3q-2)(q+4)(2q-3)}$ **63.** $\frac{9r+2}{r(r+2)(r-1)}$ **65.** $\frac{2x^2+6xy+8y^2}{(x+y)(x+y)(x+3y)}$, or $\frac{2x^2+6xy+8y^2}{(x+y)^2(x+3y)}$ **67.** $\frac{15r^2+10ry-y^2}{(3r+2y)(6r-y)(6r+y)}$ **69. (a)** $\frac{9k^2+6k+26}{5(3k+1)}$ **(b)** $\frac{1}{4}$

SECTION 7.5 (pages 523–524)

1. division **2.** identity property for multiplication **3. (a)** 6; $\frac{1}{6}$ **(b)** 12; $\frac{3}{4}$ **(c)** $\frac{1}{6} \div \frac{3}{4}$ **(d)** $\frac{2}{9}$ **5.** -6 **7.** $\frac{1}{pq}$ **9.** $\frac{1}{xy}$ **11.** $\frac{2a^2b}{3}$ **13.** $\frac{m(m+2)}{3(m-4)}$ **15.** $\frac{2}{x}$ **17.** $\frac{8}{x}$ **19.** $\frac{a^2-5}{a^2+1}$ **21.** $\frac{3(p+2)}{2(2p+3)}$ **23.** $\frac{40-12p}{85p}$ **25.** $\frac{t(t-2)}{4}$ **27.** $\frac{-k}{2+k}$ **29.** $\frac{2x-7}{3x+1}$ **31.** $\frac{3m(m-3)}{(m-1)(m-8)}$ **33.** $\frac{6}{5}$

SECTION 7.6 (pages 533–536)

1. proposed; original **2.** extraneous; extraneous **3.** expression; $\frac{43}{40}x$ **5.** equation; $\left\{\frac{40}{43}\right\}$ **7.** expression; $-\frac{1}{10}y$ **9.** $\{-6\}$ **11.** $\{-15\}$ **13.** $\{7\}$ **15.** $\{-15\}$ **17.** $\{-5\}$ **19.** $\{-6\}$ **21.** $\{5\}$ **23.** $\{12\}$ **25.** $\{2\}$ **27.** $x \neq -2, x \neq 0$ **29.** $x \neq -3, x \neq 4, x \neq -\frac{1}{2}$ **31.** $x \neq -9, x \neq 1, x \neq -2, x \neq 2$ **33.** $\left\{\frac{20}{9}\right\}$ **35.** $\emptyset$ **37.** $4(x-1)$; $\{3\}$ **39.** $\{3\}$ **41.** $\{-2, 12\}$ **43.** $\left\{-\frac{1}{5}, 3\right\}$ **45.** $\left\{-\frac{3}{5}, 3\right\}$ **47.** $\{-4\}$ **49.** $\emptyset$ **51.** $\{-1\}$ **53.** $\{-6\}$ **55.** $\left\{-6, \frac{1}{2}\right\}$ **57.** Transform the equation so that the terms with k are on one side and the remaining term is on the other. **58.** Factor out k on the left side. **59.** $F=\frac{ma}{k}$ **61.** $a=\frac{kF}{m}$ **63.** $y=mx+b$ **65.** $R=\frac{E-Ir}{I}$, or $R=\frac{E}{I}-r$ **67.** $b=\frac{2A-hB}{h}$, or $b=\frac{2A}{h}-B$ **69.** $a=\frac{2S-dnL}{dn}$, or $a=\frac{2S}{dn}-L$ **71.** $t=\frac{rs}{rs-2s-3r}$, or $t=\frac{-rs}{-rs+2s+3r}$ **73.** $c=\frac{ab}{b-a-2ab}$, or $c=\frac{-ab}{-b+a+2ab}$ **75.** $z=\frac{3y}{5-9xy}$, or $z=\frac{-3y}{9xy-5}$

SUMMARY EXERCISES Simplifying Rational Expressions vs. Solving Rational Equations (page 538)

1. expression; $\frac{10}{p}$ **2.** expression; $\frac{y^3}{x^3}$ **3.** expression; $\frac{1}{2x^2(x+2)}$ **4.** equation; $\{9\}$ **5.** expression; $\frac{x+2}{x-1}$ **6.** expression; $\frac{5k+8}{k(k-4)(k+4)}$ **7.** equation; $\{39\}$ **8.** expression; $\frac{t-5}{3(2t+1)}$ **9.** expression; $\frac{13}{3(p+2)}$ **10.** equation; $\left\{-1, \frac{12}{5}\right\}$ **11.** equation; $\left\{\frac{1}{7}, 2\right\}$ **12.** expression; $\frac{16}{3x}$ **13.** expression; $\frac{7}{12z}$ **14.** equation; $\{13\}$ **15.** expression; $\frac{3m+5}{(m+2)(m+3)(m+1)}$ **16.** expression; $\frac{k+3}{5(k-1)}$ **17.** equation; $\emptyset$ **18.** equation; $\{-7\}$

SECTION 7.7 (pages 546–549)

1. into a headwind: $(m-5)$ mph; with a tailwind: $(m+5)$ mph **2.** $\frac{D}{R} = \frac{d}{r}$ **3.** $\frac{1}{10}$ job per hr **4.** $\frac{2}{3}$ of the job **5. (a)** the amount **(b)** $5+x$ **(c)** $\frac{5+x}{6} = \frac{13}{3}$ **7.** x represents the original numerator; $\frac{x+3}{(x-4)+3} = \frac{3}{2}; \frac{9}{5}$ **9.** x represents the original numerator; $\frac{x+2}{3x-2} = 1; \frac{2}{6}$ **11.** x represents the number; $\frac{1}{6}x = x + 5$; -6 **13.** x represents the quantity; $x + \frac{3}{4}x + \frac{1}{2}x + \frac{1}{3}x = 93$; 36 **15.** 18.809 min **17.** 3.565 hr **19.** 314.248 m per min **21.** table entries: (fourth column) $\frac{8}{4-x}, \frac{24}{4+x}; \frac{8}{4-x} = \frac{24}{4+x}$ **23.** 8 mph **25.** 32 mph **27.** 3 mph **29.** table entries: (second column) $\frac{1}{2}, \frac{1}{3}$; (fourth column) $\frac{1}{2}x, \frac{1}{3}x$; $\frac{1}{2}x + \frac{1}{3}x = 1$ **31.** $2\frac{2}{5}$ hr **33.** $4\frac{8}{19}$ hr **35.** 10 hr **37.** 36 hr **39.** 10 mL

SECTION 7.8 (pages 552–554)

1. direct **2.** direct **3.** inverse **4.** inverse **5.** inverse **6.** direct **7.** direct **8.** inverse **9. (a)** increases **(b)** decreases **10.** The customers in the lower-priced seats know more about the game than those in the higher-priced seats. **11.** inverse **12.** inverse **13.** direct **14.** direct **15.** 15 **17.** 300 **19.** 4 **21.** 6 **23.** 15 in.^{2} **25.** $42\frac{2}{3}$ in. **27.** 15 ft **29.** 20 lb per ft^2 **31.** 25 kg per hr **33.** 4

Chapter 7 REVIEW EXERCISES (pages 560–563)

1. $x \neq 3$ **2.** $x \neq 0$ **3.** $m \neq -1, m \neq 3$ **4.** $k \neq -5, k \neq -\frac{2}{3}$ **5. (a)** $-\frac{4}{7}$ **(b)** -16 **6. (a)** $\frac{11}{8}$ **(b)** $\frac{13}{22}$ **7. (a)** undefined **(b)** 1 **8. (a)** undefined **(b)** $\frac{1}{2}$ **9.** $\frac{b}{3a}$ **10.** -1 **11.** $\frac{-(2x+3)}{2}$ **12.** $\frac{2p+5q}{5p+q}$

Answers may vary in Exercises 13 and 14.

13. $\frac{-(4x-9)}{2x+3}, \frac{-4x+9}{2x+3}, \frac{4x-9}{-(2x+3)}, \frac{4x-9}{-2x-3}$ **14.** $\frac{-(8-3x)}{3-6x}, \frac{-8+3x}{3-6x}, \frac{8-3x}{-(3-6x)}, \frac{8-3x}{-3+6x}$ **15.** 2 **16.** $\frac{2}{3m^6}$ **17.** $\frac{5}{8}$ **18.** $\frac{r+4}{3}$ **19.** $\frac{3}{2}$ **20.** $\frac{y-2}{y-3}$ **21.** $\frac{p+5}{p+1}$ **22.** $\frac{3z+1}{z+3}$ **23.** 96 **24.** $108y^4$ **25.** $m(m+2)(m+5)$ **26.** $(x+3)(x+1)(x+4)$ **27.** $\frac{35}{56}$ **28.** $\frac{40}{4k}$ **29.** $\frac{15a}{10a^4}$ **30.** $\frac{-54}{18-6x}$ **31.** $\frac{15y}{50-10y}$ **32.** $\frac{4b(b+2)}{(b+3)(b-1)(b+2)}$ **33.** $\frac{15}{x}$ **34.** $-\frac{2}{p}$ **35.** $\frac{4k-45}{k(k-5)}$ **36.** $\frac{28+11y}{y(7+y)}$ **37.** $\frac{-2-3m}{6}$ **38.** $\frac{3(16-x)}{4x^2}$ **39.** $\frac{7a+6b}{(a-2b)(a+2b)}$ **40.** $\frac{-k^2-6k+3}{3(k+3)(k-3)}$ **41.** $\frac{5z-16}{z(z+6)(z-2)}$ **42.** $\frac{-13p+33}{p(p-2)(p-3)}$ **43.** $\frac{a}{b}$ **44.** $\frac{4(y-3)}{y+3}$ **45.** $\frac{6(3m+2)}{2m-5}$ **46.** $\frac{(q-p)^2}{pq}$ **47.** $\frac{xw+1}{xw-1}$ **48.** $\frac{1-r-t}{1+r+t}$ **49.** $\left\{\frac{35}{6}\right\}$ **50.** $\{-16\}$ **51.** $\{-4\}$ **52.** $\emptyset$ **53.** $\{3\}$ **54.** $t = \frac{Ry}{m}$ **55.** $y = \frac{4x+5}{3}$ **56.** $t = \frac{rs}{s-r}$ **57.** $\frac{20}{15}$ **58.** $\frac{3}{18}$ **59.** 1.843 hr **60.** 800.897 m per min **61.** $7\frac{1}{2}$ min **62.** $3\frac{1}{13}$ hr **63.** $\frac{36}{5}$ **64.** 4 cm **65.** $\frac{m+7}{(m-1)(m+1)}$ **66.** $8p^2$ **67.** $\frac{1}{6}$ **68.** 3 **69.** $\frac{z+7}{(z+1)(z-1)^2}$ **70.** $\frac{x-7}{2x+3}$ **71.** $\{4\}$ **72.** $\{-2, 3\}$ **73.** $\{2\}$ **74.** $w = v - at$ **75.** 3 **76.** $7\frac{1}{2}$ hr **77.** table entries: (third column) $x+50$, $x-50$; (fourth column) $\frac{400}{x+50}$, $\frac{200}{x-50}$; 150 km per hr **78.** 10 hr **79.** 50 amps **80.** inverse

Chapter 7 TEST (pages 564–565)

1. $x \neq -2, x \neq 4$ **2. (a)** $\frac{11}{6}$ **(b)** undefined **3.** (Answers may vary.) $\frac{-(6x-5)}{2x+3}, \frac{-6x+5}{2x+3}, \frac{6x-5}{-(2x+3)}, \frac{6x-5}{-2x-3}$ **4.** $-3x^2y^3$ **5.** $\frac{3a+2}{a-1}$ **6.** $\frac{25}{27}$ **7.** $\frac{3k-2}{3k+2}$ **8.** $\frac{a-1}{a+4}$ **9.** $\frac{x-5}{3-x}$ **10.** $150p^5$ **11.** $(2r+3)(r+2)(r-5)$ **12.** $\frac{240p^2}{64p^3}$ **13.** $\frac{21}{42m-84}$ **14.** 2 **15.** $\frac{-14}{5(y+2)}$ **16.** $\frac{x^2+x+1}{3-x}$, or $\frac{-x^2-x-1}{x-3}$ **17.** $\frac{-m^2+7m+2}{(2m+1)(m-5)(m-1)}$ **18.** $\frac{2k}{3p}$ **19.** $\frac{-2-x}{4+x}$ **20.** $\left\{-\frac{1}{2}, 1\right\}$ **21.** $\left\{-\frac{1}{2}, 5\right\}$ **22.** $\left\{-\frac{1}{2}\right\}$ **23.** $D = \frac{dF-k}{F}$, or $D = d - \frac{k}{F}$ **24.** -4 **25.** 3 mph **26.** $2\frac{2}{9}$ hr **27.** 27 days **28.** 27

Chapters R–7 CUMULATIVE REVIEW EXERCISES (pages 566–567)

1. 2 **2.** $\{17\}$ **3.** $b = \frac{2A}{h}$ **4.** $\left\{-\frac{2}{7}\right\}$ **5.** $[-8, \infty)$

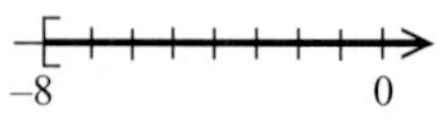

6. $(4, \infty)$

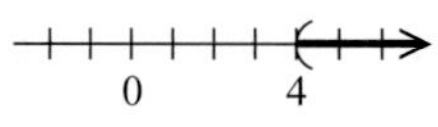

7.

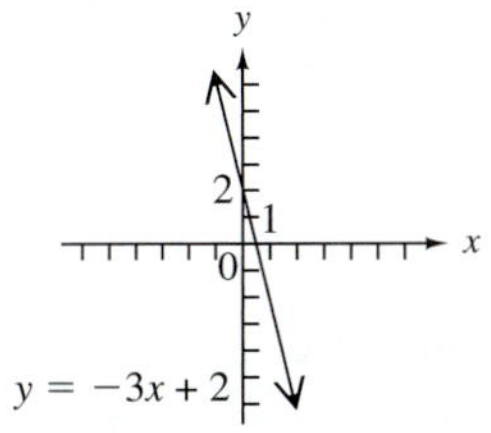

8.

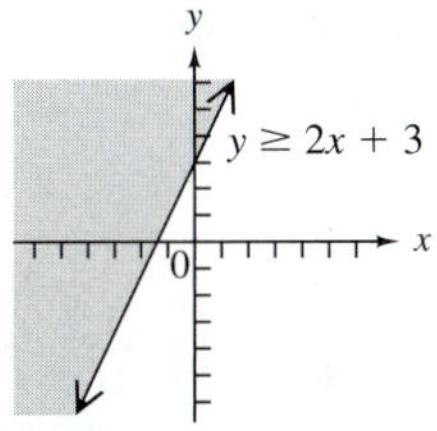

9. $\{(-1, 2)\}$ **10.** $\{(1, -2)\}$ **11.** $\frac{1}{16x^7}$ **12.** $\frac{1}{m^6}$ **13.** $\frac{q}{4p^2}$ **14.** $k^2 + 2k + 1$ **15.** $72x^6y^7$ **16.** $4a^2 - 4ab + b^2$ **17.** $3y^3 + 8y^2 + 12y - 5$ **18.** $6p^2 + 7p + 1 + \frac{3}{p-1}$ **19.** $(4t + 3v)(2t + v)$ **20.** prime **21.** $(4x^2 + 1)(2x + 1)(2x - 1)$ **22.** $\{-3, 5\}$ **23.** $\left\{-\frac{1}{2}, \frac{2}{3}, 5\right\}$ **24.** -2 or -1 **25.** 6 m **26.** $t \neq -2, t \neq 2$ **27.** A **28.** D **29.** $\frac{4}{q}$ **30.** $\frac{3r + 28}{7r}$ **31.** $\frac{7}{15(q-4)}$ **32.** $\frac{-k-5}{k(k+1)(k-1)}$ **33.** $\frac{7(2z+1)}{24}$ **34.** $\frac{195}{29}$ **35.** $\left\{\frac{21}{2}\right\}$ **36.** $\{-2, 1\}$ **37.** 150 mi **38.** $1\frac{1}{5}$ hr

CHAPTER 8 Roots and Radicals

SECTION 8.1 (pages 577–580)

1. true **2.** false; A negative number has no real square roots. **3.** false; Zero has only one square root. **4.** true **5.** true **6.** false; A positive number has just one real cube root. **7.** $-3, 3$ **9.** $-8, 8$ **11.** $-13, 13$ **13.** $-\frac{5}{14}, \frac{5}{14}$ **15.** $-30, 30$ **17.** a must be positive. **18.** a must be positive. **19.** a must be negative. **20.** a must be negative. **21.** 64; 8 **23.** 1 **25.** 7 **27.** -16 **29.** $-\frac{12}{11}$ **31.** 0.8 **33.** not a real number **35.** not a real number **37.** 100 **39.** 19 **41.** $\frac{2}{3}$ **43.** $3x^2 + 4$ **45.** 9 and 10 **46.** 6 and 7 **47.** 7 and 8 **48.** 5 and 6 **49.** 4 and 5 **50.** 3 and 4 **51.** rational; 5 **53.** irrational; 5.385 **55.** rational; -8 **57.** irrational; -17.321 **59.** not a real number **61.** irrational; 34.641 **63.** C **65.** $c = 17$ **67.** $b = 8$ **69.** $c \approx 11.705$ **71.** 24 cm **73.** 80 ft **75.** 195 ft **77.** 158.6 ft **79.** 11.1 ft **81.** 9.434 **83.** 4; 4 **85.** 5 **87.** 8 **89.** -3 **91.** -6 **93.** 2 **95.** 4 **97.** 6 **99.** not a real number **101.** -5 **103.** -4

SECTION 8.2 (pages 587–590)

1. false; $\sqrt{(-6)^2} = \sqrt{36} = 6$ **2.** true **3.** false; $2\sqrt{7}$ represents the product of 2 and $\sqrt{7}$. **4.** true **5.** $\sqrt{15}$ **7.** $\sqrt{22}$ **9.** $\sqrt{42}$ **11.** $\sqrt{13r}$ **13.** A **14.** Yes; because a prime number cannot be factored (except as the product of itself and 1), it cannot have any perfect square factors (except 1). **15.** $3\sqrt{5}$ **17.** $2\sqrt{6}$ **19.** $3\sqrt{10}$ **21.** $5\sqrt{3}$ **23.** $5\sqrt{5}$ **25.** cannot be simplified **27.** $4\sqrt{10}$ **29.** $-10\sqrt{7}$ **31.** $6\sqrt{13}$ **33.** $25\sqrt{2}$ **35.** $5\sqrt{10}$ **37.** $6\sqrt{2}$ **39.** $3\sqrt{6}$ **41.** $12\sqrt{2}$ **43.** 24 **45.** $6\sqrt{10}$ **47.** $12\sqrt{5}$ **49.** $30\sqrt{5}$ **51.** *Method 1:* $\sqrt{8} \cdot \sqrt{32} = \sqrt{8 \cdot 32} = \sqrt{256} = 16$; *Method 2:* $\sqrt{8} = 2\sqrt{2}$ and $\sqrt{32} = 4\sqrt{2}$, so $\sqrt{8} \cdot \sqrt{32} = 2\sqrt{2} \cdot 4\sqrt{2} = 8 \cdot 2 = 16$; The same answer results. Either method can be used to obtain the correct answer. **52.** *Method 1:* $\sqrt{288} = \sqrt{144 \cdot 2} = \sqrt{144} \cdot \sqrt{2} = 12\sqrt{2}$; *Method 2:* $\sqrt{288} = \sqrt{48 \cdot 6} = \sqrt{48} \cdot \sqrt{6} = \sqrt{16 \cdot 3} \cdot \sqrt{6} = 4\sqrt{3} \cdot \sqrt{6} = 4\sqrt{18} = 4\sqrt{9 \cdot 2} = 4 \cdot 3\sqrt{2} = 12\sqrt{2}$; The same answer results. Method 1 yields the answer more quickly using the greatest perfect square factor first. **53.** $\frac{4}{15}$ **55.** $\frac{\sqrt{7}}{4}$ **57.** $\frac{\sqrt{2}}{5}$ **59.** 5 **61.** $\frac{25}{4}$ **63.** $6\sqrt{5}$ **65.** m **67.** y^2 **69.** $6z$ **71.** $20x^3$ **73.** $3x^4\sqrt{2}$ **75.** $3c^7\sqrt{5}$ **77.** $z^2\sqrt{z}$ **79.** $a^6\sqrt{a}$ **81.** $8x^3\sqrt{x}$ **83.** x^3y^6 **85.** $9m^2n$ **87.** $\frac{\sqrt{7}}{x^5}$ **89.** $\frac{y^2}{10}$ **91.** $\frac{x^3}{y^4}$ **93.** $2\sqrt[3]{5}$ **95.** $3\sqrt[3]{2}$ **97.** $4\sqrt[3]{2}$ **99.** $2\sqrt[4]{5}$ **101.** $\frac{2}{3}$ **103.** $-\frac{6}{5}$ **105.** p **107.** x^3 **109.** $4z^2$ **111.** $7a^3b$ **113.** $2t\sqrt[3]{2t^2}$ **115.** $\frac{m^4}{2}$ **117.** 6 cm **119.** 6 in. **121.** D

SECTION 8.3 (pages 593–594)

1. radicand; index; like; 7; 7 **2.** like; 4; $3xy^3$ **3.** radicands; unlike **4.** indexes or roots; unlike **5.** distributive **6.** like; 1; $5\sqrt{5x}$ **7.** $7\sqrt{3}$ **9.** $-5\sqrt{7}$ **11.** $5\sqrt{17}$ **13.** $5\sqrt{7}$ **15.** $2\sqrt{6}$ **17.** These unlike radicals cannot be combined. **19.** $7\sqrt{3}$ **21.** $11\sqrt{5}$ **23.** $15\sqrt{2}$ **25.** $-6\sqrt{2}$ **27.** $17\sqrt{7}$ **29.** $-16\sqrt{2} - 8\sqrt{3}$ **31.** $20\sqrt{2} + 6\sqrt{3} - 15\sqrt{5}$ **33.** $4\sqrt{2}$ **35.** $22\sqrt{2}$ **37.** $11\sqrt{3}$ **39.** $3\sqrt{21}$ **41.** $\sqrt{2x}$ **43.** $7\sqrt{3r}$ **45.** $5\sqrt{x}$ **47.** $3x\sqrt{6}$ **49.** 0 **51.** $-20\sqrt{2k}$ **53.** $42x\sqrt{5z}$ **55.** $-\sqrt[3]{2}$ **57.** $6\sqrt[3]{p^2}$ **59.** $21\sqrt[4]{m^3}$

SECTION 8.4 (pages 599–600)

1. We are actually multiplying by 1. The identity property for multiplication justifies our result. **3.** $\frac{6\sqrt{5}}{5}$ **5.** $\sqrt{5}$ **7.** $4\sqrt{2}$ **9.** $\frac{-\sqrt{33}}{3}$ **11.** $\frac{7\sqrt{15}}{5}$ **13.** $\frac{\sqrt{30}}{2}$ **15.** $\frac{16\sqrt{3}}{9}$ **17.** $\frac{-3\sqrt{2}}{10}$ **19.** $\frac{21\sqrt{5}}{5}$ **21.** $\sqrt{3}$ **23.** $\frac{\sqrt{2}}{2}$ **25.** $\frac{\sqrt{65}}{5}$ **27.** $\frac{\sqrt{21}}{3}$ **29.** $\frac{3\sqrt{14}}{4}$ **31.** $\frac{1}{6}$ **33.** 1 **35.** $\frac{\sqrt{7x}}{x}$ **37.** $\frac{2x\sqrt{xy}}{y}$ **39.** $\frac{x\sqrt{30xz}}{6}$ **41.** $\frac{3ar^2\sqrt{7rt}}{7t}$ **43.** B **44.** yes **45.** $\frac{\sqrt[3]{4}}{2}$ **47.** $\frac{\sqrt[3]{15}}{3}$ **49.** $\frac{\sqrt[3]{196}}{7}$ **51.** $\frac{\sqrt[3]{6y}}{2y}$ **53.** $\frac{\sqrt[3]{42mn^2}}{6n}$ **55.** **(a)** $\frac{9\sqrt{2}}{4}$ sec **(b)** 3.182 sec

SECTION 8.5 (pages 605–608)

1. 13 **2.** 1 **3.** 4 **4.** 8 **5.** Because multiplication must be performed before addition, it is incorrect to add -37 and -2. $-2\sqrt{15}$ cannot be simplified further. The final answer is $-37 - 2\sqrt{15}$. **6.** **(a)** $x - y$ **(b)** 14 **7.** $\sqrt{15} - \sqrt{35}$ **9.** $2\sqrt{10} + 30$ **11.** $4\sqrt{7}$ **13.** $57 + 23\sqrt{6}$ **15.** $81 + 14\sqrt{21}$ **17.** $71 - 16\sqrt{7}$ **19.** $37 + 12\sqrt{7}$ **21.** $a + 2\sqrt{a} + 1$ **23.** 23 **25.** 1 **27.** $y - 10$ **29.** $2\sqrt{3} - 2 + 3\sqrt{2} - \sqrt{6}$ **31.** $15\sqrt{2} - 15$ **33.** $\sqrt{30} + \sqrt{15} + 6\sqrt{5} + 3\sqrt{10}$ **35.** $\sqrt{5x} - \sqrt{10} - \sqrt{10x} + 2\sqrt{5}$ **37.** **(a)** $\sqrt{5} - \sqrt{3}$ **(b)** $\sqrt{6} + \sqrt{5}$ **38.** There would still be a radical in the denominator. He should multiply by $\frac{4 - \sqrt{3}}{4 - \sqrt{3}}$. **39.** $\frac{3 - \sqrt{2}}{7}$

41. $-4-2\sqrt{11}$ **43.** $1+\sqrt{2}$ **45.** $-\sqrt{10}+\sqrt{15}$
47. $2\sqrt{5}+\sqrt{15}+4+2\sqrt{3}$ **49.** $\frac{-6\sqrt{2}+12+\sqrt{10}-2\sqrt{5}}{2}$
51. $\frac{12(\sqrt{x}-1)}{x-1}$ $(x \neq 1)$ **53.** $\frac{3(7+\sqrt{x})}{49-x}$ **55.** $\sqrt{11}-2$
57. $\frac{\sqrt{3}+5}{8}$ **59.** $\frac{6-\sqrt{10}}{2}$ **61.** $\frac{2+\sqrt{2}}{3}$ **63.** **(a)** 4 in. **(b)** 3.4 in.
65. $30+18x$ **66.** They are not like terms. **67.** $30+18\sqrt{5}$
68. They are not like radicals. **69.** Make the first term $30x$, so that $30x+18x=48x$. Make the first term $30\sqrt{5}$, so that $30\sqrt{5}+18\sqrt{5}=48\sqrt{5}$. **70.** Both like terms and like radicals are combined by adding their numerical coefficients. The variables in like terms are replaced by radicals in like radicals.

SUMMARY EXERCISES Applying Operations with Radicals (pages 609–610)

1. $-3\sqrt{10}$ **2.** $5-\sqrt{15}$ **3.** $2-\sqrt{6}+2\sqrt{3}-3\sqrt{2}$ **4.** $6\sqrt{2}$
5. $73-12\sqrt{35}$ **6.** $\frac{\sqrt{6}}{2}$ **7.** $3\sqrt[3]{2t^2}$ **8.** $4\sqrt{7}+4\sqrt{5}$ **9.** $-3-2\sqrt{2}$
10. 4 **11.** -33 **12.** $\frac{\sqrt{t}-\sqrt{3}}{t-3}$ $(t \neq 3)$ **13.** $2xyz^2\sqrt[3]{y^2}$
14. $4\sqrt[3]{3}$ **15.** $\sqrt{6}+1$ **16.** $\frac{\sqrt{6x}}{3x}$ **17.** $\frac{3}{5}$ **18.** $4\sqrt{2}$ **19.** $-2\sqrt[3]{2}$
20. $11-2\sqrt{30}$ **21.** $3\sqrt{3x}$ **22.** $52+30\sqrt{3}$ **23.** 1 **24.** $\frac{2\sqrt[3]{18}}{9}$
25. $-x^2\sqrt[4]{x}$ **26.** $2\sqrt{7}$ **27.** cannot be combined **28.** $12\sqrt{6}+6\sqrt{5}$
29. $\frac{\sqrt{15}}{10}$ **30.** $\sqrt{5}$ **31.** $20\sqrt[3]{3}$ **32.** $\frac{8(4+\sqrt{x})}{16-x}$ **33.** $2-3\sqrt[3]{4}$
34. $\sqrt{2x}$ **35.** $\frac{\sqrt{10}}{4}$ **36.** $49+14\sqrt{x}+x$ **37.** 57 species
38. 858 species

SECTION 8.6 (pages 617–620)

1. squaring; squared; original **2.** $a^2+2ab+b^2$; $a^2-2ab+b^2$
3. $\{49\}$ **5.** $\{7\}$ **7.** $\{85\}$ **9.** $\emptyset$ **11.** $\{-45\}$ **13.** $\left\{-\frac{3}{2}\right\}$
15. $\emptyset$ **17.** $\{121\}$ **19.** $\{8\}$ **21.** $\{1\}$ **23.** $\{6\}$ **25.** $\emptyset$ **27.** $\{5\}$
29. When the left side is squared, the result should be $x-1$, not $-(x-1)$. The correct solution set is $\{17\}$. **30.** $x^2-14x+49$ **31.** $\{12\}$ **33.** $\{5\}$
35. $\{0,3\}$ **37.** $\{-1,3\}$ **39.** $\{8\}$ **41.** $\{4\}$ **43.** $\{8\}$ **45.** $\{9\}$
47. $\{4,20\}$ **49.** $\{-5\}$ **51.** 21 **53.** 8 **55.** 17,616 ft^2
57. **(a)** 70.5 mph **(b)** 59.8 mph **(c)** 53.9 mph **59.** yes; 26 mi
61. 47 mi **63.** $s=13$ units **64.** $6\sqrt{13}$ sq. units **65.** $h=\sqrt{13}$ units
66. $3\sqrt{13}$ sq. units **67.** $6\sqrt{13}$ sq. units **68.** They are both $6\sqrt{13}$.

Chapter 8 REVIEW EXERCISES (pages 625–628)

1. $-7, 7$ **2.** $-9, 9$ **3.** $-14, 14$ **4.** $-11, 11$ **5.** $-16, 16$ **6.** $-27, 27$
7. 4 **8.** -0.6 **9.** -8 **10.** 3 **11.** not a real number **12.** -65
13. $\frac{12}{13}$ **14.** $-\frac{10}{9}$ **15.** 8 **16.** 48.3 cm **17.** irrational; 8.544
18. rational; 13 **19.** rational; -25 **20.** not a real number **21.** $4\sqrt{3}$
22. $-12\sqrt{2}$ **23.** $2\sqrt[3]{2}$ **24.** $5\sqrt[3]{3}$ **25.** 18 **26.** $16\sqrt{6}$ **27.** $-\frac{11}{20}$
28. $\frac{\sqrt{7}}{13}$ **29.** $\frac{\sqrt{5}}{6}$ **30.** $\frac{2}{15}$ **31.** $3\sqrt{2}$ **32.** $2\sqrt{2}$ **33.** r^9 **34.** x^5y^8
35. $9x^4\sqrt{2x}$ **36.** $\frac{6}{p}$ **37.** $a^7b^{10}\sqrt{ab}$ **38.** $11x^3y^5$ **39.** y^2 **40.** $6x^5$
41. $8\sqrt{11}$ **42.** $9\sqrt{2}$ **43.** $21\sqrt{3}$ **44.** $12\sqrt{3}$ **45.** 0 **46.** $3\sqrt{7}$
47. $2\sqrt{3}+3\sqrt{10}$ **48.** $2\sqrt{2}$ **49.** $6\sqrt{30}$ **50.** $5\sqrt{x}$ **51.** 0
52. $11k^2\sqrt{2n}$ **53.** $\frac{10\sqrt{3}}{3}$ **54.** $\frac{8\sqrt{10}}{5}$ **55.** $\sqrt{6}$ **56.** $\frac{\sqrt{10}}{5}$
57. $\sqrt{10}$ **58.** $\frac{\sqrt{42}}{21}$ **59.** $\frac{r\sqrt{x}}{4x}$ **60.** $\frac{\sqrt[3]{9}}{3}$ **61.** $-\sqrt{15}-9$
62. $3\sqrt{6}+12$ **63.** $22-16\sqrt{3}$ **64.** $2\sqrt{21}-\sqrt{14}+12\sqrt{2}-4\sqrt{3}$
65. -13 **66.** $x+4\sqrt{x}+4$ **67.** $-2+\sqrt{5}$
68. $\frac{3(1-\sqrt{x})}{1-x}$ $(x \neq 1)$ **69.** $\frac{-\sqrt{10}+3\sqrt{5}+\sqrt{2}-3}{7}$
70. $\frac{3+2\sqrt{6}}{3}$ **71.** $\frac{1+3\sqrt{7}}{4}$ **72.** $3+4\sqrt{3}$ **73.** $\emptyset$ **74.** $\{48\}$
75. $\{1\}$ **76.** $\{2\}$ **77.** $\{6\}$ **78.** $\{-3,-1\}$ **79.** $\{-2\}$ **80.** $\{4\}$
81. $\{7\}$ **82.** **(a)** B **(b)** F **(c)** D **(d)** A **(e)** C **(f)** A
83. $11\sqrt{3}$ **84.** $\frac{5-\sqrt{2}}{23}$ **85.** $\frac{2\sqrt{10}}{5}$ **86.** $3a^2b^3\sqrt[3]{2ab}$
87. $-\sqrt{10}-5\sqrt{15}$ **88.** $\frac{4r\sqrt{3rs}}{3s}$ **89.** $\frac{2+\sqrt{13}}{2}$
90. $7-2\sqrt{10}$ **91.** $166+2\sqrt{7}$ **92.** $\{7\}$ **93.** $\emptyset$ **94.** $\{8\}$
95. **(a)** $(a+b)^2$, or $a^2+2ab+b^2$ **(b)** c^2+2ab **(c)** Subtract $2ab$ from each side to get $a^2+b^2=c^2$.

Chapter 8 TEST (pages 629–630)

1. $-20, 20$ **2.** **(a)** irrational **(b)** 11.916 **3.** a must be negative.
4. 6 **5.** $-3\sqrt{6}$ **6.** $\frac{8\sqrt{2}}{5}$ **7.** $2\sqrt[3]{4}$ **8.** $4\sqrt{6}$ **9.** $9\sqrt{7}$
10. $11+2\sqrt{30}$ **11.** $4xy\sqrt{2y}$ **12.** 31 **13.** $6\sqrt{2}+2-3\sqrt{14}-\sqrt{7}$
14. $-5\sqrt{3x}$ **15.** **(a)** $6\sqrt{2}$ in. **(b)** 8.485 in. **16.** 50 ohms
17. $\frac{5\sqrt{14}}{7}$ **18.** $\frac{\sqrt{6x}}{3x}$ **19.** $-\sqrt[3]{2}$ **20.** $\frac{-3(4+\sqrt{3})}{13}$
21. $\frac{\sqrt{3}+12\sqrt{2}}{3}$ **22.** $\emptyset$ **23.** $\{3\}$ **24.** $\{1,4\}$ **25.** $\{9\}$
26. 12 is not a solution. A check shows that it does not satisfy the original equation. The solution set is $\emptyset$.

Chapters R–8 CUMULATIVE REVIEW EXERCISES (pages 631–632)

1. 54 **2.** 6 **3.** 3 **4.** $\{3\}$ **5.** $[-16, \infty)$ **6.** $(5, \infty)$
7. 2009: \$346,779; 2010: \$507,921
8.

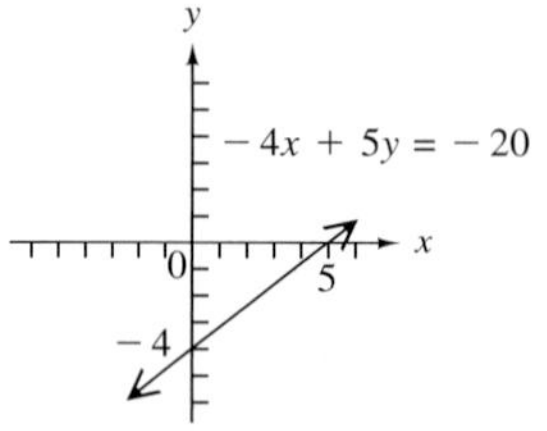

9.

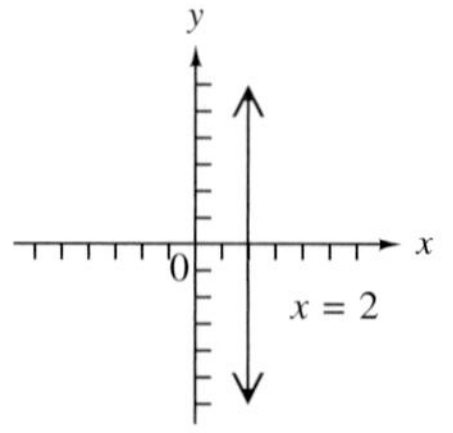

10.

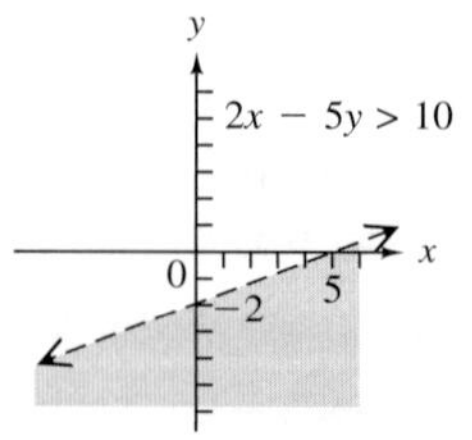

11. 19.3; The number of subscribers increased by an average of 19.3 million per yr.
12. $y=19.3x+109.5$
13. 341.1 million **14.** $\{(3,-7)\}$
15. $\{(x,y) \mid 2x-y=6\}$
16. from Chicago: 61 mph; from Des Moines: 54 mph **17.** $12x^{10}y^2$

18. $\dfrac{y^{15}}{5832}$ **19.** $3x^3 + 11x^2 - 13$ **20.** $4t^2 - 8t + 5$ **21.** 1,340,000,000 **22.** $(m + 8)(m + 4)$ **23.** $(5t^2 + 6)(5t^2 - 6)$ **24.** $(6a + 5b)(2a - b)$ **25.** $(9z + 4)^2$ **26.** $\{3, 4\}$ **27.** $\{-2, -1\}$ **28.** $\{12\}$ **29.** $\dfrac{x + 1}{x}$ **30.** $(t + 5)(t + 3)$ **31.** $\dfrac{-2x - 14}{(x + 3)(x - 1)}$ **32.** $29\sqrt{3}$ **33.** $-\sqrt{3} + \sqrt{5}$ **34.** $21 - 5\sqrt{2}$ **35.** $\emptyset$ **36.** $\{16\}$

CHAPTER 9 Quadratic Equations

SECTION 9.1 (pages 638–640)

1. C **2.** A **3.** D **4.** B **5.** According to the square root property, -9 is also a solution, so her answer was not completely correct. The solution set is $\{-9, 9\}$, or $\{\pm 9\}$. **6.** The equation $x^2 = 21$ can be solved with the square root property. Because the square of x is 21, x can be either the positive or the negative square root of 21. The solution set is $\{-\sqrt{21}, \sqrt{21}\}$, or $\{\pm\sqrt{21}\}$. **7.** $\{-7, 8\}$ **9.** $\{-11, 11\}$ **11.** $\left\{-\dfrac{5}{3}, 6\right\}$ **13.** $\{-9, 9\}$ **15.** $\{-\sqrt{14}, \sqrt{14}\}$ **17.** $\{-4\sqrt{3}, 4\sqrt{3}\}$ **19.** $\left\{-\dfrac{5}{2}, \dfrac{5}{2}\right\}$ **21.** $\emptyset$ **23.** $\{-1.5, 1.5\}$ **25.** $\{-\sqrt{3}, \sqrt{3}\}$ **27.** $\left\{-\dfrac{2\sqrt{7}}{7}, \dfrac{2\sqrt{7}}{7}\right\}$ **29.** $\{-3\sqrt{2}, 3\sqrt{2}\}$ **31.** $\{-2\sqrt{6}, 2\sqrt{6}\}$ **33.** $\left\{-\dfrac{2\sqrt{5}}{5}, \dfrac{2\sqrt{5}}{5}\right\}$ **35.** $\{-2, 8\}$ **37.** $\emptyset$ **39.** $\{8 + 3\sqrt{3}, 8 - 3\sqrt{3}\}$ **41.** $\left\{-3, \dfrac{5}{3}\right\}$ **43.** $\left\{0, \dfrac{3}{2}\right\}$ **45.** $\left\{\dfrac{5 + \sqrt{30}}{2}, \dfrac{5 - \sqrt{30}}{2}\right\}$ **47.** $\left\{\dfrac{-1 + 3\sqrt{2}}{3}, \dfrac{-1 - 3\sqrt{2}}{3}\right\}$ **49.** $\{-10 + 4\sqrt{3}, -10 - 4\sqrt{3}\}$ **51.** $\left\{\dfrac{1 + 4\sqrt{3}}{4}, \dfrac{1 - 4\sqrt{3}}{4}\right\}$ **53.** about $\dfrac{1}{2}$ sec **55.** 9 in. **57.** 5%

SECTION 9.2 (pages 647–648)

1. D **2.** yes **3.** 25; $(x + 5)^2$ **5.** 1; $(x + 1)^2$ **7.** $\dfrac{25}{4}$; $\left(p - \dfrac{5}{2}\right)^2$ **9.** $\{1, 3\}$ **11.** $\{-3, -2\}$ **13.** $\{-1 + \sqrt{6}, -1 - \sqrt{6}\}$ **15.** $\{4 + 2\sqrt{3}, 4 - 2\sqrt{3}\}$ **17.** $\{-3\}$ **19.** $\left\{\dfrac{-1 + \sqrt{5}}{2}, \dfrac{-1 - \sqrt{5}}{2}\right\}$ **21.** $\left\{-\dfrac{3}{2}, \dfrac{1}{2}\right\}$ **23.** $\left\{\dfrac{2 + \sqrt{14}}{2}, \dfrac{2 - \sqrt{14}}{2}\right\}$ **25.** $\emptyset$ **27.** $\left\{\dfrac{-7 + \sqrt{97}}{6}, \dfrac{-7 - \sqrt{97}}{6}\right\}$ **29.** $\{-4, 2\}$ **31.** $\{4 + \sqrt{3}, 4 - \sqrt{3}\}$ **33.** $\{1 + \sqrt{6}, 1 - \sqrt{6}\}$ **35.** 1 sec and 5 sec **37.** 3 sec and 5 sec **39.** 75 ft by 100 ft

SECTION 9.3 (pages 653–654)

1. $2a$ should be in the denominator for $-b$ as well. The correct formula is $x = \dfrac{-b \pm \sqrt{b^2 - 4ac}}{2a}$. **2.** There is no real number solution, so the solution set is $\emptyset$. **3.** 4; 5; -9 **5.** 3; -4; -2 **7.** 3; 7; 0 **9.** $\{-13, 1\}$ **11.** $\{2\}$ **13.** $\left\{-1, \dfrac{5}{2}\right\}$ **15.** $\left\{\dfrac{-6 + \sqrt{26}}{2}, \dfrac{-6 - \sqrt{26}}{2}\right\}$ **17.** $\{-1, 0\}$ **19.** $\emptyset$ **21.** $\left\{\dfrac{-5 + \sqrt{13}}{6}, \dfrac{-5 - \sqrt{13}}{6}\right\}$ **23.** $\left\{0, \dfrac{12}{7}\right\}$ **25.** $\{-2\sqrt{6}, 2\sqrt{6}\}$ **27.** $\left\{-\dfrac{2}{5}, \dfrac{2}{5}\right\}$ **29.** $\left\{\dfrac{6 + 2\sqrt{6}}{3}, \dfrac{6 - 2\sqrt{6}}{3}\right\}$ **31.** $\emptyset$ **33.** $\left\{-\dfrac{2}{3}, \dfrac{4}{3}\right\}$ **35.** $\left\{\dfrac{-1 + \sqrt{73}}{6}, \dfrac{-1 - \sqrt{73}}{6}\right\}$ **37.** $\emptyset$ **39.** $\left\{-1, \dfrac{5}{2}\right\}$ **41.** $\{-3 + \sqrt{5}, -3 - \sqrt{5}\}$ **43.** 3.5 ft **45.** -8, 16; Only 16 board feet is a reasonable answer.

SUMMARY EXERCISES Applying Methods for Solving Quadratic Equations (pages 655–656)

1. $\{-6, 6\}$ **2.** $\left\{\dfrac{-3 + \sqrt{5}}{2}, \dfrac{-3 - \sqrt{5}}{2}\right\}$ **3.** $\left\{-\dfrac{10}{9}, \dfrac{10}{9}\right\}$ **4.** $\left\{-\dfrac{7}{9}, \dfrac{7}{9}\right\}$ **5.** $\{1, 3\}$ **6.** $\{-2, -1\}$ **7.** $\{4, 5\}$ **8.** $\left\{\dfrac{-3 + \sqrt{17}}{2}, \dfrac{-3 - \sqrt{17}}{2}\right\}$ **9.** $\left\{-\dfrac{1}{3}, \dfrac{5}{3}\right\}$ **10.** $\left\{\dfrac{1 + \sqrt{10}}{2}, \dfrac{1 - \sqrt{10}}{2}\right\}$ **11.** $\{-17, 5\}$ **12.** $\left\{-\dfrac{7}{5}, 1\right\}$ **13.** $\left\{\dfrac{7 + 2\sqrt{6}}{3}, \dfrac{7 - 2\sqrt{6}}{3}\right\}$ **14.** $\left\{\dfrac{1 + 4\sqrt{2}}{7}, \dfrac{1 - 4\sqrt{2}}{7}\right\}$ **15.** $\emptyset$ **16.** $\emptyset$ **17.** $\left\{-\dfrac{1}{2}, 2\right\}$ **18.** $\left\{-\dfrac{1}{2}, 1\right\}$ **19.** $\left\{-\dfrac{5}{4}, \dfrac{3}{2}\right\}$ **20.** $\left\{-3, \dfrac{1}{3}\right\}$ **21.** $\{1 + \sqrt{2}, 1 - \sqrt{2}\}$ **22.** $\left\{\dfrac{-5 + \sqrt{13}}{6}, \dfrac{-5 - \sqrt{13}}{6}\right\}$ **23.** $\left\{\dfrac{2}{5}, 4\right\}$ **24.** $\{-3 + \sqrt{5}, -3 - \sqrt{5}\}$ **25.** $\left\{\dfrac{-3 + \sqrt{41}}{2}, \dfrac{-3 - \sqrt{41}}{2}\right\}$ **26.** $\left\{-\dfrac{5}{4}\right\}$ **27.** $\left\{\dfrac{1}{4}, 1\right\}$ **28.** $\left\{\dfrac{1 + \sqrt{3}}{2}, \dfrac{1 - \sqrt{3}}{2}\right\}$ **29.** $\left\{\dfrac{-2 + \sqrt{11}}{3}, \dfrac{-2 - \sqrt{11}}{3}\right\}$ **30.** $\left\{\dfrac{-5 + \sqrt{41}}{8}, \dfrac{-5 - \sqrt{41}}{8}\right\}$ **31.** $\left\{\dfrac{-7 + \sqrt{5}}{4}, \dfrac{-7 - \sqrt{5}}{4}\right\}$ **32.** $\left\{-\dfrac{8}{3}, -\dfrac{6}{5}\right\}$ **33.** $\left\{\dfrac{8 + 8\sqrt{2}}{3}, \dfrac{8 - 8\sqrt{2}}{3}\right\}$ **34.** $\left\{\dfrac{-5 + \sqrt{5}}{2}, \dfrac{-5 - \sqrt{5}}{2}\right\}$ **35.** $\emptyset$ **36.** $\emptyset$ **37.** $\left\{-\dfrac{2}{3}, 2\right\}$ **38.** $\left\{-\dfrac{1}{4}, \dfrac{2}{3}\right\}$ **39.** $\left\{-4, \dfrac{3}{5}\right\}$ **40.** $\{-3, 5\}$ **41.** $\left\{-\dfrac{2}{3}, \dfrac{2}{5}\right\}$ **42.** $\{-4, 6\}$

SECTION 9.4 (pages 661–662)

1. parabola **2.** vertex; highest **3.** axis of symmetry; mirror **4.** one; one; no **5.** $(0, 0)$

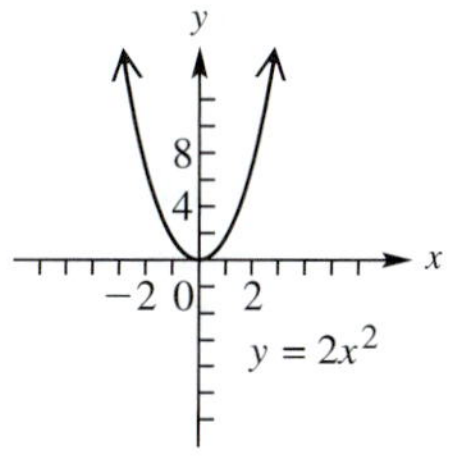

7. $(0, -4)$

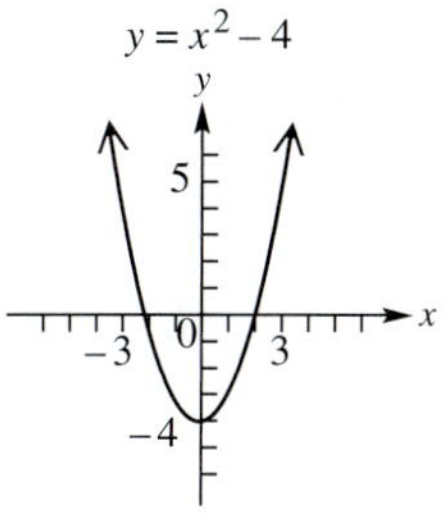

ANSWERS

9. (0, 2)

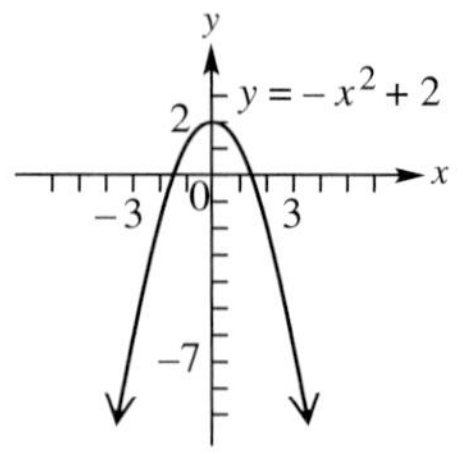

11. (−1, 0)

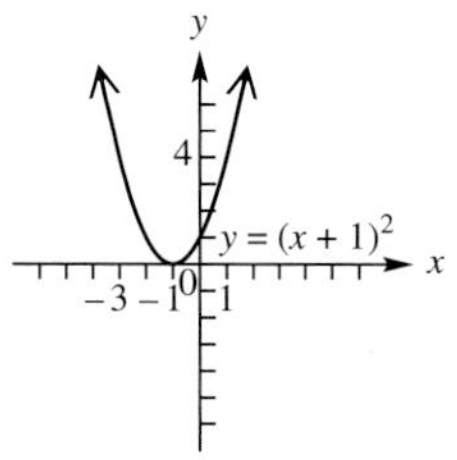

13. (−1, 2)

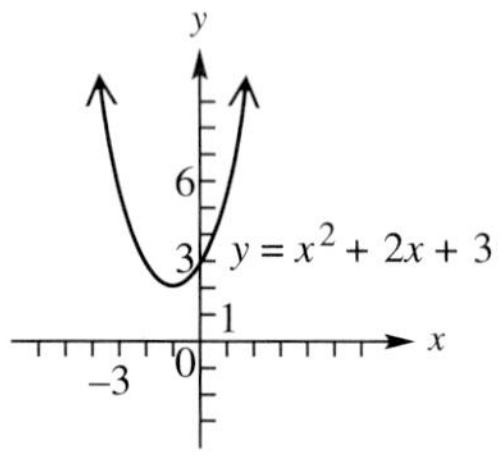

15. (3, 4)

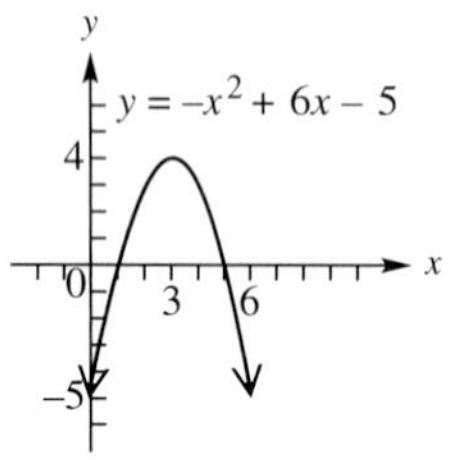

17. If $a > 0$, it opens upward, and if $a < 0$, it opens downward.

18. (a) none **(b)** one **(c)** two **19.** $y = \frac{11}{5625}x^2$

SECTION 9.5 (pages 668–669)

1. 3; 3; (1, 3) **2.** 4; 4; (2, 4) **3.** 5; 5; (3, 5) **4.** 6; 6; (4, 6) **5.** The graph consists of the five points (0, 2), (1, 3), (2, 4), (3, 5), and (4, 6). **6.** The graph is the line through the points (0, 2), (1, 3), (2, 4), (3, 5), and (4, 6). **7.** not a function; domain: {−4, −2, 0}; range: {3, 1, 5, −8} **9.** function; domain: {A, B, C, D, E}; range: {2, 3, 6, 4} **11.** not a function; domain: {−4, −2, 0, 2, 3}; range: {−2, 0, 1, 2, 3} **13.** function **15.** not a function **17.** function **19.** function **21.** not a function **23.** function **25. (a)** 11 **(b)** 3 **(c)** −9 **27. (a)** 4 **(b)** 2 **(c)** 14 **29. (a)** absolute value; 2; 2 **(b)** 0 **(c)** 3 **31.** {(1980, 14.1), (1990, 19.8), (2000, 31.1), (2010, 40.0)}; yes **33.** $g(1980) = 14.1$ (million); $g(2000) = 31.1$ (million) **35.** For the year 2008, the function gives 38.0 million foreign-born residents in the United States.

Chapter 9 REVIEW EXERCISES (pages 674–676)

1. $\{-7, 4\}$ **2.** $\{-9, -5\}$ **3.** $\left\{-5, \frac{3}{2}\right\}$ **4.** $\{-13, 13\}$ **5.** $\{-12, 12\}$

6. $\{-\sqrt{37}, \sqrt{37}\}$ **7.** $\{-8\sqrt{2}, 8\sqrt{2}\}$ **8.** $\{-7, 3\}$

9. $\{3 + \sqrt{10}, 3 - \sqrt{10}\}$ **10.** $\left\{\frac{-1 + \sqrt{14}}{2}, \frac{-1 - \sqrt{14}}{2}\right\}$ **11.** $\emptyset$

12. $\left\{-\frac{5}{3}\right\}$ **13.** $\{-5, -1\}$ **14.** $\{-2 + \sqrt{11}, -2 - \sqrt{11}\}$

15. $\{-1 + \sqrt{6}, -1 - \sqrt{6}\}$ **16.** $\left\{\frac{-4 + \sqrt{22}}{2}, \frac{-4 - \sqrt{22}}{2}\right\}$

17. $\left\{-\frac{7}{2}\right\}$ **18.** $\emptyset$ **19.** 2.5 sec **20.** 6, 8, 10 **21.** $\{-4, 9\}$

22. $\left\{0, \frac{14}{5}\right\}$ **23.** $\emptyset$ **24.** $\left\{\frac{3}{2}\right\}$ **25.** $\left\{\frac{-1 + \sqrt{29}}{4}, \frac{-1 - \sqrt{29}}{4}\right\}$

26. $\left\{\frac{2 + \sqrt{10}}{2}, \frac{2 - \sqrt{10}}{2}\right\}$ **27.** $\left\{\frac{1 + \sqrt{21}}{10}, \frac{1 - \sqrt{21}}{10}\right\}$

28. $\left\{\frac{-3 + \sqrt{41}}{2}, \frac{-3 - \sqrt{41}}{2}\right\}$ **29.** $\{-3 + \sqrt{19}, -3 - \sqrt{19}\}$

30. $\left\{\frac{2 + \sqrt{2}}{2}, \frac{2 - \sqrt{2}}{2}\right\}$

31. (0, 5)

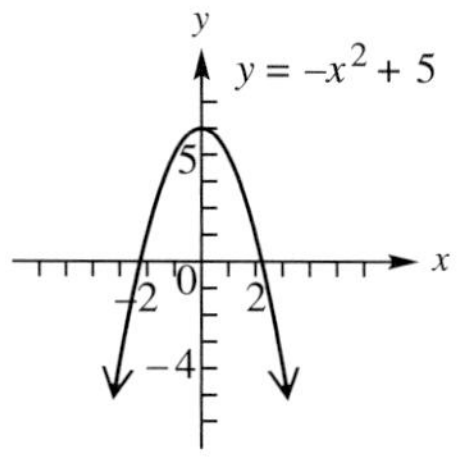

32. (1, 0)

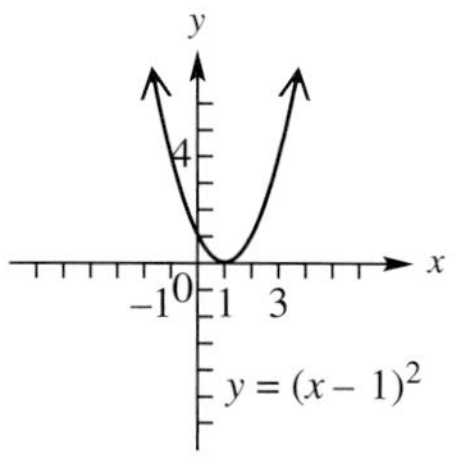

33. (1, 4)

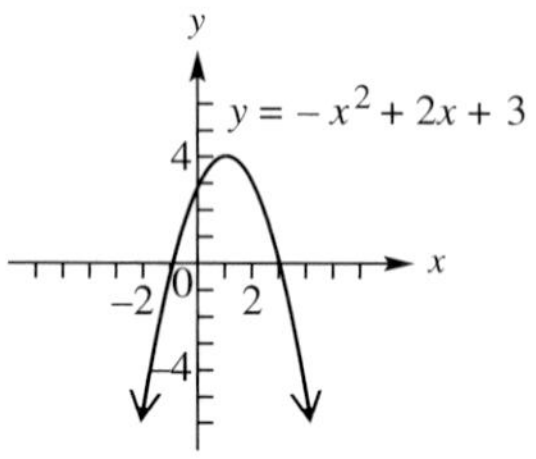

34. (−2, −2)

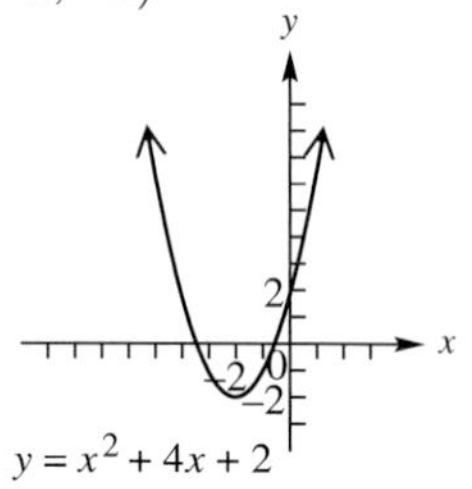

35. not a function; domain: {−2, 0, 2}; range: {4, 8, 5, 3}

36. function; domain: {8, 7, 6, 5, 4}; range: {3, 4, 5, 6, 7}

37. not a function **38.** function **39.** function **40.** function

41. (a) 8 **(b)** −1 **42. (a)** 7 **(b)** 1 **43. (a)** 5 **(b)** 2

44. (a) −3 **(b)** −6 **45.** $\left\{-\frac{11}{2}, 5\right\}$ **46.** $\left\{-\frac{11}{2}, \frac{9}{2}\right\}$

47. $\left\{\frac{-1 + \sqrt{21}}{2}, \frac{-1 - \sqrt{21}}{2}\right\}$ **48.** $\left\{-\frac{3}{2}, \frac{1}{3}\right\}$

49. $\left\{\frac{-5 + \sqrt{17}}{2}, \frac{-5 - \sqrt{17}}{2}\right\}$ **50.** $\{-1 + \sqrt{3}, -1 - \sqrt{3}\}$

51. $\emptyset$ **52.** $\left\{\frac{9 + \sqrt{41}}{2}, \frac{9 - \sqrt{41}}{2}\right\}$ **53.** $\left\{-\frac{6}{5}\right\}$

54. $\{-1 + 2\sqrt{2}, -1 - 2\sqrt{2}\}$ **55.** $\{-2 + \sqrt{5}, -2 - \sqrt{5}\}$

56. $\{-2\sqrt{2}, 2\sqrt{2}\}$ **57. (a)** $\{-3, 3\}$ **(b)** $\{-3, 3\}$ **(c)** $\{-3, 3\}$ **(d)** We will always get the same results, no matter which method of solution is used. **58.** The fraction bar should be under both $-b$ and $\sqrt{b^2 - 4ac}$. The term $2a$ should not be in the radicand. The correct formula is $x = \frac{-b \pm \sqrt{b^2 - 4ac}}{2a}$.

Chapter 9 TEST (pages 677–678)

1. $\{-\sqrt{39}, \sqrt{39}\}$ **2.** $\{-11, 5\}$ **3.** $\left\{\frac{-3 + 2\sqrt{6}}{4}, \frac{-3 - 2\sqrt{6}}{4}\right\}$

4. $\{2 + \sqrt{10}, 2 - \sqrt{10}\}$ **5.** $\left\{\frac{-6 + \sqrt{42}}{2}, \frac{-6 - \sqrt{42}}{2}\right\}$

6. $\left\{-3, \frac{1}{2}\right\}$ **7.** $\left\{\frac{3 + \sqrt{3}}{3}, \frac{3 - \sqrt{3}}{3}\right\}$ **8.** $\emptyset$

9. $\left\{\frac{5 + \sqrt{13}}{6}, \frac{5 - \sqrt{13}}{6}\right\}$ **10.** $\{1 + \sqrt{2}, 1 - \sqrt{2}\}$

11. $\left\{\frac{-1 + 3\sqrt{2}}{2}, \frac{-1 - 3\sqrt{2}}{2}\right\}$ **12.** $\left\{\frac{11 + \sqrt{89}}{4}, \frac{11 - \sqrt{89}}{4}\right\}$

13. $\{5\}$ **14.** 2 sec **15.** 12, 16, 20

16. (3, 0)

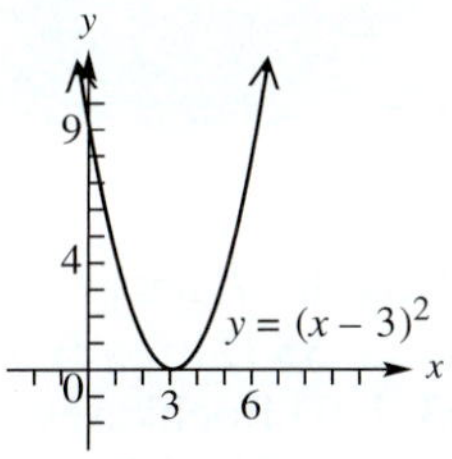

17. (−1, −3)

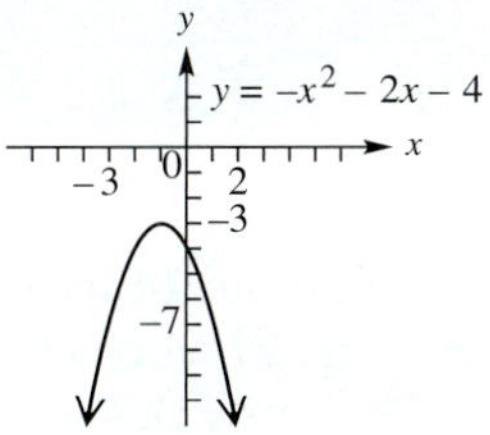

18. **(a)** not a function **(b)** function; domain: $\{0, 1, 2\}$; range: $\{2\}$
19. not a function **20.** 1

Chapters R–9 CUMULATIVE REVIEW EXERCISES (pages 679–680)

1. 15 **2.** −2 **3.** $19m - 17$ **4.** $\{3\}$ **5.** $\{5\}$ **6.** $\{2\}$ **7.** width: 50 ft; length: 94 ft **8.** 100°; 80° **9.** $L = \frac{P - 2W}{2}$, or $L = \frac{P}{2} - W$
10. $(-\infty, 4]$

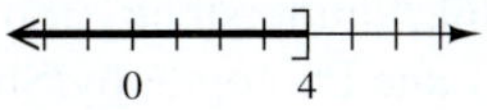

11.

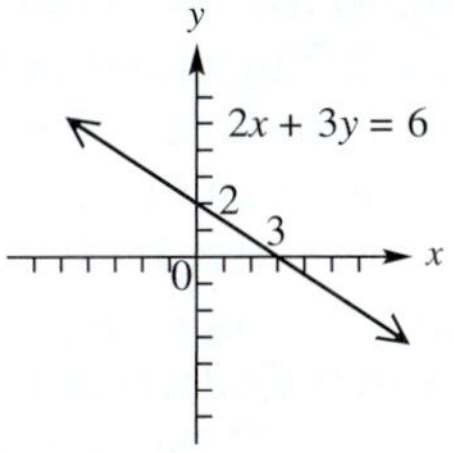

12.

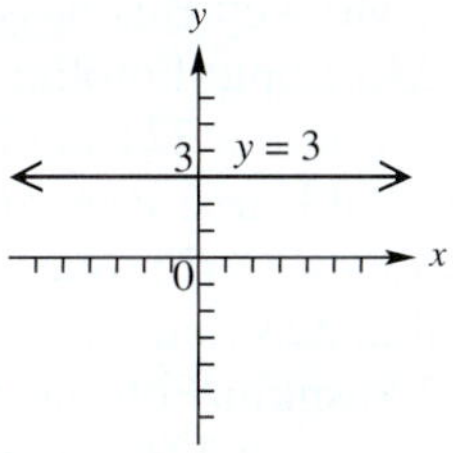

13.

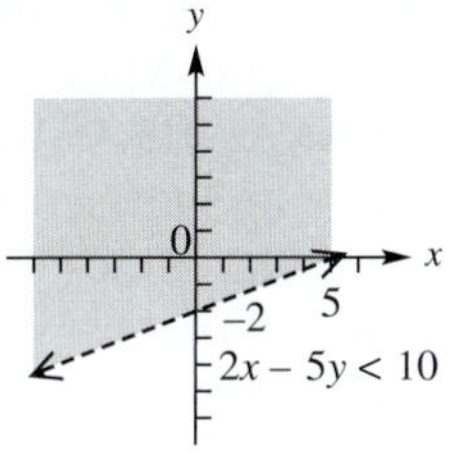

14. $-\frac{1}{3}$ **15.** $2x - y = -3$
16. $\{(-3, 2)\}$ **17.** $\emptyset$
18. Motorola Theory: \$79.99; Motorola i412: \$69.99 **19.** $\frac{x^4}{9}$
20. $\frac{b^{16}}{c^2}$ **21.** $\frac{27}{125}$

22. $8x^5 - 17x^4 - x^2$ **23.** $2x^4 + x^3 - 19x^2 + 2x + 20$ **24.** $16x^2(x - 3y)$
25. $(2a + 1)(a - 3)$ **26.** $(5m - 2)^2$ **27.** $\left\{-1, \frac{4}{3}\right\}$ **28.** 50 m
29. $\frac{4}{5}$ **30.** $\frac{-k - 1}{k(k - 1)}$ **31.** $\frac{6x + 1}{3x - 1}$ **32.** $\left\{-\frac{15}{7}, 2\right\}$ **33.** $\{7\}$
34. $\left\{\frac{1 + \sqrt{3}}{2}, \frac{1 - \sqrt{3}}{2}\right\}$ **35.** $\frac{6\sqrt{30}}{5}$ **36.** $4\sqrt{5}$ **37.** $-ab\sqrt[3]{2b}$
38. (2, −4); (0, 0) and (4, 0)

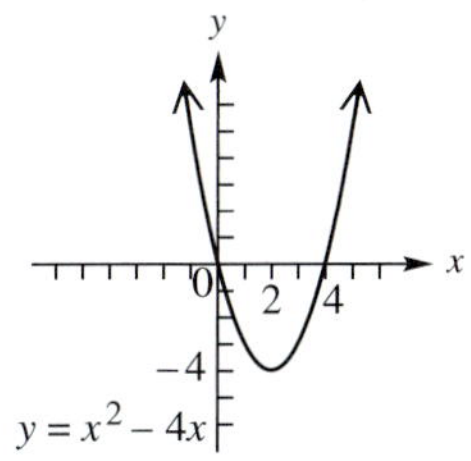

39. **(a)** yes **(b)** domain: $\{0, 1, 3\}$; range: $\{4, 2, 5\}$ **40.** 11

APPENDIX Factoring Sums and Differences of Cubes

(pages 683–684)

1. 1; 8; 27; 64; 125; 216; 343; 512; 729; 1000 **2.** 3 **3.** C, D **4.** A, D
5. $(x + 1)(x^2 - x + 1)$ **7.** $(x - 1)(x^2 + x + 1)$
9. $(p + q)(p^2 - pq + q^2)$ **11.** $(y - 6)(y^2 + 6y + 36)$
13. $(k + 10)(k^2 - 10k + 100)$ **15.** $(3x - 1)(9x^2 + 3x + 1)$
17. $(5x + 2)(25x^2 - 10x + 4)$ **19.** $(y - 2x)(y^2 + 2xy + 4x^2)$
21. $(3x - 4y)(9x^2 + 12xy + 16y^2)$
23. $(2p + 9q)(4p^2 - 18pq + 81q^2)$ **25.** $2(2t - 1)(4t^2 + 2t + 1)$
27. $5(2w + 3)(4w^2 - 6w + 9)$ **29.** $(x + y^2)(x^2 - xy^2 + y^4)$
31. $(5k - 2m^3)(25k^2 + 10km^3 + 4m^6)$ **33.** $(x^3 - 1)(x^3 + 1)$
34. $(x - 1)(x^2 + x + 1)(x + 1)(x^2 - x + 1)$
35. $(x^2 - 1)(x^4 + x^2 + 1)$ **36.** $(x - 1)(x + 1)(x^4 + x^2 + 1)$
37. The result in **Exercise 34** is completely factored. **38.** Show that $x^4 + x^2 + 1 = (x^2 + x + 1)(x^2 - x + 1)$. **39.** difference of squares
40. $(x - 3)(x^2 + 3x + 9)(x + 3)(x^2 - 3x + 9)$

Photo Credits

iii Courtesy of Janet Lial; **10** Giangi555/Fotolia; **14** Ryan McVay/Photodisc/Getty Images; **26** WavebreakMediaMicro/Fotolia; **27** Michael Flippo/Fotolia; **35** Darren Baker/ Fotolia; **44** Monkey Business/Fotolia; **59** Sirena Designs/Fotolia; **63** (left) Bochkarev Photography/Shutterstock, (top right) UryadnikovS/Fotolia, (bottom right) Vlad G/Fotolia; **71**(left) Jesse Kunerth/Fotolia, (top right) John Hornsby, (bottom right) Igor Kushnir/ Fotolia; **72** Horticulture/Fotolia; **84** Rob Byron/Fotolia; **88** Jason Stitt/Fotolia; **97** Rhona Wise/EPA/Newscom; **104** EloPaint/Fotolia; **111** Paul Maguire/Fotolia; **113** Elainekay/ Fotolia; **117** Huaxiadragon/Fotolia; **125** Alex SK/Fotolia; **149** Dimitar Dilkoff/AFP/ Getty Images/Newscom; **158** (top) Emkaplin/Fotolia, (bottom) DenisNata/Fotolia; **160** Tmass/Fotolia; **170** Roger Cracknell 01/classic/Alamy; **174** Jaimie Duplass/Fotolia; **177** GW Images/Fotolia; **183** Leah-Anne Thompson/Shutterstock; **185** (left) Yanlev/ Fotolia, (right) Pressmaster/Fotolia; **187** (left) Peter Bernik/Shutterstock, (top right) Tim Dominick/MCT/Newscom, (bottom right) Morgan Lane Photography/Shutterstock; **194** Lightpoet/Shutterstock; **208** (left) Monkey Business/Fotolia, (right) Michaeljung/ Fotolia; **209** Sababa66/Shutterstock; **213** Eppic/Fotolia; **214** Konstantin Yuganov/Fotolia; **215** Dianapaz/Fotolia; **219** KingPhoto/Fotolia; **221** Sanjay Goswami/Fotolia; **227** Igor Mojzes/Fotolia; **235** Zentilia/Fotolia; **260** Karen Roach/Fotolia; **266** Rebecca Fisher/ Fotolia ; **287** Zimmytws/Fotolia; **289** Geewhiz/Fotolia; **306** Vadim Ponomarenko/Fotolia; **313** Sashkin/Fotolia; **316** Julijah/Fotolia; **323** (left) Scott Rothstein/Shutterstock, (top right) AbleStock/Thinkstock, (bottom right) Sashkin/Fotolia; **325** (left) Everett Collection Inc./ Alamy, (right) Everett Collection Inc./Alamy; **326** Eric Gevaert/Fotolia; **327** Ints/Fotolia; **340** (left) Iophoto/Shutterstock, (right) Marty/Fotolia; **343** (top) Jezper/Fotolia, (bottom) Sara Piaseczynski; **401** Johan Swanepoel/Shutterstock; **402** (left) Carlos Santa Maria/ Fotolia, (right) Ivan Cholakov/Shutterstock; **411** JPL-Caltech/STScI/NASA; **415** World History Archive/Alamy; **453** Nickolae/Fotolia; **461** OnFilm/iStockphoto; **465** Offscreen/ Shutterstock; **467** Gino Santa Maria/Fotolia; **479** Giordano Aita/Fotolia; **482** Mikesch112/ Fotolia; **485** Matthew O'Haren/Icon SMI 394/Newscom; **544** Ernest Prim/Fotolia; **547** (left) Jason Roberts/EPA/Newscom, (right) Gero Breloer/EPA/Newscom; **548** Epantha/Fotolia; **550** John Hornsby; **553** Mary Durden/Fotolia; **568** Everett Collection, Inc.; **569** Danny E Hooks/Shutterstock; **570** Beth Anderson/Pearson Education, Inc.; **580** (left) John Hornsby, (right) Patric Martel/Fotolia; **600** Apollo 17 Crew, NASA; **608** Serhiy Kobyakov/Fotolia; **610** Kjersti/Fotolia; **616** Betacam-SP/Fotolia; **619** Timothy Large/Shutterstock; **620** (top left) Lotharingia/Fotolia, (top right) Beboy/Fotolia, (bottom left) Tmax/Fotolia, (bottom right) Sculpies/Fotolia; **630** Comstock/Thinkstock; **631**, Tyler Olson/Fotolia; **633** Badmanproduction/Fotolia; **637** Sablin/Fotolia; **654** NASA Headquarters; **664** Lisa F. Young/Shutterstock

Solutions to Selected Exercises

CHAPTER 1 The Real Number System

SECTION 1.1 (pages 33–35)

33. $\frac{1}{4}\cdot\frac{2}{3}+\frac{2}{5}\cdot\frac{11}{3}$

$$=\frac{1}{6}+\frac{22}{15} \qquad \text{Multiply; } \frac{2}{12}=\frac{1}{6}$$

$$=\frac{5}{30}+\frac{44}{30} \qquad \text{LCD} = 30$$

$$=\frac{49}{30} \qquad \text{Add.}$$

45. $\left(\frac{3}{2}\right)^2\left[\left(11+\frac{1}{3}\right)-6\right]$

$$=\frac{9}{4}\left[\left(11+\frac{1}{3}\right)-6\right] \qquad \text{Apply the exponent.}$$

$$=\frac{9}{4}\left[\left(\frac{33}{3}+\frac{1}{3}\right)-6\right] \qquad \text{LCD} = 3$$

$$=\frac{9}{4}\left(\frac{34}{3}-6\right) \qquad \text{Add inside parentheses.}$$

$$=\frac{9}{4}\left(\frac{34}{3}-\frac{18}{3}\right) \qquad \text{LCD} = 3$$

$$=\frac{9}{4}\left(\frac{16}{3}\right) \qquad \text{Subtract inside parentheses.}$$

$$=\frac{144}{12}, \text{ or } 12 \qquad \text{Multiply. Divide.}$$

65. $\frac{9(7-1)-8\cdot 2}{4(6-1)} > 3$

$$\frac{9(6)-8\cdot 2}{4(5)} > 3 \qquad \text{Work inside parentheses.}$$

$$\frac{54-16}{20} > 3 \qquad \text{Multiply.}$$

$$\frac{38}{20} > 3 \qquad \text{Subtract.}$$

$$1\frac{9}{10} > 3 \qquad \text{Write as a mixed number.}$$

This statement is *false*.

SECTION 1.2 (pages 41–44)

31. (a) $\frac{3x+y^2}{2x+3y}$

$$=\frac{3(2)+1^2}{2(2)+3(1)} \qquad \text{Let } x=2 \text{ and } y=1.$$

$$=\frac{3(2)+1}{4+3}$$

$$=\frac{6+1}{7}$$

$$=\frac{7}{7}, \text{ or } 1$$

(b) $\frac{3x+y^2}{2x+3y}$

$$=\frac{3(1)+5^2}{2(1)+3(5)} \qquad \text{Let } x=1 \text{ and } y=5.$$

$$=\frac{3+25}{2+15} \qquad \text{Multiply.}$$

$$=\frac{28}{17} \qquad \text{Add.}$$

33. (a) $0.841x^2+0.32y^2$

$$=0.841\cdot 2^2+0.32\cdot 1^2 \qquad \text{Let } x=2,\ y=1.$$

$$=0.841\cdot 4+0.32\cdot 1 \qquad \text{Apply the exponents.}$$

$$=3.364+0.32 \qquad \text{Multiply.}$$

$$=3.684 \qquad \text{Add.}$$

(b) $0.841x^2+0.32y^2$

$$=0.841\cdot 1^2+0.32\cdot 5^2 \qquad \text{Let } x=1,\ y=5.$$

$$=0.841\cdot 1+0.32\cdot 25$$

$$=0.841+8$$

$$=8.841$$

57. $\frac{z+4}{2-z}=\frac{13}{5}$

$$\frac{\frac{1}{3}+4}{2-\frac{1}{3}} \stackrel{?}{=} \frac{13}{5} \qquad \text{Let } z=\frac{1}{3}.$$

$$\frac{\frac{1}{3}+\frac{12}{3}}{\frac{6}{3}-\frac{1}{3}} \stackrel{?}{=} \frac{13}{5} \qquad \text{Write with LCD} = 3.$$

$$\frac{\frac{13}{3}}{\frac{5}{3}} \stackrel{?}{=} \frac{13}{5} \qquad \text{Add and subtract.}$$

$$\frac{13}{3}\cdot\frac{3}{5} \stackrel{?}{=} \frac{13}{5} \qquad \text{Definition of division}$$

$$\frac{13}{5}=\frac{13}{5}\ \checkmark \qquad \text{True}$$

The true result shows that $\frac{1}{3}$ is a solution.

SECTION 1.3 (pages 52–54)

45. In order to compare these two numbers, write them with a common denominator.

$$-\frac{2}{3}=-\frac{8}{12} \quad \text{and} \quad -\frac{1}{4}=-\frac{3}{12}$$

Since $\frac{8}{12}>\frac{3}{12}$,

$-\frac{2}{3}$ is farther to the left of 0 on a number line than $-\frac{1}{4}$, so $-\frac{2}{3}$ is the lesser number.

73. A decrease is represented by a negative number, and the amount of decrease is represented by its absolute value. There are three negative numbers in the table: -43.6, -16.2, and -0.6. Of these, -43.6 has the greatest absolute value, so Energy, 2008 to 2009 represents the greatest decrease.

SECTION 1.4 (pages 60–63)

39. $\left(-\frac{1}{2}+0.25\right)+\left(-\frac{3}{4}+0.75\right)$

$$=\left(-\frac{1}{2}+\frac{1}{4}\right)+\left(-\frac{3}{4}+\frac{3}{4}\right)$$

$$=\left(-\frac{2}{4}+\frac{1}{4}\right)+0$$

$$=-\frac{1}{4}, \text{ or } -0.25$$

59. "$\frac{2}{7}$ more than the sum of $\frac{5}{7}$ and $-\frac{9}{7}$" is interpreted and evaluated as follows.

$$\left[\frac{5}{7}+\left(-\frac{9}{7}\right)\right]+\frac{2}{7}$$

$$=-\frac{4}{7}+\frac{2}{7} \qquad \text{Work inside brackets.}$$

$$=-\frac{2}{7} \qquad \text{Add.}$$

SECTION 1.5 (pages 68–72)

39. $\left(-\frac{3}{8}-\frac{2}{3}\right)-\left(-\frac{9}{8}-3\right)$

$$=\left[-\frac{3}{8}+\left(-\frac{2}{3}\right)\right]-\left[-\frac{9}{8}+(-3)\right]$$

$$=\left[-\frac{9}{24}+\left(-\frac{16}{24}\right)\right]-\left[-\frac{9}{8}+\left(-\frac{24}{8}\right)\right] \qquad \text{Get a common denominator for each pair of fractions.}$$

$$=-\frac{25}{24}-\left(-\frac{33}{8}\right) \qquad \text{Add inside brackets.}$$

$$=-\frac{25}{24}-\left(-\frac{99}{24}\right) \qquad \text{Get a common denominator.}$$

$$=-\frac{25}{24}+\frac{99}{24} \qquad \text{Definition of subtraction}$$

$$=\frac{74}{24} \qquad \text{Add.}$$

$$=\frac{37}{12} \qquad \text{Lowest terms}$$

41. $[-12.25 - (8.34 + 3.57)] - 17.88$

$= [-12.25 - 11.91] - 17.88$

$= [-12.25 + (-11.91)] - 17.88$

$= -24.16 - 17.88$

$= -24.16 + (-17.88)$

$= -42.04$

59. Sum of checks:

$35.84 + $26.14 + $3.12

= $61.98 + $3.12

= $65.10

Sum of deposits:

$85.00 + $120.76 = $205.76

Final balance:

= Beginning balance

− checks + deposits

= $904.89 − $65.10 + $205.76

= $839.79 + $205.76

= $1045.55

Her account balance at the end of August was $1045.55.

SECTION 1.6 (pages 81–84)

43. $\dfrac{4(2^3 - 5) - 5(-3^3 + 21)}{3[6 - (-2)]}$

$= \dfrac{4(8 - 5) - 5(-27 + 21)}{3[6 - (-2)]}$

$= \dfrac{4(3) - 5(-6)}{3[8]}$

$= \dfrac{12 + 30}{24}$

$= \dfrac{42}{24}$, or $\dfrac{7}{4}$

49. $\left(\frac{5}{6}x + \frac{3}{2}y\right)\left(-\frac{1}{3}a\right)$

$= \left[\frac{5}{6}(6) + \frac{3}{2}(-4)\right]\left[-\frac{1}{3}(3)\right]$ Let $x = 6$, $y = -4$, and $a = 3$.

$= [5 + (-6)](-1)$

$= (-1)(-1)$

$= 1$

69. "The product of $-\frac{2}{3}$ and $-\frac{1}{5}$, divided by $\frac{1}{7}$" is interpreted and evaluated as follows.

$\dfrac{-\frac{2}{3}\left(-\frac{1}{5}\right)}{\frac{1}{7}}$

$= \dfrac{\frac{2}{15}}{\frac{1}{7}}$ Multiply in the numerator.

$= \frac{2}{15} \cdot \frac{7}{1}$ Definition of division

$= \frac{14}{15}$ Multiply.

SECTION 1.7 (pages 93–96)

73. $-3(8x + 3y + 4z)$

$= -3(8x) + (-3)(3y) + (-3)(4z)$ Distributive property

$= -24x - 9y - 12z$ Multiply.

79. $-(-3q + 5r - 8s)$

$= -1(-3q + 5r - 8s)$

$= 3q - 5r + 8s$

SECTION 1.8 (pages 101–103)

45. $-\frac{4}{3} + 2t + \frac{1}{3}t - 8 - \frac{8}{3}t$

$= \left(2t + \frac{1}{3}t - \frac{8}{3}t\right) + \left(-\frac{4}{3} - 8\right)$

$= \left(2 + \frac{1}{3} - \frac{8}{3}\right)t + \left(-\frac{4}{3} - 8\right)$

$= \left(\frac{6}{3} + \frac{1}{3} - \frac{8}{3}\right)t + \left(-\frac{4}{3} - \frac{24}{3}\right)$

$= -\frac{1}{3}t - \frac{28}{3}$

47. $-5.3r + 4.9 - (2r + 0.7) + 3.2r$

$= -5.3r + 4.9 - 2r - 0.7 + 3.2r$

$= (-5.3r - 2r + 3.2r) + (4.9 - 0.7)$

$= (-5.3 - 2 + 3.2)r + (4.9 - 0.7)$

$= -4.1r + 4.2$

CHAPTER 2 Equations, Inequalities, and Applications

SECTION 2.1 (pages 123–124)

33. $10x + 4 = 9x$

$10x + 4 - 9x = 9x - 9x$ Subtract $9x$.

$1x + 4 = 0$

$x + 4 - 4 = 0 - 4$ Subtract 4.

$x = -4$

CHECK $10x + 4 = 9x$

$10(-4) + 4 \stackrel{?}{=} 9(-4)$ Let $x = -4$.

$-36 = -36$ ✓ True

Solution set: $\{-4\}$

53. $\frac{5}{7}x + \frac{1}{3} = \frac{2}{5} - \frac{2}{7}x + \frac{2}{5}$

$\frac{5}{7}x + \frac{1}{3} = \frac{4}{5} - \frac{2}{7}x$

$\frac{5}{7}x + \frac{1}{3} + \frac{2}{7}x = \frac{4}{5} - \frac{2}{7}x + \frac{2}{7}x$ Add $\frac{2}{7}x$.

$\frac{7}{7}x + \frac{1}{3} = \frac{4}{5}$ Combine like terms.

$1x + \frac{1}{3} - \frac{1}{3} = \frac{4}{5} - \frac{1}{3}$ Subtract $\frac{1}{3}$.

$x = \frac{12}{15} - \frac{5}{15}$ LCD = 15

$x = \frac{7}{15}$ Subtract.

CHECK $\frac{5}{7}x + \frac{1}{3} = \frac{2}{5} - \frac{2}{7}x + \frac{2}{5}$

$\frac{5}{7}\left(\frac{7}{15}\right) + \frac{1}{3} \stackrel{?}{=} \frac{2}{5} - \frac{2}{7}\left(\frac{7}{15}\right) + \frac{2}{5}$ Let $x = \frac{7}{15}$.

$\frac{1}{3} + \frac{1}{3} \stackrel{?}{=} \frac{4}{5} - \frac{2}{15}$ Multiply.

$\frac{2}{3} \stackrel{?}{=} \frac{12}{15} - \frac{2}{15}$ LCD = 15

$\frac{2}{3} = \frac{10}{15}$, or $\frac{2}{3}$ ✓ True

Solution set: $\left\{\frac{7}{15}\right\}$

61. $10(-2x + 1) = -19(x + 1)$

$-20x + 10 = -19x - 19$ Distributive property

$-20x + 10 + 19x = -19x - 19 + 19x$ Add $19x$.

$-x + 10 = -19$ Combine like terms.

$-x + 10 - 10 = -19 - 10$ Subtract 10.

$-x = -29$

$x = 29$

CHECK $10(-2x + 1) = -19(x + 1)$

$10[-2(29) + 1] \stackrel{?}{=} -19(29 + 1)$ Let $x = 29$.

$10[-57] \stackrel{?}{=} -19(30)$

$-570 = -570$ ✓ True

Solution set: $\{29\}$

SECTION 2.2 (pages 130–131)

53. $\frac{2}{5}x - \frac{3}{10}x = 2$

$\frac{4}{10}x - \frac{3}{10}x = 2$ LCD = 10

$\frac{1}{10}x = 2$ Combine like terms.

$10 \cdot \frac{1}{10}x = 10 \cdot 2$ Multiply by 10.

$x = 20$

CHECK $\frac{2}{5}x - \frac{3}{10}x = 2$

$\frac{2}{5}(20) - \frac{3}{10}(20) \stackrel{?}{=} 2$ Let $x = 20$.

$8 - 6 \stackrel{?}{=} 2$

$2 = 2$ ✓ True

Solution set: $\{20\}$

61. $0.9w - 0.5w + 0.1w = -3$

$0.5w = -3$ Combine like terms.

$\frac{0.5w}{0.5} = \frac{-3}{0.5}$ Divide by 0.5.

$w = -6$

A check confirms this solution.

Solution set: $\{-6\}$

SECTION 2.3 (pages 141–144)

55. $0.02(5000) + 0.03x = 0.025(5000 + x)$

Multiply both sides by 1000, *not* 100.

$1000[0.02(5000) + 0.03x]$
$= 1000[0.025(5000 + x)]$

$20(5000) + 30x = 25(5000 + x)$

$100{,}000 + 30x = 125{,}000 + 25x$ Distributive property

$5x + 100{,}000 = 125{,}000$ Subtract $25x$.

$5x = 25{,}000$ Subtract 100,000.

$x = 5000$ Divide by 5.

CHECK Substituting 5000 for x in the original equation results in a true statement, $250 = 250$. ✓

Solution set: $\{5000\}$

59. $-(6k - 5) - (-5k + 8) = -3$

$-1(6k - 5) - 1(-5k + 8) = -3$

$-6k + 5 + 5k - 8 = -3$

$-k - 3 = -3$

$-k = 0$

$k = 0$

CHECK Substituting 0 for k in the original equation results in a true statement, $-3 = -3$. ✓

Solution set: $\{0\}$

SECTION 2.4 (pages 156–161)

21. ***Step 1***

Read the problem again.

Step 2

Let $x =$ the unknown number. Then $3x$ is three times the number, $x + 7$ is 7 more than the number, and the sum is

$3x + (x + 7)$.

$2x$ is twice the number, and

$-11 - 2x$

is the difference between -11 and twice the number.

Step 3

$3x + (x + 7) = -11 - 2x$

Step 4

$4x + 7 = -11 - 2x$

$6x + 7 = -11$ Add $2x$.

$6x = -18$ Subtract 7.

$x = -3$ Divide by 6.

Step 5

The number is -3.

Step 6

Check that -3 is the correct answer by substituting this result into the words of the original problem. The sum of three times this number and 7 more than the number is

$3(-3) + (-3 + 7) = -5$.

The difference between -11 and twice this number is

$-11 - 2(-3) = -5$.

The values are equal, so the number -3 is the correct answer.

49. ***Step 1***

Read the problem again.

Step 2

Let $x =$ the measures of angles A and B.

$x + 60 =$ the measure of angle C.

Step 3

The sum of the measures of the angles of any triangle is 180°, so

$x + x + (x + 60) = 180$.

Step 4

$3x + 60 = 180$

$3x = 120$ Subtract 60.

$x = 40$ Divide by 3.

Step 5

Angles A and B have measures of 40°, and angle C has a measure of $40 + 60 = 100°$.

Step 6

The answer checks since

$40 + 40 + 100 = 180$.

SECTION 2.5 (pages 168–173)

55. Let L represent the length of the box. The girth is $4 \cdot 18 = 72$ in. Since the length plus the girth is 108, we have

$L + 72 = 108$

$L = 36$ in. Subtract 72.

The maximum volume of the box is

$V = LWH$

$V = 36(18)(18)$ Substitute.

$V = 11{,}664$ in.^{3}.

59. The two angles are complementary, so the sum of their measures is 90°.

$(8x - 1) + 5x = 90$

$13x - 1 = 90$ Combine like terms.

$13x = 91$ Add 1.

$x = 7$ Divide by 13.

Since $x = 7$, we have

$8x - 1 = 8(7) - 1 = 56 - 1 = 55$

and $5x = 5(7) = 35$.

The angles measure 55° and 35°.

85. $M = C(1 + r)$ for r

$M = C + Cr$ Distributive property

$M - C = Cr$ Subtract C.

$\frac{M - C}{C} = \frac{Cr}{C}$ Divide by C.

$\frac{M - C}{C} = r$, or $r = \frac{M - C}{C}$

SECTION 2.6 (pages 180–185)

47. Let $x =$ the number of cups of cleaner. Set up a proportion with one ratio involving the number of cups of cleaner and the other involving the number of gallons of water.

$\frac{x \text{ cups}}{\frac{1}{4} \text{ cup}} = \frac{10\frac{1}{2} \text{ gallons}}{1 \text{ gallon}}$

$x \cdot 1 = \frac{1}{4}\left(10\frac{1}{2}\right)$ Cross products

$x = \frac{1}{4}\left(\frac{21}{2}\right)$ Write an improper fraction.

$x = \frac{21}{8}$, or $2\frac{5}{8}$

The amount of cleaner needed is $2\frac{5}{8}$ cups.

89. The increase in value was

$\$6200 - \$625 = \$5575$.

We can state the problem as follows:

"What percent of \$625 is \$5575?"

This translates into a percent equation, where p is the percent as a decimal.

$p \cdot 625 = 5575$

$p = \frac{5575}{625}$ Divide by 625.

$p = 8.92$, or 892%

The percent increase in the value of the coin was 892%.

SECTION 2.7 (pages 197–200)

53. $\frac{2}{3}(p + 3) > \frac{5}{6}(p - 4)$

$6\left(\frac{2}{3}\right)(p + 3) > 6\left(\frac{5}{6}\right)(p - 4)$ Multiply by 6, the LCD.

$4(p + 3) > 5(p - 4)$

(*continued*)

$$4p + 12 > 5p - 20$$ Distributive property

$$-p + 12 > -20$$ Subtract $5p$.

$$-p > -32$$ Subtract 12.

$$\frac{-p}{-1} < \frac{-32}{-1}$$ Divide by -1. Reverse the direction of the symbol.

$$p < 32$$

Solution set: $(-\infty, 32)$

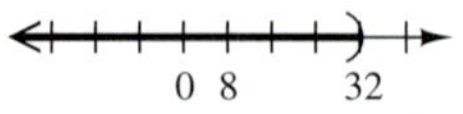

71. The Celsius temperature C must give a Fahrenheit temperature F that has *never exceeded* 104°, which translates as *is less than or equal to* 104°.

$$F \leq 104$$

$$\frac{9}{5}C + 32 \leq 104$$

$$\frac{9}{5}C \leq 72$$ Subtract 32.

$$\frac{5}{9}\left(\frac{9}{5}C\right) \leq \frac{5}{9}(72)$$ Multiply by $\frac{5}{9}$.

$$C \leq 40$$

The temperature of Providence, Rhode Island, has never exceeded 40°.

CHAPTER 3 Graphs of Linear Equations and Inequalities in Two Variables

SECTION 3.1 (pages 222–227)

29. Is $(5, -6)$ a solution of the equation $x = -6$?

Since y does not appear in the equation, we just substitute 5 for x.

$$x = -6$$

$$5 \stackrel{?}{=} -6$$ Let $x = 5$.

The result is false, so $(5, -6)$ is not a solution of the equation $x = -6$.

49. The given equation $x - 8 = 0$ may be written as $x = 8$. For any value of y, the value of x will always be 8. The completed table of values follows.

x	y	Ordered pairs
8	8 ⟶	(8, 8)
8	3 ⟶	(8, 3)
8	0 ⟶	(8, 0)

SECTION 3.2 (pages 236–241)

39. $-3y = 15$

$$y = \frac{15}{-3}, \text{ or } -5$$ Divide by -3.

For any value of x, the value of y is -5. Three ordered pairs are $(-2, -5)$, $(0, -5)$, and $(1, -5)$. Plot these points and draw a line through them. The graph is a horizontal line.

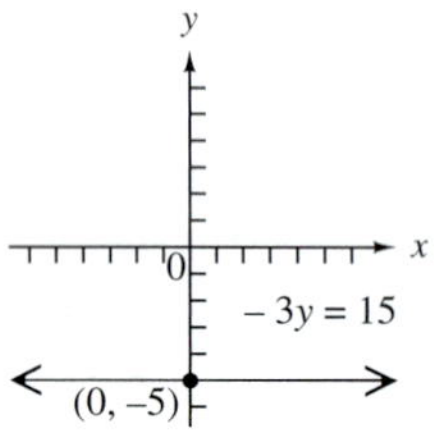

SECTION 3.3 (pages 250–253)

17. The slope (or grade) of the hill is the ratio of the rise to the run, or the ratio of the vertical change to the horizontal change. Since the rise is 32 and the run is 108, the slope is

$$\frac{32}{108} = \frac{8 \cdot 4}{27 \cdot 4} = \frac{8}{27}.$$ Lowest terms

35. First label the given points.

$$\begin{matrix} (x_1, y_1) & & (x_2, y_2) \\ \downarrow \downarrow & & \downarrow \downarrow \\ \left(-\frac{7}{5}, \frac{3}{10}\right) & \text{and} & \left(\frac{1}{5}, -\frac{1}{2}\right) \end{matrix}$$

Then apply the slope formula.

$$\text{slope } m = \frac{\text{change in } y}{\text{change in } x}$$

$$= \frac{y_2 - y_1}{x_2 - x_1}$$

$$= \frac{-\frac{1}{2} - \frac{3}{10}}{\frac{1}{5} - \left(-\frac{7}{5}\right)}$$ Substitute.

$$= \frac{-\frac{5}{10} - \frac{3}{10}}{\frac{1}{5} + \frac{7}{5}}$$ Get a common denominator in the numerator. Simplify the denominator.

$$= \frac{-\frac{8}{10}}{\frac{8}{5}}$$ Subtract in the numerator. Add in the denominator.

$$= -\frac{8}{10} \div \frac{8}{5}$$ $\frac{a}{b} = a \div b$

$$= -\frac{8}{10} \cdot \frac{5}{8}$$ Multiply by the reciprocal.

$$= -\frac{1}{2}$$ Multiply. Write in lowest terms.

SECTION 3.4 (pages 261–266)

19. The rise is 3 and the run is 1, so the slope m is given by

$$m = \frac{\text{rise}}{\text{run}} = \frac{3}{1} = 3.$$

The y-intercept is $(0, -3)$, so $b = -3$. The equation of the line, written in slope-intercept form $y = mx + b$, is

$$y = 3x - 3.$$ $m = 3, b = -3$

55. (a) First, find the slope of the line.

$$\begin{matrix} (x_1, y_1) & & (x_2, y_2) \\ \downarrow \downarrow & & \downarrow \downarrow \\ \left(\frac{1}{2}, \frac{3}{2}\right) & \text{and} & \left(-\frac{1}{4}, \frac{5}{4}\right) \end{matrix}$$

$$m = \frac{\frac{5}{4} - \frac{3}{2}}{-\frac{1}{4} - \frac{1}{2}} = \frac{\frac{5}{4} - \frac{6}{4}}{-\frac{1}{4} - \frac{2}{4}} = \frac{-\frac{1}{4}}{-\frac{3}{4}}$$

$$= -\frac{1}{4} \div \left(-\frac{3}{4}\right) = -\frac{1}{4}\left(-\frac{4}{3}\right) = \frac{1}{3}$$

Now use the point $\left(\frac{1}{2}, \frac{3}{2}\right)$ for (x_1, y_1) and $m = \frac{1}{3}$ in the point-slope form.

$$y - y_1 = m(x - x_1)$$

$$y - \frac{3}{2} = \frac{1}{3}\left(x - \frac{1}{2}\right)$$ Substitute.

$$y - \frac{3}{2} = \frac{1}{3}x - \frac{1}{6}$$ Distributive property

$$y = \frac{1}{3}x - \frac{1}{6} + \frac{9}{6}$$ Add $\frac{3}{2} = \frac{9}{6}$.

$$y = \frac{1}{3}x + \frac{4}{3}$$ Combine like terms; $\frac{8}{6} = \frac{4}{3}$

This last equation is written in slope-intercept form $y = mx + b$.

(b) Clear the fractions in the final equation in part (a) by multiplying each term by 3.

$$y = \frac{1}{3}x + \frac{4}{3}$$ From part (a)

$$3y = x + 4$$ Multiply by 3.

$$-x + 3y = 4$$ Subtract x.

$$x - 3y = -4$$ Multiply by -1.

This last equation is written in standard form $Ax + By = C$.

61. Solve the given equation for y.

$$3x = 4y + 5$$

$$3x - 4y = 5$$ Subtract $4y$.

$$-4y = -3x + 5$$ Subtract $3x$.

$$y = \frac{3}{4}x - \frac{5}{4}$$ Divide by -4.

The slope is $\frac{3}{4}$. A line parallel to this line has the same slope.

SOLUTIONS

Use the point-slope form with $m = \frac{3}{4}$ and $(x_1, y_1) = (2, -3)$.

$$y - y_1 = m(x - x_1)$$

$$y - (-3) = \frac{3}{4}(x - 2) \quad \text{Substitute.}$$

$$y + 3 = \frac{3}{4}x - \frac{3}{2}$$

$$y = \frac{3}{4}x - \frac{3}{2} - \frac{6}{2} \quad \text{Subtract } 3 = \tfrac{6}{2}.$$

$$y = \frac{3}{4}x - \frac{9}{2} \quad \text{Combine like terms.}$$

CHAPTER 4 Systems of Linear Equations and Inequalities

SECTION 4.1 (pages 295–298)

27. $2x - 3y = -6$

$y = -3x + 2$

To graph $2x - 3y = -6$, find the intercepts.

$$2x - 3(0) = -6 \quad \text{Let } y = 0.$$

$$2x = -6$$

$$x = -3$$

The x-intercept is $(-3, 0)$.

$$2(0) - 3y = -6 \quad \text{Let } x = 0.$$

$$-3y = -6$$

$$y = 2$$

The y-intercept is $(0, 2)$.

Plot the intercepts, $(-3, 0)$ and $(0, 2)$, and draw the line through them.

To graph the second line, which is in slope-intercept form, start by plotting the y-intercept, $(0, 2)$. From this point, move 3 units down and 1 unit to the right (because the slope is -3) to reach the point $(1, -1)$. Draw the line through $(0, 2)$ and $(1, -1)$.

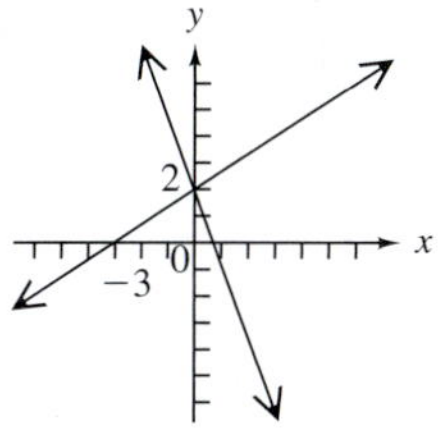

The lines intersect at their common y-intercept, $(0, 2)$.

Solution set: $\{(0, 2)\}$

35. $4x - 2y = 8$ (1)

$2x = y + 4$ (2)

Graph the line $4x - 2y = 8$ using its intercepts, $(2, 0)$ and $(0, -4)$.

Graph $2x = y + 4$ using its intercepts, $(2, 0)$ and $(0, -4)$. Since both equations have the same intercepts, they are equations of the same line.

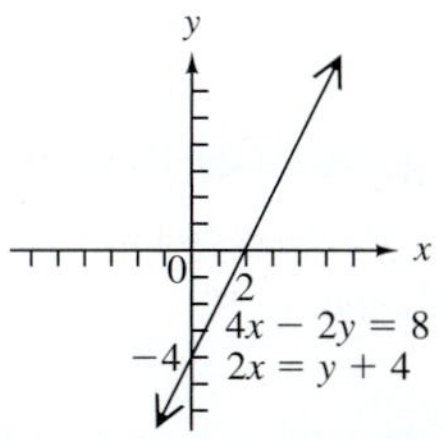

There are an infinite number of solutions, so we use set-builder notation to write the solution set. Rewrite equation (2) in standard form.

$$2x - y = 4$$

The solution set is

$$\{(x, y) \mid 2x - y = 4\}.$$

The given system consists of dependent equations.

SECTION 4.2 (pages 305–306)

21. $6x - 8y = 6$ (1)

$2y = -2 + 3x$ (2)

Solve equation (2) for y.

$$2y = -2 + 3x$$

$$y = \frac{3x - 2}{2} \quad (3)$$

Substitute $\frac{3x - 2}{2}$ for y in equation (1).

$$6x - 8y = 6 \quad (1)$$

$$6x - 8\left(\frac{3x - 2}{2}\right) = 6$$

$$6x - 4(3x - 2) = 6$$

$$6x - 12x + 8 = 6$$

$$-6x + 8 = 6$$

$$-6x = -2$$

$$x = \frac{-2}{-6}, \quad \text{or} \quad \frac{1}{3}$$

To find y, let $x = \frac{1}{3}$ in equation (3).

$$y = \frac{3x - 2}{2} = \frac{3\left(\frac{1}{3}\right) - 2}{2} = \frac{1 - 2}{2} = -\frac{1}{2}$$

Solution set: $\left\{\left(\frac{1}{3}, -\frac{1}{2}\right)\right\}$

27. $\frac{x}{5} + 2y = \frac{16}{5}$ (1)

$\frac{3x}{5} + \frac{y}{2} = -\frac{7}{5}$ (2)

Multiply each side of equation (1) by 5.

$$5\left(\frac{x}{5} + 2y\right) = 5\left(\frac{16}{5}\right)$$

$$x + 10y = 16 \quad (3)$$

Multiply each side of equation (2) by 10.

$$10\left(\frac{3x}{5} + \frac{y}{2}\right) = 10\left(-\frac{7}{5}\right)$$

$$6x + 5y = -14 \quad (4)$$

We now have an equivalent system that does not include fractions.

$$x + 10y = 16 \quad (3)$$

$$6x + 5y = -14 \quad (4)$$

Solve equation (3) for x.

$$x = 16 - 10y \quad (5)$$

Substitute $16 - 10y$ for x in equation (4).

$$6x + 5y = -14 \quad (4)$$

$$6(16 - 10y) + 5y = -14$$

$$96 - 60y + 5y = -14$$

$$-55y = -110$$

$$y = 2$$

To find x, let $y = 2$ in equation (5).

$$x = 16 - 10y \quad (5)$$

$$x = 16 - 10(2)$$

$$x = -4$$

Solution set: $\{(-4, 2)\}$

SECTION 4.3 (pages 311–313)

25. $-x + 3y = 4$ (1)

$-2x + 6y = 8$ (2)

Multiply equation (1) by -2 and add the result to equation (2).

$$2x - 6y = -8 \quad (3)$$

$$\underline{-2x + 6y = 8} \quad (2)$$

$$0 = 0 \quad \text{True}$$

The true result, $0 = 0$, indicates that the equations of the original system are dependent. To obtain an equation in standard form $Ax + By = C$, with $A > 0$, multiply equation (1) by -1.

$$x - 3y = -4$$

Solution set: $\{(x, y) \mid x - 3y = -4\}$

37. $6x + 3y = 0$ (1)

$-18x - 9y = 0$ (2)

Multiply equation (1) by 3 and add the result to equation (2).

$$18x + 9y = 0 \quad (3)$$

$$\underline{-18x - 9y = 0} \quad (2)$$

$$0 = 0 \quad \text{True}$$

The true result, $0 = 0$, means that the system has an infinite number of solutions. To obtain an equation for the solution set in which the coefficients have greatest common factor 1, divide both sides of equation (1) by 3 to obtain the equivalent equation

$$2x + y = 0.$$

Solution set: $\{(x, y) \mid 2x + y = 0\}$

SECTION 4.4 (pages 322–327)

19. $C = 85x + 900$ (1)

$R = 105x$ (2)

No more than 38 units can be sold.

(a) Since $C = R$, substitute $105x$ for C and solve for x.

$$105x = 85x + 900$$

$$20x = 900 \quad \text{Subtract } 85x.$$

$$x = 45 \quad \text{Divide by 20.}$$

The break-even quantity is 45 units.

(b) Since no more than 38 units can be sold and the break-even quantity is 45, do not produce the product $(45 > 38)$. Doing so will lead to a loss.

37. ***Step 1***

Read the problem again.

Step 2

Assign variables.

Let $x =$ the average rate of the bicycle;

$y =$ the average rate of the car.

	r	*t*	*d*
Bicycle	x	7.5	$7.5x$
Car	y	7.5	$7.5y$

Use the formula $d = rt$.

Step 3

The total distance traveled by the bicycle and the car is 471 mi.

$$7.5x + 7.5y = 471 \quad (1)$$

The car traveled 35.8 mi faster than the bicycle.

$$y = x + 35.8 \quad (2)$$

These equations form the following system.

$$7.5x + 7.5y = 471 \quad (1)$$

$$y = x + 35.8 \quad (2)$$

Step 4

Because equation (2) is already solved for y, the substitution method is a good choice. Substitute $x + 35.8$ for y in equation (1).

$$7.5x + 7.5y = 471 \quad (1)$$

$$7.5x + 7.5(x + 35.8) = 471 \quad \text{Let } y = x + 35.8.$$

$$7.5x + 7.5x + 268.5 = 471 \quad \text{Distributive property}$$

$$15x + 268.5 = 471 \quad \text{Combine like terms.}$$

$$15x = 202.5 \quad \text{Subtract 268.5.}$$

$$x = 13.5 \quad \text{Divide by 15.}$$

To find the value of y substitute 13.5 for x in equation (2).

$$y = x + 35.8 \quad (2)$$

$$y = 13.5 + 35.8 \quad \text{Let } x = 13.5.$$

$$y = 49.3$$

Step 5

The average rate of the bicycle is 13.5 mph, and the average rate of the car is 49.3 mph.

Step 6

In 7.5 hr, the total distance traveled by the bicycle and the car is

$$7.5(13.5) + 7.5(49.3) = 471 \text{ mi.}$$

Since $49.3 - 13.5 = 35.8$, the car traveled 35.8 mph faster than the bicycle, as stated.

SECTION 4.5 (pages 331–332)

19. $x \le 2y + 3$

$x + y < 0$

Graph $x = 2y + 3$ as a solid line through $(3, 0)$ and $(5, 1)$. Using $(0, 0)$ as a test point results in the true statement $0 \le 3$, so shade the region containing the origin. Graph $x + y = 0$ as a dashed line through $(0, 0)$ and $(1, -1)$. We cannot use $(0, 0)$ as a test point because it is on the boundary line. Using $(1, 0)$ results in the false statement $1 < 0$, so shade the region *not* containing $(1, 0)$.

The solution set of the system is the intersection of the two shaded regions. It includes the portion of the line $x = 2y + 3$ that bounds the region but not the portion of the line $x + y = 0$.

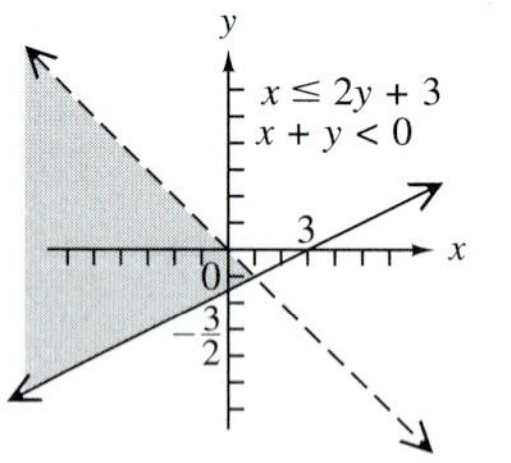

23. $3x - 2y \ge 6$

$x + y \le 4$

$x \ge 0$

$y \ge -4$

Graph $3x - 2y = 6$, $x + y = 4$, $x = 0$, and $y = -4$ as solid lines, as shown in the graph at the top of the next column. All four inequalities are true for $(2, -2)$. Shade the region bounded by the four lines, which contains the test point $(2, -2)$. The solution set is the shaded region.

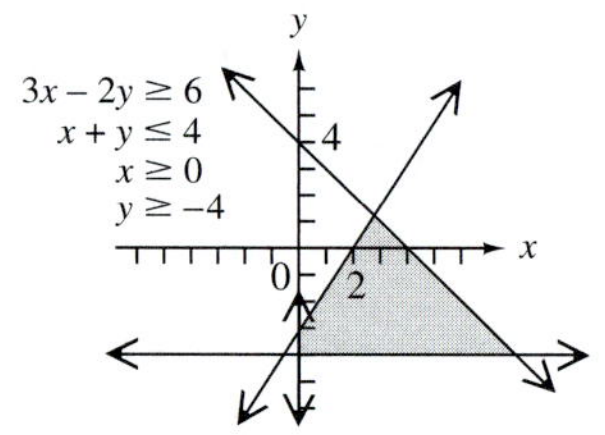

CHAPTER 5 Exponents and Polynomials

SECTION 5.1 (pages 349–352)

21. $\frac{1}{2}x^4 + \frac{1}{6}x^4$

$$= \left(\frac{1}{2} + \frac{1}{6}\right)x^4 \quad \text{Distributive property}$$

$$= \left(\frac{3}{6} + \frac{1}{6}\right)x^4 \quad \text{Write fractions with a common denominator.}$$

$$= \frac{4}{6}x^4 \quad \text{Add fractions.}$$

$$= \frac{2}{3}x^4 \quad \text{Lowest terms}$$

37. $0.8x^4 - 0.3x^4 - 0.5x^4 + 7$

$$= (0.8 - 0.3 - 0.5)x^4 + 7$$

$$= 0x^4 + 7$$

$$= 7$$

Since 7 can be written as $7x^0$, the degree of the simplified polynomial is 0. It has one term, so it is a monomial.

53. Add.

$$\begin{array}{r} \frac{2}{3}x^2 + \frac{1}{5}x + \frac{1}{6} \\ \frac{1}{2}x^2 - \frac{1}{3}x + \frac{2}{3} \\ \hline \end{array}$$

Rewrite so that the fractions in each column have a common denominator. Then add column by column.

$$\begin{array}{r} \frac{4}{6}x^2 + \frac{3}{15}x + \frac{1}{6} \\ \frac{3}{6}x^2 - \frac{5}{15}x + \frac{4}{6} \\ \hline \frac{7}{6}x^2 - \frac{2}{15}x + \frac{5}{6} \end{array}$$

71. Use the formula for the perimeter of a square, $P = 4s$, with $s = \frac{1}{2}x^2 + 2x$.

$$P = 4s$$

$$P = 4\left(\frac{1}{2}x^2 + 2x\right) \quad \text{Substitute.}$$

$$P = 4\left(\frac{1}{2}x^2\right) + 4(2x) \quad \text{Distributive property}$$

$$P = 2x^2 + 8x \quad \text{Multiply.}$$

SECTION 5.2 (pages 359–360)

43. $(-5^2)^6$

$= (-1 \cdot 5^2)^6$ Write the factor -1.

$= (-1)^6 \cdot (5^2)^6$ Power rule (b)

$= 1 \cdot 5^{2 \cdot 6}$ Power rule (a)

$= 5^{12}$ Multiply.

77. $\left(\frac{5a^2b^5}{c^6}\right)^3 \quad (c \neq 0)$

$= \frac{(5a^2b^5)^3}{(c^6)^3}$ Power rule (c)

$= \frac{5^3(a^2)^3(b^5)^3}{(c^6)^3}$ Power rule (b)

$= \frac{125a^6b^{15}}{c^{18}}$ Power rule (a)

79. $(-5m^3p^4q)^2(p^2q)^3$

$= (-1 \cdot 5m^3p^4q)^2(p^2q)^3$ Write the factor -1.

$= (-1)^2 \cdot 5^2 \cdot (m^3)^2 \cdot (p^4)^2 \cdot q^2 \cdot (p^2)^3 \cdot q^3$ Power rule (b)

$= 1 \cdot 25 \cdot m^6p^8q^2p^6q^3$ Power rule (a)

$= 25m^6p^{8+6}q^{2+3}$ Product rule

$= 25m^6p^{14}q^5$

SECTION 5.3 (pages 365–367)

57. $\left(x - \frac{2}{3}\right)\left(x + \frac{1}{4}\right)$

F O I L

$= x(x) + x\left(\frac{1}{4}\right) + \left(-\frac{2}{3}\right)x + \left(-\frac{2}{3}\right)\frac{1}{4}$

$= x^2 + \frac{1}{4}x - \frac{2}{3}x - \frac{1}{6}$

$= x^2 + \left(\frac{3}{12}x - \frac{8}{12}x\right) - \frac{1}{6}$

$= x^2 - \frac{5}{12}x - \frac{1}{6}$

SECTION 5.4 (pages 372–374)

11. $\left(x - \frac{3}{4}\right)^2$

$= x^2 - 2(x)\left(\frac{3}{4}\right) + \left(\frac{3}{4}\right)^2$

$(a - b)^2 = a^2 - 2ab + b^2$

$= x^2 - \frac{3}{2}x + \frac{9}{16}$

21. $(0.8t + 0.7s)^2$

$= (0.8t)^2 + 2(0.8t)(0.7s) + (0.7s)^2$

$(a + b)^2 = a^2 + 2ab + b^2$

$= 0.64t^2 + 1.12ts + 0.49s^2$

29. $-(4r - 2)^2$

First square the binomial.

$(4r - 2)^2$

$= (4r)^2 - 2(4r)(2) + 2^2$

$= 16r^2 - 16r + 4$

Now multiply by -1.

$-1(16r^2 - 16r + 4)$

$= -16r^2 + 16r - 4$

43. $(2x^2 - 5)(2x^2 + 5)$

$= (2x^2)^2 - 5^2$ $(a - b)(a + b) = a^2 - b^2$

$= 4x^4 - 25$ $(2x^2)^2 = 2^2x^4 = 4x^4$

SECTION 5.5 (pages 382–383)

17. $(-2)^0 - 2^0$

$= 1 - 1$ $a^0 = 1$

$= 0$ Subtract.

33. $-2^{-1} + 3^{-2}$

$= -(2^{-1}) + 3^{-2}$ In -2^{-1}, the base is 2.

$= -\frac{1}{2^1} + \frac{1}{3^2}$ $a^{-n} = \frac{1}{a^n}$

$= -\frac{1}{2} + \frac{1}{9}$ Apply the exponents.

$= -\frac{9}{18} + \frac{2}{18}$ Write fractions with a common denominator.

$= -\frac{7}{18}$ Add.

69. $\frac{(m^7n)^{-2}}{m^{-4}n^3}$

$= \frac{(m^7)^{-2}n^{-2}}{m^{-4}n^3}$ Power rule (b)

$= \frac{m^{7(-2)}n^{-2}}{m^{-4}n^3}$ Power rule (a)

$= \frac{m^{-14}n^{-2}}{m^{-4}n^3}$

$= m^{-14-(-4)}n^{-2-3}$ Quotient rule

$= m^{-10}n^{-5}$

$= \frac{1}{m^{10}n^5}$ $a^{-n} = \frac{1}{a^n}$

SECTION 5.6 (pages 388–389)

23. $\frac{-3x^3 - 4x^4 + 2x}{-3x^2}$

$= \frac{-4x^4 - 3x^3 + 2x}{-3x^2}$ Write in descending powers.

$= \frac{-4x^4}{-3x^2} - \frac{3x^3}{-3x^2} + \frac{2x}{-3x^2}$ Divide each term by $-3x^2$.

$= \frac{4x^2}{3} + x - \frac{2}{3x}$ Quotient rule; be careful with signs.

Notice how the third term $-\frac{2}{3x}$ is written with x in the denominator, which is ***not*** the same as $-\frac{2}{3}x$ or $-\frac{2x}{3}$. In $-\frac{2}{3x}$, we are *dividing* by x. In $-\frac{2}{3}x$ and $-\frac{2x}{3}$, we are *multiplying* by x. Applying the quotient rule to the term $\frac{2x}{-3x^2}$ gives the following.

$\frac{2x}{-3x^2} = \frac{2x^1}{-3x^2} = -\frac{2}{3}x^{1-2} = -\frac{2}{3}x^{-1}$

$= -\frac{2}{3}\left(\frac{1}{x}\right) = -\frac{2}{3x}$

($\frac{4}{3}x^2$ is an acceptable form for the first term, $\frac{4x^2}{3}$. Why?)

33. Use the formula for the area of a rectangle, $A = LW$, with $A = 12x^2 - 4x + 2$ and $W = 2x$.

$A = LW$

$12x^2 - 4x + 2 = L(2x)$ Substitute for A and W.

$\frac{12x^2 - 4x + 2}{2x} = L$ Divide each side by $2x$.

$\frac{12x^2}{2x} - \frac{4x}{2x} + \frac{2}{2x} = L$ Divide each term by $2x$.

$6x - 2 + \frac{1}{x} = L$ $\frac{2}{2x} = \frac{2}{2} \cdot \frac{1}{x} = 1 \cdot \frac{1}{x} = \frac{1}{x}$

$6x - 2 + \frac{1}{x}$ represents the length of the rectangle.

SECTION 5.7 (pages 394–395)

23. $(x^4 - x^2 - 2) \div (x^2 - 2)$

Use 0 as the coefficient of the missing x^3- and x-terms in the dividend and the missing x-term in the divisor.

$$\begin{array}{r} x^2 \qquad\quad + 1 \\ x^2 + \mathbf{0x} - 2 \overline{)\, x^4 + \mathbf{0x^3} - x^2 + \mathbf{0x} - 2} \\ \underline{x^4 + 0x^3 - 2x^2} \qquad\qquad \\ x^2 + 0x - 2 \\ \underline{x^2 + 0x - 2} \\ 0 \end{array}$$

The remainder is 0. The answer is the quotient, $x^2 + 1$.

29. $\dfrac{2x^5 + x^4 + 11x^3 - 8x^2 - 13x + 7}{2x^2 + x - 1}$

$$\begin{array}{r}
x^3 \qquad\quad + \;\; 6x - 7 \\
2x^2 + x - 1\overline{)2x^5 + x^4 + 11x^3 - 8x^2 - 13x + 7} \\
\underline{2x^5 + x^4 - \quad x^3} \qquad\qquad\qquad\quad \\
12x^3 - 8x^2 - 13x \quad\;\; \\
\underline{12x^3 + 6x^2 - \;\; 6x} \quad\;\; \\
-14x^2 - \;\; 7x + 7 \\
\underline{-14x^2 - \;\; 7x + 7} \\
0
\end{array}$$

The remainder is 0. The answer is the quotient, $x^3 + 6x - 7$.

SECTION 5.8 (pages 400–402)

51. $\dfrac{4 \times 10^5}{8 \times 10^2}$

$= \dfrac{4}{8} \times \dfrac{10^5}{10^2}$

$= 0.5 \times 10^3$
Divide; quotient rule

$= (5 \times 10^{-1}) \times 10^3$
Write 0.5 in scientific notation.

$= 5 \times (10^{-1} \times 10^3)$
Associative property

$= 5 \times 10^2$
Product rule

$= 500$
Write without exponents.

55. $\dfrac{(1.65 \times 10^8)(5.24 \times 10^{-2})}{(6 \times 10^4)(2 \times 10^7)}$

$= \dfrac{1.65 \times 5.24}{6 \times 2} \times \dfrac{10^8 \times 10^{-2}}{10^4 \times 10^7}$
Associative and commutative properties

$= 0.7205 \times \dfrac{10^6}{10^{11}}$ Product rule

$= 0.7205 \times 10^{-5}$ Quotient rule

$= (7.205 \times 10^{-1}) \times 10^{-5}$
Write 0.7205 in scientific notation.

$= 7.205 \times (10^{-1} \times 10^{-5})$
Associative property

$= 7.205 \times 10^{-6}$ Product rule

$= 0.000007205$
Write without exponents.

CHAPTER 6 Factoring and Applications

SECTION 6.1 (pages 423–425)

79. $18r^2 + 12ry - 3xr - 2xy$

$= (18r^2 + 12ry) + (-3xr - 2xy)$
Group the terms.

$= 6r(3r + 2y) - x(3r + 2y)$
Factor each group.

$= (3r + 2y)(6r - x)$
Factor out the common factor, $3r + 2y$.

87. $y^2 + 3x + 3y + xy$

$= y^2 + 3y + xy + 3x$
Rearrange terms.

$= (y^2 + 3y) + (xy + 3x)$
Group the terms.

$= y(y + 3) + x(y + 3)$
Factor each group.

$= (y + 3)(y + x)$
Factor out the common factor, $y + 3$.

SECTION 6.2 (pages 431–432)

41. $a^2 - 8a - 48$

Find two integers whose product is -48 and whose sum is -8. Since c is negative, one integer must be positive and one must be negative.

Factors of -48	Sums of Factors
$-1, 48$	47
$1, -48$	-47
$-2, 24$	22
$2, -24$	-22
$-3, 16$	13
$3, -16$	-13
$-4, 12$	8
$4, -12$	-8 ←
$-6, 8$	2
$6, -8$	-2

Thus,
$a^2 - 8a - 48$ factors as $(a + 4)(a - 12)$.

57. $-2x^6 - 8x^5 + 42x^4$

First, factor out the negative common factor, $-2x^4$.

$-2x^6 - 8x^5 + 42x^4$
$= -2x^4(x^2 + 4x - 21)$

Now factor $x^2 + 4x - 21$.

Factors of -21	Sums of Factors
$1, -21$	-20
$-1, 21$	20
$3, -7$	-4
$-3, 7$	4 ←

Thus,
$x^2 + 4x - 21$ factors as $(x - 3)(x + 7)$.
The completely factored form is
$-2x^6 - 8x^5 + 42x^4$
$= -2x^4(x - 3)(x + 7)$.

61. $5m^5 + 25m^4 - 40m^2$

First, factor out the GCF, $5m^2$.

$5m^5 + 25m^4 - 40m^2$
$= 5m^2(m^3 + 5m^2 - 8)$

The trinomial $m^3 + 5m^2 - 8$ cannot be factored further, so the polynomial is in completely factored form.

63. $m^3n - 10m^2n^2 + 24mn^3$

First, factor out the GCF, mn.

$m^3n - 10m^2n^2 + 24mn^3$
$= mn(m^2 - 10mn + 24n^2)$

The expressions $-6n$ and $-4n$ have a product of $24n^2$ and a sum of $-10n$. The completely factored form is

$m^3n - 10m^2n^2 + 24mn^3$
$= mn(m - 6n)(m - 4n)$.

SECTION 6.3 (pages 435–436)

29. $16 + 16x + 3x^2$

Write the trinomial in descending powers.

$3x^2 + 16x + 16$

To factor, find two integers whose product is $3(16) = 48$ and whose sum is 16. The integers are 12 and 4.

$3x^2 + 16x + 16$

$= 3x^2 + 12x + 4x + 16$

$= (3x^2 + 12x) + (4x + 16)$
Group the terms.

$= 3x(x + 4) + 4(x + 4)$
Factor each group.

$= (x + 4)(3x + 4)$
Factor out the common factor, $x + 4$.

35. $-32z^5 + 20z^4 + 12z^3$

First factor out the negative common factor, $-4z^3$.

$-32z^5 + 20z^4 + 12z^3$
$= -4z^3(8z^2 - 5z - 3)$

To factor $8z^2 - 5z - 3$, find two integers whose product is $8(-3) = -24$ and whose sum is -5. These integers are -8 and 3. Now rewrite the given trinomial and factor it.

$-32z^5 + 20z^4 + 12z^3$

$= -4z^3(8z^2 - 5z - 3)$

$= -4z^3(8z^2 - 8z + 3z - 3)$
$-5z = -8z + 3z$

$= -4z^3[(8z^2 - 8z) + (3z - 3)]$
Group the terms.

$= -4z^3[8z(z - 1) + 3(z - 1)]$
Factor each group.

$= -4z^3[(z - 1)(8z + 3)]$
Factor out the common factor, $z - 1$.

$= -4z^3(z - 1)(8z + 3)$

SECTION 6.4 (pages 441–442)

43. $6m^6n + 7m^5n^2 + 2m^4n^3$

Factor out the GCF, m^4n.

$6m^6n + 7m^5n^2 + 2m^4n^3$
$= m^4n(6m^2 + 7mn + 2n^2)$

Now factor $6m^2 + 7mn + 2n^2$ by trial and error. Possible factors of $6m^2$ are $6m$ and m or $3m$ and $2m$. Possible factors of $2n^2$ are $2n$ and n.

$$(3m + 2n)(2m + n)$$
$$= 6m^2 + 7mn + 2n^2 \quad \text{Correct}$$

The completely factored form is

$$6m^6n + 7m^5n^2 + 2m^4n^3$$
$$= m^4n(3m + 2n)(2m + n).$$

SECTION 6.5 (pages 447–448)

51. $4m^2 - \frac{9}{25}$

Because $4m^2 = (2m)^2$ and $\frac{9}{25} = \left(\frac{3}{5}\right)^2$, $4m^2 - \frac{9}{25}$ is a difference of squares.

$$4m^2 - \frac{9}{25}$$
$$= (2m)^2 - \left(\frac{3}{5}\right)^2$$
$$= \left(2m + \frac{3}{5}\right)\left(2m - \frac{3}{5}\right)$$

57. $x^2 - 1.0x + 0.25$

The first and last terms are perfect squares, x^2 and $(-0.5)^2$. The trinomial is a perfect square, since the middle term is

$$2 \cdot x \cdot (-0.5) = -1.0x.$$

$$x^2 - 1.0x + 0.25$$
$$= (x)^2 - 2(x)(0.5) + (0.5)^2$$
$$= (x - 0.5)^2$$

SECTION 6.6 (pages 459–461)

57. $9y^3 - 49y = 0$

To factor the polynomial, begin by factoring out the greatest common factor.

$$y(9y^2 - 49) = 0$$

Now factor $9y^2 - 49$ as the difference of two squares.

$$y(3y + 7)(3y - 7) = 0$$

Set each of the three factors equal to 0 and solve.

$$y = 0 \quad \text{or} \quad 3y + 7 = 0 \quad \text{or} \quad 3y - 7 = 0$$
$$3y = -7 \quad \text{or} \quad 3y = 7$$
$$y = -\frac{7}{3} \quad \text{or} \quad y = \frac{7}{3}$$

Solution set: $\left\{-\frac{7}{3}, 0, \frac{7}{3}\right\}$

59. $(2r + 5)(3r^2 - 16r + 5) = 0$

$$(2r + 5)(3r - 1)(r - 5) = 0 \quad \text{Factor } 3r^2 - 16r + 5.$$

Set each factor equal to 0 and solve.

$$2r + 5 = 0 \quad \text{or} \quad 3r - 1 = 0 \quad \text{or} \quad r - 5 = 0$$
$$2r = -5 \quad \text{or} \quad 3r = 1 \quad \text{or} \quad r = 5$$
$$r = -\frac{5}{2} \quad \text{or} \quad r = \frac{1}{3}$$

Solution set: $\left\{-\frac{5}{2}, \frac{1}{3}, 5\right\}$

65. $r^4 = 2r^3 + 15r^2$

Write with 0 on one side of the equation.

$$r^4 - 2r^3 - 15r^2 = 0 \quad \text{Subtract } 2r^3 \text{ and } 15r^2.$$
$$r^2(r^2 - 2r - 15) = 0 \quad \text{GCF} = r^2$$
$$r \cdot r \cdot (r + 3)(r - 5) = 0 \quad \text{Factor.}$$

Set each of the factors equal to 0, and solve the resulting equations. (Since r is a factor twice, it gives a double solution.)

$$r = 0 \quad \text{or} \quad r + 3 = 0 \quad \text{or} \quad r - 5 = 0$$
$$r = -3 \quad \text{or} \quad r = 5$$

Solution set: $\{-3, 0, 5\}$

SECTION 6.7 (pages 468–472)

15. Let x = the length of a side of the square painting.

Then $x - 2$ = the length of a side of the square mirror.

Since the formula for the area of a square is $A = s^2$, the area of the painting is x^2, and the area of the mirror is $(x - 2)^2$. The difference between their areas is 32.

$$x^2 - (x - 2)^2 = 32$$
$$x^2 - (x^2 - 4x + 4) = 32$$
$$x^2 - x^2 + 4x - 4 = 32$$
$$4x - 4 = 32$$
$$4x = 36$$
$$x = 9$$

The length of a side of the painting is 9 ft. The length of a side of the mirror is

$$9 - 2 = 7 \text{ ft}.$$

As a check,

$$9^2 - 7^2 = 81 - 49 = 32, \quad \text{as required.}$$

23. Let x = the least even integer.

Then $x + 2$ = the next even integer and

$x + 4$ = the third even integer.

$$x^2 + (x + 2)^2 = (x + 4)^2$$
$$x^2 + x^2 + 4x + 4 = x^2 + 8x + 16 \quad \text{Square the binomials.}$$
$$2x^2 + 4x + 4 = x^2 + 8x + 16 \quad \text{Combine like terms.}$$
$$x^2 - 4x - 12 = 0 \quad \text{Standard form}$$
$$(x + 2)(x - 6) = 0 \quad \text{Factor.}$$
$$x + 2 = 0 \quad \text{or} \quad x - 6 = 0 \quad \text{Zero-factor property}$$
$$x = -2 \quad \text{or} \quad x = 6$$

If $x = -2$, then $x + 2 = 0$ and $x + 4 = 2$.
If $x = 6$, then $x + 2 = 8$ and $x + 4 = 10$.
The integers are -2, 0, and 2 or 6, 8, and 10.

CHECK

$(-2)^2 + 0^2 \stackrel{?}{=} 2^2$	$6^2 + 8^2 \stackrel{?}{=} 10^2$
$4 + 0 \stackrel{?}{=} 4$	$36 + 64 \stackrel{?}{=} 100$
$4 = 4$ ✓	$100 = 100$ ✓
True	True

CHAPTER 7 Rational Expressions and Applications

SECTION 7.1 (pages 493–494)

9. (a) $\frac{(-3x)^2}{4x + 12}$

$$= \frac{(-3 \cdot 2)^2}{4(2) + 12} \quad \text{Let } x = 2.$$
$$= \frac{(-6)^2}{8 + 12}$$
$$= \frac{36}{20}$$
$$= \frac{9}{5}$$

(b) $\frac{(-3x)^2}{4x + 12}$

$$= \frac{[-3(-3)]^2}{4(-3) + 12} \quad \text{Let } x = -3.$$
$$= \frac{9^2}{-12 + 12}, \quad \text{or} \quad \frac{81}{0}$$

Since substituting -3 for x makes the denominator 0, the given rational expression is *undefined* for $x = -3$.

35. $\frac{12m^2 - 3}{8m - 4}$

$$= \frac{3(4m^2 - 1)}{4(2m - 1)} \quad \text{Factor out the common factors.}$$
$$= \frac{3(2m + 1)(2m - 1)}{4(2m - 1)} \quad \text{Factor again in the numerator.}$$
$$= \frac{3(2m + 1)}{4} \quad \text{Divide out the common factor.}$$

43. We factor the numerator and denominator by grouping in this rational expression.

$$\frac{zw + 4z - 3w - 12}{zw + 4z + 5w + 20}$$
$$= \frac{(zw + 4z) - (3w + 12)}{(zw + 4z) + (5w + 20)} \quad \text{Group the terms.}$$
$$= \frac{z(w + 4) - 3(w + 4)}{z(w + 4) + 5(w + 4)} \quad \text{Factor each group.}$$
$$= \frac{(w + 4)(z - 3)}{(w + 4)(z + 5)} \quad \text{Factor by grouping.}$$
$$= \frac{z - 3}{z + 5} \quad \text{Divide out the common factor.}$$

61. $LW = A$ Formula for the area of a rectangle

$W = \frac{A}{L}$ Divide by L.

$W = \frac{x^4 + 10x^2 + 21}{x^2 + 7}$ Substitute for A and L.

$W = \frac{(x^2+7)(x^2+3)}{x^2+7}$ Factor.

$W = x^2 + 3$ Divide out the common factor.

Note: If it is not apparent that A,

$$x^4 + 10x^2 + 21,$$

factors as $(x^2+7)(x^2+3)$, we can use "long division" to find the quotient $\frac{A}{L}$. Remember to insert zeros for the coefficients of the missing terms.

$$\begin{array}{r} x^2 \qquad\quad + 3 \\ x^2+0x+7\overline{)x^4+0x^3+10x^2+0x+21} \\ \underline{x^4+0x^3+\ 7x^2\qquad\qquad} \\ 3x^2+0x+21 \\ \underline{3x^2+0x+21} \\ 0 \end{array}$$

The width of the rectangle is $x^2 + 3$.

SECTION 7.2 (pages 499–500)

31. $\frac{6(m-2)^2}{5(m+4)^2} \cdot \frac{15(m+4)}{2(2-m)}$

$= \frac{6 \cdot 15(m-2)^2(m+4)}{5 \cdot 2(m+4)^2(2-m)}$ Multiply.

$= \frac{2 \cdot 3 \cdot 3 \cdot 5(m-2)(m-2)(m+4)}{5 \cdot 2(m+4)(m+4)(-1)(m-2)}$ Factor.

$= \frac{3 \cdot 3(m-2)}{(m+4)(-1)}$ Divide out the common factors.

$= \frac{9(m-2)}{-(m+4)}$, or $\frac{-9(m-2)}{m+4}$

39. $\frac{m^2+2mp-3p^2}{m^2-3mp+2p^2} \div \frac{m^2+4mp+3p^2}{m^2+2mp-8p^2}$

$= \frac{m^2+2mp-3p^2}{m^2-3mp+2p^2} \cdot \frac{m^2+2mp-8p^2}{m^2+4mp+3p^2}$

Multiply by the reciprocal.

$= \frac{(m+3p)(m-p)(m+4p)(m-2p)}{(m-2p)(m-p)(m+3p)(m+p)}$

Multiply numerators and multiply denominators. Factor.

$= \frac{m+4p}{m+p}$

Divide out the common factors.

43. The formula for the area of a rectangle is $A = LW$. Solving for W gives $W = \frac{A}{L}$, which can be written as $W = A \div L$, since the fraction bar indicates division. Substituting the given expressions for A and L gives the width.

$$\begin{array}{ccc} A & & L \\ \downarrow & & \downarrow \\ \frac{5x^2y^3}{2pq} & \div & \frac{2xy}{p} \end{array}$$

$= \frac{5x^2y^3}{2pq} \cdot \frac{p}{2xy}$ Multiply by the reciprocal.

$= \frac{5xy^2}{4q}$ Multiply. Divide out the common factors.

SECTION 7.3 (pages 505–506)

39. $\frac{-4t}{3t-6} = \frac{?}{12-6t}$

Factor each denominator. (On the right, rewrite $12 - 6t$ as $-6t + 12$, and factor.)

$$\frac{-4t}{3(t-2)} = \frac{?}{-6(t-2)}$$

The missing factor is -2, so multiply the fraction on the left by $\frac{-2}{-2}$.

$$\frac{-4t}{3(t-2)} \cdot \frac{-2}{-2} = \frac{8t}{-6(t-2)}, \text{ or } \frac{8t}{12-6t}$$

SECTION 7.4 (pages 513–516)

31. $\frac{x+1}{6} + \frac{3x+3}{9}$

Write the second fraction in lowest terms.

$\frac{3x+3}{9}$

$= \frac{3(x+1)}{9}$, or $\frac{x+1}{3}$

The sum is then as follows.
The LCD is 6.

$\frac{x+1}{6} + \frac{x+1}{3}$

$= \frac{x+1}{6} + \frac{x+1}{3} \cdot \frac{2}{2}$ Get a common denominator.

$= \frac{x+1}{6} + \frac{2(x+1)}{6}$

$= \frac{x+1+2x+2}{6}$ Add numerators; distributive property

$= \frac{3x+3}{6}$ Combine like terms.

$= \frac{3(x+1)}{3 \cdot 2}$ Factor.

$= \frac{x+1}{2}$ Lowest terms

43. $\frac{t}{t+2} + \frac{5-t}{t} - \frac{4}{t^2+2t}$

$= \frac{t}{t+2} + \frac{5-t}{t} - \frac{4}{t(t+2)}$ Factor.

$= \frac{t}{t+2} \cdot \frac{t}{t} + \frac{5-t}{t} \cdot \frac{t+2}{t+2}$

$- \frac{4}{t(t+2)}$ LCD $= t(t+2)$

$= \frac{t \cdot t + (5-t)(t+2) - 4}{t(t+2)}$

Write using the LCD.

$= \frac{t^2 + 5t + 10 - t^2 - 2t - 4}{t(t+2)}$

Multiply in the numerator.

$= \frac{3t+6}{t(t+2)}$ Combine like terms.

$= \frac{3(t+2)}{t(t+2)}$ Factor.

$= \frac{3}{t}$ Lowest terms

61. $\frac{2q+1}{3q^2+10q-8} - \frac{3q+5}{2q^2+5q-12}$

$= \frac{2q+1}{(3q-2)(q+4)} - \frac{3q+5}{(2q-3)(q+4)}$

Factor.

$= \frac{(2q+1)(2q-3)}{(3q-2)(q+4)(2q-3)}$

$- \frac{(3q+5)(3q-2)}{(2q-3)(q+4)(3q-2)}$

LCD $= (3q-2)(q+4)(2q-3)$

$= \frac{(2q+1)(2q-3) - (3q+5)(3q-2)}{(3q-2)(q+4)(2q-3)}$

Subtract numerators.

$= \frac{(4q^2-4q-3) - (9q^2+9q-10)}{(3q-2)(q+4)(2q-3)}$

Multiply in the numerator.

$= \frac{4q^2-4q-3-9q^2-9q+10}{(3q-2)(q+4)(2q-3)}$

Distributive property

$= \frac{-5q^2-13q+7}{(3q-2)(q+4)(2q-3)}$

Combine like terms.

65. $\frac{x+3y}{x^2+2xy+y^2} + \frac{x-y}{x^2+4xy+3y^2}$

$= \frac{x+3y}{(x+y)(x+y)} + \frac{x-y}{(x+3y)(x+y)}$

$= \frac{(x+3y)(x+3y)}{(x+y)(x+y)(x+3y)}$

$+ \frac{(x-y)(x+y)}{(x+3y)(x+y)(x+y)}$

LCD $= (x+y)(x+y)(x+3y)$

$= \frac{(x^2+6xy+9y^2) + (x^2-y^2)}{(x+y)(x+y)(x+3y)}$

$= \frac{2x^2+6xy+8y^2}{(x+y)(x+y)(x+3y)}$,

or $\frac{2x^2+6xy+8y^2}{(x+y)^2(x+3y)}$

There is no need to factor out 2 in the numerator, since the denominator does not have 2 as a factor. The answer is in lowest terms.

SECTION 7.5 (pages 523–524)

17. $\dfrac{\frac{1}{x}+x}{\frac{x^2+1}{8}}$ Use Method 2.

$= \dfrac{8x\left(\frac{1}{x}+x\right)}{8x\left(\frac{x^2+1}{8}\right)}$ LCD $= 8x$

$= \dfrac{8+8x^2}{x(x^2+1)}$ Distributive property

$= \dfrac{8(1+x^2)}{x(x^2+1)}$ Factor.

$= \dfrac{8}{x}$ Lowest terms

31. $\dfrac{\frac{1}{m-1}+\frac{2}{m+2}}{\frac{2}{m+2}-\frac{1}{m-3}}$ Use Method 2.

$= \dfrac{(m-1)(m+2)(m-3)\left(\frac{1}{m-1}+\frac{2}{m+2}\right)}{(m-1)(m+2)(m-3)\left(\frac{2}{m+2}-\frac{1}{m-3}\right)}$

LCD $= (m-1)(m+2)(m-3)$

$= \dfrac{(m+2)(m-3)+2(m-1)(m-3)}{2(m-1)(m-3)-(m-1)(m+2)}$

Distributive property

$= \dfrac{(m-3)[(m+2)+2(m-1)]}{(m-1)[2(m-3)-(m+2)]}$

Factor out $m-3$ in the numerator and $m-1$ in the denominator.

$= \dfrac{(m-3)[m+2+2m-2]}{(m-1)[2m-6-m-2]}$

Distributive property

$= \dfrac{3m(m-3)}{(m-1)(m-8)}$

Combine like terms.

33. $2-\dfrac{2}{2+\frac{2}{2+2}}$

$= 2-\dfrac{2}{2+\frac{2}{4}}$ Simplify the denominator.

$= 2-\dfrac{2}{\frac{5}{2}}$ $\quad 2+\frac{2}{4}=\frac{4}{2}+\frac{1}{2}=\frac{5}{2}$

$= 2-\left(2\cdot\frac{2}{5}\right)$ $\quad \dfrac{2}{\frac{5}{2}} = 2 \div \frac{5}{2} = 2\cdot\frac{2}{5}$

$= 2-\frac{4}{5}$ Multiply.

$= \frac{10}{5}-\frac{4}{5}$ $\quad 2=\frac{10}{5}$

$= \frac{6}{5}$ Subtract.

SECTION 7.6 (pages 533–536)

11. $\frac{3x}{5}-6=x$

$5\left(\frac{3x}{5}-6\right)=5(x)$ Multiply by the LCD, 5.

$5\left(\frac{3x}{5}\right)-5(6)=5x$ Distributive property

$3x-30=5x$ Multiply.

$-30=2x$ Subtract $3x$.

$-15=x$ Divide by 2.

A check confirms that -15 is the solution.

Solution set: $\{-15\}$

21. $\frac{q+2}{3}+\frac{q-5}{5}=\frac{7}{3}$

$15\left(\frac{q+2}{3}+\frac{q-5}{5}\right)=15\left(\frac{7}{3}\right)$

Multiply by the LCD, 15.

$15\left(\frac{q+2}{3}\right)+15\left(\frac{q-5}{5}\right)=5\cdot 7$

$5(q+2)+3(q-5)=35$

$5q+10+3q-15=35$

$8q-5=35$

$8q=40$

$q=5$

A check confirms that 5 is the solution.

Solution set: $\{5\}$

41. $\frac{2}{m}=\frac{m}{5m+12}$

$m(5m+12)\left(\frac{2}{m}\right)=m(5m+12)\left(\frac{m}{5m+12}\right)$

Multiply by the LCD, $m(5m+12)$.

$(5m+12)(2)=m(m)$ Divide out the common factors.

$10m+24=m^2$ Multiply.

$m^2-10m-24=0$ Standard form

$(m+2)(m-12)=0$ Factor.

$m+2=0$ or $m-12=0$

Zero-factor property

$m=-2$ or $m=12$

The proposed solutions are -2 and 12.

CHECK $\frac{2}{-2} \stackrel{?}{=} \frac{-2}{5(-2)+12}$ Let $m=-2$.

$-1=\frac{-2}{2}$ ✓ True

CHECK $\frac{2}{12} \stackrel{?}{=} \frac{12}{5(12)+12}$ Let $m=12$.

$\frac{1}{6}=\frac{12}{72}$ ✓ True

Solution set: $\{-2, 12\}$

69. $d=\frac{2S}{n(a+L)}$ for a

We need to isolate a on one side of the equation.

$d=\frac{2S}{n(a+L)}$

$d\cdot n(a+L)=\frac{2S}{n(a+L)}\cdot n(a+L)$

Multiply by $n(a+L)$.

$dn(a+L)=2S$

$dna+dnL=2S$ Distributive property

$dna=2S-dnL$ Subtract dnL.

$a=\frac{2S-dnL}{dn}$, Divide by dn.

or $a=\frac{2S}{dn}-L$

75. $9x+\frac{3}{z}=\frac{5}{y}$ for z

$yz\left(9x+\frac{3}{z}\right)=yz\left(\frac{5}{y}\right)$ Multiply by the LCD, yz.

$yz(9x)+yz\left(\frac{3}{z}\right)=yz\left(\frac{5}{y}\right)$ Distributive property

$9xyz+3y=5z$

$9xyz-5z=-3y$ Get the z-terms on one side.

$z(9xy-5)=-3y$ Factor out z.

$z=\frac{-3y}{9xy-5}$, Divide by $9xy-5$.

or $z=\frac{3y}{5-9xy}$

SECTION 7.7 (pages 546–549)

27. Let $x=$ the rate of the current of the river. Then $(12-x)$ mph is the rate upstream (against the current) and $(12+x)$ mph is the rate downstream (with the current).

Use $t=\frac{d}{r}$ to complete the table.

	d	r	t
Upstream	6	$12-x$	$\frac{6}{12-x}$
Downstream	10	$12+x$	$\frac{10}{12+x}$

The times are equal.

$\frac{6}{12-x}=\frac{10}{12+x}$ Write an equation.

$(12+x)(12-x)\frac{6}{12-x}$

$=(12+x)(12-x)\frac{10}{12+x}$

Multiply by the LCD, $(12+x)(12-x)$.

$6(12+x)=10(12-x)$

$72+6x=120-10x$

$16x=48$

$x=3$

The rate of the current of the river is 3 mph.

SOLUTIONS

35. Let x = the number of hours it would take Brenda to paint the room by herself.

	Rate	Time Working Together	Fractional Part of the Job Done When Working Together
Hilda	$\frac{1}{6}$	$3\frac{3}{4} = \frac{15}{4}$	$\frac{1}{6} \cdot \frac{15}{4} = \frac{5}{8}$
Brenda	$\frac{1}{x}$	$3\frac{3}{4} = \frac{15}{4}$	$\frac{1}{x} \cdot \frac{15}{4} = \frac{15}{4x}$

Since together Hilda and Brenda complete 1 whole job, we must add their individual fractional parts and set the sum equal to 1.

$$\frac{5}{8} + \frac{15}{4x} = 1$$

$$8x\left(\frac{5}{8} + \frac{15}{4x}\right) = 8x(1) \qquad \text{Multiply by the LCD, } 8x.$$

$$8x\left(\frac{5}{8}\right) + 8x\left(\frac{15}{4x}\right) = 8x$$

$$5x + 30 = 8x$$

$$30 = 3x$$

$$10 = x$$

It will take Brenda 10 hr to paint the room by herself.

37. Let x = the number of hours to fill the pool with both pipes left open.

	Rate	Time to Fill the Pool	Part Done by Each Pipe
Inlet Pipe	$\frac{1}{9}$	x	$\frac{1}{9}x$
Outlet Pipe	$\frac{1}{12}$	x	$\frac{1}{12}x$

Part done by inlet pipe − Part done by outlet pipe = Full pool

$$\frac{1}{9}x - \frac{1}{12}x = 1$$

$$36\left(\frac{1}{9}x - \frac{1}{12}x\right) = 36(1)$$

Multiply by the LCD, 36.

$$36\left(\frac{1}{9}x\right) - 36\left(\frac{1}{12}x\right) = 36$$

$$4x - 3x = 36$$

$$x = 36$$

It will take 36 hr to fill the pool.

SECTION 7.8 (pages 552–554)

1. For a constant time of 3 hr, if the rate of the pickup truck *increases,* then the distance traveled *increases*. Thus, the variation between the quantities is *direct.*

23. For a given base, the area A of a triangle varies directly as its height h, so there is a constant k such that $A = kh$. Find the value of k.

$$10 = k(4) \qquad \text{Let } A = 10,\ h = 4.$$

$$k = \frac{10}{4} = 2.5 \qquad \text{Solve for } k.$$

For $k = 2.5$, $A = kh$ becomes

$$A = 2.5h.$$

Now find A for $h = 6$.

$$A = 2.5(6) = 15$$

When the height of the triangle is 6 in., the area of the triangle is 15 in.^{2}

CHAPTER 8 Roots and Radicals

SECTION 8.1 (pages 577–580)

63. $\sqrt{103} \approx \sqrt{100} = 10$ Length

$\sqrt{48} \approx \sqrt{49} = 7$ Width

The best estimate for the length and width, in meters, of the rectangle is 10 by 7, choice C.

79. Let $a = 4.5$ and $c = 12.0$. Use the Pythagorean theorem.

$$c^2 = a^2 + b^2$$

$$(12.0)^2 = (4.5)^2 + b^2 \qquad \text{Substitute carefully.}$$

$$144 = 20.25 + b^2 \qquad \text{Square.}$$

$$123.75 = b^2 \qquad \text{Subtract 20.25.}$$

$$b = \sqrt{123.75} \qquad \text{Solve for } b.$$

$$b \approx 11.1 \qquad \text{Use a calculator.}$$

The distance from the base of the tree to the point where the broken part touches the ground is 11.1 ft (to the nearest tenth).

93. $-\sqrt[3]{-8} = -(-2) = 2$

SECTION 8.2 (pages 587–590)

31. $3\sqrt{52}$

$$= 3 \cdot \sqrt{4 \cdot 13} \qquad \text{Factor; 4 is a perfect square.}$$

$$= 3\sqrt{4} \cdot \sqrt{13} \qquad \text{Product rule}$$

$$= 3 \cdot 2 \cdot \sqrt{13} \qquad \sqrt{4} = 2$$

$$= 6\sqrt{13} \qquad \text{Multiply.}$$

49. $5\sqrt{3} \cdot 2\sqrt{15}$

$$= 5 \cdot 2 \cdot \sqrt{3 \cdot 15} \qquad \text{Commutative property; product rule}$$

$$= 10\sqrt{45} \qquad \text{Multiply.}$$

$$= 10\sqrt{9 \cdot 5} \qquad \text{Factor; 9 is a perfect square.}$$

$$= 10\sqrt{9} \cdot \sqrt{5} \qquad \text{Product rule}$$

$$= 10 \cdot 3 \cdot \sqrt{5} \qquad \sqrt{9} = 3$$

$$= 30\sqrt{5} \qquad \text{Multiply.}$$

57. $\sqrt{\frac{4}{50}}$

$$= \sqrt{\frac{2}{25}} \qquad \text{Write the fraction in lowest terms.}$$

$$= \frac{\sqrt{2}}{\sqrt{25}} \qquad \text{Quotient rule}$$

$$= \frac{\sqrt{2}}{5} \qquad \sqrt{25} = 5$$

119. Use the formula for the volume of a sphere.

$$V = \frac{4}{3}\pi r^3$$

Let $V = 288\pi$ and solve for r.

$$288\pi = \frac{4}{3}\pi r^3 \qquad V = 288\pi$$

$$\frac{3}{4}(288\pi) = \frac{3}{4}\left(\frac{4}{3}\pi r^3\right) \qquad \text{Multiply by } \frac{3}{4}.$$

$$216\pi = \pi r^3 \qquad \text{Simplify.}$$

$$216 = r^3 \qquad \text{Divide by } \pi.$$

$$\sqrt[3]{216} = r \qquad \text{Take cube roots.}$$

$$6 = r$$

The radius is 6 in.

121. $2\sqrt{26} \approx 2\sqrt{25} = 2 \cdot 5 = 10$ Length

$\sqrt{83} \approx \sqrt{81} = 9$ Width

Using 10 and 9 as estimates for the length and the width of the rectangle gives us

$$10 \cdot 9 = 90$$

as an estimate for the area, in square inches. Choice D is the best estimate.

SECTION 8.3 (pages 593–594)

29. $2\sqrt{8} - 5\sqrt{32} - 2\sqrt{48}$

$$= 2(\sqrt{4} \cdot \sqrt{2}) - 5(\sqrt{16} \cdot \sqrt{2}) - 2(\sqrt{16} \cdot \sqrt{3}) \qquad \text{Product rule}$$

$$= 2(2\sqrt{2}) - 5(4\sqrt{2}) - 2(4\sqrt{3}) \qquad \sqrt{4} = 2;\ \sqrt{16} = 4$$

$$= 4\sqrt{2} - 20\sqrt{2} - 8\sqrt{3} \qquad \text{Multiply.}$$

$$= -16\sqrt{2} - 8\sqrt{3} \qquad \text{Subtract like radicals.}$$

33. $\frac{1}{4}\sqrt{288} + \frac{1}{6}\sqrt{72}$

$$= \frac{1}{4}\sqrt{144 \cdot 2} + \frac{1}{6}\sqrt{36 \cdot 2}$$

$$= \frac{1}{4}(\sqrt{144} \cdot \sqrt{2}) + \frac{1}{6}(\sqrt{36} \cdot \sqrt{2}) \qquad \text{Product rule}$$

$$= \frac{1}{4}(12\sqrt{2}) + \frac{1}{6}(6\sqrt{2}) \qquad \sqrt{144} = 12;\ \sqrt{36} = 6$$

$$= 3\sqrt{2} + 1\sqrt{2} \qquad \text{Multiply.}$$

$$= 4\sqrt{2} \qquad \text{Add like radicals.}$$

35. Use the formula for the perimeter of a rectangle.

$P = 2L + 2W$

$P = 2(7\sqrt{2}) + 2(4\sqrt{2})$ Let $L = 7\sqrt{2}$ and $W = 4\sqrt{2}$.

$P = 14\sqrt{2} + 8\sqrt{2}$ Multiply.

$P = 22\sqrt{2}$ Add like radicals.

59. $5\sqrt[4]{m^3} + 8\sqrt[4]{16m^3}$

$= 5\sqrt[4]{m^3} + 8\sqrt[4]{16 \cdot m^3}$

$= 5\sqrt[4]{m^3} + 8\sqrt[4]{16} \cdot \sqrt[4]{m^3}$ Product rule

$= 5\sqrt[4]{m^3} + 8 \cdot 2 \cdot \sqrt[4]{m^3}$ $\sqrt[4]{16} = 2$

$= 5\sqrt[4]{m^3} + 16\sqrt[4]{m^3}$ Multiply.

$= 21\sqrt[4]{m^3}$ Add like radicals.

SECTION 8.4 (pages 599–600)

41. $\sqrt{\dfrac{9a^2r^5}{7t}}$

$= \dfrac{\sqrt{9a^2r^5}}{\sqrt{7t}}$ Quotient rule

$= \dfrac{\sqrt{9a^2r^4 \cdot r}}{\sqrt{7t}}$ Factor.

$= \dfrac{\sqrt{9a^2r^4} \cdot \sqrt{r}}{\sqrt{7t}}$ Product rule

$= \dfrac{3ar^2\sqrt{r}}{\sqrt{7t}}$ $\sqrt{9} = 3$; $\sqrt{a^2} = a$, since $a > 0$; $\sqrt{r^4} = r^2$

$= \dfrac{3ar^2\sqrt{r} \cdot \sqrt{7t}}{\sqrt{7t} \cdot \sqrt{7t}}$ Rationalize the denominator.

$= \dfrac{3ar^2\sqrt{7rt}}{7t}$ Product rule; $\sqrt{7t} \cdot \sqrt{7t} = 7t$

47. $\sqrt[3]{\dfrac{5}{9}}$

$= \dfrac{\sqrt[3]{5}}{\sqrt[3]{9}}$ Quotient rule

$= \dfrac{\sqrt[3]{5} \cdot \sqrt[3]{3}}{\sqrt[3]{9} \cdot \sqrt[3]{3}}$ Multiply numerator and denominator by $\sqrt[3]{3}$ to make the radicand in the denominator a perfect cube.

$= \dfrac{\sqrt[3]{15}}{\sqrt[3]{27}}$ Product rule

$= \dfrac{\sqrt[3]{15}}{3}$ $\sqrt[3]{27} = 3$

55. (a) $p = k \cdot \sqrt{\dfrac{L}{g}}$

$p = 6 \cdot \sqrt{\dfrac{9}{32}}$ Let $k = 6$, $L = 9$, $g = 32$.

$p = \dfrac{6\sqrt{9}}{\sqrt{32}}$ Quotient rule; multiply.

$p = \dfrac{6 \cdot 3}{\sqrt{16 \cdot 2}}$ $\sqrt{9} = 3$; Factor.

$p = \dfrac{18}{4\sqrt{2}}$ Multiply; product rule

$p = \dfrac{9 \cdot 2}{2 \cdot 2\sqrt{2}}$ Factor.

$p = \dfrac{9}{2\sqrt{2}}$ Divide out 2.

$p = \dfrac{9 \cdot \sqrt{2}}{2\sqrt{2} \cdot \sqrt{2}}$ Rationalize the denominator.

$p = \dfrac{9\sqrt{2}}{4}$ Multiply.

The period of the pendulum is $\dfrac{9\sqrt{2}}{4}$ sec.

(b) Using a calculator, $\dfrac{9\sqrt{2}}{4} \approx 3.182$ sec.

SECTION 8.5 (pages 605–608)

15. $(5\sqrt{7} - 2\sqrt{3})(3\sqrt{7} + 4\sqrt{3})$

$= 5\sqrt{7}(3\sqrt{7}) + 5\sqrt{7}(4\sqrt{3}) - 2\sqrt{3}(3\sqrt{7}) - 2\sqrt{3}(4\sqrt{3})$ FOIL method

$= 15 \cdot 7 + 20\sqrt{21} - 6\sqrt{21} - 8 \cdot 3$ Multiply.

$= 105 + 14\sqrt{21} - 24$ Multiply numbers. Add like radicals.

$= 81 + 14\sqrt{21}$ Subtract.

45. $\dfrac{\sqrt{5}}{\sqrt{2} + \sqrt{3}}$

$= \dfrac{\sqrt{5}(\sqrt{2} - \sqrt{3})}{(\sqrt{2} + \sqrt{3})(\sqrt{2} - \sqrt{3})}$

Multiply numerator and denominator by the conjugate of the denominator.

$= \dfrac{\sqrt{5} \cdot \sqrt{2} - \sqrt{5} \cdot \sqrt{3}}{(\sqrt{2})^2 - (\sqrt{3})^2}$

Distributive property; $(a + b)(a - b) = a^2 - b^2$

$= \dfrac{\sqrt{10} - \sqrt{15}}{2 - 3}$ Product rule; $(\sqrt{a})^2 = a$

$= \dfrac{\sqrt{10} - \sqrt{15}}{-1}$ Subtract.

$= -\sqrt{10} + \sqrt{15}$

SECTION 8.6 (pages 617–620)

23. $\sqrt{3x - 5} = \sqrt{2x + 1}$

$(\sqrt{3x - 5})^2 = (\sqrt{2x + 1})^2$ Squaring property

$3x - 5 = 2x + 1$ $(\sqrt{a})^2 = a$

$x = 6$ Subtract $2x$. Add 5.

CHECK $\sqrt{3x - 5} = \sqrt{2x + 1}$

$\sqrt{3(6) - 5} \stackrel{?}{=} \sqrt{2(6) + 1}$ Let $x = 6$.

$\sqrt{13} = \sqrt{13}$ ✓ True

Solution set: $\{6\}$

33. $\sqrt{3k + 10} + 5 = 2k$

$\sqrt{3k + 10} = 2k - 5$ Subtract 5.

$(\sqrt{3k + 10})^2 = (2k - 5)^2$ Squaring property

$3k + 10 = 4k^2 - 20k + 25$

$(\sqrt{a})^2 = a$; Square the binomial.

$4k^2 - 23k + 15 = 0$ Standard form

$(4k - 3)(k - 5) = 0$ Factor.

$k = \dfrac{3}{4}$ or $k = 5$ Zero-factor property

Checking both proposed solutions in the original equation will show that 5 is a solution, but $\frac{3}{4}$ is not.

Solution set: $\{5\}$

49. $\sqrt{2x + 11} + \sqrt{x + 6} = 2$

$\sqrt{2x + 11} = 2 - \sqrt{x + 6}$ Isolate a radical.

$(\sqrt{2x + 11})^2 = (2 - \sqrt{x + 6})^2$ Square both sides.

$2x + 11 = 4 - 4\sqrt{x + 6} + x + 6$

$(\sqrt{a})^2 = a^2$; Square the binomial.

$x + 1 = -4\sqrt{x + 6}$ Isolate the radical.

$(x + 1)^2 = (-4\sqrt{x + 6})^2$ Square both sides again.

$x^2 + 2x + 1 = 16(x + 6)$

Square the binomial; $(a\sqrt{b})^2 = a^2b$

$x^2 + 2x + 1 = 16x + 96$ Distributive property

$x^2 - 14x - 95 = 0$ Standard form

$(x + 5)(x - 19) = 0$ Factor.

$x = -5$ or $x = 19$ Zero-factor property

Checking both proposed solutions in the original equation will show that -5 is a solution, but 19 is not.

Solution set: $\{-5\}$

SOLUTIONS

CHAPTER 9 Quadratic Equations

SECTION 9.1 (pages 638–640)

27. $7x^2 = 4$

$x^2 = \frac{4}{7}$ Divide by 7.

$x = \sqrt{\frac{4}{7}}$ or $x = -\sqrt{\frac{4}{7}}$ Square root property

$x = \frac{\sqrt{4}}{\sqrt{7}} \cdot \frac{\sqrt{7}}{\sqrt{7}}$ or $x = -\frac{\sqrt{4}}{\sqrt{7}} \cdot \frac{\sqrt{7}}{\sqrt{7}}$ Rationalize denominators.

$x = \frac{2\sqrt{7}}{7}$ or $x = -\frac{2\sqrt{7}}{7}$

Solution set: $\left\{-\frac{2\sqrt{7}}{7}, \frac{2\sqrt{7}}{7}\right\}$

49. $\left(\frac{1}{2}x + 5\right)^2 = 12$

$\frac{1}{2}x + 5 = \sqrt{12}$ or $\frac{1}{2}x + 5 = -\sqrt{12}$

$\frac{1}{2}x = -5 + \sqrt{12}$ or $\frac{1}{2}x = -5 - \sqrt{12}$ Subtract 5.

$\frac{1}{2}x = -5 + 2\sqrt{3}$ or $\frac{1}{2}x = -5 - 2\sqrt{3}$

$\sqrt{12} = \sqrt{4} \cdot \sqrt{3} = 2\sqrt{3}$

$x = -10 + 4\sqrt{3}$ or $x = -10 - 4\sqrt{3}$ Multiply each term by 2.

Solution set: $\{-10 + 4\sqrt{3}, -10 - 4\sqrt{3}\}$

SECTION 9.2 (pages 647–648)

19. $x^2 + x - 1 = 0$

$x^2 + x = 1$ Add 1.

Take half of 1, the coefficient of x, square it, and add the result to each side.

$x^2 + x + \frac{1}{4} = 1 + \frac{1}{4}$ Add $\left[\frac{1}{2}(1)\right]^2 = \frac{1}{4}$.

$\left(x + \frac{1}{2}\right)^2 = \frac{5}{4}$ Factor and add.

$x + \frac{1}{2} = \sqrt{\frac{5}{4}}$ or $x + \frac{1}{2} = -\sqrt{\frac{5}{4}}$

$x + \frac{1}{2} = \frac{\sqrt{5}}{2}$ or $x + \frac{1}{2} = -\frac{\sqrt{5}}{2}$

$x = -\frac{1}{2} + \frac{\sqrt{5}}{2}$ or $x = -\frac{1}{2} - \frac{\sqrt{5}}{2}$

$x = \frac{-1 + \sqrt{5}}{2}$ or $x = \frac{-1 - \sqrt{5}}{2}$

Solution set: $\left\{\frac{-1 + \sqrt{5}}{2}, \frac{-1 - \sqrt{5}}{2}\right\}$

23. $2x^2 - 4x = 5$

$x^2 - 2x = \frac{5}{2}$ Divide by 2.

$x^2 - 2x + 1 = \frac{5}{2} + 1$ Add $\left[\frac{1}{2}(-2)\right]^2 = 1$.

$(x - 1)^2 = \frac{7}{2}$ Factor and add.

$x - 1 = \sqrt{\frac{7}{2}}$ or $x - 1 = -\sqrt{\frac{7}{2}}$

$x - 1 = \frac{\sqrt{7}}{\sqrt{2}}$ or $x - 1 = -\frac{\sqrt{7}}{\sqrt{2}}$

$x - 1 = \frac{\sqrt{14}}{2}$ or $x - 1 = -\frac{\sqrt{14}}{2}$ Rationalize denominators by multiplying by $\frac{\sqrt{2}}{\sqrt{2}}$.

$x = 1 + \frac{\sqrt{14}}{2}$ or $x = 1 - \frac{\sqrt{14}}{2}$

$x = \frac{2}{2} + \frac{\sqrt{14}}{2}$ or $x = \frac{2}{2} - \frac{\sqrt{14}}{2}$

$x = \frac{2 + \sqrt{14}}{2}$ or $x = \frac{2 - \sqrt{14}}{2}$

Solution set: $\left\{\frac{2 + \sqrt{14}}{2}, \frac{2 - \sqrt{14}}{2}\right\}$

SECTION 9.3 (pages 653–654)

23. $7x^2 = 12x$

Write the equation in standard form.

$7x^2 - 12x = 0$

Substitute $a = 7$, $b = -12$, and $c = 0$ into the quadratic formula.

$x = \frac{-b \pm \sqrt{b^2 - 4ac}}{2a}$

$x = \frac{-(-12) \pm \sqrt{(-12)^2 - 4(7)(0)}}{2(7)}$

$x = \frac{12 \pm \sqrt{144 - 0}}{14}$

$x = \frac{12 \pm 12}{14}$

$x = \frac{12 + 12}{14} = \frac{24}{14} = \frac{12}{7}$ or

$x = \frac{12 - 12}{14} = \frac{0}{14} = 0$

Solution set: $\left\{0, \frac{12}{7}\right\}$

29. $3x^2 - 2x + 5 = 10x + 1$

Write the equation in standard form.

$3x^2 - 12x + 4 = 0$

Substitute $a = 3$, $b = -12$, and $c = 4$ into the quadratic formula.

$x = \frac{-b \pm \sqrt{b^2 - 4ac}}{2a}$

$x = \frac{-(-12) \pm \sqrt{(-12)^2 - 4(3)(4)}}{2(3)}$

$x = \frac{12 \pm \sqrt{144 - 48}}{6}$

$x = \frac{12 \pm \sqrt{96}}{6}$

$x = \frac{12 \pm 4\sqrt{6}}{6}$ $\sqrt{96} = \sqrt{16} \cdot \sqrt{6} = 4\sqrt{6}$

$x = \frac{2(6 \pm 2\sqrt{6})}{2 \cdot 3}$ Factor.

$x = \frac{6 \pm 2\sqrt{6}}{3}$ Divide out 2.

Solution set: $\left\{\frac{6 + 2\sqrt{6}}{3}, \frac{6 - 2\sqrt{6}}{3}\right\}$

SECTION 9.4 (pages 661–662)

19. Because the vertex is at the origin, an equation of the parabola is of the form

$y = ax^2$.

As shown in the figure, one point on the graph has coordinates $(150, 44)$.

$y = ax^2$ General equation

$44 = a(150)^2$ Let $x = 150$, $y = 44$.

$44 = 22{,}500a$ Apply the exponent.

$a = \frac{44}{22{,}500}$ Solve for a.

$a = \frac{4 \cdot 11}{4 \cdot 5625}$, or $\frac{11}{5625}$ Lowest terms

Thus, an equation of the parabola is

$y = \frac{11}{5625}x^2$.

SOLUTIONS

Index

Geometry Formulas

Square
Perimeter: $P = 4s$
Area: $A = s^2$

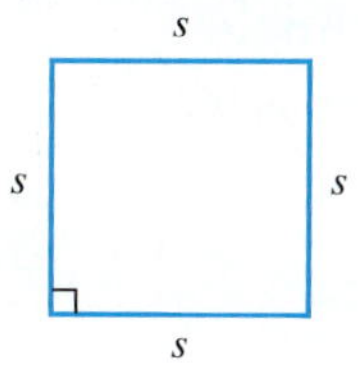

Rectangular Solid
Volume: $V = LWH$
Surface area: $A = 2HW + 2LW + 2LH$

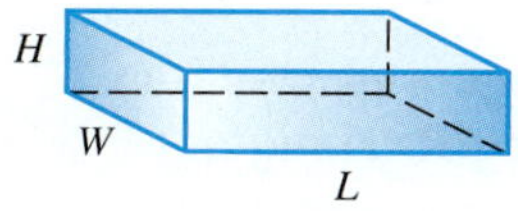

Rectangle
Perimeter: $P = 2L + 2W$
Area: $A = LW$

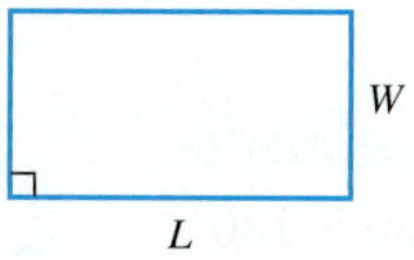

Cube
Volume: $V = e^3$
Surface area: $S = 6e^2$

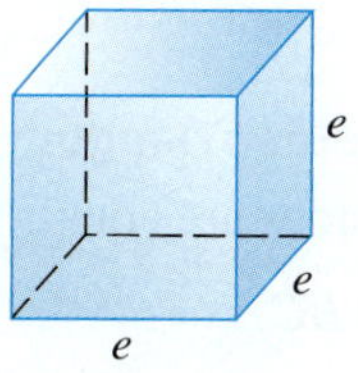

Triangle
Perimeter: $P = a + b + c$
Area: $A = \frac{1}{2}bh$

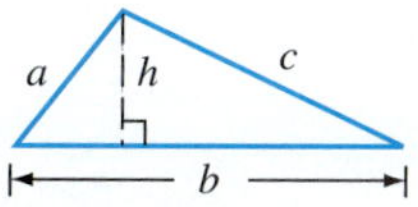

Right Circular Cylinder
Volume: $V = \pi r^2 h$
Surface area: $S = 2\pi rh + 2\pi r^2$
(Includes both circular bases)

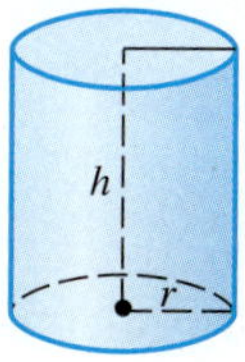

Parallelogram
Perimeter: $P = 2a + 2b$
Area: $A = bh$

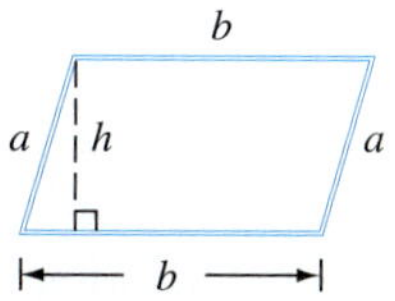

Cone
Volume: $V = \frac{1}{3}\pi r^2 h$
Surface area: $S = \pi r\sqrt{r^2 + h^2} + \pi r^2$
(Includes circular base)

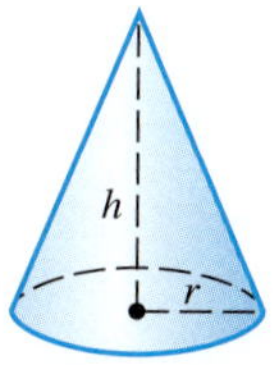

Trapezoid
Perimeter: $P = a + b + c + B$
Area: $A = \frac{1}{2}h(b + B)$

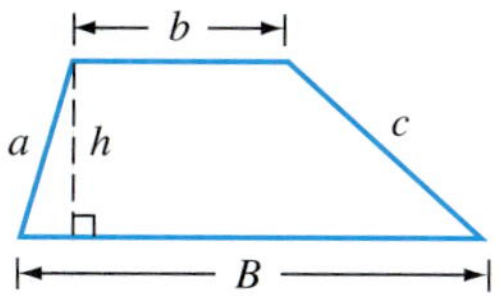

Right Pyramid
Volume: $V = \frac{1}{3}Bh$
B = area of the base

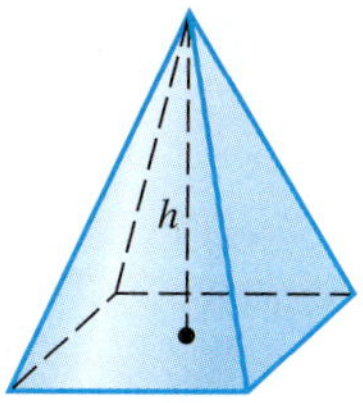

Circle
Diameter: $d = 2r$
Circumference: $C = 2\pi r$
$C = \pi d$
Area: $A = \pi r^2$

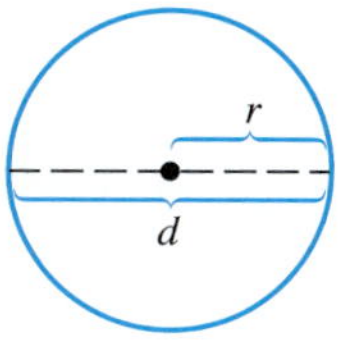

Sphere
Volume: $V = \frac{4}{3}\pi r^3$
Surface area: $S = 4\pi r^2$

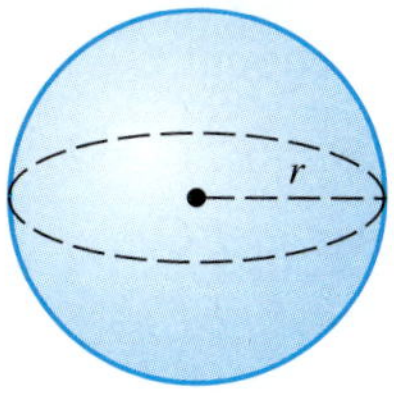

Triangles and Angles

Right Triangle
Triangle has one 90° (right) angle.

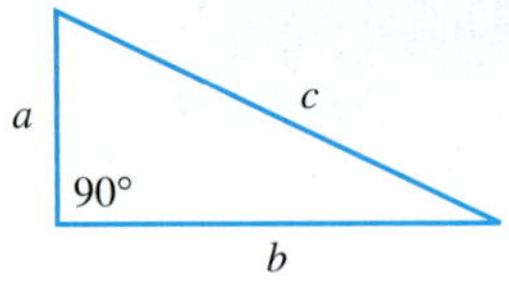

Pythagorean Theorem (for right triangles)
$a^2 + b^2 = c^2$

Isosceles Triangle
Two sides are equal.

$AB = BC$

Equilateral Triangle
All sides are equal.

$AB = BC = CA$

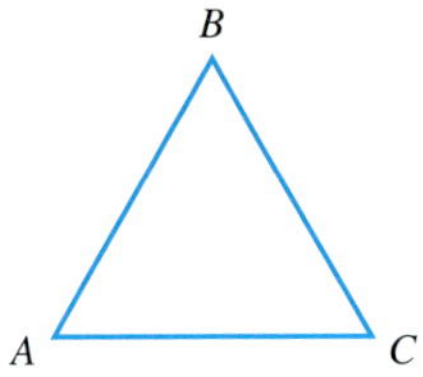

Sum of the Angles of Any Triangle
$A + B + C = 180°$

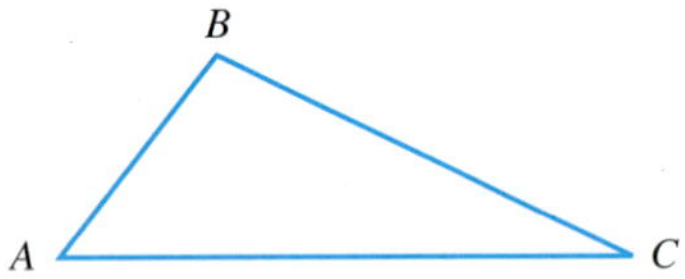

Similar Triangles
Corresponding angles are equal. Corresponding sides are proportional.

$A = D, B = E, C = F$

$$\frac{AB}{DE} = \frac{AC}{DF} = \frac{BC}{EF}$$

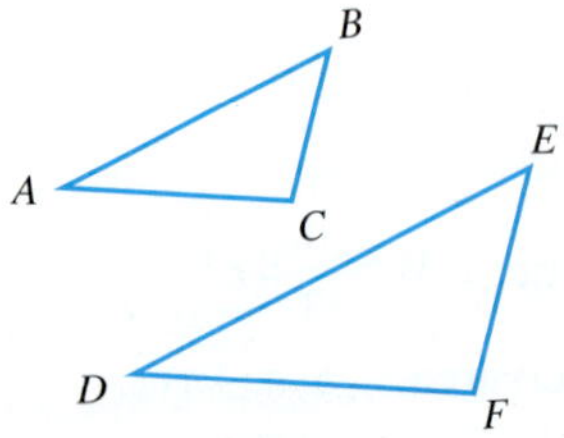

Right Angle
Measure is 90°.

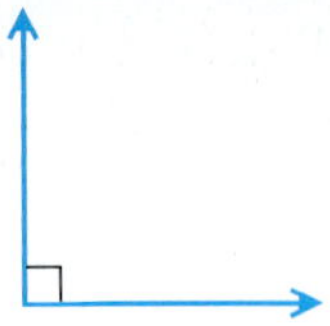

Straight Angle
Measure is 180°.

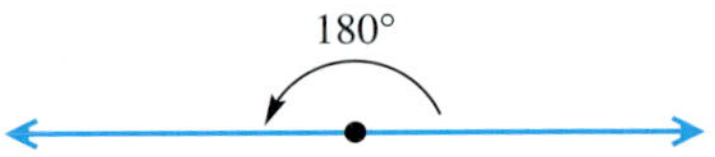

Complementary Angles
The sum of the measures of two complementary angles is 90°.

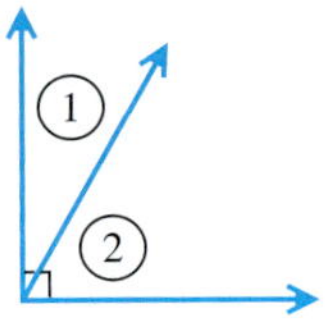

Angles (1) and (2) are complementary.

Supplementary Angles
The sum of the measures of two supplementary angles is 180°.

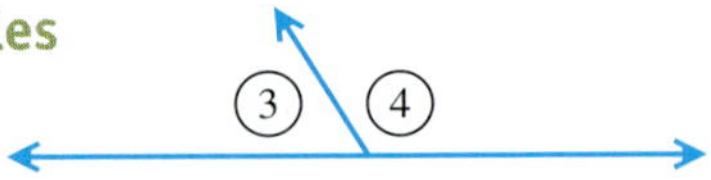

Angles (3) and (4) are supplementary.

Vertical Angles
Vertical angles have equal measures.

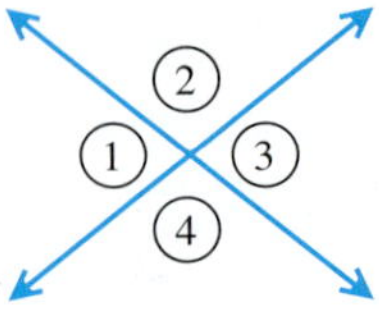

Angle (1) = Angle (3)

Angle (2) = Angle (4)